Petroleum Engineering Handbook

Petroleum Engineering Handbook

Larry W. Lake, Editor-in-Chief

  I   **General Engineering** *John R. Fanchi, Editor*

 II **Drilling Engineering** *Robert F. Mitchell, Editor*

III **Facilities and Construction Engineering** *Kenneth E. Arnold, Editor*

IV **Production Operations Engineering** *Joe Dunn Clegg, Editor*

 V **Reservoir Engineering and Petrophysics** *Edward D. Holstein, Editor*

VI **Emerging and Peripheral Technologies** *H.R. Warner Jr., Editor*

VII **Indexes and Standards**

Petroleum Engineering Handbook

Larry W. Lake, Editor-in-Chief
U. of Texas at Austin

Volume II

Drilling Engineering

Robert F. Mitchell, Editor
Landmark Graphics Corp.

Society of Petroleum Engineers

ISBN 978-1-55563-114-7 (print)
ISBN 978-1-55563-115-4 (CD)
ISBN 978-1-55563-129-1 (print and CD)
ISBN 978-1-55563-126-0 (Complete 7-Vol. Set, print)
ISBN 978-1-55563-127-7 (Complete 7-Vol. Set, CD)
ISBN 978-1-55563-135-2 (Complete 7-Vol. Set, print and CD)

06 07 08 09 10 11 12 13 14 / 9 8 7 6 5 4 3 2 1

Society of Petroleum Engineers
222 Palisades Creek Drive
Richardson, TX 75080-2040 USA

http://www.spe.org/store
service@spe.org
1.972.952.9393

Foreword

This 2006 version of SPE's *Petroleum Engineering Handbook* is the result of several years of effort by technical editors, copy editors, and authors. It is designed as a handbook rather than a basic text. As such, it will be of most benefit to those with some experience in the industry who require additional information and guidance in areas outside their areas of expertise. Authors for each of the more than 100 chapters were chosen carefully for their experience and expertise. The resulting product of their efforts represents the best current thinking on the various technical subjects covered in the *Handbook*.

The rate of growth in hydrocarbon extraction technology is continuing at the high level experienced in the last decades of the 20th century. As a result, any static compilation, such as this *Handbook*, will contain certain information that is out of date at the time of publication. However, many of the concepts and approaches presented will continue to be applicable in your studies, and, by documenting the technology in this way, it provides new professionals an insight into the many factors to be considered in assessing various aspects of a vibrant and dynamic industry.

The *Handbook* is a continuation of SPE's primary mission of technology transfer. Its direct descendents are the "Frick" *Handbook*, published in 1952, and the "Bradley" *Handbook*, published in 1987. This version is different from the previous in the following ways:

- It has multiple volumes in six different technical areas with more than 100 chapters.
- There is expanded coverage in several areas such as health, safety, and environment.
- It contains entirely new coverage on Drilling Engineering and Emerging and Peripheral Technologies.
- Electronic versions are available in addition to the standard bound volumes.

This *Handbook* has been a monumental undertaking that is the result of many people's efforts. I am pleased to single out the contributions of the six volume editors:

General Engineering—John R. Fanchi, Colorado School of Mines
Drilling Engineering—Robert F. Mitchell, Landmark Graphics Corp.
Facilities and Construction Engineering—Kenneth E. Arnold, AMEC Paragon
Production Operations Engineering—Joe D. Clegg, Shell Oil Co., retired
Reservoir Engineering and Petrophysics—Ed Holstein, Exxon Production Co., retired
Emerging and Peripheral Technologies—Hal R. Warner, Arco Oil and Gas, retired

It is to these individuals, along with the authors, the copy editors, and the SPE staff, that accolades for this effort belong. It has been my pleasure to work with and learn from them.

—Larry W. Lake

Preface

You hold in your hand the very first Drilling Engineering volume of the SPE *Petroleum Engineering Handbook*. This volume is intended to provide a good snapshot of the drilling state of the art at the beginning of the 21st century.

Obviously, the history of well drilling goes back for millennia. The history of "scientific" oilwell drilling had its beginnings at the end of Word War II. Perhaps one indication was that while Petroleum was first established as a Division of the American Inst. of Mining, Metallurgical, and Petroleum Engineers (AIME) in 1922, it was not established as a Branch until 1948. The first SPE reprinted volume of the Petroleum Branch was the 1953 *Transactions of the AIME, Petroleum Development and Technology* (Vol. 198). This volume had a total of seven papers related to drilling and completion topics, a relatively small proportion of the total of 344 pages.

The first wave of scientific drilling was an era of slide rules and hand calculations. Several references give an idea of the technology level of this era; *Developments in Petroleum Engineering* by Arthur Lubinski (1987) provides a good overview of the mechanical engineering aspects of drilling, while W.F. Rogers' *Composition and Properties of Oil Well Drilling Fluids* (first edition) gives a picture of wellbore hydraulics in 1948. The technology of this era consisted of relatively simple but effective models of very complex phenomena. Former SPE President Claude Hocott once said that any calculation that could not be summarized on a note card would not be useful, and for that era, he was correct. Today, it is difficult to appreciate the tedium of evaluating these simple formulas with a slide rule.

The next wave of scientific drilling introduced a new computational tool—the electronic computer—beginning in the 1970s. Young engineers, who had used primitive computers as part of their university education, were now ready to break Hocott's one-card rule and delve into the complexity of the phenomena of drilling. As an example of the explosion of knowledge, consider the 1980 *Transactions of the Society of Petroleum Engineers* (Vol. 269) (note the name change!). The size of the volume nearly doubled to 629 pages, and the number of drilling- and completion-related papers increased 10-fold. To get a feel for the technology level of this era, the textbook *Applied Drilling Engineering* by A.T. Bourgoyne *et al.* gives a good overview of the state of the art in 1984.

We are now beginning a third wave of scientific drilling. The days of novel computer application are reaching their twilight years, and a period of evaluation and consolidation is beginning. Computer science and numerical analysis are at a much higher level of accuracy and sophistication today than they were in the 1970s, and many of the technology developments of that era could be re-examined in light of modern techniques. Further, we all recognize that the computer can do far more than just execute numerical calculations. While we can anticipate some of these developments, I suspect we will find that we did not dream big enough.

In an effort like this, the editor quickly recognizes the multidisciplinary nature of the modern drilling process, and his own inadequacies in most of these disciplines. I would like to thank the efforts of this volume's authors, going above and beyond the call of duty with little financial reward but with, perhaps, a little fame for being the first. I also would like to thank James Bobo for getting this effort off the ground and providing guidance early in the process. I also thank the Editor-in-Chief, Larry W. Lake, for offering me this task and not allowing me to say no.

—*Robert F. Mitchell*

Contents

Foreword v

Preface vi

1   **Geomechanics Applied to Drilling Engineering** II-1
Dan Moos

2   **Drilling Fluids** II-89
Gary West, John Hall, and *Simon Seaton*

3   **Fluid Mechanics for Drilling** II-119
R.F. Mitchell and *Kris Ravi*

4   **Well Control: Procedures and Principles** II-185
Neal Adams

5   **Introduction to Roller-Cone and Polycrystalline Diamond Drill Bits** II-221
W.H. Wamsley, Jr. and *Robert Ford*

6   **Directional Drilling** II-265
David Chen

7   **Casing Design** II-287
R.F. Mitchell

8   **Introduction to Wellhead Systems** II-343
Mike Speer

9   **Cementing** II-369
Ron Crook

10   **Drilling Problems and Solutions** II-433
J.J. Azar

11   **Introduction to Well Planning** II-455
Neal Adams

12   **Underbalanced Drilling** II-519
Steve Nas

13   **Emerging Drilling Technologies** II-571
Roy C. Long

14 **Offshore Drilling Units** **II-589**
Mark A. Childers

15 **Drilling-Data Acquisition** **II-647**
Iain Dowell, Andrew Mills, Marcus Ridgway, and *Matt Lora*

16 **Coiled-Tubing Well Intervention and Drilling**

Operations **II-687**
Alex Sas-Jaworsky, II, Curtis Blount, and *Steve M. Tipton*

Author Index **II-743**

Subject Index **II-745**

Chapter 1
Geomechanics Applied to Drilling Engineering
Dan Moos, GeoMechanics Intl. Inc.

1.1 Introduction
In the early years of oil drilling and production, wells were primarily drilled on land to moderate depths and with relatively minor horizontal offsets, and an empirical understanding of the impact of geological forces and Earth material properties on required drilling practice was developed by region. Successful practices were defined by trial and (sometimes costly and spectacular) error. Once local conditions were understood, it then became possible to drill new wells with a sufficient degree of confidence to guarantee the safety and economic success of further field developments. However, techniques that were successful in one field were not necessarily successful in other fields, and, therefore, the trial-and-error learning process often had to be repeated.

Because wells have become more expensive and complex, both in terms of well geometry (reach and length) and access to deep, high-temperature, high-pore-pressure, and high-stress regimes, it has become clear that the economic success of field developments can only be assured if geology and tectonics are understood and field activities are designed with that understanding. Furthermore, constraints on engineering practice based on environmental and societal requirements necessitate specially designed mud formulations and drilling techniques. Development and application of these solutions depends critically not only on an understanding of the processes that act within the Earth, but also of the impact of these processes on drilling practice. The study of these processes, of the interactions between them and their effect on Earth materials is called geomechanics.

This section of the *Handbook* is devoted to geomechanics as applied to drilling engineering. As such, it discusses the geological and tectonic effects that can impact the design and successful completion of oil and gas and geothermal wells, and it introduces methods and techniques to characterize those processes and to make recommendations to mitigate their effects. First, we briefly review the concepts of stress and strain, pore pressure, and effective stress. We continue with a brief overview of tectonics and of the origins of forces within the Earth. The purpose is not to cover these subjects exhaustively, but rather to acquaint the reader with this subject sufficiently to understand what follows and to be informed in discussions with geomechanics experts. Then, we discuss the physical properties of Earth materials, including rock strength. With this background, we then focus on issues related to drilling, starting with the impact of far-field stresses on local conditions around the wellbore.

Stresses in 3D

Tensor Transformation
(axes rotation)

Principal Stress Tensor

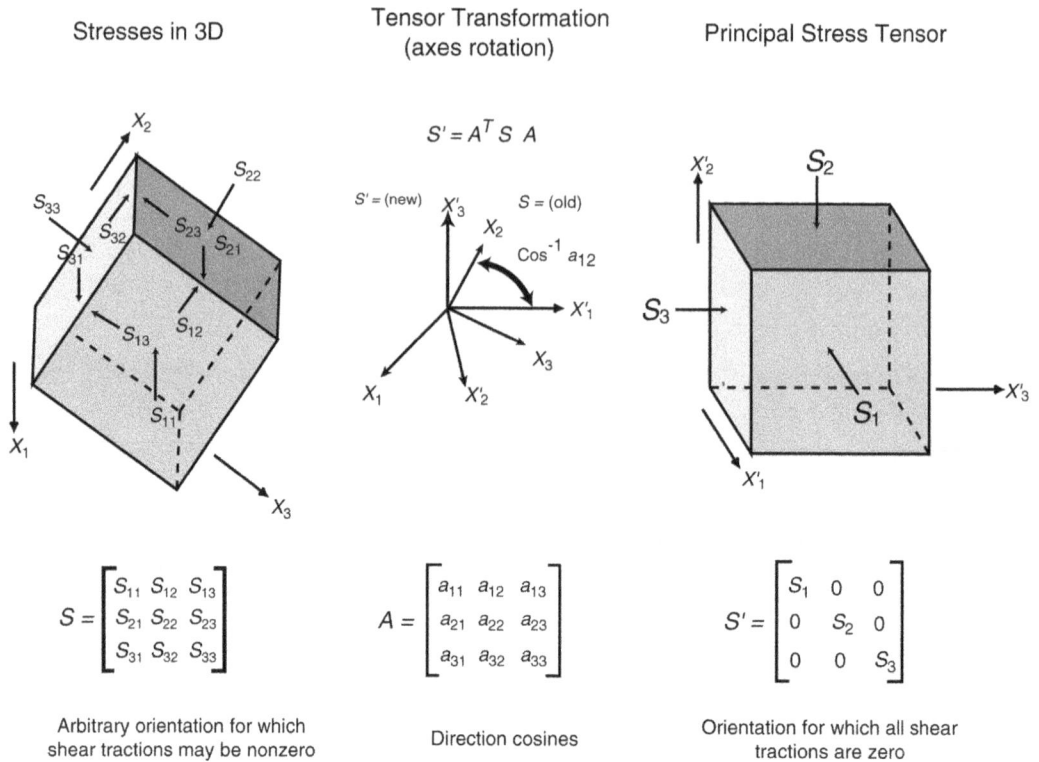

$$S' = A^T S A$$

$$S = \begin{bmatrix} S_{11} & S_{12} & S_{13} \\ S_{21} & S_{22} & S_{23} \\ S_{31} & S_{32} & S_{33} \end{bmatrix}$$

$$A = \begin{bmatrix} a_{11} & a_{12} & a_{13} \\ a_{21} & a_{22} & a_{23} \\ a_{31} & a_{32} & a_{33} \end{bmatrix}$$

$$S' = \begin{bmatrix} S_1 & 0 & 0 \\ 0 & S_2 & 0 \\ 0 & 0 & S_3 \end{bmatrix}$$

Arbitrary orientation for which
shear tractions may be nonzero

Direction cosines

Orientation for which all shear
tractions are zero

Fig. 1.1a—Definitions of the stress tensor in Cartesian coordinates, tensor transformation through direction cosines, the principal stress axes (courtesy GeoMechanics Intl. Inc.).

With this as an introduction, we then define the parameters that are required to develop the geomechanical model of a field and review the various techniques with which they can be measured or constrained. Once the geomechanical model has been developed, it can then be used in well design as part of an integrated process to minimize cost and maximize safety.

1.2 Stress, Pore Pressure, and Effective Stress

1.2.1 Definitions and Tectonic Stresses. Forces in the Earth are quantified by means of a stress tensor, in which the individual components are tractions (with dimensions of force per unit area) acting perpendicular or parallel to three planes that are in turn orthogonal to each other. The normals to the three orthogonal planes define a Cartesian coordinate system (x_1, x_2, and x_3). The stress tensor has nine components, each of which has an orientation and a magnitude (see **Fig. 1.1a**). Three of these components are normal stresses, in which the force is applied perpendicular to the plane (e.g., S_{11} is the stress component acting normal to a plane perpendicular to the x_1-axis); the other six are shear stresses, in which the force is applied along the plane in a particular direction (e.g., S_{12} is the force acting in the x_2-direction along a plane perpendicular to the x_1-axis). In all cases, $S_{ij} = S_{ji}$, which reduces the number of independent stress components to six.

At each point there is a particular stress axes orientation for which all shear stress components are zero, the directions of which are referred to as the "principal stress directions." The stresses acting along the principal stress axes are called principal stresses. The magnitudes of the principal stresses are S_1, S_2, and S_3, corresponding to the greatest principal stress, the inter-

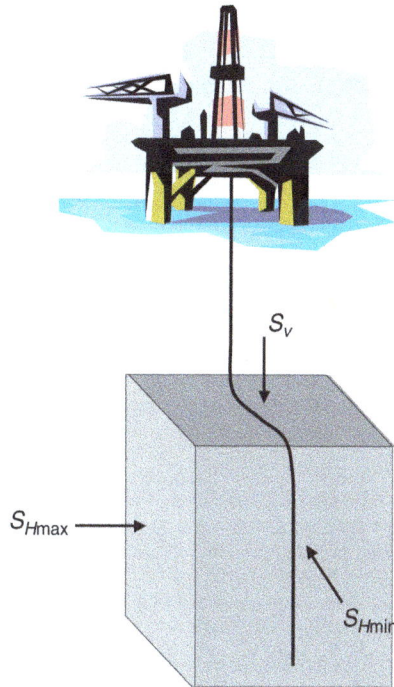

Fig. 1.1b—The vertical (S_v) and horizontal maximum and minimum stresses (S_{Hmax} and S_{Hmin}), which are usually, but need not be, principal stresses (courtesy GeoMechanics Intl. Inc.).

mediate principal stress, and the least principal stress, respectively. Coordinate transformations between the principal stress tensor and any other arbitrarily oriented stress tensor are accomplished through tensor rotations.

It has been found in most parts of the world, at depths within reach of the drill bit, that the stress acting vertically on a horizontal plane (defined as the vertical stress, S_v) is a principal stress. This requires that the other two principal stresses act in a horizontal direction. Because these horizontal stresses almost always have different magnitudes, they are referred to as the greatest horizontal stress, S_{Hmax}, and the least horizontal stress, S_{Hmin} (**Fig. 1.1b**).

The processes that contribute to the in-situ stress field primarily include plate tectonic driving forces and gravitational loading (see **Table 1.1**). Plate driving forces cause the motions of the lithospheric plates that form the crust of the Earth. Gravitational loading forces include topographic loads and loads owing to lateral density contrasts and lithospheric buoyancy. These are modified by the locally-acting effects of processes such as volcanism, earthquakes (fault slip), and salt diapirism. Human activities such as mining and fluid extraction or injection can also cause local stress changes. Because the largest components of the stress field (gravitational loading and plate driving stresses) act over large areas, stress orientations and magnitudes in the crust are remarkably uniform (**Fig. 1.2**). However, local perturbations, both natural and man-made, are important to consider for application of geomechanical analyses to drilling and reservoir engineering (**Fig. 1.3**).[1]

1.2.2 Relative Magnitudes of the Principal Stresses in the Earth. The vertical stress can be the greatest, the intermediate, or the least principal stress. In 1924, Anderson[2] developed a classification scheme to describe these three possibilities, based on the type of faulting that would occur in each case (**Table 1.2** and **Fig. 1.4**). A normal faulting regime is one in which the vertical stress is the greatest stress. When the vertical stress is the intermediate stress, a strike-

TABLE 1.1—SOURCES OF STRESS IN THE EARTH	
Plate-Driving Forces	Plate driving forces have constant orientations over wide areas. They are caused by a variety of effects, including ridge push from midocean ridges, slab pull where plates are being subducted, collision resistance forces at converging plate margins such as in Trinidad or the Himalayas, forces along transform faults where plates are moving laterally past each other such as the San Andreas fault in California, and suction above subduction zones such as the one NE of Australia.
Topographic Loads	Topographic loads can be because of large mountain chains such as the Canadian Rockies or the Himalayas or from addition or removal of loads because of ice sheets or changes in sea level. This category includes gravitational loads such as those associated with sedimentation within basins and downslope extensional loads within active depositional sequences.
Lithospheric Buoyancy	Because the lithosphere is less dense than the underlying asthenosphere, it "floats" on the underlying material, and sediment loading and lateral changes in lithospheric thickness or density cause bending forces to develop.
Flexural Forces	These are generated because of localized topographic loads and the forces acting on downgoing slabs in subduction zones.
Active Processes	Earthquakes (slip-on faults), active volcanism, and salt diapirism are all examples of processes that act to change local stresses.

slip regime is indicated. If the vertical stress is the least stress the regime is defined to be reverse. The horizontal stresses at a given depth will be smallest in a normal faulting regime, larger in a strike-slip regime, and greatest in a reverse faulting regime. In general, vertical wells will be progressively less stable as the regime changes from normal to strike-slip to reverse, and consequently will require higher mud weights to drill.

1.2.3 Pore Pressure. Pore pressure is the pressure at which the fluid contained within the pore space of a rock is maintained at depth. In the absence of any other processes, the pore pressure is simply equal to the weight of the overlying fluid, in the same way that the total vertical stress is equal to the weight of the overlying fluid and rock (**Fig. 1.5**). This pressure is often referred to as the "hydrostatic pressure." A number of processes can cause the pore pressure to be different from hydrostatic pressure. Processes that increase pore pressure include undercompaction caused by rapid burial of low-permeability sediments, lateral compression, release of water from clay minerals caused by heating and compression, expansion of fluids because of heating, fluid density contrasts (centroid and buoyancy effects), and fluid injection (e.g., waterflooding). Processes that decrease pore pressure include fluid shrinkage, unloading, rock dilation, and reservoir depletion.

Because pore pressure and horizontal stresses are interrelated, changes in pore pressure also cause similar changes in stress. While the exact relationship depends on the properties of the reservoir, it is reasonable to assume that the change in horizontal stress is approximately two-thirds of the change in pore pressure (see Eq. 1.1 and **Fig. 1.6**). This leads to a considerable reduction in leakoff pressure in a depleted reservoir and an increase in horizontal stress where pore pressure increases.

$$\Delta S_{H\max} = \Delta S_{H\min} = \alpha(1-2v)/(1-v)\Delta P_p,$$

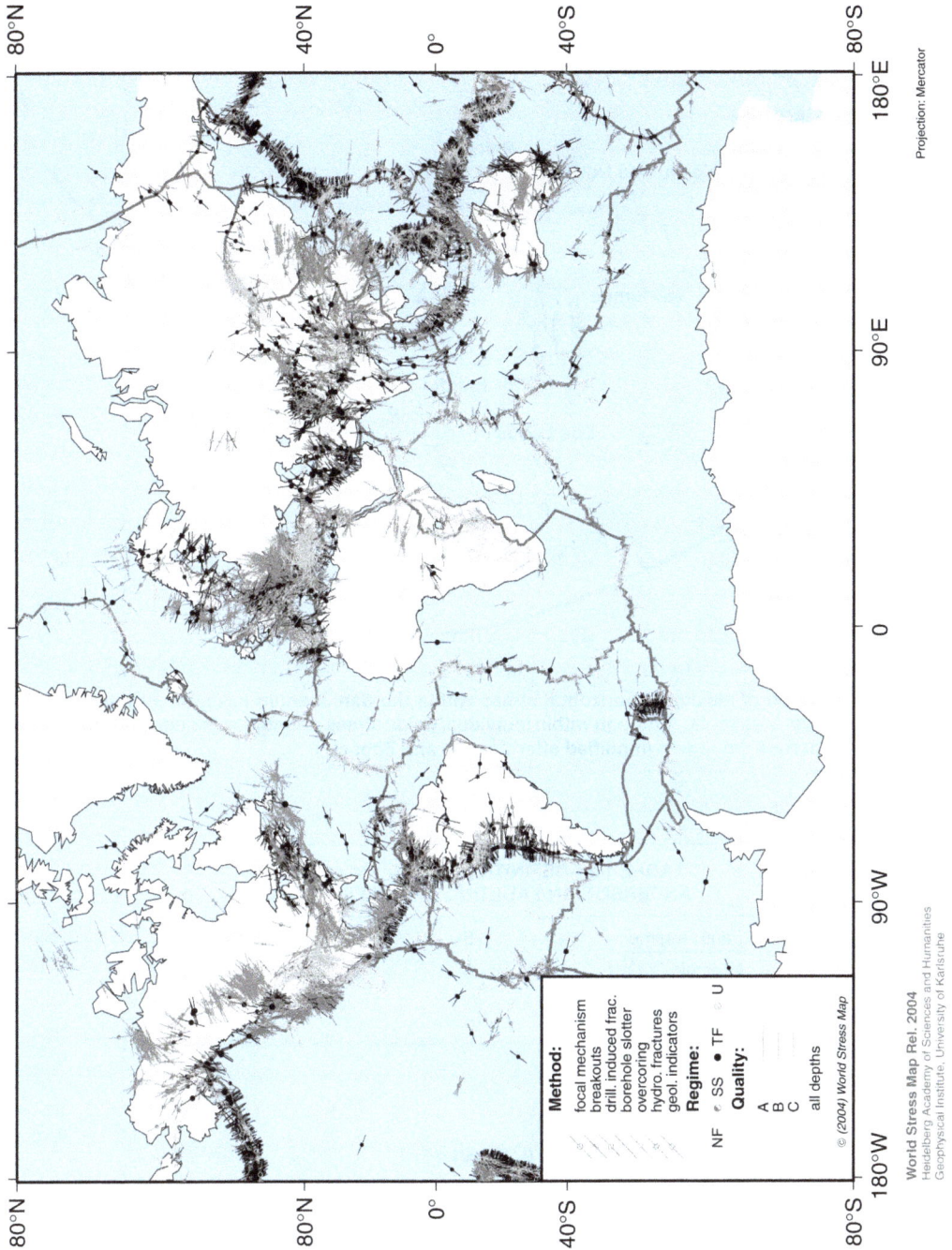

Fig. 1.2—World stress map showing orientations of the greatest horizontal stress, S_{Hmax}, where it has been measured using wellbore breakouts or inferred from earthquakes. Also shown are the boundaries of the major tectonic plates. Colors of the symbols indicate the relative stress magnitudes (light gray, normal; gray, strike-slip; black, reverse or unknown). This figure was produced using software and data available from the World Stress Map Project website.

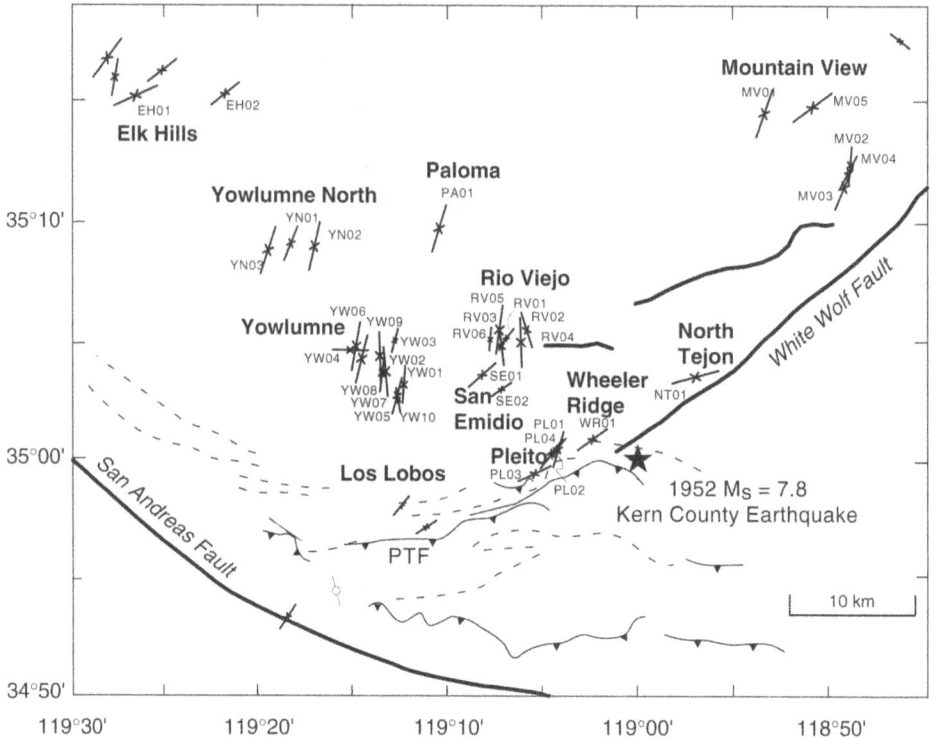

Fig. 1.3—Orientation of maximum horizontal stress within the San Joaquin basin of south-central California derived from breakouts. Although within individual fields stress orientation is quite uniform, it varies systematically across the region (modified after Castillo and Zoback[1]).

TABLE 1.2—DEFINITIONS OF S_1 AND S_3 FOR ANDERSONIAN FAULTING CLASSIFICATIONS		
Fault Regime	S_1	S_3
Normal	S_v	S_{Hmin}
Strike-slip	S_{Hmax}	S_{Hmin}
Reverse	S_{Hmax}	S_v

if

$$\nu = 0.25 \text{ and } \alpha = 1,$$

and

$$\Delta S_{H\max} = \Delta S_{H\min} = 2/3 \Delta P_p , \quad\text{...} (1.1)$$

where ν is Poisson's ratio, and α ($= 1 - K_{dry}/K_{grain}$) is the Biot poroelastic coefficient, which varies between zero for a rock that is as stiff as the minerals of which it is composed and one for most sediments, which are much softer than their mineral components. It is important to note that Eq. 1.1 cannot be used to calculate the relationship between pore pressure and stress in the Earth that develops over geological time because in that case the assumptions used to derive the equation are not valid.

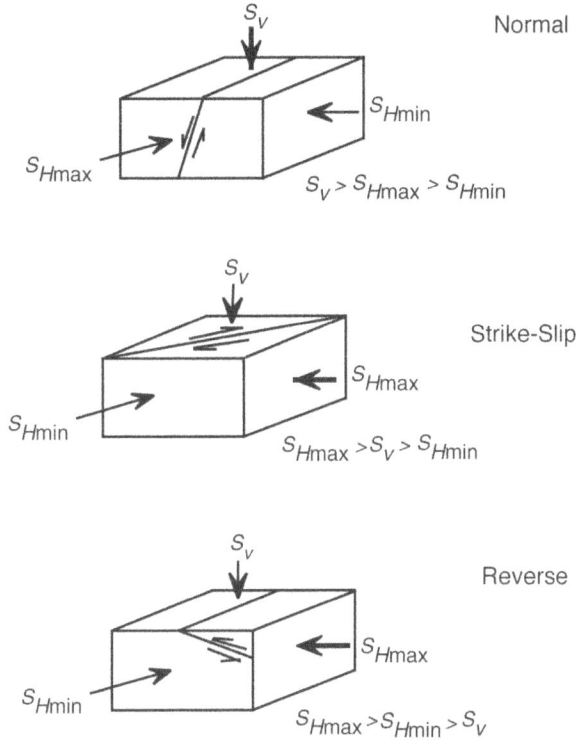

Fig. 1.4—Diagram illustrating the three faulting states based on Andersonian[2] faulting theory (courtesy GeoMechanics Intl. Inc.).

1.2.4 Effective Stress. The mathematical relationship between stress and pore pressure is defined in terms of effective stress. Implicitly, the effective stress is that portion of the external load of total stress that is carried by the rock itself. The concept was first applied to the behavior of soils subjected to both externally applied stresses and pore pressure acting within the pore volume in a 1924 paper by Terzaghi[3] as

$$\sigma_{ij} = S_{ij} - \delta_{ij} P_p, \quad\quad\quad\quad\quad\quad\quad\quad\quad\quad\quad\quad\quad (1.2)$$

where σ_{ij} is the effective stress, P_p is the pore pressure, δ_{ij} is the Kronecker delta ($\delta_{ij} = 1$, if $i = j$, $\delta_{ij} = 0$ otherwise), and S_{ij} represents the total stresses, which are defined without reference to pore pressure. While it is sometimes necessary to use a more exact effective stress law in rock ($\sigma_{ij} = S_{ij} - \delta_{ij} \, \alpha \, P_p$, where α is Biot's coefficient and varies between 0 and 1), in most reservoirs it is generally sufficient simply to assume that $\alpha = 1$. This reduces the effective stress law to its original form (Eq. 1.2). When expanded, the Terzaghi effective stress law becomes

$$\sigma_1 = S_1 - P_p,$$
$$\sigma_2 = S_2 - P_p,$$

and

$$\sigma_3 = S_3 - P_p. \quad\quad\quad\quad\quad\quad\quad\quad\quad\quad\quad\quad\quad (1.3)$$

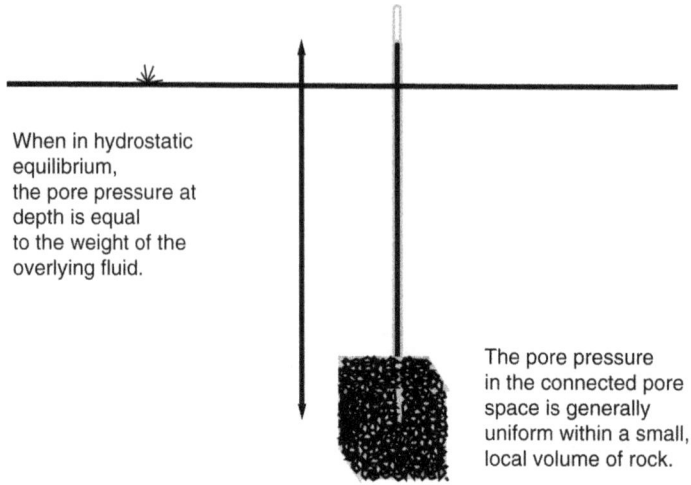

When in hydrostatic
equilibrium,
the pore pressure at
depth is equal
to the weight of the
overlying fluid.

The pore pressure
in the connected pore
space is generally
uniform within a small,
local volume of rock.

Fig. 1.5—Illustration of pore pressure in permeable rock under hydrostatic pressure (courtesy Geo-Mechanics Intl. Inc.).

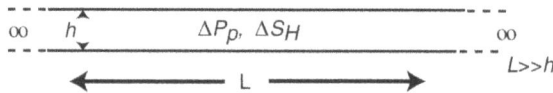

∞ h $\Delta P_p, \Delta S_H$ ∞

|← ——————— L ———————— →| $L >> h$

Fig. 1.6—In a laterally infinite reservoir where L>>h, the relationship between a change in pore pressure and the resulting change in stress is defined in Eq. 1.1 (courtesy GeoMechanics Intl. Inc.).

The concept of effective stress is important because it is well known from extensive laboratory experiments (and from theory) that properties such as velocity, porosity, density, resistivity, and strength are all functions of effective stress. Because these properties vary with effective stress, it is therefore possible to determine the effective stress from measurements of physical properties such as velocity or resistivity. This is the basis for most pore-pressure-prediction algorithms. At the same time, effective stress governs the frictional strength of faults and the permeability of fractures.

1.2.5 Constraints on Stress Magnitudes. If rock were infinitely strong and contained no flaws, stresses in the crust could, in theory, achieve any value. However, faults and fractures exist at all scales, and these will slip if the stress difference gets too large. Even intact rock is limited in its ability to sustain stress differences. It is possible to take advantage of these limits when defining a geomechanical model for a field when other data are not available.

Stress Constraints Owing to Frictional Strength. One concept that is very useful in considering stress magnitudes at depth is frictional strength of the crust and the correlative observation that, in many areas of the world, the state of stress in the crust is in equilibrium with its frictional strength. Because the Earth's crust contains widely distributed faults, fractures, and planar discontinuities at many different scales and orientations, stress magnitudes at depth (specifically, the differences in magnitude between the maximum and minimum principal effective stresses) are limited by the frictional strength of these planar discontinuities. This concept is schematically illustrated in **Figs. 1.7a and 1.7b.** In the upper part of the figure, a series of randomly oriented fractures and faults is shown. Because this is a two-dimensional (2D) illustration (for simplicity), it is easiest to consider this sketch as a map view of vertical strike-slip faults. In this case, it is the difference between σ_{Hmax} ($S_{Hmax} - P_p$) and σ_{Hmin} ($S_{Hmin} - P_p$) that is

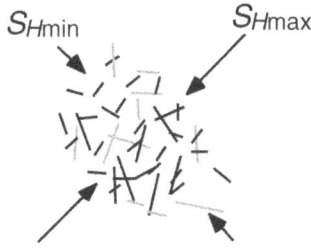

Fig. 1.7a—Map view of theoretical faults and fractures. The fractures and faults shown in gray are optimally oriented to slip in the current stress field (courtesy GeoMechanics Intl. Inc.).

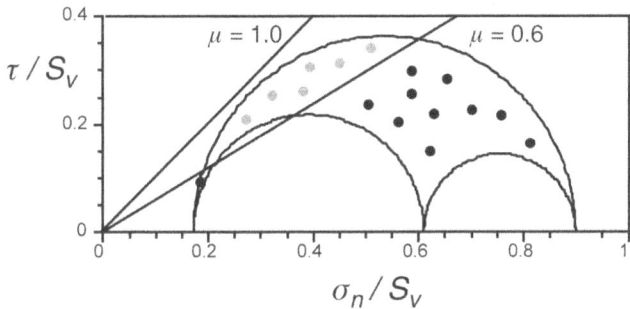

Fig. 1.7b—Mohr diagram showing poles to the critically stressed faults and fractures for $0.6 < \mu < 1.0$ (courtesy GeoMechanics Intl. Inc.).

limited by the frictional strength of these pre-existing faults. In other words, as σ_{Hmax} increases with respect to σ_{Hmin}, a subset of these pre-existing faults (shown in light gray) begins to slip as soon as its frictional strength is exceeded. Once that happens, further stress increases are not possible, and this subset of faults becomes critically stressed (i.e., just on the verge of slipping). The lower part of the figure illustrates using a three-dimensional (3D) Mohr diagram, the equivalent 3D case.

The frictional strength of faults can be described in terms of the Coulomb criterion, which states that faults will slip if the ratio of shear to effective normal stress exceeds the coefficient of sliding friction (i.e., $\tau/\sigma_n = \mu$); see **Fig. 1.8**. Because for essentially all rocks (except some shales) $0.6 < \mu < 1.0$, it is straightforward to compute limiting values of effective stresses using the frictional strength criterion.

This is graphically illustrated using a 3D Mohr diagram as shown in the lower part of Fig. 1.7. 2D Mohr diagrams plot normal stress along the x-axis and shear stress along the y-axis. Any stress state is represented by a half circle that intersects the x-axis at $\sigma = \sigma_3$ and $\sigma = \sigma_1$ and has a radius equal to $(\sigma_1 - \sigma_3)/2$. A 3D Mohr diagram plots three half circles the endpoints of which lie at values equal to the principal stresses and the radii of which are equal to the principal stress differences divided by 2. Planes of any orientation plot within and along the edges of the region between the circles at a position corresponding to the values of the shear and normal stresses resolved on the planes. Planes that contain the σ_2 plot along the largest circle are first to reach a critical equilibrium.

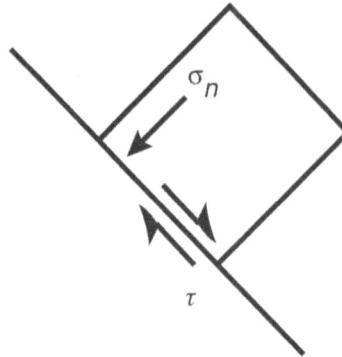

Fig. 1.8—Sliding on faults is limited by the ratio of the shear stress (τ) to the effective normal stress (σ_n) on the fault plane, as defined by the Coulomb criterion: $\tau_n = \mu$, where μ is the coefficient of sliding friction (courtesy GeoMechanics Intl. Inc.).

The critically stressed (light gray) faults in the upper part of the figure correspond to the points (also shown in light gray) in the Mohr diagram, which have ratios of shear to effective normal stress between 0.6 and 1.0. It is clear in the Mohr diagram that for a given value of σ_{Hmin}, there is a maximum value of σ_{Hmax} established by the frictional strength of pre-existing faults (the Mohr circle cannot extend past the line defined by the maximum frictional strength). The values of S_1 and S_3 corresponding to the situation illustrated in Fig. 1.7 are defined by

$$\sigma_1 / \sigma_3 = (S_1 - P_p) / (S_3 - P_p) = [(\mu^2 + 1)^{1/2} + \mu]^2 = f(\mu) \dots\dots\dots\dots\dots\dots (1.4)$$

That is, it is the effective normal stress on the fault (the total stress minus the pore pressure) that limits the magnitude of the shear stress. Numerous in-situ stress measurements have demonstrated that the crust is in frictional equilibrium in many locations around the world (**Fig. 1.9**).[4] This being the case, if one wished to predict stress differences in-situ with Eq. 1.4, one would use Anderson's faulting theory to determine which principal stress (i.e., S_{Hmax}, S_{Hmin}, or S_v) corresponds to S_1 or S_3, depending of course on whether it is a normal, strike-slip, or reverse-faulting environment, and then utilize appropriate values for S_v and P_p (the situation is more complex in strike-slip areas because S_v corresponds to neither S_1 nor S_3). Regardless of whether the state of stress in a given sedimentary basin reflects the frictional strength of pre-existing faults, the importance of the concept illustrated in Fig. 1.7 is that at any given depth and pore pressure, once we have determined the magnitude of the least principal effective stress using minifracs or leakoff tests (σ_{Hmin} in a normal or strike-slip faulting case), there is only a finite range of values that are physically possible for σ_{Hmax}.

Eq. 1.4 defines the upper limit of the ratio of effective maximum to effective minimum in-situ stress that is possible before triggering slip on a pre-existing, well-oriented fault. The in-situ effective stress ratio can never be larger than this limiting ratio. Therefore, all possible stress states must obey the relationship that the effective stress ratios must lie between 1 and the limit defined by fault slip as shown in Eq. 1.5.

$$1 \le \sigma_1 / \sigma_2 \le \sigma_1 / \sigma_3 \le [(\mu^2 + 1)^{1/2} + \mu]^2 = f(\mu). \dots\dots\dots\dots\dots\dots (1.5)$$

These equations can be used along with the Andersonian definitions of the different faulting regimes (Table 1.1) to derive a stress polygon, as shown in **Fig. 1.10**. These figures are constructed as plots at a single depth of S_{Hmax} vs. S_{Hmin}. The shaded region is the range of

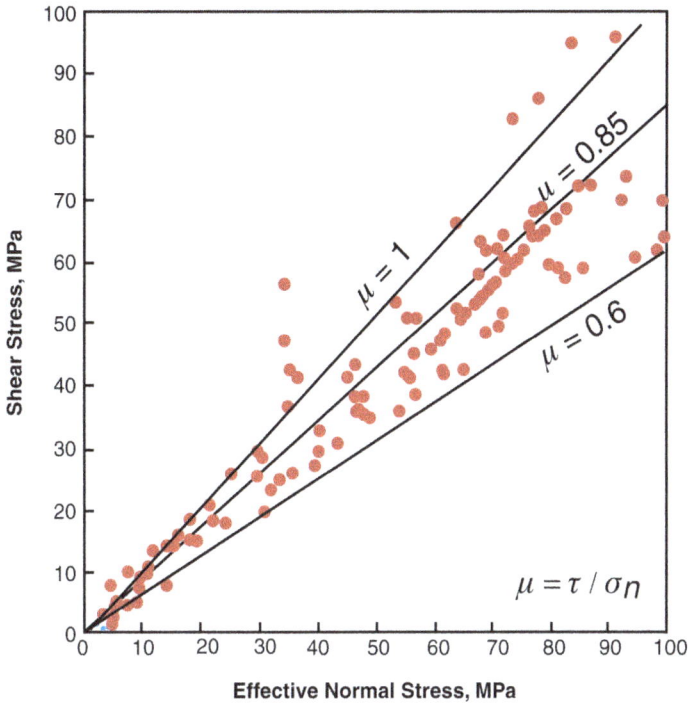

Fig. 1.9—Stress measurements made in brittle rock (dots) reveal that in most of the world, the crust is in a state of frictional equilibrium for fault slip for coefficients of sliding friction between 0.6 and 1.0 as measured in the laboratory (modified after Townend and Zoback[4]).

allowable values of these stresses. By the definitions of S_{Hmax} and S_{Hmin}, the allowable stresses lie above the line for which $S_{Hmax} = S_{Hmin}$. Along with the pore pressure, S_v, shown as the black dot on the $S_{Hmax} = S_{Hmin}$ line, defines the upper limit of S_{Hmax} [the horizontal line at the top of the polygon, for which $\sigma_{Hmax}/\sigma_v = f(\mu)$], and the lower limit of S_{Hmin} [the vertical line on the lower left of the polygon, for which $\sigma_v/\sigma_{Hmin} = f(\mu)$]. The third region is constrained by the difference in the horizontal stress magnitudes [i.e., $\sigma_{Hmax}/\sigma_{Hmin} < f(\mu)$]. The larger the magnitude of S_v, the larger the range of possible stress values; however, as the pore pressure increases, the polygon shrinks, until at the limit when $P_p = S_v$, all three stresses are equal.

It is important to emphasize that the stress limit defined by frictional faulting theory is just that—a limit—and provides a constraint only. The stress state can be anywhere within and along the boundary of the stress polygon. As discussed at length later, the techniques used for quantifying in-situ stress magnitudes are not model based, but instead depend on measurements, calculations, and direct observations of wellbore failure in already-drilled wells in the region of interest. These techniques have proved to be sufficiently robust that they can be used to make accurate predictions of wellbore failure (and determination of the steps needed to prevent failure) with a reasonable degree of confidence.

Stress Constraints Owing to Shear-Enhanced Compaction. In weak, young sediments, compaction begins to occur before the stress difference is large enough to reach frictional equilibrium. Therefore, rather than being at the limit constrained by the frictional strength of faults, the stresses will be in equilibrium with the compaction state of the material. Specifically, the porosity and stress state will be in equilibrium and lie along a compactional end cap. The physics of this process is discussed in the section on rock properties of this chapter.

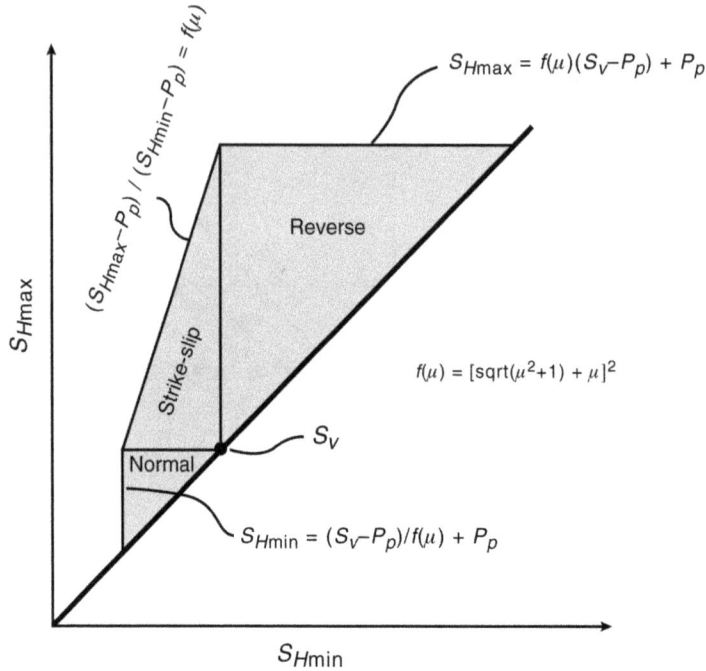

Fig. 1.10—This figure shows construction of the polygon that limits the range of allowable stress magnitudes in the Earth's crust at a fixed depth and corresponding magnitude of S_v). It is a plot of S_{Hmax} vs. S_{Hmin} as constrained by the strength of well-oriented, pre-existing faults. The limits are constrained by Eq. 1.4, with S_1 and S_3 defined by Andersonian faulting theory, as shown in Table 1.2 (courtesy GeoMechanics Intl. Inc.).

Constraints, based on compaction, define another stress polygon similar to the one shown in Fig. 1.10. It is likely that in regions such as the Gulf of Mexico, and in younger sediments worldwide where compaction is the predominant mode of deformation, this is the current in-situ condition. Unfortunately, while end-cap compaction has been studied in the laboratory for biaxial stress states ($\sigma_1 > \sigma_2 \cong \sigma_3$), there has been little laboratory work using polyaxial stresses ($\sigma_1 \neq \sigma_2 \neq \sigma_3$), and there have been relatively few published attempts to make stress predictions using end-cap models. Also, it is important to apply end-cap analyses only where materials lie along a compaction curve, and not to apply these models to overcompacted or diagenetically modified rocks. If the material lies anywhere inside the region bounded by its porosity-controlled end cap, this constraint can be used only to provide a limit on stress differences.

1.3 Rock Properties

1.3.1 Deformation of Rocks—Elasticity. To first order, most rocks obey the laws of linear elasticity. That is, for small strains, the elements of the stress and strain tensors are related through

$$\sigma_{ij} = M_{ijkl}\varepsilon_{kl},$$

where

$$\varepsilon_{kl} = \Delta l_k / l_l. \quad\quad\quad\quad\quad\quad\quad\quad\quad\quad\quad\quad (1.6)$$

In other words, the stress required to cause a given strain, or normalized length change ($\Delta l_k/l_l$), is linearly related to the magnitude of the deformation and proportional to the stiffnesses (or moduli), M_{ijkl}. Furthermore, the strain response occurs instantaneously as soon as the stress is applied, and it is reversible—that is, after removal of a load, the material will be in the same state as it was before the load was applied. A plot of stress vs. strain for a laboratory experiment conducted on rock that obeys such a law is a straight line with slope equal to the modulus. However, in real rocks, the moduli increase as a function of effective stress, particularly at low stress. Some of this increase is reversible (nonlinear elasticity), but some of it is irreversible (plasticity or end-cap compaction). To make matters even more complicated, rocks also exhibit time-dependent behavior, so that an instantaneous stress change elicits both an instantaneous and a time-dependent response. These anelastic effects can be seen in laboratory experiments, as shown in **Fig. 1.11**.[5]

At the top of Fig. 1.11 is shown the stress as a function of time applied in the laboratory to two samples of an upper Miocene turbidite. As in most experiments of this type, a cylindrical rock sample is jacketed with an impermeable soft sleeve and placed in a fluid-filled pressure cell. The fluid pressure surrounding the sample is increased slowly, and the fluid pressure (confining stress) and sample axial and circumferential strains are measured. To identify the various deformation processes that occur in this unconsolidated sand, the stress is slowly increased at a constant rate and then held constant until the sample stops deforming. Then the pressure is decreased to approximately half of the previous maximum pressure. After that, the pressure is increased at the same rate until the next pressure step is completed. This process continues until the desired maximum pressure is achieved, and then the sample is slowly unloaded and removed from the pressure cell.

All aspects of typical rock behavior can be seen in the stress-strain curve plotted on the bottom of Fig. 1.11. At low pressure, the sample is soft, and there is a rapid increase of stiffness with pressure (nonlinear elasticity) owing to crack closure, as well as an increase in stiffness caused by irreversible compaction. Once the pressure increase stops, the sample continues to deform, with deformation rate decreasing with time (time-dependent creep). The sample is stiffer during unloading than during loading, and during this phase of the experiment, it essentially behaves as a linear elastic material; the permanent strains during loading and creep that occurred through plastic/viscous deformation mechanisms are not recoverable. Reloading follows the (purely elastic) unloading path until the maximum previous pressure has been reached, after which additional plastic deformation begins to occur again as the material resumes following the compaction curve. All of these effects can be seen in situ, including the difference between the loading and unloading response.

Measurements of P-wave and S-wave velocity made on this sample during the experiment by measuring the time of flight of pulses transmitted axially along the sample were used to calculate the dynamic shear (G) and bulk (K) moduli with Eq. 1.7. The implications of the results for pore pressure prediction are discussed later in this chapter.

$$G = \rho V_s^2; \quad K = \rho V_p^2 - 4G/3. \dotfill (1.7)$$

The dynamic bulk modulus calculated from the velocity measurements is higher than the moduli computed from the slopes of the unloading/reloading curves, which in turn are larger than the modulus calculated from the slope of the loading curve. This dispersion (frequency-dependence of the moduli) also is typical of reservoir rocks, and it is the justification for empirical corrections applied to sonic log data to convert from the dynamic moduli measured by the sonic log to static moduli that are used to model reservoir response. However, it is important to realize that there are two different "static" moduli—a "compaction modulus," the slope of the loading curve, which includes plastic effects, and a "static elastic modulus" mea-

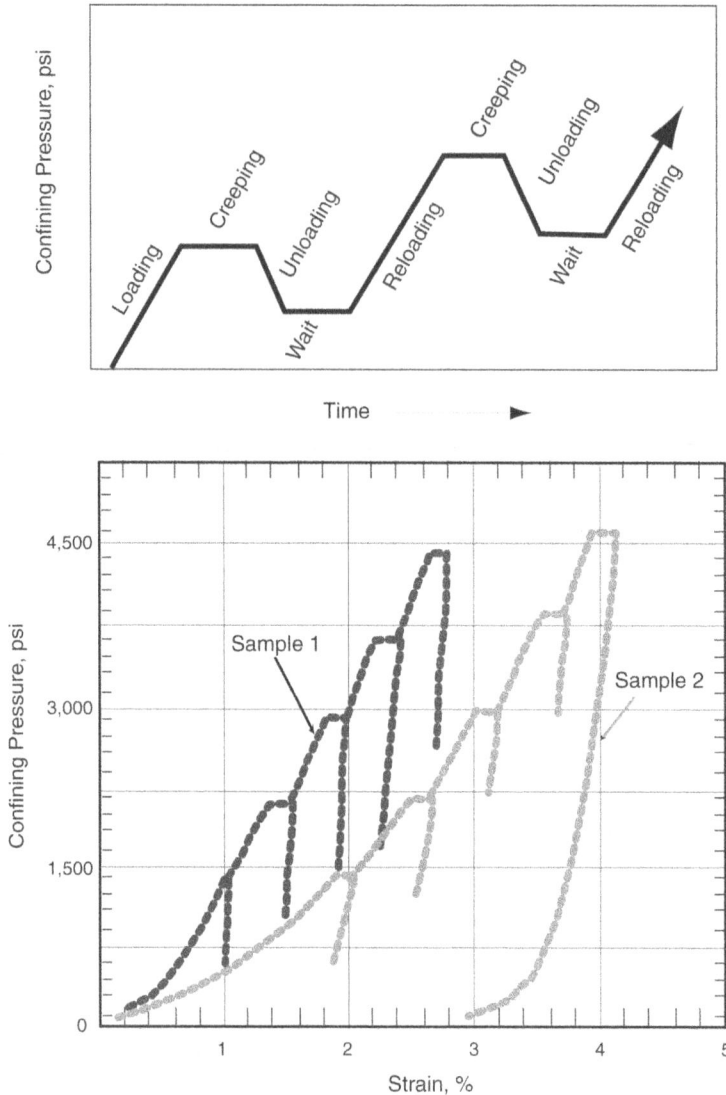

Fig. 1.11—This figure shows the loading path and the confining pressure as a function of strain recorded during compaction experiments conducted using two samples of a poorly consolidated, shaley turbiditic sand of Miocene age. Sample 1 was maintained at its saturated condition; Sample 2 was cleaned and dried before testing (modified after Moos and Chang[5]).

sured by unloading/reloading, which is truly elastic. It is critical when measuring material response in the lab to differentiate between these two and to use the appropriate one for in-situ modeling—the elastic unloading modulus when no compaction is occurring, for example when pore pressure is increased by injection during waterflooding, and the compaction modulus when modeling, for example, very large depletions in weak reservoirs.

These considerations can become very important when modeling and predicting how the wellbore will respond during and after drilling. In the discussion of wellbore stability that follows, however, we will assume that the rock is purely elastic and only briefly discuss the implications of more complicated rheological models.

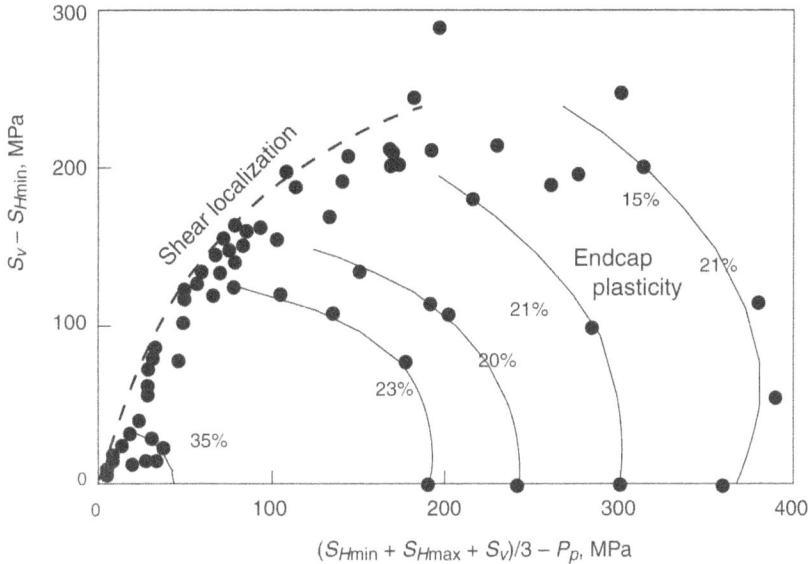

Fig. 1.12—This figure, modified from Schutjens *et al.*,[6] shows the end-cap relationship between porosity and stress for a material undergoing compaction. The *x*-axis is the mean effective stress. The *y*-axis is the difference between S_1 and S_3 (here, in a normal faulting environment, these are, respectively, S_v and S_{Hmin}). For high porosities, very little differential stress can be sustained. As compaction progresses, porosities decrease, and the rock is better able to withstand differential stress—the end-caps move to the right. The dots are laboratory data that can be used to define (1) the brittle failure line that follows a non-linear Coulomb-style failure law for shear localization, and (2) curved end caps that indicate the porosity for which the strength of the material is in equilibrium with the stress state.

1.3.2 Compaction and End-Cap Plasticity.

When rocks are loaded past a certain point, they will no longer behave elastically. If the load is approximately isotropic ($\sigma_1 \cong \sigma_2 \cong \sigma_3$), the rock will begin to compact and lose volume, primarily because of a decrease in porosity. This process is referred to as shear-enhanced compaction because, in general, the effect occurs at lower mean stress as the shear stress increases. **Fig. 1.12**[6] shows a plot of the shear stress as a function of mean stress for a variety of rocks, labeled for use in a normal faulting regime where S_v is S_1. Compaction trends are shown as arcs bounding the data from the right, and they define end caps of the stress regime within which the rock at a given porosity can exist. Values of porosity decrease as the end caps move outward, owing to material compaction that is caused by the increase in confining stress. The shapes of the end caps for any porosity depend on the form of the relationship between the mean stress at the compaction limit and the shear stress. In many studies, the shape of the end-cap is assumed to be elliptical. At any point along an end cap, the porosity is in equilibrium with the state of effective stress.

In unconsolidated materials, shear-enhanced compaction begins at zero confining stress as soon as the material begins to be loaded (see Fig. 1.11). In situ, this compaction is the primary cause of the increase in stiffness and decrease in porosity of sediments with burial. The assumption inherent in all standard pore-pressure-prediction algorithms that rock properties are uniquely related to the effective stress is equivalent to assuming that the rock in situ lies along a compaction trend defined by an end cap.

If the mean effective stress decreases (for example, because of erosion) or the pore pressure increases, the rock becomes overcompacted. When this occurs, its porosity is no longer in equilibrium with the end cap, and it will behave elastically, as occurred during the unloading stages of the experiment shown in Fig. 1.11.

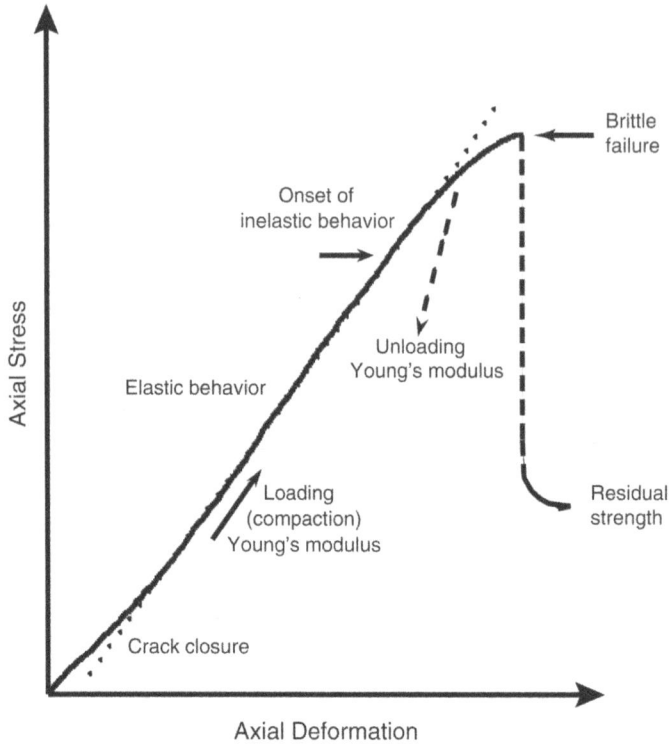

Fig. 1.13—Typical plot of axial stress vs. axial deformation during a triaxial strength test. Initially, the sample is soft, but as the axial load increases, microcracks begin to close, causing an increase in stiffness. When the axial stress is sufficiently high, inelastic behavior begins to occur. If the axial load continues to increase, the stress-strain curve will reach a maximum, followed either by catastrophic failure (as shown here) or a roll-over and continued residual strength, for which an increase in deformation can be achieved with no change in axial load (courtesy GeoMechanics Intl. Inc.).

When a differential load is applied (e.g., $\sigma_1 > \sigma_2 \cong \sigma_3$), eventually the maximum stress (σ_1) will get so large that the sample either will begin to yield through a process of distributed deformation or will fail because of shear localization and the creation of a brittle failure surface (a fault). In Fig. 1.12, the data at the left edge of the plot lie along a limit in the ratio of shear stress to mean stress that is defined by the onset of brittle failure or plastic yielding by shear localization, as discussed next.

1.3.3 Failure Models and Rock Strength. Rock strength models that define stress states at which brittle failure occurs follow stress trajectories that lie along the left edge of the data shown in Fig. 1.12. It is clear that the ability of a rock to carry differential stress increases with confining stress. To establish the exact relationship, rock strength tests are conducted at a number of confining pressures. In these tests, a jacketed, cylindrical sample is loaded into a pressure vessel, a constant confining pressure is established, and an axial load is applied by means of a hydraulic ram. The load is increased slowly by driving the ram at a constant rate, monitoring axial and circumferential strains and maintaining a constant confining pressure, until the sample fails or yields. An example of an axial stress vs. axial strain plot from a typical triaxial stress experiment is shown in **Fig. 1.13**.

One criterion to define the stress state at failure is the 2D linear Mohr-Coulomb criterion. The Mohr-Coulomb criterion defines a linear relationship between the stress difference at failure and the confining stress using two parameters: S_o, the cohesion (or C_o, the unconfined

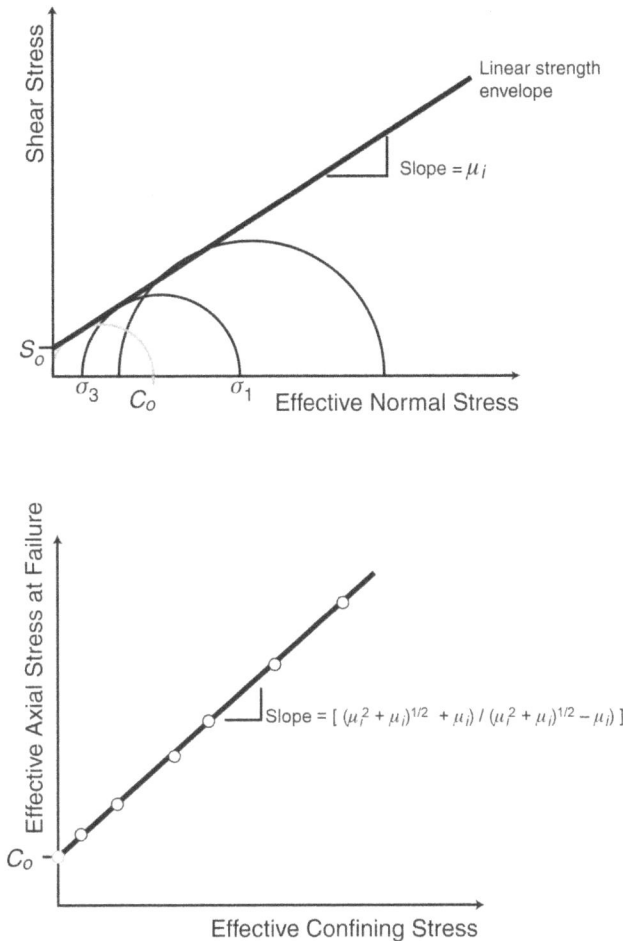

Fig. 1.14—Top is a plot of a set of Mohr circles showing the stress state at failure for a series of triaxial strength tests. The results have been fitted to a linear Mohr-Coulomb failure criterion. The lower plot shows axial load at failure vs. confining stress. S_o (or C_o) and the coefficient of internal friction, μ_i, can be derived easily from these data (courtesy GeoMechanics Intl. Inc.).

compressive strength) and Φ, the angle or μ_i, the coefficient of internal friction, where $\mu_i = \tan\Phi$. The equation that defines the 2D linear criterion is $\tau = S_o + \mu_i\sigma_n$. These parameters can be derived from triaxial strength tests on cylindrical cores by measuring the stress at failure as a function of confining pressure.

Fig. 1.14 shows graphically how the Mohr-Coulomb parameters are derived. The upper plot shows a series of Mohr circles, with x-intercepts σ_1 and σ_3 at failure and diameter $\sigma_1 - \sigma_3$, in a plot of shear stress to effective normal stress. The failure line with slope μ_i and intercept S_o that just touches each of the circles defines the parameters of the 2D linear Mohr-Coulomb strength criterion for this material. C_o is the value of σ_1 for $\sigma_3 = 0$ of the circle that just touches the failure line. The lower plot graphs σ_1 vs. σ_3 and can be used to derive C_o directly.

Some of the other strength criteria include the Hoek and Brown (HB) criterion, which, like the Mohr Coulomb criterion, is 2D and depends only on knowledge of σ_1 and σ_3, but which uses three parameters to describe a curved failure surface and, thus, can better fit Mohr envelopes than can linear criteria. The Tresca criterion is a simplified form of the linearized Mohr-Coulomb criterion in which $\mu_i = 0$. It is rarely used in rocks and is more commonly applied to metals, which have a yield point but do not strengthen with confining pressure. Other failure criteria, such as Drucker Prager (inscribed and circumscribed, both extensions of the von Mises criterion) and Weibols and Cook incorporate the dependence of rock strength on the intermediate principal stress, σ_2, but require true polyaxial rock strength measurements that have $\sigma_1 > \sigma_2 > \sigma_3$ and are difficult to carry out. The Modified Lade Criterion has considerable advantages, in that it, too, is a 3D strength criterion but requires only two empirical constants, equivalent to C_o and μ_i. Thus, it can be calibrated in the same way as the simpler 2D Mohr-Coulomb failure criterion, but because it is fully 3D, it is the preferred criterion for analysis of wellbore stability.

1.3.4 Single-Sample Testing. Because triaxial tests are so difficult and time-consuming to carry out, and because of the amount of core required and the difficulty in finding samples that are similar enough to be considered identical, it is common to attempt to reduce the number of tests requiring core preparation. One method is simply to carry out a uniaxial strength test in which the confining pressure is zero. This requires a much simpler apparatus; in fact, the sample does not even have to be jacketed, although this is recommended. By definition, the axial stress at failure in a uniaxial test is a direct measure of C_o. Unfortunately, unconfined samples can fail in a variety of ways that do not provide a good measure of C_o for use with a Mohr-Coulomb model. Furthermore, it is impossible to measure μ_i using one test unless a clearly defined failure surface is produced, the angle of which with respect to the loading axis can be measured. For these reasons, a series of triaxial tests is preferred.

An alternative method that does require testing in a triaxial cell is to carry out a series of tests on a single sample. The process proceeds by establishing a low-confining pressure and then increasing the axial stress until the sample just begins to yield. At that point, the test is stopped, the confining pressure is increased, and again the axial stress is increased until yielding occurs. In comparisons of this method against multiple triaxial tests, it is often the case that the yield stress derived from the multistage test is systematically lower, and the internal friction is also systematically lower, than the stress at failure and the internal friction derived from the triaxial tests. This is because, once the initial yielding has begun, the sample is already damaged and thus is weaker than it would be had this not occurred. However, by using this method, it may be possible to characterize the yield envelope of a plastic rock.

1.3.5 Scratch and Penetrometer Testing. A number of techniques have been developed to replace or augment triaxial tests to measure the strength properties of rocks. One such technique, which has a demonstrated ability to provide continuous, fine-scale measurements of both elastic and strength properties, is the scratch test. This test involves driving a sharp cutter across a rock surface. By monitoring the vertical and lateral forces required to maintain a certain depth of cut, it is possible to determine the uniaxial compressive strength, C_o. The Young's modulus, E, can also be estimated in some cases. **Fig. 1.15** shows a comparison of C_o derived by scratch testing to laboratory core measurements and log-derived C_o. The results are quite similar.

The advantage of scratch testing is that no special core preparation is required. This is in contrast to the extensive preparations required prior to triaxial testing. The test can be conducted either in the lab or, in principle, on the rig, almost immediately after recovery of core material. No significant damage occurs to the core, which makes this a very attractive substitute for triaxial testing when little material is available. In fact, research is now under way to evaluate the feasibility of designing a downhole tool to carry out this analysis.

In a penetrometer test, a blunt probe is pressed against the surface of a rock sample using continuously increasing pressure. The unconfined compressive strength is then computed from the pressure required to fracture the sample. As in the case of scratch testing, no special sam-

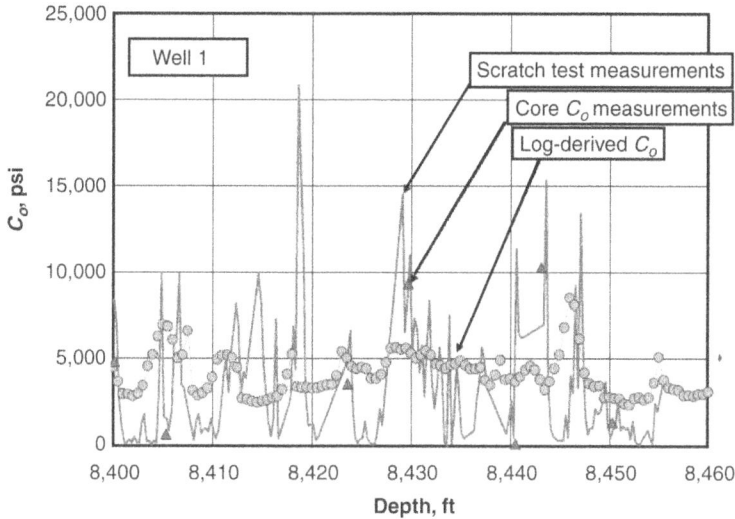

Fig. 1.15—Example plot of a comparison between log-derived C_o (gray dots), scratch-test results (solid line), and laboratory measurements of unconfined compressive strength (triangles).

ple preparation is required. In fact, any sample shape can be used for a penetrometer test, and even irregular rock fragments such as those recovered from intervals of wellbore enlargement because of compressive shear failure can be tested. Recently, methods have been developed to apply penetrometer tests to drill cuttings. Although these have not been widely used, they show considerable promise, and in the future they may become an important component of the measurement suite required to carry out wellbore stability analysis in real time.

1.3.6 Estimating Strength Parameters From Other Data. It is relatively straightforward to estimate C_o using measurements that can be obtained at the rig site. Log or logging while drilling (LWD) measurements of porosity, elastic modulus, velocity, and even gamma ray activity (GR) have all been used to estimate strength. For example, **Fig. 1.16**[7] shows a plot of C_o computed from P-wave modulus $(\rho_b V_p^2)$ for Hemlock sands (Cook Inlet, Alaska).

It is possible to develop an empirical relationship between any log parameter (even GR—see **Fig. 1.17**) and C_o or internal friction μ_i. Measurements of cation exchange capacity (CEC) and P-wave velocity have both been used for this purpose.

Because velocity, porosity, and GR can be acquired either using LWD or by measurements on cuttings carried out at the drilling rig from which CEC can also be derived, it is now possible to produce a strength log almost in real time. It is important, however, to recognize that different rock types will have very different log-strength relationships, based on their lithology (sand/shale, limestone, dolomite), age, history, and consolidation state. Therefore, it is important to be careful to avoid applying to one rock type a relationship calibrated for another.

1.4 Elastic Wellbore Stress Concentration

1.4.1 Stresses Around a Vertical Well. For a vertical well drilled in a homogeneous and isotropic elastic rock in which one principal stress (the overburden stress, S_v) is parallel to the wellbore axis, the effective hoop stress, $\sigma_{\theta\theta}$, at the wall of a cylindrical wellbore is given by Eq. 1.8.

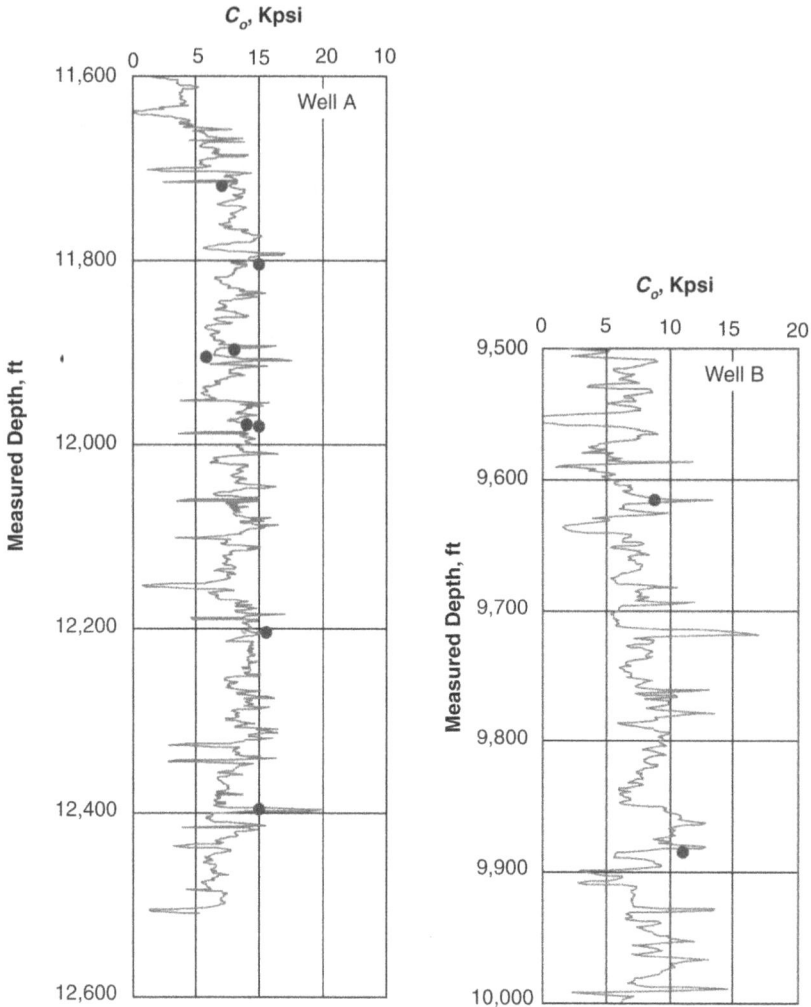

Fig. 1.16—Log-derived unconfined compressive strength for two wells drilled into sands of the Hemlock formation, Cook Inlet, Alaska, along with laboratory test data (dots). In this case, the strength log was derived from the P-wave modulus $(\rho_b V_p^2)$.[7]

$$\sigma_{\theta\theta} = S_{H\min} + S_{H\max} - 2(S_{H\max} - S_{H\min}) \cos 2\theta - 2P_p - \Delta P - \sigma^{\Delta T}$$

$$\sigma_{rr} = \Delta P. \quad\text{... (1.8)}$$

Here, θ is measured from the azimuth of the maximum horizontal stress, $S_{H\max}$ $S_{H\min}$ is the minimum horizontal stress; P_p is the pore pressure; ΔP is the difference between the wellbore pressure (mud weight) and the pore pressure, and $\sigma^{\Delta T}$ is the thermal stress induced by cooling of the wellbore by ΔT. If there is no strain in the axial direction, the effective stress acting parallel to the wellbore axis (σ_{zz}) is

$$\sigma_{zz} = S_v - 2v(S_{H\max} - S_{H\min}) \cos 2\theta - P_p - \sigma^{\Delta T}, \quad\text{................................. (1.9)}$$

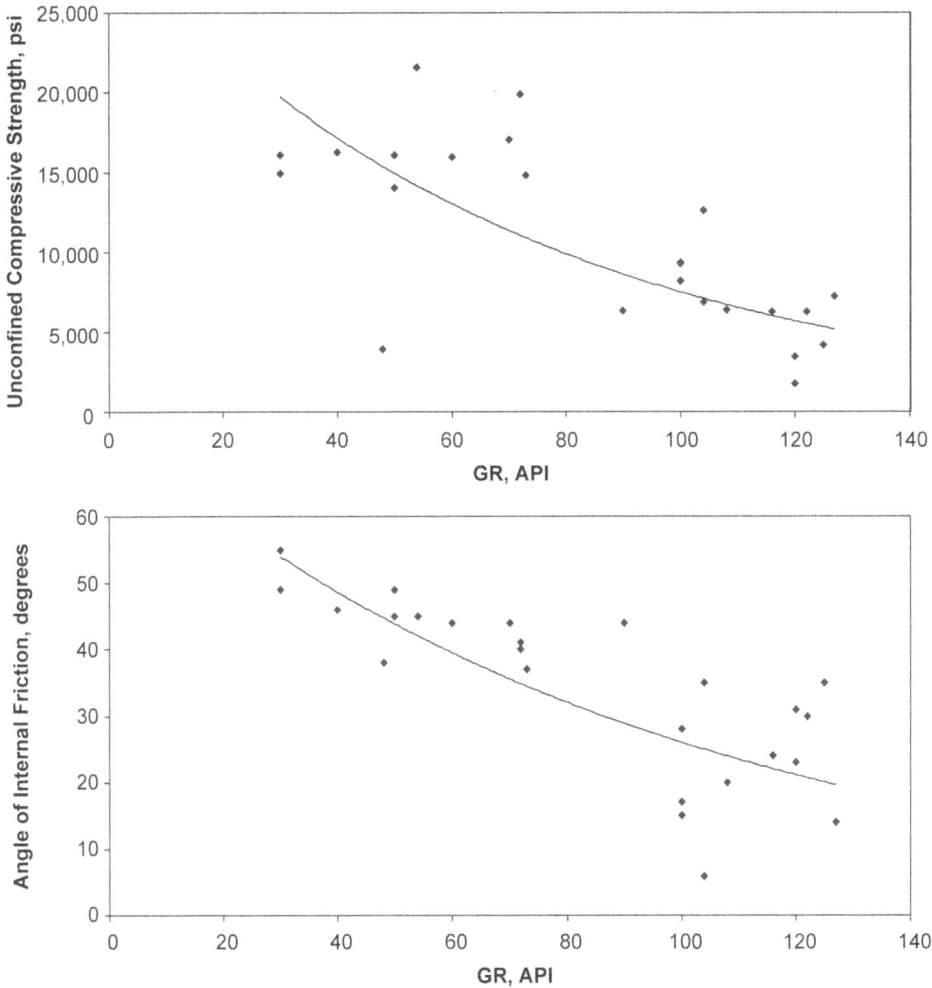

Fig. 1.17—Crossplots of unconfined compressive strength (top) and angle of internal friction (bottom) vs. GR from log inversion. Dots are laboratory test results. While there is considerable scatter, a clear relationship exists between GR and both parameters.

where v is Poisson's ratio. At the point of minimum compression around the wellbore (i.e., at $\theta = 0$, parallel to S_{Hmax}), Eq. 1.8 reduces to

$$\sigma_{\theta\theta}\text{min} = 3S_{Hmin} - S_{Hmax} - 2P_p - \Delta P - \sigma^{\Delta T}. \quad\quad\quad (1.10)$$

At the point of maximum stress concentration around the wellbore (i.e., at $\theta = 90°$, parallel to S_{Hmin}),

$$\sigma_{\theta\theta}\text{max} = 3S_{Hmax} - S_{Hmin} - 2P_p - \Delta P - \sigma^{\Delta T}. \quad\quad\quad (1.11)$$

The equations for $\sigma_{\theta\theta}$ and σ_{zz} are illustrated in **Fig. 1.18** for a strike-slip/normal faulting stress regime ($S_{Hmax} \sim S_v > S_{Hmin}$) at a depth of 5 km, where the pore pressure is hydrostatic and both ΔP and $\sigma^{\Delta T}$ are assumed to be zero for simplicity. As indicated in Eq. 1.11 and illustrated in Fig. 1.18, at the point of maximum compression around the wellbore, the maximum principal

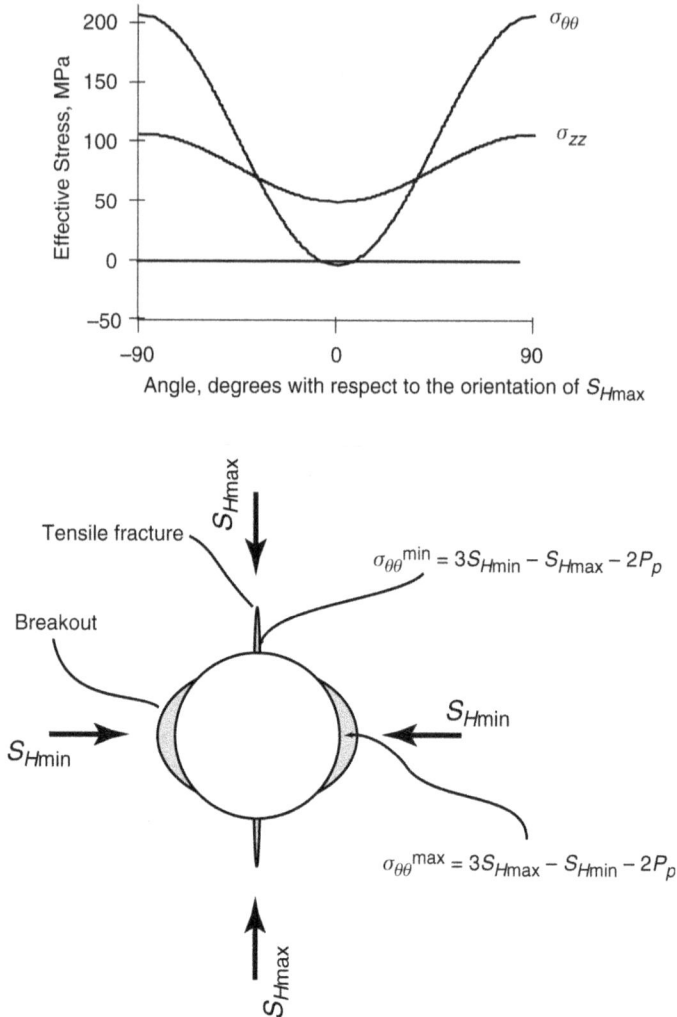

Fig. 1.18—(Top) The upper plot shows the characteristics of the wellbore stress concentration for a vertical well when the vertical stress is a principal stress. Both the circumferential ($\sigma_{\theta\theta}$) and axial (σ_{zz}) stresses are most compressive at the azimuth of the minimum far-field principal stress, leading to the formation of breakouts at that orientation if the stresses exceed the rock's compressive strength, as shown below. Both are most tensile at the azimuth of the greatest horizontal far-field principal stress, possibly leading to tensile failure at the wellbore wall 90° from the orientation of the breakouts (courtesy GeoMechanics Intl. Inc.).

horizontal stress is amplified appreciably. If the stress concentration is high enough, it can exceed the rock strength, and the rock will fail in compression. Compressive failures that form in the region around the wellbore where the stress concentration is greatest are commonly called stress-induced wellbore breakouts.

For the stress state assumed in Fig. 1.18, the stress concentration is close to zero at the azimuth of the maximum horizontal stress, S_{Hmax}. This is because a strike-slip faulting stress state was used for these calculations. It can be straightforwardly shown that in a strike-slip stress state in which the horizontal stress difference is in equilibrium with the strength of vertical strike-slip faults

$$S_{H\max} - P_p \sim 3\,(S_{H\min} - P_p). \quad\text{..} \quad (1.12)$$

Substituting this relation into Eq. 1.10 demonstrates that $\sigma_{\theta\theta}{}^{\min} \sim 0$, and it is easy for the wellbore to fail in tension, especially if ΔP and ΔT are greater than zero. Because the horizontal stress difference is smaller in a normal or reverse-faulting stress state than for a strike-slip stress state, tensile failure is less likely in these faulting regimes unless a wellbore is inclined.

To consider the potential for wellbore failure when a wellbore is inclined to the principal stresses, it is necessary to take into account the magnitudes and the orientations of the principal far-field stresses. Once these stress components are determined, in order to know whether a wellbore is likely to fail, the magnitudes of the stresses around the wellbore must be computed and the results considered in the context of a formal failure criterion. Because the equations that describe the stress concentration around a well inclined to the principal stress axes are complicated, they are usually solved using a computer application designed for the purpose.

The wellbore stress concentration decreases as a function of radial distance from the wellbore wall. Thus, the zone of failed rock will only extend to a certain depth away from the well. Once the rock has failed, however, the stresses are re-concentrated around the now broken-out wellbore, and it is possible (depending on the residual strength of the failed rock, which determines whether it can support stress) that additional failure will occur. One important thing to keep in mind is that even if the rock has failed, it may not lead to drilling difficulties.

1.4.2 Compressive Wellbore Failure. Stress-induced wellbore breakouts form because of compressive wellbore failure when the compressive strength of the rock is exceeded in the region of maximum compressive stress around a wellbore (Fig. 1.18). If the rock inside the breakout has no residual strength, the failed rock falls into the wellbore and gets washed out of the hole. The shape of these cuttings can be diagnostic of the mode of wellbore failure. Assuming (for the sake of discussion) that a Mohr-Coulomb failure criterion is appropriate for relatively brittle rocks, **Fig. 1.19**[8] shows the potential shear failure surfaces for the indicated stress field (left), and the zone of initial failure for a given cohesive strength, S_o (right). Comparison of the wellbore cross sections with the failure trajectories suggests that the surface of some breakouts is defined by a single shear fracture. It also has been demonstrated that wider and deeper breakouts will form as the maximum horizontal stress increases or as rock strength or mud weight decreases. While there is an increase in the stress concentration at the back of the breakout once it forms, any additional failure caused by that new stress concentration will result in an increase in breakout depth but will not change the width.

In a vertical well, breakouts are centered at the azimuth of minimum horizontal stress $S_{H\min}$ because this is where the compressive hoop stress is greatest. Hence, one can directly deduce the orientation of the in-situ stress tensor from the observation of breakouts. In inclined wells or in vertical wells where one principal stress axis is not parallel to the wellbore, breakout orientations are a function of both the orientations and the magnitudes of the in-situ stresses. Breakouts also may rotate in wells that intersect active shear planes. In both cases, while it is not possible to determine the stress orientation without additional information, it is often possible to determine one or more stress magnitudes.

1.4.3 Tensile Wellbore Failure. It is well known that if a vertical wellbore is pressurized, a hydraulic fracture will form at the azimuth of the maximum horizontal stress $S_{H\max}$. In some cases, the natural stress state, perhaps aided by drilling-related perturbations such as high mud weight, causes the wellbore wall to fail in tension, generating drilling-induced tensile wall fractures (DITWFs), as previously discussed for a vertical well in a strike-slip faulting environment. These fractures occur only at the wellbore wall (owing to the local stress concentration) and do not propagate any significant distance into the formation. They form 90° from the az-

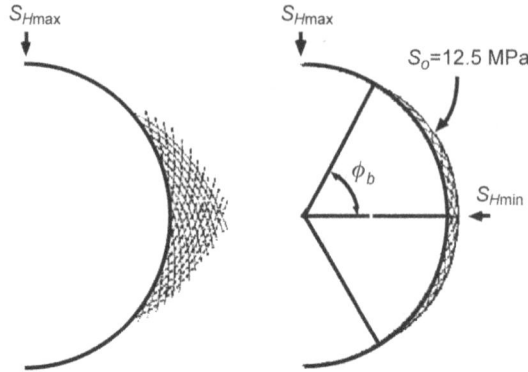

Fig. 1.19—On the left are shown the orientations of potential shear failure surfaces adjacent to a vertical wellbore for S_{Hmax} = 45 MPa, S_{Hmin} = 30 MPa, ΔP = 0, and the coefficient of internal friction μ_i = 1.0. On the right is shown the region in which failure is expected for a cohesive strength S_o= 12.5 MPa. The angle Φ_b is the half-width of wellbore breakout.[8] (After M.D. Zoback et al., "Wellbore Breakouts and In-Situ Stress," J. Geophysical Research, Vol. 90, No. B7, 5523; © American Geophysical Union; reproduced/ modified by permission of the American Geophysical Union.)

imuths of wellbore breakouts, and in vertical wells they indicate the azimuth of the maximum horizontal stress. As in the case of breakouts, tensile fractures in wells inclined to the principal stresses form at orientations that are a function of the stress magnitude as well as its orientation. In such cases, tensile fractures are inclined with respect to the wellbore axis, thus providing a clear indication that the stresses are not parallel and perpendicular to the well.

1.4.4 Detecting Wellbore Breakouts and Tensile Fractures. Wellbore breakouts were first identified by Gough and Bell[9] using 4-arm, magnetically oriented caliper logs acquired with Schlumberger dipmeters. However, to use this information for stress analysis, breakouts must be distinguished from other enlargements such as washouts (in which the entire hole is enlarged) and keyseats (caused by pipe wear or other drilling-related wellbore damage). The criteria, illustrated in **Fig. 1.20**,[10] used to distinguish stress-induced wellbore breakouts from drilling-induced features are as follows. First, when the caliper tool encounters a breakout, the tool should stop rotating in the well, because it should be engaged in the enlargement. Second, the small diameter measured by the caliper must be equal to the bit size. Third, in the case of an inclined well, the direction in which the wellbore is enlarged should not be the same as the direction of wellbore deviation. Finally, neither caliper diameter should be smaller than the bit size, as can occur in zones of keyseats owing to an associated off-centered tool. Failure to utilize criteria such as these can result in interpreting washouts and keyseats as wellbore breakouts.

While breakouts can be detected and used to determine stress orientation in many wells if 4-arm caliper data are carefully analyzed using rigorous criteria, truly unambiguous identification of breakouts requires the interactive analysis of data from full-wellbore scanning tools such as acoustic televiewers, which generate wellbore images that allow a much more detailed investigation of the wellbore wall (**Fig. 1.21**). These image data have the advantage over caliper data in that it is possible in images to

1. Study detailed variations of breakout orientation with depth.

2. Analyze the precise span of the wellbore's circumference which has failed using wellbore cross sections based on the time of flight of the acoustic pulse.

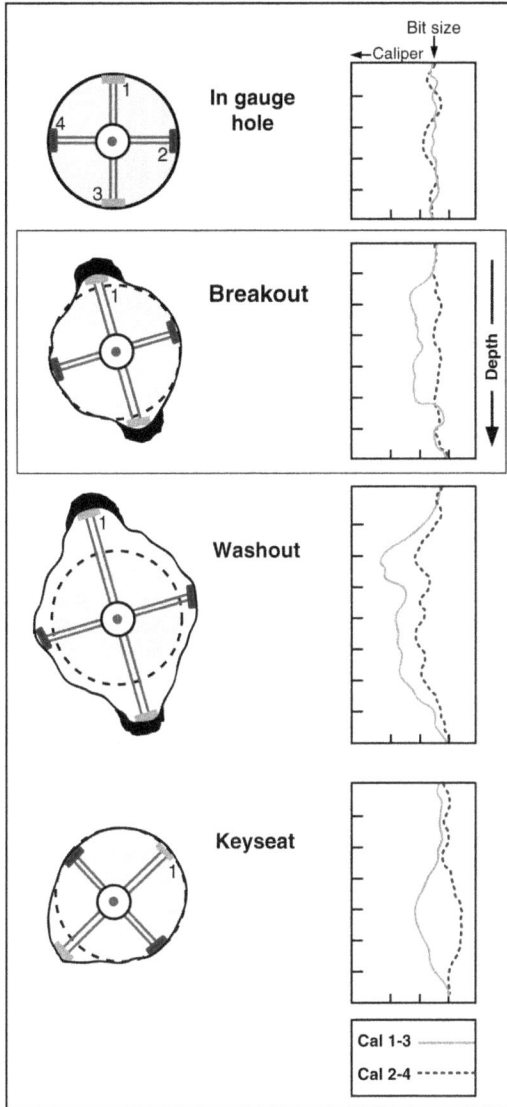

Fig. 1.20—Examples of 4-arm caliper (dipmeter) logs and common interpretations of the borehole geometry. Cal 1-3 and Cal 2-4 indicate borehole diameter as measured between perpendicular dipmeter arms. The shaded regions in the direction of enlargement represent local zones of slightly higher conductivity.[10] (After R.A. Plumb and S.H. Hickman, "Stress-Induced Borehole Elongation—A Comparison Between the Four-Arm Dipmeter and the Borehole Televiewer in the Auburn Geothermal Well," *J. Geophysical Research,* Vol. 90, B6, 5513; © American Geophysical Union; reproduced/modified by permission of the American Geophysical Union.)

3. Unambiguously distinguish stress-induced breakouts from keyseats and washouts. Although electrical image logs can also be used for wellbore failure analysis, it is more difficult to detect and characterize wellbore breakouts in electrical images than in acoustic images.

With the advent of 6-arm, oriented calipers, both those associated with electrical imaging tools and those that are run independently, it is now possible to utilize such data to define the shape of a well and identify oriented enlargements such as those caused by breakouts. To do so, however, these logs must be run in combination with orientation devices. As with 4-arm

Orientation

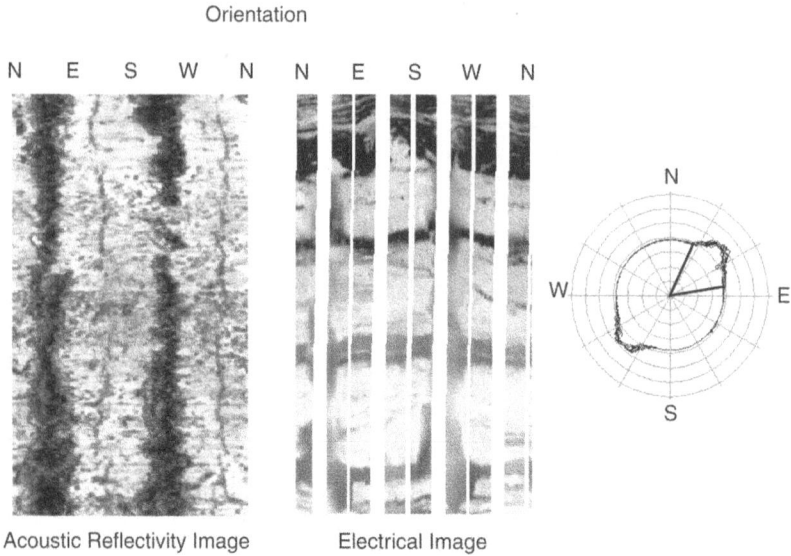

Acoustic Reflectivity Image   Electrical Image

Fig. 1.21—Example showing that it is possible to identify both tensile cracks and breakouts in acoustical wellbore images (left) and also to identify breakouts in an electrical image (center). Both images show an unwrapped view of the wellbore as a function of depth and azimuth, with azimuths starting at N on the left moving clockwise to E, S, W, and back to N on the right-hand edge. A wellbore cross section of time-of-flight data from an acoustic wellbore imaging tool is shown on the right, along with radial lines indicating the azimuthal extent of a wellbore breakout (courtesy GeoMechanics Intl. Inc.).

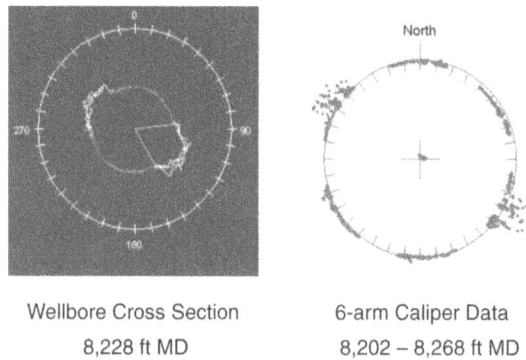

Wellbore Cross Section   6-arm Caliper Data
8,228 ft MD   8,202 – 8,268 ft MD

Fig. 1.22—Comparisons of 6-arm caliper data with wellbore cross sections from acoustic televiewer data from within the same interval reveal clearly that the 6-arm tool can be used to detect and orient breakouts.

caliper (dipmeter) data, strict criteria must be defined before using these data to determine stress orientation. An example of a case in which televiewer data was available to validate a 6-arm caliper analysis is shown in **Fig. 1.22.** Here, the wellbore cross section provided by the acoustic time-of-flight information shows enlargements in the precise orientations of the enlarged parts of the hole detected using 6-arm calipers. It is not possible to constrain breakout widths using 6-arm calipers because the orientation scatter in that data reflects only the variation in position of the centers of the caliper pads where breakouts were detected. Thus, the widths of these two cross sections have little relationship to each other.

Tensile fractures can most easily be seen in electrical image logs (see **Fig. 1.23**), whereas in acoustic images, they are most often seen when they are associated with fluid losses (e.g.,

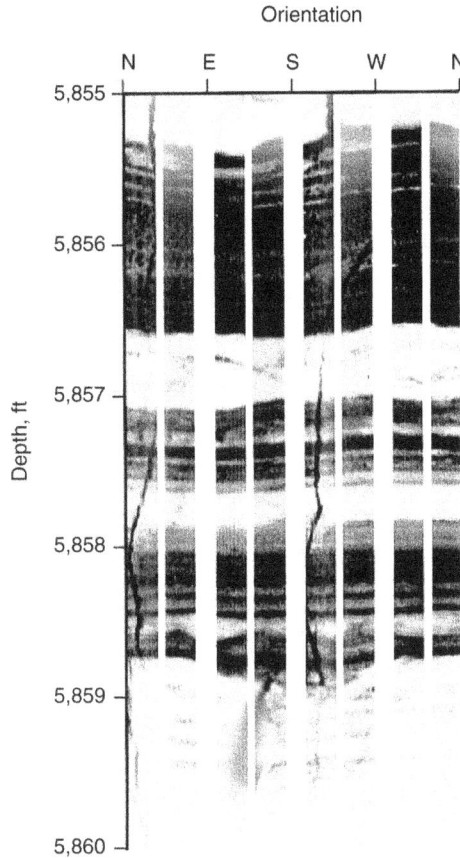

Fig. 1.23—Drilling-induced tensile fractures in Formation Micro-Imager (FMI) data. The fact that the fractures are parallel to the wellbore is an indication that the wellbore axis is parallel to a principal stress direction—in this case, the vertical stress.

Fig. 1.21). In some very rare cases, wellbores will enlarge in the direction in which tensile fractures are created by excessive amounts of wellbore cooling or extremely high mud weights. This effect has been documented using image logs in cases where stress orientations obtained from caliper logs were interpreted to indicate 90° shifts in stress orientation across bed boundaries. Even if televiewer data are available, enlargements in the direction in which tensile fractures develop can be mistaken for breakouts unless the data are studied carefully.

The cracks seen in Fig. 1.23 occur on both sides of the wellbore at the orientation of the maximum horizontal principal stress in the region, and similar cracks are seen over a ~200-m-long interval of the relatively vertical section of this well. These cracks are principally the result of the natural stress state combined with the additional effects of excess wellbore pressure and cooling, and thus the state of stress implied by the occurrence of these fractures is strike-slip ($S_{Hmax} > S_v > S_{Hmin}$).

1.4.5 Effects of Mud Weight and Temperature on the Wellbore Stress Concentration. The equations for stress around a wellbore shown above (Eqs. 1.8 and 1.9) include terms that describe thermal effects as well as the influence of the internal wellbore pressure. In the latter case, $\Delta P = P_{mud} - P_p$; in other words, the mud acts first against the pressure of the pore fluid, and any excess pressure is then applied to the rock. This assumes a reasonably efficient mud cake, and can be modified to account for its absence. If the mud weight is increased, it results

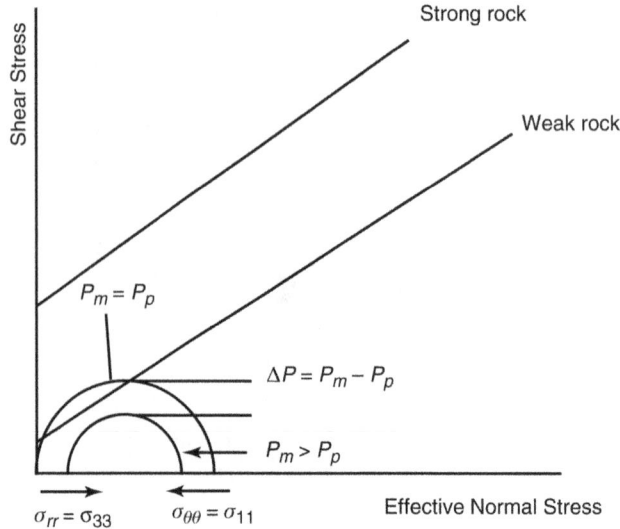

Fig. 1.24—When $P_{mud} = P_p$, $\sigma_{rr} = \Delta P = 0$, possibly leading to large amounts of failure in weak rock. When the mud weight is increased, it increases the radial stress on the wellbore wall and decreases the circumferential stress. This shrinks the Mohr circle without changing its midpoint, leading to a decreased risk of wellbore failure. The increase in effective strength can be as large in weak rock as it is in strong rock, and increases with mud weight at a rate defined by the internal friction (courtesy GeoMechanics Intl. Inc.).

in an increase in σ_{rr} and a decrease in $\sigma_{\theta\theta}$ and σ_{zz}; this usually inhibits breakout formation, which explains in part why raising mud weight can often solve wellbore instability problems (see **Fig. 1.24**). On the other hand, elevated mud pressures increase the likelihood of drilling-induced tensile wall fractures.

Thermal effects at the wellbore wall in the absence of pore fluid diffusion (that is, for purely conductive heat flow) can be described to first order by

$$\sigma_{\theta\theta}^{\Delta T} = (\alpha \, E \, \Delta T) / (1 - v). \qquad\qquad (1.13)$$

There is no effect at the wall of the hole on either σ_{rr} or σ_{zz}. Raising the temperature of the mud leads to an increase in $\sigma_{\theta\theta}$, which enhances the likelihood of breakouts and inhibits tensile fracture formation; on the other hand, cooling the mud inhibits breakouts (at least as long as the mud is kept at a temperature below the temperature of the rock) and increases the likelihood of development of tensile wall fractures. It has recently been noted that leakoff pressure can be increased by wellbore heating, which is consistent with this effect.

1.5 Determining Stress Orientation

The most reliable way to determine stress orientation is to identify features (either geological features or wellbore failures) the orientation of which is controlled by the orientations of the present-day in-situ stresses. Other methods that rely on observing the effect of stress on rock properties using oriented core have been found to be less reliable and subject to influence by factors other than in-situ stress.

1.5.1 Using Wellbore Failure. As previously discussed, wellbore breakouts occur in vertical wells at the azimuth of S_{Hmin}, and drilling-induced tensile failures occur 90° to breakouts at the azimuth of S_{Hmax} Therefore, the orientations of these stress-induced wellbore failures uniquely define the orientations of the far-field horizontal stresses when using data from vertical wells.

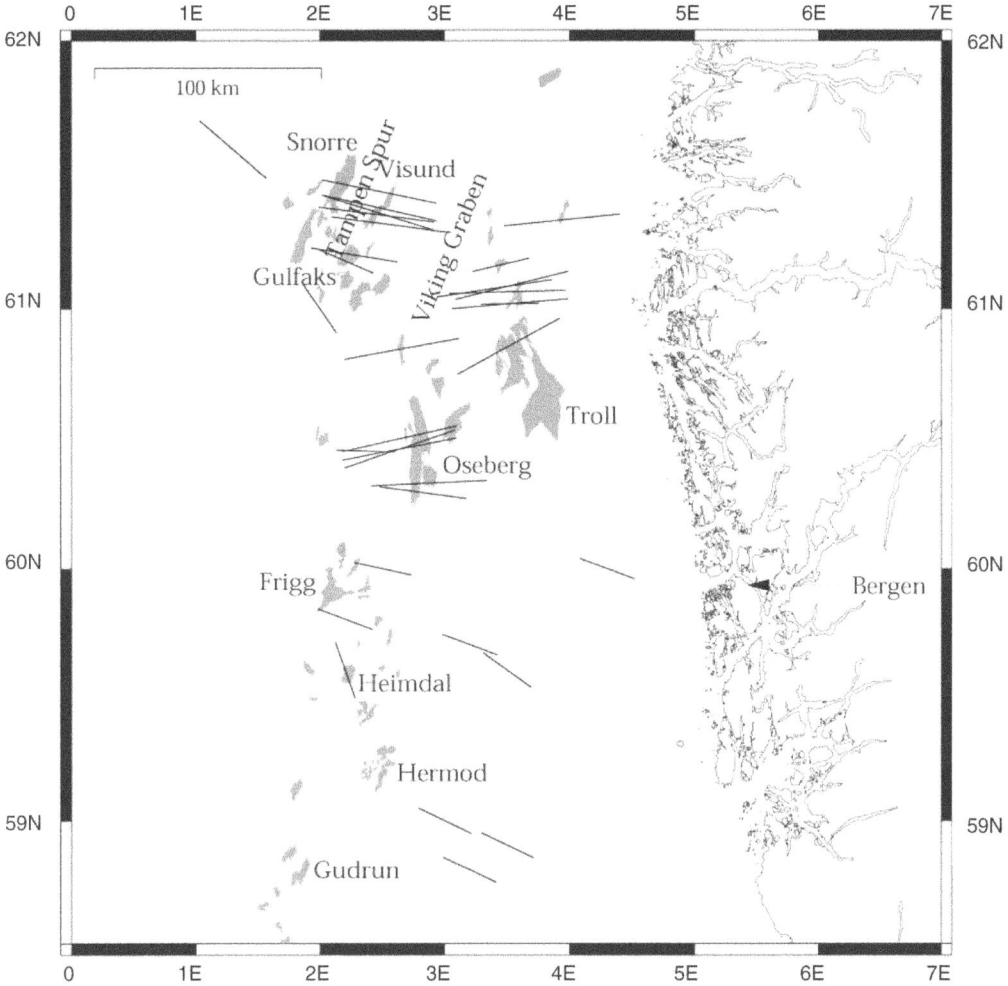

Fig. 1.25—Stress orientations derived from carefully filtered caliper logs in the North Sea (modified, after Wiprut *et al.*[11]).

This is true for breakouts whether they are detected using 4-arm- or 6-arm-oriented caliper logs or using electrical or acoustic images, whether obtained by wireline or LWD tools. In fact, with the advent of density and porosity LWD imaging tools, it is now possible to identify and orient wellbore failures while drilling.

Because mechanical calipers are still the most widely used tool in detecting breakouts, and because of the large amount of available data, a considerable amount of work has been carried out using data from these devices to identify stress orientations and their variations with depth and location. The results, when careful filtering criteria are used, indicate that stress orientations vary slowly with both depth and location. The exceptions are in cases of active faulting, rapid drawdown of compartmentalized reservoirs, and where other local stress perturbations cause changes in the stress field. **Figs. 1.25**[11] **and 1.26** illustrate that local stress orientations are quite consistent and that stresses generally do not change with depth. In addition, Fig. 1.26 illustrates the expected result that wellbore breakouts and drilling-induced tensile fractures provide similar stress orientation results.

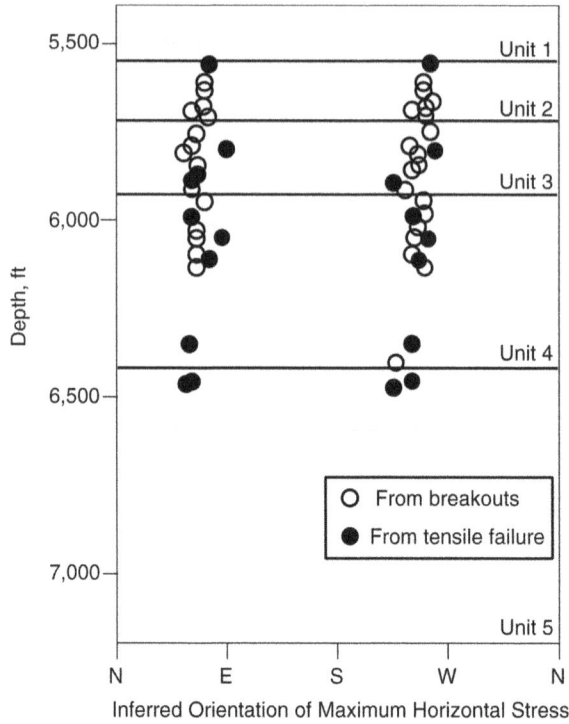

Fig. 1.26—Stress orientations in one well, from breakouts (open circles) and from drilling-induced tensile wall fractures (solid circles), are typically the same at similar depths and consistent over large depth intervals. Local variations are caused by slip on small weak faults that intersect the well and are activated by a near-wellbore pore-pressure increase caused by infiltration owing to the mud overbalance during drilling.

1.5.2 Using Seismic Anisotropy. It has long been known that elastic-wave velocities are a function of stress. That is, velocity will increase with confining pressure, as shown in **Fig. 1.27**.[12,13] It has also been demonstrated that this is owing to the presence of microcracks and pores that close in response to applied load. The amount of change of velocity with stress depends both on the number of cracks and on their compliances. As the load increases and the most compliant cracks close, the sensitivity of velocity to confining pressure decreases. Once all cracks are closed, velocities change very little with stress, and under sufficiently high confining stress, it is possible to measure the "intrinsic velocities" of a sample that are functions only of its mineralogy and morphology.

Because rocks are intrinsically anisotropic and can also be anisotropic due to structural fabric such as joints or bedding planes, it is important to differentiate between stress-induced and intrinsic or structural anisotropy. This is rarely possible, except in cases in which the geological structures can be identified and their effects quantified and removed from the data prior to analysis. It is also possible to identify stress-induced anisotropy when the characteristics of the anisotropy that is stress-induced differ from the characteristics of structural or intrinsic anisotropy.

Laboratory experiments have confirmed that, in many rocks, velocities are anisotropic (a function of direction) and are most sensitive to the stress applied in the direction of propagation or of particle motion. Thus, because in-situ stresses vary with direction in the Earth, in-situ velocities are likely to be anisotropic. In the case of compressional P-waves, the velocity depends on the direction of propagation because P-wave particle motion is parallel to propaga-

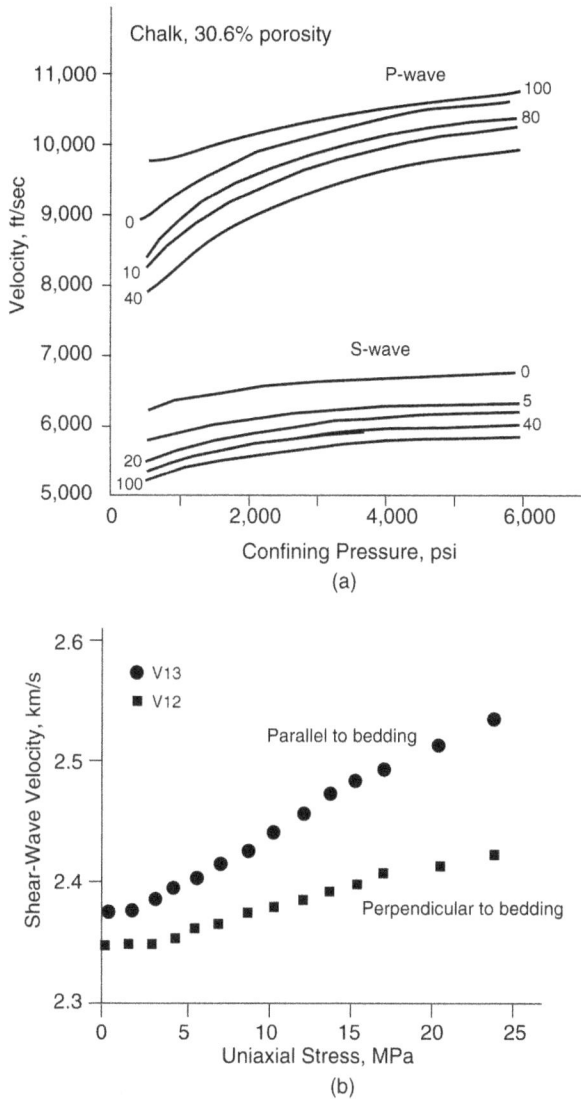

Fig. 1.27—Velocity as a function of stress derived from laboratory experiments. (a) Velocity as a function of confining pressure and saturation (after Rai and Hanson[12]). (b) Velocity as a function of direction relative to the direction of an applied uniaxial stress (after Gregory[13]).

tion direction. Under uniaxial stress, the *P*-wave velocity in the direction of applied stress increases with stress much more than the velocity in the direction perpendicular to the applied stress, as shown in Fig. 1.27b. In the case of shear *S*-waves, in which particle motion is perpendicular to propagation direction, velocities depend both on the propagation direction and on the polarization direction (i.e., the direction of particle motion). Because almost all rocks in situ have a finite porosity, this offers the opportunity to derive stress directions from in-situ seismic velocities.

Multicomponent seismic sources and receivers have been developed and deployed, and 3D multicomponent seismic surveys have been designed, to take advantage of this effect. However, while there is clear evidence that vertically and horizontally polarized shear waves have different velocities in shales, it is rare to find an appreciable amount of azimuthal anisotropy. One exception has been in cases in which anisotropy occurs owing to oriented sets of vertical fractures or joints. In such cases, however, it is not clear if the stress orientations are related in any way to the anisotropy because the joints may have been created in a quite different stress field than pertains at present.

1.5.3 Using Crossed-Dipole Sonic Logs. Modified sonic logging techniques can also be used to determine stress orientations. This is because, in a wellbore, it is possible using oriented sources and receivers to generate modes that bend the borehole in one direction. These dipole modes propagate efficiently at low frequencies and have been used to measure S-wave velocity where standard sonic logs cannot (that is, where S-wave formation velocity is lower than the acoustic velocity of fluids in the well). When velocity is stress-sensitive, dipole velocity becomes a function of the orientations of the sources and receivers, owing both to the presence of a near-wellbore stress concentration and to differences in the far-field stresses. Two modes are produced: a slow shear wave and a fast shear wave. At low frequencies, these propagate at the orientations of the least and greatest stress, respectively, perpendicular to the well. By adding an orientation device to a dipole logging tool, it is possible both to derive the velocities of the fast and slow shear waves and also to determine their directions.

Crossed-dipole logs are recorded in such a way as to allow computation of the velocities and orientations of fast and slow dipole modes. One potential benefit of these analyses is that theoretical considerations and laboratory measurements indicate that it is possible to differentiate between stress-induced and intrinsic anisotropy based on plots of velocity vs. frequency of the dipole modes. However, it is rare that field data are of sufficient quality to make this possible. In fact, reliable stress orientations are generally only possible in sands, porous limestones, and shales where the well is nearly perpendicular to bedding, where intrinsic anisotropy is low and the rocks are fairly compliant. In addition, these measurements should only be attempted when wells are drilled within approximately 20° of a principal stress direction. Finally, analyses should be believed only when very restrictive quality-assurance conditions can be met (e.g., **Fig. 1.28**).

1.5.4 Core-Based Analysis of Stress Orientation. Considerable effort has been devoted to developing and validating core-based stress analysis techniques. The one thing these have in common is the idea that post-coring deformation is dominated by expansion occurring because of removal of the core from in-situ stress conditions. The assumption is that the recovery-induced strains will have the same relative magnitudes and orientations as the original in-situ stresses. Therefore, by measuring the strains caused by removal or reloading, it is possible to constrain at least the directions of the principal stresses and their relative magnitudes.

There are basically three classes of techniques. These include measuring strain relaxation as a function of time after core removal, measuring strain as a function of orientation while reloading under isotropic conditions (possibly including monitoring for noise caused by microscopic slip events), and measuring velocities as a function of orientation under isotropic reloading. To orient stresses based on these techniques, it is necessary to know the original orientation of the core, which adds to the complexity of coring operations. Also, because the orientations of the principal strains are unknown prior to testing, to determine them it is necessary to attach more than three strain gauges to the sample.

In strain relaxation measurements, the core is recovered as quickly as possible, instrumented with strain gauges to monitor deformation as a function of time, and maintained in a constant (fixed) temperature/humidity environment. The principal strain axes are assumed to define the principal in-situ stress axes, and the relative strain magnitudes are assumed to correspond to the relative stress magnitudes. Thus, the vertical and horizontal relative stress magnitudes and the horizontal stress orientations can be derived from the principal strain orientations and magnitudes. Because most of the strain occurs during the first few minutes following removal from in-situ conditions, rapid recovery is essential to ensure accurate results.

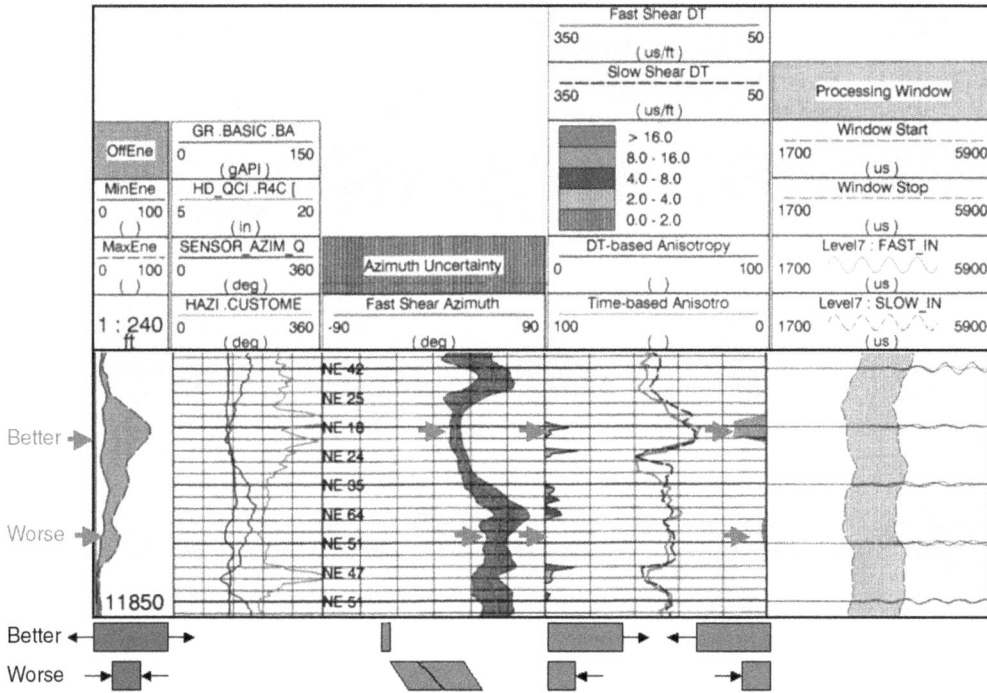

Fig. 1.28—Data and display from a Schlumberger crossed dipole log (xDSI™) showing results and quality control curves. A high-quality crossed-dipole analysis result has large maximum energy and low minimum energy (Track 1), a consistent orientation with a small uncertainty (Track 3), and large time- and velocity-domain anisotropy (Track 4). A low quality result has low maximum energy and very small anisotropy; a consistent orientation and low uncertainty are meaningless when this is the case.

The second (reloading) technique relies on the assumption that samples are much "softer" at stresses that are below their original confining stress than they are at stresses above their original confining stress. This assumption can be extended to cases in which the original stress state was anisotropic, so that the stress at which the sample gets stiffer is different in different directions. By isotropically loading a rock and monitoring strain as a function of confining stress and orientation, it is possible to determine the magnitudes of the three principal stresses by identifying the point in plots of stress vs. strain at which each curve bends over, indicating that the sample has suddenly become less sensitive to applied stress. The in-situ stress orientations and Andersonian stress state can be derived using the relative stress magnitudes at which this occurs, after resolving them into principal stress coordinates. If the sample has been instrumented to observe acoustic emissions from microscopic slip events, these will sometimes increase once the in-situ stress has been exceeded.

To derive principal stress orientations using velocity measurements, samples are instrumented with ultrasonic transmitters and receivers at a number of orientations, and the travel times of ultrasonic pulses are measured as a function of confining stress. As in the case of reloading, changes in velocity for each principal stress direction while confining pressure is below the original stress are larger than changes in velocity when confining pressure is above the original stress.

All three techniques suffer from the same limitation, which is that nearly all rocks are intrinsically anisotropic. In other words, their elastic moduli (which control the amount of strain that is caused by a given applied stress) are a function of direction. Anisotropic rocks will have

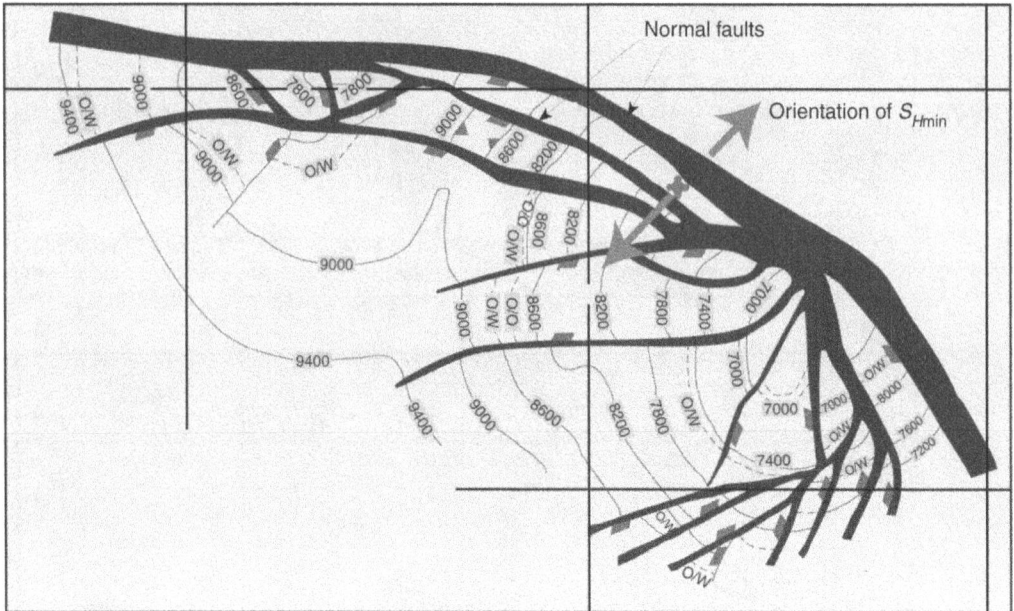

Fig. 1.29—Structure map in the South Eugene Island area of the Gulf of Mexico, showing large, active, WNW-ESE-trending normal faults. Given this fault orientation, the least stress is expected to be horizontal and oriented NE-SW. Stress orientations from breakouts confirm this (outward-facing arrows) (modified after Finkbeiner et al.[14]).

different amounts of strain in different directions, even if they are subjected to an isotropic stress state. If the intrinsic anisotropy is large enough, which it generally is in shales, laminated sands, and other finely bedded or foliated rocks, strains related to that anisotropy can mask strains caused by stress changes. Thus, while there are some situations in which these techniques work, there are many pitfalls, and the results should be used with caution.

1.5.5 Geological Indicators of Stress Orientation. In the absence of better data, it is sometimes useful to look at earthquake focal mechanisms within the region or to map local geological structures to help provide a "first look" estimate of the relative magnitudes and orientations of the current stresses. In the case of earthquake focal mechanisms, it is important to utilize data from many earthquakes within a small region to derive a "composite focal mechanism," to avoid the large uncertainties associated with individual analyses. In the case of geological structure, it is critically important to remember that many structures are inherited from older stress fields and that the only structures that do provide information are those that are currently active. **Fig. 1.29**[14] is an example, from South Eugene Island in the Gulf of Mexico, where the stress direction, confirmed by wellbore breakout analysis, is consistent with the orientation of a nearby large, active normal fault.

Salt domes can significantly perturb the local stress field because extension predominates above active salt intrusions, whereas beside the salt compression acts radially away from its walls. This is because salt is virtually unable to sustain a significant stress difference, and thus all three stresses in salt bodies are nearly equal and close to the vertical stress. This not only increases the local horizontal stresses, but also causes a rotation in the principal stress axes to be perpendicular and parallel to the salt face. Salt domes rarely have vertical walls, and thus the vertical stress may no longer be a principal stress close to their flanks.

TABLE 1.3—EMPIRICAL CONSTANTS FOR EQ. 1.15 FOR VARIOUS LITHOLOGIES (DATA FROM MAVKO et al.[14])						
Lithology	a	b	c	d	e	Valid Range, km/s
Shale	1.75	0.265	−0.0261	0.373	1.458	1.5–5.0
Sandstone	1.66	0.261	−0.0115	0.261	1.515	1.5–6.0
Limestone	1.50	0.225	−0.0296	0.461	0.963	3.5–6.4
Dolomite	1.74	0.252	−0.0235	0.390	1.242	4.5–7.1
Anhydrite	2.19	0.160	−0.0203	0.321	1.732	4.6–7.4

1.6 Building the Geomechanical Model

The elements of the geomechanical model that form the basis for analysis of wellbore stability are the state of stress (the orientations and magnitudes of the three principal stresses), the pore pressure, and the rock properties, including strength (which can be anisotropic, particularly in consolidated shales). We have already presented a number of ways to determine the rock strength and the orientations of the in-situ stresses. In this section, we outline methods for determination of the stress magnitudes and the pore pressure.

1.6.1 Overburden, S_v. Overburden pressure, or S_v, is almost always equal to the weight of overlying fluids and rock. Thus, it can be calculated by integrating the density of the materials overlying the depth of interest (see **Fig. 1.30**).

$$S_v(Z_o) = \int_0^{Z_o} \rho_b G dz. \qquad\qquad (1.14)$$

Here, G is the gravitational coefficient. The best measurement of density is derived from well logs. However, density logs are seldom acquired to ground surface or to the sea floor. If good seismic velocities are available, a velocity to density transformation can be used to estimate density where it has not been measured directly. A number of transformations from velocity to density are available (see Eq. 1.15 and **Table 1.3**).[15] In the absence of good velocity or density data, densities must be extrapolated from the surface to the depths at which they are measured. Shallow density profiles can take many forms, and thus they ideally should be calibrated against in-situ log data or measurements of sample densities. A good resource for information about shallow density profiles is the archives of the Deep Sea Drilling Project and the Ocean Drilling Program (www.oceandrilling.org). In the absence of good calibrations or data, reasonable mudline densities for clean sands are between 2.0 and 2.2 gm/cm³, and for fine-grained shales are between 1.4 and 1.8 gm/cm³.

$$\rho_b = a V_p^b,$$

and

$$\rho_b = c V_p^2 + d V_p + e, \qquad\qquad (1.15)$$

where a, b, c, d, and e are constants that vary with lithology. The values in Table 1.3 are for velocity in km/s and density in gm/cm³.

1.6.2 Pore Pressure, P_p. The only accurate way to determine pore pressure is by direct measurement. Such measurements are typically done in reservoirs at the same time fluid samples are taken with a wireline formation-testing tool. Recently, advances in while-drilling measurements make it possible to measure in-situ pore pressure while drilling. However, it is difficult

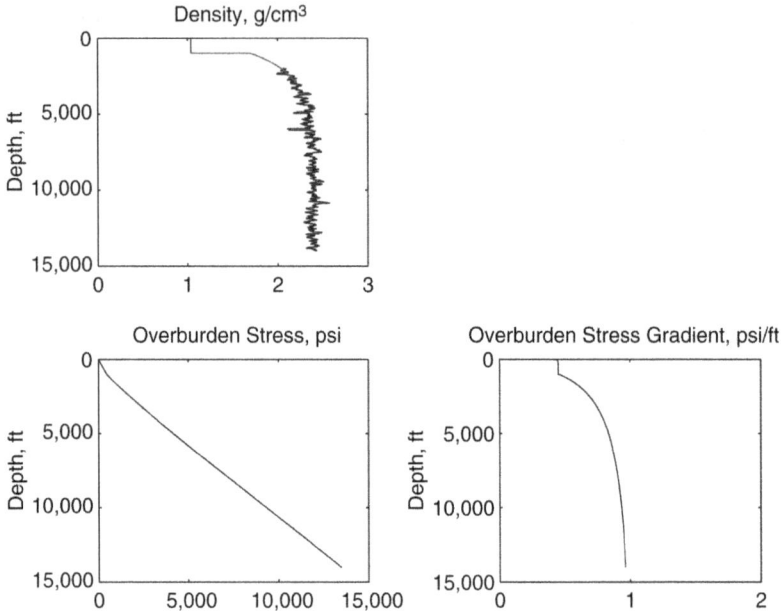

Fig. 1.30—(a) Density logs for a subsea well beneath 1,000 ft of water, extrapolated to the mud line using an exponential curve. Density within the water column is 1.04 gm/cm³. (b) Integration results in a plot of overburden (Sᵥ) vs. depth. (c) When converted to an equivalent density, overburden can be displayed in psi/ft, lbm/gal, or specific gravity (SG) (courtesy GeoMechanics Intl. Inc.).

(if not impossible) to measure pore pressure in shales because of their very low permeability and small pore volume. In addition, because of their low permeability, pore pressure in shales adjacent to permeable reservoirs may be different from pore pressures in the reservoir. However, there are a number of methods that can be used to estimate pore pressure in shales based on other measurements. Because pore pressure (and the derived fracture gradient, which will not be discussed) is often the only geomechanical parameter on which mud weights are based, we will take some time to review standard and new methods for its prediction. Keep in mind that these prediction methods are intended for use only in shales.

Pore-pressure-prediction methods fall into a few general categories. In the first category are normal compaction trend (NCT), ratio, and equivalent depth methods, which are all more or less empirical. The second category includes methods that explicitly utilize relationships between measured values and the effective stress. These first two methods assume that the in-situ material is either normally compacted or undercompacted. In the third category are models that are also applicable to overcompacted rock. All of these methods require measurement of one or more physical properties that are functions of effective stress. These include resistivity, density, and seismic or sonic velocity.

In most cases, the only measurement that is available prior to drilling is (*P*-wave) seismic velocity. After the first well has been drilled, or during drilling (using LWD tools), log data are acquired that make it possible to improve predrill pore-pressure estimates. Using LWD, and adding a pressure while drilling (PWD) measurement, pore-pressure analysis can be carried out in real time. Typically, in shale sections above a target, only LWD resistivity and gamma ray are acquired, but deeper in the well, additional measurements are often made, including density and velocity.

One additional measurement that has been used to predict pore pressure is the drilling exponent, D_c, which defines the rate of drill-bit penetration as a function of depth. Because ease of drilling is related to strength, which in turn is a function of porosity (and therefore of effective stress), the rate of penetration should be a function of effective stress, provided that it is corrected for changes in any other drilling parameters. Therefore, D_c can be used to determine pore pressure using the same analyses used to compute pore pressure from physical properties like resistivity or velocity.

Although all shale pore-pressure-prediction methods rely on the fact that rock physical properties depend on the effective stress, σ (= $S - \alpha P_p$), equivalent depth and NCT methods use depth as a proxy, and in ratio methods even depth is implicit. Effective stress methods work by

1. Measuring the total stress (S).

2. Using either an explicit relationship or an implicit function to derive the effective stress (σ) from a measured parameter.

3. Computing the pore pressure as the difference between the effective stress and the total stress, divided by alpha, the Biot coefficient [$P_p = (S - \sigma)/\alpha$].

In relatively young, unconsolidated shales $\alpha = 1$, but values of 0.9 or less may be more appropriate for older, more highly compacted sediments.

Because overburden (S_v) can be computed as the integral of the density of the rock and fluid overlying the depth of interest, pore-pressure-prediction methods were developed initially using $P_p = (S_v - \sigma)$. This is more reasonable than it might seem because properties such as velocity depend most strongly on stress in the direction of propagation, which for near-offset seismic data is nearly vertical. In some cases, the vertical stress is replaced by the mean stress. This results in adjustments to the pore pressure computations based on differences in the magnitudes of the horizontal stresses in different regions.

The relationship between the measured quantity and the effective stress is derived either using explicit functional relationships or by so-called trend-line methods. Trend-line methods require the existence of a depth section, over which the pore pressure is hydrostatic, to derive the NCT.

Equivalent Depth Methods. One example of analysis using a trend line is the equivalent depth method illustrated in **Fig. 1.31**. This method first assumes that there is a depth section over which the pore pressure is hydrostatic, and the sediments are normally compacted because of the systematic increase in effective stress with depth. When the log of a measured value is plotted as a function of depth, NCTs can be displayed as straight lines fitted to the data over the normally compacted interval. Because the value of the measured physical property is a unique function of effective stress, the pore pressure at any depth where the measured value is not on the NCT can be computed from

$$P_z = P_a + (S_z - S_a), \quad\dotfill (1.16)$$

where $P_{a,z}$ and $S_{a,z}$ are the pore pressure and the stress at z, the depth of interest and a, the depth along the normal compaction trend at which the measured parameter is the same as it is at the depth of interest. The only unique assumption required by equivalent depth methods is that effective stress is a linear function of depth.

The Ratio Method. In the ratio method, pore pressure is calculated using the assumption that, for sonic delta-t, density, and resistivity, respectively, the pore pressure is the product of the normal pressure multiplied (or divided by) the ratio of the measured value to the normal value for the same depth.

$$P_p = P_{\text{hyd}} \Delta T_{\log} / \Delta T_n,$$
$$P_p = P_{\text{hyd}} \rho_n / \rho_{\log},$$

and

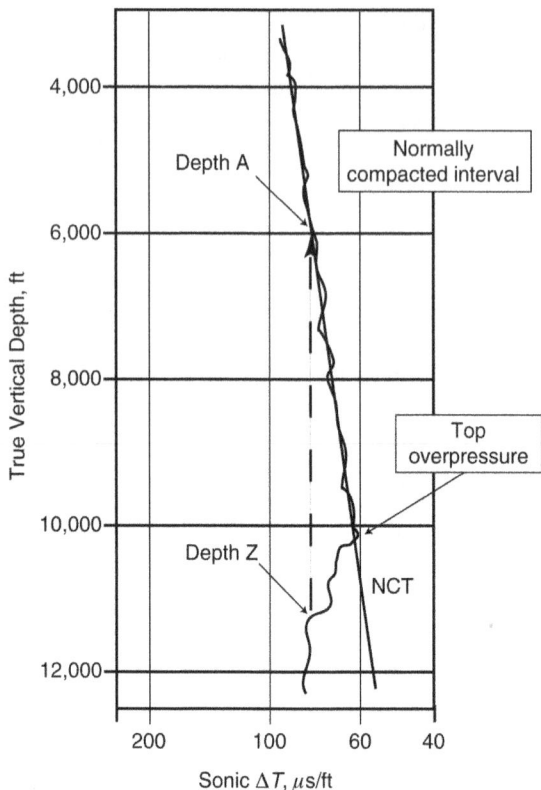

Fig. 1.31—Illustration of the equivalent depth method using sonic ΔT. The normal compaction trend (NCT) is a straight line in log-linear space that has been fitted to the decrease in slowness as a function of depth where sediments are normally compacting. The effective stress at depth Z is equal to the effective stress at depth A, and thus, the pore pressure at depth Z is simply $P_z = P_a + (S_z - S_a)$.

$$P_p = P_{hyd} R_n / R_{log}, \quad\quad\quad\quad\quad\quad\quad\quad\quad\quad\quad (1.17)$$

where the subscripts n and log refer to the normal and measured values of density, resistivity, or sonic delta-t; P_p is the actual pore pressure, and P_{hyd} is the normal hydrostatic pore pressure. Calibration of this method requires knowing the appropriate normal value of each parameter. It is important to realize that, in contrast to trend-line methods, the ratio method does not use overburden or effective stress explicitly and thus is not an effective stress method. This can lead to unphysical situations, such as calculated pore pressures that are higher than the overburden. The ratio method is also applied to analyses of pore pressure from the drilling exponent (**Fig. 1.32**).[16]

Eaton's Method. Perhaps the most widely publicized pore-pressure-estimation technique is Eaton's method, shown graphically in **Fig. 1.33**.[16] Here, stress is used explicitly in the equations

$$P_p = S - (S - P_{hyd})(R_{log} / R_n)^{1.2}$$

and

$$P_p = S - (S - P_{hyd})(\Delta T_n / \Delta T_{log})^{3.0}, \quad\quad\quad\quad\quad\quad\quad (1.18)$$

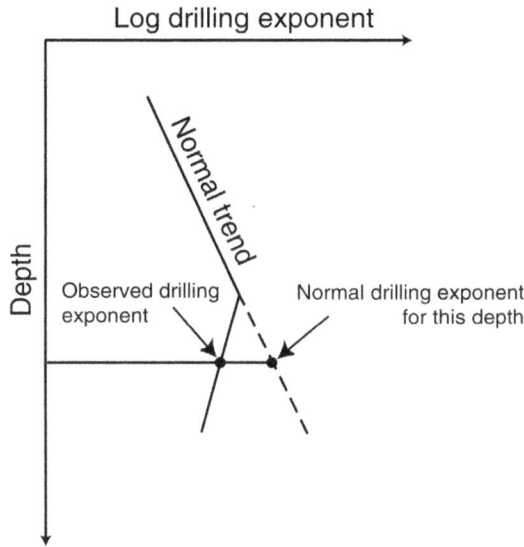

Fig. 1.32—Illustration of the ratio method. Here, d_{cn} is the expected value of the drilling exponent based on extrapolating a normal trend, and d_{co} is its measured value (modified after Mouchet and Mitchell[16]).

where P_p is pore pressure; S is the stress (typically, S_v); P_{hyd} is hydrostatic pore pressure; and the subscripts n and log refer to the normal and measured values of resistivity (R) and sonic delta-t (ΔT) at each depth. The exponents shown in Eq. 1.18 are typical values that are often changed for different regions so that the predictions better match pore pressures inferred from other data.

The major problem with all trend-line methods is that the user must pick the correct normal compaction trend. Sometimes are too few data to define the NCT. Unfortunately, if the NCT is defined over an interval with elevated pore pressure, the method will give the wrong (too low) pore pressure, leading to severe risks for drilling.

Effective Stress Methods. Methods that treat the problem correctly are often referred to as effective stress methods. The basis of the approach is summarized in **Fig. 1.34.**[17] In this example, the top set of plots shows data recorded over a normally compacted section. The mean stress and pore pressure are shown as a function of depth at the left. Because the effective stress increases with depth, the porosity decreases with depth, as shown in the middle. If the porosity-stress function is a power law, it will plot as a straight line in linear-log space. There is, in fact, no restriction on the functional form of the porosity-stress function. The lower set of plots in Fig. 1.34 show the effect of an increase in pore pressure below a certain depth, as represented by the dashed line that diverges from the hydrostatic line in the lower left plot. In this case, the pore pressure in the overpressured zone increases at the same rate as the mean stress, such that the effective stress is constant. The plot of log σ vs. Φ follows the compaction trend only until it reaches the depth of overpressure, after which there is no change in either porosity or effective stress (lower right plot). This type of profile is typical of regions in which a pore-pressure increase is caused by the inability of the pore fluids to escape during burial and compaction.

In general, effective stress methods must be calibrated, preferably using log data. However, they can also be calibrated empirically using approaches similar to those used to select trend lines, and they account explicitly for local changes in overburden and other stresses. The equation plotted in Fig. 1.34 is an example of relationships of the form

Drilling Exponent

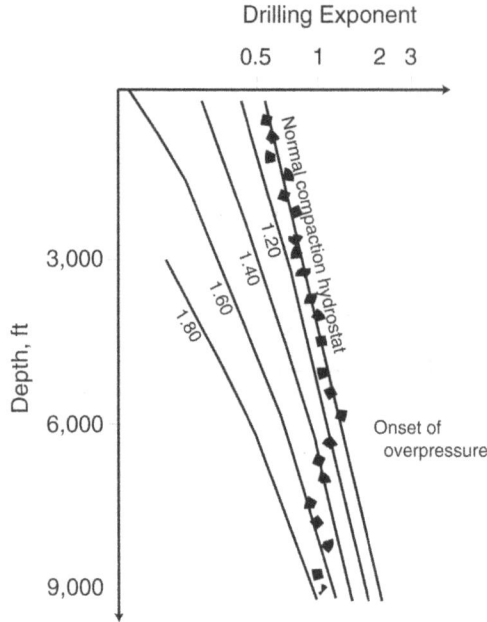

Fig. 1.33—Lines for computing pore pressure expressed as an equivalent density, calculated using Eaton's method and the drilling exponent. Notice that these "lines" are not linear in semilog space (modified after Mouchet and Mitchell[16]).

$$\sigma = \sigma_o e^{-\beta\sigma v}. \dotfill (1.19)$$

Athy[18] first proposed this type of relationship in 1930, also proposing values for the parameters. Use of Athy's original parameters is not recommended because they were based on analysis of overconsolidated shales from Oklahoma and thus are not applicable to young sediments. An appropriate algorithm for Athy's method is to solve the following set of equations.

$$P_p = S_v - [1 / \beta \ \ln \ (\varPhi_o / \varPhi)],$$

where

$$\varPhi = 1 - (\Delta t_{ma} / \Delta t)^{1/f}, \dotfill (1.20)$$

and f is the acoustic formation factor and is derived by calibration; Δt_{ma} is the matrix transit time. This type of relationship allows extension to account for effects such as cementation and thermal transformations by modifying the functional form of the exponent.

Complications. Ultimately, all pore-pressure methods must be calibrated; this is done empirically in most cases. Typical approaches rely on drilling experience to provide calibration points. These calibration points are based either on the occurrence of kicks, in which case the pore pressure in the sand producing the kick must be higher than the equivalent mud weight and lower than the kill mud weight, or on observations of instabilities in shales. In the former case, it is assumed that the pore pressures in shales and the adjacent sands are the same. In the latter case, the assumption is that instabilities occur when the mud weight has fallen below the pore pressure. In fact, wellbore instabilities that are due to compressive breakouts can occur at pressures that are higher or lower than the pore pressure. Therefore, the assumption that collapse begins to occur at a mud weight equal to the pore pressure can result either in an

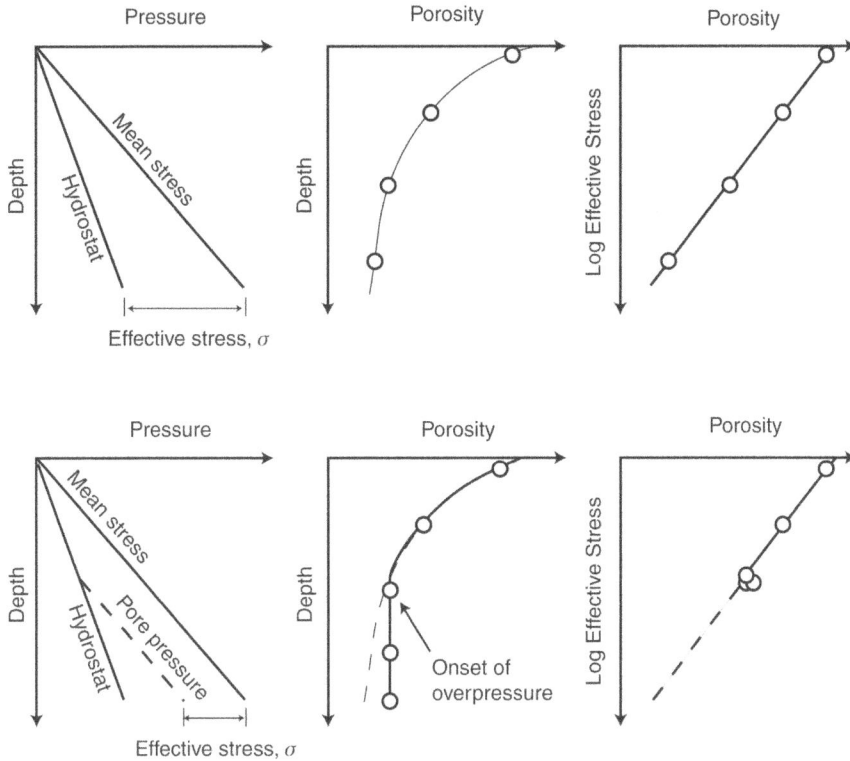

Fig. 1.34—Illustration of the effective stress method for pore pressure prediction (modified after Swarbrick[17]).

overestimate or an underestimate of P_p. If neither occurs, P_p is assumed (sometimes incorrectly) to be less than P_{mud}.

A further complication is that all of these methods require that the rock obeys a single, monotonic, compaction-induced trend, and that no other effects are operating. In reality, active chemical processes can increase cementation, leading to increased stiffness (higher velocities), which can mask high pore pressure, and increased resistance to further compaction, which can lead to erroneous prediction of the onset of pore pressure. Elevated temperatures lead to a transformation of the predominant shale mineral. For example, an increase in temperature transforms a water-bearing smectite to a relatively water-free (and more dense) illite. This transformation occurs over a range of temperatures near 110°C, but they can vary with fluid chemistry; furthermore, the depth at which this temperature is exceeded varies from basin to basin. Dutta[19] developed a method that expands the argument of the exponential relationship of Eq. 1.19 to account for temperature effects and diagenesis (cementation and other changes that occur over time).

Pore fluid properties can also have a significant effect on pore-pressure predictions. This is because resistivity and velocity are both affected by the type and properties of the pore fluid. Changes in the salinity of brines will change resistivity, because pore fluid conductivity increases with salinity; thus a salinity increase (for example, adjacent to or beneath a salt dome) could be misinterpreted as an increase in pore pressure. Fluid conductivity is also a function of temperature.

Substitution of hydrocarbons for brine will increase resistivity, because hydrocarbons do not conduct electricity; this can mask increases in pore pressure that often accompany the presence of hydrocarbons. Because hydrocarbons are more compliant and less dense than brines, compression-wave velocity will decrease and shear-wave velocity will increase as hydrocarbon saturation increases. High gas saturation or API index will amplify this affect. Because a change from water to hydrocarbon affects resistivity and compressional velocity in opposite ways, simultaneous pore-pressure analyses using both measurements can sometimes identify such zones. It is more difficult to identify and deal with changes in fluid salinity.

Undercompaction. Most shale properties are, fortunately, characterized by fairly simple and single-valued functions of effective stress while on the compaction trend. When unloading occurs and the material becomes overcompacted, they do not follow the same relationship. This is because when the effective stress decreases, porosity and other properties are less sensitive to effective stress (see Fig. 1.11). Fortunately, relationships between porosity (or density) and other properties are different for overcompacted sediments than they are for the same sediment when it is normally compacted or undercompacted, as shown in laboratory data (**Fig. 1.35**).[20] This provides a way to differentiate between undercompacted and overcompacted shales, using plots of velocity vs. density (**Fig. 1.36**). Once the domains have been separated, independent calibrations can be used to determine the pore pressure.

In highly lithified, older sediments, as in the case of overcompacted sediments, it is very difficult to use trend-line analyses to determine pore pressure. This is because, in these sediments, the sensitivity of porosity to effective stress is small. Even in such cases, however, it is sometimes possible (with accurate models derived from laboratory measurements and calibrated against in-situ direct measurements) to utilize resistivity or velocity measurements to estimate pore pressure.

Centroid and Buoyancy Effects. The previous discussion of pore-pressure-prediction algorithms applies exclusively to shales and other low-permeability materials that undergo large amounts of shear-enhanced compaction. Because these algorithms do not work very well in sands, it is often assumed that pore pressures in sands are similar to those in adjacent shales. In reality, this is often not the case because the low permeability of shale makes it possible for it to maintain a pore pressure that is quite different from that in the adjacent sand. Two active processes, both of which can lead to very much higher (or lower) sand pore pressure than shale pore pressure that can be maintained over geological time because of low shale permeability, are the centroid and buoyancy effects.

The classical centroid effect occurs when an initially flat reservoir surrounded by and in equilibrium with overpressured shale is loaded asymmetrically and tilted, leading to a hydrostatic gradient in the sand that is in equilibrium with the original pore pressure at the depth of the sand prior to tilting. At the same time, pore pressure in the shale, which has extremely low permeability after it has been compacted, changes in such a way as to maintain a constant effective stress equal to the original effective stress at the depth of the sand prior to tilting. At the depth of the centroid (usually taken to be the mean elevation of the sand), the shale and sand pore pressures are equal. This effect is shown diagrammatically in **Fig. 1.37**.[21] Because the pressure in the shale decreases upward at the same rate as the overburden (that is, proportional to the density of the shale itself), it is much lower at the top of the reservoir than is the pore pressure within the reservoir, which decreases at a slower rate that is proportional to the fluid density in the reservoir. Below the centroid, pore pressure in the sand is less than that in the adjacent shale.

Buoyancy effects occur when hydrocarbons begin to fill a tilted reservoir. The lighter hydrocarbons migrate to the top of the structure. Pressure at depth within the reservoir still follows a hydrostatic gradient (**Fig. 1.38**). The pressure in the gas at the top of the reservoir decreases upward more slowly, at a rate proportional to the density of the gas, which can be less than one-fourth the density of water. This leads to elevated pressure at the top of the structure. The

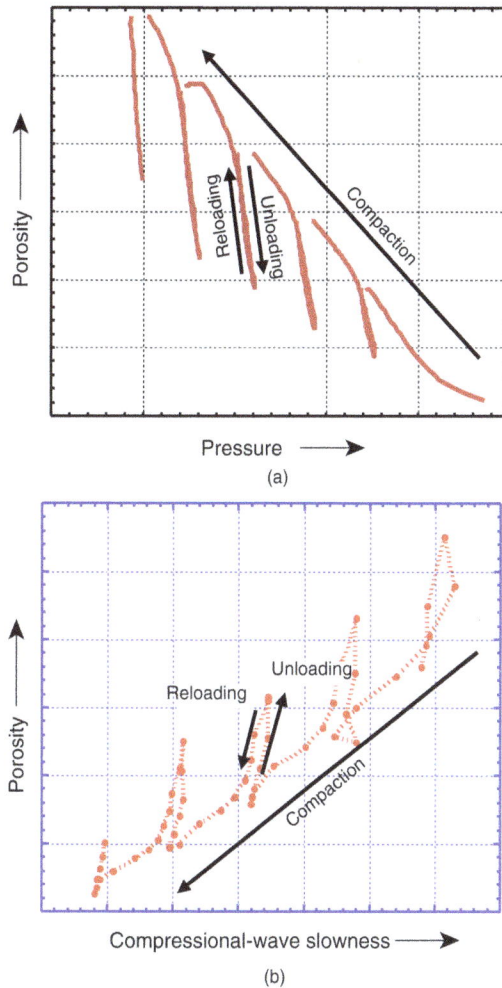

Fig. 1.35—This figure shows laboratory measurements of porosity vs. pressure (a) and porosity vs. slowness (b) along compaction trends and during reductions and subsequent increases in effective confining pressure in a poorly consolidated, shaley turbiditic sand of Miocene age. The separation of overcompacted from normally compacted or undercompacted sediments in plots of porosity vs. slowness makes it possible to use combined measurement of these parameters both to determine pore pressure and to identify the overpressure mechanism in both undercompacted and overcompacted domains (after Moos and Zwart[20]).

same process occurs when oil fills a reservoir, but since the density difference is not as large for oil, the effect is less pronounced.

The reservoir can continue to fill until the structure's sealing capacity is exceeded. In the example, seal capacity is exceeded when the pressure at the top of the reservoir is high enough to cause the sealing fault to slip. However, in extreme cases, the reservoir pressure can be close to the least principal stress in the adjacent shale.

1.6.3 Least Principal Stress, S_3. The least principal stress can be measured directly, using either extended leakoff tests or minifrac tests. These tests are similar to casing integrity tests or standard leakoff tests, except that the test procedure is slightly modified. Fluid is pumped into the wellbore to pressurize a short interval of exposed rock until the rock fractures and the frac-

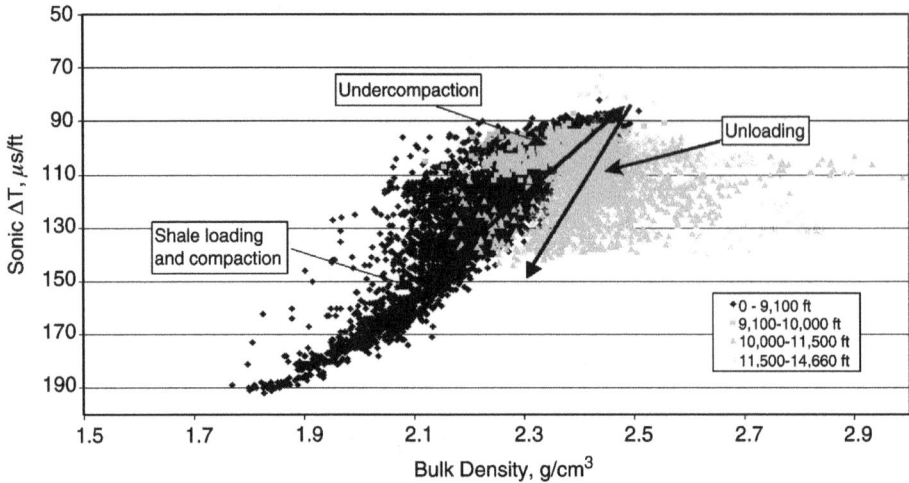

Fig. 1.36—Relationships between density and sonic ΔT can be used to distinguish overcompacted from normally compacted or undercompacted sediments, as shown in this figure.

ture is propagated a short distance away from the well by continued pumping. In either case, pumping is carried out at a constant rate, and pressure and the volume of fluid pumped are recorded as a function of time. Pressure-time curves typically look like those in **Fig. 1.39**.[22]

The theory behind using these tests to measure S_3 is that a fracture created during the test will, to minimize the energy required for its propagation, grow away from the well in an orientation that is perpendicular to the far-field least principal stress. Therefore, the pressure required to propagate the fracture will be equal to or higher than the least stress. Fracture propagation will stop when leakoff of fluid from the fracture and wellbore and into the formation occurs faster than the fluid is replaced by pumping. If pumping stops entirely, fluid leakoff will continue from the walls of the fracture until it closes, severing its connection to the wellbore. The fracture will close as soon as the pressure drops below the stress acting normal to the fracture (which is the least principal stress). The change in flow regime after pumping stops, from one in which the fracture contributes to fluid losses to one in which all fluid losses occur through the walls of the well, can be seen in pressure-time and other plots of pressure after shut-in (for example, pressure vs. the square root of time, **Fig. 1.40**).[23] The least principal stress is taken to be the pressure at which the transition in flow regime occurs. This pressure is a clear indication of the least stress regardless of whether the test created one or a multitude of subparallel fractures, as modeling suggests sometimes occurs.

Recently, and with evidence based on the ability to control pressure flowback in microfrac tests, it has been suggested that fracture closure can overestimate the least principal stress, and that choked flowback at various rates is required to determine that stress accurately. However, until such techniques become available for use in leakoff tests, the approach outlined below, based on the above theory, is the best for measuring S_3 in practice.

As for casing integrity tests or standard leakoff tests, extended leakoff tests are conducted after cementing a casing string and drilling out a short section (often between 20 and 50 ft) below the casing shoe. The conduct of the test should be as follows (refer to Fig. 1.39):

1. Perform a pretest pumping cycle with the formation isolated from flow using a very low flow rate. Pump until the pressure reaches a predefined upper limit for the casing. Record pressure and flow rate as a function of time, and draw a pressure-volume curve. This gives you a plot of pressure vs. volume (the slope of which is the system stiffness) if no fracture is initiated in the formation. The pressure vs. volume plot may initially be slightly concave up,

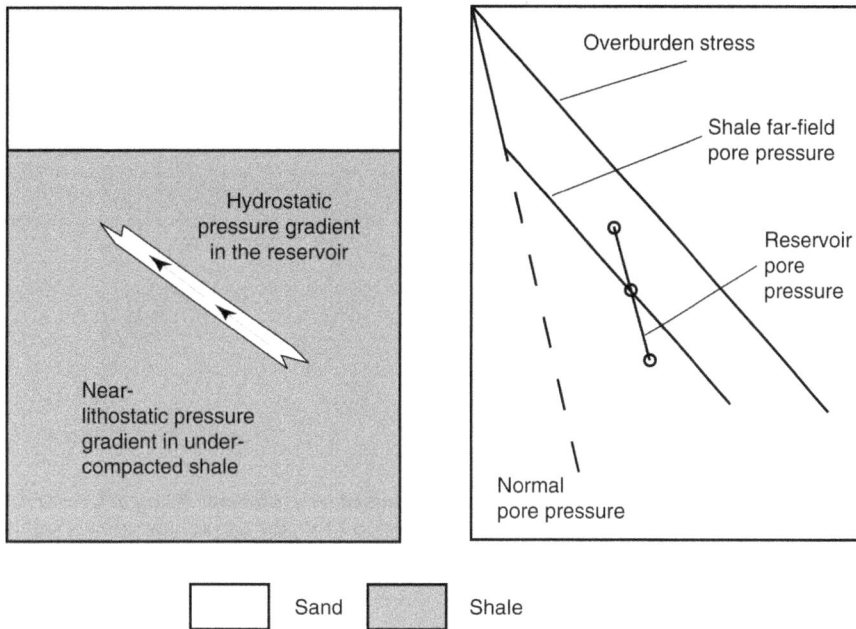

Fig. 1.37—This figure shows diagrammatically a typical centroid geometry (left) and pore pressure profiles (right) in a reservoir sand and in the surrounding shale that develop because of the centroid effect. Pressure at the top of the sand is higher than in the adjacent shale, whereas pressure at the base of the sand is lower (modified after Bruce and Bowers[21]).

indicating that the mud is compliant, possibly owing to entrained gas that is being forced into solution. Alternatively, refer to tables for the particular wellbore fluid and plot the appropriate pressure vs. volume curve by hand.

2. Open the formation to the well, pump at a low rate (¼ to ½ bbl/min), and overlay a plot of pressure vs. volume pumped on the curve from the casing test. The initial inflation should be approximately parallel to (or a little less steep than) the casing test curve. A concave down curve may be an indication of losses into the formation or a shoe integrity problem. If the former, it is possible to overcome this problem by stopping, flowing back, and starting again with a higher flow rate. If the latter, the problem must be dealt with before proceeding.

3. Pump until one of two things happens: either you have pumped a fixed volume of fluid above what was required to reach a given pressure (in which case there is a fluid loss problem to be dealt with), or the inflation pressure curve will break over, indicating the creation of a hydraulic fracture.

4. In a leakoff test, the formation would be shut in as soon as the slope of the pressure vs. volume curve begins to flatten, which is defined as the leakoff point. The fracture gradient at the shoe would be set equal to the pressure at the leakoff point. This is (in general) an upper bound on the least stress, and can, in the absence of better data, be used with caution in geomechanical models. However, to determine the least stress more accurately, continue pumping until the pressure stabilizes or begins to drop, and then shut in by stopping the pumps.

5. Record the pressure after shut-in until the pressure stabilizes. The value to which the pressure drops immediately after the pumps are shut off is typically called the instantaneous shut-in pressure (ISIP).

6. The least stress is determined using a variety of analysis methods. One method that is commonly used is to plot pressure vs. the square root of time after shut-in. The fracture closure pressure is defined by a change in the curvature of the line, as shown in Fig. 1.40.[23]

Schematic Geologic Cross Section   Pore Pressure Profile

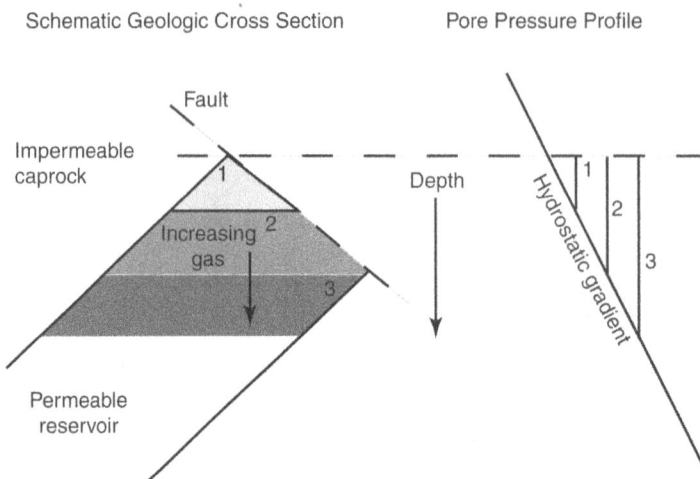

Fig. 1.38—This figure illustrates the buoyancy effect caused by systematic filling of a reservoir with low-density hydrocarbons. As filling progresses from Stage 1 to 2 to 3, the gas column grows, but the pressure is always in equilibrium with the centroid pressure at the gas-water contact, and so the pressure at the top of the reservoir increases (courtesy GeoMechanics Intl. Inc.).

7. Ideally, a second cycle should be performed. In this cycle, the previously created fracture is re-opened and extended, and then shut in again. Either or both of the following "step-rate tests" can be employed to refine the least principal stress measurement (**Fig. 1.41**).

(a) Re-open the fracture, starting with very low flow rates, and increasing the flow rate in discrete steps until the fracture opens and starts to grow. Maintain a fixed flow rate at each step until the pressure equilibrates. A plot of pressure vs. flow rate will have two slopes. At low pressure, fluid losses into the formation will result in a radial flow pattern in which flow rate increases systematically with pressure. Once the fracture is open, fracture growth and losses from the fracture walls will cause the pressure to increase much more slowly with flow rate. The intersection of lines fit to these two trends provides an upper bound on S_3. For a viscous fluid, extrapolation of the latter fit to zero flow gives a lower bound. The benefit of this procedure over 7b is that there is no need to extend the fracture as far.

(b) Open the fracture, pumping at the same rate as in the first cycle. The pressure at which a plot of pressure vs. volume deviates from the first cycle is the fracture reopening pressure. The difference between that pressure and the leakoff pressure is a measure of the formation tensile strength. Once the fracture begins to extend, decrease the pump rate in fixed increments, recording flow rate and pressure after the pressure has equilibrated at each step. Analyze this data using the same technique as described for step-rate reopening. The benefit of this procedure over 7a is that a measure of tensile strength can be obtained.

1.6.4 Estimates of Least Principal Stress From Ballooning. Ballooning is a process that occurs when wells are drilled with equivalent static mud weights close to the leakoff pressure. It occurs because during drilling, the dynamic mud weight exceeds the leakoff pressure, leading to near-wellbore fracturing and seepage loss of small volumes of drilling fluid while the pumps are on. When the pumps are turned off, the pressure drops below the leakoff pressure, and the fluid is returned to the well as the fractures close. This process has been called "breathing" or "ballooning" because it looks like the well is expanding while circulating, and contracting once the pumps are turned off. This behavior can be identified on a PWD log (**Fig. 1.42a**).[24] It can be differentiated from a small kick or gas influx (which often is used as an indication to in-

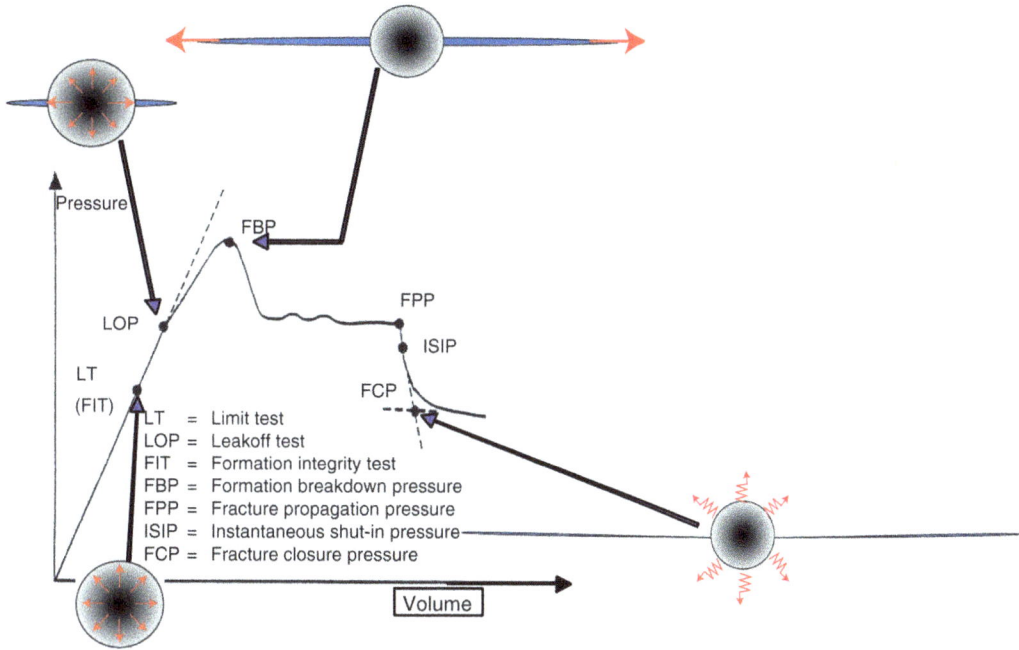

Fig. 1.39—This figure shows an idealized pressure vs. time plot for an extended leakoff test (modified after Gaarenstroom[22]) (courtesy GeoMechanics Intl. Inc.).

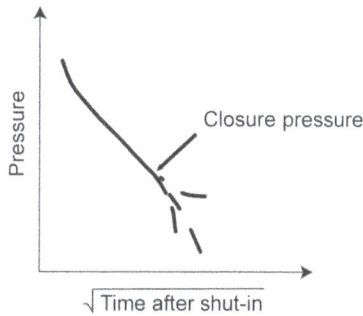

Fig. 1.40—Pressure vs. square root of time plots are often used to detect fracture closure. On these plots, the closure pressure is the pressure at the inflexion point of the pressure decay curve (modified after Nolte and Economides[23]).

crease mud weight owing to the perception that it reveals gas pressures higher than the equivalent static mud weight), as shown in Fig. 1.42b. Increasing the mud weight in a ballooning well can lead to massive lost circulation.

Ballooning is an important measure of least principal stress magnitude because it is essentially an inadvertent leakoff test conducted while drilling. The static mud weight is a lower bound on the magnitude of the least stress, and the dynamic mud weight is an upper bound. In some cases, a shut-in break can be detected, which is a very accurate measure of the least stress. The only problem is that it can be difficult to identify the depth at which the ballooning incident took place (although it is reasonable to assume that it occurred close to the bit). This is a particular problem when there is a very long openhole interval. Fortunately, it is often

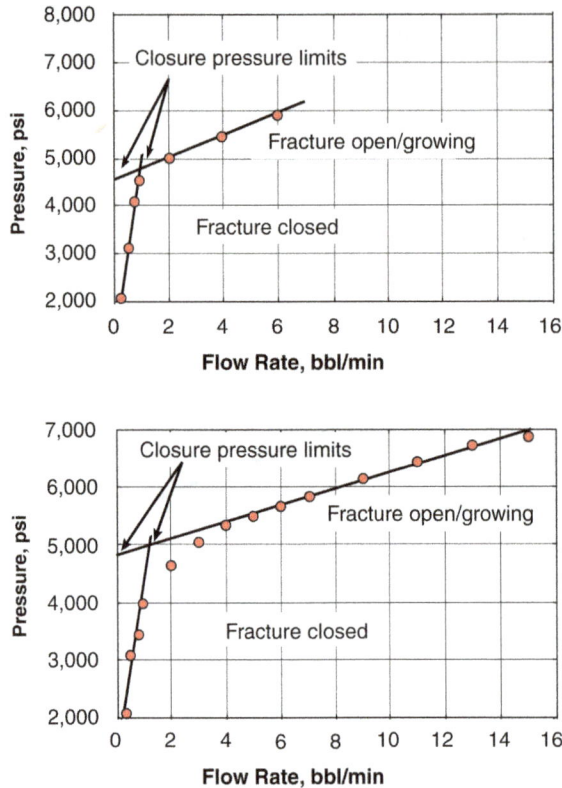

Fig. 1.41—Plots of pressure vs. flow rate showing that at low flow rates, before fracture opening, pressure increases rapidly with flow rate, but once the fracture opens, a large increase in flow rate can be accommodated with only a small increase in pressure. Sometimes the transition between the two regimes is abrupt (top); sometimes it is more gradual and requires a wide range of flow rates to delineate (bottom) (courtesy GeoMechanics Intl. Inc.).

possible to find the location of the fractures created by the ballooning incident by a change in LWD resistivity recorded before and after the event.

1.6.5 Using Wellbore Failure to Constrain the Magnitude of S_{Hmax}. Once independent knowledge of S_v and S_{Hmin} is available, S_{Hmax} can be determined from the widths of wellbore breakouts in vertical boreholes. Because the stress concentration around the well and the rock strength are equal at the point of the maximum breakout width, it is possible to re-arrange Eq. 1.8 to solve for S_{Hmax}, as shown in **Fig. 1.43**. Solving for S_{Hmax} also requires a model for rock strength and knowledge of the pore pressure and mud weight. While the equations presented here are technically accurate only for elastic, brittle rock, utilizing the results to select the appropriate mud weight for drilling future wells requires only that the same model be applied to predict wellbore stability as was used to determine the stresses.

Once breakouts have formed, they deepen but do not widen. Thus, the original width of the breakout is largely preserved, and calculations of stress magnitudes based on breakout width do not have to be adjusted for changes in the wellbore shape associated with subsequent failure (see **Fig. 1.44**).[8,25]

As previously discussed, breakout width can be determined very accurately using acoustic or electrical image logs run after the well has been drilled. With the advent of resistivity, density, and porosity LWD tools that produce an image of the borehole wall behind the bit, it is

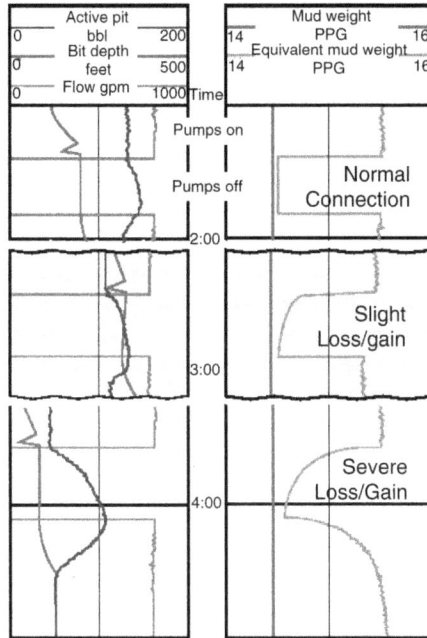

(a)

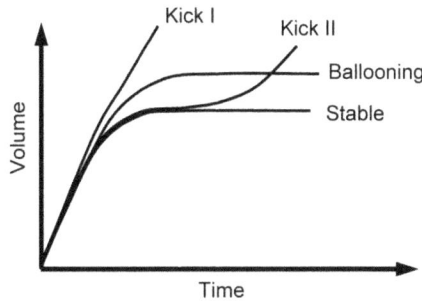

(b)

Fig. 1.42—(a) Signature of normal connections (top) and moderate and severe ballooning (middle and bottom) on a PWD log. The difference between ballooning and a potential well-control incident can be detected in a plot of mud volume vs. time after pumps-off (b) (after Ward and Beique[24]).

now possible to determine breakout widths while drilling, which then makes it possible to determine S_{Hmax} in real time. On the other hand, in the absence of borehole image data, we can only place bounds on the width of presumed breakouts if they can be detected using the electrode pads of a dipmeter tool (pad width is typically about 30° in an 8.5-in. hole). Therefore, using mechanical calipers, it is possible only to place constraints on the magnitude of S_{Hmax}.

The presence of tensile fractures in a well also gives some indication of relative stress magnitudes. This is because, as previously discussed, tensile fractures can develop at the wellbore wall only if the far-field horizontal stresses are sufficiently different. For example, when the mud weight is equal to the pore pressure, a strike-slip equilibrium state of horizontal stress is required for tensile wall fractures to develop in a vertical well.

Constraining the Magnitude of S_{Hmax} in Deviated Wellbores. In deviated wellbores, it is possible to constrain not only the orientation but also the magnitude of S_{Hmax}. This is because,

Wellbore Breakouts

$$\sigma_{\theta\theta} = S_{Hmin} - S_{Hmax} - 2(S_{Hmax} - S_{Hmin})\cos2\theta_b - 2P_p - \Delta P - \sigma^{\Delta T} = C_{eff}$$

$$\Longrightarrow \ S_{Hmax} = [C_{eff} + 2P_p + \Delta P + \sigma^{\Delta T}) - S_{Hmin}(1 + 2\cos2\theta_b)]/(1 - 2\cos2\theta_b)$$

Tensile Fractures

$$T_o = \sigma_{\theta\theta}{}^{min} = 3S_{Hmin} - S_{Hmax} - 2P_p - \Delta P - \sigma^{\Delta T}$$

$$\Longrightarrow \ S_{Hmax} = 3S_{Hmin} - 2P_p - \Delta P - T_o - \sigma^{\Delta T}$$

T_o approximately 0

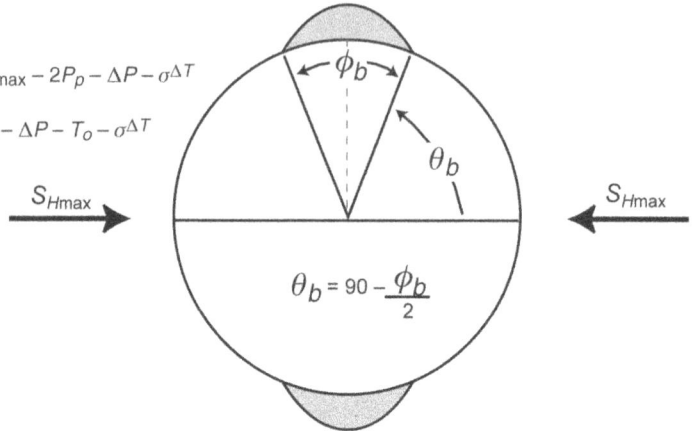

Fig. 1.43—Schematic diagram of a breakout and the Kirsch equations that are used to constrain stress magnitudes based on the widths of wellbore breakouts and the presence or absence of drilling induced tensile wall fractures. These equations apply to a vertical well when S_v is a principal stress (courtesy GeoMechanics Intl. Inc.).

in deviated wells, the position of wellbore breakouts depends on stress magnitude as well as on stress orientation (**Fig. 1.45**). It is also possible at the same time to constrain the rock strength using breakout width.

This sort of analysis can be carried out in multiple wells by use of combined analyses of tensile and compressive wellbore failures. If the wells have a sufficient number of different deviations and azimuths, a very accurate stress state can be determined using a Monte Carlo approach. Essentially, this is simply a more quantitative way of doing the same thing as creating a figure similar to Fig. 1.45 for each of the wells and overlaying the figures to identify the one stress state that comes closest to matching all of the observations. If the results for all wells are not consistent with a single stress state, then it is clear that the stress state must be different at the locations of the anomalous wells. This provides powerful evidence for reservoir compartmentalization or the influence of local sources of stress.

The constraints on in-situ stress dictated by the strength of pre-existing faults shown in Fig. 1.10 can be combined with observations of tensile and compressive wellbore failures to refine estimates of in-situ stress, as shown in **Fig. 1.46**. The frictional strength limits are as described above. Overlain on these limits are lines defining the stress states for one specific well that would cause tensile or compressive failure to occur. The near-vertical, fine lines to the left of the stress polygon represent stress states (values of S_{Hmax} and S_{Hmin}) that would create tensile fractures at the wall of this deviated well for tensile strengths of 0, 500, and 1,000 psi. If the rock has a given tensile strength and tensile cracks are found, it indicates that the stress state must lie to the left of the appropriate line. Because in most cases pre-existing flaws exist that can be opened by elevated mud weights, the effective tensile strength is often assumed to be zero. Therefore, for this example, it is apparent that if tensile failure is observed, the stress state must lie at the extreme lower left-hand corner of the strike-slip region or the extreme left side of the normal faulting region, a transitional strike-slip or normal faulting stress state for

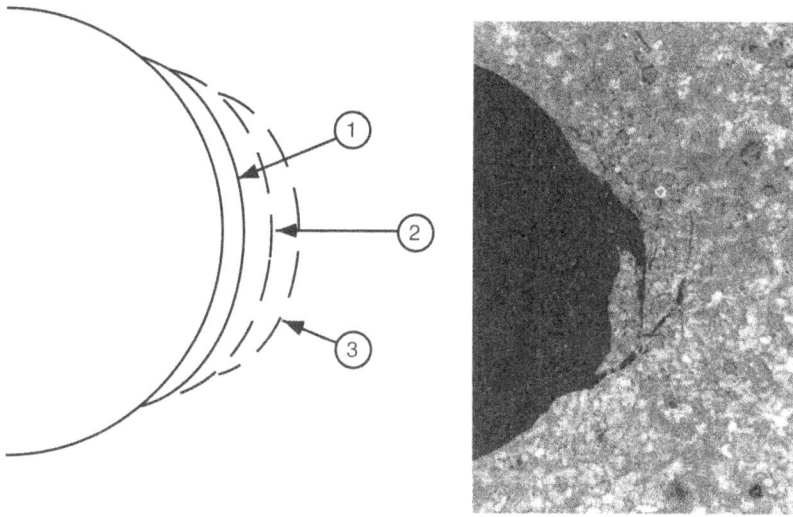

Fig. 1.44—Theoretical computations performed using boundary element methods reveal that once a breakout has formed, additional failure will occur only at the back of the breakout (left image after Ref. 8, M.D. Zoback *et al.,* "Wellbore Breakouts and In-Situ Stress," *J. Geophysical Research,* Vol. 90, No. B7, 5523; © American Geophysical Union; reproduced/modified by permission of the American Geophysical Union). Thus, the breakout may deepen with time, but not widen. Laboratory experiments reveal that breakout formation is consistent with this prediction (right image after Haimson and Herrick[25]).

which S_{Hmax} can range from 14 to as high as 38 lbm/gal. S_{Hmin} is much better constrained to between approximately 13 and 15 lbm/gal.

On the other hand, if no tensile fractures are observed and lab or log data indicate that C_o is between 10,000 and 15,000 psi, the stress state can lie anywhere to the right of the vertical lines, and within the region between the near-horizontal, light gray curves plotted for those values of C_o. In other words, S_{Hmin} can range from 13 to as high as 30 lbm/gal, and S_{Hmax} is somewhat better constrained to a range from 26 to 33 lbm/gal. If S_{Hmin} had been measured using an extended leakoff test to be approximately 20 lbm/gal, then the range of possible S_{Hmax} values would be only slightly smaller (between 28 and 33 lbm/gal). In general, in near-vertical wells, the presence of tensile cracks severely limits the magnitude of S_{Hmin} without constraining S_{Hmax}, whereas observations of breakouts provide weaker constraints on S_{Hmin} than on S_{Hmax}. Multiple observations of breakouts in strong and weak rocks can be overlain to restrict the allowable stress state to the region common to the stress states allowed by all of the observations.

When using this sort of analysis, the important thing to keep in mind is that all you are doing is providing constraints on the stress state. For example, suppose that no breakouts had formed in the well described by Fig. 1.46 and the rock strength was somewhere between 10,000 and 15,000 psi. In that case, the stress state could definitely not lie above the line corresponding to C_o = 15,000 psi, and is most likely to lie below the line corresponding to C_o = 10,000 psi (i.e., anywhere within the low-stress region, which includes the entire normal faulting stress regime). Additional observations would be required to reduce the large uncertainty in this result.

Constraining the Stress State in the Visund Field. As an example of an instance in which redundant data confirm the stress and strength values derived from combined analysis of wellbore failure and frictional constraints, consider **Fig. 1.47**,[26] prepared on the basis of data from an inclined well in the Visund field, North Sea. The frictional faulting constraints were derived from S_v and P_p calculated as described above. Breakouts were identified in caliper data, and intermittent tensile fractures were also seen in both vertical and inclined sections of the well.

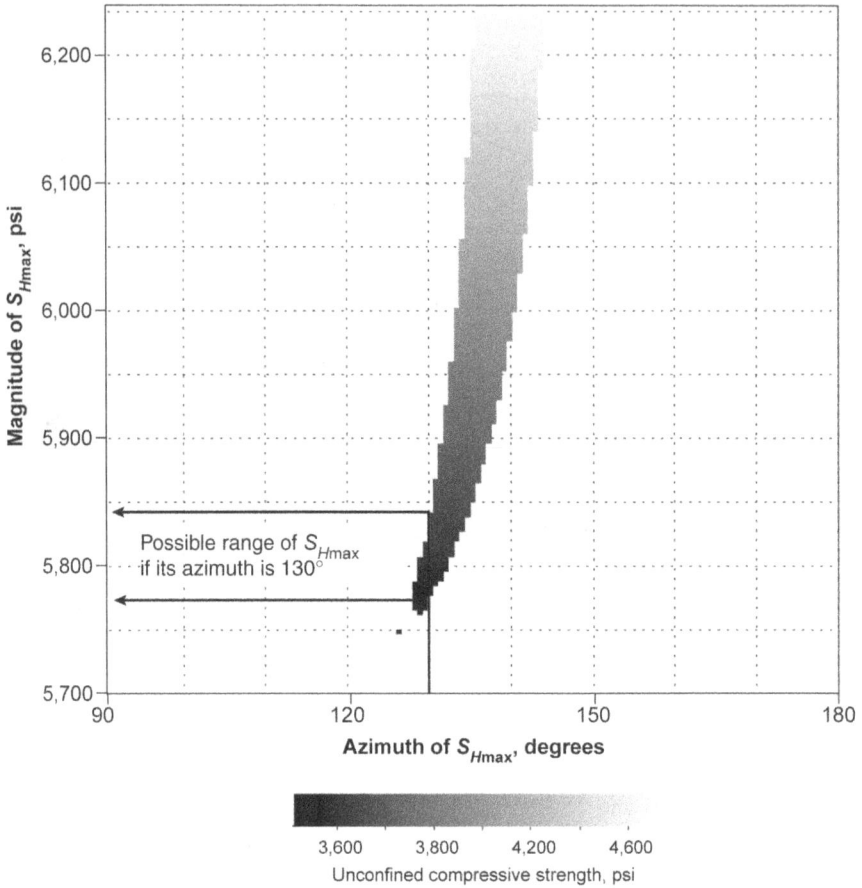

Fig. 1.45—In an inclined well, stress magnitudes can be determined simply from knowledge of the orientation of wellbore breakouts. In this case, given the magnitudes of S_v and S_{Hmin} and the orientation of S_{Hmax}, it is possible to constrain the magnitude of the maximum stress. In addition, it is possible to constrain in-situ strength using the breakout width. For example, if the azimuth of S_{Hmax} is 130°, its magnitude is 5,770 to 5,840 psi, and the in-situ unconfined compressive strength is approximately 3,600 psi.

Breakouts and tensile cracks in the vertical section provided information on the stress orientation. Based on log data, C_o ranged from 20 to 25 MPa. The light gray lines labeled 20 and 25 correspond to the stresses constrained by the breakout observations and the rock strength parameters.

Because tensile cracks are more likely to occur when circulation cools the well, it is necessary to account for that cooling in the stress constraints from their occurrence. That shifts the tensile failure line to the right. It is not necessary to include cooling in the breakout analysis, however, because the breakouts would be more likely to occur after the well temperature had equilibrated. The final constraint, based on frictional faulting theory, is that the stress state cannot lie outside the polygon.

Taken together, these observations constrain the stress state to lie in the small region bounded by the light gray lines on the top and bottom, the thin dark near-vertical line on the right, and the edge of the stress polygon on the left. This provides a very precise value for S_{Hmin} between 52.5 and 54.5 MPa, and it constrains S_{Hmax} to be between 73 and 76 MPa. A leakoff

Stress and Strength Constrained by Wellbore Failure
(strength contours in psi)

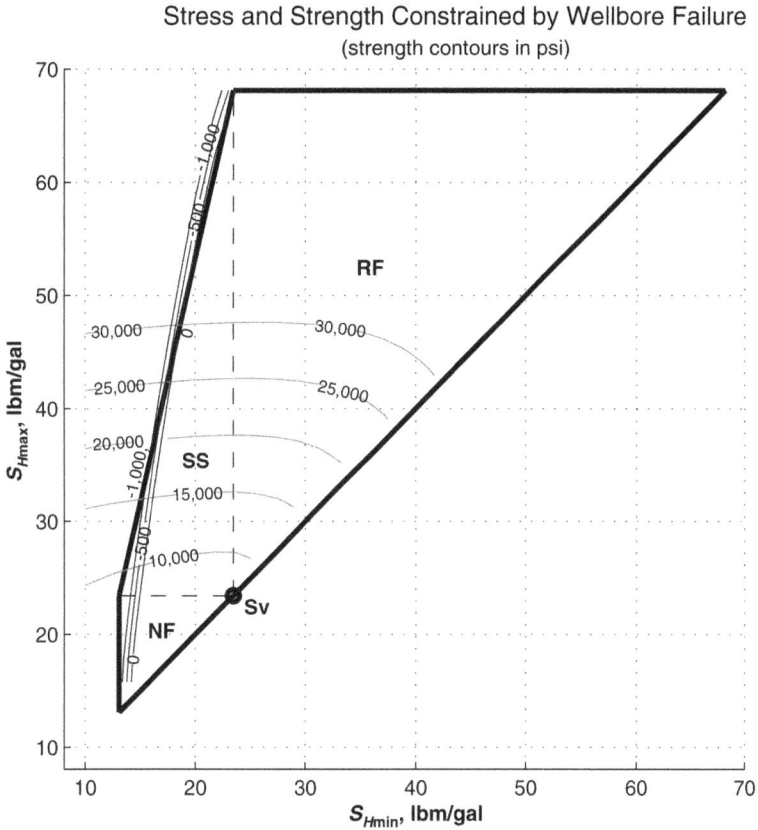

Fig. 1.46—Plots of lines corresponding to the stress magnitudes required in an inclined well for breakouts to form with the given width (in light gray), and for tensile failure to be initiated for a given tensile strength (fine dark lines), superimposed on the stress limits dictated by the strength of the crust if stresses are limited by the frictional strength of pre-existing faults. These lines correspond to equations of the form shown in Fig. 1.43.

test provided redundant information on S_{Hmin} and confirmed its value predicted from the constraints imposed by observations of failure.

1.7 Predicting Wellbore Stability

Once a geomechanical model has been developed that quantifies the principal stress magnitudes and orientations, the pore pressure, and the rock properties, it is possible to predict the amount of wellbore instability as a function of mud weight and properties. This makes it possible to reduce drilling costs by keeping lost time low and by designing wells just carefully enough to minimize problems without excessive cost. A further benefit of considering geomechanical risk is that when problems are encountered, their causes can be recognized and plans can be in place to mitigate their effects with minimal disruption of the drilling schedule.

Fig. 1.48 shows the time-depth plot of an offshore well that was designed and drilled without the use of geomechanical modeling. After setting the first string to isolate a shallow hazard, the remaining casing depth points were selected based on drilling experience in an offset block, supplemented by pore pressures predicted using seismic data. Considerable problems were experienced because of the length of the fourth casing interval. Subsequent geomechanical analysis revealed that the fourth casing interval was too long because the second and third casing strings were set too shallow (the dark dashed line on Fig. 1.48 representing AFE). A

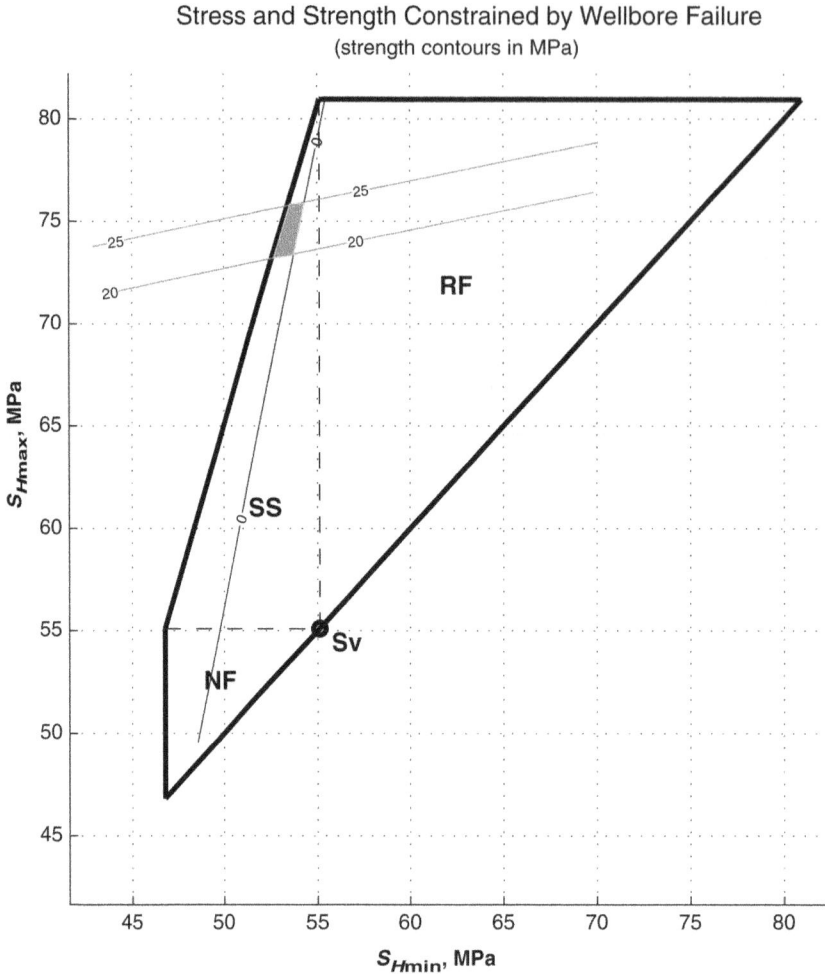

Fig. 1.47—Taken from Wiprut and Zoback,[26] this figure illustrates a case in which frictional constraints combined with observations of wellbore failure, calculated values of S_v and P_p, and measured rock strengths provided excellent constraints on the magnitudes of the horizontal stresses. A leakoff test analyzed separately from this analysis confirmed the predicted magnitude of S_{Hmin}.

new casing design was subsequently developed based on the geomechanical analysis that mitigated the problem with the fourth casing string and led to a significantly less costly well (the heavy black line on the figure).

Because the geomechanical parameters (stress, pore pressure, and strength) are largely out of our control, there are a limited number of things that can be done to minimize geomechanical stability problems. One (as illustrated in Fig. 1.48) is to optimize the locations of casing seats. Another is to optimize mud weight and drilling parameters, minimizing swab and surge while running pipe and maintaining an appropriate pumping rate to keep equivalent circulating density (ECD) low, in situations where it is necessary to maintain a close tolerance. Other options include changing the well trajectory, where that is possible, or at least identifying those trajectories that are least likely to cause drilling problems. An example in which this is particularly valuable is in drilling moderate-reach wells where there is a choice in the depth, length, and inclination of deviated hole sections. It may also be possible by use of appropriate drilling fluids to increase the pressure required to propagate hydraulic fractures, thereby reducing the

Days vs. Depth

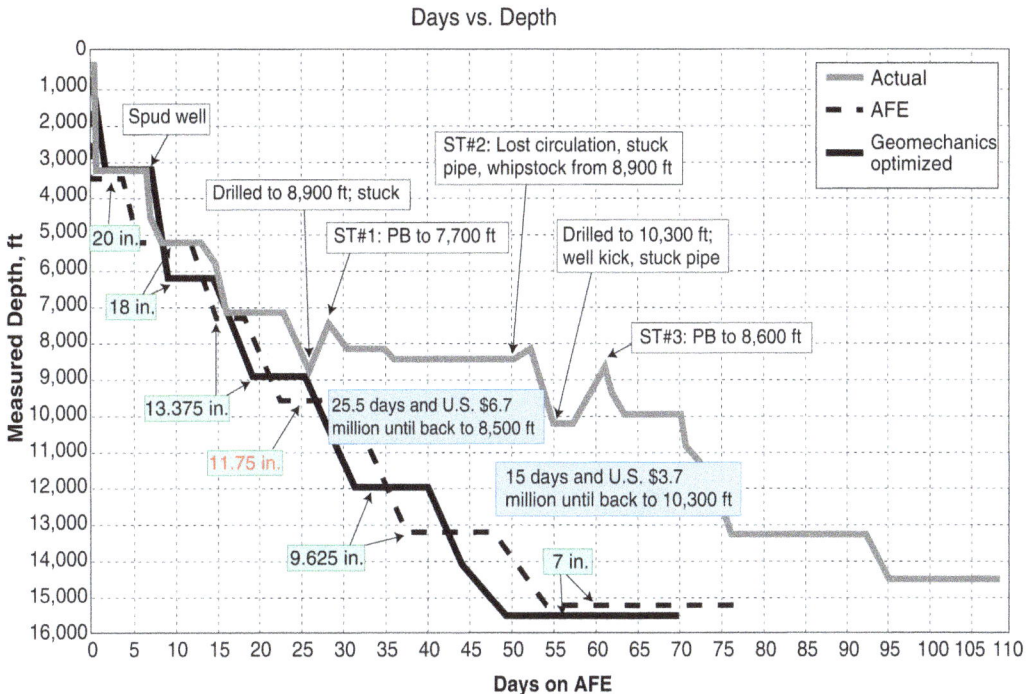

Fig. 1.48—Depth vs. time plot of an offshore well. The dashed black line is AFE. The gray line shows the initial well, drilled using the design shown in black. The solid black line represents a new well design optimized using geomechanics (courtesy GeoMechanics Intl. Inc.).

leakoff pressure, and recent developments reveal that it may be possible also to increase the leakoff pressure by changing near-wellbore conditions with use of special materials or by heating the well.

To maximize the number of options, geomechanical design constraints should be developed as early as possible in the life of a field, particularly in cases in which development will be carried out from a small number of fixed locations. This way, recovery can be maximized with the smallest number of wells drilled along risky trajectories and the lowest facilities cost.

1.7.1 Predicting Failure in Wells of Any Orientation. Fig. 1.49 shows how wellbore stability in wells of all orientations can be illustrated by a lower hemisphere projection of the likelihood of breakout formation for a single stress state at a single depth. Wells plot on the diagram at locations defined by their orientations. The deviation is represented by radial position (vertical wells plot in the center of the diagram, and horizontal wells plot at the perimeter). The well azimuth is indicated by its circumferential location in degrees clockwise from the top of the diagram; wells deviated to the north (0°) are at the top of the diagram, wells deviated to the east (90°) are on the right side, wells deviated to the south (180°) are at the bottom of the diagram, and wells deviated to the west (270°) are on the left side (Fig. 1.49a).

Fig. 1.49b illustrates the required mud weight to avoid excessive compressive wellbore failure (breakout) as a function of position defined in Fig. 1.49a. Darker gray represents orientations with higher mud weight, and lighter gray represents orientations with lower mud weight, including the strength required for a given degree of failure and mud weight and the amount of failure for a given mud weight and rock strength.

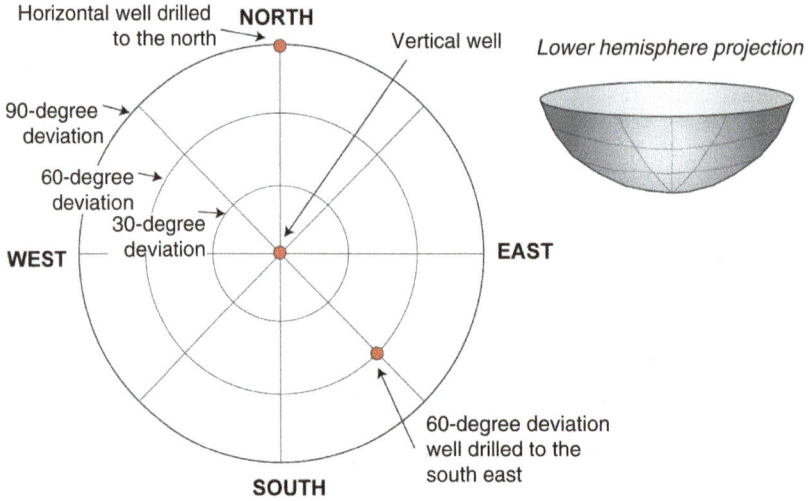

(a)

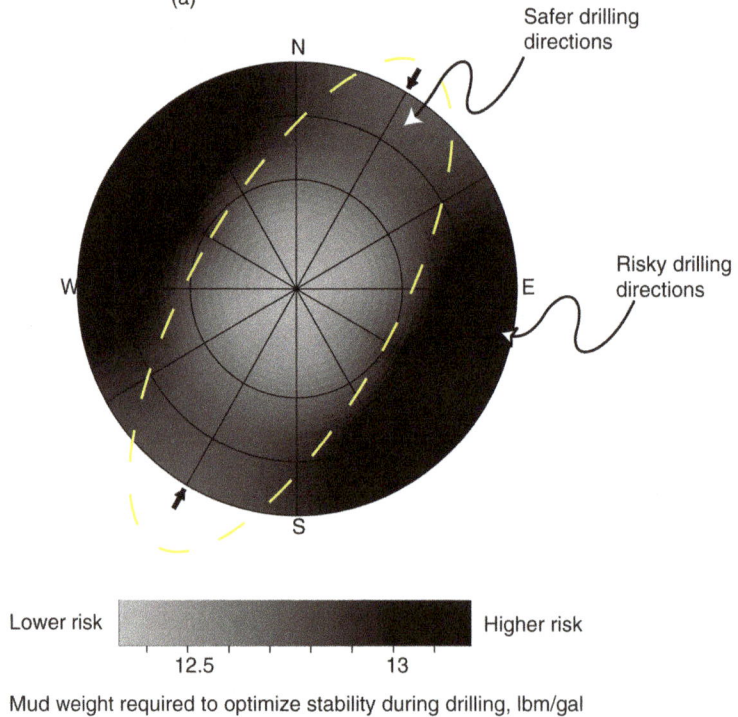

Safer drilling directions

Risky drilling directions

Lower risk ▬▬▬▬▬▬▬▬▬▬ Higher risk

12.5   13

Mud weight required to optimize stability during drilling, lbm/gal

(b)

Fig. 1.49—The risk of failure as a function of wellbore orientation can be displayed using a lower-hemisphere projection. The construction of this diagram is illustrated in (a). (b) Shows an example plot of the minimum safe mud weight to avoid excessive failure as a function of wellbore orientation as defined in (a) (courtesy GeoMechanics Intl. Inc.).

Risk of excessive rock failure around a well can be quantified in a variety of ways—for example, using the normalized radius to which the first episode of brittle failure extends (deeper is worse), or, in the case of analyses using elastoplasticity, the volume of rock that is predicted to reach the yield criterion, the depth of the yielded area, or the onset of a critical plastic strain. In the cases presented here, risk is quantified using the width at the wellbore of the failed zone or breakout. The reason breakout width is preferred is that it is easy to measure using logging data and does not change significantly with time. The same criterion should be used both to determine the magnitudes of the in-situ stresses and to calibrate stability models to improve predictions.

1.7.2 Defining the Mud Window at a Single Depth. Loosely speaking, the mud window can be defined as the range of equivalent densities or pressures that avoid drilling problems. **Fig. 1.50** shows how the mud window is defined for a single depth. The lower bound is the mud weight required to prevent excessive wellbore failure as a function of orientation (Fig. 1.50a). Similar figures can be developed to describe the risk of lost circulation, which defines the upper bound of the mud window (Fig. 1.50b). The mud window at a given depth (Fig. 1.50c) is the difference between the maximum mud weight before lost circulation occurs and the minimum mud weight to avoid excessive breakout.

In Fig. 1.50a, the variation in required mud weight to prevent excessive breakout is less than 0.9 lbm/gal. However, the lost circulation pressure (Fig. 1.50b) varies significantly. This is because to generate a lost circulation event, the wellbore pressure must be large enough to do three things: (1) create a fracture at the borehole wall, (2) propagate that fracture through the near-wellbore stress concentration, and (3) extend the fracture against the least principal far-field stress. The far-field stress, of course, is constant, and so the fracture propagation pressure is essentially independent of wellbore orientation. However, the initiation and link-up pressures are strong functions of wellbore orientation. Thus, it can be helpful to choose a wellbore orientation on the basis of maximizing the lost circulation pressure to reach a drilling objective in a low-mud window environment. Notice in this case that the mud window varies from zero for near-horizontal wells drilled to the NW or SE, to 2 lbm/gal for vertical wells, to more than 6 lbm/gal for wells drilled to the NE or SW. In this environment, wells that must be drilled to the NW or SE at this depth should have as small a deviation from vertical as possible.

What is the criterion used to establish the minimum safe mud weight? Clearly, it is one that will minimize the risk of complete hole collapse. But in addition, the volume of cuttings, the inclination of the well, and the position around the well of the breakouts can also influence this value. The cuttings volume and well inclination are important because of hole-cleaning issues. The larger the cuttings volume per unit hole length, the better hole cleaning needs to be. And, because hole cleaning is easier in vertical wells than in deviated wells, vertical wells can accommodate larger amounts of failure. Increases in pumping rate and carrying capacity, or reduced penetration rates, can mitigate the risk associated with excessive cuttings volumes. Because in deviated wells there is considerable pipe contact with the top and bottom of the well, breakouts in these locations are likely to be more problematic than breakouts on the sides of the hole. However, if the well needs to be steered, breakouts on its sides may adversely affect directional control. Because breakout width is a relatively easy measurement that is directly related to cuttings volume, and because breakout depth increases with time, we ordinarily choose the breakout width as the criterion to establish the appropriate minimum mud weight. Because breakouts have been observed that extend more than 100° on each side of a well in vertical wells drilled into some shales, this is an appropriate limit for such wells. Narrower breakouts will become problematic in more brittle rock, so in practice it is best to use a breakout width limit of 90° for breakouts on each side of a vertical well. This limit means that at least half of the wellbore circumference must be intact, a condition that has been referred to as "sufficient to maintain arch support" in sanding analyses. Because hole cleaning is more difficult in deviated wells, the maximum safe breakout width should be reduced as deviation increases.

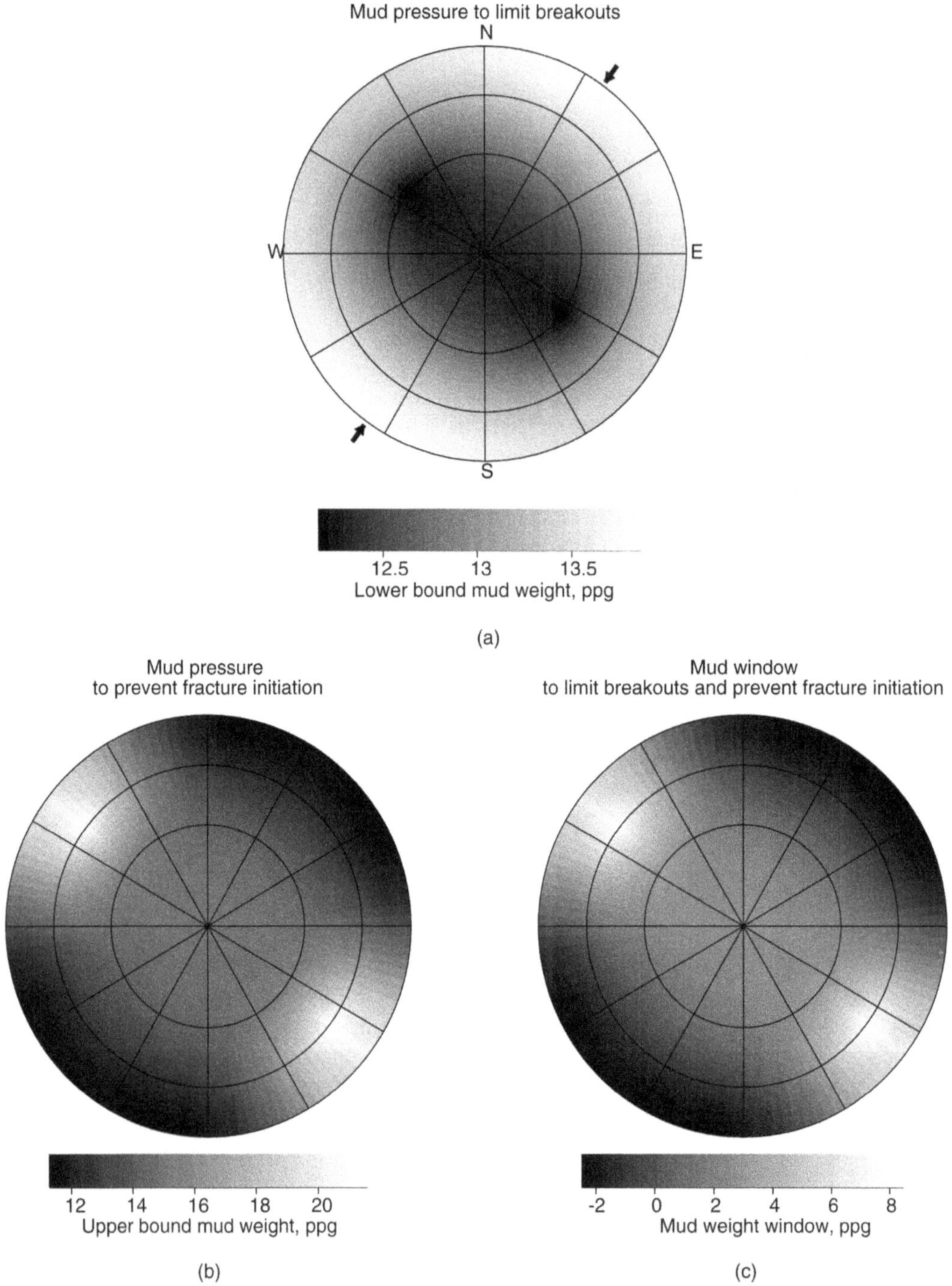

Mud pressure to limit breakouts

12.5   13   13.5
Lower bound mud weight, ppg

(a)

Mud pressure
to prevent fracture initiation

Mud window
to limit breakouts and prevent fracture initiation

12   14   16   18   20
Upper bound mud weight, ppg

-2   0   2   4   6   8
Mud weight window, ppg

(b)   (c)

Fig. 1.50—Mud weight to prevent breakouts (a), to prevent lost circulation (b), and the mud window (c), which is the difference between the two. Notice that the lost circulation pressure can be a function of wellbore orientation. This is because it depends on the fracture initiation pressure and the pressure required to propagate the fracture away from the near wellbore as well as on the pressure required to propagate the fracture in the far field.

It is important to remember that it is not necessary to completely avoid breakout formation to drill wells safely. Using such an overly restrictive criterion is not only unnecessary but will inevitably lead to recommendations for excessively high mud weights in situations in which these are not warranted.

1.7.3 Casing Seat Selection. Analyses illustrated in Figs. 1.49 and 1.50 were carried out at a single depth. However, it is necessary while drilling to maintain stability over the entire open-hole section between casing points. Therefore, analyses of stability must be carried out over that entire depth range. Using the results, the positions of casings can be adjusted to maximize wellbore stability while staying within engineering constraints. While the analysis requires knowing rock properties in detail, it is not necessary to do the calculations using every depth point. This is because although there is considerable variation in rock properties, narrow zones of severely weak rock do not, in practice, cause excessive problems. Furthermore, stresses and pore pressures generally vary slowly with depth and horizontal location. Where wells cross faults, changes in age or lithology, or fluid pressure barriers, abrupt changes in stress and pore pressure are possible. In addition, systematic changes in stress orientation and magnitude often occur adjacent to faults and salt bodies. Provided that the geomechanical model incorporates these effects, it is sufficient to use smoothly varying rock properties. The natural geological variability can be taken into account using statistical methods, as discussed next.

Fig. 1.51[27] is an illustration of the impact of geomechanics on casing selection for an offshore well. It shows plots of the equivalent densities of the pore pressure and the leakoff pressure as a function of true vertical depth for a vertical well (for deviated wells, it can be drawn as a function of measured depth). To the right of each figure is shown a casing design diagram. Superimposed on the equivalent mud weight plot are shaded rectangles that represent the limiting mud weights that are both above the minimum required mud weight (in the left plot, the pore pressure, shown in light gray) and below the maximum required mud weight (in all of these plots, the least principal stress, shown in black) at all depths within each casing interval. The upper and lower bounds on the mud weight can be selected from among several different limits. For example, in sections of underpressured sands, the upper limit may be dictated by the pressure above which differential sticking may occur. As shown in very dark gray in the center and right figures, the lower limit could be the collapse pressure computed using geomechanical analysis. And, as discussed in the context of Fig. 1.50b, the upper bound to prevent lost circulation can be the pressure required to initiate, to propagate, or to extend a hydraulic fracture.

Fig. 1.51a shows a predrill design based on offset experience and the assumption that the pore pressure and the fracture gradient are the upper and lower bounds on the mud window. When geomechanical stability is considered (Fig. 1.51b), the results indicate that over a significant portion of the well, the minimum safe mud weight required to avoid excessive breakout development (the collapse pressure) is greater than the pore pressure. One consequence is that the fourth casing section has an extremely narrow mud window. In fact, severe drilling problems developed in this section, necessitating two sidetracks and considerable lost time. On the right is shown a new well design utilizing a geomechanical model to establish safe casing points (Fig. 1.51c). This model indicates that it is possible to extend the depths of the second and third casing strings, thereby reducing the required length of the fourth. This not only increases the margin for the fourth casing string, it also makes it possible to reach the reservoir with one less casing than required by the original design.

1.7.4 Validating the Geomechanical Model. It is important when using geomechanical analysis to use prior drilling experience to validate the geomechanical model. This is possible, even when no log data are available for previous wells, by using drilling events such as mud losses, tight spots, places necessitating repeated reaming, and evidence of excessive or unusually large cuttings. If wellbore stability predictions for existing wells are capable of reproducing previous

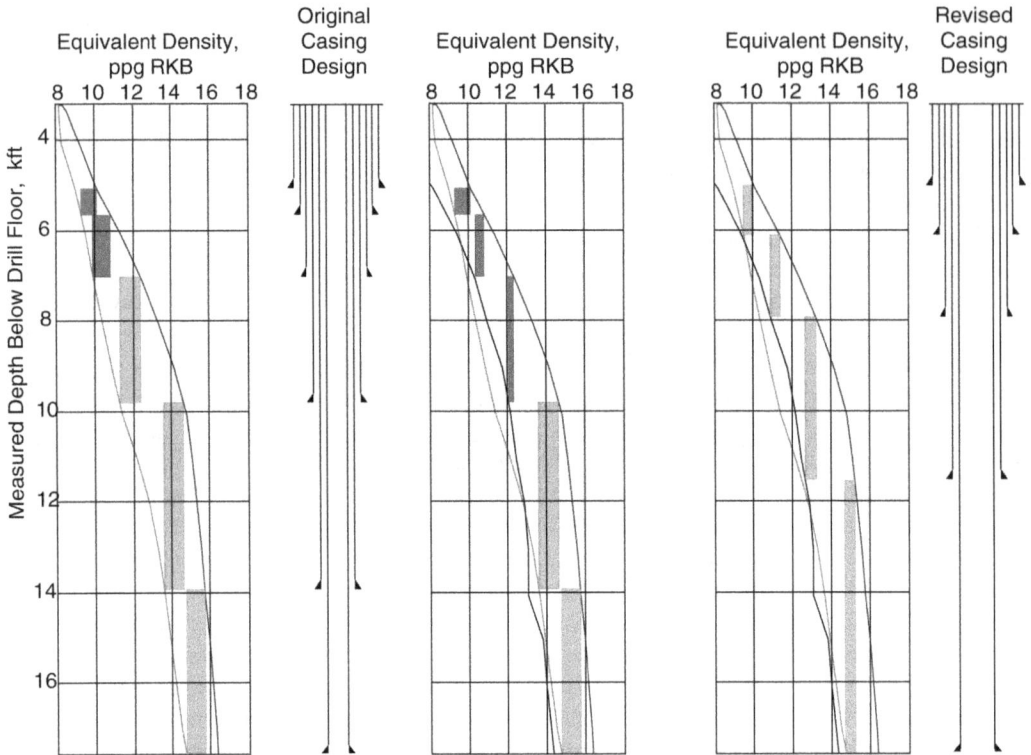

Fig. 1.51—Geomechanical analysis of two casing designs for the same well. On the left is a predrill design, made assuming that the pore pressure and the fracture gradient limit the mud window. The mud window for each casing interval is shown as a shaded rectangle. In the center is the impact of considering the collapse pressure on the predrill design. There is an extremely narrow mud window for the third casing interval. On the right is a design made utilizing geomechanics, which adjusts the positions of the first two casing seats to reduce the length of the third cased interval. Not only does this design avoid the extremely narrow mud window for the fourth casing, it also reduces the required number of casing strings.[27] (Reprinted from "Comprehensive Wellbore Stability Analysis Utilizing Quantitative Risk Assessment," Moos *et al., J. of Petroleum Science and Engineering,* Vol. 38, pages 97–109, © 2003, with permission from Elsevier.)

drilling experience, we can be confident that the geomechanical model is appropriate for use in predicting the stability of planned wells.

Fig. 1.52 shows an example prediction of the degree of wellbore instability (quantified in terms of breakout width) in a vertical well in deep water. The figure was prepared using the drilling mud program for that well and the geomechanical model developed for the field based on offset experience. The model indicates that while the section above 5,800 ft will be quite stable (no failure is predicted), below that depth, failure will progressively worsen until, at 7,400 ft, it is severe enough to cause considerable drilling problems. Although the model was not able to explain problems encountered in this well above 5,400 ft, it turned out that these problems were not caused by geomechanics because they were mitigated with no change in mud weight, and no evidence of enlargement was found in log data from this interval. In contrast, considerable drilling difficulties were encountered just above 7,800 ft in this well that were detailed in drilling reports, including several packoff and lost-circulation events. These problems required setting casing prematurely at that depth. Single-arm caliper logs subsequently revealed that this section was severely enlarged.

Predicted Breakout
Width, degrees

0   45   90   135   180

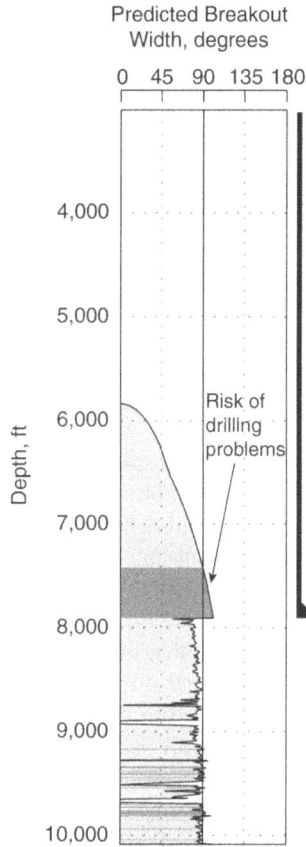

Fig. 1.52—Predicted breakout width as a function of depth, calculated using the actual mud weights used to drill the well. The black line indicates the failure width criterion (<90°) for maintaining wellbore stability. Drilling problems should be expected if the predicted failure (shaded) exceeds the failure criterion.

Below the casing point, the mud weight was increased, which reduced hole instability problems in the remaining sections of the well as predicted by the calculations. Nevertheless, there was some evidence for wellbore enlargements in caliper data in the interval below 9,200 ft, even for the higher mud weights used. These sections were those in which the predicted breakout width exceeds 90°, lending support to the validity of the geomechanical model. Subsequently, the model was used to design a number of wells, all of which reached total depth (TD) without incident.

1.7.5 Geomechanical Design With Very Little Data. It is not necessary to have a well-constrained stress state to utilize geomechanical design principals. Sometimes, knowing just the stress regime (normal, strike-slip, or reverse), it is possible to estimate relative risk as a function of wellbore deviation and determine the importance of knowing the stress orientation. If geological analysis provides information about stress orientation as well, it is also possible to determine the relative risk as a function of wellbore azimuth.

Fig. 1.53 shows relative wellbore stability as a function of wellbore orientation at 5,000 ft in normal, strike-slip, and reverse-faulting regimes. In all cases, S_{Hmax} is oriented E-W. As can be seen, the required mud weights and their variation with azimuth and deviation are quite different. The lowest mud weights are required in a normal faulting environment. Mud weight increases with deviation when both horizontal stresses are low, and there is only a small sensitivity of mud weight to drilling direction. Required mud weights are higher in the strike-slip regime, and there is a larger variation with drilling direction, especially at higher deviations. Vertical wells require the highest mud weight in this case. To drill in a reverse faulting environment, very high mud weights are necessary regardless of well orientation. The highest mud weights are required for vertical wells and for wells deviated to the North or South (the direction of the minimum horizontal stress), regardless of the amount of deviation. Mud weight decreases with increasing deviation in other directions, and the lowest mud weights are required for wells drilled with high deviations to the east and west. Based on plots similar to Fig. 1.53, it is possible, given only an indication of the stress regime and its orientation (for example, based on the orientations of currently active faults), to define the relative mud weight required for wells drilled at different orientations. If seismic data are available, and the velocity data can be inverted to constrain pore pressure and rock strength, it is possible to make approximate predictions of required mud weights for wells of all orientations.

1.7.6 Handling Uncertainty. In cases in which no wells have yet been drilled in a new exploration area, estimates of required mud weight can have considerable uncertainties. It is, however, possible to quantify those uncertainties and also to learn what measurements are required to provide the maximal improvement in prediction accuracy, using quantitative risk assessment (QRA). QRA analyses can be carried out at a single depth, or over the range of depths between casing seats. **Fig. 1.54** is an example of handling uncertainty at a single depth.

In this example, a well is being drilled at a 30° inclination to the north in the strike-slip stress state used to compute Fig. 1.53b. Based on the deterministic recommendation shown in that figure, the minimum mud weight required to drill the well without excessive instability is 14.6 lbm/gal. If at the time of analysis no well had yet been drilled, there would, however, be large uncertainties in the magnitudes of the two horizontal stresses and their orientations. It is also possible that the stress field may be inclined slightly with respect to the vertical. There may also be large uncertainties in the overburden stress, S_v, and in the rock strength and pore pressure, even if these had been estimated from seismic data. The parameter values and their uncertainties are shown in **Table 1.4.** Because QRA is carried out using a Monte Carlo approach, it is possible to allow asymmetrical distributions of the inputs, for example, for the rock strength.

The results of the analysis are shown in Fig. 1.54. This figure plots the cumulative probability of avoiding drilling problems associated with wellbore instability as a function of mud weight. The predicted likelihood of avoiding problems using the mud weight calculated deterministically is only slightly greater than 60%, as shown by the vertical dashed line. To guarantee the well's success, the mud weight would probably have to be higher.

The sensitivity of the mud weight recommendation to the parameter uncertainties is shown in **Fig. 1.55.** It can immediately be seen that the largest uncertainty is associated with the poorly constrained value of C_o. For higher values, the mud weight required to avoid instabilities is considerably reduced. In addition, the large variation in the magnitude of S_{Hmax} produces a similarly large uncertainty in the recommended mud weight. The pore-pressure uncertainty results in approximately ± 1 lbm/gal uncertainty. Uncertainties in the magnitude of the minimum stress and in the stress inclination contribute very little. Using these results, it is possible to design a data acquisition and analysis program that achieves the greatest reduction in uncertainty at the minimum cost. In this case, the most cost-effective improvement would result from a better constraint on the rock strength.

Even when rock properties and the stress model are well defined, there can be geological uncertainty based on poorly defined or unknown structure. An example of this is shown in **Figs. 1.56 and 1.57**[27] for a horizontal well drilled through hard sandstones containing an unknown distribution of intermittent shaley intervals. In such cases, the uncertainty is not caused by measurement error, but rather by the natural complexity of the structure being drilled. Fig.

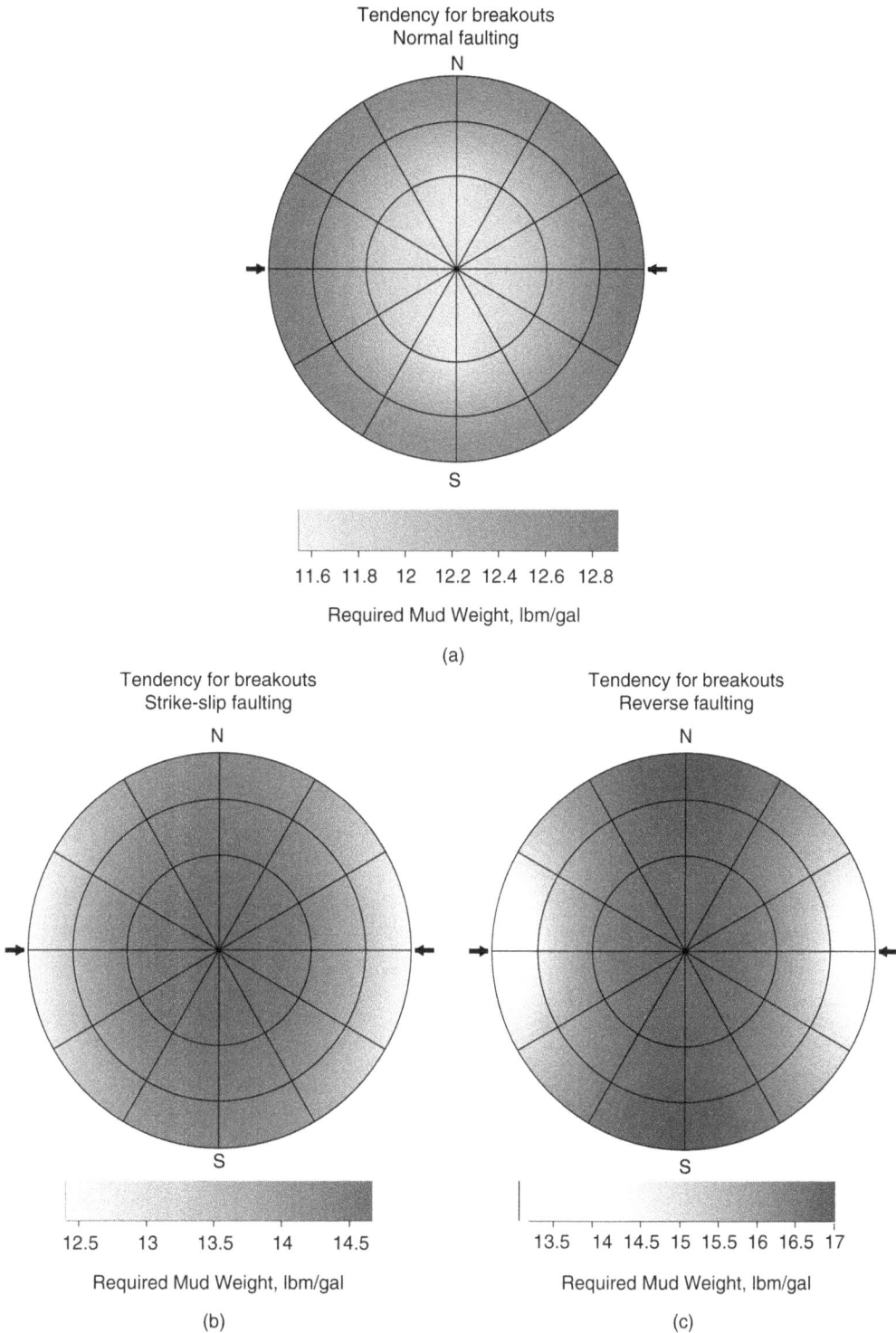

Fig. 1.53—Required mud weight to prevent excessive wellbore failure (breakouts) as a function of wellbore orientation at a depth of 5,000 ft, for normal (a), strike-slip (b), and reverse (c) faulting regimes. The gray scales are the same for all three cases.

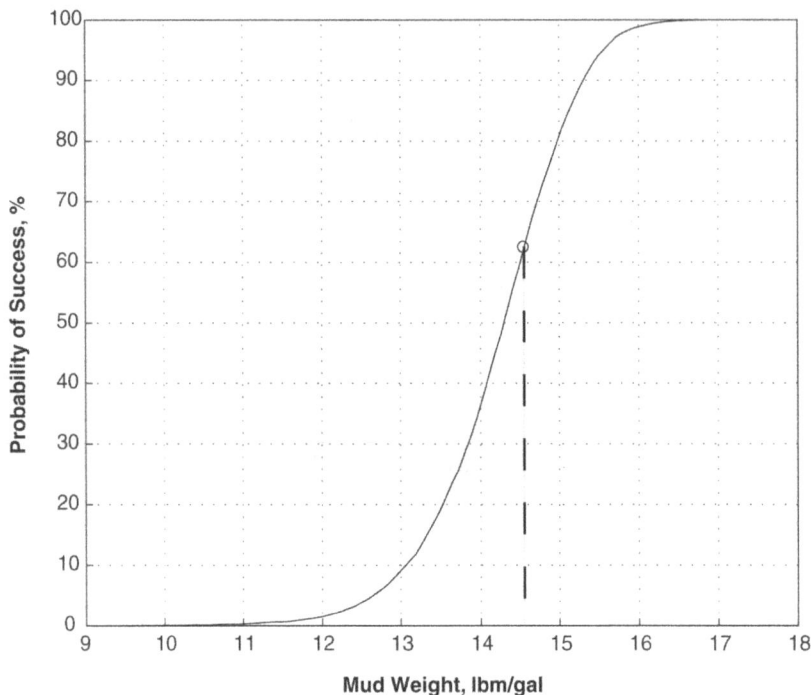

Fig. 1.54—This plot shows quantitative risk assessment (QRA) of the cumulative likelihood of avoiding excessive wellbore stability problems (in percent) as a function of mud weight, for the stress state and properties distributions shown in Table 1.4. A well drilled using a deterministic mud weight recommendation (dotted line) has a 67% likelihood to avoid wellbore instability.

TABLE 1.4—RANGE OF VALUES FOR INPUT PARAMETERS IN WELLBORE STABILITY CALCULATIONS FOR A WELL THAT IS DEVIATED 30° TO THE NORTH IN A STRIKE-SLIP ENVIRONMENT*

Parameter	−2sigma	Expected Value	+2sigma
$S_{Hmax} = S_1$, lbm/gal	20	22	24
$S_v = S_2$, lbm/gal	17	19.3	20
$S_{Hmin} = S_3$, lbm/gal	14	16	17
Deviation of S_v	−10	0	10
Azimuth of S_{Hmax}	60	90	120
P_p, lbm/gal	8.0	9.5	10
C_o, psi	1,000	2,000	5,000

*The expected values are those used to compute Fig. 1.53.

1.56a shows the distribution of log-derived strengths within this interval obtained from the pilot hole, which was drilled overbalanced. Fig. 1.56b shows the distribution used for the QRA analysis, which has a similar shape. Using this distribution, the QRA analysis of the likelihood of excessive wellbore instability is shown in Fig. 1.57. It is clear from this figure that there is little risk associated with a balanced well. However, a well drilled with a 1 lbm/gal underbalance has only a 66% likelihood of avoiding excessive wellbore failure. Because the company for which the well was drilled was risk-averse, the decision was made not to attempt underbalanced drilling.

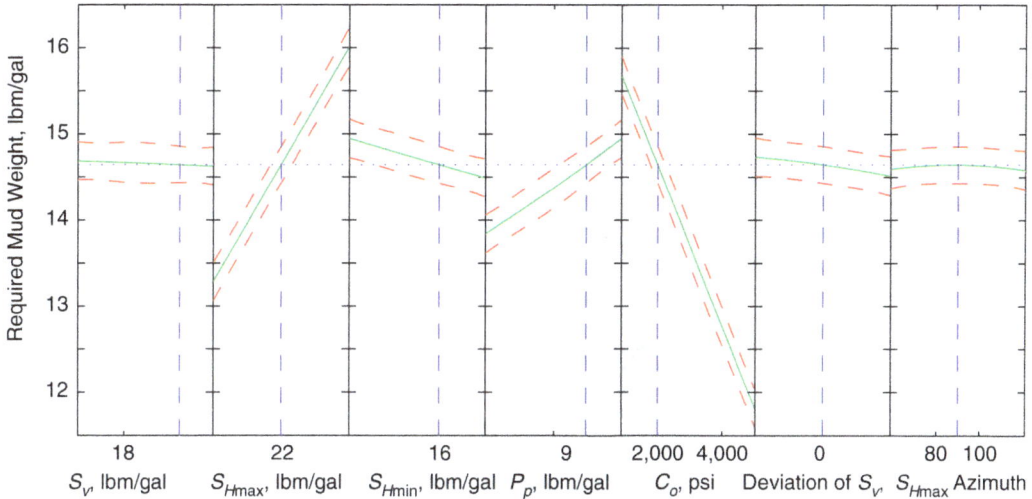

Fig. 1.55—This series of plots shows the sensitivity of the mud weight recommendations shown in Fig. 1.54 to the uncertain parameters. Plots such as these can be used to identify those parameters for which reduced data uncertainty would result in the biggest reduction in uncertainty in the recommended mud weight.

1.8 Other Models for Wellbore Stability

In many cases, wellbore stability analysis can be carried out with very simple models that are time-independent and relate stress and pore pressure only through the effective stress law. These do not account for the fact that stress changes induce pore pressure changes, and vice versa. Nor do these models account for thermal and chemical effects and their relationships to pore pressure and stress. In this section, we briefly discuss each of these issues and how they affect wellbore stability analysis. We start with a discussion of failure caused by anisotropic rock strength, which is a characteristic of consolidated shales that can cause considerable problems in wells drilled at oblique angles to bedding.

While the examples shown here demonstrate that it is possible to quantify uncertainties in the minimum safe mud weight, it is also possible to quantify uncertainties in the maximum safe mud weight. In that case, the likelihood of success decreases with increasing mud weight, and the two edges of the field defining the most stable mud weights form a possibly skewed bell-shaped curve.

1.8.1 Anisotropic Strength. In many rocks (lithified shales in particular), the elastic properties are anisotropic. In other words, they are a function of the orientation of the applied stress with respect to bedding planes (in general, shales are stiffer along the bedding planes than perpendicular to bedding). At the same time, the rock strength is also anisotropic. In both cases, the anisotropy is caused by a preferred orientation of shale particles that generally becomes more pronounced with compaction.

A well that is drilled perpendicular to shale bedding is generally not affected by bedding-parallel weakness planes. However, when a well is drilled at an oblique angle to bedding, bedding-parallel weakness planes can become very important. **Fig. 1.58** shows an acoustic wellbore image of breakouts that occur along oblique bedding planes intersecting a well, demonstrating that this mode of failure does occur. In fact, in this well failure associated with weak bedding caused severe instabilities, necessitating a sidetrack.

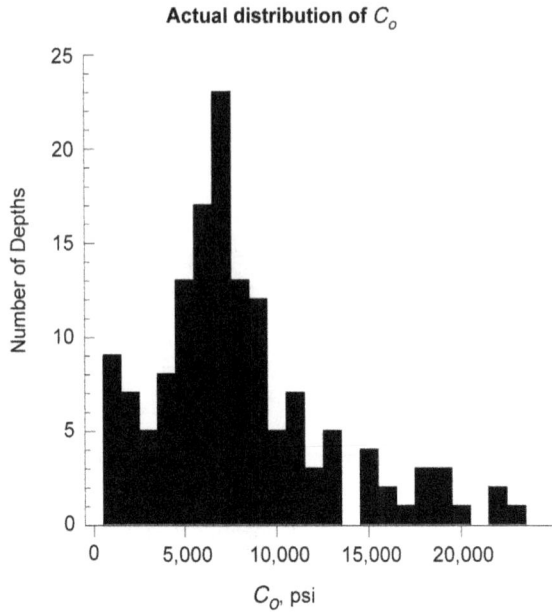

(a)

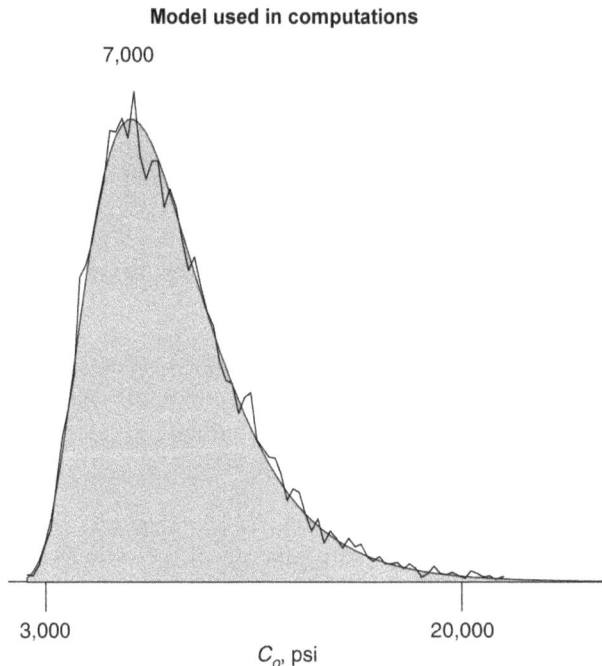

(b)

Fig. 1.56—(a) Histogram of log-derived C_o for a reservoir interval proposed for underbalanced drilling and openhole completion. (b) Log-normal probability distribution function for C_o consistent with the variation shown in the histogram in (a).[27] (Reprinted from "Comprehensive Wellbore Stability Analysis Utilizing Quantitative Risk Assessment," Moos *et al.*, *J. of Petroleum Science and Engineering,* Vol. 38, pages 97–109, © 2003, with permission from Elsevier.)

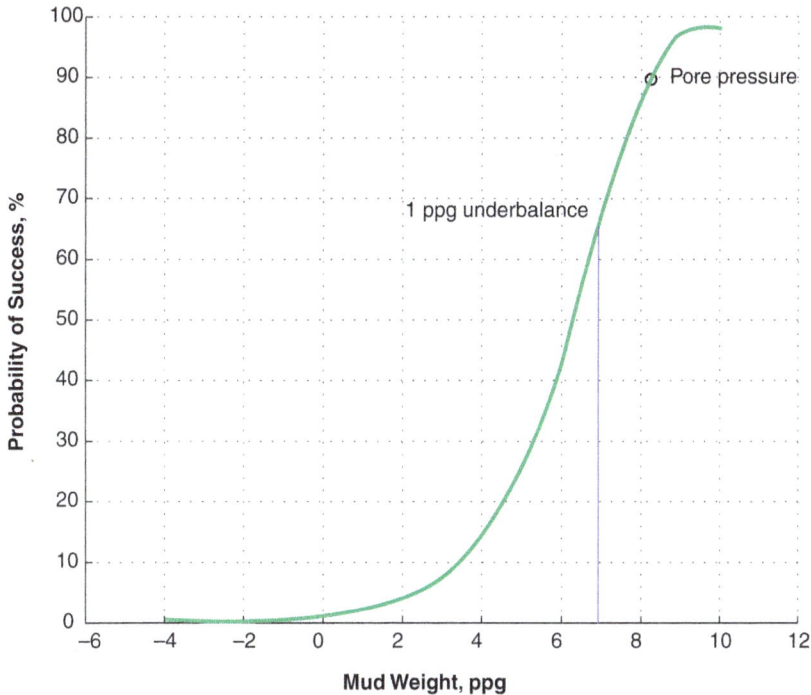

Fig. 1.57—This figure shows the cumulative likelihood of successfully drilling a well for which the uncertainty in rock strength is caused by its variability within the reservoir, as shown in Fig. 1.56. In this case, the analysis suggests that there is a 90% likelihood of success for a balanced well, but only a 64% likelihood of success when drilling with a 1 lbm/gal underbalance.[27] (Reprinted from "Comprehensive Wellbore Stability Analysis Utilizing Quantitative Risk Assessment," Moos *et al., J. of Petroleum Science and Engineering,* Vol. 38, pages 97–109, © 2003, with permission from Elsevier.)

Fig. 1.59 is an example plot that shows the required mud weight as a function of wellbore orientation for wells drilled through dipping beds that are highly anisotropic and illustrates the two angles required to define the orientation of a well with respect to bedding. The first angle is the attack angle, which is simply the angle between the well axis and the normal to the bedding plane. The larger the attack angle, the more likely it is that failure will occur because of bedding-parallel weakness planes. The second is the angle between the dip direction and the projection of the well axis onto the bedding plane. While in general it is found that wells drilled updip or downdip are more stable than those drilled along strike, the relationship depends critically on the orientations and magnitudes of the in-situ stresses.

Both of these effects can be seen in the lower right stability plot shown in Fig. 1.59, which indicates the mud weight required to maintain stability as a function of wellbore orientation. The bedding normal is shown in this plot as a white dot. The lighter gray shading close to the dot indicates that lower mud weights are required for wells drilled nearly perpendicular to the bedding planes. Darker colors show that high mud weights are required for wells drilled obliquely to bedding. The highest mud weights are required for wells drilled with moderate to high deviations to the ENE. The asymmetry in the plot is a characteristic of the effect of strength anisotropy, and it is caused by the complex interplay between the stress field concentrated around the well and the weak bedding planes. For comparison, the lower left stability plot shows recommended mud weight if there were no weak bedding planes. The difference between the two is the affect of bedding, which requires raising the mud weight if those planes are active. Where bedding planes are not active, similar mud weights are recommended. Notice

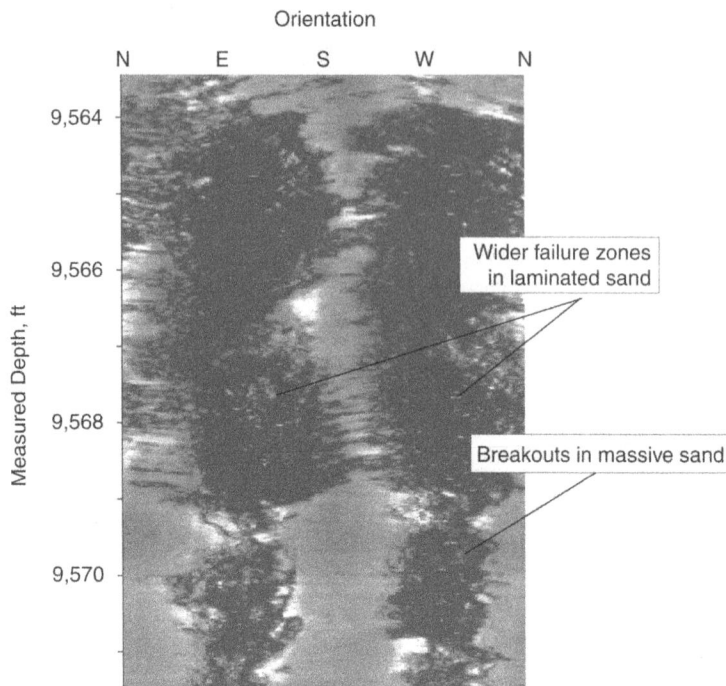

Fig. 1.58—Examples of wellbore breakouts observed in acoustic image data from a well drilled through interbedded massive and laminated sands. Wider breakouts (dark bands) can be seen in the laminated sands above 9,569.5 ft. Where laminations are less frequent, the breakouts are narrower. This pattern indicates wellbore failure in the laminated sands being exacerbated by weak bedding planes.

that the relative stability of the wells shown on the two plots is quite different when bedding weakness is taken into account from when it is not.

It is very important to realize that, contrary to cases in which wellbore instability is caused by failure of the intact rock, raising the mud weight past a certain point usually exacerbates failure in anisotropic shales. Because failure in these rocks often involves slip along discrete planes, the result is that irregular chunks of rock are often produced, and when cross-cutting fractures are present, the pieces are often spindle-shaped. Raising the mud weight when this type of failure is observed often causes an increase in fluid pressure along the weak planes, reducing their resistance to slip, thereby making failure worse. This problem is often addressed in part by adding fluid-loss-control agents to the drilling mud.

1.8.2 Poroelasticity and Thermoporoelasticity. Poroelasticity theory describes the coupling between pore pressure and stress in rocks. When pore pressure and stress are coupled, fluid diffusion plays an important role, and stability becomes time-dependent. To use the poroelasticity equations developed by Biot[28] to model this process requires knowledge of more rock properties than are required for elastic analyses. These include the elastic moduli, the porosity, the permeability, and a pore pressure-stress coupling term. Even without modeling the problem, however, it is obvious that when a well is overbalanced, fluid diffusion into the rock is likely to cause instability to increase over time. This is because diffusion causes the initial overbalance required to support the wellbore to decrease with time as the near-wellbore pore pressure increases, leading to a decrease in wellbore support and increased failure of the rock. This time-dependent weakening is reduced by development of a mud cake. Thus, it is often observed in wells with strong, brittle shales and weak, high-porosity, high-permeability sands that the

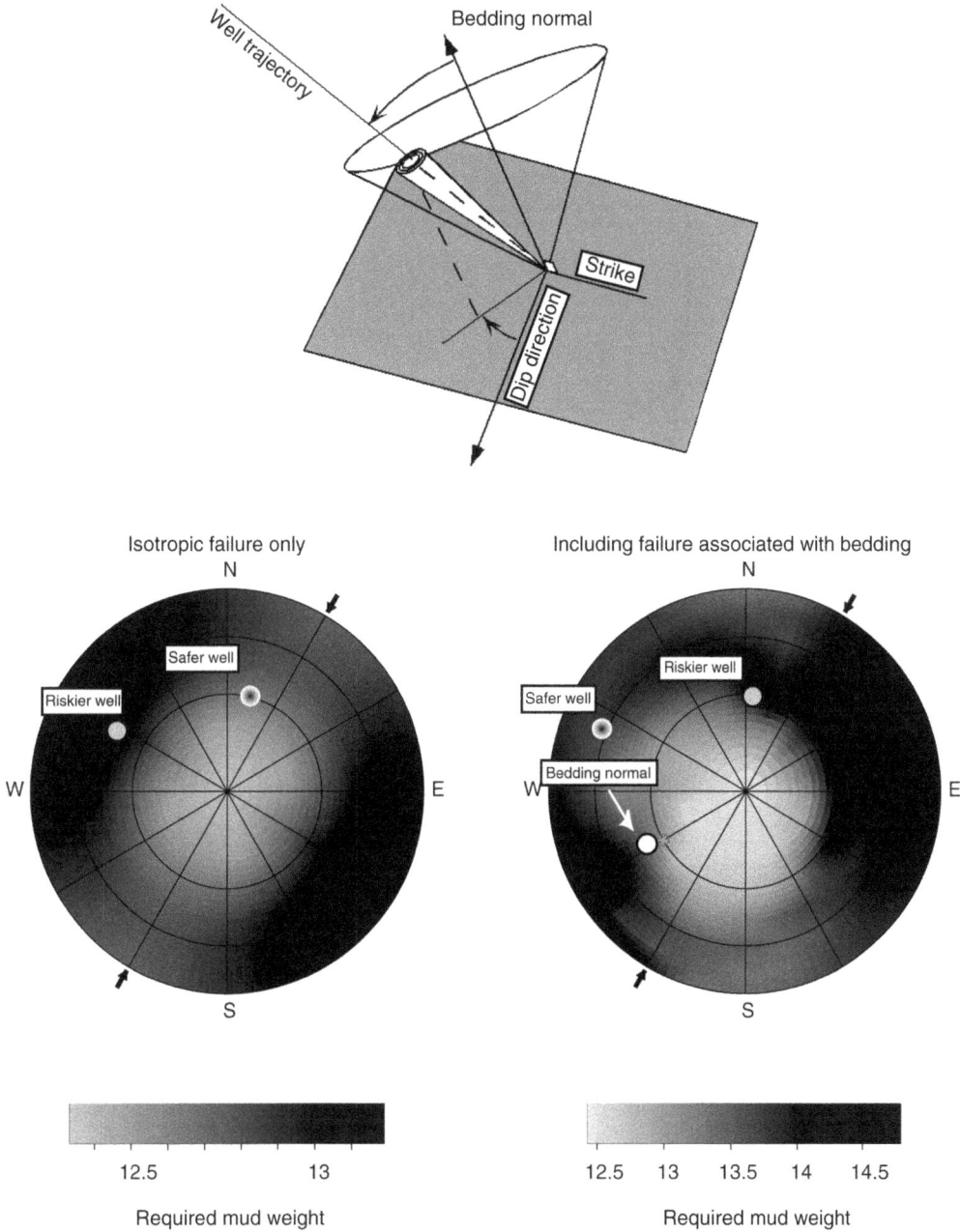

Fig. 1.59—The upper diagram defines two angles that are used to describe the orientation of a well with respect to bedding. The lower figures show the mud weight (in ppg) required to maintain stability as a function of wellbore orientation for a highly anisotropic shale if the weak bedding planes are ignored (on the left) and considered (on the right). Pale grays show that low mud weights are required for wells drilled approximately perpendicular to bedding, whereas darker grays show that higher mud weights are required for wells drilled obliquely to bedding (courtesy GeoMechanics Intl. Inc.).

strong shales break out, whereas the weaker sands appear more intact. An additional reason for the apparently anomalous stability of the sands is discussed briefly in the section on plasticity.

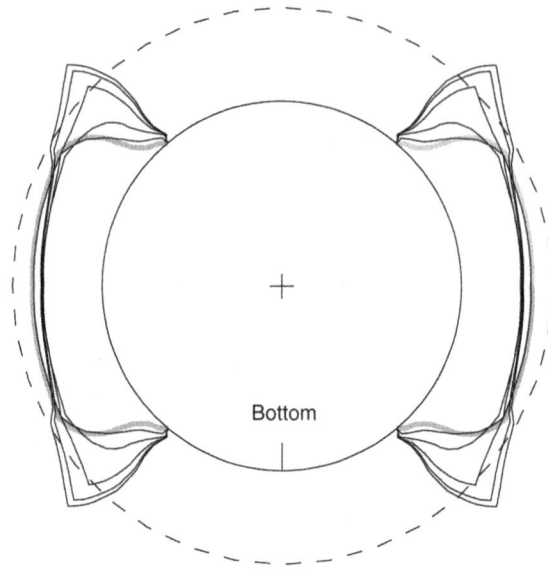

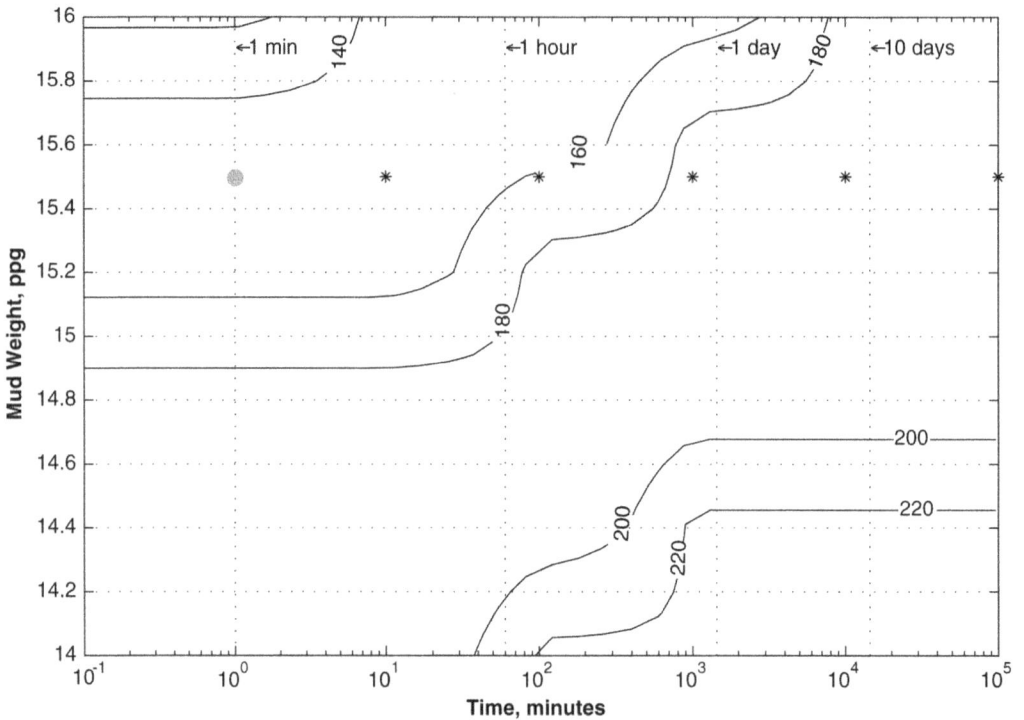

Fig. 1.60—Poroelastic analysis of failure of a horizontal well drilled through a 10% porosity gas sand with a permeability of 1 μDarcy. The amount of wellbore failure increases with time. Zones of failure shown in the wellbore cross section correspond to the positions of the stars on the lower plot. For example, the light gray line shows the extent of failure 1 minute after the well has been drilled using an equivalent mud weight of 15.5 lbm/gal. The other stars are the times corresponding to the other failure zone outlines.

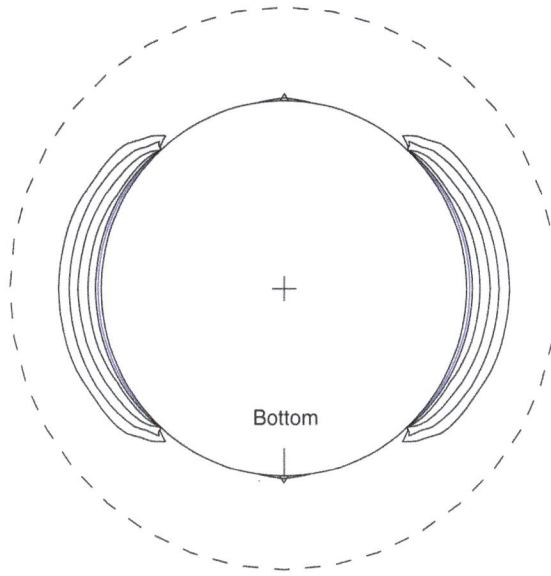

Failing fraction of wellbore circumference, degrees
at a radial distance of 1.2r

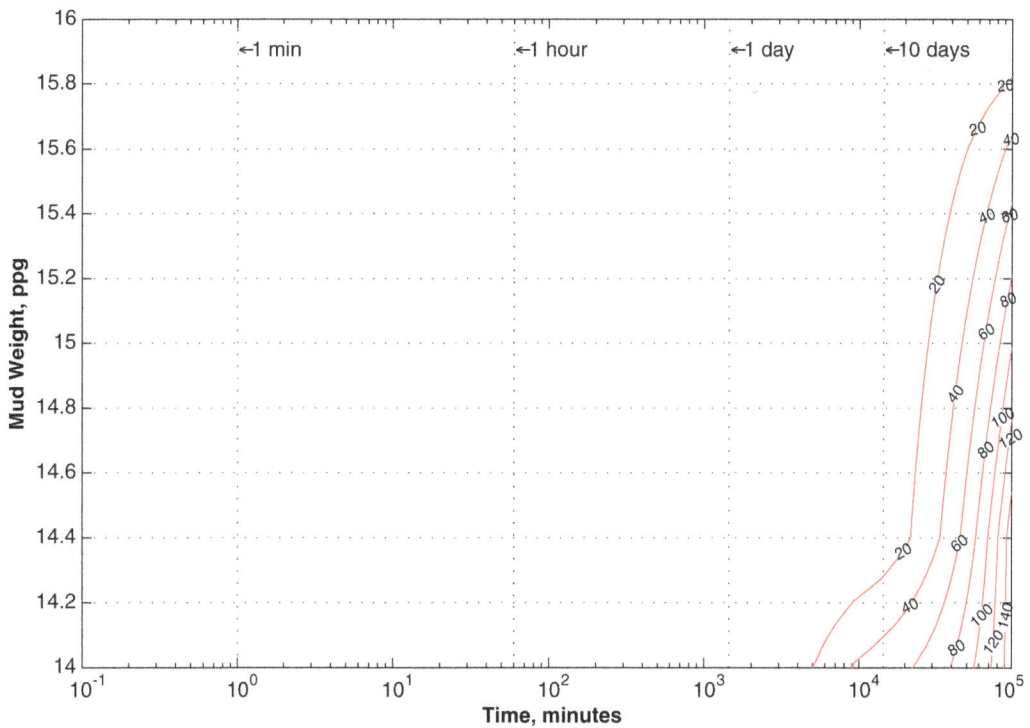

Fig. 1.61—By cooling the mud, failure of the same horizontal well as shown in Fig. 1.60 is both delayed and greatly reduced.

The effect of fluid diffusion is illustrated in **Fig. 1.60**. On the top is shown a wellbore cross section. Superimposed on the cross section is a series of contours that define the volume of rock in which the stresses exceed the rock strength as a function of time. The heavy gray curves show the boundary of the breakout zones after 1 minute, and the other curves show its shape at 10, 100, 1,000, 10,000, and 100,000 minutes. Although the amount of failure gets larger with time, the width of the failed zone at the wellbore does not change. This is because, in this example, it is assumed that no mudcake forms, and there is perfect communication between the wellbore fluid and the pore fluid. However, away from the well, the amount of failure increases with time. The lower plot shows the total angular coverage of the failure zones as a function of time and mud weight, at a radial distance from the center of the well that is 20% larger than the drilled radius. Although higher mud weights do reduce the amount of failure at short times after drilling, there is a slow but systematic increase in the amount of failure with time, regardless of the mud weight used. The stars in the lower plot show conditions corresponding to each breakout drawn on the well cross section.

Thermal energy transfer obeys the same diffusion law as does the movement of pore fluid. Hence, it is straightforward to model the time-dependent effects of wellbore cooling using the same equations as are used for poroelasticity. This is potentially quite important because cooling a well reduces the circumferential stress and thereby temporarily decreases the likelihood of breakout formation. Simply modeling the pore pressure and temperature independently is not enough, however, because thermal energy transfer occurs both by conduction (heat transfer) and by convection (motion of warm or cold fluids). Thus, a fully-coupled thermoporoelastic theory is required.

Fig. 1.61 shows analysis of the effect of a 30°F reduction in mud temperature for the same parameters used to generate Fig. 1.60. It is clear that failure is much less pronounced when the mud has been cooled. In fact, the analysis indicates that not only can cooling increase the length of time this well remains stable, it may also allow a significant decrease in mud weight. This is because of the contributions of two effects. First, cooling the wellbore reduces the circumferential stress that leads to failure. And second, cooling the fluid reduces the pore pressure, increasing the effective strength of the rock.

1.8.3 Mud/Rock Interactions. From the perspective of wellbore stability, shales are the most problematic lithologies to drill through. Evidence abounds that the shale sections of wells drilled with water-based mud are significantly more rugose than the same sections of similar wells drilled with oil-based mud. The primary reason for these observations is that chemical interactions that occur between shales and water-based drilling muds cause a significant reduction in the effective strength of the shales. Two effects contribute to this problem. The first is osmotic diffusion (the transfer of water from regions of high salinity to regions of low salinity), which causes water in low-salinity mud to diffuse across the membrane formed at the mud/rock interface. The second is chemical diffusion (the transfer of specific ions from regions of high concentration to regions of low concentration). These two effects both change the internal pressure of water in the shale and also affect its strength. Each occurs at a different rate, which in some cases can lead first to weakening and then to strengthening of a wellbore.

When the salinity of the drilling mud water phase is lower than the salinity of the pore fluid, osmotic diffusion causes shales to swell and weaken because of elevated internal pore pressure caused by uptake of water into the shale. Consequently, one solution to shale instabilities is to increase the salinity of the water phase of the mud system, and this works in some cases. However, if the salinity is increased too much, it can cause microfracturing to occur.

In calculating the magnitude of the pressure generated by osmotic diffusion, the parameter that is used to select the appropriate salinity is the water phase activity. Activity (which is explicitly the ratio of the vapor pressure above pure water to the vapor pressure above the solution being tested, and can be measured at the rig with an electrohygrometer) varies from zero to one. Typical water-based muds have activities between 0.8 and 0.9. Typical shales in situ have pore-fluid activities between 0.75 and 0.85, based on extrapolations of laboratory da-

Effect of mud chemistry on wellbore failure

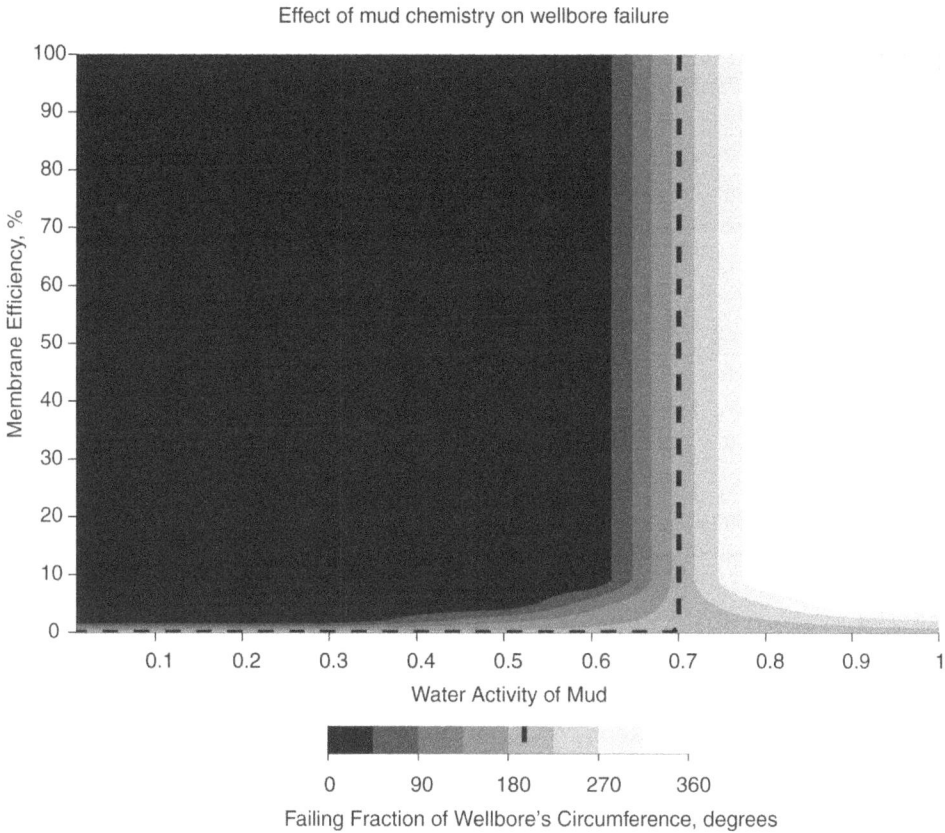

Fig. 1.62—Plot showing the amount of failure around a wellbore drilled into reactive shale as a function of the water activity of the mud and the membrane efficiency. The shale pore fluid activity is 0.7.

ta. The use of typical muds in typical shales thus causes an increase in the pore pressure within the shale, leading to shale swelling, weakening, and the development of washouts. Mody and Hale[29] published Eq. 1.21 to describe the pore pressure increase owing to a given fluid activity contrast.

$$\Delta P = E_m (RT/V) \ln (A_p / A_m). \quad\text{...} (1.21)$$

If ΔP is negative, it indicates that water will be drawn into the shale. Here, R is the gas constant; T is absolute temperature, and V is the molar volume of water (liters/mole). Decreasing the mud activity often alleviates shale swelling because ΔP is positive if A_p (the pore fluid activity) is larger than A_m (the mud activity), and water will be drawn out of the shale for this condition. The parameter E_m is the membrane efficiency, which is a measure of how close to ideal the membrane is. Explicitly, it is the pressure change across an ideal membrane owing to a fluid activity difference across the membrane, divided into the actual pressure difference across the membrane in question. Membrane efficiency is affected both by mud chemistry and by the properties of the shale. In particular, the ionic radius and the pore throat size of the shale appear to play a strong role. Oil-based mud has nearly perfect efficiency. Although water-based mud generally has very low efficiency, some recently developed water-based synthetics have been designed to have high efficiencies approaching those of oil-based mud. **Fig. 1.62** shows the relationship between membrane efficiency, mud fluid activity, and degree of failure

Failing fraction of wellbore circumference, degrees
at a radial distance of 1.25*r

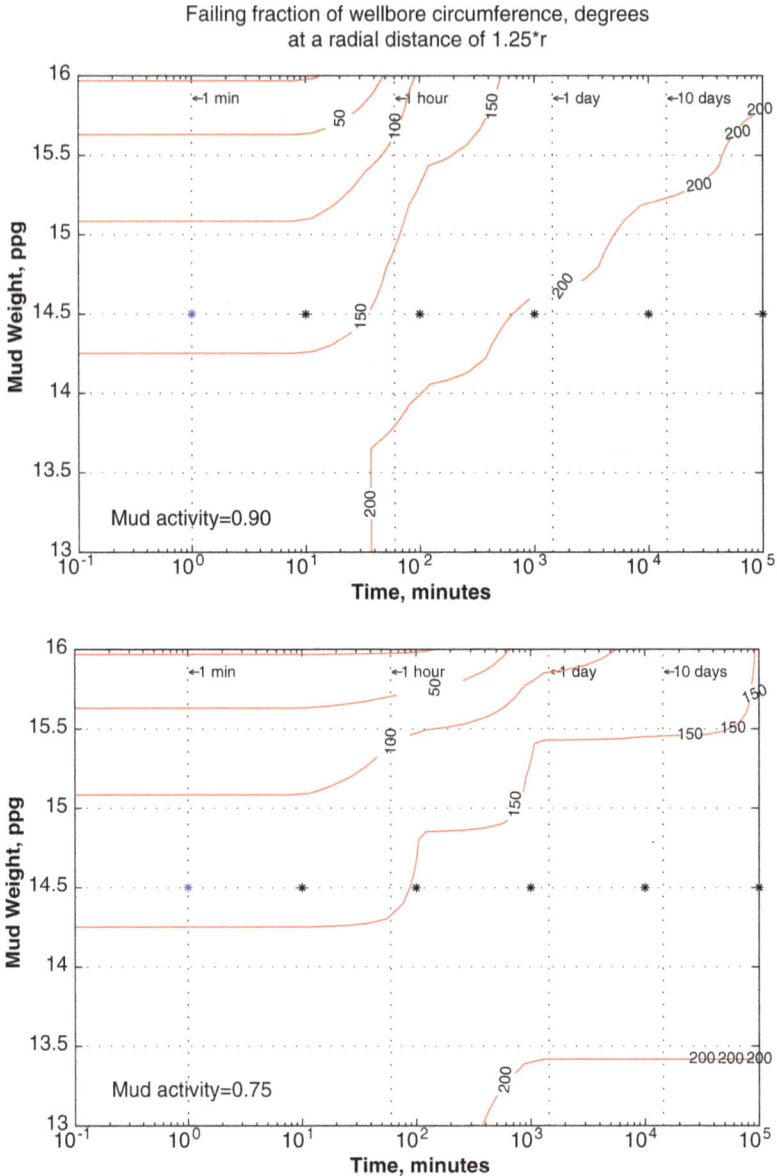

Fig. 1.63—Plot showing the amount of failure in degrees of the well's circumference as a function of time and mud weight for a shale with a pore fluid activity of 0.8, subjected to a mud with a water-phase activity of 0.9 (on the top) and 0.75 (on the bottom). When the mud activity is lower than the shale, even very high mud weights (the fracture gradient is 16 lbm/gal) only stabilize the well for less than 1 day. By lowering the mud activity, the required mud weight can be reduced while keeping failure under control and extending working time.

(quantified in terms of the widths of the failed regions) for shale with a nominal pore fluid activity of 0.7. Higher mud activities than the shale pore fluid cause an increase in breakout width, whereas predicted breakout width is less for muds with lower activities. The effect decreases for lower membrane efficiencies.

The model described by Eq. 1.21 is implicitly time-independent, and diffusion is a time-dependent process. Time-dependent models have been developed that predict variations in pore

TABLE 1.5—STRENGTH PARAMETERS USED FOR THE
ANALYSES SHOWN IN FIGS. 1.64 AND 1.65

	Onset of Yield	Peak Stress
Cohesion, S_o	943 psi	1,305 psi
UCS, C_o	3,533 psi	4,890 psi
Friction angle; μ_i	34°; 0.67	34°; 0.67

S_v=4,600 psi S_{Hmax}=4,200 psi S_{Hmin}=2,800 psi
P_p=1,700 psi P_m=1,700 psi (ΔP=0)

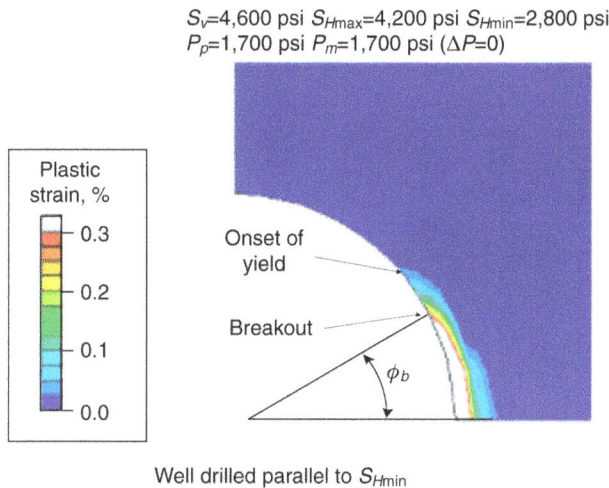

Well drilled parallel to S_{Hmin}

Fig. 1.64—Example of modeling failure in a poroelastic-plastic material. Failure is assumed to occur at a plastic strain of 3×10^{-3} (3 millistrain). This produces a breakout defined by the white area at the side of the well. The rock properties are shown in Table 1.5.

pressure due to chemical effects as a function of time and position around the hole. These are explicitly both chemo-elastic and poro-elastic (that is, they account for interactions between the pore pressure and the stress as well as the chemical effects on the pore pressure). The results allow selection of mud weights for specific mud activities, or mud activities for specific mud weights. **Fig. 1.63** (top) shows a plot of failure vs. time and mud weight for a shale with a pore-fluid activity of 0.8, for a mud activity of 0.9. As can be seen, failure gets worse over time, and even a mud weight as high as the fracture gradient of 16 lbm/gal maintains hole stability for less than one day. On the other hand, for a mud activity of 0.7 (Fig. 1.63, bottom), the time before failure begins to worsen is extended, and it is possible to select a mud weight below the fracture gradient and yet still provide several days of working time.

1.8.4 Wellbore Failure in Plastic Rock. As previously discussed, young, weak rocks that are still undergoing compaction behave plastically. The same can be said of high-porosity reservoir sands. One consequence is that these materials "fail" with only a small reduction in strength. Therefore, wellbore stability modeling can be done more accurately in young rocks using plastic models. But, because the current state of a plastic material is a function of its stress/strain history, fully 3D plastic models require numerical methods. While plastic models are not necessary for extremely brittle rocks, it is not always clear which model is the most appropriate when a rock has intermediate properties.

The simplest way to decide whether it is important to use plasticity is to look at the stress-strain curve of the rock of interest. If it has large strain at failure and it has a detectable yield

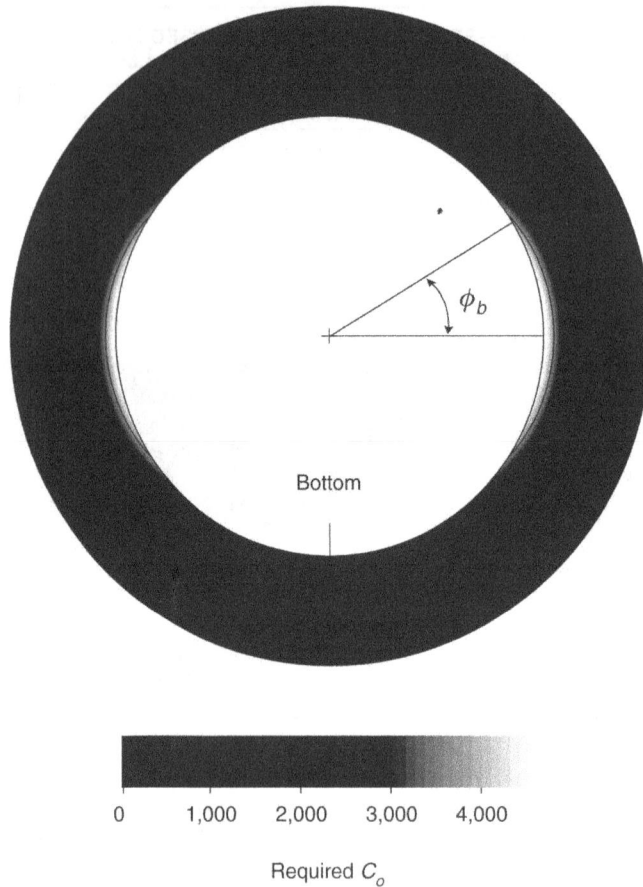

Fig. 1.65—Wellbore cross section showing the required C_o to prevent formation breakout in a poroelastic-brittle material, for the same conditions as shown in Fig. 1.64. The shape of the predicted failure zone is nearly identical to that in Fig. 1.64.

point, and/or it fails without total loss of strength, plastic models should be considered. That said, no one has yet published a definitive study that demonstrates that an elastic-plastic model is a better predictor of required mud weight for drilling than an equivalent elastic-brittle model. In fact, in cases in which both approaches are used, it is often found that predicted mud weights to avoid excess wellbore instabilities using the two techniques are within 0.1 lbm/gal.

Fig. 1.64 is the output from a strain-hardening, poroelastic-plastic analysis of failure around a balanced well. The stresses, pore pressure, and mud weight are shown in the upper right of the figure. The other parameters, obtained from measurement of the core the properties of which were used in this analysis, are shown in **Table 1.5**. With a failure model calibrated by laboratory tests, which predicts the onset of failure after 3 plastic millistrain (0.3%), a failed zone is predicted to have a half-angle of approximately 55 degrees, as shown in white on the side of the well. **Fig. 1.65** presents a similar analysis, using the same stresses, pore pressure, and mud weight, using a poroelastic-brittle model for failure. In this case, there is a remarkable similarity between the width of failure (the elastic model describes only the initial zone which will deepen with time, as discussed previously) of these two analyses, indicating that it is not necessary to use a plastic model to describe the material.

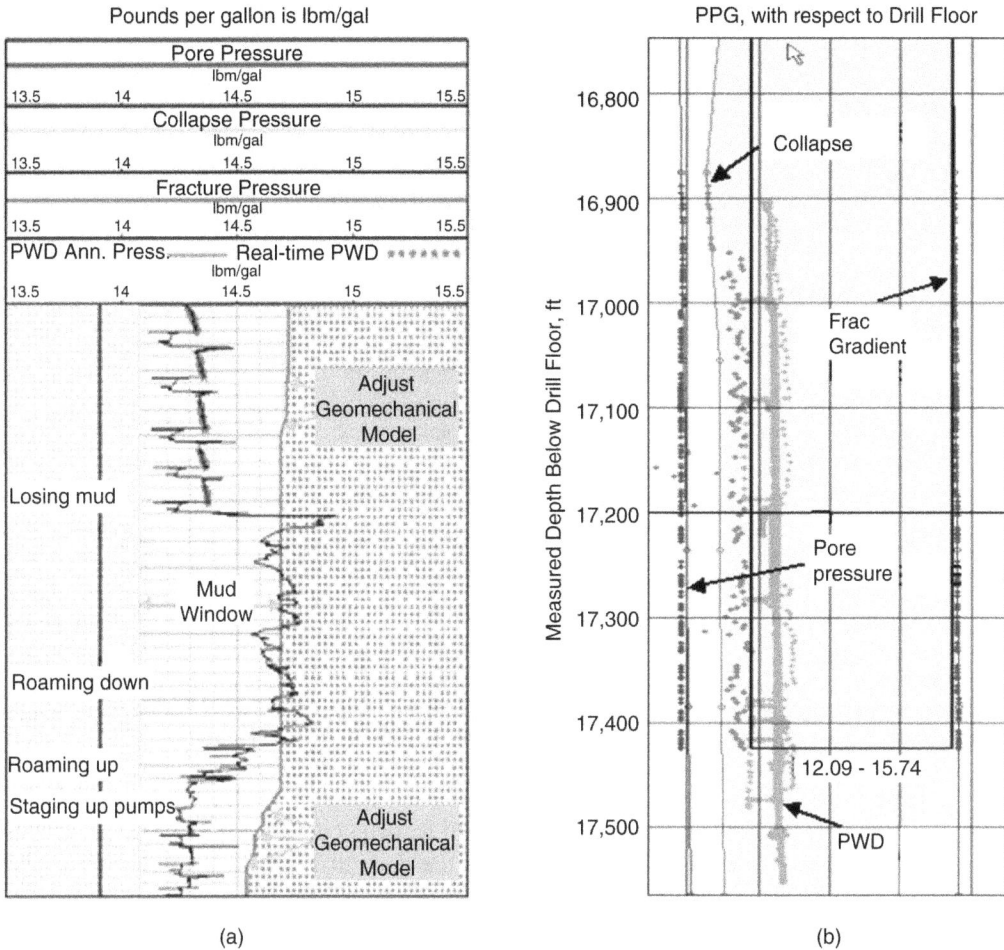

Fig. 1.66—This figure shows examples of time-based (a) and depth-based (b) displays of real-time PWD data, superimposed on event predictions of lost circulation and collapse. They are from two different depth intervals in the same well and can be produced in real time as a well is being drilled.

1.9 Making Decisions in Real Time

In situations in which predrill analysis reveals high risk but has a large uncertainty, it is possible to mitigate that risk by carrying out geomechanical analysis in real time. This can be done but requires acquisition of a variety of data while drilling. Annular pressure measurement using a PWD tool is one key component of real-time stability analysis because knowledge of the hydrostatic and circulating pressures is required to determine the magnitude of kicks, identify borehole ballooning events, and monitor hole cleaning. The measurement can also be used to show where transient pressure events such as surging and breaking the gel strength of the mud exceed fracture pressure, or where swabbing reduces the pressure below the pore or collapse pressure of the wellbore. LWD resistivity, sonic velocity, and bulk density measurements provide information for use in constraining pore pressure and rock strength. Direct pore-pressure measurements while drilling can provide critical data to calibrate pore-pressure predictions in permeable formations. Extended leakoff tests are strongly recommended. Even observations of cuttings shapes and volume can be important to identify the amount and cause of wellbore failure. Because the relationship between rock strength and log data is often poorly known, penetrometer tests and velocity measurements on cuttings are useful both to quantify the

Breakout Width, degrees

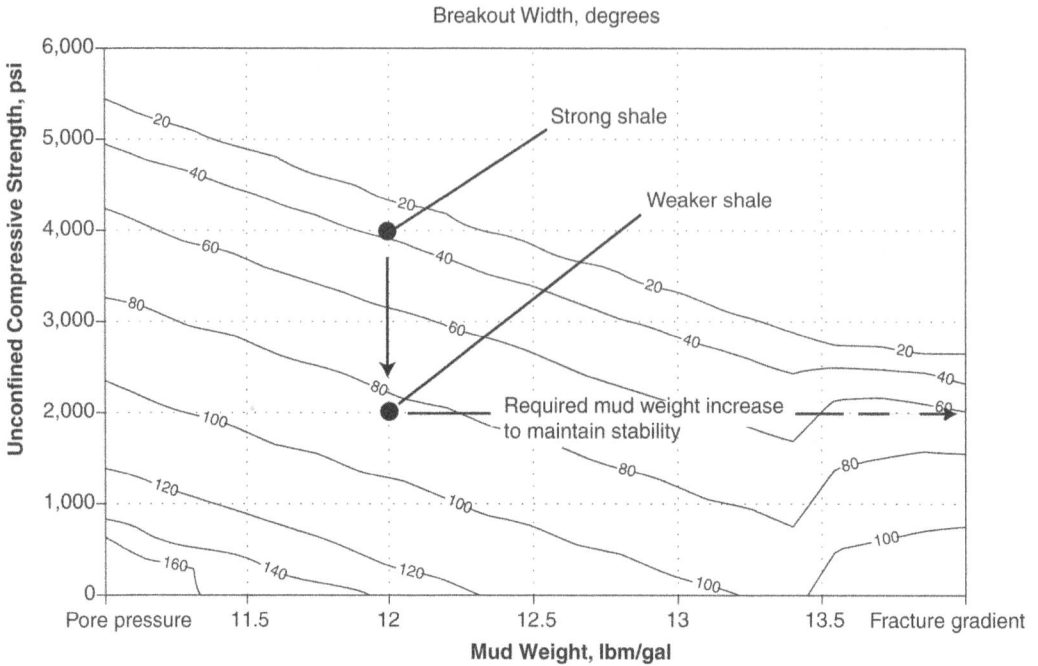

Fig. 1.67—Prediction of breakout width in reactive shale drilled using a non-reactive drilling mud as a function of rock strength and mud weight. A decrease in C_o from 4,000 to 2,000 psi necessitates an increase in mud weight from 11 to more than 14 lbm/gal to maintain the same degree of stability.

strength parameters and relate them to a measurement that can be obtained while drilling. However, the single measurement that would contribute the most to wellbore stability analysis is a wellbore image log, from which breakout characteristics can be determined.

Fig. 1.66 shows examples of displays of real-time wellbore stability pressure plots. On the left is a display as a function of time, with collapse and lost-circulation pressures predicted using the geomechanical model superimposed on the real-time and recorded data from a PWD tool. As the well was being drilled, a mud-loss episode occurred, which the analysis indicates occurred because the downhole ECD exceeded the fracture gradient. This indicated that the model was fairly accurate at that depth. Deeper in the hole, data recorded while drilling indicated that an adjustment needed to be made to the geomechanical model. Fig. 1.66b shows a plot of predicted mud weights as a function of depth on which the real-time PWD pressure data have been superimposed. In this case, the ECD is high enough to avoid drilling problems, but it appears that the static mud weight is very close to the minimum required to avoid excessive instability. Based on this observation, high cuttings volumes should be expected, and one recommendation would be to take extra care to avoid high running speeds and accelerations that might swab the hole.

A key component of real-time analysis is to provide an understanding of the origin of problems to make the right adjustment to drilling parameters to compensate. For example, when drilling through shales with inhibitive (chemically nonreactive) mud, fluid leakage may change the mud characteristics over time. When would an adjustment to the mud properties be required? A comparison of stability risk vs. mud chemistry could help make that decision. Or what happens if the rock strength suddenly decreases due to crossing from relatively strong shale into one that is much weaker? To handle this case, a crossplot of breakout width as a function of C_o and mud weight can be used to determine the amount of mud weight adjustment required.

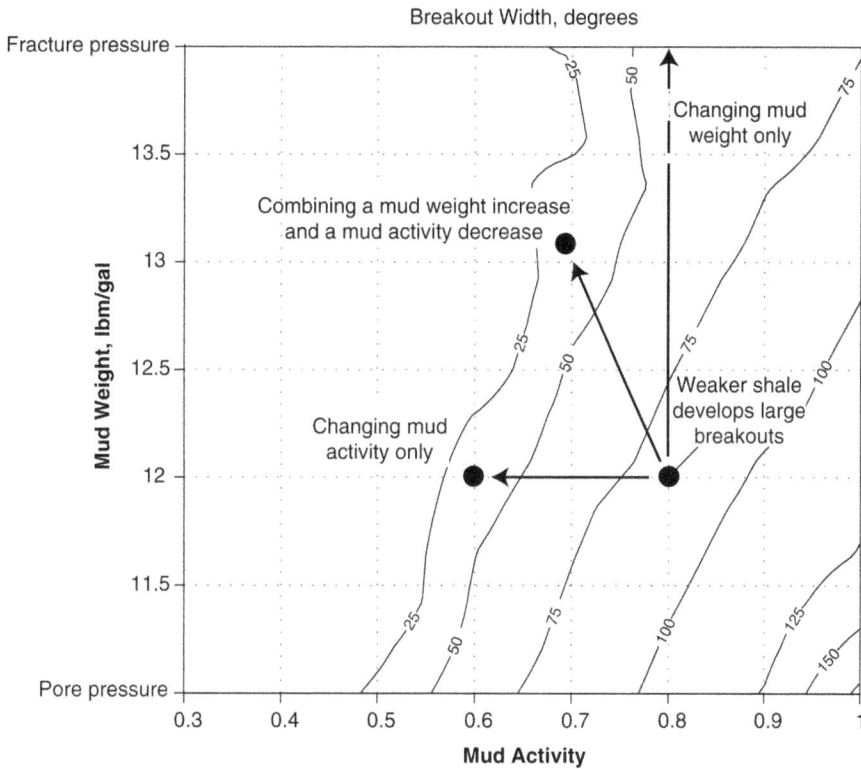

Fig. 1.68—Breakout width in reactive shale drilled using a reactive drilling mud as a function of mud weight and mud activity for the situation modeled in Fig. 1.67. The decrease in strength can be compensated for either by reducing the mud activity or by a combination of a smaller mud activity reduction and a modest increase in mud weight.

Figs. 1.67 and 1.68 show a hypothetical case of using geomechanical decision tools to help evaluate various drilling options when crossing from a strong shale into one that is substantially weaker and highly reactive. In Fig. 1.67, it is apparent for this hypothetical case that to maintain the same degree of stability if C_o decreases from 4,000 psi to 2,000 psi, it is necessary to increase the mud weight to a value that exceeds the fracture gradient. This makes it impossible to continue drilling, and the options are either to stop and set casing or to investigate other possibilities.

Plots such as Fig. 1.68 that shows breakout width as a function of mud weight and mud activity, can reveal whether it is possible to compensate for changes in strength by changing mud chemistry. Since breakout width in this particular shale is a strong function of mud activity, to compensate for the change in strength without changing mud weight it is only necessary to decrease the mud activity from its current value of 0.8 to a value of 0.6. This would achieve the desired result with no change in mud weight. It is also possible, of course, to maintain safe drilling conditions using a combination of a mud activity reduction and a mud weight increase.

The information presented in this chapter is only an introduction to the theory and application of geomechanics to drilling. It is a very young field, and rapid advances are being made. One example of this is a new approach that has been developed to model the near-wellbore behavior of fractured rock. Another is the application of uncertainty analyses to pore-pressure prediction. At the same time, new engineering techniques are being developed that provide solutions to geomechanical problems. For example, the same analysis as presented in Fig. 1.51 can be carried out for dual-gradient drilling, to quantify the potential of the technique to extend casing seats to greater depths in deep water. It is important to realize that the same geomechanical models that help improve drilling efficiency can be shared among drillers, geologists, and reservoir engineers and used to help improve operations throughout the life of a field. In fact, the best use of geomechanics is to develop initial models as early as possible and to use these models in every phase of field development, updating and refining the models as new information is obtained.

Acknowledgments

The following people contributed ideas and figures to this section: Colleen Barton, Martin Brudy, David Castillo, Balz Grollimund, Don Ritter, Thomas Finkbeiner, Wouter van der Zee, Chris Ward, David Wiprut, and Mark Zoback. Figs. 1.1, 1.4, 1.5, 1.6, 1.7, 1.8, 1.10, 1.13, 1.14, 1.18, 1.21, 1.28, 1.30, 1.39, 1.41, 1.43, 1.48, 1.49, and 1.59 were used with permission from "Reservoir Geomechanics," and Fig. 1.38 was used with permission from "Pore Pressure Prediction," both short courses taught by GeoMechanics Intl. Inc., ©M.D. Zoback and ©W. Ward, respectively. Fig. 1.64 was prepared using GMI•SandCheck™. Figs. 1.21, 1.22 (left), 1.23, and 1.58 were created using GMI•Imager™. Fig. 1.22 (right) was created using GMI•Caliper™. Figs. 1.51, 1.52, 1.67, and 1.68 were created using GMI•WellCheck™. GMI•SFIB™ was used to create Fig. 1.45 (GSTR module), Fig. 1.46 (CSTR module), Figs. 1.49, 1.50, and 1.53 (GFLR module), Figs. 1.54, 1.55, 1.56b, and 1.57 (QRA module), and Figs. 1.60, 1.61, 1.62, 1.63, and 1.65 (BSFO module). Figs. 1.51, 1.56, and 1.57 were reprinted with permission from Elsevier. Fig. 1.66 was prepared using RTWellCheck, developed jointly by Halliburton Energy Services and GeoMechanics Intl. Inc.

Nomenclature

a, b, c, d, e = constants used in Eq. 1.15 and tabulated in Table 1.4

A_m = mud activity, ratio

A_p = pore fluid activity, ratio

B = Skempton's coefficient, unitless

C_{eff} = effective strength, MPa, psi

C_o = unconfined compressive strength, MPa, psi

D_c = drilling exponent

e = base of a natural logarithm, $e = 2.718281828 \ldots$

E = Young's modulus, GPa

E_m = membrane efficiency, ratio

f = acoustic formation factor, used in Eq.1.19

G = acceleration of gravity, m/s^2

h = vertical height of a thin planar reservoir, length

K = Bulk modulus, GPa

K_{dry} = Bulk modulus of the dry frame of a porous rock, GPa

K_{grain} = Bulk modulus of the grains that make up the rock, GPa

L = horizontal length of a thin planar reservoir, length

l_k = length in the k direction

l_l = length in the l direction

M_{ijkl} = Component of the modulus tensor that relates the ij component of the stress tensor to the kl component of the strain tensor, MPa, psi

P_a = Pore pressure at depth a

P_c = confining pressure, MPa, psi

P_{hyd} = hydrostatic pressure, MPa, psi, lbm/gal

P_m = pressure of mud in a well, MPa, psi, lbm/gal

P_p = pore pressure, MPa, psi, lbm/gal

P_z = Pore pressure at depth z, MPa, psi, lbm/gal

R = gas constant, J/mol/°K

$R_{\log}$ = measured value of resistivity, ohm-m

R_n = normal value of resistivity, ohm-m

S_1 = greatest principal stress, MPa, psi

S_2 = intermediate principal stress, MPa, psi

S_3 = least principal stress, MPa, psi

S_{11} = stress component acting normal to a plane perpendicular to the x_1-axis, MPa, psi

S_{12} = stress component acting in the x_2-direction along a plane perpendicular to the x_1-axis, MPa, psi

S = total stress, MPa, psi

S_a = stress at depth a, MPa, psi, lbm/gal

$S_{H\min}$ = least horizontal stress, MPa, psi, lbm/gal

$S_{H\max}$ = greatest horizontal stress, MPa, psi, lbm/gal

S_{ij} = component of the stress tensor acting in the x_j direction on a plane perpendicular to x_i, MPa, psi

S_{ji} = component of the stress tensor acting in the x_i direction on a plane perpendicular to x_j, MPa, psi

S_n = normal stress acting on a plane, MPa, psi

S_o = cohesion, MPa, psi

S_v = vertical stress, MPa, psi

S_z = axial stress along a wellbore, MPa, psi

T = absolute temperature, °K

T_o = tensile strength, MPa, psi

V = molar volume of water, liters/mole

V_p = compressional-wave velocity, km/s

V_s = shear-wave velocity, km/s

x_1, x_2, x_3 = Cartesian coordinate system

x, y, z = Cartesian coordinate system

Z_o = depth, ft, m

α = Biot poroelastic coefficient

β = term used in equation 1.20

δ_{ij} = Kronecker delta (δ_{ij} = 1, if $i = j$; δ_{ij} = 0 otherwise)

Δ = operator indicating a change in a parameter (ΔP_p is change in P_p)

ΔP = difference between the pressure of fluid in a well and the pore pressure

Δt_{ma} = matrix transit time, μ_s/ft

ΔT = temperature difference between the fluid in a well and the adjacent rock

$\Delta T_{\log}$ = measured value of sonic transit-time at a given depth, μ_s/ft

ΔT_n = normal value of sonic transit-time at a given depth, μ_s/ft

ε_{kl} = component of strain acting in the l direction per unit length in the k direction

θ = angle around the wellbore measured from the $S_{H\max}$ direction, degrees

θ_b = angle between the $S_{H\max}$ direction and the edge of a breakout, degrees

μ = coefficient of sliding friction on a pre-existing weak plane, where $\mu = \tan\Phi$

μ_i = coefficient of internal friction, where $\mu_i = \tan\Phi$

ρ = density, gm/cm^3

ρ_b = bulk density, gm/cm^3

$\rho_{\log}$ = measured value of density, gm/cm^3

ρ_n = normal value of density, gm/cm^3

v = Poisson's ratio

σ = Terzaghi effective stress, MPa, psi

$\sigma_1, \sigma_2, \sigma_3$ = maximum, intermediate, and least effective stresses, MPa, psi

σ_{Hmin} = minimum horizontal effective stress, MPa, psi, lbm/gal

σ_{Hmax} = maximum horizontal effective stress, MPa, psi, lbm/gal

σ_{ij} = effective stress acting in the i direction on a plane perpendicular to the j direction, MPa, psi

σ_n = effective stress acting normal to a plane, MPa, psi

σ_o = mean effective stress, MPa, psi

σ_{rr} = effective normal stress acting in the radial direction, MPa, psi

σ_v = effective normal stress acting in a vertical direction, MPa, psi

σ_{zz} = effective normal stress acting on a plane perpendicular to the z direction, MPa, psi

$\sigma^{\Delta T}$ = thermal stress induced by the cooling of the wellbore by ΔT, MPa, psi

$\sigma_{\theta\theta}$ = the effective hoop stress, MPa, psi

τ = shear stress, MPa, psi

Φ = porosity

Φ_b = breakout width, degrees

Subscripts

i = index

j = index

Superscripts

β = coefficient multiplying the effective vertical stress in Athy's relationship, Eq. 1.19

σ_v = effective vertical stress in Athy's relationship, Eq. 1.19

References

1. Castillo, D.A. and Zoback, M.D.: "Systematic Stress Variations in the Southern San Joaquin Valley and Along the White Wolf Fault: Implications For The Rupture Mechanics of the 1952 M 7.8 Kern County Earthquake and Contemporary Seismicity," *J. Geophys. Res.,* **100,** No. 4, 6249.
2. Anderson, E.M.: *The Dynamics of Faulting and Dyke Formation with Applications to Britain* second edition, Oliver and Boyd, Edinburgh (1951) 206.
3. Terzaghi, K.: "Die Theorie der Hydrodynamischen Spannungserscheinungen und ihr Erdbautechnisches Anwendungsgebiet," *Proc.,* International Applied Mechanics, Delft, The Netherlands (1924) 288–294.
4. Townend, J. and Zoback, M.D.: "How Faulting Keeps the Crust Strong," *Geology,* **28,** No. 5, 399.
5. Moos, D. and Chang, C.: "Relationships between Porosity, Pressure, and Velocities in Unconsolidated Sands," A. Mitchell and D. Grauls (eds.) *Proc.,* Overpressure in Petroleum Exploration Workshop, Pau, France (1998).
6. Schutjens, P.M.T.M. *et al.:* "Compaction-Induced Porosity/Permeability Reduction in Sandstone Reservoirs: Data and Model for Elasticity-Dominated Deformation," paper SPE 71337 presented at the 2001 SPE Annual Technical Conference and Exhibition, New Orleans, 30 September–3 October.
7. Moos, D., Zoback, M.D., and Bailey, L.: "Feasibility Study of the Stability of Openhole Multilaterals, Cook Inlet, Alaska," *SPEDC* (September 2001) 140.

8. Zoback, M.D. *et al.:* "Wellbore Breakouts and In-Situ Stress," *J. Geophys. Res.,* **90,** No. B7, 5523.

9. Gough, D.I. and Bell, J.S.: "Stress Orientations from Borehole Wall Fractures with Examples From Colorado, East Texas, and Northern Canada," *Cnd. J. Earth Science,* **19,** 1358.

10. Plumb, R.A. and Hickman, S.H.: "Stress-Induced Borehole Elongation—A Comparison between the Four-Arm Dipmeter and the Borehole Televiewer in the Auburn Geothermal Well," *J. Geophys. Res.,* **90,** B6, 5513.

11. Wiprut, D.J. *et al.:* "Constraining the Full Stress Tensor from Observations of Drilling-Induced Tensile Fractures and Leak-Off Tests: Application to Borehole Stability and Sand Production on the Norwegian Margin," *Intl. J. Rock Mech. & Min. Sci.,* **37,** 317.

12. Rai, C.S. and Hanson, K.E.: "Shear-wave Velocity Anisotropy in Sedimentary Rocks: A Laboratory Study," *Geophysics,* **53,** No. 6, 800.

13. Gregory, A.R.: "Fluid Saturation Effects on Dynamic Elastic Properties of Sedimentary Rocks," *Geophysics,* **41,** No. 5, 895.

14. Finkbeiner, T. *et al.:* "Stress, Pore Pressure, and Dynamically Constrained Hydrocarbon Columns in the South Eugene Island 330 Field, Northern Gulf of Mexico," *AAPG Bull.,* **85,** No. 6, 1007.

15. Mavko, G., Mukerji, T., and Dvorkin, J.: *The Rock Physics Handbook: Tools for Seismic Analysis of Porous Media,* Cambridge U. Press, Cambridge (1998) 329.

16. Mouchet, J.P. and Mitchell, A.: *Abnormal Pressures While Drilling: Origins, Prediction, Detection, Evaluation,* Gulf Publishing Co., Houston (1989) 255.

17. Swarbrick, R.: "Challenges of Porosity-based Pore Pressure Prediction," *CSEG Recorder* (September 2002) 74.

18. Athy, L.F.: "Density, Porosity, and Compaction of Sedimentary Rocks," *Bull. American Association of Petroleum Geologists,* **14,** No. 1, 1.

19. Dutta, N.C.: *Shale Compaction, Burial Diagenesis, and Geopressures: A Dynamic Model, Solution, and Some Results in Thermal Modeling in Sedimentary Basins,* J. Burrus (ed.), Editions Technip, Paris (1986).

20. Moos, D. and Zwart, G.: "Predicting Pore Pressure from Porosity and Velocity," paper presented at the 1998 AADE Industry Forum, Pressure Regimes in Sedimentary Basins and Their Prediction, Del Lago, Texas, 2–4 September.

21. Bruce, B. and Bowers, G.: "Pore Pressure Terminology," *The Leading Edge—Special Section Pore Pressure,* Soc. of Exploration Geophysics (February 2002) **21,** No. 2, 170.

22. Gaarenstroom, L. *et al.:* "Overpressures in the Central North Sea: Implications for Trap Integrity and Drilling Safety," J.R. Parker (ed.), *Proc.,* Fourth Conference, Petroleum Geology of Northwest Europe, Geological Soc., London, 1305–1313.

23. Nolte, K.G. and Economides, M.J.: *Fracturing Diagnosis Using Pressure Analysis in Reservoir Simulation,* Prentice Hall, Englewood Cliffs, New Jersey (1989).

24. Ward, C. and Beique, M: "Pore and Fracture Pressure Information from PWD Data," paper presented at the 2000 AADE Drilling Technology Forum, Best Available Practical Drilling Technology—The Search Continues, Houston, 9–10 February.

25. Haimson, B.C. and Herrick, C.G.: "Borehole Breakouts—A New Tool for Estimating In-Situ Stress?" *Rock Stress,* O. Stephansson (ed.), Centek Publishers, Lulea, Sweden (1986) 271–280.

26. Wiprut, D.J. and Zoback, M.D.: "High Horizontal Stress in the Visund Field, Norwegian North Sea: Consequences for Borehole Stability and Sand Production," paper SPE 47244 presented at the 1998 SPE/ISRM Eurock, Trondheim, Norway, 8–10 July.

27. Moos, D. *et al.:* "Comprehensive Wellbore Stability Analysis Utilizing Quantitative Risk Assessment," B.S. Aadnoy and S. Ong (eds.) *J. Pet. Science and Engin.,* Special Issue on Wellbore Stability, **38,** No. 1, 97.

28. Biot, M.A.: "General Theory of Three-Dimensional Consolidation," *J. Appl. Phys.* (February 1941) 155.

29. Mody, F.K. and Hale, A.H.: "A Borehole Stability Model to Couple the Mechanics and Chemistry of Drilling-Fluid/Shale Interactions," paper SPE 25728 presented at the 1993 SPE/IADC Drilling Conference, Amsterdam, 23–25 February.

General References

Alixant, J.-L.: "Real-Time Effective Stress Evaluation in Shales: Pore Pressure and Permeability Estimation," PhD dissertation, Louisiana State U., Baton Rouge, Louisiana (1987).

Barton, C.A., Tesler, L., and Zoback, M.D.: "Interactive Analysis of Borehole Televiewer Data," *Automated Pattern Analysis in Petroleum Exploration,* I. Palaz and S. Sengupta (eds.) Springer-Verlag, New York City (1991) 223–248.

Barton, C.A. and Zoback, M.D.: "Stress Perturbations Associated with Active Faults Penetrated by Boreholes: Evidence for Near Complete Stress Drop and a New Technique for Stress Magnitude Measurement," *J. Geophys. Res.,* **99**, No. 5, 9373.

Biot, M.A.: "General Solutions of the Equations of Elasticity and Consolidation for a Porous Material," *J. Appl. Mech.* (1956) **23**, 91.

Bowers, G.L.: "Pore Pressure Information From Velocity Data: Accounting for Overpressure Mechanisms Besides Compaction," paper SPE 27488 presented at the 1994 IADC/SPE Drilling Conference, Dallas, 15–18 February.

Brudy, M. *et al.:* "Estimation of the Complete Stress Tensor to 8 km Depth in the KTB Scientific Drill Holes: Implications for Crustal Strength," *J. Geophys. Res.,* **102**, 18453.

Byerlee, J.: "Friction of Rocks," *Pure and Applied Geophys.,* **116**, 615.

Chenevert, M.E. and Pernot, V.: "Control of Shale Swelling Pressures Using Inhibitive Water-Based Muds," paper SPE 49263 presented at the 1998 SPE Annual Technical Conference and Exhibition, New Orleans, 27–30 September.

Colmenares, L.B. and Zoback, M.D.: "A Statistical Evaluation of Rock Failure Criteria Constrained by Polyaxial Test Data for Five Different Rocks," *Intl. J. Rock Mech. & Mining Sciences,* **39**, 695.

Detournay, E. and Cheng, A. H.-D.: "Poroelastic Response of a Borehole in a Non-hydrostatic Stress Field," *Int. J. Rock Mech. Min. & Geomech. Abstr.,* **23**, No. 3, 171.

Eaton, B.A.: "Graphical Method Predicts Geopressures Worldwide, *World Oil,* **182**, No. 6, 51.

Ewy, R.T.: "Wellbore Stability Predictions Using a Modified Lade Criterion," *JPT* (November 1998) 64.

Pepin, G. *et al.:* "Effect of Drilling Fluid Temperatures on Fracture Gradient: Field Measurements and Model Predictions," ARMA/NARMS paper 04-527 presented at the 2004 Gulf Rocks North American Rock Mechanics Symposium: Rock Mechanics Across Borders and Disciplines, Houston, 5–9 June.

Gassemi, A., Diek, A., and dos Santos, H.: "Effects of Ion Diffusion and Thermal Osmosis on Shale Deteriorization and Borehole Instability," paper AADE 01-NC-HO-40 presented at the 2001 AADE National Drilling Conference, Houston, 27–29 March.

Haimson, B.C. and Fairhurst, C.: "Initiation and Extension of Hydraulic Fractures in Rock," *SPEJ* (July 1967) 310.

Haimson, B.C. and Lee, H.: "Borehole Breakouts and Compaction Bands in Two High-Porosity Sandstones," *Int. J. Rock Mech. Min Sci.,* **41**, No. 287.

Hillis, R.R.: "Coupled Changes in Pore Pressure and Stress in Oil Fields and Sedimentary Basins," *Petroleum Geoscience* (1980) **7**, 419.

Hoek, E. and Brown, E.T.: "Empirical Strength Criterion for Rock Masses," *J. of Geotechnical Engin. Div.,* **106**, 1013.

Holbrook, P.W.: "Real-Time Pore Pressure and Fracture Gradient Evaluation in all Sedimentary Lithologies," paper SPE 26791 presented at the 1993 SPE Offshore European Conference, Aberdeen, 7–10 September.

Holbrook, P.W.: "The Use of Petrophysical Data for Well Planning, Drilling Safety and Efficiency," paper presented at the 1996 SPWLA Annual Logging Symposium, New Orleans, 16–19 June.

Horsrud, P.: "Estimating Mechanical Properties of Shale from Empirical Correlations," *SPEDC* (June 2001) 68.

Hubbert, M.K. and Willis, D.G.: "Mechanics of Hydraulic Fracturing," *Trans.*, AIME (1957) **210**, 153.

Huffman, A.R.: "The Future of Pore Pressure Prediction Using Geophysical Methods," paper OTC 13041 presented at the 2001 Offshore Technology Conference, Houston, 30 April–3 May.

Infante, E.F. and Chenevert, M.E.: "Stability of Boreholes Drilled Through Salt Formations Displaying Plastic Behavior," paper SPE 15513 presented at the 1986 SPE Annual Technical Conference and Exhibition, New Orleans, 5–8 October.

Ito, T., Zoback, M.D., and Peska, P.: "Utilization of Mud Weights in Excess of the Least Principal Stress in Extreme Drilling Environments," *SPEDC* (December 2001) 221.

Jaeger, J.C. and Cook, N.G.W.: *Fundamentals of Rock Mechanics,* third edition, Chapman and Hill, New York City (1979) 593.

Kunze, K.R. and Steiger, R.P.: "Accurate In-Situ Stress Measurements During Drilling Operations," paper SPE 24593 presented at the 1992 SPE Annual Technical Conference and Exhibition, Washington, DC, 4–7 October.

Lal, M.: "Shale Stability: Drilling Fluid Interaction and Shale Strength," paper SPE 54356 presented at the 1999 SPE Latin American and Caribbean Petroleum Engineering Conference, Caracas, 21–23 April.

Li, X., Cui, L., and Roegiers, J.C.: "Thermoporoelastic Analyses of Inclined Boreholes," paper SPE 47296 presented at the 1998 SPE/ISRM Eurock 98, Trondheim, Norway, 8–10 July.

Maury, V.: "Rock Failure Mechanisms Identification: A Key for Wellbore Stability and Reservoir Behavior Problems," paper SPE 28049 presented at the 1994 SPE/ISRM Eurock 94, Delft, The Netherlands, 29–31 August.

Mavko, G., Mukerji, T., and Dvorkin, J.: *The Rock Physics Handbook: Tools for Seismic Analysis of Porous Media,* Cambridge U. Press, Cambridge (1998) 329.

McLean, M.R. and Addis, M.A.: "Wellbore Stability Analysis: A Review of Current Methods of Analysis and Their Field Application," paper SPE 19941 presented at the 1990 IADC/SPE Drilling Conference, Houston, 27 February–2 March.

McLellan, P.J. and Hawkes, C.D.: "Application of Probabilistic Techniques for Assessing Sand Production and Wellbore Instability Risks," paper SPE 47334 presented at the 1998 SPE/ISRM Eurock 98, Trondheim, Norway, 8–10 July.

Moos, D. and Zoback, M.D.: "Utilization of Observations of Well Bore Failure to Constrain the Orientation and Magnitude of Crustal Stresses: Application to Continental, Deep Sea Drilling Project and Ocean Drilling Program Boreholes," *J. Geophys. Res.,* **95**, B6, 9305.

Morita, N. *et al.*: "Realistic Sand-Production Prediction: Numerical Approach," *SPEPE* (February 1989) 15.

Mroz, Z. and Nawrocki, P.: "Deformation and Stability of an Elasto-Plastic Softening Pillar," *Rock Mechanics and Rock Engineering,* **22**, 69.

Nur, A.M. and Wang, Z.: *Seismic and Acoustic Velocities in Reservoir Rocks,* Vol. 1, *Experimental Studies,* Geophysics Reprint Series 10, F.K. Levin (ed.), Society of Exploration Geophysics, Tulsa (1989) 405.

Okabe, T. and Hayashi, K.: "Estimation of Stress Field by Using Drilling-Induced Tensile Fractures Observed at Well TG-2 and a Study of Critically Stressed Shear Fractures Based on the Stress Field," *Proc.,* World Geothermal Congress (2000) 1533.

Okabe, T., Shinohara, N., and Takasugi, S.: "Earth's Crust Stress Field Estimation by Using Vertical Fractures Caused by Borehole Drilling," *Proc.,* Eighth International Symposium on the Observation of the Continental Crust Through Drilling, Tsukuba, Japan (1996) 265.

Ottesen, S., Zheng, R.H., and McCann, R.C.: "Wellbore Stability Assessment Using Quantitative Risk Analysis," paper SPE 52864 presented at the 1999 SPE/IADC Drilling Conference, Amsterdam, 9–11 March.

Papanastasiou, P. *et al.: Behavior and Stability Analysis of a Wellbore Embedded in an Elastoplastic Medium in Rock Mechanics,* N. Laubach (ed.), A.A. Balkema Publishers, Rotterdam, The Netherlands (1994) 209–216.

Pells, P.J.N.: "The Use of Point Load Test in Predicting the Compressive Strength of Rock Materials," *Australian Geomechanics J.,* 54.

Peska, P. and Zoback, M.D.: "Compressive and Tensile Failure of Inclined Wellbores and Determination of In-Situ Stress and Rock Strength," *J. Geophys. Res.,* **100,** No. 7, 12,791.

Raaen, A.M. and Brudy, M.: "Pump-In/Flowback Tests Reduce the Estimate of Horzontal In-Situ Stress Significantly," paper SPE 71367 presented at the 2001 SPE Annual Technical Conference and Exhibition, New Orleans, 30 September–3 October.

Santarelli, F.J. and Carminati, S.: "Do Shales Swell? A Critical Review of Available Evidence," paper SPE 29421 presented at the 1995 SPE/IADC Drilling Conference, Amsterdam, 28 February–2 March.

Schei, G. *et al.:* "The Scratch Test: An Attractive Technique for Determining Strength and Elastic Properties of Sedimentary Rocks," paper SPE 63255 presented at the 2000 SPE Annual Technical Conference and Exhibition, Dallas, 1–4 October.

Sherwood, J.D. and Bailey, L.: "Swelling of Shale Around a Cylindrical Wellbore," *Proc.,* R. Soc. London, A, 444, 161–184.

Stephens, G. and Voight, B.: "Hydraulic Fracturing Theory for Condition of Thermal Stresses," *Intl. J. Rock Mech. Min. Sci.,* **19,** 279.

Suárez-Rivera, R. *et al.:* "Continuous Rock Strength Measurements on Core and Neural Network Modeling Result in Significant Improvements in Log-Based Rock Strength Predictions Used to Optimize Completion Design and Improve Prediction of Sanding Potential and Wellbore Stability," paper SPE 84558 presented at the 2003 SPE Annual Technical Conference and Exhibition, Denver, 5–8 October.

Swarbrick, R.E. and Osborne, M.J.: "Mechanisms that Generate Abnormal Pressures: An Overview," *Abnormal Pressures in Hydrocarbon Environments,* B.E. Law, G.F. Ulmishck, and V.I. Slavin (eds.) AAPG Memoirs, **70,** 13.

Tan, C.P. and Wu, B.: "Effects of Chemical Potential Mechanism on CU Triaxial and Borehole Collapse Tests on Shales in Rock Mechanics for Industry," Amadei *et al.* (eds.) A.A. Balkema Publishers, Rotterdam, The Netherlands (1999) 225–232.

van Oort, E., Nicholson, J., and D'Agostino, J.: "Integrated Borehole Stability Studies: Key to Drilling at the Technical Limit and Trouble Cost Reduction," paper SPE 67763 presented at the 2001 SPE/IADC Drilling Conference, Amsterdam, 17 February–1 March.

Wipurt, D. and Zoback, M.D.: "Fault Reactivation and Fluid Flow Along a Previously Dormant Normal Fault in the Northern North Sea," *Geology,* **28,** No. 7, 595.

Willson, S.M. *et al.:* "Drilling in South America: A Wellbore Stability Approach for Complex Geologic Conditions," paper SPE 53940 presented at the 1999 LACPEC Conference, Caracas, 21–23 April.

Zheng, Z., Kemeny, J., and Cook, N.G.: "Analysis of Borehole Breakouts," *J. Geophys. Res.,* **94,** No. 6, 7171.

Zhou, S.: "A Program to Model Initial Shape and Extent of Borehole Breakout," *Computers and Geosciences,* **20,** No. 7–8, 1143.

Zoback, M.D. *et al.:* "Global Patterns of Tectonic Stress," *Nature,* 341, 291.

Zoback, M.D. *et al.:* "Strength of Continental Crust and the Transmission of Plate-Driving Forces," *Nature,* **365,** 633.

Zoback, M.D., and Peska, P.: "In-Situ Stress and Rock Strength in the GBRN/DOE Pathfinder Well, South Eugene Island, Gulf of Mexico," *JPT* (July 1995) 582.

Zoback, M.D., Townend, J., and Grollimund, B.: "Steady-State Failure Equilibrium and Deformation of Intraplate Lithosphere," *Intl. Geology Rev.,* **44,** 383.

SI Metric Conversion Factors

$$
\begin{array}{rl l l}
\text{bbl} \times & 1.589\ 873 & \text{E} - 01 & = \text{m}^3 \\
\text{ft} \times & 3.048^* & \text{E} - 01 & = \text{m} \\
\text{in.} \times & 2.54^* & \text{E} + 00 & = \text{cm} \\
\text{lbf} \times & 4.448\ 222 & \text{E} + 00 & = \text{N} \\
\text{lbm} \times & 4.535\ 924 & \text{E} - 01 & = \text{kg} \\
\text{psi} \times & 6.894\ 757 & \text{E} + 00 & = \text{kPa}
\end{array}
$$

*Conversion factor is exact.

Chapter 2
Drilling Fluids

Gary West, John Hall, and **Simon Seaton,** Halliburton Fluid Systems Div., Baroid Fluid Services

2.1 Introduction

The drilling-fluid system—commonly known as the "mud system"—is the single component of the well-construction process that remains in contact with the wellbore throughout the entire drilling operation. Drilling-fluid systems are designed and formulated to perform efficiently under expected wellbore conditions. Advances in drilling-fluid technology have made it possible to implement a cost-effective, fit-for-purpose system for each interval in the well-construction process.

The active drilling-fluid system comprises a volume of fluid that is pumped with specially designed mud pumps from the surface pits, through the drillstring exiting at the bit, up the annular space in the wellbore, and back to the surface for solids removal and maintenance treatments as needed. The capacity of the surface system usually is determined by the rig size, and rig selection is determined by the well design. For example, the active drilling-fluid volume on a deepwater well might be several thousand barrels. Much of that volume is required to fill the long drilling riser that connects the rig floor to the seafloor. By contrast, a shallow well on land might only require a few hundred barrels of fluid to reach its objective.

A properly designed and maintained drilling fluid performs several essential functions during well construction:

• Cleans the hole by transporting drilled cuttings to the surface, where they can be mechanically removed from the fluid before it is recirculated downhole.

• Balances or overcomes formation pressures in the wellbore to minimize the risk of well-control issues.

• Supports and stabilizes the walls of the wellbore until casing can be set and cemented or openhole-completion equipment can be installed.

• Prevents or minimizes damage to the producing formation(s).

• Cools and lubricates the drillstring and bit.

• Transmits hydraulic horsepower to the bit.

• Allows information about the producing formation(s) to be retrieved through cuttings analysis, logging-while-drilling data, and wireline logs.

The cost of the drilling fluid averages 10% of the total tangible costs of well construction; however, drilling-fluid performance can affect overall well-construction costs in several ways. A correctly formulated and well-maintained drilling system can contribute to cost containment throughout the drilling operation by enhancing the rate of penetration (ROP), protecting the reservoir from unnecessary damage, minimizing the potential for loss of circulation, stabilizing the wellbore during static intervals, and helping the operator remain in compliance with environmental and safety regulations. Many drilling-fluid systems can be reused from well to well, thereby reducing waste volumes and costs incurred for building new mud.

To the extent possible, the drilling-fluid system should help preserve the productive potential of the hydrocarbon-bearing zone(s). Minimizing fluid and solids invasion into the zones of interest is critical to achieving desired productivity rates. The drilling fluid also should comply with established health, safety, and environmental (HSE) requirements so that personnel are not endangered and environmentally sensitive areas are protected from contamination. Drilling-fluid companies work closely with oil-and-gas operating companies to attain these mutual goals.

2.2 Basic Functions of a Drilling Fluid

2.2.1 Transport Cuttings to Surface. Transporting drilled cuttings to surface is the most basic function of drilling fluid. To accomplish this, the fluid should have adequate suspension properties to help ensure that cuttings and commercially added solids such as barite weighing material do not settle during static intervals. The fluid should have the correct chemical properties to help prevent or minimize the dispersion of drilled solids, so that these can be removed efficiently at the surface. Otherwise, these solids can disintegrate into ultrafine particles that can damage the producing zone and impede drilling efficiency.

2.2.2 Prevent Well-Control Issues. The column of drilling fluid in the well exerts hydrostatic pressure on the wellbore. Under normal drilling conditions, this pressure should balance or exceed the natural formation pressure to help prevent an influx of gas or other formation fluids. As the formation pressures increase, the density of the drilling fluid is increased to help maintain a safe margin and prevent "kicks" or "blowouts"; however, if the density of the fluid becomes too heavy, the formation can break down. If drilling fluid is lost in the resultant fractures, a reduction of hydrostatic pressure occurs. This pressure reduction also can lead to an influx from a pressured formation. Therefore, maintaining the appropriate fluid density for the wellbore pressure regime is critical to safety and wellbore stability.

2.2.3 Preserve Wellbore Stability. Maintaining the optimal drilling-fluid density not only helps contain formation pressures, but also helps prevent hole collapse and shale destabilization. The wellbore should be free of obstructions and tight spots, so that the drillstring can be moved freely in and out of the hole (tripping). After a hole section has been drilled to the planned depth, the wellbore should remain stable under static conditions while casing is run to bottom and cemented. The drilling-fluid program should indicate the density and physicochemical properties most likely to provide the best results for a given interval.

2.2.4 Minimize Formation Damage. Drilling operations expose the producing formation to the drilling fluid and any solids and chemicals contained in that fluid. Some invasion of fluid filtrate and/or fine solids into the formation is inevitable; however, this invasion and the potential for damage to the formation can be minimized with careful fluid design that is based on testing performed with cored samples of the formation of interest. Formation damage also can be curtailed by expert management of downhole hydraulics using accurate modeling software, as well as by the selection of a specially designed "drill-in" fluid, such as the systems that typically are implemented while drilling horizontal wells.

2.2.5 Cool and Lubricate the Drillstring. The bit and drillstring rotate at relatively high revolutions per minute (rev/min) all or part of the time during actual drilling operations. The circulation of drilling fluid through the drillstring and up the wellbore annular space helps reduce friction and cool the drillstring. The drilling fluid also provides a degree of lubricity to aid the movement of the drillpipe and bottomhole assembly (BHA) through angles that are created intentionally by directional drilling and/or through tight spots that can result from swelling shale. Oil-based fluids (OBFs) and synthetic-based fluids (SBFs) offer a high degree of lubricity and for this reason generally are the preferred fluid types for high-angle directional wells. Some water-based polymer systems also provide lubricity approaching that of the oil- and synthetic-based systems.

2.2.6 Provide Information About the Wellbore. Because drilling fluid is in constant contact with the wellbore, it reveals substantial information about the formations being drilled and serves as a conduit for much data collected downhole by tools located on the drillstring and through wireline-logging operations performed when the drillstring is out of the hole. The drilling fluid's ability to preserve the cuttings as they travel up the annulus directly affects the quality of analysis that can be performed on the cuttings. These cuttings serve as a primary indicator of the physical and chemical condition of the drilling fluid. An optimized drilling-fluid system that helps produce a stable, in-gauge wellbore can enhance the quality of the data transmitted by downhole measurement and logging tools as well as by wireline tools.

2.2.7 Minimize Risk to Personnel, the Environment, and Drilling Equipment. Drilling fluids require daily testing and continuous monitoring by specially trained personnel. The safety hazards associated with handling of any type of fluid are clearly indicated in the fluid's documentation. Drilling fluids also are closely scrutinized by worldwide regulatory agencies to help ensure that the formulations in use comply with regulations established to protect both natural and human communities where drilling takes place. At the rigsite, the equipment used to pump or process fluid is checked constantly for signs of wear from abrasion or chemical corrosion. Elastomers used in blowout-prevention equipment are tested for compatibility with the proposed drilling-fluid system to ensure that safety is not compromised.

The upper hole sections typically are drilled with low-density water-based fluids (WBFs). Depending on formation types, downhole temperatures, directional-drilling plans, and other factors, the operator might switch to an OBF or SBF at a predetermined point in the drilling process. High-performance WBFs also are available to meet a variety of drilling challenges.

Depending on the location of the well, the drilling-fluid system can be exposed to saltwater flows, influxes of carbon dioxide and hydrogen sulfide, solids buildup, oil or gas influxes, or extreme temperatures at both ends of the scale—or all of these. Contamination also comes from contact with the spacers and cement slurries used to permanently install casing and in the course of displacing from one drilling-fluid system to another.

The drilling-fluid specialists who prepare drilling-fluid programs should be aware of the operational and environmental challenges posed by any well. Working closely with the operator, the specialist (who typically is supported by technical experts and a research staff) can plan for the scope of conditions that are likely to be encountered and generate a program that is both safe and cost-effective. The planning stage usually includes the identification of specific performance objectives and the means by which success will be measured.

Throughout the well-construction process, the drilling-fluid personnel assigned to the operation maintain accurate records of test results, fluid volumes, drilling events, product inventory, and actions related to achieving environmental compliance. The standard drilling-mud report reflects the type of information the drilling-fluid personnel (often called "mud engineers") provide at the rig site on a daily basis. These reports, often computer-generated and stored in a database, and the post-well analysis performed at the conclusion of the well serve as reference materials for future wells in the same area or wells that present similar challenges.

2.3 Types of Drilling Fluids

World Oil's annual classification of fluid systems lists nine distinct categories of drilling fluids, including freshwater systems, saltwater systems, oil- or synthetic-based systems, and pneumatic (air, mist, foam, gas) "fluid" systems.[1] Three key factors usually determine the type of fluid selected for a specific well: cost, technical performance, and environmental impact.

Water-based fluids are the most widely used systems and generally are considered less expensive than OBFs or SBFs. The OBFs and SBFs—also known as invert-emulsion systems—have an oil or synthetic base fluid as the continuous, or external, phase and brine as the internal phase. Whereas invert-emulsion systems have a higher cost per unit than most water-based fluids, they often are selected when well conditions call for reliable shale inhibition and/or excellent lubricity. Water-based systems and invert-emulsion systems can be formulated to tolerate relatively high downhole temperatures. Pneumatic systems most commonly are implemented in areas where formation pressures are relatively low and the risk of lost circulation or formation damage is relatively high. The use of these systems requires specialized pressure-management equipment to help prevent the development of hazardous conditions when hydrocarbons are encountered.

2.3.1 WBFs.

Water-based fluids are used to drill approximately 80% of all wells.[2] The base fluid may be fresh water, seawater, brine, saturated brine, or a formate brine. The type of fluid selected depends on anticipated well conditions or on the specific interval of the well being drilled. For example, the surface interval typically is drilled with a low-density water- or seawater-based mud that contains few commercial additives. These systems incorporate natural clays in the course of the drilling operation. Some commercial bentonite or attapulgite also may be added to aid in fluid-loss control and to enhance hole-cleaning effectiveness. After surface casing is set and cemented, the operator often continues drilling with a WBF unless well conditions require displacing to an oil- or synthetic-based system.

WBFs fall into two broad categories: nondispersed and dispersed. Simple gel-and-water systems used for tophole drilling are nondispersed, as are many of the advanced polymer systems that contain little or no bentonite. The natural clays that are incorporated into nondispersed systems are managed through dilution, encapsulation, and/or flocculation. A properly designed solids-control system can be used to remove fine solids from the mud system and help maintain drilling efficiency. The low-solids, nondispersed (LSND) polymer systems rely on high- and low-molecular-weight long-chain polymers to provide viscosity and fluid-loss control. Low-colloidal solids are encapsulated and flocculated for more efficient removal at the surface, which in turn decreases dilution requirements. Specially developed high-temperature polymers are available to help overcome gelation issues that might occur on high-pressure, high-temperature (HP/HT) wells.[3] With proper treatment, some LSND systems can be weighted to 17.0 to 18.0 ppg and run at 350°F and higher.

Dispersed systems are treated with chemical dispersants that are designed to deflocculate clay particles to allow improved rheology control in higher-density muds. Widely used dispersants include lignosulfonates, lignitic additives, and tannins. Dispersed systems typically require additions of caustic soda (NaOH) to maintain a pH level of 10.0 to 11.0. Dispersing a system can increase its tolerance for solids, making it possible to weight up to 20.0 ppg. The commonly used lignosulfonate system relies on relatively inexpensive additives and is familiar to most operator and rig personnel. Additional commonly used dispersed muds include lime and other cationic systems. A solids-laden dispersed system also can decrease the rate of penetration significantly and contribute to hole erosion.

Saltwater drilling fluids often are used for shale inhibition and for drilling salt formations. They also are known to inhibit the formation of ice-like hydrates that can accumulate around subsea wellheads and well-control equipment, blocking lines and impeding critical operations. Solids-free and low-solids systems can be formulated with high-density brines, such as calcium chloride, calcium bromide, zinc bromide, and potassium and cesium formate.

2.3.2 Drill-In Fluids (DIFs). Too often, drilling into a pay zone with a conventional fluid can introduce a host of previously undefined risks, all of which diminish reservoir connectivity with the wellbore or reduce formation permeability. This is particularly true in horizontal wells, where the pay zone can be exposed to the drilling fluid over a long interval. Selecting the most suitable fluid system for drilling into the pay zone requires a thorough understanding of the reservoir. Using data generated by lab testing on core plugs from carefully selected pay zone cores, a reservoir-fluid-sensitivity study should be conducted to determine the morphological and mineralogical composition of the reservoir rock. Natural reservoir fluids should be analyzed to establish their chemical makeup. The degree of damage that could be caused by anticipated problems can be modeled, as can the effectiveness of possible solutions for mitigating the risks.

A DIF is a clean fluid that is designed to cause little or no loss of the natural permeability of the pay zone and to provide superior hole cleaning and easy cleanup. DIFs can be water-based, brine-based, oil-based, or synthetic-based systems. In addition to being safe and economical for the application, a DIF should be compatible with the reservoir's native fluids to avoid causing precipitation of salts or production of emulsions. A suitable nondamaging fluid should establish a filter cake on the face of the formation, but should not penetrate too far into the formation pore pattern. The fluid filtrate should inhibit or prevent swelling of reactive clay particles within the pore throats.

Formation damage commonly is caused by pay zone invasion and plugging by fine particles, formation clay swelling, commingling of incompatible fluids, movement of dislodged formation pore-filling particles, changes in reservoir-rock wettability, and formation of emulsions or water blocks. Once a damage mechanism has diminished the permeability of a reservoir, it seldom is possible to restore the reservoir to its original condition.

2.3.3 OBFs. Oil-based systems were developed and introduced in the 1960s to help address several drilling problems: formation clays that react, swell, or slough after exposure to WBFs; increasing downhole temperatures; contaminants; and stuck pipe and torque and drag.

Oil-based fluids in use today are formulated with diesel, mineral oil, or low-toxicity linear paraffins (that are refined from crude oil). The electrical stability of the internal brine or water phase is monitored to help ensure that the strength of the emulsion is maintained at or near a predetermined value. The emulsion should be stable enough to incorporate additional water volume if a downhole water flow is encountered.

Barite is used to increase system density, and specially-treated organophilic bentonite is the primary viscosifier in most oil-based systems. The emulsified water phase also contributes to fluid viscosity. Organophilic lignitic materials are added to help control low-pressure/low-temperature (LP/LT) and HP/HT fluid loss. Oil-wetting is essential for ensuring that particulate materials remain in suspension; the surfactants used for oil-wetting also can work as thinners. Oil-based systems usually contain lime to maintain an elevated pH, resist adverse effects of hydrogen sulfide (H_2S) and carbon dioxide (CO_2) gases, and enhance emulsion stability.

Shale inhibition is one of the key benefits of using an oil-based system. The high-salinity-water phase helps to prevent shales from hydrating, swelling, and sloughing into the wellbore. Most conventional oil-based mud (OBM) systems are formulated with calcium-chloride brine, which appears to offer the best inhibition properties for most shales.

The ratio of the oil percentage to the water percentage in an oil-based system is called its oil/water ratio. Oil-based systems generally function well with an oil/water ratio of from 65/35 to 95/5, but the most commonly observed range is from 80/20 to 90/10.

The discharge of whole fluid or cuttings generated with OBFs is not permitted in most offshore-drilling areas. All such drilled cuttings and waste fluid are processed and shipped to shore for disposal. Whereas many land wells continue to be drilled with diesel-based fluids, the development of SBFs in the late 1980s provided new options to offshore operators who depend on the drilling performance of oil-based systems to help hold down overall drilling costs but require more environmentally-friendly fluids.

2.3.4 Synthetic-Based Drilling Fluids. Synthetic-based fluids were developed out of an increasing desire to reduce the environmental impact of offshore drilling operations, but without sacrificing the cost-effectiveness of oil-based systems.

Like traditional OBFs, SBFs help maximize ROPs, increase lubricity in directional and horizontal wells, and minimize wellbore-stability problems such as those caused by reactive shales. Field data gathered since the early 1990s confirm that SBFs provide exceptional drilling performance, easily equaling that of diesel- and mineral-oil-based fluids.

In many offshore areas, regulations that prohibit the discharge of cuttings drilled with OBFs do not apply to some of the synthetic-based systems. SBFs' cost per barrel can be higher, but they have proved economical in many offshore applications for the same reasons that traditional OBFs have: fast penetration rates and less mud-related nonproductive time (NPT). SBFs that are formulated with linear alphaolefins (LAO) and isomerized olefins (IO) exhibit the lower kinematic viscosities that are required in response to the increasing importance of viscosity issues as operators move into deeper waters. Early ester-based systems exhibited high kinematic viscosity, a condition that is magnified in the cold temperatures encountered in deepwater risers. However, a shorter-chain-length (C_8), low-viscosity ester that was developed in 2000 exhibits viscosity similar to or lower than that of the other base fluids, specifically the heavily used IO systems. Because of their high biodegradability and low toxicity, esters are universally recognized as the best base fluid for environmental performance.

By the end of 2001, deepwater wells were providing 59% of the oil being produced in the Gulf of Mexico.[4] Until operators began drilling in these deepwater locations, where the pore pressure/fracture gradient (PP/FG) margin is very narrow and mile-long risers are not uncommon, the standard synthetic formulations provided satisfactory performance. However, the issues that arose because of deepwater drilling and changing environmental regulations prompted a closer examination of several seemingly essential additives.

When cold temperatures are encountered, conventional SBFs might develop undesirably high viscosities as a result of the organophilic clay and lignitic additives in the system. The introduction of SBFs formulated with zero or minimal additions of organophilic clay and lignitic products allowed rheological and fluid-loss properties to be controlled through the fluid-emulsion characteristics. The performance advantages of these systems include high, flat gel strengths that break with minimal initiation pressure; significantly lower equivalent circulating densities (ECDs); and reduced mud losses while drilling, running casing, and cementing.

2.3.5 All-Oil Fluids. Normally, the high-salinity water phase of an invert-emulsion fluid helps to stabilize reactive shale and prevent swelling; however, drilling fluids that are formulated with diesel- or synthetic-based oil and no water phase are used to drill long shale intervals where the salinity of the formation water is highly variable. By eliminating the water phase, the all-oil drilling fluid can preserve shale stability throughout the interval.

2.3.6 Pneumatic-Drilling Fluids. Compressed air or gas can be used in place of drilling fluid to circulate cuttings out of the wellbore. Pneumatic fluids fall into one of three categories: air or gas only, aerated fluid, or foam.[5] Pneumatic-drilling operations require specialized equipment to help ensure safe management of the cuttings and formation fluids that return to surface, as well as tanks, compressors, lines, and valves associated with the gas used for drilling or aerating the drilling fluid or foam.

Except when drilling through high-pressure hydrocarbon- or fluid-laden formations that demand a high-density fluid to prevent well-control issues, using pneumatic fluids offers several advantages: little or no formation damage, rapid evaluation of cuttings for the presence of hydrocarbons, prevention of lost circulation, and significantly higher penetration rates in hard-rock formations.[6]

2.3.7 Specialty Products. Drilling-fluid service companies provide a wide range of additives that are designed to prevent or mitigate costly well-construction delays. Examples of these products include:
• Lost-circulation materials (LCM) that help to prevent or stop downhole mud losses into weak or depleted formations.
• Spotting fluids that help to free stuck pipe.
• Lubricants for WBFs that ease torque and drag and facilitate drilling in high-angle environments.
• Protective chemicals (e.g., scale and corrosion inhibitors, biocides, and H_2S scavengers) that prevent damage to tubulars and personnel.

LCMs. Many types of LCM are available to address loss situations. Sized calcium carbonate, mica, fibrous material, cellophane, and crushed walnut shells have been used for decades. The development of deformable graphitic materials that can continuously seal off fractures under changing pressure conditions has allowed operators to cure some types of losses more consistently. The application of these and similar materials to actually strengthen the wellbore has proved successful. Hydratable and rapid-set lost-circulation pills also are effective for curing severe and total losses. Some of these fast-acting pills can be mixed and pumped with standard rig equipment. Others require special mixing and pumping equipment.

Spotting Fluids. Most spotting fluids are designed to penetrate and break up the wall cake around the drillstring. A soak period usually is required to achieve results. Spotting fluids typically are formulated with a base fluid and additives that can be incorporated into the active mud system with no adverse effects after the pipe is freed and/or circulation resumes.

Lubricants. Lubricants might contain hydrocarbon-based materials or can be formulated specifically for use in areas where environmental regulations prohibit the use of an oil-based additive. Tiny glass or polymer beads also can be added to the drilling fluid to increase lubricity. Lubricants are designed to reduce friction in metal-to-metal contact and to provide lubricity to the drillstring in the open hole, especially in deviated wells, where the drillstring is likely to have continuous contact with the wellbore.

Corrosion Inhibitors, Biocides, and Scavengers. Corrosion causes the majority of drillpipe loss and also damages casing, mud pumps, bits, and downhole tools. As downhole temperatures increase, corrosion also increases at a corresponding rate if the drillstring is not protected by chemical treatment. Abrasive materials in the drilling fluid can accelerate corrosion by scouring away protective films. Corrosion typically is caused by one or more factors that include exposure to oxygen, H_2S, and/or CO_2; bacterial activity in the drilling fluid; high-temperature environments; and contact with sulfur-containing materials. Drillstring coupons can be inserted between joints of drillpipe as the pipe is tripped in the hole. When the pipe next is tripped out of the hole, the coupon can be examined for signs of pitting and corrosion to determine whether the drillstring components are undergoing similar damage.

H_2S and CO_2 frequently are present in the same formation. Scavenger and inhibitor treatments should be designed to counteract both gases if an influx occurs because of underbal-

TABLE 2.1—TYPICAL FIELD TESTS FOR WBFs[7]	
Drilling-fluid density	Mud weight, in ppg or equivalent unit of measure, as appropriate to the region [e.g., specific gravity (SG)].
Viscosity	Viscosity exhibited when a specific quantity of fluid is poured through a Marsh funnel (typically recorded in seconds per quart).
Rheology	Rheological properties exhibited at various rotational speeds using a viscometer (also called rheometer). The Fann* 35 viscometer can test the fluid at multiple speeds and temperatures to give a detailed viscosity profile for the fluid.
Gel strength	Suspension characteristics developed over specified time intervals.
Filtration	Surface indication of filtrate invasion into the near wellbore (also called fluid loss). Performed under LP/LT conditions and HP/HT conditions as required.
Retort analysis	Percentages of water, oil, and solids making up the active system.
Sand content	Percentage of sand in the active system.
Methylene blue capacity	Clay content in the active system [also commonly called methylene blue test (MBT)].
pH	Indication of system acidity or alkalinity.
Chemical analysis: Alkalinity/lime content Chlorides Total hardness (calcium)	Indication of variations from base-fluid formulation caused by surface treatment and/or influx or contamination from downhole formations.

*Fann is the registered name for fluids-testing equipment provided by Fann Instrument Co., Houston.

anced drilling conditions. Maintaining a high pH helps control H_2S and CO_2 and prevents bacteria from souring the drilling fluid. Bacteria also can be controlled using a microbiocide additive.

2.4 Drilling-Fluids Testing

2.4.1 Field Tests. The drilling-fluids specialist in the field conducts a number of tests to determine the properties of the drilling-fluid system and evaluate treatment needs. Although drilling-fluid companies might use some tests that are designed for evaluating a proprietary product, the vast majority of field tests are standardized according to *American Petroleum Institute Recommended Practices (API RP) 13B-1*[7] and *13B-2*,[8] for WBFs and OBFs, respectively.

Table 2.1 shows typical API-recommended field tests for WBFs. **Table 2.2** shows typical API-recommended field tests for OBFs and SBFs. Several tests are identical to those performed on WBFs. For all three fluid types, depending on the type of fluid in use, some or all tests should be performed.

2.4.2 Laboratory Tests. Extensive testing of the fluid is performed in the design phase of the fluid, either to achieve desired fluid characteristics or to determine the performance limitations of the fluid.

Laboratory testing aids in fluid design and expands the capacity to monitor and evaluate fluids when field-testing procedures prove inadequate. Some laboratory tests are identical to field-testing methods, whereas others are unique to the laboratory environment. In the laboratory

TABLE 2.2—TYPICAL FIELD TESTS FOR OBFs AND SBFs[8]	
Drilling-fluid density	Mud weight in ppg or equivalent unit of measure, as appropriate to the region (e.g., SG).
Viscosity	Viscosity exhibited when a specific quantity of fluid is poured through a Marsh funnel (typically recorded in seconds per quart).
Rheology	Rheological properties exhibited at various rotational speeds using a viscometer (also called a rheometer). The Fann* 35 viscometer can test the fluid at multiple speeds and temperatures to give a detailed viscosity profile for the fluid.
Gel strength	Suspension characteristics developed over specified time intervals.
Filtration	Surface indication of filtrate invasion into the near wellbore (also called fluid loss). Performed under LP/LT conditions and HP/HT conditions, as required.
Retort analysis	Percentages of water, oil, and solids making up the active system.
Sand content	Percentage of sand in the active system.
Chemical analysis: Alkalinity/lime content Chlorides Total hardness (calcium)	For whole-mud analysis: indication of variations from base-fluid formulation caused by surface treatment and/or influx or contamination from downhole formations.
Electrical stability (ES)	Indication of emulsion stability of the water phase of the oil- or synthetic-based system, performed with an ES meter.
Water-phase salinity	Presence of chlorides, in parts per million (ppm).
Sulfide concentration	Indication of the concentration of soluble sulfides, in ppm, performed with Garrett gas-train apparatus.
*Fann is the registered name for fluids-testing equipment provided by Fann Instrument Co., Houston.	

setting, testing and equipment are available to determine toxicity, fluid rheology, fluid loss, particle plugging, high-angle sag, dynamic high-angle sag, high-temperature fluid aging, cuttings erosion, shale stability, capillary suction, lubricity, return permeability, X-ray diffraction, and particle-size distribution (PSD).

Toxicity. The environmental and toxicity standards of the region in which the fluid is being used will require testing either of the whole drilling fluid or of its individual components. Toxicity tests generally are used for offshore applications. An approved laboratory can perform the proper testing to ensure compliance of the fluid or its components.

Fluid Rheology. Fluid rheology is an important parameter of drilling-fluid performance. For critical offshore applications with extreme temperature and pressure requirements, the viscosity profile of the fluid often is measured with a controlled-temperature and -pressure viscometer (e.g., the Fann 75). Fluids can be tested at temperatures of < 35°F to 500°F, with pressures of up to 20,000 psia. Cold-fluid rheology is important because of the low temperatures that the fluid is exposed to in deepwater risers. High temperatures can be encountered in deep wells or in geothermally heated wells. The fluid can be under tremendous pressure downhole, and its viscosity profile can change accordingly.

Fluid Loss. If fluid (or filtration) loss is excessive, formation instability, formation damage, or a fractured formation and loss of drilling fluid can occur. In the field, LP/LT and HP/HT fluid-loss tests are performed routinely. Fluid loss also can be measured under dynamic conditions using the Fann 90 viscometer, which incorporates a rotating bob to provide fluid shear in the center of a ceramic-filter core. The fluid is heated and pressurized. Fluid loss is measured radially through the entire core, giving a sophisticated simulation of the drilling fluid circulating in the wellbore.

Particle Plugging. The particle-plugging test (PPT) often is used to evaluate the ability of plugging particles added to a fluid to mitigate formation damage by stopping or slowing filtrate invasion into a core. A PPT uses an inverted HP/HT-filter-press cell that has been fitted with a ceramic disk as a filtering medium and is pressurized with a hydraulic cylinder. Ceramic disks with different mean pore-throat diameters are used to simulate a wellbore wall. A PPT typically is run with a 2,000-psi or higher differential pressure. The spurt loss and total fluid loss are measured over a 30-minute period. The cell is inverted, and fluid loss is measured from the top of the cell to eliminate the effects of fluid settling.

High-Angle Sag and Dynamic High-Angle Sag. The weighting material used to increase the density of the drilling fluid can settle at a faster rate in an angled well than in a vertical well. The high-angle sag test (HAST) and dynamic high-angle sag test (DHAST) measure density differences in the fluid as the angle of drilling changes. The HAST is used with fluids under static conditions, whereas the DHAST is used under dynamic conditions, in which the fluid can be subjected to shear or observed statically. The DHAST has temperature and pressure specifications of 350°F and 10,000 psia, respectively. Measuring the changes in density allows the fluid's propensity to undergo these changes in the drilling process to be evaluated and curtailed by modifications to the fluid design.

High-Temperature Fluid Aging. Over time, high temperatures can degrade the components of a drilling fluid and alter its performance. High-temperature aging of the fluid is conducted to assess the impact that temperatures > 250°F have on performance. Fluid can be aged statically and dynamically. In the static-aging process, the fluid is placed in a pressurized cell and allowed to stand without rolling at the desired test temperature for a desired length of time (rarely < 16 hours). This simulates the stress the fluid might be subjected to during static periods in the wellbore (e.g., logging and tripping). In dynamic aging, the fluid is rolled in a pressurized cell at the desired test temperature to simulate the fluid under drilling conditions. After undergoing aging, the fluid can be evaluated using the same tests that are applied to nonaged fluid.

Cuttings Erosion. If drilled cuttings undergo significant erosion before being removed from the drilling fluid, that fluid's colloidal content can increase and interfere with drilling performance. Also, cuttings erosion usually is accompanied by wellbore erosion, which leads to hole washout. Two tests are available to aid in designing fluids that reduce cuttings erosion. One is an API-approved method in which the user measures out a known amount of shale material that is representative of the formation to be drilled and that has been broken and sized between No. 6 and No. 12 shaker screens. The shale then is placed in a jar and exposed to the drilling fluid, where it is aged by hot-rolling it at 150°F for 16 hours. After aging, the shale is collected on the No. 12 shaker screen, carefully washed, and dried. The percent recovery then is calculated on the basis of the weight of the recovered shale vs. that of the shale originally used. Variations of this method are used for individual component testing. The aging time is varied to determine the erosion rate.

The second available test is the slake-durability tester, which measures chemical and mechanical erosion to the shale. This tester resembles the API-approved method in that it uses a known amount of test shale and in that recovery is calculated in the same way. It differs in that the shale sample is placed inside a mesh-screen cage that is immersed in the drilling fluid and rolled continually throughout the test.[9]

Shale Stability. Reactive shales cause many difficulties in a drilling operation. Fluids should be designed to mitigate these shale problems. Along with erosion testing, four other distinct tests are used to assess the interaction between the drilling fluid and shale: capillary suction time (CST), return permeability, X-ray diffraction and PSD.

CST. The CST test investigates the chemical effects of the drilling fluid on the dispersive properties of shale and active clays. The CST test measures filter-cake permeability by timing the capillary action of filtrate onto a paper medium. Changes in permeability then can be related to the inhibitive characteristics of the fluid.[10]

Return Permeability. When drilling reaches a hydrocarbon-bearing zone, of great concern is the potential to damage the formation and thereby to reduce the ability of the well to produce hydrocarbons. A return-permeability test can reveal formation damage and can be conducted using a return permeameter. The porosity and conductivity of a core sample are determined by flowing a refined mineral oil through the core. To simulate fluid and filtrate invasion into the core, drilling fluid then is placed against the outflow side of the core, and differential pressure is applied in the direction opposite that of the previous flow measurement. After contamination, mineral oil again is flowed through the core in the original direction, and the resultant porosity is compared to the original porosity to determine whether a reduction in permeability has occurred.

X-Ray Diffraction. Knowing the mineral composition of a formation to be drilled is important for determining how the drilling fluid will react with the formation and how to prevent potential drilling problems. Fluid labs use X-ray diffraction to determine the mineralogical composition of shale or cuttings. They expose a crystalline mineral sample to X-ray radiation and then compare the resultant diffraction pattern to known standards to determine which minerals are present in the sample.

PSD. Particle-size distributions are determined for various solid materials that are added to drilling fluids. A particle-size analyzer determines PSD by measuring laser-light diffraction, which then can be related to particle size. PSDs are used to determine what screen size is needed for removing particles from the fluid system for conditioning, whether the particles present are small enough to cause formation damage by becoming trapped in the formation's pores, and whether the present distribution of particle sizes will allow effective bridging of pore openings to help control fluid loss without causing excessive formation damage.

2.5 Challenges Related to Drilling Fluid

All drilling challenges relate to the fundamental objective of maintaining a workable wellbore throughout the well-construction process. A workable wellbore can be drilled, logged, cased, cemented, and completed with minimal nonproductive time. The design of the drilling-fluid system is central to achieving this objective.

Most operational problems are interrelated, making them more difficult to resolve. For example, loss of circulation into a depleted zone causes a drop in hydrostatic pressure in the wellbore. When the hydrostatic pressure falls too low to hold back formation fluids, the loss incident can be compounded by an influx of gas or water, known as a flow or (when more severe) a kick. In these circumstances, the operator should increase the fluid density to stop the kick, yet avoid exacerbating the lost-circulation problem. Furthermore, the pressure differential created at the loss zone can cause the drillstring to become embedded in the wall cake, a situation called differential sticking. The drillstring should be freed quickly by mechanical or chemical methods because the longer it remains stuck, the lower the likelihood of freeing it. Failure to free the pipe can require an expensive fishing job that cannot be undertaken until the well is under control.

Another example of interrelated problems occurs when the directional-drilling operation requires an interval of sliding, in which the drillstring is not rotated for a period of time but drilling continues by means of a downhole motor. Sliding allows better directional control, but the lack of pipe rotation can impair hole cleaning. Good hole cleaning is important in all wells, but it is critical in high-angle wells, in which cuttings might fall to the low side of the wellbore and form deep cuttings beds. Failure to remove the cuttings can lead to packing off of the drillstring—another version of stuck pipe. If the drillstring is severely packed off, attempts to circulate drilling fluid might lead to excessive pressure on the wellbore, which in turn can cause the formation below the packoff to break down.

Incidents like these are not uncommon. The drilling fluid alone cannot correct all problems, but skillful management of the drilling-fluid system by a specialist can prevent conditions that lead to wellbore instability, thereby helping the operator to achieve a workable wellbore.

2.5.1 Loss of Circulation.

A lost-circulation incident exacts a heavy cost that goes far beyond the price of products that are used to treat it. Lost circulation always causes nonproductive time that includes the cost of rig time and all the services that support the drilling operation. Losing mud into the oil or gas reservoir can drastically reduce or even eliminate the operator's ability to produce the zone. Prevention is critical, but because lost circulation is such a common occurrence, effective methods of remediation are also a high priority.

Rock mechanics and hydraulic-fracture theory indicate that it is easier to prevent fracture propagation than it is to plug the fracture later to prevent fluid from re-entering.[11] Because of the high cost of most weighted, treated drilling-fluid systems, LCM routinely is carried in the active system on many operations in which probable lost-circulation zones exist, such as in a "rubble" zone beneath salt or in a known depleted zone. Other conditions that are prone to loss of circulation include natural and induced fractures, formations with high permeability and/or high porosity, and vugular formations (e.g., limestone and chalk). Using an LCM that can be carried in the drilling fluid without significantly affecting its rheology or fluid-loss characteristics facilitates the preventive pretreatment. Pretreatment can mitigate wellbore breathing (ballooning), seepage losses, and/or potential lost circulation when drilling depleted zones.

When a loss zone is encountered, the top priority is keeping the hole full so that the hydrostatic pressure does not fall below formation pressure and allow a kick to occur. The hydrostatic pressure may be purposely reduced to stop the loss, as long as sufficient density is maintained to prevent well-control problems. Loss zones also pose a high risk of differential sticking. Rotating and reciprocating the drillstring helps reduce this risk while an LCM treatment is prepared. If the location of the loss zone is known, it might be advisable to pull the drillstring to above the affected area.

A variety of LCM is available, and combining several types and particle sizes for treatment purposes is common practice. Conventional—and relatively inexpensive—materials include sized calcium carbonate, paper, cottonseed hulls, nutshells, mica, and cellophane. Because lost circulation always has been one of the most costly issues facing the industry, a focus on healing the loss zone quickly and safely encouraged the development of proprietary materials that conform to the fracture to seal off pores, regardless of changes in annular pressure. In some cases, such deformable, expanding LCM is pumped ahead of cement jobs in which losses are expected. This type of material has a comparatively high success rate for the prevention and remediation of severe losses.

Severe lost-circulation problems that do not respond to conventional treatments might be curable by spotting a hydratable LCM pill and holding it under gentle squeeze pressure for a predetermined period. At downhole temperatures, the LCM pill expands rapidly to fill and bridge fractures, allowing drilling and cementing operations to resume quickly, sometimes in 4 hours or less. Alternatively, rapid-set LCM products are available that react quickly with the drilling fluid after being spotted across the loss zone and form a dense, flexible plug that fills the fracture and adheres to the wellbore. In some cases, this type of plug has proved so effective that the natural fracture gradient of the formation actually increased, allowing the operator to resume drilling and increase the mud weight beyond constraints established before the treatment.[12]

Leakoff Test (LOT). Conducting an accurate leakoff test is fundamental to preventing lost circulation. The LOT is performed by closing in the well and pressuring up in the open hole immediately below the last string of casing before drilling ahead in the next interval. On the basis of the point at which the pressure drops off, the test indicates the strength of the wellbore at the casing seat, typically considered one of the weakest points in any interval. However, extending an LOT to the fracture-extension stage can seriously lower the maximum mud weight that may be used to safely drill the interval without lost circulation. Consequently, stopping the test as early as possible after the pressure plot starts to break over is preferred.

Formation Integrity Test (FIT). To avoid breaking down the formation, many operators perform an FIT at the casing seat to determine whether the wellbore will tolerate the maximum mud weight anticipated while drilling the interval. If the casing seat holds pressure that is equivalent to the prescribed mud density, the test is considered successful and drilling resumes.

When an operator chooses to perform an LOT or an FIT, if the test fails, some remediation effort—typically a cement squeeze—should be carried out before drilling resumes to ensure that the wellbore is competent.

2.5.2 Stuck Pipe. Complications related to stuck pipe can account for nearly half of total well cost, making stuck pipe one of the most expensive problems that can occur during a drilling operation.[13] Stuck pipe often is associated with well-control and lost-circulation events—the two other costly disruptions to drilling operations—and is a significant risk in high-angle and horizontal wells.

Drilling through depleted zones, where the pressure in the annulus exceeds that in the formation, might cause the drillstring to be pulled against the wall and embedded in the filter cake deposited there (**Fig. 2.1**). The internal cake pressure decreases at the point where the drillpipe contacts the filter cake, causing the pipe to be held against the wall by differential pressure. In high-angle and horizontal wells, gravitational force contributes to extended contact between the drillstring and the formation. Properly managing the lubricity of the drilling fluid and the quality of the filter cake across the permeable formation can help reduce occurrences of stuck pipe.

Mechanical causes for stuck pipe include keyseating, packoff from poor hole-cleaning, shale swelling, wellbore collapse, plastic-flowing formation (i.e., salt), and bridging. Preventing stuck pipe can require close monitoring of early warning signs, such as increases in torque and drag, indications of excessive cuttings loading, encountering tight spots while tripping, and experiencing loss of circulation while drilling.

Depending on what the suspected cause of sticking is, it might be necessary to increase the drilling-fluid density (to stabilize a swelling shale) or to decrease it (to protect the depleted zone and avoid differential sticking). A drilling fluid's friction coefficient is an important factor in its effectiveness in preventing stuck pipe and/or enabling stuck pipe to be worked free. OBFs and SBFs offer the maximum lubricity; inhibitive WBFs can be treated with a lubricant (typically 1 to 5% by volume) and formulated to produce a thin, impermeable filter cake that offers increased protection against sticking. High-performance-polymer WBFs that are designed specifically to serve as alternates to OBFs and SBFs exhibit a high degree of natural lubricity and might not require the addition of a lubricant.

Lubricants for WBFs. The quality of the emulsion is important to a lubricant's performance in WBF. If the lubricant is too tightly emulsified, it no longer functions as a lubricant. If the emulsion is too loose, there is a risk that the lubricant will destabilize into a stringy, semisolid material. Overtreatment with lubricants might cause flocculation of the drilling fluid because of oil-wet solids.

Film strength is the main indicator of lubricant performance; generally, the higher the strength, the better the lubricant performance. However, environmental issues might arise with

Ideal

Embedment

Mudcake

Formation

Mud pressure

Mudcake

Mud

Resultant force

Mud

Formation

Drillpipe

Mudcake pore pressure

Pore Pressure

Time

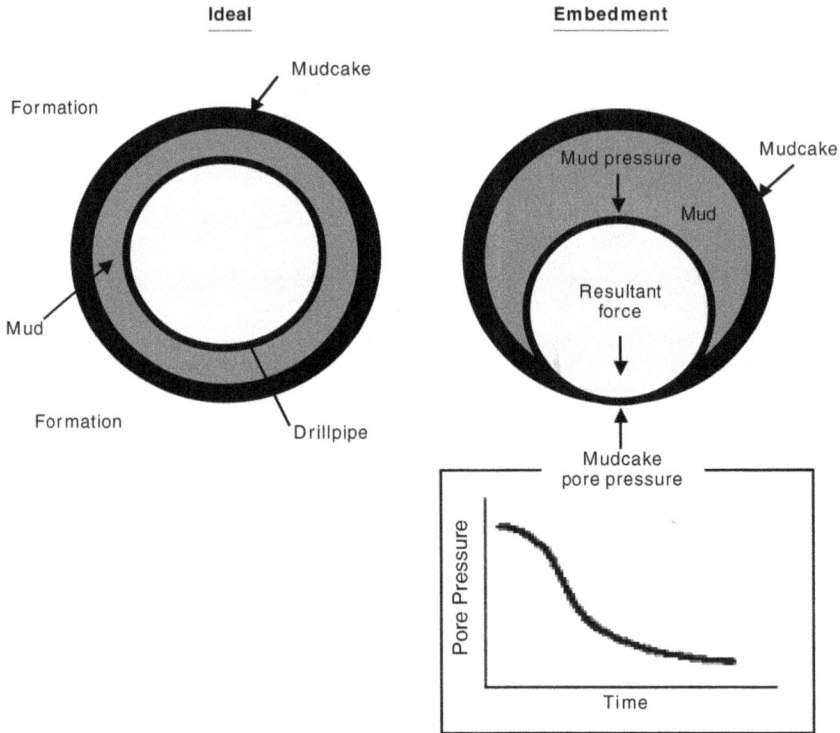

Fig. 2.1—Principle of differential sticking (modified from Ref. 14).

the use of certain high-film-strength lubricants. For this reason, alcohol/glycol-type lubricants, considered to be more environmentally friendly, have gained popularity. The alcohol/glycol lubricants also might perform better at low temperatures. The typical operating range for alcohol/glycol lubricants is 40 to 350°F.

Maximum film strength is required under high-pressure conditions and where elevated torque and drag measurements indicate a high risk of stuck pipe. Sulfurized oils (e.g., sulfurized olefin) have proved effective under these conditions. High-film-strength lubricants generally demonstrate increased thermal stability.

Other lubricant types include glass, plastic, and ceramic beads.

Spotting Fluids. Spotting fluids that are used to free stuck pipe are formulated to first crack the filter cake and then provide sufficient lubricity to allow the pipe to be worked free. Time is the key factor in successfully freeing stuck pipe. Spotting fluids routinely are included in rigsite inventory so that the spotting fluid can be applied as soon as possible after the pipe sticks, ideally within 6 hours. The length of free pipe may be estimated on the basis of drillstring stretch measurements, allowing the operator to determine the stuck point and deliver the spotting fluid as accurately as possible. If circulation is possible, decreasing the drilling-fluid density might relieve differential sticking; however, stuck pipe might be caused by a well kick combined with loss of circulation in a higher zone, which would eliminate the option of an intentional reduction in mud weight.

2.5.3 Shale Instability. Shales make up the majority of drilled formations and cause most wellbore-instability problems, ranging from washout to complete collapse of the hole. Shales are fine-grained sedimentary rocks composed of clay, silt, and, in some cases, fine sand. Shale types range from clay-rich gumbo (relatively weak) to shaly siltstone (highly cemented) and have in common the characteristics of extremely low permeability and a high proportion of clay minerals.

Drilling overbalanced through a shale formation with a WBF allows drilling-fluid pressure to penetrate the formation. Because of the saturation and low permeability of the formation, the penetration of a small volume of mud filtrate into the formation causes a considerable increase in pore-fluid pressure near the wellbore wall. The increase in pore-fluid pressure reduces the effective mud support, which can cause instability. Several polymer WBF systems have made shale-inhibition gains on OBFs and SBFs through the use of powerful inhibitors and encapsulators that help prevent shale hydration and dispersion.

2.5.4 Hole Cleaning. Effective drilling-fluid selection and management is important to the successful outcome of a high-angle or horizontal extended-reach drilling (ERD) operation. In addition to formation protection, the most important ERD challenges include the narrow margin between the pore pressure and the fracture gradient, ECD management, adequate hole cleaning, reduction of torque and drag, wellbore stability, barite sag, and loss of circulation. Years of operational data indicate that hole angles of between 30 and 60° create the most difficult hole-cleaning conditions. Good management of annular velocities, drilling-fluid viscosity, pipe-rotation speed, and pipe eccentricity can help minimize hole-cleaning problems.

Because of their reliable performance under adverse downhole conditions, invert-emulsion drilling fluids (OBFs and SBFs) usually are the first choice for ERD operations; however, the use of invert-emulsion muds is becoming more restricted because of environmental considerations, and several inhibitive WBF systems have been developed for use as alternatives.

Using hydraulics-modeling software that is programmed specifically for oilfield applications, it is possible to accurately predict drilling-fluid properties under actual downhole conditions, including static and dynamic temperature profiles, hydraulic pressures, ECD, annular pressure loss, rheological properties, cuttings loading and transport efficiency, effects of pipe eccentricity, and pressures required to break gels.[15] The modeled properties are confirmed by real-time pressure-while-drilling (PWD) data. The immediate feedback from the modeling process can allow the operator to optimize hole cleaning by several means, including:

• Adjustment of surface mud properties to meet changing downhole conditions.

• Adjustment of mechanical parameters such as penetration rate, flow rate, pipe-rotation speed, and tripping speed.

• Design and implementation of an effective sweep program.

Even while still in the well-planning stages, the casing design, bit selection, and drilling-fluid properties can be optimized to achieve the best drilling conditions, given the rig's pumps and fluid-handling capabilities. An accurate hydraulics modeling package should incorporate Bingham-plastic, power-law, and the Herschel-Bulkley rheological models.[15] Surface rheological properties are measured with a six-speed rheometer; such input allows the hydraulics-modeling software to determine actual annular shear rates at any depth in the well, taking into account the temperature and pressure regime at that depth.

The basis for the rheological modeling is either a matrix database of rheometer data or real-time data obtained from an HP/HT viscometer while drilling. A built-in routine can calculate the pressure required to break gels, allowing the operator to minimize surge pressures while tripping and running casing and to reduce the risk of breaking down the formation.

A comprehensive modeling software package also should accurately predict the cuttings loading in the annulus, the cuttings-bed height, the effect of drillstring rotary speed and pipe eccentricity, and the maximum recommended ROP for the given conditions.

These tools are useful not only for drilling extended-reach wells, but also for optimizing drilling performance in deepwater operations, HP/HT wells, and slimhole-drilling operations.

Hole-Cleaning Sweeps. High-viscosity sweeps that provide effective hole-cleaning in vertical wellbores might not be the best option for high-angle and horizontal wells because of the flow distribution around eccentric drillpipe.[16] To induce flow, the stress applied to a fluid must exceed that fluid's yield stress. In the narrow annular space created by eccentric drillpipe, it is possible that little or no flow will occur and that the cuttings bed will remain in place. Pumping a high-viscosity sweep might exacerbate this problem in a deviated well.

Applying a weighted sweep program that targets the silt bed that accumulates on the low side of the hole can mitigate hole-cleaning problems that often occur in ERD wells. As early as 1986, hole-cleaning research indicated that turbulent flow produced by relatively thin drilling fluid is more effective at silt-bed removal than is flow produced under a high-viscosity flow profile.[17] Consistent results in silt-bed removal have been achieved with fully-circulated, low-viscosity, weighted sweeps that exceed the drilling mud weight by 3 to 4 ppg and provide a 200- to 400-ft column in the annulus.[16] The guidelines for an effective weighted sweep program are:
- The sweep is pumped at regular intervals at the normal circulating rate.
- The pipe-rotation speed is ≥ 60 rev/min once the sweep has reached the bit.
- The sweep is allowed to return to the surface with continuous circulation.[16]

The additional buoyancy that a weighted sweep provides helps to reduce cuttings-settling tendency while the sweep travels up the annulus; however, the efficiency of the weighted sweep in dislodging cuttings might cause an increase in ECD while the annulus becomes loaded. If a PWD tool is used, effects on the ECD can be monitored and the pump rate reduced as needed to maintain an acceptable ECD without allowing cuttings to settle.

2.5.6 Barite Sag. Barite sag can occur in high-angle wells (possibly at 35°, but increasingly likely at ≥ 50°, then diminishing as the interval approaches 75 to 90°). The most severe sag incidents typically occur in the 45 to 65° range. Sag causes a decrease in drilling-fluid density, which is particularly noticeable when circulating bottoms up after a long noncirculating period. The barite falls to the low side of the wellbore and slides toward the bottom, creating an accumulation of weighted silt around the lower part of the drillstring.

Barite sag can lead to well-control issues and stuck pipe and can aggravate hole-cleaning problems. A well-designed sweep program can help prevent or minimize the occurrence of sag. Recent field results indicate significant success in preventing sag using properly formulated weighted sweeps. When using an emulsion-based synthetic fluid that contains no commercial clays, operators have experienced little or no detectable barite sag, based on data retrieved from downhole pressure-sampling tools and drilling-fluid-density measurements recorded while circulating bottoms up.[18,19]

2.5.7 Salt Formations and Rubble Zones. The five major problems that typically are associated with drilling of salt formations are bit-balling and packoff because of reactive shales within the salt, wellbore erosion when drilling through the salt formation and/or through shales above or below the salt formation, excessive torque and packoffs caused by salt creep, well-control issues, and excessive mud losses. The rubble zone that might lie beneath or adjacent to the salt section usually consists of a series of highly reactive shale stringers that are embedded in unconsolidated sand. The zone could be overpressured at the entry point because of a gas pocket under the salt, then underpressured for the remainder of the section.

Catastrophic mud loss below the salt is the most challenging of these problems and prevents most operators from drilling rubble zones with OBFs and SBFs. The decision about whether to use an SBF or a salt-saturated WBF usually is based on the known risk of lost returns. The SBF can provide increased drilling efficiency and a faster ROP, but a salt-saturated WBF provides adequate control over hole enlargement and might be preferable where the potential for large losses exists.[20] Seawater or undersaturated pills might facilitate ROPs without creating excessive washout, but extended use of undersaturated WBF might cause excessive hole enlargement where the longest exposure to the drilling fluid occurs.

Using hydraulics-modeling software and PWD data, a hydraulics baseline can be established while drilling through the salt formation and before reaching the rubble zone. If it is necessary to lay down the PWD tool before drilling through the rubble, this baseline can help the operator maintain a minimal ECD in the rubble zone.

2.6 Special Drilling Situations

2.6.1 Riserless Interval. Until the riser is in place, all drilling fluid returns to the seabed. A variety of drilling-fluid systems are available for drilling the top-hole riserless interval, including 1:1, 2:1, and 3:1 seawater/base-fluid blends; ballast-storable fluids; and precisely engineered "pad muds" for maintaining wellbore stability while running casing.

Typically, drilling fluid must be mixed and treated at flow rates of up to 1,000 U.S. gal/min (1,400 bbl/hr). A three-inlet, high-performance eductor often is used to blend riserless drilling fluids, which allows a third stream of $CaCl_2$ or an alternate inhibitive brine to be mixed with standard seawater and base-fluid streams.

Using a specialized, polymer-free base fluid facilitates efficient pumpoff from supply boats. The ideal fluid is thin enough to be transported by a standard centrifugal pump, yet retains the desired rheological properties when blended with seawater.

For remote locations that do not have continuous workboat support, using a ballast-storable fluid system helps to ensure that the operator has access to the required volume of drilling fluid during inclement weather. The fluid should be virtually solids-free to help avoid settling in the tanks. This type of system has been used for the riserless portion in several deepwater operations.[21]

Accurate modeling helps when formulating an optimal pad mud for each well. A properly designed pad mud helps to ensure that the wellbore will remain stable while surge pressures are minimized. Modeling-anticipated hole conditions allow the drilling-fluids engineer to predict the effects of interdependent parameters such as the maximum recommended ROP at various flow rates and cuttings concentrations; the ECD on the bottom and the annular cuttings concentration at different ROPs and flow rates; transport efficiencies and annular shear rates at different flow rates; and a comparison of proposed casing running speed with surge-pressure tolerances. The pad mud should exhibit controllable fluid-loss characteristics to minimize filter-cake buildup across permeable sands.

2.6.2 Deepwater Operations. Drilling operations in water depths of between 5,000 and 10,000 ft take place all over the world, and their success underscores the adaptability of oil-field technology and the industry's capacity to overcome significant technical challenges. The unique conditions presented by deepwater drilling require certain drilling-fluid characteristics. These are related to temperature variation, overpressured shallow water-bearing sands, narrow pore pressure/fracture gradient margins resulting from the extra fluid weight in the long drilling riser, and the potential for hydrates at the mudline.

Temperature Variation. The seafloor temperature in deepwater locations is approximately 40°F, but it can approach 32°F. The temperature downhole can exceed 300°F. The drilling fluid should exhibit the appropriate rheological properties throughout this wide range. In the riser near the mudline, the fluid is apt to thicken excessively from exposure to the cold seafloor temperature. Downhole, the fluid might become too thin as it heats up, and problems with hole cleaning and barite sag might develop. SBFs that contain little or no commercial clay appear to remain the most stable under these conditions.[18] These clay-free and low-clay systems rely on emulsion characteristics to achieve the desired rheological properties and provide sufficient barite suspension.

Shallow-Water Flow (SWF). Seismic data can help operators to predict and evaluate the risk of encountering an SWF on a given well. These water-bearing sands typically are located in the first 2,000 ft below the mudline and often are encountered while drilling in riserless mode. Stopping the flow under these circumstances is difficult. Pumping weighted mud that is cut with seawater on the fly generally is successful, but it requires pumping thousands of barrels of weighted WBF that returns to the seafloor because the riser is not yet connected to the wellbore. If the SWF is not brought under control or cased off successfully, its continued flow can undermine the structural integrity of the well and even affect neighboring wells.[22]

PP/FG. Because of the long riser that is required in deepwater operations, the hydrostatic pressure from the column of drilling fluid can approach or exceed the fracture gradient, especially when breaking circulation after a static period, tripping in the hole, or running casing. Significant loss of whole mud can occur and might lead to well-control problems. Control of the ECD, as verified by PWD data, is critical to maintaining wellbore stability. The drilling fluid that is selected should be evaluated for its demonstrated capacity to minimize or eliminate whole-mud losses. A suitable fluid will be characterized in part by a comparatively small pressure spike on PWD logs when circulation is resumed after a long static period.

Hydrates. When a WBF is used, the cold seafloor temperatures coupled with high pressures can cause the formation of hydrates, or "dirty ice." Hydrates form from hydrogen bonding between water molecules and low-molecular-weight gas. The water actually forms a crystalline cage structure around the gas and creates the risk of blocking the choke and kill lines at the blowout preventers. The four conditions required for hydrate formation are the presence of gas, the presence of water, low temperature, and high pressure. Shutting in a gas influx on a deepwater well that is drilled with WBF makes a likely scenario for hydrate formation.

Maintaining the appropriate salinity level in the WBF suppresses hydrate formation. For extreme situations, glycerine, polyglycerine, and polyglycol products might be needed to further suppress the hydrate-formation temperature.

2.6.3 HP/HT Wells. A well with a bottomhole temperature of > 350°F generally is considered to be in the high-temperature category. Where possible, these wells are drilled with OBFs or SBFs because of the thermal limitations of most WBFs. Such limitations include temperature-induced gelation, high risk of CO_2 contamination from the formation being drilled and/or from the degradation of organic mud additives, and increased solids sensitivity that is related to high temperatures.

Historically, WBFs have relied on bentonite clay for both rheology and filtration control. When tested at temperatures $\geq$ 300°F under laboratory conditions, bentonite slurries begin to thermally flocculate. Under HP/HT conditions with significantly elevated temperatures, a traditional WBF such as the lignosulfonate system might thicken so much that it no longer is usable or requires drastic and costly dilution and conditioning.

The ability to maintain bentonite and other active solids in a deflocculated state is the key to obtaining acceptable rheological and fluid-loss properties for WBFs exposed to high temperatures.[3] Bentonite can be used in relatively low concentrations if it is supplemented with a high-temperature, high-molecular-weight synthetic polymer for additional carrying capacity. This combination helps to make it possible to maintain 6% by weight of low-gravity solids and a PSD of these solids in an acceptable micron range. Adding polymeric deflocculant at depths where elevated temperatures are expected assists in rheology control.

An HP/HT viscometer typically is used to monitor the temperature stability of the drilling fluid and to evaluate its rheological properties at up to 500°F and 20,000 psia. This test is especially useful for determining whether high-temperature flocculation occurs in water-based muds. The test results can be presented graphically by plotting the change in viscosity with respect to temperature over the heating and cooling cycle, which establishes a baseline for recognizing indicators of temperature instability.

Eliminating lignite and lignite derivatives from the WBF formulation, lowering the bentonite concentration, and supplementing the high-temperature water-based system with synthetic polymers and copolymers can help minimize problems with temperature gelation.

OBFs and SBFs are subject to temperature thinning. Surface density should be corrected on the basis of downhole pressure data from a PWD tool. Hydraulics-modeling software that accurately accounts for fluid compressibility and the effect of temperature can improve the performance of the SBF system by allowing more precise surface conditioning.

2.7 Environmental Considerations

A prime objective in all drilling operations is to minimize safety and environmental risks while maintaining drilling performance. Operators and service companies alike take a proactive stance to reduce the potential for hazardous incidents and to minimize the impact of any single incident. The HSE policies of many companies are more stringent than those required by national governments and the various agencies charged with overseeing drilling operations. All personnel who take part in the well-construction process must comply with these standards to ensure their own safety and that of others. On most locations, a "zero-tolerance" policy is in effect concerning behaviors that might endanger workers, the environment, or the safe progress of the operation. Additionally, all personnel are encouraged to report potentially hazardous activities or circumstances through a variety of observational safety programs.

The packaging, transport, and storage of drilling-fluid additives and/or premixed fluid systems are closely scrutinized regarding HSE issues. Personnel who handle drilling fluid and its components are required to wear personal protective equipment (PPE) to prevent inhalation or other direct contact with potentially hazardous materials. Risk-assessed ergonomic programs have been established to reduce the potential for injuries related to lifting sacks and other materials and operating mud-mixing equipment.

When possible, drilling-fluid additives, base fluids, and whole mud are transported in bulk-tote tanks or are containerized. These transport methods help reduce packaging-related waste and minimize the risk of harming personnel, polluting the environment, and impairing operations. High-volume materials such as barite, bentonite, salt, and base fluids almost always are provided in bulk to offshore installations; onshore locations might use both bulk and packaged-unit materials, depending on the well depth and complexity.

The drilling-fluids specialist and operator representative at each location are responsible for ensuring that the available volume and properties of the drilling fluid will meet the immediate demands of a well-control situation, a loss of circulation, a tripping of a wet string, and/or a material-delivery delay caused by adverse transport conditions, and that additional volumes of drilling fluid with the appropriate properties can be mixed as needed at the rigsite or obtained in a timely manner. Published well-control guidelines recommend storing a riser volume plus 200 bbl (for pumping and line losses) in water depths $\geq$ 1,000 ft.[23] Many deepwater-drilling operations take place in water depths exceeding this and approaching 10,000 ft. Nonsettling, ballast-storable drilling fluids have been used offshore to eliminate the risk of disruptions to supply created by inclement weather and to prepare for drilling through SWFzones.[24]

2.7.1 Protecting the Environment. Keeping drilling-related accidents to "few and far between" not only provides the obvious benefit of minimizing, if not eliminating, sources of pollution and related threats to the ecosystem, but also enables the oil and gas industry more easily to obtain governmental permission to acquire and develop commercial reserves worldwide.

The drilling activities of countries that are emerging as energy producers should reflect the successful practices established by operators in well-regulated areas, as outlined in **Table 2.3**, which is modified from the *CAPP Technical Report 2001-0007*.[25] The associated technologies, procedures, financial arrangements, and records serve as project blueprints for newcomers to the industry. Developing nations can better protect their natural environments and resources by

TABLE 2.3—EXAMPLES OF WORLDWIDE REQUIREMENTS FOR DRILLING-FLUID AND CUTTINGS DISCHARGE[25]

Country	WBF and Cuttings	OBF and Cuttings	SBF and Cuttings	Monitoring Requirements
Australia	• Discharge is allowed, subject to 1% oil limit and/or 17% KCl content (exploratory drilling). Predischarge sampling is required. • Risk assessments are required. • Operators submit environmental plan(s) that are binding once accepted.	• 1% oil limit effectively eliminates discharge.	• No specific regulatory language is in place. • In Western Australia, a 10% dry weight limit on SBF-cuttings discharges applies. • Monitoring requirements for IO-cuttings discharges are determined on a case-by-case basis. • General acceptability of SBF is not resolved.	• Monitoring currently is not required. • Monitoring may be a component of the environmental plan that is submitted by the operator.
Brazil	• No specific regulatory language. • Discharge is allowed.	• No specific regulatory language. • Discharge plans require Inst. Brasiliero do Meio Ambiente e dos Recursos Naturais Renováveis (IBAMA) approval. • Approval of a low-toxicity, mineral-oil-based fluid is unlikely.	• No specific regulatory language is in place. • Discharge plans require IBAMA approval. • Industry workshop currently is formulating guidelines.	
Canada	• 1996 guidelines (under review) allow unrestricted WBF discharge. • Cuttings may be discharged if reinjection is unfeasible.	• Specific approval is required to use OBFs. • Imposes aromatic-content and toxicity limits. • Cuttings discharge is allowed if reinjection is unfeasible, subject to 15% oil-on-cuttings (OOC) limit. • Cuttings from diesel and highly aromatic oils are prohibited. • Nova Scotia has a 1% OOC limit as of 1999.	• Cuttings discharge is allowed if reinjection is unfeasible, subject to a 15% OOC limit. • SBF bulk discharges are prohibited. • Nova Scotia has a 1% OOC limit as of 1999. • The Newfoundland agency stipulates that SBF cuttings may be discharged where reinjection is unfeasible (including paraffins). The cuttings volume should be reduced to lowest possible level that allows efficient solids control.	• Offshore Waste Treatment Guidelines require environmental-effects and compliance monitoring.
China	• Discharge is allowed. • Use of oil is minimal or eliminated. • Notification and sample are required before discharge of oil-containing fluid. Fee is required for discharge of fluid with oil content of <10%. • Discharge is prohibited if oil content is >10%. • No dispersants may be added before discharge.	• Cuttings discharge is allowed, subject to OOC restrictions. Fluid discharge is prohibited. • OOC must be <10%; agency approval and discharge fee are required.	• Regulation is not established.	• No monitoring requirements are in place for exploratory drilling.

TABLE 2.3—EXAMPLES OF WORLDWIDE REQUIREMENTS FOR DRILLING-FLUID AND CUTTINGS DISCHARGE[25] (Continued)

Country	WBF and Cuttings	OBF and Cuttings	SBF and Cuttings	Monitoring Requirements
China	• No known KCl restrictions. • Discharge of residual oil, waste oil, and oil that contains waste (liquids, solids) is prohibited. • Operator must maintain antipollution record book, noting drilling fluid used, OOC, and time and volume of discharge.			
Kazakhstan	• No discharge is allowed.	• No discharge is allowed. • Cuttings are treated onshore with thermal desorption and fluid recovery.	• No discharge is allowed.	• Baseline survey is required before commencing operations. • Monitoring requirements are stated in regulation, but also are negotiated by each operator through the environmental impact assessment (EIA). • Survey required 2 years after completion of drilling.
Malaysia	• Discharge is allowed.	• Discharge is allowed. • No oil limit applies.	• Refined paraffins and low-toxicity OBFs are in use. • Cuttings discharge is allowed. • No oil limit applies.	• No drilling monitoring requirements are in place. • A one-time baseline study is required in new field areas.
Nigeria	• Proof of low toxicity is required before discharge. • WBF cuttings may be discharged offshore in deep water without treatment.	• Proof of low toxicity is required before discharge. • OBF must be recovered, reconditioned, and recycled. • OOC is limited to 1%, with a 0% goal. • On-site disposal is permitted if no sheen is on the water. • Daily cuttings sample analysis by operator required. • Point of discharge is classed as end of shunt line. • The Director of Petroleum Resources (DPR) may analyze samples at will. • Detailed sampling and analysis records must be provided to the DPR within 2 weeks of completing the well.	• SBF must be recovered, reconditioned, and recycled. • Cuttings SBF content limited to 5% (10% for ester SBF). • Some deepwater wells are granted special provision for higher retention limits.	• Post-drilling seabed survey required 9 months after five wells drilled; subsequent surveys after 18 months or ten additional wells.

TABLE 2.3—EXAMPLES OF WORLDWIDE REQUIREMENTS FOR DRILLING-FLUID AND CUTTINGS DISCHARGE[25] (Continued)

Country	WBF and Cuttings	OBF and Cuttings	SBF and Cuttings	Monitoring Requirements
Nigeria		• Operations must be open to inspection at all reasonable times. • OOC limited to 1% (effectively prohibits discharge).		
Norway	• Discharge is allowed, subject to preapproval for all fluid chemicals according to the OsloParis Convention (OSPAR) protocols. • Drilling-fluid makeup monitored and reported. • Discharge-site monitoring might be required. • No KCl limits. • Cuttings discharge calculation is based on well dimensions plus washout factor. • Daily sampling. • Preapproval is required for discharge of other drilling wastes. • Discharge permit is required for cementing and completion chemicals.		• Discharge is permitted only with developmental drilling. • SBF discharge is allowed only where use of WBF is precluded • SBF cuttings content is limited to between 8 and 18%. (The operator is required to set the limit on the basis of formation properties.) • Annual chemical monitoring of cuttings is required; biological monitoring is required every 3 years. • OSPAR protocols require preapproval testing.	• Baseline survey required before commencing production drilling. • Monitoring required every 3 years (sediment sampling, analysis for biological/chemical properties). • Guidelines for characterizing cuttings piles available from OLF.
Russia (Sakhalin Island)	• Legal basis for discharge is being clarified. • Toxicity testing is performed on mud additives, lab-formulated muds, and used muds. • Sampling is required several times while drilling. • Drilling-fluid constituents, discharge rates, etc., might be regulated.	• Regulations are likely to prohibit OBF-cuttings discharge.	• Regulation is not established.	• Regulation is not established.
United Kingdom	• Preapproval is required for drilling-fluid chemicals; includes toxicity testing per OSPAR protocols.	• OOC limited to 1% (effectively prohibits discharge). • Cuttings reinjection and onshore processing to recover oil are established practices.	• All discharge except that of ester-based-fluids cuttings is being phased out.	• Adheres to OSPAR requirements. • Seabed monitoring to follow discharge of SBF cuttings.
Vietnam	• Discharge is allowed. • No restriction on KCl. • Oil content should be <1%. • Preapproval is required for use of drilling fluids and potentially toxic or hazardous chemicals. • Drilling-fluid makeup is monitored and reported in EIA report.	• Discharge is prohibited less than three nautical miles from shore. • 1% oil limit applies for areas beyond three nautical miles. • Diesel-based fluids are prohibited.	• No restrictions currently are stipulated.	• Operator must perform environmental monitoring and supervision in accordance with Ministry of Science, Technology and Environment. • Baseline and impact studies carried out as stipulated in approved EIA.

implementing proven standards. Environmental-protection agencies and industry associations worldwide continue to study the effects on air and water quality had by drilling-fluid- and cuttings-related discharges specifically and by drilling operations generally.

Attention to international environmental issues often is channeled through the Intl. Assn. of Oil and Gas Producers (OGP), the membership of which consists of 45 oil companies, 10 industry associations, and 2 service companies operating in more than 80 countries. The OGP Environmental Quality Committee addresses drilling fluids and cuttings, environmental-performance indicators, and related regulatory issues.

The United Kingdom Offshore Operators Assn. (UKOOA) and the *Oljeindustriens Landsforening*/Norwegian Oil Industry Association (OLF), a corresponding Norwegian organization, have been formally examining the effects of drill-cuttings beds in the central and northern North Sea since June 1998.[26] Nearly U.K. £5 million was budgeted to assess the environmental impact of existing cuttings beds, compare options for accelerating degradation, and investigate the risks associated with removing the beds mechanically.

In the United States, an API report on environmental protection indicates that in 2000, the U.S. oil-and-gas industry spent $7.8 billion in this area, an amount that represents approximately 10% of the net income of the top 200 oil and natural gas companies. Since such record keeping began in 1992, the oil-and-gas industry has spent an estimated $90 billion to protect that nation's environment.[27]

Drilling-fluid companies strive to maintain an "econo-ecological" balance when choosing drilling-fluid systems and additives. An ecologically friendly drilling-fluid system that performs poorly will be used seldom because its poor performance extends drilling time and increases both the likelihood of hole problems and the cost of well construction. Conversely, using a properly managed high-performance SBF can shorten the duration of the drilling operation and/ or help maintain wellbore stability, thereby reducing opportunities for environmental damage. These and other factors must be weighed in the selection and design of any drilling-fluid system.

2.7.2 Sources of Contamination. Land and offshore drilling locations are regulated regarding the disposal of whole mud, drilled cuttings and other solids, and runoff generated by rainfall, wave action, or water used at the rigsite. Industrywide efforts to eliminate environmental hazards resulting from accidents or the negligent handling of drilling fluids and/or drilled cuttings encompass several contamination issues related to drilling fluids: formulation (chlorides, base oils, heavy metals, and corrosion inhibitors), natural sources (crude oil, salt water, or salt formation), and rigsite materials (pipe dope, lubricants, and fuel).

In some cases, reformulating drilling-fluid systems makes them environmentally more benign. For example, chrome lignosulfonate WBF is available in a chrome-free formulation. The development of SBFs stemmed from the need to replace diesel- and mineral-oil-based fluids (OBFs) because of environmental restrictions.

The discharge of conventional OBFs and drilled cuttings effectively was prohibited in the North Sea in 2000. According to the OsloParis Convention (OSPAR) Commission for the Protection of the Marine Environment of the North-East Atlantic, 98% of the total hydrocarbon discharge volume consists of produced water and drilled cuttings generated with SBFs.[28]

Cuttings that are generated by drilling with certain compliant SBFs may be discharged overboard in the western Gulf of Mexico if they comply with the retention-on-cuttings (ROC) limits introduced in 2002 by the U.S. Environmental Protection Agency (EPA). Neither traditional OBFs nor the drilled cuttings produced while using them can be discharged in the Gulf of Mexico; the rare offshore operation that uses a diesel- or mineral-based fluid must include a closed-loop process for continuously capturing all drilled cuttings and returning them to shore for regulated disposal.

2.7.3 Gulf-of-Mexico Compliance-Testing Profile. In deciding to permit the use of SBFs rather than restrict operators to WBFs, the EPA acknowledged the importance of the econo-ecological balance. The EPA felt that switching solely to WBF would cause more WBF development wells and more discharges to the ocean because WBF operations produce more

TABLE 2.4—SYNTHETIC DRILLING-FLUID STANDARDS

Testing For	Ester Standard	IO Standard
ROC	9.4 wet wt% of base fluid, measured every 500 ft. If a best-practices plan is in use, the extent of ROC monitoring is reduced. Measure one to three times per day.	6.9 wet wt% base-fluid measured every 500 ft. If a best-practices plan is in use, the extent of ROC monitoring is reduced. Measure one to three times per day.
Base-fluid degradation	Equal to or better than C8 or C12-C14 2-ethylhexanol/palm-oil ester in a 275-day test. Test annually.	Equal to or better than 65/35 blend of C16-C18 IO in a 275-day test. Test annually.
Leptocheirus plumulosus base-fluid toxicity	10-day *Leptocheirus plumulosus* LC_{50} must show no more toxicity than for a C8 or C12-C14 2-ethyl-hexanol/palm-oil ester. Test annually.	10-day *Leptocheirus plumulosus* LC_{50} must show no more toxicity than for a 65/35 blend of C16-C18 IO. Test annually.
Field-mud static sheen	Test weekly as per existing permit.	Test weekly as per existing permit.
Mysidopsis bahia field-mud toxicity	Monthly 30,000-ppm 96-hr LC_{50} suspended-particulate-phase (SPP) mysid shrimp test, as per existing permit.	Monthly 30,000-ppm 96-hr LC_{50} SPP mysid shrimp test, as per existing permit.
Leptocheirus plumulosus field-mud toxicity	*Leptocheirus plumulosus* 96-hr LC_{50} must show no more toxicity than for standard IO laboratory mud. Test once per month.	*Leptocheirus plumulosus* 96-hr LC_{50} must show no more toxicity than for standard IO laboratory mud. Test once per month.
Barite heavy-metal content	Total mercury 1 mg/kg and cadmium 3 mg/kg.	Total mercury 1 mg/kg and cadmium 3 mg/kg.
Polynuclear-aromatic-hydrocarbon (PAH) content of base fluid	10 ppm or 0.001% PAH (as phenanthrine) content as determined by high-performance-liquid chromatography with ultraviolet (UV) detector.	10 ppm or 0.001% PAH (as phenanthrine) content as determined by high-performance-liquid chromatography with UV detector.
Crude/PAH content (before drilling)	Conduct a gas-chromatography/mass-spectrometry (GC/MS) test on each batch of fluid before shipping it offshore. Fluid must contain <1% crude oil.	Conduct a GC/MS test on each batch of fluid before shipping it offshore. Fluid must contain <1% crude oil.
Crude/PAH content (field mud)	Test weekly using reverse-phase extraction (RPE) or GC/MS method. If fluid fails test, ship sample to shore for GC/MS testing.	Test weekly using RPE or GC/MS method. If fluid fails test, ship sample to shore for GC/MS testing.

waste per well than do SBF wells. WBF and OBF operations also progress more slowly than do SBF operations, leading to increased air emissions and fuel usage. Furthermore, the technical demands of drilling in deepwater locations require the use of either an SBF or OBF, and the pollution risks from a riser disconnect where OBF is in use are far greater than with SBF. Therefore, SBFs became the generally approved fluids for the Gulf of Mexico.

The toxicity of a drilling fluid and/or of cuttings that are generated with the fluid is determined by the fluid composition and is measured using a variety of testing protocols. **Table 2.4** lists tests that the 2002 modifications to the EPA General Permit[29] require to be performed where SBF is used. Discharge of whole SBF is prohibited. Unblended linear paraffin and LAO base fluids are not expected to comply with the modified requirements because of biodegradation and/or toxicity issues.

Although technical advances in the design and formulation of SBFs have been spurred mainly by offshore drilling conditions, the biodegradability and relatively low toxicity of ester- and

IO-based fluids make them suitable for onshore operations as well. Experimentation with various soil types, plant germination, and earthworm survival rates indicates that these base oils respond well to bioremediation.[30]

2.8 Solids Control and Waste Management

2.8.1 Fundamental Concepts. Contamination of drilling fluids with drilled cuttings is an unavoidable consequence of successful drilling operations. If the drilling fluid does not carry cuttings and cavings to the surface, the rig either is not "making hole" or soon will be stuck in the hole it is making. Before the introduction of mechanical solids-removal equipment, dilution was used to control solids content in the drilling fluid. The typical dilution procedure calls for dumping a portion of the active drilling-fluid volume to a waste pit and then diluting the solids concentration in the remaining fluid by adding the appropriate base fluid, such as water or synthetic oil.

Using solids-control equipment to minimize dilution has been a standard practice for the drilling industry for more than 60 years. Equipment and methods have changed over that time, but the fundamentals behind the process have not:
- Solids concentration matters—increasing solids content is detrimental to fluid performance.
- Economics matter—mechanical removal of solids costs less than dilution.
- Volume matters—the volume of waste generated is indicative of performance.
- Size matters—fine solids are the most detrimental and difficult to remove.
- Stokes' law matters—viscosity and density affect gravity separations.
- Shaker-screen selection matters—shaker screens make the only separation based on size.
- Footprint matters—the space available for equipment on rigs always is limited.

2.8.2 Solids Concentration. Increasing solids concentration in drilling fluid is a problem for the operator, the drilling contractor, and the fluids provider. It is well established that increasing solids content in a drilling fluid leads to a lower ROP. Other problems that are related to excessive solids concentration include:
- High viscosity and gel strength.
- High torque and drag.
- Lost circulation caused by higher ECD.
- Abrasion and wear on pump fluid ends.
- Production loss caused by formation damage from filtrate or solids invasion.
- Stuck pipe caused by filtrate loss.
- Poor cement jobs caused by excessive filter cakes.
- Generation of excessive drilling waste.
- Higher drilling-fluid maintenance costs.

2.8.3 Particle Size and Surface Area. From the perspective of both the drilling-fluids specialist and the solids-control technician, the effects of particle size and surface area are perhaps the most important concepts to understand. The fluids industry describes particle size in microns.

One micron (μm) is one one-thousandth of a meter and is equivalent to one inch divided by 25,400. The visual acuity of an unaided eye is approximately 35 μm, and fingertip sensitivity is approximately 20 μm. Drilled solids vary in size from < 1 μm to 15,000 μm in average particle diameter. Colloidal-sized particles are < 2 μm (average particle diameter) and will not settle out under gravitational forces. Ultrafines range from 2 to 44 μm and are unlikely to settle out of a drilling fluid unless it is centrifuged.

Solids of colloid and ultrafine size have the most adverse effect on fluid rheology. Ultrafines and colloids have pronounced effects on mud properties because both particle types have large surface-to-volume ratios. Like bentonite particles, the exposed surfaces of fine drilled

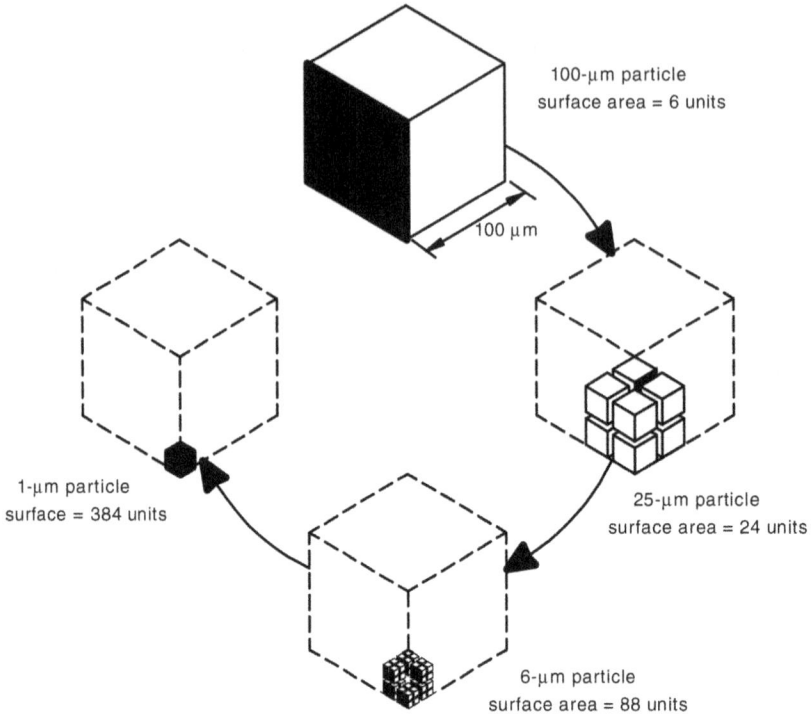

Fig. 2.2—Particle-size and surface-area comparison (courtesy of Cutpoint Inc.).

solids contain charges that increase the viscosity and gel strengths of the drilling fluid. Unlike bentonite, drilled solids do not plate out on sides of the wellbore to form a compressible and slick filter cake. The viscosity and fluid loss properties of a drilling fluid are difficult to control with high concentrations of drilled solids that are < 20 μm. The effect of drilled-solids degradation can be demonstrated by the fact that the available surface area increases almost 400 times when a particle degrades from 100 μm to 1 μm in diameter **(Fig. 2.2)**.

Fluid-technology advances have solved many of the problems that contribute to fines buildup in drilling fluids. Today, fluids are highly inhibitive and prevent cuttings dispersion like that shown in Fig. 2.2; however, the fluids cannot prevent cuttings recirculation or the inherent mechanical degradation that occurs during recirculation. Drilled solids that circulate through the mud tanks and back downhole are subject to mechanical degradation from surface pumps, drillpipe rotation, mud motors, and the bit's jet nozzles. The solids-control equipment must remove solids as far upstream as possible in the surface drilling-fluid system. When drilled solids have degraded to ultrafines or colloids, it is increasingly difficult to remove them by mechanical separation or settling.

2.8.4 Separation by Settling. Hydrocyclones, centrifuges, and settling tanks rely on settling velocity to concentrate and separate solids from slurry. Settling velocity is described mathematically by Stokes' law (Eq. 2.1), which states that the velocity at which a particle will settle in a liquid is proportional to the density difference between the particle and the liquid, the acceleration, and the square of the particle diameter. The settling velocity is inversely proportional to the viscosity of the liquid or slurry.

$$V_s \propto \frac{d^2(\rho_p - \rho_l)}{\eta} g, \quad .... (2.1)$$

where V_s = the settling velocity because of G-force, d = the particle diameter, ρ_p = the density of the particle, ρ_l = the density of the liquid, g = the acceleration or G-force, and η = the viscosity of the liquid.

Because particle diameter is squared, it has a great effect on separation efficiency; however, other factors that affect the settling rate should not be overlooked. For example, if a fluid contains both high-gravity solids (HGS) and low-gravity solids (LGS), a centrifuge that recovers barite particles in the 10-μm range also will recover LGS particles in the 15-μm range because both settle at the same rate. It also follows from Stokes' law that the settling velocity is lower in viscous and dense fluids; therefore the cut point and the capacity of centrifugal separators will be adversely affected by increasing viscosity and density.

2.8.5 Screen Selection. Obviously, shale-shaker screens are important for controlling the concentration of LGSs. What often is overlooked is the impact that proper screen selection can have on the other functions of a mud system:
• Screens are the only solids-control devices that are changed to handle changes in fluid properties or drilling conditions.
• Screens generate the bulk of drilling waste and reclaim the bulk of the mud.
• Screens must be able to handle the full circulation rate.
• Screens are the only devices on a rig that separate solids on the basis of size.

2.8.6 Waste Volumes. The combined waste volume of cuttings that are created while drilling and the excess or spent drilling fluid might be the best measure of performance and cost savings offered by a fluids system. The volume of spent mud determines what the mud-maintenance and disposal costs are and affects the long-term liabilities that are associated with waste disposal. Even under ideal situations, the volume of wet cuttings generated can easily exceed hole volume by a factor of two or more (three is a good rule of thumb). Minimizing the volume of spent mud and cuttings is the key to effective waste management. The increase in volume of the wet cuttings stems only partly from the added volume of cavings, washouts, or drilling a nongauge hole.

Cuttings are not discharged from mechanical separators as dry particulate matter. Much of the volume increase comes from the effect of surface-to-volume ratio. As drilled cuttings are ground down by the bit or dispersed from fluid interaction, they become thoroughly wetted with the drilling fluid. This fluid is known as ROC and is difficult to remove mechanically. Furthermore, a certain amount of carrier fluid usually discharges from mechanical separators with the cuttings. Unless measures are taken to dry the cuttings, the volume of drilled cuttings that is discharged will be more than double that of the theoretical gauge hole. Hydrocyclones in particular discharge very wet cuttings. The volume of spent mud that is created will depend largely on mechanical solids-removal efficiency.

2.8.7 Total Fluids Management. The importance of viewing fluids, solids control, and waste management as a process that must be designed to meet specific drilling conditions cannot be overemphasized. This process design is the key to helping improve the economics and minimizing the environmental impact of drilling activities. Many operators prefer a total fluids management approach that integrates fluids, solids control, and waste management to deliver a cost-effective wellbore in a safe and successful manner.

During the project planning stage, the following questions should be asked to help ensure successful field operations:
• What equipment best suits the drilling-fluid program and waste-disposal options?
• What are the solids loading and liquid loading that the equipment must handle?
• How much time will it take to install equipment, and who will install it?

- Are pumps, piping, chutes, conveyors, etc., adequate for the intended service?
- Is there enough power on the rig for the proposed equipment set?
- Is space available to install the proposed equipment set?
- Can the drilling-fluid program be modified to assist the mud- and cuttings-treatment system?
- What information needs to be collected and reported?
- What training needs to take place before startup?
- What safety issues need to be addressed?
- What environmental issues need to be addressed?
- What contingency or emergency operations need to be planned?

2.9 Drilling-Fluid Considerations

Drilling-fluid selection remains one of the most important components of a successful well-construction operation. Drilling-fluid service companies strive to provide the analytical tools, test equipment, stockpoint facilities, and innovative materials that will help operators to overcome the familiar issues (e.g., lost circulation) as well as the challenges that are brought on by drilling in ultradeep waters, extreme HP/HT formations, or remote environmentally sensitive areas.

The ability to simulate downhole conditions and optimize fluid design will continue to help reduce nonproductive time, and real-time management of hole conditions through data feed from downhole tools allows the operator and drilling-fluid specialist to fine-tune drilling parameters.

The demand for drilling-waste-management services that are dedicated to reducing, recovering, and recycling the volume of spent fluids and drilled cuttings continues to grow rapidly. These services and the related equipment have demonstrated their worth by helping operators achieve environmental compliance, reducing disposal costs, and returning more fluid and water for reuse in multiple applications.

Drilling-fluid services of some kind are required on every well. They encompass a broad spectrum of systems, products, software, personnel specializations, and logistical support. As wells become more complex, total drilling costs can increase dramatically. Because the drilling-fluid system comes in contact with almost every aspect of the drilling operation, proper drilling-fluid selection can help the operator minimize costs throughout the well-construction process.

Nomenclature

d = particle diameter, L, μm
g = acceleration or G-force (constant 980 cm/s^2), L/t^2
η = viscosity of the liquid, m/Lt, cp
V_s = settling velocity because of G-force, L/t, cm/s^2
ρ_l = density of the liquid = SG
ρ_p = density of the particle = SG

References

1. "World Oil 2004 Drilling, Completion and Workover Fluids," *World Oil* (June 2004) **225**, No. 6, F-1.
2. *Oilfield Market Report 2004*, Spears & Assoc. Inc., Tulsa, www.spearsresearch.com.
3. Mason, W. and Gleason, D.: "System Designed for Deep, Hot Wells," *American Oil and Gas Reporter* (August 2003) **46**, No. 8, 70.
4. "Deepwater Production Summary by Year, Gulf of Mexico Region, Offshore Information," Minerals Management Service, U.S. Dept. of the Interior, www.gomr.mms.gov/homepg/offshore/deepwatr/summary.asp.

5. Lyons, W.C., Guo, B., and Seidel, F.: *Air and Gas Drilling Manual*, McGraw-Hill, New York (2001) 1.

6. Negrao, A.F., Lage, A.C.V.M., and Cunha, J.: "An Overview of Air/Gas/Foam Drilling in Brazil," *SPEDC* (June 1999) 109.

7. *API RP 13-B1, Recommend Practice for Field Testing Water-Based Drilling Fluids,* third edition, API, Washington, DC (2003).

8. *API RP 13-B2, Recommended Practice Standard Procedure for Field Testing Oil-Based Drilling Fluids,* third edition, API, Washington, DC (1998).

9. Patel, A. *et al.:* "Designing for the Future—A Review of the Design, Development and Testing of a Novel, Inhibitive Water-Based Mud," paper AADE-02-DFWM-HO-33, 2002 AADE Annual Technology Conference Drilling and Completion Fluids and Waste Management, Houston, 2–3 April.

10. Wilcox, R.D., Fisk, J.V. Jr., and Corbett, G.E.: "Filtration Method Characterizes Dispersive Properties of Shale," *SPEDE* (June 1987) 149.

11. Whitfill, D.L. and Hemphill, T.: "All Lost-Circulation Materials and Systems Are Not Created Equal," paper SPE 84319 presented at the 2003 Annual Technical Conference and Exhibition, Denver, 5–8 October.

12. Sweatman, R., Wang, H., and Xenakis, H.: "Wellbore Stabilization Increases Fracture Gradients and Controls Losses/Flows During Drilling," paper SPE 88701 presented at the 2004 Abu Dhabi International Conference and Exhibition, Abu Dhabi, UAE, 10–13 October.

13. Isambourg, P. *et al.:* "Down-Hole Simulation Cell for Measurement of Lubricity and Differential Pressure Sticking," paper SPE 52816 presented at the 1999 SPE/IADC Drilling Conference, Amsterdam, 9–11 March.

14. Santos, H.: "Differentially Stuck Pipe: Early Diagnostic and Solution," paper SPE 59127 presented at the 2000 SPE/IADC Drilling Conference, New Orleans, 23–25 February.

15. Cameron, C.: "Drilling Fluids Design and Management for Extended Reach Drilling," paper SPE 72290 presented at the 2001 SPE/IADC Middle East Drilling Technology Conference, Bahrain, 22–24 October.

16. Sewell, M. and Billingsley, J.: "An Effective Approach to Keeping the Hole Clean in High-Angle Wells," *World Oil* (October 2002) **223,** No. 10, 35.

17. Pilehvari, A.A., Azar, J.J., and Shirazi, S.A.: "State-of-the-Art Cuttings Transport in Horizontal Wellbores," *SPEDC* (September 1999) 196.

18. Burrows, K. *et al.:* "Benchmark Performance: Zero Barite Sag and Significantly Reduced Downhole Losses With the Industry's First Clay-Free Synthetic-Based Fluid," paper SPE 87138 presented at the 2004 SPE/IADC Drilling Conference, Dallas, 2–4 March.

19. Hsia, R. and Patrickis, A.: "Case History: Zero Whole Mud Losses Achieved During Casing and Cementing Operations on Challenging Deepwater Well Drilled with Clay-Free Synthetic-Based Fluid," paper AADE-04-DF-HO-36, 2004 AADE Drilling Fluids Conference, Houston, 6–7 April.

20. Whitfill, D. *et al.:* "Drilling Salt—Effect of Drilling Fluid on Penetration Rate and Hole Size," paper SPE 74546 presented at the 2002 SPE/IADC Drilling Conference, Dallas, 26–28 February.

21. Morales, L.: "Drill Fluid Stored in Semi Ballast Tanks Controls Shallow Water Flow," *World Oil* (July 1999) **220,** No. 7, 59.

22. Eaton, L.: "Drilling Through Deepwater Shallow Water Flow Zones at Ursa," paper SPE 52780 presented at the 1999 SPE/IADC Drilling Conference, Amsterdam, 9–11 March.

23. *Deepwater Well Control Guidelines*, IADC, Houston (1998) 1–83.

24. Whitfill, D.L. *et al.:* "Fluids for Drilling and Cementing Shallow Water Flows," paper SPE 62957 presented at the 2000 Annual Technical Conference and Exhibition, Dallas, 1–4 October.

25. "Country-Specific Requirements for Discharge of Drilling Muds and Cuttings," *CAPP Technical Report 2001-0007,* Canadian Assn. of Petroleum Producers, Calgary (February 2001) I-1–I-11.

26. "Drill Cuttings Initiative Has Begun Its Second Phase Drilling," *Drilling Contractor* (May/June 2001) **57,** No. 3, 43.

27. Arafa, H.: "U.S. Petroleum Industry's Environmental Expenditures (1992–2001)," API, Washington, DC (20 February 2003) http://api-ep.api.org/filelibrary/FinalEnvExpS01%20022003.pdf.

28. "Annual Report of the OSPAR Commission, 2000–2001," OSPAR Commission 2001, London, 30.

29. *Final NPDES General Permit for New and Existing Sources and New Dischargers in the Offshore Subcategory of the Oil and Gas Extraction Category for the Western Portion of the Outer Continental Shelf of the Gulf of Mexico GMG290000,* U.S. EPA, Washington, DC (December 2004) 37.

30. Lee, B. *et al.:* "Reducing Drilling Fluid Toxicity," *Drilling* (June 2002) **75,** No. 6, 30.

SI Metric Conversion Factors

bbl ×	1.589 873	E – 01	= m³
ft ×	3.048*	E – 01	= m
ft² ×	9.290 304*	E – 02	= m²
ft³ ×	2.831 685	E – 02	= m³
°F	(°F – 32)/1.8		= °C
°F	(°F + 459.67)/1.8		= K
U.S. gal ×	3.785 412	E – 03	= m³
hp ×	7.460 43	E – 01	= kW
in. ×	2.54*	E + 00	= cm
in.² ×	6.451 6*	E + 00	= cm²
in.³ ×	1.638 706	E + 01	= cm³
lbm ×	4.535 924	E – 01	= kg
mL ×	1.0*	E + 00	= cm³
oz ×	2.957 353	E + 01	= cm³
ppm ×	1.0		= mg/kg
psi ×	6.894 757	E + 00	= kPa

*Conversion factor is exact.

Chapter 3
Fluid Mechanics for Drilling
R.F. Mitchell, Landmark Graphics and **Kris Ravi,** Halliburton

3.1 Introduction
The three primary functions of a drilling fluid—the transport of cuttings out of the wellbore, prevention of fluid influx, and the maintenance of wellbore stability—depend on the flow of drilling fluids and the pressures associated with that flow. For example, if the wellbore pressure exceeds the fracture pressure, fluids will be lost to the formation. If the wellbore pressure falls below the pore pressure, fluids will flow into the wellbore, perhaps causing a blowout. It is clear that accurate wellbore pressure prediction is necessary. To properly engineer a drilling fluid system, it is necessary to be able to predict pressures and flows of fluids in the wellbore. The purpose of this chapter is to describe in detail the calculations necessary to predict the flow performance of various drilling fluids for the variety of operations used in drilling and completing a well.

3.2 Overview
Drilling fluids range from relatively incompressible fluids, such as water and brines, to very compressible fluids, such as air and foam. Fluid mechanics problems range from the simplicity of a static fluid to the complexity of dynamic surge pressures associated with running pipe or casing into the hole. This chapter first presents a general overview of one-dimensional (1D) fluid flow so that the common features of all these problems can be studied. Next, each specific wellbore flow problem is examined in detail, starting from the simplest and progressing to the most complicated. These problems are considered in the following order:

1. Static incompressible fluids.
2. Static compressible fluids.
3. Circulation of incompressible fluids.
4. Circulation of compressible fluids.
5. General wellbore steady flow.
6. Steady-state surge pressure prediction.

Following these basic problems, a series of special topics are presented:

1. Fluid friction and rheology.
2. Dynamic wellbore pressure prediction.
3. Cuttings transport.
4. Air, mist, and foam drilling.

3.3 Governing Equations

A complete fluid mechanics analysis of wellbore flow solves the equations of mass, momentum, and energy for each flow stream and the energy equation for the wellbore and formation. In the usual treatment of these equations, the mass conservation equations are not stated explicitly, and the temperatures are given, rather than calculated from the energy equation. Here, the whole problem is set out, with appropriate assumptions made for special, simplified cases as they are considered.

The flow streams are treated as 1D constant area flow but with recognition of the effects of discontinuous area changes, such as nozzles. The assumption of 1D flow means that the flow variables, such as density, velocity, viscosity, etc., are given as their average values over the cross-sectional area of the flow stream. For instance, for flow in a tube, the frictional pressure drop is formulated in terms of the average velocity, density, and viscosity. The equations of mass and momentum conservation are solved subject to the assumption of steady flow. This assumption is that time variations of all variables are neglected in a time increment. In particular, this means that mass accumulation effects are not considered in the mass balance equation and that velocity is only a function of position in the momentum balance equation. Fully dynamic momentum equations are considered in a later section.

The balance equations are written in a control volume form. The equations are written as integrals over a specified volume with specified surface areas rather than as partial differential equations at a point. For these flow streams, the volume under consideration has cross-sectional area A and length Δz. The surface areas are the circular or annular cross-sectional areas, A, of the ends and the cylindrical lateral surface area. In the solution of these equations, only the entrance and exit values of the flow variables are calculated. To evaluate the integrals, the variation of variables between the entrance and exit of each space increment may be needed. The usual assumption used in wellbore calculations is that density, velocity, viscosity, and thermal conductivity are constant and equal to entrance conditions through the increment and that pressure and temperature vary linearly between entrance and exit. Experience has shown these assumptions to be good except for compressible flow. For most cases, the increased accuracy from other interpolation methods does not justify the computation penalty. On the other hand, none of these calculations are so numerically intensive that they cannot be done with more accurate integration methods on a personal computer, and some of these methods are mentioned later in the text.

3.3.1 Single-Phase Flow. The balance of mass for single-phase flow is given by

$$\dot{m} = \rho v A = \text{constant}, \quad\text{...} (3.1)$$

where

ρ = density, kg/m^3,

$\dot{m}$ = mass flow rate, kg/s,

v = average velocity, m/s,

and

A = area, m^2,

where steady flow (time independent flow) has been assumed. By Eq. 3.1, the mass flow rate in any flow stream is constant. In other words, the rate of fluid flow into a volume equals the rate of fluid flow out of the volume. Note that this relation does not change with area changes. However for nonsteady-state flow, we find that mass can accumulate in the volume so that flow out does not necessarily equal flow in.

The balance of momentum for single-phase flow has the form

$$\Delta P + \rho v \Delta v = \int_{\Delta z} \rho g \cos \Phi \, dz \pm \int_{\Delta z} \frac{2 f \rho v^2}{D_h} dz, \quad\text{..........................} (3.2)$$

where
  P = pressure, Pa;
  g = acceleration of gravity, m/s^2;
  Φ = angle of pipe with vertical;
  f = Fanning friction factor;
  D_{hyd} = hydraulic diameter, m;
and
  Δz = length of flow increment, m,
where steady flow has again been assumed. The Δv term is called the fluid acceleration and is important only for compressible fluids. The ρg term is the fluid weight term, which is positive for flow downward. The ρv^2 term is the fluid friction term. The Fanning friction factor f depends on the fluid density, velocity, viscosity, fluid type, and pipe roughness. Appropriate models for f, considering a variety of different fluid types, are considered in detail in the section on rheology. The sign of the friction term is counter to the flow direction (e.g., negative for flow in the positive direction). For area changes, the following relation holds

$$C_d \int_{P_o}^{P_1} \frac{dP}{\rho} + \frac{1}{2} \Delta v^2 = 0 . \quad\text{.......................................} (3.3)$$

Eq. 3.3 simplifies for incompressible flow and is written as

$$P_1 - P_o + \frac{\rho}{2 C_d}\left(v_1^2 - v_o^2\right) = 0, \quad\text{...................................} (3.4)$$

where subscript o indicates upstream properties, and subscript 1 indicates downstream properties. The quantities C_d are the discharge coefficients for the flow through an area change. Exact treatment of the effect of area change on momentum would have C_d equal to 1. However, real flow is not 1D through area changes, so a factor is needed to account for the real three-dimensional (3D) flow effects. Flow into a smaller area results in a reversible pressure drop plus an irreversible pressure drop. Flow into a larger area results in a reversible pressure increase plus an irreversible pressure drop. Thus, the values of C_d are different for flow into a restriction (reduced area) and flow out of a restriction (increased area). The following values of C_d are typical:
  C_d = .95 into reduced area.
  C_d = 1.00 into increased area.
  The basic balance of energy equation for single-phase flow is given by

$$\int_{\Delta z} \rho \dot\varepsilon A \, dz = - \int_{\Delta z} p \frac{dv}{dz} A \, dz + \Phi + Q + \Theta, \quad\text{.............................} (3.5)$$

where
  ε = internal energy, J/kg;
  Φ = viscous dissipation, W;
  Q = heat transferred into volume, W;

Θ = rate of volume energy added, W;
and

$$\dot{\varepsilon} = \frac{\partial \varepsilon}{\partial t} + v \frac{\partial \varepsilon}{\partial z} \ .$$

This equation is given in the fully transient form because temperature variation with time may be significant and because steady-state temperatures are usually not achieved in typical wellbore hydraulic operations. The viscous dissipation term Φ depends on the fluid friction model. The term Q is usually written as the total heat flux into the control volume. An example of Θ would be the heat of hydration for cement. Eq. 3.3 can be rewritten in terms of enthalpy h.

$$\int_{\Delta z} \rho \dot{h} A dz = \int_{\Delta z} v \frac{dp}{dz} A dz + \Phi + Q + \Theta, \dotfill (3.6)$$

and by choosing pressure and temperature as independent variables can be further rewritten as

$$\int_{\Delta z} \rho C_p \frac{\partial T}{\partial t} A dz + \int_{\Delta z} \rho v C_p \frac{\partial T}{\partial z} A dz - \int_{\Delta z} v \beta T \frac{\partial p}{\partial z} A dz = \Phi + Q + \Theta \ . \dotfill (3.7)$$

$C_p = \dfrac{\partial h}{\partial T}(P, T)=$ heat capacity at constant pressure, J/kg-K.

$\beta = -\dfrac{1}{\rho} \dfrac{\partial \rho}{\partial T}(P, T)$ = coefficient of thermal expansion, 1/K.

T = absolute temperature, °K.

At area changes, the following relation holds

$$\Delta h + \frac{1}{2} \Delta v^2 = 0 \ . \dotfill (3.8)$$

3.4 Key Considerations for Wellbore Hydraulic Simulation

While many applications can be done as hand calculations, more complex problems, especially involving temperature changes, require a hydraulic simulator. To address the wellbore operations of interest, a wellbore simulator should have a wide range of capabilities. These fall into four categories:

1. Transient effects.
2. Fluid models.
3. Wellbore geometry.
4. Flow types.

Many applications for operational design involve highly transient behavior where temperatures are changing rapidly. Drilling, cementing, fracturing, and production startup are all transient operations where fluid temperatures can change on the order of 100°F or more in a matter of minutes during flow in the well. Fully transient thermal response should be modeled in the flowing stream, the wellbore assembly, and the formation. The model should handle changing flow conditions, including changes in flow rate, inlet temperature and pressure, fluid type, and flow direction.

Oil and gas well operations involve fluids of many different types. The heat transfer characteristics and temperature-pressure coupling vary with fluid type. Oil- and water-based liquids and polymers behave differently from compressible systems. Multiple fluids in the wellbore, including spacers and displacement fluids, are an important consideration. Temperature dependent properties must be updated as temperatures and rheological properties change with time and depth. Even with drilling muds, the viscosity changes with temperature during the mud's circuit down the drillpipe and up the annulus, affecting the overall hydraulics of the system.

Flexibility in wellbore geometry is needed to accommodate different configurations such as deviated wells, liners, dual completions, and offshore risers. The geometry determines the cross-sectional flow area and the fluid velocity, which, in turn, governs the heat transfer. Temperatures during liner cementing are strongly influenced by the size of the liner and the annular clearance.

Flow types include production, injection, forward circulation, reverse circulation, drilling, and shut-in. Drilling is a special case of forward circulation, in which the depth of circulation and the wellbore thermal resistance change as the well is drilled and casing is set.

3.5 Static Wellbore Pressure Solutions

Static wellbore pressure solutions are the easiest to determine and are the most suitable for hand calculation. Because velocity is zero and no time dependent effects are present, we need only consider Eq. 3.2 with velocity terms deleted.

$$\Delta P = \int_{\Delta z} \rho g \cos \Phi dz \ . \ \dots\dots (3.9)$$

Temperatures are assumed to be static (often the undisturbed geothermal temperature) and known functions of measured depth.

3.5.1 Constant Density. The simplest version of Eq. 3.9 is the case of an incompressible fluid with constant density ρ.

$$\Delta P = \rho g \Delta Z (\text{TVD}), \ \dots\dots (3.10)$$

where ΔZ is the change in true vertical depth (TVD) (i.e., hydrostatic head). For constant slope Φ, ΔZ equals $\cos\Phi \ \Delta z$. For a slightly compressible fluid, such as water, Eq. 3.9 could be used for small ΔZ increments where temperature and pressure values do not vary greatly.

3.5.2 Compressible Gas. To show a somewhat more complicated static pressure solution, consider the density equation for an ideal gas: $\rho = \dfrac{P}{RT}$, where T is absolute temperature, and R is a constant. For an ideal gas, density has an explicit dependence on pressure and temperature. The solution to Eq. 3.9 for a well with constant slope Φ is

$$\Delta P(z) = P_o \left[\exp \left(\frac{g \cos \Phi}{R} \int_{\Delta z} \frac{1}{T(\varsigma)} d\varsigma \right) - 1 \right], \ \dots\dots (3.11)$$

where the initial condition for P is P_o. $T(z)$ is a given absolute temperature distribution, and z is the measured depth. For constant T, we see that the pressure of an ideal gas increases exponentially with depth, while an incompressible fluid pressure increases linearly with depth.

3.6 Flowing Wellbore Pressure Solutions

The next level of complexity in hydraulic calculations is the steady flow of the wellbore fluids. One part of this complexity is the calculation of the Fanning friction factor, f. This subject is postponed to the section on rheology, and f is assumed to be known in the following calculations.

3.6.1 Constant Density. The simplest version of Eq. 3.2 for flowing fluids is the case of an incompressible fluid with constant density ρ and rheological properties, and a constant-slope wellbore.

$$\Delta P = \left(\rho g \cos \Phi \pm \frac{2 f \rho v^2}{D_{\text{hyd}}}\right)\Delta z . \ \dots\dots\dots\dots\dots\dots\dots\dots\dots (3.12)$$

In this case, we have evaluated the Fanning friction factor from the appropriate equation in the rheological section. Note that Δv is zero through the mass conservation equation.

3.6.2 Linearly Varying Density. The next simplest solution has a linearly varying density along z. Conservation of mass requires that

$$\rho v = \rho_o v_o, \ \dots\dots\dots\dots\dots\dots\dots\dots\dots\dots\dots\dots (3.13)$$

where the o subscript indicates initial values, and no subscript indicates final values. With Eq. 3.13, we can calculate Δv in terms of $\Delta \rho$.

$$\Delta v = -v_o \Delta \rho / \rho = -v \Delta \rho / \rho_o . \ \dots\dots\dots\dots\dots\dots\dots\dots (3.14)$$

Assuming a linear variation in density, constant wellbore angle, and constant Fanning friction factor, the pressure drop equation gives

$$\Delta P - \rho_o v_o^2 \frac{\Delta \rho}{\rho} = \left[\frac{1}{2}(\rho + \rho_o)g \cos \Phi \pm \frac{f \rho_o v_o^2}{D}\frac{(\rho + \rho_o)}{\rho}\right]\Delta z . \ \dots\dots\dots\dots (3.15)$$

Eq. 3.15 may be used directly to calculate ΔP if the final density is insensitive to pressure. Otherwise, this equation must be solved numerically for the pressure. For instance, the density terms could be linearized with respect to P and Newton's method used to converge to the final pressure.

3.6.3 Compressible Fluid. The flow of a compressible fluid can often produce results that seem counter to intuition. For example, consider the steady flow of air in a constant area duct. As with all fluids, there is a pressure loss because of friction, and the pressure decreases continuously from the entrance of the duct to the exit. Unlike the flow of incompressible fluids, the fluid velocity increases from the entrance of the duct to the exit. How could friction make the fluid speed up?

Two facts account for this acceleration. First, the gas pressure is proportional to the density (as in the ideal gas law $P = \rho RT$). As the pressure of the gas decreases, the density must decrease also. Second, because the mass flow through the duct is constant, the product of density and velocity is constant. Thus, as the density decreases with the pressure, the velocity must increase to maintain the mass flow.

This example demonstrates a typical compressible flow characteristic—the interrelationship of pressure and mass flow. In air drilling, high velocities are needed at bottom of hole to remove the cuttings. High velocities result in friction pressure drops in the drillpipe and annulus, so higher standpipe pressures may be needed to keep the air flowing. Higher standpipe pressures result in higher gas densities and, hence, result in lower velocities. Fortunately, most air drilling operations do not result in the vicious circle situation previously described.

The pressure change in a flowing gas is properly a problem in gas dynamics. Gas dynamic analytic solutions are available for two cases of flow with friction: adiabatic and isothermal flow. Neither case is exactly what we need for these calculations, but the reader is directed to a gas dynamics reference for additional depth of understanding of this problem.[1] If we assume a linearly varying density and temperature, we can use the results of the previous section with the addition of a pressure/volume/temperature (PVT) relationship. For most applications, an ideal gas model is sufficiently accurate. With the relation $P = \rho RT$, we can calculate ΔP in terms of density as

$$\Delta P = P_o\left(\frac{\rho T - \rho_o T_o}{\rho_o T_o}\right). \quad\text{...} \quad (3.16)$$

Substituting Eq. 3.16 into Eq. 3.15, we derive the quadratic equation for density.

$$\left(\frac{P_o T}{\rho_o T_o} - \frac{1}{2}g\cos\Phi\Delta z\right)\rho^2 + \left(-P_o - \rho_o v_o^2 + \frac{1}{2}\rho_o g\cos\Phi\Delta z \pm f\rho_o v_o^2\Delta z\right)\rho$$

$$+\rho_o^2 v_o^2(1 \pm f\Delta z). \quad\text{...} \quad (3.17)$$

If there are two positive roots for the density, the root that gives a subsonic velocity is the correct root. The speed of sound for an ideal gas is

$$a = \sqrt{\frac{c_p RT}{c_v}}, \quad\text{...} \quad (3.18)$$

where c_p and c_v are the heat capacities at constant pressure and volume, respectively.

3.7 General Steady Flow Wellbore Pressure Solutions

We make only one assumption in this general discussion of wellbore flow modeling, and that is that the temperature distribution is given. To make any other assumption requires a general solution of the energy equation for the wellbore, which is beyond the scope of this chapter. For review, we repeat the balance of mass and momentum for 1D flow along a constant area duct.

$$\rho V A = \text{constant}. \quad\text{...} \quad (3.19)$$

$$\frac{dP}{dz} + \rho v\frac{dv}{dz} = \rho g\cos\Phi \pm \frac{2f\rho v^2}{D_{\text{hyd}}}. \quad\text{...} \quad (3.20)$$

At changes of area, the following conditions hold.

$$\rho v A = \rho_o v_o A_o, \quad\text{...} \quad (3.21)$$

and

$$C_d \int_{P_o}^{P} \frac{dP}{\rho} + \frac{1}{2}\Delta v^2 = 0. \quad\text{...} \quad (3.22)$$

Eq. 3.22 simplifies for incompressible flow.

$$P - P_o + \frac{\rho}{2C_d}(v^2 - v_o^2) = 0, \dots\dots\dots\dots\dots (3.23)$$

where subscript o indicates inlet properties.

Given that we have a means of calculating the Fanning friction factor, we need a PVT relationship relating pressure, density, and temperature, and what is often available is a pressure function $P(\rho,T)$ depending on density and temperature. When this is substituted into Eq. 3.20, with Eq. 3.19, we obtain the first-order differential equation in density.

$$\left[\frac{\partial P(\rho, T)}{\partial \rho} - \frac{\rho_o v_o}{\rho}\right]\frac{d\rho}{dz} + \frac{\partial P(\rho, T)}{\partial T}\frac{dT}{dz}$$
$$= \rho g \cos \Phi \pm \frac{2f(\rho_o v_o)^2}{\rho D_{\text{hyd}}}. \dots\dots\dots\dots (3.24)$$

Eq. 3.24 can be integrated numerically using any of several methods, such as adaptive Runge-Kutta or Bulirsh-Stoer. The reader is referred to numerical analysis books for details of these two methods.[2] Once the new density has been determined, the pressure can be calculated from the PVT relationship, and the velocity can be calculated from Eq. 3.19. Alternately, we might have density as a function of pressure and temperature: $\rho(P,T)$. In this case, the differential equation is in terms of pressure.

$$\left[1 - \frac{\rho_o v_o}{\rho(P, T)^2}\frac{\partial \rho}{\partial P}\right]\frac{dP}{dz} - \frac{\rho_o v_o}{\rho(P, T)^2}\frac{\partial \rho}{\partial T}$$
$$= \rho(P, T)g \cos \Phi \pm \frac{2f(\rho_o v_o)^2}{\rho(P, T)D_h}. \dots\dots\dots (3.25)$$

Again, this is a first-order differential equation but now in terms of pressure. Once the pressure is determined, density is determined from the PVT relationship, and the density together with Eq. 3.19 gives the velocity.

Flow-area changes may act like nozzles (area reduction) or diffusers (area increases). The actual calculation of flow-property changes is beyond the capability of a 1D flow analysis. Often, we insert coefficients into the 1D equations to account for the complexity, and then we evaluate these coefficients from experiments. One such coefficient is the discharge coefficient shown in Eqs. 3.22 and 3.23. A further comment is needed about the general Eq. 3.22. Some assumption must be made about the variation of temperature with pressure within the nozzle before the integral can be evaluated. A typical assumption is that the flow is adiabatic (i.e., negligible heat transfer takes place in the nozzle). For our purposes, adiabatic is equivalent to isentropic, and entropy functions are available for many fluid models. For isentropic flow,

$$S(P, T) = S(P_o, T_o). \dots\dots\dots\dots (3.26)$$

Or, we can derive the change of temperature with pressure.

$$\frac{dT}{dP} = -\frac{\partial S/\partial P}{\partial S/\partial T} \ . \ \text{...} \ (3.27)$$

For an ideal gas, we can solve Eq. 3.27 to get

$$p^{1-k}T^k = p_o^{1-k}T_o^{1-k}, \ \text{...} \ (3.28)$$

where $k = c_p/c_v$. Using the ideal gas law, we can eliminate T.

$$p = p_o\left(\frac{\rho}{\rho_0}\right)^k . \ \text{...} \ (3.29)$$

We can now express pressure in terms of density in Eq. 3.22 and express v in terms of density with Eq. 3.21. The resulting equation can be solved numerically for density. Then, Eq. 3.29 can be used to determine the pressure.

3.8 Calculating Pressures in a Wellbore

Assuming we can calculate ΔP for each constant area section of drillpipe or annulus and can calculate ΔP for nozzles and area changes, we are now ready to evaluate the pressures in a wellbore. A typical wellbore fluid system is illustrated in **Fig. 3.1**.[3] Summing all pressure drops give the standpipe pressure.

$$P\text{standpipe} = \Delta P(\text{pipe joints}) + \Delta P(\text{internal upsets}) + \Delta P(\text{area changes}) + \Delta P(\text{bit})$$
$$+\Delta P(\text{annulus}) + \Delta P(\text{tool joints}) + \Delta P(\text{misc.}) + \Delta P(\text{choke}) + P\text{atm} . \ \text{.................} \ (3.30)$$

In this calculation, we assume that the calculations are started from a known pressure value, most conveniently the atmospheric pressure at the exit of the annulus. This choice is particularly suitable if air or foam drilling is being considered because "choked" gas flow almost never occurs. For this choice of "boundary condition," flow calculations proceed backward from the annulus exit to the standpipe pressure. For flow in the annulus, both fluid density and fluid friction increase pressure going down the annulus. Where fluid type changes, the pressure and flow velocity are continuous.

$$P(\text{fluid}_A) = P(\text{fluid}_B),$$

and

$$v(\text{fluid}_A) = v(\text{fluid}_B) \ \text{at the interface} \ . \ \text{...} \ (3.31)$$

Notice that mass flow rate may not be continuous at the interface between two fluids because the densities may be different: $\rho_1 vA \neq \rho_2 vA$, where v and A are continuous at the interface. When calculating from the bit to the standpipe, inside the drillstring, fluid density decreases pressure and fluid friction increases pressure. Pressure changes because of internal upsets and tool joints consist of two area changes and a short flow section, as shown in **Fig. 3.2**.

Pressure drop across the bit consists of two area changes: into the nozzles and exit from the nozzles into the openhole annular area. Miscellaneous pressure drops are drops through tools, mud motors, floats, or in-pipe chokes. Sometimes, the manufacturer will have this pressure-loss information tabulated; otherwise, you will have to estimate the pressure loss through use of the tool internal dimensions.

If the standpipe pressure is given, then the flow exiting the annulus must be choked back to atmospheric pressure.

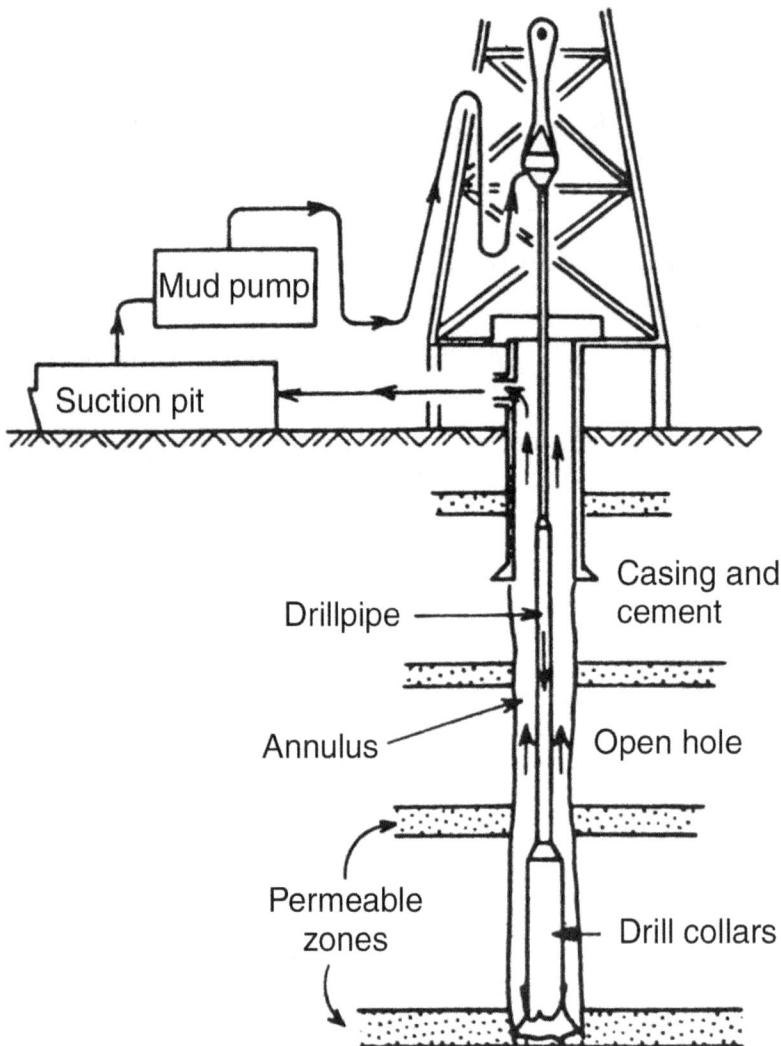

Fig. 3.1—Typical wellbore hydraulic system.[3]

$$\Delta P(\text{exit choke}) = P\text{standpipe} - \Delta P(\text{pipe joints}) - \Delta P(\text{internal upsets})$$
$$- \Delta P(\text{area changes}) - \Delta P(\text{bit}) - \Delta P(\text{annulus}) - \Delta P(\text{tool joints})$$
$$- \Delta P(\text{misc.}) - P\text{atm} . \quad\text{...} \quad (3.32)$$

3.9 Surge Pressure Prediction

3.9.1 Introduction. An exceptional flow case is the operation of running pipe or casing into the wellbore. Moving pipe into the wellbore displaces fluid, and the flow of this fluid generates pressures called surge pressures. When the pipe is pulled from the well, negative pressures are generated, and these pressures are called swab pressures. In most wells, the magnitude of the pressure surges is not critical because proper casing design and mud programs leave large enough margins between fracture pressures and formation-fluid pressures. Typically, dynamic fluid flow is not a consideration, so a steady-state calculation can be performed. A certain fraction of wells, however, cannot be designed with large surge-pressure margins. In these critical

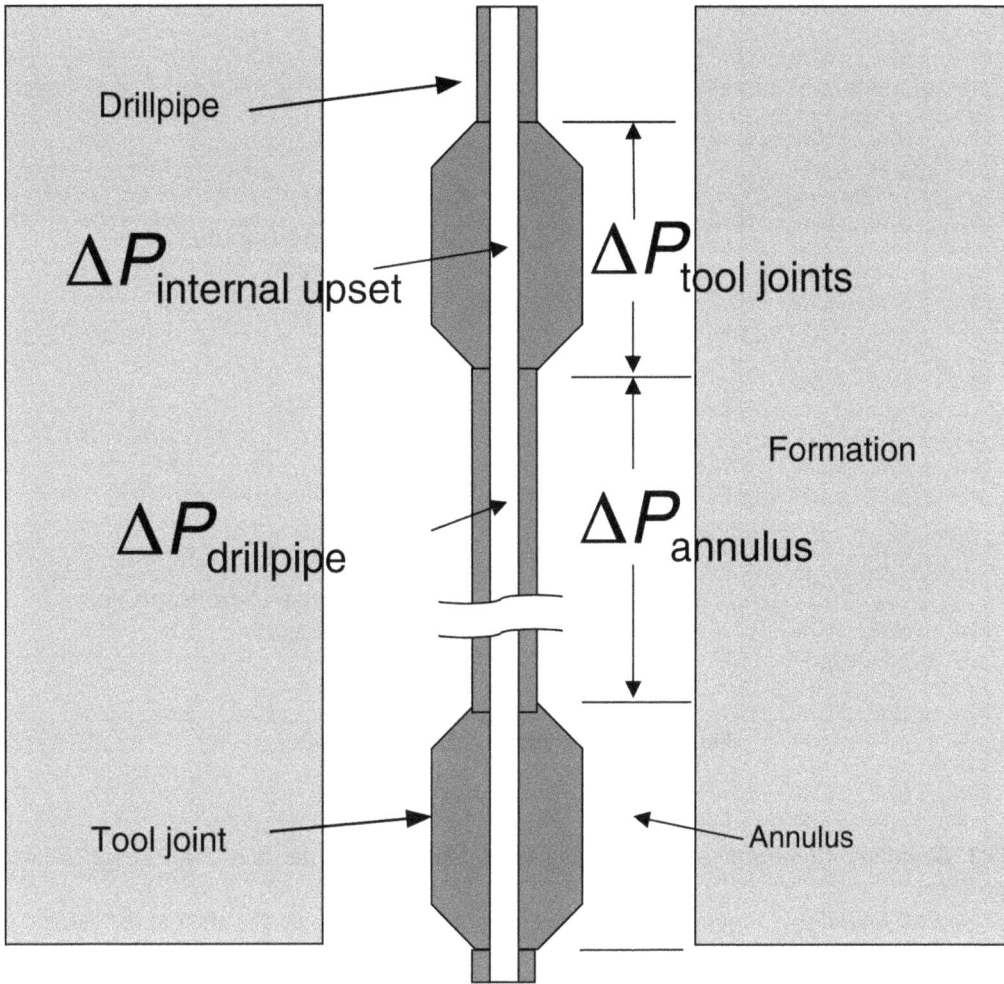

Fig. 3.2—Pressure-drop calculation sections.

wells, pressure surges must be maintained within narrow limits. In other critical wells, pressure margins may be large, but pressure surges may still be a concern. Some operations are particularly prone to large pressure surges (e.g., running of low-clearance liners in deep wells). The reader is referred to papers on dynamic surge calculations,[4,5] and a later section on dynamic pressure calculation gives a taste of this type of calculation.

The surge pressure analysis consists of two analytical regions: the pipe-annulus region and the pipe-to-bottomhole region (**Fig. 3.3**). The fluid flow in the pipe-annulus region should be solved using techniques already discussed, but with the following special considerations: frictional pressure drop must be solved for flow in an annulus with a moving pipe, and in deviated wells, the effect of annulus eccentricity should be considered. The analysis of the pipe-to-bottomhole region should consist of a static pressure analysis, with pressure boundary condition determined by the fluid flow at the bit, or pipe end if running casing. The pipe-annulus model and the pipe-to-bottomhole model then are connected through a comprehensive set of force and displacement compatibility relations.

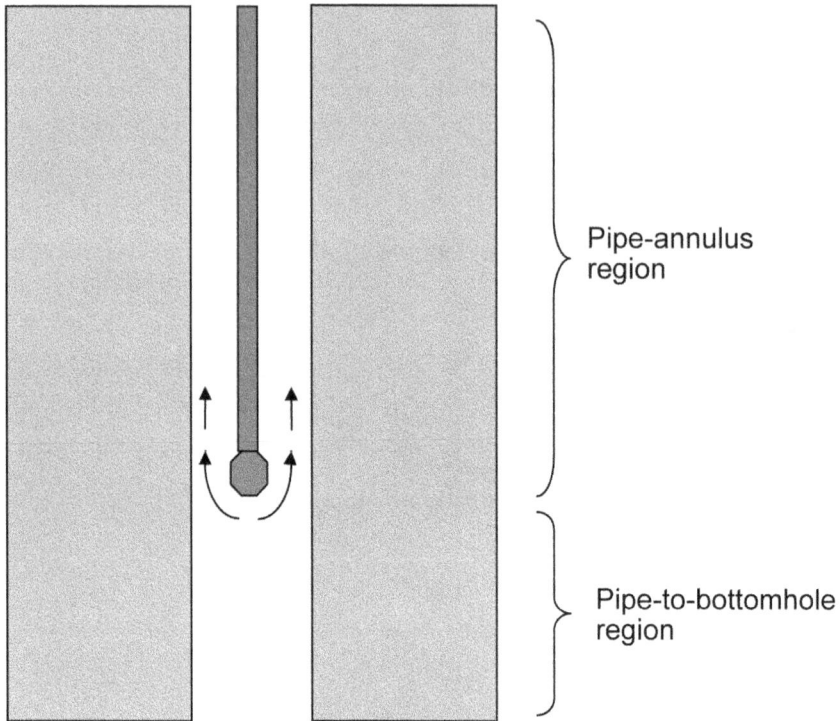

Fig. 3.3—Surge-pressure calculation regions.

3.9.2 Boundary Conditions. The following conditions describe the flow for a surge or swab operation.

Surface Boundary Conditions. There are six variables that can be specified at the surface:

P_1 = pipe pressure.

v_1 = pipe fluid velocity.

P_2 = annulus pressure.

v_2 = annulus fluid velocity.

v_3 = pipe velocity.

A maximum of three boundary conditions can be specified at the surface. For surge without circulation, the following boundary conditions hold:

P_1 = atmospheric pressure.

P_2 = atmospheric pressure.

v_3 = specified pipe velocity.

For a closed-end pipe, the following boundary conditions hold:

$v_1 = v_3$, and fluid velocity equals pipe velocity.

P_2 = atmospheric pressure.

v_3 = specified pipe velocity.

For circulation with circulation rate Q, the boundary conditions are

$v_1 = v_3 + Q/A_1$ (i.e., fluid velocity equals pipe velocity plus circulation velocity).

P_2 = atmospheric pressure.

v_3 = specified pipe velocity.

End of Pipe Boundary Conditions. There are 11 variables that can be specified at the moving pipe end (see **Fig. 3.4**):

P_1 = pipe pressure.

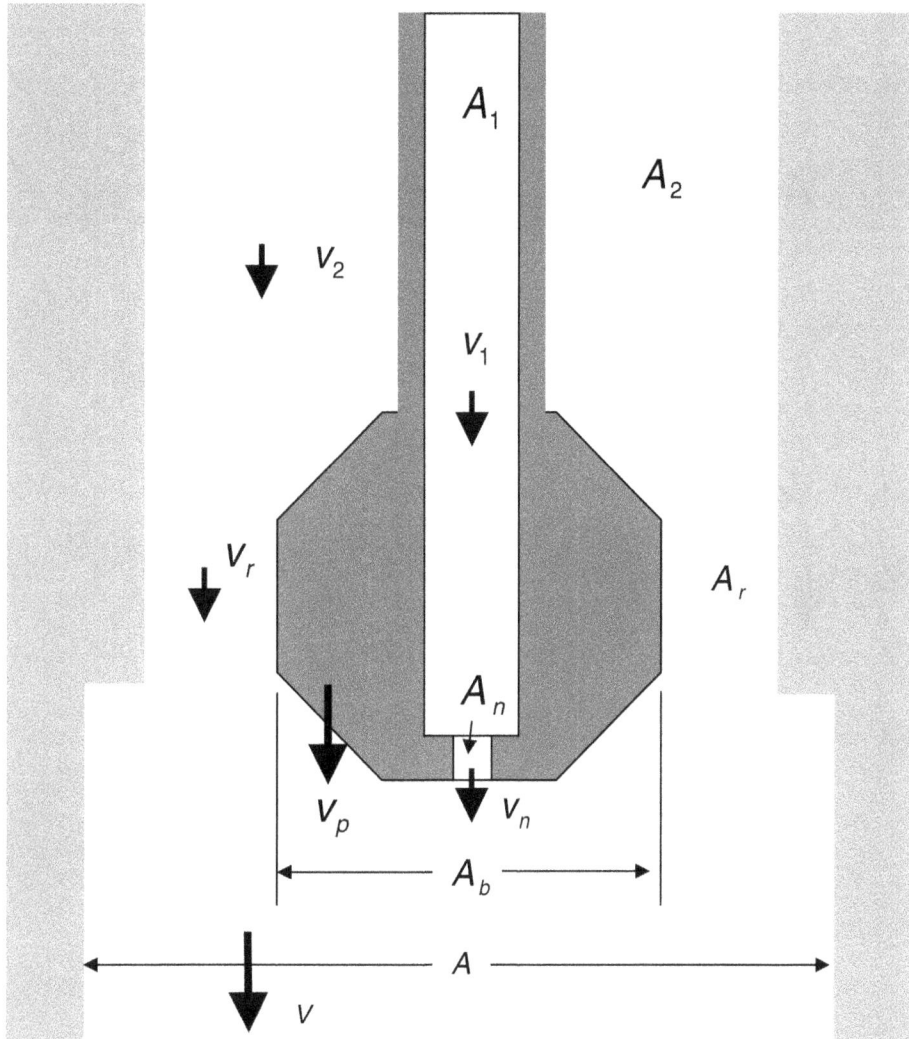

Fig. 3.4—Balance of mass at the bit.

v_1 = pipe velocity.

P_2 = pipe annulus pressure.

v_2 = pipe annulus velocity.

P_n = pipe nozzle pressure.

v_n = pipe nozzle velocity.

P_r = annulus return area pressure.

v_r = annulus return area velocity.

P = pipe-to-bottomhole pressure.

v = pipe-to-bottomhole velocity.

v_3 = pipe velocity.

A total of seven boundary conditions can be specified at the moving pipe end with bit (see **Fig. 3.5**). For the surge model, three mass balance equations and four nozzle pressure relations were used.

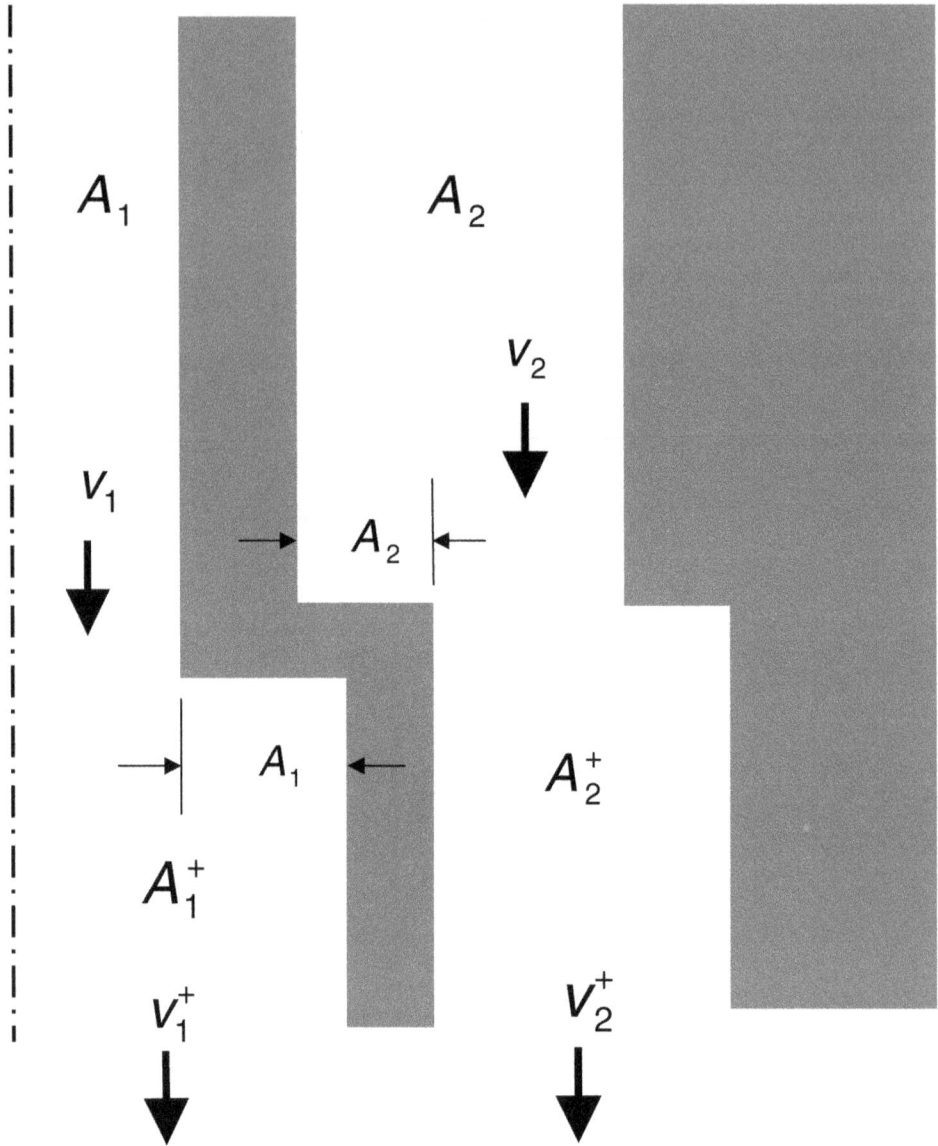

Fig. 3.5—Balance of mass for cross-sectional area change.

Pipe-to-Bottomhole Mass Balance.

$$A_r v_r + A_n v_n + A_b v_3 - A_v = 0 .$$

Pipe Annulus Mass Balance.

$$A_2 v_2 - (A_2 - A_r)v_3 - A_r v_r = 0 .$$

Pipe Mass Balance.

$$A_1 v_1 - (A_1 - A_n) v_3 - A_n v_n = 0 .$$

Pipe Nozzle Pressures.

$$P_1 - P_n = \frac{\rho}{2C_d} \left(v_n^2 - v_1^2 \right) .$$

$$P_2 - P_r = \frac{\rho}{2C_d} \left(v_r^2 - v_2^2 \right) .$$

$$P - P_n = \frac{\rho}{2C_d} \left(v_n^2 - v^2 \right) .$$

$$P - P_r = \frac{\rho}{2C_d} \left(v_r^2 - v^2 \right) .$$

Annulus Return Pressures. The boundary conditions are greatly simplified for a pipe without a bit.

$$A_1 v_1 + A_2 v_2 + A_3 v_3 - A v = 0 .$$
$$P_1 = P_2 = P_r = P_n = P .$$
$$v_1 = v_n .$$
$$v_2 = v_r .$$

The boundary condition imposed by a float is the requirement that

$$v_1 - v_3 < 0 .$$

If the solution of the boundary conditions does not satisfy this condition, the boundary conditions must be solved again with the new requirement:

$$v_1 = v_3 .$$

Change of Cross-Sectional Area. Changes in the cross-sectional area of the moving pipe generate an additional term in the balance of mass equations because of the fluid displaced by the moving pipe (see Fig. 3.5).

The following was already inserted:

$$A_1^+ v_1^+ = A_1^- v_1^- + \Delta A_1 v_3,$$
$$A_2^+ v_2^+ = A_2^- v_2^- + \Delta A_2 v_3,$$

where

$$\Delta A_1 = A_1^+ - A_1^-,$$

and

$$\Delta A_2 = A_2^+ - A_2^- .$$

The superscript – denotes upsteam properties, and the superscript + denotes downstream properties.

3.9.3 Surge Pressure Solution.
Because of the complex boundary conditions, the solution of a steady-state surge pressure is most easily solved with a computer program. For closed-pipe and circulating cases, the flow is defined so that pressures can be calculated from the annulus exit to the standpipe, as discussed previously. For open-pipe surges, the problem is finding how the flow splits between the pipe and the annulus, so that the pressures for both the pipe and the annulus match at the bit. One strategy for solving this problem is given next.

1. Calculate all pressures with all flow in the annulus. Then, check pressures at the bit; annulus pressure will be lower because of fluid friction.

2. Calculate all pressures with all flow in the pipe. Then, check pressures at the bit; pipe pressure will be lower because of fluid friction.

3. Calculate a division of flow between the pipe and annulus that will equalize the pressures at the bit.

4. Repeat Step 3 until the two pressures match within an acceptable tolerance.

The efficiency of this calculation will depend on the method chosen for Step 3. With modern computers, this is not a particularly critical problem, so a simple interval halving technique would work. For the ith iteration of Step 3, f_i is the fraction of flow in the pipe, and $(1 - f_i)$ is the fraction in the annulus. Previous steps show that f_p gives a higher annulus pressure and f_m gives a lower annulus pressure. Our new choice for f_i is $\frac{1}{2}(f_p + f_m)$. We perform the pressure calculation and find that the annulus pressure is higher, so we assign $f_p = f_i$. If the pressure difference is less than our tolerance, which we chose to be 1 psi, then the calculation is complete. Otherwise, we try another step. How do we establish f_p and f_m? The initial two steps in the solution step should give us $f_p = 0$ and $f_m = 1$, respectively. In some cases, such as small nozzles or restricted flow around the bit, fluid must flow into either the pipe or annulus, or the fluid level must fall. For these cases, f may be negative or greater than one. It may be necessary to repeat Steps 1 and 2 to establish the initial set f_m and f_p.

3.10 Fluid Friction
In the previous sections, we calculated the pressures in a flowing fluid, assuming we knew the value of the Fanning friction factor. Determination of the Fanning friction factor, in fact, may be the most difficult step in this calculation. Fluid friction is studied by the science of rheology.

3.10.1 Fluid Rheology.
The science of rheology is concerned with the deformation of all forms of matter, but has had its greatest development in the study of the flow behavior of suspensions in pipes and other conduits. The rheologist is interested primarily in the relationship between flow pressure and flow rate, and in the influence thereon of the flow characteristics of the fluid. There are two fundamentally different relationships:

1. The laminar flow regime prevails at low flow velocities. Flow is orderly, and the pressure-velocity relationship is a function of the viscous properties of the fluid.

2. The turbulent flow regime prevails at high velocities. Flow is disorderly and is governed primarily by the inertial properties of the fluid in motion. Flow equations are empirical.

The laminar flow equations relating flow behavior to the flow characteristics of the fluid are based on certain flow models, namely the Newtonian, the Bingham plastic, the pseudoplastic, the yield power-law, and the dilatant. Only the first four are of interest in drilling-fluid technology. Most drilling fluids do not conform exactly to any of these models, but drilling-fluid behavior can be predicted with sufficient accuracy by one or more of them. Flow models are usually visualized by means of consistency curves, which are plots either of flow pressure vs. flow rate or of shear stress vs. shear rate.

Shear stress is force per unit area and is expressed as a function of the velocity gradient of the fluid as

$$\tau = -\mu \frac{dv}{dr}, \quad \text{..} \quad (3.33)$$

where μ is the fluid viscosity and dv/dr is the velocity gradient. The negative sign is used in Eq. 3.33 because momentum flux flows in the direction of negative velocity gradient. That is, the momentum tends to go in the direction of decreasing velocity The absolute value of velocity gradient is called the shear rate and is defined as

$$\gamma = \left| \frac{dv}{dr} \right| . \quad \text{..} \quad (3.34)$$

Then, Eq. 3.33 can be written as

$$\tau = \mu\gamma . \quad \text{..} \quad (3.35)$$

Viscosity is the resistance offered by a fluid to deformation when it is subjected to a shear stress. If the viscosity is independent of the shear rate, the fluid is called a Newtonian fluid. Water, brines, and gases are examples of Newtonian fluid. The shear stress is linear with the shear rate for a Newtonian fluid and is illustrated by Curve A in **Fig. 3.6**. The symbol μ without any subscript is used to refer to the viscosity of Newtonian fluid. Most of the fluids used in drilling and cementing operations are not Newtonian, and their behavior is discussed next.

If the viscosity of a fluid is a function of shear stress (or, equivalently, of shear rate), such a fluid is called non-Newtonian fluid. Non-Newtonian fluids can be classified into three general categories:

1. Fluid properties are independent of duration of shear.
2. Fluid properties are dependent on duration of shear.
3. Fluid exhibits many properties that are characteristics of solids.

Time Independent. The following three types of materials are in this class.

Bingham Plastic. These fluids require a finite shear stress, τ_y; below that, they will not flow. Above this finite shear stress, referred to as yield point, the shear rate is linear with shear stress, just like a Newtonian fluid. Bingham fluids behave like a solid until the applied pressure is high enough to break the sheer stress, like getting catsup out of a bottle. The fluid is illustrated by Curve B in Fig. 3.6. The shear stress can be written as

$$\tau = \tau_y + \mu_p\gamma, \quad \text{..} \quad (3.36)$$

where τ_y is called the yield point (YP), and μ_p is referred to as the plastic viscosity (PV) of the fluid. Some water-based slurries and sewage sludge are examples of Bingham plastic fluid. Most of the water-based cement slurries and water-based drilling fluids exhibit Bingham plastic behavior. Drilling muds are often characterized with YP and PV values, but this is for historical reasons and does not necessarily imply that the Bingham fluid model is the best model for all muds.

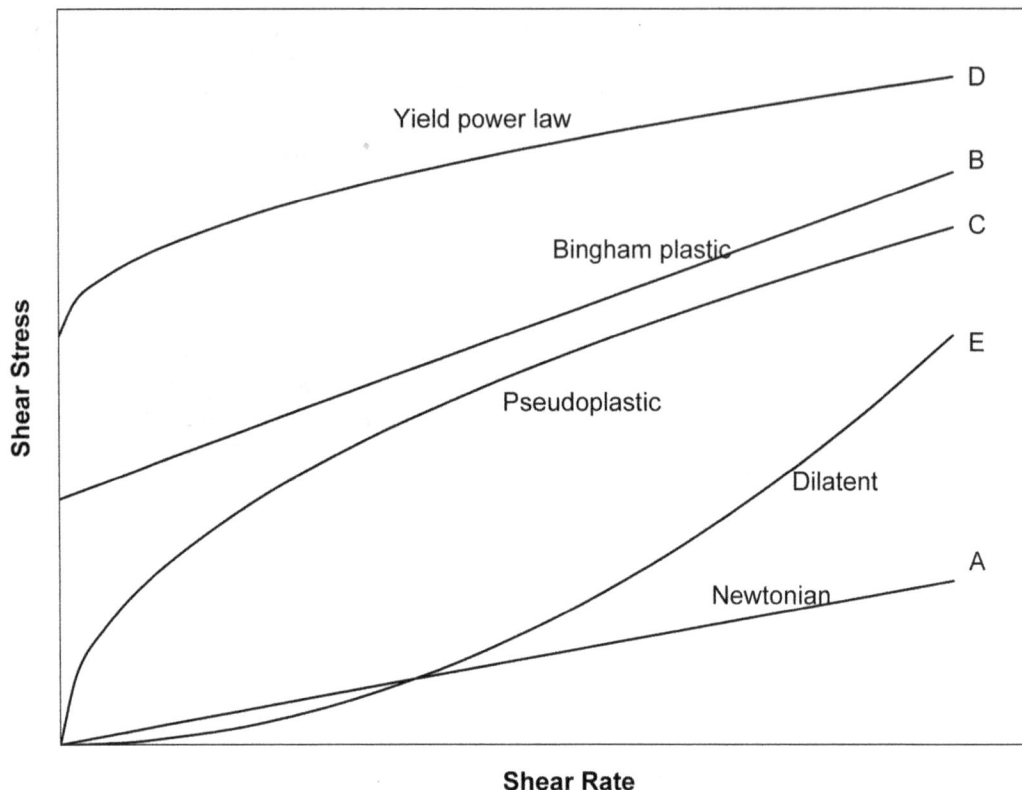

Shear Rate
Fig. 3.6—Rheology of fluids.

Pseudoplastic. These fluids exhibit a linear relationship between shear stress and shear rate when plotted on a log-log paper. This is illustrated by Curve C in Fig. 3.6. This fluid is also commonly referred to as a power-law non-Newtonian fluid. The shear stress can be written as

$$\tau = K\gamma^n \text{ and } n < 1, \dots\dots\dots\dots\dots\dots\dots (3.37)$$

where K is the consistency index, and n is the exponent, referred to as power-law index. A term μ_a is defined that is called the apparent viscosity and is

$$\mu_a = K\gamma^{n-1} . \dots\dots\dots\dots\dots\dots\dots\dots (3.38)$$

Note that apparent viscosity and effective viscosity as defined by different authors are not always defined in the sense used here, so read with caution. The apparent viscosity decreases as the shear rate increases for power-law fluids. For this reason, another term commonly used for pseudoplastic fluids is "shear thinning." Polymeric solutions and melts are examples of power-law fluid. Some drilling fluids and cement slurries, depending on their formulation, may exhibit power-law behavior.

Yield Power Law. Also known as Herschel-Bulkley fluids, these fluids require a finite shear stress, τ_y, below which they will not flow. Above this finite shear stress, referred to as yield point, the shear rate is related to the shear stress through a power-law type relationship. The shear stress can be written as

$$\tau = \tau_y + K\gamma^m, \text{...} (3.39)$$

where τ_y is called the yield point, K is consistency index, and m is the exponent, referred to as power-law index.

Dilatant. These fluids also exhibit a linear relationship between shear stress and shear rate when plotted on a log-log paper and are illustrated as Curve D in Fig. 3.6. The shear stress expression for dilatant fluid is similar to power-law fluid, but the exponent n is greater than 1. The apparent viscosity for these fluids increases as shear rate increases. For this reason, dilatant fluids are often called "shear-thickening."

Quicksand is an example of dilatant fluid. In cementing operations, it would be disadvantageous if fluids increased in viscosity as shear stress increased.

Time Dependent. These fluids exhibit a change in shear stress with the duration of shear. This does not include changes because of reaction, mechanical effects, etc. Cement slurries and drilling fluids usually do not exhibit time-dependent behavior. However, with the introduction of new chemicals on a regular basis, one should test and verify the behavior.

Solids Characteristic. These fluids exhibit elastic recovery from deformation that occurs during flow and are called viscoelastic. Most of the cement slurries and drilling fluids do not exhibit this behavior. However, as mentioned earlier, new polymers are being introduced on a regular basis, and tests should be conducted to verify the behavior.

The unit of viscosity is Pascal-second (Pa-s) in the SI system and lbf/(ft-s) in oilfield units. One Pa-s equals 10 poise (P), 1,000 centipoise (cp), or 0.672 lbf/(ft-s). The exponent n is dimensionless, and consistency index, K, has the units of Pa-s^n in the SI system and lbf/(sec^n-ft^{-2}) in oilfield units. One Pa-s^n equals 208.86 lbf/(sec^n.ft^{-2}). The yield point for Bingham fluids is often characterized in units of lbf/(1,00ft^2), and plastic viscosity is usually given in centipoise.

Viscometry. The rheology parameters of the fluids, μ and μ_p, and τ_o, K, and n, are determined by conducting tests in a concentric viscometer. This consists of concentric cylinders with one of them rotating, usually the outer one. A sample of fluid is placed between the cylinders, and the torque on the inner cylinder is measured. Assuming an incompressible fluid, with flow in the laminar flow regime, the equations of motion can be solved for τ to give

$$\tau = M_T / (2\pi r^2 L), \text{...} (3.40)$$

where
 τ = shear stress, Pa;
 M_T = torque, N-m;
 L = length, m;
and
 r = radius, m.

In a concentric viscometer, torque, M_T, is measured at a different rotational speed of the outer cylinder. Shear stress is then calculated from Eq. 3.40, and shear rate is given by

$$\gamma = \frac{4\pi\Omega_0 R_c^2}{R_c^2 - R_b^2} = \frac{4\pi\Omega_0}{1 - \kappa^2},$$

and

$$\kappa = \frac{R_b}{R_c}, \quad\text{(3.41)}$$

where

R_b = radius of inner cylinder (bob), m;

R_c = radius of outer cylinder (cup), m;

κ = ratio of radius of inner cylinder to outer cylinder;

and

Ω_0 = angular velocity of outer cylinder.

Shear stress and shear rate are then analyzed to determine the rheology model.

A number of commercially available concentric cylinder rotary viscometers are suitable for use with drilling muds. They are similar in principle to the viscometer already discussed. All are based on a design by Savins and Roper, which enables the plastic viscosity and yield point to be calculated very simply from two dial readings, at 600 and 300 rpm, respectively.[6] They are referred to in the industry as the direct-indicating viscometer and typically are called Fann viscometers.

The underlying theory is as follows: Eqs. 3.40 and 3.41 are combined to give

$$\mu = \frac{a_{vs}\theta - b_{vs}\tau_y}{\omega}, \quad\text{(3.42)}$$

where a_{vs} and b_{vs} are constants that include the instrument dimensions, the spring constant, and all conversion factors; ω is the rotor speed in revolutions per minute (rpm).

Then,

$$\mu_p = \overline{\text{PV}} = a_{vs}\left(\frac{\theta_1 - \theta_2}{\omega_1 - \omega_2}\right), \quad\text{(3.43)}$$

where θ_1 and θ_2 are dial readings taken at ω_1 and ω_2 rpm, respectively. PV is the conventional oilfield term for plastic viscosity, thus measured. Then, the yield point is determined.

$$\tau_y = \overline{\text{YP}} = \frac{a_{vs}}{b_{vs}}\left[\theta_1 - \left(\frac{\omega_1}{\omega_1 - \omega_2}\right)(\theta_1 - \theta_2)\right]. \quad\text{(3.44)}$$

YP is the conventional oilfield term for yield point, thus measured. The numerical values of a_{vs}, b_{vs}, ω_1, and ω_2 were chosen so that

$$\overline{\text{PV}} = \theta_1 - \theta_2, \quad\text{(3.45)}$$

and

$$\overline{\text{YP}} = \theta_2 - \overline{\text{PV}}. \quad\text{(3.46)}$$

Apparent viscosity μ_a may be calculated from the Savins-Roper viscometer reading as

$$1° \text{ dial reading } = 1.067 \text{ lbf}/100 \text{ ft}^2$$

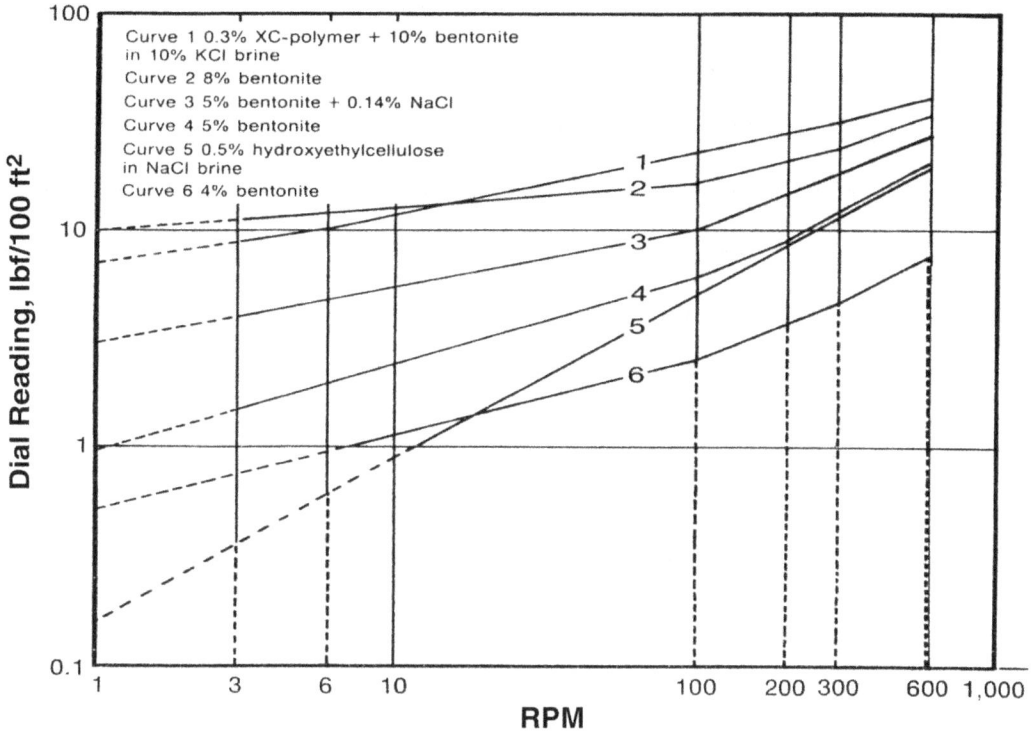

Fig. 3.7—Typical drilling fluid consistency curves.[7] (Reprinted from *Composition and Properties of Oil Well Drilling Fluids,* G.R. Gray and H.C.H. Darley, fourth edition, © 1980, with permission from Elsevier.)

$$= 5.11 \text{dynes} / \text{cm}^2 \ \text{shear stress,}$$

$$1 \ \text{rpm} = 1.703 \ \text{reciprocal seconds, shear rate,}$$

and

$$\mu_a = \tau / \gamma = 5.11 / 1.703 \ \text{poise} / \text{degree} / \text{rpm}$$

$$= 300 \ \text{cp} / \text{degree} / \text{rpm}$$

$$= 300 \ \theta / \omega, \ .... (3.47)$$

where θ is the dial reading at ω rpm. Typical viscometer results are shown in **Fig. 3.7.**[7] Notice that real fluids are not ideally any of the models shown, but generally are pretty close to one model or another. The selection of the model may be motivated by a particular fluid velocity of interest. For instance, fluid 6 in Fig. 3.7 would be modeled well by a yield-power law for rpm below about 100.

Fanning Friction Factor Correlations. Flow in pipes and annuli are typically characterized as laminar or turbulent flow. Laminar flow often can be solved analytically. Correlation for turbulent flow is usually developed empirically by conducting experiments in a flow loop. Typical data will look like those that are is shown in **Fig. 3.8**. Experimental data are usually analyzed and correlated through the use of two dimensionless numbers: f, the Fanning friction factor, and Re, the Reynolds number. The relationship between the friction factor, f, and Reynolds number for Newtonian fluids is given in **Fig. 3.9,**[8] with the pipe roughness given in **Fig. 3.10**. This figure is based on the experimental results of Colebrook.[9] The relationship between friction factor f vs. Re for pseudoplastic fluids is shown in **Fig. 3.11**. This figure is

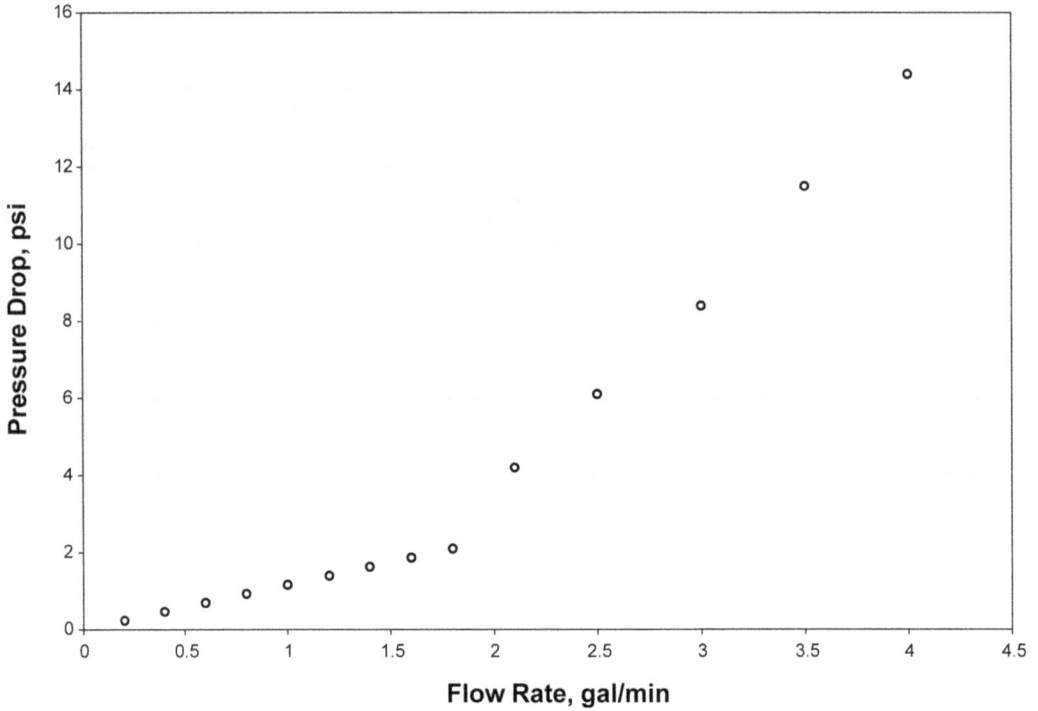

Fig. 3.8—Flow loop experimental data.

based on the experimental results of Dodge and Metzner.[10] Here non-Newtonian fluids usually assume this pseudoplastic friction factor for turbulent flow.

The pressure drop per unit length for flow through a duct is given by

$$\Delta P = \frac{2f\rho v^2}{D_{hyd}}\Delta z, \quad\text{...} (3.48)$$

where f is Fanning friction factor, Δz is the length, v is the velocity, ρ is the density, D_{hyd} is a characteristic "diameter," and ΔP is the pressure drop. The friction factor depends on Reynolds number, Re, and the roughness of the pipe. The Reynolds number, Re, is defined as

$$\text{Re} = D\rho v / \mu, \quad\text{...} (3.49)$$

where ρ is the density of the fluid, v is the average velocity, D is a characteristic length (e.g., pipe diameter), and μ is a characteristic viscosity. Correlations for friction factor, f, in both laminar and turbulent flow regime and for critical Reynolds number are available for a number of fluids and geometries. However, in critical situations, it is recommended that flow-loop tests be conducted and data compared with calculations that are based on fundamental equations for flow. For example, experimental data in laminar flow should be compared with estimated values from correlation such as Eq. 3.52. However, some solid-laden polymers are known to exhibit what is known as shear-induced diffusion, in which solids migrate away from the walls to the center of the pipe. These fluids show deviation in calculated and experimental values in laminar flow. Correlations should be modified as needed to reflect this behavior. Several poly-

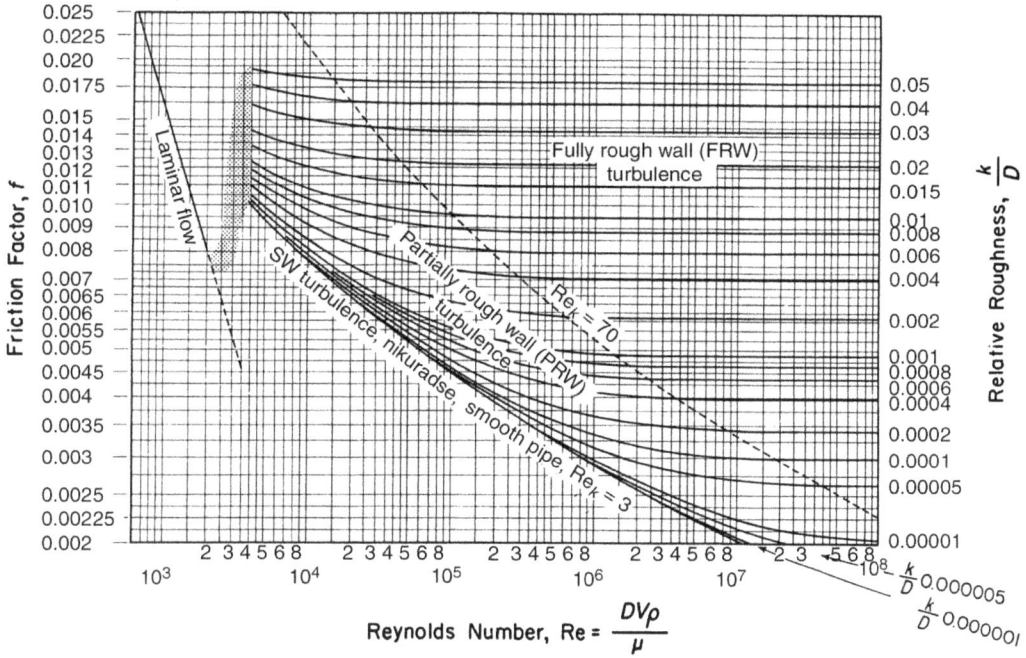

Fig. 3.9—Newtonian fluid friction factor (after Govier and Aziz).[8]

mers are known to exhibit drag reduction in turbulent flow. Theoretical prediction of polymer-flow behavior is not yet good enough, so flow-loop data are almost always needed.

Commonly used Fanning friction correlations are summarized in the next section. Correlations are provided for three geometric configurations: pipe flow, concentric annular flow, and slit flow. For each case, the ΔP and Re are defined for the specific geometry and flow model. The laminar flow equations for annular flow are approximate for Newtonian and power-law flow in annuli with low-clearance but reasonably accurate and much simpler than the exact solutions. Note that for low clearance annuli, the slit flow model provides almost as accurate a result as the concentric annular model but can also be modified to account for eccentric annuli.

Rheological Model 1: Newtonian Fluids.

Pipe Flow.

Frictional pressure drop:

$$\Delta P = \frac{2 f \rho v^2}{D_i} \Delta z \ . \ \text{..} (3.50)$$

Reynolds number:

$$\text{Re} = D_i \rho v / \mu, \text{..} (3.51)$$

where D_i is the pipe inside diameter (ID).

Laminar flow:

$$f = 16 / \text{Re} , \text{..} (3.52)$$

Pipe Diameter, *D*, ft

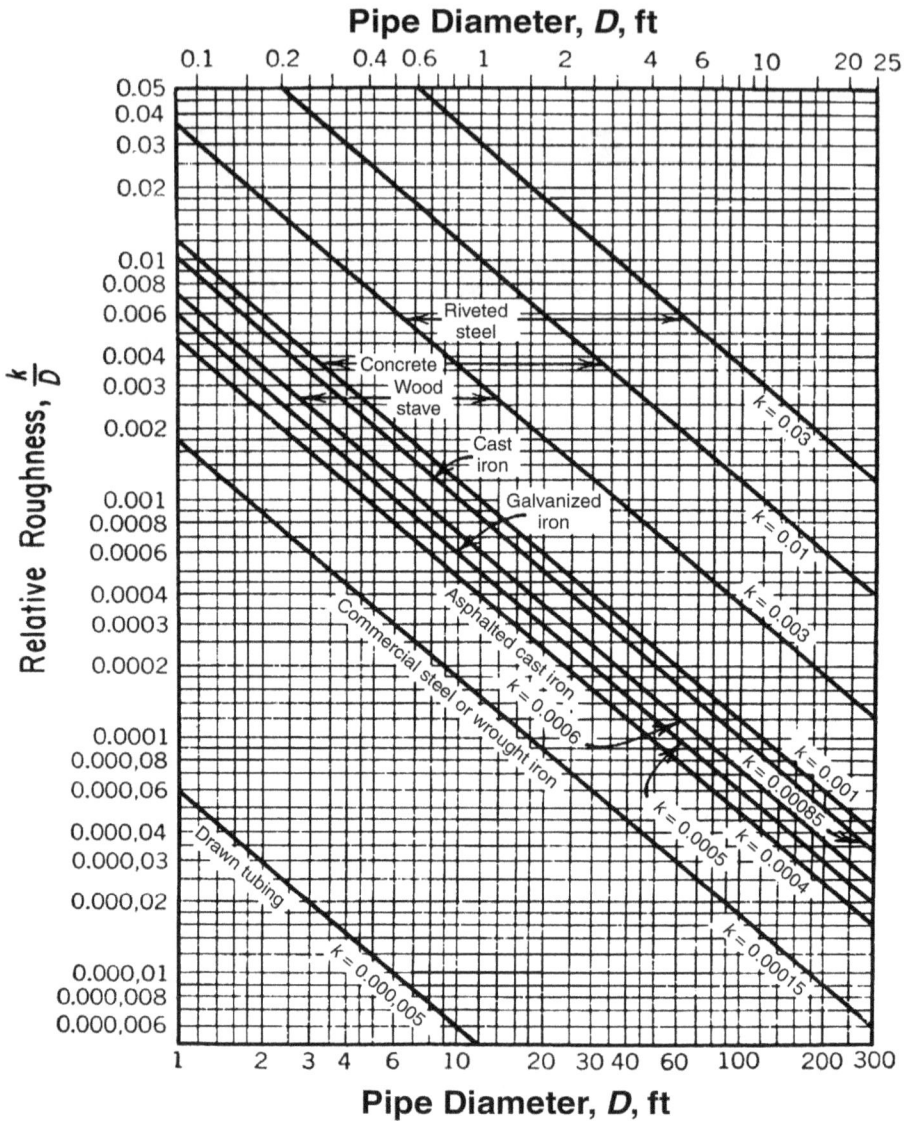

Fig. 3.10—Pipe roughness (after Govier and Aziz).[8]

for Re < 2,100.

Turbulent flow:

$$\frac{1}{\sqrt{f}} = -4 \log_{10} \left(\frac{k/D}{3.7065} + \frac{1.2613}{\text{Re} \sqrt{f}} \right), \quad \dots\dots\dots\dots\dots\dots\dots\dots (3.53)$$

for Re > 3,000, and *k* is the absolute pipe roughness in the same units as *D*.

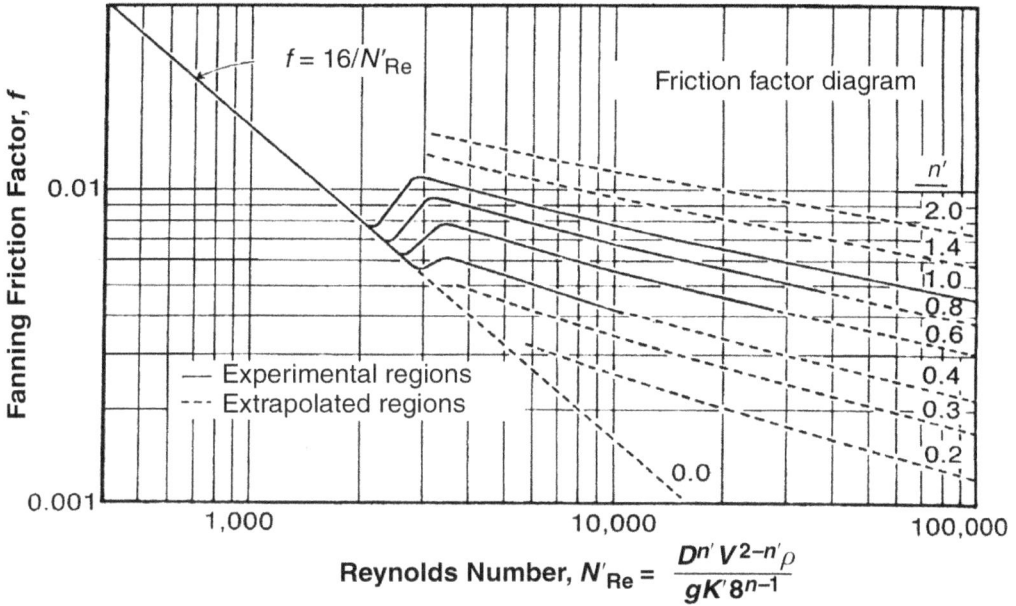

Fig. 3.11—Pseudoplastic friction factor vs. Reynolds number (Govier and Aziz).[8]

Annular Flow.
Frictional pressure drop:

$$\Delta P = \frac{2 f \rho v^2}{D_o - D_i} \Delta z \ . \ \text{..} \ (3.54)$$

Reynolds number:

$$\text{Re} \ = (D_o - D_i) \rho v / \mu, \ \text{..} \ (3.55)$$

where D_o is the annulus outside diameter (OD), and D_i is the ID.
Laminar flow:

$$f = 16 / \ \text{Re} \ , \text{(approximate)} \text{..} \ (3.56)$$

for Re < 2,100.
Turbulent flow:

$$\frac{1}{\sqrt{f}} = - 4\log_{10} \left(\frac{k / D}{3.7065} + \frac{1.2613}{\text{Re} \ \sqrt{f}} \right) \text{..} \ (3.57)$$

for Re > 3,000, and k is the absolute pipe roughness in the same units as D.

Rheological Model 2: Bingham Plastic Fluids.
Pipe Flow.
Frictional pressure drop:

$$\Delta P = \frac{2f\rho v^2}{D_i}\Delta z . \quad\text{...} \quad (3.58)$$

Reynolds number:

$$\text{Re} = D_i\rho v / \mu_p, \quad\text{...} \quad (3.59)$$

where D_i is the pipe ID, and μ_p is the plastic viscosity.
Laminar flow:

$$f = 16\{(1 / \text{Re}) + [\text{He}/(6\text{Re}^2)] - [\text{He}^4/(3f^3\text{Re}^8)]\}, \quad\text{...........................} \quad (3.60)$$

for $\text{Re} < \text{Re}_{BP1}$, where

$$\text{He} = \tau_o\rho D_i^2 / \mu_p^2,$$
$$\text{Re}_{BP1} = \text{Re}_{BP2} - 866(1 - \alpha_c),$$
$$\text{Re}_{BP2} = \text{He}[(0.968774 - 1.362439\alpha_c + 0.1600822\alpha_c^4)/(8\alpha_c)],$$
$$\alpha_c = \tfrac{3}{4}\{[(2\text{He}/24,500) + (3/4)] = [(2\text{He}/24,500) + (3/4)]^2$$
$$-[4(\text{He}/24,500)^2]^{1/2}/[2(\text{He}/24,500)]\} .$$

Turbulent flow:

$$f = A(\text{Re})^{-B}, \quad\text{...} \quad (3.61)$$

for $\text{Re} > \text{Re}_{BP2}$, where:
For $\text{He} <= 0.75 \times 10^5$, $A = 0.20656$, and $B = 0.3780$.
For $0.75 \times 10^5 < \text{He} <= 1.575 \times 10^5$, $A = 0.26365$, and $B = 0.38931$.
For $\text{He} > 0.75 \times 10^5$, $A = 0.20521$, $B = 0.35579$, and $\text{He} = \tau_o\rho D^2/\mu_p^2$.
Annular Flow.
Frictional pressure drop:

$$\Delta P = \frac{2f\rho v^2}{D_o - D_i}\Delta z . \quad\text{...} \quad (3.62)$$

Reynolds number:

$$\text{Re} = (D_o - D_i)\rho v / \mu_p, \quad\text{...} \quad (3.63)$$

where D_o is the annulus OD; D_i is the ID; and μ_p is the plastic viscosity.

Laminar flow:

$$f = 16\left\{\left[(1/\ \mathrm{Re}\ ) + [(\mathrm{He}/6\mathrm{Re}^2)] - [\mathrm{He}^4/(3\mathrm{f}^3\mathrm{Re}^8)]\right]\right\},\ \dotfill (3.64)$$

for $\mathrm{Re} < \mathrm{Re}_{\mathrm{BP1}}$, where:

$$\mathrm{He} = \tau_o\rho(D_o^2 - D_i^2)/\mu_p^2\ .$$

$$\mathrm{Re}_{\mathrm{BP1}} = \mathrm{Re}_{\mathrm{BP2}} - 866(1 - \alpha_c)\ .$$

$$\mathrm{Re}_{\mathrm{BP2}} = \mathrm{He}[(0.968774 - 1.362439\alpha_c + 0.1600822\alpha_c^4)/(8\alpha_c)]\ .$$

$$\alpha_c = \tfrac{3}{4}\left\{[(2\mathrm{He}/24{,}500) + (3/4)] - [(2\mathrm{He}/24{,}500) + (3/4)]^2\right.$$
$$\left. -[4(\mathrm{He}/24{,}500)^2]^{1/2}/[2(\mathrm{He}/24{,}500)]\right\}\ .$$

Turbulent flow:

$$f = A(\ \mathrm{Re}\ )^{-B},\ \dotfill (3.65)$$

for $\mathrm{Re} > \mathrm{Re}_{\mathrm{BP2}}$, where:

For $\mathrm{He} < = 0.75 \times 10^5$, $A = 0.20656$, and $B = 0.3780$.
For $0.75 \times 10^5 < \mathrm{He} < = 1.575 \times 10^5$, $A = 0.26365$, and $B = 0.38931$.
For $\mathrm{He} > 0.75 \times 10^5$, $A = 0.20521$, $B = 0.35579$, and $\mathrm{He} = \tau_o\rho(D_o^2 - D_i^2)/\mu_p^2$.
Slit Flow.
Frictional pressure drop:

$$\Delta P = \frac{2f\rho v^2}{D_o - D_i}\Delta z\ .\ \dotfill (3.66)$$

Reynolds number:

$$\mathrm{Re}\ = (D_o - D_i)\rho v / (1.5\mu_p),\ \dotfill (3.67)$$

where D_o is the annulus OD; D_i is the ID; and μ_p is the plastic viscosity.
Laminar flow:

$$f = 16\left\{\left[(1/\ \mathrm{Re}\ ) + [(9/8)\mathrm{He}/(6\mathrm{Re}^2)] - [\mathrm{He}^4/(3\mathrm{f}^3\mathrm{Re}^8)]\right]\right\},\ \dotfill (3.68)$$

for $\mathrm{Re} < \mathrm{Re}_{\mathrm{BP1}}$, where:

$$\mathrm{He} = \tau_o\rho(D_o^2 - D_i^2)/(1.5\mu_p)^2\ .$$

$$\mathrm{Re}_{\mathrm{BP1}} = \mathrm{Re}_{\mathrm{BP2}} - 577(1 - \alpha_c)\ .$$

$$\mathrm{Re}_{\mathrm{BP2}} = \mathrm{He}[(0.968774 - 1.362439\alpha_c + 0.1600822\alpha_c^4)/(12\alpha_c)]\ .$$

$$\alpha_c = \tfrac{3}{4}\left\{[(2\mathrm{He}/24{,}500) + (3/4)] - [(2\mathrm{He}/24{,}500) + (3/4)]^2\right.$$
$$\left. -4(\mathrm{He}/24{,}500)^2]^{1/2}/[2(\mathrm{He}/24{,}500)]\right\}\ .$$

Turbulent flow:

$$f = A(\text{Re})^{-B}, \quad\quad\quad\quad\quad\quad\quad\quad\quad\quad\quad (3.69)$$

for $\text{Re} > \text{Re}_{BP2}$, where:

For $\text{He} <= 0.75 \times 10^5$, $A = 0.20656$, and $B = 0.3780$.

For $0.75 \times 10^5 < \text{He} <= 1.575 \times 10^5$, $A = 0.26365$, and $B = 0.38931$.

For $\text{He} > 0.75 \times 10^5$, $A = 0.20521$, $B = 0.35579$, and $\text{He} = \tau_o \rho (D_o^2 - D_i^2)/(1.5\mu_p)^2$.

Rheological Model 3: Power Law Fluids.

Pipe Flow.

Frictional pressure drop:

$$\Delta P = \frac{2 f \rho v^2}{D_i} \Delta z . \quad\quad\quad\quad\quad\quad\quad\quad\quad (3.70)$$

Reynolds number:

$$\text{Re} = D_i^n v^{2-n} \rho / (8^{n-1}[(3n+1)/4n]^n K), \quad\quad\quad\quad\quad (3.71)$$

where D_i is the pipe ID.

Laminar flow:

$$f = 16/\text{Re} , \quad\quad\quad\quad\quad\quad\quad\quad\quad\quad\quad (3.72)$$

for $\text{Re} \leq 3{,}250 - 1{,}150n$.

Turbulent flow:

$$1/f^{1/2} = \{[(4.0/n^{0.75}) \log (\text{Re} f^{1-n/2})] - (0.4/n^{1.2})\}, \quad\quad\quad (3.73)$$

for $\text{Re} \geq 4{,}150 - 1{,}150n.$ [10]

Annular Flow.

Frictional pressure drop:

$$\Delta P = \frac{2 f \rho v^2}{D_o - D_i} \Delta z . \quad\quad\quad\quad\quad\quad\quad\quad\quad (3.74)$$

Reynolds number:

$$\text{Re} = (D_o - D_i)^n v^{2-n} \rho / \{8^{n-1}[(3n+1)/4n]^n K\}, \quad\quad\quad\quad (3.75)$$

where D_o is the annulus OD, and D_i is the ID.

Laminar flow:

$$f = 16/\text{Re} \text{ (approximate)}, \quad\quad\quad\quad\quad\quad\quad\quad (3.76)$$

for $\text{Re} \leq 3{,}250 - 1{,}150n$.

Turbulent flow:

$$1/f^{1/2} = \left\{ [(4.0/n^{0.75}) \log (\mathrm{Re}f^{1-n/2})] - (0.4/n^{1.2}) \right\}, \quad \text{.............................} \quad (3.77)$$

for $\mathrm{Re} \geq 4{,}150 - 1{,}150n$.
Slit Flow.
Frictional pressure drop:

$$\Delta P = \frac{2f\rho v^2}{D_o - D_i} \Delta z . \quad \text{.......................................} \quad (3.78)$$

Reynolds number:

$$\mathrm{Re} = (D_o - D_i)^n v^{2-n} \rho / (12^{n-1}[(2n+1)/3n)]^n K) . \quad \text{..............................} \quad (3.79)$$

Laminar flow:

$$f = 24 / \mathrm{Re} , \quad \text{...} \quad (3.80)$$

for $\mathrm{Re} \leq 3{,}250 - 1{,}150n$.
Turbulent flow:

$$1/f^{1/2} = \left\{ [(4.0/n^{0.75}) \log (\mathrm{Re}f^{1-n/2})] - (0.4/n^{1.2}) \right\}, \quad \text{.............................} \quad (3.81)$$

for $\mathrm{Re} \geq 4{,}150 - 1{,}150n$.
Rheological Model 4: Yield Power Law (YPL) Fluids.
Pipe Flow.
Frictional pressure drop:

$$\Delta P = \frac{2f\rho v^2}{D_i} \Delta z . \quad \text{.......................................} \quad (3.82)$$

Reynolds number:

$$\mathrm{Re}_{\mathrm{YPL}} = 8\rho v^2 / (\tau_y + K\gamma_e^m), \quad \text{.....................................} \quad (3.83)$$

where

$$\gamma_e = 8v / D_e .$$

$$D_e = \frac{4m}{3m+1} C_c D_i .$$

$$C_c = (1-x)\left[\frac{2m^2 x^2}{(1+2m)(1+m)} + \frac{2mx}{1+2m} + 1 \right].$$

$$x = \frac{\tau_y}{\tau_w}.$$

$$\tau_w = \frac{D}{4}\frac{\Delta P}{\Delta z}.$$

Laminar flow:

$$f = 16 / \text{Re}_{\text{YPL}}, \quad\text{...} \quad (3.84)$$

for $\text{Re} \leq 3{,}250 - 1{,}150n$.

Turbulent flow:

$$1 / f^{1/2} = \left\{ [(4.0 / n^{0.75}) \log (\text{Re} f^{1 - n/2})] - (0.4 / n^{1.2}) \right\}, \quad\text{............................}\quad (3.85)$$

for $\text{Re} \geq 4{,}150 - 1{,}150n$.

Slit Flow.

Frictional pressure drop:

$$\Delta P = \frac{2 f \rho v^2}{D_o - D_i} \Delta z. \quad\text{...}\quad (3.86)$$

Reynolds number:

$$\text{Re}_{\text{YPL}} = 12 \rho v^2 / (\tau_y + K \gamma_e^m), \quad\text{...}\quad (3.87)$$

where

$$\gamma_e = 12 v / D_e,$$

$$D_e = \frac{3m}{2m + 1} C_c (D_o - D_i),$$

$$C_c = (1 - x) \left(\frac{mx}{1 + m} + 1 \right),$$

$$x = \frac{\tau_y}{\tau_w},$$

and

$$\tau_w = \frac{(D_o - D_i)}{4} \frac{\Delta P}{\Delta z}. \quad\text{...}\quad (3.88)$$

Laminar flow:

$$f = 24 / \text{Re} \quad\text{...}\quad (3.89)$$

for $\text{Re} \leq 3{,}250 - 1150n$.

Turbulent flow:

$$1/f^{1/2} = \{[(4.0/n^{0.75}) \log (Re f^{1-n/2})] - (0.4/n^{1.2})\}, \quad\text{(3.90)}$$

for $Re \geq 4,150 - 1,150n$.

Frictional Pressure Drop in Eccentric Annulus. The frictional pressure drop in an eccentric annulus is known to be less than the frictional pressure drop in a concentric annulus. For laminar flow of Newtonian fluids, the pressure drop in a fully eccentric annulus is half the pressure drop in a concentric annulus. For turbulent flow, the difference is about 10%. For non-Newtonian fluids, the effect is less but still significant. In deviated wells, the drillpipe should be fully eccentric over much of the deviated wellbore, resulting in reduced fluid friction.

Define the correction factor for eccentricity.

$$C_e = \left(\frac{dp}{dz}\right)_e \bigg/ \left(\frac{dp}{dz}\right)_c, \quad\text{(3.91)}$$

where subscript e denotes eccentric, and subscript c denotes concentric.

C_e for laminar flow is determined based on the methods used by Uner et al.[11] The flow rate through a concentric annulus is given by

$$q_c = \frac{\pi r_o^3}{2} \frac{n}{2n+1} \left| \frac{dp}{dz} \frac{r_o}{2K} \right|^{1/n} (1+R_r)(1-R_r)^{2+1/n}, \quad\text{(3.92)}$$

where $R_r = r/r_o$. The flow rate through an eccentric annulus was determined to be

$$q_e = \frac{\pi r_o^3}{2} \frac{n}{2n+1} \left| \frac{dp}{dz} \frac{r_o}{2K} \right|^{1/n} \frac{(1-R_r^2)}{(2E - \pi R_r)} F(f, n, R_r), \quad\text{(3.93)}$$

where

$$F(f, n, R_r) = \int_0^\pi \left(\sqrt{1 - f^2 \sin^2 \vartheta} + f \cos \vartheta - R_r\right)^{2+1/n} d\vartheta, \quad\text{(3.94)}$$

$$E = \int_0^{\pi/2} \sqrt{1 - f^2 \sin^2 \vartheta}\, d\vartheta, \quad\text{(3.95)}$$

and

$$f = \delta_r(1 - R_r)/(r_o - r_i), \quad\text{(3.96)}$$

where δ_r is the distance between centers of the inside and outside pipes (e.g., $\delta_r = 0$ for concentric pipes). The geometry of the eccentric annulus is illustrated in **Fig. 3.12**.

The function E may be evaluated using a six-coefficient approximation. The function F must be evaluated using numerical methods (e.g., a seven-point Newton-Cotes numerical integration formula). Setting q_a and q_e equal, then

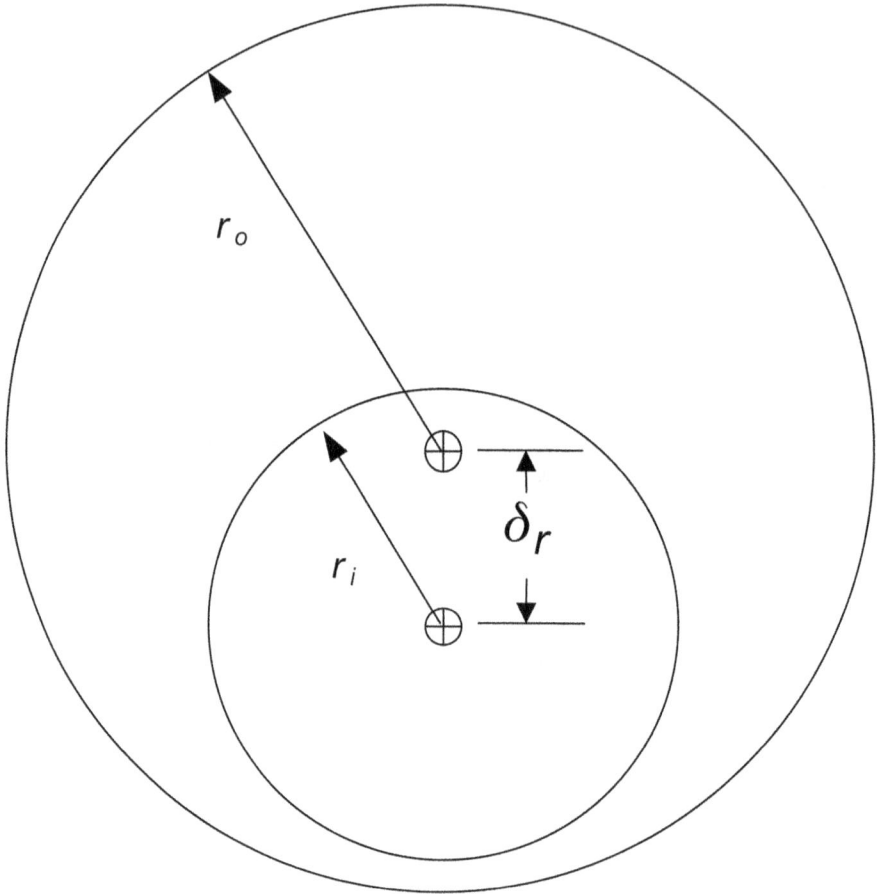

Fig. 3.12—Annulus eccentricity.

$$C_e = \left(1 - R_r\right)^{n+1} \left(\frac{2E - \pi R_r}{F(f,\, n,\, R_r)}\right)^n . \quad\text{... (3.97)}$$

Because C_e depends only on f, n, and R_r, C_e need be calculated only once, then used for all future frictional pressure drop calculations, as long as the property n does not vary.

C_e for turbulent flow is determined by applying the same techniques to the turbulent velocity profile determined by Dodge and Metzner.[10]

$$v/v^* = f_1 \ln |y| + f_2 \ln (v^*) + f_3, \quad\text{.. (3.98)}$$

where

$$f_1 = 2.456 n^{.25} ; \quad\text{... (3.99)}$$

$$f_2 = 2.458(2 - n) n^{-.75} ; \quad\text{.. (3.100)}$$

$$f_3 = 2.458 \ln (\rho/K) n^{-.75} + 3.475 n^{-.75}[1.960 + 0.815 n - .707 n \ln (3 + 1/n)] ; \quad\text{......... (3.101)}$$

and

$$v^* = \sqrt{\tau / \rho} . \qquad (3.102)$$

The volume flow rate through the concentric annulus is given by

$$q_c = v_c^* \int_0^w \int_{-h/2}^{h/2} \left[f_1 \ln y + f_2 \ln \left(v_c^* \right) + f_3 \right] dy \, dx, \qquad (3.103)$$

where
$$h = r_o - r_i,$$
and
$$w = \pi (r_o + r_i).$$
Integrating Eq. 3.67 gives

$$q_c = v_c^* A \left\{ f_1 [\ln (h/2) - 1] + f_2 \ln \left(v_c^* \right) + f_3 \right\}, \qquad (3.104)$$

where A is the flow area.

The equivalent integral to Eq. 3.103 for eccentric flow is given by

$$q_e = v_e^* \left\{ \left[B \int_0^{2\pi} \int_{-h/2}^{h/2} f_1 \ln y \; dy \; d\vartheta \right] + A \left[f_2 \ln \left(u_e^* \right) + f_3 \right] \right\}, \qquad (3.105)$$

where

$$B = \frac{\pi r_o}{2} \frac{\left(1 - R_r^2 \right)}{2E - \pi R_r}, \qquad (3.106)$$

and

$$h(\vartheta) = r_o \left(\sqrt{1 - f^2 \sin^2 \vartheta} + f \cos \vartheta - R_r \right) . \qquad (3.107)$$

The integral in Eq. 3.105 must be evaluated numerically (e.g., by a seven-point Newton-Cotes numerical integration. C_e can be determined by setting Eq. 3.104 equal to Eq. 3.105 and noting that

$$v_c^* = \sqrt{C_e} v_e^*, \qquad (3.108)$$

where v_c^* is determined from the concentric solution given by Dodge and Metzner.[10] The resulting nonlinear equation must be solved for C_e numerically (e.g., by using Newton's method). Because C_e depends only on f_1, f_2, f_3, n, and R_r, C_e need be calculated only once, then used for all future frictional pressure-drop calculations, as long as the properties ρ, K, and n do not vary.

3.11 Dynamic Pressure Prediction

3.11.1 Introduction. Calculating dynamic pressures in a wellbore are significantly more difficult than calculating steady-state flowing conditions. In a dynamic calculation, there are two effects not considered in steady flow: fluid inertia and fluid accumulation. In steady-state mass conservation, flow of fluid into a volume was matched by an equivalent flow out of the volume. In the dynamic calculation, there may not be equal inflow and outflow, but instead, fluid may accumulate within the volume. For fluid accumulation to occur, either the fluid must compress or the wellbore must expand. When considering the momentum equation, the fluid at rest must be accelerated to its final flow rate. The fluid inertia resists the change in velocity.

Typically, dynamic fluid flow is not a consideration. One exception is the operation of running pipe or casing into the wellbore, where dynamic pressure variation may be as important as pressures because of fluid friction. A second area of interest might be water-hammer effects during production startup.

3.11.2 Governing Equations—Dynamic Pressure Prediction. The fluid pressures and velocities in open hole are determined by solving two coupled partial differential equations: the balance of mass and the balance of momentum.

Balance of Mass.

$$\left(\frac{1}{A}\frac{dA}{dP} + \frac{1}{K_b}\right)\frac{dP}{dt} + \frac{\partial v}{\partial z} = 0, \quad\ldots\ldots\ldots\ldots\ldots\ldots\ldots\ldots (3.109)$$

where
A = cross-sectional area, m^2;
P = pressure, Pa;
K_b = fluid bulk modulus, Pa;
and
v = fluid velocity, m/s.
The term

$$\left(\frac{1}{A}\frac{dA}{dP} + \frac{1}{K}\right) = C \ldots\ldots\ldots\ldots\ldots\ldots\ldots\ldots\ldots\ldots (3.110)$$

is the compressibility, C, of the wellbore/fluid system (i.e., the change in wellbore volume per unit change in pressure). The balance of mass consists of three effects: the expansion of the hole because of internal fluid pressure, the compression of the fluid because of changes in fluid pressure, and the influx or outflux of the fluid. The expansion of the hole is governed by the elastic response of the formation and any casing cemented between the fluid and the formation. The fluid volume change is given by the bulk modulus K. For drilling muds, K varies as a function of composition, pressure, and temperature. The reciprocal of the bulk modulus is called the compressibility.

Balance of Momentum.

$$\rho\frac{dv}{dt} = -\frac{\partial P}{\partial z} + \frac{2f\rho v^2}{D_h} + \rho g \cos \Phi, \quad\ldots\ldots\ldots\ldots\ldots\ldots\ldots (3.111)$$

where
ρ = fluid density, kg/m^3;
f = Fanning friction factor;
D_h = wellbore diameter, m;
g = gravitational constant, m/s^2;
Φ = angle of inclination from the vertical;
and

$$\frac{d}{dt} = \frac{\partial}{\partial t} + v \frac{\partial}{\partial z} .$$

The balance of momentum equation consists of four terms. The first term in Eq. 3.111 represents the inertia of the fluid [i.e., the acceleration of the fluid (left side of Eq. 3.111) equals the sum of the forces on the fluid (right side of Eq. 3.111)]. The last three terms are the forces on the fluid. The first of these terms is the pressure gradient. The second is the drag on the fluid because of frictional or viscous forces. The friction factor f is a function of the type of fluid and the velocity of the fluid. Frictional drag is discussed in the section on rheology. The last force is the gravitational force.

The balance equations for flow with a pipe in the wellbore are similar to the equations for the openhole model with two important differences. First, the expansivity terms in the balance of mass equations depend on the pressures both inside and outside the pipe. For instance, increased annulus pressure can decrease the cross-sectional area inside the pipe, and increased pipe pressure can increase the cross-sectional area because of pipe elastic deformation. The second major difference is the effect of pipe speed on the frictional pressure drop in the annulus, as discussed in the steady-state surge article. Consult papers on dynamic surge pressures for more detail concerning the wellbore/pipe problem.[4,5]

3.11.3 Borehole Expansion. The balance of mass equation contains a term that relates the flow cross-sectional area to the fluid pressures. This section discusses the application of elasticity theory to the determination of the coefficients in the balance of mass equation. If we assume that the formation outside the wellbore is elastic, then the displacement of the borehole wall because of change in internal pressure is given by the elastic formula.

$$u = \frac{D_h}{2E_f}(1 + v_f)\Delta P, \quad\text{...} (3.112)$$

where
  u = radial displacement, m;
  v_f = Poisson's ratio for the formation;
and
  E_f = Young's modulus for the formation, Pa.
  The cross-sectional area of the annulus is given by

$$A = \pi(\tfrac{1}{2} D + u)^2 . \quad\text{...} (3.113)$$

If we assume u is small compared to D, we can calculate the following formula from Eqs. 3.112 and 3.113.

$$\frac{1}{A}\frac{dA}{dP} = \frac{2(1 + v)}{E_f} . \quad\text{...} (3.114)$$

Using typical values of formation elastic modulus, the borehole expansion term is the same order of magnitude as the fluid compressibility and cannot be neglected.

3.11.4 Solution Method—Fluid Dynamics. The method of characteristics is the method most commonly used to solve the dynamic pressure-flow equations. This method has been extensively used in the analysis of dynamic fluid flow. However, applying the method of characteristics to realistic wellbore flow problems has the following difficulties:

• Iteration may be necessary to solve for characteristics and flow variables when properties and geometry vary in space.

• Multiple coordinate systems must be computed and related to a fixed coordinate system.

• Interpolation is necessary when characteristic curves do not intersect the spatial point of interest.

• Moving coordinate systems must be continually updated so that only points within the fixed-coordinate system are computed.

These difficulties can be reduced or eliminated by using the following approach:

• Adopt a fixed spatial grid.

• For a given time step, integrate the characteristic curves and flow equations from each gridpoint. Note that the flow equations are now evaluated at the new spatial point obtained from the characteristic curves.

• Interpolate the flow equations back to the fixed grid and solve for the flow variables.

This method eliminates the moving coordinate systems and replaces them with a set of interpolation factors. Because the grid is fixed, fluid properties and well geometry are known at each gridpoint, and no iteration is necessary. Most of the equations can be "presolved" so that they only need to be evaluated at each timestep. The disadvantages of this method are that the fluid variables must be evaluated at each gridpoint rather than only at points of interest, and that a maximum timestep size is required for stability.

The characteristic equations are developed using the methods given in Chap. 1 of Ref. 12. For the open hole below the moving pipe, the fluid motion is governed by the system of equations shown in Eq. 3.115.

$$
\begin{vmatrix} 1 & 0 & 0 & C \\ 0 & \rho & 1 & 0 \\ a & 1 & 0 & 0 \\ 0 & 0 & a & 1 \end{vmatrix}
\begin{vmatrix} \partial v / \partial z \\ \partial v / \partial t \\ \partial p / \partial z \\ \partial p / \partial t \end{vmatrix}
=
\begin{vmatrix} 0 \\ h \\ dv / dt \\ dp / dt \end{vmatrix} , \dotfill (3.115)
$$

where the first two equations are the balance of mass, with C equal to the wellbore-fluid compressibility, and the balance of momentum, with friction and gravitation terms lumped together as h.

$$
h = \frac{2 f \rho v^2}{D} + \rho g \cos \Phi . \dotfill (3.116)
$$

The last two equations describe the variation of p and v along the characteristic curve $\xi = z \pm a_t$, where a is the acoustic velocity. Subscripts here denote partial derivatives (e.g., $v_z = \partial v / \partial z$). This system of equations is overdetermined; that is, there are more equations than unknowns. For this system to have a solution, the following condition must hold.

$$
\det \begin{vmatrix} 1 & 0 & 0 & C \\ 0 & \rho & 1 & 0 \\ a & 1 & 0 & 0 \\ 0 & 0 & a & 1 \end{vmatrix} = 0 . \dotfill (3.117)
$$

Evaluating the determinant (Eq. 3.117) defines the acoustic velocity.

$$a^2 = 1/(\rho C) . \qquad (3.118)$$

The second condition that the equations have a solution requires

$$\det \begin{vmatrix} 1 & 0 & 0 & 0 \\ 0 & \rho & 1 & h \\ a & 1 & 0 & dv/dt \\ 0 & 0 & a & dp/dt \end{vmatrix} = 0 . \qquad (3.119)$$

This determinant produces the following differential equations along the characteristic curve.

$$\frac{dp}{dt} \pm \rho a \frac{dv}{dt} = \pm ah . \qquad (3.120)$$

The characteristic equations are solved to give $p(x,t)$ and $v(x,t)$ in the following way. Eq. 3.120 is integrated along the characteristics for time step Δt.

$$p(x, \Delta t) + \rho a v(x, \Delta t) = p(x - a\Delta t, 0) + \rho a v(x - a\Delta t, 0) + \int_0^{\Delta t} ah d\xi = c^+(x, \Delta t), \qquad (3.121)$$

and

$$p(x, \Delta t) - \rho a v(x, \Delta t) = p(x + a\Delta t, 0) - \rho a v(x + a\Delta t, 0) - \int_0^{\Delta t} ah d\xi = c^-(x, \Delta t) . \qquad (3.122)$$

Eqs. 3.121 and 3.122 can be solved simultaneously to give

$$p(x, \Delta t) = (c^+ + c^-)/2, \qquad (3.123)$$

and

$$v(x, \Delta t) = (c^+ - c^-)/(2\rho a) . \qquad (3.124)$$

Generally, c^+ and c^- must be interpolated to give values at the points of interest.[13]

3.12 Cuttings Transport

3.12.1 Introduction. Of the many functions that are performed by the drilling fluid, the most important is to transport cuttings from the bit up the annulus to the surface. If the cuttings cannot be removed from the wellbore, drilling cannot proceed for long. In rotary drilling operations, both the fluid and the rock fragments are moving. The situation is complicated further by the fact that the fluid velocity varies from zero at the wall to a maximum at the center of pipe. In addition, the rotation of the drillpipe imparts centrifugal force on the rock fragments, which affects their relative location in the annulus. Because of the extreme complexity of this flow behavior, drilling personnel have relied primarily on observation and experience for determining the lifting ability of the drilling fluid. In practice, either the flow rate or effective viscosity of the fluid is increased if problems related to inefficient cuttings removal are encountered. This has resulted in a natural tendency toward thick muds and high annular velocities. However, increasing the mud viscosity or flow rate can be detrimental to the cleaning action beneath the bit and cause a reduction in the penetration rate. Thus, there may be a considerable economic penalty associated with the use of a higher flow rate or mud viscosity than necessary. Transport is usually not a problem if the well is near vertical. However, considerable difficulties can occur when the well is being drilled directionally, because cuttings may accumulate either in a stationary bed at hole angles above about 50° or in a moving, churning bed at lower hole angles. Drilling problems that may result include stuck pipe, lost circulation, high torque and drag, and poor cement jobs. The severity of such problems depends on the amount and location of cuttings distributed along the wellbore.

Vertical Wells. The problem of cuttings transport in vertical wells has been studied for many years, with the earliest analysis of the problem being that of Pigott.[14] Several authors have conducted experimental studies of drilling-fluid carrying capacity. Williams and Bruce were among the first to recognize the need for establishing the minimum annular velocity required to lift the cuttings.[15] In 1951, they reported the results of extensive laboratory and field measurements on mud carrying capacity. Before their work, the minimum annular velocity generally used in practice was about 200 ft/min. As a result of their work, a value of about 100 ft/min gradually was accepted. More recent experimental work by Sifferman and Becker indicates that while 100 ft/min may be required when the drilling fluid is water, a minimum annular velocity of 50 ft/min should provide satisfactory cutting transport for a typical drilling mud.[16,17]

The transport efficiency in vertical wells is usually assessed by determining the settling velocity, which is dependent on particle size, density and shape; the drilling fluid rheology and velocity; and the hole/pipe configuration. Several investigators have proposed empirical correlations for estimating the cutting slip velocity experienced during rotary-drilling operations. While these correlations should not be expected to give extremely accurate results for such a complex flow behavior, they do provide valuable insight in the selection of drilling-fluid properties and pump-operating conditions. The correlations of Moore, Chien, and Walker and Mayes have achieved the most widespread acceptance.[18]

Deviated Wells. Since the early 1980s, cuttings transport studies have focused on inclined wellbores. And an extensive body of literature on both experimental and modeling work has developed. Experimental work on cuttings transport in inclined wellbores has been conducted using flow loops at the U. of Tulsa and elsewhere. Different mechanisms, which dominate within different ranges of wellbore angle, determine cuttings bed heights and annular cuttings concentrations as functions of operating parameters (flow rate and penetration rate), wellbore configuration (depth, hole angle, hole size or casing ID, and pipe size), fluid properties (density and rheology), cuttings characteristics (density, size, bed porosity, and angle of repose), and pipe eccentricity and rotary speed.

Laboratory experience indicates that the flow rate, if high enough, will always remove the cuttings for any fluid, hole size, and hole angle. Unfortunately, flow rates high enough to transport cuttings up and out of the annulus effectively cannot be used in many wells because of limited pump capacity and/or high surface or downhole dynamic pressures. This is particularly true for high angles with hole sizes larger than 12¼ in. High rotary speeds and backreaming are often used when flow rate does not suffice.

3.12.2 Particle Slip Velocity. The earliest analytical studies of cuttings transport considered the fall of particles in a stagnant fluid, with the hope that these results could be applied to a moving fluid with some degree of accuracy. Most start with the relation developed by Stokes for creeping flow around a spherical particle.[19]

$$F_d = 3\pi\mu d_s v_{sl}, \quad\quad\quad\quad\quad\quad\quad\quad\quad\quad\quad\quad (3.125)$$

where
  μ = Newtonian viscosity of the fluid, Pa-s;
  d_s = particle diameter, m;
  v_{sl} = particle slip velocity, m/s;
and
  F_d = total viscous drag force on the particle, N.
  When the Stokes drag is equated to the buoyant weight of the particle W,

$$W = \frac{\pi}{6}\left(\rho_s - \rho_f\right)g d_s^3 . \dotfill (3.126)$$

Then, the slip velocity is given by

$$v_{sl} = \frac{d_s^2 g\left(\rho_s - \rho_f\right)}{18\mu}, \dotfill (3.127)$$

where
  ρ_s = solid density, kg/m³;
  ρ_f = fluid density, kg/m³;
and
  g = acceleration of gravity, m/s².
  Stokes' law is accurate as long as turbulent eddies are not present in the particle's wake. The onset of turbulence occurs for

$$\mathrm{Re}_p > 0.1, \dotfill (3.128)$$

where the particle Reynolds number is given by

$$\mathrm{Re}_p = \frac{\rho_f v_{sl} d_s}{\mu} . \dotfill (3.129)$$

For turbulent slip velocities, the drag force is given by

$$F_d = \frac{\pi}{8} f \rho_f v_{sl}^2 d_s, \dotfill (3.130)$$

where f is an empirically determined friction factor. The friction factor is a function of the particle Reynolds number and the shape of the particle given by Ψ, the sphericity. **Table 3.1** gives the sphericity of various particle shapes.
  The friction factor/Reynolds number relationship is shown in **Fig. 3.13** for a range of sphericity. The particle slip velocity for turbulent flow is given by

$$v_{sl} = \frac{2}{3}\sqrt{\frac{3 g d_s\left(\rho_s - \rho_f\right)}{f \rho_f}} . \dotfill (3.131)$$

If we define a laminar friction factor, $f = 24/\mathrm{Re}_p$, then Eq. 3.131 is valid for all Reynolds numbers.

TABLE 3.1—SPHERICITIES OF VARIOUS PARTICLE SHAPES		
Shape	Aspect Ratio	Sphericity, Ψ
Sphere		1.00
Octahedron		0.85
Prism	$L \times L \times L$	0.81
	$L \times L \times 2L$	0.77
	$L \times 2L \times 2L$	0.76
	$L \times 2L \times 3L$	0.73
Cylinder	$H = R/15$	0.25
	$H = R/10$	0.32
	$H = R/3$	0.59
	$H = R$	0.83
	$H = 2 \times R$	0.87
	$H = 3 \times R$	0.96
	$H = 10 \times R$	0.69
	$H = 20 \times R$	0.58

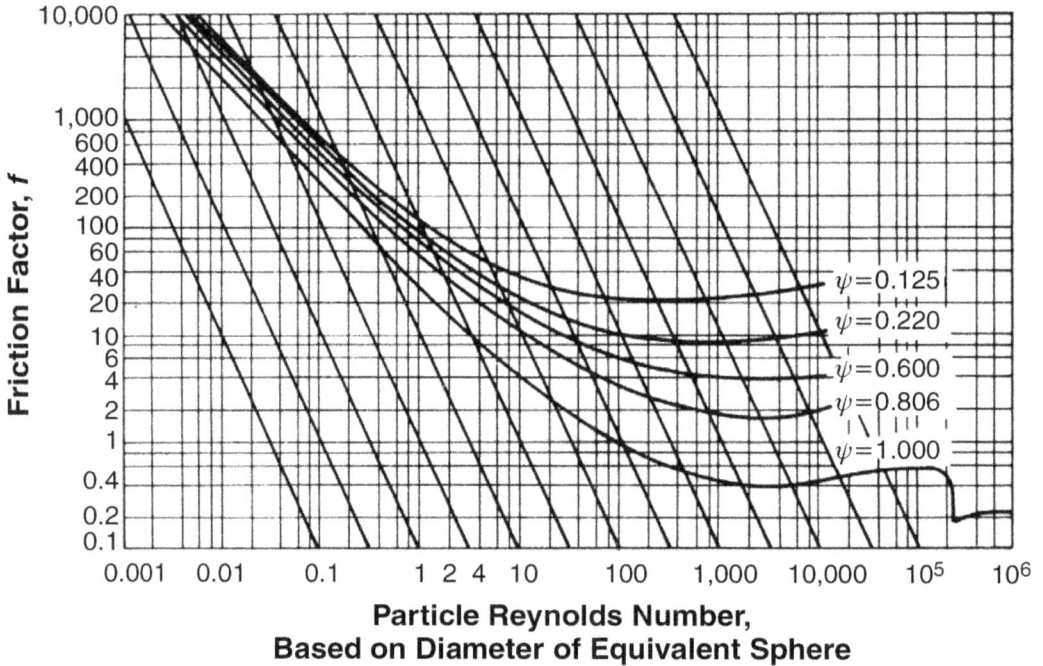

Particle Reynolds Number, Based on Diameter of Equivalent Sphere

Fig. 3.13—Particle slip velocity friction factor (Bourgoyne).[3]

Non-Newtonian fluids introduce new factors into particle-settling calculations. For a Bingham fluid, the particle will remain suspended with no settling if

$$\tau_y \geq \frac{d_s}{6}\left(\rho_s - \rho_f\right), \quad\quad\quad (3.132)$$

where τ_y is the fluid YP. Otherwise, because no other analytic solutions exist, an "apparent" or "equivalent" viscosity is determined from the non-Newtonian fluid parameters. For example, Moore used the apparent viscosity proposed by Dodge and Metzner for a pseudoplastic fluid.[17]

$$\mu_a = \frac{K}{144} \left(\frac{D_o - D_i}{v} \right)^{1-n} \left(\frac{2n+1}{.0208n} \right)^n, \dots\dots\dots\dots\dots\dots\dots (3.133)$$

where
μ_a = apparent viscosity, Pa-s;
K = consistency index for pseudoplastic fluid, Pa-s^n;
n = power law index;
D_o = annulus OD, m;
D_i = annulus ID, m;
and
v = annulus average flow velocity.
Chien determines apparent viscosity for a Bingham plastic fluid shown in Eq. 3.134.[20]

$$\mu_a = \mu_p + 5\frac{\tau_y d_s}{v}, \dots\dots\dots\dots\dots\dots\dots\dots (3.134)$$

where μ_p is the plastic viscosity. The apparent viscosity models with most widespread acceptance are those of Moore.[21]

3.12.3 Carrying Capacity of a Drilling Fluid for Vertical Wells.
The cuttings slip velocity is used to specify the minimum flow rate needed to clean the wellbore. This determination is not as straightforward as one might expect. In rotary-drilling operations, both the fluid and the rock fragments are moving. The situation is complicated further by the fact that the fluid velocity varies from zero at the wall to a maximum at the center of annulus. In addition, the rotation of the drillpipe imparts centrifugal force to the rock fragments, which affects their relative location in the annulus. Because of the extreme complexity of this flow behavior, drilling personnel have relied primarily on observation and experience for determining the lifting ability of the drilling fluid. In practice, either the flow rate or effective viscosity of the fluid is increased if problems related to inefficient cuttings removal are encountered. This has resulted in a natural tendency toward thick muds and high annular velocities. However, increasing the mud viscosity or flow rate can be detrimental to the cleaning action beneath the bit and cause a reduction in the penetration rate. Thus, there may be a considerable economic penalty associated with the use of a higher flow rate or mud viscosity than necessary.

As stated in the previous section, several investigators have proposed empirical correlations for estimating the cutting slip velocity experienced during the drilling process. While these correlations are not extremely accurate, they do give useful qualitative information about the cuttings transport process in vertical wells.

3.12.4 Five Percent Maximum Concentration Model for Vertical Wells.
The following model was taken from Clark and Bickham.[22] For vertical well conditions, **Fig. 3.14** shows a schematic of the cuttings transport process in a YPL fluid under laminar flow conditions. The area open to flow is characterized as a tube instead of an annulus. This simplifies the wellbore geometry. The tube diameter is based on the hydraulic diameter for pressure-drop calculations.

Because drilling mud often exhibits a yield stress, there may be a region, near the center of the cross section, where the shear stress is less than the yield stress. There, the mud will move as a plug (i.e., rigid body motion). The plug velocity is v_p. The average cuttings concentration and velocity in the plug are c_p and v_{cp}, respectively. In the annular region around the plug, the mud flows with a velocity gradient and behaves as a viscous fluid. The average annular velocity of the mud in this region is v_a. In addition, for the cuttings in this region, the average concentration and velocity are C_a and v_{ca}, respectively.

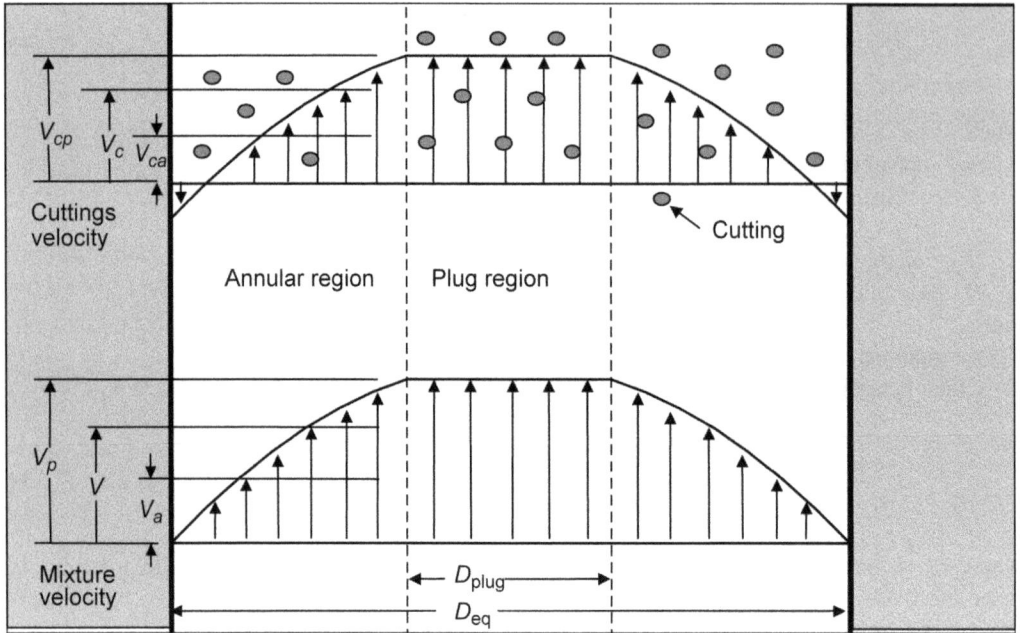

Fig. 3.14—Cuttings velocity profiles: YPL fluid.

Cross-Sectional Geometry. First, let us define the basic wellbore geometry. The hydraulic diameter is defined as four times the flow area divided by the length of the wetted perimeter; namely,

$$D_{\text{hyd}} = \frac{4 \times \text{ cross-sectional area}}{\text{wetted perimeter}} \cdot \quad \dots\dots\dots\dots\dots\dots\dots\dots\dots\dots\dots\dots \text{(3.135)}$$

For the wellbore annulus, the hydraulic diameter of the wellbore cross section is

$$D_{\text{hyd}} = D_h - D_p, \quad \dots\dots\dots\dots\dots\dots\dots\dots\dots\dots\dots\dots\dots\dots\dots \text{(3.136)}$$

where D_h is the wellbore diameter, and D_p is the drillpipe OD. The equivalent diameter is defined as

$$D_{\text{eq}} = \sqrt{4A/\pi}, \quad \dots\dots\dots\dots\dots\dots\dots\dots\dots\dots\dots\dots\dots\dots\dots \text{(3.137)}$$

where A is the area open to flow. For the wellbore annulus, the equivalent diameter is

$$D_{\text{eq}} = \sqrt{D_h^2 - D_P^2}, \quad \dots\dots\dots\dots\dots\dots\dots\dots\dots\dots\dots\dots\dots\dots \text{(3.138)}$$

The plug diameter ratio is

$$\lambda_p = D_{\text{plug}}/D_{\text{eq}} \cdot \quad \dots\dots\dots\dots\dots\dots\dots\dots\dots\dots\dots\dots\dots\dots\dots \text{(3.139)}$$

Flow Conditions. The mixture velocity is

$$v_{mix} = \frac{Q_c + Q_m}{A}, \quad \text{...} \quad (3.140)$$

where Q_m is the volumetric flow rate of the mud and Q_c is the volumetric flow rate of the cuttings, which depends on the bit size and the penetration rate. In addition, the mixture velocity can be calculated from the average plug and annulus velocities in the equivalent pipe; namely,

$$v_{mix} = v_a(1 - \lambda_P^2) + v_p\lambda_P^2 . \quad \text{.....................................} \quad (3.141)$$

Cuttings Concentration. The feed concentration is defined as

$$c_o = \frac{Q_c}{Q_c + Q_m} . \quad \text{...} \quad (3.142)$$

The average concentration, c, of cuttings in a short segment with length, Δz, and cross-sectional area, A, can be calculated as

$$c = c_a(1 - \lambda_P^2) + c_p\lambda_P^2 . \quad \text{.....................................} \quad (3.143)$$

The cuttings concentrations in the plug and annular regions are assumed equal. This means that the suspended cuttings are uniformly distributed across the area open to flow. Obviously, this assumption has a major impact, and the actual distribution is probably a function of wellbore geometry, mud properties, cuttings properties, and operating conditions. Thus, we obtain

$$v_{mix} = \frac{cv_s(1 - c)}{c - c_o}, \quad \text{...} \quad (3.144)$$

where

$$v_s = v_{sa}(1 - \lambda_P^2) - v_{sp}\lambda_P^2 . \quad \text{...} \quad (3.145)$$

is the average settling velocity in the axial direction. The components of the settling velocities in the axial direction are

$$v_{sa} = v'_{sa}(1 - c)^n . \quad \text{...} \quad (3.146)$$

and

$$v_{sp} = v'_{sp}(1 - Y_a^{0.94}), \quad \text{...} \quad (3.147)$$

where

$$v'_{sa} = \sqrt{\frac{4d_sg(\rho_s - \rho)}{3\rho C_D}},$$

$$n = \exp(0.0811y - 1.19),$$

$$y = - \text{sgn} \, (x)(0.0001 + 0.865 \mid x \mid^{-6})^{-1/6},$$
$$x = 1.24 \ln (\text{Re}_p) - 4.59,$$

$$v_{sp}' = \cos \alpha \sqrt{\frac{4}{\rho C_D} \left\{ \frac{d_s g(\rho_s - \rho)}{3} - \pi \tau_y \right\}},$$

$$\text{Re}_p = \frac{d_s v \rho}{\mu_a},$$

and

$$Y_a = \frac{3\tau_y}{d_s g(\rho_s - \rho)} . \quad\dotfill (3.148)$$

C_D is the drag coefficient of a sphere, τ_y is the yield stress of the mud, and μ_a is the apparent viscosity of the mud at a shear rate resulting from the settling cutting.

The value calculated using Eq. 3.144 is the minimum acceptable mixture velocity required for a cuttings concentration, c. Pigott[14] recommended that the concentration of suspended cuttings be a value less than 5%. With this limit ($c = 0.05$), Eq. 3.144 becomes

$$v_{\text{mix}} = \frac{0.0475 v_s}{0.05 - c_o}, \quad\dotfill (3.149)$$

where $c_o < 0.05$. This implies that the penetration rate must be limited to a rate that satisfies this equality.

For near-vertical cases, the critical mud-cuttings mixture velocity equals the value of Eq. 3.149. If the circulation rate exceeds this value, the suspended cuttings concentration will remain less than 5%. However, if the mud circulation velocity is less than the cuttings' settling velocity, the cuttings will eventually build up in the wellbore and plug it.

3.12.5 Cuttings Transport in Deviated Wells. A comprehensive cuttings transport model should allow a complete analysis for the entire well, from surface to the bit. The different mechanisms which dominate within different ranges of wellbore angle should be used to predict cuttings bed heights and annular cuttings concentrations as functions of operating parameters (flow rate and penetration rate), wellbore configuration (depth, hole angle, hole size or casing ID, and pipe size), fluid properties (density and rheology), cuttings characteristics (density, size, bed porosity, and angle of repose), pipe eccentricity, and rotary speed. Because of the complexity, extensive experimental data were necessary to help formulate and validate the new cuttings transport models.

New Experimental Data. Large-scale cuttings transport studies in inclined wellbores were initiated at the Tulsa U. Drilling Research Projects (TUDRP) in the 1980s with the support of major oil and service companies. A flow loop was built that consisted of a 40-ft length of 5-in. transparent annular test section and the means to vary and control
- The angles of inclination between vertical and horizontal.
- Mud pumping flow rate.
- Drilling rate.
- Drillpipe rotation and eccentricity.

Tomren *et al.*[23] found marked difference between the cuttings transport in inclined wellbores and that of vertical wellbores. A cuttings bed was observed to form at inclination angles of more than 35° from vertical, and this bed could slide back down for angles up to 50°. Eccentricity, created by the drillpipe lying on the low side of the annulus, was found to worsen the situation. Analysis of annular fluid flow showed that eccentricity diverts most of the mud flow away from the low side of the annulus, where the cuttings tend to settle, to the more open area above the drillpipe. Okrajni and Azar[24] investigated the effect of mud rheology on hole cleaning. They observed that removing a cuttings bed with a high-viscosity mud, a remedy for the hole-cleaning problem in vertical wells, may in fact be detrimental in high-angle wellbores (assuming a zero to low drillpipe rotation) and that a low-viscosity mud that can promote turbulence is more helpful. On the basis of this finding and on the previous study, hole cleaning was found to depend on the angle of inclination, hydraulics, mud rheological properties, drillpipe eccentricity, and rate of penetration. Becker *et al.*[25] then showed that the cuttings transport performance of the muds tested correlated best with the low-end-shear-rate viscosity, particularly the 6-rpm Fann V-G viscometer dial readings.

By the mid-1980s, a general qualitative understanding of the hole-cleaning problem in highly inclined wellbores had been gained. Because more directional and horizontal wells with longer lateral reaches were being drilled, the need for more and new experimental data created a demand for additional flow loops. In partnership with Chevron, Conoco, Elf Aquitaine, and Philips, TUDRP built a new and larger flow loop, with a 100-ft-long test section of 8-in. annulus, while construction of new flow loops was also done at Heriot-Watt U., BP, Southwest Research, M.I. Drilling Fluids, and the Inst. Français du Pétrole. All the flow loops had a transparent part of the annular test section that allowed observation of the cuttings transport mechanism. These flow loops provided the necessary tools for collecting the badly needed experimental data.

Because of the new flow loops, a significant amount of experimental data was collected on the effect of different parameters on cuttings transport under various conditions. The observations made and subsequent analysis of the data collected provided the basis for work toward formulating correlations/models.

Larsen conducted extensive studies on cuttings transport, totaling more than 700 tests with the TUDRP's 5-in. flow loop. Tests were performed for angles from vertical to horizontal under critical as well as subcritical flow conditions. Critical flow corresponds to the minimum annular average fluid velocity that would prevent stationary accumulation of cuttings bed. Subcritical flow refers to the condition where a stationary cuttings bed forms. Analysis of the experimental data shows that when the fluid velocity is below the critical value, a cuttings bed starts to form and grows in thickness until the fluid velocity above the bed reaches the critical value. The critical velocity was reported in the range of 3 to 4 ft/sec, depending on the value of various parameters, such as the mud rheology, drilling rate, pipe eccentricity, and rotational speed. There were several new findings:

- Under subcritical flow conditions, a medium-rheology mud (PV = 14 and YP = 14) consistently resulted in slightly smaller cuttings beds than those obtained with the low-rheology (PV = 7 and YP = 7) or the high-rheology (PV = 21 and YP = 21) muds. Calculation of the Reynolds number for the tests suggests that the flow regime for this mud is neither turbulent nor laminar but in the transition range.

- The small cuttings size used (0.1 in.) in the study was more difficult to clean than the medium (0.175 in.) and the large (0.275 in.) sizes (drillpipe rpm 0 to 50). The small cuttings formed a more packed and smooth bed.

- The height of the cuttings bed between 55 and 90° remained about the same, but there was a slight increase at about 65 to 70°.

- Significant backsliding of the cuttings bed was observed for angles from 35 to 55°.

Seeberger *et al.*[26] reported that elevating the low shear rate viscosities enhances the cuttings-transport performance of oil muds. Sifferman and Becker[17] conducted a series of hole-cleaning experiments in an 8-in. flow loop. Statistical analysis of the data showed interaction among various parameters; thus, simple relationships could not be derived. For example, the effect of drillpipe rotation on cuttings transport depended also on the size of the cuttings and the mud rheology. The effect of rotation was more pronounced for smaller particles and for more viscous muds. Bassal[27] completed a study of the effect of drillpipe rotation on cuttings transport in inclined wellbores. The variables considered in this work were drillpipe rotary speed, hole inclination, mud rheology, cuttings size, and mud flow rate. Results have shown that drillpipe rotation has a significant effect on hole cleaning in directional well drilling. The level of enhancement in cuttings removal as a result of rotary speed is a function of a combination of mud rheology, cuttings size, mud flow rate, and the manner in which the drillstring behaves dynamically.

New Cuttings Transport Models. Larsen *et al.*[28] developed a model for highly inclined (50 to 90° angle) wellbores. The model predicts the critical velocity as well as the cuttings-bed thickness when the flow rate is below that of the critical flow. Hemphill and Larsen[29] showed that oil-based muds with comparable rheological properties performed about the same. Jalukar *et al.*[30] modified this model with a scaleup factor to correlate with the data obtained with the 8-in. TUDRP flow loop.

Zamora and Hanson,[31] on the basis of laboratory observations and field experience, compiled 28 rules of thumb to improve high-angle hole cleaning. Luo and Bern[32] presented charts to determine hole-cleaning requirements in deviated wells. These empirical charts were developed on the basis of the data collected with the BP 8-in. flow loop, and they predicted the critical flow rates required for prevention of cuttings-bed accumulation. The predictions have also been compared with some field data.

Mechanistic Modeling. The existing cuttings-transport correlations and/or models have a few empirical coefficients, determined based on laboratory and/or field data. There is a need for developing comprehensive cuttings transport mechanistic models that can be verified with experimental data. Different levels of the mechanistic approach are possible and can be built on gradually. Ideally, a fluid/solids interaction model, which would be coupled and integrated with a fluid-flow model to simulate the whole cuttings-transport process, is needed. Campos *et al.*[33] recently made such an attempt, but much more work is needed to develop a comprehensive solids/liquid flow model.

Ford *et al.*[34] published a model for the prediction of minimum transport velocity for two modes: cuttings suspension and cuttings rolling. The predictions were compared with laboratory data.

Gavignet and Sobey[35] presented a cuttings transport model based on physical phenomena, similar to that published by Wilson,[36] for slurry flow in pipelines that is known as the double-layer model. The model has many interrelated equations and a substantial number of parameters, a few of which are difficult to determine. Martin *et al.*[37] developed a numerical correlation based on the cuttings-transport data that they had collected in the laboratory and in the field.

Clark and Bickham[22] presented a cuttings-transport model based on fluid mechanics relationships, in which they assumed three cuttings-transport modes: settling, lifting, and rolling—each dominant within a certain range of wellbore angles. Predictions of the model were compared with critical and subcritical flow data they had collected with the TUDRP's 5- and 8-in.flow loops. A prediction of the model was also used to examine several situations in which poor cuttings transport had been responsible for drilling problems.

Campos *et al.*[33] developed a mechanistic model for predicting the critical velocity as well as the cuttings-bed height for subcritical flow conditions. Their work was based on earlier work by Oraskar and Whitmore[38] for slurry transport in pipes. The model's predictions are good for thin muds, but the model needs to be further refined to account for thick muds and pipe rotation.

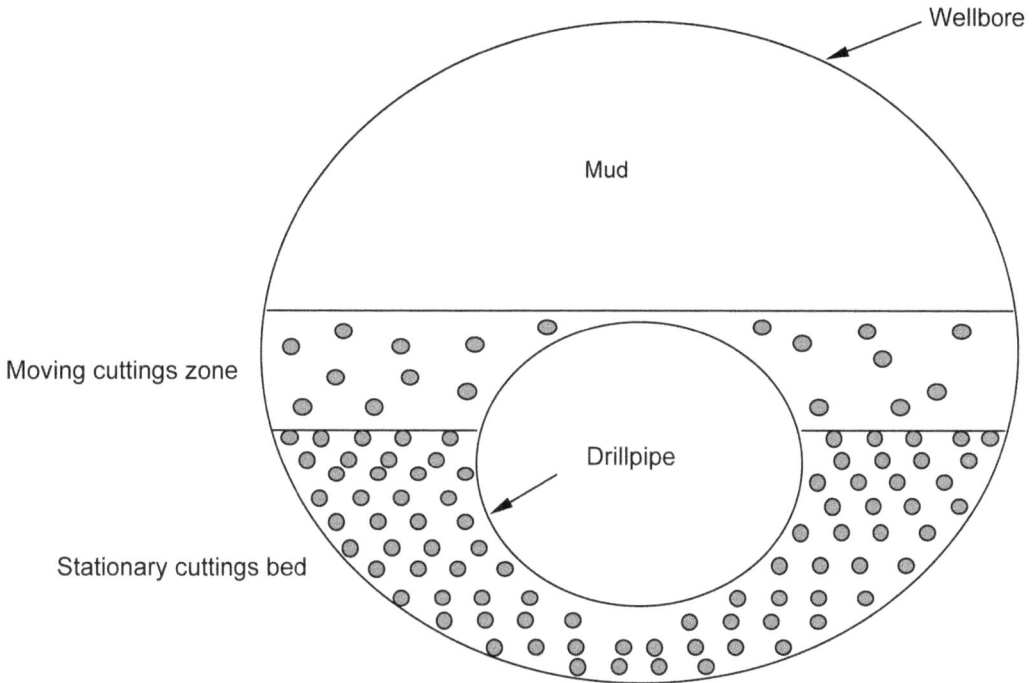

Fig. 3.15—Wellbore cross section with a cuttings bed.

Kenny *et al.*[39] defined a lift factor that they used as an indicator of cuttings-transport performance. The lift factor is a combination of the fluid velocity in the lower part of the annulus and the mud-settling velocity determined by Chien's correlation.[20]

Fig. 3.15 illustrates the basic flow configuration for mechanistic cuttings transport modeling. There are three distinct zones in this model: stationary cuttings bed, moving cuttings zone, and "cuttings free" mud-flow zone.

The cuttings-free mud flow creates a shear force at the interface with the moving cuttings bed, which drags the moving cuttings zone along with it. In the moving cuttings zone, gravity forces tend to make the cuttings fall onto the fixed cuttings bed, while aerodynamic and gel forces tend to keep the cuttings suspended. At the interface between the moving cuttings zone and the stationary cuttings bed, fluid friction is trying to strip off cuttings, which are held by gravity and cohesive forces. The balance of these forces determines whether the cuttings bed increases or decreases in depth. The critical flow rate for cuttings transport leaves the cuttings bed unchanged. For effective hole cleaning, the desired flow rate exceeds the critical flow rate.

Field Application. When the results of cuttings transport research and field experience are integrated into a drilling program, hole-cleaning problems are avoided, and excellent drilling performance follows. This has certainly been the case when engineers achieved two new world records in extended-reach drilling.

Guild and Hill[40] presented another example of integration of hole-cleaning research into field practice. They reported trouble-free drilling in two extended-reach wells after they lost one well because of poor hole cleaning. Their program was designed to maximize the footage drilled between wiper trips and eliminate hole-cleaning backreaming trips before reaching the casing point. They devised a creative way to avoid significant cuttings accumulation by carefully monitoring the pickup weight, rotating weight, and slackoff weight as drilling continued. They observed that cuttings accumulation in the hole caused the difference between the pickup weight and the slackoff weight to keep increasing, while cleaning the hole decreased the difference. By observing the changes in these parameters and by the use of other readily available information, they were able to closely monitor hole cleaning and control the situation.

3.12.6 Air, Mist, and Foam Drilling. Air and mist drilling have several advantages over conventional drilling fluids. The principle advantages are higher penetration rates, longer bit life, and no lost-circulation problems. The usual disadvantages are control of fluid influx and control of high-pressure zones.

To realize these advantages, it is important to maintain adequate circulation. Determining the required volume flow rate to maintain this "adequate" circulation has always been difficult. The best available technique has been the chart developed by R.R. Angel.[41] This chart allows the estimation of volume circulation rates for various hole sizes, drillpipe sizes, and penetration rates,

One difficulty with Angel's result is that the equation giving the volume flow rate must be solved by trial and error. This difficulty is avoided by using the charts prepared by Angel, provided the case of interest is tabulated or can be estimated from similar cases. A second difficulty is that the drill cuttings are assumed to travel at the same velocity as the air. Angel notes that this is not a conservative assumption, and the analysis by Mitchell[42] demonstrates that the predicted flow rates are 20 to 30% too low. The downhole temperatures used for Angel's chart are assumed to be 80°F at the surface, increasing 1°F per 100 ft of depth. There is no convenient way to convert to other temperatures. A final consideration is that the Angel charts do not apply to mist drilling. The addition of water to the air requires increases in both the volume flow rate and standpipe pressures to maintain the same penetration rate.

Compressible Flow. The flow of a compressible fluid can often produce results that seem to go counter to common sense. For instance, consider the steady flow of air in a constant area duct. As with all fluids, there is a pressure loss because of friction, and the pressure decreases continuously from the entrance of the duct to the exit. Unlike the flow of incompressible fluids, the fluid velocity increases from the entrance of the duct to the exit. How could friction make the fluid speed up?

Two facts account for this acceleration. First, the gas pressure is proportional to the density (as in the ideal gas law $P = \rho RT$). As the pressure of the gas decreases, the density must decrease also. Second, because the mass flow through the duct is constant, the product of density and velocity is constant. Thus, as the density decreases with the pressure, the velocity must increase to maintain the mass flow.

This example demonstrates a typical compressible flow characteristic—the interrelationship of pressure and mass flow. In air drilling, high velocities are needed at bottomhole to remove the cuttings. High velocities result in friction-pressure drops in the drillpipe and annulus, so higher standpipe pressures may be needed to keep the air flowing. Higher standpipe pressures result in higher gas densities, and, hence, result in lower velocities. Fortunately, most air-drilling operations do not result in the vicious circle situation previously described.

Cuttings Transport and Mist Flow in Vertical Wells. The addition of the effect of cuttings and mist to the equations already developed require two changes. First, the effect of the cuttings and mist on momentum equation must be accounted for; and second, the forces exerted on the cuttings and mist must be determined. The principles of multiphase flow can be applied to both of these effects.

Two basic ideas are sufficient to develop the modified momentum equation. First, the mass flow rate of the cuttings is easy to determine; it is the product of the penetration rate, the hole area, and the density of the rock. Assuming that the cuttings velocity is known, a "density" for the cuttings mass flow rate can be determined.

$$\bar{\rho}_s V_s A = \dot{m}_s = G_s A . \quad\text{...} \quad (3.150)$$

This density represents the total mass of cuttings in a volume of the duct divided by the volume of the duct. The ratio of this density to the actual density of the rock is the volume fraction of the cuttings,

$$h = \bar{\rho}_s / \rho_s . \quad\text{...} \quad (3.151)$$

The remainder of the volume is filled by the air, with an air in-mixture density defined as

$$\bar{\rho} = (1 - h)\rho . \quad\text{...} \quad (3.152)$$

With these definitions, the cuttings transport equivalents to the single-phase flow equations can be written as

$$\bar{\rho} V A = \dot{m} = GA, \quad\text{..} \quad (3.153)$$

$$\Delta P + G\Delta V + G_s \Delta V_s + F + W = 0, \quad\text{...} \quad (3.154)$$

$$F = \frac{2f}{D} \Big|_{\Delta z} (GV + G_s V_s) dz, \quad\text{..} \quad (3.155)$$

and

$$W = \Big|_{\Delta z} (\bar{\rho} + \bar{\rho}_s) g \, \cos \, \Phi dz . \quad\text{..} \quad (3.156)$$

Note that G and G_s are constant. The final missing piece is the relationship between the velocity of the air and the velocity of the cuttings. There is a large body of literature on the data necessary to determine this relationship. For example, in the petroleum engineering literature, there is the work of Gray.[43] There is also a large amount of literature on terminal settling velocities for solid particles (see Ref. 3, Chap. 1, Sec. 3, pages 172–174). The equation is given by Gray as Eq. 15 of the Appendix to Ref. 43. Rewritten in terms of flow variables previously defined, this equation becomes

$$G_s \Delta v_s + W_s - \hat{P} = 0, \quad\text{..} \quad (3.157)$$

where

$$W_s = \Big|_{\Delta z} (\rho_s - \rho) H_s g \, \cos \, \Phi dz, \quad\text{..} \quad (3.158)$$

and

$$\hat{P} = \frac{C_D}{\delta} \Big|_{\Delta z} \rho (V - V_s)^2 dz . \quad\text{..} \quad (3.159)$$

The term W_s is the buoyant weight of the cuttings. The term P is the aerodynamic force exerted on the cuttings by the air, with C_D the drag coefficient and δ the ratio of the average particle volume to its cross-sectional area. Values of C_D can be found for various types of rock in Ref. 3, pages 172–174. The term δ was evaluated for an average cutting diameter of $^3/_8$ in. This size is considered to be typical of cuttings at the bit. Higher up the hole, these cuttings get broken into smaller pieces. Because there is no way of predicting the change in average particle size as the cuttings move up the annulus, the average diameter is held fixed at $^3/_8$ in. This assumption causes the model to overpredict the relative velocity between the air and the cuttings. This assumption is conservative because higher air velocities are now needed to lift the cuttings. The assumption used by Angel[41] is that the particle velocity and the air velocity are equal, and he notes that this is not conservative.

The addition of mist to the flowing equations is much simpler than adding the cuttings. The water droplets in a mist are very small, and, as a result, the relative velocity between the air and the mist droplets is small. The usual assumption used in two-phase flow analysis is that the air and mist move at the same velocity, and simulations using Eq. 3.22 verify this. Eqs. 3.17 through 3.19 are suitable to model mist flow with the following changes: the mass flow and density of the mist replace those for the cuttings, and the velocities of the mist and the air are set equal.

This cuttings model predicts higher-volume flow rates than Angel's model, which was expected because of the conservative nature of the cuttings model. The cuttings model also shows, however, that the flow rates specified by Angel are adequate to clean the bole, even though they do not satisfy the 3,000 ft/min requirement. The predicted temperatures are reasonably near the undisturbed geothermal temperature, which justifies the temperature assumptions used by Angel.

Foams. Foams are being used in a number of petroleum industry applications that exploit their high viscosity and low liquid content. Some of the earliest applications for foam dealt with its use as a displacing agent in porous media and as a drilling fluid. Following these early applications, foam was introduced as a wellbore circulating fluid for cleanout and workover applications. In the mid-1970s, N_2-based foams became popular for both hydraulic fracturing and fracture-acidizing stimulation treatments. In the late 1970s and early 1980s, foamed cementing became a viable service, as did foamed gravel packing. Most recently, CO_2 foams have been found to exhibit their usefulness in hydraulic fracturing stimulation.

Regardless of why they are applied, these compressible foams are structured, two-phase fluids that are formed when a large internal phase volume (typically 55 to 95%) is dispersed as small discrete entities through a continuous liquid phase. Under typical formation temperatures of 90°F (32.2°C) encountered in stimulation work, the internal phases N_2 or CO_2 exist as a gas and, hence, are properly termed foams in their end-use application. However, the formations of such fluids at typical surface conditions of 75°F (23.9°C) and 900 psi [6205 kPa] produce N_2 as a gas but CO_2 as a liquid. A liquid/liquid two-phase structured fluid is classically called an emulsion. The end-use application of the two-phase fluid, however, normally is above the critical temperature of CO_2 at which only a gas can exist, so we consider the fluids together as foams. The liquid phase typically contains a surfactant and/or other stabilizers to minimize phase separation (or bubble coalescence).

These dispersions of an internal phase within a liquid can be treated as homogeneous fluids, provided bubble size is small in comparison to flow geometry dimensions. Volume percent of the internal phase within a foam is its quality. The degree of internal phase dispersion is its texture. At a fixed quality, foams are commonly referred to as either fine or coarse textured. Fine texture denotes a high level of dispersion characterized by many small bubbles with a narrow size distribution and a high specific surface area, and coarse texture denotes larger bubbles with a broad size distribution and a lower specific surface area.

Because foams exhibit shear-rate-dependent viscosities in laminar flow, they are classified as non-Newtonian fluids. In addition to shear rate, their apparent viscosities also appear to be dependent on quality, texture, and liquid-phase rheological properties. Measured laminar-flow apparent viscosities generally are larger than those of either constituent phase at all shear rates. When the liquid phase is thickened by addition of solids, soluble high-molecular-weight polymers, or other viscosifying agents, we see production of even larger foam viscosities. While laminar flow is characterized by strictly viscous energy dissipation, turbulent flow is characterized more by kinetic than viscous energy dissipation. Density and velocity are the factors that establish kinetic energy, and reduced foam density may outweigh an increased viscosity contribution and produce a turbulent-flow friction loss less than liquid-phase friction loss. Soluble high-molecular-weight polymers produce a form of turbulent drag reduction that is analogous to that which occurs in a nonfoamed liquid. In this case, a substantial drag-reduction effect is evident when one compares the turbulent-flow friction loss of foams with and without a gelled liquid phase.

Interactions between forces caused by surface tension, viscosity, inertia, and buoyancy produce a variety of effects observable in foams. These effects include different bubble shapes and sizes. Anomalous effects have been attributed to slippage as well as bubble size or texture. Buoyancy and inertia forces act on the foam and tend to destroy the discrete bubble structure, which makes the foam dynamically unstable. However, when work is performed on foam, as is the case when foam flows in a pipe, the bubble structure is being destroyed dynamically and then rebuilt, making the foam macroscopically act as a homogeneous fluid.

Beyer et al.[44] developed foam flow equations from data collected in horizontal pipes. They observed slippage, applied Mooney's method for flow data correction, and correlated the data with a Bingham plastic flow model. Blauer et al.[45] concluded that foam behaves as a Bingham plastic without slippage in laminar flow. They equated the Buckingham-Reiner equation to the Hagen-Poiseuille equation to determine an expression for effective viscosity for use in conventional Newtonian fluid laminar and turbulent-flow friction-loss relationships.[8] A critical Reynold's number of 2,100 was used to denote transition from laminar to turbulent flow. To the best of our knowledge, they present the only experimental turbulent foam flow data in the literature. Sanghani and Ikoku[46] studied the rheological properties of foam flowing in an annulus and concluded that foam rheological behavior was best represented as a pseudoplastic fluid. They also stated that their data could be represented by a Bingham plastic model and a yield pseudoplastic model without large errors.

Earlier investigators noted drag-reduction effects in turbulent two-phase flows when drag-reducing additives were introduced into the liquid phase. These investigators, however, did not deal directly with foam flow but with such diverse two-phase flow regimes as slug, plug, and annular mist.

The importance of foam rheological properties has been recognized by investigators; however, very little agreement exists among them. Foam has been characterized as a Bingham plastic, a pseudoplastic, and a yield pseudoplastic. Slippage has been observed in some, but not all, cases. Unexplained anomalous effects were observed in many cases. Bubble size and shape have been considered and neglected. All these vastly different observations indicate that foam is a very complex fluid that could exhibit a number of characteristics. All investigators agree, however, that a rheological dependence on quality exists. That foams in general exhibit a yield stress is also well supported. Investigations have been conducted primarily with water as the liquid phase.

3.12.7 Summary Guidelines for Efficient Hole Cleaning. Based on the results of many laboratories' research and various field experiences and observations, the following general guidelines are recommended.
• Design the well path so that it avoids critical angles, if possible.

• Use top-drive rigs, if possible, to allow pipe rotation while tripping.
• Maximize fluid velocity, while avoiding hole erosion, by increasing pumping power and/or using large-diameter drillpipes and drill collars.
• Design the mud rheology so that it enhances turbulence in the inclined/horizontal sections while maintaining sufficient suspension properties in the vertical section.
• In large-diameter horizontal wellbores, where turbulent flow is not practical, use muds with high-suspension properties and muds with high meter-dial readings at low shear rates.
• Select bits, stabilizers, and bottomhole assemblies (BHAs) with minimum cross-sectional areas to minimize plowing of cuttings while tripping.
• Use various hole-cleaning monitoring techniques including a drilled cuttings retrieval rate, a drilled cuttings physical appearance, pressure while drilling, and a comparison of pickup weight, slackoff weight, and rotating weight.
• Perform wiper trips as the hole condition dictates.

3.13 Sample Calculations

3.13.1 Introduction. The most important consideration in making hydraulic calculations is the use of consistent units. Unfortunately, oilfield units are rarely consistent; in some cases they are unique to the industry. The universal set of consistent units is the SI Metric System of Units. The Society of Petroleum Engineers (SPE) has available a publication: "The SI Metric System of Units and SPE Metric Standard" that contains every conversion factor necessary. Whenever there is a question of units, the safest solution is to convert all units to SI units, solve the problem, and then convert the answer back to the common engineering units.

Sample Problem. Geometry. A deviated well kicks off at 3,000 ft and is drilled to total depth (TD) at an angle of 30° to the vertical. The well's total measured depth is 11,000 ft. The well is cased with 72-ppf 13⅜-in. casing (13.375 × 12.347 in.) set at 3,000 ft. The drillstring consists of 900 ft of 8-in. 147-ppf drill collars (8 × 3 in.), 19½-ppf drillpipe (5 × 4.206 in.), a 9⅝-in. bit with 3 × ¹³⁄₃₂-in. nozzles. The undisturbed temperature is 70°F at the surface with a 1.4°F/100-ft gradient. We will neglect the build section and assume the well trajectory is vertical to 3,000 ft measured depth, and deviated at 30° to the vertical from 3,000 ft measured depth to 11,000 ft measured depth. We will assume the open hole is gauge (9.625 in.).

True Vertical Depth. For measured depth < 3,000 ft, $Z = z$.

For measured depth > 3,000 ft, $Z = 3,000$ ft $+ (z - 3,000) \sin(30) \cong 402 + 0.866z$, where z is measured depth in feet, Z is true vertical depth in feet.

Hydrostatic Pressure. 1. Assume the wellbore is filled with 8.34 lbm/gal fluid (fresh water). What is the pressure at TD? True vertical depth at TD is $402 + 0.866 \times 11,000 = 9,928$ ft. Using Eq. 3.10 and converting to SI units:

$$\rho = 8.34 \text{ lbm/gal} \times 119.8264 (\text{kg/m}^3)/(\text{lbm/gal}) = 999.4 \text{ kg/m}^3 .$$
$$\Delta Z = 9,928 \text{ ft} \times 0.3048 \text{ m/ft} = 3026 \text{ m} .$$
$$\Delta P = \rho g \Delta Z = 999.4 \text{ kg/m}^3 \times 9.80665 \text{ m/s}^2 \times 3026 \text{ m} = 29.66 \times 10^6 \text{ Pa}$$
$$= 29.66 \times 10^6 \text{ Pa}/(6,894.757 \text{ Pa/psi}) = 4301.4 \text{ psi} .$$

This pressure is gauge pressure at TD. For absolute pressure, add atmospheric pressure, 14.7 psi:

$$P_{TD} = 4,301.4 \text{ psi} + 14.7 \text{ psi} = 4,316.1 \text{ psi} .$$

2. Assume a layered wellbore with 14 lbm/gal mud from surface to 5,000 ft (measured depth) and 9 lbm/gal mud from 5,000 ft to TD. What is the pressure at TD?

For layer 1:

$$\rho = 14 \ \text{lbm/gal} \times 119.8264(\text{kg/m}^3)/(\text{lbm/gal}) = 1677.6 \ \text{kg/m}^3 \ .$$
$$\Delta Z = (402 + 0.866 \times 5,000) \ \text{ft} \ \times 0.3048 \ \text{m/ft} = 1,442.3 \ \text{m} \ .$$
$$\Delta P1 = \rho g \Delta Z = 1677.6 \ \text{kg/m}^3 \times 9.80665 \ \text{m/s}^2 \times 1442.3 \ \text{m} \ = 23.728 \times 10^6 \ \text{Pa}$$
$$= 23.728 \times 10^6 \ \text{Pa}/(6,894.757 \ \text{Pa/psi}) = 3,441.5 \ \text{psi} \ .$$

For layer 2:

$$\rho = 9 \ \text{lbm/gal} \times 119.8264(\text{kg/m}^3)/(\text{lbm/gal}) = 1078.4 \ \text{kg/m}^3 \ .$$
$$\Delta Z = 3026 \ \text{m} - 1,442.3 \ \text{m} = 1583.7 \ \text{m}$$
$$\Delta P2 = \rho g \Delta Z = 1078.4 \ \text{kg/m}^3 \times 9.80665 \ \text{m/s}^2 \times 1583.7 \ \text{m} = 16.748 \times 10^6 \ \text{Pa}$$
$$= 16.748 \times 10^6 \ \text{Pa}/(6,894.757 \ \text{Pa/psi}) = 2,429.2 \ \text{psi} \ .$$
$$P_{TD} = 14.7 \ \text{psi} \ (\text{surface}) + 3,441.5 \ \text{psi} \ (\text{layer 1}) + 2,429.2 \ \text{psi} \ (\text{layer 2}) = 5,885.4 \ \text{psi} \ .$$

3. Assume the wellbore is filled with nitrogen with a surface pressure of 2,000 psi. What is the pressure at TD? This problem is much more difficult because the gas density and temperature vary over the length of the wellbore. The pressure change is given by

$$\Delta P(z) = P_o \left[\exp \left(\frac{g \cos \Phi}{R} \int_{\Delta z} \frac{1}{T(\varsigma)} d\varsigma \right) - 1 \right] .$$

The temperature distribution is given by $T(Z) = 70 + .014 \ Z$, where Z is true vertical depth. Because we need absolute temperature, in Kelvin: $(T \ °F + 459.67)/1.8 = T \ °K$,

$$T(Z) = 294.3 \ \text{K} + 0.025518 \ \text{K/m} \ Z(\text{meters}) \ .$$

The integral of $1/T$ with respect to Z is

$$\ln \left[(294.4 \ \text{K} + 0.025518 \ \text{K/m} \ \times 3026 \ \text{m})/(294.4 \text{K}) \right]/0.025518 \ \text{K/m} \right] = 9.128 \ \text{m/K},$$
$$\Delta P = 2,000 \ \text{psi} [\exp (9.80665 \ \text{m/s}^2 \times 9.128 \ \text{m/K} \ /296.8 \ \text{m}^2/\text{s}^2 - \text{K}) - 1]$$
$$= 2000 \ \text{psi} \ \times 0.3520 = 704 \ \text{psi},$$

and

$$P_{TD} = 2,000 \ \text{psi} \ + 704 \ \text{psi} \ = 2704 \ \text{psi} \ .$$

Frictional Pressure Loss. 4. Assume fresh water is being circulated at 600 gal/min. What is the pressure change inside a single vertical 30-ft joint of drillpipe? Assume the density is 8.34 lbm/gal and the viscosity is 1 cp.

The pipe flow area $= A_i = \frac{1}{4} \pi(4.206)^2 \ \text{in.}^2 \times 6.4516 \times 10^{-4} \ \text{m}^2/\text{in.}^2 = 0.008964 \ \text{m}^2$.

The flow rate $= Q = 600 \ \text{gal/min} \times 6.30902 \times 10^{-5} \ \text{m}^3/\text{s-gal/min} = 0.037854 \ \text{m}^3/\text{s}$.

$$\text{Fluid velocity} = Q / A_i = (0.37854 \ \text{m}^3/\text{s})/(0.008964 \ \text{m}^2) = 4.222 \ \text{m/s}.$$

$$\text{Reynolds number} = \rho V D / \mu$$

$$= (8.34 \ \text{lbm/gal})(119.8265)(\text{kg/m}^3/(\text{lbm/gal})(4.222 \ \text{m/s})$$

$$\times (4.206 \ \text{in.})(0.0254 \ \text{m/in.})/(0.001 \ \text{Pa-s/cp}) = 4{,}509.$$

This Reynolds number indicates turbulent flow. To determine the friction factor, first determine the relative roughness k/D. From Fig. 3.10, the relative roughness is about .0004 for commercial steel. The friction factor is about .011 from Fig. 3.9. Friction pressure drop is given by

$$\Delta P_{fr} = 2 f \rho V^2 / \text{D} = 2(0.011)(8.34)(119.8264)(4.222)^2/(4.206)/(0.0254)$$

$$= 1.834 \ \text{kPa/m} = 0.267 \ \text{psi/ft}.$$

The hydrostatic pressure change per foot is

$$\Delta P_{\text{hyd}} = (8.34)(119.8264)(9.80665) = 9.800 \ \text{KPa/m} = 0.433 \ \text{psi/ft}.$$

Total pressure change per length of pipe for flow downward is

$$\Delta P_{\text{tot}} = \Delta P_{\text{hyd}} - \Delta P_{fr} = 0.433 - 0.267 = 0.166 \ \text{psi/ft}.$$

The total pressure change in a 30-ft pipe joint is $0.166 \times 30 = 4.98$ psi.

5. Assume a 10-lbm/gal mud is being circulated at 100 gal/min. What is the frictional pressure change in the annulus outside a single 30-ft joint of drillpipe? Use the Bingham plastic model and assume the plastic viscosity is 40 cp and the YP is 15 lbf/100 ft^2.

$$\text{Flow area} = \tfrac{1}{4} \pi(9.625^2 - 5^2) \times 6.4516 \times 10^{-4} \text{m}^2 = 0.03427 \text{m}^2.$$

$$\text{Flow velocity} = 100 \ \text{gal/min} \times 6.30902 \times 10^{-5} \ \text{m}^3/\text{s-gal/min}/(0.03427 \ \text{m}^2) = 0.1841 \ \text{m/s}.$$

$$\text{Plastic viscosity} = 40 \ \text{cp} \times 0.001 \ \text{Pa-s/cp} = 0.040 \ \text{Pa-s}.$$

$$\text{Yield point} = 15 \ \text{lbf/100 ft}^2 \times 47.88026 \ \text{Pa/(lbf/ft}^2)/100 = 7.182 \ \text{Pa}.$$

$$\text{Density} = (10 \ \text{lbm/gal})(119.8264 \ \text{kg/m}^3/(\text{lbm/gal})) = 1198.3 \ \text{kg/m}^3.$$

$$\text{Reynolds number} = (1198.3 \ \text{kg/m}^3)(9.625 - 5)\text{in.} \ (0.0254 \ \text{m/in.})(0.1841 \ \text{m/s})/0.04 \ \text{Pa-s} = 648.$$

$$\text{Hedstrom number} = \text{He} = \tau_o \rho(D_o^2 - D_i^2)/\mu_p^2$$

$$= (7.182)(1198.3)(9.625^2 - 5^2)(0.0254)^2/(0.04)^2$$

$$= 234{,}729.$$

$$N = 2\text{He}/24{,}500 = 19.16$$

$$\alpha_c = \tfrac{3}{4} + 3/16(3 - \text{sqrt}\,(24N + 9)N/) = 0.5675.$$

$$\text{Re}_{\text{BP2}} = \text{He}(0.968774 - 1.362439\alpha_c + 0.1600822\alpha_c^4)/(8\alpha_c) = 10{,}973.$$

$$\text{Re}_{\text{BP1}} = \text{Re}_{\text{BP2}} - 866(1 - \alpha_c) = 10{,}973 - 866(1 - 0.5675) = 10{,}599.$$

$$\text{Laminar flow because Re} = 648 < \text{Re}_{\text{BP1}} = 10{,}599 \ (\text{Eq. 3.64}).$$

$$f = 16(1/\text{ Re } + 1/6 \text{ He}/\text{Re}^2 - 1/3 \text{ He}^4/\text{Re}^8/f^3) = 1.5154 - 0.52079/f^3$$
$$= 1.0101 \text{ approximately (Eq. 3.64)}.$$
$$\Delta P_{fr} = 2f\rho V^2/(D_o - D_i) = 2(1.0101)(1,198.3 \text{ kg}/\text{m}^3)(0.1841 \text{ m}/\text{s})^2/(9.625 - 5)/0.0254$$
$$= 698.5 \text{ Pa}/\text{m }(0.3048 \text{ m}/\text{ft})/(6894.757 \text{ Pa}/\text{psi}) = 0.03088 \text{ psi}/\text{ft (Eq. 3.62)}.$$
$$\Delta P_{\text{joint}} = 0.03088 \text{ psi}/\text{ft }(30 \text{ ft}) = 0.926 \text{ psi}.$$

6. Repeat Calculation 5, but assume the fluid is a power-law fluid. Remember that PV and YP were determined from the 300-rpm and 600-rpm readings of the Fann viscometer. The equivalent shear stresses are

$$\tau_{300} = \tau_o + \mu_p(300 \times 1.703 \text{ s}^{-1}) = 7.182 \text{ Pa} + 0.040(\text{Pa-s})(510.9 \text{ s}^{-1}) = 27.62 \text{ Pa}.$$
$$\tau_{600} = \tau_o + \mu_p(600 \times 1.703 \text{ s}^{-1}) = 7.182 \text{ Pa} + 0.040(\text{Pa-s})(1021.8 \text{ s}^{-1}) = 44.05 \text{ Pa}.$$
$$n = \ln(\tau_{600}/\tau_{300})/\ln(600/300) = 0.7988.$$
$$K = \tau_{300}/(510.9)^n = 0.1896 \text{ Pa-s}^n.$$
$$\text{Reynolds number} = (D_o - D_i)^n V^{2-n}\rho/[(3n+1)/4n]^n/8^n/K$$
$$= [(9.625 - 5)(0.0254)]^n(0.1841)^{2-n}(1198.3)/(0.1896)/8^{n-1}/[(3n+1)/4n]^n$$
$$= 216.5.$$
$$\text{Laminar flow, because Re} = 216.5 < 3,250 - 1150n = 2,331.$$
$$\text{Friction factor} = 16/\text{ Re } = 0.73903.$$
$$\Delta P_{fr} = 2f\rho V^2/(D_o - D_i) = 2(0.073903)(1198.3 \text{ kg}/\text{m}^3)(0.1841 \text{ m}/\text{s})^2/(9.625 - 5)/0.0254$$

$$= 51.11 \text{ Pa}/\text{m }(0.3048 \text{ m}/\text{ft})/(6894.757 \text{ Pa}/\text{psi}) = 0.002259 \text{ psi}/\text{ft (Eq. 3.62)}.$$
$$\Delta P_{\text{joint}} = 0.002259 \text{ psi}/\text{ft }(30 \text{ ft}) = 0.06778 \text{ psi}.$$

7. For a flow rate of 600 gal/min, what is the fluid pressure in the bit nozzles? The mud density is 12 lbm/gal. What is the pressure recovery in the annulus?

$$\text{The flow rate} = Q = 600 \text{ gal}/\text{min} \times 6.30902 \times 10^{-5} \text{ m}^3/\text{s-gal}/\text{min} = 0.037854 \text{ m}^3/\text{s}.$$
$$\text{Fluid velocity in pipe} = V_o = Q/A_i = (0.037854 \text{ m}^3/\text{s})/(0.008964 \text{ m}^2) = 4.222 \text{ m}/\text{s}.$$
$$\text{Total nozzle area} = 3 \text{ ¼ } \pi(13/32 \times 0.0254)^2 = 0.0002509 \text{ m}^2.$$
$$\text{Nozzle velocity} = 150.9 \text{ m}/\text{s}.$$
$$\text{Density} = 12 \text{ lbm}/\text{gal }[119.8264 \text{ kg}/\text{m}^3/(\text{lbm}/\text{gal})] = 1437.9 \text{ kg}/\text{m}^3.$$
$$\Delta P_{\text{nozzle}} = -\text{ ½ }\rho(V^2 - V_o^2)/C_D = \text{ ½ }(1,437.9 \text{ kg}/\text{m}^3)[(150.9 \text{ m}/\text{s})^2 - (4.222)^2]/0.98$$
$$= -16,692 \text{ kPa}$$
$$= -2,421 \text{ psi}.$$
$$\text{Annulus flow area} = \text{ ¼ }\pi(9.625^2 - 5^2) - 6.4516 - 10^{-4} \text{ m}^2 = 0.03427 \text{ m}^2.$$
$$\text{Flow velocity} = 600 \text{ gal}/\text{min} - 6.30902 - 10^{-5} \text{ m}^3/\text{s-gal}/\text{min}/(0.03427 \text{ m}^2) = 1.105 \text{ m}/\text{s}.$$

$$\Delta P_{\text{annulus}} = -\text{ ½ }\rho(V_a^2 - V^2)/C_D = \text{ ½ }(1,437.9 \text{ kg}/\text{m}^3)[(1.105)2 - (150.9 \text{ m}/\text{s})^2]$$
$$= 16,370 \text{ kPa}$$
$$= 2,375 \text{ psi}.$$

The net pressure change is $-2421 + 2375 = -46.7$ psi.

Nomenclature

a = acoustic velocity, m/s

a_{vs}, b_{vs} = constants that include the viscometer dimensions, the spring constant, and all conversion factors

A = flow area (see subscripts), m^2

c = average concentration of cuttings overall

c_a = cuttings concentration in annular region

c_o = feed concentration of cuttings

c_p = cuttings concentration in plug region

C = compressibility

C_d = discharge coefficients for the flow through an area change, dimensionless

C_D = drag coefficient, dimensionless

C_e = pressure drop correction factor for pipe eccentricity, dimensionless

C_p = heat capacity at constant pressure, J/kg-K

C_v = heat capacity at constant volume, J/kg-K

d_s = particle diameter, m

dv/dr = velocity gradient, s^{-1}

dv/dt = total derivative of velocity with respect to time, Pa/s

D = characteristic length in Reynolds number, m

D_e = special equivalent diameter for yield power law fluid, m

D_{eq} = equivalent diameter, m

D_{hyd} = hydraulic diameter, m

D_h = wellbore diameter, m

D_i = inside diameter, m

D_o = outside diameter, m

D_p = drillpipe outside diameter, m

D_{plug} = plug diameter, m

E_f = Young's modulus for the formation, Pa

$E(k)$ = complete elliptic integral of the second kind, parameter k

f = Fanning friction factor, dimensionless

f_1, f_2, f_3 = turbulent flow velocity profile parameters, dimensionless

f_i = the fraction of flow in the pipe, ith iteration

f_m = the fraction of flow in the pipe, lower annular pressure

f_p = the fraction of flow in the pipe, higher annular pressure

$F(f,n,R_r)$ = eccentric flow function

F_d = total viscous drag force on the particle, N

g = acceleration of gravity, m/s^2

G = mass flow rate density of mixture, kg/m^{3-s}

G_s = mass flow rate density of solids, kg/m^{3-s}

h = specific enthalpy, J/kg

h = total friction pressure drop, Pa/m

He = Hedstrom number

H_s = holdup of solid particles, volume fraction of solids

k = absolute pipe roughness, m

k = c_p/c_v

K = consistency index for pseudoplastic fluid, Pa-s^n
K_b = elastic bulk modulus, Pa
L = length of viscometer bob, m
m = power-law exponent for Herschel-Bulkley fluids
$\dot{m}$ = mass flow rate, kg/s
$\dot{m}_s$ = mass flow rate of solid, kg/s
M_T = torque measured by viscometer, N-m
n = power law exponent for pseudoplastic fluids
p_n = pressure in bit nozzle, Pa
p_r = pressure in bit annular area, Pa
P = pressure, Pa
p_{atm} = atmospheric pressure, Pa
$\hat{P}$ = aerodynamic force exerted on the cuttings by the air, N
q_a = total volumetric flow rate, m^3/s
q_c = volumetric flow rate through concentric annulus, m^3/s
q_e = volumetric flow rate through eccentric annulus, m^3/s
Q = heat transferred into volume, W
Q_c = volumetric flow rate of the cuttings, m^3/s
Q_m = volumetric flow rate of the mud, m^3/s
r_i = inside radius of annulus, m
r_o = outside radius of annulus, m
R = ideal gas constant, m^{3}Pa/kg-K
R_b = radius of inner cylinder (bob) of viscometer, m
R_c = radius of outer cylinder (cup) of viscometer, m
Re = Reynolds number
Re$_p$ = particle Reynolds number
R_r = r_i/r_o
S = entropy, J/K
t = time, s
T = absolute temperature, °K
u = radial displacement, m
v^* = characteristic velocity for turbulent flow calculations, m/s
v = average velocity, m/s
v_a = average annulus velocity, m/s
v_{mix} = mixture velocity, m/s
v_n = velocity in bit nozzle, m/s
v_p = plug velocity, m/s
v_r = velocity in bit annular area, m/s
v_s = average settling velocity
v_{sa} = average cuttings velocity in annular region, m/s
v_{sl} = particle slip velocity, m/s
v_{sp} = average cuttings velocity in plug, m/s
W = buoyant weight or particle, N
W_s = buoyant weight of the cuttings, N
x = parameter in settling velocity equation
y = parameter in settling velocity equation
Y_a = parameter in settling velocity equation

z = measure depth, ft

Z = true vertical depth, ft

α_c = parameter in Bingham fluid friction factor

β = $-\dfrac{1}{\rho}\dfrac{\partial \rho}{\partial T}(P, T)$ = coefficient of thermal expansion, 1/K

γ = shear stress, s^{-1}

γ_e = equivalent shear stress, s^{-1}

δ = ratio of the average particle volume to its cross-sectional area

δ_r = the distance between centers of the inside and outside pipes, m

ΔP = pressure drop, Pa

Δt = time increment, s

Δv = change in velocity, m/s

Δz = length of flow increment, m

ε = internal energy, J/kg

ζ = measured depth integration variable, m

θ = viscometer reading, degrees

ϑ = integration variable

κ = ratio of radius of inner cylinder to outer cylinder

λ_P = $D_{\text{plug}}/D_{\text{eq}}$, the plug diameter ratio

μ = Newtonian viscosity of the fluid, Pa-s

μ_a = apparent viscosity, Pa-s

μ_p = plastic viscosity, centipoise

ξ = integration variable corresponding to depth z, m

ρ = fluid density, kg/m^3

$\bar{\rho}$ = fluid in-mixture density, kg/m^3

ρ_f = fluid density in solid/fluid mixture, kg/m^3

ρ_s = solid density in solid/fluid mixture, kg/m^3

$\bar{\rho}s$ = solid in-mixture density, kg/m^3

τ = shear stress, Pa

τ_w = wall shear stress, Pa

τ_y = yield point, Pa

υ_f = Poisson's ratio for the formation

Φ = angle of inclination from the vertical

Φ = viscous dissipation, W

Ψ = sphericity

ω = rotor speed, rev/min

Ω_o = angular velocity of outer cylinder

Subscripts

1 = properties inside pipe, surge calculations

2 = properties inside annulus, surge calculations

3 = properties of moving pipe, surge calculations

c = concentric

e = eccentric

n = properties in bit nozzle, surge calculations

o = upstream, initial, or inlet

r = properties in annulus outside bit, surge calculations

Superscripts

− = upsteam properties

+ = downstream properties

References

1. Zucrow, M.J. and Hoffman, J.D.: *Gas Dynamics,* Vol. I, John Wiley & Sons Inc., New York City (1976).
2. Press, W.H. *et al.: Numerical Recipes in Fortran 77, The Art of Scientific Computing,* second edition, Cambridge U. Press, New York City (1992).
3. Bourgoyne, A.T. *et al.: Applied Drilling Engineering,* SPE Textbook Series, Vol. 2, Society of Petroleum Engineers, Richardson, Texas (1991).
4. Lubinski, A., Hsu, F.H., and Nolte, K.G.: "Transient Pressure Surges Because of Pipe Movement in an Oil Well," *Revue de l'Inst. Fran. du Pet.* (May/June 1977) 307.
5. Mitchell, R.F.: "Dynamic Surge/Swab Pressure Predictions," *SPEDE* (September 1988) 325.
6. Savins, J.G. and Roper, W.F.: "A Direct Indicating Viscometer for Drilling Fluids," *API Drill. & Prod. Prac.* (1954) 7.
7. Gray, G.R. and Darley, H.C.H.: *Composition and Properties of Oilwell Drilling Fluids,* fourth edition, Gulf Publishing Co., Houston (1980).
8. Govier, G.W. and Aziz, K.: *The Flow of Complex Mixtures in Pipes*, Robert E. Krieger Publishing Co., Huntington, New York (1987).
9. Colebrook, C.F.: "Turbulent Flow in Pipes, with Particular Reference to the Transition Region Between the Smooth and Rough Pipe Laws," *J. Inst. Civil Eng.* (1939) **11**, 133.
10. Dodge, D.W. and Metzner, A.B.: "Turbulent Flow of Non-Newtonian Systems," *AIChE. J.*, **5**, No. 2, 189.
11. Uner, D., Ozgen, C., and Tosun, I.: "An Approximate Solution for Non-Newtonian Flow in Eccentric Annulus," *Ind. Eng. Chem. Res.***27**, No. 4, 698.
12. Lapidus, L. and Pindar, G.F.: *Numerical Solution of Partial Differential Equations in Science and Engineering,* John Wiley & Sons Inc., New York City (1982) 1–26.
13. Streeter, V.L.: *Fluid Mechanics*, McGraw-Hill Book Co. Inc., New York City (1962).
14. Pigott, R.J.S.: "Mud Flow in Drilling," *API Drill. & Prod. Prac.* (1941) 91.
15. Williams, C.E. and Bruce, G.H.: "Carrying Capacity of Drilling Muds," *Trans.*, AIME (1951) **192**, 111–120.
16. Sifferman, T.R. *et al.:* "Drill Cutting Transport in Full-Scale Vertical Annuli," *JPT* (November 1974) 1295.
17. Sifferman, T.R. and Becker, T.E.: "Hole Cleaning in Full-Scale Inclined Wellbores," *SPEDE* (June 1992) 115; *Trans.,* AIME, **293.**
18. Walker, R.E. and Mayes, T.M.: "Design of Muds for Carrying Capacity," *JPT* (July 1975) 893; *Trans.,* AIME, **259.**
19. Stokes, G.G.: *Transactions of the Cambridge Philosophical Society* (1845, 1851) **8**, 9.
20. Chien, S.F.: "Annular Velocity for Rotary Drilling Operations," *Proc.,* SPE Fifth Conference on Drilling and Rock Mechanics, Austin, Texas (1971) 5–16.
21. Moore, P.L.: *Drilling Practices Manual,* Petroleum Publishing Co., Tulsa (1974)
22. Clark, R.K. and Bickham, K.L.: "A Mechanistic Model for Cuttings Transport," paper SPE 28306 presented at the 1994 SPE Annual Technical Conference and Exhibition, New Orleans, 23–26 September.
23. Tomren, P.H., Iyoho, A.W., and Azar, J.J.: "Experimental Study of Cuttings Transport in Directional Wells," *SPEDE* (February 1986) 43.
24. Okrajni, S.S. and Azar, J.J.: "The Effects of Mud Rheology on Annular Hole Cleaning in Directional Wells," *SPEDE* (August 1986) 297; *Trans.* AIME, **285.**
25. Becker, T.E., Azar, J.J., and Okrajni, S.S.: "Correlations of Mud Rheological Properties With Cuttings-Transport Performance in Directional Drilling," *SPEDE* (March 1991) 16; *Trans.*, AIME, **291.**
26. Seeberger, M.H., Matlock, R.W., and Hanson, P.M.: "Oil Muds in Large-Diameter, Highly Deviated Wells: Solving the Cuttings Removal Problem," paper SPE/IADC 18635 presented at the 1989 SPE/IADC Drilling Conference, New Orleans, 28 February–3 March.

27. Bassal, A.A.: "A Study of the Effect of Drill Pipe Rotation on Cuttings Transport in Inclined Wellbores," MS thesis, U. of Tulsa, Tulsa, Oklahoma (1995).

28. Larsen, T.I., Pilehvari, A.A., and Azar, J.J.: "Development of a New Cuttings Transport Model for High-Angle Wellbores Including Horizontal Wells," paper SPE 25872 presented at the 1993 SPE Rocky Mountain Regional/Low Permeability Reservoir Symposium, Denver, 12–14 April.

29. Hemphill, T. and Larsen, T.I.: "Hole-Cleaning Capabilities of Oil-Based and Water-Based Drilling Fluids: A Comparative Experimental Study," paper SPE 26328 presented at the 1993 SPE Annual Technical Conference and Exhibition, Houston, 3–6 October.

30. Jalukar, L.S. *et al.:* "Extensive Experimental Investigation of Hole Size Effect on Cuttings Transport in Directional Well Drilling," paper presented at the 1993 ASME Fluids Engineering Division Annual Summer Meeting, San Diego, California, 7–12 July.

31. Zamora, M. and Hanson, P.: "More Rules of Thumb to Improve High Angle Hole Cleaning," *Pet. Eng. Intl.* (February 1991) 22.

32. Luo, Y. and Bern, P.A.: "Flow-Rate Predictions for Cleaning Deviated Wells," paper SPE/IADC 23884 presented at the 1992 SPE/IADC Drilling Conference, New Orleans, 18–21 February.

33. Campos, W. *et. al.:* "A Mechanistic Modeling of Cuttings Transport in Highly Inclined Wells," *Liquid-Solid Flows,* ASME *FED* Vol. 189 (June 1994) 145–155.

34. Ford, J.T. *et al.:* "Experimental Investigation of Drilled Cuttings Transport in Inclined Boreholes," paper SPE 20421 presented at the 1990 SPE Annual Technical Conference and Exhibition, New Orleans, 23–26 September.

35. Gavignet, A. and Sobey, I.: "Model Aids Cuttings Transport Prediction," *JPT* (September 1989) 916; *Trans.*, AIME, **287.**

36. Wilson, K.C.: "Slip Point of Beds in Solid-Liquid Pipeline Flow," *J. Hydrol. Division of ASCE* (1970) **96,** HYI, 1.

37. Martin, M. *et al.:* "Transport of Cuttings in Directional Wells," paper SPE/IADC 16083 presented at the 1987 SPE/IADC Drilling Conference, New Orleans, 15–18 March.

38. Oraskar, A.D. and Whitmore, R.L.: "The Critical Velocity in Pipeline Flow of Slurries," *AIChE. J.*, **26,** No. 4, 550.

39. Kenny, P., Sunde, E., and Hemphill, T.: "Hole Cleaning Modeling: What's 'n' Got to Do with It?" paper SPE/IADC 35099 presented at the 1996 SPE/IADC Drilling Conference, New Orleans, 12–15 March.

40. Guild, G.J. and Hill, T.H.: "Hole Cleaning Program for Extended Reach Wells," paper SPE 29381 presented at the 1995 SPE/IADC Drilling Conference, Amsterdam, The Netherlands, 28 February–5 March.

41. Angel, R.R.: "Volume Requirements for Air and Gas Drilling," *Trans.,* AIME, **210,** 325.

42. Mitchell, R.F.: "The Simulation of Air and Mist Drilling for Geothermal Wells," *JPT* (November 1983) 2120.

43. Gray, K.E.: "The Cutting Carrying Capacity of Air at Pressures above Atmospheric," AIME technical paper 874-G, Dallas (October 1957).

44. Beyer, A.H., Millhone, R.S., and Foote, R.W.: "Flow Behavior of Foam as a Well Circulating Fluid," paper SPE 3986 presented at the 1972 SPE Annual Fall Meeting, San Antonio, Texas, 2–5 October.

45. Blauer, R.D., Mitchell, B.J., and Kohlhaas, C.A.: "Determination of Laminar, Turbulent and Transitional Foam-Flow Friction Losses in Pipes," paper SPE 4885 presented at the 1974 SPE Annual California Regional Meeting, San Francisco, 4–5 April.

46. Sanghani, V. and Ikoku, C.U.: "Rheology of Foam and Its Implications in Drilling and Cleanout Operations," U.S. DOE Report DOE/BC/10079-47, U.S. DOE, Washington, DC (June 1982).

General References

Bikerman, J.J.: *Foams: Theory and Industrial Applications*, Reinhold Publishing Corp., New York City (1953) 124–125.

Bird, R.B. *et al.: Dynamics of Polymeric Fluids,* Vol. 1, *Fluid Mechanics,* John Wiley & Sons Inc., New York City (1987).

Bird, R.B., Stewart, W.E., and Lightfoot, E.N.: *Transport Phenomenon*, John Wiley & Sons Inc., New York City (1960).

Blauer, R.E. and Kohlhaas, C.A.: "Formation Fracturing with Foam," paper SPE 5003 presented at the 1974 SPE Annual Fall Meeting, Houston, 6–9 October.

Brown, N.P. and Bern, P.A.: "Cleaning Deviated Holes: New Experimental and Theoretical Studies," paper SPE/IADC 18636 presented at the 1989 SPE/IADC Drilling Conference, New Orleans, 28 February–3 March.

Burkhardt, J.A.: "Wellbore Pressure Surges Produced by Pipe Movement," *JPT* (June 1961) 595; *Trans.*, AIME, **222.**

Caldwell, J.: "The Hydraulic Mean Depth as a Basis for Form Comparison in the Flow of Fluids in Pipes," *J. Roy. Tech. College,* Glasgow (January 1930) **2,** No. 2, 203.

Cannon, G.E.: "Changes in Hydrostatic Pressure Because of Withdrawing Drill Pipe from the Hole," *Drill. & Prod. Prac.,* API (1934) 42.

Cartalos, U. *et al.:* "Field Validated Hydraulic Model Predictions Give Guidelines for Optimal Annular Flow in Slim Hole Drilling," paper SPE 35131 presented at the 1996 IADC/SPE Drilling Conference, New Orleans, 12–15 March.

Cawiezel, K.E. and Niles, T.D.: "Rheological Properties of Foam Fracturing Fluids Under Downhole Conditions," paper SPE 16191 presented at the 1987 SPE Hydrocarbon Economics and Evaluation Symposium, Dallas, 2–3 March.

Churchill, S.W.: "Friction Factor Equation Spans All Fluid Flow Regimes," *Chem. Eng.* (7 November 1977) 91.

Clark, R.K. and Fontenot, J.E.: "Field Measurements of the Effect of Drillstring Velocity, Pump Speed, and Lost Circulation Material on Downhole Pressures," paper SPE 4970 presented at the 1974 SPE Annual Meeting, Houston, 6–9 October.

David, A. and Marsden, S.S. Jr.: "The Rheology of Foam," paper SPE 2544 presented at the 1969 SPE Annual Meeting, Denver, 28 September–1 October.

Eckel, J.R. and Bielstein, W.J.: "Nozzle Design and Its Effect on Drilling Rate and Pump Operations," *Drill. & Prod. Prac.* (1951) 28.

Economides, M.J., Watters, L.T., and Dunn-Norman, S.: *Petroleum Well Construction*, John Wiley & Sons Inc., New York City (1998).

Eirich, F.R.: *Rheology,* Vol. 3, Academic Press, San Diego, California (1960).

Fontenot, J.E. and Clark, R.E.: "An Improved Method for Calculating Swab and Surge Pressures and Circulating Pressures in a Drilling Well," *SPEJ* (October 1974) 451.

Ford, J.T. *et al.:* "Experimental Investigation of Drilled Cuttings Transport in Inclined Boreholes," paper SPE 20421 presented at the 1990 SPE Annual Technical Conference and Exhibition, New Orleans, 23–26 September.

Franco V., and Verduzco, M.B.: "Transition Critical Velocity In Pipes Transporting Slurries with Non-Newtonian Behavior," *Proc.,* 14th Miami U., Coral Gables, Multiphase Transport & Particulate Phenomena International Symposium, Miami Beach, Florida (1988) Vol. 4, 289–299.

Fredrickson, A.G. and Bird, R.B.: "Non-Newtonian Flow in Annuli," *Ind. & Eng. Chem.* (1958) **50,** No. 3, 347.

Goins, W.C. *et al.:* "Down-The-Hole Pressure Surges and Their Effect on Loss of Circulation," *Drill. & Prod. Prac.*, API (1951) 125.

Grundmann, S.R. and Lord, D.L.: "Foam Stimulation," *JPT* (March 1983) 597.

Gucuyener, H.L. and Mehmetoglulu,T.: "Flow of Yield-Pseudoplastic Fluids through a Concentric Annulus," *AIChE J.* (July 1992) **38,** No. 7, 1139.

Haciislamoglu, M. and Langlinais, J.: "Non-Newtonian Flow in Eccentric Annuli," *J. of Energy Resources Tech.* (September 1990) **112,** 163.

Hanks, R.W.: "The Laminar-Turbulent Transition In Non-isothermal Flow of Pseudoplastic Fluids In Tubes," *AIChE J.* (September 1962) **8**, No. 4, 467.

Hanks, R.W.: "The Laminar-Turbulent Transition for Fluids with a Yield Stress," *AIChE J.* (May 1963) **9**, No. 3, 306.

Hanks, R.W. and Pratt, D.R.: "On the Flow of Bingham Plastic Slurries in Pipes and Between Parallel Plates," *SPEJ* (December 1967) 342; *Trans.,* AIME, **240.**

Hanks, R.W.: "A Theory of Laminar Flow Stability," *AIChE J.* (January 1969) **15,** No. 1, 25.

Hanks, R.W.: "On the Prediction of Non-Newtonian Flow Behavior in Ducts of Noncircular Cross Section," *Ind. Eng. Fundam.* (1974) **13,** No. 1, 62.

Hanks, R.W. and Ricks, B.L.: "Laminar-Turbulent Transition in Flow of Pseudoplastic Fluids with Yield Stresses," *AIAA J. of Hydronautics* (October 1974) **8,** No. 4, 163.

Hanks, R.W.: "Low Reynolds Number Turbulent Pipeline Flow of Pseudo-homogeneous Slurries," *Proc.,* Fifth International Conference on the Hydraulic Transport of Solids In Pipes, British Hydromechanics Research Association (May 1978) paper C2.

Hanks, R.W.: "The Axial Laminar Flow of Yield-Pseudoplastic Fuids in a Concentric Annulus," *Ind. Eng. Chem. Process Des. Dev.* (1979) **18,** No. 3, 488.

Hanks, R.W.: "The Not so 'Generalized' Reynolds Number," *Proc.,* Fourth International Technical Conference on Slurry Transport, Slurry Transport Association, Washington, DC (1979) 91–98.

Hanks, R.W.: "Critical Reynolds Numbers for Newtonian Flow in Concentric Annuli," *AIChE J.* (January 1980) **26,** No. 1, 152.

Hanks, R.W. and Peterson, J.M.: "Complex Transitional Flows to Concentric Annuli," *AIChE J.* (September 1982) **28,** No. 5, 800.

Hansen S.A.: "Laminar Non-Newtonian Flow in an Eccentric Annulus, a Numerical Solution," paper presented at the 1993 Annual Meeting of the Nordic Rheology Society, Gothenburg, Sweden, 19–20 August.

Hansen S.A. and Sterri, N.: "Drill Pipe Rotation Effects on Frictional Pressure Losses in Slim Annuli," paper SPE 30488 presented at the 1995 SPE Annual Technical Conference and Exhibition, Dallas, 22–25 October.

Harris, P.C., Haynes, R.J., and Egger, J.P.: "The Use of CO_2-Based Fracturing Fluids in the Red Fork Formation in the Anadarko Basin, Oklahoma," *JPT* (June 1984) 1003.

Harris, P.C.: "Effects of Texture on Rheology of Foam Fracturing Fluids," paper SPE 14257 presented at the 1985 SPE Annual Technical Conference and Exhibition, Las Vegas, 22–25 September.

Holcomb, D.L.: "Foamed Acid as a Means of Providing Extended Retardation,'" paper SPE 6376 presented at the 1977 SPE Permian Basin Oil and Gas Recovery Conference, Midland, Texas, 10–11 March.

Honore Jr., R.S., and Tarr, B.A.: "Cementing Temperature Predictions Based on Both Downhole Measurements and Computer Predictions: A Case History," paper SPE 25436 presented at the 1993 SPE Productions Operations Symposium, Oklahoma City, Oklahoma, 21–23 March.

Jalukar, L.S.: "Study of Hole Size Effect on Critical and Subcritical Drilling Fluid Velocities in Cuttings Transport for Inclined Wellbores," MS thesis, U. of Tulsa, Tulsa, Oklahoma (1993).

Jensen, T.B. and Sharma, M.P.: "Study of Friction Factor Equivalent Diameter Correlations for Annular Flow of Non- Newtonian Drilling Fluids," *ASME J. Energy Resources Tech.* (December 1987) **109,** 200.

Jones, O.C. Jr. and Leung, J.C.M.: "An Improvement in the Calculation of Turbulent Friction in a Smooth Concentric Annuli," *ASME J. of Fluids Eng.* (December 1981) **103,** 615.

Kendal, W.A. and Goins, W.C.: "Design and Operation of Jet Bit Programs for Maximum Hydraulic Horsepower, Impact Force, or Jet Velocity," *Trans.,* AIME (1960) **219,** 238.

Lal, M.: "Surge and Swab Modeling for Dynamic Pressures and Safe Trip Velocities," paper SPE 11412 presented at the 1983 SPE/IADC Drilling Technology Conference, New Orleans, 20–23 February.

Lamb, H.: *Hydrodynamics*, sixth edition, Dover Publications, New York City (1945).

Larsen, T.I.: "A Study of the Critical Fluid Velocity in Cuttings Transport," MS thesis, U. of Tulsa, Tulsa, Oklahoma (1990).

Larsen, T.I., Pilehvari, A.A., and Azar, J.J.: "Development of a New Cuttings Transport Model for High-Angle Wellbores Including Horizontal Wells," paper SPE 25872 presented at the 1993 SPE Rocky Mountain Regional/Low Permeability Reservoir Symposium, Denver, 12–14 April.

Leinan, A.G. and Harris, P.C.: "Carbon Dioxide/Aqueous Mixtures as Fracturing Fluids in Western Canada," paper CIM No. 83-34-36 presented at the 1983 Annual Meeting of the Petroleum of CIM, Banff, Alberta, 10–13 May.

Lockett, T.J., Richardson, S.M., and Worraker, W.J.: "The Importance of Rotation Effects for Efficient Cuttings Removal During Drilling," paper SPE 25768 presented at the 1993 SPE/IADC Drilling Conference, Amsterdam, 23–25 February.

Lord, D.L.: "Analysis of Dynamic and Static Foam Behavior," *JPT* (January 1981) 39.

Martins, A.L. *et al.:* "Foam Rheology Characterization as a Tool for Predicting Pressures While Drilling Offshore Wells in UBD Conditions," paper SPE 67691 presented at the 2001 SPE/IADC Drilling Conference, Amsterdam, 27 February–1 March.

McCann, R.C. *et al.:* "Effects of High-Speed Pipe Rotation on Pressures in Narrow Annuli," paper SPE 26343 presented at the 1993 SPE Annual Technical Conference and Exhibition, Houston, 3–6 October.

Metzner, A.B. and Reed, J.C.: "Flow of Non-Newtonian Fluids—Correlation of the Laminar, Transition, and Turbulent-flow Regions," *AIChE J.* (December 1955) **1**, 434.

Metzner, A.B.: "Non-Newtonian Fluid Flow," *Ind. Eng. Chem.* (September 1957) **49**, No. 9, 1429.

Miller, C.: "Predicting Non-Newtonian Flow Behavior in Ducts of Unusual Cross Section," *Ind. Eng. Chem. Fund.* (1972) **11**, No. 4, 524.

Mishra, P. and Tripathi, G.: "Transition from Laminar to Turbulent Flow of Purely Viscous Non-Newtonian Fluids in Tubes," *Chem. Eng. Sci.* (1971) **26**, 915.

Mitchell, B.J. "Viscosity of Foam," PhD dissertation, U. of Oklahoma, Oklahoma City, Oklahoma (1969).

Moran, L.K. and Lindstrom, K.O.: "Cement Spacer Fluid Solids Settling," paper IADC/SPE 19936 presented at the 1990 IADC/SPE Drilling Conference, Houston, 27 February–2 March.

Okafor, M.N. and Evers, J.F.: "Experimental Comparison of Rheology Models for Drilling Fluids," paper SPE 24086 presented at the 1992 SPE Western Regional Meeting, Bakersfield, California, 30 March–1 April.

Oraskar, A.R. and Turian, R.M.: "The Critical Velocity in Pipeline Flow of Slurries," *AIChE J.* (July 1980) 550.

Ozbayoglu, M. *et al.:* "A Comparative Study of Hydraulic Models for Foam Drilling," paper SPE 65489 presented at the 2000 SPE/PS-CIM International Conference on Horizontal Well Technology, Calgary, Alberta, 6–8 November.

Perry, R.: *Perry's Chemical Engineers' Handbook*, McGraw-Hill Book Co. Inc., New York City (1985).

Piercy, N.A.V., Hooper, M.S., and Winney, H.F.: "Viscous Flow Through Pipes with Cores," *Proc. of the Royal Society*, London (1933).

Princen, H.M.: "Rheology of Foams and Highly Concentrated Emulsions," *J. Colloid Interface Sci.* (January 1983) **91**, No. 1, 160.

Ramsey, M.S. *et al.:* "Bit Hydraulics: Net Pressure Drops Are Lower Than You Think," *World Oil* (October 1983) 65.

Rasi, M.: "Hole Cleaning in Large, High-Angle Wellbores," paper IADC/SPE 27464 presented at the 1994 IADC/SPE Drilling Conference, Dallas, 15–18 February.

Ravi, K.: "New Rheological Correlation for Cement Slurries as a Function of Temperature," paper presented at the 1991 AIChE Annual Conference, Los Angeles, 13–19 November.

Ravi, K. and Sutton, D.L.: "New Rheological Correlation for Cement Slurries as a Function of Temperature," paper presented at the 1990 SPE Annual Technical Conference and Exhibition, New Orleans, 23–26 September.

Ravi, K. and Weber, L.: "Drill Cutting Removal in a Horizontal Wellbore for Cementing," paper IADC/SPE 35081 presented at the 1996 IADC/SPE Drilling Conference, New Orleans, 12–15 March.

Raza, S.H. and Marsden, S.S.: "The Streaming Potential and the Rheology of Foam," *SPEJ* (April 1967) 359.

Reed, T.D. and Pilehvari, A.A.: "A New Model for Laminar, Transitional, and Turbulent Flow of Drilling Muds," paper SPE 25456 presented at the 1993 SPE Production Operations Symposium, Oklahoma City, Oklahoma, 21–23 March.

Reidenbach, V.G. *et al.:* "Rheological Study of Foam Fracturing Fluids Using Nitrogen and Carbon Dioxide," *SPEPE* (January 1986) 31.

Reynolds, O.: "An Experimental Investigation of the Circumstances Which Determine Whether the Motion of Water Shall Be Direct or Sinuous, and the Laws of Resistance in Parallel Channels," *Trans.,* Royal Society, London (1883) 174.

Sabins, F.L.: "The Relationship of Thickening Time, Gel Strength, and Compressive Strengths of Oilwell Cements," *SPEPE* (March 1986) 143.

Sample, K.J. and Bourgoyne, A.T.: "Development of Improved Laboratory and Field Procedures for Determining the Carrying Capacity of Drilling Fluids," paper SPE 7497 presented at the 1978 SPE Annual Technical Conference and Exhibition, Houston, 1–4 October.

Sassen, A. *et al:* "Monitoring of Barite Sag Important in Deviated Drilling," *Oil & Gas J.* (26 August 1991) 43.

Savins, F.J.: "Generalized Newtonian (Pseudoplastic) Flow in Stationary Pipes and Annuli," *Trans.,* AIME (1958) **213,** 325.

Schuh, F.J.: "Computer Makes Surge Pressure Calculations Useful," *Oil & Gas J.* (3 August 1964) 96.

Shah, R.K. and London, A.L.: *Laminar Flow Forced Convection in Ducts,* Academic Press, New York City (1978).

Shah, S.N. and Lord, D.L.: "Hydraulic Fracturing Slurry Transport in Horizontal Pipes," *SPEDE* (September 1990) 225.

Shah, S.N. and Sutton, D.L.: "New Friction Correlation for Cements from Pipe and Rotational-Viscometer Data," *SPEPE* (November 1990) 415.

Skelland A.H.P.: *Non-Newtonian Flow and Heat Transfer,* John Wiley & Sons Inc., New York City (1967).

Smith, T.R. and Ravi, K.: "Investigation of Drilling Fluid Properties to Maximize Cement Displacement Efficiency," paper SPE 22775 presented at the 1991 SPE Annual Technical Conference and Exhibition, Dallas, 6–9 October.

Stevenic, B.C.: "Design and Construction of a Large-Scale Wellbore Simulator and Investigation of Hole Size Effects on Critical Cuttings Transport Velocity in Highly Inclined Wells," MS thesis, U. of Tulsa, Tulsa, Oklahoma (1991).

Sutton, D.L. and Ravi, K.: "New Method for Determining Downhole Properties that Affect Gas Migration and Annular Sealing," paper SPE 19520 presented at the 1989 SPE Annual Technical Conference and Exhibition, San Antonio, Texas, 8–11 October.

Sylvester, N.D. and Brill, J.P.: "Drag Reduction in Two-Phase Annular-Mist Flow of Air and Water," *AIChE J.* (May 1976) **22,** No. 3.

Tilghman, S.E., Benge, O.G., and George, C.R.: "Temperature Data for Optimizing Cementing Operations," paper IADC/SPE 19939 presented at the 1990 IADC/SPE Drilling Conference, Houston, 27 February–March 3.

Timoshenko, S.P. and Goodier, J.N.: *Theory of Elasticity,* McGraw-Hill Book Co. Inc., New York City (1970).

Tosun, I.: "Axial Laminar Flow in an Eccentric Annulus: An Approximate Solution," *AIChE J.* **30,** No. 5, 877.

Vieira, P. *et al.:* "Minimum Air and Water Flow Rates Required for Effective Cuttings Transport in High Angle and Horizontal Wells," paper SPE 74463 presented at the 2002 IADC/SPE Drilling Conference, Dallas, 26–28 February.

Walton, I.C. and Bittleston, S.H.: "The Axial Flow of a Bingham Plastic in a Narrow Eccentric Annulus," *J. Fluid Mech.* (1991) **222,** 39.

Warnock, W.E. Jr., Harris, P.C., and King, D.S.: "Successful Field Applications of CO_2 Foam Fracturing Fluids in the Arkansas-Louisiana-Texas Region," *JPT* (January 1985) 80.

Wedelich, H., and Galate, J.W.: "Key Factors That Affect Cementing Temperatures," paper IADC/SPE 16133 presented at the 1987 IADC/SPE Drilling Conference, New Orleans, 15–18 March.

Wenzel, H.G. Jr. *et al.:* "Flow of High Expansion Foam in Pipes," *Proc.,* American Society of Civil Engineers, *J. of the Engineering Mechanics Division* (December 1976) 153.

White, F.M.: *Fluid Mechanics,* McGraw-Hill Book Co. Inc., New York City (1979).

Whittaker, A.: *Theory and Application of Drilling Fluid Hydraulics,* EXLOG staff (eds.) International Human Resources Dev. Corp., Boston (1985) 102.

Wilson, N.W. and Azad, R.S.: "A Continuous Prediction Method for Fully Developed Laminar, Transitional, and Turbulent Flow In Pipes," *ASME J. of Applied Mechanics* (March 1975) 51.

Wooley, G.R. *et al.:* "Cementing Temperatures for Deep-Well Production Liners," paper SPE 13046 presented at the 1984 SPE Annual Technical Conference and Exhibition, Houston, 16–19 September.

Wylie, E.B. and Streeter, V.L.: *Fluid Transients,* corrected edition 1983, FEB Press, Ann Arbor, Michigan (1982) 31–63.

Zamora, M. and Bleier, R.: "Prediction of Drilling Mud Rheology Using a Simplified Herschel-Bulkley Model," paper presented at the 1976 ASME International Joint Conference, Mexico City, 19–24 September.

SI Metric Conversion Factors

cp	$\times$	1.0*	E – 03	= Pa·s
ft	$\times$	3.048*	E – 01	= m
ft²	$\times$	9.290 304*	E – 02	= m²
°F		(°F– 32)/1.8		= °C
gal	$\times$	3.785 412	E – 03	= m³
in.	$\times$	2.54*	E + 00	= cm
in.²	$\times$	6.451 6*	E +00	= cm²
lbf	$\times$	4.448 222	E + 00	= N
lbm	$\times$	4.535 924	E – 01	= kg

*Conversion factor is exact.

Chapter 4
Well Control: Procedures and Principles
Neal Adams, Neal Adams Services

Well control and blowout prevention have become particularly important topics in the hydrocarbon production industry for many reasons. Among these reasons are higher drilling costs, waste of natural resources, and the possible loss of human life when kicks and blowouts occur. One concern is the increasing number of governmental regulations and restrictions placed on the hydrocarbon industry, partially as a result of recent, much-publicized well-control incidents. For these and other reasons, it is important that drilling personnel understand well-control principles and the procedures to follow to properly control potential blowouts.

This chapter discusses the key elements that can be used to control kicks and prevent blowouts. These steps are based on the work of a blowout specialist and are briefly presented below:

• Quickly shut in the well.

• When in doubt, shut down and get help. Kicks occur as frequently while drilling as they do while tripping out of the hole. Many small kicks turn into big blowouts because of improper handling.

• Act cautiously to avoid mistakes—take your time to get it right the first time. You may not have another opportunity to do it correctly.

These and other well-control details are presented in detail throughout this chapter. Unusual problems occurring during kick killing are discussed in other referenced sources.

4.1 Introduction to Kicks
Various drilling problems confront operators daily. Among these are lost circulation, stuck pipe, deviation control, and well control. The drilling problem specifically examined in this chapter is well control. Other drilling problems will be presented as they relate to aspects of well control. One of the most pervasive problems with well control is the "kick."

A kick is a well control problem in which the pressure found within the drilled rock is higher than the mud hydrostatic pressure acting on the borehole or rock face. When this occurs, the greater formation pressure has a tendency to force formation fluids into the wellbore. This forced fluid flow is called a kick. If the flow is successfully controlled, the kick is considered to have been killed. An uncontrolled kick that increases in severity may result in what is known as a "blowout."

4.1.1 Factors Affecting Kick Severity. Several factors affect the severity of a kick. One factor, for example, is the "permeability" of rock, which is its ability to allow fluid to move through the rock. Another factor affecting kick severity is "porosity." Porosity measures the amount of space in the rock containing fluids. A rock with high permeability and high porosity has greater potential for a severe kick than a rock with low permeability and low porosity. For example, sandstone is considered to have greater kick potential than shale because sandstone, in general, has greater permeability and greater porosity than shale.

And yet another factor affecting kick severity is the "pressure differential" involved. Pressure differential is the difference between the formation fluid pressure and the mud hydrostatic pressure. If the formation pressure is much greater than the hydrostatic pressure, a large negative differential pressure exists. If this negative differential pressure is coupled with high permeability and high porosity, a severe kick may occur.

4.1.2 Kick Labels. A kick can be labeled in several ways, including one that depends on the type of formation fluid that entered the borehole. Known kick fluids include gas, oil, salt water, magnesium chloride water, hydrogen sulfide (sour) gas, and carbon dioxide. If gas enters the borehole, the kick is called a "gas kick." Furthermore, if a volume of 20 bbl (3.2 m^3) of gas entered the borehole, the kick could be termed a 20-bbl (3.2-m^3) gas kick.

Another way of labeling kicks is by identifying the required mud weight increase necessary to control the well and kill a potential blowout. For example, if a kick required a 0.7-lbm/gal (84-kg/m^3) mud weight increase to control the well, the kick could be termed a 0.7-lbm/gal (84-kg/m^3) kick. It is interesting to note that an average kick requires approximately 0.5 lbm/gal (60 kg/m^3), or less, mud weight increase.

4.1.3 Other Factors Affecting Well Control. Another important consideration in well control is the pressure the formation rock can withstand without sustaining an induced fracture. This rock strength is often called the "fracture mud weight," or "gradient," and is usually expressed in lbm/gal of equivalent mud weight.

The "equivalent mud weight" is the summation of pressures exerted on the borehole wall and includes mud hydrostatic pressure, pressure surges caused by pipe movement, friction pressures applied against the formation as a result of pumping the drilling fluid, or any casing pressure caused by a kick. For example, if the fracture mud weight of a formation is determined to be 16.0 lbm/gal, the well can withstand any combination of the above-mentioned pressures that yield the same pressure as a column of 16.0-lbm/gal (1920-kg/m^3) mud to the desired depth. This combination could be 16.0-lbm/gal (1920-kg/m^3) mud, 15.0-lbm/gal (1800-kg/m^3) mud and some amount of casing pressure, 15.5-lbm/gal (1860-kg/m^3) mud and a smaller amount of casing pressure, or other combinations.

4.1.4 Causes of Kicks. Kicks occur as a result of formation pressure being greater than mud hydrostatic pressure, which causes fluids to flow from the formation into the wellbore. In almost all drilling operations, the operator attempts to maintain a hydrostatic pressure greater than formation pressure and, thus, prevent kicks; however, on occasion the formation will exceed the mud pressure and a kick will occur. Reasons for this imbalance explain the key causes of kicks:
   • Insufficient mud weight.
   • Improper hole fill-up during trips.
   • Swabbing.
   • Cut mud.
   • Lost circulation.
   Insufficient Mud Weight. Insufficient mud weight is the predominant cause of kicks. A permeable zone is drilled while using a mud weight that exerts less pressure than the formation

TABLE 4.1—ABNORMAL PRESSURE INDICATORS	
Qualitative Methods	Quantitative Methods
Lithology	Shale density
Offset well-log analysis	d exponent
Temperature anomaly	Normalized penetration rate
Gas counting	Other drilling equations
Mud or cuttings resistivity	
Cutting character	
Hole condition	

pressure within the zone. Because the formation pressure exceeds the wellbore pressure, fluids begin to flow from the formation into the wellbore and the kick occurs.

These abnormal formation pressures are often associated with causes for kicks. Abnormal formation pressures are greater pressures than in normal conditions. In well control situations, formation pressures greater than normal are the biggest concern. Because a normal formation pressure is equal to a full column of native water, abnormally pressured formations exert more pressure than a full water column. If abnormally pressured formations are encountered while drilling with mud weights insufficient to control the zone, a potential kick situation has developed. Whether or not the kick occurs depends on the permeability and porosity of the rock. A number of abnormal pressure indicators can be used to estimate formation pressures so that kicks caused by insufficient mud weight are prevented (some are listed in **Table 4.1**).

An obvious solution to kicks caused by insufficient mud weights seems to be drilling with high mud weights; however, this is not always a viable solution. First, high mud weights may exceed the fracture mud weight of the formation and induce lost circulation. Second, mud weights in excess of the formation pressure may significantly reduce the penetration rates. Also, pipe sticking becomes a serious consideration when excessive mud weights are used. The best solution is to maintain a mud weight slightly greater than formation pressure until the mud weight begins to approach the fracture mud weight and, thus, requires an additional string of casing.

Improper Hole Fill-Up During Trips. Improperly filling up of the hole during trips is another prominent cause of kicks. As the drillpipe is pulled out of the hole, the mud level falls because the pipe steel no longer displaces the mud. As the overall mud level decreases, the hole must be periodically filled up with mud to avoid reducing the hydrostatic pressure and, thereby, allowing a kick to occur.

Several methods can be used to fill up the hole, but each must be able to accurately measure the amount of mud required. It is not acceptable—under any condition—to allow a centrifugal pump to continuously fill up the hole from the suction pit because accurate mud-volume measurement with this sort of pump is impossible. The two acceptable methods most commonly used to maintain hole fill-up are the trip-tank method and the pump-stroke measurements method.

The trip-tank method has a calibration device that monitors the volume of mud entering the hole. The tank can be placed above the preventer to allow gravity to force mud into the annulus, or a centrifugal pump may pump mud into the annulus with the overflow returning to the trip tank. The advantages of the trip-tank method include that the hole remains full at all times, and an accurate measurement of the mud entering the hole is possible.

The other method of keeping a full hole—the pump-stroke measurement method—is to periodically fill up the hole with a positive-displacement pump. A flowline device can be

TABLE 4.2—SWAB PRESSURES (psig) FOR A 14.0-ppg MUD, 4¹/₂-in. PIPE WITH VARIOUS HOLE SIZES AND SEVERAL PULLING SPEEDS

Hole Size (in.)	Pulling Speeds (sec/stand)					
	15	22	30	45	68	75
8¹/₂	267	167	124	98	84	75
6¹/₂	589	344	256	192	159	140
5³/₄	921	524	294	289	231	200

installed with the positive-displacement pump to measure the pump strokes required to fill the hole. This device will automatically shut off the pump when the hole is full.

Swabbing. Pulling the drillstring from the borehole creates swab pressures. Swab pressures are negative and reduce the effective hydrostatic pressure throughout the hole and below the bit. If this pressure reduction lowers the effective hydrostatic pressure below the formation pressure, a potential kick has developed. Variables controlling swab pressures are pipe pulling speed, mud properties, hole configuration, and the effect of "balled" equipment. Some swab pressures can be seen in **Table 4.2.**

Cut Mud. Gas-contaminated mud will occasionally cause a kick, although this is rare. The mud density reduction is usually caused by fluids from the core volume being cut and released into the mud system. As the gas is circulated to the surface, it expands and may reduce the overall hydrostatic pressure sufficient enough to allow a kick to occur.

Although the mud weight is cut severely at the surface, the hydrostatic pressure is not reduced significantly because most gas expansion occurs near the surface and not at the hole bottom.

Lost Circulation. Occasionally, kicks are caused by lost circulation. A decreased hydrostatic pressure occurs from a shorter mud column. When a kick occurs from lost circulation, the problem may become severe. A large volume of kick fluid may enter the hole before the rising mud level is observed at the surface. It is recommended that the hole be filled with some type of fluid to monitor fluid levels if lost circulation occurs.

4.1.5 Warning Signs of Kicks. Warning signs and possible kick indicators can be observed at the surface. Each crew member has the responsibility to recognize and interpret these signs and take proper action. All signs do not positively identify a kick; some merely warn of potential kick situations. Key warning signs to watch for include the following:
 • Flow rate increase.
 • Pit volume increase.
 • Flowing well with pumps off.
 • Pump pressure decrease and pump stroke increase.
 • Improper hole fill-up on trips.
 • String weight change.
 • Drilling break.
 • Cut mud weight.

Each is identified below as a primary or secondary warning sign, relative to its importance in kick detection.

Flow Rate Increase (Primary Indicator). An increase in flow rate leaving the well, while pumping at a constant rate, is a primary kick indicator. The increased flow rate is interpreted as the formation aiding the rig pumps by moving fluid up the annulus and forcing formation fluids into the wellbore.

Pit Volume Increase (Primary Indicator). If the pit volume is not changed as a result of surface-controlled actions, an increase indicates a kick is occurring. Fluids entering the wellbore displace an equal volume of mud at the flowline, resulting in pit gain.

Flowing Well With Pumps Off (Primary Indicator). When the rig pumps are not moving the mud, a continued flow from the well indicates a kick is in progress. An exception is when the mud in the drillpipe is considerably heavier than in the annulus, such as in the case of a slug.

Pump Pressure Decrease and Pump Stroke Increase (Secondary Indicator). A pump pressure change may indicate a kick. Initial fluid entry into the borehole may cause the mud to flocculate and temporarily increase the pump pressure. As the flow continues, the low-density influx will displace heavier drilling fluids and the pump pressure may begin to decrease. As the fluid in the annulus becomes less dense, the mud in the drillpipe tends to fall and pump speed may increase.

Other drilling problems may also exhibit these signs. A hole in the pipe, called a "washout," will cause pump pressure to decrease. A twist-off of the drillstring will give the same signs. It is proper procedure, however, to check for a kick if these signs are observed.

Improper Hole Fill-Up on Trips (Primary Indicator). When the drillstring is pulled out of the hole, the mud level should decrease by a volume equivalent to the removed steel. If the hole does not require the calculated volume of mud to bring the mud level back to the surface, it is assumed a kick fluid has entered the hole and partially filled the displacement volume of the drillstring. Even though gas or salt water may have entered the hole, the well may not flow until enough fluid has entered to reduce the hydrostatic pressure below the formation pressure.

String Weight Change (Secondary Indicator). Drilling fluid provides a buoyant effect to the drillstring and reduces the actual pipe weight supported by the derrick. Heavier muds have a greater buoyant force than less dense muds. When a kick occurs, and low-density formation fluids begin to enter the borehole, the buoyant force of the mud system is reduced, and the string weight observed at the surface begins to increase.

Drilling Break (Secondary Indicator). An abrupt increase in bit-penetration rate, called a "drilling break," is a warning sign of a potential kick. A gradual increase in penetration rate is an abnormal pressure indicator and should not be misconstrued as an abrupt rate increase.

When the rate suddenly increases, it is assumed that the rock type has changed. It is also assumed that the new rock type has the potential to kick (as in the case of a sand), whereas the previously drilled rock did not have this potential (as in the case of shale). Although a drilling break may have been observed, it is not certain that a kick will occur, only that a new formation has been drilled that may have kick potential.

It is recommended when a drilling break is recorded that the driller should drill 3 to 5 ft (1 to 1.5 m) into the sand and then stop to check for flowing formation fluids. Flow checks are not always performed in tophole drilling or when drilling through a series of stringers in which repetitive breaks are encountered; unfortunately, many kicks and blowouts have occurred because of this lack of flow checking.

Cut Mud Weight (Secondary Indicator). Reduced mud weight observed at the flow line has occasionally caused a kick to occur. Some causes for reduced mud weight are core volume cutting, connection air, or aerated mud circulated from the pits and down the drillpipe. Fortunately, the lower mud weights from the cuttings effect are found near the surface (generally because of gas expansion) and do not appreciably reduce mud density throughout the hole. **Table 4.3** shows that gas cutting has a very small effect on bottomhole hydrostatic pressure.

An important point to remember about gas cutting is that if the well did not kick within the time required to drill the gas zone and circulate the gas to the surface, only a small possibility exists that it will kick. Generally, gas cutting indicates that a formation has been drilled that contains gas. It does not mean that the mud weight must be increased.

TABLE 4.3—EFFECT OF GAS-CUT MUD ON THE BOTTOMHOLE HYDROSTATIC PRESSURE			
	Pressure reduction (psi)		
Depth (ft)	10 lbm/gal cut to 5 lbm/gal	18 lbm/gal cut to 16.2 lbm/gal	18.0 lbm/gal cut to 9 lbm/gal
1,000	51	31	60
5,000	72	41	82
10,000	86	48	95
20,000	97	51	105

4.2 Kick Detection and Monitoring With MWD Tools

Measurement while drilling (MWD) systems monitor mud properties, formation parameters, and drillstring parameters during circulation and drilling operations. The system is widely used for drilling, but it also has applications for well control, including the following:

• Drilling-efficiency data, such as downhole weight on bit and torque, can be used to differentiate between rate of penetration changes caused by drag and those caused by formation strength. Monitoring bottomhole pressure, temperature, and flow with the MWD tool is not only useful for early kick detection, but can also be valuable during a well-control kill operation. Formation evaluation capabilities, such as gamma ray and resistivity measurements, can be used to detect influxes into the wellbore, identify rock lithology, and predict pore pressure trends.

• The MWD tool enables monitoring of the acoustic properties of the annulus for early gas-influx detection. Pressure pulses generated by the MWD pulser are recorded and compared at the standpipe and the top of the annulus. Full-scale testing has shown that the presence of free gas in the annulus is detected by amplitude attenuation and phase delay between the two signals. For water-based mud systems, this technique has demonstrated the capacity to consistently detect gas influxes within minutes before significant expansion occurs. Further development is currently under way to improve the system's capability to detect gas influxes in oil-based mud.

• Some MWD tools feature kick detection through ultrasonic sensors. In these systems, an ultrasonic transducer emits a signal that is reflected off the formation and back to the sensor. Small quantities of free gas significantly alter the acoustic impedance of the mud. Automatic monitoring of these signals permits detection of gas in the annulus. It should be noted that these devices only detect the presence of gas at or below the MWD tool.

The MWD tool offers kick-detection benefits if the response time is less than the time it takes to observe the surface indicators. The tool can provide early detection of kicks and potential influxes as well as monitor the kick-killing process. Tool response time is a function of the complexity of the MWD tool and the mode of operation. The sequence of data transmission determines the update times of each type of measurement. Many MWD tools allow for reprogramming of the update sequence while the tool is in the hole. This feature can enable the operator to increase the update frequency of critical information to meet the expected needs of the section being drilled. If the tool response time is longer than required for surface indicators to be observed, the MWD only serves as a confirmation source.

4.3 Shut-In Procedures

When one or more warning signs of kicks are observed, steps should be taken to shut in the well. If there is any doubt that the well is flowing, shut it in and check the pressures. It is important to remember that there is no difference between a small-flow well and a full-flowing well, because both can very quickly turn into a big blowout.

In the past, there has been some hesitation to close in a flowing well because of the possibility of sticking the pipe. It can be proven that, for common types of pipe sticking (e.g., differential pressure, heaving, or sloughing shale), it is better to close in the well quickly, re-

TABLE 4.4—THE EFFECT OF CONTINUOUS INFLUX ON THE CASING PRESSURE RESULTING FROM FAILURE TO CLOSE IN THE WELL	
Volume of Gas Gained (bbl)	Casing Pressure (psi)
20	1,468
30	1,654
40	1,796

duce the kick influx, and, thereby, reduce the chances of pipe sticking. The primary concern at this point is to kill the kick safely; when feasible, the secondary concern is to avoid pipe sticking.

There is some concern about fracturing the well and creating an underground blowout resulting from shutting in the well when a kick occurs. If the well is allowed to flow, it will eventually become necessary to shut in the well, at which time the possibility of fracturing the well will be greater than if the well had been shut in immediately after the initial kick detection. **Table 4.4** shows an example of higher casing pressures resulting from continuous flow because of failure to close in the well.

4.3.1 Initial Shut-In. Considerable discussion has occurred regarding the merits of "hard" shut-in procedures vs. "soft" shut-in procedures. In the hard shut-in procedure, the annular preventer(s) are closed immediately after the pumps are shut down. In soft shut-in procedures, the choke is opened before closing the preventers, and then, once the preventers are closed, the choke is closed. Some arguments in favor of soft shut-in procedures are that they avoid a "water-hammer" effect caused by abruptly stopping the fluid flow, and they provide an alternate means of well control (i.e., the low-choke-pressure method) if the casing pressure becomes excessive. But, the water-hammer effect has no proven substance, and the low-choke-pressure method is an unreliable procedure. The primary argument against the soft shut-in procedure is that a continuous influx is permitted while the procedures are executed. For these reasons, only the hard shut-in procedures are presented in this chapter.

Hard shut-in procedures for well control depend on the type of rig and the drilling operation occurring when the kick is taken, such as the following:
- Drilling—land or bottom-supported offshore rig.
- Tripping—land or bottom-supported offshore rig.
- Drilling—floating rig.
- Tripping—floating rig.
- Diverter procedures—all rigs (when surface pipe is not set).

Drilling—Land or Bottom-Supported Offshore Rig. These rigs do not move during normal drilling operations. They include land-and-barge rigs, jack-ups, and platform rigs.

Shut-In Procedures. When a primary kick warning sign has been observed, do the following immediately:
1. Raise the kelly until a tool joint is above the rotary table.
2. Stop the mud pumps.
3. Close the annular preventer.
4. Notify company personnel.
5. Read and record the shut-in drillpipe pressure, the shut-in casing pressure, and the pit gain.

Raising the kelly is an important procedure. With the kelly out of the hole, the valve at the bottom of the kelly can be closed if necessary. Also, the annular-preventer members can attain a more secure seal on the pipe than a kelly.

Tripping—Land or Bottom-Supported Offshore Rig. A high percentage of well-control problems occur when a trip is being made. The kick problems may be compounded when the rig crew is preoccupied with the trip mechanics and fails to observe the initial warning signs of the kick.

Shut-In Procedures. When a primary warning sign of a kick has been observed, do the following immediately:

1. Set the top tool joint on the slips.
2. Install and make up a full-opening, fully opened safety valve on the drillpipe.
3. Close the safety valve and the annular preventer.
4. Notify company personnel.
5. Pick up and make up the kelly.
6. Open the safety valve.
7. Read and record the shut-in drillpipe pressure, shut-in casing pressure, and pit gain.

Installing a fully opened, full-opening safety valve in preference to an inside blowout preventer (BOP), or float, valve is a prime consideration because of the advantages offered by the full-opening valve. If flow is encountered up the drillpipe as a result of a trip kick, the fully opened, full-opening valve is physically easier to stab. Also, a float-type inside-BOP valve would automatically close when the upward-moving fluid contacts the valve.

If wireline work, such as drillpipe perforating or logging, becomes necessary, the full-opening valve will accept logging tools approximately equal to its inside diameter, whereas the float valve may prohibit wireline work altogether. After the kick is shut in, an inside-BOP float valve may be stabbed on the full-opening valve to allow stripping operations.

Drilling—Floating Rig. A floating rig moves during normal drilling operations. The primary types of floating vessels are semisubmersibles and drillships.

Several differences in shut-in procedures apply to floating rigs. Drillstring movement can occur, even with a motion compensator in operation. Also, the BOP stack is on the sea floor. To solve the problem of possible vessel and drillstring movement, and the resulting wear on the preventers, a tool joint may be lowered on the closed pipe rams. The string weight is hung on these rams. This procedure may not be necessary if the rig has a functional motion compensator.

When the stack is located a considerable distance from the rig floor, the problem is to ensure that a tool joint does not interfere with closing the preventer elements. A spacing-out procedure should be executed when the BOP is tested. After running the BOP stack, close the rams, slowly lower the drillstring until a tool joint contacts the rams, and record the position of the kelly at that point. Space out should occur so that a tool joint and lower-kelly valve are above the rotary table. Spacing should be correlated to tide-measuring equipment on the rig floor.

The following procedure could be altered in emergency situations to use the annular preventer and motion compensator for cases in which the shut-in drillpipe pressure and shut-in casing pressure are low and near the same value (indicating oil or water), or the "kick volume" is less than 20 to 30 bbl and the time to kill the well is less than 2 to 3 hours. The closing pressure on the annular preventer must be reduced to the range recommended by the manufacturer for this situation to avoid annular element failure.

Shut-In Procedures. When a primary warning sign of a kick has been observed, do the following immediately:

1. Raise the kelly to the level previously designated during the spacing-out procedure (tide adjusted).
2. Stop the mud pumps.
3. Close the annular preventer.
4. Notify company personnel.
5. Close the upper set of pipe rams.
6. Reduce the hydraulic pressure on the annular preventer.
7. Lower the drillpipe until the pipe is supported entirely by the rams.

8. Read and record the shut-in drillpipe pressure, shut-in casing pressure, and pit gain.

Tripping—Floating Rig. The procedures for kick closure during a tripping operation on a floater is a combination of floating drilling procedures and immobile rig-tripping procedures.

Shut-In Procedures. When a primary warning sign of a kick has been observed, do the following immediately:

1. Set the top tool joint on the slips.
2. Install and make up a full-opening, fully opened safety valve in the drillpipe.
3. Close the safety valve and the annular preventer.
4. Notify company personnel.
5. Pick up and make up the kelly.
6. Reduce the hydraulic pressure on the annular preventer.
7. Lower the drillpipe until the rams support it.
8. Read and record the shut-in drillpipe pressure, shut-in casing pressure, and pit gain.

Diverter Procedures—All Rigs. When a kick occurs in a well with insufficient casing to safely control a kick, a blowout will occur. Because a shallow underground blowout is difficult to control and may cause the loss of the rig, an attempt is usually made to divert the surface blowout away from the rig. This is common practice on land or offshore rigs that are not mobile. Special attention must be given to this procedure so that the well is not shut in until after the diverter lines are opened.

Shut-In Procedures. When a primary warning sign of a kick has been observed, do the following immediately:

1. Raise the kelly until a tool joint is above the rotary table.
2. Increase the pump rate to maximum output.
3. Open the diverter line valve(s).
4. Close the diverter unit (or annular preventer).
5. Notify company personnel.

Recent experiences show that shallow gas flows are difficult to control, but the industry philosophy is improving, and new handling procedures are being developed.

4.3.2 Crewmember Responsibilities for Shut-In Procedures. Each crewmember has different responsibilities during shut-in procedures. These responsibilities follow and are listed according to job classification.

Floorhand (Roughneck). These responsibilities for shut-in procedures belong to the floorhand:

1. Notify the driller of any observed kick-related warning signs.
2. Assist in installing the full-opening safety valve if a trip is being made.
3. Initiate well-control responsibilities after shut-in.

Derrickman. These responsibilities for shut-in procedures belong to the derrickman:

1. Notify the driller of any observed kick-related warning signs.
2. Initiate well-control responsibilities.
3. Begin mud-mixing preparations.

Driller. These responsibilities for shut-in procedures belong to the driller:

1. Immediately shut in the well if any of the primary kick-related warning signs are observed.
2. If a kick occurs while making a trip, set the top tool joint on the slips and direct the crews in the installation of the safety valve before closing the preventers.
3. Notify all proper company personnel.

4.4 Obtaining and Interpreting Shut-In Pressures

"Shut-in pressures" are defined as pressures recorded on the drillpipe and on the casing when the well is closed. Although both pressures are important, the drillpipe pressure will be used almost exclusively in killing the well. The shut-in drillpipe pressure is shown as p_{sidp}. Shut-in casing pressure is p_{sic}. (At this point, assume that the drillpipe does not contain a float valve.)

4.4.1 Reading Pressures. During a kick, fluids flow from the formation into the wellbore. When the well is closed to prevent a blowout, pressure builds at the surface because of formation fluid entry into the annulus, as well as because of the difference between the mud hydrostatic pressure and the formation pressure.

Because this pressure imbalance cannot exist for long, the surface pressures will finally build so that the surface pressure, plus the mud and influx hydrostatic pressures in the well, are equal to the formation pressures. Eqs. 4.1 and 4.2 express this relationship for the drillpipe and the annular side, respectively:

$$p_{form} = p_{sidp} + p_{dph}, \quad\dotfill (4.1)$$

where p_{sidp} = shut-in drillpipe pressure, psi; p_{dph} = drillpipe hydrostatic pressure, psi; p_{form} = formation pressure, psi; and

$$p_{form} = p_{sic} + p_{ah} + p_i, \quad\dotfill (4.2)$$

where p_{sic} = shut-in casing pressure, psi; p_{ah} = annular-hydrostatic pressure, psi; and p_i = influx-hydrostatic pressure, psi.

Example 4.1 and **Fig. 4.1** show how the shut-in pressures are read.

Example 4.1 While drilling at 15,000 ft, the driller observed several primary warning signs of kicks and proceeded to shut in the well. After the shut-in was completed (note: the well was shut in at 6 a.m.), he called company personnel and began recording the pressures and pit gains in **Table 4.5.**

After 15 minutes, the final shut-in pressures were recorded as follows:

p_{sidp}= 780 psi
p_{sic}= 1,040 psi
Pit gain = 20 bbl

4.4.2 Interpreting Recorded Pressures. An important basic principle can be seen in Fig. 4.1. It shows that formation pressure (p_{form}) is greater than the drillpipe hydrostatic pressure by an amount equal to the p_{sidp}. The drillpipe pressure gauge is the bottomhole pressure gauge. The casing pressure cannot be considered a direct bottomhole pressure gauge because of generally unknown amounts of formation fluid in the annulus.

4.4.3 Constant-Bottomhole-Pressure Concept. Fig. 4.1 can be used to illustrate another important basic principle. It was stated that the 780 psi (5.4 MPa) observed on the drillpipe gauge was the amount necessary to balance mud pressure at the hole bottom with the pressure in the gas sand at 15,000 ft (4600 m). Formation fluids travel from areas of high pressure to areas of lower pressures only. They do not travel between areas of equal pressures, assuming gravity segregation is neglected.

If the drillpipe pressure is controlled so that the total mud pressure at the hole bottom is slightly greater than formation pressure, then there will be no additional kick influx entering the well. This concept is the basis of the constant-bottomhole-pressure method of well control, in which the pressure at the hole bottom is constant and at least equal to formation pressure.

4.4.4 Effects of Time. In Example 4.1, 15 minutes were used to obtain shut-in pressures. The purpose of this time is to allow pressures to reach equilibrium sufficient to balance formation

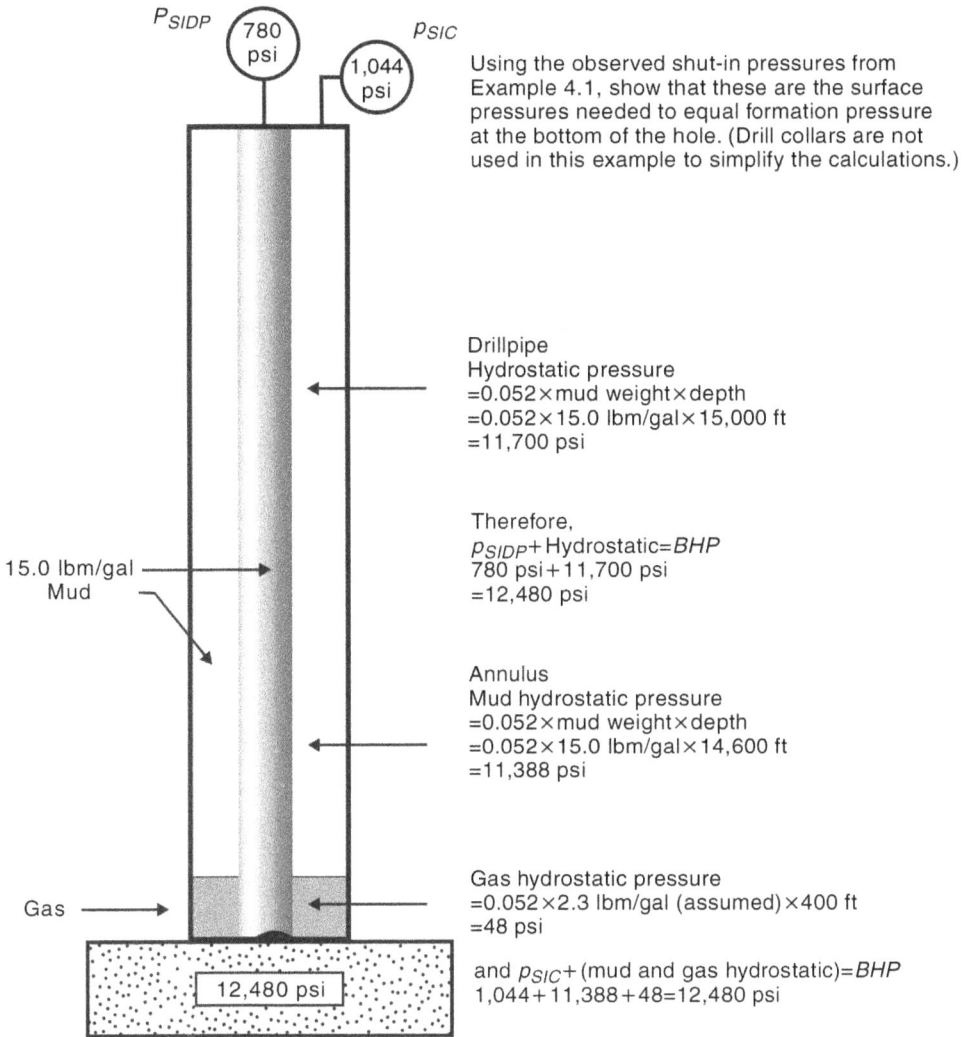

P_{SIDP} 780 psi

p_{SIC} 1,044 psi

Using the observed shut-in pressures from Example 4.1, show that these are the surface pressures needed to equal formation pressure at the bottom of the hole. (Drill collars are not used in this example to simplify the calculations.)

Drillpipe
Hydrostatic pressure
$=0.052 \times$ mud weight $\times$ depth
$=0.052 \times 15.0$ lbm/gal $\times 15,000$ ft
$=11,700$ psi

Therefore,
$p_{SIDP}+$ Hydrostatic$=BHP$
780 psi$+11,700$ psi
$=12,480$ psi

15.0 lbm/gal
Mud

Annulus
Mud hydrostatic pressure
$=0.052 \times$ mud weight $\times$ depth
$=0.052 \times 15.0$ lbm/gal $\times 14,600$ ft
$=11,388$ psi

Gas hydrostatic pressure
$=0.052 \times 2.3$ lbm/gal (assumed) $\times 400$ ft
$=48$ psi

Gas

and $p_{SIC}+$ (mud and gas hydrostatic)$=BHP$
$1,044+11,388+48=12,480$ psi

12,480 psi

NOTE: In practical situations, the amount or type of influx will not be (exactly) known and, therefore, the annulus pressure should not be used to calculate formation pressures.

Fig. 4.1—Pressure relationships at shut-in conditions.

pressures. The required time will depend on variables such as the rock type, permeability, porosity, and the original amount of pressure underbalance. This may take a few minutes to several hours. The required time depends on conditions surrounding the kick.

Several other factors affect the time allowed for pressures to stabilize. Gas migration is the movement of low-density fluids up the annulus. It tends to build pressure at the surface, if time is allowed for migration. Also, the influx may have a tendency to deteriorate the hole stability and cause either stuck pipe or hole bridging. These problems must also be considered when reading the shut-in pressures.

4.4.5 Trapped Pressure. "Trapped pressure" is any pressure recorded on the drillpipe or annulus greater than the amount needed to balance the bottomhole pressure. Pressure can be trapped

TABLE 4.5—SHUT-IN PRESSURES AND PIT GAIN			
Shut-In Time	Shut-In Drillpipe (psi)	Shut-In Casing Pressure (psi)	Pit Gain (bbl)
6:00 a.m.	50	950	20
6:05 a.m.	750	1,000	20
6:10 a.m.	775	1,040	20
6:15 a.m.	780	1,040	20

TABLE 4.6—GUIDELINES TO CHECK FOR TRAPPED PRESSURE
1.　Bleed from the casing side only when checking for trapped pressure. The reasons for this are that the choke is located on the casing side, it avoids contamination of the mud in the drillpipe, and because it avoids the possibility of plugging the bits jets.
2.　Use the drillpipe pressure as a guide because it is a direct bottomhole pressure indicator.
3.　Bleed small amounts ($^1/_4$ to $^1/_2$ bbl) of mud at a time.
4.　Close the choke after bleeding and observe the pressure on the drillpipe.
5.　Continue to alternate the bleeding and observing the pressure as long as the drillpipe pressure continues to decrease. When it ceases to fall, stop bleeding and record the true shut-in drillpipe and casing pressure.
6.　If the drillpipe pressure should decrease to zero during this procedure, continue to bleed and check pressures on the casing side as long as the casing pressure decreases.*
*This step is not normally necessary.

in the system in several ways. Common ways include gas migrating up the annulus and tending to expand, or closing the well in before the mud pumps have stopped running. Using pressure readings containing trapped pressure results in erroneous kill calculations.

Guidelines help when releasing trapped pressure. If they are not properly executed, the well will be much more difficult to kill. These guidelines are listed and explained in **Table 4.6.**

Because trapped pressure is greater than the amount needed to balance bottomhole pressure, trapped pressure can be bled without allowing any additional influx into the well. However, after the trapped pressure is bled off, and if the bleeding is continued, more influx will be allowed into the well, and the surface pressures will begin to increase.

Although bleeding procedures can be implemented at any time, it is advisable to check for trapped pressure when the well is shut in initially and to recheck it when the drillpipe is displaced with kill mud if any pressure remains on the shut-in drillpipe.

4.4.6 Drillpipe Floats. A kick can occur when a drillpipe float valve is used. Because a float valve prevents fluid and pressure movement up the drillpipe, there will not be a drillpipe pressure reading after the well is shut in. Several procedures can obtain the drillpipe pressure, and each depends on the amount of information known when the kick occurs.

Table 4.7 describes the procedures to obtain the drillpipe pressure if the slow pumping rate (kill rate) is known. **Table 4.8** outlines the procedure if the kill rate is not known; **Fig. 4.2** illustrates this process.

Table 4.7 is important in other applications. For example, assume a kick was taken, the shut-in drillpipe pressure was known (no float valve), and a kill rate had not yet been established. Step 6 of this table (Eq. 4.3),

$$p_{sidp} = p_\Sigma - p_{kr}, \quad\quad\quad\quad\quad\quad\quad\quad\quad\quad (4.3)$$

TABLE 4.7—PROCEDURES TO ESTABLISH SHUT-IN DRILLPIPE PRESSURE IF THE KILL RATE IS KNOWN
1. Shut in the well, record the shut-in casing pressure (p_{sic}), and obtain the kill rate, either from the driller or the daily tour report.
2. Instruct the driller to start the pumps and maintain the pumping rate at the kill rate (strokes).
3. As the driller starts the pumps, use the choke to regulate the casing pressure at the same pressure that was originally recorded at shut-in conditions.
4. After the pumps are running at the kill rate, with the casing pressure properly regulated at shut-in pressure, record the pressure on the drillpipe while pumping.
5. Shut down the pumps and close the choke.
6. The shut-in drillpipe pressure equals the total pumping pressure minus the kill rate pressure, as shown in Eq. 4.3.

TABLE 4.8—PROCEDURES TO ESTABLISH THE SHUT-IN DRILLPIPE PRESSURE IF THE KILL RATE IS NOT KNOWN
1. Shut in the well.
2. Line up a low-volume, high-pressure reciprocating pump on the standpipe.
3. Start pumping and fill up all of the lines.
4. Gradually increase the torque on the pumps until they begin to move fluid down the drillpipe.
The shut-in drillpipe pressure is the amount of pressure required to initiate the fluid movement. This is assumed to be the amount needed to overcome the pressure acting against the bottom side of the valve.

could be modified to

$$p_{kr} = p_{\Sigma} - p_{sidp}, \qquad\qquad\qquad\qquad\qquad\qquad (4.4)$$

where p_{kr} = pump pressure at kill rate, psi, and p_{Σ} = total pressure, psi.

The procedures in Table 4.7 remain the same, with the exception that Eq. 4.4 would be substituted for Eq. 4.3.

Establishing the shut-in drillpipe pressure becomes more complex if the kill rate is not previously determined and a float valve in the string prohibits pressure readings at the surface. Table 4.8 must be used initially to determine the p_{sidp}, after which Eq. 4.4 and Table 4.7 must be implemented to establish the kill rate.

4.5 Kick Identification

When a kick occurs, note the type of influx (gas, oil, or salt water) entering the wellbore. Remember that well-control procedures developed here are designed to kill all types of kicks safely. The formula required to make this kick influx calculation is as follows:

$$g_i = g_{mdp} - (p_{sic} - p_{sidp}) / h_i, \qquad\qquad\qquad\qquad (4.5)$$

where g_i = influx gradient, psi/ft; g_{mdp} = mud gradient in drillpipe, psi/ft; and h_i = influx height, ft. The influx gradient can be evaluated using the guidelines in **Table 4.9.**

Although p_{sidp} and p_{sic} can be determined accurately for Eq. 4.5, it is difficult to determine the influx height. This requires knowledge of the pit gain and the exact hole size. Example 4.2, described later, illustrates Eq. 4.5.

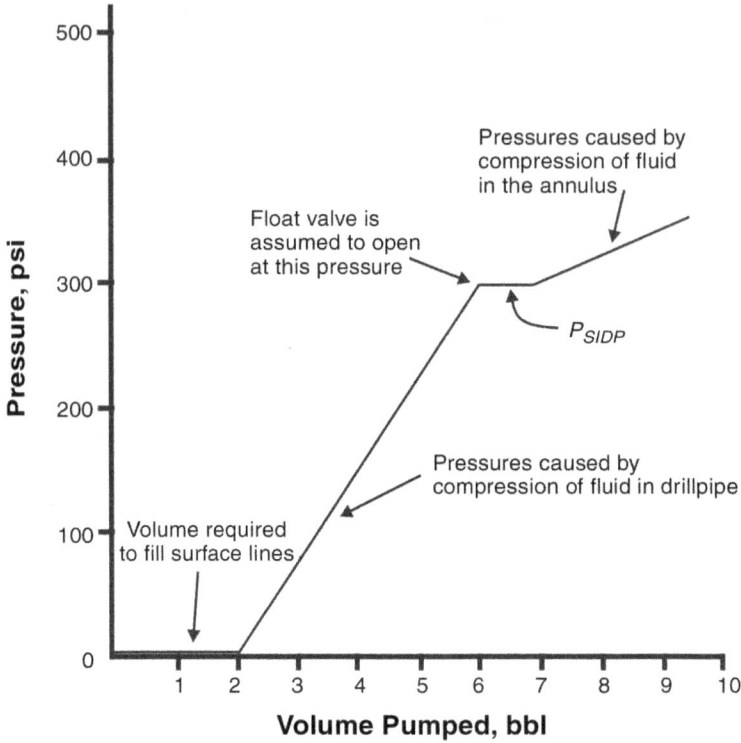

Fig. 4.2—Procedure to establish the shut-in drillpipe pressure when the kill rate is not known.

4.6 Kill-Weight Mud Calculation

It is necessary to calculate the mud weight needed to balance bottomhole formation pressure. "Kill-weight mud" is the amount of mud necessary to exactly balance formation pressure. It will be shown in later sections that it is safer to use the exact required mud weight without variation.

Because the drillpipe pressure has been defined as a bottomhole pressure gauge, the p_{sidp} can be used to calculate the mud weight necessary to kill the well. The kill mud formula follows:

$$\rho_{kw} = 19.23 p_{sidp} / D_{tv} + \rho_o, \quad \text{.. (4.6)}$$

where ρ_{kw} = kill-mud weight, lbm/gal; 19.23 = conversion constant; D_{tv} = true vertical-bit depth, ft; and ρ_o = original mud weight, lbm/gal.

Because the casing pressure does not appear in Eq. 4.6, a high casing pressure does not necessarily indicate a high kill-weight mud. The same is true for pit gain because it does not appear in Eq. 4.6. Example 4.2 uses the kill-weight mud formula.

Example 4.2 What will the kill-weight mud density be for the kick data given below?
D_{tv} = 11,550 ft
ρ_o = 12.1 lbm/gal
p_{sidp} = 240 psi
p_{sic} = 1,790 psi
Pit gain = 85 bbl

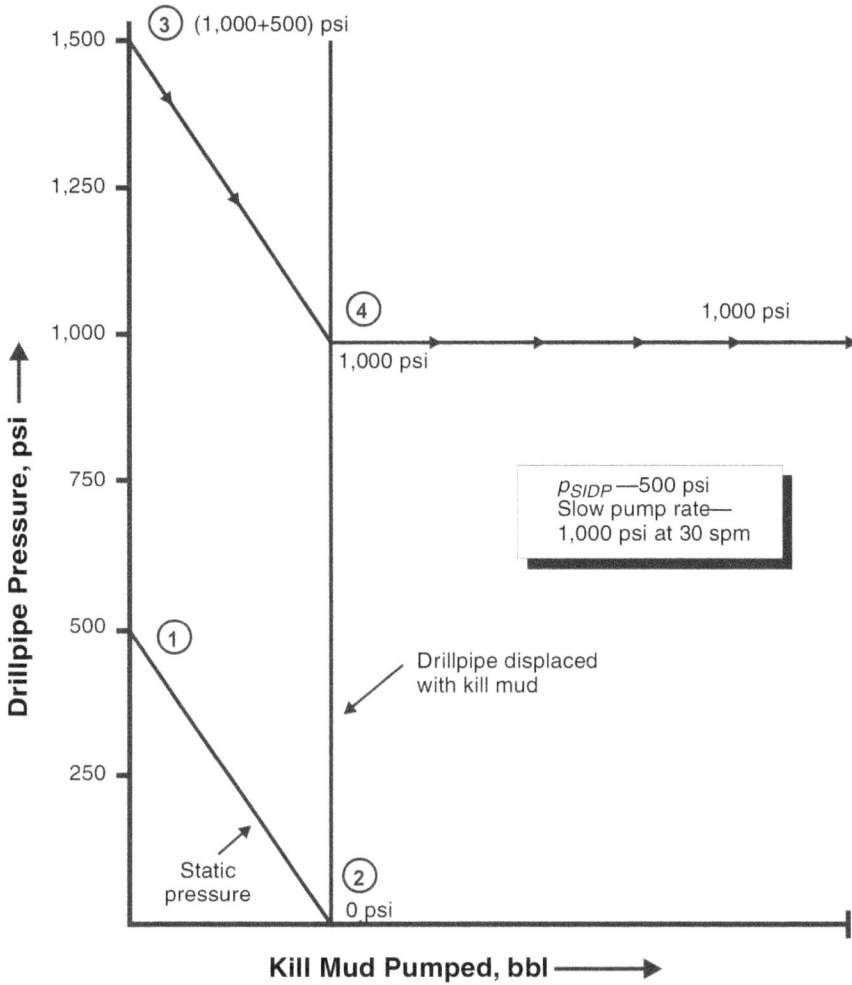

Fig. 4.3—Static drillpipe pressure of the one-circulation method of well control.

TABLE 4.9—INFLUX GRADIENT EVALUATION GUIDELINES	
Gradient (psi/ft)	Influx Type
0.05–0.2	Gas
0.2–0.4	Probable combination of gas, oil, and/or salt water
0.4–0.5	Probable oil or salt water

Solution.

$\rho_{kw} = p_{sidp} \times 19.23/D_{tv} + \rho_o$

$= 240 \text{ psi} \times 19.23/11{,}550 \text{ ft} + 12.1 \text{ lbm/gal}$

$= 0.4 \text{ lbm/gal} + 12.1 \text{ lbm/gal}$

$= 12.5 \text{ lbm/gal}$

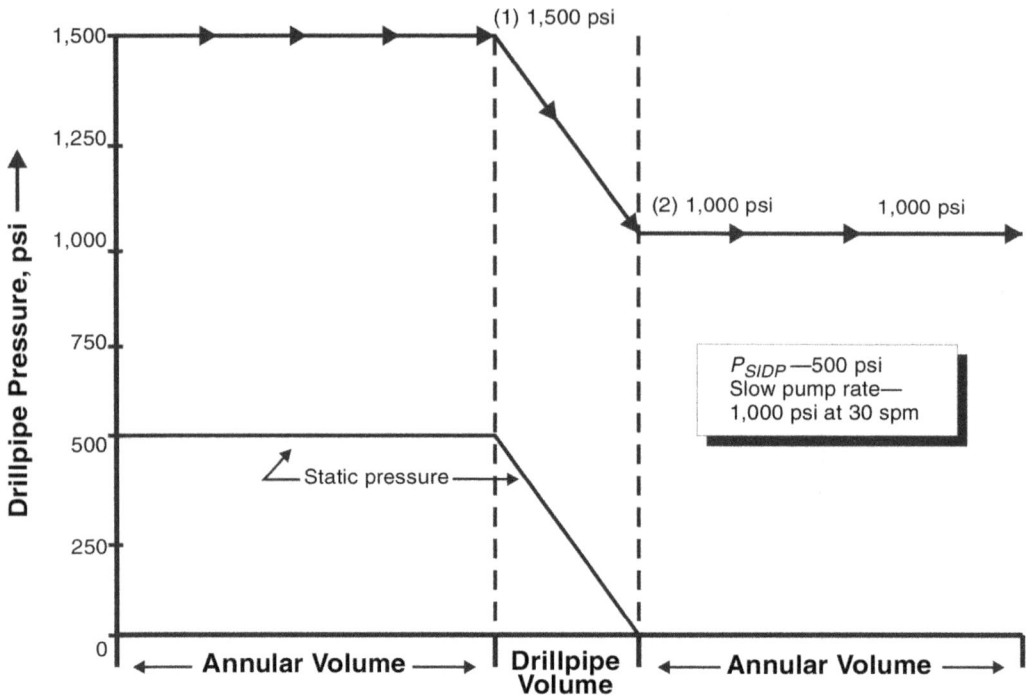

Fig. 4.4—Static drillpipe pressure of the two-circulation method of well control.

4.7 Well-Control Procedures

Many well-control procedures have been developed over the years. Some have used systematic approaches, while others are based on logical, but perhaps unsound, principles. The systematic approaches will be presented here.

In previous sections, the constant-bottomhole-pressure concept was introduced. With this concept, the total pressures (e.g., mud hydrostatic pressure and casing pressure) at the hole bottom are maintained at a value slightly greater than the formation pressures to prevent further influxes of formation fluids into the wellbore. And, because the pressure is only slightly greater than the formation pressure, the possibility of inducing a fracture and an underground blowout is minimized. This concept can be implemented in three ways:

• One-Circulation, or Wait-and-Weight, Method. After the kick is shut in, weight the mud to kill density and then pump out the kick fluid in one circulation using the kill mud. (Another name often applied to this method is "the engineer's method.")

• Two-Circulation, or Driller's, Method. After the kick is shut in, the kick fluid is pumped out of the hole before the mud density is increased.

• Concurrent Method. Pumping begins immediately after the kick is shut in and pressures are recorded. The mud density is increased as rapidly as possible while pumping the kick fluid out of the well.

If applied properly, each method achieves constant pressure at the hole bottom and will not allow additional influx into the well. Procedural and theoretical differences make one procedure more desirable than the others.

4.7.1 One-Circulation Method. Fig. 4.3 depicts the one-circulation method. At Point 1, the shut-in drillpipe pressure is used to calculate the kill-weight mud. The mud weight is increased to kill density in the suction pit. As the kill mud is pumped down the drillpipe, the static drillpipe

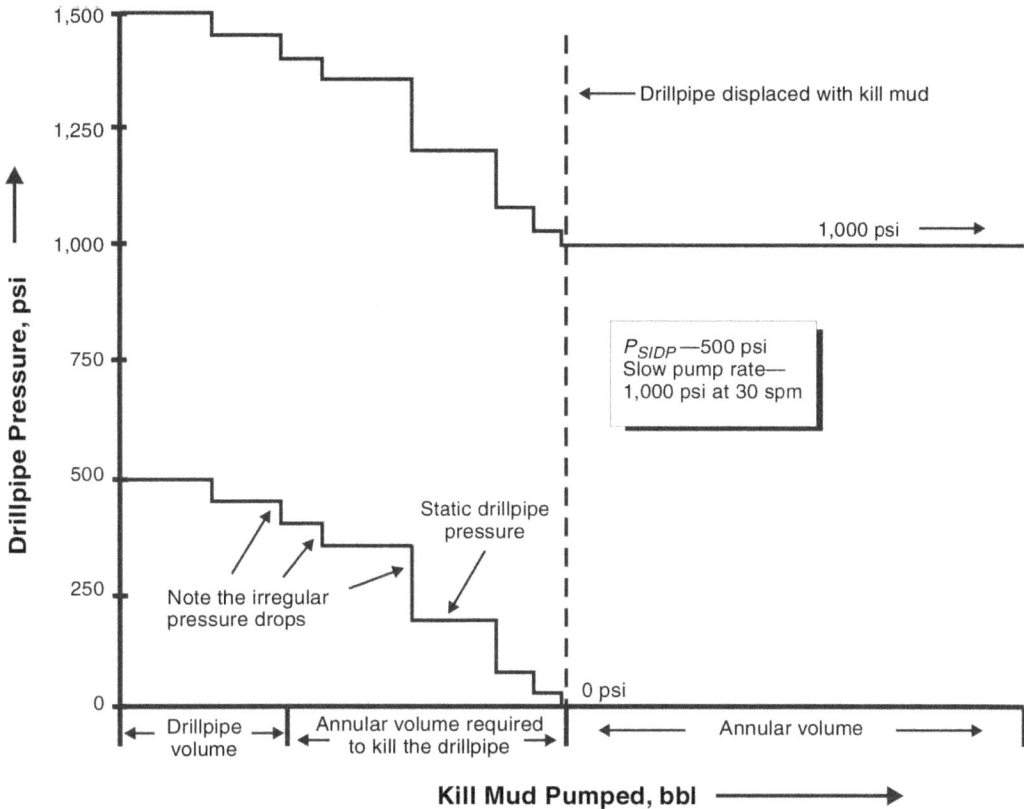

Fig. 4.5—Static drillpipe pressure of the concurrent method.

pressure is controlled to decrease linearly until at Point 2, the drillpipe pressure is zero. The heavy mud has killed the drillpipe pressure. Point 3 shows that the initial pumping pressure on the drillpipe is the total of p_{sidp} plus the kill-rate pressure. While pumping kill mud down the pipe, the circulating pressure decreases until, at Point 4, only the pumping pressure remains. From the time kill mud is at the bit until it reaches the flow line, the choke is used to control the drillpipe pressure at the final circulating pressure. The driller ensures the pump remains at the kill speed.

4.7.2 Two-Circulation Method. In the two-circulation method, the circulation is started immediately. Kill mud is not added in the first circulation. As seen in **Fig. 4.4**, the drillpipe pressure will not decrease during the first circulation. The purpose is to remove the kick fluid from the annulus.

In the second circulation, the mud weight increases, but causes a decrease from the initial pumping pressure at Point 1, to the final circulating pressure at Point 2. This pressure is held constant while the annulus is displaced with kill mud.

4.7.3 Concurrent Method. This method is the most difficult to execute properly (see **Fig. 4.5**). As soon as the kick is shut in and the pressures are read, pumping immediately begins. The mud density is increased as rapidly as rig facilities will allow. The difficulty is determining the mud density being circulated and its relative position in the drillpipe. Because this position determines the drillpipe pressures, the rate of pressure decrease may not be as consistent as in the other two methods. As a new density arrives at the bit, or a predetermined depth,

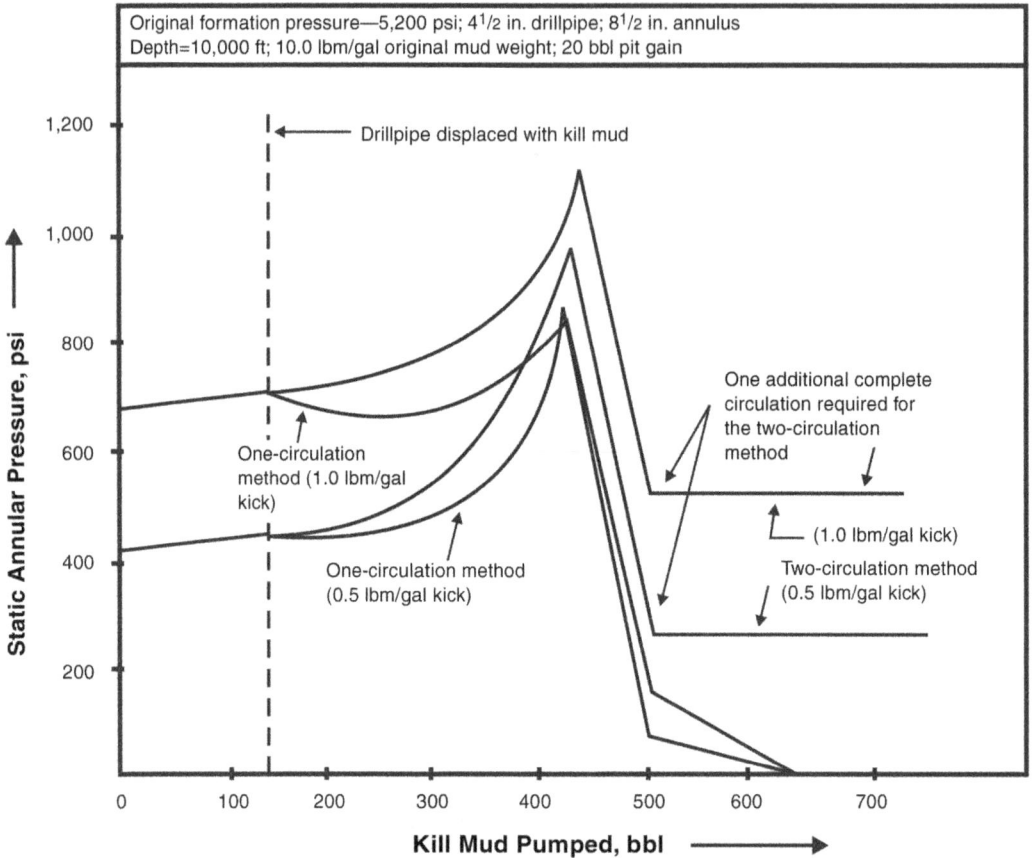

Fig. 4.6—Static annular pressures for one-circulation method vs. two-circulation method in a 10,000-ft well.

the drillpipe pressure is decreased by an amount equal to the hydrostatic pressure of the new mud-weight increment. When the drillpipe is displaced with kill mud, the pumping pressure is maintained constant until kill mud reaches the flow line.

4.8 Choosing the Best Method

Determining the best well-control method for most situations involves several considerations including the time required to execute the kill procedure, the surface pressures from the kick, the complexity relative to the ease of implementation, and the downhole stresses applied to the formation during the kick-killing process. All points must be analyzed before a procedure can be selected. The following list briefly summarizes the general opinion in the industry regarding these methods:

• The one-circulation method should be used in most cases.

• The two-circulation method should be used if a good casing shoe exists and there is going to be a delay in weighting up the system.

The concurrent method should be used only in rare cases, such as for a severe (1.5 lbm/gal or greater) kick with a large influx and a potential problem with developing lost circulation. In this case, the pump rate should be kept to a minimum to allow the weight to be raised continuously. In an analysis of kick-killing procedures, emphasis is placed on the one- and two-circulation methods (i.e., the wait-and-weight method and the driller's method, respectively).

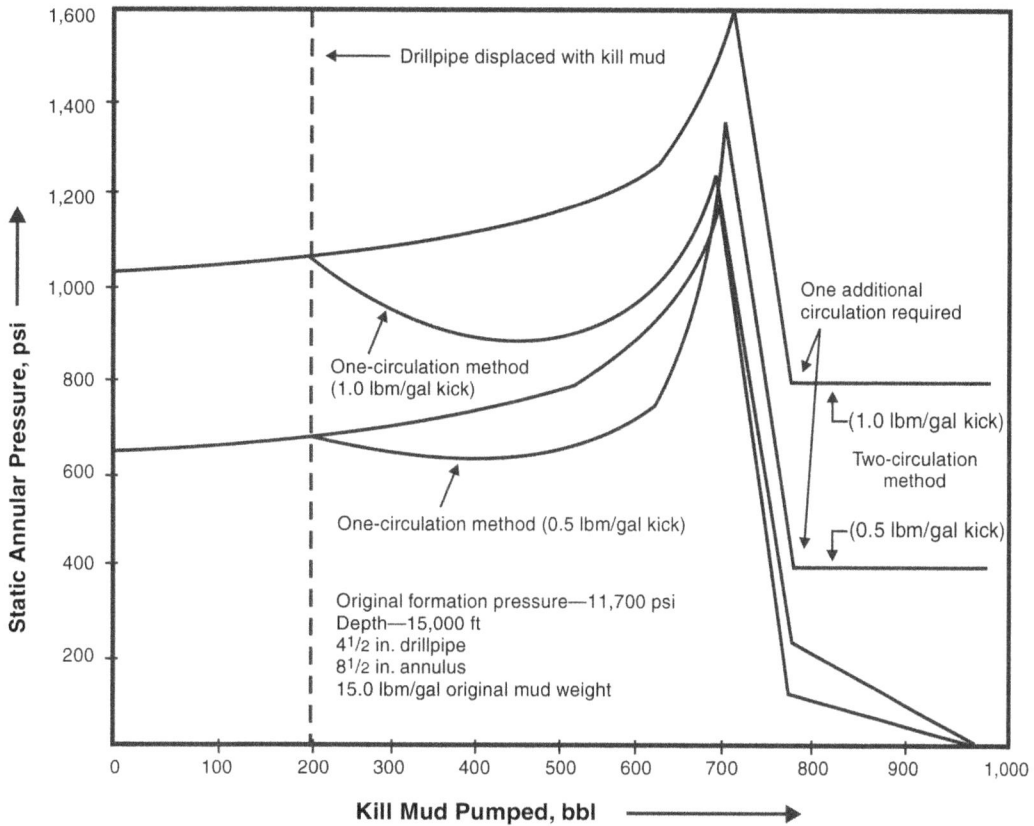

Fig. 4.7—Static annular pressure for one-circulation method vs. two-circulation method in a 15,000-ft well.

Inspection of the procedures will show that these are opposite approaches, while the concurrent method falls somewhere in between.

4.8.1 Time. Two important considerations relative to time are required for the kill procedure: initial wait time and overall time required. The first concern with time is the amount required to increase the mud density from the original weight to the final kill-weight mud. Because some operators are very concerned with pipe sticking during this time, the well-control procedure that minimizes the initial wait time is often chosen. These are the concurrent method and the two-circulation method. In both procedures, pumping begins immediately after the shut-in pressures are recorded.

The other important time consideration is the overall time required for the complete procedure to be implemented. Fig. 4.3 shows that the one-circulation method requires one complete fluid displacement (i.e., within the drillpipe and the annulus), while the two-circulation method (Fig. 4.4) requires the annulus to be displaced twice, in addition to the drillpipe displacement. In certain situations, extra time for the two-circulation method may be extensive with respect to hole stability or preventer wear.

4.8.2 Surface Pressures. During the course of well killing, surface pressures may approach alarming heights. This may be a problem in gas-volume expansion near the surface. The kill procedure with the least surface pressure required to balance the bottomhole formation pressure is important.

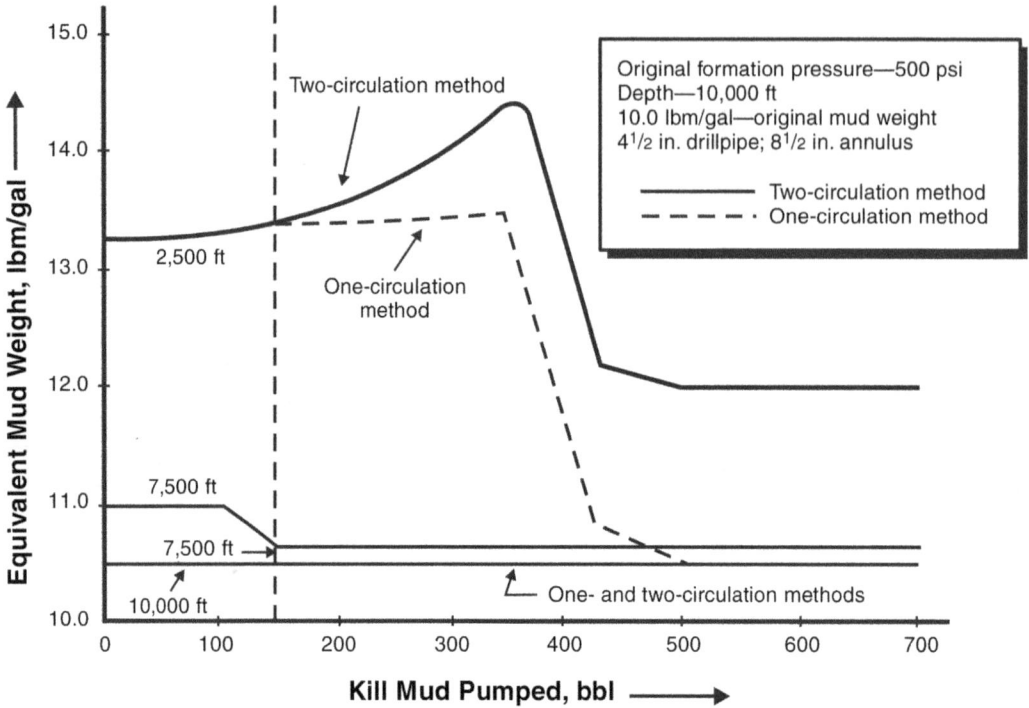

Fig. 4.8—Equivalent mud-weight comparison for the one-circulation vs. the two-circulation kill procedure (0.5-lbm/gal kick at 10,000 ft).

Figs. 4.6 and 4.7 show the different surface-pressure requirements for several kick situations. The first major difference is noted immediately after the drillpipe is displaced with kill mud. The amount of casing pressure required begins to decrease because of the increased kill-mud hydrostatic pressure during the one-circulation procedure. This decrease is not seen in the two-circulation method because this procedure does not circulate kill mud initially. In fact, in the two-circulation method, the casing pressure increases as the gas-bubble expansion displaces mud from the hole.

The second difference in pressure occurs as the gas approaches the surface. The two-circulation procedure has higher pressures resulting from the lower-density original mud weight. It is interesting to note these high casing pressures that are necessary to suppress the gas expansion to a small degree result in a later arrival of gas at the surface.

4.8.3 Procedure Complexity. Process suitability partially depends on the ease with which the procedure can be executed. The same principle holds true for well control. If a kick-killing procedure is difficult to comprehend and implement, its reliability diminishes.

The concurrent method is less reliable because of its complexity. To perform this procedure properly, the drillpipe pressure must be reduced according to the mud weight being circulated and its position in the pipe. This implies that the crew will inform the operator when a new mud weight is being pumped, that the rig facilities can maintain this increased mud-weight increment, and that the mud-weight position in the pipe can be determined by counting pump strokes. Many operators have stopped using this complex method entirely.

One- and two-circulation methods are used more prominently because of their ease of application. In both procedures, the drillpipe pressure remains constant for long intervals of time. In

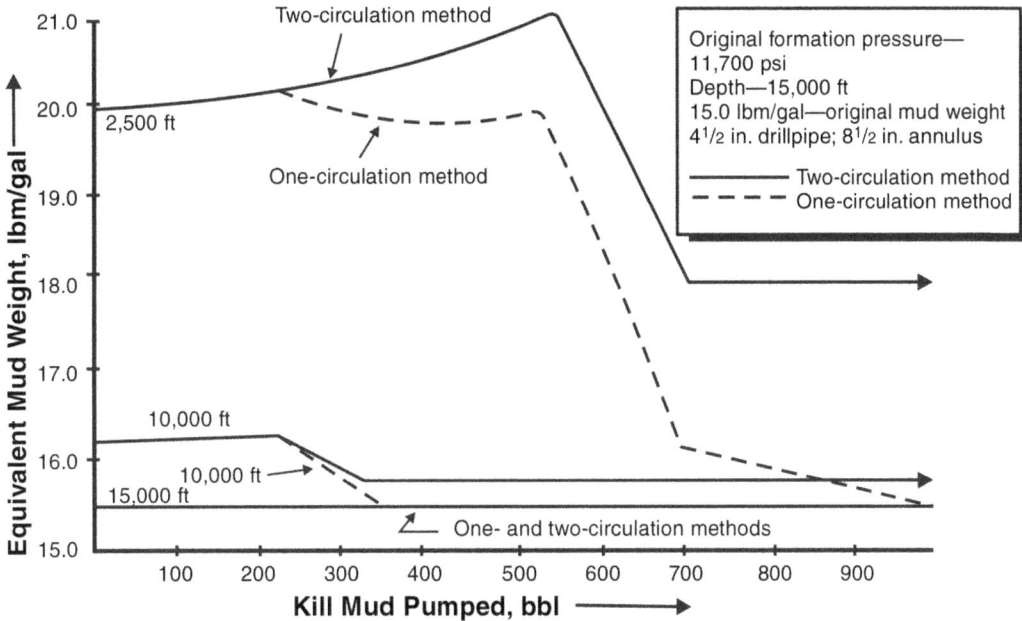

Fig. 4.9—Equivalent mud-weight comparison for the one-circulation vs. the two-circulation kill procedure (0.5-lbm/gal kick at 15,000 ft).

addition, while displacing the drillpipe with kill mud, the drillpipe pressure decrease is virtually a straight-line relationship, not staggered, as in the concurrent method (Fig. 4.5).

4.8.4 Downhole Stresses. Although all considerations for choosing the best method are important, the primary concern should always be the stresses imposed on the borehole wall. If the kick-imposed stresses are greater than the formation can withstand, an induced fracture occurs, creating the possibility of an underground blowout. The procedure that imposes the least downhole stress while maintaining constant pressures on the kicking zone is considered the most conducive to safe kick killing.

One way to measure downhole stresses is by use of "equivalent mud weights," or the total pressures to a depth converted to lbm/gal mud weight. For example,

$$\rho_e = 19.23 p_\Sigma / D_e, \quad\quad\quad\quad\quad\quad\quad\quad\quad\quad\quad\quad (4.7)$$

where ρ_e = equivalent mud weight, lbm/gal.

The equivalent mud weights for the systems in Figs. 4.6 and 4.7 are presented in **Figs. 4.8 and 4.9**. The one-circulation method has consistently lower equivalent mud weights throughout the killing process after the drillpipe has been displaced. The procedures generally exhibit the same maximum equivalent mud weights. They occur from the time the well is shut in until the drillpipe is displaced.

Figs. 4.8 and 4.9 illustrate an important principle: maximum stresses occur very early in circulation for the deeper depth, not at the maximum casing pressure intervals. The maximum lost-circulation possibilities will not occur at the gas-to-surface conditions, as might seem logical. If a fracture is not created at shut-in, it probably will not occur throughout the remainder of the process. A full understanding of this behavior may calm operators' concerns about formation fracturing as the gas approaches the surface.

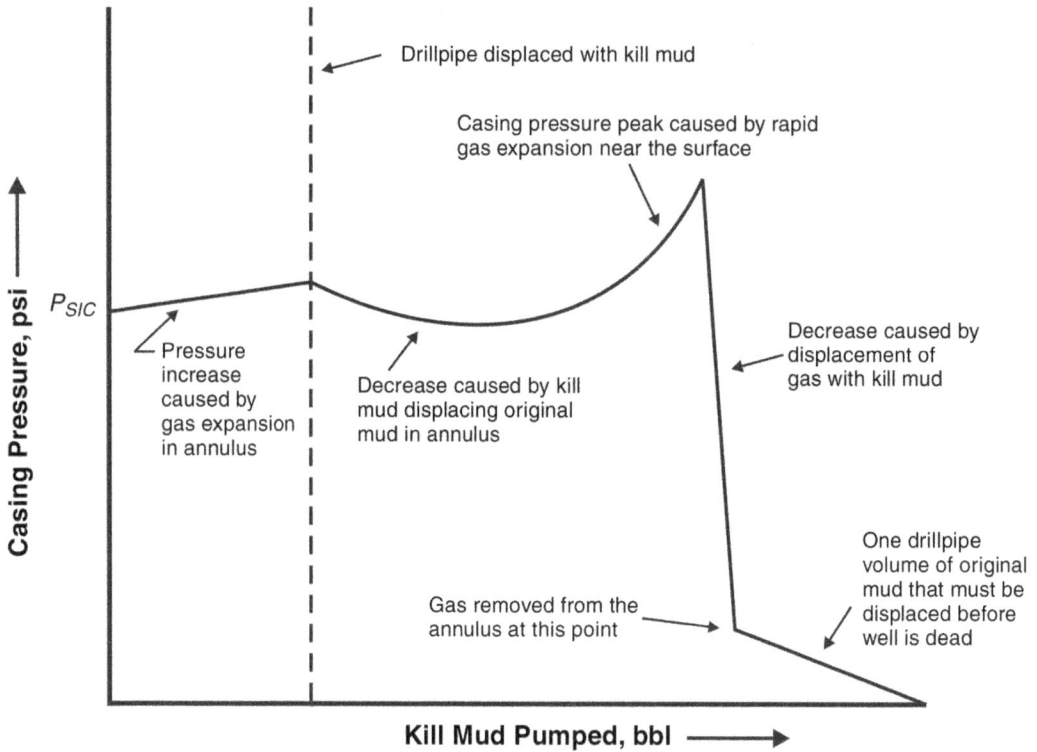

Fig. 4.10—Typical gas-kick casing-pressure curve for the one-circulation method.

4.9 Variables Affecting Kill Procedures

Although variables that affect kick-killing do not necessitate a change in the basic procedural structure, they may cause unexpected behaviors that can mislead an operator into choosing the wrong procedure. The one-circulation method will be used in this section to demonstrate the effect of these variables.

4.9.1 Influx Type. The influx type entering the wellbore plays a key role in casing-pressure behavior. The influx can range from heavy oil to fresh water. The most common is gas or salt water; each has a pronounced casing pressure curve and different downhole effects.

Gas Kicks. Gas kicks are generally more dramatic than other influx types. Reasons for this include the rate at which gas enters the wellbore, the high casing pressures resulting partially from the low-density fluid, gas expansion as it approaches the surface, fluid migration up the wellbore, and fluid flammability. A typical gas-kick casing-pressure curve is shown in **Fig. 4.10**.

Gas expanding from a decrease in confining pressures while the fluid is pumped up the wellbore affects the kick-killing process (Figure 4.10). As the gas begins to expand, the previously decreasing casing pressure begins to increase at an accelerating rate. This higher casing pressure may give the false impression that another kick influx is entering the well. Immediately after the gas-to-surface conditions, the casing pressure decreases rapidly, which may give the impression that lost circulation has occurred. Both casing pressure changes are expected behaviors and do not indicate an additional influx or lost circulation. The possibility of lost circulation is smaller at the gas-to-surface conditions than at the initial shut-in conditions (Figs. 4.8 and 4.9).

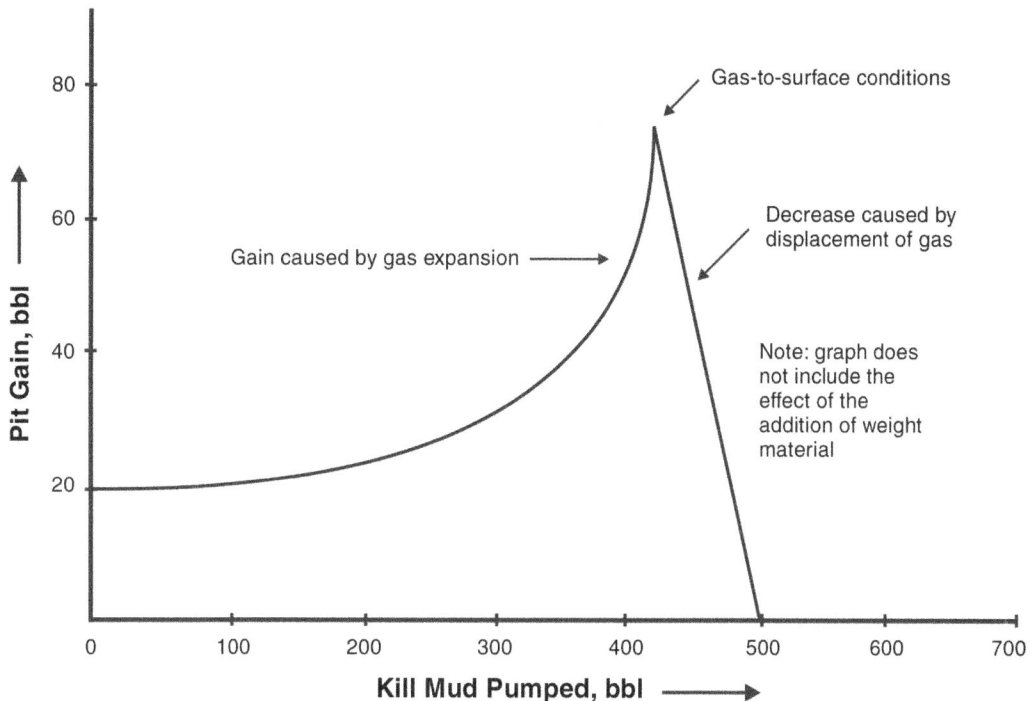

Fig. 4.11—Pit gain for the 1.0-lbm/gal kick in Fig. 4.6.

When gas expands, the increased gas volume displaces fluid from the well, resulting in a pit gain. **Fig. 4.11** shows the pit gain for the problem illustrated in Fig. 4.6. This pit gain is in addition to the volume increase from weight materials. Because the pit gains in volume, the flow rate exiting the well increases **(Fig. 4.12)**.

Gas migration may cause special problems. There have been numerous recent studies of gravity-segregation phenomena in an effort to quantify a migration rate. Field data from one professional well-killing corporation suggests a rate of 7 to 15 ft/min in mud systems. Regardless of the rate, the migration effect must be considered because of the potential for gas expansion. If the fluid is not allowed to expand properly during the migration period, trapped pressure will be generated at the surface. If unnecessary expansion occurs, additional formation gas will enter the well. Example 4.3 illustrates the gas-migration phenomenon with an actual field case.

Example 4.3 While drilling a development well from an offshore platform, a kick was taken. The p_{sidp} was 850 psi, and the p_{sic} was 1,100 psi. Storm conditions forced the tender (barge) to be towed away from the platform to avoid damage to the tender or platform legs. The removal of the tender caused all support services to the platform to be severed, including the mud and pumps.

The engineer on the platform knew the kick would become a problem from gas migration up the annulus. To rectify the situation, he allowed the migration to build pressure on the drillpipe, up to 900 psi, which he used as a 50-psi safety margin. Thereafter, the migration was allowed to build the p_{sidp} up to 950 psi before he bled a small volume of mud from the annulus to reduce the drillpipe pressure down to 900 psi. Because bottomhole pressure was still 50 psi more than formation pressure, no additional influx occurred. This procedure was continued until the gas reached the surface, at which time the pressures ceased to increase and remained

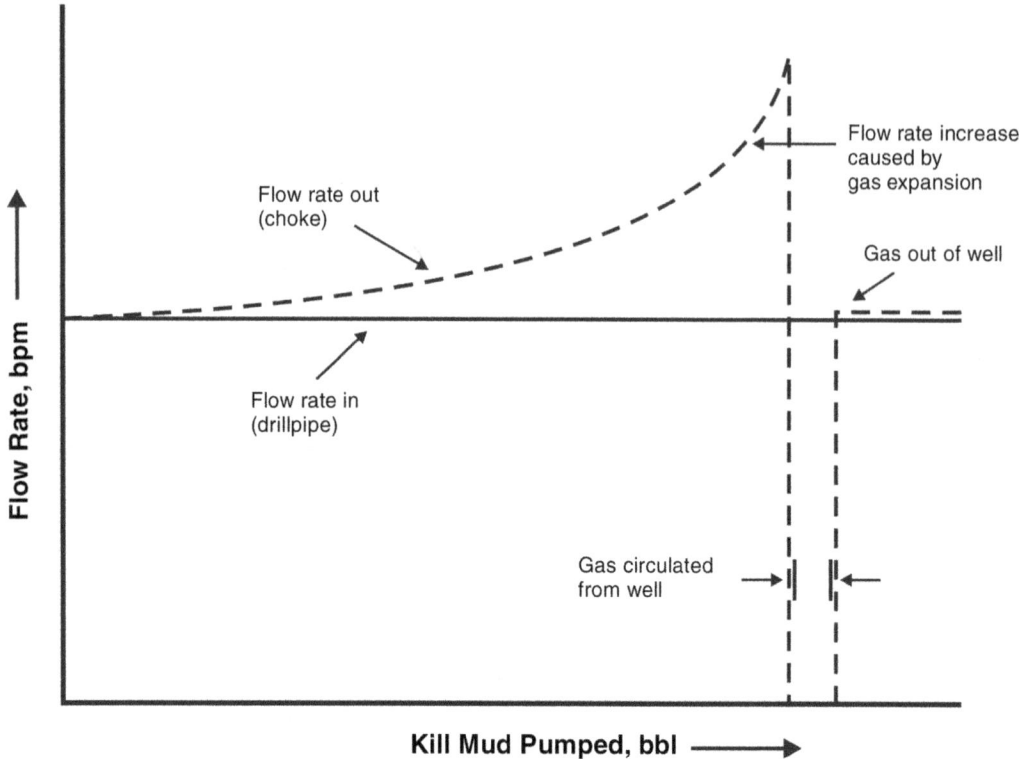

Fig. 4.12—Typical representation of flow rates in and flow rates out of the well during a kick-killing operation.

at 900 psi. After support services were restored to the rig, the gas was pumped from the well, and kill procedures were initiated.

This example points out the manner in which gas migration can be safely controlled with the drillpipe pressure acting as a bottomhole pressure indicator.

Saltwater Kicks. Saltwater-kick problems differ from gas-kick problems. Volume expansion does not occur. Because salt water is more dense than gas, casing pressures are lower than for a comparable volume of gas (**Fig. 4.13**). Shut-in pressures for the 50-bbl (7.9-m³) saltwater kick are approximately the same as those seen in Fig. 4.6 for a 20-bbl (3.2-m³) gas kick under the same conditions.

Hole stability and pipe sticking are generally more severe with a saltwater kick than a gas kick. The saltwater fluid causes a freshwater mud-filter cake to flocculate and create pipe-sticking tendencies and unstable hole conditions. The severity increases with large kick volumes and extended waiting periods before the fluid is pumped from the hole.

4.9.2 Volume of Influx. The fluid volume entering the well is a variable controlling the casing pressure throughout the kill process. Increased influx volumes give rise to higher initial p_{sic}, as well as greater pressure differences at the gas-to-surface conditions. **Fig. 4.14** depicts the importance of quick closure over closure with hesitation.

4.9.3 Kill-Weight Increment Variations. The original mud density must be increased in most kick situations to kill the well. The incremental density increase has some effect on casing

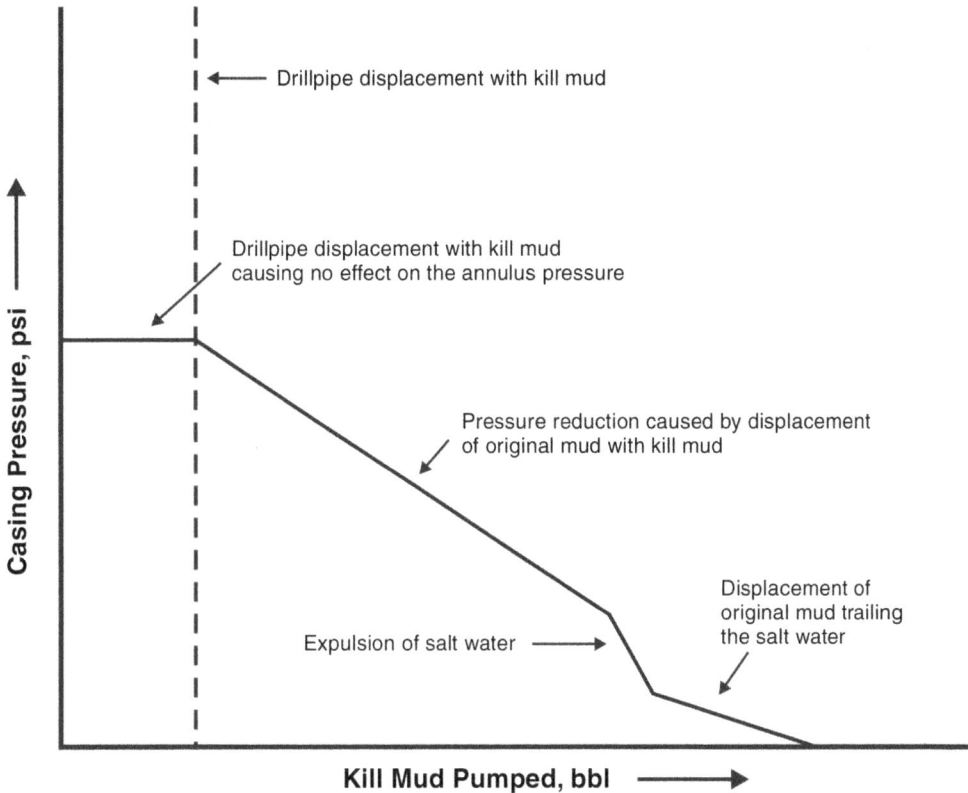

Fig. 4.13—Typical salt water-kick casing-pressure curve.

pressure behavior. In **Fig. 4.15**, the gas-to-surface pressure conditions are higher than the original shut-in pressures for 0.5-lbm/gal (60-kg/m^3) and 1.0-lbm/gal (120-kg/m^3) kicks. The 2.0-lbm/gal (240-kg/m^3) and 3.0-lbm/gal (360-kg/m^3) mud weight increases do not show this tendency. The 3.0-lbm/gal (360-kg/m^3) kick has a lower gas-to-surface pressure than at the initial closure. This is caused by suppressed gas expansion, which minimizes the associated pressures. This is generally observed in kicks requiring greater than a 2.0-lbm/gal (240-kg/m^3) incremental increase.

An important mud-weight variation is the difference between the kill-mud weight necessary to balance bottomhole pressure and the mud weight actually circulated. If the circulated mud is less than the kill-mud weight, the casing pressure is higher than if kill mud had been used because it was necessary to maintain a balanced pressure at the hole bottom (Figs. 4.6 and 4.7). The equivalent mud weights will then be greater, increasing formation fracture possibility.

Circulated mud weights greater than the calculated kill mud weight do not decrease the casing pressure. The situation is synonymous with mud-weight safety factors and is termed "overkill." As the extra-heavy mud is pumped down the drillpipe, the U-tube effect (**Fig. 4.16**) causes the casing pressure to increase (**Fig. 4.17**). The U-tube principle states that the pressures on each side of the tube must be equal. These higher casing pressures have associated downhole stresses that increase formation fracture potential.

Several attempts have been made to achieve the benefits of "safety factors" while avoiding the ill effects of high casing pressures caused by the U-tube effect. The most common attempt at this effort is to subtract the hydrostatic pressure supplied by the extra mud-weight increment from the final circulating pressure, creating a net-zero effect from the added mud weight.

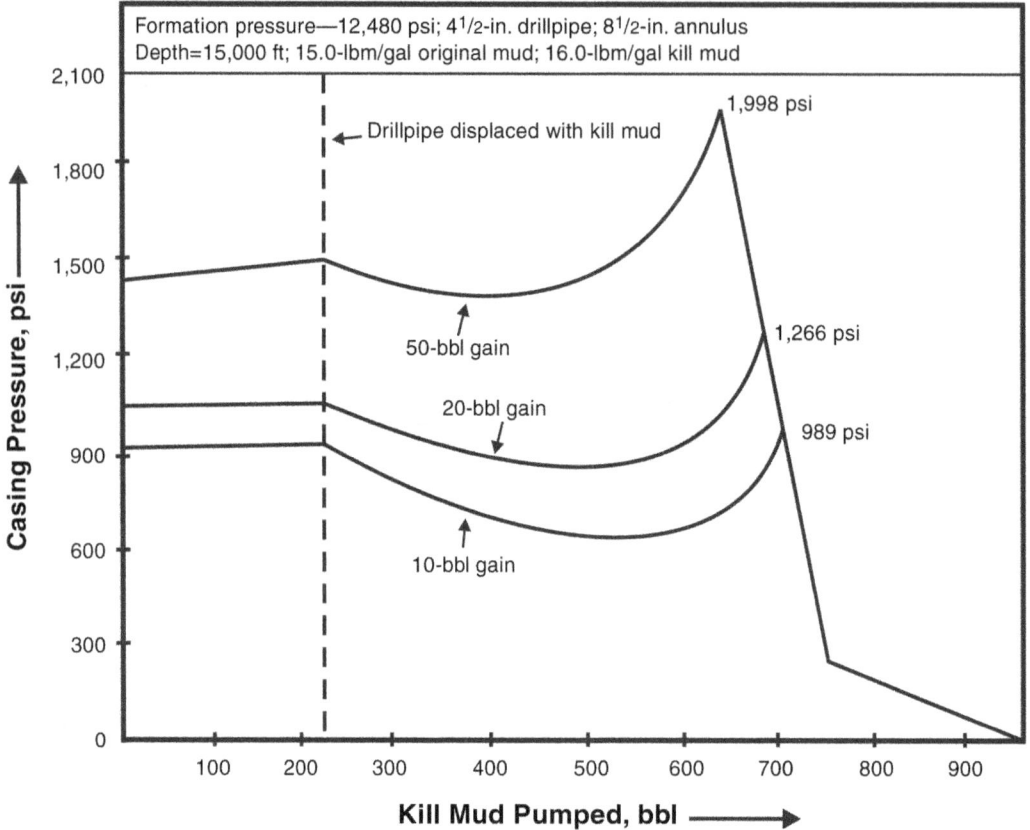

Fig. 4.14—Comparison of casing-pressure curves for 10-, 20-, and 50-bbl kick volumes.

In a static situation, the casing pressure is reduced by an amount equal to the safety-factor hydrostatic pressure, which results in a zero net effect. From a theoretical standpoint, the approach is based on sound principles; however, field experience has shown that this procedure is not practical because of its complexity. This procedure is not necessary for proper well control, and only experienced well-control engineers should use it.

4.9.4 Hole Geometry Variations. In practical kick-killing situations, hole- and drillstring-size changes cause the kick fluid geometry to be altered. This is particularly a problem in deep tapered holes in which several pipe and hole sizes are used. The influx may occupy a large vertical space at the hole bottom, creating a high casing pressure. As the fluid is pumped into the larger annular spaces, the vertical height is decreased, thus increasing the mud column height and resulting in lower casing pressures. **Figs. 4.18a through 4.18c** show a typical tapered hole and the associated casing and drillpipe pressure curves.

4.10 Implementation of the One-Circulation Method

To implement the one-circulation method, certain guidelines must be followed to ensure a safe kick-killing exercise. Although the procedure is relatively simple, its mastery demands basic knowledge of the practical steps taken during the process. Checkpoints indicate potential problems.

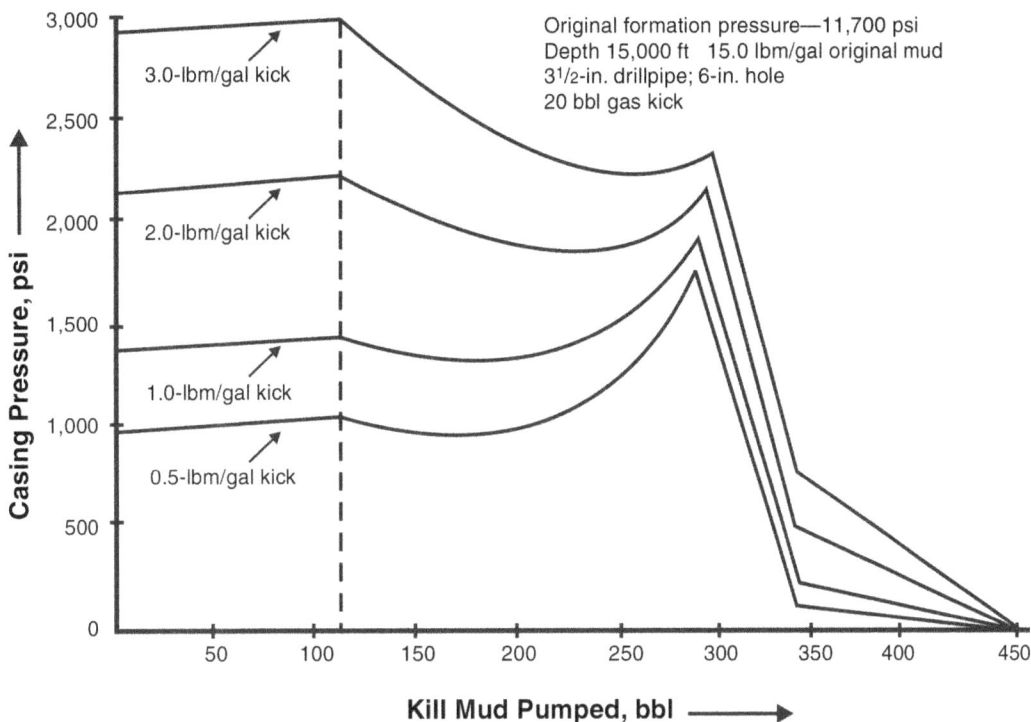

Original formation pressure—11,700 psi
Depth 15,000 ft   15.0 lbm/gal original mud
3¹/₂-in. drillpipe; 6-in. hole
20 bbl gas kick

3.0-lbm/gal kick

2.0-lbm/gal kick

1.0-lbm/gal kick

0.5-lbm/gal kick

Fig. 4.15—Comparison of different required kill-mud weight increments.

A kill sheet is normally used during conventional operations. It contains prerecorded data, formulas for the various calculations, and a graph—or other means—for determining the required pressures on the drillpipe as the kill mud is pumped. Although many operators have complex kill sheets, only the basic required kick-killing data is necessary. A kill sheet is shown in the example problem in the following section.

A summary of the steps involved in proper kick killing follows. The sections not directly applicable to deepwater situations are noted. When a kick occurs, shut in the well using the appropriate shut-in procedures. Once the pressures have stabilized, follow these steps to kill the kick:

1. Read and record the shut-in drillpipe pressure, the shut-in casing pressure, and the pit gain. If a float valve is in the drillpipe, use the established procedures to obtain the shut-in drillpipe pressure.

2. Check the drillpipe for trapped pressure.

3. Calculate the exact mud weight necessary to kill the well and prepare a kill sheet.

4. Mix the kill mud in the suction pit. It is not necessary to weight up the complete surface-mud volume, initially. First pump some mud into the reserve pits.

5. Initiate circulation after the kill mud has been mixed, by adjusting the choke to hold the casing pressure at the shut-in value, while the driller starts the mud pumps. (Not applicable in deep water.)

6. Use the choke to adjust the pumping pressure according to the kill sheet while the driller displaces the drillpipe with the exact kill-mud weight at a constant pump rate (kill rate).

7. Consider shutting down the pumps and closing the choke to record pressures when the drillpipe has been displaced with kill mud. (Note: If the kill mud is highly weighted up, set-

U-Tube—The pressure at the bottom of the casing side must always balance the pressure at the bottom of the drillpipe.

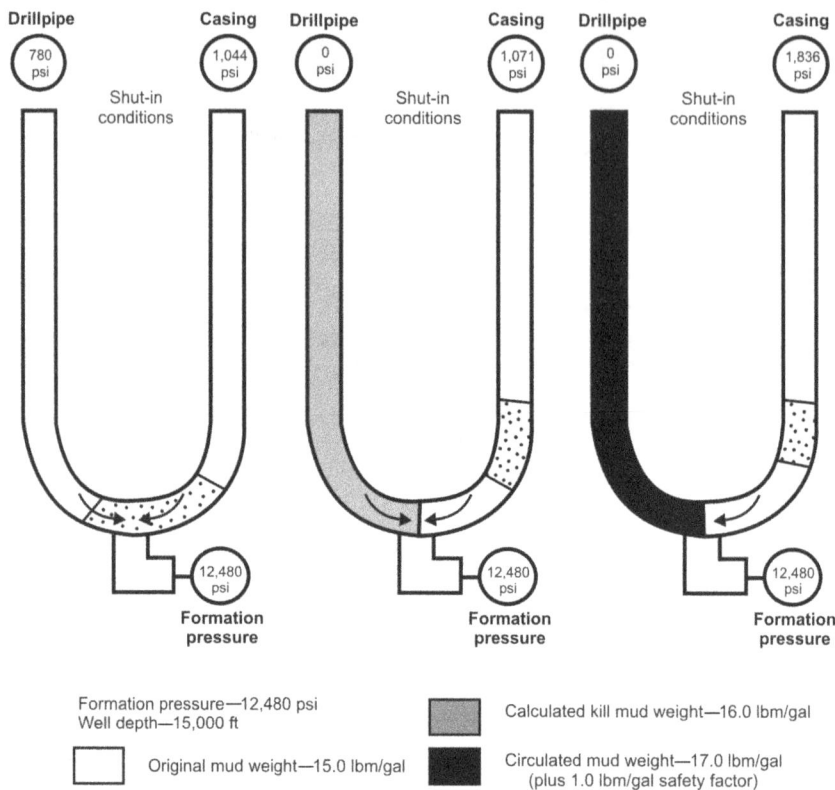

Fig. 4.16—The effect of safety factors (1.0 lbm/gal, in this example) causes higher casing pressure than the proper calculated kill-mud density.

tling and plugging may occur.) The drillpipe pressure should be zero, and the casing should have pressure remaining. If the pressure on the drillpipe is not zero, execute the following steps:

• Check for trapped pressure using the established procedures. If the drillpipe pressure is still not zero, pump an additional 10 to 20 bbl (1.5 to 3 m³) to ensure that kill mud has reached the bit. The pump efficiency may be reduced at the low circulation rate.

• If pressure remains on the drillpipe, recalculate the kill mud weight, prepare a new kill sheet, and return to the first steps of this procedure.

8. Maintain the drillpipe pumping pressure and pumping rate constant to displace the annulus with the kill mud by using the choke to adjust the pressures, as necessary.

9. Shut down the pumps and close the choke after the kill mud has reached the flow line. The well should be dead. If pressure remains on the casing, continue circulation until the annulus is dead.

10. Open the annular preventers, circulate and condition the mud, and add a trip margin when the pressures on the drillpipe and casing are zero. In subsea applications, the trapped gas under the annular is circulated out by pumping down the kill line and up the choke line with the ram preventer below the annular closed. The riser must then be circulated with kill mud by reverse circulation, down the choke line and up the riser, before the preventers can be opened.

Well-control learning experiences are often best accomplished by observing an actual kick problem. Example 4.4 has been provided for this purpose.

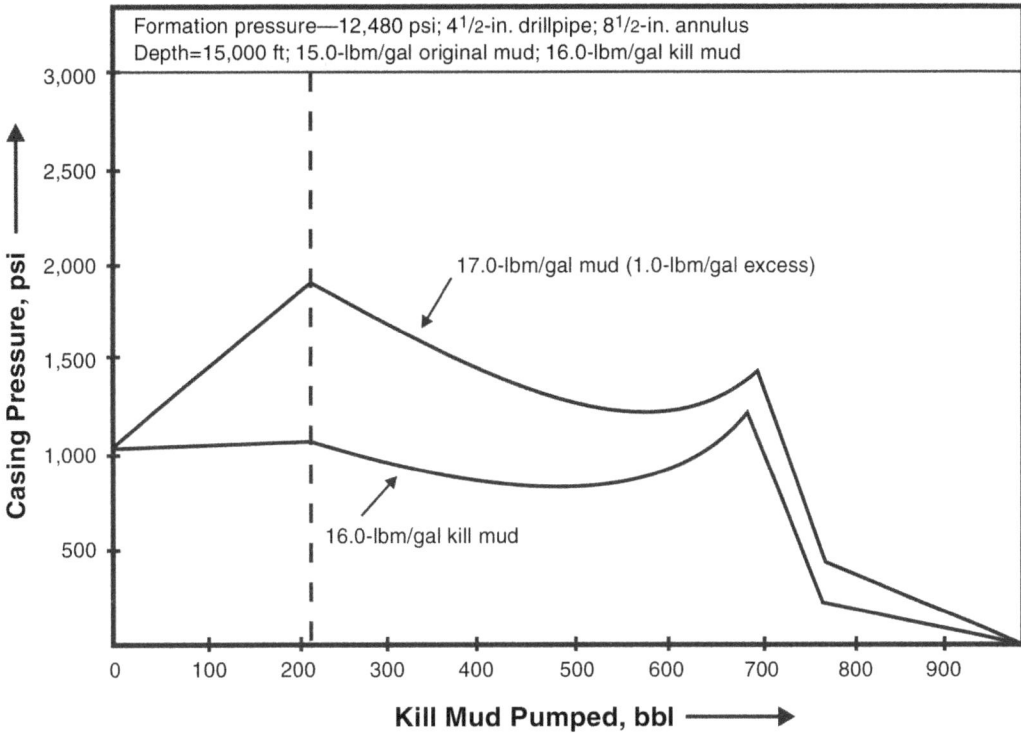

Fig. 4.17—Effect of excess mud weight on annulus pressure.

Example 4.4 Prekick Considerations. While drilling the R.B. Texas No. 1 in the Louisiana Gulf Coast offshore area, a company representative carried out his normal drilling responsibilities related to well control in the event that a kick should occur. Some items that the representative did are listed below:

• Read the appropriate MMS orders and complied with the provisions.

• Checked the barite supplies to ensure that a sufficient amount of barite was on board to kill a 1.0-lbm/gal kick, if necessary.

• Recorded on the driller's book that the kill rate was 21 spm and 800-psi pump pressure.

• Calculated the drillstring volume as follows:

4½-in. drillpipe to 14,000 ft.

6½×2-in. drill collars to 15,000 ft.

4½-in., 16.6-lbf pipe capacity

= 0.01422 bbl/ft×14,000 ft =199 bbl

6½×2-in. collar capacity

=0.0039 bbl/ft×1,000 ft = 3.9 bbl

Total = 199 + 3.9 = 202.9 bbl

Shut-In and Weight-Up Procedures. The drillers on the rig had just changed tours when a drilling break was observed. The well was checked for flow. A flow was recorded with the pumps off, and the following steps were taken:

1. The kelly was raised until a tool joint cleared the floor. (A jackup rig was in use.)

2. The pumps were shut down.

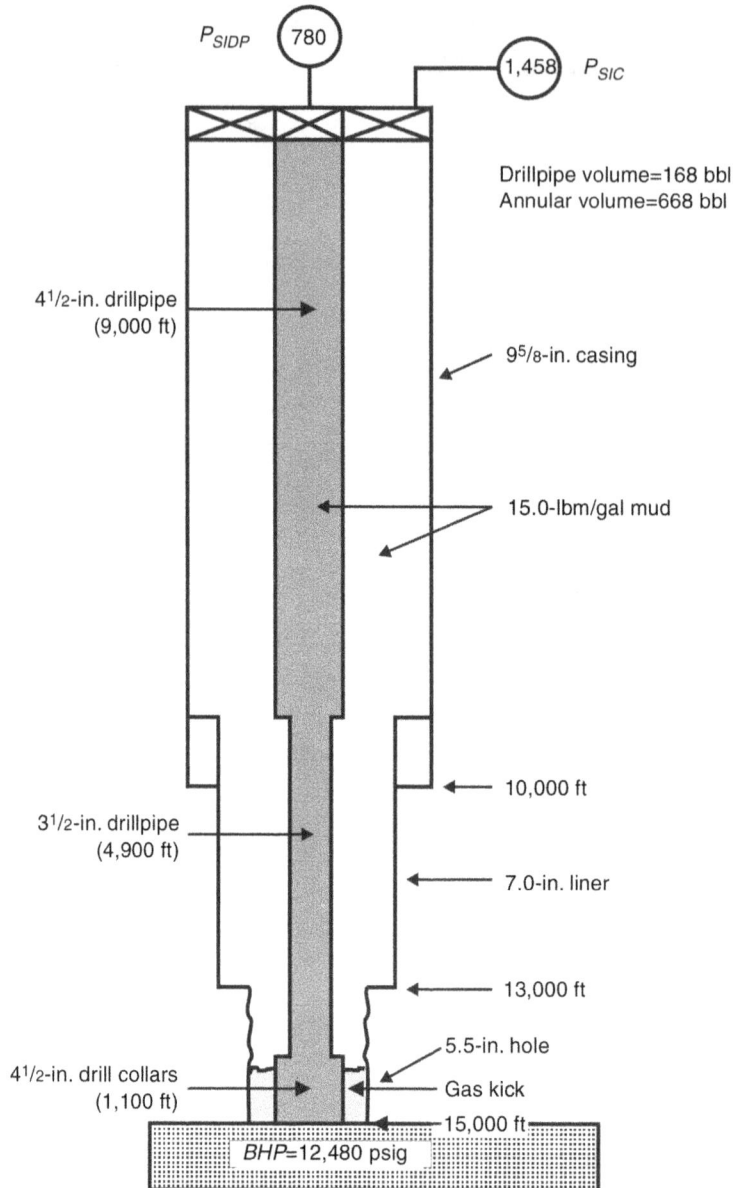

Fig. 4.18a—Tapered hole diagram.

3. The annular preventer was closed.

4. The company representative was notified that the well was shut in.

5. The driller told his crew in the mudroom to stand by in case the mud weight had to be increased. Then, the company representative went to the floor and read the pressures as follows:

p_{sidp} = 240 psi

p_{sic} = 375 psi

Pit gain = 31 bbl

After checking for trapped pressures, he recorded the information on his kill sheet. From the kill sheet, he calculated that he needed to raise the mud weight from the 13.1-lbm/gal origi-

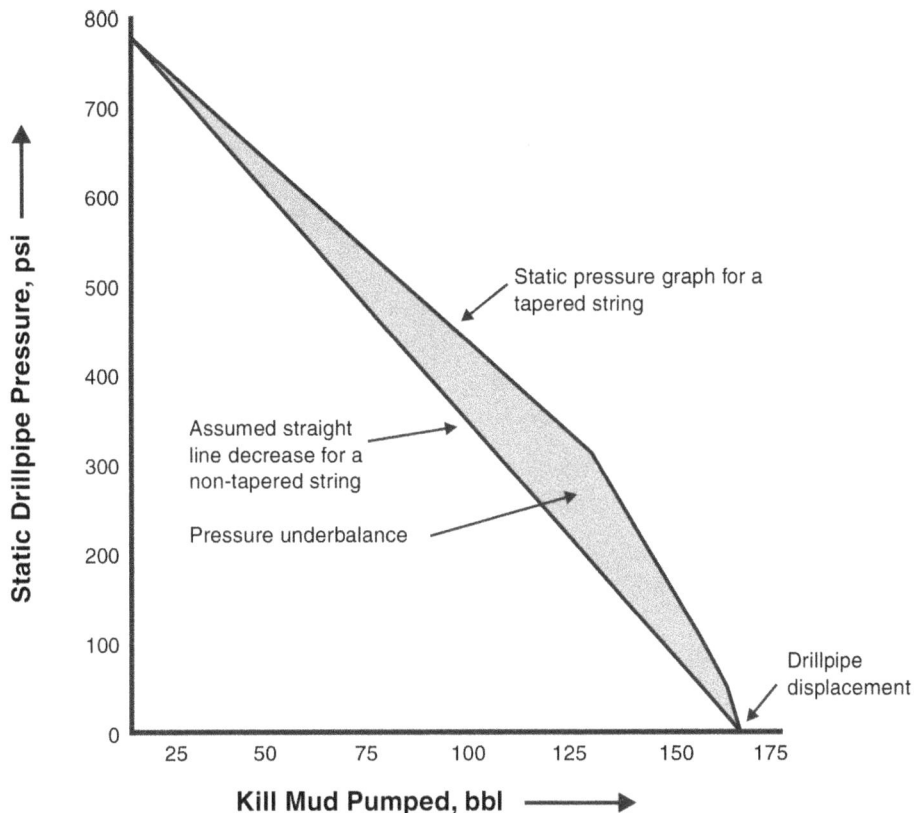

Fig. 4.18b—Static drillpipe pressure for a typical tapered string.

nal weight to 13.4 lbm/gal. He was walking to the mudroom, to tell the derrickman that he needed 13.4-lbm/gal kill mud, when he noticed the pits were almost full. He knew the needed barite would raise the mud level, so he instructed the derrickman to pump off a foot of mud, section off the suction pit, and increase the weight to 13.4 lbm/gal. The representative judged that it would be better to pump off the mud at that time, rather than after the killing operation was started.

Pump Rates. The pump output was read from the mud engineer's report as 5.2 strokes/bbl for the 6×18-in. duplex mud pump. The volumetric output at 21 spm was 0.1916 stroke/bbl×21 spm = 4.0 bbl/min. The representative knew he could cripple his pumps according to the chart previously provided to him but felt that 4.0 bbl/min was not much more than the recommended 1 to 3 bbl/min as a kill rate.

Kill Sheet Preparation. The representative prepared his kill sheet as shown in **Fig. 4.19**.

Working the Pipe. While the mud weight was increased and the kill sheet was being prepared, the driller was instructed to work the pipe every 10 minutes by moving it up and down. He was also instructed not to move a tool joint through the annular preventer.

Displacing the Drillpipe. After the mud was weighted to 13.4 lbm/gal, the representative was ready to displace the drillpipe. He instructed the driller to start his pumps and run them at 21 spm. Then, he cracked open the choke slightly and held his casing pressure at 375 psi until the driller had the pumps at the kill rate. The choke was used to control the drillpipe pressure to decrease it gradually according to values on his kill sheet. The pressures were maintained as shown in **Table 4.10.**

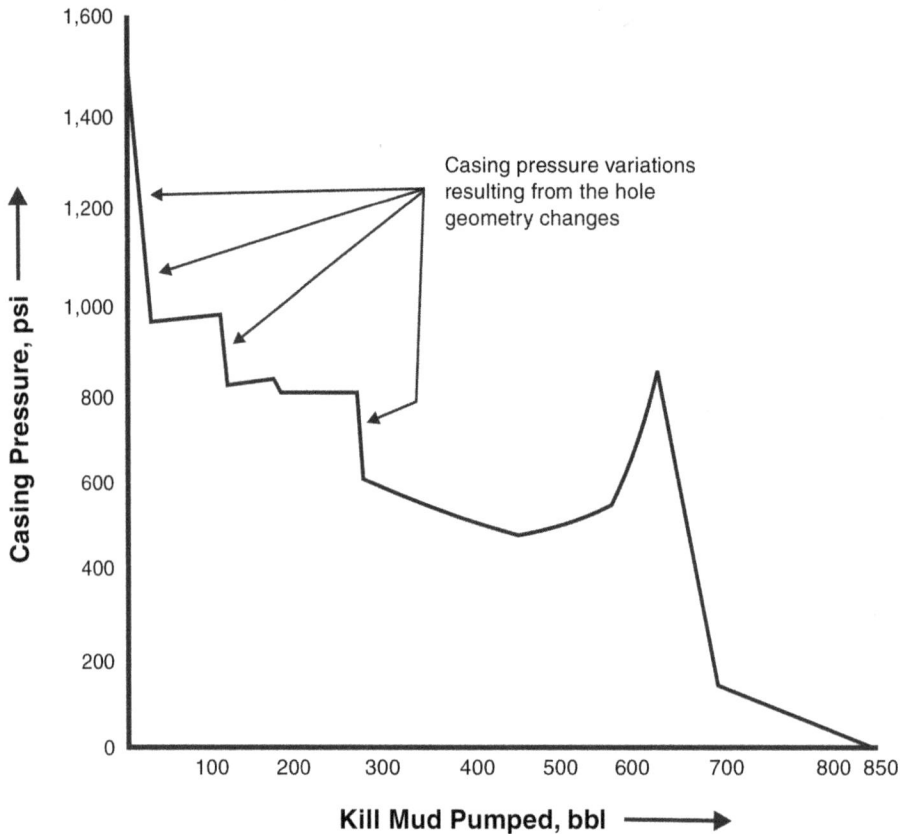

Fig. 4.18c—Effect of hole size changes on casing pressure.

When the drillpipe had been displaced, the pump was shut down and the choke was closed. The pressures were then as follows:

p_{sidp} = 0 psi

p_{sic} = 350 psi

The pressure on the drillpipe told the representative that the heavier kill-mud weight was sufficient to kill the well. If it had not been of sufficient density, some pressure would have remained on the drillpipe.

Displacing the Annulus. The representative was now ready to displace the annulus with kill mud. He initiated pumping by adjusting his choke to maintain 350 psi on the casing while the driller started the pumps. After the pumps were running at 21 spm, he used the choke to maintain the drillpipe pressure constant at the final circulating pressure of 820 psi. He held this pressure until a 13.4-lbm/gal mud weight was observed at the shaker, at which time he closed in the well. The drillpipe and casing had zero pressure. The choke and the annular preventer were opened. The well was dead.

Post-Kick Considerations. There are several items that the representative considered after the well was dead to ensure that the procedure was complete. He circulated and conditioned the mud in the hole and added a trip margin to the mud weight so that he could make a short trip. Additional barite was ordered from the mud company to resupply the bulk tank. He also took time to inspect his equipment to identify any damage sustained from the kick.

Prerecorded Data

Original mud weight = __13.1__ lbm/gal
Slow pump rate = __21__ spm at __800__ psi
Drillpipe volume = __203__ bbl
Annulus volume = __700__ bbl
Pump output = __.1916__ bbl/strokes

$$\text{Drillpipe strokes} = \frac{\text{Drillpipe volume (bbl)}}{\text{Pump output (bbl/stroke)}}$$

$$= \frac{203 \text{ bbl}}{.1916 \text{ bbl/stroke}}$$

$$= \underline{\quad 1060 \quad} \text{ strokes}$$

Kick Data

P_{SIDP} = __240__ psi
P_{SIC} = __375__ psi
Pit gain = __31__ bbl
True vertical depth = __15,000__ ft

Kill Mud Data

$$\text{Mud weight increase} = \frac{P_{SIDP} \times 19.23}{Depth} = \frac{240 \times 19.23}{15,000'} = 0.3 \text{ lbm/gal}$$

Kill mud weight = original weight + increase = __13.1__ lbm/gal + __0.3__ lbm/gal = __13.4__ lbm/gal

Pump Pressure

Initial drillpipe pressure = P_{SIDP} + slow pump pressure
$$= \underline{\quad 240 \quad} \text{ psi} + \underline{\quad 800 \quad} \text{ psi} = \underline{\quad 1040 \quad} \text{ psi}$$

$$\text{Final drillpipe pressure} = \frac{\text{Kill mud weight} \times \text{slow pump pressure}}{\text{Original mud weight}}$$

$$= \frac{13.4 \text{ lbm/gal} \times 800 \text{ psi}}{13.1 \text{ lbm/gal}}$$

$$= \underline{\quad 820 \quad} \text{ psi}$$

Pressure profile

Strokes	Pressure, psi
200	1,000
400	960
600	915
800	870
1,000	830
1,055	820

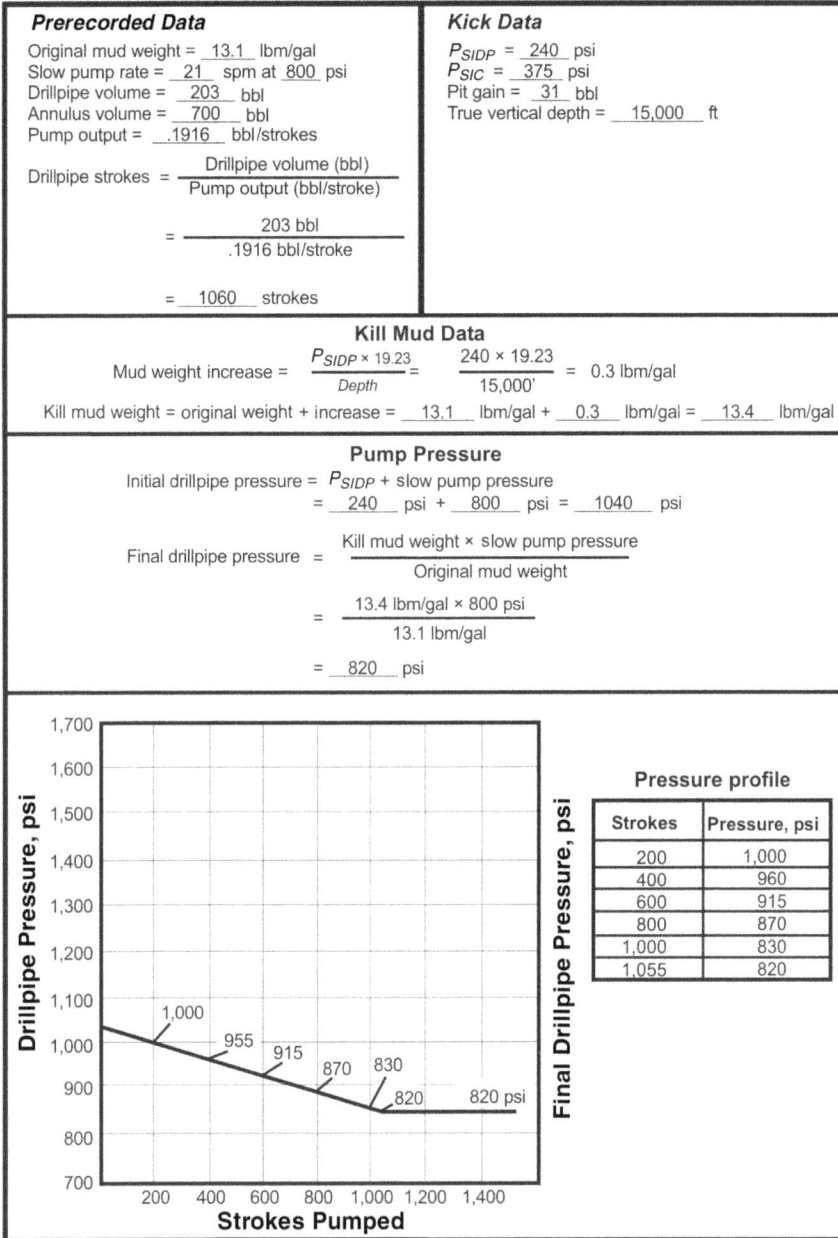

Fig. 4.19—Kill sheet.

4.11 Nonconventional Well-Control Procedures

Many attempts have been made to develop well-control procedures based on principles other than the constant-bottomhole-pressure concept. These procedures may be based on specific problems peculiar to a geological area. One example is low-permeability, high-pressured formations contiguous to structurally weak rocks that cannot withstand hydrostatic kill pressures. Often, nonconventional procedures are used to overcome problem situations that result from poor well design.

TABLE 4.10—PUMPING PRESSURE vs. STROKES PUMPED	
Strokes Pumped (bbl)	p_{sidp} (psi)
200	1,000
400	960
600	915
800	870
1,000	830
1,055	820

Nomenclature

d = other drilling equations
D_e = depth equivalent, ft
D_{tv} = true vertical depth, bit depth, ft
g_i = influx gradient, psi/ft
g_{mdp} = mud gradient in drillpipe, psi/ft
h_i = influx height, ft
p_{ah} = annular hydrostatic pressure, psi
ρ_e = equivalent mud weight, lbm/gal
p_{dph} = drillpipe hydrostatic pressure, psi
p_{form} = formation pressure, psi
p_i = influx-hydrostatic pressure, psi
p_{kr} = pump pressure at kill rate, psi
ρ_{kw} = kill mud weight, lbm/gal
ρ_o = original mud weight, lbm/gal
p_{sic} = shut-in casing pressure, psi
p_{sidp} = shut-in drillpipe pressure, psi
p_Σ = total pressure, psi

General References

Able, L.W., Bowden, J.R., and Campbell, P.J.: *Firefighting and Blowout Control,* Wild Well Control, Tulsa (1994).

Adams, N.J.: "Low Choke Pressure Method of Well Control Generally Not Recommended," *Oil & Gas J.* (29 September 1975) **73**, 109.

Adams, N.J.: "Deep Water Poses Unique Well Control Problems," *Petroleum Engineer* (May 1977).

Adams, N.J.: "Pressures Define Well Control Objectives," *Oil and Gas J.* (8 October 1979).

Adams, N.J.: "Kicks Give Clear Warning Signs," *Oil and Gas J.* (15 October 1979).

Adams, N.J.: "Drilling Variables Fix Kick Control Method," *Oil and Gas J.* (29 October 1979).

Adams, N.J.: "Variables Can Affect Kill Procedure," *Oil and Gas J.* (12 November 1979).

Adams, N.J.: "How to Implement the One-Circulation Method," *Oil and Gas J.* (26 November 1979).

Adams, N.J.: "Deepwater Kick—Control Methods Take Special Planning," *Oil and Gas J.* (4 February 1980).

Adams, N.J.: "Nonconventional Control Methods May Offer Emergency Solutions," *Oil and Gas J.* (18 February 1980).

Adams, N.J.: *Well Control Problems and Solutions,* Petroleum Publishing Co., Tulsa (1980).

Adams, N.J.: *Workover Well Control,* Petroleum Publishing Co., Tulsa (1981).

Adams, N.J.: *Drilling Engineering: A Well Planning Approach,* PennWell Publishing Co., Tulsa (1985).

Adams, N.J.: *Kicks & Blowout Control,* second edition, PennWell Publishing Co., Tulsa (1994).

Grace, R.D.: *Advanced Blowout and Well Control,* Gulf Publishing Co., Houston (1994).

Holland, P.: *Offshore Blowouts,* Gulf Publishing Co., Houston (1997).

Westergaard, R.H.: *All About Blowout,* Senter for Industriforskning, Oslo, Norway (1987).

SI Metric Conversion Factors

bbl ×	1.589 873	E–01	= m^3
ft ×	3.048*	E–01	= m
ft^3 ×	2.831 685	E–02	= m^3
ft/min ×	5.080*	E–03	= m/sec
in ×	2.54*	E+00	= cm
kg/m ×	1.488 164	E+00	= lbm/ft
lbf ×	4.448 222	E+00	= N
lbm/gal ×	9.977 633	E+01	= kg/m^3
MPa ×	1.378 951	E+01	= Pa
psi ×	6.894 757	E+00	= kPa
psi/ft ×	2.262 059	E+01	= Pa/m

*Conversion factor is exact.

Chapter 5
Introduction to Roller-Cone and Polycrystalline Diamond Drill Bits

W.H. Wamsley, Jr. and **Robert Ford,** Smith Intl. Inc.

5.1 Introduction

Rotary drilling uses two types of drill bits: roller-cone bits and fixed-cutter bits. Roller-cone bits are generally used to drill a wide variety of formations, from very soft to very hard. Milled-tooth (or steel-tooth) bits are typically used for drilling relatively soft formations. Tungsten carbide inserts bits (TCI or button bits) are used in a wider range of formations, including the hardest and most abrasive drilling applications (see **Fig. 5.1**). Fixed-cutter bits, including polycrystalline diamond compact (PDC), impregnated, and diamond bits, can drill an extensive array of formations at various depths. The following material outlines design considerations and general product characteristics for the two types.

5.2 Roller-Cone Drill Bits

Wide varieties of roller-cone bit designs are available. They provide optimum performance in specific formations and/or particular drilling environments. Manufacturers meticulously collect information on the operation of their bits to enhance future production efficiency. Modern drill bits incorporate significantly different cutting structures and use vastly improved materials compared with those used in the recent past. As a result, bit efficiency has improved systematically through the years. Variations in operating practices, types of equipment used, and hole conditions commonly require design adjustments, and manufacturers usually work closely with drilling companies to ensure that opportunities for design improvement are expeditiously identified and implemented.

5.2.1 Roller-Cone Bit Design. Roller-cone bit design goals expect the bit to do the following:
- Function at a low cost per foot drilled.
- Have a long downhole life that minimizes requirements for tripping.
- Provide stable and vibration-free operation at the intended rotational speed and weight on bit (WOB).
- Cut gauge accurately throughout the life of the bit.

To achieve these goals, bit designers consider several factors. Among these are the formation and drilling environment, expected rotary speed, expected WOB, hydraulic arrangements, and anticipated wear rates from abrasion and impact. The bit body, cone configurations, and

Fig. 5.1—Different bit types (from left to right: PDC, TCI, and milled-tooth bits).

cutting structures are design focal points, as are metallurgical, tribological, and hydraulic considerations in engineering bit design solutions. (Tribology is a science that deals with the design, friction, wear, and lubrication of interacting surfaces in relative motion.)

Basic Design Principles. Drill-bit performance is influenced by the environment in which it operates. Operating choices such as applied WOB, rotary speed, and hydraulic arrangements all have important implications in both the way that bits are designed and their operating performance.

Environmental factors, such as the nature of the formation to be drilled, hole depth and direction, characteristics of drilling fluids, and the way in which a drill rig is operated, are also of critical importance in bit performance and design. Engineers consider these factors for all designs, and every design should begin with close cooperation between the designer and the drilling company to ensure that all applicable inputs contribute to the design.

Design activities are focused principally on four general areas: material selection for the bit body and cones, geometry and type of cutting structure to be used, mechanical operating requirements, and hydraulic requirements. The dimensions of a bit at the gauge (outside diameter) and pin (arrangement for attachment to a drillstem) are fixed, usually by industry standards, and resultant design dimensions always accommodate them (**Fig 5.2**).

For roller-cone bits, steels must have appropriate yield strength, hardenability, impact resistance, machineability, heat treatment properties, and ability to accept hard facing without damage. Cutting structure designs provide efficient penetration of the formation(s) to be drilled and accurately cut gauge. The importance of bearing reliability in roller-cone bits cannot be understated. In an operational sense, bearings, seals, and lubrication arrangements function as a unit, and their designs are closely interrelated. Bearing systems must function normally under high loads from WOB, in conditions of large impact loads, while immersed in abrasive- and chemical-laden drilling fluids, and in some cases, in relatively high-temperature environments. Hydraulic configurations are designed to efficiently remove cuttings from cutting structure and bottomhole and then evacuate cuttings to the surface.

Design Methods and Tools. *How Teeth and Inserts "Drill."* To understand design parameters for roller-cone bits, it is important to understand how roller-cone bits drill. Two types of drilling action take place at the bit. A crushing action takes place when weight applied to the bit forces inserts (or teeth) into the formation being drilled (WOB in **Fig. 5.3**). In addition, a

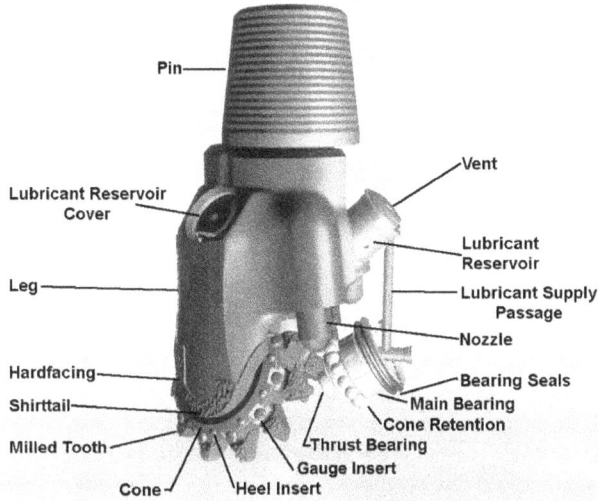

Fig. 5.2—Roller-cone bit general nomenclature.

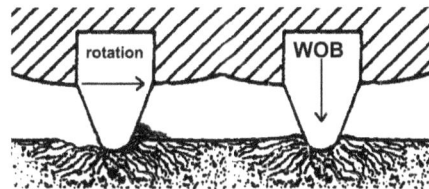

Fig. 5.3—Cutting actions for roller-cone bits.

skidding, gouging type of action results partly because the designed axis of cone rotation is slightly angled to the axis of bit rotation (rotation in Fig. 5.3). Skidding and gouging also take place because the rotary motion of a bit does not permit a penetrated insert to rotate out of a crushed zone it has created without causing it to exert a lateral force at the zone perimeter. Both effects contribute to cutting action (Fig. 5.3).

Bit Design Method. The bit geomety and cutting structure engineering method of Bentson has since 1956 been the root from which most roller-cone bit design methods have been designed (see also Reference 1). Although modern engineering techniques and tools have advanced dramatically from those used in 1956, Bentson's method is the heritage of modern design and continues to be useful for background explanation.

• Bit diameter/available space. Well diameter and the bit diameter required to achieve it influence every design feature incorporated into every efficient bit. The first consideration in the physical design of a roller-cone bit is the permissible bit diameter or, in the words of the designer, available space. Every element of a roller-cone bit must fit within a circle representative of the required well diameter. The API has issued specifications establishing permissible tolerances for standard bit diameters.[2] The sizes of journals, bearings, cones, and hydraulic and lubrication features are collectively governed by the circular cross section of the well. Individually, the sizing of the various elements can, to an extent, be varied. Repositioning or altering the size or shape of a single component nearly always requires subsequent additional changes in one or more of the other components. In smaller bits, finding good compromises can be difficult because of a shortage of space.

• Journal angle. "Journal angle" describes an angle formed by a line perpendicular to the axis of a bit and the axis of the bit's leg journal. Journal angle is usually the first element in a roller-cone bit design. It optimizes bit insert (or tooth) penetration into the formation being drilled; generally, bits with relatively small journal angles are best suited for drilling in softer formations, and those with larger angles perform best in harder formations.

• Cone offset. To increase the skidding-gouging action, bit designers generate additional working force by offsetting the centerlines of the cones so that they do not intersect at a common point on the bit. This "cone offset" is defined as the horizontal distance between the axis of a bit and the vertical plane through the axis of its journal. Offset forces a cone to turn within the limits of the hole rather than on its own axis. Offset is established by moving the centerline of a cone away from the centerline of the bit in such a way that a vertical plane through the cone centerline is parallel to the vertical centerline of the bit. Basic cone geometry is directly affected by increases or decreases in either journal or offset angles, and a change in one of the two requires a compensating change in the other. Skidding-gouging improves penetration in soft and medium formations at the expense of increased insert or tooth wear. In abrasive formations, offset can reduce cutting structure service life to an impractical level. Bit designers thus limit the use of offset so that results just meet requirements for formation penetration.

Teeth and Inserts. Tooth and insert design is governed primarily by structural requirements for the insert or tooth and formation requirements, such as penetration, impact, and abrasion. With borehole diameter and knowledge of formation requirements, the designer selects structurally satisfactory cutting elements (steel teeth or TCIs) that provide an optimum insert/tooth pattern for efficient drilling of the formation.

Factors that must be considered to design an efficient insert/tooth and establish an advantageous bottomhole pattern include bearing assembly arrangement, cone offset angle, journal angle, cone profile angles, insert/tooth material, insert/tooth count, and insert/tooth spacing. When these requirements have been satisfied, remaining space is allocated between insert/tooth contour and cutting structure geometry to best suit the formation.

In general, the physical appearance of cutting structures designed for soft, medium, and hard formations can readily be recognized by the length and geometric arrangement of their cutting elements.

Design as Applied to Cutting Structure. Application of design factors produces diverse results (**Fig. 5.4**). The cutting structure on the left is designed for the softest formation types; that on the right, for formations that are harder.

The action of bit cones on a formation is of prime importance in achieving a desirable penetration rate. Soft-formation bits require a gouging-scraping action. Hard-formation bits require a chipping-crushing action. These actions are governed primarily by the degree to which the cones roll and skid. Maximum gouging-scraping (soft-formation) actions require a significant amount of skid. Conversely, a chipping-crushing (hard-formation) action requires that cone roll approach a "true roll" condition with very little skidding. For soft formations, a combination of small journal angle, large offset angle, and significant variation in cone profile is required to develop the cone action that skids more than it rolls. Hard formations require a combination of large journal angle, no offset, and minimum variation in cone profile. These will result in cone action closely approaching true roll with little skidding.

Inserts/Teeth and the Cutting Structure. Because formations are not homogeneous, sizable variations exist in their drillability and have a large impact on cutting structure geometry. For a given WOB, wide spacing between inserts or teeth results in improved penetration and relatively higher lateral loading on the inserts or teeth. Closely spacing inserts or teeth reduces loading at the expense of reduced penetration. The design of inserts and teeth themselves depends large-

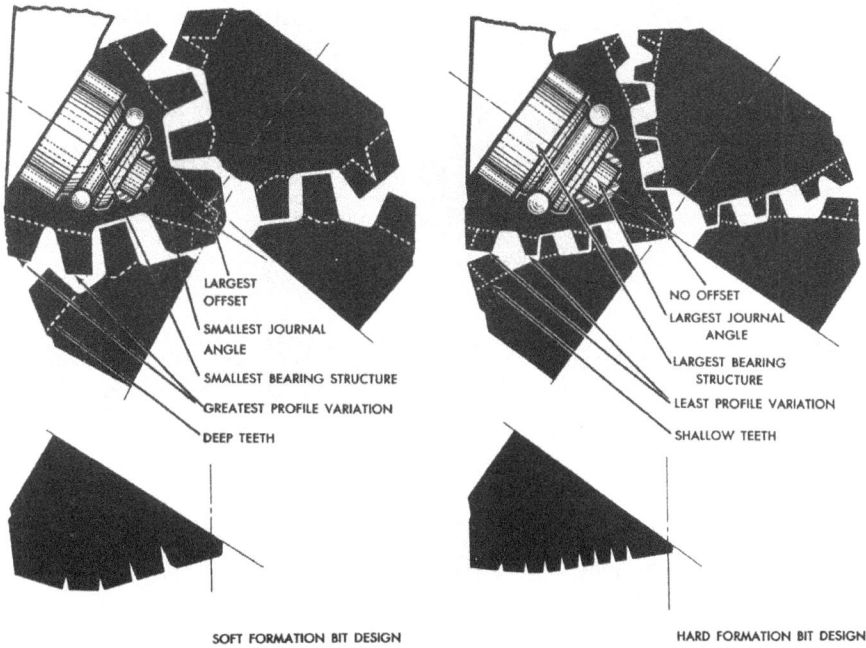

Fig. 5.4—Cutting structure for soft (left) and hard (right) formations.

TABLE 5.1—INTERRELATIONSHIP BETWEEN INSERTS, TEETH, HYDRAULIC REQUIREMENTS, AND THE FORMATION				
Formation Characteristics	Insert/Tooth Spacing	Insert/Tooth Properties	Penetration and Cuttings Production	Cleaning/Hydraulic Flow Rate Requirement
Soft	Wide	Long and sharp	High	High
Medium	Relatively wide	Shorter and stubbier	Relatively high	Relatively high
Hard	Close	Short and rounded	Relatively low	Relatively lower

ly on the hardness and drillability of the formation. Penetration of inserts and teeth, cuttings production rate, and hydraulic requirements are interrelated, as shown in **Table 5.1.**

Formation and cuttings removal influence cutting structure design. Soft, low-compressive-strength formations require long, sharp, and widely spaced inserts/teeth. Penetration rate in this type of formation is partially a function of insert/tooth length, and maximum insert/tooth depth must be used. Limits for maximum insert/tooth length are dictated by minimum requirements for cone-shell thickness and bearing-structure size. Insert/tooth spacing must be sufficiently large to ensure efficient fluid flows for cleaning and cuttings evacuation.

Requirements for hard, high-compressive-strength formation bits are usually the direct opposite of those for soft-formation types. Inserts are shallow, heavy, and closely spaced. Because of the abrasiveness of most hard formations and the chipping action associated with drilling of hard formations, the teeth must be closely spaced (**Fig. 5.5**). This close spacing distributes loading widely to minimize insert/tooth wear rates and to limit lateral loading on individual teeth. At the same time, inserts are stubby and milled tooth angles are large to withstand the heavy

Fig. 5.5—Comparison of softer IADC 427y (left) and harder 837Y (right) cutting structures.

WOB loadings required to overcome the formation's compressive strength. Close spacing often limits the size of inserts/teeth.

In softer and, to some extent, medium-hardness formations, formation characteristics are such that provisions for efficient cleaning require careful attention from designers. If cutting structure geometry does not promote cuttings removal, bit penetration will be impeded and force the rate of penetration (ROP) to decrease. Conversely, successful cutting structure engineering encourages both cone shell cleaning and cuttings removal.

Materials Design. Materials properties are a crucial aspect of roller-cone bit performance. Components must be resistant to abrasive wear, erosion, and impact loading. Metallurgical characteristics, such as heat treatment properties, weldability, capacity to accept hard facing without damage, and machineability, all figure into the eventual performance and longevity results for a bit.

Physical properties for bit components are contingent on the raw material from which a component is constructed, the way the material has been processed, and the type of heat treatment that has been applied. Steels used in roller-cone bit components are all melted to exacting chemistries, cleanliness, and interior properties. All are wrought because of grain structure refinements obtained by the rolling process. Most manufacturers begin with forged blanks for both cones and legs because of further refinement and orientation of microstructure that result from the forging process.

Structural requirements and the need for abrasion and erosion resistance are different for roller-cone bit legs and cones. Predictably, the materials from which these components are constructed are normally matched to the special needs of the component. Furthermore, different sections of a component often require different physical properties. Leg journal sections, for example, require high hardenabilities that resist wear from bearing loads, whereas the upper portion of legs are configured to provide high tensile strengths that can support large structural loads.

Roller-cone bit legs and cones are manufactured from low-alloy steels. Legs are made of a material that is easily machinable before heat treatment, is weldable, has high tensile strength, and can be hardened to a relatively high degree. Cones are made from materials that can be easily machined when soft, are weldable when soft, and can be case hardened to provide higher resistance to abrasion and erosion.

Inserts and Wear-Resistant Hard-Facing Materials. Tungsten carbide is one of the hardest materials known. Its hardness makes it extremely useful as a cutting and abrasion-resisting material for roller-cone bits. The compressive strength of tungsten carbide is much greater than its tensile strength. It is thus a material whose usefulness is fully gained only when a design maximizes compressive loading while minimizing shear and tension. Tungsten carbide is the most popular material for drill-bit cutting elements. Hard-facing materials containing tungsten carbide grains are the standard for protection against abrasive wear on bit surfaces.

When most people say "tungsten carbide," they do not refer to the chemical compound (WC) but rather to a sintered composite of tungsten carbide grains embedded in, and metallurgically bonded to, a ductile matrix or binder phase. Such materials are included in a family of materials called ceramic metal, or "cermets." Binders support tungsten carbide grains and provide tensile strength. Because of binders, cutters can be formed into useful shapes that orient tungsten carbide grains so that they will be loaded under compression. Tungsten carbide cermets can also be polished to very smooth finishes that reduce sliding friction. Through the controlled grain size and binder content, hardness and strength properties of tungsten carbide cermets are tailored for specific cutting or abrasion resistances.

The most common binder metals used with tungsten carbide are iron, nickel, and cobalt. These materials are related on the periodic table of elements and have an affinity for tungsten carbide (cobalt has the greatest affinity). Tungsten carbide cermets normally have binder contents in the 6% to 16% (by weight) range. Because tungsten carbide grains are metallurgically bonded with binder, there is no porosity at boundaries between the binder and grains of tungsten carbide, and the cermets are less susceptible to damage by shear and shock.

Properties of Tungsten Carbide Composites. The process of "designing" cermet properties makes it possible to exactly match a material to the requirements for a given drilling application. Tungsten carbide particle size (normally 2 to 6 μm), shape, and distribution, together with binder content (as a weight percent), affect composite material hardness, toughness, and strength. As a generalization, increasing binder content for a given tungsten carbide grain size will cause hardness to decrease and fracture toughness to increase. Conversely, increasing tungsten carbide grain size affects both hardness and toughness. Smaller tungsten carbide particle size and less binder content produce higher hardness, higher compressive strength, and better wear resistance. In general, cermet grades are developed in a range in which hardness and toughness vary oppositely with changes in either particle size or binder content. In any case, subtle variations in tungsten carbide content, size distribution, and porosity can markedly affect material performance (**Fig. 5.6**).

TCI Design. TCI design takes the properties of tungsten carbide materials and the geometric efficiency for drilling of a particular rock formation into account. As noted, softer materials require geometries that are long and sharp to encourage rapid penetration. Impact loads are low, but abrasive wear can be high. Hard formations are, on the other hand, drilled more by a crushing and grinding action than by penetration. Impact loads and abrasion can be very high. Tough materials, such as carbonates, are drilled by a gouging action and can sustain high impact loads and high operating temperatures. Variations in the way that drilling is accomplished and rock formation properties govern the shape and grade of the correct TCIs to be selected.

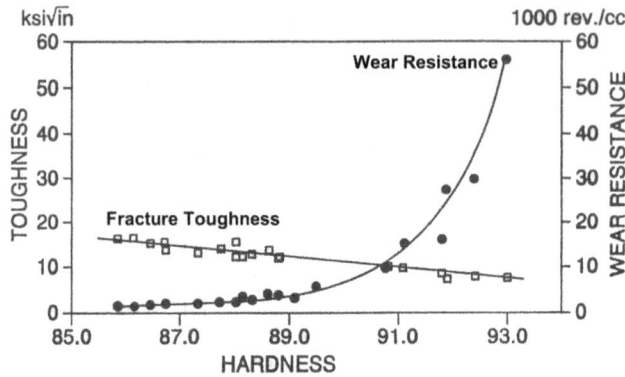

Fig. 5.6—Hardness, toughness, and wear resistance of cemented tungsten carbide.

Fig. 5.7—Typical insert types (height ≈¾ in. but varies with bit size).

The shape and grade of TCIs are influenced by their respective location on a cone. Inner rows of inserts function differently from outer rows. Inner rows have relatively lower rotational velocities about both the cone and bit axes. As a result, they have a natural tendency to gouge and scrape rather than roll. Inner insert rows thus generally use softer, tougher insert grades that best withstand crushing, gouging, and scraping actions. Gauge inserts are commonly constructed of harder, more wear-resistant tungsten carbide grades that best withstand severe abrasive wear. It is thus seen that requirements at different bit locations dictate different insert solutions. A large variety of insert geometries, sizes, and grades through which bit performance can be optimized are available to the designer (**Fig. 5.7**) (see also Ref. 3).

Gauge Cutting Structure. The most critical cutting structure feature is the gauge row. Gauge cutting structures must cut both the hole bottom and its outside diameter. Because of the severity of gauge demands on a bit, both milled tooth and insert type bits can use either tungsten carbide or diamond-enhanced inserts on the gauge. Under abrasive conditions, severe wear or gauge rounding is common, and at high rotary speeds, the gauge row can experience temperatures that lead to heat checking, chipping, and eventually breakage.

Diamond-Enhanced TCIs. Diamond-enhanced inserts are frequently used to prevent wear in the highly loaded, highly abraded gauge area of bits and in all insert positions for difficult drilling conditions. They are made up of PDC, which is chemically bonded, synthetic diamond grit supported in a matrix of tungsten carbide cermet. PDC has higher compressive strength and higher hardness than tungsten carbide. In addition, diamond materials are largely unaffected by chemical interactions and are less sensitive to heat than tungsten carbides. These properties make it possible for diamond-enhanced materials to function normally in drilling environments in which tungsten carbide grades deliver disappointing or unsatisfactory results (**Table 5.2**) (see also Refs. 4 through 6).

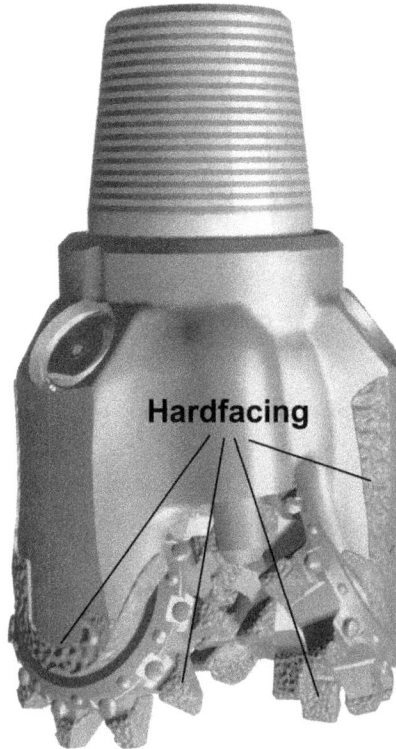

Fig. 5.8—Typical hard-facing applications on a milled-tooth bit.

TABLE 5.2—COMPARISON OF DIAMOND, PDC, AND TUNGSTEN CARBIDE MATERIALS					
	Knoop Hardness, kg/mm^2	Coefficient of Friction	Coefficient of Thermal Expansion, 10^{-4}/°C	Density, g/cm^3	Fracture Toughness, K_{IC}
Tungsten carbide (6% Co)	1,475	0.2	4.3–5.6	14.95	10.8
PDC	5,000–8,000	0.08–0.15	1.5–3.8	3.80–4.10	6.1–8.9
Natural diamond	6,000–9,000	0.05–0.10	0.8–4.8	3.52	3.4

When diamond-enhanced inserts are designed, higher diamond densities increase impact resistance and ability to economically penetrate abrasive formations. Increased diamond density increases insert cost, however. In the past, diamond-enhanced inserts have been available only in symmetrical shapes. The first of these was the semiround top insert. Today, some manufacturers have developed processes that make it possible to produce complex diamond-enhanced insert shapes.

Tungsten Carbide Hard Facing. Hard-facing materials are designed to provide wear resistance (abrasion, erosion, and impact) for the bit (**Fig. 5.8**). To be effective, hard facing must be resistant to loss of material by flaking, chipping, and bond failure with the bit. Hard facing

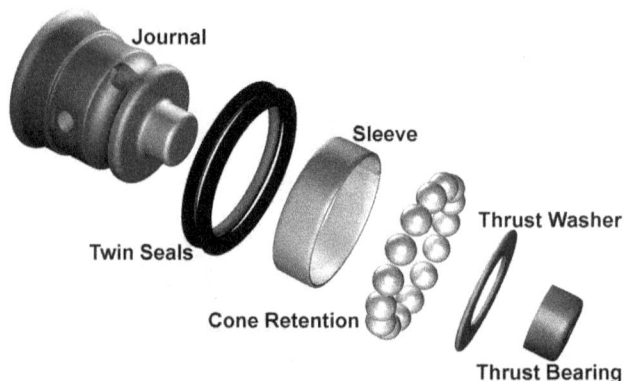

Fig. 5.9—Exploded view of seal and bearing components.

provides wear protection on the lower (shirttail) area of all roller-cone bit legs and as a cutting structure material on milled-tooth bits (**Fig. 5.9**).

Hard facing is commonly installed manually by welding. A hollow steel tube containing appropriately sized grains of tungsten carbide is held in a flame until it melts. The resulting molten steel bonds, through surface melting, with the bit feature being hard faced. In the process, tungsten carbide grains flow as a solid, with molten steel from the rod, onto the bit. The steel then solidifies around the tungsten carbide particles, firmly attaching them to the bit.

Hydraulic Features. *Nozzles and Flow Tubes.* Drilling fluids circulate through a drillstring to nozzles at the bit and back to the surface via the system annulus. They provide three crucial functions to drilling: cleaning of the cutting structure, cuttings removal from the hole bottom, and efficient cuttings evacuation to the surface. the hydraulic energy that causes fluid circulation is one of only three variable energy inputs (wob, rotary speed, and hydraulic flow) available on a drill rig for optimization of drilling performance.

Many roller-cone bit options, such as nozzle selection, flow tubes, vectored flow tubes, and center nozzle ports, help optimize hydraulic performance. These features provide alternatives for precise placement of hydraulic energy according to well bottom needs.

Generating cuttings is the first step needed to achieve high ROPs; cleaning those cuttings from the cone and hole bottom and lifting them through the annulus to the rig surface is the remaining part of a hydraulic solution. Computer modeling supported by laboratory testing is the most common approach to development and verification of hydraulic designs. Efficient velocity profiles deliver hydraulic energy to the most needed points, even in cases for which drilling flow rates are compromised.

Normally, several different nozzles can be used interchangeably on a particular bit. Nozzles are commonly classified into standard, extended, and diverging categories. Extended nozzles release the flow at a point closer than standard to the hole bottom. Diverging nozzles release the flow in a wider-than-normal, lower-velocity stream. They are designed primarily for use in center jet installations (see also Ref. 7).

Asymmetric Nozzle Configurations and Crossflow. A bit has a symmetric nozzle configuration when three nozzles of the same size and type, at the same level on the periphery of a bit, are installed 120° from each other. An asymmetric nozzle configuration has two or more different nozzle sizes and/or types.

When the fluid from a nozzle impinges on the well bottom, it moves away from the point of impingement in a 360°, fan-like, spray. A boundary forms at which fluids from two different jets meet. Fluids at these boundaries create stagnant zones known as dead zones. In the

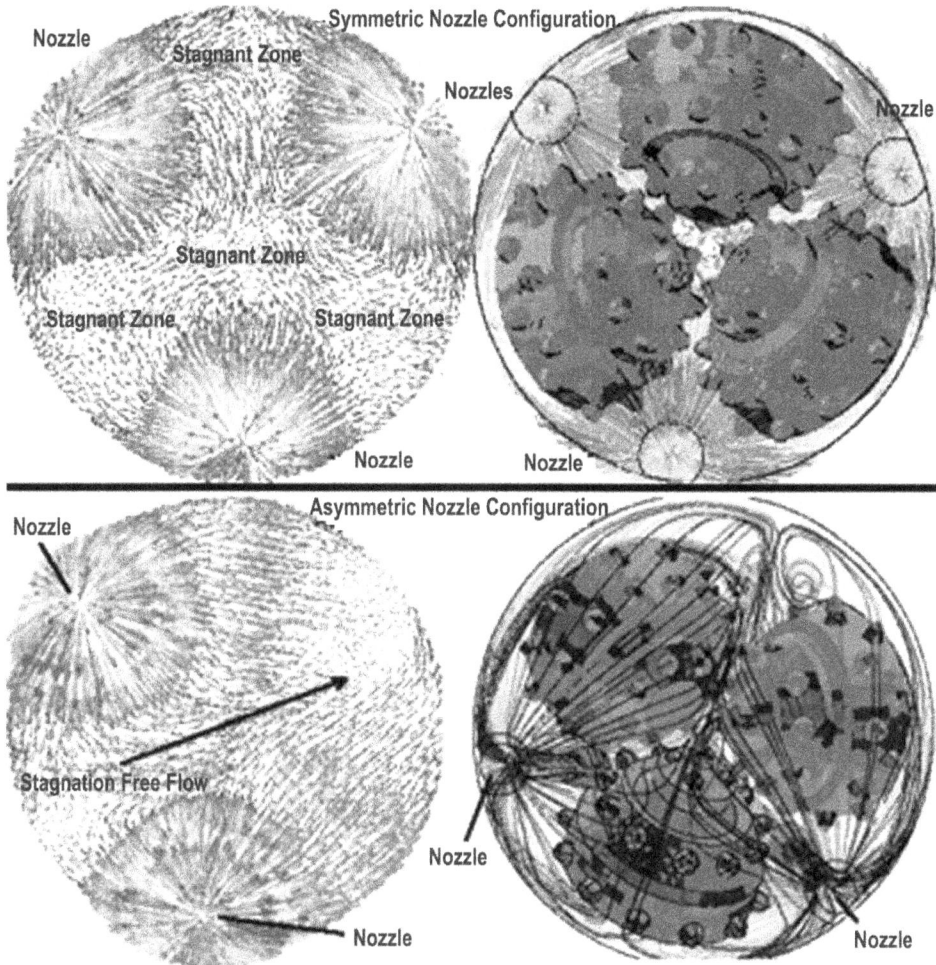

Fig. 5.10—Symmetric and asymmetric flow.

case of a symmetric nozzle configuration, dead zones occur under the middle part of the cone's asymmetric nozzle configurations; dead zones are moved away from the impingement zone of the larger jet and toward that of the smaller jet (i.e., away from the middle of the cone). Asymmetric flows resist entrapment of cuttings under a bit and help prevent the inefficiencies of regrind, lower ROPs, and erosive wear on the bit. **Fig. 5.10** shows typical flow patterns.

Crossflow is a subset of asymmetric nozzle sizing in which one jet is blocked by nozzle blank. The blanked side of the bit leaves a natural exit path for the fluid from the opposing two jets. The flow from the two jets sweeps under two of the cones to improve bottomhole cleaning and chip removal.

Practical Hydraulic Guidelines. **Table 5.3** is a summary of accepted starting hydraulics configurations for roller-cone bits.

5.2.2 Roller-Cone Bit Components. *Bearing, Seal, and Lubrication Systems.* Roller cone bearing systems are designed to be in satisfactory operating condition when the cutting structure of the bit is worn out. To achieve this standard of bearing performance, modern goals for seal and bearing system life are 1 million or more revolutions of a bit without failure, as opposed to

TABLE 5.3—RULES OF THUMB FOR OPTIMIZATION OF ROLLER-CONE BIT HYDRAULIC PERFORMANCE

1. Use mini-extended nozzles except in severe bit balling conditions.

2. Use vectored (or slant) nozzles in bit-balling conditions.

3. Run center jets in formations that have bit-balling tendencies.

4. Use diffusing center jets to minimize cone shell erosion and fluid wash.

5. Re-evaluate and optimize hydraulics as applications and equipment change.

6. Design the hydraulic program to address requirements of key lithologies in the bit run.

≈300,000 or fewer in the recent past. To achieve this goal, research into bearing, seal, and lubricant designs and into materials that improve seal and bearing life is ongoing.

Roller-cone bits primarily use two types of bearings: roller bearings and journal bearings, sometimes called friction bearings. Each type is normally composed of a number of separate components, including primary bearings, secondary bearings, seal system, features that resist thrust loading, cone retention balls, and a lubrication system (Fig. 5.9).

Primary bearings are as large as possible within the limits of available space. Secondary bearings are smaller, reduced-diameter bearings located adjacent to the interior apex area in a cone. Secondary bearings provide supplemental load-bearing capability. Primary and secondary bearings can individually be either roller bearings or journal bearings. It is not uncommon for a bearing system to be made up of combinations of the two.

Seals prevent cuttings and drilling fluids from entering the bearing system and prevent lubricant from escaping the bearing system. Thrust washers are located on the end of leg journals and between the primary and secondary bearing surfaces to resist axial loading.

Most roller-cone bits incorporate what appears to be a ball-type bearing. This is the cone retention feature. The balls prevent cones from separating from their journals. Finally, the lubrication system contains the lubricant that, throughout the life of the bearing system, provides lubrication to bearings and seals. These features are described below.

Roller-Bearing Systems. Roller bearings are a common bit-bearing system because they can reliably support large loads and generally perform well in the drilling environment (**Fig. 5.11**). They are typically used on larger-diameter bits (> 14 in.), which have more physical space to accommodate the rollers. To enhance bearing life, leading manufacturers continually research bearing materials, sizing, and shape.

Journal-Bearing Systems. Journal bearings consist of at least one rotating surface separated from the journal by a film of lubricant. The surfaces are specially designed so that the film of lubricant separates them; were they to touch, mating bearing components would gall or possibly fuse. As long as satisfactory lubrication is provided and loading remains within design limits, journal bearings are extremely efficient. **Fig. 5.12** compares roller-bearing and journal-bearing assemblies.

Design of Journal Bearings. Journal bearings must provide a balanced bearing geometry and adequate journal strength and must maximize the thickness of the high-pressure lubricating film developed during hydrodynamic lubrication. Surface areas, journal and cone diameters, and clearances between journal and cone all affect the thickness of lubricating films (**Fig. 5.13**). Manufacturing tolerances must be precise so that surfaces run true. Roundness of journal and cone surfaces is important, and if any part of a bearing is out of round, the effectiveness of the lubrication regime will be adversely affected.

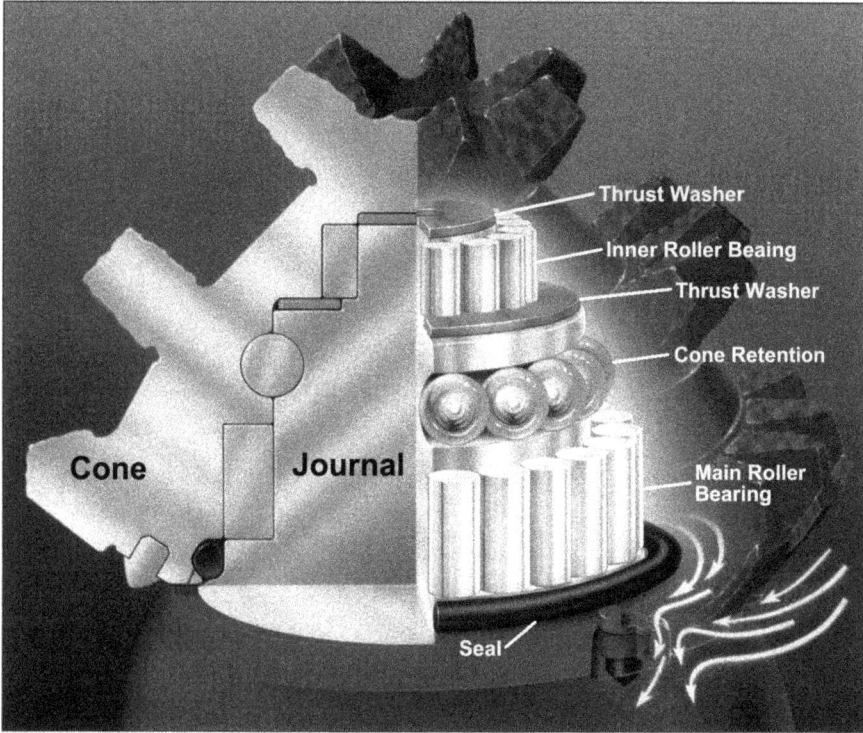

Fig. 5.11—Typical roller-bearing arrangement.

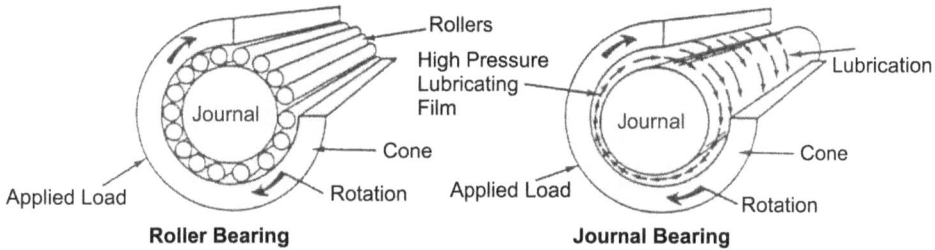

Fig. 5.12—Comparison of journal- and roller-bearing arrangements.

The metallurgy of a bearing must be balanced to minimize heat generation during boundary lubrication. Cone-bearing surfaces are steel. Soft, silver-plated sleeves are installed on the journal. Silver polishes easily, and minor surface irregularities from machining are quickly smoothed. This smoothness ensures low-friction operation and uniform lubricant flow over the bearing surface (**Fig. 5.14**).

Open Bearing Systems. Nonsealed roller bearings, referred to as open bearing systems, are typically used in large-diameter (> 20 in.) bits. These bits are often used to drill from surface to relatively shallow depths with a simple drilling fluid system (e.g., seawater). This drilling application does not necessitate the use of seals in the bits. They rely on the drilling fluid for cooling, cleaning, and lubrication of the bearings.

Seal Systems. In general, seal systems are classified as either static or dynamic. Roller-cone bits use both types of seals. Dynamic seals involve sealing across surfaces that are moving in

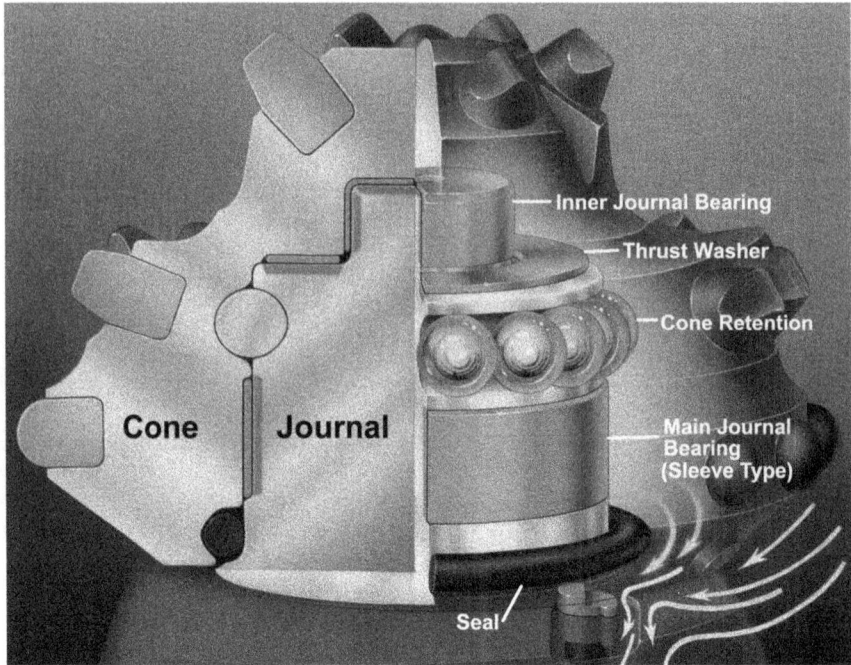

Fig. 5.13—Typical journal-bearing system.

Fig. 5.14—Cutaway of typical journal-bearing sleeve.

relationship to one another, as would be the case for a bearing seal. Seal parts or surfaces that do not move in relationship to one another during bit operation, such as the seal between a hydraulic nozzle and a bit to prevent leakage around the joint, are static seals.

Bearing Seals. Roller-cone bearing seals operate in an exceptionally harsh environment. Drilling mud and most cuttings are extremely abrasive. Drilling fluids often contain chemicals, and operating temperatures can be sufficiently high to break down the elastomers from which seals are made. Pressure pulses often occur in downhole drilling fluids that apply lateral loading on seals that must be resisted.

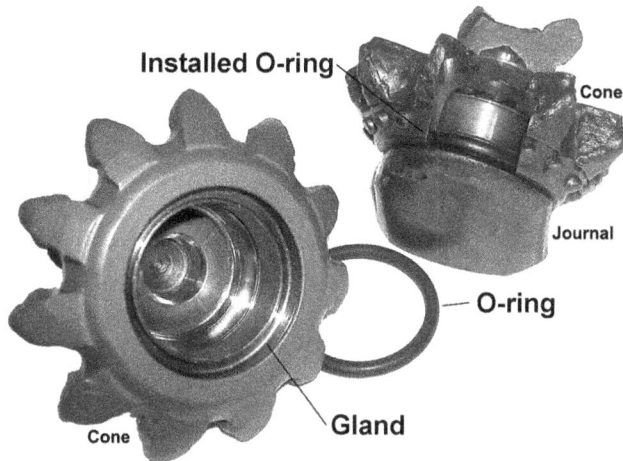

Fig. 5.15—Cone cutaway showing O-ring and gland.

On a purely practical level, bearing seals have two functions: to prevent foreign materials, such as mud, cuttings, chemicals, and water, from entering the bearings and to prevent bearing lubricant from escaping the bit.

Visualize the difference in the nature of these two duties. On the interior side, the seal is excluding clean, functional lubricant from escaping the bit, while on the exterior side, the seal is excluding dirt and chemicals from penetrating the bit. The separation of these two extremely different functions takes place at a small point between the two sides of a seal. If either of these functions breaks down, the bearings and the bit could be destined for failure.

Seal Definitions. In a rotating bearing, the two working sides of a seal are called the static energizer and the dynamic wear face. These two parts are directly opposite each other, with the energizing portion bearing on the gland and the dynamic wear face bearing against the rotating unit. For the energizing portion of the seal to function properly, it must have a surface against which to react. This is provided by a channel-shaped groove called a seal gland.

The wearing portion of the seal must have the capability to withstand the heat and abrasion generated as the rotating surface passes over it. The energizer, when functioning correctly, is not a high-wear area. Ideally, it simply bears against the gland and provides the pushing energy that maintains firm contact between the wear surface and rotating cone.

O-Rings. Donut-shaped O-rings are used in many roller-cone bit applications. O-rings are manufactured from elastomers (synthetic rubbers) that withstand the temperatures, pressures, and chemicals encountered in drilling environments. They are a traditional, but still consistently reliable, seal system.

An O-ring is installed in a seal gland to form a seal system. The gland holds the O-ring in place and is sized so that the O-ring is compressed between the gland and the bearing hub at which sealing is required. It presses the interior wall of the O-ring against the hub and the exterior diameter of the O-ring against the gland. These latter forces prevent the seal from turning in the gland and experiencing wear on the outer surfaces by rotational contact with the gland (**Fig. 5.15**).

Lubrication of Seals. Seals must be lubricated to prevent high wear rates and excessive temperaratures that could lead to seal material failures. Lubricant for the bearings also lubricates the seals.

Lubrication Systems and Lubricants. Lubricants play a vital role in bearing performance. They provide lubrication for both bearings and seals, and they provide a medium for heat transfer away from the bearings. To achieve these functions, lubricants are specially engineered and continually improved. Lubrication systems are engineered to provide reserve storage, positive delivery to the bearing system, capacity for thermal expansion, and pressure equalization with fluids on the bit exterior.

Lubrication systems include a resupply reservoir large enough to ensure availability of lubricant for all lubrication functions throughout the life of the bit. A small positive pressure differential in the system ensures flow from reservoir to bearings. The system is vented to equalize internal and external reservoir pressures. Without equalization, a pressure differential between bit exterior and interior could be sufficient to cause seal damage, leading to bearing failure.

Lubricants. High drilling temperatures and high pressures in the lubrication system, together with the potential of exposure to water and chemicals, require high performance from lubricants. Most bit lubricants are specially formulated. Leading bit manufacturers employ scientists to develop and test lubricants. Better drill-bit lubricants are stable to temperatures > 300°F, and many function normally at temperatures down to ≈0°F. They are hydrophobic (repel water) and retain their stability if water penetrates the bit. Quality lubricants are also resistant to chemicals commonly found in drilling fluids, are environmentally safe, and do not contain the lead additives that have traditionally helped resist high pressures.

•Lubricant supply. Roller-cone bits typically contain one lubricant reservoir in each leg (**Fig. 5.16**). Thus, for a three-cone bit, there are three reservoirs. Each must have the capacity for sufficient reserves of lubricant for operation of the bearing assembly it serves throughout the bit's life.

•Pressure equalization and relief. A column of drilling fluids and cuttings contained in a well exerts very high pressures on a bit operating at the well bottom. These high pressures are applied to the seal system and must be resisted by lubricant in the seal and bearing system. At installation, lubricant is at atmospheric pressure and cannot provide significant resistance to well-bottom pressures. Accordingly, internal lubrication system pressures equalize themselves with external bit pressures to prevent seal failure caused by differential pressure. Equalization is accomplished by a small relief valve installed in the lubricant reservoir system.

5.2.3 Special-Purpose Roller-Cone Bit Designs. *Monocone Bits.* Monocone bits were first used in the 1930s. The design has several theoretical advantages but has not been widely used. Bit researchers, encouraged by advances in cutting structure materials, continue to keep this concept in mind because it has the room for extremely large bearings and has very low cone rotation velocities, which suggest a potential for long bit life. While of a certain general interest, monocone bits are potentially particularly advantageous for use in small-diameter bits in which bearing sizing presents significant engineering problems.

Monocone bits drill differently from three-cone bits. Drilling properties can be similar to both the beneficial crushing properties of roller-cone bits and the shearing action of PDC bits. Cutting structure research thus focuses partly on exploitation of both mechanisms encouraged by the promise of efficient shoe drillouts and drilling in formations with hard stingers interrupting otherwise "soft" formations. Modern ultrahard cutter materials properties can almost certainly extend insert life and expand the range of applications in which this design could be profitable. The design also provides ample space for nozzle placements for efficient bottomhole and cutting structure cleaning.

Two-Cone Bits. The origin of two-cone bit designs lies in the distant past of rotary drilling. The first roller-cone patent, issued in August 1909, covered a two-cone bit. As with monocone bits, two-cone bits have available space for larger bearings and rotate at lower speeds than three-cone bits. Bearing life and seal life for a particular bit diameter are greater than for comparable three-cone bits. Two-cone bits, although not common, are available and perform well in special applications (**Fig 5.17**). Their advantages cause this design to persist, and designers have never completely lost interest in them.

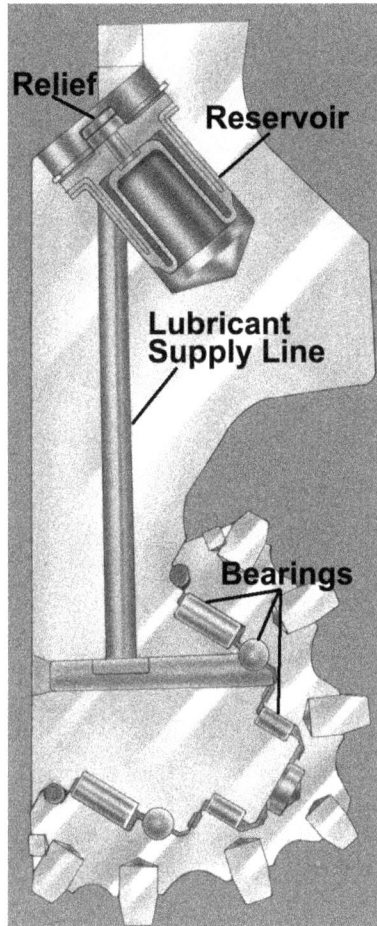

Fig. 5.16—Typical roller-cone bit lubricant reservoir system.

The cutting action of two-cone bits is similar to that of three-cone bits, but fewer inserts simultaneously contact the hole bottom. Penetration per insert is enhanced, providing particularly beneficial results in applications in which capabilities to place WOB are limited.

The additional space available in two-cone designs has several advantages. It is possible to have large cone offset angles that produce increased scraping action at the gauge. Space also enables excellent hydraulic characteristics through room for placement of nozzles very close to bottom. It also allows the use of large inserts that can extend bit life and efficiency.

Two-cone bits have a tendency to bounce and vibrate. This characteristic is a concern for directional drilling. Because of this concern and advances in three-cone bearing life and cutting structures, two-cone bits do not currently have many clear advantages. As with many roller-cone bit designs, however, modern materials and engineering capabilities may resolve problems and again underscore their recognized advantages.

5.2.4 Roller-Cone Bit Nomenclature. Roller-cone bits are generally classified as either TCI bits or milled-tooth bits. To assist in comparison of similar products from various manufacturers, the International Association of Drilling Contractors (IADC) has established a unified bit classification system for the naming of drill bits.

Fig. 5.17—Two-cone bit.

IADC Roller-Cone Bit Classification Method. The IADC Roller-Cone Bit Classification Method is an industry-wide standard for the description of milled-tooth and insert-type roller-cone bits. This coding system is based on key design- and application-related criteria. The currently used version was introduced in 1992 and incorporates criteria cooperatively developed by drill bit manufacturers under the auspices of the Society of Petroleum Engineers (see also Reference 8).

IADC Classification. The IADC classification system is a four-character design- and application-related code. The first three characters are always numeric; the last character is always alphabetic. The first digit refers to bit series, the second to bit type, the third to bearings and gauge arrangement, and the fourth (alphabetic) character to bit features.

•Series. Series, the first character in the IADC system, defines general formation characteristics and divides milled-tooth and insert-type bits. Eight series or categories are used to describe roller-cone rock bits. Series 1 through 3 apply to milled-tooth bits; series 4 through 8 apply to insert-type bits. The higher the series number is, the harder or more abrasive the rock type is. Series 1 represents the softest (easiest drilling applications) for milled-tooth bits; series 3 represents the hardest and most abrasive applications for milled-tooth bits. Series 4 represents the softest (easiest drilling applications) for insert-type bits, and series 8 represents very hard and abrasive applications for insert-type bits.

Unfortunately, rock hardness is not clearly defined by the IADC system. The meanings of "hard" sandstone or "medium-soft" shale, for example, are subjective and open to a degree of interpretation. Thus, information should be used only in a descriptive sense; actual rock hardness will vary considerably, depending on such factors as depth, overbalance pressure, porosity, and others that are difficult to quantify.

•Type. The second character in the IADC categorization system represents bit type, insert or milled tooth, and describes a degree of formation hardness. Type ranges from 1 through 4.

•Bearing design and gauge protection. The third IADC character defines both bearing design and gauge protection. IADC defined seven categories of bearing design and gauge protection: (1) nonsealed roller bearing (also known as open bearing bits); (2) air-cooled roller bearing (designed for air, foam, or mist drilling applications); (3) nonsealed roller bearing, gauge protected; (4) sealed roller bearing; (5) sealed roller bearing, gauge protected; (6) sealed friction bearing; and (7) sealed friction bearing, gauge protected. Note that "gauge protected" indicates only that a bit has some feature that protects or enhances bit gauge. It does not specify the nature of the feature. As examples, it could indicate special inserts positioned in the heel row location (side of the cone) or diamond-enhanced inserts on the gauge row.

•Included features. The fourth character used in the system defines features available. IADC considers this category optional. This alphabetic character is not always recorded on bit records but is commonly used within bit manufacturers' catalogs and brochures. IADC categorization assigns and defines 16 identifying features.

Only one alphabetic feature character can be used under IADC rules. Bit designs, however, often combine several of these features. In these cases, the most significant feature is usually listed.

5.3 PDC Drill Bits

5.3.1 PDC Materials and Bit Design. PDC is one of the most important material advances for oil drilling tools in recent years. Fixed-head bits rotate as one piece and contain no separately moving parts. When fixed-head bits use PDC cutters, they are commonly called PDC bits. Since their first production in 1976, the popularity of bits using PDC cutters has grown steadily, and today they are nearly as common as roller-cone bits in many drilling applications.

PDC bits are designed and manufactured in two structurally dissimilar styles: matrix-body bit and steel-body bits (**Figs. 5.18 and 5.19**). The two provide significantly different capabilities, and because both types have certain advantages, a choice between them would be decided by the needs of the application.

"Matrix" is a very hard, rather brittle composite material comprising tungsten carbide grains metallurgically bonded with a softer, tougher, metallic binder. Matrix is desirable as a bit material because its hardness is resistant to abrasion and erosion. It is capable of withstanding relatively high compressive loads but, compared with steel, has low resistance to impact loading.

Matrix is relatively heterogeneous because it is a composite material. Because the size and placement of the particles of tungsten carbide it contains vary (by both design and circumstances), its physical properties are slightly less predictable than steel.

Steel is metallurgically opposite of matrix. It is capable of withstanding high impact loads but is relatively soft and without protective features would quickly fail by abrasion and erosion. Quality steels are essentially homogeneous with structural limits that rarely surprise their users.

Design characteristics and manufacturing processes for the two bit types are, in respect to body construction, different because of the nature of the materials from which they are made. The lower impact toughness of matrix compared with steel limits some matrix-bit features, such as blade height. Conversely, steel is ductile, tough, and capable of withstanding greater impact loads. This makes it possible for steel-body PDC bits to be relatively larger than matrix bits and to incorporate greater height into features such as blades.

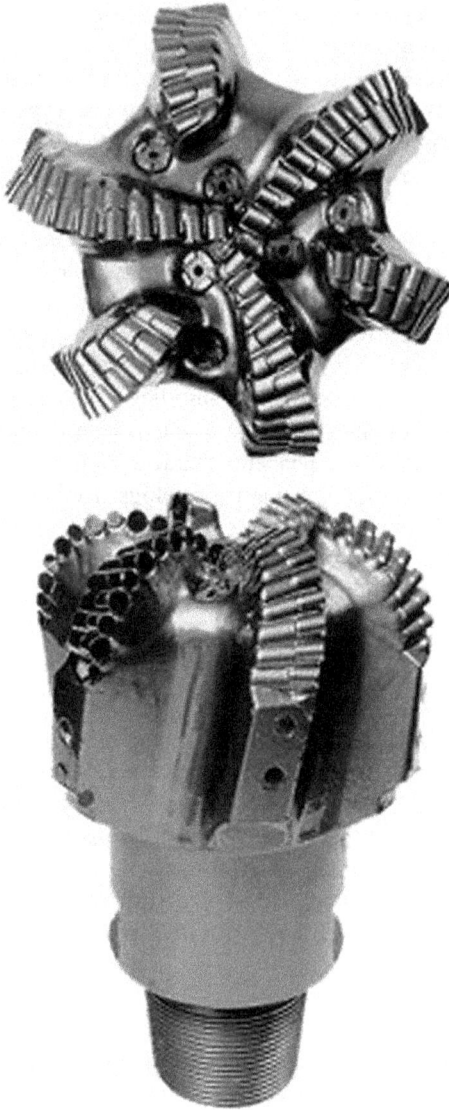

Fig. 5.18—Matrix-body bit.

Matrix-body PDC bits are commonly preferred over steel-body bits for environments in which body erosion is likely to cause a bit to fail. For diamond-impregnated bits, only matrix-body construction can be used.

The strength and ductility of steel give steel-bit bodies high resistance to impact loading. Steel bodies are considerably stronger than matrix bodies. Because of steel material capabilities, complex bit profiles and hydraulic designs are possible and relatively easy to construct on a multi-axis, computer-numerically-controlled milling machine. A beneficial feature of steel bits is that they can easily be rebuilt a number of times because worn or damaged cutters can be replaced rather easily. This is a particular advantage for operators in low-cost drilling environments.

Fortunately, both steels and matrix are rapidly evolving, and their limitations are diminishing. As hard-facing materials improve, steel bits are becoming extremely well protected with

Fig. 5.19—Steel-body bit.

materials that are highly resistant to abrasion and erosion. At the same time, the structural and wear-resisting properties of matrix materials are also rapidly improving, and the range of economic applications suitable for both types is growing.

Today's matrix has little resemblance to that of even a few years ago. Tensile strengths and impact resistance have increased by at least 33%, and cutter braze strength has increased by ≈80%. At the same time, geometries and the technology of supporting structures have improved, resulting in strong, productive matrix products. **Fig. 5.20** describes PDC bit nomenclature.

PDC Cutters. Diamond is the hardest material known. This hardness gives it superior properties for cutting any other material. PDC is extremely important to drilling because it aggregates tiny, inexpensive, manmade diamonds into relatively large, intergrown masses of randomly oriented crystals that can be formed into useful shapes called diamond tables. Diamond tables are the part of a cutter that contacts a formation. Besides their hardness, PDC diamond tables have an essential characteristic for drill-bit cutters: They efficiently bond with tungsten carbide materials that can, in turn, be brazed (attached) to bit bodies. Diamonds by themselves will not bond together, nor can they be attached by brazing.

Synthetic Diamond. Diamond grit is commonly used to describe tiny grains (≈0.00004 in.) of synthetic diamond used as the key raw material for PDC cutters. In terms of chemicals and properties, manmade diamond is identical to natural diamond. Making diamond grit involves a chemically simple process: ordinary carbon is heated under extremely high pressure and temperature. In practice, however, making diamond is far from easy.

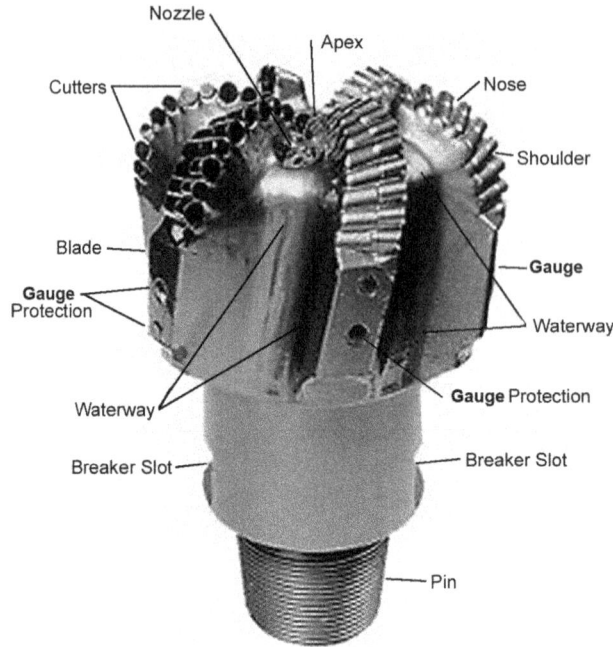

Fig. 5.20—PDC bit nomenclature.

Individual diamond crystals contained in diamond grit are diversely oriented. This makes the material strong, sharp, and, because of the hardness of the contained diamond, extremely wear resistant. In fact, the random structure found in bonded synthetic diamond performs better in shear than natural diamonds because natural diamonds are cubic crystals that fracture easily along their orderly, crystalline boundaries.

Diamond grit is less stable at high temperatures than natural diamond, however. Because metallic catalyst trapped in the grit structure has a higher rate of thermal expansion than diamond, differential expansion places diamond-to-diamond bonds under shear and, if loads are high enough, causes failure. If bonds fail, diamonds are quickly lost, so PDC loses its hardness and sharpness and becomes ineffective. To prevent such failure, PDC cutters must be adequately cooled during drilling.

Diamond Tables. To manufacture a diamond table, diamond grit is sintered with tungsten carbide and metallic binder to form a diamond-rich layer. They are wafer-like in shape, and they should be made as thick as structurally possible because diamond volume increases wear life. Highest-quality diamond tables are ≈2 to 4 mm, and technology advances will increase diamond table thickness. Tungsten carbide substrates are normally ≈0.5 in. high and have the same cross-sectional shape and dimensions as the diamond table. The two parts, diamond table and substrate, make up a cutter (**Fig. 5.21**).

Forming PDC into useful shapes for cutters involves placing diamond grit, together with its substrate, in a pressure vessel and then sintering at high heat and pressure.

PDC cutters cannot be allowed to exceed temperatures of 1,382°F [750°C]. Excessive heat produces rapid wear because differential thermal expansion between binder and diamond tends to break the intergrown diamond grit crystals in the diamond table. Bond strengths between the diamond table and tungsten carbide substrate are also jeopardized by differential thermal expansion.

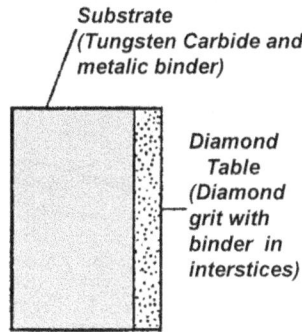

Fig. 5.21—PDC cutter construction.

5.3.2 Basic PDC Bit Design Principles. Four considerations primarily influence bit design and performance: mechanical design parameters, materials, hydraulic conditions, and properties of the rock being drilled.

Geometric Parameters of PDC Bit Design. Geometric considerations include bit shape or profile, which is predicated on cutter geometry, cutter placements, cutter density, and hydraulic requirements, along with the abrasiveness and strength of the formations to be drilled and well geometry. Each of these factors must be considered on an application-to-application basis to ensure achievement of ROP goals during cooling, cleaning the bit, and removing cuttings efficiently. During design, all factors are considered simultaneously.

Cutting Structure Characteristics. Cutting structures must provide adequate bottomhole coverage to address formation hardness, abrasiveness, and potential vibrations and to satisfy productive needs.

Early (1970s) PDC bits incorporated elementary designs without waterways or carefully engineered provisions for cleaning and cooling. By the late 1980s, PDC technology advanced rapidly as the result of new understanding of bit vibrations and their influence on productivity. Today, cutting structures are recognized as the principal determinant of force balancing for bits and for ROP during drilling.

Cutting Mechanics. The method in which rock fails is important in bit design and selection. Formation failure occurs in two modes: brittle failure and plastic failure. The mode in which a formation fails depends on rock strength, which is a function of composition and such downhole conditions as depth, pressure, and temperature.

Formation failure can be depicted with stress-strain curves (**Fig. 5.22**). Stress, applied force per unit area, can be tensile, compressive, torsional, or shear. Strain is the deformation caused by the applied force. Under brittle failure, the formation fails with very little or no deformation. For plastic failure, the formation deforms elastically until it yields, followed by plastic deformation until rupture.

PDC bits drill primarily by shearing. Vertical penetrating force from applied drill collar weight and horizontal force from the rotary table are transmitted into the cutters (**Fig. 5.23**). The resultant force defines a plane of thrust for the cutter. Cuttings are then sheared off at an initial angle relative to the plane of thrust, which is dependent on rock strength.

Formations that are drillable with PDC bits fail in shear rather than compressive stress typified by the crushing and gouging action of roller-cone bits. Thus, PDC bits are designed primarily to drill by shearing. In shear, the energy required to reach plastic limit for rupture is significantly less than by compressive stress. PDC bits thus require less WOB than roller-cone bits.

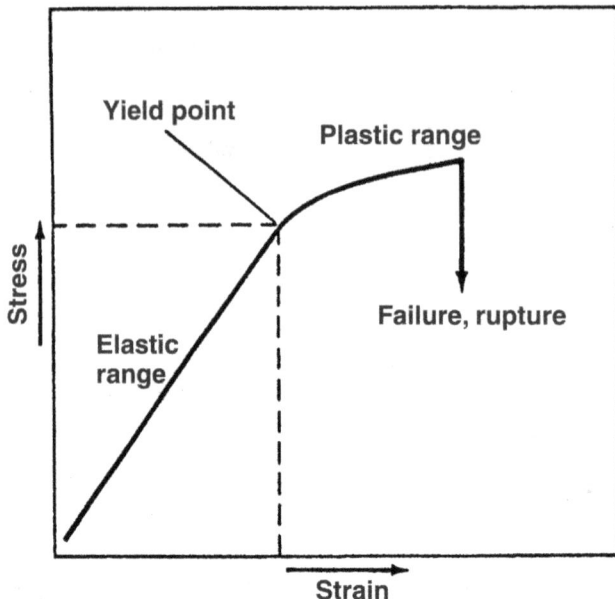

Fig. 5.22—Formation failure from stress and strain.

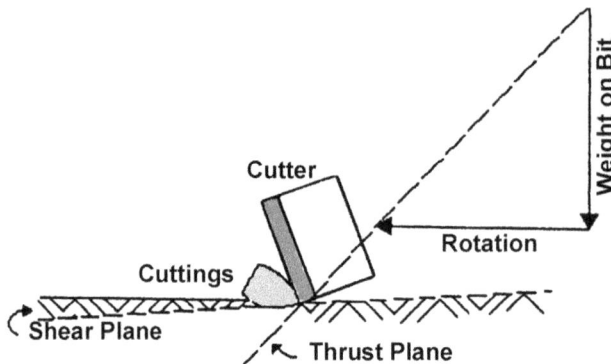

Fig. 5.23—Shear and thrust on a cutter.

Thermally stable PDC cutters are designed to plow or grind harder formations because of their thermal stability and wear resistance. This grinding action breaks cementing materials bonding individual grains of rock.

Cutters. Cutters are expected to endure throughout the life of a bit. To perform well, they must receive both structural support and efficient orientation from bit body features. Their orientation must be such that they are loaded only by compressive forces during operation. Then, to prevent loss, cutters must be retained by braze material that has adequate structural capabilities and has been properly deposited during manufacturing.

Cutter Density. Cutters are strategically placed on a bit face to ensure complete bottomhole coverage. "Cutter density" refers to the number of cutters used in a particular bit design. PDC bit cutter density is a function of profile shape and length and of cutter size, type, and quantity. If there is a redundancy of cutters, it generally increases from the center of the bit to the

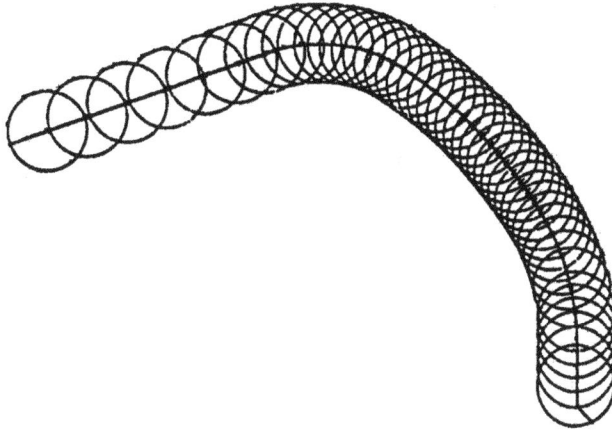

Fig. 5.24—Planar representation of cutter density increase with radial position.

outer radii because of increasing requirements for work as radial distance from the bit center-line increases. Cutters nearer to the gauge must travel farther and faster and remove more rock than cutters near the centerline. Regional cutter density can be examined by rotating each cutter's placement onto a single radial plane (**Fig. 5.24**).

If the number of cutters on a bit face is reduced, the depth of cut increases, ROP increases, and higher torque results, but life is shortened. Conversely, if cutter density is increased, ROP and cutting structure cleaning efficiency decrease, but bit life increases.

Cutter density has been increased in the "outward" radial direction from the bit centerline for the bit depicted in Fig. 5.24. Note that planar cutter strike pattern inscribes an image of bit profile.

Cutter Orientation. PDC cutters are set into bits to achieve specific rake (attack) angles relative to the formation. Back rake angle has a major effect on the way in which a bit inter-acts with a formation. Back rake is the angle between a cutter's face and a line perpendicular to the formation being drilled (**Fig 5.25**). This angle contributes to bit performance by influenc-ing cleaning efficiency, increasing bit aggressiveness, and prolonging cutter life. Back rake causes the cuttings to curl away from the cutting element, and as the back rake angle is in-creased, the tendency for cuttings to stick to the bit face is reduced.

Back rake is the amount, if any, that a cutter in a bid is tilted in the direction of bit rota-tion. It is a key factor in defining the aggressiveness or depth of cut by a cutter. Aggressive-ness is increased by decreasing back-rake angle. This increases depth of cut and results in increased ROP. Smaller back-rake angles are thus used to maximize ROP when softer forma-tions are drilled. Increased back-rake angles reduce depth of cut and thus ROP and bit vibration. It increases cutter life. An increase in angle also reduces cutter breakage from impact loading when harder formations are encountered. Harder formations require greater back rake angles to give durability to the cutting structure and reduce "chatter" or vibration. Individual cutters normally have different back-rake angles that vary with their position between the bit center and gauge.

5.3.3 PDC Bit Profile. The shape of a PDC bit body is called its profile. Bit profile has a direct influence on the following bit qualities:
- Stability (tendency to vibrate or drill laterally away from bit centerline).
- Steerability.
- Cutter density.

Fig. 5.25—Back-rake angles.

- Durability.
- ROP.
- Cleaning efficiency.
- Prevention of thermal damage to cutters by cooling.

Elements of PDC Bit Profile. A profile governs hydraulic efficiency, cutter and/or diamond loading, and wear characteristics across the bit face. It is also the principal influence on bit productivity and stability. The geometry established by the profile contributes to hydraulic flow efficiency across the bit face. Hydraulic flows directly influence ROP through the cuttings removal they provide. If cuttings are removed as rapidly as they are produced, ROP will be relatively higher. If a bit is capable of generating cuttings faster than they can be removed, however, penetration is restricted by the cuttings, and achievement of optimal ROP is impeded. Hydraulic flows also cool bit cutting elements and prevent thermal damage to them. Cutter life influences bit life and the economic efficiency of a bit investment. **Fig. 5.26** describes the nomenclature of various PDC bit profiles. Starting at the centerline of the bit and moving outward to the gauge, profile is broken into five zones: cone, nose, shoulder, taper, and gauge.

Profile Categories. Profile shape is one of the most important characteristics of fixed-cutter bits, having direct influence on possibilities for cutter placement and densities and on hydraulic layouts. Operationally, bit stability, the rotational speeds at which the bit can be run, directional characteristics, permissible WOB, and bit durability are also affected by profile.

There are four general categories of PDC bit profiles. These range from long, parabolic curves to flat shapes with narrow-radius, compressed curves. The types are described as flat profiles, short parabolic profiles, medium parabolic profiles, or long parabolic profiles (**Fig. 5.27**).

Parabolic profiles are considerably more aggressive than flatter profiles and produce higher ROPs at the expense of accelerated rates of abrasive wear. As bit profile becomes more parabolic, cutter wear on the inner radii around the nose increases. Parabolic profiles are susceptible to cutter breakage by impact, particularly if insufficient cutter density exists in the nose area.

When harder formations are drilled, flat profiles and high cutter loading are required. Flatter profiles uniformly place high loading on individual cutters and increase penetration. If

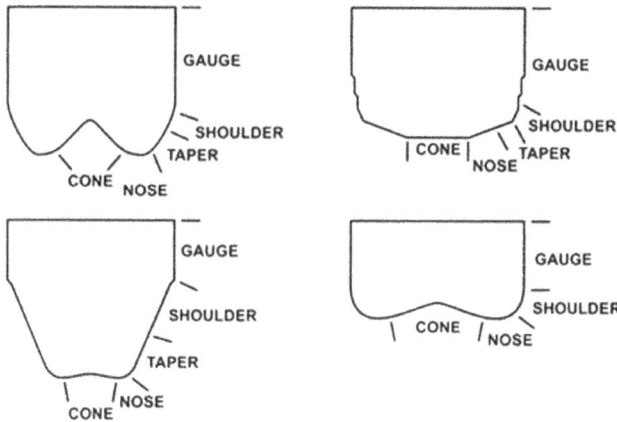

Fig. 5.26—Nomenclature of a PDC bit profile.

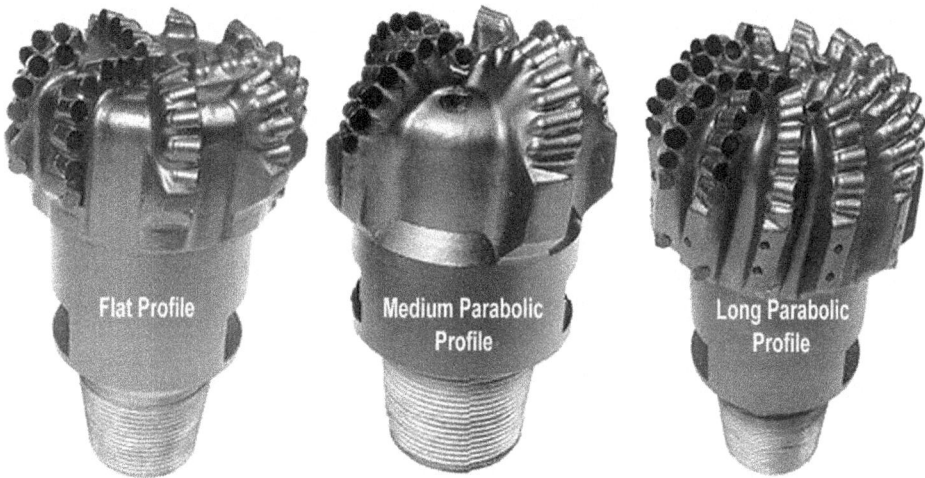

Fig. 5.27—PDC bit profile types.

abrasive wear is predominant, however, parabolic profiles enable the higher cutter densities that limit penetration but increase resistance to abrasion.

PDC bits most frequently incorporate large shoulder radii and primarily use either short or medium parabolic profiles. Cone angles are sufficient to stabilize the bit from unwanted deviation without hindering steerability. Such designs give bits the versatility to drill efficiently either by conventional rotary drilling or with downhole motors.

Flat and long parabolic profiles are less commonly used designs. Flat profiles have a single radius on the shoulder and are less aggressive than parabolic profiles. Long parabolic profiles are made up of a series of curves beginning at the cone-to-nose intersection and continuing to the outside-diameter radius and gauge intersection.

5.3.4 Cutter Design. *PDC Cutters.* PDC cutters are made up of a working component, the diamond table, and a supporting component called the substrate (**Fig. 5.28**).

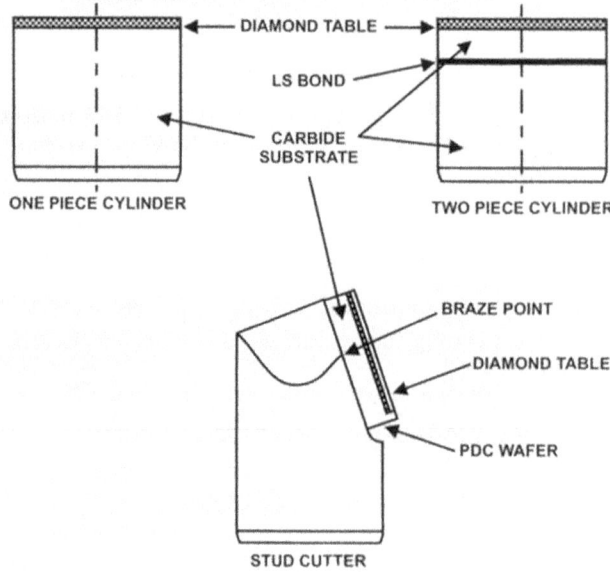

Fig. 5.28—Construction of a PDC cutter.

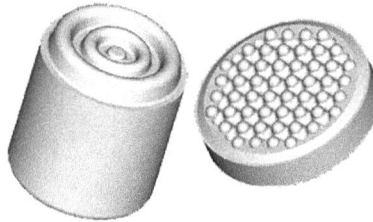

Fig. 5.29—Typical geometries between diamond tables and substrates.

Substrate. Substrates are a composite material made up of tungsten carbide grains bonded by metallic binder. This material bonds efficiently with diamond tables but is very hard and thus capable of impeding erosive damage to a working cutter.

Cutter geometry, at the interface between diamond table and substrate, seeks to enhance bonding between the two. Generally, geometries that increase interface surface area improve bonding (**Fig. 5.29**). Geometries also attempt to control stresses at the bond to the lowest possible level.

Diamond Table. The shape of a diamond table is governed by two design objectives. It must include the highest possible diamond volume and total diamond availability to its working features. It must also ensure the lowest possible stress level within the diamond table and at the substrate bond.

Geometric features of an interface between a diamond table and substrate can significantly improve the ability of the diamond table to withstand impact (**Fig. 5.30**).

Diamond Table Bonds. High stress concentrations in a diamond table can result in delamination failures between diamond tables and substrates or in diamond table edge and corner chipping. Poor bonds between grains of diamond grit can lead to cracking in a diamond layer and eventually to diamond table and substrate failure.

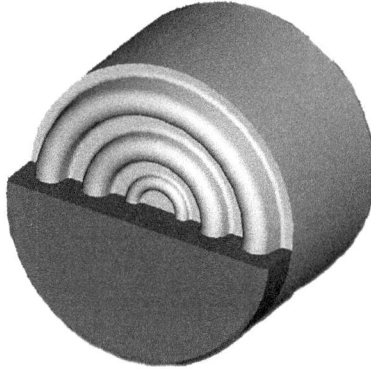

Fig. 5.30—Cutter section showing diamond table and interface style.

Thermally Stable PDC. As described earlier, the maximum safe operating temperature for PDC materials is 750°C [1382°F]. Higher temperature resistance can be achieved in a diamond table, however, by removing residual cobalt catalyst from the manufacturing process. The resulting material is called thermally stable polycrystalline diamond (TSP). When cobalt is removed, problems related to differential thermal expansion between the binder and diamond are removed, making TSP stable to ≈1200°C [2192°F].

TSP is formed like PDC and, except for thermal properties, behaves like PDC with one important exception. Because cobalt contained in PDC plays a key role in bonding PDC diamond tables to tungsten carbide substrates, attachment of TSP cutters to a bit is relatively difficult. Therefore, TSP is generally used only in applications in which bit operating temperature cannot be reliably controlled.

Cutter Optimization. To achieve cutter durability and reliable bonds between diamond tables and substrates, design engineers use a variety of application-specific cutter options. These include cutter diameter options between ≈6 and 22 mm, optimized total diamond volumes in diamond table designs, special diamond table blends, a variety of nonplanar interface shapes that increase bond area and reduce internal stresses between the diamond table and substrate, and a variety of external cutter geometries designed to improve performance in particular drilling environments.

Cutter Shape. The most common PDC shape is the cylinder, partly because cylindrical cutters can be easily arranged within the constraint of a given bit profile to achieve large cutter densities. Electron wire discharge machines can precisely cut and shape PDC diamond tables (**Fig. 5.31**). Nonplanar interface between the diamond table and substrate reduces residual stresses. These features improve resistance to chipping, spalling, and diamond table delamination. Other interface designs maximize impact resistance by minimizing residual stress levels.

Certain cutter designs incorporate more than one diamond table. The interface for the primary diamond table is engineered to reduce stress. A secondary diamond table is located in the high-abrasion area on the ground-engaging side of the cutter. This two-tier arrangement protects the substrate from abrasion without compromising structural capability to support the diamond table.

Highly specialized cutters are designed to increase penetration in tough materials such as carbonate formations. Others include engineered relief in the tungsten carbide substrate that increases penetration and reduces requirement for WOB and torque, or beveled diamond tables that reduce effective cutter back rake and lower bit aggressiveness for specific applications.

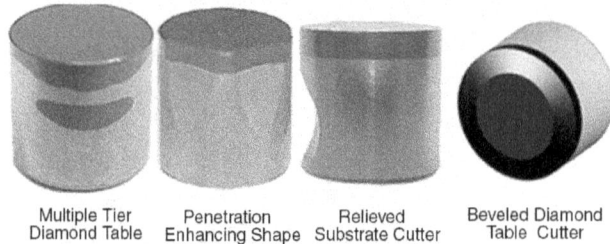

Multiple Tier   Penetration   Relieved   Beveled Diamond
Diamond Table   Enhancing Shape   Substrate Cutter   Table Cutter

Fig. 5.31—Examples of special-purpose and extreme service cutters.

5.3.5 Special Bit Configurations. *Diamond Bits.* The term "diamond bit" normally refers to bits incorporating surface-set natural diamonds as cutters. This bit type, which has been used for many years, was the predecessor to PDC bits and continues to be used in certain drilling environments. Diamond bits are used in abrasive formations. They drill by a high-speed plowing action that breaks the cementation between rock grains. Fine cuttings are developed in low volumes per rotation. To achieve satisfactory ROPs with diamond bits, they must, accordingly, be rotated at high speeds.

Diamond bits are described in terms of the profile of their crown, the size of diamond stones (stones per carat), total fluid area incorporated into the design, and fluid course design (radial or cross flow).

Diamonds do not bond with other materials. They are held in place by partial encapsulation in a matrix bit body. Diamonds are set in place on the drilling surfaces of bits (**Fig. 5.32**).

Impregnated Bits. Impregnated bits are a PDC bit type in which diamond cutting elements are fully imbedded within a PDC bit body matrix (**Fig. 5.33**).

Impregnated bit bodies are PDC matrix materials that are similar to those used in cutters. The working portions of impregnated bits are unique, however: matrix impregnated with diamonds.

Both natural and synthetic diamonds are prone to breakage from impact. When embedded in a bit body, they are supported to the greatest extent possible and are less susceptible to breakage. However, because the largest diamonds are relatively small, cut depth must be small and ROP must be achieved through increased rotational speed. Thus, impregnated bits do not perform well in rotary drilling because of relatively low rotary speeds. They are most frequently run in conjunction with turbodrills and high-speed positive displacement motors that operate at several times normal rotational velocity for rotary drilling (500 to 1500 rpm).

Impregnated bits use combinations of natural diamond, synthetic diamond, PDC, and TSP for cutting and gauge protection purposes. They are designed to provide complete diamond coverage of the well bottom with only diamonds touching the formation. Variations in diamond size and the ratio of diamond to matrix volumes allow optimization of performance in terms of aggressiveness and durability. Varying diamond distribution also affects the ratio of diamond to matrix with similar effects on aggressiveness and durability.

During drilling, individual diamonds in a bit are exposed at different rates. Sharp, fresh diamonds are always being exposed and placed into service.

Dual-Diameter Bits. Dual-diameter bits have a unique geometry that allows them to drill and underream. To achieve this, the bits must be capable of passing through the ID of a well casing and then drilling an oversized (larger than casing diameter) hole. State-of-the-art dual-diameter bits are similar to conventional PDC drill bits in the way that they are manufactured. They typically incorporate a steel body construction and a variety of PDC and/or diamond-enhanced cutters. They are unitary and have no moving parts (**Fig. 5.34**).

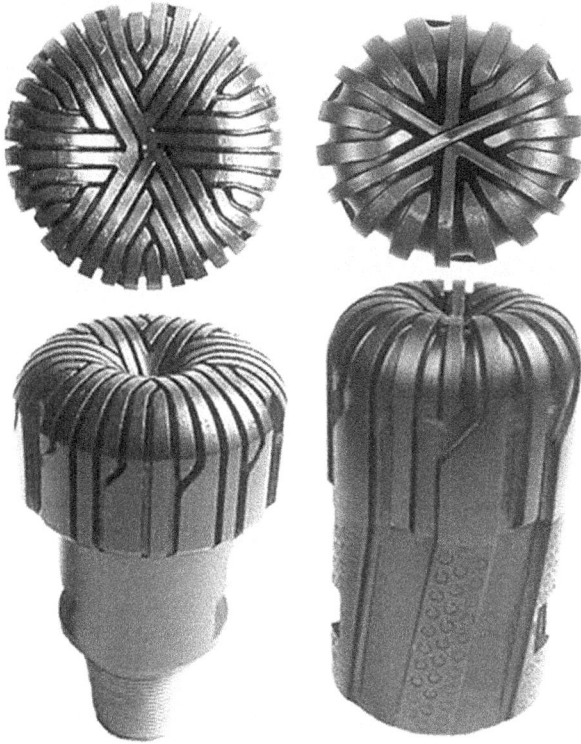

Fig. 5.32—Diamond bit examples.

Dual-diameter bits can provide drilling flexibility through well diameter control, directional aptitude, and reduction of drop tendencies. They are functional in vertical and directional wells and in a wide range of formations. Maximum benefit is realized in swelling or flowing formations in which the risk of sticking pipe can be reduced by drilling an oversized hole. They are commonly used in conjunction with applications requiring increased casing, cement, and gravel-pack clearance; they also can eliminate the need for extra trips and avoid the risk of moving part failure in mechanical underreamers in high-cost intervals. When a well is deepened below existing casing, they reduce the need for additional underreamer runs and increase clearance for smooth casing run in curve sections. With this flexibility, they are also useful in exploratory wells, in which they provide for maximum casing diameters.

Dual-Diameter Bit Drilling Method. **Fig. 5.35** shows the maximum dual-diameter bit diameter that can be tripped through casing without problems (left). On the opposite side of the reaming section, the pilot section is significantly removed from the casing pass-through diameter, and the centerline of the bit is similarly to the left (in the image) of casing/hole centerline. The only contact area between the bit and the casing pass-through diameter is at the small side of the reaming section. There is no cutter contact with casing during drillout, and neither the casing nor the bit cutting structure is damaged by tripping.

During drilling, the bit is centered on the hole, and the large sides of the reaming and pilot sections are in contact with the hole (right).

Dual-diameter bits are possible largely because of sophisticated modern engineering. Because of the unique geometry of dual-diameter bits, many obstacles and challenges must be overcome. To drill properly, this type of bit must be stable. If a conventional PDC bit becomes unstable during drilling, it will drill an oversized hole. If, on the other hand, a dual-diameter

Fig. 5.33—Impregnated bits.

bit becomes unstable during drilling, the pilot section will drill an oversized hole that will, in turn, cause the reaming section to drill undersize, and hole diameter goals will not be achieved. To drill with optimal hole-opening ability, a dual-diameter bit must rotate purely around the bit axis. Stability is achieved with bit features and through careful engineering. The bits are force and mass balanced. Without care, dual-diameter bits could have a large turning moment between the pilot and reamer sections because of the axial separation of loading. Excessive torque contributes to poor bit stability, which adversely affects hole condition.

Designs must ensure that gauge cutters are prevented from contacting the casing, even in extreme applications. Dual-diameter designs must perform similarly to conventional PDC drill bits and produce a high-quality, larger hole.

Dual-diameter bits are often configured for drillout. Drillout cutting structures are more aggressive than those that will eventually serve rotating and sliding modes but are generally more durable during drillout than most bits, even though the bits eventually perform other functions besides drillout.

Dual-diameter bit hydraulics requires special attention. Fluids provided to the pilot must fully clean the pilot section. Much of the total flow must, however, be reserved for and directed to the full-gauge section from which a much higher volume of cuttings removal is required. Excessive flow to the pilot risks washout to the hole bottom, whereas insufficient flows to the gauge cutting area will provide inadequate cuttings removal and poor ROP or even binding of the drillstring. Dual-diameter bits require a special geometric relationship between reamer nozzles and the bit profile that minimizes flow scatter and maximizes available hydraulic energy across the reamer cutters. This layout works with pilot section hydraulics and deep junk slots to ensure high overall cleaning efficiency.

5.3.6 IADC PDC Bit Classification. *IADC Fixed-Cutter Bit Classification System.* The IADC Fixed-Cutter Bit Classification System seeks to classify fixed-cutter PDC and diamond drill bits effectively so that they can be efficiently selected and used by the drilling industry. IADC classification codes for each bit are generated by placing the bit style into the category that best describes it so that similar bit types are grouped within a single category. The version currently used was introduced in 1992 using criteria that were cooperatively developed by drill-

Fig. 5.34—Dual-diameter bit.

bit manufacturers under the auspices of SPE.[10,11] The system leaves a rather broad latitude for interpretation and is not as precise or useful as the IADC Classification System for Roller-Cone Bits.[8]

The system is composed of four characters that designate body material, cutter density, cutter size or type, and bit profile. It does not consider hydraulic features incorporated into a bit and does not attempt to give a detailed description of body style beyond basic classification of the overall length of the bit cutting face. Special designs incorporating unconventional use and densities of gauge cutters are not considered for classification.

Bit Body Material. The first digit in the IADC Fixed-Cutter Bit Classification describes the material from which the bit body is constructed: M or S for matrix- or steel-body construction, respectively.

Cutter Density. The second IADC classification character is a digit that represents the density of cutting elements. Densities for PDC cutter and surface set diamond bits are described separately through use of numerals 1 through 4 for PDC bits and 6 through 8 for surface-set diamond bits. Numerals 0, 5, and 9 are not defined. Specifically, for PDC bits, density classification relates to cutter count; for surface-set bits, it relates to diamond size. Because heavier cutter densities generally correspond to tougher drilling applications, the density classification digit implies an applications aspect as it increases.

•PDC Bit Cutter Density. PDC bit cutter density represents total cutter count, usually including gauge cutter count. A designation of 1 represents a light cutter density; 4 represents a heavy density. Within the classification rules, a density of 1 refers to ≤ 30 cutters; a density of 2 refers to 30 to 40; density 3 indicates 40 to 50; and density 4 refers to ≥ 50 cutters.

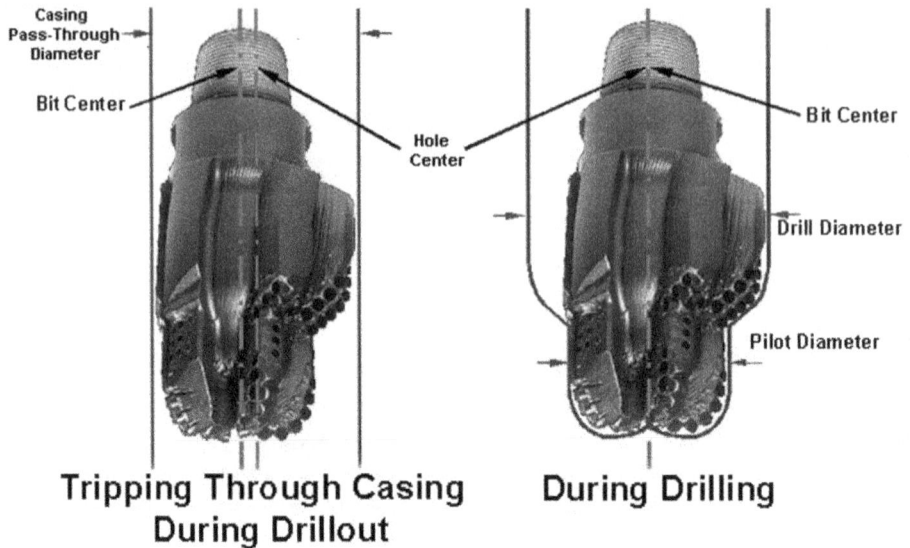

Fig. 5.35—Dual-diameter drilling and tripping.

Manufacturers classify their PDC bits within these four numeric categories, depending on a manufacturer's internal criteria for cutter density. Bits that are "borderline" are placed into a higher or lower density category, depending on manufacturer preference.

•Surface-Set Diamond Bit Density. Surface-set diamond density, numerals 6 through 8, categorize variations in the size of the cutter material. The numeral 6 represents diamond sizes > 3 stones per carat; 7 represents diamond sizes from 3 to 7 stones per carat; and 8 represents diamond sizes < 7 stones per carat. Thus, diamond size becomes smaller as the density classification increases. This generally corresponds to what would be expected in surface-set bit designs intended for harder or more abrasive formations.

Cutter Size or Type. The third character in the IADC classification designates the "size" or "type" of cutter. This again differs for PDC and surface-set diamond bits. For PDC cutter bits, the third character is a digit that represents cutter size: 1 indicates PDC cutters > 24 mm in diameter; 2 represents cutters from 14 to 24 mm in diameter; 3 indicates PDC cutters < 14 but > 8 mm; and 4 is used for cutters < 8 mm.

For surface-set bits, the third character represents diamond type, with 1 indicating natural diamonds, 2 referring to TSP material, 3 representing combinations such as mixed diamond and TSP materials, and 4 indicating impregnated diamond bits.

Bit Profile. The final (fourth) character describes the basic appearance of the bit based on overall length of the cutting face. "Fishtail"-type PDC bits are an exception as bits; for this type of bit, the ability to clean in fast-drilling, soft formations is thought to be a more important body feature than profile. The numeral 1 represents fishtail PDC bits and "flat" TSP and natural diamond bits; 2, 3, and 4 indicate increasingly longer bit profiles of both types (a virtually flat PDC bit would be identified by 2, whereas a long-flanked "turbine style" bit would be categorized as 4). In lieu of developing a formula relating overall bit face length (depth) to bit diameter, each manufacturer classifies its own product profiles using these rules.

5.3.7 IADC Bit Dull Grading. The IADC, in conjunction with SPE, has established a systematic method for communication of bit failures The intent of the system is to facilitate and accelerate product and operational development based on accurate recording of bit experiences. This system is called dull grading. The IADC Dull Grading Protocol evaluates eight roller-cone

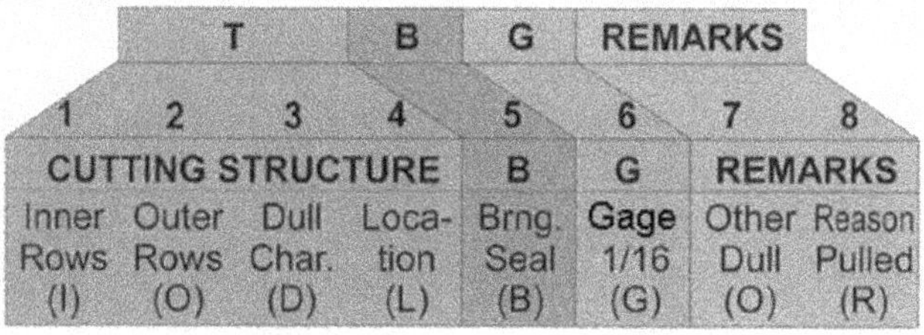

T				B	G	REMARKS	
1	2	3	4	5	6	7	8
CUTTING STRUCTURE				B	G	REMARKS	
Inner Rows (I)	Outer Rows (O)	Dull Char. (D)	Loca- tion (L)	Brng. Seal (B)	Gage 1/16 (G)	Other Dull (O)	Reason Pulled (R)

Fig. 5.36—IADC dull grading categories.

or seven PDC bit areas, provides a mechanism for systematically evaluating the reasons for removal of a bit from service, and establishes a uniform method for reporting.[12,13]

Partly because of dull analyses, bit design processes and product operating efficiencies evolve rapidly. Engineers identify successful design features that can be reapplied and unsuccessful features that must be corrected or abandoned; manufacturing units receive feedback on product quality; sales personnel migrate performance gains and avoid duplication of mistakes between similar applications, and so forth. All bit manufacturers require collection of dull information for every bit run.

IADC dull grading is closely associated with its bit classification systems, and the general formats for fixed-cutter bit and roller-cone bit dull grading are similar. There are important differences that must be taken into account, however, and the two approaches are not interchangeable. The following explains IADC Dull Grading and points out the differences between diamond/PDC and roller-cone bit rules.

IADC Dull Grading System. IADC dull grading reviews four general bit wear categories: cutting structure (T), bearings and seals (B), gauge (G), and remarks. These and their subcategories are outlined in **Fig. 5.36**.[10]

Cutting Structure Wear Grading (T). For dull grading purposes, cutting structures are subdivided into four subcategories: inner rows, outer rows, major dull characteristic of the cutting structure, and location on bit face where the major dull characteristic occurs. **Fig. 5.37** illustrates the dull grading system.

•Roller Cone Cutting Structure Evaluation. Dull grading begins with evaluation of wear on the inner rows of inserts/teeth (i.e., with the cutting elements not touching the wall of the hole bore). Grading involves measurement of combined inner row structure reduction caused by loss, wear, and/or breakage with the measurement method described above. Outer rows of inserts/teeth are those that touch the wall of the hole bore. Grading involves measurement of combined outer row teeth/insert structure reduction caused by loss, wear, and/or breakage with the measurement method described above.

•Roller-Cone Cutter or Insert/Tooth Wear Measurement. Measurement of roller-cone cutting structure condition requires evaluation of bit tooth/insert wear status. Wear is reported by use of an eight-increment wear scale in which no wear is represented by "0" and completely worn (100%) is represented by "8" (**Fig. 5.38**).

•PDC Bit Cutter Wear Evaluation. Cutter wear is graded with a 0 to 8 scale in which 0 represents no wear and 8 indicates that no usable cutting surface remains (**Fig. 5.39**). PDC cutter wear is measured across the diamond table, regardless of the cutter shape, size, type, or exposure. The location of cutter wear is categorized as either the inner two-thirds or outer third of the bit radius (**Fig. 5.40**).

•PDC Bit Inner and Outer Row Cutter Wear Measurement. For both PDC and surface-set diamond bits, a value is given to cutter wear with the method described above. To obtain average wear for the inner rows of cutters depicted in Fig. 5.40, the six included cutters must be individually graded, summed as a group, and averaged to obtain the inner row wear grade,

IADC DULL BIT GRADING

	Cutting Structure							
Inner	Outer	Dull Char.	Location	Bearings/ Seals	Gage	Other Dull Char.	Reason Pulled
1	2	3	4	5	6	7	8

Inner Cutting Structure (1) (All Inner Rows)
(For fixed cutter bits, use the inner 2/3 of the bit radius)

Outer Cutting Structure (2) (Gage Row Only)
(For fixed cutter bits, use the outer 1/3 of the bit radius)

In columns 1 and 2, a linear scale from 0 to 8 is used to describe the condition of the cutting structure according to the following:

Steel Tooth Bits
A measure of lost tooth height due to abrasion and/or damage
0 - No Loss of Tooth Height
8 - Total Loss of Tooth Height

Inner Bits
A measure of total cutting structure reduction due to lost, worn and/or broken inserts
0 - No Lost, Worn and/or Broken Inserts
8 - All Inserts Lost, Worn and/or Broken

Fixed Cutter Bits
A measure of lost, worn and/or broken cutting structure
0 - No Lost, Worn and/or Broken Cutting Structure
8 - All of Cutting Structure Lost, Worn and/or Broken

Dull Characteristics (3)
(Use only cutting structure related codes)

*BC - Broken Cone	OC - Off Center Wear
BF - Bond Failure	PB - Pinched Bit
BT - Broken Teeth/ Cutters	PN - Plugged Nozzle/ Flow Passage
BU - Balled Up Bit	RG - Rounded Gage
*CC - Cracked Cone	RO - Ring Out
*CD - Coned Dragged	SD - Shirttail Damage
CI - Cone Interference	SS - Self Sharpening Wear
CR - Cored	TR - Tracking
CT - Chipped Teeth/ Cutters	WO - Washed Out Bit
ER - Erosion	WT - Worn Teeth/ Cutters
FC - Flat Crested Wear	NO - No Dull Characteristics
HC - Heat Checking	
JD - Junk Damage	* Show cone number(s) under location(4)
*LC - Lost Cone	
LN - Lost Nozzle	
LT - Lost Teeth/Cutters	

Location (4)

Roller Cone
N - Nose Row
M - Middle Row
G - Gage Row
A - All Rows
Cone #
1   2   3

Fixed Cutter
C - Cone
N - Nose
T - Taper
S - Shoulder
G - Gage
A - All Areas

Bearings/Seals (5)

Non-Sealed Bearings
A linear scale estimating bearing life used
0 - No Life Used
8 - All Life Used. i.e., no bearing life remaining

Sealed Bearings
E - Seals Effective
F - Seals Failed
N - Not Able to Grade
X - Fixed Cutter Bit (Bearingless)

Gage (6)

Measure to nearest
1/16 of an inch
I - In Gage
1 - 1/16" Out of Gage
2 - 2/16" Out of Gage
4 - 4/16" Out of Gage

Other Dull Characteristics (7)

Refer to column 3 codes

Reason Pulled or Run Terminated (8)

BHA - Change Bottom-Hole Assembly	LH - Left in Hole
	HR - Hours on Bit
DMF - Downhole Motor Failure	LOG - Run Logs
DTF - Downhole Tool Failure	PP - Pump Pressure
DSF - Drillstring Failure	PR - Penetration Rates
DST - Drill Stem Test	RIG - Rig Repair
DP - Drill Plug	TD - Total Depth/Casing Depth
CM - Condition Mud	
CP - Core Point	TW - Twist Off
FM - Formation Change	TQ - Torque
HP - Hole Problems	WC - Weather Conditions

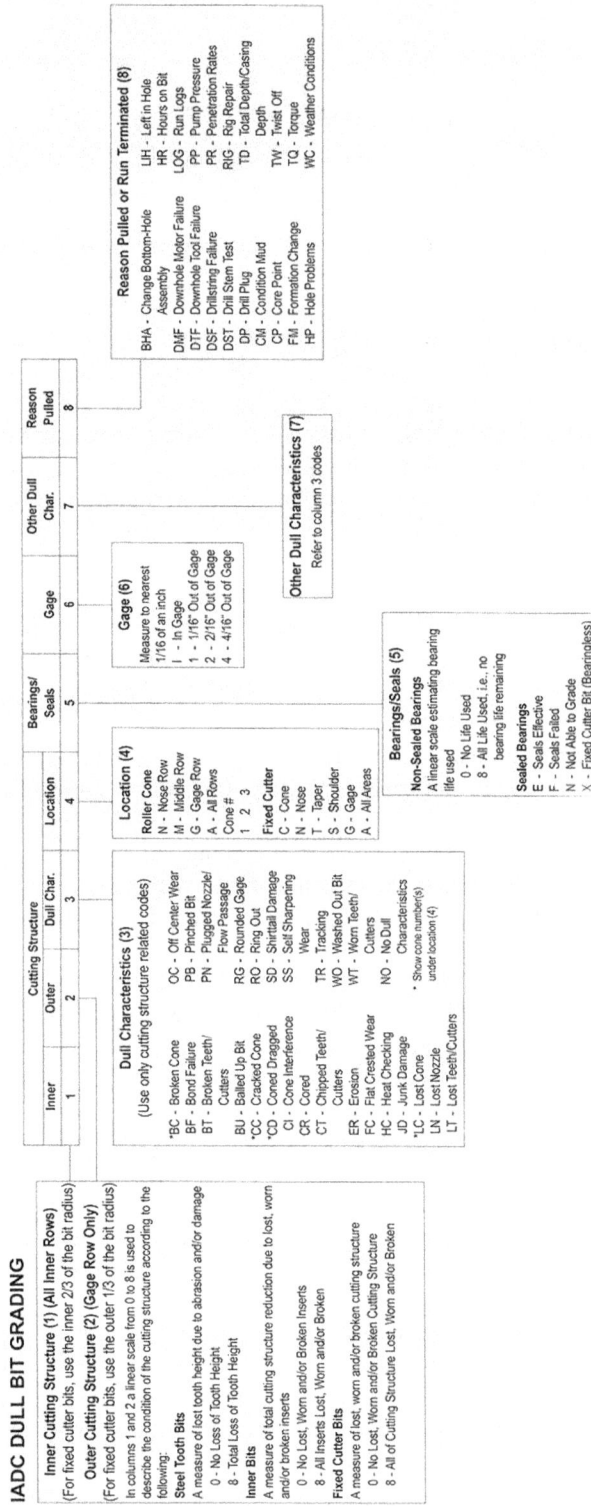

Fig. 5.37—IADC Dull Grading System.

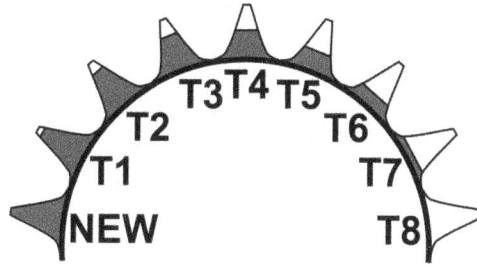

Fig. 5.38—Tooth height measurement.

Degrees of Cutter Wear

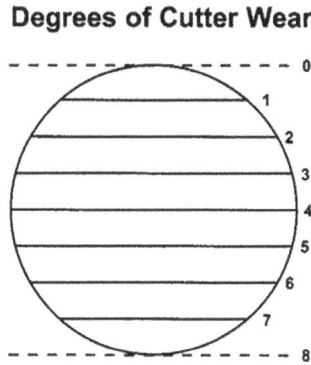

Fig. 5.39—Cutter wear convention (zero is no wear).

$(a + b + c + d + e + f)/6$. This analysis is repeated for each blade, and blade results are summed and averaged for the final result. A similar analysis is made for the seven cutters used in the outer bit rows, and the two results are recorded in the first two spaces of the dull grading form.

•Dull Characteristic (D). The cutting structure dull characteristic is the observed characteristic most likely to limit further use of the bit in the intended application. A two-letter code is used to indicate the major dull characteristics of the cutting structure.

The primary cutter dull characteristic, the third cutting structure subcategory, is recorded in the third space on the dull grading record. (Note that noncutting structure or "other" dull characteristics that a bit might exhibit are noted in the seventh grading category.) Category 3 defines only primary cutter wear, whereas Category 7 can be used to describe either secondary cutting structure wear or wear characteristics that relate to the bit as a whole and are unrelated to cutting structure. Grading codes for the other dull characteristics category are the same as those listed above.

•Roller-Cone Bit Dull Location (L). A two-letter code is used to indicate the location of the wear or failure that necessitated removal of the bit from service. These codes are listed in Fig. 5.37.

•PDC Bit Cutting Structure Dull Location. The last of the cutting structure-related wear grades, dull location, indicates the location of the primary dull characteristic. Possible locations include the cone (C), nose (N), taper (T), shoulder (S), gauge (G), all areas (A), middle row (M), and heel row (H). Location grades are reported in the fourth space on the dull grading form.

Bearing and Seal Criteria (Not Used for PDC Bits). IADC provides separate protocols for estimation of bearing and seal wear in nonsealed and sealed bearing assemblies. Seal and bearing grading applies only to roller-cone bits. It is always marked "X" for PDC bits.

Inner 2/3 Radius   Outer 1/3 Radius

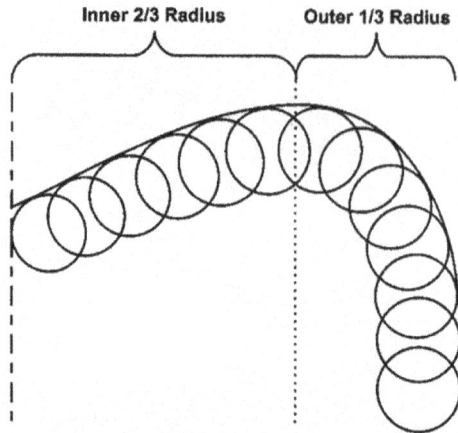

Inner / Outer Body Designation
PDC Bits / Impregnated Bits

Fig. 5.40—Inner/outer body designation for PDC and impregnated bits.

TABLE 5.4—SEAL/BEARING EVALUATION CHECKLIST	
Evaluation/Description	Acceptable Condition
Ability to rotate cone	Rotates normally
Cone springback	Springback exists
Seal squeak	Seal squeak exists
Internal sounds	No internal noises exist
Weeping grease	No lubricant leaks exist
Shale burn	No shale burn exists
Shale packing	If packing exists, remove before measuring
Gaps at the back face or throat	No bearing gaps exist
Inner or outer bearing letdown	No bearing letdown exists

•Estimating Wear on Nonsealed Bearings. For nonsealed bearings, wear is estimated on a linear scale of 0 to 8: 0 is new, 8 is 100% expended.

•Estimating Wear on Sealed Bearings. A checklist for the seal and bearing system condition is provided in **Table 5.4**. The grading protocol is as follows:

• If no seal problems are encountered, use the grading code E.

• If any component in the assembly has failed, use the grading code F.

• If any portion of the bearing is exposed or missing, it is considered an ineffective assembly; again, use the grading code F.

• Use the grading code N if it is not possible to determine the condition of both the seal and the bearing.

• Grade each seal and bearing assembly separately by cone number. If grading all assemblies as one, report the worst case.

Gauge Grading (G). The gauge category of the Dull Bit Grading System is used to report an undergauge condition for cutting elements intended to touch the wall of the hole bore. For

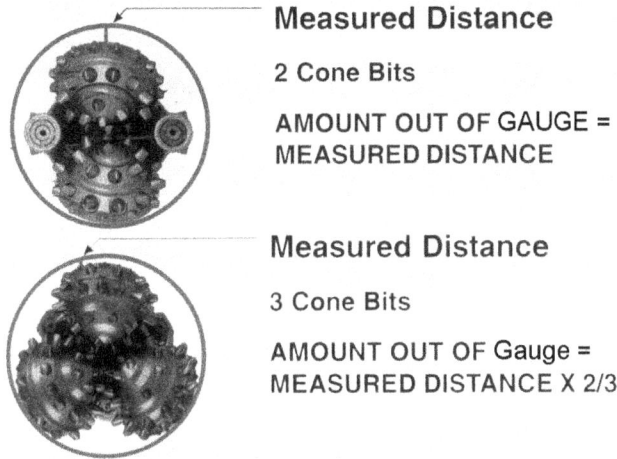

Measured Distance

2 Cone Bits

AMOUNT OUT OF GAUGE =
MEASURED DISTANCE

Measured Distance

3 Cone Bits

AMOUNT OUT OF Gauge =
MEASURED DISTANCE X 2/3

Fig. 5.41—Measuring out of gauge.

diamond and PDC bits only, gauge is measured with an API-specified ring gauge. (API specifi-cations for ring gauges for roller-cone bits have not been issued.)

•Roller-Cone Bit Gauge Grading. For three-cone bits, the "two-thirds rule" is applied to mea-suring the gauge condition. The amount out of gauge, as measured by the ring gauge, is multiplied by two-thirds to give the true gauge condition.

For two-cone bits, gauge is the measured distance from either the gauge or heel elements, whichever is closer to gauge.

Measurements are taken at either the gauge or heel cutting elements, whichever is closer to gauge (**Fig. 5.41**). Undergauge increments of $\frac{1}{16}$ in. are reported. If a bit is $\frac{1}{16}$ in. undergauge, the gauge report is 1. If a bit is $\frac{1}{8}$ in. ($\frac{2}{16}$ in.) undergauge, the gauge report is 2. If, a bit is $\frac{3}{16}$ in. undergauge, the gauge report is 3, and so forth. Round to the nearest $\frac{1}{16}$ in. Gauge rules apply to cutting structure elements only.

•PDC Bit Gauge Grading. For diamond and PDC bits, gauge is measured with a nominal ring gauge. Use of an "IN" code indicates that the bit remains in gauge. Undergauge incre-ments of $\frac{1}{16}$ in. are reported. If a bit is $\frac{1}{16}$ in. undergauge, the gauge report is 1. If a bit is $\frac{1}{8}$ in. ($\frac{2}{16}$ in.) undergauge, the gauge report is 2, and so forth. Round to nearest $\frac{1}{16}$ in. Gauge rules apply to cutting structure elements only. Measurements are taken at the gauge cutting elements.

Roller Cone and PDC Bit Remarks. The "remarks" category allows explanation of dull char-acteristics that do not correctly fit into other categories and is the category in which the reason a bit was removed from service is recorded.

•Roller-Cone Bit Other Dull Characteristics (O). Dull characteristics can be used to report dull characteristics other than those reported under cutting structure dull characteristics (D). Evidence of secondary bit wear is reported in the seventh grading category. Such evidence could relate to cutting structure wear, as recorded in the third space, or may report identifiable wear, such as erosion, for the bit as a whole. The secondary dull characteristic often identifies the cause of the dull characteristic noted in the third space.

•Roller-Cone and PDC Bit Reason Pulled (R). The eighth dull grading category reports the reason why a bit was pulled.

5.3.8 Bit Hydraulics. *Hydraulic Energy.* Energy is the rate of doing work. A practical aspect of energy is that it can be transmitted or transformed from one form to another (e.g., from an electrical form to a mechanical form by a motor). A loss of energy always occurs during transformation or transmission. In drilling fluids, energy is called hydraulic energy or commonly hydraulic horsepower.

The basic equation for hydraulic energy is

$$H = (pq)/1{,}714,$$

where H = hydraulic horsepower, p = pressure (psi or kPa), q = flow rate (gal/min or L/min), and 1,714 is the conversion of (psi-gal/min) to hydraulic horsepower [or (kPa·L/min) = 44 750].

Rig pumps are the source of hydraulic energy carried by drilling fluids. This energy is commonly called the total hydraulic horsepower or pump hydraulic horsepower:

$$H_1 = (p_1 q)/1{,}714,$$

where H_1 = total hydraulic energy (hydraulic horsepower) and p_1 = actual or theoretical rig pump pressure (psi). (See prior equation for metric conversion.) Note that the rig pump pressure (p_1) is the same as the total pressure loss or the system pressure loss. H_1 is the total hydraulic energy (rig pump) required to counteract all friction energy (loss) starting at the Kelly hose (surface line) and Kelly, down the drillstring, through the bit nozzles, and up the annulus at a given flow rate (q).

Bit hydraulic energy, H_b, is the energy needed to counteract frictional energy (loss) at the bit or can be expressed as the energy expended at the bit:

$$H_b = (p_b q)/1{,}714.$$

See prior equation for metric conversion.

Fluid Velocity. The general formula for fluid velocity is

$$v = q/A,$$

where v = velocity (ft/min or m/min), q = flow rate (gal/min or L/min), and A = area of flow (ft^2 or m^2).

The average velocity of a drilling fluid passing through a bit's jet nozzles is derived from the fluid velocity equation:

$$v_j = (0.32086q)/A_n,$$

where v_j = average jet velocity of bit nozzles (ft/sec or m/s) and A_n = total bit nozzle area (in.^{2} or cm^2).

Nozzle sizes are expressed in $\frac{1}{32}$-in. (inside diameter) increments. Examples are $\frac{9}{32}$ and $\frac{12}{32}$ in. The denominator is not usually mentioned; the size is understood to be in 32nds of an inch. For example, $\frac{9}{32}$- and $\frac{12}{32}$-in. nozzles are expressed as sizes 9 and 12.

The impact force of the drilling fluid at velocity v_{j1} can be derived from Newton's Second Law of Motion: force equals mass times acceleration. Assuming that all the fluid momentum is transferred to the bottomhole,

$$I_j = 0.000518 W q v_j,$$

where I_j = impact force of nozzle jets (lbf or kPa), W = mud weight (lbm/gal or kg/L), q = flow rate (gal/min or L/min), and v_j = average jet velocity from bit nozzles (ft/sec or m/s).

System Pressure Loss. Pressure losses inside the drillstring result from turbulent conditions. Viscosity has very little effect on pressure losses in turbulent flow. At higher Reynold's numbers, a larger variation results in only a small variation in friction factor. The calculated pressure loss equations are based on turbulent flow and are corrected for mud weight instead of viscosity:

$$\Delta p = pb = (Wq)^2 / (10,858 A_n)^2,$$

where A_n = total combined area of the bit nozzles (in.^{2} or cm^2), W = mud weight (lb/gal or kg/L), p_b = bit nozzle jets pressure loss (psi or kPa), and q = flow rate (gal/min or L/min).

5.3.9 Bit Economics. Regardless of how good a new product or method may be to a drilling operation, the result is always measured in terms of cost per foot or meter. Lowest cost per foot indicates to drilling engineers and supervisors which products to use most advantageously in each situation. Reduced costs lead directly to higher profits or, in some cases, to the difference between profit and loss.

For those in administration, engineering, manufacturing, and sales, cost calculations are used to evaluate the effectiveness of any product or method, new or old. Because drilling costs are so important, everyone involved should know how to make a few simple cost calculations.

For example, the cost of a PDC bit can be up to 20 times the cost of a milled-tooth bit and up to 4 times the cost of a TCI bit. The choice of a PDC bit, a milled-tooth bit, or an insert roller-cone bit must be economically justified by its performance. Occasionally, this performance justification is accomplished by simply staying in the hole longer. In such cases, the benefits of using it are intangible.

The main reason for using a bit, however, is that it saves money on a cost-per-foot basis. To be economical, a PDC bit must make up for its additional cost by either drilling faster or staying in the hole longer. Because the bottom line on drilling costs is dollars and cents, bit performance is based on the cost of drilling each foot of hole.

Breakeven analysis of a bit is the most important aspect of an economic evaluation. A breakeven analysis is necessary to determine whether the added bit cost can be justified for a particular application.

The breakeven point for a bit is simply the footage and hours needed to equal the cost-per-foot that would be obtained on a particular well if the bit were not used. To break even, a good offset well must be used for comparative purposes.

If the bit record in **Table 5.5** were used, we could determine whether a bit would be economical.

Example 5.1 *Economic Analysis.*
Total rotating time = 212.5 hr
Total trip time = 54.3 hr
Rig operating cost = $300/hr
Total bit cost = $16,148
Total footage = 3,380 ft
Note: Tripping rate is computed at 1,000-ft/hr average. This rate will vary, depending on rig type and operation. Therefore, the offset cost per foot for this interval (8,862 to 12,242 ft) is calculated with the standard cost-per-foot equation:

$$C = [R(t + t_d) + C_b] / F,$$

TABLE 5.5—BIT RECORD						
Bit Size, in.	Type	Bit Cost, $	Depth Out, ft	Footage Drilled, ft	Time, hr	ROP
8¹/₂	FDS+	2,820	9,618	756	26	29.1
8¹/₂	FDT	1,848	10,271	653	24	27.2
8¹/₂	FDG	1,848	10,699	428	22	19.5
8¹/₂	F2	4,816	11,614	915	71.5	12.8
8¹/₂	F2	4,816	12,242	628	69	9.1
Total		16,148		3,380	212.5	
Average						15.9

where C = drilling cost per foot ($/ft), R = rig operating cost (plus add-on equipment, such as downhole motor) ($/hr), t = trip time (hr), t_d = drilling time (hr), C_b = bit cost ($), and F = footage drilled (ft).

From the data provided in the example above, the cost per foot is

$$C = [300(212.5 + 54.3) + 16,148]/3,380 = 28.46 \ \text{ft/hr}.$$

In determinations of whether an application is suitable for a bit, the offset performances are given, but bit performance must be estimated. Thus, we must assume either the footage the bit will drill or the ROP it will obtain. If the footage is assumed, then we use the following equation to calculate the break-even ROP:

$$\text{Breakeven ROP} = C_r/[C_o(Rt + C_b)]/F,$$

where C_r = rig operating cost ($/hr), C_o = offset cost per foot ($), t = trip time of bit (hr), C_b = bit cost ($), and F = assumed bit footage (ft). Therefore, in the above example,

$$\text{Breakeven ROP} = 300/\{28.46[(300 \times 12) + 18,300]\}/3,380 = 13.7 \ \text{ft/hr}.$$

The bit must drill the 3,380 ft at an ROP of 13.7 ft/hr to equal the offset cost per foot of $28.46 for the same 3,380 ft.

If an ROP is assumed, use the following equation to calculate the breakeven footage:

$$\text{Breakeven footage} = [(C_r t) + C_b]/[C_o - (r/\text{ROP})].$$

Thus, in the above example, if we assume an ROP of 20 ft/hr, we have

$$\text{Breakeven footage} = [(300 \times 12) + 18,300]/[28.46(300/20)] = 1,627 \ \text{ft}.$$

In this case, the bit must drill 1,627 ft to attain the breakeven point.

5.3.10 Bit Selection and Operating Practices. *Rules of Thumb for Bit Selection.*
- Shale has a better drilling response to drill speed.
- Limestone has a better drilling response to bit weight.

- Bits with roller bearings can be run at a higher speed than bits with journal bearings.
- Bits with sealed bearings have a longer life than bits with open bearings.
- Bits with journal bearings can be run at higher weights than bits with roller bearings.
- Diamond product bits can run at higher speeds than three-cone bits.
- Bits with high offset may wear more on gauge.
- Cost-per-foot analysis can help you decide which bit to use.
- Examination of dulls can also help you decide which bit to use.

Tripping Can Ruin a New Bit.
- Make the bit up to proper torque.
- Hoist and lower the bit slowly through ledges and doglegs.
- Hoist and lower the bit slowly at liner tops.
- Avoid sudden stops. Drillpipe stretch can cause a bit to hit the hole bottom.
- If reaming is required, use a light weight and low speed.

Establish a Bottomhole Pattern.
- Rotate the bit and circulate mud when approaching bottom. This will prevent plugged nozzles and clear out fill.
- Lightly tag bottom with low speed.
- Gradually increase speed and then gradually increase weight.

Use a Drill-Off Test To Select Best WOB and Speed.
- Select speed.
- Select bit weight. Depending on bit selected, refer to appropriate manufacturer's recommended maximum speed and WOB.
- Lock brake.
- Record drill-off time for 5,000-lbm increments of weight indicator decrease.
- Repeat this procedure for different speeds.
- Drill at the weight and speed that give the fastest drill-off time.

The Bit Is Not Always To Blame for Low ROP.
- Mud weight may be too high with respect to formation pressure.
- Mud solids may need to be controlled.
- Pump pressure or pump volume may be too low.
- Formation hardness may have increased.
- Speed and weight may not be the best for bit type and formation. Use drill-off test.
- Bit may not have adequate stabilization.
- Bit may be too hard for the formation.

Acknowledgments

Figures and tables in this chapter are courtesy of Smith Intl. Inc.

References

1. Bentson, H.G., and Smith Intl. Inc: Roller-Cone Bit Design, API Division of Production, Pacific Coast District, Los Angeles (May 1956).
2. *Spec. 7,* Specification for Rotary Drilling Equipment, 37th edition, Section 7, API, Washington, DC (August 1990).
3. Portwood G. *et al.:* "Improved Performance Roller-Cone Bits for Middle Eastern Carbonates," paper SPE 72298 presented at the 2001 SPE Middle East Drilling Technology Conference, Bahrain, 22–24 October.
4. Keshavan, M.K. *et al.:* "Diamond-Enhanced Insert: New Compositions and Shapes for Drilling Soft-to-Hard Formations," paper SPE 25737 presented at the 1993 SPE/IADC Drilling Conference, Amsterdam, 23–25 February.

5. Salesky, W.J. and Payne, B.R.: "Preliminary Field Test Results of Diamond-Enhanced Inserts for Three Cone Rock Bits," paper SPE 16115 presented at the 1987 SPE/IADC Drilling Conference, New Orleans, 15–18 March.

6. Salesky, W.J. *et al.*: "Offshore Tests of Diamond-Enhanced Rock Bits," paper SPE 18039 presented at the 63rd SPE Annual Technical Conference and Exhibition, Houston, 2–5 October.

7. Chia, R. and Smith, R.: "A New Nozzle System to Achieve High ROP Drilling," paper SPE 15518 presented at the 1986 SPE Annual Technical Conference, New Orleans, 5–8 October.

8. McGehee, D.Y. *et al.*: "Roller-Cone Bit Classification System," paper SPE 23937 presented at the 1992 IADC/SPE Drilling Conference, New Orleans, 18–21 February.

9. Clark, D.A. *et al.*: "Application of the New IADC Dull Grading System for Fixed-cutter bits," paper 16145 presented at the 1987 SPE/IADC Drilling Conference, New Orleans, 15–18 March.

10. Brandon, B.D. *et al.*: "Development of a New IADC Fixed-Cutter Drill Bit Classification System," paper SPE 23940 presented at the 1992 IADC/SPE Drilling Conference, New Orleans, 18–21 February.

11. Brandon, B.D. *et al.*: "IADC Fixed-cutter bit Classification System," paper SPE 16142 presented at the 1987 SPE/IADC Drilling Conference, New Orleans, 15–18 March.

12. Brandon, B.D. *et al.*: "First Revision to the IADC Fixed Cutter Dull Grading System," paper SPE 23939 presented at the 1992 IADC/SPE Drilling Conference, New Orleans, 18–21 February.

13. McGehee, D.Y. *et al.*: "The IADC Roller Bit Dull Grading System," paper SPE 23938 presented at the 1992 SPE/IADC Drilling Conference, New Orleans, 18–21 February.

SI Metric Conversion Factors

ft	× 3.048*	E–01	= m
gal	× 3.785 412	E–03	= m^3
in.	× 2.54*	E–02	= cm
lbf	× 4.448 222	E+00	= N
lbm	× 4.535 924	E–01	= kg
psi	× 6.894 757	E+00	= kPa
sq in.	× 6.451 6*	E–04	= m^2

*Conversion factor is exact.

Chapter 6
Directional Drilling
David Chen, Halliburton

6.1 Introduction to Directional Drilling

Directional drilling is defined as the practice of controlling the direction and deviation of a wellbore to a predetermined underground target or location. This section describes why directional drilling is required, the sort of well paths that are used, and the tools and methods employed to drill those wells.

6.1.1 Applications.

• Multiple wells from a single location. Field developments, particularly offshore and in the Arctic, involve drilling an optimum number of wells from a single platform or artificial island. Directional drilling has helped by greatly reducing the costs and environmental impact of this application.

• Inaccessible surface locations. A well is directionally drilled to reach a producing zone that is otherwise inaccessible with normal vertical-drilling practices. The location of a producing formation dictates the remote rig location and directional-well profile. Applications like this are where "extended-reach" wells are most commonly drilled.

• Multiple target zones. A very cost-effective way of delivering high production rates involves intersecting multiple targets with a single wellbore. There are certain cases in which the attitudes (bed dips) of the producing formations are such that the most economical approach is a directional well for a multiple completion. This is also applicable to multiple production zones adjacent to a fault plane or beneath a salt dome.

• Sidetrack. This technique may be employed either to drill around obstructions or to reposition the bottom of the wellbore for geological reasons. Drilling around obstructions, such as a lost string of pipe, is usually accomplished with a blind sidetrack. Oriented sidetrack is required if a certain direction is critical in locating an anticipated producing formation.

• Fault drilling. It is often difficult to drill a vertical well through a steeply inclined fault plane to reach an underlying hydrocarbon-bearing formation. Instead, the wellbore may be deflected perpendicular or parallel to the fault for better production. In unstable areas, a wellbore drilled through a fault zone could be at risk because of the possibility of slippage or movement along the fault. Formation pressures along fault planes may also affect hole conditions.

• Salt-dome exploration. Producing formations can be found under the hard, overhanging cap of salt domes. Drilling a vertical well through a salt dome increases the possibility of drilling problems, such as washouts, lost circulation, and corrosion.

• Relief-well drilling. An uncontrolled (wild) well is intersected near its source. Mud and water are then pumped into the relief well to kill the wild one. Directional control is extremely exacting for this type of application.

• River-crossing applications. Directional drilling is employed extensively for placing pipelines that cross beneath rivers, and has even been used by telecommunication companies to install fiber-optic cables.

6.2 Directional-Well Profiles

A directional well can be divided into three main sections—the surface hole, overburden section, and reservoir penetration. Different factors are involved at each stage within the overall constraints of optimum reservoir penetration.

6.2.1 Surface-Hole Section.
Most directional wells are drilled from multiwell installations, platforms, or drillsites. Minimizing the cost or environmental footprint requires that wells be spaced as closely as possible. It has been found that spacing on the order of 2 m (6 ft) can be achieved. At the start of the well, the overriding constraint on the well path is the presence of other wells. Careful planning is required to assign well slots to bottomhole locations in a manner that avoids the need for complex directional steering within the cluster of wells. At its worst, the opportunity to reach certain targets from the installation can be lost if not carefully planned from the outset. Visualizing the relative positions of adjacent wells is important for correct decisions to be made about placing the well path to minimize the number of adjacent wells that must be shut in as a safety precaution against collisions. The steel in nearby wells requires that special downhole survey techniques be used to ensure accurate positioning. This section is generally planned with very low curvatures to minimize problems in excessive torque and casing wear resulting from high contact forces between drillstrings and the hole wall.

Many directional wells are drilled from surface pads and offshore locations. Close surface locations always have the potential for collisions near the surface. Planning proper surface-hole surveying strategies to prevent collisions is critical in well planning. Gyro surveys (single/multiple shots) are often used to eliminate problems associated with close wellbore spacing. Modern well-planning software has used the survey uncertainty model in the anticollision calculations.

Perhaps the most important technique in collision avoidance is the traveling-cylinder diagram (TCD). The TCD provides an effective means of portraying the actual position of the well being drilled relative to its planned course and to adjacent wells. It also allows complex, 3D interwell tolerances on the allowable position of the borehole trajectory to be presented in a simple and unambiguous form. Because the original hand-drawn version was developed in 1968, various algorithms have been devised to produce the TCD in the well-planning software package. Among the three versions of the diagram commonly available, the normal-plane TCD is the most efficient tool. The normal-plane projection displays the intersection of wells with a plane constructed in space to be normal to the direction of the planned well at the point of interest. Because of its clear and simple presentation of a complex situation, the normal-plane TCD has recently been used at the wellsite to assist the simple go/no-go decision and the visualization of collision potential without making any interpretive judgments on well convergence or survey error values.[1-5]

6.2.2 Overburden Section.
Having steered away from the congestion of the surface section, the main part of the well path through the overburden is specifically designed to put the well in the best possible position for penetrating the reservoir. There are three different overall

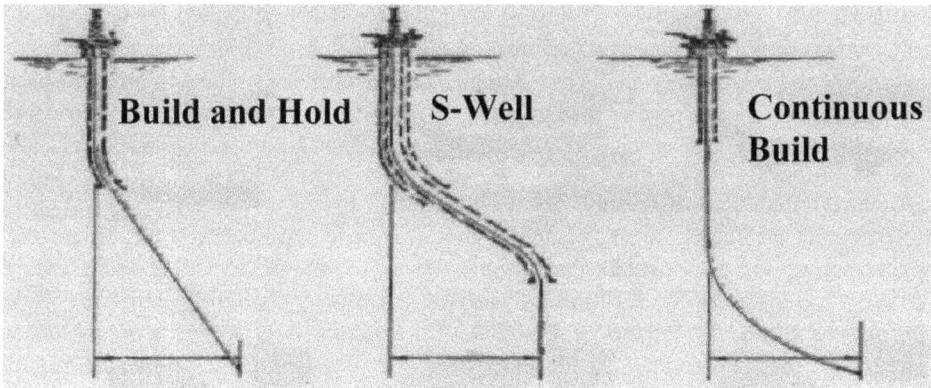

Fig. 6.1—Schematic of wellbores through overburdens.

shapes of the well, depending on the penetration requirements. These are build-and-hold, S-shaped, and continuous build. In practice, these generic shapes will be modified by local conditions. Getting the right well path through the overburden is a multidisciplinary task in which geologists advise the designer about the presence of faults, the precise shape of salt formations, mud diapirs, and other subsurface hazards. Understanding the interaction between the 3D well trajectory and the formation stresses, particularly in overthrust areas, is vital to ensuring that the well can be drilled safely and efficiently. See **Fig. 6.1** for an illustration of these wellbores.

In general, a build-and-hold profile is planned so that the initial deflection angle is obtained at a shallow depth, and from that point on the angle is maintained as a straight line to the target zone. Once the angle and deflection are obtained, casing may be set through the deviated section and cemented. In general, the build-and-hold profile is the basic building block of extended-reach wells. These profiles can usually be employed in two distinct depth programs. These profiles can be used for moderate-depth drilling in areas where intermediate casing is not required and where oil-bearing strata are a single horizon. They can also be used for deeper wells requiring a large lateral displacement. In this case, an intermediate-casing string can be set to the required depth, and then the angle and direction can be maintained after drilling out below the string.

The main reasons for drilling an S-shaped well are completion requirements for the reservoir; for example, when a massive stimulation operation is required during the completion. An S-shaped well also sets the initial deflection angle near the surface. After the angle is set, drilling continues on this line until the appropriate lateral displacement is attained. The hole is then returned to vertical or near vertical and drilled until the objective depth is reached. Surface casing is set through the upper deviated section and cemented. The wellbore is then continued at the desired angle until the lateral displacement has been reached and then returns to vertical. Intermediate casing is set through the lower vertical-return section. Drilling then continues below the intermediate casing in a vertical hole. The S-shaped well is often employed with deep wells in areas where gas troubles, saltwater flows, etc. dictate the setting of intermediate casing. It permits more-accurate bottomhole spacing in a multiple-pay area. The deflection angle may be set in surface zones in which drilling is fast and round-trip costs can be held to a minimum.

A continuous-build well starts its deviation well below the surface. The angle is usually achieved with a constant build to the target point. The deflection angles may be relatively high, and the lateral distances from vertical to the desired penetration point are relatively shorter than other well types. Typical applications would be in exploring a stratigraphic trap or obtaining additional geological data on a noncommercial well. Because deflection operations take place deep in the hole, trip time for such operations is high, and the deflected part of the hole is not normally protected by casing. The continuous-build profile may also commonly be found in old fields in which development of bypassed oil is carried out by means of sidetracks from existing wells that have ceased to produce economically from the original completion.

6.2.3 Reservoir-Penetration Section. The penetration of the reservoir is the realization of the whole purpose of drilling the well, whether a producer or an exploration well. Therefore, correct placement of the well within the target zone is of utmost importance. Designing the penetration is clearly a major multidisciplinary task involving not only the drilling team but also geologists and reservoir engineers. As indicated previously, for some wells, a simple straight-line penetration may suffice to provide an economical flow. Sometimes the path should be brought back to vertical to assist in stimulation operations or to keep the well within a fault block. Increasingly, though, target penetrations can be very complex undertakings in high-cost, Arctic, onshore extended-reach, and offshore-platform operations. At its most basic is the horizontal well; at the other extreme is the designer well.

There are two important aspects of reservoir penetration. First is allowing for the effects of a wellbore position error on defining the target location; the other is placing the wellbore within the formation for maximum production efficiency. The surveys used to calculate the well position always contain some errors. These errors result in a difference between the apparent location of the well, as derived from the survey data, and the actual location, which, by definition, is never known. The likely size of these errors can be quantified for different well locations, surveying methods, and wellbore shapes. These errors must be taken into account when defining the boundaries, or tolerances, around the target location. In extreme cases, such as extended-reach wells in the Arctic, the errors can be much greater than the size of the target unless special surveying techniques are employed. Under these conditions and even though the apparent position of the well is within the target, the actual location may be outside. When undetected, this misleading information can have a significant impact on understanding the geological model and can result in substantial losses of reserves. If detected, there may be the need to undertake a costly sidetracking operation to place the well correctly.

6.2.4 Horizontal Wells. Horizontal wells are high-angle wells (with an inclination of generally greater than 85°) drilled to enhance reservoir performance by placing a long wellbore section within the reservoir. This contrasts with an extended-reach well, which is a high-angle directional well drilled to intersect a target point. There was relatively little horizontal drilling activity before 1985. The Austin Chalk play is responsible for the boom in horizontal drilling activity in the U.S. Now, horizontal drilling is considered an effective reservoir-development tool.[6-9]

The advantages of horizontal wells include:

1. Reduced water and gas coning because of reduced drawdown in the reservoir for a given production rate, thereby reducing the remedial work required in the future.

2. Increased production rate because of the greater wellbore length exposed to the pay zone.

3. Reduced pressure drop around the wellbore.

4. Lower fluid velocities around the wellbore.

5. A general reduction in sand production from a combination of Items 3 and 4.

6. Larger and more efficient drainage pattern leading to increased overall reserves recovery.

Horizontal wells are normally characterized by their buildup rates and are broadly classified into three groups that dictate the drilling and completion practices required, as shown in **Table 6.1.**

The "build rate" is the positive change in inclination over a normalized length (e.g., 3°/100 ft.) A negative change in inclination would be the "drop rate." A long-radius horizontal well is characterized by build rates of 2 to 6°/100 ft, which result in a radius of 3,000 to 1,000 ft.

TABLE 6.1—HORIZONTAL-WELL CLASSIFICATIONS			
Well Type	Build Rate (ft)	Radius (m)	Radius (ft)
Long radius	2 to 6°/100 ft	900 to 290	3,000 to 1,000
Medium radius	6 to 35°/100 ft	290 to 50	1,000 to 160
Short radius	5 to 10°/3 ft	12 to 6	40 to 20

This profile is drilled with conventional directional-drilling tools, and lateral sections of up to 8,000 ft have been drilled. This profile is well suited for applications in which a long, horizontal displacement is required to reach the target entry point. The use of rotary-steerable systems (RSSs) may be required to drill an extra-long lateral section because slide drilling may not be possible with the conventional steerable motors.

Medium-radius horizontal wells have build rates of 6 to 35°/100 ft, radii of 1,000 to 160 ft, and lateral sections of up to 8,000 ft. These wells are drilled with specialized downhole mud motors and conventional drillstring components. Double-bend assemblies are designed to build angles at rates up to 35°/100 ft. The lateral section is often drilled with conventional steerable motor assemblies. This profile is common for land-based applications and for re-entry horizontal drilling. In practical terms, a well is classified as medium radius if the bottomhole assembly (BHA) cannot be rotated through the build section at all times. At the upper end of the medium radius, drilling the maximum build rate is limited by the bending and torsional limits of API tubulars. Smaller holes with more-flexible tubulars have a higher allowable maximum dogleg severity (DLS).

Short-radius horizontal wells have build rates of 5 to 10°/3 ft (1.5 to 3°/ft), which equates to radii of 40 to 20 ft. The length of the lateral section varies between 200 and 900 ft. Short-radius wells are drilled with specialized drilling tools and techniques. This profile is most commonly drilled as a re-entry from any existing well.

6.2.5 Multilateral Wells. Multilateral wells are new evolution of horizontal wells in which several wellbore branches radiate from the main borehole. In 1997, Technology Advancement for Multi-Laterals (TAML), an industry consortium of operators and service companies, was formed to categorize multilateral wells by their complexity and functionality. The designated categories (levels) are as follows.
- Level 1—openhole junction.
- Level 2—cased-hole exit.
- Level 3—junction with connection but no seal.
- Level 4—sealed junction.
- Level 5—mechanical sealed junction with reduced inside diameter (ID).
- Level 6—mechanical sealed junction with full ID.
- Level 6S—downhole splitter.

Multilateral technology has advanced dramatically in recent years to assist in recovering hydrocarbons, particularly in heavy-oil applications.[9–11]

6.2.6 Extended-Reach Wells. An extended-reach well is one in which the ratio of the measured depth (MD) vs. the true vertical depth (TVD) is at least 2:0. The current world record is Brintnell Well 2-10 (Amoco in Canada), with the highest MD/TVD ratio of 8.00. The top four wells are as follows.
- Amoco Brintnell 2-10 (Wabasca): MD/TVD = 8.00.
- Amoco Brintnell 1-18 (Wabasca): MD/TVD = 7.39.
- Maersk Qatar BA-26 (Al Shaheen): MD/TVD = 7.04.

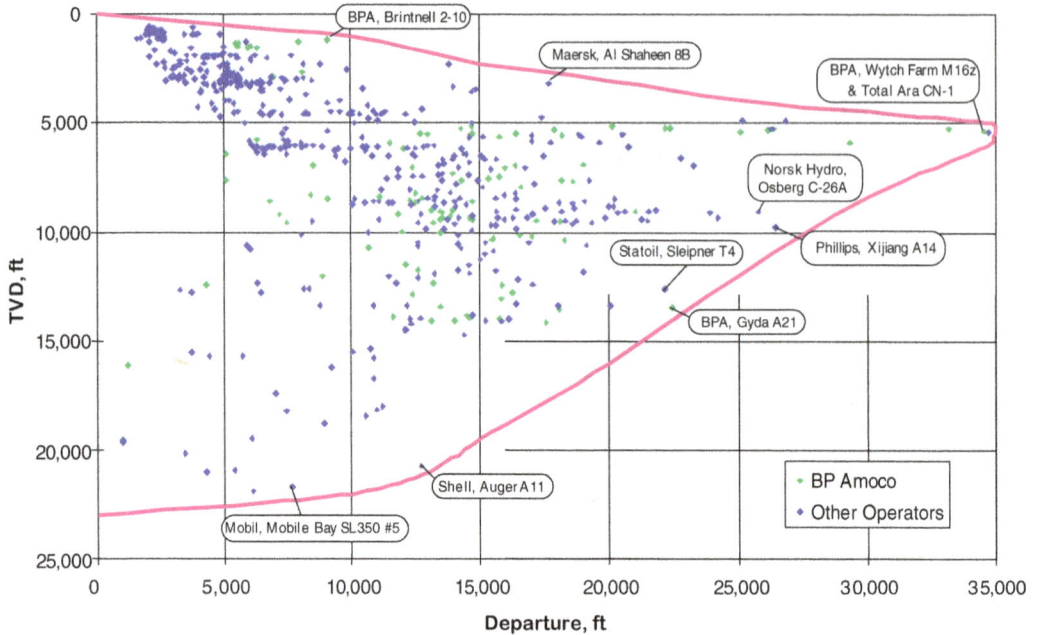

Fig. 6.2—Extended-reach wells drilled to date.

- BP Wytch Farm M16z: MD/TVD = 6.89.

BP Wytch Farm Well M16z still holds the world record MD and horizontal departure (MD = 37,001 ft, and departure = 35,197 ft).

Other notable extended-reach-drilling (ERD) achievements in pushing the horizontal departure limit are:

- 9,000 to 10,000 ft TVD: Phillips (Xijiang, China) = 8 km, Norsk Hydro (Oseberg, Norway) = 7.8 km, Woodside (Goodwyn, Australia) = 7.4 km, and Statoil (Sleipner, Norway) = 7.4 km.
- 13,000 to 15,000 ft TVD: Woodside (Goodwyn, Australia) = 7.4 km, and Statoil (Sleipner, Norway) = 7.4 km.
- More than 20,000 ft TVD: Shell (Auger, U.S.A.) = 3.9 km.

Extended-reach wells are expensive and technically challenging.[12–15] However, they can add value to drilling operations by making it possible to reduce costly subsea equipment and pipelines, by using satellite field development, by developing near-shore fields from onshore, and by reducing the environmental impact by developing fields from pads.

There have been more than 1,700 extended-reach wells drilled to date, as shown in **Fig. 6.2**.

Fig. 6.3 shows the evolution of extended-reach wells in the 1990s and the future. With new technology, the goal is to see the TVD push to 30,000 ft and the horizontal departure to 50,000 ft in the 21st century. As of this printing, we are still waiting to see that goal achieved.

6.2.7 Design Wells. Today, most directional-well planning is done on the computer. Modern computer technologies, such as 3D visualization and 3D earth models, have provided geoscientists and engineers with integrated and interactive tools to create, visualize, and optimize well paths through reservoir targets, as shown in **Fig. 6.4**. Furthermore, recently developed geosteering systems and RSSs allow more-complex directional-well trajectories that are designed to drain more of the reservoir. The future is the real-time integration of the drilling and logging-

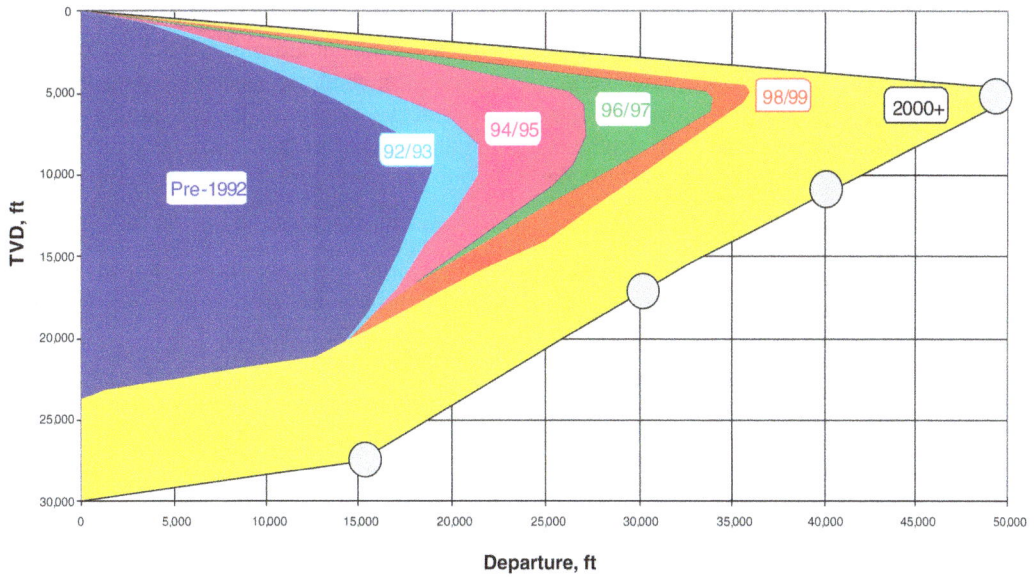

Departure, ft

Fig. 6.3—Evolution of departure distance in ERD.

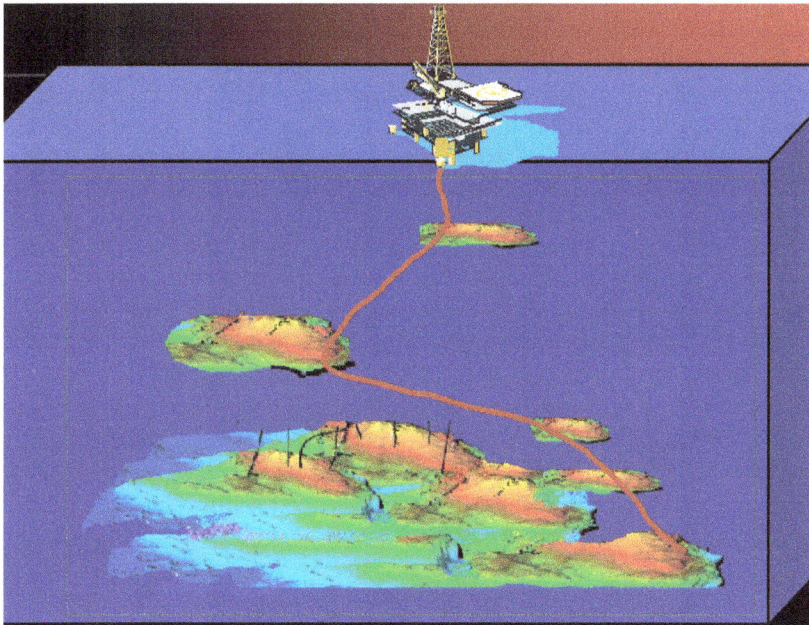

Fig. 6.4—Illustration of the visualization of a well path.

while-drilling (LWD) data with geosteering and the earth model. The 3D visualization of real-time data, together with the earth model, would allow integrated knowledge management and real-time decision making.

6.3 Directional Survey

The method used to obtain the measurements needed to calculate and plot the 3D well path is called directional survey. Three parameters are measured at multiple locations along the well path—MD, inclination, and hole direction. MD is the actual depth of the hole drilled to any point along the wellbore or to total depth, as measured from the surface location. Inclination is the angle, measured in degrees, by which the wellbore or survey-instrument axis varies from a true vertical line. An inclination of 0° would be true vertical, and an inclination of 90° would be horizontal. Hole direction is the angle, measured in degrees, of the horizontal component of the borehole or survey-instrument axis from a known north reference. This reference is true north, magnetic north, or grid north, and is measured clockwise by convention. Hole direction is measured in degrees and is expressed in either azimuth (0 to 360°) or quadrant (NE, SE, SW, NW) form.

Each recording of MD, inclination, and hole direction is taken at a survey station, and many survey stations are obtained along the well path. The measurements are used together to calculate the 3D coordinates, which can then be presented as a table of numbers called a survey report. Surveying can be performed while drilling occurs or after it has been completed.

The purposes of directional survey are to:
• Determine the exact bottomhole location to monitor reservoir performance.
• Monitor the actual well path to ensure the target will be reached.
• Orient deflection tools for navigating well paths.
• Ensure that the well does not intersect nearby wells.
• Calculate the TVD of the various formations to allow geological mapping.
• Evaluate the DLS, which is the total angular inclination and azimuth in the wellbore, calculated over a standard length (100 ft or 30 m).
• Fulfill requirements of regulatory agencies, such as the Minerals Management Service (MMS) in the U.S.

6.3.1 Survey Instruments. Survey instruments can be set up in several different variations, depending on the intended use of the instrument and the methods used to store or transmit survey information. Basically, there are two types of survey instruments: magnetic and gyroscopic. Depending on the method used to store the data, there are film and electronic systems. Survey systems can also be categorized by the methods used to transmit the data to the surface, such as wireline or measurement while drilling (MWD).

Magnetic Sensors. Magnetic sensors must be run within a nonmagnetic environment [i.e., in uncased hole either in a nonmagnetic drill collar(s) or on a wireline]. In any case, there must not be any magnetic interference from adjacent wells. Magnetic sensors can be classified into two categories—mechanical and electronic compasses.

A mechanical compass uses a compass card that orients itself to magnetic north, similar to a hiking-compass needle. Inclination is measured by means of a pendulum or a float device. In the pendulum device, the pendulum is either suspended over a fixed grid or along a vernier scale and is allowed to move as the inclination changes. The float device suspends a float in fluid that allows the instrument tube to move around it independently as the inclination changes. The only advantage of mechanical compasses is the low cost, while several disadvantages have limited them from being used widely in directional surveys. The drawbacks are high maintenance costs, a need to choose inclination range, limited temperature capability, the possibility of human error in reading film, and the inability to use them in MWD tools.

The electronic compass system is a solid-state, self-contained, directional-surveying instrument that measures the Earth's magnetic and gravitational forces. Inclination is measured by gravity accelerometers, which measure the Earth's gravitational field in the x, y, and z planes. The z plane is along the tool axis, x is perpendicular to z and in line with the tool's reference slot, and y is perpendicular to both x and z. From this measurement, the vector components can be summed to determine inclination. Hole direction is measured by gravity accelerometers and fluxgate magnetometers. Fluxgate magnetometers measure components of the Earth's magnetic field orthogonally (i.e., in the same three axes as the accelerometers). From this measurement, the vector components can be summed to determine hole direction.

Depending on the packaging of the electronic sensors, the electronic-compass system can be employed in different modes, such as single-shot, multishots, and MWD, in which data are sent to surface in real time through the mud-pulse telemetry system.

The electronic magnetic single-shot records a single survey record while drilling the well. The sensors measure the Earth's magnetic and gravitational forces with fluxgate magnetometers and gravity accelerometers, respectively. The components of this survey system include the probe and a battery stack that supplies power to the probe. The raw data are stored downhole in the memory and retrieved at the surface to calculate the hole direction, inclination, and tool face. The electronic magnetic multishot uses the same components as the electronic single-shot; the only difference is that electronic multishots record multiple survey records. The MWD acquires downhole information during drilling operations that can be used to make timely decisions about the drilling process. The magnetic survey information is obtained with an electronic compass, but, unlike previous systems that stored the information, the MWD encodes the survey data in mud pulses that are sent up and decoded at the surface. The real-time survey information enables the drillers to make directional-drilling decisions while drilling. The sensors used in MWD tools are the same design as those used in electronic magnetic single-shot and multishots (i.e., gravity accelerometers and fluxgate magnetometers).

The Geomagnetic Field. Both types of magnetic sensors rely upon detecting the Earth's magnetic field to determine hole direction. The Earth can be imagined as having a large bar magnet at its center, laying (almost) along the north/south spin axis (see **Fig. 6.5**). The normal lines of the magnetic field will emanate from the bar magnet in a pattern such that at the magnetic north and south poles, the lines of force (flux lines) will lay vertically, or at 90° to the Earth's surface, while at the magnetic equator, the lines of force will be horizontal, or at 0° to the Earth's surface. At any point on the Earth, a magnetic field can be observed having a strength and a direction (vector). The strength is called magnitude and is measured in units of tesla. Usual measurements are approximately 60 microtesla at the magnetic north pole and 30 microtesla at the magnetic equator. The direction is always called magnetic north. However, although the direction is magnetic north, the magnitude will be parallel to the surface of the Earth at the equator and point steeply into the Earth closer to the north pole. The angle that the vector makes with the Earth's surface is called the dip.

The prevailing models used to estimate the local magnetic field are provided by the British Geological Survey (BGS) or, alternatively, by the U.S. Geological Soc. (USGS). These models carry out a high-order spherical harmonic expansion of the Earth's magnetic field and provide a very accurate global calculation of the magnetic field rising from the Earth's core and mantle. The models are based on measurements from hundreds of magnetic stations on the surface of the earth, airborne magnetic surveys, and magnetic-field data gathered by satellites. Because even the field of the Earth's core and mantle varies with time, these models are updated on an approximately annual basis. Note that these models include neither effects from materials near the surface of the earth (termed "crustal anomalies"), which can be quite significant, nor separate effects from various electrojets in the Earth's atmosphere,* the effects of solar storms, or the diurnal variation in the earth's magnetic field. At high latitudes,** these effects can be quite significant. A way of getting around this problem is to make magnetic-observatory-quality measurements directly at the wellsite; however, this is rarely possible. A very useful alternative is to interpolate the field at a given location and time, as measured by at least three nearby magnetic observatories, the triangle of which preferably includes the wellsite being surveyed. This is referred to as interpolated in-field referencing. Scientists who use this technique

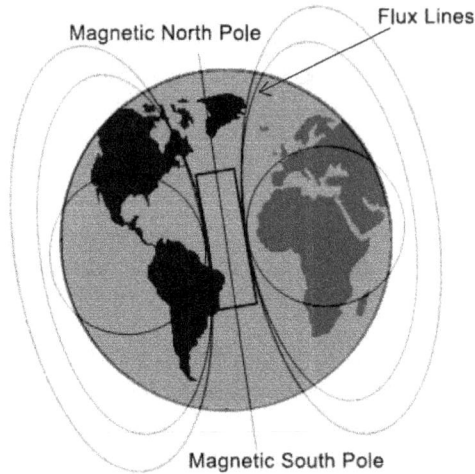

Fig. 6.5—The Earth's magnetic field.

on surveys taken at high latitudes and with axial magnetic interference report achieving an accuracy approaching that otherwise attainable only with gyros.

Gyroscopic Sensors. Gyroscopic surveying instruments are used when the accuracy of a magnetic survey system may be corrupted by extraneous influences, such as cased holes, production tubing, geographic location, or nearby existing wells. A rotor gyroscope is composed of a spinning wheel mounted on a shaft, is powered by an electric motor, and is capable of reaching speeds of greater than 40,000 rev/min. The spinning wheel (rotor) can be oriented, or pointed, in a known direction. The direction in which the gyro spins is maintained by its own inertia; therefore, it can be used as a reference for measuring azimuth. An outer and inner gimbal arrangement allows the gyroscope to maintain its predetermined direction, regardless of how the instrument is positioned in the wellbore.

Gyroscopic systems (gyros) can be classified into three categories—free gyros, rate gyros, and inertial navigation systems.

• Free gyros. There are three types of free gyros: tilt scale, level rotor, and stable platform. The tilt scale and level rotor are film systems, while the stable platform uses the electronic system, which has shorter run time, faster data processing, and monitors continuously. Thus, most free gyros are the stable-platform type, which uses a two-gimbal gyro system like the level-rotor gyro, but the gimbals remain perpendicular to each other, even when the instrument is tilted during use. The inner gimbal remains perpendicular to the tool axis (platform) instead of perpendicular to the horizon.

• Rate gyros (north-seeking gyros). These use the horizontal component of the Earth's rotational rate to determine north. The Earth rotates 360° in 24 hours, or 15° in 1 hour. The horizontal component of the Earth's rate decreases with the cosine of latitude; however, a true-north reference will always be resolved at a latitude of less than 80° north or south. Therefore, the rate gyro does not have to rely on a known reference direction for orientation. Inclination is measured by a triaxial gravity-accelerometer package. Rate gyros have a very precise drift rate that is small compared to the Earth's spin rate. The Earth's spin rate becomes less at higher latitudes, affecting the gyro's ability to seek north. This effect also increases the time required to seek north accurately and decreases the accuracy of the north reference.

• Inertial navigation systems. This is the most accurate surveying method. Inertial navigation systems use groups of gyros to orient the system to north. It can measure movement in the

TABLE 6.2—COMPARISON OF RESULTS OF THE FIVE COMMONLY USED SURVEY METHODS		
Calculation Methods	TVD Error (ft)	Displacement Error (ft)
Tangential	−4.76	+14.99
Balanced tangential	−0.11	−0.03
Average angle	0.00	−0.25
Curvature radius	−0.04	−0.31
Minimum curvature	—	—

Well Specifications: well = Sperry-Sun test hole, total survey depth = 6,023 ft, maximum angle = 26°, survey interval = approximately 62 ft, and well profile = vertical hole to 4,064 ft building to 26° at 6,023 ft.

x, y, and z axes of the wellbore with gyros and gravity accelerometers. Because of the sensor design, this instrument can survey in all latitudes without sacrificing accuracy.

6.3.3 Calculation Methods. There are several known methods of computing directional survey. The five most commonly used are: tangential, balanced tangential, average angle, curvature radius, and minimum curvature (most accurate).[16]

• Tangential. This method uses the inclination and hole direction at the lower end of the course length to calculate a straight line representing the wellbore that passes through the lower end of the course length. Because the wellbore is assumed to be a straight line throughout the course length, it is the most inaccurate of the methods discussed and should be abandoned completely.

• Balanced Tangential. Modifying the tangential method by taking the direction of the top station for the first half of the course length, then that of the lower station for the second half can substantially reduce the errors in that method. This modification is known as the balanced-tangential method. This method is very simple to program on hand-held calculators and in spreadsheets and gives accuracy comparable to the minimum-curvature method.

• Average Angle. The method uses the average of the inclination and hole-direction angles measured at the upper and lower ends of the course length. The average of the two sets of angles is assumed to be the inclination and the direction for the course length. The well path is then calculated with simple trigonometric functions.

• Curvature Radius. With the inclination and hole direction measured at the upper and lower ends of the course length, this method generates a circular arc when viewed in both the vertical and horizontal planes. Curvature radius is one of the most accurate methods available.

• Minimum Curvature. Like the curvature-radius method, this method, the most accurate of all listed, uses the inclination and hole direction measured at the upper and lower ends of the course length to generate a smooth arc representing the well path. The difference between the curvature-radius and minimum-curvature methods is that curvature radius uses the inclination change for the course length to calculate displacement in the horizontal plane (the TVD is unaffected), whereas the minimum-curvature method uses the DLS to calculate displacements in both planes. Minimum curvature is considered to be the most accurate method, but it does not lend itself easily to normal, hand-calculation procedures.

The survey results are compared against those from the minimum-curvature method, as shown in **Table 6.2.** Large errors are seen in the tangential method for only approximately 1,900 ft of deviation. This demonstrates that the tangential method is inaccurate and should be abandoned completely. The balanced-tangential and average-angle methods are more practical for field calculations and should be used when sophisticated computational equipment or expertise may not be available. These should be noted as "Field Results Only."

TABLE 6.3—SOURCES OF TOOL-MISALIGNMENT ERRORS	
Misalignment Source	Cause
Sensor to instrument	Manufacturing tolerance or poor calibration
Instrument to drillstring/casing	Lack of or poor centralization
Drillstring to wellbore	BHA deflection

6.3.4 Sources of Errors in Directional Survey. *Survey Instruments.* The survey instrument's performance depends on the package design elements, calibration performance, and quality control during operation. System performance will functionally depend on the borehole inclination, azimuth, geomagnetic-field vector, and geographical position. Because of the dependency on sensing Earth's spin rate, the performance of gyro compassing tools is inversely proportional to the cosine of the latitude of wellbore location. The sensor systems' performance generally degrades as the inclination increases, especially in an east/west direction at higher latitudes. For magnetic tools, high latitudes result in weaker horizontal components of Earth's field. For two-axis gyro tools, the approach to east/west at high inclinations places the sensor axes increasingly parallel to Earth's spin. With magnetic tools, errors increase at high east/west inclinations because of the progressive difficulty in compensating for the effect of drillstring magnetism.

Gyros suffer from the additional problem of time-related drift uncertainty. The time component may be significant for gyro systems, particularly in horizontal wells and possibly in east/west orientation. The survey duration inevitably extends beyond the average survey-duration period. Long survey duration means larger drift uncertainty and more exposure to the wellbore environment, which may potentially reduce the accuracy of directional data.

The ability of the tool to freefall into the well will decrease substantially at approximately 60°. Consequently, gyro performance degrades at 60°, and most gyros cannot be used to survey at greater than 70°.

Tool Misalignment. The misalignment of the survey instrument with the wellbore results in errors in measuring wellbore-axis direction and inclination. (Note: inclination and azimuth are affected.) Sources of this kind of error are detailed in **Table 6.3.** Sensor-to-instrument error is independent of inclination, which is an important variable for both instrument to drillstring/casing and drillstring to wellbore. Misalignments have long been recognized as significant error sources in directional surveying.

MD Error. Sources of depth error depend on the type of survey system used. Drillpipe-conveyed tools (MWD, multishots, and single-shot) suffer from errors in the physical measurement of drillpipes and the differential effects of drillstring compression and stretch. Because of wellbore friction, drillstring compression and stretch are not easily calculated, particularly in inclined wells. Depth errors can account for the relatively large angular errors frequently observed when comparing overlapping, high-accuracy surveys in deviated wells.

Wireline survey tools generally have smaller depth errors than drillstring-conveyed tools, provided adequate quality-control measures have been taken. Errors on the order of 1/1,000 for gyroscopic tools and 2/1,000 for drillstring tools are commonly quoted. However, this may not apply to horizontal-well situations.

Magnetic Interference. Magnetic interference may be defined as corruption of the geomagnetic field by a field from an external source. This can cause serious errors in measuring hole direction (azimuth). Potential sources of magnetic interference are:
- Drillstrings.
- Adjacent wells.
- Casing shoes.
- Magnetic formations.

• "Hot spots" in nonmagnetic drill collars.

Although all the previous error sources may compromise the magnetic survey's quality, drill-string (axial) interference is probably the most common and frequent cause of errors in hole direction. The drillstrings may be regarded as a steel-bar, dipole magnet. The normal approach for magnetic survey tools is to place the survey sensor within sufficient quantity of nonmagnetic drill collars in the BHA. Azimuth measurement errors are minimized by virtue of their distance from the interference source. Magnetic interference diminishes proportionally with the inverse of the square of the distance from the source. The bar-dipole-magnet analogy is simplistic. There is evidence that downhole drillstring magnetism may be much more complex, even dynamic in nature. In practice, it may be hard to remove interference completely. The magnitude of the effect of magnetic interference depends on the strength of the interference field as well as the inclination and direction of the wellbore and its geographical latitude. Highly deviated wells drilled in an east/west direction are likely to suffer greater magnetic-interference errors, especially in higher latitudes.[17–18]

There are several techniques to correct the effects of magnetic interference. These tend to be proprietary, but at least two are based upon a common hypothesis. The corrupted sensor measurements can be replaced with values calculated from a model of the local geomagnetic parameters, which allows azimuth estimation without interference errors. The techniques have been proved to be sound, in theory. In practice, the available geomagnetic models are imperfect, resulting in potentially significant errors in the calculated azimuth. If good geomagnetic-field information is available, then these correction routines can provide accurate azimuth data. In some cases, the hole direction's (azimuth's) accuracy has approached gyro quality.[19]

Cross-Axial Interference. This can arise from hot spots or from close proximity to magnetic elements in the drillstring. Cross-axial magnetic interference can cause significant survey errors, especially when the well being surveyed is in an east/west direction or approximately horizontal. A few companies have devised means for dealing with this type of interference, and at least one company combines this with an axial interference correction. These techniques also rely on knowing the magnitude of the local magnetic field and the dip angle. As with axial interference corrections, performance can be affected significantly by the imperfections in commonly available geomagnetic models.

6.3.5 Wellbore Position Error. The survey errors described previously must be translated into positional errors so that geoscientists can assess the impact of those errors on their understanding of the subsurface model and behavior of the well when on production. In extreme cases, these errors, if not recognized, can result in a well missing its target completely. Thus, the wellbore position error is a multidisciplinary problem and should be considered during well planning. Also, drillstring magnetic interference affects hole-direction measurements most severely at high inclinations when the well is traveling close to east or west. Planning the drainage of a field with wells oriented not along east or west greatly improves the accuracy of the directional MWD surveys.

To quantify the effects of the instrument errors described previously on bottomhole location, Walstrom et al.[20] introduced the concept of survey uncertainty by generating 2D ellipses of uncertainty. An ellipse is used because the greatest survey errors are usually azimuth errors rather than inclination ones. The ellipse is expressed as an ellipsoid with the long axis at right angles to the wellbore direction. These calculations were based on the assumption that most survey errors were random. It later became evident that the calculated ellipses were too small and generally would not overlap.

In 1981, Wolff and de Wardt[21] introduced an alternative method of determining wellbore uncertainty by suggesting that most survey errors were systematic rather than random. This method, or similar ones that also use systematic error sources, has become the accepted method of computing error source. While work was done in this area during the late 1980s, there was little standardization in computational technique. This caused many problems within the industry. To address these issues, a number of individuals created the Industry Steering Committee on Wellbore Survey Accuracy (ISCWSA) in 1995. The committee's objective is to produce and maintain standards relating to wellbore-survey accuracy for the industry. ISCWSA published a paper describing in detail how errors in sensor bias, scale factor, and misalignment propagate into errors in measured inclination and azimuth. Readers interested in survey accuracy and error models should contact ISCWSA for more information.[22]

6.3.6 Survey Quality Control. The nature of downhole directional surveying is that it can never be independently verified. It is very difficult to go down the well to check if the bottom is located where the calculations claim. Practically, the best way of verifying survey results is to have surveys obtained from two different sources, preferably from two different sensor types, such as a magnetic MWD survey checked by a rate gyro or inertial navigation system.

Survey-tool performance often is dependent on how it is run. Regardless of the system or sensor type, the quality of the survey is controlled by the surveyor. The surveyor must follow the procedures and verification checks specified by the survey company and possibly even apply additional procedures and checks, as specified by the operating company, to ensure the best possible survey. Unless the proper procedures and checks are adhered to, the quality of the survey is questionable. These checks should include pre- and post-job calibration checks and paperwork and procedures verification by someone other than the original surveyor.

6.4 BHA Design for Directional Control

6.4.1 Design Principles—Bit Side Force and Tilt. The BHA is a portion of the drillstring that affects the trajectory of the bit and, consequently, of the wellbore. In general, the factors that determine the drilling tendency of a BHA are bit side force, bit tilt, hydraulics, and formation dip. The BHA design objective for directional control is to provide the directional tendency that will match the planned trajectory of the well.

The bit side force is the most important factor affecting the drilling tendency. The direction and magnitude of the bit side force determine the build, drop, and turn tendencies. A drop assembly is defined as when the bit side force acts toward the low side, whereas a build assembly is when the bit side force acts toward the high side of the hole. A hold assembly is when the inclination side force at the bit is zero. The bit tilt angle is the angle between the bit axis and the hole axis and affects the drilling direction because a drill bit is designed to drill parallel to its axis.

6.4.2 Rotary Assemblies. Rotary assemblies are designed to build, drop, or hold angle. The behavior of any rotary assembly is governed by the size and placement of stabilizers within the first 120 ft from the bit. Additional stabilizers run higher on the drillstring will have limited effect on the assembly's performance. Rotary assemblies are not "steerable"; first, the azimuth behavior (right/left turn) of a rotary assembly is nearly uncontrollable. Second, each rotary assembly has its own unique build/drop tendency that cannot be adjusted from the surface. Thus, tripping for the assembly change is required to correct the wellbore course.

Commonly used stabilizer types are sleeve, welded blade, and integral blade. For long wear life, geology is the most important consideration when selecting one type of stabilizer vs. another. Sleeve stabilizers are most economical, but ruggedness often is an issue. Welded-blade stabilizers are best suited to large holes in soft formations. Integral-blade stabilizers are the most expensive but very rugged, making them the ideal choice in hard and abrasive formations. Roller reamers are sometimes used with stabilizers to open the hole to full gauge, extend bit life, and prevent possible sticking problems.

Building Assemblies: Fulcrum Principle. Building assemblies use the fulcrum principle—a near-bit stabilizer, closely placed above the bit, creates a pivot point wherein the bending drill collars force the near-bit stabilizer to the low side of the hole and create a lateral force at the bit to the high side of the hole. Experience has shown that the more limber the portion of the assembly just above the fulcrum, the faster the increase in angle.

A typical build assembly uses two to three stabilizers. The first (near-bit) stabilizer usually connects directly to the bit. If a direct connection is not possible, the distance between the bit and the first stabilizer should be less than 6 ft to ensure it remains an angle-building assembly. The second stabilizer is added to increase the control of side force and to alleviate other problems. Build rates can be increased by increasing the distance between the first and second stabilizers. When the distance between the stabilizers increases enough to cause the drill collar sag to touch the low side of the hole, the bit side force and bit tilt reach their maximum build rate for the assembly. Generally, the drill collars will sag to touch the borehole wall when the distance between the stabilizers is greater than 60 ft. The amount of sag will also depend on the hole and collar sizes, inclination, stabilizer gauge, and weight on bit (WOB).

Other important factors for the fulcrum assemblies are inclination, WOB, and rotary speed. The build rate of a fulcrum assembly increases as inclination increases because the larger component of the collar's own weight causes them the bend. Increasing the WOB will bend the drill collars behind the near-bit stabilizer even more, increasing the build rate. A higher rotary speed tends to straighten out the drill collars, thus reducing the build rate. Therefore, low rotary speeds (70 to 100 rev/min) are generally used with fulcrum assemblies. Sometimes, in soft formations, a high flow rate can lead to formation washout, resulting in decreased stabilizer contacts and, thus, a reduced build tendency.

Holding Assemblies: Packed Hole. The packed-hole assemblies contain three to five stabilizers properly spaced to maintain the angle. The increased stiffness on the BHA from the added stabilizers keeps the drillstring from bending or bowing and forces the bit to drill straight ahead. The assembly may be designed for slight build or drop tendency to counteract formation tendencies.

Dropping Assemblies: Pendulum Principle. The pendulum effect is produced by removing the stabilizer just above the bit while retaining the upper ones. While the remaining stabilizers hold the bottom drill collar away from the low side of the wall, gravity acts on the bit and the bottom drill collar and tends to pull them to the low side of the hole, thus decreasing the hole angle. Pendulum assemblies sometimes can be run slick (without stabilizers). Although a slick assembly is simple and economical, it is difficult to control and maintain the drop tendency.

A dropping assembly usually contains two stabilizers. As the distance between the bit and the first stabilizer increases, gravity pulls the bit to the low side of the hole, increasing the downward bit tilt and bit side force. If the distance between the bit and the first stabilizer is too large, the bit will begin to tilt upward, and the drop rate will reach a maximum. With a higher WOB, the drop assembly could even start building angle. Generally, the distance between the bit and the first stabilizer will be approximately 30 ft. The second stabilizer is added to increase control of the side force.

Initially, low WOB should be used to avoid bending the pendulum toward the low side of the hole. Once a dropping trend has been established, moderate WOB can be used to achieve a higher penetration rate.

6.4.3 Deviation Tools. The most common deviation tools for directional drilling are steerable motor assemblies (or so-called positive-displacement motors [PDMs]) and RSSs. Adjustable-gauge stabilizers, known as "2D rotary systems," have become quite popular to run with the rotary and PDM assemblies to control inclination. Whipstocks, especially casing whipstocks, are used routinely to sidetrack out of cased wellbores. Other tools, such as turbines, are used mainly in Russia, and jetting bits are seldom used today.

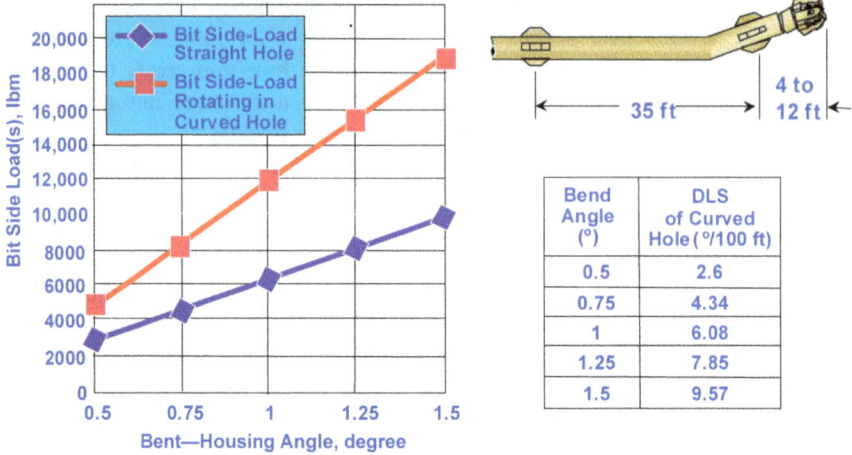

Fig. 6.6—The effect of the bent-housing angle on build rates and bit side load.

Steerable Motor Assemblies or PDMs. The most important advancements in trajectory control are the steerable motor assemblies, which contain PDMs with bent subs or bent housing. The PDM is based on the Moineau principle. The first commercial PDM was introduced to the petroleum industry in the late 1960s. Since then, PDM use has been accelerated greatly for directional-drilling applications. Steerable motor assemblies are versatile and are used in all sections of directional wells, from kicking off and building angle to drilling tangent sections and providing accurate trajectory control. Among the PDM assemblies, the most commonly used deviation tool today is the bent-housing mud motor.

The bent sub and bent housing use bit tilt (misalignment of bit face away from the drillstring axis) and bit side force to change the hole direction and inclination. Bent housing is more effective than the bent sub because of a shorter bit-to-bend distance, which reduces the bit offset and creates a higher build rate for a given bend size. A shorter bit-to-bend distance also reduces the moment arm, which, in turn, reduces the bending stress at the bend. As a result, the bent-housing PDM is easier to orient and allows for a long rotation period. Larger hole sizes (22 to 26 in.) are the only application for a bent sub. The requirement for bent subs is obsolete in most applications, particularly with the introduction of the adjustable bent housing.

Before the personal computer become widely available, the simple "three-point curvature" calculation was used to predict the build rates of the motor assemblies as

$$r_b = 200 \frac{\theta}{L_1 + L_2}, \quad\text{..} \quad (6.1)$$

in which r_b = build rate in degrees/100 ft, θ = bend angle in degrees, L_1 = distance from the first contact point (bit) to the second (bend) in ft, and L_2 = distance from the second contact point to the third (motor top stabilizer) in ft.

For more-accurate results, a BHA-analysis program is often used to calculate the build/drop/turn rates of the motor assemblies. **Fig. 6.6** shows the expected DLS and the bit side forces for a two-stabilizer motor assembly.

Bent-Housing Motor Components. A typical bent-housing motor contains the following four sections: dump sub, power unit, transmission/bent-housing unit, and bearing section.

• Dump sub. Located on top of the motor assembly, the dump sub contains a valve that is ported to allow fluid flow between the drillstrings and the annulus. This allows the drillstrings to fill when tripping in the hole and empty when tripping out of the hole. The dump sub also permits low-rate circulation bypassing of the motor, if required.

• Power unit. Most motor assemblies use the Moineau pump principle to convert hydraulic energy to mechanical energy: a rotor/stator pair converts the hydraulic energy of the pressurized circulating fluid to mechanical energy for a rotating shaft. The rotor and stator are of lobed design. Both rotor- and stator-lobe profiles are similar, with the steel rotor having one less lobe than the elastomeric stator. The rotor and stator lobes are helical in nature, with one stage equating to the linear distance of a full "wrap" of the stator helix. Power units may be categorized with respect to the number of lobes and the effective stages. The speed and torque of a power section is linked directly to the number of lobes on the rotor and stator. The greater the number of lobes, the greater the torque and the lower the rotary speed. Typical rotor/stator configurations range from 1:2 to 9:10 lobe, as shown in **Fig. 6.7.** The power section should be matched to the bit and the formation being drilled for best performance.

• Transmission bent-housing/unit. The universal couplings inside the transmission/bent-housing unit eliminate all eccentric rotor motion and accommodate the misalignment motion of the bent housing while transmitting torque and down thrust to the drive shaft, which is held concentrically by the bearing assembly.

• Bearing section. The bearing assembly consists of multiple thrust-bearing cartridges, radial bearings, a flow restrictor, and a drive shaft. The thrust bearings support the down thrust of the rotor, the hydraulic down thrust from bit pressure loss, and the reactive upward thrust from the applied WOB. For larger-diameter motors, the thrust bearings are usually of multistack ball- and track-design. Small-diameter motors use carbide/diamond-enhanced friction bearings. Metallic and nonmetallic radial bearings are employed above and below the thrust bearings to absorb lateral side loading of the drive shaft. The flow restrictor allows approximately 5 to 8% of the circulating fluid to flow through the bearing section to cool and lubricate the bearing assembly. On the basis of planned bit hydraulics, the type of flow restrictor used is preselected and set in the motor shop; it cannot be changed at the rig site. The drive shaft transmits both axial load and torque to the bit. The drive shaft is a forged component designed such that fatigue, axial, and torque strengths are maximized. It has a threaded connection at the bottom to facilitate connection to the drill bit.

• Power delivered to the bit. Eqs. 6.2 and 6.3 are used to calculate the horsepower delivered to the bit from the motor. Note T (torque) is in the unit of lbf-ft (not lbf/ft).

$$h_p = \frac{2\pi T N}{33,000}, \dotfill (6.2)$$

$$\text{or, more simply, } h_p = \frac{T \times N}{5,252}, \dotfill (6.3)$$

in which h_p = horsepower, T = torque in lb-ft, and N = speed of rotation in rev/min.

PDM Applications in Directional Drilling. *Sidetrack.* The most common method of sidetracking out of casing, especially when considerable drilling is to follow, is milling a length of casing with a section mill, then diverting the trajectory with a bent-housing motor assembly. The assembly usually contains a stabilizer on the motor and possibly another one above the motor.

Steerable Drilling and Kickoff. The essential requirement for a steerable drilling system is that it be capable of making both inclination and azimuth changes. Thus, this is the most commonly used configuration because:

• An average planned curvature can be adhered to by a combination of orienting and rotating.

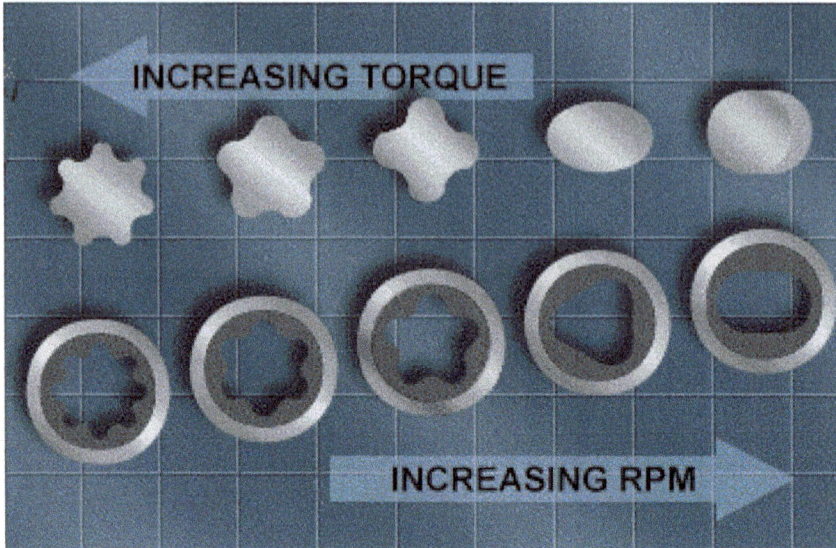

Fig. 6.7—Typical rotor/stator configurations.

• After completing the buildup, the assembly can be rotated ahead to hold the angle with minor corrections to inclination and azimuth as necessary.

• Extended intervals can be drilled through different formations without tripping for assembly changes.

• Drilling performance is maximized by efficiently delivering the torque and horsepower at the bit.

This system usually consists of a bent-housing motor and a stabilizer on the bearing housing. To enhance the motor's sliding capability, the stabilizer has wide, straight blades that are tapered at either end and is undergauge relative to the hole size (typically ⅛ to ½ in.). Depending on the application, additional stabilizers may be used above the motor. Although these stabilizers are generally spiral, the blades should be tapered and undergauge.

The overall design of this steerable assembly will depend on its application. The important considerations are as follows:

1. The expected build rate in oriented mode should be slightly greater than (typically 1 to 2°/100 ft) that required to guarantee the planned build rate.

2. The number of stabilizers used should be kept to a minimum to reduce drag in the oriented mode.

3. If the drillstring is rotated in a curved section, bending stresses around the bent housing should be checked to ensure that they are less than the endurance limit.

Medium-Radius Applications (6 to 15°/100 ft DLS). The vast majority of medium-radius drilling is undertaken in hole sizes of 12¼ in. and less with 8-in. (and less) -diameter motors for build rates of 6 to 15°/100 ft. There are a number of motor configurations used to drill medium-radius wells, each with its own merits—single bent-housing motor, single bent housing with offset pad, double-bend motor, bent-housing motor with bent sub positioned on top of the motor and aligned with the bend, and double bent-housing motor.

Intermediate- and Short-Radius Applications. Intermediate-radius drilling systems are used to achieve build rates from 15 to 65°/100 ft. The build and lateral sections are drilled with a short-bearing pack motor. When the build rate exceeds 45°/100 ft, an articulated motor and flexed MWD tool should be used. Both system types can be used for new or re-entry wells.

Two types of motors are used to drill the short-radius wells with build rates ranging from 65 to 125°/100 ft: a "build" articulated motor used to drill the build section and a "hybrid lateral" motor for the horizontal lateral section. The articulated MWD tool is used on both the build and lateral sections.

RSS. The RSS is an evolution in directional-drilling technology that overcomes the drawbacks in steerable motors and in conventional rotary assemblies. To initiate a change in the wellbore trajectory with steerable motors, the drilling rotation is halted in such a position that the bend in the motor points in the direction of the new trajectory. This mode, known as the sliding mode, typically creates higher frictional forces on the drillstring. In extreme ERD, the frictional force builds to the point at which no axial weight is available to overcome the drag of the drillstring against the wellbore, and, thus, further drilling is not possible. To overcome this limitation in steerable motor assemblies, the RSS was developed in the early 1990s to respond to this need from ERD. The first RSS was used in BP plc's Wytch Farm (U.K.) extended-reach wells.

RSSs allow continuous rotation of the drillstring while steering the bit. Thus, they have better penetration rate, in general, than the conventional steerable motor assemblies. Other benefits include better hole cleaning, lower torque and drag, and better hole quality. RSSs are much more complex mechanically and electronically and are, therefore, more expensive to run compared to conventional steerable motor systems. This economic penalty tends to limit their use to highly demanding extended-reach wells or the very complex profiles associated with designer wells. Additionally, the technology is still very new. As a result, the current generation of systems (2002) is climbing a very steep learning curve in regard to run length, performance, and mechanical reliability.

There are two steering concepts in the RSS—point the bit and push the bit. The point-the-bit system uses the same principle employed in the bent-housing motor systems. In RSSs, the bent housing is contained inside the collar, so it can be oriented to the desired direction during drillstring rotation.[23] Point-the-bit systems claim to allow the use of a long-gauge bit to reduce hole spiraling and drill a straighter wellbore.[24] The push-the-bit system uses the principle of applying side force to the bit, pushing it against the borehole wall to achieve the desired trajectory. The force can be hydraulic pressure[25] or in the form of mechanical forces.[26] In general, either a point-the-bit or a push-the-bit RSS allows the operator to expect a maximum build rate of approximately 6 to 8°/100 ft for the 8½-in.-hole-sized tool.

Adjustable-Gauge Stabilizers (AGSs). In the late 1980s, the industry developed AGSs, the effective blade outer diameter (OD) of which could be changed while the tool was downhole. With AGSs, the drillers could change the stabilizer OD without making time-consuming and costly trips out of the hole. AGSs run in rotary assemblies were often placed near the bit or positioned approximately 15 to 30 ft from the bit. In these positions, changes in their gauge could effectively control the build or drop tendency of the assembly. Because they could control inclination while in the rotary mode, these assemblies became known as "2D rotary systems." AGSs can also be run with steerable motor systems. Running AGSs with the steerable motor assemblies makes it possible to control inclination with the stabilizer while drilling in the rotary mode. If the wellbore requires a change in azimuth, one would have to revert to a sliding mode.

AGSs have been widely used recently, particularly in drilling the horizontal section with a geological steering or pay-zone steering device that usually consists of an LWD tool. With its deep investigation depth, a resistivity sensor can detect a geological change many feet before the bit penetrates that boundary. This ability may allow the drilling assembly to be held in the reservoir and steered away from either an upper or lower boundary.[27]

Whipstocks. Openhole whipstocks are the first type of deflection tool used to change the wellbore trajectory but are seldom used today. Bent-housing motors have replaced openhole whipstocks as the most commonly used deviation tool in openhole sidetracking. Casing whipstocks, on the other hand, are routinely used to sidetrack out of cased wellbores. Whipstocks can be either retrievable or nonretrievable. Retrievable ones are ideal for drilling multiple laterals from a single wellbore. A typical casing-sidetracking operation involves multiple trips to set the cement plug and the whipstock, start the window mill, complete the mill, and clean up. To save time, recently developed systems can accomplish all these tasks in one trip. Note that the gyro survey is usually required for setting the whipstock and for initial tool orientations.

Turbines. Turbines, commonly known as turbodrills, are powered by a turbine motor, which has a series of rotors/stators (stages) connected to a shaft. As the drilling fluid is pumped through the turbine, the stators deflect the fluid against the rotors, forcing the rotors to rotate the drive shaft to which they are connected. Turbines are designed to run on high speed and low torque; thus, they are suited for running with diamond or polycrystalline-diamond compact bits. Turbines are not only less flexible and efficient than PDMs but are also more expensive, so they are not as widely used, except in Russia.

Jetting Bits. Jetting bits can be used to change the trajectory of a borehole, with the hydraulic energy of the drilling fluid used to erode a pocket out of the bottom of the borehole. The tricone bit with one large nozzle is oriented to the desired hole direction to create a pocket. The drilling assembly is forced into the jetted pocket for a short distance. This procedure continues until the desired trajectory change is achieved. Jetting is seldom used today because of its slow penetration rate and its limitations in soft formations.

Nomenclature

h_p = horsepower
L_1 = distance from the first contact point to the second, ft
L_2 = distance from the second contact point to the third, ft
N = speed of rotation, rev/min
r_b = build rate, ft
T = torque, lbf-ft
θ = bend angle, degrees

Acknowledgments

The author is deeply grateful to John Thorogood for his guidance and inspiration during the preparation of this work. He would also like to thank BP plc for providing the ERD database as well as his colleagues, Alpar Cseley, Steve D'aunoy, and Paul Rodney, for their review and comments.

References

1. Lyons E.P. and Mecham, O.E.: "Design and Implementation of Directional Drilling Programs: Thums Offshore Islands Development Wells, East Wilmington Field," paper 801-44M presented at the 1968 Spring Meeting, Pacific Coast District, Production Div., API, Bakersfield, California, 14–16 May.
2. Thorogood, J.L.: "How BNOC Controls Directional Drilling," *Petroleum Engineer Intl.* (May 1980) 26–44.
3. Hodgeson, H. and Varnado, S.G.: "Computerized Well Planning for Directional Wells," paper SPE 12071 presented at the 1983 SPE Annual Technical Conference and Exhibition, San Francisco, 5–8 October.
4. Hauck, M.: "Planning Platform Wells: The Below Ground Structure," *Ocean Industry* (May 1989) 36–40.
5. Thorogood, J.L. and Sawaryn, S.J.: "The Traveling-Cylinder Diagram: A Practical Tool for Collision Avoidance," *SPEDE* (March 1991) 31.

6. Burgess, T. *et al.:* "Horizontal Drilling Comes of Age," *Oil Field Review* (1991) **2,** No. 3.

7. Fisher, E.K. and French, M.R.: "Drilling the First Horizontal Well in the Gulf of Mexico: A Case History of East Cameron Block 278 Well B-12," *SPEDE* (June 1992) 86.

8. Gust, D.A. and MacDonald, R.R.: "Rotation of a Long Liner in a Shallow Long Reach Well," *JPT* (April 1989) 401; *Trans.,* AIME, **287.**

9. Smith, R.C. *et al.:* "The Lateral Tie-Back System: The Ability To Drill and Case Multiple Wells," paper IADC/SPE 27436 presented at the 1994 IADC/SPE Drilling Conference, Dallas, 15–18 February.

10. Bosworth, S. *et. al.:* "Key Issues in Multilateral Technology," *Oil Field Review* (Winter 1998).

11. Pasicznyk, A.: "Evolution Toward Simpler, Less Risky Multilateral Wells," paper IADC/SPE 67825 presented at the 2001 IADC/SPE Drilling Conference, Amsterdam, 27 February–1 March.

12. Scott, P.W.: "Increasing Reach from 3000 m to 5000 m," paper SPE/IADC 21983 presented at the 1991 SPE/IADC Drilling Conference, Amsterdam, 11–14 March.

13. Payne, M. *et al.:* "Drilling Dynamic Problems and Solutions for Extended Reach Operations," paper presented at the 1995 ASME Energy-Sources Technology Conference and Exhibition, Houston, 29 January–1 February.

14. Modi, S. *et al.*: "Meeting the 10km Drilling Challenge," paper SPE 38583 presented at the 1997 SPE Annual Technical Conference and Exhibition, San Antonio, Texas, 5–8 October.

15. Guild, J. *et al.:* "Designing and Drilling Extended Reach Wells, Part 2," *Petroleum Engineer Intl.* (January 1995).

16. *Bull. D20, Directional Drilling Survey Calculation Methods and Terminology,* API (December 1985).

17. Russell, A.W. and Roesler, R.F.: "Reduction of Nonmagnetic Drill Collar Length Through Magnetic Azimuth Correction Technique," paper SPE/IADC 13476 presented at the 1985 SPE/IADC Drilling Conference, New Orleans, 6–8 March.

18. Cheatham, C.A. *et al.:* "Effects of Magnetic Interference on Directional Surveys in Horizontal Wells," paper IADC/SPE 23852 presented at the 1992 IADC/SPE Drilling Conference, New Orleans, 18–21 February.

19. Russell, J., Shiells, G., and Kerridge, D.J.: "Reduction of Well-Bore Position Uncertainty Through Application of a New Geomagnetic In-Field Referencing Technique," paper SPE 30452 presented at the 1995 SPE Annual Technical Conference and Exhibition, Dallas, 22–25 October.

20. Walstrom, J.E., Brown, A.A., and Harvey, R.P.: "An Analysis of Uncertainty in Directional Surveying," *JPT* (April 1969) 515; *Trans.,* AIME, **271.**

21. Wolff, C.J.M. and de Wardt, J.P.: "Borehole Position Uncertainty – Analysis of Measuring Methods and Derivation of Systematic Error Model," *JPT* (December 1981) 2339; *Trans.,* AIME, **271.**

22. Williamson, H.S.: "Accuracy Prediction for Directional MWD," paper SPE 56702 presented at the 1999 SPE Annual Technical Conference and Exhibition, Houston, 3–6 October.

23. Schaaf, S., Pafitis, D., and Guichemerre, E.: "Application of a Point the Bit Rotary Steerable System in Directional Drilling Prototype Well-Bore Profile," paper SPE 62519 presented at the 2000 SPE/AAPG Western Regional Meeting, Long Beach, California, 19–23 June.

24. Yonezawa, T. *et al.:* "Robotic Controlled Drilling: A New Rotary Steerable Drilling System for the Oil and Gas Industry," paper IADC/SPE 74458 presented at the 2002 IADC/SPE Drilling Conference, Dallas, 26–28 February.

25. Barr, J.D., Clegg, J.M., and Russell, M.K.: "Steerable Rotary Drilling With an Experimental System," paper SPE/IADC 29382 presented at the 1995 SPE/IADC Drilling Conference, Amsterdam, 28 February–2 March.

26. Gruenhagen, H., Hahne, U., and Alvord, G.: "Application of New Generation Rotary Steerable System for Reservoir Drilling in Remote Areas," paper IADC/SPE 74457 presented at the 2002 IADC/SPE Drilling Conference, Dallas, 26–28 February.

27. Lawrence, L. *et al.:* "Adjustable-Gauge Stabilizer in Motor Provides Greater Inclination Control," *Oil and Gas J.* (February 2002) 37–41.

SI Metric Conversion Factors

ft	× 3.048*	E–01	= m
in.	× 2.54*	E+00	= cm
hp	× 7.46*	E–01	= kW
lbm	× 4.535 924	E–01	= kg
lbf	× 4.448 222	E+00	= N
lbf-ft	× 1.356*	E+00	= N-m

*Conversion factor is exact.

Chapter 7
Casing Design
R.F. Mitchell, Landmark Graphics

7.1 Introduction

Casing and tubing strings are the main parts of the well construction. All wells drilled for the purpose of oil/gas production (or injecting materials into underground formations) must be cased with material with sufficient strength and functionality. Therefore, this chapter provides the basic knowledge for practical casing and tubing strength evaluation and design.

7.2 Casing

Casing is the major structural component of a well. Casing is needed to maintain borehole stability, prevent contamination of water sands, isolate water from producing formations, and control well pressures during drilling, production, and workover operations. Casing provides locations for the installation of blowout preventers, wellhead equipment, production packers, and production tubing. The cost of casing is a major part of the overall well cost, so selection of casing size, grade, connectors, and setting depth is a primary engineering and economic consideration.

7.2.1 Casing Strings. There are six basic types of casing strings. Each is discussed next.

Conductor Casing. Conductor casing is the first string set below the structural casing (i.e., drive pipe or marine conductor run to protect loose near-surface formations and to enable circulation of drilling fluid). The conductor isolates unconsolidated formations and water sands and protects against shallow gas. This is usually the string onto which the casing head is installed. A diverter or a blowout prevention (BOP) stack may be installed onto this string. When cemented, this string is typically cemented to the surface or to the mudline in offshore wells.

Surface Casing. Surface casing is set to provide blowout protection, isolate water sands, and prevent lost circulation. It also often provides adequate shoe strength to drill into high-pressure transition zones. In deviated wells, the surface casing may cover the build section to prevent keyseating of the formation during deeper drilling. This string is typically cemented to the surface or to the mudline in offshore wells.

Intermediate Casing. Intermediate casing is set to isolate unstable hole sections, lost-circulation zones, low-pressure zones, and production zones. It is often set in the transition zone from normal to abnormal pressure. The casing cement top must isolate any hydrocarbon zones.

Some wells require multiple intermediate strings. Some intermediate strings may also be production strings if a liner is run beneath them.

Production Casing. Production casing is used to isolate production zones and contain formation pressures in the event of a tubing leak. It may also be exposed to injection pressures from fracture jobs, downcasing, gas lift, or the injection of inhibitor oil. A good primary cement job is very critical for this string.

Liner. Liner is a casing string that does not extend back to the wellhead but instead is hung from another casing string. Liners are used instead of full casing strings to reduce cost, improve hydraulic performance when drilling deeper, allow the use of larger tubing above the liner top, and not represent a tension limitation for a rig. Liners can be either an intermediate or a production string. Liners are typically cemented over their entire length.

Tieback String. Tieback string is a casing string that provides additional pressure integrity from the liner top to the wellhead. An intermediate tieback is used to isolate a casing string that cannot withstand possible pressure loads if drilling is continued (usually because of excessive wear or higher than anticipated pressures). Similarly, a production tieback isolates an intermediate string from production loads. Tiebacks can be uncemented or partially cemented. An example of a typical casing program that illustrates each of the specified casing string types is shown in **Fig. 7.1**.

7.3 Tubing

Tubing is the conduit through which oil and gas are brought from the producing formations to the field surface facilities for processing. Tubing must be adequately strong to resist loads and deformations associated with production and workovers. Further, tubing must be sized to support the expected rates of production of oil and gas. Clearly, tubing that is too small restricts production and subsequent economic performance of the well. Tubing that is too large, however, may have an economic impact beyond the cost of the tubing string itself because the tubing size will influence the overall casing design of the well.

7.4 Properties of Casing and Tubing

The American Petroleum Inst. (API) has formed standards for oil/gas casing that are accepted in most countries by oil and service companies. Casing is classified according to five properties: the manner of manufacture, steel grade, type of joints, length range, and the wall thickness (unit weight).

Almost without exception, casing is manufactured of mild (0.3 carbon) steel, normalized with small amounts of manganese. Strength can also be increased with quenching and tempering. API has adopted a casing "grade" designation to define the strength of casing steels. This designation consists of a grade letter followed by a number, which designates the minimum yield strength of the steel in ksi (10^3 psi). **Table 7.1** summarizes the standard API grades.

The yield strength, for these purposes, is defined as the tensile stress required to produce a total elongation of 0.5% of the length. However, the case of P–110 casing is an exception where yield is defined as the tensile stress required to produce a total elongation of 0.6% of the length. There are also proprietary steel grades widely used in the industry, which do not conform to API specifications. These steel grades are often used in special applications requiring high strength or resistance to hydrogen sulfide cracking. **Table 7.2** gives a list of commonly used non-API grades.

7.5 Pipe Strength

To design a reliable casing string, it is necessary to know the strength of pipe under different load conditions. Burst strength, collapse resistance, and tensile strength are the most important mechanical properties of casing and tubing.

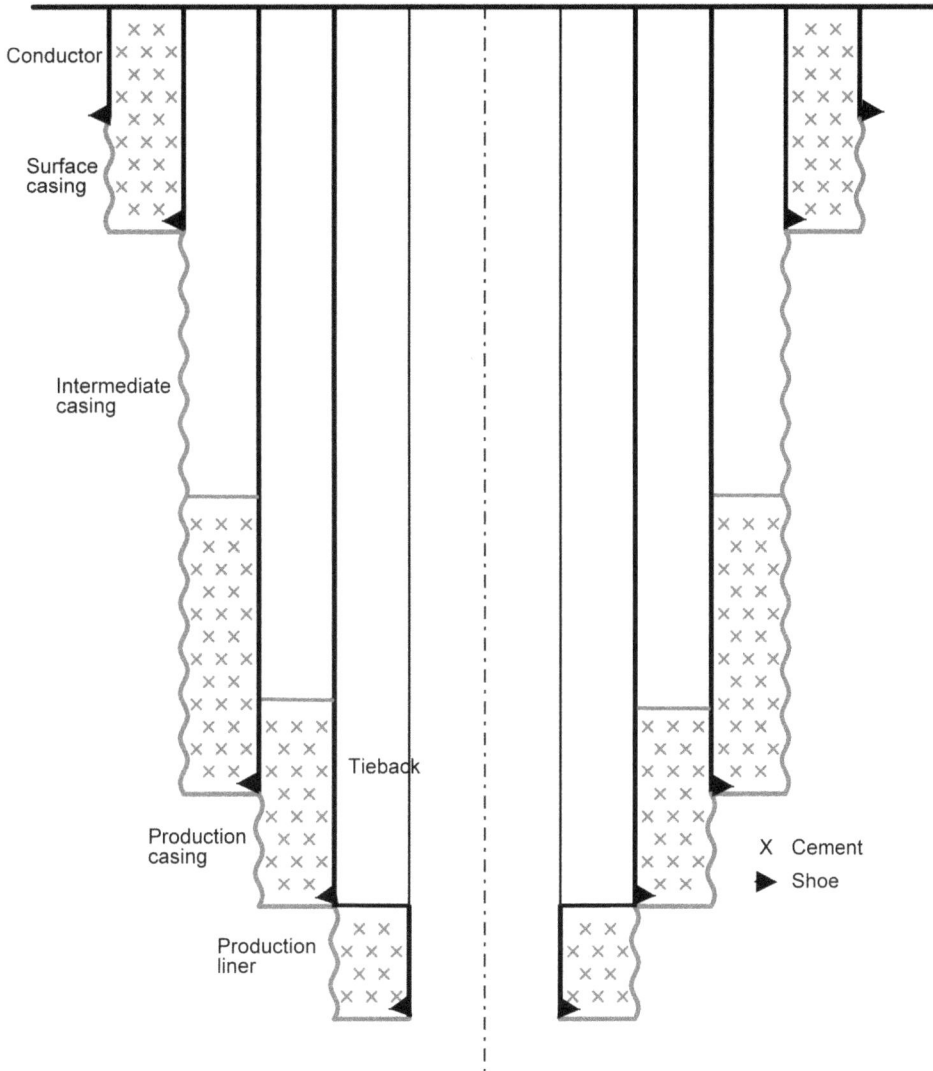

Fig. 7.1—Typical casing program.

7.5.1 Mechanical Properties. Each mechanical property of casing and tubing is discussed next.

Burst Strength. If casing is subjected to internal pressure higher than external, it is said that casing is exposed to burst pressure. Burst pressure conditions occur during well control operations, integrity tests, and squeeze cementing. The burst strength of the pipe body is determined by the internal yield pressure formula found in API *Bull. 5C3, Formulas and Calculations for Casing, Tubing, Drillpipe, and Line Pipe Properties.*[1]

$$P_B = 0.875 \left[\frac{2Y_{pt}}{D} \right], \quad \text{..} \quad (7.1)$$

where

P_B = minimum burst pressure, psi,

Y_p = minimum yield strength, psi,

TABLE 7.1—API STEEL GRADES				
API Grade	Yield Stress, psi		Minimum Ult. Tensile, psi	Minimum Elongation, %
	Minimum	Maximum		
H–40	40,000	80,000	60,000	29.5
J–55	55,000	80,000	75,000	24.0
K–55	55,000	80,000	95,000	19.5
N–80	80,000	110,000	100,000	18.5
L–80	80,000	95,000	95,000	19.5
C–90	90,000	105,000	100,000	18.5
C–95	95,000	110,000	105,000	18.5
T–95	95,000	110,000	105,000	18.0
P–110	110,000	140,000	125,000	15.0
Q–125	125,000	150,000	135,000	18.0

TABLE 7.2—NON-API STEEL GRADES					
Non-API Grade	Manufacturers	Yield Stress, psi		Minimum Ult. Tensile, psi	Minimum Elongation, %
		Minimum	Maximum		
S–80	Lone Star	75,000	–	75,000	20.0
	Longitudinal	55,000	–	–	–
modN–80	Mannesmann	80,000	95,000	100,000	24.0
C–90	Mannesmann	90,000	105,000	120,000	26.0
SS–95	Lone Star	95,000	–	95,000	18.0
	Longitudinal	75,000	–	–	–
SOO–95	Mannesmann	95,000	110,000	110,000	20.0
S–95	Lone Star	95,000	–	110,000	16.0
	Longitudinal	92,000	–	–	–
SOO–125	Mannesmann	125,000	150,000	135,000	18.0
SOO–140	Mannesmann	140,000	165,000	150,000	18.0
V–150	U.S. Steel	150,000	180,000	160,000	14.0
SOO–155	Mannesmann	155,000	180,000	165,000	20.0

t = nominal wall thickness, in.,

and

D = nominal outside pipe diameter, in.

This equation, commonly known as the Barlow equation, calculates the internal pressure at which the tangential (or hoop) stress at the inner wall of the pipe reaches the yield strength (YS) of the material. The expression can be derived from the Lamé equation for tangential stress by making the thin-wall assumption that $D/t \gg 1$. Most casing used in the oilfield has a D/t ratio between 15 and 25. The factor of 0.875 appearing in the equation represents the allowable manufacturing tolerance of –12.5% on wall thickness specified in API *Bull. 5C2, Performance Properties of Casing, Tubing, and Drillpipe.*[2]

Because a burst failure will not occur until after the stress exceeds the ultimate tensile strength (UTS), using a yield strength criterion as a measure of burst strength is an inherently conservative assumption. This is particularly true for lower-grade materials such H-40, K-55, and N-80 whose UTS/YS ratio is significantly greater than that of higher-grade materials such as P-110 and Q-125. The effect of axial loading on the burst strength is discussed later.

TABLE 7.3—YIELD COLLAPSE PRESSURE FORMULA RANGE	
Grade*	Maximum D/t*
H–40	16.40
–50	15.24
J–K–55	14.81
–60	14.44
–70	13.85
C–75 & E	13.60
L–N–80	13.38
C–90	13.01
C–T–95 & X	12.85
–100	12.70
P–105 & G	12.57
P–110	12.44
–120	12.21
Q–125	12.11
–130	12.02
S–135	11.92
–140	11.84
–150	11.67
–155	11.59
–160	11.52
–170	11.37
–180	11.23

*Grades indicated without a letter designation are not API grades but are grades in use or grades being considered for use. They are shown for information purposes.

Collapse Strength. If external pressure exceeds internal pressure, the casing is subjected to collapse. Such conditions may exist during cementing operations or well evacuation. Collapse strength is primarily a function of the material's yield strength and its slenderness ratio, D/t. The collapse strength criteria, given in API *Bull. 5C3, Formulas and Calculations for Casing, Tubing, Drillpipe, and Line Pipe Properties*,[1] consist of four collapse regimes determined by yield strength and D/t. Each criterion is discussed next in order of increasing D/t.

Yield Strength Collapse. Yield strength collapse is based on yield at the inner wall using the Lamé thick wall elastic solution. This criterion does not represent a "collapse" pressure at all. For thick wall pipes ($D/t < 15\pm$), the tangential stress exceeds the yield strength of the material before a collapse instability failure occurs.

$$P_{Yp} = 2Y_p \left[\frac{(D/t) - 1}{(D/t)^2} \right] \dots\dots\dots\dots\dots\dots\dots\dots\dots\dots\dots (7.2)$$

Nominal dimensions are used in the collapse equations. The applicable D/t ratios for yield strength collapse are shown in **Table 7.3**.

Plastic Collapse. Plastic collapse is based on empirical data from 2,488 tests of K-55, N-80, and P-110 seamless casing. No analytic expression has been derived that accurately models collapse behavior in this regime. Regression analysis results in a 95% confidence level that 99.5% of all pipes manufactured to API specifications will fail at a collapse pressure higher than the plastic collapse pressure. The minimum collapse pressure for the plastic range of collapse is calculated by Eq. 7.3.

TABLE 7.4—FORMULA FACTORS AND *D/t* RANGES FOR PLASTIC COLLAPSE				
	Formula Factor			
Grade*	A	B	C	*D/t* Range
H–40	2.950	0.0465	754	16.40–27.01
–50	2.976	0.0515	1,056	15.24–25.63
J–K–55	2.991	0.0541	1,206	14.81–25.01
–60	3.005	0.0566	1,356	14.44–24.42
–70	3.037	0.0617	1,656	13.85–23.38
C–75 & E	3.054	0.0642	1,806	13.60–22.91
L–N–80	3.071	0.0667	1,955	13.38–22.47
C–90	3.106	0.0718	2,254	13.01–21.69
C–T–95 & X	3.124	0.0743	2,404	12.85–21.33
–100	3.143	0.0768	2,553	12.70–21.00
P–105 & G	3.162	0.0794	2,702	12.57–20.70
P–110	3.181	0.0819	2,852	12.44–20.41
–120	3.219	0.0870	3,151	12.21–19.88
Q–125	3.239	0.0895	3,301	12.11–19.63
–130	3.258	0.0920	3,451	12.02–19.40
S–135	3.278	0.0946	3,601	11.92–19.18
–140	3.297	0.0971	3,751	11.84–18.97
–150	3.336	0.1021	4,053	11.67–18.57
–155	3.356	0.1047	4,204	11.59–18.37
–160	3.375	0.1072	4,356	11.52–18.19
–170	3.412	0.1123	4,660	11.37–17.82
–180	3.449	0.1173	4,966	11.23–17.47

*Grades indicated without a letter designation are not API grades but are grades in use or grades being considered for use. They are shown for information purposes.

$$P_p = Y_p \left[\frac{A}{D/t} - B \right] - C \quad \dotfill \quad (7.3)$$

The factors *A, B,* and *C* and applicable *D/t* range for the plastic collapse formula are shown in **Table 7.4**.

Transition Collapse. Transition collapse is obtained by a numerical curve fit between the plastic and elastic regimes. The minimum collapse pressure for the plastic-to-elastic transition zone, P_T, is calculated with Eq. 7.4.

$$P_T = Y_p \left[\frac{F}{D/t} - G \right] \quad \dotfill \quad (7.4)$$

The factors *F* and *G* and applicable *D/t* range for the transition collapse pressure formula, are shown in **Table 7.5**.

Elastic Collapse. Elastic Collapse is based on theoretical elastic instability failure; this criterion is independent of yield strength and applicable to thin-wall pipe (*D/t* > 25±). The minimum collapse pressure for the elastic range of collapse is calculated with Eq. 7.5.

$$P_E = \frac{46.95 \times 10^6}{(D/t)[(D/t) - 1]^2} \quad \dotfill \quad (7.5)$$

The applicable *D/t* range for elastic collapse is shown in **Table 7.6**.

Most oilfield tubulars experience collapse in the "plastic" and "transition" regimes. Many manufacturers market "high collapse" casing, which they claim has collapse performance prop-

TABLE 7.5—FORMULA FACTORS AND *D/t* RANGE FOR TRANSITION COLLAPSE			
	Formula Factor		
Grade*	F	G	*D/t* Range
H–40	2.063	0.0325	27.01–42.64
–50	2.003	0.0347	25.63–38.83
J–K–55	1.989	0.0360	25.01–7.21
–60	1.983	0.0373	24.42–5.73
–70	1.984	0.0403	23.38–33.17
C–75 & E	1.990	0.0418	22.91–32.05
L–N–80	1.998	0.0434	22.47–1.02
C–90	2.017	0.0466	21.69–29.18
C-T–95 & X	2.029	0.0482	21.33–28.36
–100	2.040	0.0499	21.00–27.60
P–105 & G	2.053	0.0515	20.70–26.89
P–100	2.066	0.0532	20.41–26.22
–120	2.092	0.0565	19.88–25.01
Q–125	2.106	0.0582	19.63–24.46
–130	2.119	0.0599	19.40–23.94
S–135	2.133	0.0615	19.18–23.44
–140	2.146	0.0632	18.97–22.98
–150	2.174	0.0666	18.57–22.11
–155	2.188	0.0683	18.37–21.70
–160	2.202	0.0700	18.19–21.32
–170	2.231	0.0734	17.82–20.60
–180	2.261	0.0769	17.47–19.93

*Grades indicated without a letter designation are not API grades but are grades in use or grades being considered for use. They are shown for information purposes.

erties that exceed the ratings calculated with the formulae in API *Bull. 5C3, Formulas and Calculations for Casing, Tubing, Drillpipe, and Line Pipe Properties.*[1] This improved performance is achieved principally by using better manufacturing practices and stricter quality assurance programs to reduce ovality, residual stress, and eccentricity. High collapse casing was initially developed for use in the deeper sections of high-pressure wells. The use of high collapse casing has gained wide acceptance in the industry, but its use remains controversial among some operators. Unfortunately, all manufacturers' claims have not been substantiated with the appropriate level of qualification testing. If high collapse casing is deemed necessary in a design, appropriate expert advice should be obtained to evaluate the manufacturer's qualification test data such as lengths to diameter ratio, testing conditions (end constraints), and the number of tests performed.

Equivalent Internal Pressure. If the pipe is subjected to both external and internal pressures, the equivalent external pressure is calculated as

$$p_e = p_o - \left[1 - \frac{2}{D/t}\right]p_i = \Delta p + \left(\frac{2}{D/t}\right)p_i, \dots \dots \dots \dots (7.6)$$

where
   p_e = equivalent external pressure,
   p_o = external pressure,
   p_i = internal pressure,
and
   $\Delta p = p_o - p_i$.

TABLE 7.6—*D/t* RANGE FOR ELASTIC COLLAPSE	
Grade*	Minimum *D/t* Range
H–40	42.64
–50	38.83
J–K–55	37.21
–60	35.73
–70	33.17
C–75 & E	32.05
L–N–80	31.02
C–90	29.18
C–T–95 & X	28.36
–100	27.60
P–105 & G	26.89
P–110	26.22
–120	25.01
Q–125	24.46
–130	23.94
S–135	23.44
–140	22.98
–150	22.11
–155	21.70
–160	21.32
–170	20.60
–180	19.93

*Grades indicated without a letter designation are not API grades but are grades in use or grades being considered for use. They are shown for information purposes.

To provide a more intuitive understanding of the sense of this relationship, Eq. 7.6 can be rewritten as

$$p_e D = p_o D - p_i d, \quad\dots\dots\dots\dots\dots\dots\dots\dots\dots\dots\dots\dots\dots (7.7)$$

where
D = nominal outside diameter,
and
d = nominal inside diameter.
In Eq. 7.7, we can see the internal pressure applied to the internal diameter and the external pressure applied to the external diameter. The "equivalent" pressure applied to the external diameter is the difference of these two terms.

Axial Strength. The axial strength of the pipe body is determined by the pipe body yield strength formula found in API *Bull. 5C3, Formulas and Calculations for Casing, Tubing, Drillpipe, and Line Pipe Properties.*[1]

$$F_y = \frac{\pi}{4}\left(D^2 - d^2\right)Y_p, \quad\dots\dots\dots\dots\dots\dots\dots\dots\dots\dots\dots\dots (7.8)$$

where
F_y = pipe body axial strength (units of force),
Y_p = minimum yield strength,
D = nominal outer diameter,
and

d = nominal inner diameter.

Axial strength is the product of the cross-sectional area (based on nominal dimensions) and the yield strength.

Combined Stress Effects. All the pipe-strength equations previously given are based on a uniaxial stress state (i.e., a state in which only one of the three principal stresses is nonzero). This idealized situation never occurs in oilfield applications because pipe in a wellbore is always subjected to combined loading conditions. The fundamental basis of casing design is that if stresses in the pipe wall exceed the yield strength of the material, a failure condition exists. Hence, the yield strength is a measure of the maximum allowable stress. To evaluate the pipe strength under combined loading conditions, the uniaxial yield strength is compared to the yielding condition. Perhaps the most widely accepted yielding criterion is based on the maximum distortion energy theory, which is known as the Huber-Hencky-Mises yield condition or simply the von Mises stress, triaxal stress, or equivalent stress.[3] Triaxial stress (equivalent stress) is not a true stress. It is a theoretical value that allows a generalized three-dimensional (3D) stress state to be compared with a uniaxial failure criterion (the yield strength). In other words, if the triaxial stress exceeds the yield strength, a yield failure is indicated. The triaxial safety factor is the ratio of the material's yield strength to the triaxial stress.

The yielding criterion is stated as

$$\sigma_{VME} = \frac{1}{\sqrt{2}}\sqrt{(\sigma_z - \sigma_\theta)^2 + (\sigma_\theta - \sigma_r)^2 + (\sigma_r - \sigma_z)^2} \geq Y_p, \quad\quad\quad (7.9)$$

where

Y_p = minimum yield stress, psi,

σ_{VME} = triaxial stress, psi,

σ_z = axial stress, psi,

σ_θ = tangential or hoop stress, psi,

and

σ_r = radial stress, psi.

The calculated axial stress, σ_z, at any point along the cross-sectional area should include the effects of self-weight, buoyancy, pressure loads, bending, shock loads, frictional drag, point loads, temperature loads, and buckling loads. Except for bending/buckling loads, axial loads are normally considered to be constant over the entire cross-sectional area.

The tangential and radial stresses are calculated with the Lamé equations for thick-wall cylinders.

$$\sigma_\theta = \frac{\left(1 + r_o^2/r^2\right)}{\left(r_o^2 - r_i^2\right)}r_i^2 p_i - \frac{\left(1 + r_i^2/r^2\right)}{\left(r_o^2 - r_i^2\right)}r_o^2 p_o \quad\quad\quad (7.10)$$

and

$$\sigma_r = \frac{\left(1 - r_o^2/r^2\right)}{\left(r_o^2 - r_i^2\right)}r_i^2 p_i - \frac{\left(1 - r_i^2/r^2\right)}{\left(r_o^2 - r_i^2\right)}r_o^2 p_o \quad\quad\quad (7.11)$$

where

p_i = internal pressure,

p_o = external pressure,

r_i = inner wall radius,

r_o = outer wall radius,

and

r = radius at which the stress occurs.

The absolute value of σ_θ is always greatest at the inner wall of the pipe and that for burst and collapse loads, where $|p_i - p_o| \gg 0$, then $|\sigma_\theta| \gg |\sigma_r|$. For any p_i and p_o combination, the sum of the tangential and radial stresses is constant at all points in the casing wall. Substituting Eq. 7.10 and Eq. 7.11 into Eq. 7.9, after rearrangements, yields

$$\sigma_{VME} = \sqrt{(f_1 f_2)^2 + f_3^2}, \dots\dots\dots\dots\dots\dots\dots\dots\dots\dots\dots (7.12)$$

in which

$$f_1 = \left(\frac{r_i}{r}\right)^2 \frac{\sqrt{3}}{2}(p_o - p_i)$$

$$f_2 = \frac{1}{2}\frac{\left(\frac{D}{t}\right)^2}{\frac{D}{t} - 1},$$

and

$$f_3 = \sigma_z - \frac{r_i^2 p_i - r_o^2 p_o}{r_o^2 - r_i^2}$$

where

D = outside pipe diameter,

and

t = wall thickness.

Eq. 7.12 calculates the equivalent stress at any point of the pipe body for any given pipe geometry and loading conditions. To illustrate these concepts, let us consider a few particular cases.

Combined Collapse and Tension. Assuming that $\sigma_z > 0$ and $\sigma_\theta \gg \sigma_r$ and setting the triaxial stress equal to the yield strength results in the next equation of an ellipse.

$$Y_p = \left[\sigma_z^2 - \sigma_z \sigma_\theta + \sigma_\theta^2\right]^{1/2} \dots\dots\dots\dots\dots\dots\dots\dots\dots\dots (7.13)$$

This is the biaxial criterion used in API *Bull. 5C3, Formulas and Calculations for Casing, Tubing, Drillpipe, and Line Pipe Properties,*[1] to account for the effect of tension on collapse.

$$Y_{pa} = \left[\sqrt{1 - 0.75\left(\frac{S_a}{Y_p}\right)^2} - 0.5\frac{S_a}{Y_p}\right]Y_p, \dots\dots\dots\dots\dots\dots\dots\dots (7.14)$$

where

S_a = axial stress based on the buoyant weight of pipe,

and

Y_p = yield point.

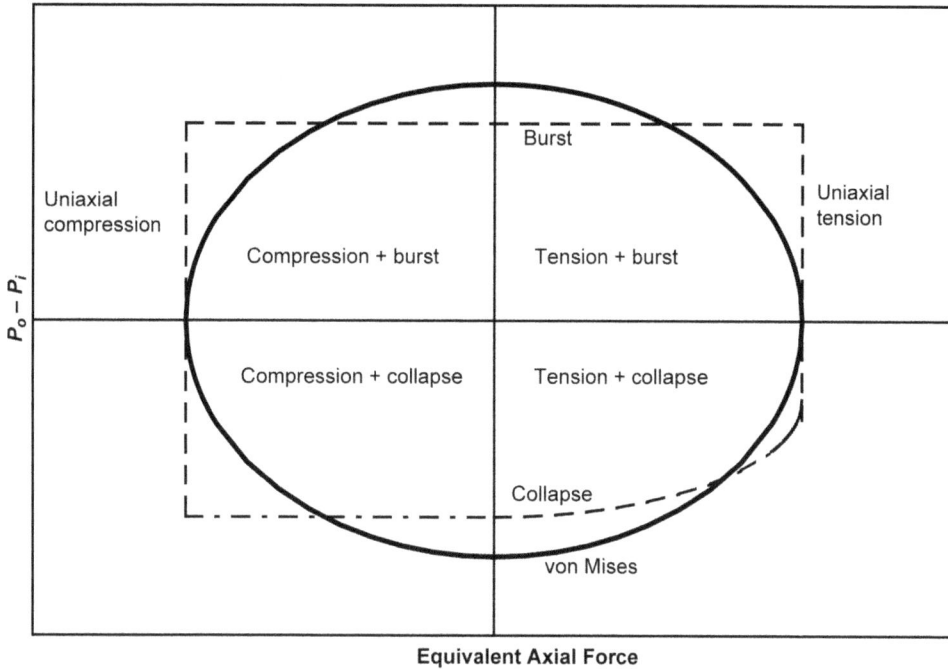

Equivalent Axial Force
Fig. 7.2—Casing failure criteria.

It is clearly seen that as the axial stress S_a increases, the pipe collapse resistance decreases. Plotting this ellipse, **Fig. 7.2** allows a direct comparison of the triaxial criterion with the API ratings. Loads that fall within the design envelope meet the design criteria. The curved lower right corner is caused by the combined stress effects, as described in Eq. 7.14.

Combined Burst and Compression Loading. Combined burst and compression loading corresponds to the upper left-hand quadrant of the design envelope. This is the region where triaxial analysis is most critical because reliance on the uniaxial criterion alone would not predict several possible failures. For high burst loads (i.e., high tangential stress and moderate compression), a burst failure can occur at a differential pressure less than the API burst pressure. For high compression and moderate burst loads, the failure mode is permanent corkscrewing (i.e., plastic deformation because of helical buckling). This combined loading typically occurs when a high internal pressure is experienced (because of a tubing leak or a buildup of annular pressure) after the casing temperature has been increased because of production. The temperature increase, in the uncemented portion of the casing, causes thermal growth, which can result in significant increases in compression and buckling. The increase in internal pressure also results in increased buckling.

Combined Burst and Tension Loading. Combined burst and tension loading corresponds to the upper right-hand quadrant of the design envelope. This is the region where reliance on the uniaxial criterion alone can result in a design that is more conservative than necessary. For high burst loads and moderate tension, a burst yield failure will not occur until after the API burst pressure has been exceeded. As the tension approaches the axial limit, a burst failure can occur at a differential pressure less than the API value. For high tension and moderate burst loads, pipe body yield will not occur until a tension greater than the uniaxial rating is reached.

Taking advantage of the increase in burst resistance in the presence of tension represents a good opportunity for the design engineer to save money while maintaining wellbore integrity. Similarly, the designer might wish to allow loads between the uniaxial and triaxial tension ratings. However, great care should be taken in the latter case because of the uncertainty of what burst pressure might be seen in conjunction with a high tensile load (an exception to this is the green cement pressure test load case). Also, connection ratings may limit your ability to design in this region.

Use of Triaxial Criterion for Collapse Loading. For many pipes used in the oil field, collapse is an inelastic stability failure or an elastic stability failure independent of yield strength. The triaxial criterion is based on elastic behavior and the yield strength of the material and, hence, should not be used with collapse loads. The one exception is for thick-wall pipes with a low D/t ratio, which have an API rating in the yield strength collapse region. This collapse criterion along with the effects of tension and internal pressure (which are triaxial effects) result in the API criterion being essentially identical to the triaxial method in the lower right-hand quadrant of the triaxial ellipse for thick-wall pipes.

For high compression and moderate collapse loads experienced in the lower left-hand quadrant of the design envelope, the failure mode may be permanent corkscrewing because of helical buckling. It is appropriate to use the triaxial criterion in this case. This load combination typically can occur only in wells that experience a large increase in temperature because of production. The combination of a collapse load that causes reverse ballooning and a temperature increase acts to increase compression in the uncemented portion of the string.

Most design engineers use a minimum wall for burst calculations and nominal dimensions for collapse and axial calculations. Arguments can be made for using either assumption in the case of triaxial design. Most importantly, more so than the choice of dimensional assumptions, is that the results of the triaxial analysis should be consistent with the uniaxial ratings with which they may be compared.

Triaxial analysis is perhaps most valuable when evaluating burst loads. Hence, it makes sense to calibrate the triaxial analysis to be compatible with the uniaxial burst analysis. This can be done by the appropriate selection of a design factor. Because the triaxial result nominally reduces to the uniaxial burst result when no axial load is applied, the results of both of these analyses should be equivalent. Because the burst rating is based on 87.5% of the nominal wall thickness, a triaxial analysis based on nominal dimensions should use a design factor that is equal to the burst design factor multiplied by 8/7. This reflects the philosophy that a less conservative assumption should be used with a higher design factor. Hence, for a burst design factor of 1.1, a triaxial design factor of 1.25 should be used.

Final Triaxial Stress Considerations. **Fig. 7.3** graphically summarizes the triaxial, uniaxial, and biaxial limits that should be used in casing design along with a set of consistent design factors.

Because of the potential benefits (both cost savings and better mechanical integrity) that can be realized, a triaxial analysis is recommended for all well designs. Specific applications include saving money in burst design by taking advantage of the increased burst resistance in tension; accounting for large temperature effects on the axial load profile in high-pressure, high-temperature wells (this is particularly important in combined burst and compression loading); accurately determining stresses when using thick-wall pipe ($D/t < 12$) (conventional uniaxial and biaxial methods have imbedded thin-wall assumptions); and evaluating buckling severity (permanent corkscrewing occurs when the triaxial stress exceeds the yield strength of the material).

While it is acknowledged that the von Mises criterion is the most accurate method of representing elastic yield behavior, use of this criterion in tubular design should be accompanied by a few precautions.

First, for most pipe used in oilfield applications, collapse is frequently an instability failure that occurs before the computed maximum triaxial stress reaches the yield strength. Hence,

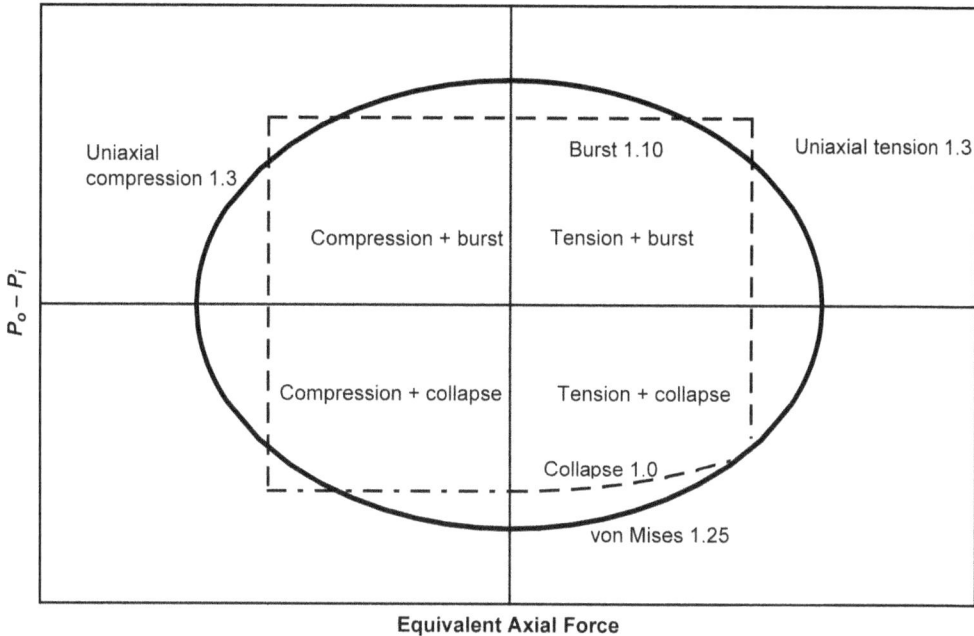

Fig. 7.3—Design factors for casing failure criteria.

triaxial stress should not be used as a collapse criterion. Only in thick-wall pipe does yielding occur before collapse.

Second, the accuracy of triaxial analysis is dependent upon the accurate representation of the conditions that exist both for the pipe as installed in the well and for the subsequent loads of interest. Often, it is the change in load conditions that is most important in stress analysis. Hence, an accurate knowledge of all temperatures and pressures that occur over the life of the well can be critical to accurate triaxial analysis.

7.6 API Connection Ratings
While a number of joint connections are available, the API recognizes three basic types: coupling with rounded thread (long or short); coupling with asymmetrical trapezoidal thread buttress; and extreme-line casing with trapezoidal thread without coupling.

Threads are used as mechanical means to hold the neighboring joints together during axial tension or compression. For all casing sizes, the threads are not intended to be leak resistant when made up. API *Spec. 5C2, Performance Properties of Casing, Tubing, and Drillpipe,*[2] provides information on casing and tubing threads dimensions.

7.6.1 Coupling Internal Yield Pressure. The internal yield pressure is the pressure that initiates yield at the root of the coupling thread.

$$P_{CIY} = Y_c\left(\frac{W - d_1}{W}\right), \dots\dots\dots\dots\dots\dots\dots\dots\dots\dots\dots (7.15)$$

where
P_{CIY} = coupling internal yield pressure, psi,
Y_c = minimum yield strength of coupling, psi,
W = nominal outside diameter of coupling, in.,
and

d_1 = diameter at the root of the coupling thread in the power tight position, in.

This dimension is based on data given in API *Spec. 5B, Threading, Gauging, and Thread Inspection of Casing, Tubing, and Line Pipe Threads,*[4] and other thread geometry data. The coupling internal yield pressure is typically greater than the pipe body internal yield pressure. The internal pressure leak resistance is based on the interface pressure between the pipe and coupling threads because of makeup.

$$P_{ILR} = \frac{ETN p_t \left(W^2 - E_s^2 \right)}{2 E_S W^2}, \dotfill (7.16)$$

where

P_{ILR} = coupling internal pressure leak resistance, psi,

E = modulus of elasticity,

T = thread taper, in.,

N = a function of the number of thread turns from hand-tight to power-tight position, as given in API *Spec. 5B, Threading, Gauging, and Thread Inspection of Casing, Tubing, and Line Pipe Threads,*[4]

p_t = thread pitch, in.,

and

E_s = pitch diameter at plane of seal, as given in API *Spec. 5B, Threading, Gauging, and Thread Inspection of Casing, Tubing, and Line Pipe Threads.*[4]

This equation accounts only for the contact pressure on the thread flanks as a sealing mechanism and ignores the long helical leak paths filled with thread compound that exist in all API connections.

In round threads, two small leak paths exist at the crest and root of each thread. In buttress threads, a much larger leak path exists along the stabbing flank and at the root of the coupling thread. API connections rely on thread compound to fill these gaps and provide leak resistance. The leak resistance provided by the thread compound is typically less than the API internal leak resistance value, particularly for buttress connections. The leak resistance can be improved by using API connections with smaller thread tolerances (and, hence, smaller gaps), but it typically will not exceed 5,000 psi with any long-term reliability. Applying tin or zinc plating to the coupling also results in smaller gaps and improves leak resistance.

7.6.2 Round-Thread Casing-Joint Strength. The round-thread casing-joint strength is given as the lesser of the fracture strength of the pin and the jump-out strength. The fracture strength is given by

$$F_j = 0.95 A_{jp} U_p. \dotfill (7.17)$$

The jump-out strength is given by

$$F_j = 0.95 A_{jp} L \left[\frac{0.74 D^{-0.59} U_p}{0.5 L + 0.14 D} + \frac{Y_p}{L + 0.14 D} \right], \dotfill (7.18)$$

where

F_j = minimum joint strength, lbf,

A_{jp} = cross-sectional area of the pipe wall under the last perfect thread, in.^{2},

= $\pi/4 [(D - 0.1425)^2 - d^2]$,

D = nominal outside diameter of pipe, in.,

d = nominal inside diameter of pipe, in.,

L = engaged thread length, in., as given in API *Spec. 5B, Threading, Gauging, and Thread Inspection of Casing, Tubing, and Line Pipe Threads,*[4]

Y_p = minimum yield strength of pipe, psi,

and

U_p = minimum ultimate tensile strength of pipe, psi.

These equations are based on tension tests to failure on 162 round-thread test specimens. Both are theoretically derived and adjusted using statistical methods to match the test data. For standard coupling dimensions, round threads are pin weak (i.e., the coupling is noncritical in determining joint strength).

7.6.3 Buttress Casing Joint Strength. The buttress thread casing joint strength is given as the lesser of the fracture strength of the pipe body (the pin) and the coupling (the box). Pipe thread strength is given by

$$F_j = 0.95 A_p U_p [1.008 - 0.0396(1.083 - Y_p/U_p)D]. \dots\dots\dots\dots\dots (7.19)$$

Coupling thread strength is given by

$$F_j = 0.95 A_c U_c, \dots\dots\dots\dots\dots\dots\dots\dots (7.20)$$

where

U_c = minimum ultimate tensile strength of coupling, psi,

A_p = cross-sectional area of plain-end pipe, in.^{2},

and

A_c = cross-sectional area of coupling, in.,

= $\pi/4(W^2 - d_1^2)$.

These equations are based on tension tests to failure on 151 buttress-thread test specimens. They are theoretically derived and adjusted using statistical methods to match test data.

7.6.4 Extreme-Line Casing-Joint Strength. Extreme-line casing-joint strength is calculated as

$$F_j = A_{cr} U_p, \dots\dots\dots\dots\dots\dots\dots\dots (7.21)$$

where

F_j = minimum joint strength, lbf,

and

A_{cr} = critical section area of box, pin, or pipe, whichever is least, in.^{2}.

When performing casing design, it is very important to note that the API joint-strength values are a function of the ultimate tensile strength. This is a different criterion from that used to define the axial strength of the pipe body, which is based on the yield strength. If care is not taken, this approach can lead to a design that inherently does not have the same level of safety for the connections as for the pipe body. This is not the most prudent practice, particularly in light of the fact that most casing failures occur at connections. This discrepancy can be countered by using a higher design factor when performing connection axial design with API connections.

The joint-strength equations for tubing given in API *Bull. 5C3, Formulas and Calculations for Casing, Tubing, Drillpipe, and Line Pipe Properties,*[1] are very similar to those given for

TABLE 7.7—TENSILE PROPERTY REQUIREMENTS			
Grade	Y_p, psi	U_p, psi	U_p/Y_p
H–40	40,000	60,000	1.50
J–55	55,000	75,000	1.36
K–55	55,000	95,000	1.73
N–80	80,000	100,000	1.25
L–80	80,000	95,000	1.19
C–90	90,000	100,000	1.11
C–95	95,000	105,000	1.11
T–95	95,000	105,000	1.11
P–110	110,000	125,000	1.14
Q–125	125,000	135,000	1.08

round-thread casing except they are based on yield strength. Hence, the *UTS/YS* discrepancy does not exist in tubing design.

If API casing connection joint strengths calculated with the previous formulae are the basis of a design, the designer should use higher axial design factors for the connection analysis. The logical basis for a higher axial design factor (DF) is to multiply the pipe body axial design factor by the ratio of the minimum ultimate tensile strength, U_p, to the minimum yield strength, Y_p.

$$DF_{connection} = DF_{pipe} \times \left(\frac{U_p}{Y_p} \right) \dotfill (7.22)$$

Tensile property requirements for standard grades are given in API *Spec. 5C2, Performance Properties of Casing, Tubing, and Drillpipe,*[2] and are shown in **Table 7.7** for reference along with their ratio.

7.6.5 Proprietary Connections. Special connections are used to achieve gas-tight sealing reliability and 100% connection efficiency (joint efficiency is defined as a ratio of joint tensile strength to pipe body tensile strength) under more severe well conditions. Severe conditions include high pressure (typically > 5,000 psi); high temperature (typically > 250°F); a sour environment; gas production; high-pressure gas lift; a steam well; and a large dogleg (horizontal well). Also, efficiency in flush joint, integral joint or other special clearance applications improves connections; a large diameter (> 16 in.) pipe improves the stab-in and makeup characteristics; galling should be reduced (particularly in CRA applications and tubing strings that will be re-used); and connection failure under high torsional loads (e.g., while rotating pipe) should be prevented.

The improved performance of many proprietary connections results from one or more of these features not found in API connections: more complex thread forms; resilient seals; torque shoulders; and metal-to-metal seals. The "premium" performance of most proprietary connections comes at a "premium" cost. Increased performance should always be weighed against the increased cost for a particular application. As a general rule, it is recommended to use proprietary connections only when the application requires them. "Premium" performance may also be achieved using API connections if certain conditions are met. Those conditions are tighter dimensional tolerance; plating applied to coupling; use of appropriate thread compound; and performance verified with qualification testing.

The performance of a proprietary connection can be reliably verified by performing three steps: audit the manufacturer's performance test data (sealability and tensile load capacity under combined loading); audit the manufacturer's field history data; and require additional performance testing for the most critical applications. When requesting tensile performance data, make sure that the manufacturer indicates whether quoted tensile capacities are based on the ultimate tensile strength (i.e., the load at which the connection will fracture, commonly called the "parting load") or the yield strength (commonly called the "joint elastic limit"). If possible, it is recommended to use the joint elastic limit values in the design so that consistent design factors for both pipe-body and connection analysis are maintained. If only parting load capacities are available, a higher design factor should be used for connection axial design.

7.7 Connection Failures
Most casing failures occur at connections. These failures can be attributed to improper design or exposure to loads exceeding the rated capacity; failure to comply with makeup requirements; failure to meet manufacturing tolerances; damage during storage and handling; and damage during production operations (corrosion, wear, etc.).

Connection failure can be classified broadly as leakage; structural failure; galling during makeup; yielding because of internal pressure; jump-out under tensile load; fracture under tensile load; and failure because of excessive torque during makeup or subsequent operations. Avoiding connection failure is not only dependent upon selection of the correct connection but is strongly influenced by other factors, which include manufacturing tolerances; storage (storage thread compound and thread protector); transportation (thread protector and handling procedures); and running procedures (selection of thread compound, application of thread compound, and adherence to correct makeup specifications and procedures).

The overall mechanical integrity of a correctly designed casing string is dependent upon a quality assurance program that ensures damaged connections are not used and that operations personnel adhere to the appropriate running procedures.

7.8 Connection Design Limits
The design limits of a connection are not only dependent upon its geometry and material properties but are influenced by surface treatment; phosphating; metal plating (copper, tin, or zinc); bead blasting; thread compound; makeup torque; use of a resilient seal ring (many companies do not recommend this practice); fluid to which connection is exposed (mud, clear brine, or gas); temperature and pressure cycling; and large doglegs (e.g., medium- or short-radius horizontal wells).

7.9 Casing and Tubing Buckling

7.9.1 Introduction. As installed, casing usually hangs straight down in vertical wells or lays on the low side of the hole in deviated wells. Thermal or pressure loads might produce compressive loads, and if these loads are sufficiently high, the initial configuration will become unstable. However, because the tubing is confined within open hole or casing, the tubing can deform into another stable configuration, usually a helical or coil shape in a vertical wellbore or a lateral S-shaped configuration in a deviated hole. These new equilibrium configurations are what we mean when we talk about buckling in casing design. In contrast, conventional mechanical engineering design considers buckling in terms of stability (i.e., the prediction of the critical load at which the original configuration becomes unstable).

Accurate analysis of buckling is important for several reasons. First, buckling generates bending stresses not present in the original configuration. If the stresses in the original configuration were near yield, this additional stress could produce failure, including permanent plastic deformation called "corkscrewing." Second, buckling causes tubing movement. Coiled tubing is shorter than straight tubing, and this is an important consideration if the tubing is not fixed. Third, tubing buckling causes the relief of compressive axial loads when the casing is fixed.

This effect is not as recognized as the first two buckling effects but is equally important. The axial compliance of buckled tubing is much less than the compliance of straight tubing. Casing movement, because of thermal expansion or ballooning, can be accommodated with a lower increase in axial load for a buckled casing.

The accuracy and comprehensiveness of the buckling model is important for designing tubing. The most commonly used buckling solution is the model developed by Lubinski in the 1950s. This model is accurate for vertical wells but needs modification for deviated wells. Tubing bending stress, because of buckling, will be overestimated for deviated wells using Lubinski's formula. However, Lubinski's solution, applied to deviated wells, will also overpredict tubing movement. This solution overestimates tubing compliance, which might greatly underestimate the axial loads, resulting in a nonconservative design.

7.9.2 Casing Buckling in Oilfield Operations. Buckling should be avoided in drilling operations to minimize casing wear. Buckling can be reduced or eliminated by applying a pickup force before landing the casing; holding pressure, while weighing on cement (WOC), to pretension the string (subsea wells); raising the top of cement; using centralizers; and increasing pipe stiffness.

In production operations, casing buckling is not normally a critical design issue. However, a large amount of buckling can occur because of increased production temperatures in some wells. A check should be made to ensure that plastic deformation or corkscrewing will not occur. This check is possible using triaxial analysis and including the bending stress because of buckling. Corkscrewing occurs only if the triaxial stress exceeds the yield strength of the material.

7.9.3 Tubing Buckling in Oilfield Operations. Buckling is typically a more critical design issue for production tubing than for casing. Tubing is typically exposed to the hottest temperatures during production. Pressure/area effects in floating seal assemblies can significantly increase buckling. Tubing is less stiff than casing, and annular clearances can be quite large. Buckling can prevent wireline tools from passing through the tubing. Buckling can be controlled by tubing-to-packer configuration (latched or free, seal bore diameter, allowable movement in seals, etc.); slackoff or pickup force at surface; cross-sectional area changes in tubing; packer fluid density; pipe stiffness; centralizers; and hydraulic set pressure. As in casing design, a triaxial check should be made to ensure that plastic deformation or corkscrewing will not occur.

7.9.4 Buckling Models and Correlations. Buckling occurs if the buckling force, F_b, is greater than a threshold force, F_p, known as the Paslay buckling force. The buckling force, F_b, is defined as

$$F_b = -F_a + p_i A_i - p_o A_o , \qquad (7.23)$$

where
  F_b = buckling force, lbf,
  F_a = axial force (tension positive), lbf,
  p_i = internal pressure, psi,
  $A_i = r_i^2$, where r_i is the inside radius of the tubing, in.^{2},
  p_o = external pressure, psi,
and
  $A_o = r_o^2$, where r_o is the outside radius of the tubing, in.^{2}
  The Paslay buckling force, F_p, is defined as

TABLE 7.8—BUCKLING CRITERIA	
Buckling Force Magnitude	**Result**
$F_b < F_p$	No buckling
$F_p < F_b < \sqrt{2}F_p$	Lateral (S-shaped) buckling
$\sqrt{2}F_p < F_b < 2\sqrt{2}F_p$	Lateral or helical buckling
$2\sqrt{2}F_p < F_b$	Helical buckling

$$F_p = \sqrt{\frac{EIw_c}{r}}$$

$$w_c = \sqrt{\left(w_e \sin \Phi + F_b \frac{d\Phi}{dz}\right)^2 + \left(F_b \sin \Phi \frac{d\Theta}{dz}\right)^2}, \quad\text{................................ (7.24)}$$

where

F_p = Paslay buckling force, lbf,

w_c = casing contact load, lbf/in.,

w_e = distributed buoyed weight of casing, lbf/in.,

Φ = wellbore angle of inclination, radians,

Θ = wellbore azimuth angle, radians,

EI = pipe bending stiffness, lbf-in.^{2},

and

r = radial annular clearance, in.

Table 7.8 gives the relationship between the buckling force F_b, the Paslay buckling force F_p, and the type of buckling expected for the tubing.

An increase in internal pressure acts on the buckling force in two ways. It increases F_a because of ballooning, which tends to decrease buckling, and increases the p_iA_i term, which tends to increase buckling. The second effect is much greater; hence, an increase in internal pressure will result in an increase in buckling.

A temperature increase results in a reduction in the axial tension (or increase in the compression). This reduction in tension results in an increase in buckling. The onset and type of buckling is a function of hole angle. Because of the stabilizing effect of the lateral distributed force of a casing lying on the low side of the hole in an inclined wellbore, a greater force is required to induce buckling. In a vertical well, $F_p = 0$, and helical buckling occurs at any $F_b > 0$. For production tubing that is free to move in a seal assembly, the upward force, because of pressure/area effects in the seal assembly, will decrease F_a, which, in turn, increases buckling.

In order to give the correlations for tubing stresses and movement, definitions are made. The lateral displacements of the tubing, shown in **Fig. 7.4**, are given by

$$u_1 = r \cos \theta, \quad\text{... (7.25)}$$

and

$$u_2 = r \sin \theta, \quad\text{... (7.26)}$$

where θ is the helix angle. The quantity θ', where $'$ denotes d/dz, is important and appears often in the next analysis. It can be related to the more familiar quantity, pitch through Eq. 7.27.

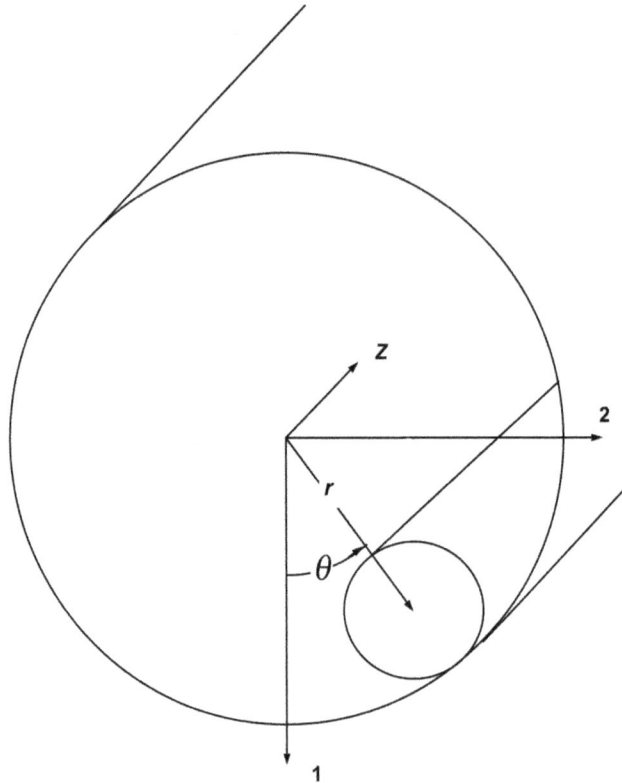

Fig. 7.4—Coordinate system for buckling analysis.

$$P_{hel} = 2\pi / \theta', \dotfill (7.27)$$

where P_{hel} = pitch of helically buckled pipe, in.

Other important quantities, such as pipe curvature, bending moment, bending stress, and tubing length change are proportional to the square of θ'. Nonzero θ' indicates that the pipe is curving, while zero θ' indicates that the pipe is straight.

7.9.5 Correlations for Maximum Buckling Dogleg. The correlation for the maximum value of θ' for lateral buckling, with $2.8\,F_p > F_b > F_p$, can be expressed by

$$\theta'_{max} = \frac{1.1227}{\sqrt{2EI}} F^{0.04}(F_b - F_p)^{0.46} \dotfill (7.28)$$

For $F_b > 2.8\,F_p$, the corresponding helical buckling correlation is

$$\theta'_{max} = \pm \sqrt{\frac{F_b}{2EI}} \dotfill (7.29)$$

The region $2.8\,F_p > F_b > 1.4\,F_p$ may be either helical or lateral; however, $2.8\,F_p$ is believed to be the lateral buckling limit on loading, while $1.4\,F_p$ is believed to be the helical buckling limit on unloading from a helical buckled state. An important distinction between Eq. 7.28 and Eq. 7.29 is that Eq. 7.28 is the maximum value of θ', while Eq. 7.29 is the actual value of θ'.

The equation for a dogleg curvature for a helix is

$$\kappa = r(\theta')^2, \quad\text{..}\quad (7.30)$$

assuming θ'' is negligible. The dogleg unit for Eq. 7.30 is radians per inch. To convert to the conventional unit of degrees per 100 ft, multiply the result by 68,755.

7.9.6 Correlations for Bending Moment and Bending Stress. Given the tubing curvature, the bending moment is determined by

$$M = EI\kappa = EIr(\theta')^2 . \quad\text{..}\quad (7.31)$$

The corresponding maximum bending stress is

$$\sigma_b = \frac{Md_o}{2I} = \frac{Ed_o r(\theta')^2}{2}; \quad\text{..}\quad (7.32)$$

where d_o is the outside diameter of the pipe. The following correlations can be derived with Eqs. 7.28 and 7.29. $M = 0$, for $F_b < F_p$;

$$M = .6302 r F_b^{.08}(F_b - F_p)^{0.92}, \text{ for } 2.8F_p > F_b > F_p. \quad\text{................................}\quad (7.33)$$

$$M = .5 \ r \ F_b, \text{ for } F_b > 2.8 \ F_p. \quad\text{..}\quad (7.34)$$

$\sigma_b = 0$, for $F_b < F_p$;

$$\sigma_b = .3151 \frac{d_o r}{I} F_b^{.08}(F_b - F_p)^{0.92}, \quad \text{for } 2.8F_p > F > F_p. \quad\text{................................}\quad (7.35)$$

$$\sigma_b = .2500 \frac{d_o r}{I} F_b, \text{ for } F_b > 2.8 \ F_p. \quad\text{..}\quad (7.36)$$

7.9.7 Correlations for Buckling Strain and Length Change. The buckling "strain," in the sense of Lubinski, is the buckling length change per unit length. The buckling strain is given by

$$e_b = -\tfrac{1}{2}(r\theta')^2 \quad\text{..}\quad (7.37)$$

For the case of lateral buckling, the actual shape of the θ' curve was integrated numerically to determine the relationship,

$$e_{\text{bavg}} = -.7285 \frac{r^2}{4EI} F_b^{.08}(F_b - F_p)^{0.92}, \quad\text{..}\quad (7.38)$$

for $2.8 \ F_p > F_b > F_p$, which compares to the helical buckling strain given by

$$e_b = -\frac{r^2}{4EI}F_b, \quad\quad\quad\quad\quad\quad\quad\quad\quad\quad\quad\quad\quad\quad (7.39)$$

for $F_b > 2.8\ F_p$. The lateral buckling strain is roughly half the conventional helical buckling strain. To determine the buckling length change, ΔL_b, we must integrate Eqs. 7.38 and 7.39 over the appropriate length interval, which is written as

$$\Delta L_b = \int_{z_1}^{z_2} e_b dz, \quad\quad\quad\quad\quad\quad\quad\quad\quad\quad\quad\quad\quad (7.40)$$

where z_1 and z_2 are defined by the distribution of the buckling force, F. For the general case of arbitrary variation of F_b over the interval $\Delta L = Z_2 - Z_1$, Eq. 7.40 must be numerically integrated. However, there are two special cases that are commonly used. For the case of constant force, F_b, such as in a horizontal well, Eq. 7.40 is easily integrated.

$$\int_{z_1}^{z_2} e_b\ dz = e_b\ \Delta L, \quad\quad\quad\quad\quad\quad\quad\quad\quad\quad\quad (7.41)$$

where e_b is defined by either Eq. 7.28 or Eq. 7.29. The second special case is for a linear variation of F_b over the interval.

$$F_b(z) = wz + c \quad\quad\quad\quad\quad\quad\quad\quad\quad\quad\quad\quad\quad\quad (7.42)$$

The length change is given for this case by Eqs. 7.43 and 7.44.

$$\Delta L_b = \frac{-r^2}{4E\ Iw}(F_2 - F_p)[.3771\ F_2 - .3668\ F_p], \quad\quad\quad\quad\quad (7.43)$$

for $2.8\ F_p > F_2 > F_p$.

$$\Delta L_b = -\frac{r^2}{8E\ Iw}[F_2^2 - F_1^2], \quad\quad\quad\quad\quad\quad\quad\quad\quad (7.44)$$

for $F > 2.8\ F_p$.

7.9.8 Correlations for Contact Force. From equilibrium considerations only, the average contact force for lateral buckling is

$$W_n = w_e g_c \sin\theta \quad\quad\quad\quad\quad\quad\quad\quad\quad\quad\quad\quad\quad (7.45)$$

The average contact force for the helically buckled section is

$$W_n = \frac{rF^2}{4EI}\ . \quad\quad\quad\quad\quad\quad\quad\quad\quad\quad\quad\quad\quad\quad (7.46)$$

TABLE 7.9—BUCKLING FORCES			
Deviation Angle, degrees	Minimum Lateral	Buckling Forces, lbf Maximum Lateral	Minimum Helical
0	0	0	0
5	2,201	6,226	3,113
10	3,107	8,788	4,394
15	3,793	10,729	5,365
30	5,272	14,913	7,456
60	6,939	19,626	9,813
90	7,456	21,090	10,545

When the buckling mode changes from lateral to helical, the contact force increases substantially.

7.9.9 Sample Buckling Calculations. The basis of the sample calculations is the buckling of tubing ($2\frac{7}{8}$ in., 6.5 lbm/ft) inside of casing (7 in., 32 lbm/ft). The tubing is submerged in 10-lbm/gal packer fluid with no other pressures applied. The effect of the packer fluid is to reduce the tubing weight per unit length through buoyancy. $w_e = w + A_i \gamma_i - A_o \gamma_o$, where w_e is the effective weight per unit length of the tubing, A_i is the inside area of the tubing, γ_i is the density of the fluid inside the tubing, A_o is the outside area of the tubing, and γ_o is the density of the fluid outside the tubing. The calculation gives

$$w_e = 6.5 \ \text{lbm/ft} + (4.68 \ \text{in.}^2)(.052 \ \text{psi/ft/lbm/gal})(10.0 \ \text{lbm/gal})$$

$$-(6.49 \ \text{in.}^2)(.052 \ \text{psi/ft/lbm/gal})(10.0 \ \text{lbm/gal})$$

$$= 5.56 \ \text{lbm/ft} = 0.463 \ \text{lbm/in.}$$

Other information useful for the buckling calculations are radial clearance $= r = 1.61$ in.; moment of inertia $= I = 1.611$ in.^{4}, and Young's modulus $= 30 \times 10^6$ psi.

Sample Buckling Length Calculations. From Eq. 7.24, we can calculate the Paslay force for a variety of inclinations. First, we calculate the value for a horizontal well, which is written as

$$F_p = \sqrt{\left[4(0.463 \ \text{lbm/in.})(30 \times 10^6 \ \text{psi})(1.611 \ \text{in.}^4)/(1.61 \ \text{in.})\right]} = 7,456 \ \text{lbf}.$$

This means that the axial buckling force must exceed 7,500 lbf before the tubing will buckle. We can evaluate other angles by multiplying the horizontal Fp by the square root of the sine of the inclination angle. **Table 7.9** was developed with this procedure. Of particular notice in Table 7.9 is how large these buckling forces are for relatively small deviations from vertical. For a well 10° from the vertical, the buckling forces are nearly half of the horizontal well buckling forces.

With Table 7.9, the total buckled length of the tubing can be calculated, as well as maximum and minimum lateral buckling or helical buckling. Assume an applied buckling force of 30,000 lbf is applied at the end of the tubing in a well with a 60° deviation from vertical. The tubing will buckle for any force between 6,939 lbf and 30,000 lbf. The axial force will vary as $w_e \cos\Phi$ (i.e., $w_a = w_e \cos(60) = 5.56 \ \text{lbf/ft} \ (0.50) = 2.78 \ \text{lbf/ft}$). Therefore, the total buckled length, L_{bkl}, is $L_{bkl} = (30,000 - 6939)\text{lbf}/(2.78 \ \text{lbf/ft}) = 8,295$ ft. The maximum helically buckled length, L_{helmax}, is $L_{helmax} = (30,000 - 9,813)\text{lbf}/(2.78 \ \text{lbf/ft}) = 7,262$ ft. The minimum helically buckled length, L_{helmin}, is $L_{helmin} = (30,000 - 19,626)\text{lbf}/(2.78 \ \text{lbf/ft}) = 3,732$ ft.

Sample Buckling Bending Stress Calculations. The maximum bending stress, because of buckling, can be evaluated with Eq. 7.38. $\sigma_b = .25(2.875 \ \text{in.})(1.61 \ \text{in.})(30,000 \ \text{lbf})/(1.611 \ \text{in.}^4)$

= 21,550 psi. This stress is fairly large compared to tubing yield strengths of about 80,000 psi, so buckling bending stresses can be important for casing and tubing design. At the buckling load of 19,626 lbf, both helical and lateral buckling can occur. The lateral bending stress is given by Eq. 7.35. σ_b = .3151 (2.875 in.)(1.61 in.)/(1.611 in.^{4}) (6,939 lbf)$^{.08}$(19,626 – 6,939 lbf)$^{0.92}$ = 10,945 psi. The equivalent calculation for helical buckling gives σ_b = .25(2.875 in.) (1.61 in.)(19,626 lbf)/(1.611 in.^{4}) = 14,097 psi, so helical buckling produces approximately 29% higher stresses than lateral buckling. This indicates that determination of buckling type can be important in casing design where casing strength is marginal.

Sample Buckling Length Change Calculations—Tubing Movement. Tubing length change calculations involve two calculations for this case, tubing movement because of lateral buckling and tubing movement because of helical buckling. Eqs. 7.43 and 7.44 are used to calculate tubing movement, and these equations assume the minimum amount of helical buckling. A third calculation is made to show the movement because of pure helical buckling. The lateral buckling tubing movement is given by

$$\Delta L_b = -(1.61 \text{ in.})^2(19,626 - 6,939 \text{ lbf})$$
$$\times [.3771 \ (19,626) - .3668(6,939) \text{ lbf}]$$
$$/[(4)(30 \times 10^6 \text{ psi})(1.611 \text{ in.}^4)(2.78 \text{ lbf}/\text{ft})] = 0.297 \text{ ft.}$$

The helical buckling tubing movement is given by

$$\Delta L_b = -(1.61 \text{ in.})^2(30,000^2 - 19,626^2 \text{ lbf}^2)$$
$$/[(8)(30 \times 10^6 \text{ psi})(1.611 \text{ in.}^4)(2.78 \text{ lbf}/\text{ft})] = \quad \text{ft.}$$

The total tubing movement is 0.297 ft plus 1.242 ft, which equals 1.539 ft. Pure helical buckling produces the length change,

$$\Delta L_b = -(1.61 \text{ in.})^2(30,000^2\text{lbf}^2)$$
$$/[8)(30 \times 10^6 \text{ psi})(1.611 \text{ in.}^4)(2.78 \text{ lbf}/\text{ft})] = 2.170 \text{ ft.}$$

Tubing movement is a design consideration for packer selection. Seal length is an important criterion for tubing well completion design. The use of pure helical buckling produces a 41% error in the calculation of tubing movement. When designing seal length in a deviated well, use of pure helical buckling can produce significant error.

7.10 Loads on Casing and Tubing Strings

In order to evaluate a given casing design, a set of loads is necessary. Casing loads result from running the casing, cementing the casing, subsequent drilling operations, production and well workover operations. Casing loads are principally pressure loads, mechanical loads, and thermal loads. Pressure loads are produced by fluids within the casing, cement and fluids outside the casing, pressures imposed at the surface by drilling and workover operations, and pressures imposed by the formation during drilling and production. Mechanical loads are associated with casing hanging weight, shock loads during running, packer loads during production and workovers, and hanger loads. Temperature changes and resulting thermal expansion loads are induced in casing by drilling, production, and workovers, and these loads might cause buckling (bending stress) loads in uncemented intervals.

Next, we discuss casing loads that are typically used in preliminary casing design. However, each operating company usually has its own special set of design loads for casing, based on their experience. If you are designing a casing string for a particular company, this load information must be obtained from them. Because there are so many possible loads that must be evaluated, most casing design today is done with computer programs that generate the appropriate load sets (often custom tailored for a particular operator), evaluate the results, and sometimes even determine a minimum-cost design automatically.

7.11 External Pressure Loads

7.11.1 Pressure Distributions. Pressure distributions are typically used to model the external pressures in cemented intervals.

Mud/Cement Mix-Water. Fluid pressure is given by the mud gradient above the top-of-cement (TOC) and by the cement gradient below TOC.

Permeable Zones: Good Cement. Again, fluid pressure is given by the mud gradient above TOC and by the cement gradient below TOC. The exception is that formation pore pressure is imposed over the permeable zone interval. This pressure profile is discontinuous.

Permeable Zones: Poor Cement, High Pressure. In this case, the formation pore pressure is felt at the surface through the poor cement. This pressure profile is continuous with depth.

Permeable Zones: Poor Cement, Low Pressure. In this case, the mud surface drops so that the mud pressure equals the formation pressure. This pressure profile is continuous with depth.

Openhole Pore Pressure: TOC Inside Previous Shoe. In this case, fluid pressure is given by mud gradient above TOC, cement gradient to the shoe, and the minimum equivalent mud weight gradient of the openhole below the shoe. This pressure profile is not continuous with depth; it is discontinuous at the previous shoe.

Openhole Pore Pressure: TOC Below Previous Shoe, Without Mud Drop. In this case, fluid pressure is given by the mud gradient above TOC and by the minimum equivalent mud weight gradient of the openhole below the shoe. This pressure profile is not continuous with depth but is discontinuous at TOC.

TOC Below Previous Shoe, With Mud Drop. In this case, the mud surface drops so that the mud pressure equals the minimum equivalent mud weight gradient of the openhole at the TOC. This pressure profile is continuous with depth.

Above/Below TOC External Pressure Profile. In this case, fluid pressure is given by mud gradient above TOC, cement gradient to the shoe, and a specified pressure profile below a specified depth. This external pressure distribution may be discontinuous at the specified depth. If a pressure gradient is specified, the pressure profile may also be continuous at the specified depth.

7.12 Internal Pressure Loads

7.12.1 Pressure Distributions. Pressure distributions are typically used to model the internal pressures. These pressure distributions are discussed next.

Burst: Gas Kick. This load case uses an internal pressure profile, which is the envelope of the maximum pressures experienced by the casing while circulating out a gas kick using the driller's method. It should represent the worst-case kick to which the current casing can be exposed while drilling a deeper interval. Typically, this means taking a kick at the total depth (TD) of the next openhole section. If the kick intensity or volume cause the fracture pressure at the casing shoe to exceed, the kick volume is often reduced to the maximum volume that can be circulated out of the hole without exceeding the fracture pressure at the shoe. The maximum pressure experienced at any casing depth occurs when the top of the gas bubble reaches that depth.

Burst: Displacement to Gas. This load case uses an internal pressure profile consisting of a gas gradient extending upward from a formation pressure in a deeper hole interval or from the fracture pressure at the casing shoe. This pressure physically represents a well control situation, in which gas from a kick has completely displaced the mud out of the drilling annulus from the surface to the casing shoe. This is the worst-case drilling burst load that a casing string could experience, and if the fracture pressure at the shoe is used to determine the pressure profile, it ensures that the weak point in the system is at the casing shoe and not the surface. This, in turn, precludes a burst failure of the casing near the surface during a severe well-control situation.

Burst: Maximum Load Concept. This load case is a variation of the displacement-to-gas load case that has wide usage in the industry and is taught in several popular casing design schools. It has been used historically because it results in an adequate design (though typically quite conservative, particularly for wells deeper than 15,000 ft), and it is simple to calculate. The load case consists of a gas gradient extending upward from the fracture pressure at the shoe up to a mud/gas interface and then a mud gradient to the surface. The mud/gas interface is calculated in a number of ways—the most common being the "fixed endpoint" method. The interface is calculated on the basis of surface pressure typically equal to the BOP rating and the fracture pressure at the shoe and assuming a continuous pressure profile.

Burst: Lost Returns With Water. This load case models an internal pressure profile, which reflects pumping water down the annulus to reduce surface pressure during a well-control situation in which lost returns are occurring. The pressure profile represents a freshwater gradient applied upward from the fracture pressure at the shoe depth. A water gradient is used, assuming that the rig's barite supply has been depleted during the well-control incident. This load case typically dominates the burst design when compared to the gas-kick load case. This is particularly the case for intermediate casing.

Burst: Surface Protection. This load case is less severe than the displacement-to-gas criteria and represents a moderated approach to preventing a surface blowout during a well-control incident. It is not applicable to liners. The same surface pressure calculated in the "lost returns with water" load case is used, but in this load case, a gas gradient from this surface pressure is used to generate the rest of the pressure profile. This load case represents no actual physical scenario; however, when used with the gas-kick criterion, it ensures that the casing weak point is not at the surface. Typically, the gas-kick load case will control the design deep, and the surface-protection load case will control the design shallow, leaving the weak point somewhere in the middle.

Burst: Pressure Test. This load case models an internal pressure profile, which reflects a surface pressure applied to a mud gradient. The test pressure typically is based on the maximum anticipated surface pressure determined from the other selected burst load cases plus a suitable safety margin. For production casing, the test pressure is typically based on the anticipated shut-in tubing pressure. This load case may or may not dominate the burst design depending on the mud weight in the hole at the time the test occurs. The pressure test is normally performed prior to drilling out the float equipment.

Collapse: Cementing. This load case models an internal and external pressure profile, which reflects the collapse load imparted on the casing after the plug has been bumped during the cement job and the pump pressure bled off. The external pressure considers the mud hydrostatic column and different densities of the lead and tail cement slurries. The internal pressure is based on the gradient of the displacement fluid. If a light displacement fluid is used, the cementing collapse load can be significant.

Collapse: Lost Returns With Mud Drop. This load case models an internal pressure profile, which reflects a partial evacuation or a drop in the mud level because of the mud hydrostatic column equilibrating with the pore pressure in a lost-circulation zone. The heaviest mud weight used to drill the next openhole section should be used along with a pore pressure and depth that result in the largest mud drop. Many operators make the conservative assumption that the lost-circulation zone is at the TD of the next openhole section and is normally pressured. A partial evacuation of more than 5,000 ft, because of lost circulation during drilling, is normally not seen. Many operators use a partial evacuation criterion in which the mud level is assumed to be a percentage of the openhole TD.

Collapse: Other Load Cases. Full Evacuation. This load case should be considered when drilling with air or foam. It may also be considered for conductor or surface casing where shallow gas is encountered. This load case would represent all of the mud being displaced out of the wellbore (through the diverter) before the formation bridged off.

Water Gradient. For wells with a sufficient water supply, an internal pressure profile consisting of a freshwater or seawater gradient is sometimes used as a collapse criterion. This assumes a lost-circulation zone that can only withstand a water gradient.

Burst: Gas Migration (Subsea Wells). This load case models bottomhole pressure applied at the wellhead (subject to fracture pressure at the shoe) from a gas bubble migrating upward behind the production casing with no pressure bleedoff at the surface. The pressure is the minimum of the fracture pressure at the shoe and the reservoir pressure plus the mud gradient. The load case has application only to the intermediate casing in subsea wells where the operator has no means of accessing the annulus behind the production casing.

Burst: Tubing Leak. This load case applies to both production and injection operations and represents a high surface pressure on top of the completion fluid because of a tubing leak near the hanger. A worst-case surface pressure is usually based on a gas gradient extending upward from reservoir pressure at the perforations. If the proposed packer location has been determined when the casing is designed, the casing below the packer can be assumed to experience pressure, based on the produced fluid gradient and reservoir pressure only.

Burst: Injection Down Casing. This load case applies to wells that experience high-pressure annular injection operations such as a casing fracture stimulation job. The load case models a surface pressure applied to a static fluid column. This is analogous to a screenout during a frac job.

Collapse Above Packer: Full Evacuation. This severe load case has the most application in gas lift wells. It is representative of a gas filled annulus that loses injection pressure. Many operators use the full evacuation criterion for all production casing strings regardless of the completion type or reservoir characteristics.

Collapse Above Packer: Partial Evacuation. This load case is based on a hydrostatic column of completion fluid equilibrating with depleted reservoir pressure during a workover operation. Some operators do not consider a fluid drop but only a fluid gradient in the annulus above the packer. This is applicable if the final depleted pressure of the formation is greater than the hydrostatic column of a lightweight packer fluid.

Collapse Below Packer: Common Load Cases. Full Evacuation. This load case applies to severely depleted reservoirs, plugged perforations, or a large drawdown of a low-permeability reservoir. It is the most commonly used collapse criterion.

Fluid Gradient. This load case assumes zero surface pressure applied to a fluid gradient. A common application is the underbalanced fluid gradient in the tubing before perforating (or after if the perforations are plugged). It is a less conservative criterion for formations that will never be drawn down to zero.

Collapse: Gas Migration (Subsea Wells). This load case models bottomhole pressure applied at the wellhead (subject to fracture pressure at the prior shoe) from a gas bubble migrating upward behind the production casing with no pressure bleedoff at the surface. The pressure distribution is the minimum of the following two pressure distributions. The load case has application only in subsea wells where the operator has no means of accessing the annulus behind the production casing. An internal pressure profile consisting of a completion fluid gradient is typically used.

Collapse: Salt Loads. If a formation that exhibits plastic behavior, such as a salt zone, is to be isolated by the current string, then an equivalent external collapse load (typically taken to be the overburden pressure) should be superimposed on all of the collapse load cases from the top to the base of the salt zone.

Annular Pressure Buildup. In offshore wells with sealed annuli, increases in fluid temperatures caused by production will cause fluid expansion, resulting in increased fluid pressures. For instance, for water at 100°F, a 1°F increase in temperature will produce a pressure increase of 38 ksi in a rigid container. Fortunately, the casing and formation are sufficiently elastic to greatly reduce this pressure. The equilibrium pressure produced by thermal expansion must be calculated to balance fluid volume change with annular volume change. Nevertheless, the annular pressure change produced by thermal expansion has proved to be a serious design consideration, especially in the North Sea and in deep water.

7.13 Mechanical Loads

7.13.1 Changes in Axial Load. In tubing and over the free length of the casing above TOC, changes in temperatures and pressures will have the largest effect on the ballooning and temperature load components. The incremental forces, because of these effects, are given here.

$$\Delta F_{bal} = 2v\left(\Delta p_i A_i - \Delta p_o A_o\right) + vLg_c\left(\Delta \rho_i A_i - \Delta \rho_o A_o\right), \quad \text{..........................} \quad (7.47)$$

where

ΔF_{bal} = incremental force because of ballooning, lbf,

v = Poisson's ratio (0.30 for steel),

g_c = gravity constant, = 1 lbf/lbm,

Δp_i = change in surface internal pressure, psi,

Δp_o = change in surface external pressure, psi,

A_i = cross-sectional area associated with casing inside diameter (ID), in.,

A_o = cross-sectional area associated with casing outside diameter (OD), in.,

L = free length of casing, in.,

$\Delta \rho_i$ = change in internal fluid density, lbm/in.^{3},

and

$\Delta \rho_o$ = change in external fluid density, lbm/in.^{3}.

$$\Delta F_{temp} = -\alpha E A_s \Delta T, \quad \text{...} \quad (7.48)$$

where

ΔF_{temp} = incremental force because of temperature change, lbf,

α = thermal expansion coefficient (6.9 × 10^{-6} °F^{-1} for steel), °F^{-1},

E = Young's modulus (3.0 × 10^7 psi for steel), psi,

A_s = cross-sectional area of pipe, in.^{2},

and

ΔT = average change in temperature over free length, °F.

7.13.2 Axial: Running in Hole. This installation load case represents the maximum axial load that any portion of the casing string experiences when running the casing in the hole. It can include effects such as: self-weight; buoyancy forces at the end of the pipe and at each cross-sectional area change; wellbore deviation; bending loads superimposed in dogleg regions;

shock loads based on an instantaneous deceleration from a maximum velocity [this velocity is often assumed to be 50% greater than the average running speed (typically 2 to 3 ft/sec)]; and frictional drag (typically, the maximum axial load experienced by any joint in the casing string is the load when the joint is picked up out of the slips after being made up).

7.13.3 Axial: Overpull While Running. This installation load case models an incremental axial load applied at the surface while running the pipe in the hole. Casing designed using this load case should be able to withstand an overpull force applied with the shoe at any depth if the casing becomes stuck while running in the hole. Certain effects must be considered, such as self-weight; buoyancy forces at the end of the pipe and at each cross-sectional area change; wellbore deviation; bending loads superimposed in dogleg regions; frictional drag; and the applied overpull force.

7.13.4 Axial: Green Cement Pressure Test. This installation load case models applying surface pressure after bumping the plug during the primary cement job. Because the cement is still in its fluid state, the applied pressure will result in a large piston force at the float collar and often results in the worst-case surface axial load. The effects that should be considered are self-weight; buoyancy forces at the end of the pipe and at each cross-sectional area change; wellbore deviation; bending loads superimposed in dogleg regions; frictional drag; and piston force because of differential pressure across float collar.

7.13.5 Axial: Other Load Cases. *Air Weight of Casing Only.* This axial load criterion has been used historically because it is an easy calculation to perform, and it normally results in adequate designs. It still enjoys significant usage in the industry. Because a large number of factors are not considered, it is typically used with a high axial design factor (e.g., 1.6+).

Buoyed Weight Plus Overpull Only. Like the air weight criterion, this load case has wide usage because it is an easy calculation to perform. Because a large number of factors are not considered, it is typically used with a high axial design factor (e.g., 1.6+).

7.13.6 Axial: Shock Loads. Shock loads can occur if the pipe hits an obstruction or the slips close while the pipe is moving. The maximum additional axial force, because of a sudden deceleration to zero velocity, is given by the equation,

$$F_{shock} = \frac{v_{run} A_s}{12} \sqrt{\frac{E \rho_s}{g_c}}, \dots\dots\dots\dots\dots\dots\dots\dots\dots\dots\dots (7.49)$$

where
F_{shock} = shock loading axial force, lbf,
v_{run} = running speed, ft/sec,
A_s = pipe cross-sectional area, in.^{2},
E = Young's modulus for pipe, lbf/in.^{2},
ρ_s = density of pipe, lbm/ft^3,
and
g_c = gravity constant, ft/sec^2.
The shock load equation is often expressed as

$$F_{shock} = \frac{w_a}{g_c} v_{run} v_{sonic}, \dots\dots\dots\dots\dots\dots\dots\dots\dots\dots\dots (7.50)$$

where

w_a = pipe weight per unit length in air, lbm/ft,

and

v_{sonic} = speed of sound in pipe, ft/sec,

$$= \sqrt{\frac{144 E g_c}{\rho_s}}. \text{ (For steel, } v_{sonic} \text{ is 16,800 ft/sec.)}$$

For practical purposes, some operators specify an average velocity in this equation and multi-ply the result by a factor that represents the ratio between the peak and average velocities (typically 1.5).

7.13.7 Axial: Service Loads. For most wells, installation loads will control axial design. How-ever, in wells with uncemented sections of casing and where large pressure or temperature changes will occur after the casing is cemented in place, changes in the axial load distribution can be important because of effects such as self-weight; buoyancy forces; wellbore deviation; bending loads; changes in internal or external pressure (ballooning); temperature changes; and buckling.

7.13.8 Axial: Bending Loads. Stress at the pipe's OD because of bending can be expressed as

$$\sigma_b = \frac{ED}{2R}, \quad\text{..} \quad (7.51)$$

where

σ_b = stress at the pipe's outer surface, psi,

E = modulus of elasticity, psi,

D = nominal outside diameter, in.,

and

R = radius of curvature, in.

This bending stress can be expressed as an equivalent axial force as

$$F_{bnd} = \frac{E\pi}{360} D\left(\frac{\alpha}{L}\right) A_s, \quad\text{..} \quad (7.52)$$

where

F_{bnd} = axial force because of bending, lbf,

α/L = dogleg severity (°/unit length),

and

A_s = cross-sectional area, in.^{2}.

The bending load is superimposed on the axial load distribution as a local effect.

7.14 Thermal Loads and Temperature Effects

In shallow normal-pressured wells, temperature will typically have a secondary effect on tubu-lar design. In other situations, loads induced by temperature can be the governing criteria in the design. Next, we discuss how temperature can affect tubular design.

7.14.1 Temperature Effects on Tubular Design. *Annular Fluid Expansion Pressure.* Increases in temperature after the casing is landed can cause thermal expansion of fluids in sealed annuli and result in significant pressure loads. Most of the time, these loads need not be included in the design because the pressures can be bled off. However, in subsea wells, the outer annuli cannot be accessed after the hanger is landed. The pressure increases will also influence the axial load profiles of the casing strings exposed to the pressures because of ballooning effects.

Tubing Thermal Expansion. Changes in temperature will increase or decrease tension in the casing string because of thermal contraction and expansion, respectively. The increased axial load, because of pumping cool fluid into the wellbore during a stimulation job, can be the critical axial design criterion. In contrast, the reduction in tension during production, because of thermal expansion, can increase buckling and possibly result in compression at the wellhead.

Temperature Dependent Yield. Changes in temperature not only affect loads but also influence the load resistance. Because the material's yield strength is a function of temperature, higher wellbore temperatures will reduce the burst, collapse, axial, and triaxial ratings of the casing.

Sour Gas Well Design. In sour environments, operating temperatures can determine what materials can be used at different depths in the wellbore.

Tubing Internal Pressures. Produced temperatures in gas wells will influence the gas gradient inside the tubing because gas density is a function of temperature and pressure.

7.15 Casing Design
To design a casing string, one must know the purpose of the well, the geological cross section, available casing and bit sizes, cementing and drilling practices, rig performance, as well as safety and environmental regulations. To arrive at the optimal solution, the design engineer must consider casing as a part of a whole drilling system. A brief description of the elements involved in the design process is presented next.

7.16 Design Objectives
The engineer responsible for developing the well plan and casing design is faced with a number of tasks that can be briefly characterized.

• Ensure the well's mechanical integrity by providing a design basis that accounts for all the anticipated loads that can be encountered during the life of the well.

• Design strings to minimize well costs over the life of the well.

• Provide clear documentation of the design basis to operational personnel at the well site. This will help prevent exceeding the design envelope by application of loads not considered in the original design.

While the intention is to provide reliable well construction at a minimum cost, at times failures occur. Most documented failures occur because the pipe was exposed to loads for which it was not designed. These failures are called "off-design" failures. "On-design" failures are rather rare. This implies that casing-design practices are mostly conservative. Many failures occur at connections. This implies that either field makeup practices are not adequate or the connection design basis is not consistent with the pipe-body design basis.

7.17 Design Method

7.17.1 Phases of Design Process. The design process can be divided into two distinct phases.

Preliminary Design. Typically the largest opportunities for saving money are present while performing this task. This design phase includes data gathering and interpretation; determination of casing shoe depths and number of strings; selection of hole and casing sizes; mud-weight design; and directional design. The quality of the gathered data will have a large impact on the appropriate choice of casing sizes and shoe depths and whether the casing design objective is successfully met.

Detailed Design. The detailed design phase includes: Selection of pipe weights and grades for each casing string. Connection selection. The selection process consists of comparing pipe ratings with design loads and applying minimum acceptable safety standards (i.e., design factors). A cost-effective design meets all the design criteria with the least expensive available pipe.

7.18 Required Information

The items listed next are a checklist, which is provided to aid the well planners/casing designers in both the preliminary and detailed design.

  • Formation properties: pore pressure; formation fracture pressure; formation strength (borehole failure); temperature profile; location of squeezing salt and shale zones; location of permeable zones; chemical stability/sensitive shales (mud type and exposure time); lost-circulation zones, shallow gas; location of freshwater sands; and presence of H_2S and/or CO_2.

  • Directional data: surface location; geologic target(s); and well interference data.

  • Minimum diameter requirements: minimum hole size required to meet drilling and production objectives; logging tool OD; tubing size(s); packer and related equipment requirements; subsurface safety valve OD (offshore well); and completion requirements.

  • Production data: packer-fluid density; produced-fluid composition; and worst-case loads that might occur during completion, production, and workover operations.

  • Other: available inventory; regulatory requirements; and rig equipment limitations.

7.19 Preliminary Design

The purpose of preliminary design is to establish casing and corresponding drill-bit sizes, casing setting depths and, consequently, the number of casing strings. Casing program (well plan) is obtained as a result of preliminary design. Casing program design is accomplished in three major steps. First, mud program is prepared; second, the casing sizes and corresponding drill-bit sizes are determined; and next, the setting depths of individual casing strings are found.

7.19.1 Mud Program. The most important mud program parameter used in casing design is the "mud weight." The complete mud program is determined from: pore pressure; formation strength (fracture and borehole stability); lithology; hole cleaning and cuttings transport capability; potential formation damage, stability problems, and drilling rate; formation evaluation requirement; and environmental and regulatory requirements.

7.19.2 Hole and Pipe Diameters. Hole and casing diameters are based on the requirements discussed next.

Production. The production equipment requirements include tubing; subsurface safety valve; submersible pump and gas lift mandrel size; completion requirements (e.g., gravel packing); and weighing the benefits of increased tubing performance of larger tubing against the higher cost of larger casing over the life of the well.

Evaluation. Evaluation requirements include logging interpretation and tool diameters.

Drilling. Drilling requirements include a minimum bit diameter for adequate directional control and drilling performance; available downhole equipment; rig specifications; and available BOP equipment.

These requirements normally impact the final hole or casing diameter. Because of this, casing sizes should be determined from the inside outward starting from the bottom of the hole. Usually the design sequence is as described next.

Based upon reservoir inflow and tubing intake performance, proper tubing size is selected. Then, the required production casing size is determined considering completion requirements. Next, the diameter of the drill bit is selected for drilling the production section of the hole considering drilling and cementing stipulations. Next, one must determine the smallest casing through which the drill bit will pass, and the process is repeated. Large cost savings are possible by becoming more aggressive (using smaller clearances) during this portion of the preliminary design phase. This has been one of the principal motivations in the increased popularity of slimhole drilling. Typical casing and rock bit sizes are given in **Table 7.10**.

Casing Shoe Depths and the Number of Strings. Following the selection of drillbit and casing sizes, the setting depth of individual casing strings must be determined. In conventional

TABLE 7.10—COMMONLY USED BIT SIZES THAT WILL PASS THROUGH API CASING				
Casing Size, OD, in.	Weight/ft, lbm/ft	ID, in.	Drift Diameter, in.	Commonly Used Bit Sizes, in.
4 1/2	9.5	4.090	3.965	3 7/8
	10.5	4.052	3.927	
	11.6	4.000	3.875	
	13.5	3.920	3.795	3 3/4
5	11.5	4.560	4.435	4 1/4
	13.0	4.494	4.369	
	15.0	4.408	4.283	
	18.0	4.276	4.151	3 7/8
5 1/2	13.0	5.044	4.919	4 3/4
	14.0	5.012	4.887	
	15.5	4.950	4.825	
	17.0	4.892	4.764	
	20.0	4.778	4.653	4 5/8
	23.0	4.670	4.545	4 1/4
6 5/8	17.0	6.135	6.010	6
	20.0	6.049	5.924	5 5/8
	24.0	5.921	5.796	
	28.0	5.791	5.666	
	32.0	5.675	5.550	4 3/4
7	17.00	6.538	6.413	6 1/4
	20.00	6.456	6.331	
	23.00	6.366	6.241	
	26.00	6.276	6.151	6 1/8
	29.00	6.184	6.059	6
	32.00	6.094	5.969	
	35.00	6.006	5.879	
	38.00	5.920	5.795	5 5/8
7 5/8	20.00	7.125	7.000	6 3/4
	24.00	7.025	6.900	
	26.40	6.969	6.844	
	29.70	6.875	6.750	
	33.70	6.765	6.640	6 1/2
	39.00	6.625	6.500	
8 5/8	24.00	8.097	7.972	7 7/8
	28.00	8.017	7.892	
	32.00	7.921	7.796	6 3/4
	36.00	7.825	7.700	
	40.00	7.725	7.600	
	44.00	7.625	7.500	
	49.00	7.511	7.386	

rotary drilling operations, the setting depths are determined principally by the mud weight and the fracture gradient, as schematically depicted in **Fig. 7.5**, which is sometimes called a well plan. Equivalent mud weight (EMW) is pressure divided by true vertical depth and converted to units of lbm/gal. EMW equals actual mud weight when the fluid column is uniform and static. First, pore and fracture gradient lines must be drawn on a well-depth vs. EMW chart. These are the solid lines in Fig. 7.5. Next, safety margins are introduced, and broken lines are drawn, which establish the design ranges. The offset from the predicted pore pressure and fracture gradient nominally accounts for kick tolerance and the increased equivalent circulating density (ECD) during drilling. There are two possible ways to estimate setting depths from this figure.

Bottom-Up Design. This is the standard method for casing seat selection. From Point A in Fig. 7.5 (the highest mud weight required at the total depth), draw a vertical line upward to Point B. A protective 7⅝-in. casing string must be set at 12,000 ft, corresponding to Point B,

TABLE 7.10—COMMONLY USED BIT SIZES THAT WILL PASS THROUGH API CASING (continued)				
$9^5/_8$	29.30	9.063	8.907	$8^3/_4$, $8^1/_2$
	32.30	9.001	8.845	
	36.00	8.921	8.765	
	40.00	8.835	8.679	$8^5/_8$, $8^1/_2$
	43.50	8.755	8.599	
	47.00	8.681	8.525	$8^1/_2$
	53.50	8.535	8.379	$7^7/_8$
$10^3/_4$	32.75	10.192	10.036	$9^7/_8$
	40.50	10.050	9.894	
	45.50	9.950	9.794	$9^5/_8$
	51.00	9.850	9.694	
	55.00	9.760	9.604	
	60.70	9.660	9.504	$8^3/_4$, $8^1/_2$
	65.37	9.560	9.404	$8^3/_4$, $8^1/_2$
$11^3/_4$	38.00	11.154	10.994	11
	42.00	11.084	10.928	$10^5/_8$
	47.00	11.000	10.844	
	54.00	10.880	10.724	
	60.00	10.772	10.616	
$13^3/_8$	48.00	12.715	12.559	$12^1/_4$
	54.50	12.615	12.459	
	61.00	12.515	12.359	
	68.00	12.415	12.259	
	72.00	12.347	12.191	11
16	55.00	15.375	15.188	15
	65.00	15.250	15.062	
	75.00	15.125	14.939	$14^3/_4$
	84.00	15.010	14.822	
	109.00	14.688	14.500	
$18^5/_8$	87.50	17.755	17.567	$17^1/_2$
20	94.00	19.124	18.936	$17^1/_2$

to enable safe drilling on the section AB. To determine the setting depth of the next casing, draw a horizontal line BC and then a vertical line CD. In such a manner, Point D is determined for setting the 9⅝-in. casing at 9,500 ft. The procedure is repeated for other casing strings, usually until a specified surface casing depth is reached.

Top-Down Design. From the setting depth of the 16-in. surface casing (here assumed to be at 2,000 ft), draw a vertical line from the fracture gradient dotted line, Point A, to the pore pressure dashed line, Point B. This establishes the setting point of the 11¾-in. casing at about 9,800 ft. Draw a horizontal line from Point B to the intersection with the dotted frac gradient line at Point C; then, draw a vertical line to Point D at the pore pressure curve intersection. This establishes the 9⅝-in. casing setting depth. This process is repeated until bottom hole is reached.

There are several things to observe about these two methods. First, they do not necessarily give the same setting depths. Second, they do not necessarily give the same number of strings. In the top-down design, the bottomhole pressure is missed by a slight amount that requires a short 7-in. liner section. This slight error can be fixed by resetting the surface casing depth. The top-down method is more like actually drilling a well, in which the casing is set when necessary to protect the previous casing shoe. This analysis can help anticipate the need for additional strings, given that the pore pressure and fracture gradient curves have some uncertainty associated with them.

In practice, a number of regulatory requirements can affect shoe depth design. These factors are discussed next.

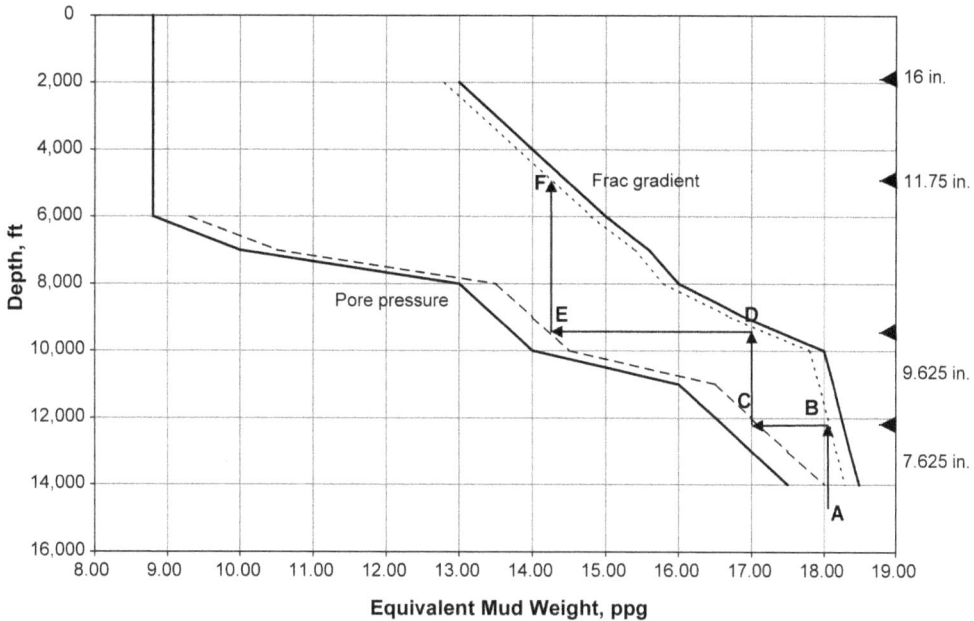

Fig. 7.5—Casing setting depths—bottom-up design.

Hole Stability. This can be a function of mud weight, deviation and stress at the wellbore wall, or can be chemical in nature. Often, hole stability problems exhibit time-dependent behavior (making shoe selection a function of penetration rate). The plastic flowing behavior of salt zones must also be considered.

Differential Sticking. The probability of becoming differentially stuck increases with increasing differential pressure between the wellbore and formation, increasing permeability of the formation, and increasing fluid loss of the drilling fluid (i.e., thicker mudcake).

Zonal Isolation. Shallow freshwater sands must be isolated to prevent contamination. Lost-circulation zones must be isolated before a higher-pressure formation is penetrated.

Directional Drilling Concerns. A casing string is often run after an angle building section has been drilled. This avoids keyseating problems in the curved portion of the wellbore because of the increased normal force between the wall and the drillpipe.

Uncertainty in Predicted Formation Properties. Exploration wells often require additional strings to compensate for the uncertainty in the pore pressure and fracture gradient predictions.

Another approach that could be used for determining casing setting depths relies on plotting formation and fracturing pressures vs. hole depth, rather than gradients, as shown in **Fig. 7.6** and Fig. 7.5. This procedure, however, typically yields many strings and is considered to be very conservative. See the chapter on geoscience principles in this volume of the handbook.

The problem of choosing the casing setting depths is more complicated in exploratory wells because of shortage of information on geology, pore pressures, and fracture pressures. In such a situation, a number of assumptions must be made. Commonly, the formation pressure gradient is taken as 0.54 psi/ft for hole depths less then 8,000 ft and taken as 0.65 psi/ft for depths greater than 8,000 ft. Overburden gradients are generally taken as 0.8 psi/ft at shallow depth and as 1.0 psi/ft for greater depths.

TOC Depths. TOC depths for each casing string should be selected in the preliminary design phase because this selection will influence axial load distributions and external pressure profiles used during the detailed design phase. TOC depths are typically based on zonal isola-

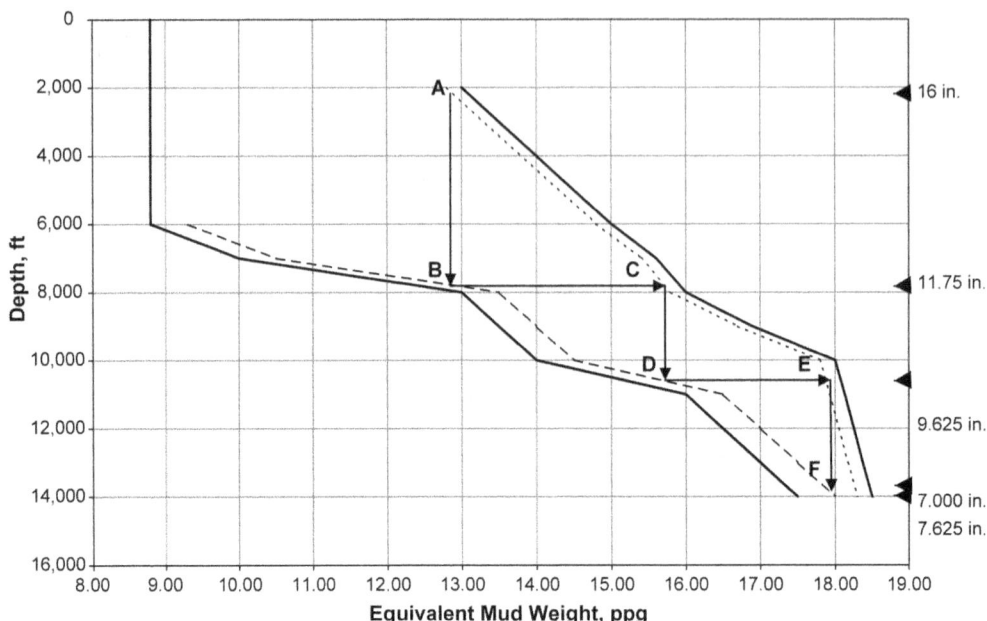

Fig. 7.6—Casing setting depths—top-down design.

tion; regulatory requirements; prior shoe depths; formation strength; buckling; and annular pressure buildup in subsea wells. Buckling calculations are not performed until the detailed design phase. Hence, the TOC depth may be adjusted, as a result of the buckling analysis, to help reduce buckling in some cases.

Directional Plan. For casing design purposes, establishing a directional plan consists of determining the wellpath from the surface to the geological targets. The directional plan influences all aspects of casing design including mud weight and mud chemistry selection for hole stability, shoe seat selection, casing axial load profiles, casing wear, bending stresses, and buckling. It is based on factors that include geological targets; surface location; interference from other wellbores; torque and drag considerations; casing wear considerations; bottomhole assembly [(BHA) an assembly of drill collars, stabilizers, and bits]; and drill-bit performance in the local geological setting.

To account for the variance from the planned build, drop, and turn rates, which occur because of the BHAs used and operational practices employed, higher doglegs are often superimposed over the wellbore. This increases the calculated bending stress in the detailed design phase.

7.20 Detailed Design

7.20.1 Load Cases. In order to select appropriate weights, grades, and connections during the detailed design phase using sound engineering judgment, design criteria must be established. These criteria normally consist of load cases and their corresponding design factors that are compared to pipe ratings. Load cases are typically placed into categories that include burst loads; drilling loads; production loads; collapse loads; axial loads; running and cementing loads; and service loads.

7.20.2 Design Factors. In order to make a direct graphical comparison between the load case and the pipe's rating, the DF must be considered.

$$DF = SF_{min} \leq SF = \frac{pipe\ rating}{applied\ load}, \quad\text{.. (7.53)}$$

where
  DF = design factor (the minimum acceptable safety factor), and
  SF = safety factor.
  It follows that

$$DF \times (applied\ load) \leq pipe\ rating \quad\text{.. (7.54)}$$

Hence, by multiplying the load by the DF, a direct comparison can be made with the pipe rating. As long as the rating is greater than or equal to the modified load (which we will call the design load), the design criteria have been satisfied.

7.20.3 Other Considerations. After performing a design based on burst, collapse and axial considerations, an initial design is achieved. Before a final design is reached, design issues (connection selection, wear, and corrosion) must be addressed. In addition, other considerations can also be included in the design. These considerations are triaxial stresses because of combined loading (e.g., ballooning and thermal effects)—this is often called "service life analysis"; other temperature effects; and buckling.

7.21 Sample Design Calculations
In the examples that are discussed next, burst, collapse, and uniaxial tension failure criteria are examined. Triaxial stresses are calculated for a variety of load situations to demonstrate how the casing strength formulas and the load formulas are actually used.

Example 7.1: Sample Burst Calculation With Triaxial Comparison. Assume that we have a 13⅜-in., 72-lbm/ft N-80 intermediate casing set at 9,000 ft and cemented to surface. The burst differential pressure for this casing is given by Eq. 7.1.

$$\Delta P = 0.875(2)(80,000\ psi)(0.515\ in.)/(13.375\ in) = 5,380\ psi.$$

The load case we will test against is the burst displacement-to-gas case, with formation pressure of 6,000 psi, formation depth at 12,000 ft, and gas gradient equal to 0.1 psi/ft.

  Surface internal pressure = 6,000 psi − 0.1 psi/ft (12,000 ft = 4,800 psi.
  Surface external pressure = 0.
  Net pressure differential = 4,800 psi.

According to this calculation, the casing is strong enough to resist this burst pressure. As an additional test, let us calculate the von Mises stress associated with this case. Surface axial stress is the casing weight divided by the cross-sectional area (20.77 in.^{2}) less pressure loads when cemented (assume 15 lbm/gal cement).

$$\sigma_z = (72\ lbm/ft)(9,000\ ft)/(20.77\ in.^2) - (15\ lbm/gal)(.052\ psi/lbm/gal)(9,000\ ft)$$

  $$= 24,182\ psi\ (tensile\ stresses\ are\ positive\ by\ convention).$$

The radial stresses for the internal and external radii are the internal and external pressures.

σ_{ri} = $-$4,800 psi (pressures are compressive stresses and negative by convention).

σ_{ro} = 0 psi.

The hoop stresses are calculated by the Lamé formula (Eq. 7.10).

$\sigma_{\theta i}$ = (4,800 psi[(6.688 in.)^2 + (6.174 in.)^2]/[(6.688 in.)^2 − (6.174 in.)^2] = 60,152 psi.

$\sigma_{\theta o}$ = (4,800 psi)(2)(6.174 in.)^2/[(6.688 in.)^2 − (6.174 in.)^2] = 55,352 psi.

The von Mises equivalent stress or triaxial stress is given as Eq. 7.9. Evaluating Eq. 7.9 at the inside radius and at the outside radius, we have

$$\sigma_{VMI} = \sqrt{\{[(0 - 24,182 \text{ psi})^2 + (24,182 - 60,152 \text{ psi})^2 + (60,152 - 0 \text{ psi})^2]/2\}}$$
$$= 52,426 \text{ psi,}$$

and

$$\sigma_{VMO} = \sqrt{\{[(-4,800 - 24,182 \text{ psi})^2 + (24,182 - 55,352 \text{ psi})^2 + (55,352 + 4,800 \text{ psi})^2]/2\}}$$
$$= 47,905 \text{ psi.}$$

The maximum von Mises stress is at the inside of the 13 3/8-in. casing with a value that is 66% of the yield stress. In the burst calculation, the applied pressure was 89% of the calculated burst pressure. Thus, the burst calculation is conservative compared to the von Mises calculation for this case.

Example 7.2: Sample Collapse Calculation. For the sample collapse calculation, we will test the collapse resistance of a 7-in., 23-lbm/ft P-110 liner cemented from 8,000 to 12,000 ft. Comparing the 7-in. liner properties against the various collapse regimes, it was found that transition collapse was predicted for this liner. The collapse pressure for this liner is calculated from Eq. 7.4 with the following values for F and G, as taken from Table 7.5.

$$F = 2.066, \quad \text{and} \quad G = 0.0532.$$

The collapse pressure is then given by

$$p_c = (110,000 \text{ psi})(2.066/(22.08) - (0.0532) = 4,440 \text{ psi.}$$

To evaluate the collapse of this liner, we need internal and external pressures. Internal pressure is determined with the full evacuation above packer.

$$p_i = 0.1 \text{ psi/ft} \ (12,000 \text{ ft}) = 1,200 \text{ psi.}$$

The external pressure is based on a fully cemented section behind the 7-in. liner. The external pressure profile is given by the mud/cement mix-water external pressure profile where the liner is assumed to be cemented in 10-lbm/gal mud with an internal mix-water pressure gradient of 0.45.

$$P_o = (10 \text{ lbm/gal}) (.052 \text{ psi/ft/lbm/gal ft}) + 0.45 \text{ psi/ft} (12,000 - 8,000 \text{ ft})$$
$$= 5,960 \text{ psi.}$$

An equivalent pressure is calculated from p_i and p_o for comparison with the collapse pressure, p_c, through use of Eq. 7.6.

$$p_e = 5,960 \text{ psi} - (1 - 2/22.08)(1,200 \text{ psi}) = 4,869 \text{ psi.}$$

Because p_e exceeds p_c (4,440 psi), the liner is predicted to collapse. It is not appropriate to calculate a von Mises stress for collapse in this case because collapse in the transitional region is not strictly a plastic yield condition.

Example 7.3: Sample Uniaxial Tension Calculation. For this example, consider a 9⅝-in. 43.5-lbm/ft N-80 production casing in an 11,000-ft vertical well, with TOC at 8,000 ft. The casing is run in 11-lbm/gal water-based mud. The hanging weight in air for the casing is

$$F_{air} = 43.5 \text{ lbm/ft} (11,000 \text{ ft}) = 478,500 \text{ lbm.}$$

The casing stress at the surface is F_{air} divided by the cross-sectional area of the casing, less the hydrostatic pressure at the bottom of the casing when cemented. If we assume 15-lbm/gal cement and 11-lbm/gal displaced mud, this bottomhole pressure is

$$p_{bh} = 11 \text{ lbm/gal} (.052 \text{ psi/ft/lbm/gal}) (8,000 \text{ ft})$$
$$+(15 \text{ lbm/gal}) (.052 \text{ psi/ft/lbm/gal}) (11,000 - 8,000 \text{ ft})$$
$$= 6,916 \text{ psi.}$$

Therefore, the surface hanging stress is

$$\sigma_z = 478,500 \text{ lbm}/(12.56 \text{ in.}^2) - 6,916 \text{ psi} = 31,181 \text{ psi.}$$

For N-80 casing, a stress of 31,181 psi leaves a large margin of safety. Next, consider the effects of a stimulation treatment on this surface stress. Assume that the average temperature change in the 0–8,000-ft interval is –50°F. The change in axial stress, because of this temperature increase, is given by Eq. 7.48.

$$\Delta\sigma_z = -\alpha E \Delta T,$$

where α is the coefficient of thermal expansion ($6.9 \times 10^6/°F$ for steel) and E is Young's modulus (30×10^6 psi for steel). The net surface stress in the casing is

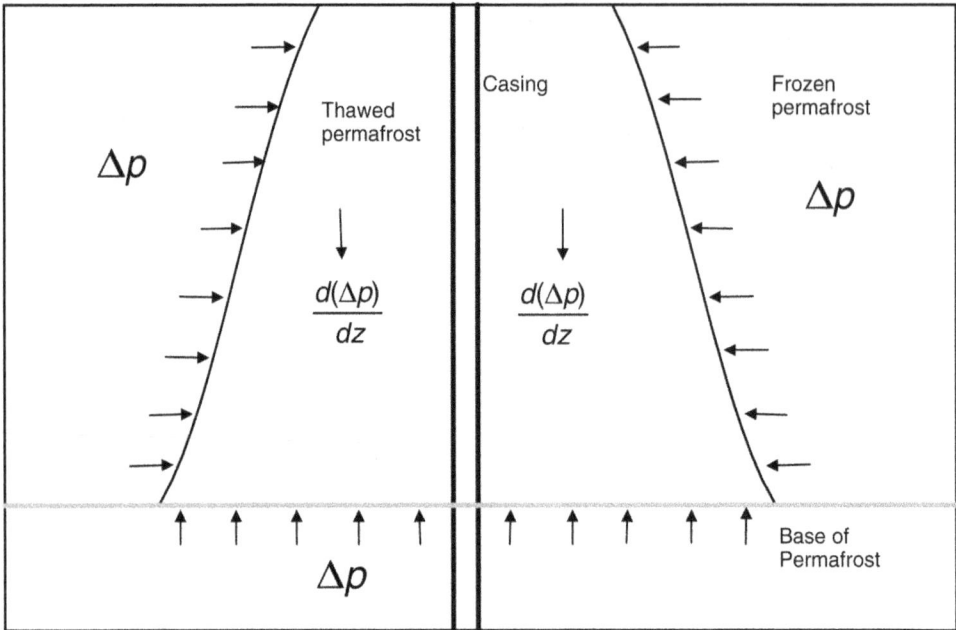

Fig. 7.7—Loading mechanisms in thawed permafrost.

$$\sigma_z = 31,181 \text{ psi} - (6.9 \times 10^6 / {}^\circ\text{F})(30 \times 10^6 \text{ psi})(-50{}^\circ\text{F})$$
$$= 41,531 \text{ psi}.$$

7.22 Arctic Well Completions

The surface formations in the Arctic, called permafrost, may be frozen to depths in excess of 2,000 ft. In addition to addressing concerns about the freezing of water-based fluids and cement, the engineer must also design surface casing for the unique loads generated by the thawing and refreezing of the permafrost. There are also road and foundation design problems, associated with ice-rich surface permafrost, that are not addressed here.

The following is a qualitative description of the loading mechanism in permafrost. If we consider a block of permafrost before thaw, the overburden and lateral earth pressures surrounding this block are balanced by the intergranular stresses between the soil panicles and the pore pressure in the ice. Upon thaw, the ice changes to water; the volume of the pore fluid decreases by about 9%; and the pore pressure decreases. To maintain equilibrium, the soil compacts, increasing intergranular forces until a new stress state is reached that balances the surrounding earth pressures.

The loading of the permafrost is the pore-pressure change caused by the phase change of the pore ice, illustrated in **Fig. 7.7.** The pore pressure is discontinuous at the thaw boundary and equal to Δp. Associated with the thaw is a body force or "gravity like" loading caused by the gradient of the pore-pressure change. This loading is equivalent to the loss of the buoyant pressure of the ice on the soil particles.

The mechanical response of the permafrost to the pore-pressure loads determines the casing loads. Experiments on simulated deep-frozen permafrost show that it can be characterized as a linear, isotropic elastic material with coefficients corresponding to the compressibility, C, and shear modulus, G. These moduli are functions of the mean normal effective stress, the soil type, and the degree of consolidation of the soil.

Determining the pore-pressure loading requires knowledge of the pore pressure before and after thaw. Thaw subsidence and freeze-back field tests at Prudhoe Bay suggest that the initial pore pressure is hydrostatic. The following mechanisms influence the final pore pressure. First, water may flow into the thawed zone from the surface, the base of the permafrost, or horizontally through the permafrost. Second, water may flow from one part of the thaw zone to another. Third, dissolved or trapped gases within the frozen ice may evolve and maintain some pressure upon thaw. Finally, the soil may compact so that the pore spaces are no longer undersaturated. If the compaction is sufficient to remove voidage and recompress the pore water, then the pressure within the pore space will rise. This limiting compaction is particularly important near the base of the permafrost, where the permafrost contains initially unfrozen water. Unfrozen water leads to a smaller amount of voidage upon thaw; hence, compaction and repressurization occur at lower soil strains. Unfrozen water may occur as a result of the effects of salinity and of adsorption in fine-grained materials. These effects not only depress the initial freezing point but also cause freezing to occur over a range of several degrees.

7.22.1 Internal Freeze-Back. Most of the discovery wells in the Prudhoe Bay field were lost because of the freezing of annular fluids. This failure mode is called internal freeze-back, to distinguish it from the refreezing of the permafrost, called external freeze-back. The solution to internal freeze-back is to replace freezeable fluids in the annuli with nonfreezeable fluids, such as oil-based fluids or alcohol-based fluids, such as glycol. Complete displacement of water-based fluids is essential for successful mitigation of internal freeze-back.

7.22.2 Permafrost Cementing. Experience has shown that a cement system used for permafrost cementing must meet a minimum set of requirements:
- Provide an ample thickening time.
- The ability to set at bottomhole temperatures without requiring external heat.
- The ability to set with a low heat of hydration.
- Provide an acceptable WOC time.
- The ability to set without freezing.
- The ability to attain adequate compressive strength for the well conditions.
- Provide stability to freeze/thaw cycling.

Other desirable qualities of a permafrost cement system include:
- The ability to be bulk blended and easily handled by field equipment and personnel.
- Provide controlled rheology.
- Provide the ability to be easily mixed in a continuous process at Arctic temperatures.
- Have no free water.

As with any cementing system, once the slurry is in place, the major consideration of system design becomes long-term performance of the cement. In permafrost cementing, considerations are compressive strength development and stability to freeze/thaw cycling.

Experience with permafrost cementing has shown the value of using high-alumina cements for this application. A high-alumina cement marketed under the name of Ciment Fondu has been used extensively in Arctic/North Slope operations.

Through the use of chemical extenders and freeze depressants, a high-alumina cement can be used to make a permafrost cement system. The system exhibits heat of hydration high enough to enhance the setting process. However, the large quantity of water in an extended system absorbs heat generated during hydration, eliminating the need for fly ash.

A high-alumina cement cannot be blended with Portland cement because blending the two causes extreme acceleration of the high-alumina cement, resulting in severe gelation or "flash" setting. Operators must use extreme caution to prevent contamination of a high-alumina cement system with Portland cement. The chance of contamination can be minimized with astringent cleaning of field bins, bulk trucks, and storage facilities before and after each job using a high-alumina cement system. However, under normal operations, it becomes almost impossible to eliminate the chance of alumina cement and Portland cement contacting each other.

A permafrost cementing system using Portland cement and appropriate cement additives eliminates the chance of this problem occurring. An extended Class G permafrost cement may offer the same performance as the high-alumina cement except that it is compatible with conventional permafrost tail-in cement systems, whereas the high-alumina cement is not. Another feature of extended Class G permafrost cement is superior compressive strength after freeze/thaw cycling. The extended Class G system eliminates the storage and handling problems previously associated with a high-alumina cement system. These attributes make an extended system using Class G Portland cement more cost effective than a high-alumina cement system.

7.22.3 External Freeze-Back. Drilling and production in the Arctic thaws the permafrost. If thawed permafrost is allowed to freeze back, significant collapse loads near the bottom of the permafrost will be generated and must be considered in casing design. The loading mechanism is associated with the phase-change expansion of pore water in the thawed permafrost. The magnitude of the pressure buildup depends on the mechanical response of the frozen permafrost.

The following analytic model was effective in predicting freeze-back pressures. The permafrost is initially thawed to radius r_b and then allowed to freeze back to radius r_a. These two radii serve to determine the amount of phase-change expansion at each instant in time. At the beginning of freeze-back, the thawed permafrost is nearly saturated because of vertical drainage, water influx from drilling fluids, and compaction of the soil structure. The freeze-back process occurs in three stages: relief of effective stress, elastic behavior, and elastic-yield behavior (see **Fig. 7.8**). In the first stage, as the pore water freezes, the ice expands into the fluid-filled pores, increasing the porosity and, at the same time, compressing the pore water. The grain size and permeability of the solids and the pressure conditions on the solids and fluids determine which of the two situations will occur. In either case, however, the pressure in the thawed zone will increase until the effective stress between grains is relieved and the material is fluidized (can no longer support shear). The freeze-back radius at which this occurs is denoted by r_e. Further freezing generates a zone of excess ice between r_e and r_a together with higher pressures within this zone. The second stage of the freeze-back process then occurs, as the frozen permafrost outside r_e is loaded and responds elastically. Elastic behavior continues until the third stage, when the stress in the permafrost reaches the yield point. A yielded region between r_p and r_e is created, as shown in Fig. 7.8 and grows as freeze-back proceeds. In the model, each of the three stages of freeze-back is treated as a separate boundary-value problem. The model predicts pressures along the entire length of casing through the permafrost at any instant in time during the freeze-back process.

This analytical freeze-back model and its correlation with freeze-back field-test data from Prudhoe Bay yielded the conclusions that are listed next.

• The 13-in., 72-lbm/ft N-80 casing used in the field test and commonly used at Prudhoe Bay can safely withstand the maximum freeze-back pressures.

• For freeze-back from large thaw radii (50 ft of production thaw), the maximum pressure is not significantly greater than that for freeze-back from small radii (3 ft of drilling thaw).

• The maximum freeze-back pressure depends on the elastic and yield properties of permafrost but is most sensitive to the Young's modulus of frozen permafrost.

• Based on laboratory studies and supported by field-test data, the creep or viscoelastic behavior of permafrost subject to freeze-back is negligible compared with the purely elastic and yield behavior.

• To limit the freeze-back pressure, the model is useful in the design of methods to limit the amount of initial thaw or to limit the extent of freeze-back.

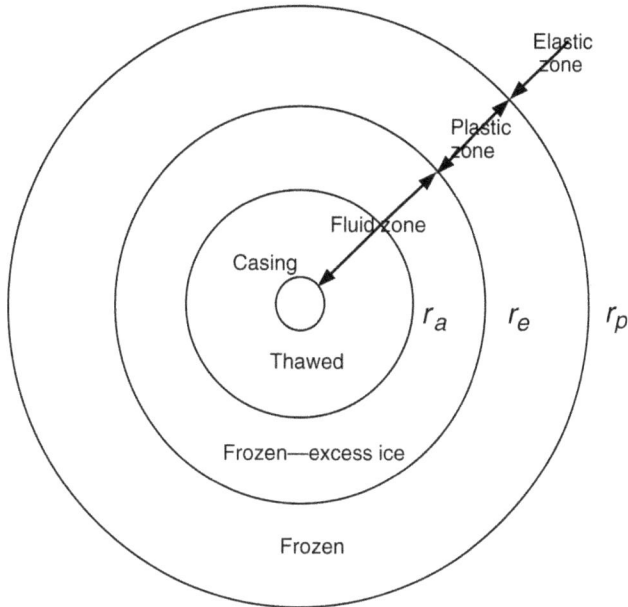

Fig. 7.8—Freeze-back model formulation.

7.22.4 Thaw Subsidence. Thaw subsidence is the soil compaction resulting from the thawing of permafrost by a producing oil well. Thaw subsidence should be considered in well design because of the strains induced on well casing by this compaction. Thaw-subsidence effects are influenced considerably by the geometry of the thawed zone. A typical thaw zone is roughly cylindrical and, even after 20 years of production, the radius of this cylinder is less than 2% of the length. The consequences of this geometry are that one-dimensional, vertical compaction is not applicable and that the full 3D geometry must be considered in the analysis. Further, the permafrost loading illustrated in Fig. 7.7 shows radial inward loading applied to the surface of the thawed zone. Thus, any resulting compaction of the permafrost should be predominantly in the radial direction with the gravity like loads carried by the arching support of the surrounding permafrost.

The lateral loading produced some very interesting effects in the thaw-subsidence field test. From 400 to 1,300 ft, the measured strains along the casing alternated between compression and tension. In **Fig. 7.9**, the alternating strain behavior is explained in terms of layering in the permafrost. A sand layer is bounded above and below by a fine silt layer. As the pore pressure decreases in the thawed zone, the thawed/frozen interface moves inward and the sand layer, which is relatively incompressible compared with the silt layers, elongates along the casing, at the expense of the compressible silts, which contract. The casing experiences tension adjacent to the elongating sands and compression opposite the contracting silts.

Another interesting effect occurs at the base of the permafrost. Below the base, the casing experiences tension, while above the base, the casing experiences compression; this indicates uplifting of the permafrost base. The decrease in pore pressure (as shown in Fig. 7.7) not only causes the thawed/frozen interface to move inward but also causes the permafrost base to move upward.

Thaw-subsidence strain is the most difficult arctic well design quantity to evaluate. The problem is complex and very dependent on lithology and permafrost mechanical properties. On the basis of numerous sensitivity studies, Prudhoe Bay operators developed "bounding curves" for tensile and compressive thaw-subsidence strains. At Prudhoe Bay, for single wells assuming

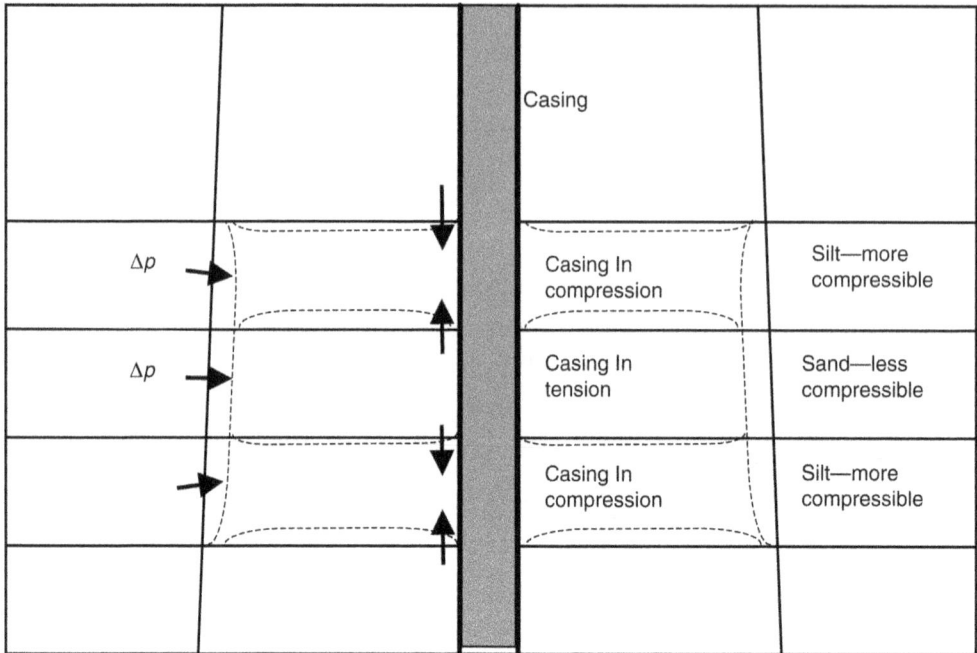

Fig. 7.9—Layering produces alternating tension and compression.

no thaw interference from adjacent wells, calculations give upper-bound tensile strains of 0.5% and upper-bound compressive strains of 0.7 to 0.9%, depending on production variables.

Calculated maximum strains are much higher than those measured in the ARCO/Exxon field test. Maximum field-test strains are 0.08% tension and 0.13% compression. The principal reason for this difference is that the field test did not have a worst-case lithology near the permafrost base where loading mechanisms are greatest. Recall that sand/silt layering is required for maximum strain generation, which was not present at depth in the field test. These values are considerably less than 13⅜-in. L-80 buttress-casing strain limits. Safety factors are 2.3 in compression and 8.8 in tension.

7.23 Risk-Based Casing Design

7.23.1 Introduction. Oilfield tubulars have been traditionally designed using a deterministic working stress design (WSD) approach, which is based on multipliers called safety factors (SFs). The primary role of a safety factor is to account for uncertainties in the design variables and parameters, primarily the load effect and the strength or resistance of the structure. While based on experience, these factors give no indication of the probability of failure of a given structure, as they do not explicitly consider the randomness of the design variables and parameters. Moreover, the safety factors tend to be rather conservative, and most limits of design are established using failure criteria based on elastic theory. In contrast, reliability-based approaches are probabilistic in nature and explicitly identify all the design variables and parameters that determine the load effect and strength of the structure. Moreover, they use a limit-states approach to the design of tubulars, rather than elasticity-based initial yield criteria to predict structural failure. Such probabilistic design methodologies allow either the computation of the probability of failure (P_f) of a given structure or the design of a structure that meets a target probability of failure.

Reliability-based techniques have been formally applied to the design of load-bearing structures in several disciplines. However, their application to the design of oilfield tubulars is relatively new. Two different reliability-based approaches have been considered: the more fundamental quantitative risk assessment (QRA) approach and the more easily applied load and resistance factor design (LRFD) format. Comparison of SF to the estimated design reliability offers a reliability-based interpretation of WSD and gives insight into the design reliabilities implicit in WSD.

7.23.2 Background. In all design procedures, a primary goal is to ensure that the total load effect of the applied loads is lower than the strength of the tubular to withstand that particular load effect, given the uncertainty in the estimate of the load effect, resistance, and their relationship.

The load effect is related to the resistance of the tubular by means of a relationship, often known as the "failure criterion," which is thought to represent the limit of the tubular under that particular load effect. Thus, the failure criterion is specific to the response of the tubular to that load effect. Three conventional design procedures are considered: WSD, QRA, and LRFD.

Clearly, the relationship between the load effect and resistance and the means of ensuring safety or reliability are different in each of these procedures. In what follows, z_i are the variables and parameters (such as tension, pressure, diameter, yield stress, etc.) that determine the load effect and resistance; Q is the total load effect; and R is the total resistance in response to the load effect, Q.

7.23.3 Working Stress Design. WSD is the conventional casing design procedure, as discussed earlier in this chapter, that is, the familiar deterministic approach to the design of oilfield tubulars. In WSD, the load effect is separated from the resistance by means of an arbitrary multiplier, the SF. The estimated load effect is often the worst-case load, Q_w, based on deterministic design values for the parameters, z_i, that determine the load effect. The estimated resistance is often the minimum resistance, R_{min}, based on deterministic design values for the parameters that determine the resistance. The design values chosen in formulating the resistance are such that the resulting resistance is a minimum. In most cases, the limits of design are established using failure criteria based on elastic theory. In some cases, such as collapse, WSD employs empirical failure criteria. In general, the design procedure can be represented by the relationship

$$\text{SF} \times Q_w(z_i) \le R_{min}(Z_i) \dots\dots\dots\dots\dots\dots\dots\dots\dots\dots\dots\dots\dots\dots\dots (7.55)$$

The ratio R_{min}/SF is called the safe working stress of the structure, hence, the name of the procedure.

The role of the SF is to account for uncertainties in the design variables and parameters, primarily the load effect and the strength or resistance of the structure. The magnitude of the SF is usually based on experience, though little documentation exists on their origin or impact. Different companies use different acceptable SFs for their tubular design. SFs give little indication of the probability of failure of a given structure, as they do not explicitly consider the randomness of the design variables and parameters. Some other limitations of this approach are listed in brief next.

• WSD designs to worst-case load, with no regard to the likelihood of occurrence of the load.

• WSD mostly uses conservative elasticity-based theories and minimum strength in design (though this is not a requirement of WSD).

• WSD gives the engineer no insight into the degree of risk or safety (though the engineer assumes that it is acceptably low), thus making it impossible to accurately assess the risk-cost balance.

• SFs are based on experience and not directly computed from the uncertainties inherent in the load estimate (though these uncertainties are implicit in the experience).

• WSD sometimes makes the design engineer change loading or accept smaller SFs to fit an acceptable WSD, without giving him the means to evaluate the increased risk.

7.23.4 Reliability-Based Design Approaches. Both QRA and LRFD are reliability-based approaches. The general principles of reliability-based design are given in *ISO 2394, International Standard for General Principles on Reliability of Structures*,[5] and a detailed discussion of the underlying theory is given by Kapur and Lamberson.[6] In reliability-based approaches, the uncertainty and variability in each of the design variables and parameters is explicitly considered. In addition, a limit-states approach is used rather than elasticity-based criteria. Thus, the "failure criterion" of WSD is replaced by a limit state that represents the true limit of the tubular for a given load effect. Such probabilistic design approaches allow the estimation of a probability of failure of the structure, thus giving better risk-consistent designs.

Quantitative Risk Assessment. In QRA, the limit state is considered directly. The limit state is the relationship between the load effect and resistance that represents the true limit of the tubular. Conceptually, the limit state $G(Z_i)$ is written as

$$G(Z_i) = R(Z_i) - Q(Z_i), \quad\text{...} (7.56)$$

where Z_i are the random variables and parameters that determine the load effect and resistance for the given limit state. $G(Z_i)$ is known as the limit-state function (LSF). In Eq. 7.56, the upper case Z is used to represent the parameters to remind us that the parameters are treated as random variables in QRA. The LSF usually represents the ultimate limit of load-bearing capacity or serviceability of the structure, and the functional relationship depends upon the failure mode being considered. $G(Z_i) < 0$ implies that the limit state has been exceeded (i.e., failure). The probability of failure can be estimated if the magnitude and uncertainty of each of the basic variables, Z_i, is known and the mechanical models defining $G(Z_i)$ are known through the use of an appropriate theory. The uncertainty in $Q(Z_i)$ and $R(Z_i)$ is calculated from the uncertainty in each of the basic variables and parameters, Z_i, through an appropriate uncertainty propagation model, such as Monte Carlo simulation. **Fig. 7.10** illustrates the concept, with the load effect and resistance being shown as random variables. The shaded region shows the interference area, which is indicative of P_f, the probability of failure. It is the area where the loads exceed the strength, hence, this is the area of failure. The interference area can be estimated using reliability theory.

Thus, the probability that any given design may fail can be estimated, given an appropriate limit state and estimated magnitude and uncertainty of each of the basic variables and a reliability analysis tool. The approach previously mentioned, although simple in concept, is usually difficult to implement in practice. First, the LSF is not always a manageable function and is often cumbersome to use. Second, the uncertainty in the load and resistance parameters must be estimated each time a design is attempted. Third, the probability of failure must be estimated with an appropriate reliability analysis tool. It is tempting to treat each of the parameters, Z_i, as normal variates and use a first-order approach to the propagation of uncertainty. However, such an analysis would be in error because the variables are usually not normal, and first-order propagation gives reliable information only on the central tendencies of the resultant distributions and is erroneous in estimating the tail probabilities.

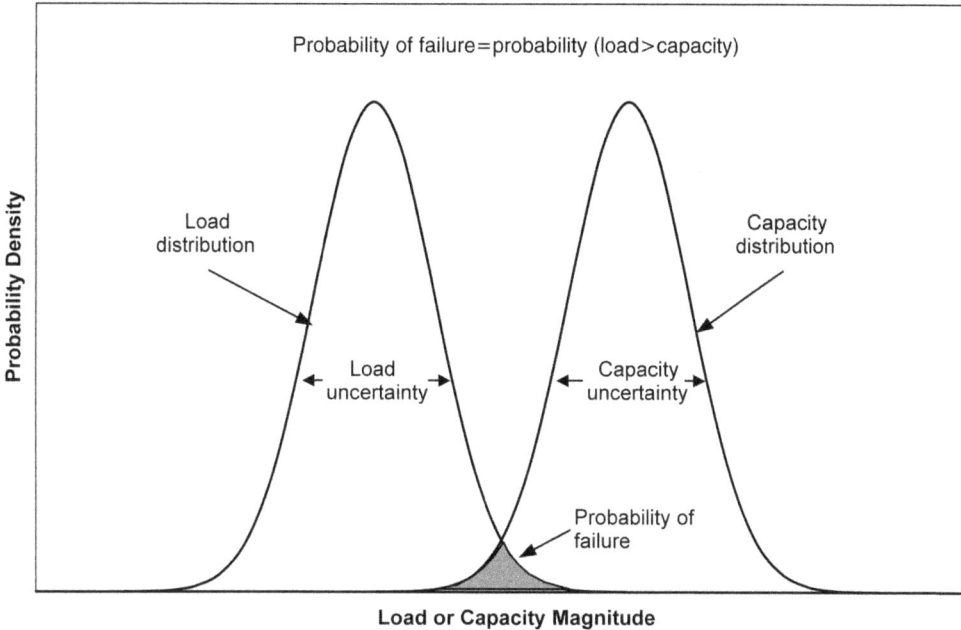

Fig. 7.10—Reliability-based design.

From Fig. 7.10, it is clear that it is the tail probabilities that are of interest in our work. Therefore, it is important to do a full Monte Carlo simulation to estimate the probability of failure of any real design with real variables. However, this too, is not easy because to obtain probability of failure information of the order 10^{-n}, the simulation has to go through 10^{n+2} iterations. Clearly, this is a computer-intensive effort. See the chapter on risk assessment in *Emerging and Peripheral Technologies*, Vol. VI of this *Handbook*, for more discussion on the Monte Carlo method.

Load and Resistance Factor Design. The load and resistance factor design approach is a reliability-based approach that captures the reliability information characteristic of quantitative risk assessment and presents it in a design format far more amenable to routine use, just like WSD. The limit state is the same as the one considered by QRA. However, the design approach is simplified by the use of a design check equation (DCE).

LRFD allows the designer to check a design using a simplified DCE. The DCE is usually chosen to be a simple and familiar equation (for instance, the von Mises criterion in tubular design). Appropriate characteristic values of the design parameters are used in the DCE, along with partial factors that account for the uncertainties in the load and resistance and the difference between the DCE and the actual limit state. Thus, if $Q_{\text{char}}(z_i)$ and $R_{\text{char}}(z_i)$, respectively, represent the characteristic value of the load effect and of resistance, with z_i being the characteristic values of each of the parameters and variables, the DCE can be represented by the inequality

$$\text{LF} \times Q_{\text{char}}(z_i) \leq \text{RF} \times R_{\text{char}}(z_i), \dots\dots\dots\dots\dots\dots\dots\dots\dots\dots\dots\dots\dots\dots \text{(7.57)}$$

where load factor (LF) and resistance factor (RF) are the partial factors required. In the literature, LF and RF are usually referred to as the load factor and resistance factor, respectively. The LF takes into account the uncertainty and variability in load effect estimation, while the RF takes into account the uncertainty and variability in the determination of resistance, as well as any difference between the LSF and DCE. Any design that satisfies Eq. 7.57 is a valid design. The design check equation can be functionally identical to the LSF, or the functional relationship can be a simple formula specified by the design code or familiar WSD formulas. Note that Eq. 7.57 is merely a conceptual representation. In practice, it might not be possible to separate the load effects and resistance in the way suggested by Eq. 7.57. Moreover, several load effects and resistance terms may be present in the DCE, with varying uncertainties, requiring the use of several partial factors.

Similarity to WSD. We observe, from Eq. 7.57 that the partial factors are, in a sense, similar to the SF used in WSD. Comparing Eq. 7.57 to Eq. 7.55, we notice that both equations are based on deterministic values, and the SF in Eq. 7.55 is replaced by two partial factors. Indeed, the ratio LF/RF is analogous to the SF used in WSD, if the DCE happens to be identical to the WSD failure criterion. Thus, in concept, it may be said that

$$LF/RF => SF \dots\dots\dots\dots\dots\dots (7.58)$$

Despite these similarities, however, there are three crucial differences. First, the loads and resistances are estimated using a set methodology. Second, the load effect and the resistance are treated separately, thus allowing the partial factors to separately account for the uncertainties in each. And third, the magnitude of loads and resistances is based on reliability, rather than being arbitrarily set.

Partial factors are chosen through a process of calibration, where the deterministic DCE with partial factors is calibrated against the probabilistic LSF. Partial-factor values are chosen such that their use in the DCE results in a design that has a preselected target reliability or target probability of failure, as determined from the LSF using reliability analysis. For the partial factors to do so, the calibration process should prescribe a scope of the application of LRFD, and the values of the partial factors should be optimized to ensure a uniform reliability across the scope. The objective is to obtain a set of factors that results in designs of this target probability. In brief, the procedure may be summarized as follows. First, choose a desired target probability of failure. Second, identify the characteristic values of each of the parameters, and the uncertainty and variability about these values. Third, for an assumed set of load and resistance factors, generate a set of "passed" designs from the DCE, across the scope of the structure, for all possible load magnitudes. In other words, all designs that pass the DCE are valid designs. The passing of a design is, of course, controlled by the assumed value of the load and resistance factors. Fourth, for each of the passed designs, estimate the probability of failure from the LSF, taking into account the uncertainty in each of the variables. Fifth, determine the statistical minimum reliability assured by the assumed set of load and resistance factors. This is the reliability (or equivalently, probability of failure) that results from the use of these partial factors. In other words, the probability of failure of any design that results from the use of these partial factors in the DCE will, statistically, be less than or equal to the probability of failure. Sixth, repeat until the set of partial factors results in the desired target probability of failure.

At the end of the process, we have a set of partial factors and their corresponding design reliability. If several target reliabilities are to be aimed for, the procedure is repeated, until a new set of partial factors is obtained.

It must be noted that this is a very brief summary of the approach. Calibration is usually the most time-consuming and rigorous step in devising an LRFD procedure. Several reliability-theory and statistical details such as uncertainty estimation, preprocessing of high-reliability designs, zonation, uniformity of reliability, multiple partial factor calibration, etc. have been omitted for brevity.

7.24 Critique of Risk-Based Design

WSD has been used successfully for many years to design casing. It is a simple system, understood by the average drilling engineer, of comparing a calculated worst-case load against the rating of the casing. The safety factors used may neither be based on strict logic nor be the same across industry, but the concept is simple and the numbers are similar. Generally, the system has served the industry well. Risk-based design advocates criticize WSD because the failure models do not always use the ultimate load limit as the failure criterion, but this is not inherent to WSD. In an ideal world, where casing is always within specification, using average safety factors and worst-case estimates of loads, the casing should always be overdesigned.

However, WSD makes no allowance for casing manufactured below minimum specification. The SF used may or may not compensate for the fact that a below-strength joint is in a critical location. The risks cannot be quantified, so there is no way of comparing the relative risks of different designs. It can also lead to a situation in which it is impossible to produce a practical design under extreme downhole conditions. There would be a temptation in this case either to try to justify a reduction in the SF, perhaps by relying on improved procedures, or to re-estimate the loads downward. Also, the system does not usually consider low levels of H_2S, causing brittle failure in burst. Improvements such as better quality control, more accurate failure equations, and considering brittle burst could be utilized within a WSD system.

It is reasonable for the nonstatistician to accept that the strengths of joints of casing of the same weight and grade from the same mill will vary symmetrically around a mean value. The product is manufactured from nominally the same materials and by the same process, with the aim of producing identical properties. The predictability of the "resistance" side of the equation has been confirmed by large-scale examination and testing of the finished product.

The "load" side of the equation, such as formation pressures and kick volumes, may not be so predictable. There is also a much smaller data bank available for estimating probabilities. Further, human factors may influence the size of a kick by such things as speed of reaction in closing the well in and choosing the correct choke pressures when killing a kick.

The designer using risk-based casing design, thus has the same problem that the WSD user has—namely, which loads to consider in the design. The risk-based designer has an additional task, the assignment of probabilities to these loads. One could argue further that these loads should be weighted according to the severity of the resulting failure.

If risk-based designs are used to justify thinner/lower-grade casing and pipe manufactured to the same quality standards as used as with WSDs, the wells will not be safer. If risk-based design systems are used by people who do not understand the system, or only use partial factors rather than the full system, wells will not be safer. If the load data have been underestimated, the wells will not be safer, especially in high-temperature/high-pressure wells.

A risk-based design system with more accurate failure equations; account taken of brittle fracture in low levels of H_2S; improved quality control of tubulars and connections; accurate load data; engineers who understand the system and the well; and a full training and competence assurance program may produce wells that are as safe as those designed using WSD.

Nomenclature

A = constant in plastic collapse equation, dimensionless
A_c = cross-sectional area of coupling, in.^{2}
A_{cr} = critical section area of box, pin, or pipe, whichever is least, in.^{2}
A_i = the inside area of the tubing, πr_i^2, in.^{2}
A_{jp} = cross-sectional area of the pipe wall under the last perfect thread,
$\pi/4[(D - 0.1425)^2 - d^2]$, in.^{2}
A_o = the outside area of the tubing, πr_o^2, in.^{2}
A_p = cross-sectional area of plain-end pipe, in.^{2}

A_s = cross-sectional area of pipe, in.^{2}

B = constant in plastic collapse equation, dimensionless

C = constant in plastic collapse equation, psi

d_1 = diameter at the root of the coupling thread in the power tight position, in.

d = nominal inside diameter of pipe, in.

D = nominal outside pipe diameter, in.

D/t = slenderness ratio, dimensionless

e_b = buckling strain, in./in.

e_{bavg} = average buckling strain, in./in.

E = Young's modulus (3.0×10^7 psi for steel)

E_s = pitch diameter at plane of seal, in.

f_1, f_2, f_3 = terms in combined stress effects for collapse, psi

F = constant in transition collapse equation, dimensionless

F_a = axial force (tension positive), lbf

F_b = buckling force (compression positive), lbf

F_{bnd} = bending stress equivalent force, lbf

F_j = minimum joint strength, lbf

F_p = Paslay buckling force, lbf

F_y = pipe-body axial strength, lbf

g_c = gravity constant, 32.2 ft/sec^2

G = constant in transition collapse equation, dimensionless

G = shear modulus, psi

$G(Z_i)$ = the limit state function

I = moment of inertia, in.^{4}

L = engaged thread length, in.

L_{bkl} = buckled length of tubing, ft

L_{helmax} = maximum helically buckled length, ft

L_{helmin} = minimum helically buckled length, ft

M = bending moment, lbf-ft

N = API-defined thread-turns from Ref. 4, dimensionless

p_e = equivalent external pressure, psi

p_i = internal pressure, psi

p_o = external pressure, psi

p_t = thread pitch, in.

P_B = minimum burst pressure, psi

P_{CIY} = coupling internal yield pressure, psi

P_E = elastic collapse pressure, psi

P_f = probability of failure, dimensionless

P_{hel} = pitch of helically buckled pipe, ft

P_{ILR} = coupling internal leak resistance pressure, psi

P_P = plastic collapse pressure, psi

P_{Yp} = yield strength collapse pressure, psi

P_T = transition collapse pressure, psi

Q_{char} = characteristic value for the load effect, lbf

r = radial annular clearance, in.

r_a = permafrost fluid-zone radius, ft

r_e = permafrost excess-ice-zone radius, ft

r_i = inside radius of the pipe, in.

r_o = outside radius of the pipe, in.

r_p = permafrost plastic-zone radius, ft

R = radius of curvature

R_{char} = characteristic value of the resistance, lbf

S_a = axial stress based on the buoyant weight of pipe, psi

t = nominal wall thickness, in.

T = thread taper, in./in.

u = tubing buckling displacement, in.

U_c = minimum ultimate tensile strength of coupling, psi

U_p = minimum ultimate tensile strength of pipe, psi

w = distributed buoyed weight of casing, lbm/in.

w_a = weight per unit length of pipe in air, lbm/ft

w_e = the effective (buoyant) weight per unit length of the tubing, lbm/ft

W = nominal outside diameter of coupling, in.

W_n = lateral contact force, lbf/in.

Y_c = minimum yield strength of coupling, psi

Y_p = minimum yield stress of pipe, psi

Z_i = the random variables and parameters that determine the load effect and resistance for the given limit state

α = thermal-expansion coefficient (6.9×10^{-6} °F^{-1} for steel), °F^{-1}

α/L = dogleg severity (°/unit length)

γ_i = the density of the fluid inside the tubing, lbm/ft^3

γ_o = the density of the fluid outside the tubing, lbm/ft^3

ΔF_{bal} = incremental force caused by ballooning, lbf

ΔF_{temp} = incremental force caused by temperature change, lbf

ΔL_b = buckling length change, ft

Δp = $p_o - p_i$, psi

Δp_i = change in surface internal pressure, psi

Δp_o = change in surface external pressure, psi

ΔT = average change in temperature over free length, °F

$\Delta \rho_i$ = change in internal fluid density, lbm/ft^3

$\Delta \rho_o$ = change in external fluid density, lbm/ft^3

θ' = rate of change of helix angle with respect to pipe length, radians/ft

θ = helix angle, radians

κ = curvature, radians/ft

σ_b = stress at the pipe's outer surface, psi

σ_r = radial stress, psi

σ_{VME} = triaxial stress, psi

σ_z = axial stress, psi

σ_θ = tangential or hoop stress, psi

υ = Poisson's ratio (0.30 for steel), dimensionless

φ = wellbore angle with the vertical, radians

References

1. *Bull. 5C3, Bulletin for Formulas and Calculations for Casing, Tubing, Drillpipe, and Line Pipe Properties,* fourth edition, API, Dallas (1985).
2. *Bull. 5C2, Bulletin for Performance Properties of Casing, Tubing, and Drillpipe,* eighteenth edition, API, Dallas (1982).
3. Crandall, S.H. and Dahl, N.C.: *An Introduction to the Mechanics of Solids,* McGraw-Hill Book Company, New York City (1959).
4. *Spec. 5B, Specification for Threading, Gauging, and Thread Inspection of Casing, Tubing, and Line Pipe Threads,* fourteenth edition, API, Dallas (1996).
5. *ISO 2394, International Standard for General Principles on Reliability of Structures,* second edition, International Organization for Standardization, Geneva (1986).
6. Kapur, K.C. and Lamberson, L.R.: *Reliability in Engineering Design,* John Wiley & Sons, New York City (1977).

General References

Aadnoy, B.S.: *Modern Well Design,* Balkema Publications, Rotterdam, The Netherlands (1996).

Adams, A.J. and Glover, S.B.: "An Investigation into the Application of QRA in Casing Design," paper SPE 48319 presented at the 1998 SPE Applied Technology Workshop on Risk-Based Design of Well Casing and Tubing, The Woodlands, Texas, 7–8 May.

Adams, A.J. and Hodgson, T.: "Calibration of Casing/Tubing Design Criteria by Use of Structural Reliability Techniques," *SPEDC* (March 1999).

Adams, A.J. and MacEachran, A.: "Impact on Casing Design of Thermal Expansion of Fluids in Confined Annuli," *SPEDE* (September 1994).

Adams, A.J. *et al.:* "Casing System Risk Analysis Using Structural Reliability," paper SPE/IADC 25693 presented at the 1993 SPE/IADC Drilling Conference, Amsterdam, 23–25 February.

Adams, A.J., Warren, A.V.R., and Masson, P.C.: "On the Development of Reliability-Based Design Rules for Casing Collapse," paper SPE 48331 presented at the 1998 SPE Applied Technology Workshop on Risk-Based Design of Well Casing and Tubing, The Woodlands, Texas, 7–8 May.

Adams, A.J.: "Quantitative Risk Analysis (QRA) in Casing/Tubing Design," paper presented at the Offshore Drilling Technology Conference, Aberdeen (November 1993).

Adams, A.J.: "QRA for Casing/Tubing Design," paper presented at the Seminar of Norwegian HPHT Programme, Stavanger (January 1995).

Ang, A. H.-S. and Tang, W.H.: *Probability Concepts in Engineering Planning and Design, Volume II: Decision, Risk and Reliability,* John Wiley & Sons Inc., New York City (1984).

Banon, H., Johnson, D.V., and Hilbert, L.B.: "Reliability Considerations in Design of Steel and CRA Production Tubing Strings," paper SPE 23483 presented at the 1991 International Conference on Health, Safety and Environment, The Hague, The Netherlands, 10–14 November.

Beach, H.J.: "Cementing Through Permafrost Environment," paper presented at the 1997 ASME Energy Technology Conference and Exhibit, Houston, Texas.

Benge, O.G. *et al.:* "A New Low-Cost Permafrost Cementing System," paper SPE 10757 presented at the 1982 California Regional Meeting of the SPE, San Francisco, California, 24–28 March.

Brand, P.R., Whitney, W.S., and Lewis, D.B.: "Load and Resistance Factor Design Case Histories," paper OTC 7937 presented at the 1995 Offshore Technology Conference, Houston, Texas, 1–4 May.

Bull. D7, Bulletin for Casing Landing Recommendations, first edition, API, Dallas (1955).

Chen, Y., Lin, Y., and Cheatham, J.B.: "Tubing and Casing Buckling in Horizontal Wells," *JPT* (February 1990).

CIRIA. 6, Construction Industry Research and Information Association: Rationalisation of Safety and Serviceability Factors in Structural Codes, SW1P 3AU, CIRIA, Storey's Gate, London (1977).

Cunningham, W.C. and Smith, D.W.: "Cementing Through the Permafrost," paper 77-Pet-37 presented at the 1977 ASME Energy Technology Conference and Exhibit, Houston, 18–22 September.

Cunningham, W.C., Fehrenbach, J.R., and Maier L.F.: "Arctic Cements and Cementing," *J. Cdn. Pet. Tech.* (1972).

Davies, B.E. and Boorman, R.D.: "Field Investigation of Effect of Thawing Permafrost Around Wellbores at Prudhoe Bay," paper SPE 4591 presented at the 1973 SPE Annual Meeting, Las Vegas, Nevada, 30 September–3 October.

Dawson, R. and Paslay, P.R.: "Drillpipe Buckling in Inclined Holes," *JPT* (October 1984).

Det Norske Veritas, Rules for the Design, Construction and Inspection of Offshore Structures, DNV, Hovik, Norway (1981).

Economides, M.J., Waters, L.T., and Dunn-Norman, S.: *Petroleum Well Construction,* John Wiley & Sons, New York City (1998).

EUROCODE 3, Common Unified Rules for Steel Structures, Commission of the European Communities (1984).

Fowler, E.D. and Taylor, T.E.: "How to Select and Test Materials for −75°F," *World Oil* (1976).

Fowler, E.D. and Taylor, T.E.: "Materials for Wellheads and Christmas Trees for Cold Climates," paper ASME 75-Pet-17 presented at the Petroleum Mechanical Engineering Conference, Tulsa, Oklahoma (September 1975).

Galambos, T.V. and Ravindra, M.K.: "Properties of Steel for Use in LRFD," *Proc.,* ASCE (1978) **104.**

Galambos, T.V. *et al.:* "Probability-based Load Criteria: Assessment of Current Design Practice," *J. of the Structural Division* (1982); *Trans.,* ASCE, **108.**

Goodman, M.A.: "A New Look at Permafrost Completions," *Pet. Eng. Intl.* (1977).

Goodman, M.A.: "Loading Mechanisms in Thawed Permafrost around Arctic Wells," paper presented at the ASME Energy Technology Conference and Exhibition, Houston (September 1977).

Goodman, M.A.: "Mechanical Properties of Simulated Deep Permafrost," *J. of Engineering for Industry* (May 1975); *Trans.,* ASME, **97.**

Goodman, M.A.: World Oil's *Handbook of Arctic Well Completions,* Gulf Publishing Co., Houston (1978).

Goodman, M.A. and Wood, D. B.: "A Mechanical Model for Permafrost Freeze-Back Pressure Behavior," *SPEJ* (August 1975).

Halal, A.S. and Mitchell, R.F.: "Casing Design for Trapped Annular Pressure Buildup," *SPEDE* (June 1994).

Halal, A.S. and Mitchell, R.F.: "Multistring Casing Design with Wellhead Movement," paper SPE 37443 presented at the 1997 SPE Production Operations Symposium, Oklahoma City, Oklahoma, 9–11 March.

Hammerlindl, D.J.: "Movement, Forces, and Stresses Associated With Combination Tubing Strings Sealed in Packers," *JPT* (February 1977).

Hammerlindl, D.J.: "Packer-to-Tubing Forces for Intermediate Packers," *JPT* (March 1980).

Health and Safety Executive: A Guide to the Offshore Installations (Safety Case) Regulations 1992, first edition, HMSO, London (1992).

Health and Safety Executive: A Guide to the Wells Aspects of the Offshore Installations and Wells (Design and Construction, etc.) Regulations 1996, first edition, HMSO, London (1996).

Hinton, A.: "Will Risk Based Casing Design Mean Safer Wells," paper SPE 48326 presented at the 1998 SPE Applied Technology Workshop on Risk-Based Design of Well Casing and Tubing, The Woodlands, Texas, 7–8 May.

Howell, E.P., Seth, M.S., and Perkins, T.K.: "Temperature Calculations for Wells, which are Completed Through Permafrost," paper SPE 3969 presented at the 1972 SPE Annual Meeting, San Antonio, Texas, 8–11 October.

Howitt, F.: "Permafrost Geology at Prudhoe Bay," *World Petroleum* (September 1971).

Hoyer, W. A. *et al.:* "Evaluation of Permafrost with Logs," *Trans.,* SPWLA 1975 Logging Symposium, 4–7 June.

Kelly, I.D. and Rabia, H.: "Applying Quantitative Risk Assessment to Casing Design," paper IADC/SPE 35038 presented at the 1996 IADC/SPE Drilling Conference, New Orleans, Louisiana, 12–15 March.

Kendall, M.G. and Stuart, A.: *The Advanced Theory of Statistics, Vol. 1: Distribution Theory,* Charles Griffen & Co., London (1958).

Klementich, E.F.: "A Rational Characterization of Proprietary High Collapse Casing Grades," paper SPE 30526 presented at the 1995 SPE Annual Technical Conference and Exhibition, Dallas, Texas, 22–25 October.

Klementich, E.F. and Jellison, M.J.: "A Service Life Model for Casing Strings," *SPEDE* (April 1986).

Kljucec, N.M., and Telford, A.S.: "Thermistor Cables Monitor Well Temperatures Effectively Through Permafrost," *J. Cdn. Pet. Tech.* (1972).

Kljucec, N.M., Telford, A.S., and Bombardier, C.C.: "Gypsum-Cement Blend Works Well in Permafrost," *World Oil* (1973).

Lewis, D.B. *et al.:* "Load and Resistance Factor Design for Oil Country Tubular Goods," paper OTC 7936 presented at the 1995 Offshore Technology Conference, Houston, 1–4 May.

Lin, C.J. and Wheeler, J.D.: "Simulation of Permafrost Thaw Behavior at Prudhoe Bay," *JPT* (March 1978).

Load and Resistance Factor Design Specification for Structural Steel Buildings, American Institute of Steel Construction, Chicago (1986).

Lubinski, A., Althouse, W.S., and Logan, J.L.: "Helical Buckling of Tubing Sealed in Packers," *JPT* (June 1962).

MacEachran, A. and Adams, A.J.: "Impact on Casing Design of Thermal Expansion of Fluids in Confined Annuli," paper SPE/IADC 21911 presented at the 1991 IADC/SPE Drilling Conference, Amsterdam, 11–14 March.

Mackay, J.R.: "The Origin of Massive Ice Beds in Permafrost, Western Arctic Coast, Canada," *Canadian J. of Earth Sciences* (1971) **8,** No. 397.

Madsen, H.O., Krenk, S., and Lind, N.C.: *Methods of Structural Safety,* Prentice-Hall Inc., Englewood Cliffs, New Jersey (1986).

Maes, M.A. *et al.:* "Reliability-Based Casing Design," *ASME J. of Energy Resources Tech.,* **117** (1995).

Maes, M.A., Breitung, K., and Dupuis, D.J.: "Asymptotic Importance Sampling," *Structural Safety,* **12** (1993).

Maier, L.F. *et al.:* "Cementing Materials for Cold Environments," *JPT* (October 1971).

Mann, N.R., Schafer, R.E., and Singpurwalla, N. D.: *Methods for Statistical Analysis of Reliability and Life Data,* John Wiley & Sons Inc., New York City (1974).

Manual for Steel Construction, Load and Resistance Factor Design, American Institute of Steel Construction, Chicago (1986).

Merriam, R. *et al:* "Insulated Hot Oil-Producing Wells in Permafrost," *JPT* (March 1975).

Miller, R.A.: "Real World Implementation of QRA Methods in Casing Design," paper SPE 48325 presented at the 1998 SPE Applied Technology Workshop on Risk-Based Design of Well Casing and Tubing, The Woodlands, Texas, 7–8 May.

Minimum Design Loads for Buildings and Other Structures, American National Standards Institute, New York City (1982) A58.1.

Miska, S. and Cunha J.C.: "An Analysis of Helical Buckling of Tubulars Subjected to Axial and Torsional Loading in Inclined Wellbores," paper SPE 29460 available from SPE, Richardson, Texas (1995).

Mitchell, R.F.: "A Mechanical Model for Permafrost Thaw-Subsidence," *J. of Pressure Vessel Technology* (February 1977); *Trans.,* ASME, **99,** Series J, No. 1.

Mitchell, R.F.: "Buckling Analysis in Deviated Wells: A Practical Method," *SPEDC* (March 1999).

Mitchell, R.F.: "Forces on Curved Tubulars due to Fluid Flow," *SPEPF* (February 1996).

Mitchell, R.F.: "Loading Mechanisms in Thawed Permafrost Around Arctic Wells," *J. of Pressure Vessel Technology* (August 1978); *Trans.,* ASME, **100,** 320.

Mitchell, R.F.: "New Concepts for Helical Buckling," *SPEDE* (September 1988).

Mitchell, R.F. and Goodman, M.A.: "Permafrost Thaw-Subsidence Casing Design," *JPT* (November 1977).

Morgenstern, N.R. and Nixon, J.F.: "One-Dimensional Consolidation of Thawing Soils," *Cdn. Geotechnical J.* (1976) **8,** 558.

Morris E.F.: "Evaluation of Cement Systems for Permafrost," paper SPE 2824 presented at the 1970 AIME Annual Meeting, Denver, 15–19 February.

National Building Code of Canada, Associate Committee on the National Building Code, Ottawa, Ontario (1980).

Nixon, J.F. and Morgenstern, N.R.: "Practical Extensions to a Theory of Consolidation for Thawing Soils," Permafrost: Second International Conference, Yakutsk, U.S.S.R. (1973).

Parfitt, S.H.L. and Thorogood, J.L.T.: "Application of QRA Methods to Casing Seat Selection," paper SPE 28909 presented at the 1994 SPE European Petroleum Conference, London, 25–27 October.

Payne, M.L. and Swanson, J.D.: "Application of Probabilistic Reliability Methods to Tubular Designs," *SPEDE* (December 1990).

Perkins, T.K. *et al.:* "Prudhoe Bay Field Permafrost Casing and Well Design for Thaw-Subsidence Protection," report to State of Alaska, Atlantic Richfield Co., North American Producing Div., Dallas (May 1975).

Perkins, T.K., Rochon, J.A., and Knowles, C.R.: "Studies of Pressures Generated Upon Refreezing of Thawed Permafrost Around a Wellbore," *JPT* (1974).

Prentice, C.M.: "Maximum Load Casing Design," *JPT* (July 1970).

Pui, N.K. and Kljucec, N.M.: "Temperature Simulation While Drilling Permafrost," paper CIM 75-14 presented at the CIM Annual Technical Meeting of the Petroleum Society, Banff, Alberta (1975).

Rabia, H.: *Fundamentals of Casing Design,* Graham & Trotman, London (1987).

Rackvitz, R. and Fiessler, B.: "Structural Reliability Under Combined Random Load Processes," *Computers and Structures* (1978) **9,** 489.

Raney, J.B., Suryanarayana, P.V.R., and Maes, M.A.: "A Comparison of Deterministic and Reliability-Based Design Methodologies for Production Tubing," paper SPE 48322 presented at the 1998 SPE Applied Technology Workshop on Risk-Based Design of Well Casing and Tubing, The Woodlands, Texas, 7–8 May.

Raney, J.B., Suryanarayana, P.V.R., and Maes, M.A.: "Implementation of a Reliability-Based Design Procedure for Production Tubing," paper 1897 presented at the Offshore Mediterranean Conference, Ravenna, Italy (March 1997).

"Recommendations for Loading and Safety Regulations for Structural Design," report no. 36, Nordic Committee on Building Regulations, NKB, Copenhagen (1978).

Reeves, T.B., Parfitt, S.H.L., and Adams, A.J.: "Casing System Risk Analysis Using Structural Reliability," *Proc.*, SPE/IADC Drilling Conference, Amsterdam (1993).

Rogers, J.C. and Sackinger, W.M.: "Investigation of Arctic Offshore Permafrost Near Prudhoe Bay," paper ASME 76-Pet-97 presented at the 1976 Petroleum Mechanical Engineering and Pressure Vessels and Piping Conference, Mexico City, 19–24 September.

RP2A-LRFD, Recommended Practice for Planning, Design and Construction of Fixed Offshore Platforms, first edition, API, Washington, D.C. (1995).

Ruedrich, R.A. *et al.:* "Casing Strain Resulting from Thawing of Prudhoe Bay Permafrost," *JPT* (March 1978).

Ruedrich, R.A., Perkins, T.K., and O'Brien, D.E.: "Precise Joint Length Determination Using a Multiple Casing Collar Locator Tool," paper SPE 5087 presented at the 1974 SPE Annual Meeting, Houston, 6–9 October.

Smith, R.E. and Clegg, M.W.: "Analysis and Design of Production Wells Through Thick Permafrost," *Proc.*, Eighth World Pet. Cong., Moscow (1971).

Smith, W.S., Nair, K., and Smith, R.E.: "Sample Disturbance and Thaw Consolidation of a Deep Sand Permafrost," *Proc.*, Permafrost Second International Conference, Yakutsk, U.S.S.R. (1973).

Thoft-Christensen, P. and Baker, M.J.: *Structural Reliability Theory and its Applications,* Springer-Verlag Inc., New York City (1982).

Timoshenko, S.P. and Goodier, J.N.: *Theory of Elasticity,* third edition, McGraw-Hill Book Co., New York City (1961).

Turner, R.C.: "Partial Factor Calibration for North Sea Adaptation of API RP2A-LRFD," *Proc.*, Institution of Civil Engineers, Water Maritime and Energy, London (1993) **101.**

Weiner, P.D. *et al.:* "Casing Strain Tests of 13 3/8" N-80 Buttress Connections," *JPT* (November 1976).

White, F.L.: "Setting Cements in Below Freezing Conditions, *Pet. Eng. Intl.* (1952).

Wooley, G.R., Christman, S.A., and Crose, J.G.: "Strain Limit Design of 13 3/8-in. N-80 Buttress Casing," *JPT* (April 1977).

SI Metric Conversion Factors

$$
\begin{array}{llll}
\text{ft} & \times & 3.048^* & \text{E}-01 & = \text{m} \\
\text{ft/sec} & \times & 3.048^* & \text{E}-01 & = \text{m/s} \\
°\text{F} & & (°\text{F}-32)/1.8 & & = °\text{C} \\
\text{gal} & \times & 3.785\ 412 & \text{E}-03 & = \text{m}^3 \\
\text{in.} & \times & 2.54^* & \text{E}+00 & = \text{cm} \\
\text{in.}^2 & \times & 6.451\ 6^* & \text{E}+00 & = \text{cm}^2 \\
\text{ksi} & \times & 6.894\ 757 & \text{E}+03 & = \text{kPa} \\
\text{lbf} & \times & 4.448\ 222 & \text{E}+00 & = \text{N} \\
\text{lbm} & \times & 4.535\ 924 & \text{E}-01 & = \text{kg} \\
\text{psi} & \times & 6.894\ 757 & \text{E}+00 & = \text{kPa}
\end{array}
$$

*Conversion factor is exact.

Chapter 8
Introduction to Wellhead Systems
Mike Speer, Dril-Quip Inc.

The objective of this chapter is to provide a brief overview of the types of wellhead systems and equipment commonly found on wells drilled in today's oil and gas industry. First, we discuss two broad categories of surface wellhead systems: onshore and offshore. Then, we discuss wellhead systems used in subsea and ultradeepwater applications.

8.1 Drilling a Well on Land

When a well is drilled on land, an interface is required between the individual casing strings and the blowout preventer (BOP) stack. This interface is required for four main reasons:
- To contain pressure through the interface with the BOP stack.
- To allow casing strings to be suspended so that no weight is transferred to the drilling rig.
- To allow seals to be made on the outside of each casing string to seal off the individual annulus.
- To provide annulus access to each intermediate casing string and the production casing string.

We will address each of these points in turn and describe in more detail how this is achieved with the wellhead.

8.1.1 Pressure Containment. When drilling a well on land, a spool wellhead system is traditionally used, as shown in **Fig. 8.1**. This wellhead is considered a "build as you go" wellhead system that is assembled as the drilling process proceeds. The spool system consists of the following main components:
- Starting casing head.
- Intermediate casing spools.
- Slip casing hanger and seal.
- Tubing spool (if well is to be tested and/or completed).
- Studs, nuts, ring gaskets, and associated accessories required to assemble the wellhead.

Starting Casing Head. The starting casing head (see **Figs. 8.2 and 8.3**) attaches to the surface casing (conductor) by either welding or threading on to the conductor. The top of the starting casing head has a flange to mate with the bottom of the BOP. The flange must meet both size and pressure requirements. The starting casing head has a profile located in the inside diameter (ID) that will accept a slip-and-seal assembly to land and support the next string of

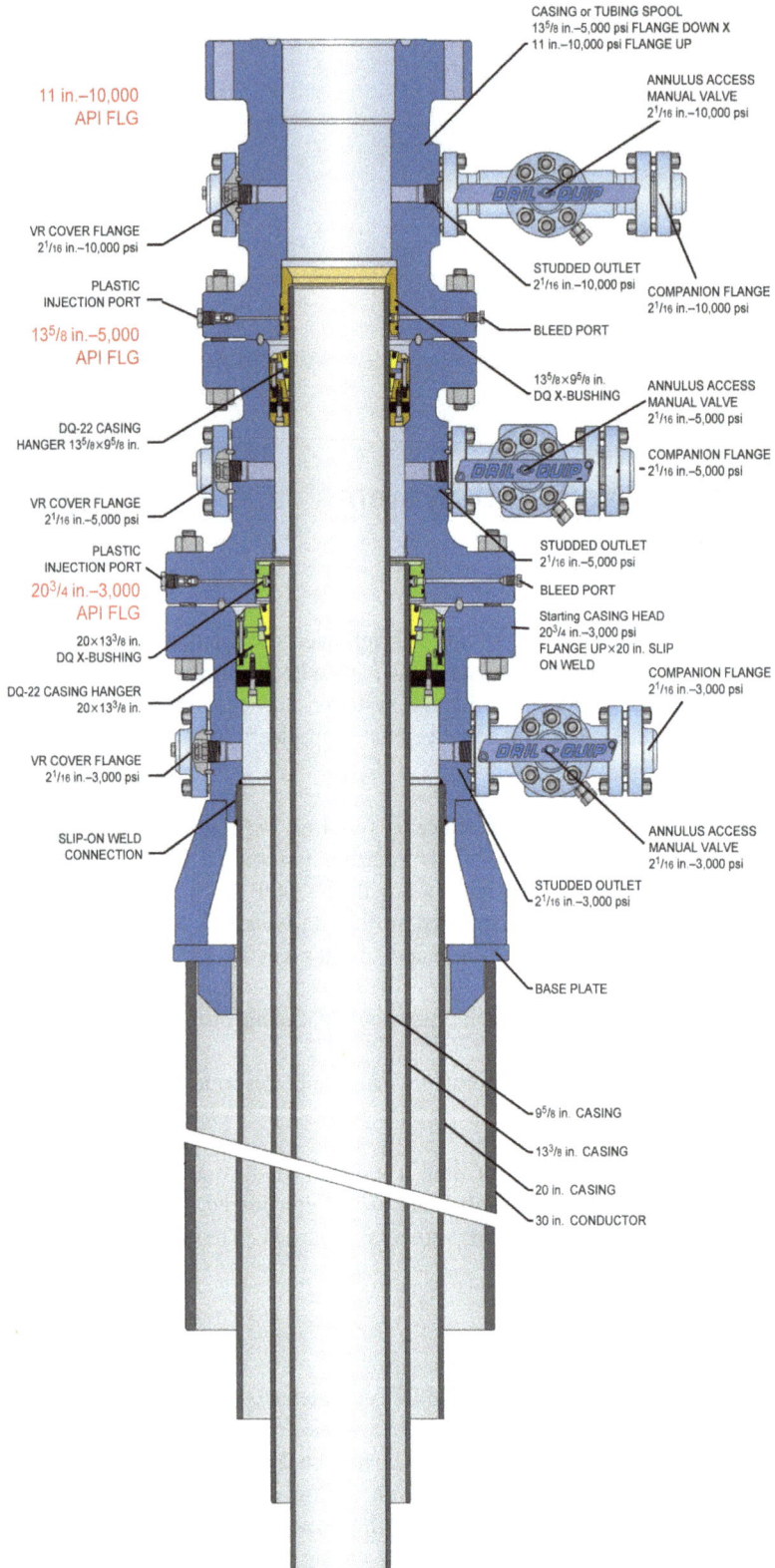

11 in.–10,000
API FLG

CASING or TUBING SPOOL
13⁵/₈ in.–5,000 psi FLANGE DOWN X
11 in.–10,000 psi FLANGE UP

ANNULUS ACCESS
MANUAL VALVE
2¹/₁₆ in.–10,000 psi

VR COVER FLANGE
2¹/₁₆ in.–10,000 psi

STUDDED OUTLET
2¹/₁₆ in.–10,000 psi

COMPANION FLANGE
2¹/₁₆ in.–10,000 psi

PLASTIC
INJECTION PORT

13⁵/₈ in.–5,000
API FLG

BLEED PORT

13⁵/₈×9⁵/₈ in.
DQ X-BUSHING

ANNULUS ACCESS
MANUAL VALVE
2¹/₁₆ in.–5,000 psi

DQ-22 CASING
HANGER 13⁵/₈×9⁵/₈ in.

COMPANION FLANGE
2¹/₁₆ in.–5,000 psi

VR COVER FLANGE
2¹/₁₆ in.–5,000 psi

PLASTIC
INJECTION PORT

STUDDED OUTLET
2¹/₁₆ in.–5,000 psi

20³/₄ in.–3,000
API FLG

BLEED PORT

20×13³/₈ in.
DQ X-BUSHING

Starting CASING HEAD
20³/₄ in.–3,000 psi
FLANGE UP×20 in. SLIP
ON WELD

DQ-22 CASING HANGER
20×13³/₈ in.

COMPANION FLANGE
2¹/₁₆ in.–3,000 psi

VR COVER FLANGE
2¹/₁₆ in.–3,000 psi

SLIP-ON WELD
CONNECTION

ANNULUS ACCESS
MANUAL VALVE
2¹/₁₆ in.–3,000 psi

STUDDED OUTLET
2¹/₁₆ in.–3,000 psi

BASE PLATE

9⁵/₈ in. CASING

13³/₈ in. CASING

20 in. CASING

30 in. CONDUCTOR

Fig. 8.1—Illustration of a typical land wellhead system and casing program (all figures in this chapter are courtesy of Dril-Quip).

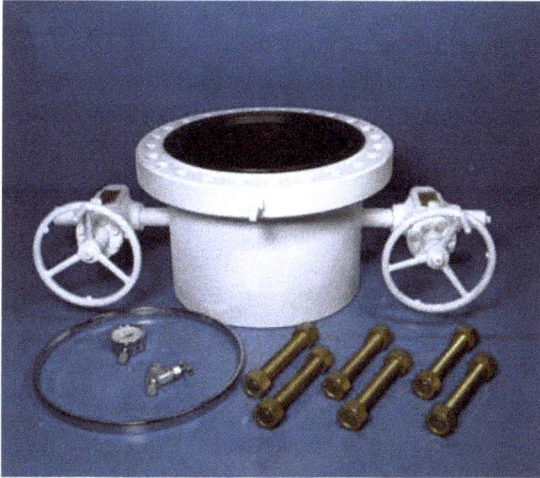

Fig. 8.2—Photo of a starting casing head and installation components. This casing head is typical of a thread-on or weld-on configuration used in land drilling operations.

casing. The slip-and-seal assembly transfers all of the casing weight to the conductor while energizing a weight-set elastomeric seal.

Intermediate Casing Spools. The intermediate casing spool is typically a flanged-by-flanged pressure vessel with outlets for annulus access (see **Fig. 8.4**). The intermediate casing spool (or spools) is installed after each additional casing string has been run, cemented, and set. The bottom section of each intermediate casing spool seals on the outside diameter (OD) of the last casing string that was installed. The bottom flange will mate with the starting casing head or the previous intermediate casing spool. The top flange will have a pressure rating higher than the bottom flange to cope with expected higher wellbore pressures as that hole section is drilled deeper.

The intermediate casing spool also incorporates a profile located in the ID, which accepts a slip-and-seal assembly similar to the one installed in the starting casing head. This slip and seal will be sized in accordance with the casing program.

Tubing Spool. The tubing spool, as shown in **Fig. 8.5**, is the last spool installed before the well is completed. The tubing spool differs from the intermediate spool in one way: it has a profile for accepting a solid body-tubing hanger with a lockdown feature located around the top flange. The lockdown feature ensures that the tubing hanger cannot move because of pressure or temperature. The flange sizes vary in accordance with pressure requirements.

8.1.2 Load-Carrying Components. Casing weight is transferred to the starting casing head and intermediate spools with two different types of hanger systems:
- A slip-and-seal casing-hanger assembly.
- A mandrel-style casing hanger.

The slip-and-seal casing-hanger assembly (**Fig. 8.6**) has an OD profile that mates with the internal profile of the starting casing head and intermediate casing spools. Integral to this casing-hanger assembly is a set of slips with a tapered wedge-type back and serrated teeth that bite into the OD of the casing being suspended.

When the casing has been run and cemented, the BOP is disconnected from the casing spool and lifted up to gain access to the spool bowl area. After the slip-and-seal casing-hanger assembly is installed, the traveling block will lower the casing and set a predetermined amount of casing load onto the slip-and-seal casing-hanger assembly. The teeth on the slips will en-

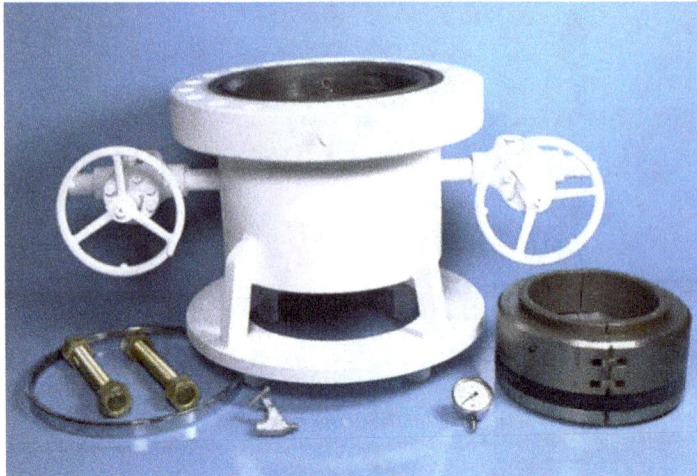

Fig. 8.3—Photo of a starting casing head and slip-and-seal assembly with installation components. This casing head has a gusseted base plate typically seen in jackup drilling operations.

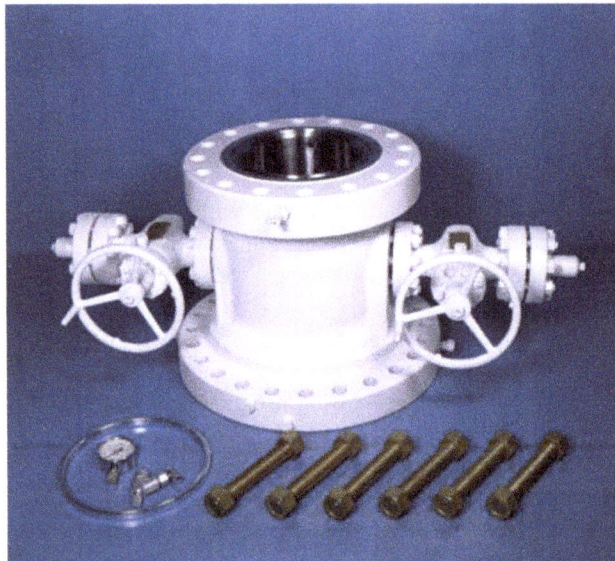

Fig. 8.4—Photo of a typical intermediate casing head and additional components required to assemble it during the drilling operation.

gage the pipe OD and transfer the suspended weight of the casing to the starting casing head. As the slips travel down, they are forced in against the casing, applying greater and greater support capacity. As the slips continue to engage the pipe, a load is placed on the automatic weight-set elastomeric seal assembly, sealing the annulus between the casing and the casing head. This installation creates a pressure barrier and isolates the annular pressure below the slip-and-seal casing hanger from the wellbore.

Traditionally, mandrel hangers (**Fig. 8.7**) are used only to suspend tubing from the tubing head. Occasionally, they can also be used in intermediate casing spools as an alternative to the slip-and-seal casing-hanger assembly. The mandrel hanger is a solid body with a through-bore

Fig. 8.5—Photo of a typical tubing head with installation components.

Fig. 8.6—Photo of a typical weight-set slip-and-seal assembly with casing-head installation components.

ID similar to that of the tubing or casing run below, and it also has penetrations for downhole safety valve line(s) and temperature and pressure gauges, if required. Traditionally in spool wellheads, elastomeric seals are used to seal the annulus between the casing-spool body and the casing or tubing hanger.

8.1.3 Annulus Seals. The seals used on spool wellhead systems are traditionally elastomeric. This is primarily because the seal must be energized against the casing-bowl ID and must also seal against the rough finish of the casing OD.

This elastomeric sealing system is used for the slip-and-seal assembly as well as the bottom of the intermediate casing or tubing spools. The slip-and-seal assembly (**Fig. 8.8**) provides a primary annulus seal, while the elastomeric seal in the bottom of each casing and tubing spool also provides a seal. The casing-spool flange connection becomes a secondary seal for both annulus and wellbore pressure. The elastomeric seals are manufactured using different materials to allow for various pressures, produced fluids, and other environmental conditions. The exception is the seal between each flange face, which is a metal-to-metal sealing ring gasket that

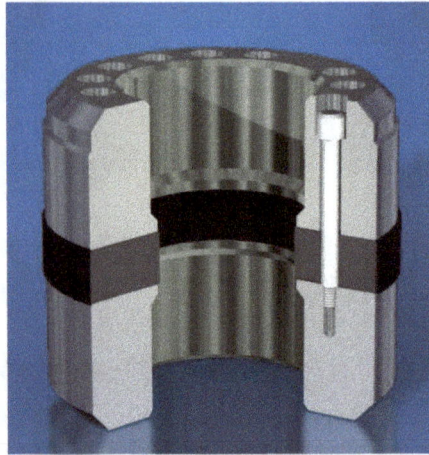

Fig. 8.7—Illustration (cutaway) of a mandrel-type tubing hanger.

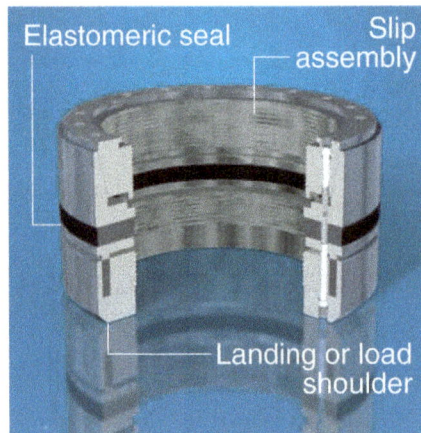

Fig. 8.8—Illustration of a weight-set slip-and-seal casing-hanger assembly.

provides a pressure-tight seal between each of the spool flanges. Ring gaskets are also used between the wellhead and the BOP stack, as well as the valves used for annulus access.

While drilling the well, it is required that the seal bores in each of the intermediate casing spools and tubing spools be protected. A series of wear bushings (**Fig. 8.9**) are supplied to protect the seal areas discussed during the drilling operation. The wear bushings are run on a drillpipe tool (**Fig. 8.10**) with J-lugs located on the OD that interface with J-slots located in the top ID section of the wear bushing.

It is also required that the flanged connections between each spool and the BOP be tested during the drilling and completion phases. The tools required are available from the equipment supplier. The tool used for testing the BOP is typically a plug type with a heavy-duty elastomer seal.

8.1.4 Annulus Access. For onshore wells, during the drilling operation, access to each annulus is required for the following reasons:

• To provide a flow-by area for returns during cementing of casing strings.

Fig. 8.9—Photos of the wear bushings for a typical land drilling wellhead system.

Fig. 8.10—Photo of the wear bushing running tools. These tools are also used to test the BOP stack.

- To provide access for possible well kill operations.
- To monitor the annulus for pressure below the slip-and-seal assembly.

8.1.5 Product Material Specifications. When ordering wellhead equipment, the following should be considered:

- All surface wellhead equipment and gate valves should be manufactured to the latest edition of the American Petroleum Inst. (API) and Intl. Organization for Standardization (ISO) standards. These standards define equipment specifications as follows:
- Material class: based on produced fluids; AA, BB, CC, DD, EE, FF, and HH (please see the example for gate-valve trims, shown in **Fig. 8.11**).
- Temperature range: 75 to +350°F.
- Please review the relevant API specifications for your application or consult your equipment supplier for further information.

8.2 Drilling a Well Offshore From a Jackup Drilling Rig Using Mudline Suspension Equipment

From a historic point of view, as jackup drilling vessels drilled in deeper water, the need to transfer the weight of the well to the seabed and provide a disconnect-and-reconnect capability became clearly beneficial. This series of hangers, called mudline suspension equipment, pro-

Gate Valve Sizes

2,000 psi	3,000 psi	5,000 psi	10,000 psi	15,000 psi
—	—	—	$1^{13}/_{16}$ in.	$1^{13}/_{16}$ in.
$2^1/_{16}$ in.	$2^1/_{16}$ in.	$2^1/_{16}$ in.	$2^1/_{16}$ in.	$2^1/_{16}$ in.
$2^9/_{16}$ in.	$2^9/_{16}$ in.	$2^9/_{16}$ in.	$2^9/_{16}$ in.	$2^9/_{16}$ in.
$3^1/_8$ in.	$3^1/_8$ in.	$3^1/_8$ in.	$3^1/_8$ in.	$3^1/_8$ in.
$4^1/_{16}$ in.	$4^1/_{16}$ in.	$4^1/_{16}$ in.	$4^1/_{16}$ in.	$4^1/_{16}$ in.
$5^1/_8$ in.	$5^1/_8$ in.	$5^1/_8$ in.	$5^1/_8$ in.	—
$6^3/_8$ in.	$6^3/_8$ in.	$6^3/_8$ in.	$6^3/_8$ in.	—

API Material Classifications

Valve Part Description	API Class AA (Trim AA) Regular Trim	API Class BB (Trim BB) Alloy Body w/ Stainless Internal Trim	API Class CC (Trim CC) Full Stainless Trim	API Class DD (Trim DD) Alloy Body w/Alloy Internal Controlled Hardness	API Class EE (Trim EE) H₂S/CO₂ Service Alloy Body w/Stainless Internal Trim Controlled Hardness	API Class FF (Trim FF) H₂S/CO₂ Service Full Stainless Controlled Hardness	API Class HH (Trim HH) H₂S/CO₂ Service CRA Cladded Controlled Hardness
Body and Bonnet	Low Alloy Forging	Low Alloy Forging	410 Stainless Steel or F6NM SS Forging	Low Alloy Forging	Low Alloy Forging	410 Stainless Steel or F6NM SS Forging	Low Alloy Forging w/625 Cladding
Stem	Low Alloy Steel w/Xylan Coating	410 Stainless Steel or Alloy K-500 w/Xylan Coating	410 Stainless Steel or Alloy K-500 w/Xylan Coating	Low Alloy Steel w/Xylan Coating	410 Stainless Steel or Alloy K-500 w/Xylan Coating	410 Stainless Steel or Monel K-500 w/Xylan Coating	Inconel 718, Xylan Coated
Gate and Seat	Low Alloy Steel w/Hard Facing	F6NM Stainless Steel w/Hard Facing	F6NM Stainless Steel w/Hard Facing	Low Alloy Steel w/Hard Facing	F6NM Stainless Steel w/Hard Facing	F6NM Stainless Steel w/Hard Facing	Inconel 718 or 625 w/Hard Facing
Lift Nut	410 Stainless Steel	410 Stainless Steel	410 Stainless Steel	410 Stainless Steel	410 Stainless Steel	410 Stainless Steel	Monel K-500 or Inconel 625
Stem and Seat Packing	Teflon Jacket Lip Seal Energized by an Elgiloy Spring	Teflon Jacket Lip Seal Energized by an Elgiloy Spring	Teflon Jacket Lip Seal Energized by an Elgiloy Spring	Teflon Jacket Lip Seal Energized by an Elgiloy Spring	Teflon Jacket Lip Seal Energized by an Elgiloy Spring	Teflon Jacket Lip Seal Energized by an Elgiloy Spring	Teflon Jacket Lip Seal Energized by an Elgiloy Spring
Seat Face Seal	Solid Teflon	Solid Teflon	Solid Teflon	Solid Teflon	Solid Teflon	Solid Teflon	Solid Teflon
Bonnet Gasket	300 Series Stainless Steel	300 Series Stainless Steel	300 Series Stainless Steel	300 Series Stainless Steel	300 Series Stainless Steel	300 Series Stainless Steel	Inconel 825
Gate Guide and Seat Ring	300 Series Stainless Steel	300 Series Stainless Steel	300 Series Stainless Steel	300 Series Stainless Steel	300 Series Stainless Steel	300 Series Stainless Steel	F6NM or CA6NM Stainless Steel
Seat Spring	X-750	X-750	X-750	X-750	X-750	X-750	X-750
Packing Junk Ring	410 Stainless Steel w/Xylan Coating	410 Stainless Steel w/Xylan Coating	410 Stainless Steel w/Xylan Coating	410 Stainless Steel w/Xylan Coating	410 Stainless Steel w/Xylan Coating	410 Stainless Steel w/Xylan Coating	410 Stainless Steel w/Xylan Coating
Packing Nut, Bonnet Cap and Bearing Thrust Ring	Low Alloy Steel	Low Alloy Steel	Low Alloy Steel	Low Alloy Steel	Low Alloy Steel	Low Alloy Steel	Low Alloy Steel

Fig. 8.11—Tables showing typical gate-valve sizes (above) and trims (below). These trims are also applicable to surface wellheads.

vides landing rings and shoulders to transfer the weight of each casing string to the conductor and the sea bed.

The mudline hanger system (shown in **Fig. 8.12**) consists of the following components:
- Butt-weld sub.
- Shoulder hangers.
- Split-ring hangers.
- Mudline hanger running tools.
- Temporary abandonment caps and running tool.
- Tieback tools.
- Cleanout tools.

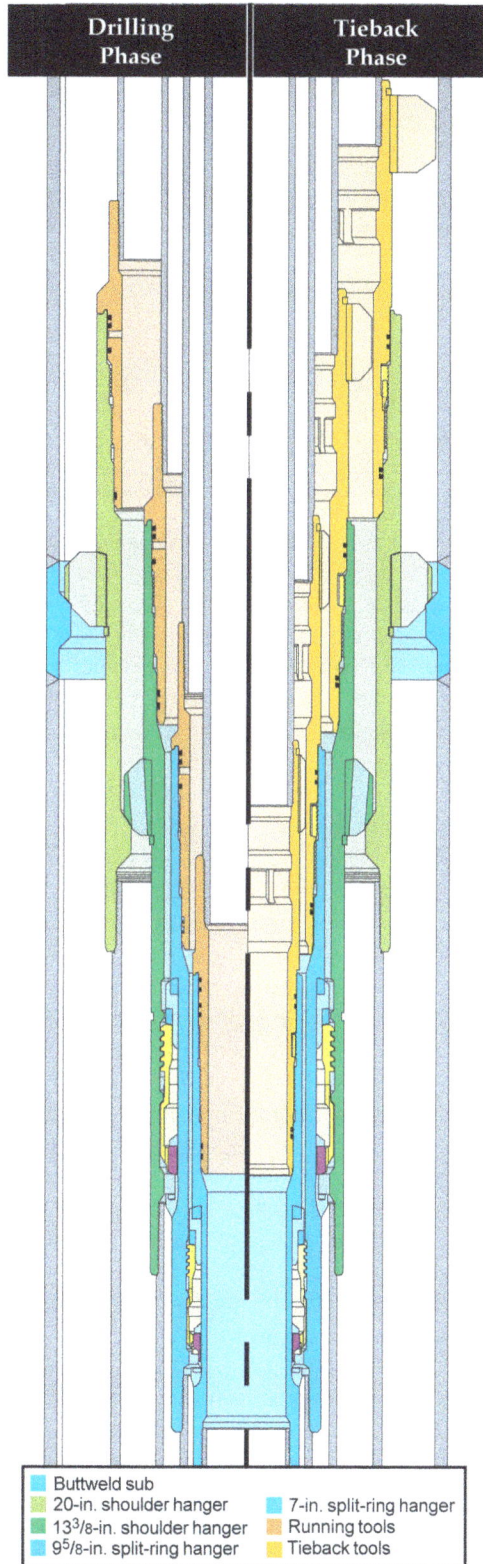

Fig. 8.12—Illustration of a typical mudline suspension system showing running tools on the left side and tieback tools on the right side.

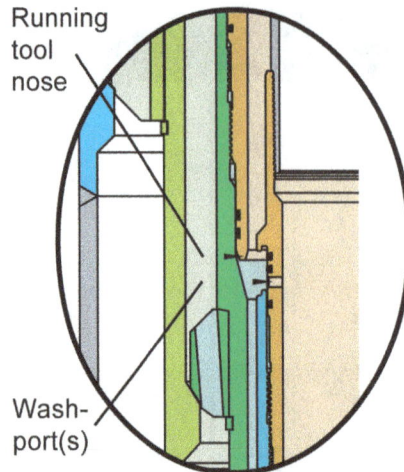

Fig. 8.13—All mudline hangers should stack down to provide washout efficiency. Washout efficiency is supplied by a series of wash ports located in the running tool that (when opened for washing out) are positioned below the running tool attached to the previously run mudline hanger.

8.2.1 Mudline Hangers. Each mudline hanger landing shoulder and landing ring centralizes the hanger body and establishes concentricity around the center line of the well. Concentricity is important when tying the well back to the surface. In addition, each hanger body stacks down relative to the previously installed hanger for washout efficiency. Washout efficiency is necessary to clean the annulus area of the previously run mudline hanger and running tool (**Fig. 8.13**). This ensures that cement and debris cannot hinder disconnect and retrieval of each casing riser to the rig floor upon abandonment of the well.

As each hole section is drilled and each casing string and mudline hanger is run, the hanger is positioned in the casing string to land on a landing shoulder inside the mudline hanger that was installed with the previous casing string. Each of the mudline hangers have casing and a mudline hanger running tool made up to it. These running tools are released through right-hand rotation to allow disconnect from the well. The threads on the mudline hanger used by the running tool can be used to install temporary abandonment caps (**Fig. 8.14**) into selected hangers to temporarily "suspend" drilling operations at the conclusion of the well.

The main difference between the wellheads used in the land drilling application and the jackup drilling application (with mudline) is the slip-and-seal assembly (**Fig. 8.15**). Because the weight of the well now sits at the seabed, a weight-set slip-and-seal assembly is not used. Instead, a mechanical set (energizing the seal by hand) is used, in which cap screws are made up with a wrench against an upper compression plate on the slip-and-seal assembly to energize the elastomeric seal.

8.2.2 Temporarily Abandoning the Well. The mudline suspension system also allows the well to be temporarily abandoned (disconnected) when "TD" is achieved (when drilling is finished at total depth). When this occurs, the conductor is normally cut approximately 5 to 6 ft above the mudline and retrieved to the surface. After each casing string is disconnected from the mudline suspension hanger and retrieved to the rig floor in the reverse order of the drilling process, threaded temporary abandonment caps or stab-in temporary abandonment caps (both of which makeup into the threaded running profile of the mudline hanger; see Fig. 8.14) are installed in selected mudline hangers before the drilling vessel finishes and leaves the location. The temporary abandonment caps can be retrieved with the same tool that installed them.

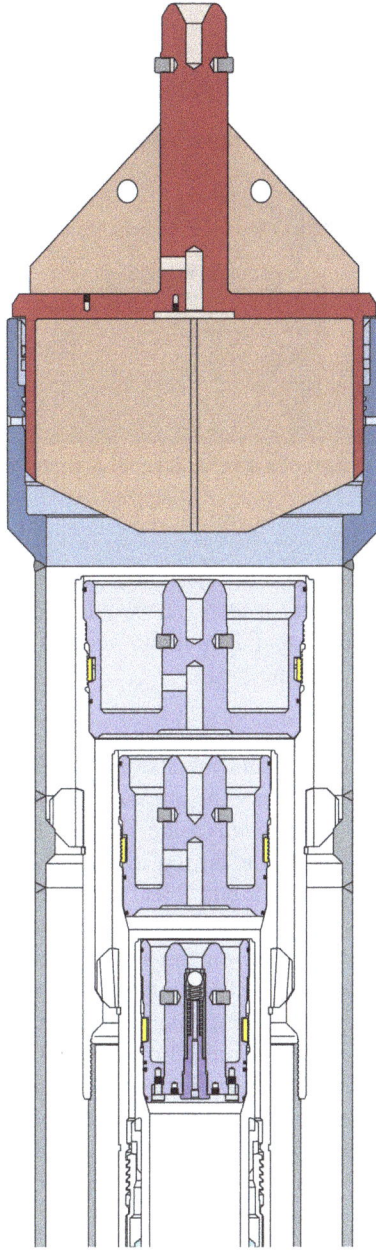

Fig. 8.14—Illustration of a mudline suspension system with temporary abandonment caps installed after the well is drilled.

8.2.3 Reconnecting to the Well. A mudline suspension system also incorporates tieback tools to reconnect the mudline hanger to the surface for re-entry and/or completion. These tieback tools can be of two types: threaded and stab-in (see **Fig. 8.16**). The tieback tools are different from the running tools in that they makeup into their own dedicated right-hand makeup threaded profile. The stab-in tieback tool offers a simple, weight-set, rotation-lock design that provides an easy way to tie the well back to the surface. A surface wellhead system is installed, and the well is completed similarly to the method used on land drilling operations.

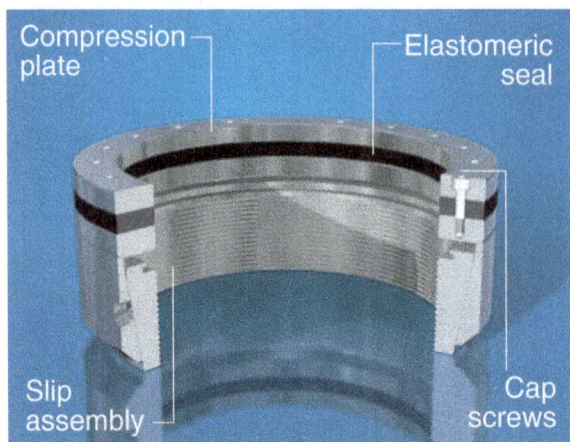

Fig. 8.15—Illustration of a mechanical-set slip-and-seal assembly.

The mudline suspension system has been designed to accommodate tying the well back to the surface for surface completion, and it also can be adapted for a subsea production tree. A tieback tubing head can be installed to the mudline suspension system at the seabed, and a subsea tree can be installed on this tubing head.

8.3 The Unitized Wellhead

The unitized wellhead is very different from the spool wellhead system because it incorporates different design characteristics and features. The unitized wellhead, shown in **Fig. 8.17**, is a one-piece body that is typically run on $13^{3}/_{8}$-in. casing through the BOP and lands on a landing shoulder located inside the starting head or on top of the conductor itself. The casing hangers used are threaded and preassembled with a pup joint. This way, the threaded connection can be pressure tested before leaving the factory, ensuring that the assembly will have pressure-containing competence. Gate valves are installed on the external outlet connections of the unitized wellhead to enable annulus access to each of the intermediate and the production casing strings.

After the next hole section is drilled, the casing string, topped out with its mandrel hanger, is run and landed on a shoulder located in the ID of the unitized wellhead. A seal assembly is run on a drillpipe tool to complete the casing-hanger and seal-installation process. Each additional intermediate casing string and mandrel hanger is run and landed on top of the previously installed casing hanger without removing the BOP stack. Besides saving valuable rig time, the other advantage of the unitized wellhead system over spool wellhead systems is complete BOP control throughout the entire drilling process.

The unitized wellhead (**Fig. 8.18**) consists of the following components:
- Unitized wellhead body.
- Annulus gate valves.
- Mandrel casing hangers.
- Mandrel tubing hangers.
- Metal-to-metal sealing for the annulus seals.

8.3.1 Mandrel Casing Hangers. The mandrel casing hangers (see **Fig. 8.19**) are a one-piece construction and are manufactured to meet the casing and thread types specified by the customer. The mandrel casing hanger has a 4° tapered sealing area on its OD. The mandrel hanger still also incorporates running threads and seal-assembly threads to facilitate installation. The hanger carries a lock ring that locks the hanger down when the seal assembly is installed. The

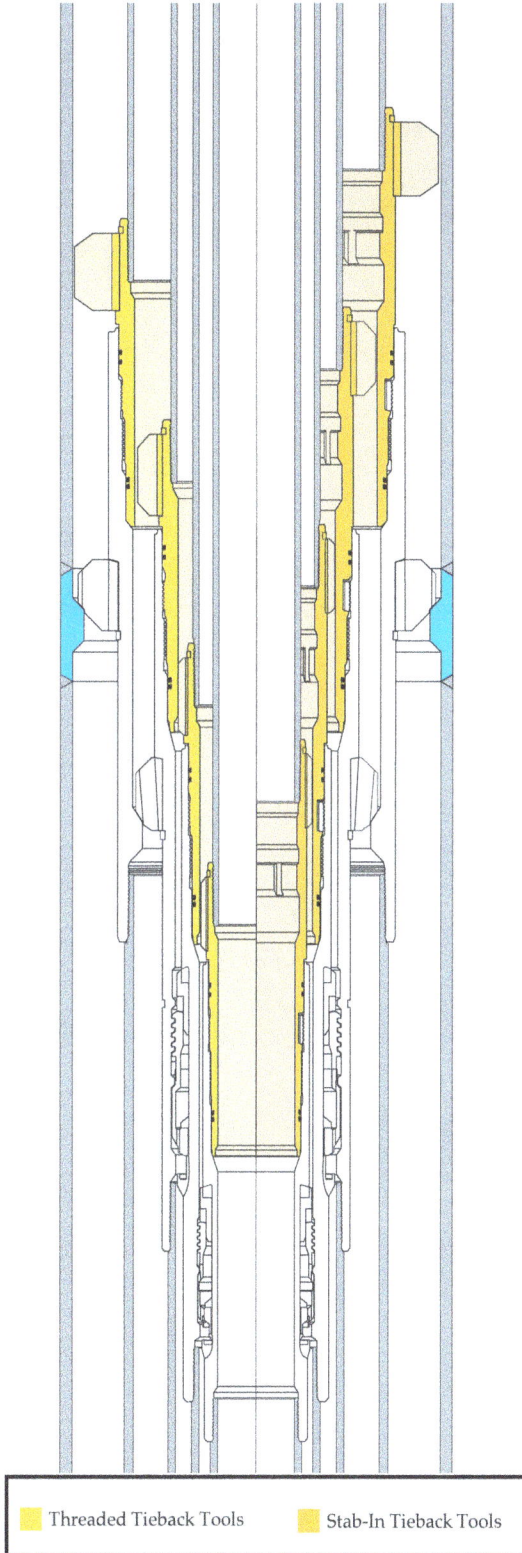

	Threaded Tieback Tools		Stab-In Tieback Tools

Fig. 8.16—Illustration of a mudline suspension system with threaded tieback tools installed (left) and "stab-in" tieback tools installed (right) in each hanger body.

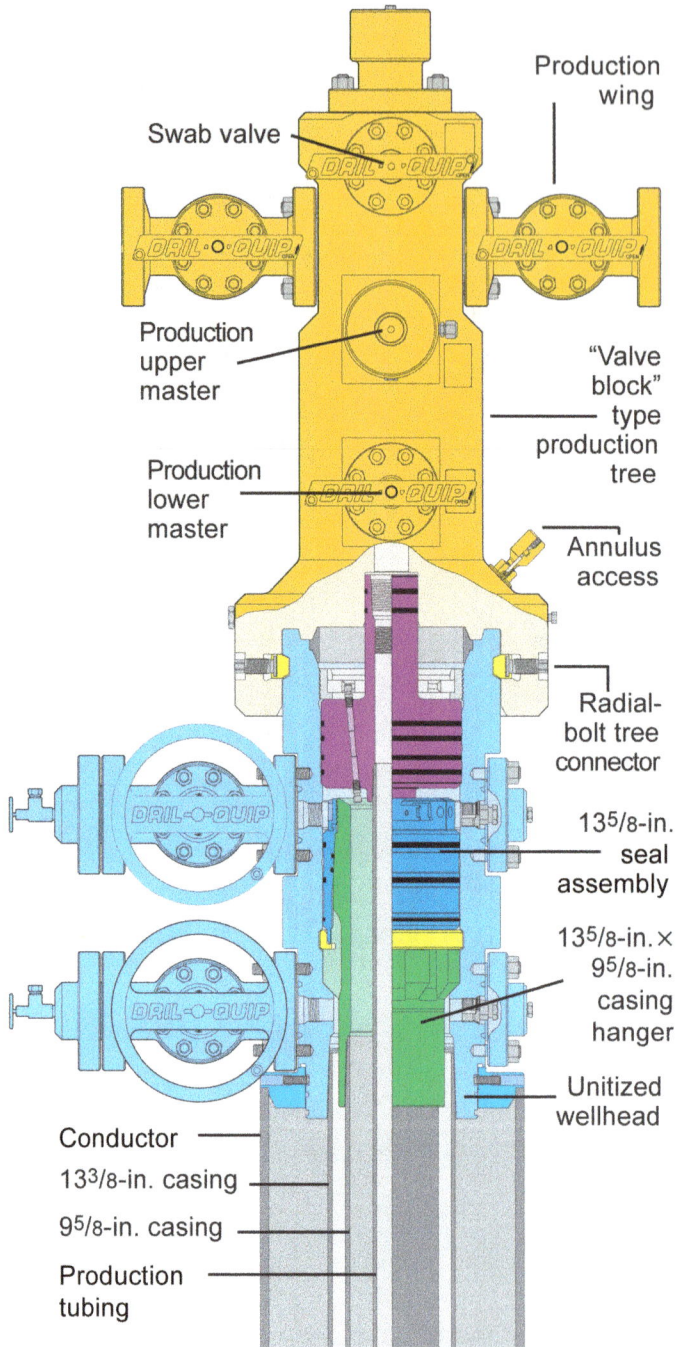

Fig. 8.17—Illustration of a typical unitized wellhead system for land applications. Offshore unitized wellhead systems are typically similar but include the use of a metal-to-metal seal assembly.

mandrel casing hanger lands on either the shoulder located in the bottom of the unitized wellhead body or on top of the previous casing hanger.

Fig. 8.18—Photo of a 13⅝-in. 5,000-psi unitized wellhead system and its major components.

Fig. 8.19—Photo of a unitized wellhead mandrel hanger.

8.3.2 Seal Assembly. The seal assembly incorporates a metal-to-metal or elastomeric seal (**Fig. 8.20**), which is run on a running tool through the BOP stack once the casing has been cemented. The seal assembly seals off the pressure from above and below and isolates the annulus from the wellbore. The annulus can still be monitored through the outlets on the unitized wellhead body and the gate valves mounted to them.

There is a full range of tools available for the unitized wellhead system:
• Wellhead-housing running tool.
• BOP test tool.
• Casing-hanger running tool.
• Seal-assembly running and retrieving tool.
• Wear-bushing running and retrieving tool.

The unitized wellhead is more often used with platform-development projects than with exploration applications.

Fig. 8.20—Photo of a seal assembly for the mandrel hanger. This seal assembly features elastomeric seals, but it also can feature true metal-to-metal seals.

8.4 Drilling a Well Subsea

The subsea wellhead system (**Fig. 8.21**) is a pressure-containing vessel that provides a means to hang off and seal off casing used in drilling the well. The wellhead also provides a profile to latch the subsea BOP stack and drilling riser back to the floating drilling rig. In this way, access to the wellbore is secure in a pressure-controlled environment. The subsea wellhead system is located on the ocean floor and therefore must be installed remotely with running tools and drillpipe.

The subsea wellhead ID is designed with a landing shoulder located in the bottom section of the wellhead body. Subsequent casing hangers land on the previous casing hanger installed. Casing is suspended from each casing-hanger top and accumulates on the primary landing shoulder located in the ID of the subsea wellhead. Each casing hanger is sealed off against the ID of the wellhead housing and the OD of the hanger itself with a seal assembly that incorporates a true metal-to-metal seal. This seal assembly provides a pressure barrier between casing strings, which are suspended in the 18¾-in. wellhead.

Once drilling is complete, the wellhead will provide an interface for the production tubing string and the subsea production tree or, if required, a point to tie back to a platform. The design objective of the subsea wellhead system is twofold: first, to provide the operator with the latest equipment technology incorporating reliable solutions for the well conditions to be encountered, as well as maximum strength and capacities; and second, to provide a system that is easy to install and requires a minimal amount of handling and rig time.

A standard subsea wellhead system will typically consist of the following:
• Drilling guide base.
• Low-pressure housing.
• High-pressure wellhead housing (typically 18¾ in.).
• Casing hangers (various sizes, depending on casing program).
• Metal-to-metal annulus sealing assembly.
• Bore protectors and wear bushings.
• Running and test tools.

8.4.1 Drilling Guide Base. The drilling guide base (**Fig. 8.22**) provides a means for guiding and aligning the BOP onto the wellhead. Guide wires from the rig are attached to the guide-

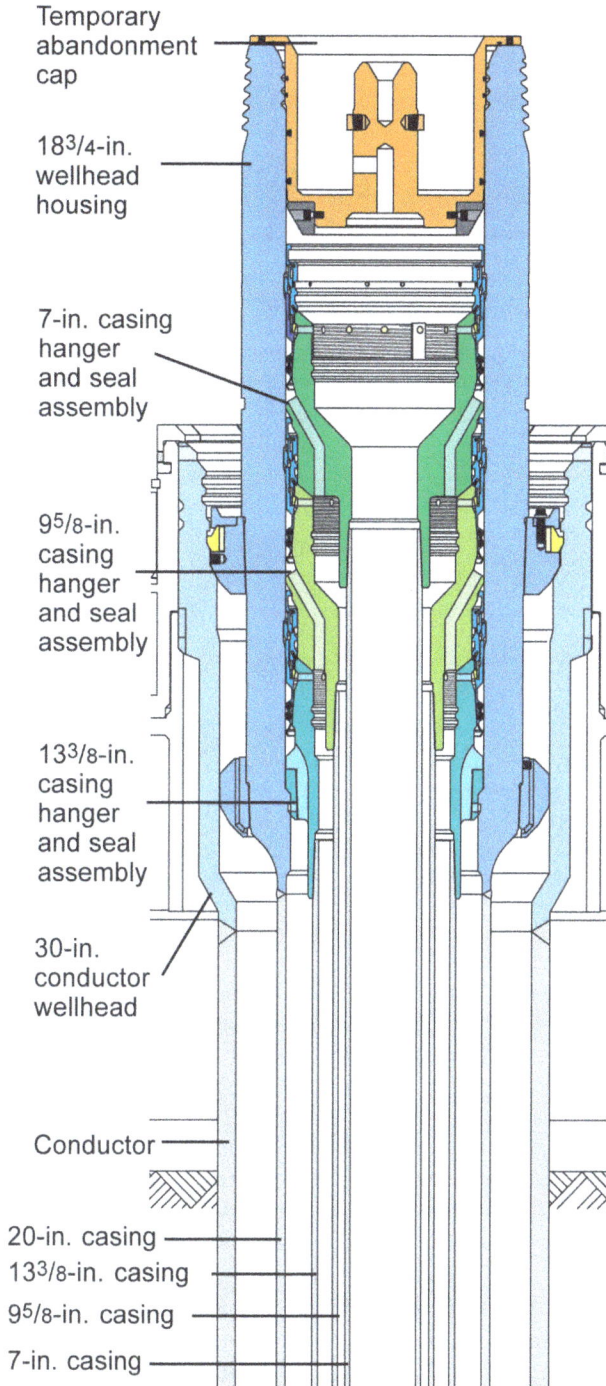

Fig. 8.21—Illustration of a typical subsea wellhead system with temporary abandonment cap installed. This illustration also shows the wellhead configuration with a 30 × 20 × 13⅜ × 9⅝ × 7-in. casing program.

posts of the base, and the wires are run subsea with the base to provide guidance from the rig down to the wellhead system.

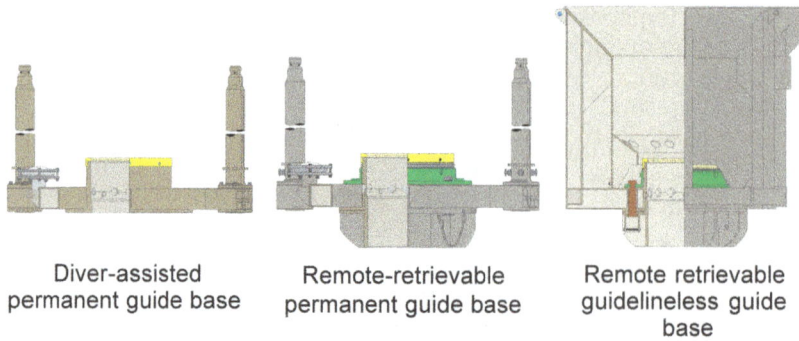

Diver-assisted
permanent guide base

Remote-retrievable
permanent guide base

Remote retrievable
guidelineless guide
base

Fig. 8.22—Illustration of typical guide bases, both guidelined and guidelineless. Each guide base can incorporate customer-specified features, such as remote-retrievable capabilities and special flow-by features.

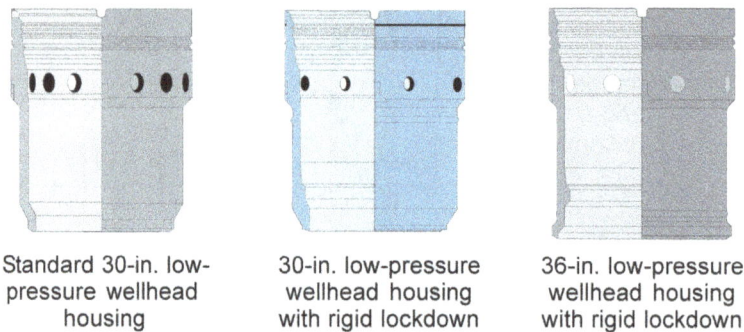

Standard 30-in. low-
pressure wellhead
housing

30-in. low-pressure
wellhead housing
with rigid lockdown

36-in. low-pressure
wellhead housing
with rigid lockdown

Fig. 8.23—Illustration of typical low-pressure wellhead housings. Each low-pressure housing can also incorporate various features based on the particular application and drilling environment.

8.4.2 Low-Pressure Housing. The low-pressure housing (typically 30 or 36 in.; see **Fig. 8.23**) provides a location point for the drilling guide base and provides an interface for the 18¾-in. high-pressure housing. It is important for this first string to be jetted or cemented in place correctly because this string is the foundation for the rest of the well.

8.4.3 High-Pressure Housing. The subsea high-pressure wellhead housing (typically 18¾ in.) is effectively a unitized wellhead with no annulus access. It also provides an interface between the subsea BOP stack and the subsea well. The subsea wellhead is the male member to a large-bore connection, as shown in **Fig. 8.24** (the female counterpart is the wellhead connector on the bottom of the BOP stack) that will be made up in a remote subsea, ocean-floor environment. The 18¾-in. wellhead will house and support each casing string by way of a mandrel-type casing hanger. The ID of the 18¾-in. wellhead provides a metal-to-metal sealing surface for the seal assembly when it is energized around the casing hanger. The wellhead provides a primary landing shoulder in the bottom ID area to support the combined casing loads and will typically accommodate two or three casing hangers and a tubing hanger. The minimum ID of the wellhead is designed to let a 17½-in. drilling bit pass through.

8.4.4 Casing Hangers. All subsea casing hangers are mandrel type, as shown in **Fig. 8.25**. The casing hanger provides a metal-to-metal sealing area for a seal assembly to seal off the annulus between the casing hanger and the wellhead. The casing weight is transferred into the wellhead

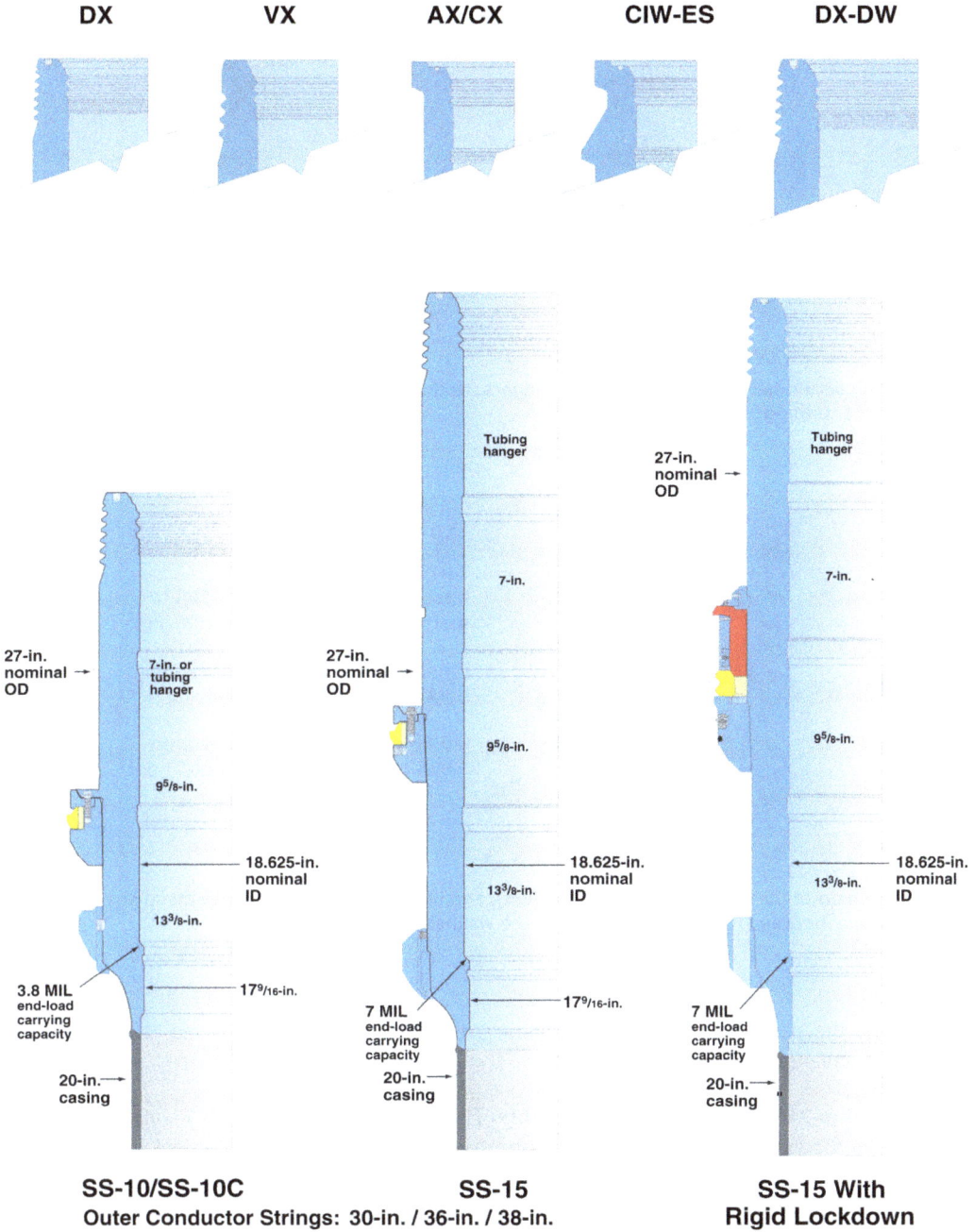

DX   VX   AX/CX   CIW-ES   DX-DW

27-in. nominal OD

7-in. or tubing hanger

9⁵/₈-in.

13³/₈-in.

3.8 MIL end-load carrying capacity

20-in. casing

17⁹/₁₆-in.

SS-10/SS-10C
Outer Conductor Strings: 30-in. / 36-in. / 38-in.

Tubing hanger

7-in.

27-in. nominal OD

9⁵/₈-in.

13³/₈-in.

18.625-in. nominal ID

7 MIL end-load carrying capacity

20-in. casing

17⁹/₁₆-in.

SS-15

27-in. nominal OD

Tubing hanger

7-in.

9⁵/₈-in.

13³/₈-in.

18.625-in. nominal ID

7 MIL end-load carrying capacity

20-in. casing

SS-15 With Rigid Lockdown

Fig. 8.24—18¾-in. wellheads are manufactured with several different locking profiles to mate with the wellhead connector located on the bottom of the BOP stack or subsea production tree. The wellhead systems are usually rated for 10,000 or 15,000 psi and can be installed with a standard lock ring or a rigid lockdown mechanism, which is the preferred choice for deepwater operations.

by means of the casing hanger/wellhead landing shoulder. Each casing hanger stacks on top of another and, consequently, all casing loads are transferred through each hanger to the landing shoulder at the bottom of the subsea wellhead. Each casing hanger incorporates flow-by slots

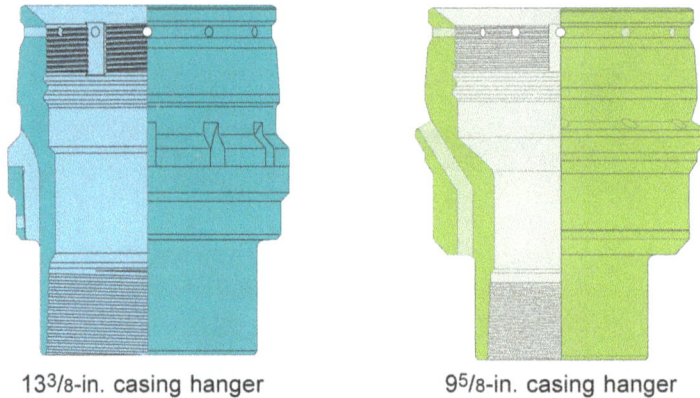

13³/₈-in. casing hanger   9⁵/₈-in. casing hanger

Fig. 8.25—Illustrations of the subsea casing hangers. Notice features to accommodate casing, a seal assembly, and a running tool.

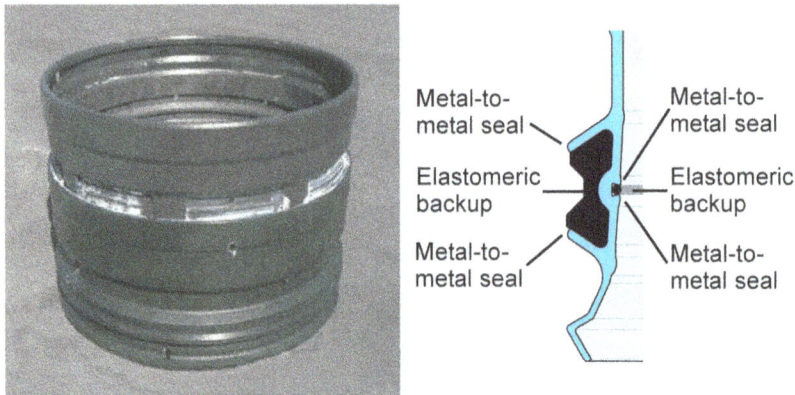

Fig. 8.26—Photo of the 18³/₄-in. seal assembly (left), and illustration of the metal-to-metal seal that seals off the annulus between the casing hanger and the wellhead.

to facilitate the passage of fluid while running through the drilling riser and BOP stack and during the cementing operation.

8.4.5 Metal-to-Metal Annulus Seal Assembly. The seal assembly (**Fig. 8.26**) isolates the annulus between the casing hanger and the high-pressure wellhead housing. The seal incorporates a metal-to-metal sealing system that today is typically weight-set (torque-set seal assemblies were available in earlier subsea wellhead systems). During the installation process, the seal is locked to the casing hanger to keep it in place. If the well is placed into production, then an option to lock down the seal to the high-pressure wellhead is available. This is to prevent the casing hanger and seal assembly from being lifted because of thermal expansion of the casing down hole.

8.4.6 Bore Protectors and Wear Bushings. Once the high-pressure wellhead housing and the BOP stack are installed, all drilling operations will take place through the wellhead housing. The risk of mechanical damage during drilling operations is relatively high, and the critical landing and sealing areas in the wellhead system need to be protected with a removable bore protector and wear bushings, as shown in **Fig. 8.27**.

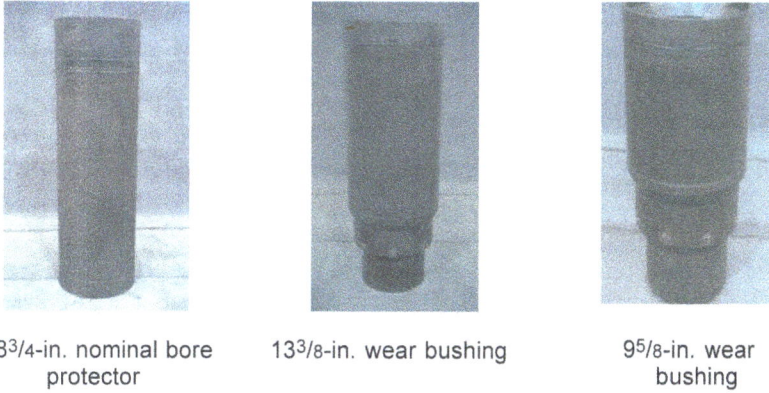

18³/₄-in. nominal bore protector   13³/₈-in. wear bushing   9⁵/₈-in. wear bushing

Fig. 8.27—Photos of the nominal bore protector, 13³/₈-in. wear bushing, and 9⁵/₈-in. wear bushing. These wellhead components are run on a multipurpose tool.

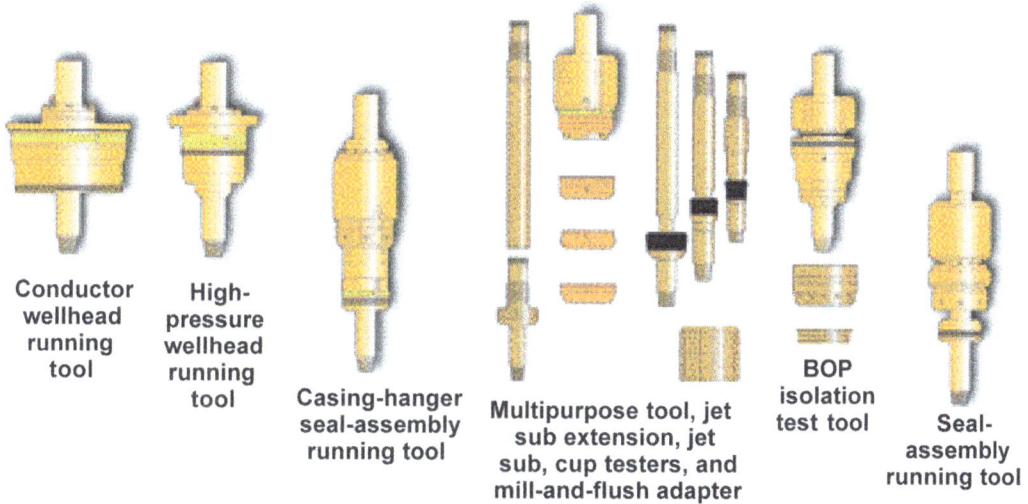

Conductor wellhead running tool

High-pressure wellhead running tool

Casing-hanger seal-assembly running tool

Multipurpose tool, jet sub extension, jet sub, cup testers, and mill-and-flush adapter

BOP isolation test tool

Seal-assembly running tool

Fig. 8.28—Illustration of the SS-15 Subsea Wellhead System running tool family.

8.4.7 Running and Test Tools. The standard subsea wellhead system will include typical running, retrieving, testing, and reinstallation tools (see **Fig. 8.28**). These tools include:

• *Conductor Wellhead Running Tool.* The conductor wellhead running tool runs the conductor casing, conductor wellhead, and guide base. This tool can be used for jetting in the conductor or cementing the conductor into a predrilled hole. The tool is a cam-actuated tool that minimizes any high torque that may be encountered during operations.

• *High-Pressure Wellhead Running Tool.* The high-pressure wellhead running tool operates just like the conductor wellhead running tool, but it runs the high-pressure wellhead and 20-in. casing. It is a cam-actuated tool that minimizes any high torque that may be encountered during operations.

• *Casing-Hanger Seal-Assembly Running Tool.* The casing-hanger seal-assembly running tool runs the casing, casing hanger, and seal assembly in one trip. It also allows testing of the seal assembly (after installation) and the BOP stack, and it has the additional benefit of bringing back the seal assembly if debris is in the way and the seal assembly cannot be installed.

• *Multipurpose Tool and Accessories.* The multipurpose tool runs and retrieves the nominal bore protector and all wear bushings. A jet sub and/or jet sub extension can be attached to the multipurpose tool so that wellhead washout can occur during the retrieval process. The multipurpose tool also retrieves the seal assembly and becomes a mill-and-flush tool by attaching the mill-and-flush adapter.

• *BOP Isolation Test Tool.* The BOP isolation test tool allows testing of the BOP stack without allowing pressure to be applied against the casing-hanger seal assembly. The BOP isolation test tool can land on the casing hangers or wear bushings.

• *Seal-Assembly Running Tool.* The seal-assembly running tool is used in the event that a second seal assembly needs to be run. The seal-assembly running tool is a weight-set tool and, like the casing-hanger seal-assembly running tool, it allows testing of the BOP stack and recovers the seal assembly if it cannot be installed (because of debris in the sealing area of the annulus).

8.5 Big Bore Subsea Wellhead Systems

As the offshore oil and gas industry has continued to explore for oil and gas in deeper and deeper waters, the requirements for well components have changed as a result of the challenges associated with deepwater drilling. Ocean-floor conditions in deep and ultradeep water can be extremely mushy and unconsolidated, which creates well-foundation problems that require development of new well designs to overcome the conditions. Second, underground aquifers in deep water have been observed in far greater frequency than in shallower waters, and it quickly became clear that these zones would have to be isolated with a casing string. Cementing requirements changed, and wellhead equipment designs would also have to change to accommodate the additional requirements.

With subsea wellhead systems, conductor and intermediate casing strings can be reconfigured to strengthen and stiffen the upper section of the well (for higher bending capacities) and overcome the challenges of an unconsolidated ocean floor at the well site. But each "water flow" zone encountered while drilling requires isolation with casing and, at the same time, consumes a casing-hanger position in the wellhead. It became obvious that more casing strings and hangers were required to reach the targeted depth than the existing wellhead-system designs would accommodate.

The 18¾-in. Big Bore Subsea Wellhead System (**Fig. 8.29**) was designed for wells that will be installed in unconsolidated ocean-floor conditions and will penetrate shallow water-flow zones. These well conditions require additional casing strings. The wellhead system incorporates an 18¾-in. high-pressure wellhead housing designed for 15,000 psi and 7 million pounds end-load carrying capacity. Unlike conventional subsea wellhead systems, the big-bore high-pressure wellhead housing (**Fig. 8.30**) is run atop 22-in. pipe (as opposed to 20-in. pipe) and has a large minimum ID bore to pass 18-in. casing. The wellhead system incorporates a rigid lockdown mechanism to preload the connection between the high-pressure wellhead and the conductor wellhead. A supplemental hanger adapter is installed in the 22-in. casing to provide a landing shoulder and seal area for the 18-in. and 16-in. supplemental hangers and their testable, retrievable seal assemblies.

Optional 28-in., 26-in., and 24-in. supplemental casing-hanger systems can be incorporated into the design to accommodate a secondary conductor string and thereby increase the overall bending capacity of the upper section of the well and/or provide an additional barrier for a water-flow zone. All casing hangers and seal assemblies are run, set, and tested on drillpipe in a single trip. These subsea wellhead systems can easily accommodate alternative casing programs and can be configured to address any deepwater (and shallow-water) drilling application.

8.6 Summary

As has been discussed in this chapter, wellhead systems (whether the application is surface wellheads on land, jackups or offshore production platforms, or subsea wellheads) serve as the

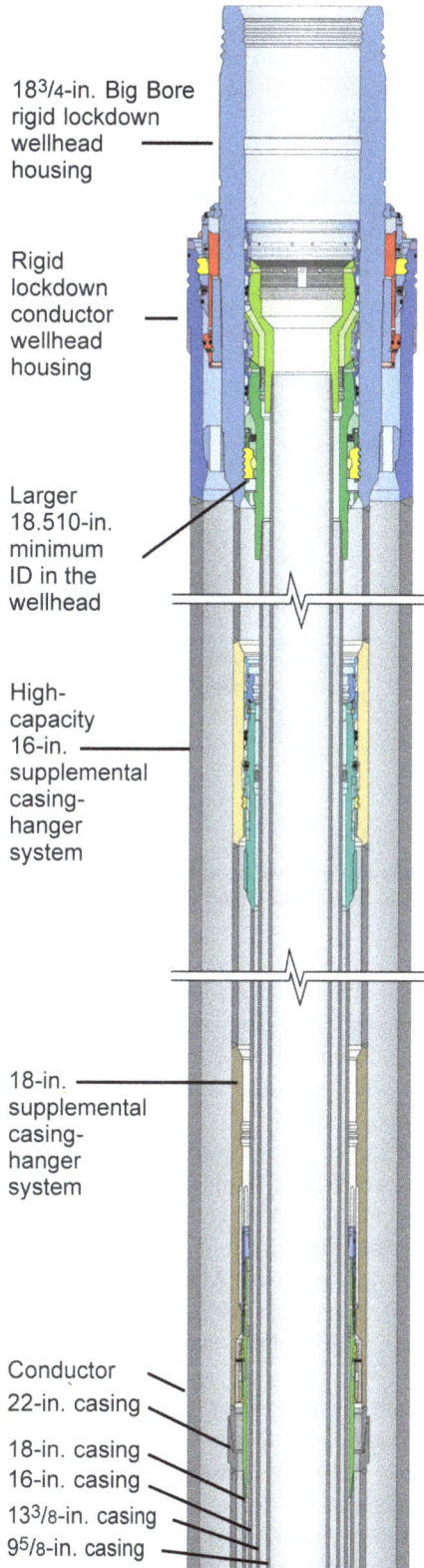

18³/4-in. Big Bore rigid lockdown wellhead housing

Rigid lockdown conductor wellhead housing

Larger 18.510-in. minimum ID in the wellhead

High-capacity 16-in. supplemental casing-hanger system

18-in. supplemental casing-hanger system

Conductor 22-in. casing

18-in. casing
16-in. casing
13³/8-in. casing
9⁵/8-in. casing

Fig. 8.29—Illustration of the Big Bore Subsea Wellhead System with the 18-in. and 16-in. supplemental casing-hanger systems.

Fig. 8.30—Deepwater subsea wellhead, designed specifically to meet the requirements of higher-strength and pressurized shallow-zone water flows associated with ultradeepwater drilling in the Gulf of Mexico.

termination point of casing and tubing strings. As such, these systems control pressure and provide access to the main bore of the casing or tubing or to the annulus. This pressure-controlled access allows drilling and completion activities to take place safely and with minimal environmental risk. Multiple barriers are used, such as primary and secondary seals, to reduce risk in case of equipment failure.

Land wellhead systems, offshore surface wellhead systems, and subsea wellhead systems have been discussed. Offshore wellhead systems are normally more sophisticated in design to handle ocean currents, bending loads, and other loads induced by the environment during the life of the well. Some of these loads are cyclic in nature, so fatigue-resistant designs are desirable, particularly for deepwater developments. Material specifications play an important role in equipment performance; organizations such as API, the American Soc. of Mechanical Engineers (ASME), and NACE Intl. offer helpful standards to provide cost-effective solutions to technical challenges.

In certain applications such as deepwater platforms, spars, and tension-leg platforms (TLPs), surface wellheads and subsea wellheads are used together to safely produce hydrocarbons. In water depths of 500 to 1,400 ft, subsea wellheads are used to explore and develop offshore fields. Deepwater production platforms can be placed over these wells and tied back to the subsea wellheads; the top termination of the tieback at the platform will typically use surface unitized wellheads with solid block Christmas trees (which have fewer leak paths) as pressure-controlled access points to each well. Spars and TLPs are floating vessels used in deep water up to 4,500 ft. The wells are drilled using subsea wellheads, which are then tied back to the production deck of the spar or TLP, again using unitized wellheads and solid block trees to safely control and produce the well. For these special applications, it is recommended to contact your equipment supplier for more detailed information.

General References

Aldridge, D. and Dodd, P.: "Meeting the Challenges of Deepwater Subsea Completion Design," paper SPE 36991 presented at the 1996 SPE Asia Pacific Oil and Gas Conference, Adelaide, Australia, 28–31 October.

Anchaboh, L. *et al.:* "Conductor Sharing Wellheads—More For Less," paper SPE 68699 presented at the 2001 SPE Asia Pacific Oil and Gas Conference and Exhibition, Jakarta, 17–19 April.

Andersen, J.N., Rosine, R.S., and Marshall, M.: "Full-Scale High-Pressure Stripper/Packer Testing With Wellhead Pressure to 15,000 psi," paper SPE 60699 presented at the 2000 SPE/ICoTA Coiled Tubing Roundtable, Houston, 5–6 April.

Bazile, D.J. II and Kluck, L.M.: "New Wellhead Equipment for Old Oilfields," paper SPE 16122 presented at the 1987 SPE/IADC Drilling Conference, New Orleans, 15–18 March.

Burman, S.S. and Norton, S.J.: "Mensa Project: Well Drilling and Completion," paper OTC 8578 presented at the 1998 Offshore Technology Conference, Houston, 4–7 May.

Cort, A.J.C. and Ford, J.T.: "The Design and Testing of Subsea Production Equipment: Current Practice and Potential for the Future," paper SPE 30675 presented at the 1995 SPE Annual Technical Conference and Exhibition, Dallas, 22–25 October.

Dupal, K. and Flodberg, K.D.: "Auger TLP: Drilling Engineering Overview," paper SPE 22543 presented at the 1991 SPE Annual Technical Conference and Exhibition, Dallas, 6–9 October.

Eaton, L.F.: "Drilling Through Deepwater Shallow Water Flow Zones at Ursa," paper SPE/IADC 52780 presented at the 1999 SPE/IADC Drilling Conference, Amsterdam, 9–11 March.

Gordy, C.A., Combes, J.F., and Childers, M.A.: "Case History of a 22,000-ft Deepwater Wildcat," paper SPE 16084 presented at the 1987 SPE/IADC Drilling Conference, New Orleans, 15–18 March.

Hadj-Moussa, N.: "Rig Equipment Planning for a Deep, Deviated High-Pressure/High Temperature Khuff Well," paper SPE 57553 presented at the 1999 SPE/IADC Middle East Drilling Technology Conference, Abu Dhabi, UAE, 8–10 November.

Harms, D.A.: "Coiled-Tubing Completion Procedure Reduces Cost and Time for Hydraulically Fractured Wells," paper SPE 27892 presented at the 1994 SPE Western Regional Meeting, Long Beach, California, 23–25 March.

Heijnen, W.H.P.M.: "A New Wellhead Design Concept," paper SPE 19252 presented at the 1989 SPE Offshore Europe Conference, Aberdeen, 5–8 September.

Holand, P.: "Reliability of Deepwater Subsea Blowout Preventers," *SPEDC* (March 2001).

Huntoon, G.G. II: "A Systems Approach to Completing Hostile Environment Reservoirs," paper SPE 28738 presented at the 1994 SPE International Petroleum Conference and Exhibition of Mexico, Veracruz, Mexico, 10–13 October.

Kenda, W.P., Allen, T.J., and Herbel, R.R.: "Offline Subsea Wellhead MODU Operations Provide Significant Time Savings," paper SPE 79834 presented at the 2003 SPE/IADC Drilling Conference, Amsterdam, 19–21 February.

Landeck, C.R.: "Application of Stacked Template Structures Offshore Indonesia," paper SPE 37033 presented at the 1996 SPE Asia Pacific Oil and Gas Conference, Adelaide, Australia, 28–31 October.

Mcleod, S. and Hartley, F.: "Drilling Technology: Part I: Poseidon 29,000-Ft. Well in 4,800-Ft. Depths Harbinger of Future for US Gulf," *Offshore Magazine* (April 2001).

Mcleod, S. and Hartley, F.: "Drilling Technology: Part II: Positive and Negative Events in Drilling of Poseidon 29,750-Ft. Well," *Offshore Magazine* (May 2001).

Nikravesh, M. *et al.*: "Design of Smart Wellhead Controllers for Optimal Fluid Injection Policy and Producibility in Petroleum Reservoirs: A Neuro-Geometric Approach," paper SPE 37557 presented at the 1997 SPE International Thermal Operations and Heavy Oil Symposium, Bakersfield, California, 10–12 February.

Schulz, R.R., Stehle, D.E., and Murall, J.: "Completion of a Deep, Hot, Corrosive East Texas Gas Well," *SPERE* (1988).

Tuah, J.B. *et al.*: "Triple Wellhead Technology in Sarawak Operations," paper SPE 64277 presented at the 2000 SPE Asia Pacific Oil and Gas Conference and Exhibition, Brisbane, Australia, 16–18 October.

Watson, P. *et al.*: "An Innovative Approach to Development Drilling in Deepwater Gulf of Mexico," paper SPE 79809 presented at the 2003 SPE/IADC Drilling Conference, Amsterdam, 19–21 February.

Wei, J.: "Technique for Close Cluster Well Head Concentrated High Quality and High Speed Drilling Cementing Surface Interval in Sz36-1d," paper SPE 62776 presented at the 2000 IADC/SPE Asia Pacific Drilling Technology Conference, Kuala Lumpur, 11–13 September.

SI Metric Conversion Factors

$$
\begin{array}{lll}
\text{ft} \times 3.048^* & E-01 & = \text{m} \\
°\text{F} \quad (°\text{F}-32)/1.8 & & = °\text{C} \\
\text{in.} \times 2.54^* & E+00 & = \text{cm} \\
\text{psi} \times 6.894\ 757 & E+00 & = \text{kPa}
\end{array}
$$

*Conversion factor is exact.

Chapter 9
Cementing
Ron Crook, Halliburton

9.1 Cementing Operations

Cementing operations can be divided into two broad categories: primary cementing and remedial cementing. The objective of primary cementing is to provide zonal isolation. Cementing is the process of mixing a slurry of cement and water and pumping it down through casing to critical points in the annulus around the casing or in the open hole below the casing string. The two principal functions of the cementing process are to restrict fluid movement between the formations and to bond and support the casing.

If this is achieved effectively, the economic, liability, safety, government regulations, and other requirements imposed during the life of the well will be met. Zonal isolation is not directly related to production; however, this necessary task must be performed effectively to allow production or stimulation operations to be conducted. Thus, the success of a well depends on this primary operation. In addition to isolating oil-, gas-, and water-producing zones, cement also aids in (1) protecting the casing from corrosion, (2) preventing blowouts by quickly forming a seal, (3) protecting the casing from shock loads in deeper drilling, and (4) sealing off zones of lost circulation or thief zones.

Remedial cementing is usually done to correct problems associated with the primary cement job. The most successful and economical approach to remedial cementing is to avoid it by thoroughly planning, designing, and executing all drilling, primary cementing, and completion operations. The need for remedial cementing to restore a well's operation indicates that primary operational planning and execution were ineffective, resulting in costly repair operations. Remedial cementing operations consist of two broad categories: squeeze cementing and plug cementing.

In general, there are five steps required to obtain successful cement placement and meet the objectives previously outlined.

1. Analyze the well parameters; define the needs of the well, and then design placement techniques and fluids to meet the needs for the life of the well. Fluid properties, fluid mechanics, and chemistry influence the design used for a well.

2. Calculate fluid (slurry) composition and perform laboratory tests on the fluids designed in Step 1 to see that they meet the needs.

3. Use necessary hardware to implement the design in Step 1; calculate volume of fluids (slurry) to be pumped; and blend, mix, and pump fluids into the annulus.

4. Monitor the treatment in real time; compare with Step 1, and make changes as necessary.

5. Evaluate the results; compare with the design in Step 1, and make changes as necessary for future jobs.

9.2 Well Parameters

Along with supporting the casing in the wellbore, the cement is designed to isolate zones, meaning that it keeps each of the penetrated zones and their fluids from communicating with other zones. To keep the zones isolated, it is critical to consider the wellbore and its properties when designing a cement job.

9.2.1 Depth. The depth of the well influences the amount of wellbore fluids involved, the volume of wellbore fluids, the friction pressures, the hydrostatic pressures, the temperature, and, thus, the cement slurry design. Wellbore depth also controls hole size and casing size. Extremely deep wells have their own distinct design challenges because of high temperatures, high pressures, and corrosive fluids.

9.2.2 Wellbore Geometry. The geometry of the wellbore is important in determining the amount of cement required for the cementing operation. Hole dimensions can be measured using a variety of methods, including acoustic calipers, electric-log calipers, and fluid calipers. Openhole geometry can indicate adverse (undesirable) conditions such as washouts. Wellbore geometry and casing dimensions determine the annular volume and the amount of fluid necessary.

The hole shape also determines the clearance between the casing and the wellbore. This annular space influences the effectiveness of drilling-fluid displacement. A minimum annular space of 0.75 to 1.5 in. (hole diameter 2 to 3 in. greater than casing diameter) is recommended. Annular clearances that are smaller restrict the flow characteristics and generally make it more difficult to displace fluids.

Another aspect of hole geometry is the deviation angle. The deviation angle influences the true vertical depth and temperatures. Highly deviated wellbores can be challenging because the casing is not as likely to be centered in the wellbore, and fluid displacement becomes difficult.

Problems created by geometry variations can be overcome by adding centralizers to the casing. Centralizers help to center the casing within the hole, leaving equal annular space around the casing.

9.2.3 Temperature. The temperatures of the wellbore are critical in the design of a cement job. There are basically three different temperatures to consider: the bottomhole circulating temperature (BHCT), the bottomhole static temperature (BHST), and the temperature differential (temperature difference between the top and bottom of cement placement). The BHCT is the temperature to which the cement will be exposed as it circulates past the bottom of the casing. The BHCT controls the time that it takes for the cement to setup (thickening time). BHCT can be measured using temperature probes that are circulated with the drilling fluid. If actual wellbore temperature cannot be determined, the BHCT can be estimated using the temperature schedules of American Petroleum Inst. (API) *RP10B*.[1] The BHST considers a motionless condition where no fluids are circulating and cooling the wellbore. BHST plays a vital role in the strength development of the cured cement.

The temperature differential becomes a significant factor when the cement is placed over a large interval and there are significant temperature differences between the top and bottom cement locations. Because of the different temperatures, commonly, two different cement slurries may be designed to better accommodate the difference in temperatures.

The bottomhole circulating temperature affects slurry thickening time, rheology, fluid loss, stability (settling), and set time. BHST affects compressive-strength development and cement integrity for the life of the well. Knowing the actual temperature that the cement will encounter during placement allows operators to optimize the slurry design. The tendency to overestimate the amount of materials required to keep the cement in a fluid state for pumping and the amount of pumping time required for a job often results in unnecessary cost and well-control problems. Most cement jobs are completed in less than 90 minutes.

To optimize cost and displacement efficiency, the guidelines discussed next are recommended. Design the job on the basis of actual wellbore circulating temperatures. A downhole temperature subrecorder can be used to measure the circulating temperature of the well. A subrecorder is a memory-recorder device that can either be lowered by wireline or dropped into the drillpipe and measures the temperature downhole during the circulating operation before cementing. The memory recorder is then retrieved from the drillpipe and the BHCT is measured. This allows for accurate determination of the downhole temperature.

• If determining the actual wellbore circulating temperature is not possible, use API *RP10B* to estimate the BHCT.[1]

• Do not "pad" the actual downhole temperatures measured, and do not exceed the amount of dispersants, retarders, etc. recommended for the temperature of the wellbore. When determining the amount of retarder required for a specific application, consider the rate at which the slurry will be heated.

9.2.4 Formation Pressures. When a well is drilled, the natural state of the formations is disrupted. The wellbore creates a disturbance where only the formations and their natural forces existed before. During the planning stages of a cement job, information about the formations' pore pressure, fracture pressure, and rock characteristics must be known. Generally, these factors will be determined during drilling. The density of the drilling fluids in a properly balanced drilling operation can be a good indication of the limitations of the wellbore.

To maintain the integrity of the wellbore, the hydrostatic pressure exerted by the cement, drilling fluid, etc. must not exceed the fracture pressure of the weakest formation. The fracture pressure is the upper safe pressure limitation of the formation before the formation breaks down (the pressure necessary to extend the formation's fractures). The hydrostatic pressures of the fluids in the wellbore, along with the friction pressures created by the fluids' movement, cannot exceed the fracture pressure, or the formation will break down. If the formation does break down, the formation is no longer controlled, and lost circulation results. Lost circulation, or fluid loss, must be controlled for successful primary cementing. Pressures experienced in the wellbore also affect the strength development of the cement.

9.2.5 Formation Characteristics. The composition of formations can present compatibility problems. Shale formations are sensitive to fresh water and can slough off if special precautions, such as increasing the salinity of the water, are not taken. Other formation and chemistry considerations, such as swelling clays and high-pH fluids, should be taken into consideration. Some formations may also contain flowing fluids, high-pressure fluids, corrosive gases, or other complex features that require special attention.

9.3 Cement-Placement Design

9.3.1 Primary Cementing. Most primary cement jobs are performed by pumping the slurry down the casing and up the annulus; however, modified techniques can be used for special situations. These techniques are cementing through pipe and casing (normal displacement technique), stage cementing (for wells with critical fracture gradients), inner-string cementing through tubing (for large-diameter pipe), outside or annulus cementing through tubing (for surface pipe or large casing), reverse-circulation cementing (for critical formations), delayed-set cementing (for critical formations and to improve placement), and multiple-string cementing (for small-diameter tubing).

9.3.2 Cementing Through Pipe and Casing. Conductor, surface, protection, and production strings are usually cemented by the single-stage method, which is performed by pumping cement slurry through the casing shoe and using top and bottom plugs. There are various types of heads for continuous cementing, as well as special adaptors for rotating or reciprocating casing.

9.3.3 Stage Cementing. Stage cementing is used to ensure annular fill and seal across selected intervals whenever a continuous single-stage, lead and tail, or lightweight (foamed, ceramic spheres, etc.) cementing application cannot be performed. Stage-cementing tools, or differential valve (DV) tools, are used to cement multiple sections behind the same casing string, or to cement a critical long section in multistages. Stage cementing may reduce mud contamination and lessens the possibility of high filtrate loss or formation breakdown caused by high hydrostatic pressures, which is often a cause for lost circulation.

Stage tools are installed at a specific point in the casing string as casing is being run into the hole. The first (or bottom) cement stage is pumped through the tool to the end of the casing and up the annulus to the calculated-fill volume (height). When this stage is completed, a shutoff or bypass plug can be dropped or pumped in the casing to seal the stage tool. A freefalling plug or pumpdown dart is then used to hydraulically set the stage tool and open the side ports, allowing the second cement stage (top stage) to be displaced above the tool. A closing plug is used to close the sliding sleeve over the side ports at the end of the second stage and serves as a check valve to keep the cement from U-tubing above and back through the tool.

The displacement stage-cementing method is used when the cement is to be placed in the entire annulus from the bottom of the casing up to or above the stage tool. The displacement method is often used in deep or deviated holes in which too much time is needed for a freefalling plug to reach the tool.

Fluid volumes (mud, spacer, cement) must be accurately calculated and prepared on locations and densities closely measured to prevent over- or underdisplacement of the first stage. Overdisplacement can result in improper opening of the tool to apply the second (upper) stage, resulting in excess pressures or job failure. Underdisplacement creates a gap (void) in the cement column at the stage tool, which results in poor zonal isolation.

Two-stage cementing is the most widely used multiple-stage cementing technique. However, when a cement slurry must be distributed over a long column and hole conditions will not allow circulation in one or two stages, a three-stage method can be used. The same steps are involved as in the two-stage methods, except that there is an additional stage. Obviously, the more stages used in the application, the more complicated the job will become. Although stage cementing was very popular many years ago, new foamed-cement and nonfoamed-ultralightweight-cement technologies have successfully reduced the need for multistage cementing in many operations.

9.3.4 Inner-String Cementing. When large-diameter pipe is cemented, tubing or drillpipe is commonly used as an inner string to place the cement. This procedure reduces the cementing time and the volume of cement required to bump the plug. The technique uses modified float shoes, guide shoes, or baffle equipment, with sealing adaptors attached to small-diameter pipe. Cementing through the inner string permits the use of small-diameter cementing plugs. If the casing is equipped with a backpressure valve or latchdown baffle, the inner string can be disengaged and withdrawn from the casing as soon as the plug is seated, while preparations are made to drill deeper.

9.3.5 Outside or Annulus Cementing. A method commonly used on conductor or surface casing to bring the top of the cement to the surface consists of pumping cement through tubing or small-diameter pipe run between casings or between the casing and the hole. This method is sometimes used for remedial work. Casing can suffer damage when gas sands become charged with high pressure from surrounding wells. In such instances, cementing the annulus between strings through a casinghead connection can repair the casing.

9.3.6 Reverse-Circulation Cementing. The reverse-circulation cementing technique involves pumping the slurry down the annulus and displacing the drilling fluid back up through the casing. The float equipment, differential fill-up equipment, and wellhead assembly must be modified. This method is used when the cement slurry cannot be pumped in turbulent flow without breaking down the weak zones above the casing shoe. Reverse circulation allows for a wider range in slurry compositions, so heavier or more-retarded cement can be placed at the lower portion of casing, and lighter or accelerated cement can be placed at the top of the annulus. Caliper surveys should be made before the casing is run, to determine the necessary volume of cement and minimize overplacement.

9.3.7 Delayed-Set Cementing. Delayed-set cementing involves placing a retarded cement slurry containing a filtration-control additive in a wellbore before running the casing. This method can help to obtain a more uniform sheath of cement around the casing than may be possible with conventional methods. The cement is placed by pumping it down the drillpipe and up the annulus. The drillpipe is then removed from the well, and casing or liner is sealed at the bottom and lowered into the unset cement slurry. After the cement slurry is set, the well can be completed with conventional methods.

This technique has been used in tubingless-completion wells by placing the slurry down one string and lowering multiple tubing strings into the unset cement. When the casing is run into the cement slurry, drilling fluid left in the annulus mixes with the cement slurry. Although not ideal, this development is preferred to leaving the drilling fluid in the annulus as a channel or pocket. The delayed-set cement slurry allows protracted reciprocation of the casing string, which is more likely to ensure a uniform cement sheath.

A disadvantage to delayed-set cementing is the increased water/oil-contact (WOC) time, which could be expensive if a drilling rig is kept on location while the cement sets and gains strength. If the drilling rig can be moved off location and a workover rig can complete the well, the cost can be reduced.

9.3.8 Multiple-String Cementing. Multiple-casing completions are used when single or conventional completions are not economically attractive. When multiple strings are placed in a well, each string is usually run independently, and the longest string is landed first. The first string is set in the hanger and is circulated before the second string is run. After the second string is landed in the hanger, it is circulated while the third string is run. In areas where lost circulation is a known problem, cement can be placed through the longest casing string. Once the cement fill-up has been established, the remainder of the hole is filled with cement slurry through a shorter string.

Centralizers are frequently used, one per joint from 100 ft above to 100 ft below productive zones. Other casing equipment in these small-diameter holes includes landing collars for cement wiper plugs, full-opening guide shoes, and limited-rotating scratchers for single completions. All float equipment, centralizers, and scratchers should be able to pass the hanger assembly in the casinghead.

Other factors considered in the design of cement slurry are similar to those considered in the design of slurry for a single string of pipe. The cement is usually pumped down the longest strings simultaneously, although this is not mandatory. The idle strings may be pressured to 1,000 to 2,000 psi during cementing to safeguard against leakage, thermal buckling, or collapse.

9.3.9 Cementing of High-Pressure/High-Temperature Wells. Recent technological advances have allowed the production of reservoirs that were once considered too expensive and risky to be commercially viable. Designs for these wells must withstand high temperatures and pressures, as well as frequently encountered corrosive gases such as H_2S and CO_2. Completions performed in high-pressure/high-temperature (HP/HT) reservoirs are some of the most expensive in the industry. High completion costs make it a necessity to successfully cement the well casing on the primary cementing job and eliminate the need for remedial cementing. HP/HT reservoirs are characterized by reservoir depths greater than 15,000 ft, reservoir pressure greater than 15,000 psi, and reservoir-fluid temperatures from 300 to 500°F.

To provide optimum zonal isolation, one should consider not only the primary cementing job, but also the long-term, post-placement effects of various operations that can place stress on the set cement. In the initial cementing, the job should be designed to displace the drilling fluid completely and to prevent gas migration and fluid loss. Once the initial cement job is completed, the effects of stress throughout the well's life will determine the cement sheath's future viability.

In most wells, the liner or production string is the most important component. In HP/HT wells, the conductor string can be placed under greater loading and all sections of the well can be exposed to formation, temperature, and pressure changes that are greater than normal; therefore, the well should be examined from the whole-well perspective.

A well's characteristics determine the cement-slurry properties and performance. A careful and thorough review of these characteristics is essential for designing an effective cement slurry and ensuring correct placement. Engineers should combine individual variables to develop a total-cement-job design.

Guidelines for improving cementing results are:

• Condition the drilling fluid to break its gel structure, thereby reducing its viscosity and improving its mobility.

• Use pipe movement to dislodge pockets of gelled, immobile drilling fluid.

• Use mechanical scratchers and wall cleaners to maximize pipe-movement effectiveness, which can erode excess drilling fluid.

• Centralize pipe in and near "critical" zones. A minimum of 70% casing standoff is recommended. Good pipe standoff helps increase drilling-fluid removal, thereby equalizing forces exerted by cement flowing up the annulus.

• Use the highest possible pump rates to get the greatest displacement efficiency.

• Use spacers and/or flushes to isolate dissimilar fluids and prevent potential contamination problems.

• Use a drilling fluid with a rheology that allows efficient drilling-fluid removal without raising the equivalent circulating density (ECD) to an unacceptable level.

• Use enough spacer and/or flush to allow adequate contact time (7 to 10 minutes contact and 500 to 1,000 ft of annulus).

9.4 Remedial Cementing

9.4.1 Squeeze Cementing. *Introduction.* Remedial cementing requires as much technical, engineering, and operational experience, as primary cementing but is often done when wellbore conditions are unknown or out of control, and when wasted rig time and escalating costs force poor decisions and high risk. Squeeze cementing is a "correction" process that is usually only necessary to correct a problem in the wellbore. Before using a squeeze application, a series of decisions must be made to determine (1) if a problem exists, (2) the magnitude of the problem, (3) if squeeze cementing will correct it, (4) the risk factors present, and (5) if economics will support it. Most squeeze applications are unnecessary because they result from poor primary-cement-job evaluations or job diagnostics.

Squeeze cementing is a dehydration process. A cement slurry is prepared and pumped down a wellbore to the problem area or squeeze target. The area is isolated, and pressure is applied from the surface to effectively force the slurry into all voids. The slurry is designed specifically to fill the type of void in the wellbore, whether it is a small crack or micro-annuli, casing split or large vug, formation rock or another kind of cavity. Thus, the slurry design and rate of dehydration or fluid loss designed into the slurry is critical, and a poor design may not provide a complete fill and seal of the voids.

9.4.2 Techniques. The following techniques are the six commonly recognized squeeze applications.

Running Squeeze. A running squeeze is any squeeze operation in which continuous pumping is used to force the cement into the squeeze interval. This technique is sometimes referred to as a "walking squeeze" when low pump rates and minimal graduating pressure is used. Although the running squeeze is easier to design and apply, it is probably the most difficult to control because the rate of pressure increase and final squeeze pressure are difficult to determine. As running-squeeze pressure builds, the pump rate should be reduced, creating a walking squeeze. Running squeezes may be applied whenever the wellbore can be circulated at a reasonable pump rate (approximately 2 bbl/min). When applied correctly, most running squeezes are low-pressure applications; however, they often turn into high-pressure applications because of unknown formation characteristics, the quality of slurry used, or lack of job control.

Hesitation Squeeze. This technique is often used when a squeeze pressure cannot be obtained using a running technique because of the size of the void, lack of filtrate control, or when the squeeze must be performed below a critical wellbore pressure. During a hesitation squeeze, the pumping sequence is started and stopped repeatedly, while the pressure is closely monitored on the surface. Cement is deposited in waves into the squeeze interval, and the slurry is designed to increase resistance (gel-strength development and fluid-leakoff rate) until the final squeeze pressure is reached. Operators must thoroughly design and test the cement slurry to understand how its properties will change with frequent shutdowns and to safely approximate the shutdown period between pumping cycles. The slurry volume should be clear of all downhole tools before the hesitation cycles begin. For many otherwise large and expensive conventional squeeze applications, a hesitation squeeze can be a safer, less expensive, and effective technique.

High-Pressure Squeeze. A high-pressure squeeze is an application performed above formation fracturing pressures when fracturing is necessary to displace the cement and seal off formations or establish injection points between channels and perforations. Slurry volumes and leakoff vary with the size of the interval. "Block" squeezing is the process of squeezing off permeable sections above and below a production zone, which requires isolation of the zone with a packer and retainer, using high pressure to force cement slurry (fracture) into the zone. Cement slurry will not invade a formation unless it is fractured away, creating a large crack to accommodate the entire slurry. Otherwise, dehydration occurs and only the filtrate enters the zone. High pressure is usually required to force all wellbore fluids into the formations ahead of the cement slurry. This technique is often referred to as "bullheading."

Low-Pressure Squeeze. A low-pressure squeeze, the most common technique, is any squeeze application conducted below the fracturing pressure. This method can be applied whenever clean wellbore fluids can be injected into a formation, such as permeable sand, lost-circulation interval, fractured limestone, vugs, or voids. Filtrate from the cement slurry is easily displaced at low pressures, and the dehydrated cement is deposited in the void. Whole cement slurries will not invade most formations unless a fracture is readily open or is created during the squeeze process.

Packer/Retainer Squeeze. Squeeze tools are often used to isolate the squeeze interval and place the cement as close to the squeeze target as possible before applying pressure. Retainers or bridge plugs are used to create a false bottom and are set just below the squeeze target inside the casing or tubing. This procedure seals off the open wellbore below the target (which may be several thousands of feet) and reduces the volume of cement needed for the squeeze. A packer can be run into the wellbore and set above the squeeze interval, between two intervals, or below an interval. Packers allow circulation of the wellbore until the cement slurry is pumped; then the packer is set, which seals off the annulus so the cement can be squeezed through tubing below the packer or down the backside between the tubing/casing annulus above the packer. Cement volumes, squeeze pressures, and squeeze targets can be more accurately determined and controlled using squeeze tools.

Bradenhead Squeeze. This technique is often applied when the problem occurs during drilling (lost circulation) or soon after a primary cement job (weak casing shoe). A Bradenhead squeeze is performed when squeeze tools are unavailable or cannot be run in the hole, or when the operator feels he can successfully control the problem without pulling the drillstring, tubulars, etc. out of the wellbore. Whether during drilling or completion, a Bradenhead is performed by circulating cement slurry down to the squeeze interval, then pulling the workstring above the top of the cement column. The backside of the wellbore is closed in, and pressure is applied through the workstring to force cement into the squeeze interval. A hesitation squeeze is sometimes used to more effectively pack off the cement into all voids. Most coiled-tubing (CT) squeeze applications are performed using this technique.

9.4.3 Plug Cementing.
In oil-gas-well construction, a plug must prevent fluid flow in a wellbore, either between formations or between a formation and the surface. As such, a competent plug must provide a hydraulic and mechanical seal. Each plugging operation presents a common problem in that a relatively small volume of plugging material, usually a cement slurry, is placed in a large volume of wellbore fluid. Wellbore fluids can contaminate the cement, and even after a reasonable WOC time, the result is a weak, diluted, nonuniform or unset plug. In addition, plugging situations frequently present unique issues that require sound engineering design and judgment. For these reasons, both mechanical and chemical technologies are necessary for successful plugging.

Displacement efficiency, slurry stability, fluid compatibilities, and all the issues that are normally considered for a primary cement job must be carefully considered for a plug job. Plugging operations are difficult because the work string from a heavier balanced cement plug must be removed from its position above a lighter wellbore fluid. Some of the varied reasons for performing plugging operations are discussed next.

Abandonment. To seal off selected intervals of a dry hole or a depleted well, operators can place a cement plug at the required depth to help prevent zonal communication and migration of any fluids that might infiltrate underground freshwater sources.

Directional Drilling/Sidetracking. When sidetracking a hole around a non-retrievable fish, such as a stuck bottomhole assembly (BHA) or changing the direction of drilling for geological reasons, it is often necessary to place a cement plug at the required depth to change the wellbore direction or to help support a mechanical whipstock, so the bit can be guided in the desired direction.

Lost-Circulation Control. When mud circulation is lost during drilling, lost returns can sometimes be restored by spotting a cement plug across the thief (lost-circulation) zone and then drilling back through the plug. Efforts should be made to identify the source and reason for lost returns when planning a plugging job. Drilling-induced fractures, chemically induced formation instability, natural fractures, vugs, and high permeability can contribute to lost circulation.

Well Control. Plugs, typically made of cement, are sometimes placed in a wellbore when the well has reached a critical state in which no margins remain between pore and frac pressures and no other options exist. In fact, the drillstring is sometimes intentionally cemented in place because it cannot be pulled without risk of inducing an uncontrolled flow to the surface or a crossflow from a high-pressured zone into a weak or low-pressured zone.

Zonal Isolation/Conformance. One of the more common reasons for plugging is to isolate a specific zone. The purpose may be to shut off water, to recomplete a zone at a shallower depth, or to protect a low-pressure zone in an openhole before squeezing. In a well with two or more producing intervals, abandoning a depleted or unprofitable producing zone may be beneficial. A permanent cement plug is used to isolate the zone, helping to prevent possible production losses into another interval or fluid migration from another interval. The integrity of such plugs is frequently enhanced mechanically by placing them above bridge plugs or through and above squeeze retainers. Other methods involve combining the spotting of plugging fluids with the remedial squeeze process of injecting a polymer plugging material into the formation matrix, followed by a small volume of cement slurry to shut off perforations.

Formation Testing. Plugs are occasionally placed in the open hole below a zone to be tested that is a considerable distance off-bottom, where other means of isolating the interval are not possible or practical. Although cement is the most commonly used plug material, barite, sand, and polymers may also serve as plugging agents.

Wellbore Stability. At times during drilling, placing a plugging material across an unstable formation can be beneficial. Polymer, resins, cements, or combinations of these materials can be used to consolidate formations and alter the near-wellbore stresses and formation integrity. A balanced cement plug is sometimes placed to simply "backfill" a severely washed out or elliptical hole section. In such cases, the plug is subsequently drilled out, leaving a cement sheath in place to reduce or prevent further wellbore enlargement and to help return the wellbore to its original diameter and circular shape for improved annular velocities.

9.4.4 Placement Techniques. This section describes some common placement techniques in basic terms, but these techniques can be custom-designed for specific situations.

Dump Bailer. Dump bailers are used for placing very small volumes of plugging material precisely and economically. Different types of dump bailers, including gravity and positive-displacement bailers, are shown in **Fig. 9.1**. These are generic dump bailers and are indications of various types. Generally, any company in the business or setting wireline plugs (both slickline and e-line) will have some type of dump-bailer service.

These tools can be run on wireline, slickline, or sandline, depending on the tool. Both through-tubing and through-casing sizes are also available. Sand, barite, polymers, thermal-set resins, plastics, and cement slurries are all placed with this technique. The use of dump bailers for spotting materials that thermally depend on set times (such as polymers, resins, and cement slurries) has historically been limited to shallow depths because of temperature concerns; but modern polymer and retarder technology allows for broader use.

A limit plug, cement basket, permanent bridge plug, or sand pill is often placed below the desired plugging location to provide a solid bottom in the wellbore. The dump bailer, containing a measured quantity of plugging material, is lowered to the desired depth. The bailer is opened either electronically by the wireline operator or mechanically by tagging the bridge plug and then raised to release the plugging material at this location. At times, the job is performed either with a lubricator on the wellhead or under overbalanced conditions so that the plugging fluid may achieve limited forced entry through gravel packs, perforations, and other passages into formation matrix.

Coiled Tubing. Probably the most technically efficient way to spot fluid in a wellbore is to lay it in with CT; but it is not always the most economical or logistically efficient way. The process consists of placing the end of the CT at the bottom of the planned plug depth, and while the cement or other plugging slurry exits, the nozzle at the end of the coil slowly extracts the coil so that the pull-out rate matches the fluid-pump rate and keeps the end of the coil just below the top of the slurry. This placement method results in a volume of plugging fluid with little or no contamination in the wellbore. After placement, the operator may wash

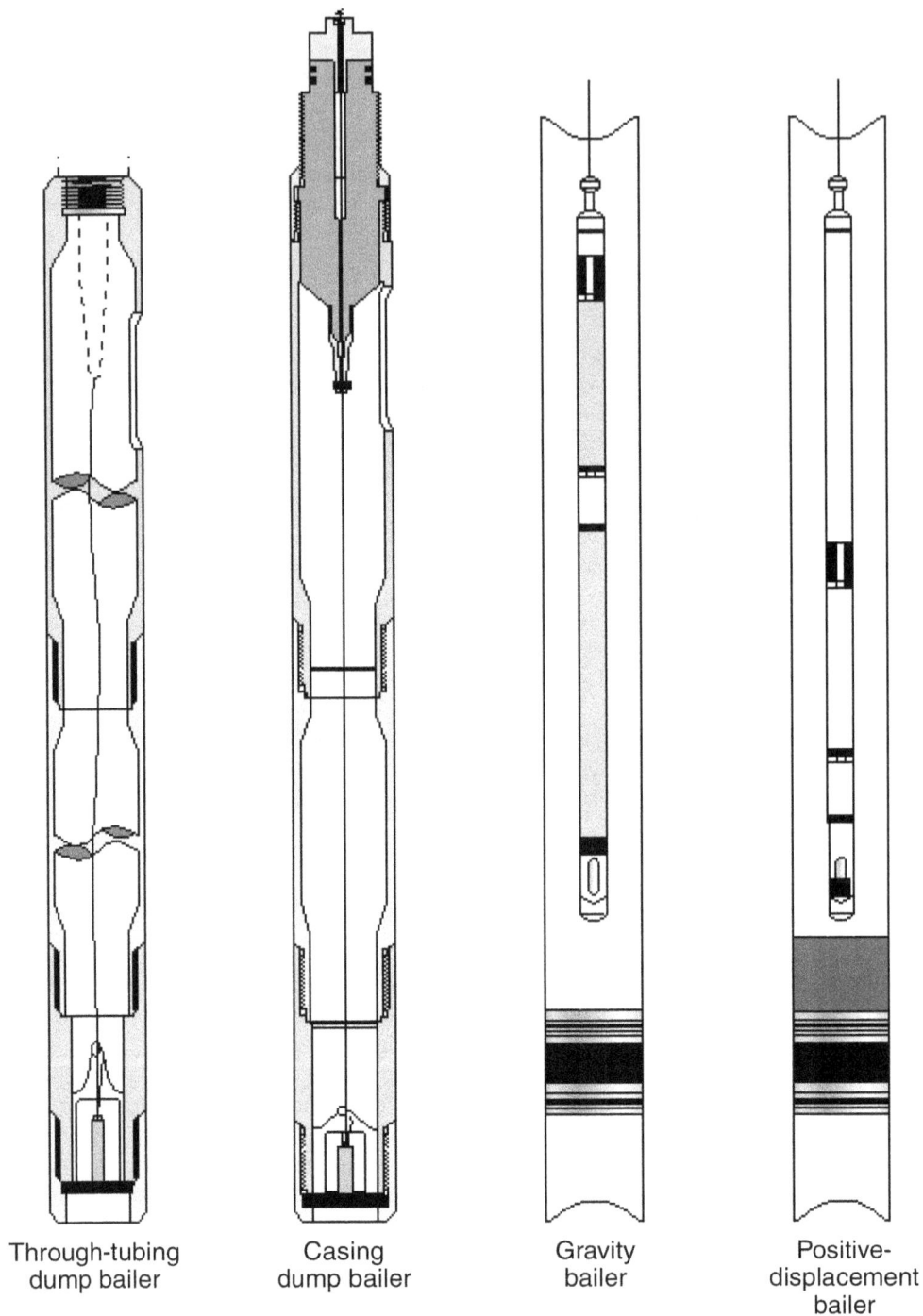

Fig. 9.1—Various types of dump bailers.

out the wellbore above the plug to establish a very accurate top of cement, or apply squeeze pressure in some prescribed manner.

Bullheading or Bradenhead Placement. The bullheading or Bradenhead placement method consists of injecting a plugging slurry into a formation with the intent to leave some portion of the plugging material in the wellbore. Typically applied in cases of well control, lost circulation, or abandonment, this method is probably one of the less accurate placement methods because of the uncertainty of the fluid path. The general assumption is that the fluids will follow the path of least resistance, but that is not always reliable. Consequently, when a slurry goes into the annulus rather than down into a lower portion of the wellbore, work strings are sometimes unintentionally cemented in the wellbore. Despite the uncertainty involved, this plugging method has been used successfully when it is executed with caution.

Balanced Plugs. Probably the most common technique in both drilling and abandonment operations, the balanced-plug method involves pumping the slurry through drillpipe or tubing until the level outside is equal to that inside the string. The volume and hydrostatics of wellbore fluids, preflushes, spacers, and plugging fluids must be carefully calculated to ensure that the system is being correctly balanced in the hole. The pipe or tubing is then pulled slowly from the plugging material before it sets, leaving the plug in place. The method is simple in theory, but depending on wellbore conditions, the fluid mechanics can be extremely complex. Wellbore, fluid, and hardware constraints must all be considered during the design and execution of the job.

Preparing a drilling fluid for cementing can be difficult before a primary cementing operation; it can be next to impossible when preparing to spot a balanced cement plug because of time, economic, and technical constraints. When these conditions exist, a simple way to ensure maximum mud removal is to wash across the interval where the plug will be placed, typically with a diverter tool (such as a drillpipe, CT, or a specialized tailpipe assembly) on the end of the work string. This approach ensures that the wellbore fluids are as close as possible to being 100% mobilized across this critical interval.

In cases where well control is a concern, such as placement of thixotropic slurries, short slurry-thickening time, or other instances with a high risk of compromising plug stability/integrity when the work string is removed, the operator should consider running a sacrificial tailpipe. This tailpipe can be released by either shearing it off at the end of the job or leaving the work string in place until the plugging material has set and cutting or backing off the pipe at the first free connection. In extreme situations, a sacrificial string may include the bottomhole drilling assembly, but if the pipe can be tripped, releasable tailpipe assemblies can be quickly fabricated out of locally available tubing and hardware. If necessary, such assemblies can be constructed out of any drillable material, such as aluminum.

9.5 Hole Preparation

The predominant cause of cementing failure appears to be channels of gelled drilling fluid remaining in the annulus after the cement is in place. If drilling-fluid channels are eliminated, any number of cementing compositions will provide an effective seal.

In evaluating factors that affect the displacement of drilling fluid, it is necessary to consider the flow pattern in an eccentric annulus (i.e., where the pipe is closer to one side of the hole than the other). Flow velocity in an eccentric annulus is not uniform, and the highest velocity occurs in the side of the hole with the largest clearance.

If the casing is close to the wall of the hole, it may not be possible to pump the cement at a rate high enough to develop uniform flow throughout the entire annulus (**Fig. 9.2**). To reduce the chances for eccentricity, centralizers should be used to maintain the pipe in the center of the annulus.

9.5.1 Standoff. *100% Standoff.* This shows a hole with casing that is exactly centralized in the hole. The shaded areas are the cement and it shows the cement level is the same on both sides of the casing.

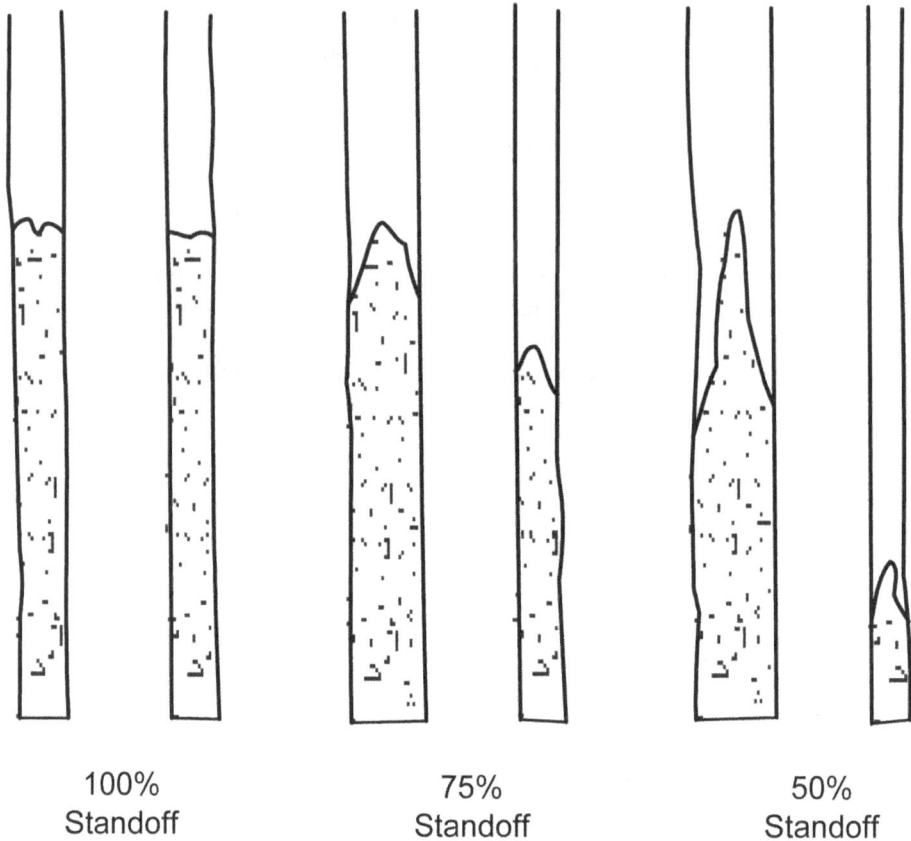

100%   75%   50%
Standoff   Standoff   Standoff

Fig. 9.2—Effect of eccentricity on cement flow in the annulus.

75% Standoff. This shows a hole with the casing decentralized to a 75% standoff, and it shows that as you decentralize the casing the flow is higher up the wide side of the hole compared to the narrow side.

50% Standoff. Same as the 75% standoff only more pronounced with the cement height.

9.5.2 Circulatability Treatment. The condition of the drilling fluid is the most important variable in achieving good displacement during cementing. Regaining and maintaining good fluid mobility is the key. An easily displaced drilling fluid will have low gel strengths and low fluid loss. Pockets of gelled fluid, which commonly exist following drilling, make displacement difficult and should be broken up. Circulating or conditioning the drilling mud for at least two hole volumes, prior to cementing, is preferred. Varying pump rates during the conditioning process enhances hole cleaning. Pipe movement (reciprocation or rotation) helps to break up mud gels for greater displacement efficiency. Performing the steps discussed next conditions the drilling fluid for a cement job.

• Determine the volume of the circulatable hole, and evaluate the percentage of the hole that is being circulated. Good fluid returns do not reliably indicate the mobility of fluid in the annular space.

• When the casing is on bottom and before displacement begins, circulate the drilling fluid to help break the gel structure of the fluid, decreasing its viscosity and increasing its mobility. Condition the drilling fluid until equilibrium is achieved.

• Never allow the drilling fluid to remain static for extended periods, especially at elevated temperatures. When the drilling fluid is well conditioned (the drilling-fluid properties going in equal the drilling-fluid properties at the outlet), continue circulating it until the displacement program begins.

• Modify the flow properties of the drilling fluid to optimize its mobility and drill-cuttings removal.

• Measure gel strengths at 10 seconds, 10 minutes, 30 minutes, and 4 hours to examine the gel-strength profile of the drilling fluid. This testing would typically be performed during the job-planning stage. During conditioning just before the job, readings taken at 10 seconds, 10 minutes, and 30 minutes are typically sufficient. An optimum drilling fluid has flat, nonprogressive gel strengths (e.g., gel-strength values of 1, 3, and 7). Note: Gel strength is measured using a rotational viscometer. The unit of measure is lbf/100 ft^2. The test procedure is outlined in API *RP13B-1, Recommended Practice Standard Procedure for Field Testing Water-based Drilling Fluids.*[2]

• Measure gel-strength development of drilling fluid to be left in the well at downhole conditions of temperature and pressure. At elevated temperatures and pressure, some drilling fluids gel to a consistency that prohibits removal. These increased gel strengths are not detectable at surface conditions. This testing should take place during the planning stages.

In deviated wellbores, a drilling fluid that has a higher viscosity at low shear rates may be required to help prevent drilling-fluid or wellbore solids from settling on the low side of the wellbore. The presence of large drill cuttings may also necessitate higher-viscosity fluids. This testing should take place during the planning stages.

9.5.3 Spacers and Flushes. Spacers and flushes are effective displacement aids because they separate incompatible fluids such as cement and drilling fluid. A spacer is a fluid used to separate drilling fluids and cementing slurries. A spacer can be designed for use with either water-based or oil-based drilling fluids and prepares both pipe and formation for the cementing operation. Spacers are typically densified with insoluble-solid weighting agents.

For example, a spacer is a volume of fluid injected ahead of the cement but behind the drilling fluid. It can also enhance the removal of gelled drilling fluid, allowing a better cement bond. Spacers can be designed to serve various needs. For example, weighted spacers can help with well control, and reactive spacers can provide increased benefits for removing drilling fluids. The drilling-fluid/spacer interface and the spacer/cement-slurry interface must be compatible. The use of the compatibility procedures outlined in API *RP10B*[1] is highly recommended. Parameters governing the effectiveness of a spacer include flow rate, contact time, and fluid properties. To achieve maximum drilling-fluid displacement, consider these guidelines: pump the spacer fluid at an optimized rate; provide a contact time (10-minute minimum) and volume of spacer that will remove the greatest possible amount of drilling fluid; make sure that the viscosity, yield point, and density of the spacer and the cement slurry are at least the same as the drilling fluid; and when an oil-based or synthetic-based drilling fluid is used, the spacer package should be formulated to thoroughly water-wet the surface of the pipe and the formation. To achieve a high level of water-wettability, test the spacer system using a newly developed API apparent-wettability testing technique. This technique is highly recommended for customizing the spacer/surfactant package to help ensure water-wetting.

Flushes are used to thin and disperse drilling-fluid particles and are used to separate drilling fluids and cementing slurries. They can be designed for use with either water-based or oil-based drilling fluids. Flushes prepare both the pipe and formation for the cementing operation and are not typically densified with insoluble-solid weighting agents. They go into turbulent flow at low rates. Flushes are also referred to as washes and preflushes.[1]

9.5.4 Contact Time. Contact time is the period of time that a fluid flows past a particular point in the annular space during displacement. Studies indicate that a contact time of 10 minutes or longer provides excellent removal of most drilling fluids. The volume of fluid needed to provide a specific contact time is

$$V_t = t_c \times q_d \times 5.615, \quad\dots\dots\dots\dots\dots\dots\dots\dots\dots\dots\dots\dots (9.1)$$

where
V_t = volume of fluid (turbulent flow), ft^3;
t_c = contact time, min;
q_d = displacement rate, bbl/min;
and
5.615 = conversion between ft^3 and bbl.

The calculation is simple because only two readily available factors are required, and the calculation is independent of casing and hole size. The equation holds as long as all of the fluid passes the point of interest.[3]

9.5.5 Sweep Pill Design and Analysis. The most important factor in a sweep program is to carry it out in a proactive manner. It is much easier to keep the hole clean than it is to try to clean it up after solids buildup has occurred. Hole cleaning depends on fluid type. When wells are drilled with invert oil emulsion systems, cuttings tend to be harder, more competent, and better defined than in water-based mud (WBM). This method allows the cuttings to be removed from the wellbore more readily. Even highly inhibitive, high-performance WBM systems do not generate cuttings of the same high level of integrity as inverts. Hole cleaning may also be compromised by the effect of WBM fluids on the nature of the borehole, which will often exhibit rugosity or out-of-gauge hole, thereby reducing annular flow velocities. Large washouts tend to require more frequent hole-conditioning trips. Silicate, CaCl$_2$, and some cationic polymer WBM systems produce near-gauge holes in formations of moderate or low chemical reactivity, but in the Gulf of Mexico (GOM), even these WBM types can fall short, and poor hole cleaning and packing off are very real risks. This problem is also manifested at the shakers, which usually require extra attention to keep the screens clean and handle the larger dilution volumes needed to maintain WBM properly.

Cuttings transport in deviated wellbores is more challenging than in vertical wells. Correct well planning, drilling practices, and sweep procedures can have a positive influence on "on-bottom" drilling times. Primary factors contributing to hole-cleaning challenges include drillpipe eccentricity, the need for sliding while maintaining hole direction, and the resultant flow-path changes in the annulus. A factor that compounds the situation is that cuttings settle toward the low side of the deviated hole. This situation, shown in **Fig. 9.3**,[4] is known as the Boycott Effect.

Regardless of drilling-fluid rheology, it is almost impossible to clean a highly deviated wellbore without drillpipe rotation. Drillpipe rotation agitates the settled cuttings back into the flow stream, so they can be transported to the surface.

9.6 Cement Composition

9.6.1 Manufacture of Cement. Almost all drilling cements are made of Portland cement, a calcined (burned) blend of limestone and clay. A slurry of Portland cement in water is used in wells because it can be pumped easily and hardens readily, even under water. It is called Portland cement because its inventor, Joseph Aspdin, thought the solidified cement resembled stone

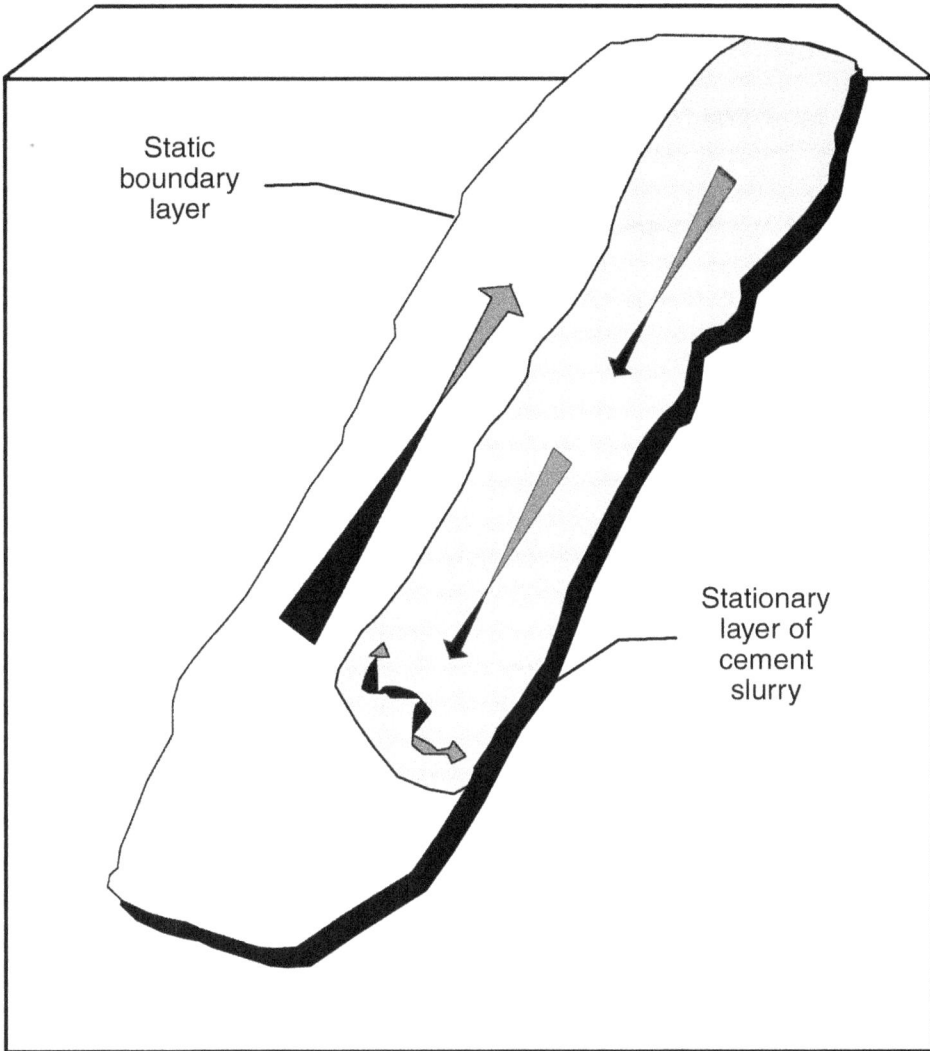

Fig. 9.3—Boycott effect: the tendency for cuttings to settle along the low side of a deviated well.[4]

quarried on the Isle of Portland off the coast of England. Portland cements can be modified easily, depending on the raw materials used and the process used to combine them.

Proportioning of the raw materials is based on a series of simultaneous calculations that take into consideration the chemical composition of the raw materials and the type of cement to be produced: American Soc. for Testing and Materials (ASTM) Type I, II, III, or V white cement, or API Class A, C, G, or H.[5,6]

9.6.2 Classification of Cement. The basic raw materials used to manufacture Portland cements are limestone (calcium carbonate) and clay or shale. Iron and alumina are frequently added if they are not already present in sufficient quantity in the clay or shale. These materials are blended together, either wet or dry, and fed into a rotary kiln, which fuses the limestone slurry at temperatures ranging from 2,600 to 3,000°F into a material called cement clinker. After it cools, the clinker is pulverized and blended with a small amount of gypsum to control the setting time of the finished cement.

TABLE 9.1—TYPICAL MILL RUN ANALYSIS OF PORTLAND CEMENT

Oxide	Class G, wt%	Class H, wt%
Silicon dioxide, SiO_2	21.7	21.9
Calcium oxide, CaO	62.9	64.2
Aluminum oxide, Al_2O_3	3.2	4.2
Iron oxide, Fe_2O_3	3.7	5
Magnesium oxide, MgO	4.3	1.1
Sulfur trioxide, SO_3	2.2	2.4
Sodium oxide, Na_2O		0.09
Potassium oxide, K_2O		0.66
Total alkali as Na_2O	0.54	0.52
Loss on ignition	0.74	1.1
Insoluble residue	0.14	0.21
Phase Composition		
C_3S	58	52
C_2S	19	24
C_3A	2	3
C_4AF	11	15
Physical Properties		
% passing 325 mesh	87	70
Blaine fineness, cm^2/gm	3,470	2,610
Physical Requirements		
Thickening time, min, Sch 5	1:40	1:38
B_c at 30 min	14	15
8 hr compressive strength, 110°F (38°C)	928 psi (6.4 MPa)	650 psi (4.5 MPa)
8 hr compressive strength, 140°F (60°C)	2,247 psi (15.5 MPa)	1,650 psi (11.4 MPa)
Free fluid, mL[13]	4.4	4.0

TABLE 9.2—TYPICAL COMPOSITION AND PROPERTIES OF API CLASSES OF PORTLAND CEMENT

API Class	Compounds, %				Wagner Fineness, cm^2/g
	C_3S	C_2S	C_3A	C_4AF	
A	53	24	8+	8	1,500 to 1,900
B	47	32	5–	12	1,500 to 1,900
C	58	16	8	8	2,000 to 2,800
G & H	50	30	5	12	1,400 to 1,700

Property	How Achieved
High early strength	By increasing the C_3S
Better retardation	By controlling C_3S and C_3A
Low heat of hydration	By limiting the C_3S and C_3A content
Resistance to sulfate attack	By limiting the C_3A content

When these clinkers hydrate with water in the setting process, they form four major crystalline phases, as shown in **Tables 9.1 and 9.2**.[7] The chemical formulas and standard designations of these phases are discussed later in this chapter.

Portland cements are usually manufactured to meet certain chemical and physical standards that depend upon their application. In some cases, additional or corrective components must be added to produce the optimum compositions. Examples of such additives are sand, siliceous loams, pozzolans, diatomaceous earth (DE), iron pyrites, and alumina. Calculations also take into account argillaceous or siliceous materials that may be present in high proportions in some limestones, as well as from the ash produced when coal is used to fire the kiln. Minor impurities in the raw material also must be taken into account, as they can have a significant effect on cement performance. In the U.S., there are several agencies that study and write specifications for the manufacture of Portland cement. Of these groups, the best known to the oil industry are ASTM, which deals with cements for construction and building use, and API, which writes specifications for cements used only in wells.

The ASTM *Spec. C150*[5] provides for eight types of Portland cement: Types I, IA, II, IIA, III, IIIA, IV, and V, where the "A" denotes an air-entraining cement. These cements are designed to meet the varying needs of the construction industry. Cements used in wells are subjected to conditions not encountered in construction, such as wide ranges in temperature and pressure. For these reasons, different specifications were designed and are covered by API specifications. API currently provides specifications covering eight classes of oilwell cements, designated Classes A through H. API Classes G and H are the most widely used.

Oilwell cements are also available in either moderate sulfate-resistant (MSR) or high sulfate-resistant (HSR) grades. Sulfate-resistant grades are used to prevent deterioration of set cement downhole caused by sulfate attack by formation waters.

9.6.3 API Classifications. The oil industry purchases cements manufactured predominantly in accordance with API classifications as published in API *Spec. 10A*.[8] The different classes of API cements for use at downhole temperatures and pressures are defined next.

Class A. This product is intended for use when special properties are not required. [Available only in ordinary, *O*, grade (similar to ASTM *Spec. C150*, Type I)].[5]

Class B. This product is intended for use when conditions require moderate or high sulfate resistance. Available in both MSR and HSR grades (similar to ASTM *Spec. C150*, Type II).[5]

Class C. This product is intended for use when conditions require high early strength. Available in ordinary, *O*, MSR, and HSR grades (similar to ASTM *Spec. C150*, Type III).[5]

Class G. No additions other than calcium sulfate or water, or both, shall be interground or blended with the clinker during manufacture of Class G well cement. This product is intended for use as a basic well cement. Available in MSR and HSR grades.

Class H. No additions other than calcium sulfate or water, or both, shall be interground or blended with the clinker during manufacture of Class H well cement. This product is intended for use as a basic well cement. Available in MSR and HSR grades.

9.6.4 Properties of Cement Covered by API Specifications. Chemical properties and physical requirements are summarized in **Tables 9.3 and 9.4**, respectively.[7] Typical physical requirements of the various API classes of cement are shown in **Table 9.5**.[7]

Although these properties describe cements for specification purposes, oilwell cements should have other properties and characteristics to provide for their necessary functions downhole. API *RP10B* provides standards for testing procedures and special apparatus used for testing oilwell cements and includes slurry preparation, slurry density, compressive-strength tests and nondestructive sonic testing, thickening-time tests, static fluid-loss tests, operating free fluid tests, permeability tests, rheological properties and gel strength, pressure-drop and flow-regime calculations for slurries in pipes and annuli, arctic (permafrost) testing procedures, slurry-stability test, and compatibility of wellbore fluids.[1]

9.6.5 Specialty Cements. A number of cementitious materials, used very effectively for cementing wells, do not fall into any specific API or ASTM classification. These materials include pozzolanic Portland cements, pozzolan/lime cements, resin or plastic cements, gypsum cements,

TABLE 9.3—CHEMICAL REQUIREMENTS FOR API CEMENTS

	Cement Class				
	A	B	C	G	H
Ordinary Grade, O					
Magnesium oxide, MgO, maximum, %	6.0	—	6.0	—	—
Sulfur trioxide, SO_3, maximum, %	3.5[1]	—	4.5	—	—
Loss on ignition, maximum, %	3.0	—	3.0	—	—
Insoluble residue, maximum, %	0.75	—	0.75	—	—
Tricalcium aluminate, $3CaO \cdot Al_2O_3$, maximum, %	—	—	15	—	—
Moderate-Sulfate-Resistant Grade, MSR					
Magnesium oxide, MgO, maximum, %	—	6.0	6.0	6.0	6.0
Sulfur trioxide, SO_3, maximum, %	—	3.0	3.5	3.0	3.0
Loss on ignition, maximum, %	—	3.0	3.0	3.0	3.0
Insoluble residue, maximum, %	—	0.75	0.75	0.75	0.75
Tricalcium silicate, C_3S maximum, %	—	—	—	58[2]	58[2]
minimum, %	—	—	—	48[3]	48[3]
Tricalcium aluminate, C_3A, maximum, %[2]	—	8	8	8	8
Total alkali content expressed as sodium oxide, Na_2O, equivalent, maximum, %[3]	—	—	—	0.75	0.75
High-Sulfate-Resistant Grade (HSR)					
Magnesium oxide, MgO	—	6.0	6.0	6.0	6.0
Sulfur trioxide, SO_3,maximum, %	—	3.0	3.5	3.0	3.0
Loss on ignition, maximum, %	—	3.0	3.0	3.0	3.0
Insoluble residue, maximum, %	—	0.75	0.75	0.75	0.75
Tricalcium silicate, C_3S, maximum, %	—	—	—	65[2]	65[2]
minimum, %	—	—	—	48[2]	48[2]
Tricalcium aluminate, C_3A, maximum, %[2]	—	3	3	3	3
Tetracalcium aluminoferrite, C_4AF, plus twice the tricalcium aluminate, C_3A, maximum, %[2]	—	24	24	24	24
Total alkali content expressed as sodium oxide, Na_2O, equivalent, maximum, %[3]	—	—	—	0.75	0.75

[1]When the tricalcium aluminate content (expressed as C_3A) of the Class A cement is 8% or less, the maximum SO_3 content shall be 3%.
[2]The expressing of chemical limitations by means of calculated assumed compounds does not necessarily mean that the oxides are actually or entirely present as such compounds. When the ratio of the percentages of Al_2O_3 to Fe_2O_3 is 0.64 or less, the C_3A content is zero. When the Al_2O_3 to Fe_2O_3 ratio is greater than 0.64, the compounds shall be calculated as $C_3A = (2.65 \times \% \, Al_2O_3) - (1.69 \times \% \, Fe_2O_3)$, $C_4AF = 3.04 \times \% \, Fe_2O_3$, $C_3S = (4.07 \times \% \, CaO) - (7.60 \times \% \, SiO_2) - (6.72 \times \% \, Al_2O_3) - (1.43 \times \% \, FeO_3) - (2.85 \times \% \, SO_3)$. When the ratio of Al_2O_3 to Fe_2O_3 is less than 0.64, the C_3S shall be calculated as $C_3S = (4.07 \times \% \, CaO) - (7.60 \times \% \, SiO_2) - (4.48 \times \% \, Al_2O_3) - (2.86 \times \% \, Fe_2O_3) - (2.85 \times \% \, SO_3)$.
[3]The sodium oxide equivalent (expressed as Na_2O equivalent) shall be calculated by Na_2O equivalent $= (0.658 \times \% \, K_2O) + \% \, Na_2O$.

microfine cements, expanding cements, refractory cements, latex cements, cements for permafrost environments, Sorel cements, and cements for carbon dioxide (CO_2) resistance.

9.6.6 Pozzolanic Cements. Pozzolanic materials include any natural or industrial siliceous or silico-aluminous material, which will combine with lime in the presence of water at ordinary temperatures to produce strength-developing insoluble compounds similar to those formed from hydration of Portland cement. Typically, pozzolanic material is categorized as natural or artificial, and can be either processed or unprocessed. The most common sources of natural pozzolanic materials are volcanic materials and DE. Artificial pozzolanic materials are produced by partially calcining natural materials such as clays, shales, and certain siliceous rocks, or are more usually obtained as an industrial byproduct. Artificial pozzolanic materials include metakaolin, fly ash, microsilica (silica fume), and ground granulated blast-furnace slag.

TABLE 9.4—PHYSICAL REQUIREMENTS FOR API CEMENTS

Well cement class:			A	B	C	G	H
Mix water, wt% of well cement:			46	46	56	44	38
Fineness tests (alternative methods):							
Turbidimeter (specified surface, minimum, m_2/kg):			150	160	220	—	—
Air permeability (specified surface, minimum, m_2/kg):			280	280	400	—	—
Free-fluid content, maximum, mL:			—	—	—	3.5	3.5

Compressive-strength test, 8-hour curing time	Schedule number, Table 7	Curing temp., °F (°C)	Curing pressure, psi (kPa)	Minimum Compressive Strength, psi (MPa)				
				250 (1.7)	200 (1.4)	300 (2.1)	300 (2.1)	300 (2.1)
	—	100 (38)	Atmos.				1,500	1,500
	—	140 (60)	Atmos.	—	—	—	(10.3)	(10.3)

Compressive-strength test, 24-hour curing time	Schedule number, Table 7	Final curing temp., °F (°C)	Final curing pressure, psi (kPa)	Minimum Compressive Strength, psi (MPa)				
				1,800 (12.4)	1,500 (10.3)	2,000 (18.8)	—	—
	—	100 (38)	Atmos.					

Pressure/ temperature thickening-time test	Specification test schedule number, Table 10	Maximum consistency, 15 to 30 min stirring period, B_c	Minimum Thickening Time, min				
	4	30	90	90	90	—	—
	5	30	—	—	—	90	90
	5	30	—	—	—	120 max.	120 max.

B_c = Bearden units of consistency, obtained on a pressurized consistometer, as defined in Sec. 9 of API *Spec. 10A* and calibrated as per the same section.[8]

Pozzolanic oilwell cements are typically used to produce lightweight slurries. Because the specific gravity of the pozzolanic material is lower than that of the cement, a pozzolan slurry has a lighter weight than a corresponding Portland cement slurry of similar consistency. The lighter weight keeps the formation from breaking down. It is important not to exceed the fracture pressure of the formation while cementing.

Some pozzolanic materials also have a high water demand that effectively gives a higher yield and lighter slurry. They also tend to improve compressive strength over time. The additional binding material also reduces permeability and minimizes attack from formation waters. In most cases, pozzolanic materials can also reduce the effect of sulfate attack, though this is to a certain degree dependent on the slurry design.

Commercial cements such as TXI Lightweight™ for use in oil wells are a special formulation composed of Portland-cement clinker interground with lightweight siliceous aggregate to produce, in effect, a pozzolanic cement.

9.6.7 Pozzolan/Lime Cements. Pozzolan/lime or silica/lime cements are usually blends of fly ash (silica), hydrated lime, and small quantities of calcium chloride. At low temperatures, the initial reactions of these cements are slower than similar reactions in Portland cements, and therefore, they are generally recommended for primary cementing at temperatures greater than

TABLE 9.5—PHYSICAL REQUIREMENTS OF VARIOUS TYPES OF CEMENTS

Properties of API Classes of Cement	Class A	Class C	Classes G and H
Specific gravity, average	3.14	3.14	3.15
Surface area, range, cm^2/g	1,500	2,000 to 2,800	1,400 to 1,700
Weight per sack, lbm	94	94	94
Bulk volume, ft^3/sk	1	1	1
Absolute volume, gal/sk	3.6	3.6	3.58

Properties of Neat Slurries	Portland	High Early Strength	API Class G	API Class H
Water, gal/sk, API	5.19	6.32	4.97	4.29
Slurry weight, lbm/gal	15.6	14.8	15.8	16.5
Slurry volume, ft^3/sk	1.18	1.33	1.14	1.05

Temperature, °F	Pressure, psi	Typical Compressive Strength, psi at 24 hours			
60	0	615	780	440	325
80	0	1,470	1,870	1,185	1,065
95	800	2,085	2,015	2,540	2,110
110	1,600	2,925	2,705	2,915	2,525
140	3,000	5,050	3,560	4,200	3,160
170	3,000	5,920	3,710	4,830	4,485
200	3,000	*	*	5,110	4,575

Temperature, °F	Pressure, psi	Typical Compressive Strength, psi at 72 hours			
60	0	2,870	2,535	—	—
80	0	4,130	3,935	—	—
95	800	4,670	4,105	—	—
110	1,600	5,840	4,780	—	—
140	3,000	6,550	4,960	—	7,125
170	3,000	6,210	4,460	5,685	7,310
200	3,000	*	*	7,360	9,900

Depth, ft	Temperature, °F		High-Pressure Thickening Time, hr:min			
	Static	Circulation				
2,000	110	91	4:00+	4:00+	3:00+	3:57
4,000	140	103	3:26	3:10	2:30	3:20
6,000	170	113	2:25	2:06	2:10	1:57
8,000	200	125	1:40*	1:37*	1:44	1:40

*Not generally recommended at this temperature.

284°C (140°F). The merits of this type of cement are ease of retardation, light weight, economy, and strength stability at high temperatures.

9.6.8 Gypsum Cements. Gypsum cement is a blend of API Class A, C, G, or H cement and a hemihydrate form of gypsum ($CaSO_4 \cdot \frac{1}{2}H_2O$). Gypsum cements are commonly used in low-temperature applications for primary cementing or remedial cementing work. This combination is particularly useful in shallow wells to minimize fallback after placement. A high-gypsum-content cement has increased ductility, thixotropy, and acid solubility. It is usually used in situations of high lateral stress or in temporary plugging applications. A 50:50 gypsum cement is frequently used in fighting lost circulation, to form a permanent insoluble plugging material. These blends should be used cautiously because they have very rapid setting properties and could set prematurely during placement. A limitation of gypsum cements is that they are slowly soluble, and they are not stable in contact with external sources of water. This would be a fatal error for an oilfield cement.

9.6.9 Microfine Cements. Microfine cements are composed of very finely ground sulfate-resisting Portland cements, Portland cement blends with ground granulated blast-furnace slag, and alkali-activated ground granulated blast-furnace slag. Such cements have a high penetrability and are ultrarapid-hardening. Applications for such cements are in consolidation of unsound formations and in repairing casing leaks in squeeze operations, particularly "tight" leaks that are inaccessible to conventional cement slurries because of their penetrability. Ultrafine alkali-activated ground blast-furnace slag is the product used in the mud-to-cement technology, in which water-based drilling mud is converted to cement.

9.6.10 Expanding Cements. Expansive cements are available for the primary purpose of improving the bond of cement to pipe and formation. If expansion is properly restrained, its magnitude will be reduced and a prestress will develop. Expansion can also be used to compensate for the effects of shrinkage in normal Portland cement.

At this time, there is no test procedure or specifications in the API standards for measuring the expansion forces in cement. Most laboratories use the expansive bar test, employing a molded $1 \times 1 \times 10$-in. cement specimen. Ring molds are also available, though they are not as commonly used. The expansive force is measured soon after the cement sets for a base reference and then at various time intervals until the maximum expansion is reached. Hydraulic bonding tests have also been used to evaluate the growth of expanding cements.

9.6.11 Calcium Aluminate Cements. High-alumina cement (HAC) is used in well-cementing operations at both temperature extremes in permafrost zones with temperatures at 32°F or below; in-situ combustion well's (fireflood) where temperatures may range from 750 to 2,000°F, and thermal-recovery wells where temperatures can exceed 1,300°F and temperature fluctuations can be high.

A number of HACs have been developed with alumina contents between 35 and 90%, and there is a move to term these collectively as calcium aluminate cements (CACs) because the reactive phase in all cases is calcium aluminate.

It is the standard type (e.g., Ciment Fondu) that is mostly used in well cementing. These cements can be accelerated or retarded to fit individual-well conditions, but the retardation characteristics will differ from those of Portland cements. The addition of Portland cement to refractory cement causes a flash set; therefore, when both are handled in the field, they should be stored separately.

9.6.12 Latex Cement. Latex cement, although sometimes identified as a special cement, is actually a blend of API Class A, G, or H with latex. In general, a latex emulsion contains only 50% latex by weight of solids and is usually stabilized by an emulsifying surface-active agent. Latexes impart elasticity to the set cement and improve the bonding strength and filtration control of the cement slurry. Latex in powdered form can be dry-blended with the cement before it is transported to the wellsite and is not susceptible to freezing.

9.6.13 Permafrost Cement. It is normally desirable to use a quick-setting, low-heat-of-hydration cement that will not melt the permafrost. API *RP10B,* Sec. 14,[1] gives special cementing procedures for simulating arctic conditions and cementing in such environments.

Two cement systems that have been used successfully are calcium aluminate cement blends and gypsum cement blends. Fly ash or natural pozzolan is normally blended (at about 50% by weight) with calcium aluminate cements to lower the heat of hydration, thus preventing permafrost damage. Gypsum-cement blends can be accelerated or retarded and will set at 15°F below freezing. For surface pipe, these slurries are normally designed for 2 to 4 hours of pumpability, yet their strength development is quite rapid and varies little at temperatures between 20 and 80°F.

9.6.14 Resin or Plastic Cements. Resin and plastic cements are specialty materials used for selectively plugging open holes, squeezing perforations, and cementing waste disposal wells, especially in highly aggressive, acidic environments. They are usually mixtures of water, liquid resins, and a catalyst blended with an API Class A, B, G, or H cement.

When pressure is applied to the slurry, the resin phase may be squeezed into a permeable zone to form a seal within the formation. These specialty cements are used in wells in relatively small volumes. They are effective at temperatures from 140 to 392°C (60 to 200°F).

9.6.15 Cements for CO_2 Resistance. The hydration products of Portland cement are susceptible to carbonation in the presence of moisture. Carbonation is the attack resulting from dissolved CO_2 in formation waters or as a result of CO_2-injection processes. The CO_2 dissolves in the aqueous pore solution of the hydrated cement, ultimately producing calcium carbonate ($CaCO_3$).

Carbonation can be minimized by the use of a specially formulated calcium phosphate cement, ThermaLock™, that is resistant to both CO_2 and acid. This cement can be used at temperatures typically ranging from 140°F (60°C) to 700°F (371°C). ThermaLock™ is an ideal cement for environments in which high concentrations of CO_2 are anticipated. The one disadvantage is that it is more expensive than Portland cement; however, it greatly reduces concerns on the long-term affects of CO_2; saves on remedial operations, abandonments, and redrilling or recompletion; and it does not require special cementing equipment or techniques.

9.7 Cement Hydration

The reactions involved when cement is mixed with water are complex. Each phase hydrates by a different reaction mechanism and at different rates (**Fig. 9.4**).

The reactions, however, are not independent of each other because of the composite nature of the cement particle and proximity of the phases. In all, five distinct stages have been identified: (1) pre-induction, (2) "dormant" (induction) period, (3) acceleration, (4) deceleration, and (5) steady state. In cementing operations, the most important of these are Stages 1 through 3. Stage 1 dictates the initial mixability of the cement and is attributed primarily to the aluminate and ferrite phase reactions. Stage 2 relates to the pumpability time, while Stage 3 gives an indication on setting properties and gel-strength development.

9.7.1 Hydration of Pure Mineral Phases. During hydration, the cement forms four major crystalline phases.

Tricalcium Silicate ($3CaO \cdot SiO_2 = C_3S$). C_3S on reaction with water produces C-S-H and calcium hydroxide, CH, (also known as Portlandite). The hyphens used in the C-S-H formula are to depict its variable composition: CSH would imply a fixed composition of $CaO.SiO_2.H_2O$. C/S ratios in C-S-H vary from 1.2 to 2.0, and H/S ratios vary between 1.3 and 2.1.

Dicalcium Silicate ($2CaO \cdot SiO_2 = C_2S$). The kinetics and hydration mechanism for C_2S are similar to those of C_3S except that the rate of reaction is much slower. The hydration products are the same except that the proportion of CH produced is about one-third of that obtained on hydration of C_3S.

Tricalcium aluminate ($3CaO \cdot Al_2O_3 = C_3A$). The initial reaction of C_3A with water in the absence of gypsum is vigorous and can lead to "flash set" caused by the rapid production of

Cement Powder	Reaction Rate	Reaction Products	Effect on Performance

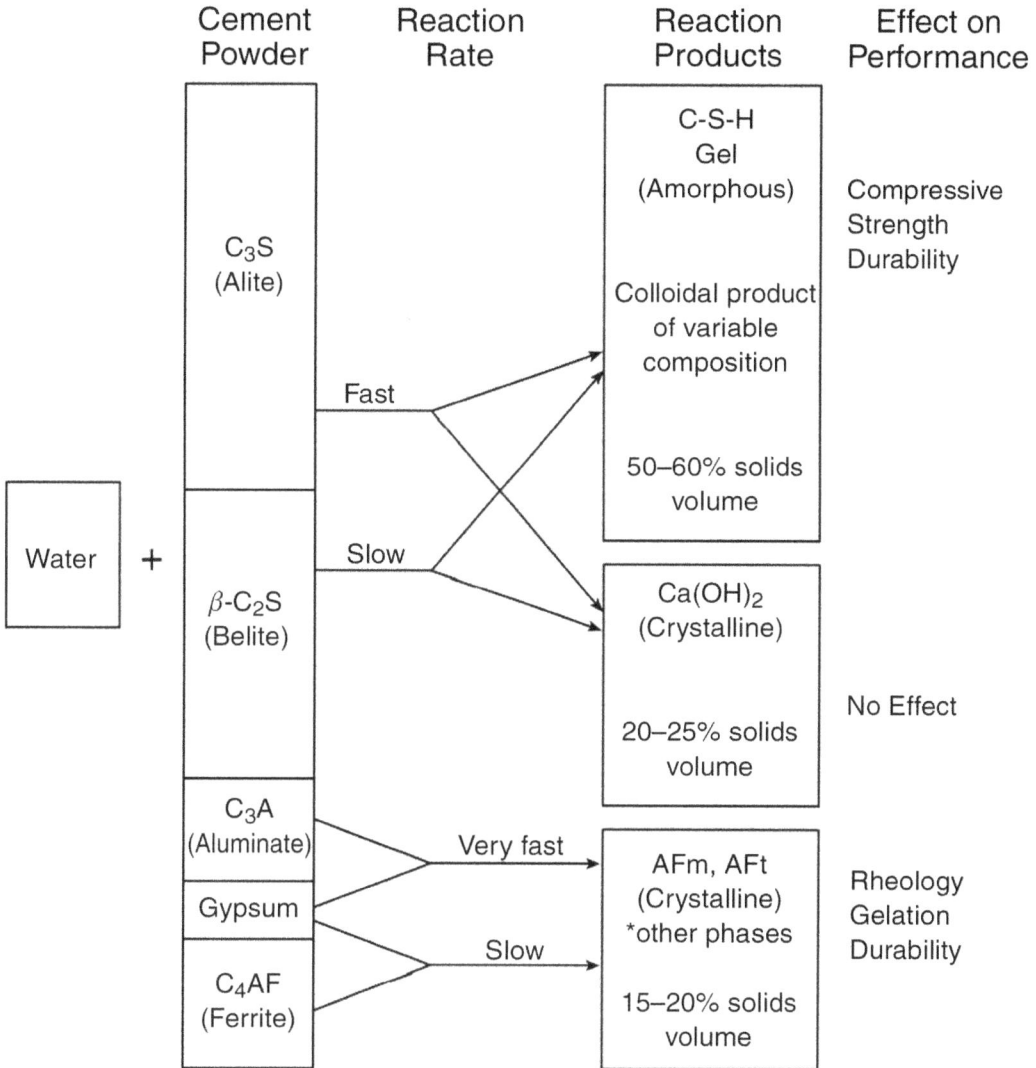

Fig. 9.4—Schematic of cement-hydration reactions (courtesy of Halliburton).

the hexagonal crystal phases, C_2AH_8 (H = H_2O) and C_4AH_{19}. Sufficient strength is developed to prevent continued mixing. The C_2AH_8 and C_4AH_{19} subsequently convert to cubic C_3AH_6 (hydrogarnet), which is the thermodynamically stable phase at ambient temperature. Typically, gypsum is added to retard this reaction, though other chemical additives can be used.

The reaction products formed on reaction of C_3A in the presence of gypsum depend primarily on the supply of sulfate ions available from the dissolution of gypsum. The primary phase formed is ettringite ($C_6A\bar{S}_3H_{32}$) ($\bar{S} = SO_3$). Ettringite is the stable phase only as long as there is an adequate supply of soluble sulfate. A second reaction takes place if all of the soluble sulfate is consumed before the C_3A has completely reacted. In this reaction, the ettringite formed initially reacts with the remaining C_3A to form a tetracalcium aluminate monosulfate-12- hydrate known as monosulfate or monosulfoaluminate ($C_4A\bar{S}H_{12}$).

Tetracalcium Aluminoferrite (4CaO·Al$_2$O$_3$·Fe$_2$O$_3$ = C$_4$AF). Hydration of C_4AF gives hydration products that are similar in many respects to those formed from C_3A under comparable

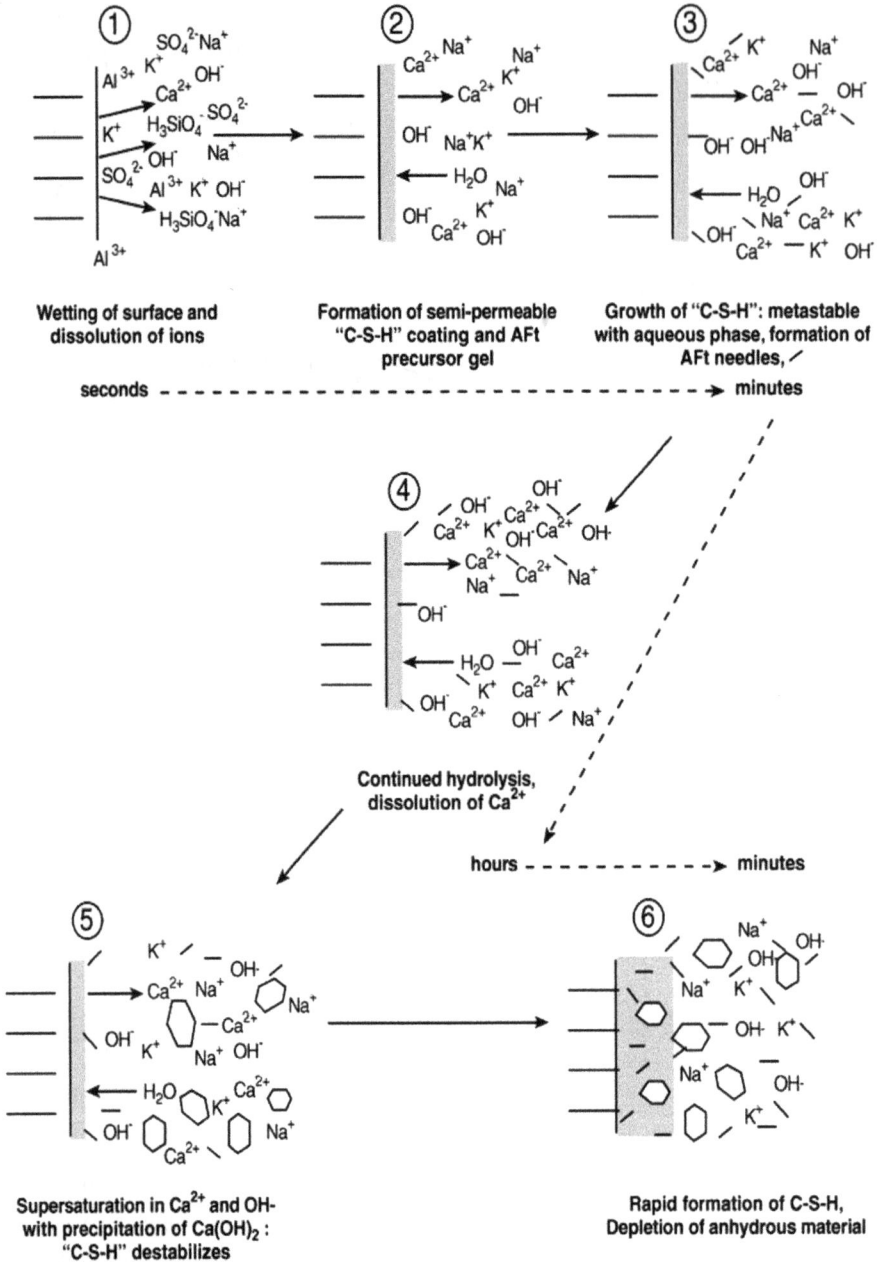

Fig. 9.5—Cement hydration from mixing to setting (courtesy of Halliburton).

conditions, though typically they contain Fe^{3+} as well as Al^{3+}. An iron (III) hydroxide gel and calcium ferrite gel are also possible products of C_4AF hydration. The reactivity of the pure C_4AF is, in general, much slower than that of the C_3A.

9.7.2 Hydration of Cement Phases. Although the basic reaction mechanisms and theories on the hydration of the pure phases pertain to the phases in cement, there are some significant differences. A schematic of the initial hydration reactions up to the time of set is illustrated in **Fig. 9.5**.

Alkalis. The alkalis, primarily sodium and potassium, are impurities that arise from shales, clays, or the fuel used in the manufacture of the cement. Although present in small amounts, < 1%, they have a significant effect on the hydration. Typically, they are present as sulfates, in the form of K_2SO_4, Na_2SO_4, $Na_2SO_4 \cdot 3K_2O$ (aphthitalite), and/or $2CaSO_4 \cdot K_2SO_4$ (calcium langbeinite), and they are usually deposited on the surface of the cement particles. The alkali sulfates dissolve almost immediately on contact with water, and alkalis can also be present as impurities in the cement phases, with sodium preferentially in the aluminate (C_3A) phase and potassium more widely distributed in both calcium and aluminate phases. API *Spec. 10A* for Class G and H cements limits the alkali to 0.75% as Na_2O_4 to allow adequate thickening times to be achieved downhole.[8]

In cements high in K_2SO_4, reaction between K_2SO_4 and gypsum in the presence of water can produce syngenite, KCS_2H. This can cause lumpiness on storage of the dry cement powder under high-humidity conditions (> 90% relative humidity) because the KCS_2H acts as an effective binder to the dry cement particles. Precipitation of KCS_2H during cement hydration can cause false or even flash setting.

Calcium Sulfates. Gypsum ($CaSO_4 \cdot H_2O = C\bar{S}H_2$) is added to the cement primarily to retard the hydration of the aluminate and ferrite phases. The effectiveness of the gypsum depends on the rate at which the relevant ionic species dissolve and come in contact with each other. Thus, interground gypsum is far more effective than interblending the same proportion because intergrinding brings the gypsum particles into closer contact with the cement particles and produces a shorter diffusion distance between the two. Temperature and humidity in the grinding mill can cause the gypsum to dehydrate, resulting in the formation of hemihydrate ($C\bar{S}H_{0.5}$) and/or soluble anhydrite (γ-$C\bar{S}$). Hemihydrate or soluble anhydrite can rehydrate to give "secondary" gypsum, causing a rapid set, known as "false set." Pumpability can be regained on further mixing or addition of water, assuming the quantity of secondary gypsum is not too great.

The reactivity and performance of cement is a culmination of the effect of the different impurities on the number of defects and morphology of the crystal structure of the different phases. This is why cement can vary not only from one source to another but also between batches from the same source.

9.7.3 Effects of Temperature on Hydration. The rate of hydration of the cement phases, however, will increase with increasing temperature, and the resulting thickening and setting times will, consequently, decrease. Above 230°F (110°C), the hydration products formed differ considerably from those obtained at lower temperatures. Alite and belite phases hydrate to give crystalline α-C_2SH rather than amorphous C-S-H. α-C_2SH is a relatively dense crystalline material that is porous and weak and is deleterious in that it provides high permeability and low compressive strength. Formation of α-C_2SH can be prevented, or at least minimized, by the addition of finely ground silica, such as silica flour, to the cement.

Normally, in oilwell cementing, ~ 35% silica in the form of silica flour is used to prevent strength retrogression that can occur at temperatures above ~ 248°F (120°C). This percentage of added silica gives an effective C/S ratio in the cement blend of approximately 1.0. Generally, over time, the permeability increases slightly, and the compressive strength decreases as the phases increase in crystallinity.

Fly ash has often been considered as a potential source of silica for hydrothermal systems. There is considerable variability in the alumina/silica ratio of fly ashes from different sources, as well as in the reactivity of the aluminosilicate glass, and this clearly has an impact on the phases formed and their stability fields. The influence of this variability in composition and reactivity is that the fly ash, if used as a source of silica, can give properties that range from good to deleterious.

9.7.4 Sulfate Attack. Sulfate attack is normally a problem only where BHSTs are below approximately 60°C (140°F), where ettringite is present. Some formation waters contain high concentrations of sulfate. These sulfates attack the cement, and, as a result, the cement will crumble with time.

Resistance to sulfate attack is increased on modifying the cement powder by replacing the aluminate with ferrite, which reduces the amount of ettringite formed on hydration, and also by lowering the amount of free lime. Addition of pozzolanic materials, such as fly ash, also reduce sulfate attack because they react with the CH in cement and render it unavailable for reaction.

9.8 Slurry Design

The properties of Portland cements must often be modified to meet the demands of a particular well application. These modifications are accomplished by the admixing of chemical compounds commonly referred to as additives that effectively alter the hydration chemistry. An overview of the most common cementing additives is given in **Table 9.6.**[7]

The table also includes an indication of the primary uses and benefits, along with the cements that they can be used with. The primary effects of the cement admixtures on the physical properties of the cement, either as a slurry or set, are presented in **Table 9.7.**[7] This is a quick reference, and individual additives in a given category may not agree in total with the effects as given. It is also typically defined for individual additives, the properties and effects of which can be modified when additive combinations are used.

Many chemical compounds have proved to be effective in modifying the properties of Portland-cement slurries. These compounds, when used alone, will have a primary effect upon the cement slurry that is considered to be beneficial. They will also exhibit at least one secondary characteristic that may be either beneficial or detrimental to the cement-slurry performance properties. The effects of the additives are reduced or enhanced by modifying the additive or by using additional additives. For most downhole requirements, more than one additive is needed. This give-and-take relationship between additives is the basis of cement-slurry design.

The reaction of these additives with the cement and the interaction between them is not well defined chemically. What is actually known are the physical effects of these additives on the slurry performance properties. The slurry performance properties that are measured include: thickening time, compressive strength, rheology, fluid loss, free fluid, and slurry stability.

Cement manufactured to API depth and temperature requirements can be purchased in most oil-producing areas of the world. Any properly made Portland cement (consistent from batch to batch) can be used at temperatures up to 570°F. For example, Class H cement with the proper additives has routinely been used at depths up to 20,000 ft.

In addition to the cement, other factors, such as the correct BHCT, should be considered when designing a cement slurry to meet well requirements. In formulating a cement slurry, the designer must consider not only the temperature but also the other downhole conditions, such as permeability and water-sensitive formations.

A slurry should be designed for its specific application, with good properties to allow placement in a normal period. The ideal cement slurry should have no measurable free water, provide adequate fluid-loss control, contain adequate retarder to help ensure proper placement, and maintain a stable density to ensure hydrostatic control. Do not add dispersants or retarders in excess of the amounts indicated by wellbore conditions, and provide just enough fluid-loss control to place the cement before it gels.

Slurry design is affected by the following criteria: well depth, quality of mix water, BHCT, fluid-loss control, BHST, flow regime, drilling fluid's hydrostatic pressure, settling and free water, type of drilling fluid, quality of cement, slurry density, dry or liquid additives, lost circulation, strength development, gas-migration potential, quality of the cement testing, pumping time, and laboratory and equipment.

TABLE 9.6—SUMMARY OF OILWELL CEMENTING ADDITIVES

Type of Additive	Use	Chemical Composition	Benefit	Type of Cement
Accelerators	Reducing WOC time	Calcium chloride	Accelerated setting	All API classes
	Setting surface pipe	Sodium chloride	High early strength	Pozzolans
	Setting cement plugs	Gypsum		Diacel systems
	Combating lost circulation	Sodium silicate		
		Dispersants		
		Seawater		
Retarders	Increasing thickening time for placement	Lignosulfonates	Increased pumping time	API Classes D, E, G, and H
	Reducing slurry viscosity	Organic acids	Better flow properties	
		CMHEC		Pozzolans
		Modified lignosulfonates		Diacel systems
Weight-reducing additives	Reducing weight	Bentonite/attapulgite	Lighter weight	All API classes
	Combating lost circulation	Gilsonite	Economy	Pozzolans
		Diatomaceous earth	Better fill-up	Diacel systems
		Perlite	Lower density	
		Pozzolans		
		Microspheres (glass spheres)		
		Nitrogen (foam cement)		
Heavyweight additives	Combating high pressure	Hematite	Higher density	API Classes D, E, G, and H
	Increasing slurry weight	Limenite		
		Barite		
		Sand		
		Dispersants		
Additives for controlling lost circulation	Bridging	Gilsonite	Bridged fractures	All API classes
	Increasing fill-up	Walnut hulls	Lighter fluid columns	Pozzolans
	Combating lost circulation	Cellophane flakes	Squeezed fractured zones	Diacel systems
	Fast-setting systems	Gypsum cement		
		Bentonite/diesel oil	Treating lost circulation	
		Nylon fibers		
		Thixotropic additives		
Filtration-control additives	Squeeze cementing	Polymers	Reduced dehydration	All API classes
	Setting long liners	Dispersants	Lower volume of cement	Pozzolans
	Cementing in water-sensitive formations	CMHEC		Diacel systems
		Latex	Better fill-up	

TABLE 9.6—SUMMARY OF OILWELL CEMENTING ADDITIVES (continued)

Type of Additive	Use	Chemical Composition	Benefit	Type of Cement
Dispersants	Reducing hydraulic horsepower	Organic acids	Thinner slurries	All API classes
		Polymers	Decreased fluid loss	Pozzolans
	Densifying cement slurries for plugging	Sodium chloride	Better mud removal	Diacel systems
		Lignosulfonates	Better placement	
	Improving flow properties			
Special cements or additives				
Salt	Primary cementing	Sodium chloride	Better bonding to salt, shales, sands	All API classes
Silica flour	High-temperature cementing	Silicon dioxide	Stabilized strength	All API classes
Radioactive tracers	Tracing flow patterns Locating leaks	$_{53}I^{131}$, $_{77}Ir^{192}$		All API classes
Pozzolan lime	High-temperature cementing	Silica-lime reactions	Lighter weight Economy	
Silica lime	High-temperature cementing	Silica-lime reactions	Lighter weight	
Gypsum cement	Dealing with special conditions	Calcium sulfate Hemihydrate	Higher strength Faster setting	
Latex cement	Dealing with special conditions	Liquid or powdered latex	Better bonding Controlled filtration	API Classes A, B, G, and H
Thixotropic additives	Covering lost-circulation zones Preventing gas migration	Organic additives Inorganic additives	Fast setting and/or gelation Less fallback Reduces lost circulation	All API Classes

When estimating job time, include the mixing time on the surface, especially if the job is going to be batch-mixed. Calculate the actual job time, using the slurry volume and average displacement rate; then, limit the amount of trouble time to 1 to 1.5 hours. To calculate the approximate thickening time for slurry design, add 1 to 1.5 hours to the job time.

9.9 Additives
The additives used to modify the properties of cement slurries for use in oilfield well-cementing applications fall into the following broad categories: accelerators, retarders, extenders, weighting agents, dispersants, fluid-loss control agents, lost-circulation agents, strength-retrogression prevention agents, free-water/free-fluid control, expansion agents, and special additives.

TABLE 9.7—EFFECTS OF CEMENT ADMIXTURES ON THE PHYSICAL PROPERTIES OF CEMENT

		Accelerator (Calcium Chloride)	Bentonite	Pozzolan (Fly Ash)	Heavyweight Hematite	Retarders	Friction Reducers (Dispersants)	Filtration Additives	Lost-Circulation Additives	Sand	Salt, 10 to 20%	Silica Flour	Seawater
Water Requirements	Increases		Y	X				X	X				
	Decreases						X						
Density	Increases				Y					X	X	X	X
	Decreases		Y	Y									
Viscosity	Increases		Y		X							X	
	Decreases	X		Y			Y	X	Y	X	X		
Thickening Time	Accelerates	Y						X					X
	Retards		X		X	Y	Y			X	X		
Fluid Loss of Slurry	Increases		X										
	Decreases						X	Y					
Early Strength	Increases	Y											
	Decreases		X	X		X							
Final Strength	Increases	Y		X	X			X		X		X	X
	Decreases		X			X	X						
Durability	Increases			Y		X						Y	
	Decreases		X										
Types of Cementing Job Applications (where mostly used)	Conductor casing	X					X		X		X		X
	Surface casing	X		X			X		X		X		X
	Intermediate casing		X	X			X		X		X		X
	Production casing		X	X	X	X	X	X	X			X	
	Liners				X	X	X	X				X	
	Squeezing					X	X			X			
	Plugging												

For temperature 230°F; X = Minor Effects; Y = Major Effects

The demand for new additives with special properties and tuned performance continues to increase. These demands include such factors as density range of application, temperature stability, economics, viscosity range, singular function, multifunction, rate of solubility, synergism with co-additives, and resistance to cement variability.

9.9.1 Accelerators. Accelerators speed up or shorten the reaction time required for a cement slurry to become a hardened mass. In the case of oilfield cement slurries, this indicates a reduction in thickening time and/or an increase in the rate of compressive-strength development of the slurry. Acceleration is particularly beneficial in cases where a low-density (e.g., high-water-content) cement slurry is required or where low-temperature formations are encountered.

Calcium Chloride (CaCl$_2$). Of the chloride salts, CaCl$_2$ is the most widely used, and in most applications, it is also the most economical. The exception is when water-soluble polymers such as fluid-loss-control agents are used. The major benefits of the use of CaCl$_2$ are the significant reduction in thickening time achieved and that, regardless of concentration, it always acts as an accelerator. The normal concentration range of use for CaCl$_2$ is 1 to 4% by weight of cement (BWOC). Above a concentration of 6% BWOC, the results will become unpredictable and gelation can occur.

Sodium Chloride (NaCl). NaCl is the second most widely used of the chloride salts. NaCl, common table salt, is the most versatile of the chloride salts. Depending on the concentration of use, NaCl can act as an accelerator or a retarder, and it acts a mild dispersant at all concentrations. Some additional uses for NaCl are to improve bonding to pipe, stabilize reactive formations (e.g., shale and gumbo), enhance bonding to salt formations, reduce the permeability of set cement, improve the durability of set cement in contact with saltwater-containing formations, and increase slurry density without the use of dispersants or a reduction in water content. In general, NaCl acts as an accelerator at concentrations from 1 to 10% by weight of water (BWOW), although the most commonly used concentration of NaCl as an accelerator is 3% BWOW.

Potassium Chloride (KCl). The acceleration performance of KCl is similar to that of NaCl. KCl has two advantages over other accelerators: its stabilizing effect on shale or active clay-containing formations and its minimal effect on the performance of fluid-loss additives. As an accelerator, KCl may be used at concentrations up to 5% BWOW; for formation stabilization, concentrations of 3% BWOW are effective.

Sodium Silicate (Na$_2$SiO$_3$). Sodium silicate is normally considered to be a chemical extender, although it is also functional as an accelerator. The effectiveness depends on the concentration and molecular weight. The low-molecular-weight form may be used at concentrations of 1% BWOC or less to accelerate normal-density slurries. The high-molecular-weight form is an effective accelerator at concentrations up to 4% BWOC. Sodium-meta-silicate also provides excellent lost-circulation control when used with cement or CaCl$_2$ brines.

Seawater. Seawater is a naturally occurring mixture of alkali chloride salts, including magnesium chloride. The composition of seawater varies widely around the world. For example, the equivalent chloride salt content can vary from 2.7 to 3.8% BWOW.

Alkali Hydroxides [Ca(OH)$_2$, NaOH]. Alkali hydroxides are commonly used in pozzolan-extended cements. They accelerate both the pozzolanic and the cement component by altering the aqueous chemistry.

Mono-Calcium Aluminate (CaO·Al$_2$O$_3$ = CA). Calcium aluminate is used as an accelerator in pozzolan- and gypsum-extended cements.

9.9.2 Retarders. The commonly used cements in well applications are API Class A, C, G, and H. These cements, as produced in accordance with API Spec. *10A*[8] do not have a sufficiently long fluid life (thickening time) for well applications above 38°C (100°F) BHCT. To extend the thickening time beyond that obtained with a neat (cement and water without additives or minerals) API-class cement slurry, additives known as retarders are required.

Lignosulfonates. Of the chemical compounds that have been identified as retarders, lignosulfonates are the most widely used. A lignosulfonate is a metallic sulfonate salt derived from the lignin recovered from processing wood waste. The common lignosulfonates are calcium and sodium lignosulfonate.

Three grades of lignosulfonate are available for the retardation of cement slurries. Each grade is available as calcium/sodium or sodium salts. The three grades are filtered, purified, and modified.

The filtered grade calcium or sodium salt is typically used at a temperature of 200°F BHCT or less at a concentration of 0.6% BWOC or less. It may be used at higher temperatures but will normally be limited by economic considerations.

The purified grade represents a class of lignosulfonates in which the sugar content has been reduced. The calcium/sodium salt is typically used at a BHCT of 200°F or lower and at a concentration of 0.5% BWOC or less.

The modified grade represents lignosulfonates that have been blended or reacted with a second component. The compounds most commonly used as blend components are boric acid and the hydroxycarboxylic acids, or their salts. Blended materials are available as calcium or sodium salts. The modified lignosulfonates are typically used at a BHCT of 200°F or above. They are more effective than the purified grade at temperatures greater than 250°F. The advantages, whether a blend or reacted product, are their improved high-temperature stability above 300°F BHCT, increased dispersing activity, and synergism with fluid-loss additives.

Cellulose Derivatives. Two cellulose polymers are used in well-cementing applications. They are hydroxyethyl cellulose (HEC) and carboxymethyl hydroxyethyl cellulose (CMHEC). HEC is commonly considered as a fluid-loss additive. Although as a possible option, it is worth noting that at BHCT of 125°F or less, the thickening time can be extended by approximately two hours in a freshwater slurry. Traditionally, the only cellulose that is considered as a retarder is CMHEC. This is largely because it is functional as a retarder up to approximately 230°F BHCT at the same concentrations as calcium lignosulfonate, but it also provides good fluid-loss control.

Hydroxycarboxylic Acids. The hydroxycarboxylic acids are well known for their antioxidant and sequestering properties that benefit cement-slurry performance. The antioxidant property improves the temperature stability of soluble compounds such as fluid-loss additives. Commonly used hydroxycarboxylic acids and their derivatives are citric acid, tartaric acid, gluconic acid, glucoheptonate, and glucono-delta-lactone. The commonly used hydroxycarboxylic acids are generally derived from naturally occurring sugars.

Organophosphonates. Organophosphonates, with a few exceptions, are the most powerful retarders used in cement. These materials are not widely used in well-cementing applications because of the low concentration required, difficulty of accurate measurement, and sensitivity to concentration. The advantage of organophosphate retarders is their effectiveness in ultrahigh-temperature wells (> 450°F) or in applications where extended thickening times of 24 hours or greater are desired.

Synthetic Retarders. The term synthetic retarder is a misnomer in that the previously mentioned retarding compounds are all, in effect, man-made. However, the term synthetic retarder has been applied to a family of low-molecular-weight copolymers. These retarders are based on the same function groups as those of conventional retarders (e.g., sulfonate, carboxylic acid, or an aromatic compound). Two common synthetic retarders are maleic anhydride and 2-Acrylamido-2-methylpropanesulfonic acid (AMPS) copolymers.

Inorganic Compounds. The retardation mechanism of inorganic compounds on cement hydration is different from that for the previously discussed retarders. Inorganic compounds, commonly used as cement retarders, are borax ($Na_2B_4O_7 \cdot 10H_2$) and other borates such as boric acid (H_3BO_3) and its sodium salt and zinc oxide (ZnO).

Borates are commonly used as a retarder aid for high-temperature retarders at BHCT of 300°F (149°C) and greater. At higher temperatures, the borate is a less-powerful retarder than at lower temperatures; however, it exerts a synergistic effect with other retarders such as ligno-sulfonates, whereby the combination provides better retardation than either retarder alone. ZnO is a strong retarder when used alone. It is normally used for the retardation of chemically extended cements.

Salt as a Retarder. Water containing salt concentrations of greater than 20% BWOW has a retarding effect on cement. The gelation is evident in the thickening-time viscosity profile of saturated salt slurries by a sudden increase in Bearden units of consistency that then levels off before set. Saturated salt slurries are useful for cementing through salt domes. They also help protect shale sections from sloughing and heaving during cementing and aid in preventing annular bridging and the lost circulation that could result. Saturated salt cements are also dispersed, and salt reduces the effectiveness of fluid-loss additives.

9.9.3 Lightweight Additives/Extenders.
Neat cement slurries, when prepared from API Class A, C, G, or H cements using the amount of water recommended in API *Spec. 10A*[8] will have slurry weights in excess of 15 lbm/gal. In many parts of the world, severe lost circulation and weak formations with low fracture gradients are common. These situations require the use of low-density cement systems that reduce the hydrostatic pressure of the fluid column during cement placement. Consequently, lightweight additives (also known as extenders) are used to reduce the weight of the slurry. There are several different types of materials that can be used. These include physical extenders (clays and organics), pozzolanic extenders, chemical extenders, and gases.

Any material with a specific gravity lower than that of the cement will act as an extender. These materials, in general, decrease the density of cement slurries by one of three means. The pozzolanic and inert organic materials have a lower density than cement and can be used to partially replace cement, therefore lowering the density of the solid material in the slurry. In the case of the physical and chemical extenders, they not only have a lower density but also absorb water, thus allowing more water to be added to the slurry without producing free fluid or particle segregation. The gases behave differently in that they are used to produce foamed cements that have exceptionally low density with acceptable compressive strengths.

In many lightweight slurries, it is common to use a combination of the different types of material. For example, pozzolanic and chemical extenders are, or can be, used with physical extenders and/or gases. Pozzolan slurry designs almost always incorporate bentonite, and gases generally have a chemical extender to stabilize the foam. Lightweight additives also increase the slurry yield and can result in an economical slurry.

Physical Extenders. These are particulate materials that function as cement extenders by increasing the water requirements or by reducing the average specific gravity of the dry mix. There are two general classes of materials that fall into this category: clays and inert organic materials. The most commonly used clay material is bentonite, although attapulgite is also used. The commonly used inert organic materials are perlite, gilsonite, ground coal, and ground rubber.

Bentonite (Gel). This extender is a colloidal clay mineral composed predominately of sodium montmorillonite [$NaAl_2(AlSi_3O_{10}) \cdot 2OH$]. The montmorillonite content of bentonite is the controlling factor in its effectiveness as an extender; hence, it is one of two extenders that are covered by an API specification. Bentonite can be added to any API class of cement and is commonly used in conjunction with other extenders. Bentonite is used to prevent solids separation, reduce free water, reduce fluid loss, and increase slurry yield.

Bentonite is typically used at concentrations of 1 to 16% BWOC. It may be dry-blended with the cement or prehydrated in the mixing water. In prehydrating, the effect of 1% BWOC prehydrated is approximately equal to 3.5% BWOC dry-blended, but the yield point is much higher. For best results, the prehydrated bentonite/water mixture should be used for mixing the cement slurry shortly after prehydration has been completed. Laboratory testing is advised to determine the proper gel concentration and mixing procedure for prehydrated bentonite. Tech grade or "mud gel" should not be substituted for cement-grade bentonite. Lignosulfonate is commonly used as a dispersant and retarder in high-gel cements to reduce the slurry viscosity.

Attapulgite (Salt Gel). This is a more effective extender than bentonite in seawater or high-salt slurries, but it is not regulated or does not have a specification. Attapulgite, $(Mg,Al)_2$ $(OH/Si_4O_{10}) \cdot 12H_2O$, is composed of clusters of fibrous needles that require high shear to be dispersed in water. It produces many of the same effects as bentonite, except that it does not reduce fluid loss. A disadvantage of attapulgite is that because of the similarity of the fibers to those of asbestos, its use has been prohibited in some countries. Granular forms are available that may be permitted as a replacement.

Expanded Perlite. Expanded Perlite is a siliceous volcanic glass that is heat-processed to form a porous particle that contains entrained air. It is a highly buoyant product that requires the addition of 2 to 6% BWOC bentonite to prevent separation from the slurry. Because of its low crush strength, the water requirement for perlite-containing slurries must be increased to allow for slurry compressibility under downhole conditions. Volume loss must also be taken into effect in fill-volume calculation.

Gilsonite. This is an asphaltic material, or solid hydrocarbon, found only in Utah and Colorado. It is one of the purest naturally occurring bitumens. Gilsonite can be used with slurry densities as low as 11 lbm/gal at a normal concentration of 5 to 25 lbm/sack (sk) of cement, and it will plug float equipment and bridge tight annuli. The low densities obtainable with gilsonite result from its low density (1.07 g/cm^3). Because gilsonite is an organic material, it is highly buoyant and will float out of the slurry unless inhibited. Bentonite is commonly added at a concentration of 2 to 6% to prevent bridging in the wellbore.

Crushed Coal. Crushed coal is used for the same purposes as gilsonite (i.e., for light weight and lost-circulation control). It is commonly used at concentrations up to 50 lbm/sk of cement. Its density is slightly higher (1.3 g/cm^3), requiring a slight increase in water content. The addition of bentonite to prevent separation is normally not required.

Ground Rubber. This is a low-cost alternative to gilsonite and may be used in similar applications. The density of ground rubber is slightly higher (1.14 g/cm^3). The physical properties are more variable than gilsonite and are dependent upon material source. One major advantage of ground rubber is its low cost. At present, there are no environmental issues with ground rubber when utilized in a cement system.

9.9.4 Pozzolanic Extenders. A number of pozzolanic materials are available for use in producing lightweight cement slurries. These can be either natural or artificial and include fly ash, DE, microsilica, metakaolin, and granulated blast-furnace slag. In comparison with other additives, pozzolanic materials are usually added in large volumes. Fly ash, for example, can be mixed with cement in ratios of fly ash to cement that range from 20:80 to 80:20, based on an "equivalent sack" weight (that is, where a sack of fly ash has the same absolute volume as that of a sack of cement). Pozzolanic materials have a lower specific gravity than that of cement, and it is this lower specific gravity that gives a pozzolanic-Portland-cement slurry a lower density than a Portland-cement slurry of similar consistency. Depending on the density, pozzolanic cements also tend to give a set cement that is more resistant to attack by formation waters.

Fly Ash. Fly ash is by far the most widely used of the pozzolanic materials. According to ASTM *Standard C618,*[9] there are two types of fly ash: Class F and Class C; Class N refers to natural pozzolanic materials. There is, however, a need for a third category, based on the performance of different fly ashes. ASTM *Standard C618,*[9] classifies fly ashes on the basis of the combined percentages of SiO_2 + Al_2O_3 + Fe_2O_3—Class F having a minimum of > 90% and Class C, 50%. In reality, there is a much greater relationship between CaO content and performance. The CaO content ranges from 2 or 3% to 30% by weight of the fly ash. The "true" Class F fly ash has a CaO content of less than 10%, whereas a "true" Class C has CaO greater than 20%. Fly ashes having CaO between 10 and 20% behave somewhat differently from either the true Class F or Class C. Fly ashes are generally composed of amorphous glassy particles that are spherical in shape.

The ASTM Class F fly ash is the most common used in oilwell cementing. It is this fly ash that is covered by the API specifications. The major advantages of the Class F fly ash are its low cost and abundance worldwide. The performance characteristics of a Class F fly ash vary little from batch-to-batch from a given source. However, the differences between sources can be considerable because the composition can vary from the true low CaO to 10 to 20% CaO. This produces significant variations in performance characteristics, and because of this, different sources of Class F fly ashes should be tested before use. Specific gravities also must be determined. Some power plants produce Class F fly ashes with a high-carbon content because of poor burning. These should be avoided for oilwell cementing because they can cause severe gelation problems.

The use of Class C fly ash, as an extender for well cementing, is relatively limited. This is, in part, because of the limited availability of Class C fly ash and the considerable variability that exists not only between sources but also to a large extent between batches from a given source.

Microspheres. Microspheres are used when slurry densities from 8.5 to 11 lbm/gal are required. They are hollow spheres obtained as a byproduct from power generating plants or are specifically formulated. The byproduct microspheres are essentially hollow fly-ash glass spheres. They are present, typically, in Class F fly ashes, but usually in small amounts. However, they are obtained in substantial quantities when excess fly ash is disposed of in waste lagoons. The low-density hollow spheres float to the top and are separated by a flotation process. These hollow spheres are composed of silica-rich aluminosilicate glasses typical of fly ash and are generally filled with a mixture of combustion gases such as CO_2, NO_x, and SO_x. The synthetic hollow spheres are manufactured from a soda-lime borosilicate glass and are formulated to provide a high strength-to-weight ratio—they are typically filled with nitrogen. The synthesized microspheres provide a more consistent composition and exhibit better resistance to mechanical shear and hydraulic pressure.

The primary disadvantage of most microspheres is their susceptibility to crushing during mixing and pumping and when exposed to hydrostatic pressures above the average crush strength. This can lead to increased slurry density, increased slurry viscosity, decreased slurry volume, and premature slurry dehydration.

However, crushing effects can be minimized by the suitable choice of microspheres. These effects can be predicted and can be taken into account in slurry design calculations to produce a slurry having the required characteristics for the well conditions. Lightweight systems incorporating microspheres can provide excellent strength development and can help control fluid loss, settling, and free water.

Microsilica. Microsilica, also known as silica fume, is a finely divided, high-surface-area silica that can be obtained as a liquid or powder. In the powder form, it can be either in its original state, densified, or pelletized. The bulk density of the densified microsilica is 400 to 500 kg/m^3. Microsilica typically has a specific gravity of approximately 2.2.

Microsilica is composed primarily of vitreous silica and has a SiO_2 content of 85 to 95%, which makes it considerably purer than the other pozzolanic materials. Microsilica particles are also considered to impart beneficial physical properties to the slurry. Because of their fineness, they are believed to fill in the voids between the larger cement particles, resulting in a dense, solid matrix, even before any chemical reaction between the cement particles has occurred. Rheological properties tend to be improved with addition of microsilica because the tiny spheres can act as very small ball bearings and/or they displace some of the water present between the flocculated cement grain, thereby increasing the amount of available fluid. Concentrations of microsilica can range from 3 to 30% BWOC, depending on the slurry and properties required.

The physical and chemical properties of the microsilica make it very useful for a variety of applications other than as an extender. These include compressive-strength enhancement for low-temperature lightweight cement, thixotropic properties for squeeze cementing, lost-circulation, gas migration, and a degree of fluid-loss control.

The one disadvantage of microsilica is the cost. Originally considered to be a waste product, with its increased usage in the construction industry over the last decade, it has become more of a specialty chemical. Also, with fluctuations of supply and demand, there is a question of having a constant supply of a good source of the product.

Diatomaceous Earth. DE is a natural pozzolan composed of the skeletons of microorganisms (diatoms) that were deposited in either fresh water or seawater.

9.9.5 Chemical Extenders. Several materials are effective as chemical extenders. In general, any material that can predictably accelerate and increase the concentration of the initial hydration products is effective as a chemical extender.

Sodium Silicate. This is the most commonly used chemical extender for cement slurries. Sodium silicate is five to six times as effective as bentonite on an equivalent concentration basis. Unlike the physical or pozzolanic extenders, sodium silicate is highly reactive with the cement.

Sodium silicate is available in both dry and liquid forms, making it readily adaptable to onshore and offshore applications. The solid form is sodium metasilicate (Na_2SiO_3), and it is typically dry-blended with the cement at a concentration of 1 to 3.5% BWOC at densities of 14.2 to 11.5 lbm/gal. It is not as effective if dissolved directly in the mix water unless $CaCl_2$ is dissolved in the water first. If a liquid system is desired, it is better to use the liquid form. Liquid sodium silicate is normally used in seawater applications at a concentration of 0.1 to 0.8 gal/sk of cement at densities of 14.2 to 11.5 lbm/gal. The two main advantages of sodium silicates as extenders are their high yield and low concentration of use.

Gypsum. The hemihydrate form of calcium sulfate ($CaSO_4 \cdot 0.5\ H_2O$) is typically used as an extender. It is normally used at concentrations of 15% BWOC or less for the preparation of thixotropic slurries for use in applications where there are severe lost-circulation problems or where expansion properties are desired to improve bonding. Typical slurry compositions for lost-circulation applications, BHCT $\leq$ 125°F (52°C), contain from 8 to 12% BWOC gypsum with good expansion properties (0.2 to 0.4%). For improved bonding applications, where increased expansion (0.4 to 1%) is desired, NaCl is used ($\geq$ 10% BWOW).

9.9.6 Foamed Cement. It is possible to make slurries ranging in density from 4 to 18 lbm/gal using foamed cement. Foamed cement is a mixture of cement slurry, foaming agents, and a gas. Foamed cement is created when a gas, usually nitrogen, is injected at high pressure into a base slurry that incorporates a foaming agent and foam stabilizer. Nitrogen gas can be considered inert and does not react with or modify the cement-hydration-product formation. Under special circumstances, compressed air can be used instead of nitrogen to create foamed cement. In general, because of the pressures, rates, and gas volumes involved, nitrogen-pumping equipment provides a more reliable gas supply. The process forms an extremely stable, lightweight slurry that looks like gray shaving foam. When foamed slurries are properly mixed and sheared, they contain tiny, discrete bubbles that will not coalesce or migrate. Because the bubbles that form are not interconnected, they form a low-density cement matrix with low permeability and relatively high strength.

Virtually any oilwell-cementing job can be considered a candidate for foamed cementing, including primary and remedial cementing functions onshore and offshore, and in vertical or horizontal wells. Although its design and execution can be more complex than standard jobs, foamed cement has many advantages that can overcome these concerns. Foamed cement is lightweight, provides excellent strength-to-density ratio, is ductile, enhances mud removal, expands, helps prevent gas migration, improves zonal isolation, imparts fluid-loss control, is applicable for squeezing and plugging, insulates, stabilizes at high temperatures, is compatible with non-Portland cements, simplifies admix logistics, enhances volume, has low permeability, is stable to crossflows, and forms a synergistic effect with some additives, which enhances the property of the additive. The disadvantage of foamed cement is the need for specialized cementing equipment both for field application and for laboratory testing.

9.9.7 Weighting Agents. Weighting agents or heavyweight additives are used to increase slurry density for control of highly pressured wells. Weighting agents are normally required at densities greater than 17 lbm/gal where dispersants or silica is no longer effective. The main requirements for weighting agents are that the specific gravity is greater than the cement, the particle size distribution is consistent, they have a low water requirement, they are chemically inert in the cement slurry, and they do not interfere with logging tools.

Hematite (Fe_2O_3). This is the most commonly used weighting agent. Hematite is a brick-red, naturally occurring mineral with a dull metallic luster. It contains approximately 70% iron. The specific gravity of hematite ranges from 4.9 to 5.3, depending on purity, and it has a Mohs hardness of approximately 6.

Ilmenite ($FeO \cdot TiO_2$). This is not as commonly used as hematite, although it has some advantages over hematite. Ilmenite is a black to dark brownish-black, naturally occurring mineral with a submetallic luster that contains approximately 37% iron. It resembles magnetite in appearance but has only a slightly magnetic character. The specific gravity ranges from 4.5 to 5, depending on the purity, and it has a Mohs hardness of 5 to 6.

Hausmannite (Mn_3O_4). Hausmannite is being used increasingly because of its unique properties that address many of the disadvantages encountered with the other weighting agents. Hausmannite is a dark brownish-black material that is a byproduct mineral from the processing industry. The specific gravity range or Mohs hardness has not been well established. Because of its particle size and unique wetting characteristics, the material can suspended in the mix water at up to 40 wt% with a minimum of agitation, providing a liquid weighting agent. Because the average particle size of hausmannite is much smaller than that of cement, it allows the material to fit within the cement pore matrix, displacing entrained water, resulting in a lower viscosity and significantly more-stable slurry. The main disadvantage is that it is not readily available in all geographic regions, so the additional shipping cost can make it cost-prohibitive.

Barite ($BaSO_4$). Barite is not normally used in cementing as a weighting agent because of its high surface area and high water demand. It is a soft, light gray, naturally occurring nonmetallic material. The specific gravity ranges from approximately 4.0 to 4.5, depending on purity, and it has a Mohs hardness of 2.5 to 3.5.

9.9.8 Dispersants. Dispersants, also known as friction reducers, are used extensively in cement slurries to improve the rheological properties that relate to the flow behavior of the slurry. Dispersants are used primarily to lower the frictional pressures of cement slurries while they are being pumped into the well. Converting frictional pressure of a slurry, during pumping, reduces the pumping rate necessary to obtain turbulent flow for specific well conditions, reduces surface pumping pressures and horsepower required to pump the cement into the well, and reduces pressures exerted on weak formations, possibly preventing circulation losses.

Another advantage of dispersants is that they provide slurries with high solids-to-water ratios that have good rheological properties. This factor has been used in designing high-density slurries up to approximately 17 lbm/gal without the need for a weighting additive. The concept can also be used to design low-density slurries in which the high-solids contents include lightweight extenders.

Dispersants have been extensively studied. It is generally agreed that the dispersants minimize or prevent flocculation of cement particles because the dispersant adsorbs onto the hydration cement particle, causing the particle surfaces to be negatively charged and repel each other. Water that otherwise would have been entrained in the flocculated system also becomes available to further lubricate the slurry.

Polysulfonated Naphthalene (PNS). This is the most common dispersant; it is available as a calcium and/or sodium salt and can be obtained in both solid and liquid form. The commercial liquid form typically has a solids content of approximately 40%.

The benefit of using PNS is that improved rheological properties can be obtained, and slurries can be pumped with reduced frictional pressures. PNS can also allow higher solids-to-water ratio slurries to be designed with improved properties.

Hydroxycarboxylic Acids. These acids, such as citric acid, may be used as the primary dispersant in freshwater slurries at higher temperatures (BHCT ≥ 200°F). This is typically advantageous with cements that have a high free alkali (> 0.75%) content to offset their retarding properties. Citric acid is also used as a dispersant in salt- and seawater cement slurries. The concentration of use is limited by the temperature and thickening time desired, although concentrations of 0.5 to 1.0% BWOC are usually sufficient.

9.9.9 Fluid-Loss-Control Additives (FLAs). FLAs are used to maintain a consistent fluid volume within a cement slurry to ensure that the slurry performance properties remain within an acceptable range. The variability of each of these parameters is dependent upon the water content of the slurry. For example, if the water content is greater than intended, the following will normally occur: thickening time, fluid loss, free fluid, sedimentation, permeability, and porosity will be increased; and density, viscosity, and compressive strength will be decreased. If the water content is less than intended, the opposite will normally occur. The magnitude of change is directly related to the amount of fluid lost from the slurry.

Because predictability of performance is typically the most important parameter in a cementing operation, considerable attention has been paid to mechanical control of slurry density during the mixing of the slurry to assure reproducibility. Of equivalent importance is the slurry density during displacement, which is directly related to fluid-loss control.

Cement slurries are colloidal suspensions consisting of distinct solid and liquid phases. During the cementing operation, there are several opportunities for the fluid phase to separate from the cement slurry. This can occur when the slurry is passing through small orifices or ports, and within the annulus. When the slurry is passing through orifices, the fluid phase can be accelerated, resulting in particle bridging. In a wellbore annulus, fluid can be displaced from the slurry while it is passing though constricted areas, or to the formation, resulting in an increase in the ECD, which can lead to formation fracture (lost circulation) or flash set (dehydration). After placement, the fluid phase will filter to permeable formations, resulting in a reduction in the slurry volume and effective hydrostatic pressure, creating the potential for the migration of formation fluid into and through the cement column. FLAs are, therefore, used to prevent solids segregation during placement and to control the rate of fluid leakoff in the static state.

Neat cement slurries normally exhibit an uncontrolled API fluid loss of at least 1,500 $cm^3/30$ min. This value is excessive for most cementing operations, where permeable formations are encountered or where long columns of cement will be used. The amount of fluid-loss control required for a particular operation varies widely and is largely dependent upon the slurry density, the water content, the formation properties, and annular clearance.

Several materials are effective as FLAs. The materials that are currently in use can be loosely categorized in two groups according to their solubility characteristics: water-insoluble and water-soluble. With the exception of bentonite, the water-insoluble materials are polymer resins. All of the water-insoluble materials function as permeability reducers. The water-soluble materials are modified natural polymers, cellulosics, and vinylinic-based polymers. The polymeric materials, whether water-insoluble or -soluble, are all synthetic (manmade) materials. The action of FLAs depends on their solubility. The water-insolubles function by reducing the permeability of the filter cake developed.

9.10 Water-Insoluble Materials

9.10.1 Bentonite. Bentonite is not typically used as the primary fluid-loss agent in normal-density slurries. In low-density slurries, where higher concentrations can be used, it may provide sufficient fluid-loss control (400 to 700 $cm^3/30$ min) for safe placement in noncritical well applications. Fluid-loss control, obtained through the use of bentonite, is achieved by the reduction of filter-cake permeability by pore-throat bridging. Fluid-loss rates can be erratic because of the concentration of use at a given density, variations in platelet disassociation caused by shear, and stacking arrangement in the filter cake.

Microsilica. Microsilica imparts a degree of fluid-loss control to cement slurries because of its small particle size of less than 5 microns. The small particles reduce the pore-throat volume within the cement matrix through a tighter packing arrangement, resulting in a reduction of filter-cake permeability.

Polyvinyl Alcohol (PVA). PVA is a white to cream-colored powder with a density range of 1.27 to 1.31 g/cm^3. It is a water-soluble polymer derived from polyvinyl acetate and is chemically reactive with acids and alkalis. It is not listed in the water-soluble polymers section because it loses solubility in alkaline environments such as the aqueous phase of a cement slurry. PVA also provides gas-migration control and enhances cement bonding and acid resistance.

Synthetic Latex. This is an oil-in-water emulsion system consisting of a dispersed phase of a water-insoluble elastomer, surfactants, and a water exterior phase. These emulsions are characterized by their milky-white appearance. Their density is typically approximately 1 g/cm^3. The most common emulsion used is styrene-butadiene rubber (SBR), which provides exceptionally low fluid-loss control, gas-migration control, and acid-solubility resistance.

The surfactant system plays a key role in the use of latex in well-cementing applications. In cement slurries, the emulsion system readily disperses and exhibits time-, shear-, and temperature-dependent stability. The emulsion stability can be improved by the addition of additional surfactant, and depending on the surfactant type and concentration, the emulsion stability may be controlled to above 300°F (149°C) BHCT. The surfactant system also acts as a dispersant in the cement slurry, resulting in low slurry viscosity. Control of emulsion stability is critical to slurry performance because the rate of inversion of the emulsion controls slurry viscosity and thickening time. Inversion of the emulsion system results in an almost instantaneous conversion to a rubberized mass (set) that is reported as the pumping time for the slurry.

Latex is typically used at a concentration of $\geq$ 0.8 gal/sk (~ 3.5% BWOC dry-weight equivalent) to obtain a fluid loss of less than 100 $cm^3/30$ min. Fluid-loss values of less than 20 $cm^3/30$ min are possible at 1.5 gal/sk in nonsilica slurries and 2 to 3 gal/sk with 35% BWOC silica slurries. Fluid loss is controlled in the cement slurry by particle plugging.

9.11 Water-Soluble Materials

9.11.1 Derivatized Cellulose. Two forms of derivatized cellulose have been found useful in well-cementing applications. They are the single-derivatized HEC and twice-derivatized CMHEC. The usefulness of the two materials depends on their retardational character and thermal stability limits.

Hydroxyethyl Cellulose. This is commonly used at temperatures up to approximately 82°C (180°F) for fluid-loss control and may be used at temperatures up to approximately 110°C (230°F) BHCT, depending on the co-additives used and slurry viscosity limitations. Above 110°C (230°F), HEC is not thermally stable. HEC is typically used at a concentration of 0.4 to 3.0% BWOC, densities ranging from 16.0 to 11.0 lbm/gal, and temperatures ranging from 27 to 66°C (80 to 150°F) BHCT to achieve a fluid loss of less than 100 cm^3/30 min.

Carboxymethyl Hydroxyethyl Cellulose. This is commonly used at temperatures up to 300°F for fluid-loss control and may be used at temperatures up to approximately 350°F, depending on degree of substitution, the co-additives used, and slurry viscosity limitations. CMHEC is more thermally stable than HEC and is not as susceptible to oxidative attack.

9.11.2 Synthetic Polymers. Since the 1970s, a significant amount of work has been performed concerning synthetic copolymers for use in cement slurries. Most of this work has centered on copolymers of acrylamide and/or acrylamide derivatives and their salts; however, several nonacrylamide-based monomers have also been reviewed.

Polyvinyl Pyrrolidone (PVP). This is a nonionic polymer that is typically used as a fluid-loss enhancer in conjunction with sodium naphthalene sulfonate condensed with formaldehyde (SNFC) to improve the performance of other polymers. When used alone, PVP is not very effective as an FLA. However, when PVP is used in conjunction with SNFC, the fluid loss is improved through improved particle orientation. PVP/SNFC is particularly advantageous when used in densified cements for both dispersion and fluid-loss control. The use of PVP/SNFC, in conjunction with HEC or CMHEC, results in significant improvement in fluid-loss control. Surfactants are surface-active agents that may be used to modify the interfacial tension between two liquids or between a liquid and a solid. Low-molecular-weight polymers such as SNFC and lignosulfonate are surfactants. The choice of the proper surfactant can have a significant effect on the FLA itself and its interaction with cement particles. Surfactants can be used to accelerate or retard the solubility or wettability of polymers.

9.11.3 Lost-Circulation Additives. Cement slurries can be lost to the formation and not circulated back to the surface during completion of a wellbore. This is defined as lost circulation. It should not be confused with the volume decrease resulting from fluid-loss filtration. Lost circulation tends to occur in three basic formation types:

• Unconsolidated or highly permeable. It is considered that the particles of a cement slurry can enter an unconsolidated or highly-permeable formation only if the permeability is greater than 100 darcies.

• Fractured, induced or natural. Induced fractures occur in highly incompetent zones (e.g., shale) that break down at relatively low hydrostatic pressures. Natural fractures can be encountered anywhere.

• Cavernous or vuggy. These are usually formed by erosion of the formation caused by the action of subsurface waters and are discovered unexpectedly.

In many cases, lost circulation occurs during drilling with loss of drilling fluids, and actions can be taken at that time to combat the lost circulation. At other times, difficulties may be encountered during drilling, indicating potential lost-circulation problems, and measures can be taken to prevent their occurrence during cementing. Typically, there are two steps in combating lost circulation: reducing slurry density and adding a bridging or plugging material.

Additives for prevention of lost circulation can be separated into three basic groups: bridging materials, rapid-setting or thixotropic cements, and lightweight cementing systems.

Bridging materials physically bridge over and/or plug the lost-circulation zone and are typically available in fibrous, flake, or granular form. Most bridging materials are considered to be chemically inert with respect to cement hydration.

Fibrous materials are, in general, used for controlling lost circulation in highly permeable formations where the fibers form a mat over the surface.

The most common flake material is cellophane. Cellophane flakes act by forming mats or bridges over very narrow fractures. Concentration range of cellophane is usually from 0.125 to 0.5 lbm/sk.

Granular materials are most frequently used and include gilsonite, perlite, and coal. These coarse particles are typically used for large fractures and cavernous or vuggy lost-circulation formations. As the cement slurry enters the formation, these large granular particles, in principle, become trapped and block off the opening. Concentrations vary according to the material used and are typically, 5 to 50 lbm/sk for gilsonite, 0.5 to 1.0 ft³/sk for perlite, and 1 to 10 lbm/sk for coal.

Rapid-setting and thixotropic cements are the preferred means for lost-circulation control in large cavernous or vuggy formations where bridging materials are no longer effective. These cements are usually designed to set up in the lost-circulation zone, ultimately plugging it off.

Rapid-setting cements include both quick- and flash-setting formulations. These cements generally give thin slurries but have very rapid setting times. The quick-setting cements will set up while being displaced or shortly after entering the lost-circulation zone, whereas the flash-setting cements form semisolid materials when mixed with water or water-based drilling fluids.

Thixotropic cements have a low viscosity during mixing and placing, but when they enter the formation and are no longer subjected to shear, they gel and become self-supporting. There are a number of thixotropic formulations that include gypsum cement, gypsum Portland cement, aluminum sulfate/iron (II) sulfate, clay-based systems, and crosslinked polymer systems.

It is often more effective to solve lost circulation by combining the bridging materials with rapid-setting or lightweight systems. The choice of system and the bridging material depends on the type of formation, the size of the lost-circulation zone, the fracture pressure gradient, and the downhole temperatures and pressure, as well as economics.

9.11.4 Strength-Retrogression Inhibitors. Strength retrogression is a normal phenomenon that occurs with all Portland cements at temperatures approximately 230 to 248°F (110 to 120°C) and is usually accompanied by a loss in impermeability. The use of 35 to 40% SiO_2 (sand or flour) is used to combat strength retrogression.

9.11.5 Free-Water Control. In well-cementing applications, the maintenance of a consistent column of cement is critical to assure proper zonal isolation. Because of rheological demands and the need for silica or weighting agents in some applications, this is not always possible with conventional materials. It is necessary, therefore, that an additional additive be incorporated into the cement slurry to address the potential problem of particle sedimentation. This group of additives is known as free-water-control additives.

Sodium Silicate. Sodium silicate may be used to control free water in normal- and low-density cement slurries. Typically, approximately 0.15 to 0.5% BWOC is sufficient to provide free-fluid control.

Biopolymers. Biopolymers impart the unique characteristics of thinning at higher shear rates and viscosifying at lower shear rates. This yields slurries that will more readily go into turbulent or upper laminar flow yet have sufficient low shear to prevent sedimentation. Xanthan gum and Welan gum both provide these characteristics and are typically used at an active concentration of approximately 0.2% BWOC.

Synthetic Polymers. Synthetic polymers of high molecular weight, which are resistant to alkaline hydrolysis, have been found to be effective as free-fluid-control additives at temperatures where sodium silicate and biopolymers are not effective. They are typically used at an active concentration of approximately 0.1 to 0.2% BWOC.

9.11.6 Expansive Cements. Expansive cements are used primarily for obtaining effective zonal isolation by improving the bond between the cement and the pipe and the cement and the annulus. Good zonal isolation is essential to prevent loss of production, control gas migration, provide protection from corrosive formation waters, reduce water production, and improve confinement of stimulation treatments. Poor bonding of cement to pipe and/or annulus is most often a result of a combination of effects from a variety of factors. The root causes are usually associated with drilling-fluid properties and displacement mechanics, casing expansion and contraction caused by thermal stresses or internal pressures, fluid loss from the cement, and hydration volume reduction during setting of cement. The resultant effect of poor bonding is the formation of "microannuli" or small gaps at the cement/casing or cement/formation interface. Expansive cements expand slightly after the cement has set and fill in the void spaces. Because of the restraints imposed by the casing and formation, any additional expansion will occupy the space provided by the internal cement porosity, resulting in a reduction in porosity. The two principal types of expansive additive or cement are post-set crystalline growth (or chemical expansion) and in-situ gas generation.

Crystalline-Growth Additives. The expansion mechanism is the growth of the crystals within the solid cement matrix. These crystals have a greater bulk volume than the original solids from which they form and, as such, cause a wedging action because of the internal pressure of crystalline growth, forcing the solid matrix apart. Crystal-growth expansion is unilateral in that restraint in one direction does not increase expansion in other directions. The amount of expansion is dependent on a number of factors that include amount of additive, curing time and temperature, and, in some cases, cement-slurry composition.

Cement slurries containing high concentrations of salt (NaCl, KCl, or $CaCl_2$) have a long reputation for contributing to expansion. Expansion is caused by the crystal growth of calcium chloroaluminate hydrate ($3CaO \cdot Al_2O_3 \cdot CaCl_2 \cdot H_2O$) from reaction of the chloride ions with the aluminate phase in cement. There are indications that the temperature limitation for calcium chloroaluminate hydrate is around 51°C (125°F), although salts are reported to be effective, expanding additives up to 204°C (400°F), depending on the system. Salt also contributes to bond improvement by preventing dissolution of the salt formation.

In-Situ Gas Generation. Expansion resulting from in-situ gas-generating additives occurs before set while the cement is still in the plastic state. The most common in-situ gas-generating additive is aluminum powder, although zinc, iron, and magnesium are possible alternatives. The expansion is caused by the reaction with alkali and water present in the cement aqueous phase to produce microsized bubbles of H_2 gas. Expansive forces that are a direct function of the gas generated compensate for any volume losses caused by hydration volume reduction or fluid loss and increase the pressure of the cement against the pipe and formation. In-situ gas-generating additives can be used at temperatures from 16 to 204°C (60 to 400°F). Because of the compressibility of the gas, the amount required is more dependent on the hydrostatic pressure of the slurry than on the downhole temperature. Concentrations generally range from 0.15 to 0.6%, although they can be higher.

9.11.7 Miscellaneous Additives. Several additives are used that do not fit in any of the preceding categories. These additives can be used frequently (as in antifoam additives) or in more-specialized cases, such as mud decontaminants, radioactive tracers, dyes, fibers, and cement for CO_2 resistance.

9.11.8 Antifoam Additives. Antifoam additives are frequently used to decrease foaming and minimize air entrainment during mixing. Foaming is a secondary effect, often caused by a number of additives. Excessive foaming can result in an underestimation of the density downhole and cavitation in the mixing system.

Slurry density is usually measured with a densitometer during mixing to proportion the solids and water to obtain the desired density. When a slurry foams, the entrapped air is also included in the density measurement, and because air compresses under pressure, the actual density downhole becomes greater than that measured on the surface. Another effect of foaming is that if severe, it can cause cavitation of the pumps and ultimately lead to loss of hydrostatic pressure.

Antifoam additives, in general, modify the surface tension and/or dispersion of solids in the slurry so that foaming is prevented or the foam breaks up. The concentration of foaming additive required to be effective is very small, typically less than 0.1% BWOW. Antifoam additives consist primarily of polyglycol ethers or silicones or a mixture of both, and may also include additional surfactants.

Polypropylene glycol is the most common polyglycol ether used and is favored for its low cost. It is effective in most situations, although, typically, it has to be added before mixing. In some cases, it can interact with other additives and cause increased foaming. The silicone antifoam additives are a suspension of very fine particles of silica dispersed in a silicone base and can also exist as an oil-in-water emulsion. They can be used both before and during mixing and are highly effective as antifoam additives.

9.11.9 Mud-Decontaminant Additives. Paraformaldehyde or a blend of paraformaldehyde and sodium chromate is sometimes used to minimize the cement retarding effects of various drilling-mud chemicals in the event a cement slurry becomes contaminated by intermixing with the drilling fluids. A mud decontaminant consisting of a 60:40 mixture of paraformaldehyde and sodium chromate neutralizes certain mud-treating chemicals. It is effective against tannins, lignins, starch, cellulose, lignosulfonate, ferrochrome lignosulfonate, chrome lignin, and chrome lignite. Mud decontaminants are used primarily in openhole plugback jobs and liner jobs and for squeeze cementing and tailing out on primary-casing jobs.

9.11.10 Radioactive Tracers. Radioactive tracers are added to cement slurries as markers that can be detected by logging devices. They were originally used to determine the location of fill-up or cement top and the location and disposition of squeeze cement, although, now, temperature surveys and cement-bond logs fulfill this function. Radioactive tracers are still occasionally used in remedial cementing to locate the slurry after placement, if required, and for tracing lost circulation. Radioisotopes are controlled and licensed by the U.S. Nuclear Regulatory Commission and various state agencies and cannot be used indiscriminately.

9.11.11 Dyes. Small amounts of indicator dye can be used to identify a cement of a specific API classification or an additive blended in a cementing composition. When the dyes are used downhole, however, dilution and mud contamination may dim and cloud the colors, rendering them ineffective. Naturally occurring mineral oxides and/or synthetically produced color pigments may be substituted for the dye indicator.

9.11.12 Fibers. Conventional Portland cement, mixed at normal density, has low ductility, making it somewhat brittle. This makes it susceptible to post-cementing stresses. Synthetic fibrous materials are frequently added to make the cement more ductile and to reduce the effects of shattering or partial destruction from perforation, drill-collar stress, or other downhole forces. Fibrous materials transmit localized stresses more evenly throughout the cement and, thus, improve the resistance to impact and shattering. Nylon, with fiber lengths varying up to 1 in., has commonly been used because it is resilient and imparts high shear, impact, and tensile strength. Particulated rubber also acts to improve the ductility of cement and improve on the flexural strength, and it is usually used in concentrations up to 5% BWOC. More recently, aluminum silicate and/or fibrous calcium silicates have been reported to enhance the compressive, flexural, and tensile strengths.

9.12 Slurry-Design Testing

9.12.1 Performance Testing. When determining a slurry's characteristics and performance, these testing procedures are recommended:

   • *Temperature.* Test to the highest simulated BHCT with a variety of retarders, densities, and temperatures.

   • *Pressure.* Test to the actual bottomhole pressure (BHP) thickening time. [Note: The slurry to be tested should include surface time required (if batch mixed) and calculated time to bottom.]

   • *Compressive Strength at the Following Top-of-liner (TOL).* Ensure certain conditions are met: simulated temperature and pressure, lowest simulated BHCT used with longest thermal recovery, ultrasonic cement analyzers set for simulated temperature recovery and calculated pressure not API minimum (3,000 psi).

   • *Mixing Effects.* Investigate and standardize order of addition, time taken to add, holding of mix water, time to mix at surface, surface mixing temperature/shear effects, slurry stability, sedimentation test, and HP/HT rheology (where available).

The methods of testing cement for downhole application are based on performance testing. Testing methods are usually performed according to API specifications, though specifically designed and engineered equipment or tests are also used. The choice of additives and testing criteria is dictated primarily by the specific parameters of the well to be cemented. Performance testing has proved to be the most effective in establishing how a slurry will behave under specific well conditions. There is no direct means of predicting cement performance from the properties of cement, and no technique has yet been established, or is likely to be in the near future, that would correlate cement composition and cement/additive interaction with performance.

9.12.2 Diagnostic Testing. Performance testing is not adequate in troubleshooting downhole problems where the integrity of the cement blend is in question. There are diagnostic analyses that can be performed to evaluate the cement powder, but there are no definitive tests for chemically analyzing the composition of a cement once it has been mixed with additives, either as a dry blend, a slurry, or a set cement. The primary reason for this is the low concentration of additives used in the slurry or set cement. This concentration in set cement can be even lower than that of the original slurry if the additive is consumed and/or modified during the cement hydration reaction. The content of samples taken from downhole is often questionable in that it is not clear exactly where they were obtained or if they were contaminated with drilling fluid, formation waters, or during retrieval. Many of the techniques used for understanding the chemistry of cement are designed for laboratory-prepared specimens and applications and are not applicable to field samples. However, depending on the sample and the concentration of additives, some qualitative analysis can sometimes be achieved.

Analysis of dry-blended samples is somewhat different from that of the slurry or set cement. If sufficient quantity is available for performance testing, this would be the most appropriate to compare the actual blend with that designed. If this is not the case, then the blend would require dissolution in an extracting solvent. This usually includes water and, inevitably, cement hydration will occur, with some of the additive component being removed by the hydration products. As the contact time is less, more additive should be extracted and will more likely be detectable through one of the methods previously discussed. After cement and additives are blended, it is usually not possible to separate the additive from the dry sample unless it has a significantly greater particle size or heavier density than that of the cement.

9.13 Cementing Hardware

Floating equipment, cementing plugs, stage tools, centralizers, and scratchers are mechanical devices commonly used in running pipe and in placing cement around casing.

9.13.1 Floating Equipment. Floating equipment is commonly used on the lower section of the well casing to reduce the strain on the derrick during placement of the casing in the wellbore; help guide the casing past ledges and sidewall cavings, as the casing passes through deviated sections of the hole; provide a backpressure valve to prevent re-entry of cement into the casing inner diameter (ID) after it is pumped into the casing/wellbore annulus; and provide a landing point for cementing plugs pumped in front of and behind the cement slurry. Some basic types of floating and guiding equipment are the guide shoe, with or without a hole through the guide nose; the float shoe containing a float valve and a guide nose; and the float shoe and float collar containing an automatic fill-up valve.

The simplest guide shoe is an open-end collar, with or without a molded nose. It is run on the first joint of casing and simply guides the casing past irregularities in the hole. Circulation is established down the casing and out the open end of the guide shoe or through side ports designed to create more agitation as the cement slurry is circulated up the annulus. If the casing rests on bottom or is plugged with cuttings, circulation can be achieved through the side ports.

A modified guide or float shoe with side ports may aid in running the casing into a hole where obstructions are anticipated. This tool has side ports above and a smaller opening through the rounded nose. The smaller opening ensures that approximately 60% of the fluid is pumped through the existing side ports. These ports help wash away obstructions that may be encountered and also aid in getting the casing to bottom, if some of the cuttings have settled in the bottom of the hole.

The jetting action of the side-port tool types aids in removing the cuttings and helps provide a cleaner wellbore with increased turbulence during circulation and cementing. It also aids in the uniform distribution of the slurry around the shoe.

The combination guide or float shoe usually incorporates a ball or spring-loaded backpressure valve. The outside body is made of steel of the same strength as that of the casing. The backpressure valve is enclosed in plastic and high-strength concrete. The valve, which is closed by a spring or by hydrostatic pressure from the fluid column in the well, prevents fluids from entering the casing while pipe is lowered into the hole. After the casing has been run to the desired depth, circulation is established through the casing and float valve and up through the annulus. When the cement job is completed, the backpressure valve prevents cement from flowing back into the casing.

Float collars are usually placed one to three joints above the float or guide shoe in the casing string and serve the same basic functions as the float shoe (**Fig. 9.6**). They contain a backpressure valve similar to the one in the float shoe and provide a smooth surface or latching device for the cementing plugs. Float collars are also available with nonrotating (NR) inserts. When cementing plugs with matching inserts are used during cementing operations, the plugs are locked to the float collar, preventing spinning of the plugs during drillout. This equipment may reduce drillout time of the "shoe track" by 80%. The space between the float collar and the guide shoe traps contaminated cement or mud that may accumulate from the wiping action of the top cementing plug. The contaminated cement is, thus, kept away from the shoe, where the best bond is required.

When the cement plug sits at the float collar (**Fig. 9.7**), it shuts off fluid flow and prevents overpumping of the cement. A pressure buildup at the surface indicates that cement placement

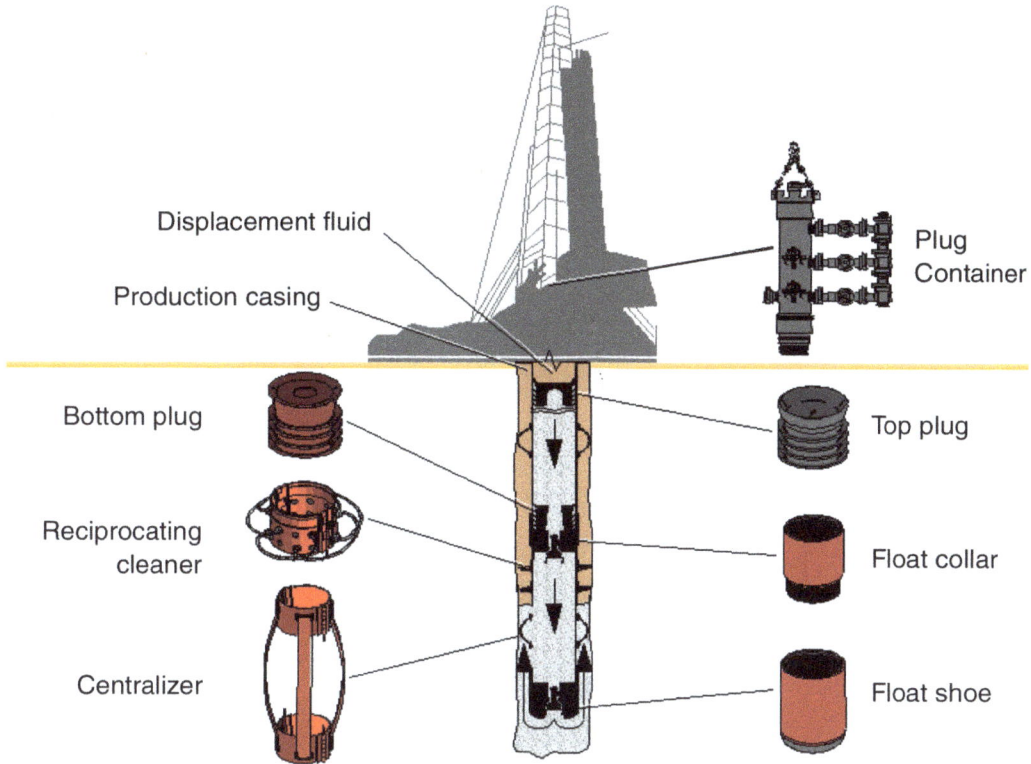

Fig. 9.6—Typical primary-cementing equipment (courtesy of Halliburton).

is complete. For larger casing, float collars or shoes may be obtained with a special stab-in device that allows the cement to be pumped through tubing or drillpipe. (This method of placement is often called inner-string cementing.) Such a device eliminates the need for large cementing plugs and oversize plug containers.

For reasons of economy, a simple insert flapper valve and seat may be installed in the casing string one or two joints above the guide shoe. This insert valve is designed for use in shallow wells for pressures less than the collapse pressure of J-55 casing in the particular weight range being used. Insert flapper-valve-equipment may be run with an orifice tube holding the flapper valve in the open position to allow the casing to automatically fill as it is being run in the wellbore. The opening through the fill tube may be varied to allow heavy concentrations of lost-circulation material to pass through the tube. After the casing has been landed at the desired depth, a weighted plastic ball is dropped in the casing to shear out the orifice tube and allow the flapper valve to close. The insert flapper valve, like the float collar, provides a space for isolating contaminated cement. It also provides a surface for landing the cement plug.

Differential-fill-up and automatic-fill-up float collars and float shoes permit a controlled amount of fluid to enter the bottom of the casing while the casing is being run in the hole (**Fig. 9.8**). They operate on the principle that hydrostatic pressure in the annulus will tend to balance the hydrostatic pressure proportionally inside the casing. A restricted area allows a controlled amount of fluid to enter the casing through the bottom of the float shoe while the casing is being run, thereby shortening running time and reducing pressure surges against the formation. The backpressure valve in automatic-fill-up equipment is held out of service until it is released by a predetermined flow rate applied from the surface through the float equipment. The rate of flow into the casing is usually low enough to hold the fluid level within 10 to 300

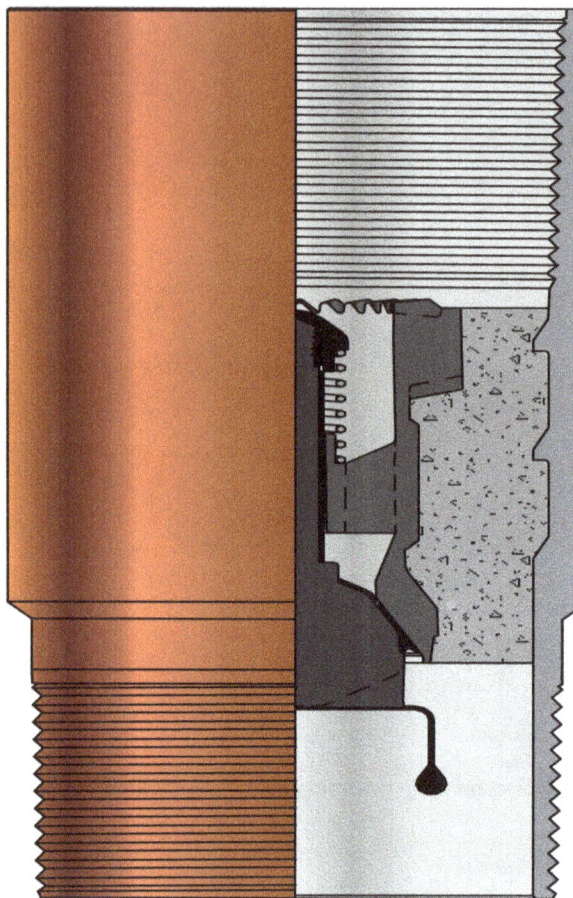

Fig. 9.7—Float collar (courtesy of Halliburton).

ft of the surface. When purchasing floating equipment, it is important to specify the outer diameter (OD), the threading, the material grade, and the pipe weight.

9.13.2 Plug Containers. Plug containers hold the top and bottom cementing plugs and come in two different versions: continuous cementing head and quick-change container. A cementing head is designed to attach to the top joint of well casing during cementing operations. The head allows cementing plugs to be released ahead of and behind the cement slurry to isolate the cement slurry from wellbore fluids ahead of the cement and displacing fluids pumped behind the slurry. Cementing heads may house one, two, or more cementing plugs. A single cementing head is used when it is not necessary to have continuous pumping of the cementing slurry. When a single cementing head is used, the bottom plug may be loaded in the head and pumped in the casing with a small volume of fluid or inserted by hand into the top of the casing and then the head installed to the top casing joint. The top plug is loaded in the cementing head for release after the cement slurry is mixed.

A double cementing head (two plugs) or multiple cementing head (three or more) allows the cementing plugs to be loaded before the cement slurry is mixed. During cementing operations, plugs can be released from the head without interrupting the pumping.

Plug containers are equipped with valves and connections for attaching cementing lines for circulation and displacement. The cement usually falls down the casing on a vacuum before the

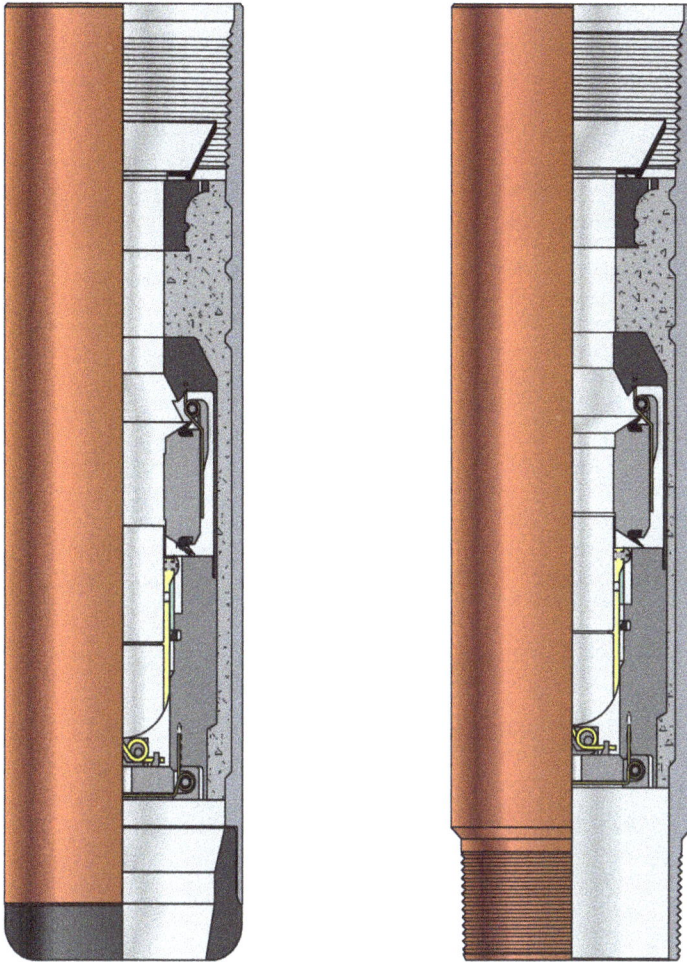

Fig. 9.8—Float collars with differential fill-up and automatic fill-up (courtesy of Halliburton).

plug is released; therefore, displacing fluid can be siphoned into the casing below the top plug if the valve to the supply source is not kept closed. Because the fluid can be siphoned through the cementing pump, the valve should not be opened until the top plug has been released. Cementing heads, with an internal swivel or a swivel between the top casing collar and the cementing head, make it possible to rotate the casing during cementing operations. Quick-connect couplings on the cementing heads allow fast connection of the cementing head to the casing when the last joint is landed so that circulation can be started immediately.

For ease of operation, the cementing head should be as near the level of the rig floor as possible. A typical plug container (**Fig. 9.9**) allows a bottom plug to be inserted through the container into the casing ahead of the cement slurry. The top plug is loaded into the plug container, where it rests on a support bar. It is released by retracting the support bar after the cement is mixed. A lever on some types of plug containers indicates the passage of the plug as it leaves the container.

9.13.3 Cementing Plugs. Cementing plugs are highly recommended to separate drilling fluid, cement, and displacing fluid. Unless a well is drilled with air or gas, the casing and hole are usually filled with drilling fluid before cementing. To minimize contamination of the interface

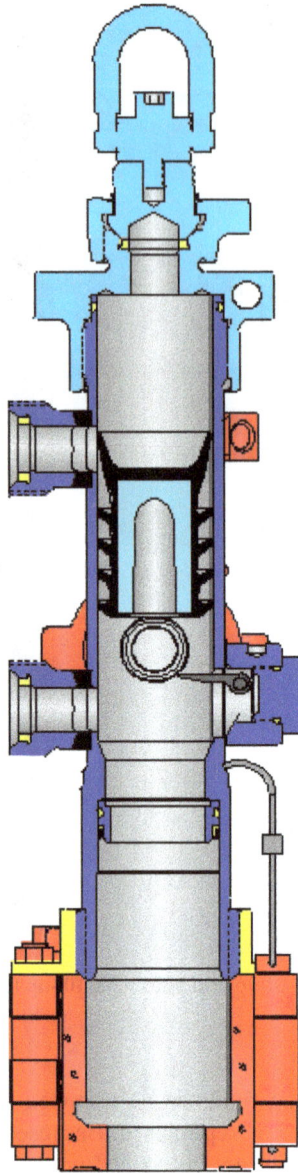

Fig. 9.9—Plug container (courtesy of Halliburton).

between the mud and the cement in the casing, a bottom plug is pumped ahead of the cement slurry. This plug wipes the mud from the casing ID as it moves down the pipe. When it reaches the float collar, differential pressure ruptures a diaphragm on top of the plug, allowing the cement slurry to flow through the plug and the floating equipment and up the annular space between the pipe and the hole (**Fig. 9.10**). The top cementing plug, pumped behind the cement slurry, is pumped to a shutoff on the float collar, causing a pressure increase at the surface, signaling that the cement has been displaced. Top and bottom plugs are similar in outward appearance but are always different colors. The top plug (black) has a solid insert with rubber wipers molded to the insert. The bottom plug (red, orange, and yellow) has a cylinder-type insert with molded wipers and a plastic or molded rubber diaphragm designed to rupture at 200 to 400 psi. Inserts are manufactured of plastic or aluminum. Aluminum inserts increase the

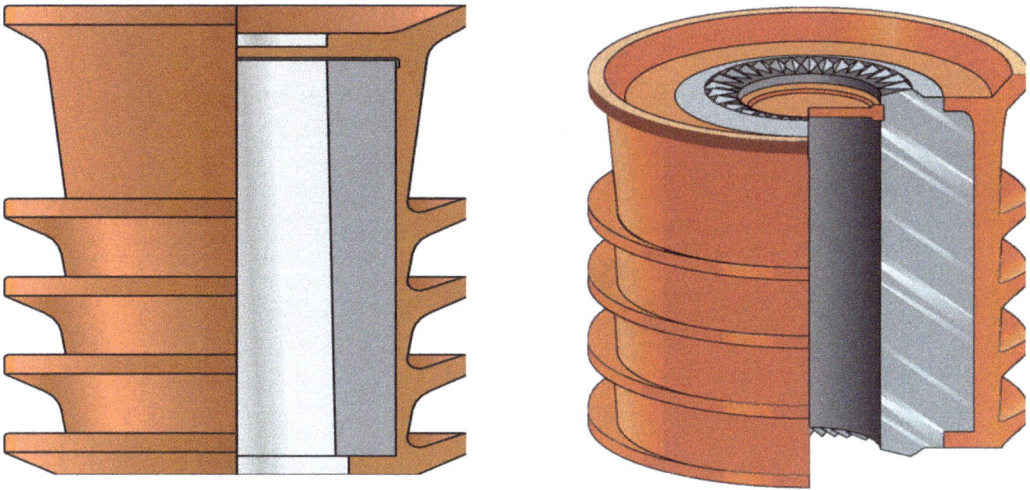

Fig. 9.10—Five-wiper (left) and nonrotating (right) cementing plugs (courtesy of Halliburton).

strength and temperature ratings of the cementing plug. Aluminum-inserted plugs should be used when the BHCT exceeds 300°F and should be drilled out with conventional tricone rock bits. The recommended landing pressures for aluminum-insert plugs vary, depending on casing size, but are normally higher than the recommended landing pressures for wiper plugs with plastic inserts. Plastic-insert plugs can be used in wells with a BHCT below 300°F and can be drilled out with tri-cone rock or polycrystalline-diamond compact (PDC) bits.

Nonrotating five-wiper cementing plugs (Fig. 9.10) are manufactured with locking teeth on both the top and bottom plug to land on an NR float collar with similar locking teeth. These locking teeth lock the plug to the float collar, preventing spinning during drillout, which reduces drillout times and associated rig costs. NR plugs use plastic inserts that allow easy drillout with either PDC bits or tricone rock bits. High-strength NR plugs and float collars can be used to pressure-test casing immediately after cementing operations are completed.

There are times, however, when a bottom plug should not be used; for example, when the cement contains large amounts of lost-circulation material or when the casing being used is badly rusted or scaled. Under such conditions, a bottom plug could cause bridging and plugging of the casing. In some cases, water or a chemical flush should precede the cement slurry to clean the casing of the mud solids. This is not as effective as the mechanical wiping action of the bottom plug, but it will reduce the amount of contaminated slurry. The top plug follows the cement slurry, wiping it from the casing wall.

Although the conventional wiper plugs are the most widely used, there are other designs available for primary cementing: balls, wooden plugs, subsea plugs, and teardrop or latch-down devices (**Fig. 9.11**). The latch-down casing plug and baffle may be used with most conventional floating equipment, but they are most commonly used in small-diameter tubing for inner-string cementing. This type of plug system, supplementing the float valve, prevents fluid from re-entering the casing string. When all the cement has been pumped, the latch-down plug allows surface pressure to be released immediately and also prevents the cement and plug from being backed up into the casing by air compressed below the plug. If completions are made fairly close to the float collar, the latch-down plug system eliminates the need to drill out the cement.

Subsea completions and conventional liner jobs can be cemented with the standard two-plug cementing techniques. They require the cement slurry to be pumped through a string of drillpipe that is smaller than the casing string being cemented. The downhole release system

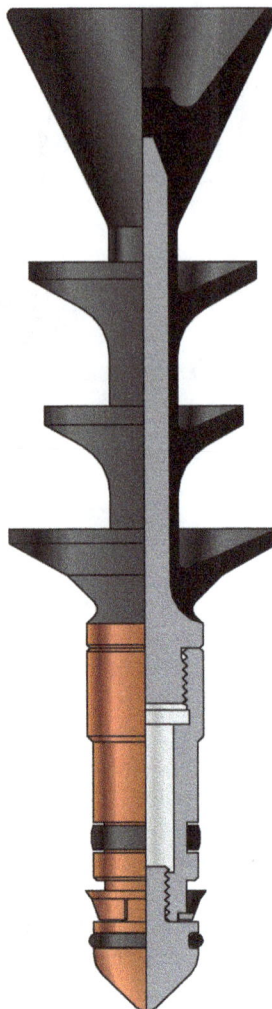

Fig. 9.11—Latch-down casing plug assembly (courtesy of Halliburton).

can wipe both the drillpipe and the casing, and can separate the cement slurry and displacing fluid.

The downhole release plugs are attached to an installation tool in the top of the casing to be cemented. The bottom plug is fastened to the top plug, which, in turn, is fastened to the installation tool. These tools use a ball or a releasing plug to release the bottom plug from the top plug by pressuring to a predetermined amount and shearing some pins. This allows the bottom plug to be pumped ahead of the cement slurry while wiping mud solids off the casing and separating the cement slurry from the wellbore fluid. A top-plug-releasing dart is pumped behind the cement slurry to separate the cement and displace fluid in the drillpipe. The top-plug-releasing dart will latch into the top wiper plug in the casing. A predetermined amount of pressure releases the top wiper plug, which is then pumped down as a solid plug through the casing behind the cement slurry.

When the top plug is to be displaced by drilling fluid or water, the volume of the displacing fluid should be measured as the cement pumps and compared with the volume measured in the water or mud tanks. Where there is a flowmeter, it can be used to crosscheck. When the top plug lands on the bottom plug, a pressure increase is indicated at the surface because no fluid can be pumped through the floating equipment. If the top plug does not "bump" (i.e., seat at the float collar) causing a pressure increase at the calculated displacement volume, the pumping should be stopped so that cement slurry will not be displaced out of the casing.

A cementing manifold is commonly used with a discharge line to the pit for flushing the cement truck. It is assembled to permit pumping the plug out of the cementing head with the displacement fluid.

If casing movement is employed, it should be continued throughout the mixing cycle. Frequently, movement is continued while plugs are released and until the top plug bumps, although it is not uncommon to stop while either or both plugs are being inserted.

9.13.4 Multiple-Stage Cementing Tools. Stage cementing usually reduces mud contamination and lessens the possibility of formation breakdown, which is often a cause of lost circulation. Stage tools are installed at a specific point in the casing string as casing is being run into the hole. When it is desirable to cement two or three separate sections behind the same casing string or to cement a long section in two or three stages, multiple-stage cementing tools are used. During multiple-stage cementing, cement slurry is placed at predetermined points around the casing string in several cementing stages. Multiple-stage cementing tools can be used for these applications: cementing wells with low formation pressures that will not withstand the hydrostatic pressure of a full column of cement; cementing to isolate only certain sections of the wellbore; placing different blends of cement in the wellbore; and cementing deep, hot holes where limited cement pump times restrict full-bore cementing of the casing string in a single stage.

Two types of multiple-stage cementing tools are available: hydraulically opened or plug-opened types. The type selected depends on well conditions. After cement has been placed around the bottom of the well casing, in the conventional manner, the multiple-stage tool may be opened, either hydraulically, by applying casing pressure (hydraulically opened tool), or with a free-fall opening plug dropped down the casing ID (plug-operated). When the tool is opened, fluid, such as cement, can be circulated through its outside ports. When all the cement slurry has been placed, a closing plug pumped down the casing behind the cement closes a sleeve over the side port.

Because the multiple-stage cementing tool contains sliding internal or external sleeves, certain precautions must be taken when it is installed into the casing string. The casing tongs should be placed only on the upper and lower 6 in. of the tool. The tongs should never be placed on the midsection of the casing. This could deform the casing, causing the tool to be inoperative.

Bending forces, resulting from hole deviation or casing deflection, will not damage the tool unless the yield strength of the casing itself is exceeded. Because the OD of the tool is larger than the casing OD, doglegs and key seats, encountered when going into the hole, may cause the tool to stick. Casing centralizers should be installed on the casing as close as possible to each end of the tool to guide it and to provide clearance with the sides of the hole.

The plug-operated free-fall stage-cementing method is used when the first-stage cement is not required to fill the annulus from the bottom of the casing all the way to the stage tool or when the distance between the tool and the casing shoe is fairly long. The primary advantage of this method is that the shutoff plug used in the first stage prevents overdisplacement of the first-stage cement.

The time for the free-falling plug to reach the tool must be estimated because there will be no surface indication when it lands. Many factors, including the viscosity and density of the fluid in the casing and large deviations of the hole from vertical, affect the plug's falling rate and must be considered when waiting time is estimated. A good rule of thumb is to allow 1 minute for each 200 to 400 ft of depth. The maximum deviation that the plug can reasonably be expected to fall is 30°. A deviation greater than 30° will probably cause the plug to hang

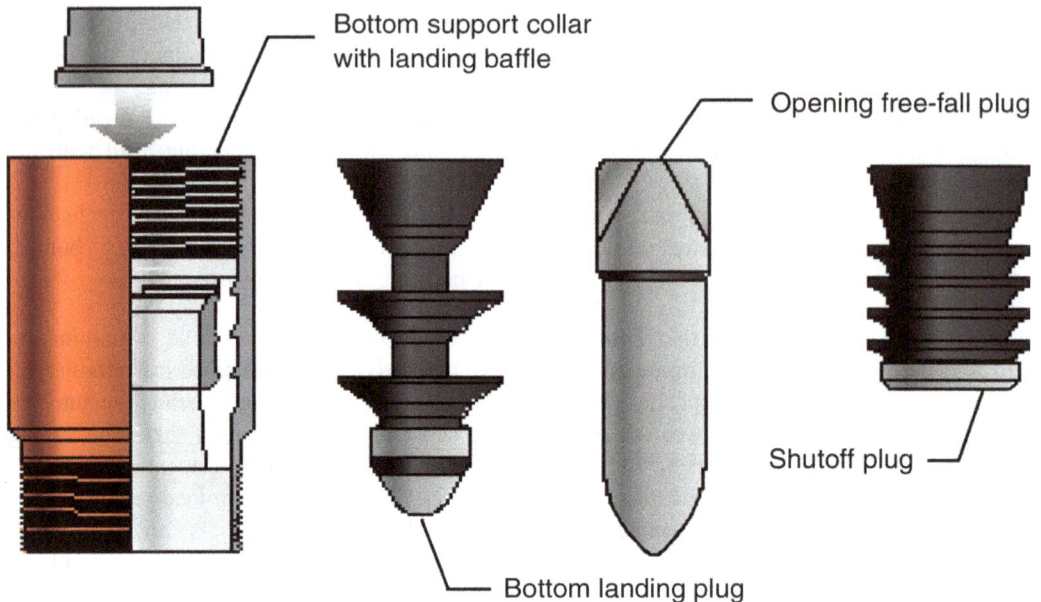

Bottom support collar
with landing baffle

Opening free-fall plug

Shutoff plug

Bottom landing plug

Fig. 9.12—Cement plugs (courtesy of Halliburton).

up at a collar, thus, requiring the plug to be pushed to the tool by a wireline sinker bar or work string (**Fig. 9.12**).

The hydraulically opened or displacement stage-cementing tool is used when the cement is to be placed in the entire annulus from the bottom of the casing up to or above the stage tool. The displacement method is often used in deep or deviated holes in which too much time is needed for a free-falling plug to reach the tool. Fluid volumes must be calculated accurately and measured carefully to prevent overdisplacement or underdisplacement of the first stage.

Two-stage cementing is the most widely used multiple-stage cementing technique. However, when a cement slurry must be distributed over a long column and hole conditions will not allow circulation in one or two stages, a three-stage method can be used. The same steps are involved as in the two-stage method, except that there is an additional stage. Most multiple-stage cementing tools are designed with drillable seats that must be drilled out after cementing operations are completed. These drillable seats allow drillout with either standard tri-cone rock bits or PDC bits.

9.13.5 Casing Centralizers.
The uniformity of the cement sheath around the pipe determines, to a great extent, the effectiveness of the seal between the wellbore and the casing. Because holes are rarely straight, the pipe is generally in contact with the wall of the hole at several places. Hole deviation may vary from zero to, in offshore directional holes, as much as 70 to 90°. Such severe deviation greatly influences the number and spacing of centralizers (**Fig. 9.13**).

A great deal of effort has been expended to determine the relative success of running casing strings with and without centralizers. Although experts differ on the proper approach to an ideal cement job, they generally agree that success hinges on the proper centralization of casing. Centralizers are among the few mechanical aids covered by API specifications.

Centralizing the casing with mechanical centralizers across the intervals to be isolated helps optimize drilling-fluid displacement. In poorly centralized casing, cement will bypass the drilling fluid by following the path of least resistance; as a result, the cement travels down the wide side of the annulus, leaving drilling fluid in the narrow side. When properly installed in

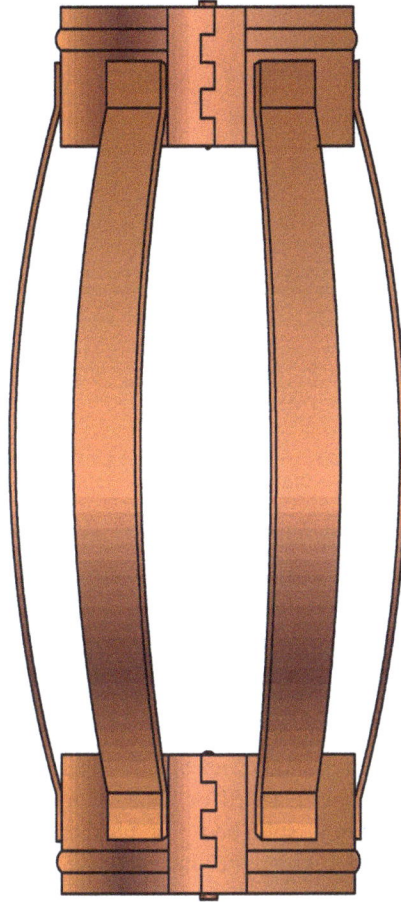

Fig. 9.13—Welded bow-spring centralizer (courtesy of Halliburton).

gauge sections of a hole, centralizers prevent drag, while pipe is run into the hole; center the casing in the wellbore; minimize differential sticking, thus, helping to equalize hydrostatic pressure in the annulus; and reduce channeling and aid in mud removal.

Two general types of centralizers are spring-bow and rigid. The spring-bow type has a greater ability to provide a standoff where the borehole is enlarged. The rigid type provides a more positive standoff where the borehole is close to gauge. Positive-type centralizers are ¼ to ½ in. smaller in diameter than the hole size where they are to be run and, therefore, have no drag forces with the wellbore. Rigid-type centralizers are commonly run in horizontal wellbores because of their positive standoff. Both spring-bow and rigid centralizers are available in almost any casing/hole size. The important design considerations are positioning, method of installation, and spacing. Centralizers should be positioned on the casing through intervals requiring effective cementing, on the casing adjacent to (and sometimes passing through) the intervals where differential-sticking is a hazard, and occasionally on the casing passing through doglegs where key seats may exist.

Good pipe standoff helps ensure a uniform flow pattern around the casing and helps equalize the force that the flowing cement exerts around the casing, increasing drilling-fluid removal. In a deviated wellbore, standoff is even more critical to help prevent a solids bed from accumulating on the low side of the annulus. The preferred standoff should be developed from computer modeling and will vary with well conditions. Under optimum rates, the best drilling-

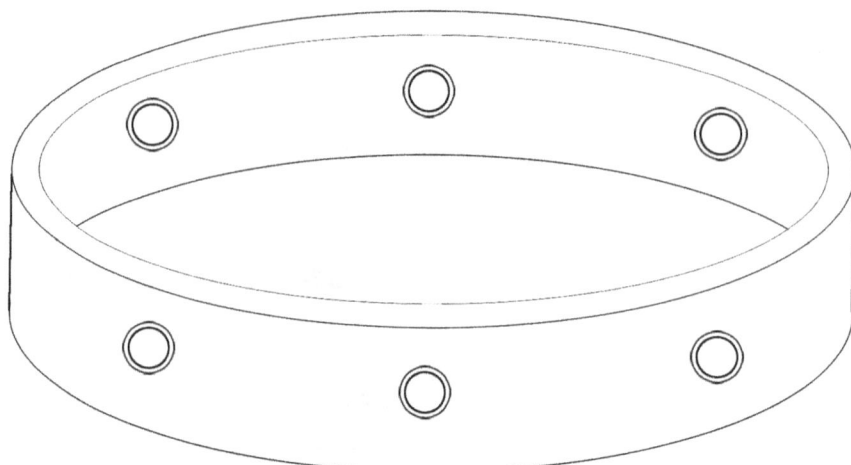

Fig. 9.14—Clamp (courtesy of Halliburton).

fluid displacement is achieved when annular tolerances are approximately 1 to 1.5 in. Effective cementing is important through the production intervals and around the lower joints of the surface and intermediate casing strings to minimize the likelihood of joint loss.

Centralizers are held in their relative position on the casing either by casing collars or mechanical stop collars. The restraining device (collar or stop collar) should always be located within the bow-spring-type centralizer, so the centralizer will be pulled, not pushed, into the hole. The bow-spring-type centralizer should not be allowed to ride free on a casing joint.

All casing attachments should be installed or fastened to the casing by some method, depending on the type (i.e., solid body, split body, or hinged). If they are not installed over a casing collar, a clamp must be used to secure or limit the travel of the various casing attachments.

There are a number of different types of clamps. One type is simply a friction clamp that uses a setscrew to keep the clamp from sliding. Another type uses spiral pins driven between the clamp and the casing to supply the holding force (**Fig. 9.14**). Others have dogs (or teeth) on the inside that actually bite into the casing. Any clamp that might scar the surface of the casing should not be used where corrosion problems exist.

Most service companies offer computer programs on the proper placement of centralizers, based on casing load, hole size, casing size, and hole deviation. All computer spacing programs are based on a standoff of 66% used in API *Spec. 10D*.[10] The computer programs determine placement of the centralizers on the casing string, depending on the well data entered into the program. The programs are based on the equations published in API *Spec. 10D*.[10]

The design of centralizers varies considerably, depending on the purpose and the vendor. For this reason, the API specifications cover minimum performance requirements for standard and close-tolerance spring-bow casing centralizers.

Definitions in API *Spec. 10D*[10] cover starting force, running force, and restoring force. The starting force is the maximum force required to start a centralizer into the previously run casing. The maximum starting force for any centralizer should be less than the weight of 40 ft of medium-weight casing. The maximum starting force should be determined for a centralizer in its new, fully assembled condition as delivered to the end user.

The running force is the maximum force required to move a centralizer through the previously run casing. The running force is proportional to and always equal to or less than the starting force. It is a practical value that gives the maximum "running drag" produced by a centralizer in the smallest specified hole size.

The restoring force is the force exerted by a centralizer against the casing to keep it away from the borehole wall. The restoring force required from a centralizer to maintain adequate standoff is small in a vertical hole but substantial for the same centralizer in a deviated hole. Centralizing smaller annuli is difficult, and pipe movement and displacement rates may be severely restricted. Larger annuli may require extreme displacement rates to generate enough flow energy to remove the drilling fluid and cuttings. Centralizers and other mechanical cementing aids that are commonly used in the industry may also serve as inline laminar-flow mixers, changing the flow pattern of the fluids, which can promote better drilling-fluid removal and greater displacement.

9.13.6 Scratchers. Scratchers, or wall cleaners, are devices that attach to the casing to remove loose filter cake from the wellbore. They are most effective when used while the cement is being pumped. Like centralizers, scratchers help to distribute the cement around the casing. There are two general types of scratchers: those that are used when the casing is rotated and those that are used when the casing is reciprocated.

The rotating scratcher is either welded to the casing or attached with limit clamps. The scratcher claws are high-strength-steel wires with angled ends that cut and remove the mudcake during rotation. The claws may have a coil spring at the base to reduce breaking or bending when the casing is run into the hole. When the pipe must be set at a precise depth, rotating scratchers should be used, but there must be assurance that the pipe can be freely rotated. Because rotating scratchers are damaged by excessive torque on the casing, they are generally not used where the risk of excessive torque is high, such as in deep or deviated wells.

Reciprocating scratchers (**Fig. 9.15**), also constructed of steel wires or cables, are installed on the casing with either an integral or a separate clamping device. When the desired depth is reached, reciprocating the casing (working it up and down) cleans the wellbore on the upstroke by removing mud and filter cake. Reciprocating scratchers are more effective where there is no depth limitation in setting casing and where the pipe can be either rotated or reciprocated after it is landed.

9.13.7 Special Equipment. Mud-diverter equipment is designed for use with a drillpipe when liners are being run or in subsea completions where the wellhead is located on the ocean floor. It allows a fluid flowpath from the drillpipe ID into the annulus above the liner. A drag-spring system on the outer case of the tool causes the drillpipe movement that opens and closes the mud-diverter-equipment ports.

This equipment is used for liner applications where small annular clearances prevent mud-flow between the liner being run and the previous casing string. Such conditions cause high mud loss into formations in the openhole section of the wellbore. Reliable automatic-fill equipment, installed on the lower end of the liner, can allow the wellbore fluid to enter the liner freely, and the drillpipe diverter equipment can allow the fluid to exit the drillpipe immediately above the liner. This arrangement helps reduce the pressure drop and the surge pressure on the formation while the liner is being run, which helps reduce costly mud loss into the formation. A mud-saver system that includes the diverter can be used on a liner or subsea completion. The use of this diverter equipment can eliminate the need to take returns at the surface.

Bridge plugs are devices that are set in open hole or casing as temporary, retrievable plugs or permanent, drillable plugs. They cannot be pumped through and are used to prevent fluid or gas from moving in the wellbore. Bridge plugs are also used to isolate a lower zone, while an upper section is being tested; establish a bridge above or below a perforated section that is to be squeezed, cemented, or fractured; provide a pressure seal for casing that is to be tested or for wells that are to be abandoned; seal off zones to be abandoned to allow the upper casing to be recovered; and plug casing, while surface equipment is being repaired.

Reciprocating wall cleaners (RWCs)

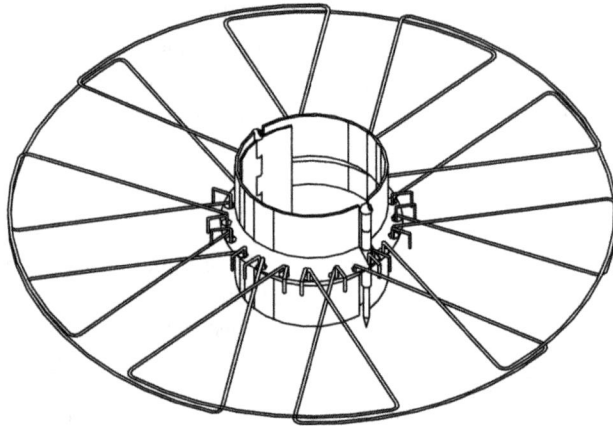

RWC wall cleaner (tubing)

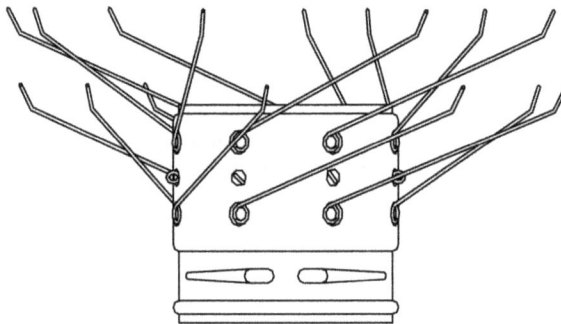

RWC wall cleaner (casing)

Fig. 9.15—Reciprocating scratchers (courtesy of Halliburton).

Cement baskets and external packers (**Fig. 9.16**) are used with casing or liner at points where porous or weak formations require help in supporting the cement column until it takes its initial set. Baskets may be installed by slipping them over the casing and using either the collars or limit clamps to hold them in place. External packers are placed in the casing string as it is run in the well. They are expanded before cementing begins.

9.13.8 Pumping Equipment. High-energy displacement rates are most effective in ensuring good displacement. Turbulent flow around the full circumference of the casing is most desirable, but it is not required. When turbulent flow is not a viable option for a formation or wellbore configuration, use the highest pump rate that is feasible for the wellbore conditions. The best results are obtained when the spacer and/or cement is pumped at maximum energy; the spacer or flush is appropriately designed to remove the drilling fluid; and a good, competent cement is used.

Cement pumping units may be mounted on a truck, trailer, skid, or waterborne vessel. They are usually powered by either internal-combustion engines or electric motors and are operated

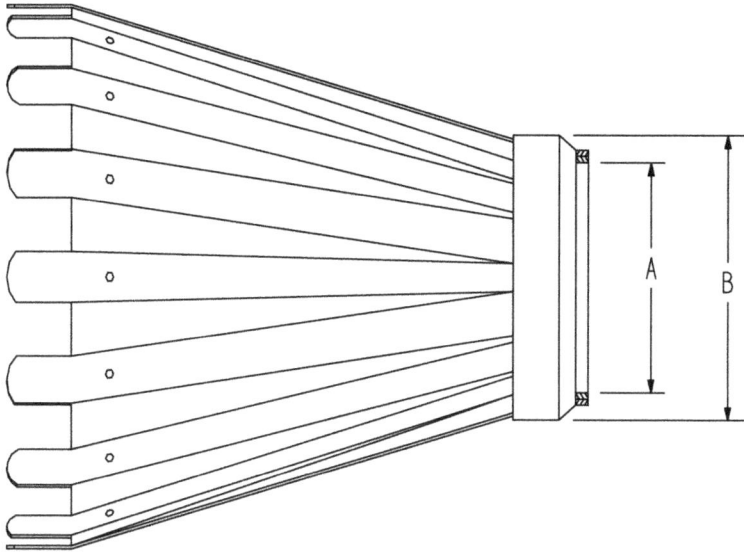

Fig. 9.16—Cement basket (courtesy of Halliburton).

intermittently at high pressures and at varying rates. Pumping units must be capable of providing a wide range of pressures and rates to facilitate the requirements of modern cementing practices, and yet, have the lowest practical weight-to-horsepower ratio to facilitate transportation.

Cementing units are normally equipped with two positive-displacement pumps. On a high-pressure system, one pump mixes while the other displaces. On a low-pressure system, a centrifugal pump mixes, and two positive-displacement pumps are available for displacement. For recirculating mixing, one centrifugal pump supplies water to the mixing jet, and another centrifugal pump recirculates slurry back through the mixing jet. As with a low-pressure system, two positive-displacement pumps are available to pick up the slurry and pump it down the well.

Nearly all cementing pumps are positive-displacement and are either duplex double-acting piston pumps or single-acting triplex plunger pumps. Either is satisfactory within its design limits. For heavy-duty pumping, triplex pumps discharge more smoothly and can usually handle higher horsepower and greater pressure than duplex pumps. Most cementing work involves a maximum pressure of less than 5,000 psi, but pressures as high as 20,000 psi are not uncommon. Because of widely varying operating conditions, the cementing pump and its power train are designed for the maximum rather than the average expected pressures.

For a given job, the number of trucks used to mix cement depends on the volume of cement, well depth, and expected pressures. For surface and conductor strings, one truck is usually adequate, whereas for intermediate or production casing, as many as three units may be required. On jobs requiring more than 1,000 sk or where high pressures are expected, two or sometimes three mixing trucks are used. A separate mixing system is used for each truck, with each unit tied to a common pumping manifold. If the pipe is to be reciprocated, the mixing trucks are tied into a temporary standpipe, which supports a flexible line leading to the cementing head.

Field slurries are usually mixed and pumped into the casing at the highest feasible rate, which varies from 20 to 50 sk/min depending on the capacity of each mixing unit. As a result, the first sack of cement on a primary cement job reaches bottomhole conditions rather quickly.

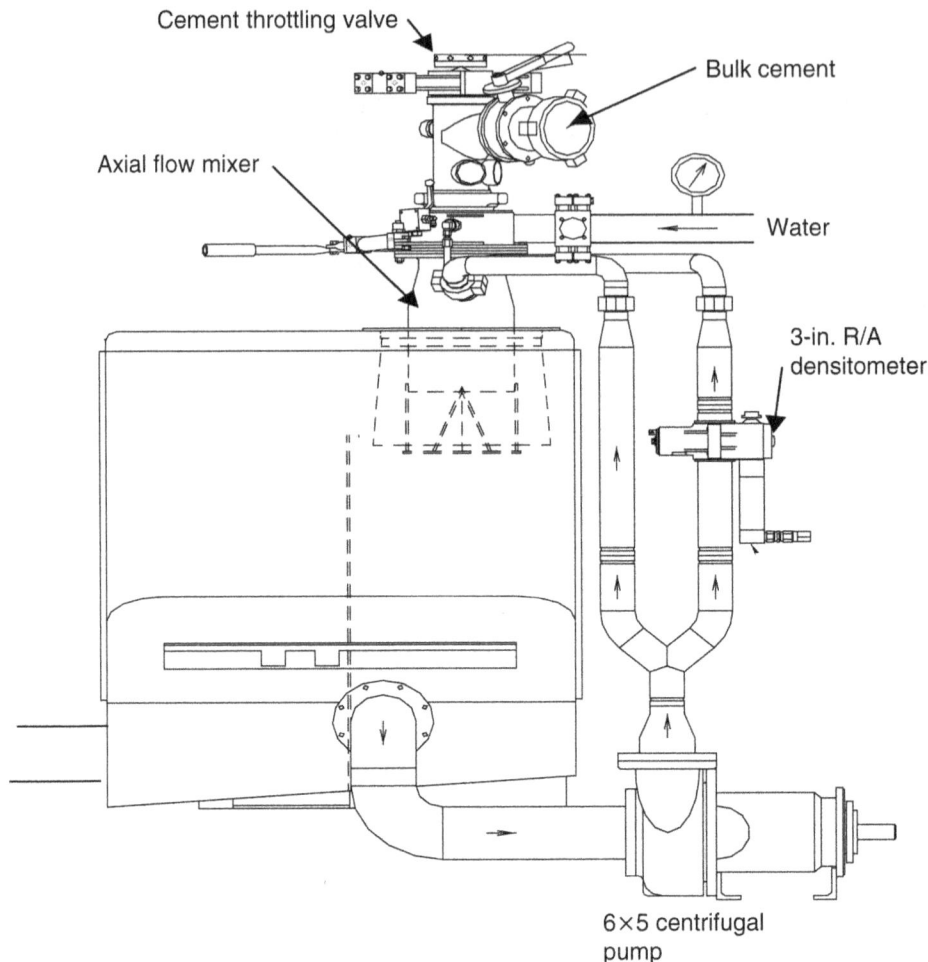

Fig. 9.17—Reciprocating mixer (courtesy of Halliburton).

9.13.9 Cement Mixing. The mixing system proportions and blends the dry cementing composition with the carrier fluid (water), supplying to the wellhead a cementing slurry with predictable properties.

The recirculating mixer, designed for mixing more-uniform homogeneous slurries, is a pressurized jet mixer with a large tub capacity (**Fig. 9.17**). It uses recirculated slurry and mixing water to partially mix and discharge the slurry into the tub. The recirculating pump provides additional shear, and agitation paddles or jets provide additional energy and improve mixing. The result is a uniform cement slurry, with a density as high as 22 lbm/gal, which can be pumped as slowly as 0.5 bbl/min.

Batch mixing is used to blend a cement slurry at the surface before it is pumped into the well. The batch mixer is not part of the cement pumping unit; it is a separate piece of equipment. The batch mixer is used when a specified volume of cement is required. The mixing tank in the batch mixer is filled with enough water for a specified amount of cement. The mixing turbine circulates the water, as dry cement is added until the desired slurry consistency and volume are obtained. A prehydrator is used to wet the dry cement to prevent dust problems. Primary disadvantages of a batch mixer are volume limitations and the need to use an additional piece of equipment. However, units with multiple mixing tanks may be used for continuous cementing to provide precise slurry consistency and volume. For mixing densified or heavyweight slurries to be pumped at rates of less than 5 bbl/min, a recirculating mixer produces a more uniform slurry.

Nomenclature

B_c = Bearden units of consistency
O = ordinary
q_d = displacement rate, bbl/min
t_c = contact time, min
V_t = volume of fluid (turbulent flow), ft^3

References

1. *RP10B1, Recommended Practice for Testing Well Cements,* 22nd edition, API, Washington, DC (1997).
2. *RP13B, Recommended Practice Standard Procedure for Field Testing Water-based Drilling Fluids,* second edition, API, Washington, DC (1997).
3. Brice, J. and Holmes, B.C.: "Engineered Casing Cementing Programs Utilizing Turbulent Flow Techniques," paper SPE 742 presented at the 1963 SPE Annual Meeting, New Orleans, 6–9 October.
4. Calvert, D.G., Heathman, J., and Griffith, J.: "Plug Cementing: Horizontal to Vertical Conditions," paper SPE 30514 presented at the 1995 SPE Annual Technical Conference and Exhibition, Dallas, 22–25 October.
5. *Standard C150-97a, Standard Specification for Portland Cement,* Vol. 04.01, ASTM, West Conshohocken, Pennsylvania (2000) 134–138.
6. *Standard C114-97a, Standard Methods for Chemical Analysis of Hydraulic Cement,* Vol. 04.01, ASTM, West Conshohocken, Pennsylvania (2000) 94–123.
7. Smith, D.K.: *Cementing,* Monograph Series, SPE, Richardson, Texas (1990) **4,** Chaps. 2 and 3.
8. *Spec. 10A, Specification for Cements and Materials for Well Cementing,* 23rd edition, ANSI/API 10A/ISO 10426-1-2001, Washington, DC (2002).
9. *Standard C618, Standard for Testing and Materials,* Vol. 04.01, ASTM, West Conshohocken, Pennsylvania (2000).
10. *Spec. 10D, Specification for Casing Centralizers,* third edition, API, Dallas (1986).

General References

Abdul-Maula, S. and Odler, I.: "Effect of Oxidic Composition on Hydration and Strength Development of Laboratory-Made Portland Cements," *World Cement,* **13,** No. 5, 216.

Anderson, P.J., Roy, D.M., and Gaidis, J.M.: "The Effects of Adsorption of Superplasticizers on the Surface of Cement," *Cement & Concrete Research,* **17,** No. 5, 805.

Annual Book of ASTM Standards, "Cement, Lime, Gypsum," ASTM, West Conshohocken, Pennsylvania (2000) **04.01,** Sec. 4.

Beach, H.J. and Goins, W.C. Jr.: "A Method of Protecting Cements Against Harmful Effects of Mud Contamination," *Trans.,* AIME (1957) **210,** 148.

Beirute, R. and Tragesser, A.: "Expansive and Shrinkage Characteristics of Cements Under Actual Well Conditions," *JPT* (August 1973) 905.

Beirute, R.M.: "The Phenomenon of Free Fall During Primary Cementing," paper SPE 13045 presented at the 1984 SPE Annual Technical Conference and Exhibition, Houston, 16–19 September.

Bensted, J.: "Calcium Aluminate Cements: Highlights from a Recent Symposium," *World Cement,* **21,** No. 10, 452.

Bensted, J.: "Microfine Cements," *World Cement*, **25**, No. 8, 45.

Brice, J.W. Jr. and Holmes, R.C.: "Engineered Casing Cementing Programs Using Turbulent Flow Techniques," *JPT* (May 1964) 503.

Brown, P.W.: "Early Hydration of Tetracalcium Aluminoferrite in Gypsum and Lime-Gypsum Solutions," *J. American Ceramic Soc.*, **70**, No. 7, 493.

Bull. 5C2, Casing, Tubing, and Drillpipe, 19th edition, API, Dallas (October 1984).

Calvert, D.G. *et al.*: "Plug Cementing: Horizontal to Vertical Conditions," paper SPE 30514 presented at the 1995 SPE Annual Technical Conference and Exhibition, Dallas, 22–25 October.

Calvert, D.G. *et al.*: "Study Reveals Variables that Affect Cement Plug Stability," *Oil and Gas J.*, **98**, No. 8.

Carter, L.G., Slagle, K.A., and Smith, D.K.: "Resilient Cement Decreases Perforating Damage," paper presented at the API Mid-Continent Dist. Div. of Production Meeting, Amarillo, Texas (April 1968).

Carter, L.G., Waggoner, H.F., and George, C.R.: "Expanding Cements for Primary Cementing," *JPT* (May 1966) 551.

Childs, J., Sabins, F., and Taylor, M.J.: "Method of Using Thixotropic Cements for Combating Lost Circulation Problems," U.S. Patent No. 4,515,216 (1985).

Childs, J.D. and Burkhalter, J.F.: "Fluid Loss Reduced Cement Compositions," U.K. Patent No. 2,247,234 (1992).

Collepardi, M. *et al.*: "Influence of Sulphonated Naphthalene on the Fluidity of Cement Pastes," seventh edition, ICCC, Paris (1980) **3**, No. VI-20–VI-25.

Cowan, K.M., Hale, A.H., and Nahm, J.J.: "Conversion of Drilling Fluids to Cements With Blast Furnace Slag: Performance Properties and Applications for Well Cementing," paper SPE 24575 presented at the 1992 SPE Annual Technical Conference and Exhibition, Washington, DC, 4–7 October.

Craft, B.C. and Hawkins, M.F.: *Applied Petroleum Reservoir Engineering*, Prentice-Hall Inc., Englewood Cliffs, New Jersey (1959) 319.

Crook, R.J. *et al.*: "Eight Steps Ensure Successful Cement Placement," *Oil and Gas J.*, **99**, No. 27, 37.

Dahl, J., Harris, K., and McKown, K.: "Uses of Small Particles Size Cement in Water and Hydrocarbon Based Slurries," *Proc.*, Ninth Kansas U. Tertiary Oil Recovery Conference, Wichita, Kansas (1991) 25–29.

Economides, M.J., Watters, L.T., and Dunn-Norman, S.: *Petroleum Well Construction*, John Wiley & Sons Inc., New York City (1998).

Encyclopedia of Polymer Science and Engineering, Vol. 11, John Wiley & Sons Inc., New York City (1988)

Enloe, J.R.: "Amerada Finds Using Multiple Casing Strings Can Cut Costs," *Oil and Gas J.*, **65**, No. 24, 76.

Gerke, R.R. *et al.*: "A Study of Bulk Cement Handling and Testing Procedures," paper SPE 14196 presented at the 1985 SPE Annual Technical Conference and Exhibition, Las Vegas, Nevada, 22–25 September.

Goins, W.C. Jr.: "Lost Circulation Problems Whipped with BDO (Bentonite Diesel Oil) Squeeze," *Drilling*, **15**, No. 11, 83.

Goins, W.C. Jr.: "Selected Items of Interest in Drilling Technology—An SPE Distinguished Lecture," *JPT* (July 1971) 857.

Golding, B: "*Polymers and Resins, Their Chemistry and Chemical Engineering*," D. Van Nostrand Co., Princton, NJ (1959).

Greminger, G.K.: "Hydraulic Cement Compositions for Wells," U.S. Patent No. 2,844,480 (1958).

Grulke, E.A.: *Polymer Process Engineering*, Prentice-Hall Inc., Englewood Cliffs, New Jersey (1994).

Hale, A.H. and Cowan, K.M.: "Solidification of Water-Based Muds," U.S. Patent No. 5,058,679 (1991).

Halliburton Oil Well Cement Manual, Halliburton Co., Duncan, Oklahoma (1983).

Hanna, E. *et al.*: "Rheological Behaviour of Portland Cement in the Presence of a Superplasticizer," Third CANMET/ACI Intl. Conference on Superplasticizers and Other Chemical Admixtures in Concrete, SP 119-9, 171–188 (1989).

Hansen, W.C.: "Oil-Well Cements," paper presented at the 1952 Intl. Symposium on the Chemistry of Cement, London, 15–19 September.

Heathman, J. *et al.*: "Removing Subjective Judgment From Wettability Analysis Aids-Displacement," paper SPE 59135 presented at the 2000 IADC/SPE Drilling Conference, New Orleans, 23–25 February.

Heathman, J.F. *et al.*: "Quality Management Alliance Eliminates Plug Failures," paper SPE 28321 presented at the 1994 SPE Annual Technical Conference and Exhibition, New Orleans, 25–28 September.

Heathman, J.F.: "Advances in Cement Plug Procedures" *JPT* (September 1996) 825.

Hemphill, A.T. *et al.*: "Field Applications of ERD Hole Cleaning Modeling," *SPEDC* (December 1999).

Hewlett, P.C.: *Lea's Chemistry of Cement and Concrete,* fourth edition, Arnold Publishers Ltd., London (1998).

Hills, J.O.: "A Review of Casing-String Design Principles and Practices," *Drill. & Prod. Prac.,* API (1951) 91.

Hook, F.E., Morris, E.F., and Rosene, R.B.: "Silica-Lime Systems for High Temperature Cementing Applications," paper SPE 3447 presented at the 1971 SPE Annual Meeting, New Orleans, 3–6 October.

Hook, F.E.: "Aqueous Cement Slurry and Method of Use," U.S. Patent No. 3,483,007 (1969).

Howard, Q.C. and Scott, P.P. Jr.: "An Analysis and Control of Lost Circulation," *Trans.,* AIME (1951) **192,** 171–182.

Huber, T.A. and Corley, C.B. Jr.: "Permanent-Type Multiple Tubingless Completions," *Pet. Eng.* (February/March 1961).

Kopp, K. *et al.*: "Foamed Cement vs. Conventional Cement for Zonal Isolation—Case Histories," paper SPE 62895 presented at the 2000 SPE Annual Technical Conference and Exhibition, Dallas, 1–4 October.

Lee, H.K., Smith, R.C., and Tighe, R.E.: "Optimal Spacing for Casing Centralizers," paper SPE 13043 presented at the 1984 SPE Annual Technical Conference and Exhibition, Houston, 16–19 September.

Ludwig, N.D.: "Portland Cements and Their Application in the Oil Industry," *Drill. & Prod. Prac.,* API (1953) 183.

Maier, L.F. *et al.*: "Cementing Practices in Cold Environments," *JPT* (October 1971) 1215.

McCarthy, G.J. *et al.*: "Use of a Database of Chemical, Mineralogical and Physical Properties on North American Fly Ash to Study the Nature of Fly Ash and Its Utilization as a Mineral Admixture in Concrete," *Proc.,* Materials Research Soc. Symposium, Boston (1990) **178,** 3–34.

Meek, J.W. and Harris, K.: "Repairing Casing Leaks Using Small Particle Size Cement," paper 3.2 presented at the 1992 Seminar of Well Cementing, Caracas, 18–20 March.

Melmick, W.E. and Longley, A.J.: "Pressure-Differential Sticking of Drillpipe and How It Can Be Avoided or Relieved," *Drill. & Prod. Prac.,* API (1957) 55.

Messenger, J.U. and McNeil, J.S. Jr.: "Lost Circulation Corrective: Time Setting Clay Cement," *Trans.,* AIME (1952) **195,** 171–182.

Mueller, D.T., Virgillio, G., and Dickerson, J.P.: "Stress Resistant Cement Compositions and Methods Using Same," U.S. Patent 6,230,804 (2001).

Myers, G.M. and Sutko, A.A.: "The Development of a Method for Calculating the Forces on Casing Centralizers," paper presented at the 1968 API Mid-Continent Meeting, Amarillo, Texas, 3–5 April.

O'Brien, T.B.: "Buckled Casing: Three Ways to Avoid It," *World Oil*, **199,** No. 5, 60.

O'Brien, T.B.: "Why Some Casing Failures Happen," *World Oil*, **198,** No. 7, 143.

Owsley, W.D.: "Improved Casing Cementing Practices in the United States." *Oil & Gas J.*, **48,** No. 32, 76.

Parcevaux, P.A. *et al.*: "Cement Compositions for Cementing Wells, Allowing Pressure Gas-Channeling in the Cemented Annulus to be Controlled," U.S. Patent No. 4,537,918 (1985).

Plowman, C. and Cabrera, J.C.: "Mechanism and Kinetics of Hydration of C_3A and C_4AF Extracted from Cement," *Cement & Concrete Research*, **14,** No. 2, 238.

Plowman, C. and Cabrera, J.G.: "The Use of Fly Ash to Improve the Sulfate Resistance of Concrete," *Waste Management*, **16,** No. 1–3, 145.

Powers, D.J. *et al.*: "Drilling Practices and Sweep Selection for Efficient Hole Cleaning in Deviated Wellbores," paper SPE 62794 presented at the 2000 IADC/SPE Asia Pacific Drilling Technology Conference, Kuala Lumpur, 11–13 September.

Pratt, P.L. and Jensen, H.-U.: "The Development of Microstructure During the Setting and Hardening of Cement Pastes," *Proc.*, RILEM Hydration and Setting of Cement, A. Nonat and J.C. Mutin (eds.) Dijon, France (1992) **16,** 353–360.

Ramachandran, V.S., Seeley, R.C., and Polomark, G.M.: "Free and Combined Chloride in Hydrating Cement and Cement components," *Materials & Structures*, **17,** No. 100, 285.

Ravi, K.M, Beirute, R.M., and Covington, R.L: "Erodability of Partially Dehydrated Gelled Drilling Fluid and Filter Cake," paper SPE 24571 presented at the 1992 SPE Annual Technical Conference and Exhibition, Washington, DC, 4–7 October.

Rike, J.I. and McGlamery, R.G.: "Recent Innovations in Offshore Completion and Workover Systems," *JPT* (January 1970) 17.

Saunders, K.J.: *Organic Polymer Chemistry,* Chapman and Hall, London (1973).

Scott, P.O. Jr., Lummus, J.L., and Howard, G.C.: "Methods for Sealing Vugular and Cavernous Formations," *Drilling Contractor*, **9,** No. 10, 70.

Shaughnessy, R. III and Clark, P.E.: "The Rheological Behavior of Fresh Cement Pastes," *Cement & Concrete Research*, **18,** No. 2, 327.

Shell, F.J. and Wynner, R.A.: "Applications of Low-Water-Loss Cement Slurries," paper API 875-12-1, API, Denver, 21–23 April.

Smith, R.C. and Calvert, D.G.: "The Use of Sea Water in Well Cementing," *JPT* (June 1975) 759.

Spangle, L.B.: "Expandable Cement Composition," Eur. Patent No. 254,342 (1988).

Suman, G.O. Jr. and Ellis, R.C.: "Cementing Oil and Gas Wells: Part 5—Guidelines for Downhole Equipment Use, Stage Cementing Methods, New Concepts for Cementing Large Diameter Casing," *World Oil's Cementing Handbook,* Gulf Publishing Co., Houston (1977).

Tang, F.J. and Gartner, E.M.: "Influence of Sulphate Source on Portland Cement Hydration," *Adv. Cem. Res.*, **1,** No. 2, 67.

Tausch, G.H. and Kenneday, J.W.: "Permanent-Type Dual Completions and Wireline Workovers," *Pet. Eng.*, **28,** No. 3, B24.

Technical Sales Catalog, Baker Oil Tools Inc., Houston (1983).

Technical Sales Catalog, BJ Services, Arlington, Texas (1983).

Technical Service Catalog, Halliburton Services, Duncan, Oklahoma (1985) No. 42.

Teplitz, A.J. and Hassebroek, W.E.: "An Investigation of Oil Well Cementing," *Drill. & Prod. Prac.*, API (1946) 76.

Training Literature: Fundamentals of Cementing Practices, Halliburton Energy Inst., Duncan, Oklahoma (1989).

Tumidajski, P.J. and Thomson, L.: "Influence of Cadmium on the Hydration of C_3A," *Cement & Concrete Research,* **24,** No 8, 1359.

Underwood, D., Broussard, P., and Walker, W.: "Long Life Cementing Slurries," paper presented at the 1965 API Southwestern Dist. Div. of Production Meeting, Dallas, 10–12 March.

Weisend, C.F.: "Cement Additive Containing Polyvinyl Pyrrolidone and a Condensate of Sodium Naphthalenesulfonate with Formaldehyde," U.S. Patent No. 3,359,225 (1967).

Weisend, C.F.: "Composition Comprising Hydroxyethyl Cellulose, Polyvinyl Pyrrolidone, and Organic Sulfonate, Cement Slurry Prepared Therefrom and Method of Cementing Well Therewith," U.S. Patent No. 3,132,693 (1964).

Well Completion Service and Equipment Catalog, B&W Inc., Houston (1983).

White, F.L.: "Setting Cements in Below Freezing Conditions," *Pet. Eng.,* **24,** No. 9 Part 1, B7.

White, R.J.: "Lost Circulation Materials and Their Evaluation," *Drill. & Prod. Prac.,* API (1956) 352.

White, W.S. *et al.*: "A Laboratory Study of Cement and Resin Plugs Placed With Thru-Tubing Dump Bailers," paper SPE 24574, presented at the 1992 SPE Annual Technical Conference and Exhibition, Washington, DC, 4–7 October.

SI Metric Conversion Factors

bbl	$\times$ 1.589 873	E – 01	= m^3
ft	$\times$ 3.048*	E – 01	= m
ft^2	$\times$ 9.290 304*	E – 02	= m^2
ft^3	$\times$ 2.831 685	E – 02	= m^3
°F	($°F$ – 32)/1.8		= °C
gal	$\times$ 3.785 412	E – 03	= m^3
in.	$\times$ 2.54*	E + 00	= cm
lbf	$\times$ 4.448 222	E + 00	= N
lbm	$\times$ 4.535 924	E – 01	= kg
psi	$\times$ 6.894 757	E + 00	= kPa

*Conversion factor is exact.

Chapter 10
Drilling Problems and Solutions
J.J. Azar, U. of Tulsa

10.1 Introduction
It is almost certain that problems will occur while drilling a well, even in very carefully planned wells. For example, in areas in which similar drilling practices are used, hole problems may have been reported where no such problems existed previously because formations are nonhomogeneous. Therefore, two wells near each other may have totally different geological conditions.

In well planning, the key to achieving objectives successfully is to design drilling programs on the basis of anticipation of potential hole problems rather than on caution and containment. Drilling problems can be very costly. The most prevalent drilling problems include pipe sticking, lost circulation, hole deviation, pipe failures, borehole instability, mud contamination, formation damage, hole cleaning, H_2S-bearing formation and shallow gas, and equipment and personnel-related problems.

Understanding and anticipating drilling problems, understanding their causes, and planning solutions are necessary for overall-well-cost control and for successfully reaching the target zone. This chapter addresses these problems, possible solutions, and, in some cases, preventive measures.

10.2 Pipe Sticking
During drilling operations, a pipe is considered stuck if it cannot be freed and pulled out of the hole without damaging the pipe and without exceeding the drilling rig's maximum allowed hook load. Differential pressure pipe sticking and mechanical pipe sticking are addressed in this section.

10.2.1 Differential-Pressure Pipe Sticking. Differential-pressure pipe sticking occurs when a portion of the drillstring becomes embedded in a mudcake (an impermeable film of fine solids) that forms on the wall of a permeable formation during drilling. If the mud pressure, p_m, which acts on the outside wall of the pipe, is greater than the formation-fluid pressure, p_{ff}, which generally is the case (with the exception of underbalanced drilling), then the pipe is said to be differentially stuck (see **Fig. 10.1**). The differential pressure acting on the portion of the drillpipe that is embedded in the mudcake can be expressed as

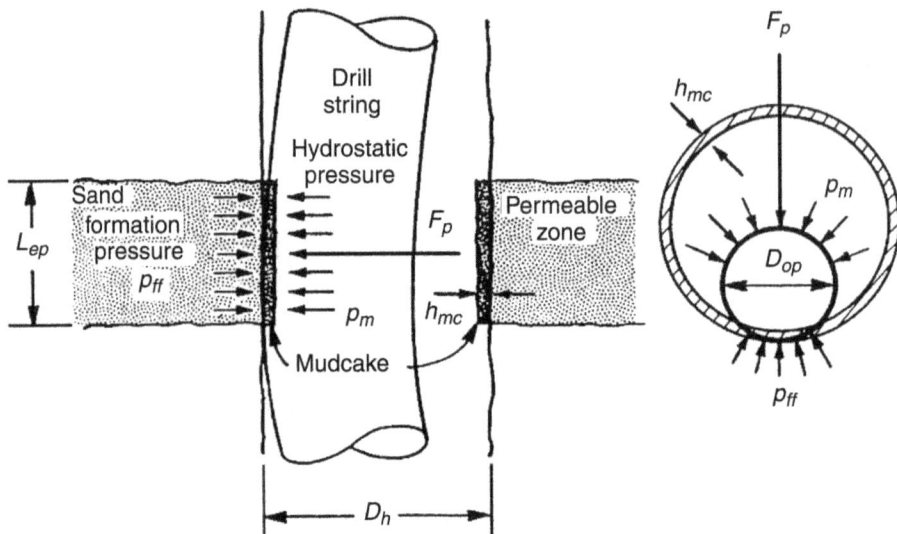

Fig. 10.1—Differential-pressure sticking.

$$\Delta p = p_m - p_{ff} \quad \text{...} \quad (10.1)$$

The pull force, F_p, required to free the stuck pipe is a function of the differential pressure, Δp; the coefficient of friction, f; and the area of contact, A_c, between the pipe and mudcake surfaces.

$$F_p = f \Delta p A_c \quad \text{...} \quad (10.2)$$

From Ref. 1,

$$A_c = 2L_{ep}\{(D_h/2 - h_{mc})^2 - [D_h/2 - h_{mc}(D_h - h_{mc})/(D_h - D_{op})]^2\}^{0.5}, \quad \text{............} \quad (10.3)$$

where

$$D_{op} \leq (D_h - h_{mc}). \quad \text{...} \quad (10.4)$$

In this formula, L_{ep} is the length of the permeable zone, D_{op} is the outside diameter of the pipe, D_h is the diameter of the hole, and h_{mc} is the mudcake thickness. The dimensionless coefficient of friction, f, can vary from less than 0.04 for oil-based mud to as much as 0.35 for weighted water-based mud with no added lubricants.

Eqs. 10.2 and 10.3 show controllable parameters that will cause higher pipe-sticking force and the potential inability of freeing the stuck pipe. These parameters are unnecessarily high differential pressure, thick mudcake (high continuous fluid loss to formation), low-lubricity mud-cake (high coefficient of friction), and excessive embedded pipe length in mudcake (delay of time in freeing operations).

Although hole and pipe diameters and hole angle play a role in the pipe-sticking force, they are uncontrollable variables once they are selected to meet well design objectives. However, the shape of drill collars, such as square, or the use of drill collars with spiral grooves and external-upset tool joints can minimize the sticking force.

Some of the indicators of differential-pressure-stuck pipe while drilling permeable zones or known depleted-pressure zones are an increase in torque and drag; an inability to reciprocate the drillstring and, in some cases, to rotate it; and uninterrupted drilling-fluid circulation. Differential-pressure pipe sticking can be prevented or its occurrence mitigated if some or all of the following precautions are taken:

• Maintain the lowest continuous fluid loss adhering to the project economic objectives.

• Maintain the lowest level of drilled solids in the mud system, or, if economical, remove all drilled solids.

• Use the lowest differential pressure with allowance for swab and surge pressures during tripping operations.

• Select a mud system that will yield smooth mudcake (low coefficient of friction).

• Maintain drillstring rotation at all times, if possible.

Differential-pressure-pipe-sticking problems may not be totally prevented. If sticking does occur, common field practices for freeing the stuck pipe include mud-hydrostatic-pressure reduction in the annulus, oil spotting around the stuck portion of the drillstring, and washing over the stuck pipe. Some of the methods used to reduce the hydrostatic pressure in the annulus include reducing mud weight by dilution, reducing mud weight by gasifying with nitrogen, and placing a packer in the hole above the stuck point.

10.2.2 Mechanical Pipe Sticking. The causes of mechanical pipe sticking are inadequate removal of drilled cuttings from the annulus; borehole instabilities, such as hole caving, sloughing, or collapse; plastic shale or salt sections squeezing (creeping); and key seating.

Drilled Cuttings. Excessive drilled-cuttings accumulation in the annular space caused by improper cleaning of the hole can cause mechanical pipe sticking, particularly in directional-well drilling. The settling of a large amount of suspended cuttings to the bottom when the pump is shut down or the downward sliding of a stationary-formed cuttings bed on the low side of a directional well can pack a bottomhole assembly (BHA), which causes pipe sticking. In directional-well drilling, a stationary cuttings bed may form on the low side of the borehole (see **Fig. 10.2**). If this condition exists while tripping out, it is very likely that pipe sticking will occur. This is why it is a common field practice to circulate bottom up several times with the drill bit off bottom to flush out any cuttings bed that may be present before making a trip. Increases in torque/drag and sometimes in circulating drillpipe pressure are indications of large accumulations of cuttings in the annulus and of potential pipe-sticking problems.

Borehole Instability. This topic is addressed in Sec. 10.6; however, it is important to mention briefly the pipe-sticking issues associated with the borehole-instability problems. The most troublesome issue is that of drilling shale. Depending on mud composition and mud weight, shale can slough in or plastically flow inward, which causes mechanical pipe sticking. In all formation types, the use of a mud that is too low in weight can lead to the collapse of the hole, which can cause mechanical pipe sticking. Also, when drilling through salt that exhibits plastic behavior under overburden pressure, if mud weight is not high enough, the salt has the tendency of flowing inward, which causes mechanical pipe sticking. Indications of a potential pipe-sticking problem caused by borehole instability are a rise in circulating drillpipe pressure, an increase in torque, and, in some cases, no fluid return to surface. **Fig. 10.3** illustrates pipe sticking caused by wellbore instability.

Key Seating. Key seating is a major cause of mechanical pipe sticking. The mechanics of key seating involve wearing a small hole (groove) into the side of a full-gauge hole. This groove is caused by the drillstring rotation with side force acting on it. **Fig. 10.4** illustrates pipe sticking caused by key seating. This condition is created either in doglegs or in undetect-

Cuttings bed during drilling

Cuttings jamming the drill bit during tripping out

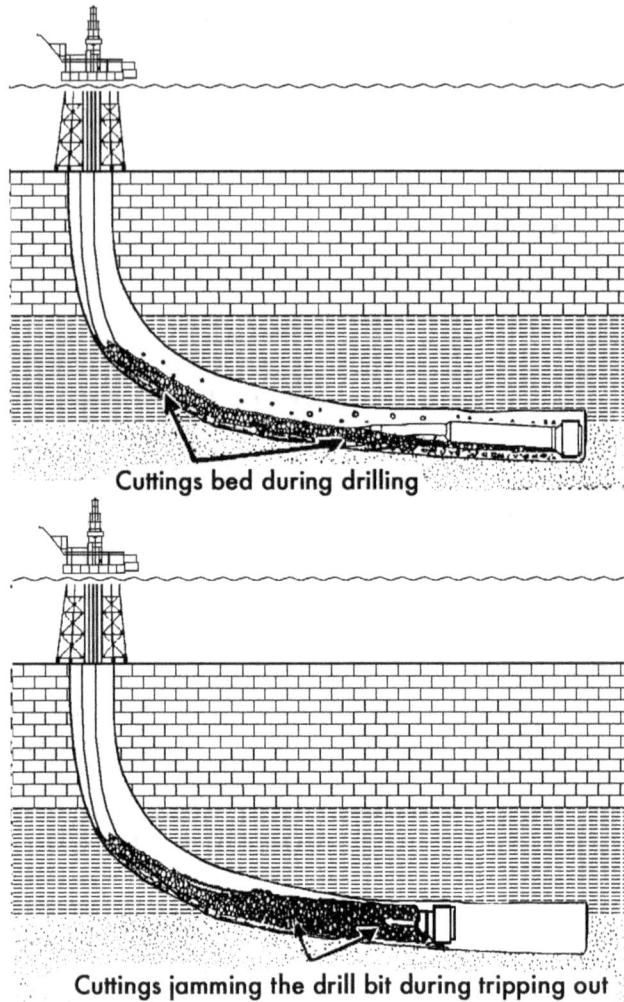

Fig. 10.2—Mechanical pipe sticking caused by drilled cuttings: (a) cuttings bed during drilling, and (b) cuttings jamming the drill bit during tripping out.

ed ledges near washouts. The lateral force that tends to push the pipe against the wall, which causes mechanical erosion and thus creates a key seat, is given by

$$F_l = T \sin \theta_{dl}, \quad\quad\quad\quad\quad\quad\quad\quad\quad\quad\quad (10.5)$$

where F_l is the lateral force, T is the tension in the drillstring just above the key-seat area, and θ_{dl} is the abrupt change in hole angle (commonly referred to as dogleg angle).

Generally, long bit runs can cause key seats; therefore, it is common practice to make wiper trips. Also, the use of stiffer BHAs tends to minimize severe dogleg occurrences. During tripping out of hole, a key-seat pipe-sticking problem is indicated when several stands of pipe have been pulled out, and then, all of a sudden, the pipe is stuck.

Freeing mechanically stuck pipe can be undertaken in a number of ways, depending on what caused the sticking. For example, if cuttings accumulation or hole sloughing is the suspected cause, then rotating and reciprocating the drillstring and increasing flow rate without

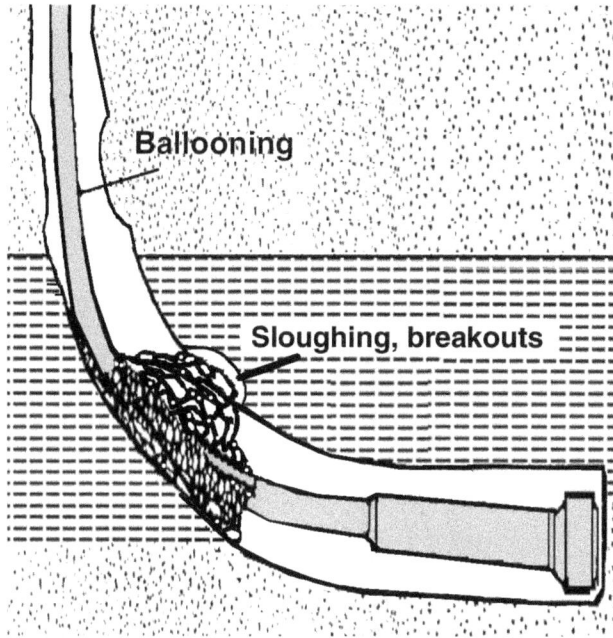

Fig. 10.3—Pipe sticking caused by wellbore instability.

exceeding the maximum allowed equivalent circulating density (ECD) is a possible remedy for freeing the pipe. If hole narrowing as a result of plastic shale is the cause, then an increase in mud weight may free the pipe. If hole narrowing as a result of salt is the cause, then circulating fresh water can free the pipe. If the pipe is stuck in a key-seat area, the most likely successful solution is backing off below the key seat and going back into the hole with an opener to drill out the key section. This will lead to a fishing operation to retrieve the fish. The decision on how long to continue attempting to free stuck pipe vs. back off, plug back, and then sidetrack is an economic issue that generally is addressed by the operating company.

10.3 Loss of Circulation

10.3.1 Definition. Lost circulation is defined as the uncontrolled flow of whole mud into a formation, sometimes referred to as thief zone. **Fig. 10.5** shows partial and total lost-circulation zones. In partial lost circulation, mud continues to flow to surface with some loss to the formation. Total lost circulation, however, occurs when all the mud flows into a formation with no return to surface. If drilling continues during total lost circulation, it is referred to as blind drilling. This is not a common practice in the field unless the formation above the thief zone is mechanically stable, there is no production, and the fluid is clear water. Blind drilling also may continue if it is economically feasible and safe.

10.3.2 Lost-Circulation Zones and Causes. Formations that are inherently fractured, cavernous, or have high permeability are potential zones of lost circulation. In addition, under certain improper drilling conditions, induced fractures can become potential zones of lost circulation. The major causes of induced fractures are excessive downhole pressures and setting intermediate casing, especially in the transition zone, too high.

Induced or inherent fractures may be horizontal at shallow depth or vertical at depths greater than approximately 2,500 ft. Excessive wellbore pressures are caused by high flow rates

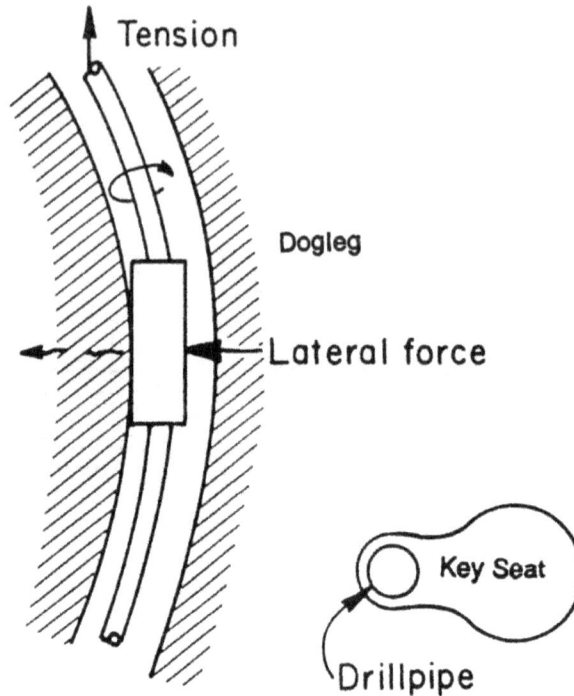

Fig. 10.4—Pipe sticking caused by key seat.

(high annular-friction pressure loss) or tripping in too fast (high surge pressure), which can lead to mud ECD. In addition, improper annular hole cleaning, excessive mud weight, or shutting in a well in high-pressure shallow gas can induce fractures, which can cause lost circulation. Eqs. 10.6 and 10.7 show the conditions that must be maintained to avoid fracturing the formation during drilling and tripping in, respectively.

$$\lambda_{eq} = \lambda_{mh} + \Delta\lambda_{af} < \lambda_{\text{frac}}, \quad\text{...} \quad (10.6)$$

$$\text{and} \quad \lambda_{eq} = \lambda_{mh} + \Delta\lambda_{s} < \lambda_{\text{frac}}, \quad\text{..} \quad (10.7)$$

where λ_{mh} = static mud weight, $\Delta\lambda_{af}$ = additional mud weight caused by friction pressure loss in annulus, $\Delta\lambda_{s}$ = additional mud caused by surge pressure, λ_{frac} = formation-pressure fracture gradient in equivalent mud weight, and λ_{eq} = equivalent circulating density of mud.

Cavernous formations are often limestones with large caverns. This type of lost circulation is quick, total, and the most difficult to seal. High-permeability formations that are potential lost-circulation zones are those of shallow sand with permeability in excess of 10 darcies. Generally, deep sand has low permeability and presents no loss-of-circulation problems. In noncavernous thief zones, mud level in mud tanks decreases gradually and, if drilling continues, total loss of circulation may occur.

10.3.3 Prevention of Lost Circulation. The complete prevention of lost circulation is impossible because some formations, such as inherently fractured, cavernous, or high-permeability zones, are not avoidable if the target zone is to be reached. However, limiting circulation loss is possible if certain precautions are taken, especially those related to induced fractures. These precautions include maintaining proper mud weight, minimizing annular-friction pressure losses

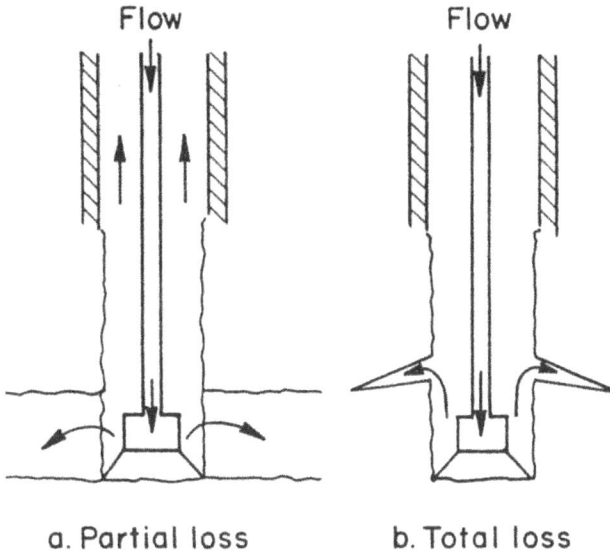

Fig. 10.5—Lost-circulation zones.

during drilling and tripping in, adequate hole cleaning, avoiding restrictions in the annular space, setting casing to protect upper weaker formations within a transition zone, and updating formation pore pressure and fracture gradients for better accuracy with log and drilling data. If lost-circulation zones are anticipated, preventive measures should be taken by treating the mud with lost-circulation materials (LCMs).

10.3.4 Remedial Measures. When lost circulation occurs, sealing the zone is necessary unless the geological conditions allow blind drilling, which is unlikely in most cases. The common LCMs that generally are mixed with the mud to seal loss zones may be grouped as fibrous, flaked, granular, and a combination of fibrous, flaked, and granular materials.

These materials are available in course, medium, and fine grades for an attempt to seal low-to-moderate lost-circulation zones. In the case of severe lost circulations, the use of various plugs to seal the zone becomes mandatory. It is important, however, to know the location of the lost-circulation zone before setting a plug. Various types of plugs used throughout the industry include bentonite/diesel-oil squeeze, cement/bentonite/diesel-oil squeeze, cement, and barite. Squeeze refers to forcing fluid into the lost-circulation zone.

10.4 Hole Deviation

10.4.1 Definition. Hole deviation is the unintentional departure of the drill bit from a preselected borehole trajectory. Whether drilling a straight or curved-hole section, the tendency of the bit to walk away from the desired path can lead to higher drilling costs and lease-boundary legal problems. **Fig. 10.6** provides examples of hole deviations.

10.4.2 Causes. It is not exactly known what causes a drill bit to deviate from its intended path. It is, however, generally agreed that one or a combination of several of the following factors may be responsible for the deviation:
- Heterogeneous nature ·of formation and dip angle.
- Drillstring characteristics, specifically the BHA makeup.
- Stabilizers (location, number, and clearances).

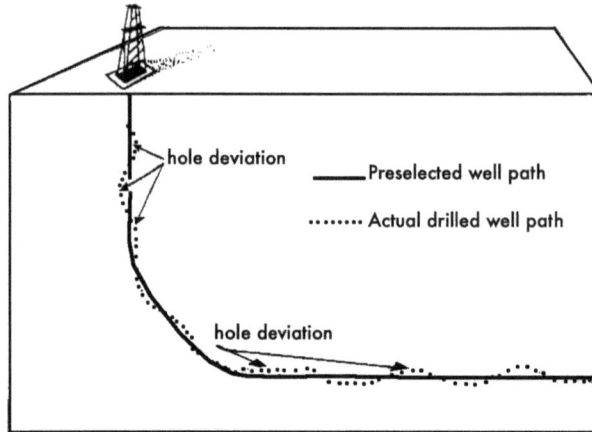

Fig. 10.6—Example of hole deviations.

- Applied weight on bit (WOB).
- Hole-inclination angle from vertical.
- Drill-bit type and its basic mechanical design.
- Hydraulics at the bit.
- Improper hole cleaning.

It is known that some resultant force acting on a drill bit causes hole deviation to occur. The mechanics of this resultant force is complex and is governed mainly by the mechanics of the BHA, rock/bit interaction, bit operating conditions, and, to some lesser extent, by the drilling-fluid hydraulics. The forces imparted to the drill bit because of the BHA are directly related to the makeup of the BHA (i.e., stiffness, stabilizers, and reamers). The BHA is a flexible, elastic structural member that can buckle under compressive loads. The buckled shape of a given designed BHA depends on the amount of applied WOB. The significance of the BHA buckling is that it causes the axis of the drill bit to misalign with the axis of the intended hole path, thus causing the deviation. Pipe stiffness and length and the number of stabilizers (their location and clearances from the wall of the wellbore) are two major parameters that govern BHA buckling behavior. Actions that can minimize the buckling tendency of the BHA include reducing WOB and using stabilizers with outside diameters that are almost in gauge with the wall of the borehole.

The contribution of the rock/bit interaction to bit deviating forces is governed by rock properties (cohesive strength, bedding or dip angle, internal friction angle); drill-bit design features (tooth angle, bit size, bit type, bit offset in case of roller-cone bits, teeth location and number, bit profile, bit hydraulic features); and drilling parameters (tooth penetration into the rock and its cutting mechanism). The mechanics of rock/bit interaction is a very complex subject and is the least understood in regard to hole-deviation problems. Fortunately, the advent of downhole measurement-while-drilling tools that allow monitoring the advance of the drill bit along the desired path makes our lack of understanding of the mechanics of hole deviation more acceptable.

10.5 Drillpipe Failures

Drillpipe failures can be put into one of the following categories: twistoff caused by excessive torque; parting because of excessive tension; burst or collapse because of excessive internal pressure or external pressure, respectively; or fatigue as a result of mechanical cyclic loads with or without corrosion.

10.5.1 Twistoff. Pipe failure as a result of twistoff occurs when the induced shearing stress caused by high torque exceeds the pipe-material ultimate shear stress. In vertical-well drilling, excessive torques are not generally encountered under normal drilling practices. In directional and extended-reach drilling, however, torques in excess of 80,000 lbf-ft are common and easily can cause twistoff to improperly selected drillstring components.

10.5.2 Parting. Pipe-parting failure occurs when the induced tensile stress exceeds the pipe-material ultimate tensile stress. This condition may arise when pipe sticking occurs, and an overpull is applied in addition to the effective weight of suspended pipe in the hole above the stuck point.

10.5.3 Collapse and Burst. Pipe failure as a result of collapse or burst is rare; however, under extreme conditions of high mud weight and complete loss of circulation, pipe burst may occur.

10.5.4 Fatigue. Fatigue is a dynamic phenomenon that may be defined as the initiation of microcracks and their propagation into macrocracks as a result of repeated applications of stresses. It is a process of localized progressive structural fractures in material under the action of dynamic stresses. It is well established that a structural member that may not fail under a single application of static load may very easily fail under the same load if it is applied repeatedly. Failure under cyclic (repeated) loads is called fatigue failure.

Drillstring fatigue failure is the most common and costly type of failure in oil/gas and geothermal drilling operations. The combined action of cyclic stresses and corrosion can shorten the life expectancy of a drillpipe by thousand folds. Cyclic stresses are induced by dynamic loads caused by drillstring vibrations and bending-load reversals in curved sections of hole and doglegs caused by rotation. Pipe corrosion occurs during the presence of O_2, CO_2, chlorides, and/or H_2S. H_2S is the most severely corrosive element to steel pipe, and it is deadly to humans. Regardless of what may have caused pipe failure, the cost of fishing operations and the sometimes unsuccessful attempts to retrieve the fish out of the hole can lead to the loss of millions of dollars in rig downtime, loss of expensive tools downhole, or abandonment of the already-drilled section below the fish.

In spite of the vast amount of work that has been dedicated to pipe fatigue failure, it is still the least understood. This lack of understanding is attributed to the wide variations of statistical data in determining type of service and environment of the drillstring, magnitude of operating loads and frequency of occurrence (load history), accuracy of methods in determining the stresses, quality control during manufacturing, and the applicability of material fatigue data.

10.5.5 Pipe-Failure Prevention. Although pipe failure cannot be eliminated totally, there are certain measures that can be taken to minimize it. Fatigue failures can be mitigated by minimizing induced cyclic stresses and insuring a noncorrosive environment during the drilling operations. Cyclic stresses can be minimized by controlling dogleg severity and drillstring vibrations. Corrosion can be mitigated by corrosive scavengers and controlling the mud pH in the presence of H_2S. The proper handling and inspection of the drillstring on a routine basis are the best measures to prevent failures.

10.6 Borehole Instability

10.6.1 Definition and Causes. Borehole instability is the undesirable condition of an openhole interval that does not maintain its gauge size and shape and/or its structural integrity. The causes can be grouped into the following categories: mechanical failure caused by in-situ stresses, erosion caused by fluid circulation, and chemical caused by interaction of borehole fluid with the formation.

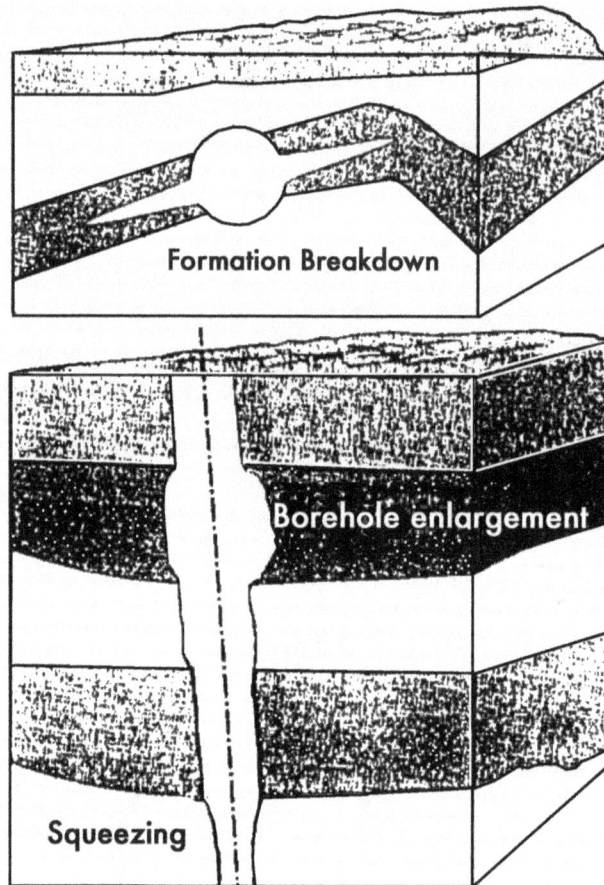

Fig. 10.7—Types of hole instability problems.

10.6.2 Types and Associated Problems. There are four different types of borehole instabilities: hole closure or narrowing, hole enlargement or washouts, fracturing, and collapse. **Fig. 10.7** illustrates hole-instability problems.

Hole Closure. Hole closure is a narrowing time-dependent process of borehole instability. It sometimes is referred to as creep under the overburden pressure, and it generally occurs in plastic-flowing shale and salt sections. Problems associated with hole closure are an increase in torque and drag, an increase in potential pipe sticking, and an increase in the difficulty of casings landing.

Hole Enlargement. Hole enlargements are commonly called washouts because the hole becomes undesirably larger than intended. Hole enlargements are generally caused by hydraulic erosion, mechanical abrasion caused by drillstring, and inherently sloughing shale. The problems associated with hole enlargement are an increase in cementing difficulty, an increase in potential hole deviation, an increase in hydraulic requirements for effective hole cleaning, and an increase in potential problems during logging operations.

Fracturing. Fracturing occurs when the wellbore drilling-fluid pressure exceeds the formation-fracture pressure. The associated problems are lost circulation and possible kick occurrence.

Collapse. Borehole collapse occurs when the drilling-fluid pressure is too low to maintain the structural integrity of the drilled hole. The associated problems are pipe sticking and possible loss of well.

10.6.3 Principles of Borehole Instability.

Before drilling, the rock strength at some depth is in equilibrium with the in-situ rock stresses (effective overburden stress, effective horizontal confining stresses). While a hole is being drilled, however, the balance between the rock strength and the in-situ stresses is disturbed. In addition, foreign fluids are introduced, and an interaction process begins between the formation and borehole fluids. The result is a potential hole-instability problem. Although a vast amount of research has resulted in many borehole-stability simulation models, all share the same shortcoming of uncertainty in the input data needed to run the analysis. Such data include in-situ stresses, pore pressure, rock mechanical properties, and, in the case of shale, formation and drilling-fluids chemistry.

10.6.4 Mechanical Rock-Failure Mechanisms.

Mechanical borehole failure occurs when the stresses acting on the rock exceed the compressive or the tensile strength of the rock. Compressive failure is caused by shear stresses as a result of low mud weight, while tensile failure is caused by normal stresses as a result of excessive mud weight.

The failure criteria that are used to predict hole-instability problems are the maximum-normal-stress criterion for tensile failure and the maximum strain energy of distortion criterion for compressive failure. In the maximum-normal-stress criterion, failure is said to occur when, under the action of combined stresses, one of the acting principal stresses reaches the failure value of the rock tensile strength. In the maximum of energy of distortion criterion, failure is said to occur when, under the action of combined stresses, the energy of distortion reaches the same energy of failure of the rock under pure tension.

10.6.5 Shale Instability.

More than 75% of drilled formations worldwide are shale formations. The drilling cost attributed to shale-instability problems is reported to be in excess of one-half billion U.S dollars per year. The cause of shale instability is two-fold: mechanical (stress change vs. shale strength environment) and chemical (shale/fluid interaction—capillary pressure, osmotic pressure, pressure diffusion, borehole-fluid invasion into shale).

Mechanical Instability. As stated previously, mechanical rock instability can occur because the in-situ stress state of equilibrium has been disturbed after drilling. The mud in use with a certain density may not bring the altered stresses to the original state; therefore, shale may become mechanically unstable.

Chemical Instability. Chemical-induced shale instability is caused by the drilling-fluid/shale interaction, which alters shale mechanical strength as well as the shale pore pressure in the vicinity of the borehole walls. The mechanisms that contribute to this problem include capillary pressure, osmotic pressure, pressure diffusion in the vicinity of the borehole walls, and borehole-fluid invasion into the shale when drilling overbalanced.

Capillary Pressure. During drilling, the mud in the borehole contacts the native pore fluid in the shale through the pore-throat interface. This results in the development of capillary pressure, p_{cap}, which is expressed as

$$p_{cap} = 2\sigma \cos \theta / r, \quad\dots\dots\dots\dots\dots\dots\dots\dots\dots\dots\dots\dots\dots\dots\dots\dots \text{(10.8)}$$

where σ is the interfacial tension, θ is the contact angle between the two fluids, and r is the pore-throat radius. To prevent borehole fluids from entering the shale and stabilizing it, an increase in capillary pressure is required, which can be achieved with oil-based or other organic low-polar mud systems.

Osmotic Pressure. When the energy level or activity in shale pore fluid, a_s, is different from the activity in drilling mud, a_m, water movement can occur in either direction across a semipermeable membrane as a result of the development of osmotic pressure, p_{os}, or chemical potential, μ_c. To prevent or reduce water movement across this semipermeable membrane that has certain efficiency, E_m, the activities need to be equalized or, at least, their differentials minimized. If a_m is lower than a_s, it is suggested to increase E_m and vice versa. The mud activity can be reduced by adding electrolytes that can be brought about through the use of mud systems such as seawater, saturated-salt/polymer, KCl/NaCl/polymer, and lime/gypsum.

Pressure Diffusion. Pressure diffusion is a phenomenon of pressure change near the borehole walls that occurs over time. This pressure change is caused by the compression of the native pore fluid by the borehole-fluid pressure, p_{wfl}, and the osmotic pressure, p_{os}.

Borehole-Fluid Invasion Into Shale. In conventional drilling, a positive differential pressure (the difference between the borehole-fluid pressure and the pore-fluid pressure) is always maintained. As a result, borehole fluid is forced to flow into the formation (fluid-loss phenomenon), which may cause chemical interaction that can lead to shale instabilities. To mitigate this problem, an increase of mud viscosity or, in extreme cases, gilsonite is used to seal off microfractures.

10.6.6 Wellbore-Stability Analysis. Several models in the literature address wellbore-stability analysis.[2] These include very-simple to very-complex models such as linear elastic, nonlinear, elastoplastic, purely mechanical, and physicochemical. Regardless of the model, the data needed include rock properties (Poisson ratio, strength, modulus of elasticity); in-situ stresses (overburden, horizontal); pore-fluid pressure and chemistry; and mud properties and chemistry.

Other than the mud data, the data are often compounded with problems of availability and/ or uncertainties. However, sensitivity analysis can be conducted by assuming data for the many variables to establish safety windows for mud selection and design.

10.6.7 Borehole-Instability Prevention Total prevention of borehole instability is unrealistic because restoring the physical and chemical in-situ conditions of the rock is impossible. However, the drilling engineer can mitigate the problems of borehole instabilities by adhering to good field practices. These practices include proper mud-weight selection and maintenance, the use of proper hydraulics to control the ECD, proper hole-trajectory selection, and the use of borehole fluid compatible with the formation being drilled. Additional field practices that should be followed are minimizing time spent in open hole; using offset-well data (use of the learning curve); monitoring trend changes (torque, circulating pressure, drag, fill-in during tripping); and collaborating and sharing information.

10.7 Mud Contamination

10.7.1 Definition. A mud is said to be contaminated when a foreign material enters the mud system and causes undesirable changes in mud properties, such as density, viscosity, and filtration. Generally, water-based mud systems are the most susceptible to contamination. Mud contamination can result from overtreatment of the mud system with additives or from material entering the mud during drilling.

10.7.2 Common Contaminants, Sources, and Treatments. The most common contaminants to water-based mud systems are solids (added, drilled, active, inert); gypsum/anhydrite (Ca^{++}); cement/lime (Ca^{++}); makeup water (Ca^{++}, Mg^{++}); soluble bicarbonates and carbonates (HCO_3^-, CO_3^{--}); soluble sulfides (HS^-, S^{--}); and salt/salt water flow (Na^+, Cl^-).

Solids Contamination. Solids are materials that are added to make up a mud system (bentonite, barite) and materials that are drilled (active and inert). Excess solids of any type are the most undesirable contaminant to drilling fluids. They affect all mud properties. It has been shown that fine solids, micron and submicron sized, are the most detrimental to the overall drilling efficiency and must be removed if they are not a necessary part of the mud makeup. The removal of drilled solids is achieved through the use of mechanical separating equipment (shakers, desanders, desilters, and centrifuges). Shakers remove solids in the size of cuttings (approximately 140μ or larger). Desanders remove solids in the size of sand (down to 50μ). Desilters remove solids in the size of silt (down to 20μ). When solids become smaller than the cutoff point of desilters, centrifuges may have to be used. Chemical flocculants are sometimes used to flocculate fine solids into a bigger size so that they can be removed by solids-removal equipment. Total flocculants do not discriminate between various types of solids, while selective flocculants will flocculate drilled solids but not the added barite solids. As a last resort, dilution is sometimes used to lower solids concentration.

Calcium-Ions Contamination. The sources of calcium ions are gypsum, anhydrite, cement, lime, seawater, and hard/brackish makeup water. The calcium ion is a major contaminant to freshwater-based sodium-clay treated mud systems. The calcium ion tends to replace the sodium ions on the clay surface through a base exchange, thus causing undesirable changes in mud properties such as rheology and filtration. It also causes added thinners to the mud system to become ineffective. The treatment depends on the source of the calcium ion. For example, sodium carbonate (soda ash) is used if the source is gypsum or anhydrite. Sodium bicarbonate is the preferred treatment if the calcium ion is from lime or cement. If treatment becomes economically unacceptable, break over to a mud system, such as gypsum mud or lime mud, that can tolerate the contaminant.

Bicarbonate and Carbonate Contamination. The contaminant ions (CO_3^{--}, HCO_3^-) are from drilling a CO_2-bearing formation, thermal degradation of organics in mud, or over treatment with soda ash and bicarbonate. These contaminants cause the mud to have high yield and gel strength and a decrease in pH. Treating the mud system with gypsum or lime is recommended.

Hydrogen Sulfide Contamination. The contaminant ions (HS^-, S^{--}) generally are from drilling an H_2S-bearing formation. Hydrogen sulfide is the most deadly ion to humans and is extremely corrosive to steel used during drilling operations. (It causes severe embrittlement to drillpipe.) Scavenging of H_2S is done by use of zinc, copper, or iron.

Salt/Saltwater Flows. The ions, Na^+Cl^-, that enter the mud system as a result of drilling salt sections or from formation saltwater flow cause a mud to have high yield strength, high fluid loss, and pH decrease. Some actions for treatment are dilution with fresh water, the use of dispersants and fluid-loss chemicals, or conversion to a mud that tolerates the problem if the cost of treatment becomes excessive.

10.8 Producing Formation Damage

10.8.1 Introduction. Producing formation damage has been defined as the impairment of the unseen by the inevitable, causing an unknown reduction in the unquantifiable. In a different context, formation damage is defined as the impairment to reservoir (reduced production) caused by wellbore fluids used during drilling/completion and workover operations. It is a zone of reduced permeability within the vicinity of the wellbore (skin) as a result of foreign-fluid invasion into the reservoir rock. **Fig. 10.8** illustrates formation skin damage.

10.8.2 Borehole Fluids. Borehole fluids are classified as drilling fluids, completion fluids, or workover fluids. Drilling fluids are categorized as mud, gas, or gasified mud. There are two types of mud: water-based (pure polymer, pure bentonite, bentonite/polymer) and oil-based (invert emulsion, oil). Completion and workover fluids are mostly brines and are solids free.

10.8.3 Damage Mechanisms. Formation damage is a combination of several mechanisms including solids plugging, clay-particle swelling or dispersion, saturation changes, wettability reversal, emulsion blockage, aqueous-filtrate blockage, and mutual precipitation of soluble salts in wellbore-fluid filtrate and formation water.

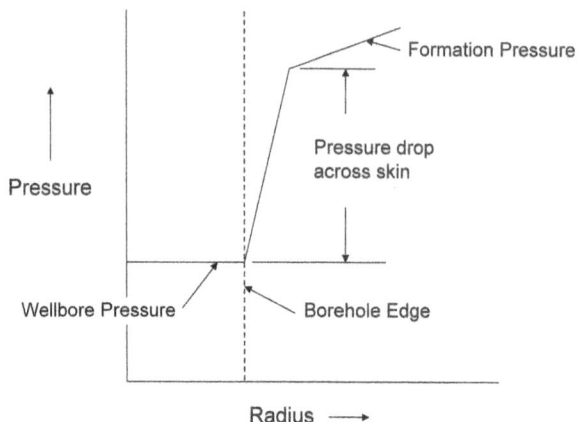

Fig. 10.8—Formation skin damage.

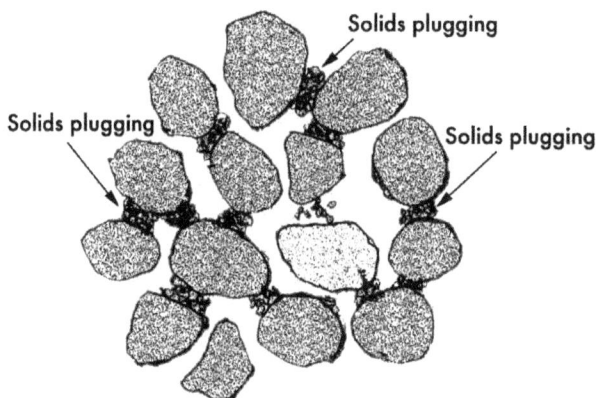

Fig. 10.9—Formation damage caused by solids plugging.

Solids Plugging. **Fig. 10.9** shows that the plugging of the reservoir-rock pore spaces can be caused by the fine solids in the mud filtrate or solids dislodged by the filtrate within the rock matrix. To minimize this form of damage, minimize the amount of fine solids in the mud system and fluid loss.

Clay-Particle Swelling. This is an inherent problem in sandstone that contains water-sensitive clays. When a fresh-water filtrate invades the reservoir rock, it will cause the clay to swell and thus reduce or totally block the throat areas.

Saturation Change. Production is predicated on the amount of saturation within the reservoir rock. When a mud-system filtrate enters the reservoir, it will cause some change in water saturation and, therefore, potential reduction in production. **Fig. 10.10** shows that high fluid loss causes water saturation to increase, which results in a decrease of rock relative permeability. See the chapter on transport properties in the General Engineering volume of this *Handbook* for additional information.

Wettability Reversal. Reservoir rocks are water-wet in nature. It has been demonstrated that while drilling with oil-based mud systems, excess surfactants in the mud filtrate that enter the rock can cause wettability reversal. It has been reported from field experience and demonstrated in laboratory tests that as much as 90% in production loss can be caused by this mecha-

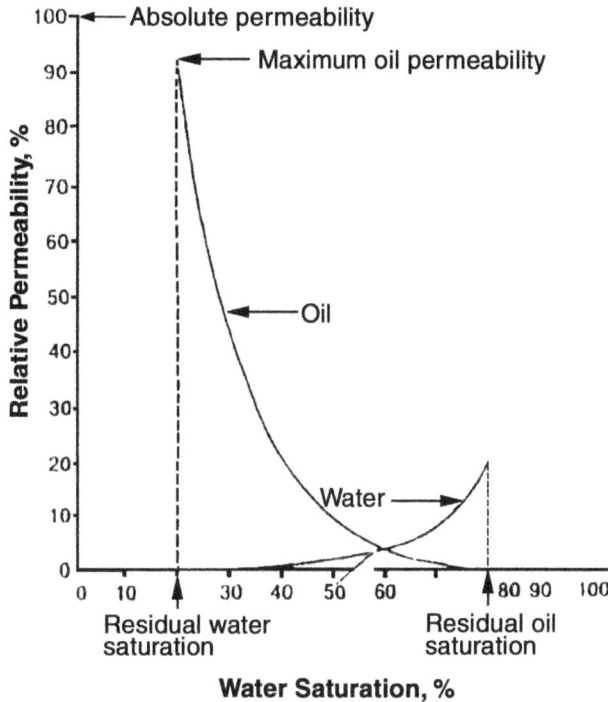

Fig. 10.10—Formation damage caused by saturation.

nism. Therefore, to guard against this problem, the amount of excess surfactants used in oil-based mud systems should be kept at a minimum.

Emulsion Blockage. Inherent in oil-based mud systems is the use of excess surfactants. These surfactants enter the rock and can form an emulsion within the pore spaces, which hinders production through emulsion blockage.

Aqueous-Filtrate Blockage. While drilling with water-based mud, the aqueous filtrate that enters the reservoir can cause some blockage that will reduce the production potential of the reservoir.

Precipitation of Soluble Salts. Any precipitation of soluble salts, whether from the use of salt mud systems or from formation water or both, can cause solids blockage and hinder production. For more information, see the Formation Damage chapter in the Production Operations Engineering volume of this *Handbook*.

10.9 Hole Cleaning

10.9.1 Introduction. Throughout the last decade, many studies have been conducted to gain understanding on hole cleaning in directional-well drilling. Laboratory work has demonstrated that drilling at an inclination angle greater than approximately 30° from vertical poses problems in cuttings removal that are not encountered in vertical wells. **Fig. 10.11** illustrates that the formation of a moving or stationary cuttings bed becomes an apparent problem if the flow rate for a given mud rheology is below a certain critical value.

Inadequate hole cleaning can lead to costly drilling problems such as mechanical pipe sticking, premature bit wear, slow drilling, formation fracturing, excessive torque and drag on drillstring, difficulties in logging and cementing, and difficulties in casings landing. The most

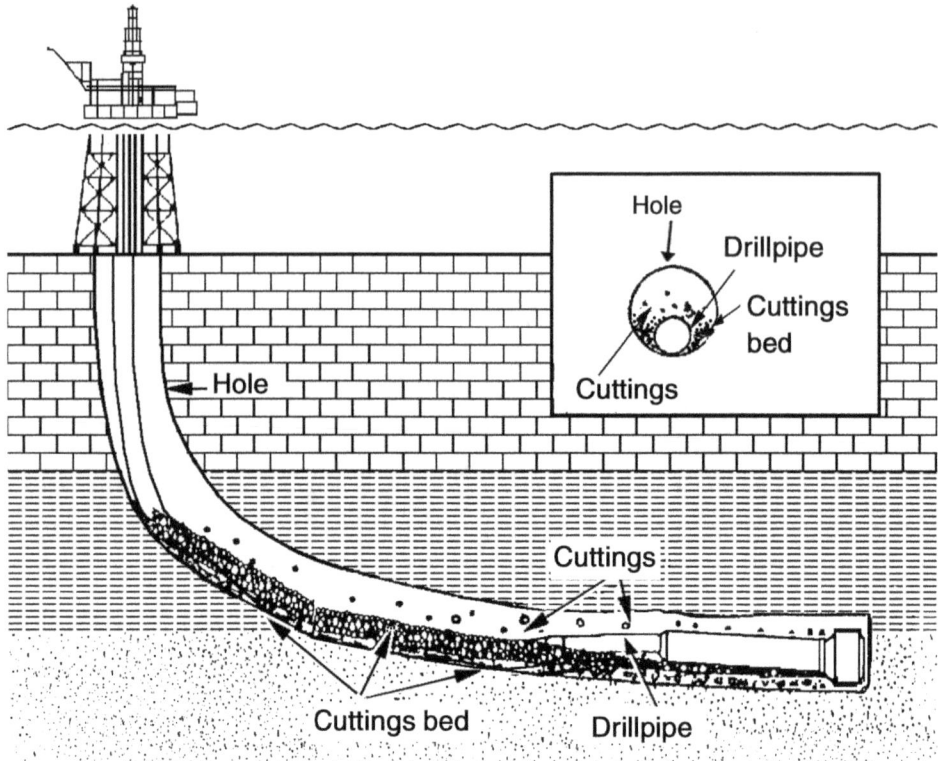

Fig. 10.11—Cuttings-bed buildup in directional wells.

prevalent problem is excessive torque and drag, which often leads to the inability of reaching the target in high-angle/extended-reach drilling.

10.9.2 Factors in Hole Cleaning. *Annular-Fluid Velocity.* Flow rate is the dominant factor in cuttings removal while drilling directional wells. An increase in flow rate will result in more efficient cuttings removal under all conditions. However, how high a flow rate can be increased may be limited by the maximum allowed ECD, the susceptibility of the openhole section to hydraulic erosion, and the availability of rig hydraulic power.

Hole Inclination Angle. Laboratory work has demonstrated that when hole angle increases from zero to approximately 67° from vertical, hole cleaning becomes more difficult, and therefore, flow-rate requirement increases. The flow-rate requirements reach a maximum at approximately 65 to 67° and then slightly decrease toward the horizontal. Also, it has been shown that at 25 to approximately 45°, a sudden pump shutdown can cause cuttings sloughing to bottom and may result in a mechanical pipe-sticking problem. Although, hole inclination can lead to cleaning problems, it is mandated by the needs of drilling inaccessible reservoir, offshore drilling, avoiding troublesome formations, and side tracking and to drill horizontally into the reservoir. Objectives in total field development (primary and secondary production), environmental concerns, and economics are some of the factors that intervene in hole angle selection.

Drillstring Rotation. Laboratory studies have shown and field cases have reported that drillstring rotation has moderate to significant effects in enhancing hole cleaning. The level of enhancement is a combined effect of pipe rotation, mud rheology, cuttings size, flow rate, and, very importantly, the string dynamic behavior. It has been proved that the whirling motion of the string around the wall of the borehole when it rotates is the major contributor to hole cleaning enhancement. Also, mechanical agitation of the cuttings bed on the low side of the hole and exposing the cuttings to higher fluid velocities when the pipe moves to the high side of the hole are results of pipe whirling action.

Although there is a definite gain in hole cleaning caused by pipe rotation, there are certain limitations to its implementation. For example, during angle building with a downhole motor (sliding mode), rotation cannot be induced. With the new steering rotary systems, this is no longer a problem. However, pipe rotation can cause cyclic stresses that can accelerate pipe failures due to fatigue, casing wear, and, in some cases, mechanical destruction to openhole sections. In slimhole drilling, high pipe rotation can cause high ECDs due to the high annular-friction pressure losses.

Hole/Pipe Eccentricity. In the inclined section of the hole, the pipe has the tendency to rest on the low side of the borehole because of gravity. This creates a very narrow gap in the annulus section below the pipe, which causes fluid velocity to be extremely low and, therefore, the inability to transport cuttings to surface. As **Fig. 10.12** illustrates, when eccentricity increases, particle/fluid velocities decrease in the narrow gap, especially for high-viscosity fluid. However, because eccentricity is governed by the selected well trajectory, its adverse impact on hole cleaning may be unavoidable.

Rate of Penetration. Under similar conditions, an increase in the drilling rate always results in an increase in the amount of cuttings in the annulus. To ensure good hole cleaning during high-rate-of-penetration (ROP) drilling, the flow rate and/or pipe rotation have to be adjusted. If the limits of these two variables are exceeded, the only alternative is to reduce the ROP. Although a decrease in ROP may have a detrimental impact on drilling costs, the benefit of avoiding other drilling problems, such as mechanical pipe sticking or excessive torque and drag, can outweigh the loss in ROP.

Mud Properties. The functions of drilling fluids are many and can have unique competing influences. The two mud properties that have direct impact on hole cleaning are viscosity and density. The main functions of density are mechanical borehole stabilization and the prevention of formation-fluid intrusion into the annulus. Any unnecessary increase in mud density beyond fulfilling these functions will have an adverse effect on the ROP and, under the given in-situ stresses, may cause fracturing of the formation. Mud density should not be used as a criterion to enhance hole cleaning.

Viscosity, on the other hand, has the primary function of the suspension of added desired weighting materials such as barite. Only in vertical-well drilling and high-viscosity pill sweep is viscosity used as a remedy in hole cleaning.

Cuttings Characteristics. The size, distribution, shape, and specific gravity of cuttings affect their dynamic behavior in a flowing media. The specific gravity of most rocks is approximately 2.6; therefore, specific gravity can be considered a nonvarying factor in cuttings transport. The cuttings size and shape are functions of the bit types (roller cone, polycrystalline-diamond compact, diamond matrix), the regrinding that takes place after they are generated, and the breakage by their own bombardment and with the rotating drillstring. It is impossible to control their size and shape even if a specific bit group has been selected to generate them. Smaller cuttings are more difficult to transport in directional-well drilling; however, with some viscosity increase and pipe rotation, fine particles seem to stay in suspension and, therefore, are easier to transport.

10.10 Hydrogen-Sulfide-Bearing Zones and Shallow Gas

Drilling H_2S-bearing formations poses one of the most difficult and dangerous problems to humans and equipment. If it is known or anticipated, there are very specific requirements to abide by in accordance with Intl. Assn. of Drilling Contractors rules and regulations. Shallow gas may be encountered at any time in any region of the world. The only way to combat this problem is to never shut in the well; divert the gas flow through a diverter system instead. High-

Fluid velocity profile   **Fluid velocity profile**   **Fluid velocity profile**

Concentric annulus   25% Eccentric   75% Eccentric

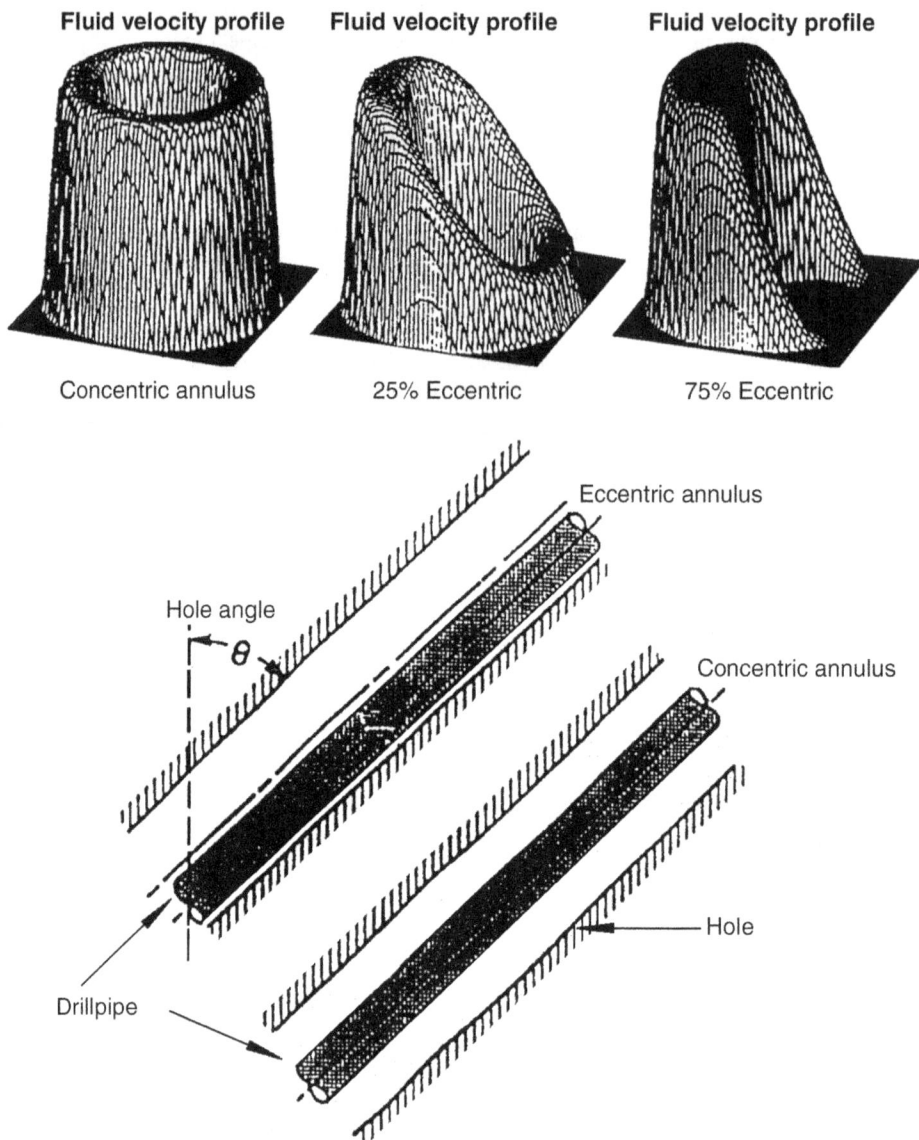

Fig. 10.12—Fluid velocity profile in eccentric annulus (after Hzouz et al.[3]).

pressure shallow gas can be encountered at depths as low as a few hundred feet where the formation-fracture gradient is very low. The danger is that if the well is shut in, formation fracturing is more likely to occur, which will result in the most severe blowout problem, underground blow.

10.11 Equipment and Personnel-Related Problems

10.11.1 Equipment. The integrity of drilling equipment and its maintenance are major factors in minimizing drilling problems. Proper rig hydraulics (pump power) for efficient bottom and annular hole cleaning, proper hoisting power for efficient tripping out, proper derrick design loads and drilling line tension load to allow safe overpull in case of a sticking problem, and well-control systems (ram preventers, annular preventers, internal preventers) that allow kick control under any kick situation are all necessary for reducing drilling problems. Proper monitoring and recording systems that monitor trend changes in all drilling parameters and can retrieve drilling data at a later date, proper tubular hardware specifically suited to accommodate all anticipated drilling conditions, and effective mud-handling and maintenance equipment that will ensure that the mud properties are designed for their intended functions are also necessary.

10.11.2 Personnel. Given equal conditions during drilling/completion operations, personnel are the key to the success or failure of those operations. Overall well costs as a result of any drilling/completion problem can be extremely high; therefore, continuing education and training for personnel directly or indirectly involved is essential to successful drilling/completion practices.

Nomenclature

a_m = activity in drilling mud, dimensionless

a_s = activity in shale pore fluid, dimensionless

A_c = area of contact, L^2, in.^{2}

D_h = diameter of the hole, L, in.

D_{op} = outside diameter of the pipe, L, in.

E_m = efficiency, dimensionless

f = coefficient of friction, dimensionless

F_l = lateral force, F, lbf

F_p = pull force, F, lbf

h_{mc} = mudcake thickness, L, in.

L_{ep} = length of the permeable zone, L, in.

p_{cap} = capillary pressure, F/L^2, psi

p_{ff} = formation-fluid pressure, F/L^2, psi

p_m = mud pressure, F/L^2, psi

p_{os} = osmotic pressure, F/L^2, psi

r = pore-throat radius, L, in.

T = tension in the drillstring just above the key-seat area, F, lbf

Δp = differential pressure, F/L^2, psi

$\Delta\lambda_{af}$ = additional mud weight caused by friction pressure loss in annulus, F/L^3, lbm/gal

$\Delta\lambda_s$ = additional mud weight caused by surge pressure, F/L^3, lbm/gal

θ = contact angle between the two fluids, degrees

θ_{dl} = abrupt change in hole angle, degrees

λ_{eq} = equivalent mud circulating density, F/L^3, lbm/gal

λ_{frac} = formation-pressure fracture gradient in equivalent mud weight, F/L^3, lbm/gal

λ_{mh} = static mud weight, F/L^3, lbm/gal

μ_c = chemical potential, dimensionless

σ = interfacial tension, F/L, lbf/in.

References

1. Bourgoyne, A.T. *et al.: Applied Drilling Engineering,* Textbook Series, SPE, Richardson, Texas (1986).
2. McLean, M.R. and Addis, M.A.: "Wellbore Stability Analysis: A Review of Current Methods and Their Field Applications," paper SPE 19947 presented at the 1990 IADC/SPE Drilling Conference, Houston, 27 February–2 March.

3. Hzouz, I. *et al.:* "Numerical Simulation of Laminar Flow of Yield-Power-Law Fluids in Conduits of Arbitrary Cross-Section," *Trans. of ASME* (December 1993) 710.

General References

Aadnoy, B.S.: "Modeling of the Stability of Highly Inclined Boreholes in Anisotropic Rock Formations," *SPEDE* (1988) 259.

Aadnoy, B.S.: *Modern Well Design,* Gulf Publishing Co., Houston (1997).

Abrams, A.: "Mud Design To Minimize Rock Impairment Due To Particle Invasion," *JPT* (May 1977) 586.

Annis, M.R. and Monoghan, P.H.: "Differential Pressure Sticking—Laboratory Studies of Friction Between Steel and Mud Filter Cake," *Petroleum Trans.* (May 1962).

Azar, J.J.: *Drilling Engineering,* U. of Tulsa Petroleum Engineering, Tulsa (1978).

Azar, J.J. and Lummus, J.L.: "The Effect of Drilling Fluid pH on Drill Pipe Corrosion Fatigue Performance," paper SPE 5516 presented at 1975 SPE Annual Technical Conference and Exhibition, Dallas, 28 September–1 October.

Azar, J.J. and Sanches, R.A.: "Important Issues in Cuttings Transport for Drilling Directional Wells," paper SPE 39020 presented at the 1997 SPE Latin American and Caribbean Petroleum Engineering Conference and Exhibition, Rio de Janeiro, 30 August–3 September.

Becker, T.E, Azar, J.J., and Okrajni, S.S.: "Correlations of Mud Rheological Properties With Cuttings-Transport Performance in Directional Drilling," *SPEDE* (March 1991) 16.

Billingston, S.A.: "Practical Approach to Circulation Problems," *Drilling Contractor* (July–August 1963) 52.

Bourgoyne, A.T. Jr., Caudle, B.H., and Kimbler, O.K.: "The Effect of Interfacial Films on the Displacement of Oil by Water in Porous Media," *SPEJ* (February 1972) 60.

Bradley, W.B.: "Deviation Forces from the Wedge Penetration Failure of Anisotropic Rock," *ASME Trans.,* (1974) **95,** Series B, No. 4, 1093.

Bradley, W.B.: "Factors Affecting the Control of Borehole Angle in Straight and Directional Wells," *JPT* (June 1975) 679.

Bradley, W.B.: "Failures of Inclined Boreholes," *J. of Energy Resources Tech.* (December 1979) 232.

Bradley, W.B. *et al.:* "A Task Force Approach to Reducing Stuck Pipe Costs," paper SPE 21999 presented at the 1991 SPE/IADC Drilling Conference, Amsterdam, 11–14 March.

Brakel, J.D. and Azar, J.J.: "Prediction of Wellbore Trajectory Considering Bottomhole Assembly and Drill-Bit Dynamics," *SPEDE* (June 1989) 109.

Cagle W.S. and Mathews, H.D.: "An Improved Lost Circulation Slurry Squeeze," *Petroleum Engineer* (July 1977) 26.

Chenevert, M.E.: "Shale Alteration By Water Adsorption," *JPT* (1970) 1141.

Civan, F., Knapp, R.M., and Henry, O.A.: "Alteration of Permeability by Fine Particle Processes," *J. of Petroleum Science and Engineering* (1989).

Clancy, L.W. and Boudreau, M. Jr.: "High-Water Loss High-Solids Slurry Stops Lost Circulation with Oil Mud," *Oil & Gas J.* (January 1981) 99.

Clark, D.A.: "An Experimental Investigation of the Mechanics of Bit-Teeth Rock Interaction," MS thesis, U. of Tulsa, Tulsa (1982).

da Fontoura, S.A.B. and dos Santos, H.M.R.: "In-Situ Stresses, Mud Weight and Modes of Failure Around Oil Wells," *Proc.,* Fifteenth Canadian Rock Mechanics Symposium (1989) 235.

Detournay, E. and Cheng, A.H.: "Poroelastic Response of a Borehole in a Non-hydrostatic Stress Field," *Intl. J. of Rock Mechanics, Mineral Science and Geomechanics* (1988) No. 25, 171.

Dunbar, M.E., Warren, T.M., and Kadaster, A.G.: "Bit Sticking Caused by Borehole Deformation," *SPEDE* (December 1986) 417.

Dykstra, M.W. *et al.:* "Drillstring Component Mass Imbalance: A Major Source of Downhole Vibrations," *SPEDC* (December 1996) 234.

Gill, J.A.: "How Borehole Ballooning Alters Drilling Responses," *Oil & Gas J.* (March 1989) 43.

Gnirk, P.F.: " The Mechanical Behavior of Uncased Wellbores Situated in Elastic/Plastic Media Under Hydrostatic Stress," *SPEJ* (February 1972) 49.

Goins, W.C. Jr.: "How to Combat Circulation Loss," *Oil & Gas J* (June 1952) 71.

Hale, A.H., Mody, F.K., and Salisbury, D.P.: "The Influence of Chemical Potential on Wellbore Stability," *SPEDC* (1993) 207.

Hanshford, J.E. and Lubinski, A.: "Cumulative Fatigue Damage of Drill Pipe in Dog-Legs," *JPT* (March 1966) 359.

Helmick, W.E. and Longley, A.J.: "Pressure Differential Sticking of Drill Pipe," *Oil & Gas J.* (June 1957) 132.

Hempkins, W.B. *et al.:* "Multivariate Statistical Analysis of Stuck Drill Pipe Situations," *SPEDE* (September 1987) 237.

Howard, G.C. and Scott, P.P. Jr.: "An Analysis of Lost Circulation," *Trans.,* AIME (1951) 171.

Hsiao, C.: "A Study of Horizontal Wellbore Failure," *SPEPE* (1988) 489.

Hussaini, S.M. and Azar, J.J.: "Experimental Study of Drilled Cuttings Transport Using Common Drilling Muds," *SPEJ* (February 1983) 11.

Krueger, R.F.: "An Overview of Formation Damage and Well Productivity in Oilfield Operations," *JPT* (February 1986) 131.

Larsen, T.I., Pilehvari, A., and Azar, J.J.: "Development of New Model For Cleaning High Angle Wells Including Horizontal," *SPEDE* (May 1997) 98.

Lubinski, A.: "Maximum Permissible Dog-legs in Rotary Boreholes," *JPT* (February 1961) 194.

Lubinski, A. and Woods, H.B.: "Factors Affecting the Angle of Inclination and Doglegging in Rotary Boreholes," *Drill. & Prod. Prac.* (1953), 222

Lummus, J. and Azar, J.J.: *Drilling Fluids Optimization—A Practical Field Approach,* PennWell Publishing Co., Tulsa (1986).

Ma, D. and Azar, J.J.: "A Study of Rock Bit Interaction and Wellbore Deviation," *ASME J. of Energy Resources Tech.* (September 1986) 228.

Ma, D. and Azar, J.J.: "Dynamics of Roller Cone Bits," *ASME J. of Energy Resources Tech.* (December 1985) **107,** 543.

McKinney, L.K.: "Formation Damage Due to Oil-Base Drilling Fluids at Elevated Temperature and Pressure," MS thesis, U. of Tulsa, Tulsa (1986).

McLamore, R.T.: "The Role of Rock Strength Anisotropy in Natural Hole Deviation," *JPT* (November 1971) 1313.

Miller, T.W. and Cheatham, J.B. Jr.: "Rock/Bit Tooth Interaction for Conical Bit Teeth," *SPEJ* (June 1971) 162.

Moore, P.L.: *Drilling Practices Manual,* second edition, PennWell Publishing Co., Tulsa (1986).

Morita, N. and Gray, K.E.: "A Constitutive Equation for Nonlinear Stress-Strain Curves in Rocks and Its Application to Stress Analysis Around a Borehole During Drilling," paper SPE 9328 presented at the 1980 SPE Annual Technical Conference and Exhibition, Dallas, 21–24 September.

Muecke, T.W.: "Formation Fines and Factors Controlling Their Movement in Porous Media," *JPT* (February 1979) 144.

Nicholson, R.W.: "Acceptable Dogleg Severity Limits," *Oil & Gas J.* (April 1974) 73.

Nyland, T. *et al.:* "Additive Effectiveness and Contaminant Influence on Fluid-Loss Control in Water-Based Muds," *SPEDE* (June 1988) 195.

O'Brien, D.E. and Chenevert, M.E.: "Stabilizing Sensitive Shales With Inhibited, Potassium-Based Drilling Fluids," *JPT* (September 1973) 1089.

Okrajni, S.S. and Azar, J.J.: "The Effect of Mud Rheology on Annular Hole Cleaning in Directional Wells," *SPEDE* (August 1986) 297.

Outmans, H.D.: "Mechanics of Differential Pressure Sticking of Drill Collars," *Petroleum Trans.* (1958) **213**, No. 8, 265.

Peterson, C.R.: "Roller Cutter Forces," *SPEJ* (March 1970) 57.

Placido, J.C. *et al.:* "Drillpipe Fatigue Life Prediction Model Based on Critical Plane Approach," paper OTC 7569 presented at the 1994 Offshore Technology Conference, Houston, 2–5 May.

Rabia, H.: *Oil Well Drilling Engineering,* Graham and Trotman Ltd., London (1985).

Roegiers, J.C. and Detournay, E.: "Consideration on Failure Initiation in Inclined Boreholes," *Key Questions in Rock Mechanics,* Cundall (ed.), Balkema, Rotterdam, The Netherlands (1988) 461.

Rollins, H.M.: "Drill Stem Failures Due to H_2S," *Oil & Gas J.* (January 1966).

Sanchez, R.A. *et al.:* "Effect of Drillpipe Rotation on Hole Cleaning During Directional-Well Drilling," *SPEJ* (June 1999) 101.

Sanner, D.: "Effect of Drilling Fluid Filtrates on Flow Properties of Various Rocks," MS thesis, U. of Tulsa, Tulsa (1989).

Sharma, M.M. and Wunderlich, R.W.: "The Alteration of Rock Properties Due to Interactions with Drilling Fluid Components," *J. of Petroleum Science and Engineering* (1987) 127.

Storli, C.: "Formation Damage of Primary and Secondary Emulsifiers and Amine Compounds at Elevated Temperatures Using Various Concentrations," MS thesis, U. of Tulsa, Tulsa (1987).

Thomas, R.P., Azar, J.J., and Becker, T.E.: "Drillpipe Eccentricity Effect on Drilled Cuttings Behavior in Vertical Wells," *JPT* (September 1982); *Trans.,* AIME, 273.

Tomren, P.H., Iyoho, A.W., and Azar, J.J.: "An Experimental Study of Cuttings Transport in Directional Wells," *SPEDE* (February 1986) 43.

Tovar, J.: "Formation Damage Studies Using Whole Drilling Muds in Simulated Boreholes," MS thesis, U. of Tulsa, Tulsa (1990).

Veeken, C.A.M. *et al.:* "Use of Plasticity Models for Predicting Borehole Stability," *Proc.,* Intl. Symposium IRSM-SPE, Pau, France (August 1989) 835.

Wolfson, L.: "Three Dimensional Analysis of Constrained Directional Drilling Assemblies in a Curved Hole," MS thesis, U. of Tulsa, Tulsa (1974).

Woods, H.B. and Lubinski, A.: "Use of Stabilizers in Controlling Hole Deviation," *Drill. & Prod. Prac.* (1955) 165.

Zoback, M.D. *et al.:* "Well Bore Breakouts and in Situ Stress," *J. of Geophysical Research* (June 1985) 5523.

SI Metric Conversion Factors

ft $\times$ 3.048*	E $-$ 01	= m
gal $\times$ 3.785 412	E $-$ 03	= m^3
in. $\times$ 2.54*	E $+$ 00	= cm
in.^{2} $\times$ 6.451 6*	E $+$ 00	= cm^2
lbf $\times$ 4.448 222	E $+$ 00	= N
lbm $\times$ 4.535 924	E $-$ 01	= kg
psi $\times$ 6.894 757	E $+$ 00	= kPa

*Conversion factor is exact.

Chapter 11
Introduction to Well Planning
Neal Adams, Neal Adams Services

Well planning is perhaps the most demanding aspect of drilling engineering. It requires the integration of engineering principles, corporate or personal philosophies, and experience factors. Although well planning methods and practices may vary within the drilling industry, the end result should be a safely drilled, minimum-cost hole that satisfies the reservoir engineer's requirements for oil/gas production.

The skilled well planners normally have three common traits. They are experienced drilling personnel who understand how all aspects of the drilling operation must be integrated smoothly. They utilize available engineering tools, such as computers and third-party recommendations, to guide the development of the well plan. And they usually have an investigative characteristic that drives them to research and review every aspect of the plan in an effort to isolate and remove potential problem areas.

11.1 Well Planning

11.1.1 Objective. The objective of well planning is to formulate from many variables a program for drilling a well that has the following characteristics: safe, minimum cost, and usable. Unfortunately, it is not always possible to accomplish these objectives on each well because of constraints based on geology, drilling equipment, temperature, casing limitations, hole sizing, or budget.

Safety. Safety should be the highest priority in well planning. Personnel considerations must be placed above all other aspects of the plan. In some cases, the plan must be altered during the course of drilling the well when unforeseen drilling problems endanger the crew. Failure to stress crew safety has resulted in loss of life and burned or permanently crippled individuals.

The second priority involves the safety of the well. The well plan must be designed to minimize the risk of blowouts and other factors that could create problems. This design requirement must be adhered to rigorously in all aspects of the plan. Example 11.1 illustrates a case in which this consideration was neglected in the earliest phase of well planning, which is data collection.

Example 11.1 A turnkey drilling contractor began drilling a 9,000-ft well in September 1979. The well was in a high-activity area where 52 wells had been drilled previously in a township (approximately 36 sq miles). The contractor was reputable and had a successful history.

The drilling superintendent called a bit company and obtained records on two wells in the section where the prospect well was to be drilled. Although the records were approximately 15 years old, it appeared that the formation pressures would be normal to a depth of 9,800 ft. Because the prospect well was to be drilled to 9,000 ft, pressure problems were not anticipated. The contractor elected to set 10¾-in. casing to 1,800 ft and use a 9.5-lbm/gal mud to 9,000 ft in a 9⅞-in. hole. At that point, responsibility would be turned over to the oil company.

Drilling was uneventful until a depth of 8,750 ft was reached. At that point, a severe kick was taken. An underground blowout occurred that soon erupted into a surface blowout. The rig was destroyed and natural resources were lost until the well was killed three weeks later.

A study was conducted that yielded the following results:

• All wells in the area appeared to be normal pressured until 9,800 ft.

• However, 4 of the 52 wells in the specific township and range had blown out in the past five years. It appeared that the blowouts came from the same zone as the well in question.

• A total of 16 of the remaining 48 wells had taken kicks or severe gas cutting from the same zone.

• All problems appeared to occur after a 1973 blowout taken from a 12,200-ft abnormal-pressure zone.

Conclusion.

• The drilling contractor did not research the surrounding wells thoroughly in an effort to detect problems that could endanger his well or crews.

• The final settlement by the insurance company was more than U.S. $16 million. The incident probably would not have occurred if the contractor had spent U.S. $800 to $1,000 to obtain proper drilling data.

Minimum Cost. A valid objective of the well-planning process is to minimize the cost of the well without jeopardizing the safety aspects. In most cases, costs can be reduced to a certain level as additional effort is given to the planning (**Fig. 11.1**). It is not noble to build "steel monuments" in the name of safety if the additional expense is not required. On the other hand, funds should be spent as necessary to develop a safe system.

Usable Holes. Drilling a hole to the target depth is unsatisfactory if the final well configuration is not usable. In this case, the term "usable" implies the following:

• The hole diameter is sufficiently large so an adequate completion can be made.

• The hole or producing formation is not irreparably damaged.

This requirement of the well planning process can be difficult to achieve in abnormal-pressure, deep zones that can cause hole-geometry or mud problems.

11.1.2 Well-Type Classification. The drilling engineer is required to plan a variety of well types, including: wildcats, exploratory holes, step-outs, infills, and re-entries. Generally, wildcats require more planning than the other types. Infill wells and re-entries require minimum planning in most cases.

Wildcats are drilled where little or no known geological information is available. The site may have been selected because of wells drilled some distance from the proposed location but on a terrain that appeared similar to the proposed site. The term "wildcatter" was originated to describe the bold frontiersman willing to gamble on a hunch.

Rank wildcats are seldom drilled in today's industry. Well costs are so high that gambling on wellsite selection is not done in most cases. In addition, numerous drilling prospects with

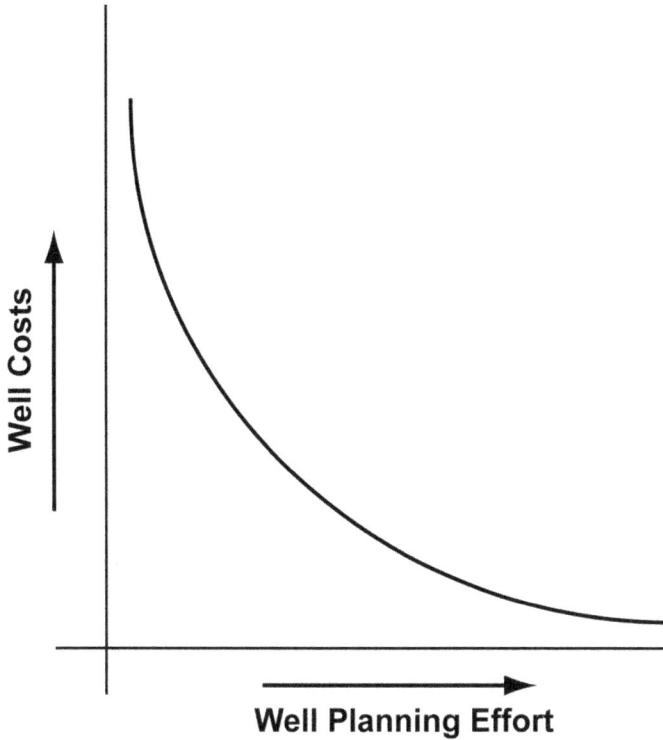

Fig. 11.1—Well costs can be reduced dramatically if proper well planning is implemented.

TABLE 11.1—WELL-TYPE CHARACTERISTICS	
Well Type	Characteristics
Wildcat	No known (or little) geological foundation for site selection.
Exploratory	Site selection based on seismic data, satellite surveys, etc.; no drilling data in the prospective horizon.
Step-out	Delineates the reservoir's boundaries; drilled after the exploratory discovery(s); site selection usually based on seismic data. Also, known as delineation well.
Infill	Infill drills the known productive portions of the reservoir; site selection usually based on patterns, drainage radius, etc.
Re-entry	Existing well re-entered to deepen, sidetrack, rework, or recomplete; various amounts of planning required, depending on purpose of re-entry.

reasonable productive potential are available from several sources. However, the romantic legend of the wildcatter will probably never die. Characteristics of various well types are shown in **Table 11.1**.

11.1.3 Formation Pressure. The formation, or pore, pressure encountered by the well significantly affects the well plan. The pressures may be normal, abnormal (high), or subnormal (low).

Normal-pressure wells generally do not create planning problems. The mud weights are in the range of 8.5 to 9.5 lbm/gal. Kicks- and blowout-prevention problems should be minimized but not eliminated altogether. Casing requirements can be stringent even in normal-pressure wells deeper than 20,000 ft because of tension/collapse design constraints.

Subnormal-pressure wells may require setting additional casing strings to cover weak or low-pressure zones. The lower-than-normal pressures may result from geological or tectonic factors or from pressure depletion in producing intervals. The design considerations can be demanding if other sections of the well are abnormal pressured.

Abnormal pressures affect the well plan in many areas, including: casing and tubing design, mud-weight and type-selection, casing-setting-depth selection, and cement planning. In addition, the following problems must be considered as a result of high formation pressures: kicks and blowouts, differential-pressure pipe sticking, lost circulation resulting from high mud weights, and heaving shale. Well costs increase significantly with geopressures.

Because of the difficulties associated with well planning for high-pressure exploratory wells, many design criteria, publications, and studies have been devoted to this area. The amount of effort expended is justified. Unfortunately, the drilling engineer still must define the planning parameters that can be relaxed or modified when drilling normal-pressure holes or well types such as step-outs or infills.

11.1.4 Planning Costs. The costs required to plan a well properly are insignificant in comparison to the actual drilling costs. In many cases, less than U.S. $1,000 is spent in planning a U.S. $1 million well. This represents 1/10 of 1% of the well costs.

Unfortunately, many historical instances can be used to demonstrate that well planning costs were sacrificed or avoided in an effort to be cost conscious. The end result often is a final well cost that exceeds the amount required to drill the well if proper planning had been exercised. Perhaps the most common attempted shortcut is to minimize data-collection work. Although good data can normally be obtained for small sums, many well plans are generated without the knowledge of possible drilling problems. This lack of expenditure in the early stages of the planning process generally results in higher-than-anticipated drilling costs.

11.1.5 Overview of the Planning Process. Well planning is an orderly process. It requires that some aspects of the plan be developed before designing other items. For example, the mud density plan must be developed before the casing program because mud weights have an impact on pipe requirements (**Fig. 11.2**).

Bit programming can be done at any time in the plan after the historical data have been analyzed. The bit program is usually based on drilling parameters from offset wells. However, bit selection can be affected by the mud plan [i.e., the performance of polycrystalline-diamond (PCD) bits in oil muds]. Casing-drift-diameter requirements may control bit sizing.

Casing and tubing should be considered as an integral design. This fact is particularly valid for production casing. A design criterion for tubing is the drift diameter of the production casing, whereas the packer-to-tubing forces created by the tubing's tendencies for movement can adversely affect the production casing. Unfortunately, these calculations are complex and often neglected.

The completion plan must be visualized reasonably early in the process. Its primary effect is on the size of casing and tubing to be used if oversized tubing or packers are required. In addition, the plan can require the use of high-strength tubing or unusually long seal assemblies in certain situations.

```
┌─────────────────────────────┐
│    Prospect development      │
└─────────────┬───────────────┘
              ▼
┌─────────────────────────────┐
│      Data collection         │
└─────────────┬───────────────┘
              ▼
┌─────────────────────────────┐
│    Pore-pressure analysis    │
└─────────────┬───────────────┘
              ▼
┌─────────────────────────────┐
│     Fracture prediction      │
└─────────────┬───────────────┘
              ▼
┌─────────────────────────────┐
│ Pipe setting-depth selection │
└─────────────┬───────────────┘
              ▼
┌─────────────────────────────┐
│   Hole-geometry selection    │
└─────────────┬───────────────┘
              ▼
┌─────────────────────────────┐
│     Completion planning      │
└─────────────┬───────────────┘
              ▼
┌─────────────────────────────┐
│          Mud plan            │
└─────────────┬───────────────┘
              ▼
┌─────────────────────────────┐
│        Cement plan           │
└─────────────┬───────────────┘
┌──────────────┐ ┄┄┄┄┄►        │
│ Bit program  │               ▼
└──────────────┘ ┌─────────────────────────────┐      ┌──────────────┐
                 │       Casing design          │─────►│  Drill time  │
                 └─────────────┬───────────────┘      │  projections │
                               ▼                       └──────┬───────┘
                 ┌─────────────────────────────┐             ▼
                 │       Tubing design          │      ┌──────────────┐
                 └─────────────┬───────────────┘      │ Cost estimation│
                               ▼                       └──────────────┘
                 ┌─────────────────────────────┐
                 │      Drillstring design      │
                 └─────────────┬───────────────┘
                               ▼
                 ┌─────────────────────────────┐
                 │    Rig sizing and selection  │
                 └─────────────────────────────┘
```

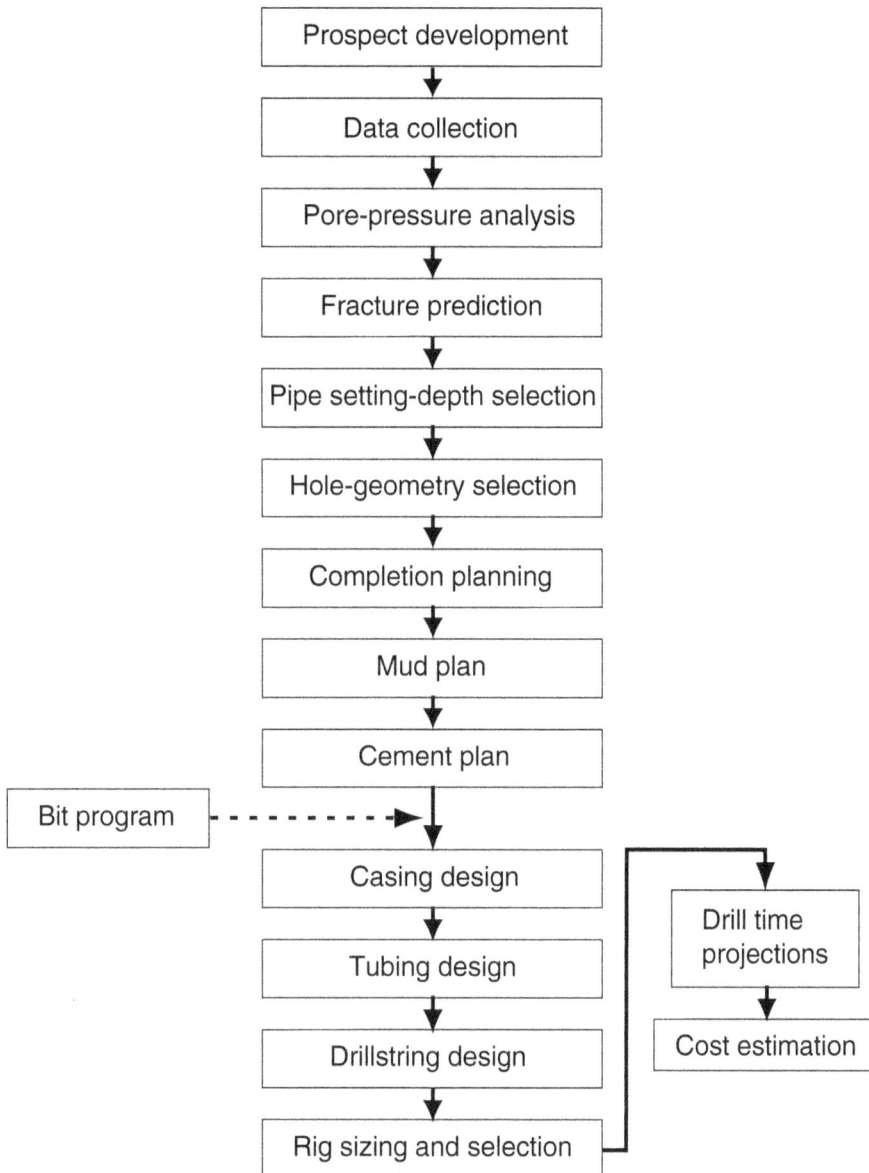

Fig. 11.2—Flow path for well planning.

Fig. 11.2 defines an orderly process for well planning. This process must be altered for various cases. The flow path in this illustration will be followed, for the most part, throughout this text.

11.2 Data Collection

The most important aspect of preparing the well plan, and subsequent drilling engineering, is determining the expected characteristics and problems to be encountered in the well. A well cannot be planned properly if these environments are unknown. Therefore, the drilling engineer must initially pursue various types of data to gain insight used to develop the projected drilling conditions.

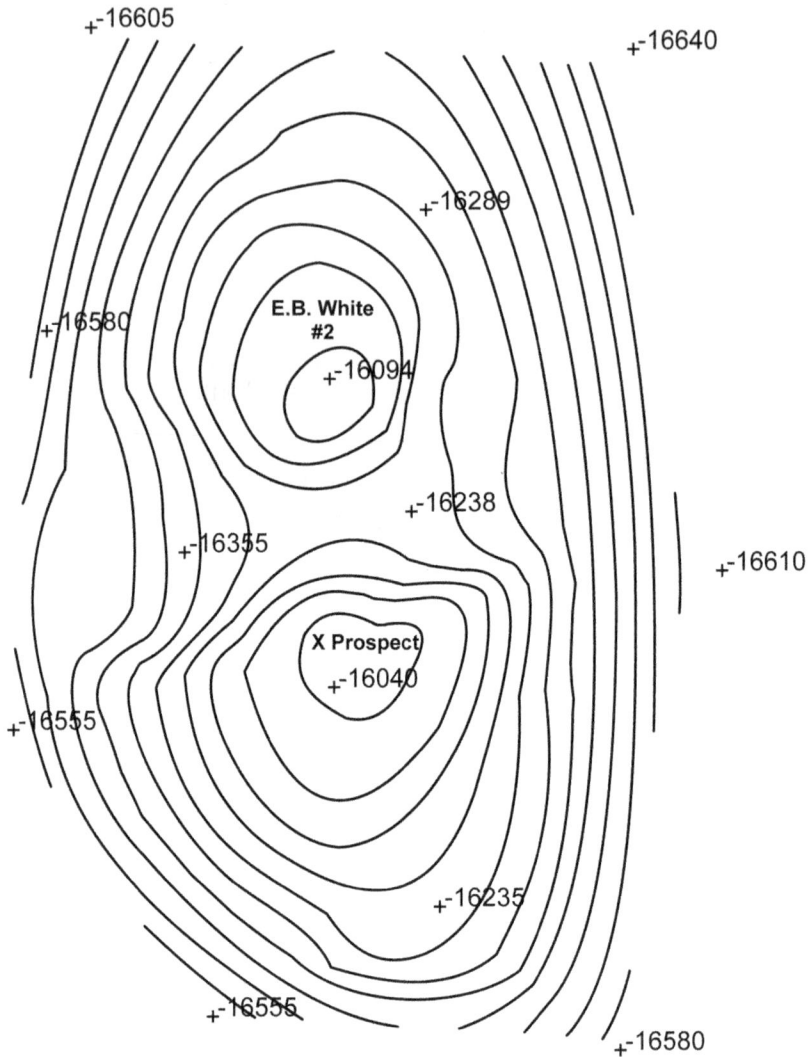

Fig. 11.3—Contour map.

11.2.1 Offset-Well Selection. The drilling engineer is usually not responsible for selecting well sites. However, he must work with the geologist for the following reasons:

• Develop an understanding of the expected drilling geology.

• Define fault-block structures to help select offset wells similar in nature to the prospect well.

• Identify geological anomalies as they may be encountered in drilling the prospect well.

A close working relationship between drilling and geology groups can be the difference between a producer and an abandoned well.

An example of geological information that the drilling group may receive is shown in **Fig. 11.3**. The geologists have found significant production from E.B. White #2. Contouring the pay zones produces the map in Fig. 11.3. The prospect well should encounter the producing structure at the approximate depth as the E.B. White #2.

Maps showing the surface location of offset wells are available from commercial cartographers (**Fig. 11.4**). These maps normally provide the well location relative to other wells,

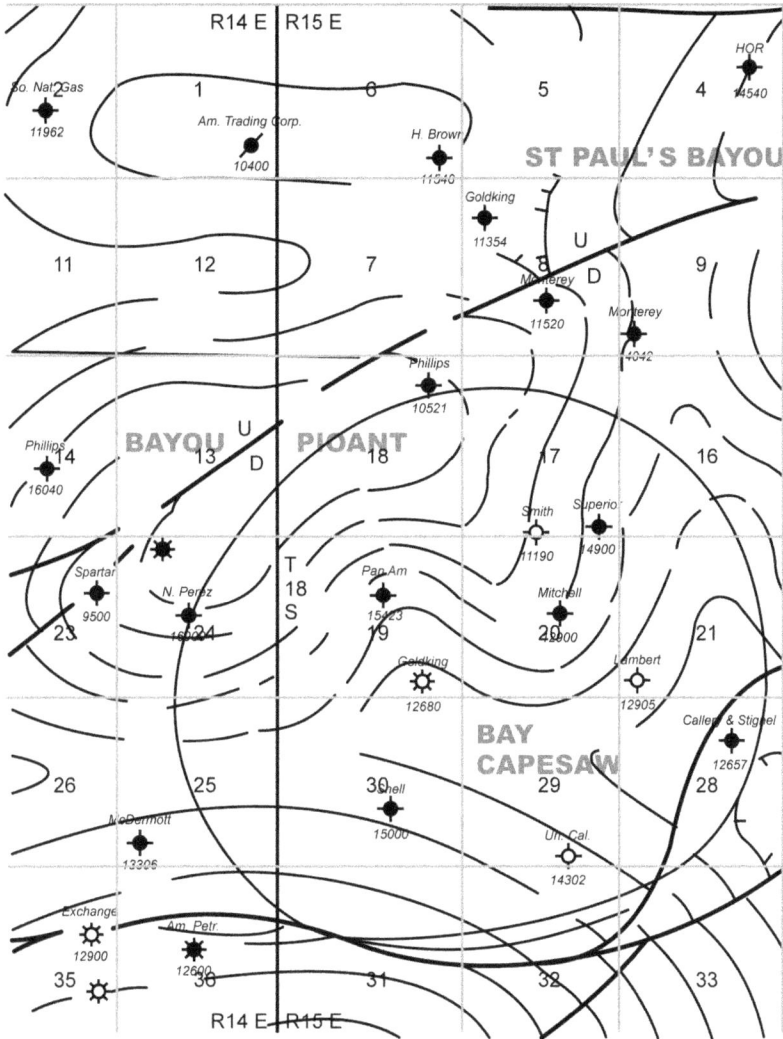

Fig. 11.4—Section map illustrating townships, ranges, and sections.

operator, well name, depth, and type of produced fluids. In addition, some maps contour regional formation tops.

The map in Fig. 11.4 is defined according to a United States land grant system using townships, ranges, and sections. Important terms used with this system are defined next.

• Section: Basic unit of the system—a square tract of land 1 × 1 mile containing 640 acres.

• Township: 36 sections arranged in a 6 × 6 array measuring 6 × 6 miles. Sections are numbered beginning with the northeast-most section, proceeding west to 6, then south along the west edge of the township and then back to the east.

• Range: Assigned to a township by measuring east or west of a principle meridian.

• Range Lines: North-to-south lines that mark township boundaries.

• Township Lines: East to west lines that mark township boundaries.

• Principal Meridian: Reference or beginning point for measuring east or west ranges.

• Base Line: Reference or beginning point for measuring north or south townships.

Fig. 11.5—Texas map illustrating abstracts.

In rare cases, a specific township and range may have several hundred sections. This scheme is used throughout the United States except in a few states including Texas, where the location is described in terms of trees, streams, rocks, and neighboring landowners (**Fig. 11.5**).

The latitude/longitude mapping system is widely used worldwide, except in the United States. This approach is more orderly and easily allows the wells to be located in relation to other known wells or landmarks. The "lat/long" system is now being introduced in the United States in conjunction with the township/range scheme.

Selecting offset wells to be used in data collection is important. Using Fig. 11.4 as an example, assume that a 13,000-ft prospect is to be drilled in the northeast corner of Section 30, T18S, R15E. The best candidates for offset analyses are shown in **Table 11.2**. Although these wells were selected for control analysis, available data from any well in the area should be analyzed.

11.2.2 Data Sources. Data sources should be available for virtually every well drilled in the United States. Drilling costs prohibit the rank wildcatting that occurred years ago. Although wildcats are currently being drilled, seismic data, as a minimum, should be available for pore-pressure estimation.

Common data types used by the drilling engineer are listed next:
- Bit, mud, mud-logging, and operator's drilling records.
- Drilling reports from operators or the Intl. Assn. of Drilling Contractors (IADC).
- Scout tickets.
- Log headers.

TABLE 11.2—OFFSET WELLS FROM FIG. 11.4	
Operator	Section (T18S, R15E)
Shell: 15,000 ft	30
Union of California: 14,562 ft	29
Huber: 12,521 ft	21
Exchange: 12,685 ft	19
Houston Oil and Minerals: 17,493 ft	19

- Production history.
- Seismic studies.
- Well surveys.
- Geological contours.
- Databases of service company files.

Each record contains data that may not be available with other sources. For example, log headers and seismic work are useful, particularly if these data are the only available sources.

Many data sources exist in the industry. Some operators consider the records confidential, when in fact the important information, such as well-testing and production data, becomes public domain a short time after the well is completed. The drilling engineer must assume the role of "detective" to define and locate the required data.

Data sources include bit manufacturers and mud companies who regularly record pertinent information on well recaps. Bit and mud companies usually make these data available to the operator. Log libraries provide log headers and scout tickets. Internal company files often contain drilling reports, IADC reports, and mud logs. Many operators share old offset information if they have no further leasing interest.

11.2.3 Bit Records. An excellent source of offset drilling information is the bit record. It contains data relative to the actual on-bottom drilling operation. A typical record for a relatively shallow well is shown in **Fig. 11.6**.

The heading of the bit record provides information such as the operator, contractor, rig number, well location, drillstring characteristics, and pump data. In addition, the bit heading provides dates for spudding, drilling out from under the surface casing, intermediate-casing depth, and reaching the hole bottom.

The main body of the bit record provides the number and type of bits, jet sizes, footage and drill rates per bit, bit weight and rotary operating conditions, hole deviation, pump data, mud properties, dull-bit grading, and comments. The vertical deviation is useful in detecting potential dogleg problems.

Comments throughout the various bit runs are informative. Typical notes such as "stuck pipe" and washout in drillstring can explain drilling times greater than expected. Drilling engineers often consider the comments section on bit (and mud) records to be as important as the information in the main body of the record.

Bit-grading data can be valuable if the operator assumes the observed data are correct and representative of the actual bit condition. The bit grades assist in preparation of a bit program identifying the most (and least) successful bits in the area. Bit running problems such as broken teeth, gauge wear, and premature failures can be observed, and preventive measures can be formulated for the new well.

Drilling Analysis. Bit records can provide additional useful data if the raw information is analyzed. Plots can be prepared that detect lithology changes and trends. Cost-per-foot analyses can be made. Crude, but often useful, pore-pressure plots can be prepared.

MADE IN U.S.A.

COUNTY	ALLEN	FIELD	REEVES
STATE	LA	SECTION	27
TOWNSHIP	6S	RANGE	7W
OPERATOR	I.T.R. Petroleum Inc	FILE NUMBER	234636
CONTRACTOR	GREAT Southern Oil	RIG NO.	37
X(12)	X	LOCATION	LAKE Charles
WELL NO.	1	SPUD	8-20-82
TOTAL DEPTH DATE	8-28-82	TOOL PUSHER	LeBlanc LaGrange
DULL BITS GIVE SIZE AND TYPE	1. 3 1/2" IF	O.D.	4 1/2
DRILL NO.	18	PUMPS	1. EMSCO F-800 LINER 5 1/2
DRAWWORKS AND POWER	IDECO	FUEL	Diesel
WATER	Well		
COMPANY MAN	MR Nick Perez	AREA	Eastern U.S. Zone 1 HB 31

NO.	SIZE	MAKE	TYPE	SERIAL	JET 32ND IN	DEPTH OUT	FEET	HOURS	SHOCK TOOL	FT/HR	ACCUM. DRLG. HRS	WT. 1000 LBS.	RPM	VERT DEV.	PUMP PRESS	SPM 1	MUD WT	VIS	DULL COND. T	B	OTHER	REMARKS
A	10 5/8	R-1		Penn	3-14	145	1415	13	N	109	13	5/10	145	1/2°	400	109	8.80		5	6		8-24-82
1	7 7/8	X3A		W881	3-10	3310	1895	27 1/2	V	69	45 1/2	24	120	1/2°	1450	109	9'40		4	5 F		
2	7 7/8	X3A		TP650	3-10	4696	1386	16 1/2	V	84	57	20	150	1°	1750	106	9'40		5	5 F		
3	7 7/8	X3A		RR76	3-10	6434	1738	25	V	76	82	20	145	1/2°	2000	105	9'40		6	5 F		
4	7 7/8	J-1		TZ61	3-11	7435	1001	25 1/2	V	39	107 1/2	24	130	1	1900	104	10'41		6	5 Q	#2	
5	7 7/8	J-1		RK188	3-11	8100	665	26	V	26	133 1/2	24	85	-	2000	104	13'40		6	5 F		8-28-82

Reported 9-20-82 — N. Perez

Fig. 11.6—Bit record for a shallow well.

Drill-Rate-vs.-Depth Plot

Well   : J.D. Sittig No. 1
Operator : Stone Oil Company
State   : LA Township: 7S   Range: 1W   Section 28

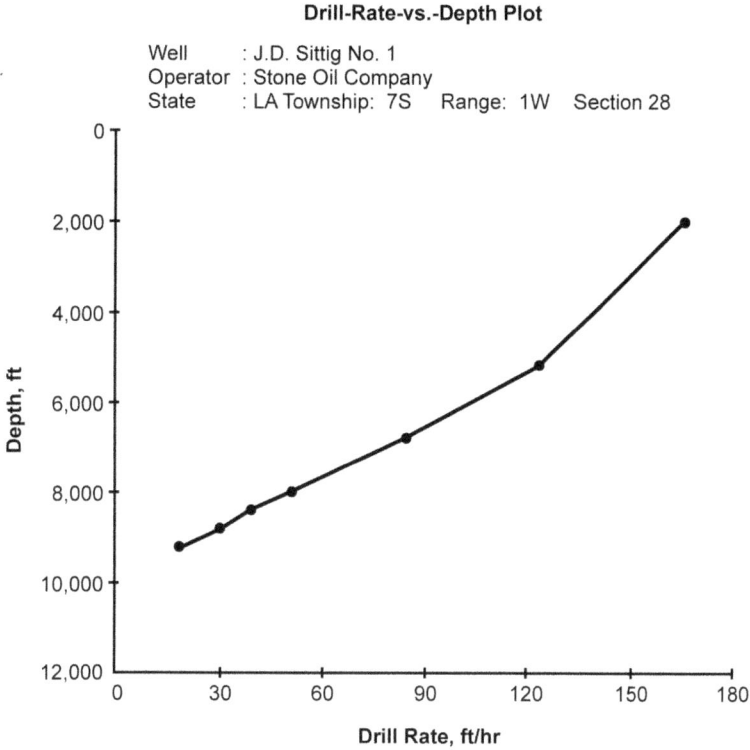

Drill Rate, ft/hr

Fig. 11.7—Raw drill rate from a south Louisiana well.

Raw drill-rate data from a well and an area can detect trends and anomalies. **Fig. 11.7** shows drill-rate data from a well in south Louisiana. A drill rate that decreases with depth is expected as shown.

Changes in the trend might suggest an anomaly, as in **Fig. 11.8**. This illustration is the composite drill rates for all wells in a south Louisiana township and range. The trend change at approximately 10,000 ft was later defined as the entrance into a massive shale section.

Cost-per-foot studies are useful in defining optimum, minimum-cost drilling conditions. A cost comparison of each bit run on all available wells in the area will identify bits and operating conditions for minimum drilling costs. The drilling engineer provides his expected rig costs, bit costs, and assumed average trip times. The cost-per-foot calculations are completed with Eq. 11.1.

$$\$ / \text{ft} = \frac{C_B + C_R T_T + C_R T_R}{Y}, \quad \dots\dots\dots\dots\dots\dots\dots\dots\dots\dots\dots\dots (11.1)$$

where
$\$/\text{ft}$ = cost per foot, U.S dollars;
C_B = bit cost, U.S. dollars;
C_R = rig cost, U.S. dollars;
T_R = rotating time, hours;
T_T = trip time, hours;
and
Y = footage per bit run.

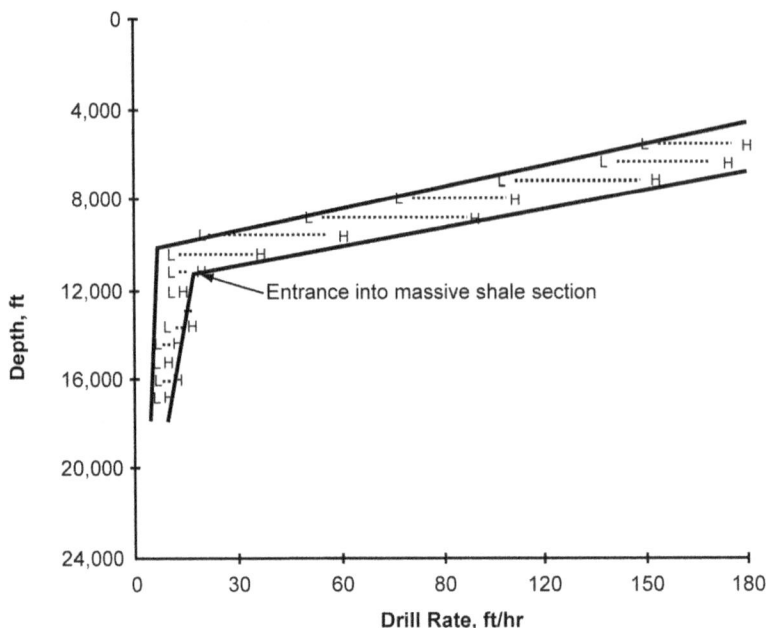

Fig. 11.8—Composite drill-rate data for a south Louisiana region.

A cost-per-foot analysis for Fig. 11.6 is shown in **Fig. 11.9**.

Trip times should be averaged for various depth intervals. Several operators have conducted field studies to develop trip-time relationships (**Table 11.3**). The most significant factors affecting trip time include depth and hole geometry (i.e., number and size of collars, and downhole tools). Table 11.3 can be used in the cost-per-foot equation (Eq. 11.1).

Example 11.2 Calculate the cost per foot and the cumulative section costs for the following data. Assume a rig cost of U.S. $12,000/day.

	Depth In, ft	Depth Out, ft	Rotating Time, hours	Bit Cost, U.S dollars
Well A	6,000	7,150	23	1,650
	7,150	8,000	20	1,650
Well B	6,000	8,000	42	2,980

Determine which drilling conditions, Well A or B, should be followed in the prospect well. Use a 9.875-in. bit.

Solution.

1. The hourly rig cost is U.S. $500. Trip times from 7,150 and 8,000 ft are 6.0 and 6.5 hours, respectively.

2. The cost per foot for Bit 1 on Well A (6,000 to 7,150 ft) is

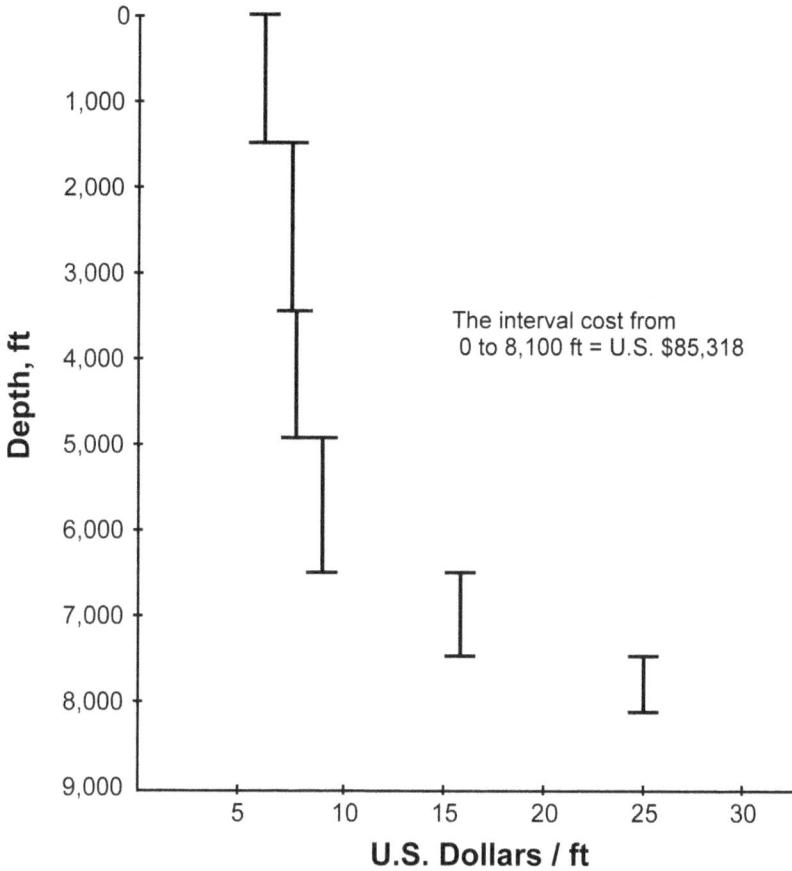

The interval cost from
0 to 8,100 ft = U.S. $85,318

Fig. 11.9—Cost per foot for the bit run in Fig. 11.6.

$$\$ \, / \, \text{ft} = \frac{C_B + C_R T_R + C_R T_T}{Y}$$

$$= \frac{1{,}650 + (500)(23) + (500)(6.0)}{1{,}150}$$

$$= \$ \, 14.04 \, / \, \text{ft.}$$

For Bit 2,

$$\$ \, / \, \text{ft} = \frac{1{,}650 + (500)(20) + (500)(6.50)}{850}$$

$$= \$ \, 17.53 \, / \, \text{ft.}$$

3. The cumulative cost for Well A is

TABLE 11.3—AVERAGE TRIP TIMES			
	Hole (Bit) Size, in.		
Depth, ft	Small (<8.75)	Medium (8.75 to 9.875)	Large (>9.875)
2,000	1.5	3.0	4.5
4,000	2.5	4.2	5.75
6,000	3.5	5.4	7.0
8,000	4.7	6.5	8.0
10,000	5.8	7.25	9.0
12,000	7.0	8.25	10.25
14,000	8.25	9.25	11.50
16,000	9.75	10.25	12.50
18,000	11.00	11.25	13.75
20,000	11.8	12.25	15.0

$$
\begin{array}{llllll}
\text{Bit} \ \#1 & \$\ 14.04/\text{ft} & \times & 1{,}150\ \text{ft} & = & \$\ 16{,}146.00. \\
\text{Bit} \ \#2 & \$\ 17.53/\text{ft} & \times & 850\ \text{ft} & = & \$\ 14{,}900.50. \\
& & & \text{Total} & = & \$\ 31{,}046.50.
\end{array}
$$

4. The cost per foot for Well B is

$$
\$\ /\text{ft} \ = \ \frac{2{,}980 + (50)(42) + (500)(6.5)}{2{,}000}
$$

$$
= \ \$\ 13.62/\text{ft}.
$$

The section cost is $27,230.

5. Because the cost per foot is lower in Well B, drilling conditions for Well B should be implemented.

11.2.4 Mud Records. Drilling-mud records describe the physical and chemical characteristics of mud system. The reports are usually prepared daily. In addition to the mud data, hole and drilling conditions can be inferred. Most personnel believe this record is important and useful.

Mud engineers usually prepare a daily mud-check report form. Copies are distributed to the operator and drilling contractor. The form contains current drilling data such as well depth, bit size and number, pit volume, pump data, solids-control equipment, and drillstring data. The report also contains mud-properties data such as mud weight; pH; funnel viscosity; plastic viscosity; yield point; gel strength; chloride, calcium, and solids content; cation-exchange capacity; and fluid loss.

An analysis of these characteristics taken in the context of the drilling conditions can provide clues to possible hole problems or changes in the drilling environment. For example, an unusual increase in the yield point, water loss, and chloride content suggests that salt (or salt water) has contaminated a freshwater mud. If kick-control problems had not been encountered, it is probable that salt zones were drilled.

A composite mud recap form is usually prepared. It contains a daily properties summary. It may also include comments pertaining to hole problems.

Drilling Analysis. Daily reports prepared by the mud engineer are useful in generating depths-vs.-days plots (**Fig. 11.10**). These plots are as important to well-cost estimating as pore

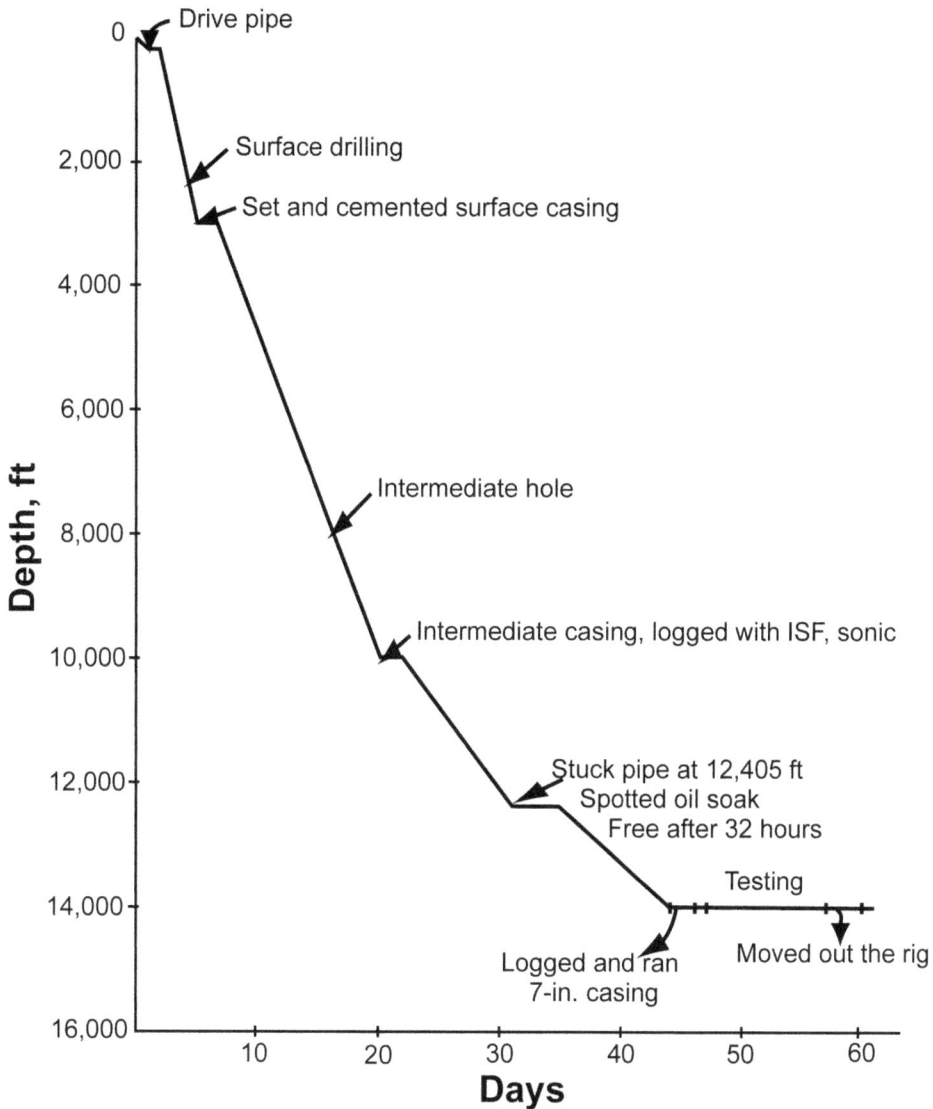

Fig. 11.10—Depth-vs.-days plot developed from a mud record.

pressures are to the overall well plan. Other types of records (i.e., bit records and log headers) do not provide sufficient daily detail to construct the plot as accurately as mud records.

An analysis of the plots in the offset area surrounding the prospect well can provide:

• Expected drilling times for various intervals.

• Identification of improved operating conditions by examining the lowest drilling times in the offset wells.

• Location of potential problem zones by comparing common difficulties in the wells.

After the offset wells have been analyzed, a projected-depth vs. days plot is prepared for the prospect well.

11.2.5 IADC Reports. The drilling contractor maintains a daily log of the drilling operations recorded on the standard IADC-API report. It contains hourly reports for drilling operations, drillstring characteristics, mud properties, bit performance, and time breakdowns for all opera-

PARISH SER:	ACADIA 1156172 & 159825	FIELD API#	CROWLEY (NORTH) 17001-20678	Sec. 9 T9S-R1E

OPR:	Amarillo O. Co. (Fmly Dixie Petro of La Inc) RESULT: Flowing Dual Gas Well
WELL:	Houssiere #1 & 1-D
LOCK:	9-9S-1E 747' FSL & 953' FEL of sec (13500' test-Nod RH) Elev: 20' RKB-CHF & 27' Grd.
SPUD:	2-9-78 COMP: 6-2-78 PBTD: 13303' TD: 14008'

CSG:	16" @ 112', 10 ¾" @ 2801' w/1665sx, 7 ⅝" @ 10803' w/1200sx, 5½" Inr @ 13370' w/200sx, 2⅜" tbg @ 13154', pkr @ 13157' 2⅜" tbg on pkr @ 10498'.
LOGS:	IEL, ISF-SONL.
PERFS:	13500-14200', PB @ 13301' w/sand, perf for prod 13093-104' & 13110-113' w/2 holes per foot, perf 13200-204' & 13210-217'.
IP:	(#1) 62 BOPD, 2109 MCFD, ¹⁰⁄₆₄" ch, TP 4208#, CP pkr, GOR 33,798-1, BS&W .1%, BHP (SI) 6171#, GR 51.2, Prod Int: 13093-13113' (Nodosaria 1RH SUB).
IP:	(#2) 82 BCPD, 2620 MCFD, ¹⁰⁄₆₄" ch, TP 4208#, CP pkr, BS&W .1%, GOR 31,943-1, BHP (SI) 5967#, GR 50.2, Prod Int: 13200-13217' (Nod 18 RB SUA)
TOPS:	Nodosaria 1: 13092', Nodosaria 2: 13200'.
	REPUBLISHED TO SHOW DUAL COMPLETION
	REPORT DATE: 6-28-78 CARD#1

Fig. 11.11—Scout ticket.

tions. These reports are usually unavailable to other contractors or operators and, as a result, cannot be obtained for offset-well analysis without the operator's cooperation.

11.2.6 Scout Tickets. Scout tickets have been available as a commercial service for many years. The tickets were originally prepared by oil company representatives who "scouted" operations of other oil companies. Current scout tickets contain a brief summary of the well (**Fig. 11.11**). The data usually include:
- Well name, location, and operator.
- Spud and completion dates.
- Casing geometries and cement volumes.
- Production-test data.
- Completion information.
- Tops of various geological zones.

The data source for scout tickets are the state or federal report forms filed by oil companies during the course of drilling the well.

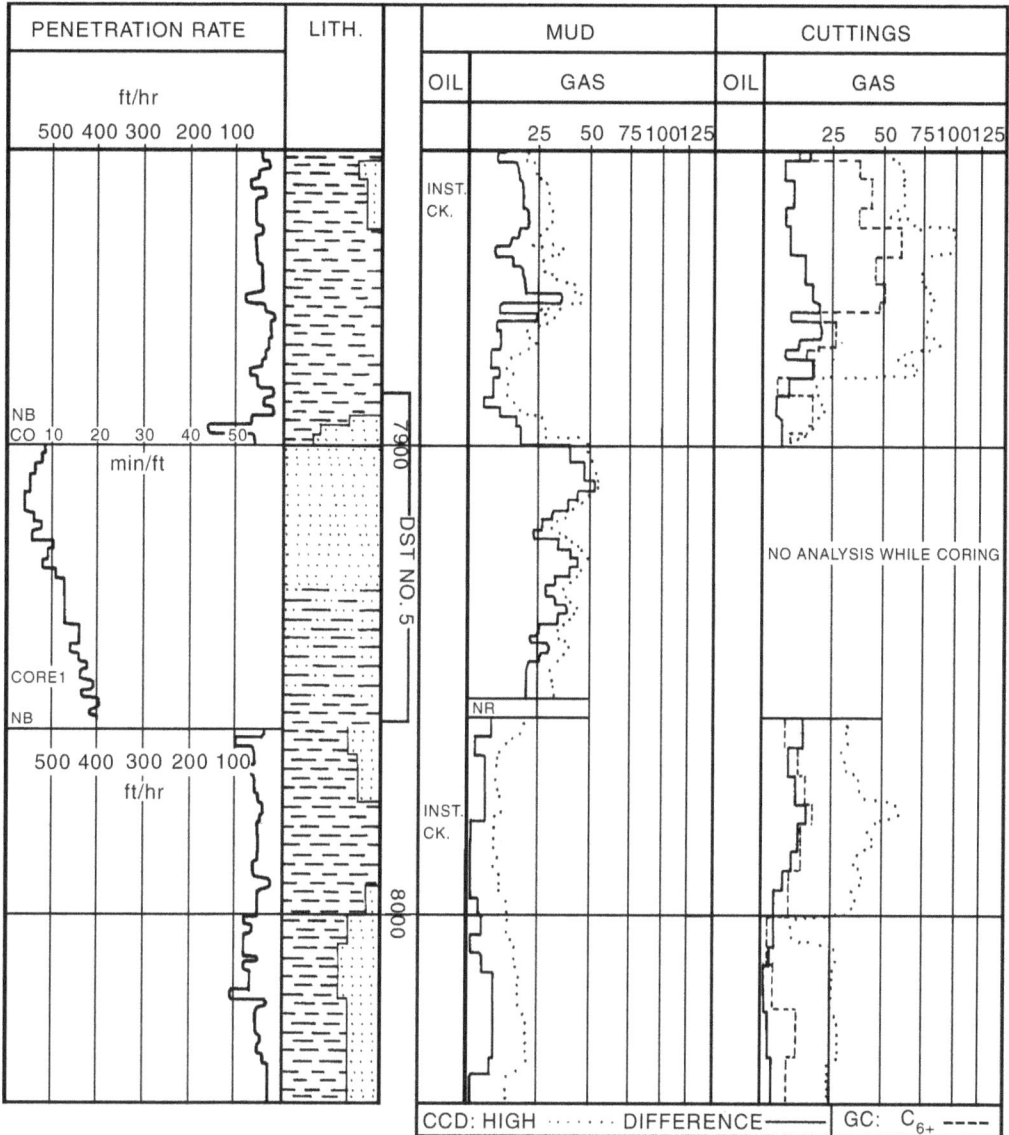

Fig. 11.12—Section of a mud log.

11.2.7 Mud-Logging Records. A mud log is a foot-by-foot record of drilling, mud, and formation parameters. Mud-logging units are often used on high-pressure or troublesome wells. Many engineers consider the mud log to be the best source of penetration-rate data. Mud logging records are seldom available to groups other than the well operators.

A section of a mud log is shown in **Fig. 11.12**. Drilling parameters normally included are penetration rate, bit weight and rotary speed, bit number and type, and rotary torque.

Mud-logging scales are often arranged so the drill-rate curve can be compared to the spontaneous potential (SP) or gamma ray curve on offset logs. The mud log may contain drilling-related parameters such as mud temperatures; chlorides; gas content in the mud and cuttings, usually measured in 'units'; lithology; and pore-pressure analysis. The pore pressure can be computed from models such as the *d*-exponent or other proprietary equations or can be measured by drillstem tests.

11.2.8 Log Headers. Drilling records similar to the previously described information are not available on all offset wells. In these cases, log headers can yield useful drilling data. Easily attainable data from the log headers include logging depths, mud weight and viscosity at each logging depth, bit sizes, inferred casing sizes, and actual setting depths. If enough logging runs were made, a useful depth-vs.-days plot can be constructed.

11.2.9 Production History. Production records in the offset area can provide clues to problems that may be encountered in the prospect well. Oil/gas production can reduce the formation pressure and cause differential pipe sticking. Production records provide pressure data from the flowing zones. Unfortunately, pressures in the over- and underlying formations will not change appreciably. This obscures detection with drilling parameters.

Example 11.3 A prospect well has the Concordia B sand as its intermediate target zone. Production records indicate the original bottomhole pressure (BHP), before production from the B sand, was 5,389 psia at 9,890 ft true vertical depth (TVD). Currently, the producing BHP is 3,812 psia, and the product is gas. A 10.7-lbm/gal mud was required to drill the intermediate shale sections contiguous to the Concordia sand. A 12.1-lbm/gal mud is required to drill to 12,050 ft. If a maximum pressure of 2,000 psi is used as the upper differential limit, can the well be drilled with the Concordia sand exposed or must the casing be set below the sand before reaching 12,050 ft? (Convert all mud hydrostatic pressures to absolute pressure by adding 15 psia for atmospheric conditions.)

Solution.

1. The mud required to balance the Concordia sand is 10.7 lbm/gal, which exerts a hydrostatic pressure of

$$p_h = 0.052 \times 9{,}890 \text{ ft } \times 10.7 \text{ lbm/gal}$$
$$= 5{,}502 \text{ psig.}$$

2. The differential pressure with 10.7 lbm/gal is

$$5{,}517 \text{ psia} - 3{,}812 \text{ psia} = 1{,}705 \text{ psia.}$$

Therefore, pipe sticking should not be a problem with the 10.7-lbm/gal mud.

3. A 12.1-lbm/gal mud is required to reach 12,050 ft. This mud weight will create a hydrostatic pressure at 9,890 ft of

$$0.052 \times 12.1 \text{ lbm/gal} \times 9{,}890 \text{ ft} = 6{,}222 \text{ psig.}$$

The differential pressure will be

$$6{,}237 \text{ psia} - 3{,}812 \text{ psia} = 2{,}425 \text{ psia.}$$

4. A casing string, or liner, must be set below 9,890 ft because the 12.1 lbm/gal required at the bottom creates a differential pressure at the Concordia B sand in excess of the 2,000-psi upper limit.

11.2.10 Seismic Studies. Wildcat wells are seldom drilled without preliminary seismic work being done in the area. Analysis of seismic reflections can eliminate the "wildcat" status of the well by predicting pore pressures. Several authors have shown that good agreement on the pore pressures can be attained with seismic and sonic-log data.

11.3 Casing Setting-Depth Selection

The first design task in preparing the well plan is selecting depths that the casing will be run and cemented. The drilling engineer must consider geological conditions such as formation pressures and fracture mud weights, hole problems, internal company policies, and, in many cases, a variety of government regulations. The program results should allow the well to be drilled safely without the necessity of building "a steel monument" of casing strings. Unfortunately, many well plans give significant considerations to the actual pipe design, yet give only cursory attention to the pipe setting depth.

The importance of selecting proper depths for setting casing cannot be overemphasized. Many wells have been engineering or economic failures because the casing program specified setting depths too shallow or deep. Applying a few basic drilling principles combined with a basic knowledge of the geological conditions in an area can help determine where casing strings should be set to ensure that drilling can proceed with minimum difficulty.

11.3.1 Types of Casing and Tubing. Drilling environments often require several casing strings to reach the total desired depth. Some of the strings are drive, or conductor; structural; surface; intermediate (also known as protection pipe); liners; production (also known as an oil string); and tubing (flow string). **Fig. 11.13** shows the relationship of some of these strings. In addition, the illustration shows some problems and drilling hazards the strings are designed to control.

All wells will not use each casing type. The conditions encountered in each well must be analyzed to determine types and amount of pipe necessary to drill it. The general functions of all casing strings are listed next.

• Segregate and isolate various formations to minimize drilling problems or maximize production.

• Furnish a stable well with a known diameter through which future drilling and completion operations can be executed.

• Provide a secure means to which pressure-control equipment can be attached.

Drive Pipe or Conductor Casing. The first string run or placed in the well is usually the drive pipe, or conductor casing. Depths range from 40 to 300 ft. In soft-rock areas such as southern Louisiana or most offshore environments, the pipe is hammered into the ground with a large diesel hammer. Hard-rock areas require that a large-diameter, shallow hole be drilled before running and cementing the pipe. Conductor casing can be as elaborate as heavy-wall steel pipe or as simple as a few old oil drums tacked together.

A primary purpose of this string is to provide a fluid conduit from the bit to the surface. Very shallow formations tend to wash out severely and must be protected with pipe. In addition, most shallow formations exhibit some type of lost-circulation problem that must be minimized.

An additional function of the pipe is to minimize hole-caving problems. Gravel beds and unconsolidated rock may continue to fall into the well if not stabilized with casing. Typically, the operator is required to drill through these zones by pumping viscous muds at high rates.

Structural Casing. Occasionally, drilling conditions will require that an additional string of casing be run between the drive pipe and surface casing. Typical depths range from 600 to 1,000 ft. Purposes for the pipe include solving additional lost-circulation or hole-caving problems and minimizing kick problems from shallow gas zones.

Surface Casing. Many purposes exist for running surface casing including:

• Cover freshwater sands.

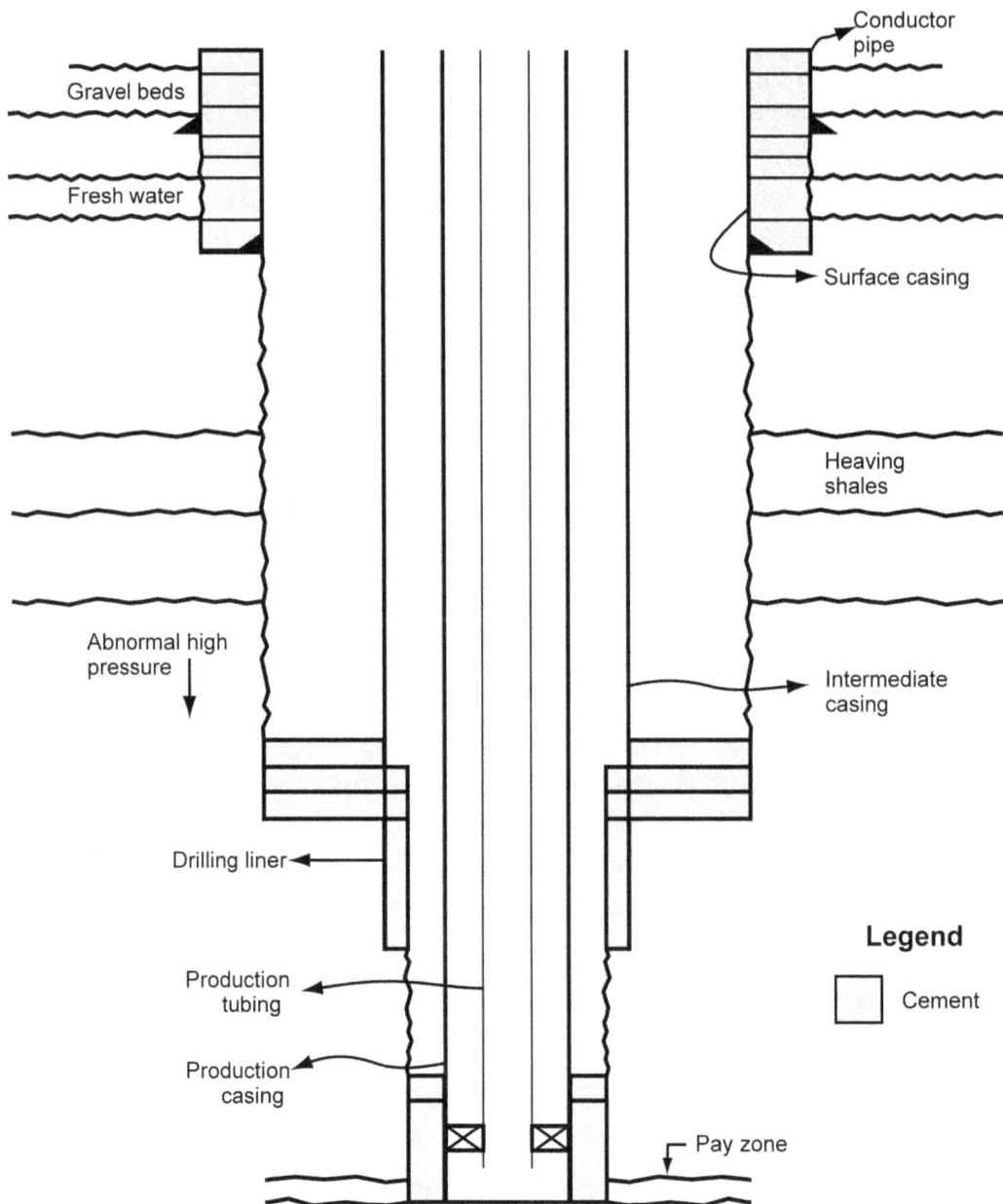

Fig. 11.13—Typical string relationships.

- Maintain hole integrity by preventing caving.
- Minimize lost circulation into shallow, permeable zones.
- Cover weak incompetent zones to control kick-imposed pressures.
- Provide a means for attaching the blowout preventers.
- Support the weight of all casing strings (except liners) run below the surface pipe.

Intermediate Casing. The primary applications of intermediate casing involve abnormally high formation pressures. Because higher mud weights are required to control these pressures, shallower weak formations must be protected to prevent lost circulation or stuck pipe. Occasionally, intermediate pipe is used to isolate salt zones or zones that cause hole problems, such as heaving and sloughing shales.

Liners. Drilling liners are used for the same purpose as intermediate casing. Instead of running the pipe to the surface, an abbreviated string is used from the bottom of the hole to a shallower depth inside the intermediate pipe. Usually, the overlap between the two strings is 300 to 500 ft. In this case, the intermediate pipe is exposed to the same drilling considerations as the liner (Fig. 11.13).

Drilling (and production) liners are used frequently as a cost-effective method to attain pressure or fracture-mud-weight control without the expense of running a string to the surface. When a liner is used, the upper exposed casing, usually intermediate pipe, must be evaluated with respect to burst and collapse pressures for drilling the open hole below the liner. Remember that a full string of casing can be run to the surface instead of a liner if required (i.e., two intermediate strings).

Production Casing. The production casing is often called the oil string. The pipe may be set at a depth slightly above, midway through, or below the pay zone. The pipe has the following purposes:
- Isolate the producing zone from the other formations.
- Provide a work shaft of known diameter to the pay zone.
- Protect the production-tubing equipment.

Tieback String. The drilling liner is often used as part of the production casing rather than running an additional full string of pipe from the surface to the producing zone. The liner is tied back or connected to the surface by running the amount of pipe required to connect to the liner top. This procedure is particularly common when producing hydrocarbons are behind the liner and the deeper section is not commercial.

11.3.2 Setting-Depth Design Procedures. Casing-seat depths are affected by geological conditions. In some cases, the prime criterion for selecting casing seats is to cover exposed, lost-circulation zones. In others, the seat may be based on differential-sticking problems, perhaps resulting from pressure depletion in a field. In deep wells, however, the primary consideration is usually based on controlling abnormal formation pressures and preventing their exposure to weaker shallow zones. This criterion of controlling formation pressures generally applies to most drilling areas.

Selecting casing seats for pressure control starts with knowing geological conditions such as formation pressures and fracture mud weights. This information is generally available within some degree of accuracy. Prespud calculations and the actual drilling conditions determine the exact locations for each casing seat.

The principle used to determine setting-depth selection can be adequately described by the adage, "hindsight is 20/20." The initial step is to determine the formation pressures and fracture mud weights that will be penetrated. After these have been established, the operator must design a casing program based on the assumption that he already knows the behavior of the well even before it is drilled.

This principle is used extensively for infill drilling where the known conditions dictate the casing program. Using these guidelines, the operator can select the most effective casing program that meets the necessary pressure requirements and minimize the casing cost.

Setting-Depth Selection for Intermediate and Deeper Strings. Setting-depth selection should be made for the deepest strings to be run in the well and successively designed from the bottom to surface. Although this procedure may appear at first to be reversed, it avoids several time-consuming iterative procedures. Surface-casing design procedures are based on other criteria.

The first criterion for selecting deep casing depths is for mud weight to control formation pressures without fracturing shallow formations. This procedure is implemented bottom to top. After these depths have been established, differential-pressure-sticking considerations are made

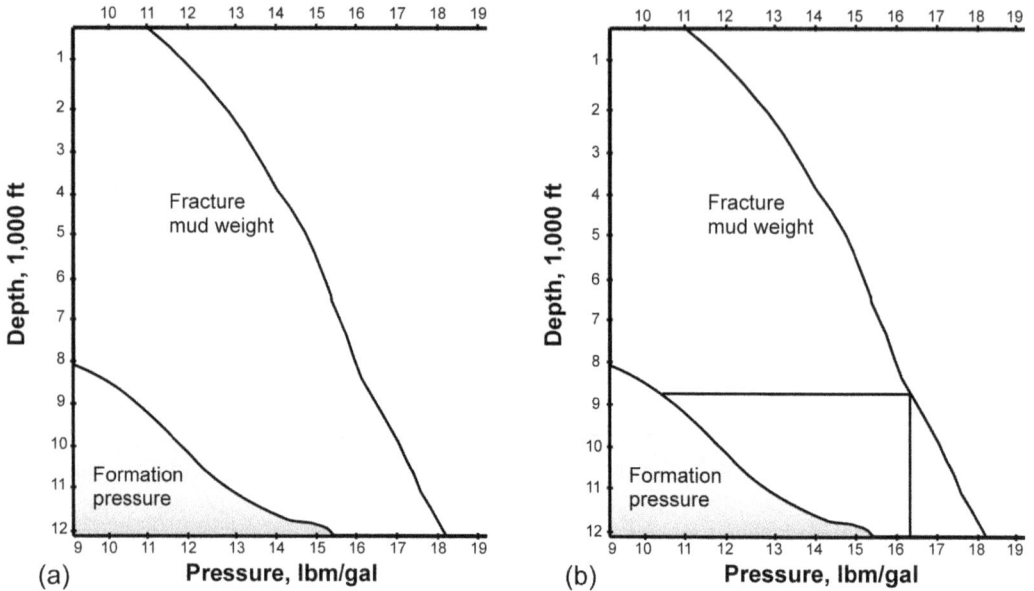

Fig. 11.14—Projected formation and fracture mud weights (a) and selection of the tentative intermediate-pipe setting depth for Example 11.4 (b).

to determine if the casing string will become stuck when running it into the well. These considerations are made from top to bottom, the reverse from the first selection criterion.

The initial design step is to establish the projected formation pressures and fracture mud weights. In **Fig. 11.14**, a 15.6-lbm/gal (equivalent) formation pressure exists at the hole bottom. To reach this depth, wellbore pressures greater than 15.6 lbm/gal are necessary and must be taken into account.

The pressures that must be considered include a trip margin of mud weight to control swab pressures, an equivalent-mud-weight increase because of surge pressures associated with running the casing, and a safety factor. These pressures usually range from 0.2 to 0.3 lbm/gal, respectively, and may vary because of mud viscosity and hole geometry. Therefore, the actual pressures at the bottom of the well include the mud weight required to control the 15.6-lbm/gal pore pressure and the 0.6- to 0.9-lbm/gal (equivalent) mud weight increases from the swab, surge, and safety factor considerations. As a result, formations exhibiting fracture mud weights 16.5 lbm/gal or less (15.6 lbm/gal + 0.9 lbm/gal) must be protected with casing. The depth at which this fracture mud weight is encountered becomes the tentative intermediate-pipe setting depth.

The next step is to determine if pipe sticking will occur when running the casing. Pipe sticking generally occurs where the maximum differential pressures are encountered. In most cases, this depth is the deepest normal-pressure zone (i.e., at the transition into abnormal pressures).

Field studies have been used to establish general values for the amount of differential pressure that can be tolerated before sticking occurs:

Normal-pressure zones 2,000 to 2,300 psi

Abnormal-pressure zones 2,500 to 3,000 psi

These values are recommended as reasonable guides. Their accuracy in day-to-day operations depends on the general attention given to mud properties and drillstring configuration.

The tentative intermediate-pipe setting depth becomes the actual setting depth if the differential pressure at the deepest normal zone is less than 2,000 to 2,300 psi. If the value is greater than this limit, the depth is redefined as the shallowest liner setting depth required to drill the well. In this case, an additional step is necessary to determine the intermediate-pipe depth.

An example problem illustrates this procedure. The section following the example shows the case in which differential pressure considerations require the additional step to select the intermediate pipe depth.

Example 11.4 Use Fig. 11.14a to determine the proper setting depth for intermediate pipe. Assume a 0.3-lbm/gal factor for swab and surge and a 0.2-lbm/gal safety factor. Use a maximum limit of 2,200-psi differential pressure for normal-pressure zones.

Solution.

1. Evaluate the maximum pressures (equivalent mud weights) at the total depth of the well.

Amount, lbm / gal	Purpose	Types of Pressure
15.6	Formation pressure	Actual mud weight
0.3	Trip margin	Actual mud weight
0.3	Surge pressure	Equivalent mud weight
0.2	Safety factor	Equivalent mud weight
16.4		

2. Determine formations that cannot withstand 16.4-lbm/gal pressures (i.e., those formations that must be protected with casing). Construct a vertical line from 16.4 lbm/gal to an intersection of the fracture-mud-weight line (Fig. 11.14 Part B). The depth of intersection is the tentative intermediate casing setting depth, or 8,600 ft in this example.

Check the tentative depth to determine if differential pipe sticking will be a problem when running the casing to 8,600 ft. The mud required to reach 8,600 ft is

10.4 lbm / gal	Formation pressure
0.3 lbm / gal	Trip margin
10.7 lbm / gal	Total required mud weight

Differential-sticking potential is evaluated at the deepest normal-pressure (9.0 lbm/gal) zone, 8,000 ft.

$$(10.7 \text{ lbm / gal} - 9.0 \text{ lbm / gal})(0.052)(8,000 \text{ ft}) = 707 \text{ psi.}$$
$$707 \text{ psi} < 2,200 \text{ psi.}$$

Because pipe can be run to 8,600 ft without differential sticking, the depth is redefined as the actual intermediate setting depth rather than the tentative depth, as defined in Step 2.

3. Check the interval from 8,600 to 12,000 ft to determine if the differential pressure exceeds the 3,000- to 3,300-psi range. In this case, pressure ≈ 2,700 psi at 8,600 ft.

Example 11.4 illustrated the case in which the vertical line from 16.4 lbm/gal intersected the fracture-mud-weight curve in an abnormal-pressure region. A calculation was performed to determine if the casing would stick when run into the well. If the pressures had been greater than the limit of 2,200 psi, procedures in the following sections would be implemented. Cases arising when the vertical line intersects the fracture-mud-weight curve in the normal-pressure region are discussed later.

Altering the tentative intermediate-casing setting depth because of potential differential-sticking problems is required in many cases. The previously defined tentative intermediate-pipe setting depth is redefined as the shallowest liner depth. The procedure must now be worked from the top to the bottom of the high-pressure zone rather than the reverse approach used to establish the tentative intermediate depth. The new intermediate depth is established using sticking criteria. The deepest liner-setting depth is determined from formation-pressure/fracture-mud-weight guidelines. After the deepest liner depth is established, the operator must determine the exact liner-setting depth between the previously calculated shallowest and deepest possible depths. The final liner depth can be established from criteria such as minimizing the amount of small hole that must be drilled below the liner and preventing excessive amounts of open hole between the intermediate-liner section or the liner pay-zone section.

Eqs. 11.2 and 11.3 can be used to help determine the new intermediate depth if sticking is a concern.

$$\Delta p = (\rho - 9)(0.052 D_n),$$

or

$$\frac{\Delta p}{0.052 \ D_n} + 9 = \rho, \quad\text{...} \quad (11.2)$$

where
  ρ = mud weight, lbm/gal;
  D_n = deepest normal zone, ft;
and
  Δp = differential pressure, psi.

A limit of 2,000 to 2,300 psi is normally used for Δp. The mud weight, ρ, from Eq. 11.2 can be used to locate the depth where the Δp value will exist.

$$\rho - \Delta \rho = p_{form}, \quad\text{...} \quad (11.3)$$

where
  ρ = mud weight, lbm/gal;
  $\Delta \rho$ = trip margin, lbm/gal;
and
  p_{form} = formation pressure, lbm/gal.

The depth at which the formation pressure, p_{form}, occurs is defined as the new intermediate-pipe depth.

The deepest liner setting depth is established from the intermediate setting depth's fracture mud weight. Using procedures reversed from those presented in Example 11.4, subtract the swab, surge, and safety factors from the fracture mud weight to determine the maximum allowable formation pressure in the deeper sections of the hole. The depth at which this pressure is encountered becomes the deepest liner depth. The establishment of a setting depth between the shallowest and deep depths generally depends on operator preference and the geological conditions.

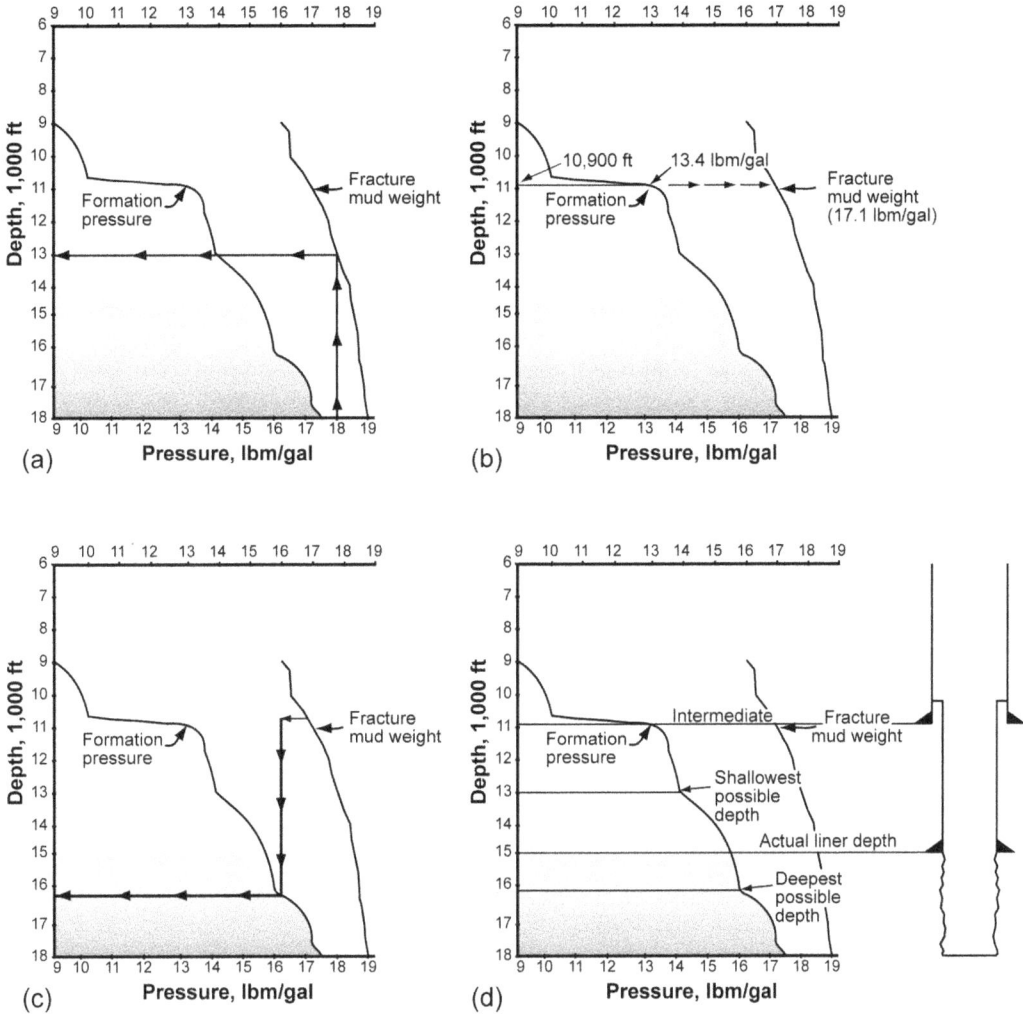

Fig. 11.15—Projected formation pressure and fracture mud weights for Example 11.5.

Example 11.5 Use **Fig. 11.15** to select liner and intermediate setting depths. Assume a differential-pressure limit of 2,200 psi. Use the following design factors:

$$\text{Swab} = 0.3 \text{ lbm / gal.}$$
$$\text{Surge} = 0.3 \text{ lbm / gal.}$$
$$\text{Safety} = 0.2 \text{ lbm / gal.}$$

Solution.

1. From Fig. 11.15, the maximum equivalent mud weight that can be seen at the bottom of the well can be calculated.

Amount, lbm / gal	Purpose
17.1	Formation pressure
0.3	Trip margin
0.3	Surge factor
0.2	Safety factor
17.9	Formation pressure (equivalent)

2. Construct a vertical line to intersect the fracture-mud-weight curve (Fig. 11.15). The depth of intersection, 13,000 ft, is the tentative intermediate casing setting depth. All shallower formations must be protected with casing because their respective fracture mud weights are less than the maximum projected requirements (18.0 lbm/gal) at the bottom of the well.

3. Evaluate the tentative depth for differential sticking by assuming that 14.3-lbm/gal mud will be required to drill the formation at 13,000 ft:

$$(9,000 \text{ ft})(0.052)(14.2 \text{ lbm / gal} - 9.0 \text{ lbm / gal}) = 2,480 \text{ psi}.$$

Because 2,480 psi > 2,200 psi, intermediate pipe cannot safely be run to 13,000 ft. The depth of 13,000 ft is redefined as the shallowest liner depth.

4. The intermediate-pipe depth is defined with Eqs. 11.2 and 11.3.

$$\Delta p = (\rho - 9)(0.052)(D);$$
$$2,200 \text{ psi} = (\rho - 9)(0.052)(9,000 \text{ ft});$$
$$\rho_e = 13.7 \text{ lbm / gal};$$

and

$$\rho - \Delta \rho = \rho_e;$$
$$13.7 \text{ lbm / gal} - 0.3 \text{ lbm / gal} = \rho_e;$$
$$\rho_e = 13.4 \text{ lbm / gal}.$$

From Fig. 11.15b, a 13.4-lbm/gal formation pressure occurs at 10,900 ft.

5. The deepest possible setting depth for the liner is determined by evaluating the fracture mud weight at 10,900 ft. What is the maximum formation pressure below 10,900 ft that can be safely controlled with a fracture mud weight of 17.1 lbm/gal?

Amount, lbm / gal	Purpose
17.1	Fracture mud weight
−0.3	Swab pressure
−0.3	Surge factor
−0.2	Safety factor
16.3	Formation pressure

From Fig. 11.15c, a 16.3-lbm/gal formation pressure occurs at 16,300 ft. The depth is defined as the deepest allowable depth for setting the liner.

6. The shallow and deep liner depths are based on formation-pressure/fracture-mud-weight considerations at the hole bottom (18,000 ft) and the intermediate-pipe depth (10,900 ft), respectively. Any depth between the 13,000- to 16,000-ft range is satisfactory. A depth selection can be based on (1) minimizing small-diameter sections below the liner, (2) minimizing the open-hole length and thereby reducing pipe costs, or (3) other considerations as specified by the operator.

As an example, assume that a depth of 15,000 ft is selected. It reduces the small-diameter hole to a 3,000-ft segment (15,000 to 18,000 ft) while allowing only 4,100 ft of open hole (10,900 to 15,000 ft) (Fig. 11.15d).

Examples 11.4 and 11.5 illustrated the cases in which the initial formation pressure/fracture mud weight at the bottom required pipe depths in the abnormal-pressure regions. Different techniques must be used if the tentative pipe-setting depth is in a normal pressure region.

The initial step is to evaluate differential-sticking possibilities at the deepest normal pressure zone. If the mud weight required at the bottom of the well does not create differential pressures in excess of some limit (2,000 to 2,300 psi), a deep surface-casing string is satisfactory. Eqs. 11.2 and 11.3 must be used when the differential pressures exceed the allowable limit.

Surface-Casing Depth Selection. Shallow casing strings, such as surface casing, are often imposed to equivalent mud weights more severe than the considerations used to select the setting depths for intermediate casing and liner. These pressures usually result from kicks inadvertently taken when drilling deeper sections. As a result, surface setting depths are selected to contain kick pressures rather than the previously described procedures for intermediate casing. This philosophy differs for the intermediate hole because the kick pressures are usually lower than the previously discussed swab/surge/safety-factor logic for deep strings.

Kick-imposed equivalent mud weights are the cause for most underground blowouts. When a kick occurs, the shut-in casing pressure added to the drilling-mud hydrostatic pressure exceeds the formation fracture pressure and results in an induced fracture. The objective of a seat-selection procedure that avoids underground blowouts would be to choose a depth that can competently withstand the pressures of reasonable kick conditions.

Determination of kick-imposed pressures can be difficult. However, a procedure that estimates the values has been proved in field applications to be quick and effective. **Fig. 11.16** represents a well whose pumps and blowout preventers have simulated a kick. Eq. 11.4 describes the pressure relationships.

$$\rho_{e\text{kick}} = \left(\frac{D}{D_i}\right)\Delta\rho + \rho_o, \dots\dots\dots\dots\dots\dots\dots\dots\dots\dots\dots\dots\dots\dots\dots\dots (11.4)$$

where
  $\rho_{e\text{kick}}$ = equivalent mud weight at the depth of interest, lbm/gal;
  D = deepest interval, ft;
  D_i = depth of interest, ft;
  $\Delta\rho$ = incremental kick mud-weight increase, lbm/gal;
and
  ρ_o = original mud weight, lbm/gal.

Eq. 11.4 can be used iteratively along with a suitable theoretical fracture-mud-weight calculation to determine a surface-pipe depth with sufficient strength to resist kick pressures. Initially, a shallow depth is chosen for which the fracture mud weight and equivalent mud

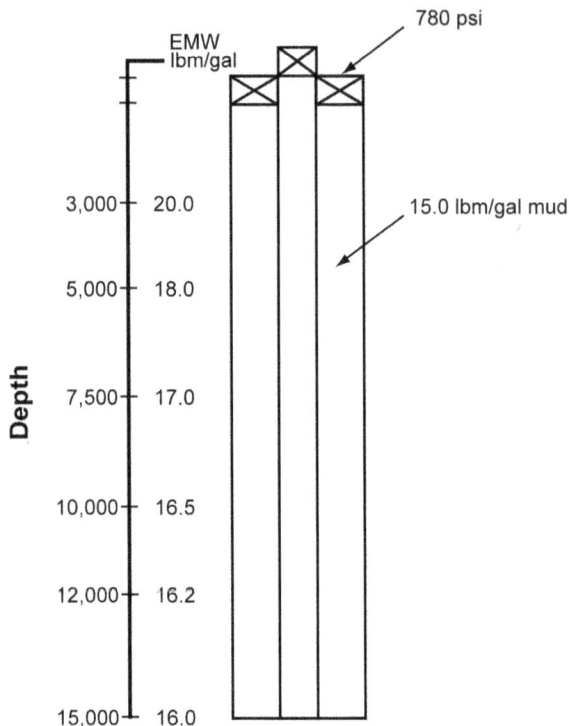

Fig. 11.16—Kick-pressure/equivalent-mud-weight (EMW) relationships.

weights are calculated. If the equivalent mud weight is greater than the fracture mud weight, a deeper interval must be selected and the calculations repeated. This procedure is iterated until the fracture mud weight exceeds the equivalent mud weights. When this occurs, a depth has been selected that will withstand the designed kick pressures. Example 11.6 illustrates the procedure.

Example 11.6 Using **Fig. 11.17**, select a surface-casing depth and, if necessary, setting depths for deeper strings. Use the following design factors:

0.3 = swab, surge factor, lbm/gal.
0.2 = safety factor, lbm/gal.
0.5 = kick factor, lbm/gal.
2,200 = maximum allowable differential pressure, psi.
Solution.
1. Evaluate the maximum pressures anticipated at the bottom of the well.

Amount, lbm / gal	Purpose
12.0	Formation pressure
0.3	Trip (swab) margin
0.3	Surge factor
0.2	Safety factor

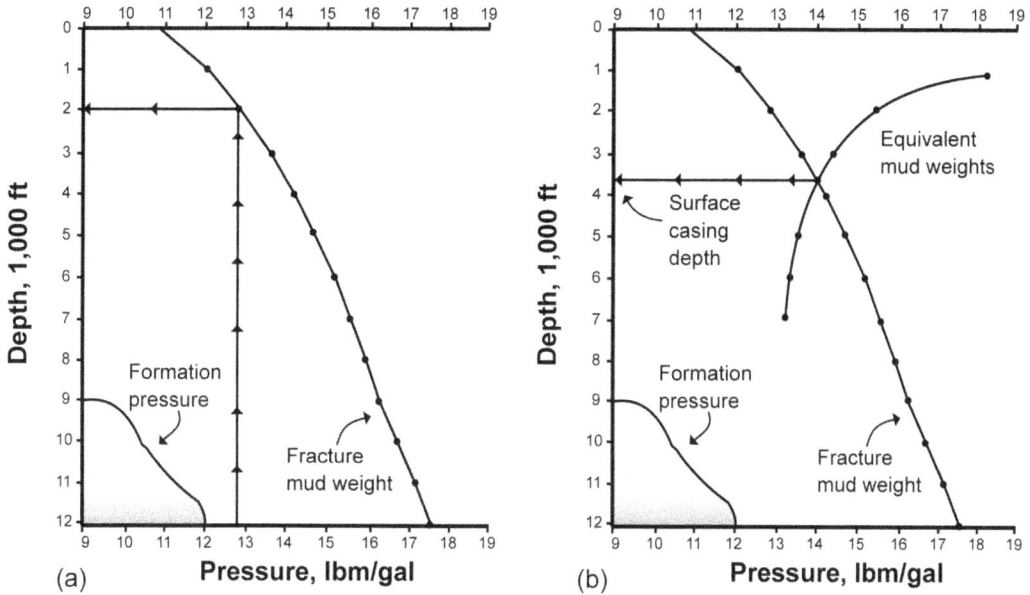

Fig. 11.17—Intermediate-casing evaluation for Example 11.6 (a) and equivalent-mud-weight/fracture-mud-weight relationship (b).

A vertical line from 12.8 lbm/gal intersects the fracture mud weight in a normal region, which indicates that intermediate casing will not be required unless differential sticking is a problem.

2. Assume that 12.3 lbm/gal will be used at the bottom of the well and determine if differential sticking may occur.

$$(12.3 - 9.0 \ \text{lbm} / \text{gal})(0.052)(9,000 \ \text{ft}) = 1,544 \ \text{psi}.$$

Because 1,544 psi is less than the arbitrary limit of 2,200 psi, intermediate casing will not be used for pipe-sticking considerations. Only surface casing is required.

3. Use Eq. 11.4 and the fracture-mud-weight curve to determine the depth at which the fracture mud weight exceeds the kick loading mud weight. Perform a trial calculation at 1,000 ft.

$$\rho_{e\text{kick}} = \left(\frac{12,000}{1,000} \right)(0.5) + 12.3$$
$$= 18.3 \ \text{lbm} / \text{gal}.$$

The fracture mud weight at 1,000 ft is 12.0 lbm/gal. Because the kick loading is greater than the rock strength, a deeper trial depth must be chosen.

Results from several iterations are given next and plotted on Fig. 11.17.

Depth, ft	EMW$_{kick}$, lbm/gal
1,000	18.3
2,000	15.3
3,000	14.3
3,500	14.0
4,000	13.8
4,500	13.6
5,000	13.5
6,000	13.3
7,000	13.2

4. A setting depth of 3,600 ft is selected.

The value of 0.5 lbm/gal used in Example 11.6 for the kick incremental mud-weight increase is widely accepted. It represents the average (maximum) mud-weight increase necessary to kill a kick. Using this variable in Eq. 11.4 allows the operator to (inadvertently) drill a formation in which the pressure is in excess of 0.5 lbm/gal greater than the original calculated value and still safely control the kick. In fact, if the original mud-weight variable is 0.3 to 0.4 lbm/gal greater than the anticipated formation pressure, the equation would account for formation-pressure calculation errors of 0.8 to 0.9 lbm/gal. If necessary, an operator may alter the 0.5-lbm/gal variable to whatever is deemed most suitable for the drilling environment.

A valid argument can be raised concerning Eq. 11.4 and its representation of field circumstances. In actual kick situations, the equivalent mud weights are controlled to a certain degree by casing pressure, which is not directly taken into account in the equation. An inspection of casing pressure shows the two components in the pressure are (1) the degree of underbalance between the original mud and the formation pressure and (2) the degree of underbalance between the influx fluid and the formation pressure.

The first of these components is taken into account in the equation by the incremental mud-weight-increase term, while the latter is not considered. In most kick situations, the average value of the second component will range from 100 to 300 psi. If an operator believes the second component is significant enough to alter the equation, he can change the incremental mud-weight-increase term to a higher value.

The considerations are illustrated in Fig. 11.16 and **Figs. 11.18 and 11.19**. Figs. 11.16 and 11.18 represent a 1.0-lbm/gal kick in simple and actual hole geometries, respectively. Fig. 11.18 shows the shut-in well with a 20-bbl kick at the bottom. Fig. 11.19 shows the equivalent mud weights for both cases. If an operator is concerned about the difference shown in Fig. 11.19, Eq. 11.4 should be modified, or a different equation should be used.

Drive Pipe and/or Conductor Casing. Pipe setting depths above the surface casing are usually determined from various government regulations or localized drilling problems. For example, an area may have severe lost-circulation problems at 75 to 100 ft that can be solved by placing drive pipe below the zone. Other drilling conditions that may affect setting depths include water-bearing sands, unconsolidated formations, or shallow gas. An evaluation of local drilling records will normally identify these conditions. Most governments require that freshwater sands be cased.

SIDPP = 780 psi

SICP = 1,370 psi

15.0-lbm/gal mud

7.0-in. drill collars
(1,000 ft)

8.5-in. hole

7.0-in. drill collars
(1,000 ft)

14,115 ft – gas top
kick – 20 bbl volume
15,000 ft

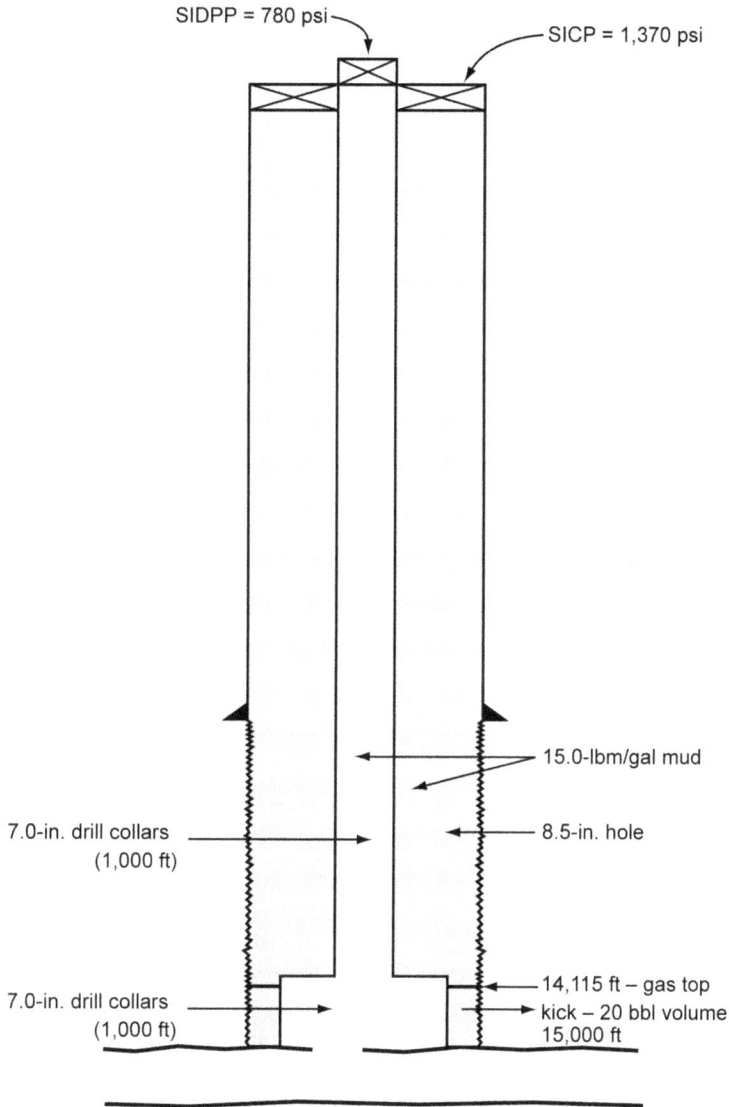

Fig. 11.18—A 20-bbl kick.

11.4 Hole-Geometry Selection

Bit- and casing-size selection can mean the difference between a well that must be abandoned before completion and a well that is an economic and engineering success. Improper size selection can result in holes so small that the well must be abandoned because of drilling or completion problems. The drilling engineer (and well planner) is responsible for designing the hole geometry to avoid these problems.

However, a successful well is not necessarily an economic success. For example, a well design that allows for satisfactory, trouble-free drilling and completion may be an economic failure because the drilling costs are greater than the expected return on investment. Hole-geometry selection is a part of the engineering plan that can make the difference between economic and engineering failure or success.

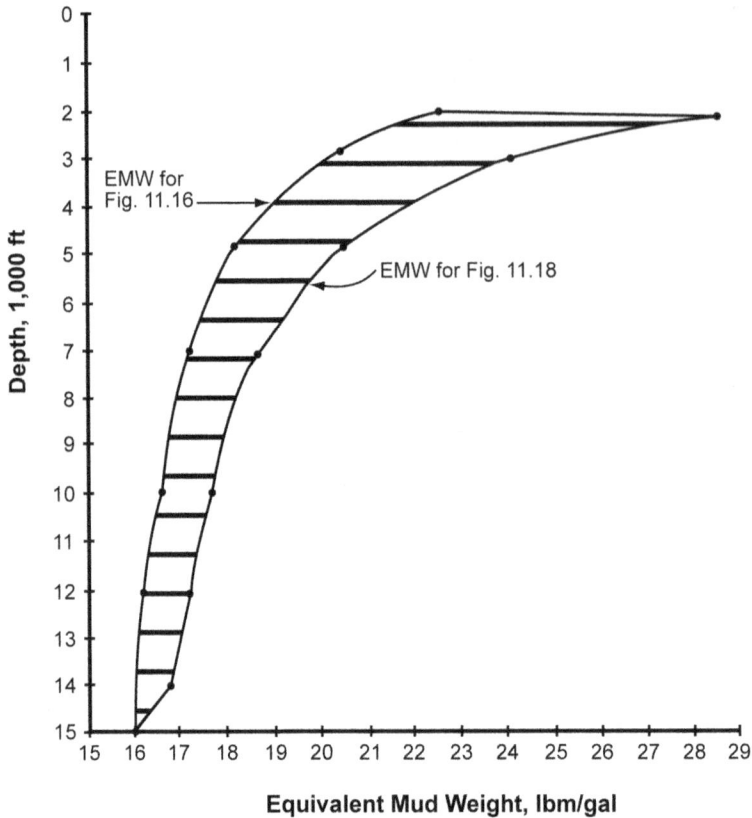

Fig. 11.19—Comparison of equivalent mud weights for rule of thumb and actual situations.

11.4.1 General Design Procedures. The drilling industry's experience has developed several commonly used hole-geometry programs. These programs are based on bit- and casing-size availability as well as the expected drilling conditions.

Deep, high-pressure wells often require deviations from common geometries. Reasons include:
• Prolific production rates requiring large tubing strings.
• Drilling problems requiring the use of an intermediate string and one or more liners.
• Tension design problems because thick-walled pipe must be used to control burst or collapse.
• Rig limitations in running heavy strings of pipe.

Because deep, high-pressure wells are being drilled with increasing frequency, careful attention must be given to hole sizing.

Bottom-to-Top Approach. The highest priority in well planning should be developing a design that provides for economic production from the pay zone. Even in exploratory drilling for geological investigations, a large hole may be necessary for thorough formation evaluation. The pay zone should be analyzed with respect to its flow potential and the drilling problems that will be encountered in reaching it.

Flow-String Sizing. The flow, or tubing, string must be given consideration relative to its ability to conduct oil/gas to surface at economical rates. Small-diameter tubing restricts, or chokes, flow rates because of high friction pressures.

Completion problems can be more complicated with small tubing and casing. The reduced radial clearances make tool placement and operations more difficult, and workover activities are more complicated.

Typical well designs are shown in **Fig. 11.20**. The geometries in parts (a) and (c) use large-diameter tubing. The small tubing string (b) will probably restrict the fluid flow from the producing zone. In addition, the design in (b) will probably require special clearance couplings, whereas parts (a) and (c) could use standard-diameter couplings.

Planning for Problems. Geological uncertainties may make it difficult to predict the expected drilling environment. For example, crossing a fault into a high-pressure region may necessitate a drilling liner, whereas an intermediate string may be satisfactory if the fault is not encountered. Hole geometries are often selected to allow the option for an additional casing string if required (**Fig. 11.21**).

11.4.2 Size-Selection Problems. Many interrelated size-selection problems must be considered before the final hole geometry is established. These problems primarily relate to casing size and openhole considerations, and they are interrelated with casing design. A working knowledge of casing-design problems influences pipe-size selection.

Casing Design. The large flow string in Fig. 11.20 resulted in a 13⅜-in. intermediate string and a 20-in. surface casing. However, these strings may be difficult to design if high formation pressures are encountered. **Table 11.4** shows the pipe required for various conditions on the intermediate string, assuming that a single weight and grade will be used.

Tension designs become critical in cases similar to Table 11.4. The in-air hook load of the string is 887,700 lbf for the worst case shown in the table. If a design factor of 1.5 is used to assess rig requirements, the design weight will be 1,331,550 lbf for derrick and substructure selection. It should be apparent that pipe yield, connector strength, and rig ratings affect casing and sizing selection.

Casing-to-Hole Annulus. Cementing problems may occur if the casing-to-hole annulus is small. Small clearances around the pipe and couplings may cause premature dehydration of the cement and result in a cement bridge. Cement companies report that this bridging occurs more frequently in deeper, hot wells. These companies suggest a minimum annular clearance of 0.375 to 0.50 in. on each side of the pipe, with 0.750 in. preferable.

Drillstring/Hole Annulus. The area between the drillstring and the hole creates problems if too large or small. Inadequate hole cleaning may occur if the hole is large. High friction pressures and turbulent erosion may occur in small holes. Large holes normally occur in the shallow depths, and small holes are found in the bottom sections.

Hole cleaning describes the ability of the drilling fluid to remove cuttings from the annulus. The important factors are mud viscosity, cuttings settling velocity, and annular mud flow rate. The annular mud velocity, Eq. 11.5, is usually considered the most important aspect.

$$v = \frac{24.50 \; Q}{d_H^2 d_{DS}^2}, \quad \text{.. (11.5)}$$

where

v = annular velocity, ft/min;

Q_m = mud flow rate, gal/min;

d_H = hole diameter, in.;

and

d_{DS} = drillstring diameter, in.

Mud engineers often use other forms of an annular velocity equation.

$$v = 100Q/V_a, \quad \text{.. (11.6)}$$

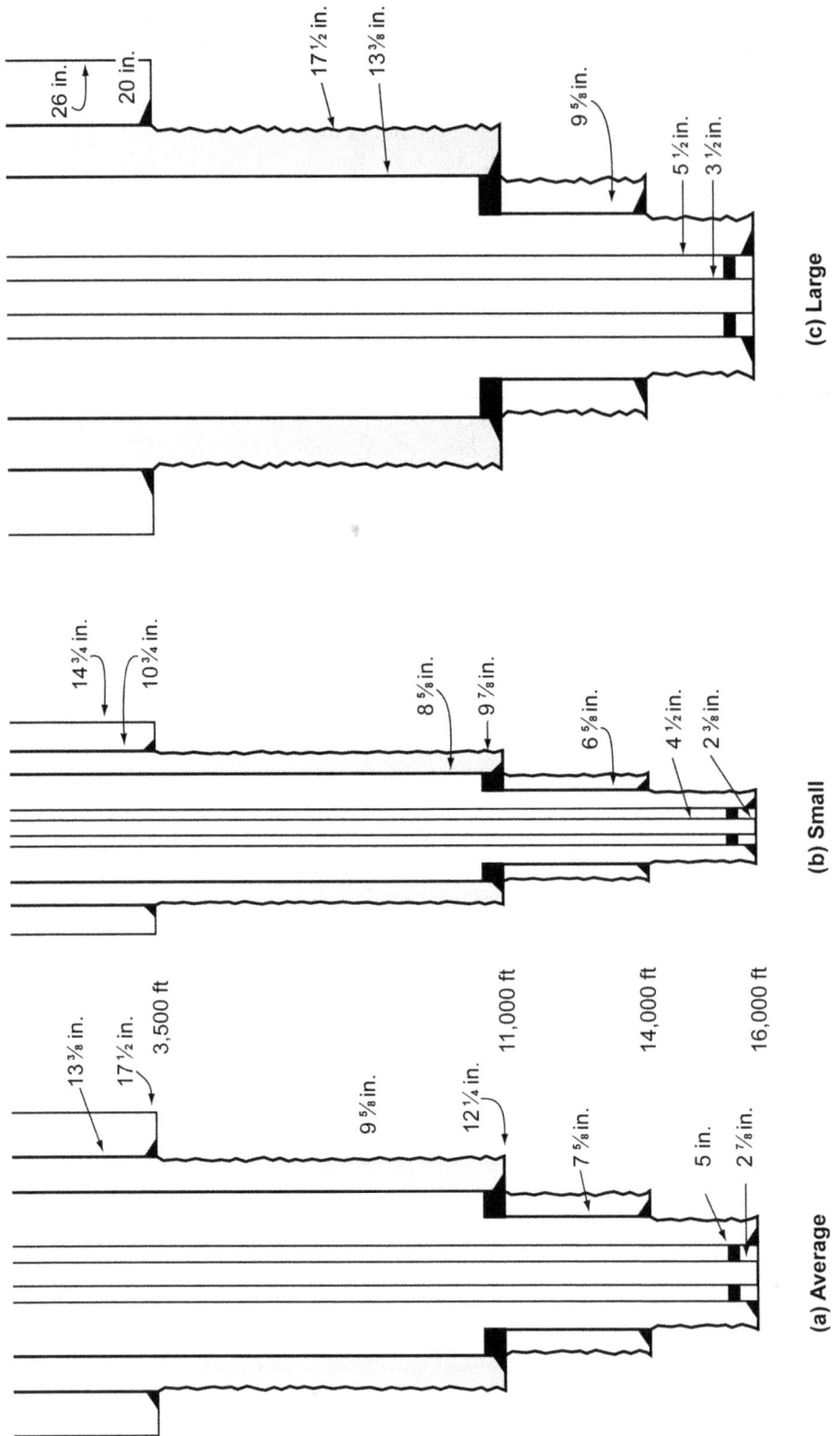

Fig. 11.20—Three hole-size combinations for a well.

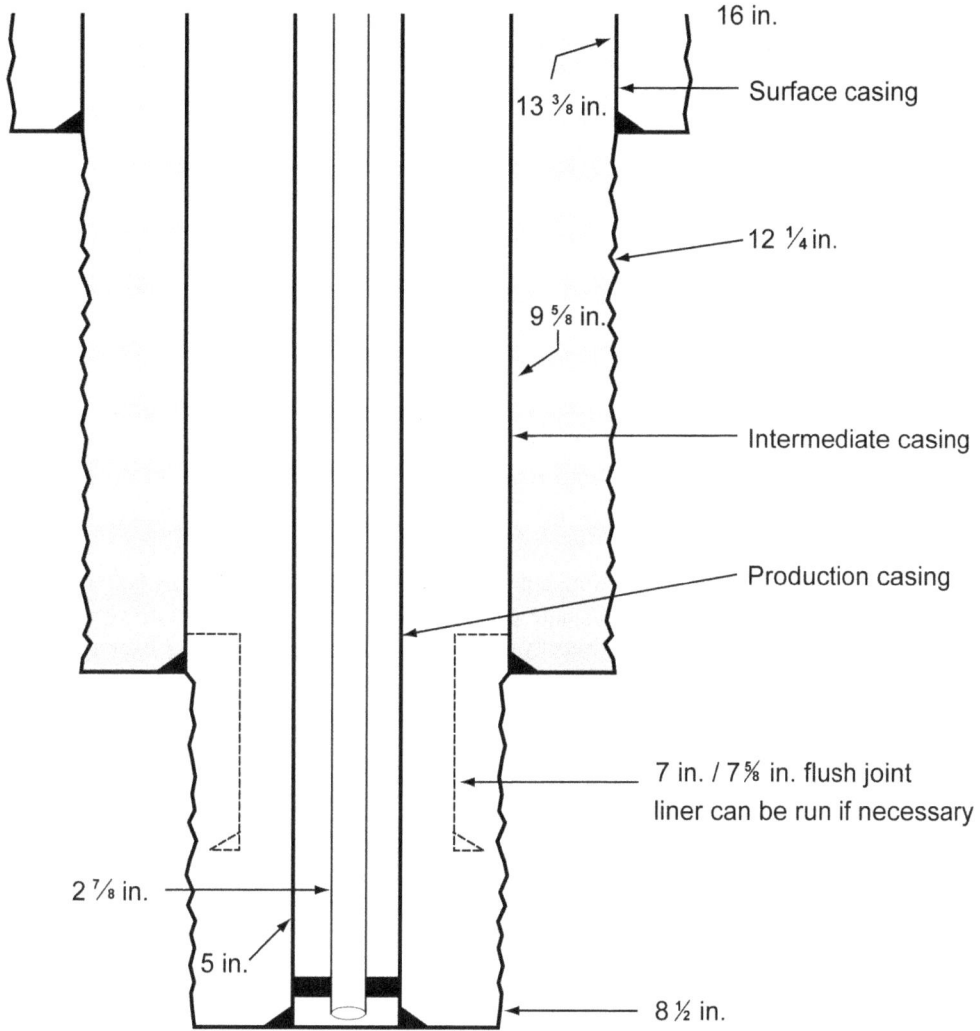

Fig. 11.21—Planning for a hole geometry that allows for liner usage if needed.

where

   v = annular velocity, ft/min;

   Q_p = pump output, bbl/min;

and

   V = annular volume, bbl/1,000 ft.

The annular volume, in bbl/1,000 ft, can be estimated from the rule-of-thumb guide in Eq. 11.7.

$$V_a = d_H^2 - d_{DS}^2, \text{..} (11.7)$$

where

   d_H and d_{DS} = hole and drillstring diameter, in.

| TABLE 11.4—CASING DESIGN REQUIREMENTS FOR 13³/₈-IN. CASING IN FIG. 11.20 ||||
| Drilling Conditions* || Casing ||
Maximum Mud Wt., lbm/gal	Surface Pressure, psi	Weight, lbm/ft	Grade
13	3,000	72.0	S-95
	5,000	72.0	L-125
14	3,000	80.7	S-95
	5,000	80.7	L-125
15	3,000	80.7	L-125
	5,000	80.7	L-125
16	3,000	80.7	L-125
	5,000	80.7	L-125
	7,500	80.7	L-125
	10,000	**	**

*Liner fracture mud weight: 16.5 lbm/gal.
Mud weight casing set in: 10.0 lbm/gal.
15.6-lbm/gal cement to 8,000 ft.
** Requires special pipe.

As an example, an 8½ × 4½-in. annulus has approximately 52 bbl/1,000 ft of annulus. Many drilling rigs do not have adequate pump horsepower to clean the surface regions of the hole and, as such, rely on high-viscosity-gel plugs to clean the annulus. Example 11.7 illustrates the hole-cleaning problem.

Example 11.7 Use the hole geometries in Fig. 11.20 to determine the required flow rate to achieve an annular velocity of 75 ft/min. In addition, determine the surface horsepower required if the pump pressure is limited to 2,500 psi. Use 5-in. drillpipe for A and C and 4½-in. pipe for B.

Solution.

1. From Fig. 11.20, the annular geometries in the largest hole sections are

Fig. 11.20	d_H, in.	d_{DS}, in.
A	17.5	5
B	14.75	4.5
C	26	5

2. Use Eq. 11.5 to determine the required pump rate for A.

$$v = \frac{24.50 \ Q}{d_H^2 - d_{DS}^2}.$$

$$75 = \frac{24.50 \ Q.}{17.5^2 - 5^2}$$

$$Q = 860 \ \text{gal/min}.$$

Likewise, for B, $Q = 640$ gal/min;

and for C, $Q = 1,992$ gal/min.

3. Determine the surface horsepower (HP) requirements if the pressure is limited to 2,500 psi. For A,

$$HP = \frac{p_p Q}{1,714}$$

$$= \frac{(2,500 \ psi)(860 \ gal/min)}{1,714}$$

$$= 1,254.$$

Likewise, for B,   HP = 880;

and for C,   HP = 2,905.

Based on results from Example 11.7, hole geometry C will be difficult to clean because many rigs are unable to deliver 2,905 hp under continuous service. Poor hole cleaning is a common cause of annular solids buildup, plugging, and lost circulation.

Most rigs are HP limited when drilling surface hole. Even though a pump may be rated to 3,000 psi, the maximum flow rate usually will be reached before achieving 3,000-psi surface pressure. Typical pressures for surface hole may be 600 to 1,500 psi even when using two pumps. If the pumps are unable to adequately clean the annulus, well-planning provisions must be made for periodic high-viscosity slurries to sweep the annulus.

Small-diameter holes create problems from turbulent erosion and hydraulics. The resultant problems can be cementing difficulties and poor hole cleaning in the enlarged area.

Hydraulics are complicated in the downhole, small-diameter sections. High friction pressures reduce the available hydraulic cleaning action at the bit and increase the chip-holddown effect on the cuttings. Swab and surge pressures can be large and range from 0.3 to 1.0-lbm/gal equivalent mud weight in small holes when heavy muds are used.

Underreaming. This technique enlarges the hole size in excess of the amount attainable with a drill bit. The underreamer tool has expandable arms with bit cones that can be activated with pump pressure. The important negative aspect of underreaming is that the tool arms are frequently damaged or lost in the hole. It is difficult to retrieve a lost underreamer arm.

This technique does have applications in some areas. One important application involves running a liner in an open hole that might be considered too small without underreaming. For example, a 7⅝-in. flush-joint liner run in an 8½-in. hole may be considered unacceptable (by some companies) without underreaming. A 7.0-in. liner may be an alternative, which would result in pipe-size restrictions in deeper sections.

11.4.3 Casing- and Bit-Size Selection. A casing- and bit-size program must consider the problems described in the previous section in addition to the actual casing- and bit-size characteristics. These include casing inner and outer diameter, drift and coupling diameter, and bit size. A working knowledge of these variables is important for selection of a viable geometry program.

Pipe Selection. Casing availability is a priority consideration in hole geometry selection. High-strength casing, often required for deep wells, may have a small (drift) diameter that will influence subsequent casing- and bit-size selection. Unfortunately, supply-and-demand cycles in the pipe industry may control the pipe design rather than engineering considerations.

The casing outer diameter (OD) is available in numerous sizes. The drift diameter, which is smaller than the inner diameter (ID), controls the bit selection for the open hole below the casing. As heavier-weight pipe is required to meet design specifications, the available drift diameter is reduced. A rule-of-thumb that has proved satisfactory in most field cases is to allow

TABLE 11.5—CLEARANCES FOR API AND VARIOUS
PREMIUM PROPRIETARY COUPLINGS

Pipe Size, in.	Coupling Size, in.			
	LTC (API)[1]	SFJ[2]	VAM[3]	IJ-4S[4]
4.5	5.0	4.59	5.106	5.150
5	5.563	5.09	5.391	5.875
5.5	6.050	5.625	5.891	6.375
6.625	7.390	6.75	7.390	7.390
7.625	8.5	7.75	8.504	8.50
8.625	9.625	8.75	9.625	9.625
9.625	10.625	9.75	10.625	10.625
10.75	11.750	10.875	11.748	—

1   Long thread and coupling (same diameter for STC, BTC)
2   Hydril
3   Vallourec
4   Atlas Bradford

1 in. of wall thickness to achieve a suitable design without resorting to the use of ultrahigh-strength pipe. As an example, 9⅞-in. casing can usually be designed properly if 8⅝-in. drift diameters are allowed.

Hole-geometry-selection approach may dictate the casing drift diameter as the controlling criterion. The options are as follows:

• Try to design the pipe under the specific drift and OD conditions.
• Use high-strength materials.
• Use special drift pipe available from some manufacturers.
• As a last resort, pipe manufacturers can prepare a special pipe design based on minimum drift requirements by enlarging the wall thickness and OD.

The fourth option is occasionally required in hydrogen sulfide environments where low-strength metals must be used.

Coupling Selection. Pipe couplings are generally designed to satisfy requirements such as burst, collapse, tension, and sealing effectiveness. However, coupling diameters may be a design guideline in some wells. **Table 11.5** shows the OD of various types of couplings and pipe sizes. American Petroleum Inst. (API) couplings are normally 1 in. larger than the pipe in sizes greater than 7⅝ in.

Advantages are provided by using premium couplings. These couplings usually have clearances less than comparable API connections and occasionally allow the use of smaller pipe in a well. In many cases, more-expensive premium couplings can reduce the total well cost by allowing smaller pipe and hole geometries. In Fig. 11.20b, the hole geometry would not be difficult to achieve if premium couplings were used, whereas clearances might be unacceptable if API couplings were used.

Bit-Size Selection. Sizing the bit program is dependent on the required casing sizes. Bits are available in almost any desired size range. However, nonstandard bits or unusual sizes may not possess all of the desirable features, such as center-jet or gauge-protection characteristics. In addition, bit selection and availability become more difficult in odd or small bit sizes (less than 6.5 in.).

Table 11.6 illustrates size availability for Hughes insert-tooth bits. Bit sizes less than 6½ in. restrict bit-type selection. In addition, bit selection is restricted for sizes greater than 12¼ in.

11.4.4 Standard Bit/Casing Combinations. Fig. 11.22 can be used to select casing and bit sizes required to fulfill many drilling programs. To use the chart, determine the casing or liner

TABLE 11.6—SIZE AVAILABILITY FOR VARIOUS INSERT-TOOTH, JOURNAL-BEARING BITS					
	IADC Designation				
Bit Size, in.	5,1,7	5,3,7	6,1,7 6,3,7	7,3,7	8,3,7
4.75			X		
5.875		X	X		
6		X	X		
6.125		X	X		
6.25		X	X		
6.5		X	X	X	X
6.75			X		
7.875	X	X	X	X	X
8.375		X			
8.5	X	X	X	X	X
8.75	X	X	X	X	X
9.5		X	X	X	
9.785	X	X	X	X	X
10.625		X	X		
11			X		
12.25	X	X	X	X	X
17.5		X			

size for the last size of pipe to be run. The flow of the chart indicates hole sizes that may be required to set that size of pipe (i.e., 5-in. liner inside 6⅛- or 6¼-in. hole).

Solid lines indicate commonly used bits for that size pipe that can be considered to have adequate clearance to run and cement the casing or liner (i.e., 5½-in. casing in a 7⅞-in. hole). The broken lines indicate less-commonly-used hole sizes. The selection of one of these broken paths requires that special attention be given to the connection, mud weight, cementing, and doglegs. Bicentered bits provide more flexibility in bit and hole size.

11.5 Preparation of Authority for Expenditures (AFE)
Preparing cost estimates for a well and getting management approval in the form of an AFE is the final step in well planning. The AFE is often accompanied by a projected payout schedule or revenue forecast. Although an essential part of well planning, the cost estimate is often the most difficult to obtain with any degree of reliability.

A properly prepared well cost estimate may require as much engineering work as the well design. The costs should address dry holes and completed wells. In addition, accounting considerations such as tangible and intangible items must be taken into account. Unfortunately, many cost "guestimates" are the "back of the napkin" type, with only a small amount of engineering work used in the process.

The cost estimate is the last item to be considered in the well plan because it is heavily dependent on the technical aspects of the projected well. After the technical aspects are established, the expected time required to drill the well must be determined. The actual well cost is obtained by integrating expected drilling and completion times with the well design.

11.5.1 Projected Drilling Time. The time required to drill the well has a significant impact on many items in the cost estimate. These items include drilling rig, mud, offshore transportation, rental tools, and support services. The effect of these items on the overall well cost is dependent on the actual unit cost (i.e., U.S. $15,000/day for a land rig vs. U.S. $250,000/day for a drillship, and the amount of drilling time).

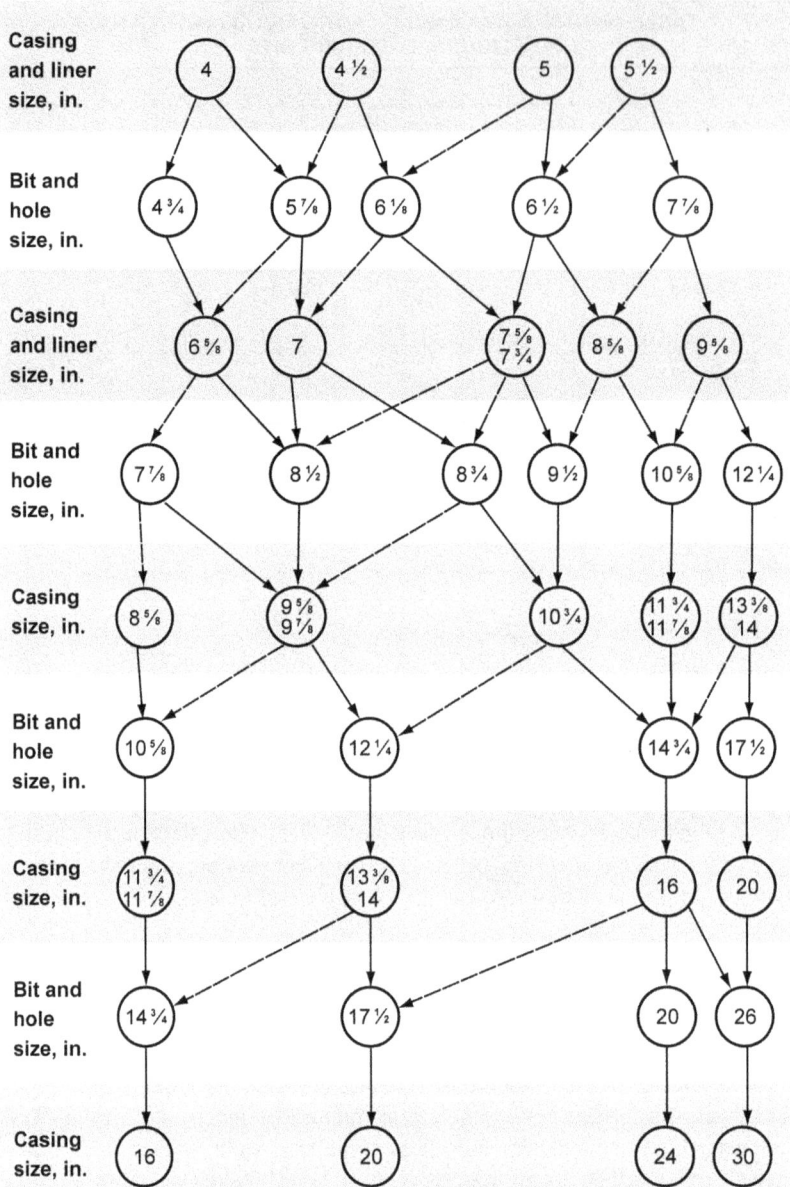

Fig. 11.22—Casing- and bit-size selection chart (courtesy of *Oil & Gas Journal*).

Consider the well in **Fig. 11.23**. Assume the well will be drilled in east Texas. **Table 11.7** summarizes the projected times for the well in three cases and illustrates the cost differences. The worst case has a 21% greater cost than the best drilling times. This example illustrates the importance of preparing accurate projections for drilling time, or "depth vs. days," as it is often termed. A typical depth-vs.-days plot is shown in **Fig. 11.24**.

Drilling-Time Information. Numerous sources are available to estimate drilling times for a well. These include bit and mud records, log header information, and operator's well histories.

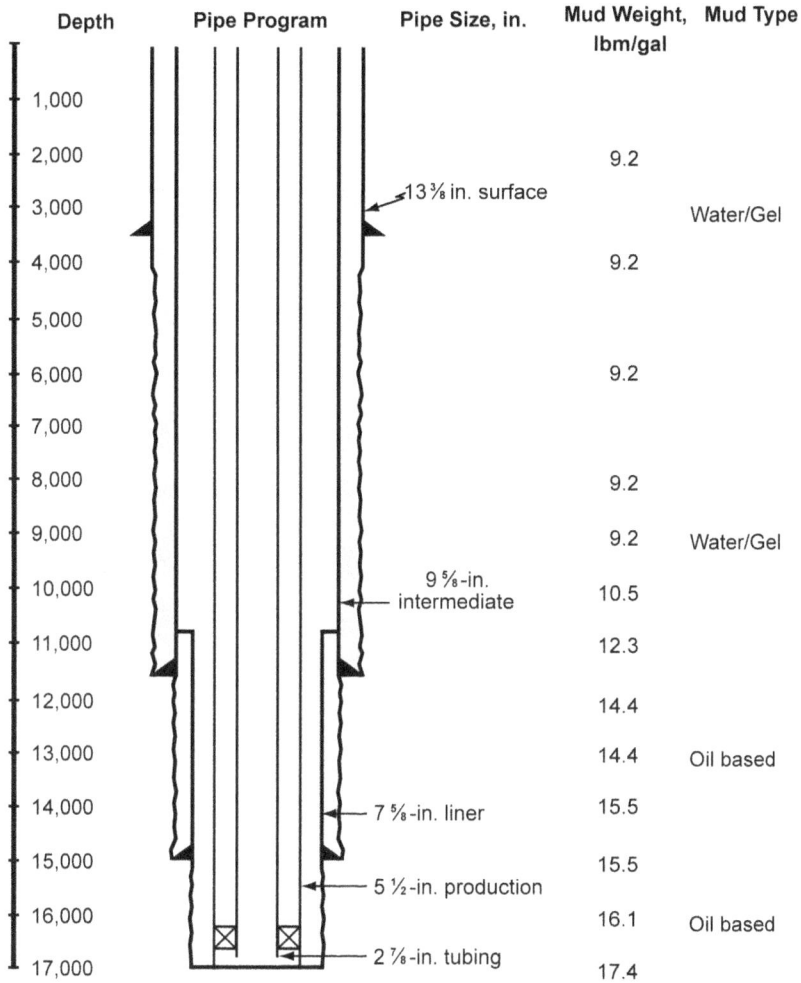

Depth	Pipe Program	Pipe Size, in.	Mud Weight, lbm/gal	Mud Type

Fig. 11.23—Example hole configuration.

Other items such as scout tickets and production histories provide information that will affect the time projections.

Bit records are valuable sources to estimate drilling time. Although few bit manufacturers incorporate a column for dates in the depth-record forms, most drilling engineers who routinely complete the forms make notes in the remarks column as to the time or date the bits were run. In addition, most records contain the dates for well spudding, completion, and pipe setting. Additional inferences can be made from the individual bit-life hours and the cumulative drilling time for each well.

Mud records usually provide the most authoritative information about the drilling-time data. These records are maintained daily and usually contain remarks about the time required for each drilling activity. In addition, time allocated to hole problems can be evaluated to determine if the same amount of time should be included in the upcoming well. For example, hole sloughing may be an expected occurrence in an area, while kicks and twist-offs are unusual activities.

Log header data contain some drilling-time information and dates for each successive logging run. In addition, scout tickets attached to some logs include spud and completion dates.

TABLE 11.7—DRILLING TIMES AND ASSOCIATED COSTS FOR FIG. 11.23			
Item	Time, days		
	1	2	3
Move in and out	8	8	8
Drill			
0 to 200 ft	1	1	1
200 to 3,580 ft	2	3	4
3,580 to 11,600 ft	8	12	18
11,600 to 15,000 ft	15	22	32
15,000 to 17,000 ft	12	18	24
Running casing/cementing	8	8	8
Logging	5	5	5
Completion	7	7	7
Total	66	84	107
Well Cost	U.S. $4,175,000	U.S. $4,565,000	U.S. $5,045,000

The operator well histories provide a comprehensive evaluation of drilling times and offset wells. Although not generally available to noncompany personnel, the histories should contain all previously described sources of information as well as geological and production data. These operator records, when available, should be the basis for the drilling-time projections on the prospect.

Scout tickets and production histories can be valuable to supplement depth-vs.-days projections. Significant production from a zone may significantly reduce formation pressures, which can induce pipe sticking or lost-circulation problems. Infill drilling or drilling adjacent to two producing wells or fields must include this factor in the time estimate for the new well.

11.5.2 Time Categories. Drilling times are usually categorized for dry holes and completed wells. These categories are important as the management decision guide to evaluate potential risk vs. production economics. The dry hole assumes that all casing strings had been run except for production casing and tubing. Dry holes must include time allotments for setting several cased and openhole plugs and the possible retrieval of some casing. Completed wells normally include all well-completion operations up to the point of building production facilities. Well testing is usually included in the time for completion.

11.5.3 Time Considerations. Several factors affect the amount of time spent in drilling a well.
- Drill rate.
- Trip time.
- Hole problems.
- Casing running.
- Directional drilling.
- Completion type.
- Move-in and move-out with the rig.
- Weather.

Each factor may vary with geology, geographical location, operator philosophy and efficiency.

Drill Rate. The cumulative drilling time spent on a well depends primarily on rock type and bit selection. Hard-rock drilling usually needs significantly more drilling time than soft-rock drilling. In addition, the wide variety of bits available to the industry makes bit selection an important factor in drilling hard and soft formations. Other items that usually affect the drill

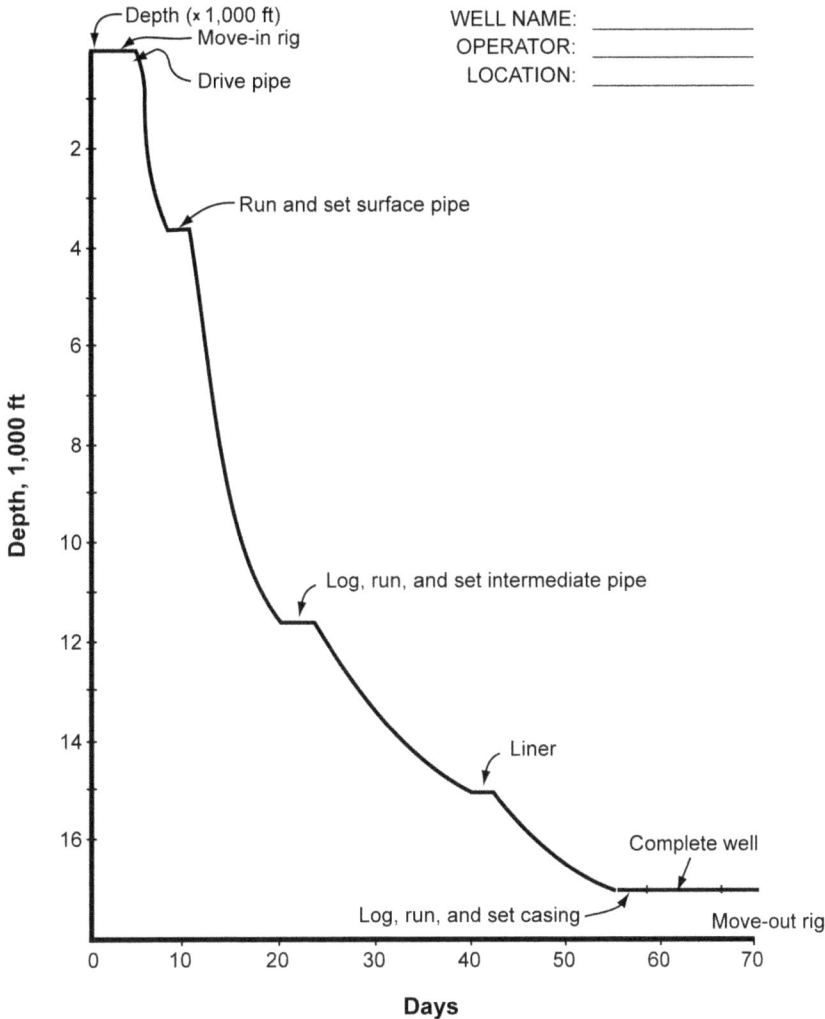

Fig. 11.24—Depth-vs.-days projection.

rate are proper selection of weight and rotary speeds for optimum drilling, mud type, and differential pressure.

Trip Time. Pulling and running the drillstring is an important item in estimating total rotating time. In many cases, it is equal to or exceeds the on-bottom drilling time. Trip time is dependent on well depth, amount of mud trip margin, hole problems, rig capacity, and crew efficiency. A rule-of-thumb for trip-time estimations is 1 hr/1,000 ft of a well.

Long bit runs from 50 to 200 hours often require a short trip of several thousand feet out of and back into the hole. The purpose of the short trip is to remove or reduce any buildup of filter cake that significantly increases the swabbing tendencies of the drillstring. Short trips are dependent on company philosophy, mud type, and bit life.

Hole Problems. Various hole problems are routinely addressed in the drilling-time projections, while others are considered improbable. For example, severe kicks and blowouts are usually unlikely if the operator devotes sufficient attention to drilling activities. The geological conditions and drilling histories and the area of the prospect well will often define other pertinent hole problems.

The type of problems often regarded as standard are hole sloughing, lost circulation, and slow drilling rates. Many operators have encountered formations that slough or heave into the wellbore regardless of the amount of attention given to the mud systems or well plan. Lost circulation will occur in some formations even if the mud density is approximately equal to that of freshwater. Slow drilling rates will usually occur in environments with high differential pressures, such as the case of formation-pressure regressions while maintaining consistent mud weights. However, these hole problems can be eliminated or mitigated in most areas by exercising good engineering judgment in preparing the well plan.

Casing Running. The time required to run casing into the well is dependent on casing size and depth, hole conditions, crew efficiency, and use special equipment such as pickup machines or electric stabbing boards. A heavy casing string may require that the drillstring be laid down rather than setback in the derrick. In addition, nippling-up the blowout preventers and testing the casing and formation must be considered.

Directional Drilling. Directional control of a well requires increases in the drilling time. These increases apply to (1) attempting to drill a well directionally, or (2) maintain vertical control of a well that has deviation tendencies. The increases in drilling time usually result from obtaining surveys and from the inability to apply desired weights or rotary speeds. Many operators increase the expected drilling time in a directional well by a factor two.

Well Completions. Completion systems vary in complexity and, as a result, have a significant variation in time to implement the system. A standard single, perforated completion can be finished in 6 to 8 days. Dual-completed wells usually require an additional 2 to 3 days. Gravel packs, acidizing, fracturing, and other forms of well treatments must be evaluated on a case-by-case basis. Needless to say, the efficiency of all associated personnel and their experiences with a particular type of completion has a major impact on the required time.

Rig Move-in and Move-out. Rig moving affects several areas of the cost estimate and must be considered in the time projections. Move-in and rig-up occur before spudding the well. Rig-down and move-out occur after well completion. If a completion rig is used rather than the drilling rig for the completion work, an additional rig move must be considered from both a cost and time standpoint.

A rule of thumb for estimating rig moving times is based on the IADC rig hydraulics code of 1, 2, 3, or 4, where the higher numbers represent larger rigs. Codes 1 and 2 can usually move in and out in 4 days because they are frequently mobile and truck mounted. Codes 3 and 4 require approximately 8 days to move in, rig up, and move out. These time estimates affect the move-in cost, supervision time, and overhead allocations.

Weather. The affect of weather on the projected time is not considered in most well plans. As an example, hurricanes and tornadoes cannot be routinely expected. However, weather problems such as those that routinely occur in the North Sea must be considered in the plan.

11.5.4 Cost Categories. The well cost estimate should be divided into several categories for engineering and accounting purposes. Engineering considerations include dry-hole and completed costs, logical grouping such as completion equipment or tubular goods, and convenience groupings such as rental equipment. Accounting considerations include tangible, intangible, and contingency items. The sample AFE summary in **Fig. 11.25** illustrates several cost categories.

11.5.5 Tangible and Intangible Costs. Accounting and tax principles treat tangible and intangible costs in different ways. As a result, they must be segregated in the cost estimate. Although intangible costs are difficult to define precisely, they include expenditures incurred by the operator for labor, fuel, repairs, hauling, and supplies used in (1) drilling, perforating, and cleaning wells, (2) preparing the surface site prior to drilling, and (3) construction derricks, tanks, pipelines, and other structures erected in connection with drilling, but not including the cost of

Operator:			Date:	
Lease:			Field:	
Sec. T R	County:		State:	
EXPENDITURE			Dry Hole (24.5 Days)	Completed (32.5 Days)
Intangible Costs			(U.S.$)	(U.S.$)
100	Location Preparation		30,000.00	65,000.00
200	Drilling Rig and Tools		298,185.75	366,612.94
300	Drilling Fluids		113,543.19	116,976.37
400	Rental Equipment		77,896.37	133,784.75
500	Cementing		49,534.68	54,368.73
600	Support Services		152,285.44	275,647.50
700	Transportation		70,200.00	83,400.00
800	Supervision and Administration		23,282.00	30,790.50
	Subtotal		814,927.94	1,126,581.00
Tangible Costs				
900	Tubular Equipment		406,100.87	846,529.44
1000	Wellhead Equipment		16,864.00	156,201.00
1100	Completion Equipment		00	15,717.00
	Subtotal		422,964.87	1,018,447.44
	Subtotal		1,237,893.00	2,145,028.00
		Contingency (15%)	185,683.94	321,754.25
		Total	142,357.00	2,466,782.00

Fig. 11.25—Summary of the authority for expenditure.

the materials themselves. The fundamental test is defining the salvage value of the item. If the item does not have a salvage value, it is an intangible.

Intangible drilling and development costs do not include the following:

• Tangible property ordinarily considered as having salvage value.

• Wages, fuel, repairs, hauling, supplies, etc., in connection with equipment facilities or structures not incident to or necessary for the drilling of wells, such as structures for storing oil.

• Casing, even though required by state law.

• Installation of production facilities.

• Oilwell pumps, separators, or pipelines.

Detailed Cost Analysis. It is usually desirable to provide more cost detail than the general summary in Fig. 11.25. A sample of a detailed summary is shown in **Fig. 11.26**. Engineers wishing to evaluate detailed cost analysis worksheets should refer to Ref. 1.

Factors considered in the detailed cost analysis will be presented in the next section. The cost divisions presented in Fig. 11.25 will be used. These factors are heavily dependent on company drilling philosophy and, as such, may not apply to all companies.

11.5.6 Location Preparation. Preparing the location to accept the rig is an important cost factor and perhaps the most difficult to quantify. It includes a legal cost, surveying the location site, physical location preparation, and post-drilling cleanup. These costs are affected by the rig type, rig size, and well location.

AFE Detailed Summary

Code	EXPENDITURE	Dry Hole (24.5 Days)	Completed (32.5 Days)
100	Location Preparation		
110	Permit	500.00	2,500.00
120	Survey	2,500.00	7,500.00
130	Right of Way, Special Permits	2,000.00	2,000.00
140	Physical Location Preparation	20,000.00	48,000.00
150	Cleanup	5,000.00	5,000.00
	Category Total	30,000.00	65,000.00
200	Drilling Rig and Tools		
210	Move in and out	57,135.37	57,135.37
220	Footage Bid	.00	.00
230	Straight Day Work Bid	182,327.06	241,862.44
240	Fuel	32,915.79	41,018.13
250	Water	5,000.00	5,000.00
260	Bits	20,807.50	21,597.00
270	Completion	.00	.00
	Category Total	298,185.75	366,612.94
300	Drilling Fluids		
310	Drilling Fluids	113,543.19	113,543.19
320	Packer Fluids	.00	3,433.16
330	Completion Fluids	.00	.00
	Category Total	113,543.19	116,976.37
400	Rental Equipment		
410	Well Control Equipment	29,852.00	43,262.00
420	Rotary Tools and Accessories	6,794.22	22,425.67
430	Mud Related Equipment	19,475.00	33,856.87
440	Casing Tools	21,775.16	44,240.16
450	Miscellaneous	.00	.00
	Category Total	77,896.37	133,784.75
500	Cementing		
510	Conductor Casing	.00	.00
520	Surface Casing	20,121.85	20,120.85
530	Intermediate Casing	15,619.91	15,619.91
540	First Liner	.00	.00
550	Second Liner	.00	.00
560	Production Casing	.00	18,626.97
570	Squeezes	.00	.00
580	Plugs	13,792.92	
	Category Total	49,534.68	54,368.73
600	Support Services		
610	Casing Crews	11,759.15	23,536.71
620	Logging	.00	.00
621	Mud Logging	18,000.00	18,000.00
623	Wireline	77,656.56	109,083.94
624	Logging	.00	11,447.00
625	Perforating	14,480.00	14,480.00
626	Testing	.00	33,597.00
627	Completion	.00	.00
630	Tubular Inspection	.00	.00
631	Surface Casing	4,896.45	4,896.45
632	Intermediate Casing	14,643.30	14,643.30
633	First Liner	.00	.00
635	Production Liner	.00	18,213.00
636	Tie Back String	.00	.00
637	Tubing	.00	13,960.10
638	Miscellaneous	.00	.00
640	Galley	10,850.00	13,790.00
650	Welding, labor	.00	.00
660	Formation Test	.00	.00
670	Consultants	.00	.00
680	Stimulation	.00	.00
690	Miscellaneous	.00	.00
	Category Total	152,285.44	275,647.50
700	Transportation		
710	Trucking	70,200.00	83,400.00
720	Marine	.00	.00
730	Air	.00	.00
	Category Total	70,200.00	83,400.00
800	Supervision and Administration		
810	Field Supervision	16,250.00	20,250.00
820	Office Supervision	7,032.00	10,540.50
830	Insurances, Bonds	.00	.00
	Category Total	23,282.50	30,790.50
900	Tubular Equipment		
905	Drive Pipe	7,498.00	7,498.00
910	Conductor Casing	.00	.00
915	Surface Casing	71,006.56	71,006.56
920	Intermediate Casing	321,156.31	321,156.31
925	First Liner	.00	.00
930	Second Liner	.00	.00
935	Production Casing	.00	325,291.06
940	Tie Back String	.00	.00
950	Tubing	.00	113,048.50
960	Casing Equipment	.00	.00
961	Drive Pipe	230.00	230.00
962	Conductor Casing	.00	.00
963	Surface Casing	3,500.00	3,500.00
964	Intermediate Casing	2,710.00	2,710.00
965	First Liner	.00	.00
966	Second Liner	.00	.00
967	Production Casing	.00	2,089.00
	Category Total	406,100.87	846,529.44
1000	Well Head Equipment		
1010	Casing Head	3,220.00	3,220.00
1020	Intermediate Spool	13,644.00	13,644.00
1030	Tubing Spool	.00	55,465.00
1040	Tree	.00	83,872.00
1050	Miscellaneous	.00	.00
	Category Total	16,864.00	156,201.00
1100	Completion Equipment		
1105	Packers	.00	2,059.00
1110	Blast Joint and Landing Nipples	.00	3,955.00
1115	Special Liners	.00	.00
1120	Safety Joints	.00	796.00
1125	Subsurface Safety Devices	.00	4,388.00
1130	Seal Assembly	.00	4,519.00
1135	Gaslift Equipment	.00	.00
1140	Gravel Packing Equipment	.00	.00
1145	Miscellaneous	.00	15,717.00
	Category Total		

Fig. 11.26—AFE detailed summary.

Location costs include only those variables actually involved with a rig move-in. These costs do not include lease fees or bidding cost. Individual companies must determine appropriate methods for handling these costs in the well cost estimate.

Permits, or "permitting" the well, are required in virtually every drilling area in the world. Some permit procedures are as simple as preparing a few fill-in-the-blank documents, while others may require extensive, time-consuming efforts such as environmental and economic impact statements. Some well permits must be granted from federal or national authorities, while

TABLE 11.8—WELL COST FOR VARIOUS RIG TYPES			
Case	Rig Type	Rig Cost, U.S. dollars/day	Completed Well, U.S. dollars
1	Land	8,500	4,620,000
2	Land	15,000	5,262,000
3	Land	22,500	5,978,000
4	Jackup	35,000	8,354,600

others may be obtained quickly from local agencies. Permitting a well is primarily a legal matter that often requires significant consultation with legal groups.

"Spotting" the well involves surveying the wellsite and determining its exact location. Land sites can be spotted by professional surveyors with the use of local, known markers. Offshore sites are spotted from offset platforms in the area. Satellite surveys can be used when spotting a well in an area, particularly in offshore environments where marker sites such as existing platforms are not available.

Right-of-way from a public access road to the actual drilling site for land wells must be considered. If the off-road distance is small or through single owner land, the permit may be obtained quite easily in some cases. Difficulties may arise for distant locations, multiple landowners, or public access areas. As in the case of obtaining permits, right-of-ways are often a matter for the legal departments.

Preparing the location to accept the rig depends on the rig type and size, as well as the location. Land rigs may require the construction of a board road and location if the soil is too soft to support transport vehicles and the rig. Sometimes pilings are required under the substructure. The size of the turnaround and the number of board plys will increase with larger rigs. Mountainous locations may need a road built to the site. In addition, factors such as the size of the mud reserve pit and the chemicals storage area depend on drilling times, mud types, and mud weights.

Marsh areas usually require that a canal or channel be dredged to the site. The depth and width of the canal must be coordinated with the size of the rig. The rigsite at the end of the canal is a larger area that must be dredged. Shell pads for a rig foundation may be required in marshy areas if the water depth is sufficiently deep to prevent the direct use of a barge rig or if the seabed is very soft or erodes because of subsea currents.

Offshore sites often require the least amount of location preparation. If surveys of the seafloor show that no obstructions are present, the rig can be moved to the site with no additional efforts. Floating rigs are seldom troubled with soft subsurface formations that may hamper settling of the legs for jackup rigs.

Location cleanup after drilling has been completed is currently undergoing close scrutiny by regulatory bodies. Most sites must be restored to a predrilling condition that may involve site leveling, trucking, and in some cases replanting wildlife vegetation. Offshore sites usually are required to ensure that no remaining obstructions will hamper commercial fishing operations.

11.5.7 Drilling Rig and Tools. The cost for drilling and completion rigs plus the associated drilling tools can be a substantial fraction of the total drilling costs. Consider drilling and completing the well in Fig. 11.24 in 75 days and use the rig costs shown in **Table 11.8** for purposes of this example.

The first three cases used the same well design criteria and equipment (i.e., casing, mud, and logging—with the exception of the rig cost). Case 4 uses the same well in an offshore environment, resulting in the need for a jackup rig. As a result, it is easily seen that careful attention must be given to defining cost for the drilling rig and tools.

Move-in and Move-out. Moving the rig into the location before drilling the well and out of the location after it is completed can be a substantial cost item. Jack-up rigs require a fleet of tugboats, while drillships may be able to move themselves onto the location. Many states have published tariffs that specify the allowable trucking charges for various types of moves. Large land rigs are normally transported by truck to the location. Generally, IADC Type 3 and 4 rigs are sufficiently large that they must be transported piecewise by truck. Types 1 and 2 are usually truck-mounted rigs, which reduces the moving time and associated trucking requirements.

Procedures for estimating rig cost can be developed with the rig cost and average moving times. A survey of numerous drilling contractors showed that Type 1 and 2 rigs usually require approximately 4 days for move-in, rig-up, rig-down, and move-out. Type-3 and -4 rigs required 8 days for land and offshore rates, although the elements of this time value are different (i.e., land rigs are transported by truck while jackups are towed by boat.

The cost for move-in and move-out is estimated as the standby rig rate over the moving time (4 or 8 days). The standby rate is slightly less than the day rate for drilling and may include support services such as crewboats that would be required for normal drilling operations. This method for estimating the rig moving costs is effective and reasonably accurate. It is not useful, however, in unusual circumstances such as overseas rig moves and drillsites requiring helicopter transportation.

Footage Bid. Many operators prefer to drill on a footage or turnkey basis. The drilling contractor provides a bid to drill the well to a certain depth, or until a certain event, such as encountering a particular formation, kickoff point, or geopressure. Footage contracts may call for drilling and casing a certain size hole through or to the expected pay zone. Contract clauses may allow reversion to day work (flat rate per day) if a marked increase in drilling hazards (loss of circulation, kick, etc.) occurs. For example, ABC Oil Co. may contract XYZ Drilling Co. to drill a well to 10,000 ft for a flat fee of U.S. $27.50/ft. The drilling company is responsible for all well operations until the contracted depth is reached.

The footage contract defines cost responsibilities for both parties. The operator may pay for all pipe, cement, logging, and mud cost. The contractor is responsible for all rig-associated costs such as move-in and move-out, drilling time, and bits. At the target depth or operation, all costs and operational responsibilities revert to the operator.

This contract arrangement can offer significant advantages to both parties. Operators are not required to staff a drilling department for drilling a single well or a few wells. The drilling contractor, with proper bid preparation and efficient drilling practices, can gain a greater profit than while on straight day-work rates. Possible problem areas for the drilling contractor include mechanical breakdowns creating unexpected costs, poor well planning, geological anomalies, or "force majeure" situations.

Day-Work Bid. Perhaps the most common drilling contract is the day-work rate. The contractor furnishes the rig at a contracted cost per day. The operator directs all drilling activities and is responsible for the well-being of the hole. The rig may be with or without crews or drillpipe. In addition, options such as high-pressure blowout preventers (BOPs) or sophisticated solids-control equipment required by the operator must be furnished at his own expense.

Rig selection and cost depend on the well. Although rigs are often rated by their capability to drill to a certain depth, the controlling criterion is usually the casing-running capability (i.e., derrick and substructure capacity). A rig rated for 18,000 ft of drilling may not be capable of running 15,000 ft of heavy 9⅝-in. casing. Therefore, the well plan must be developed and analyzed before rig selection.

Rig costs vary considerably and are dependent on items such as supply and demand, rig characteristics, and standard items found on the rig. Results of a study to compare U.S.-operated rig costs are shown in **Fig. 11.27.** The guidelines were the rig's derrick and structure capacity and disregarded items such as optional equipment that might otherwise be rented for

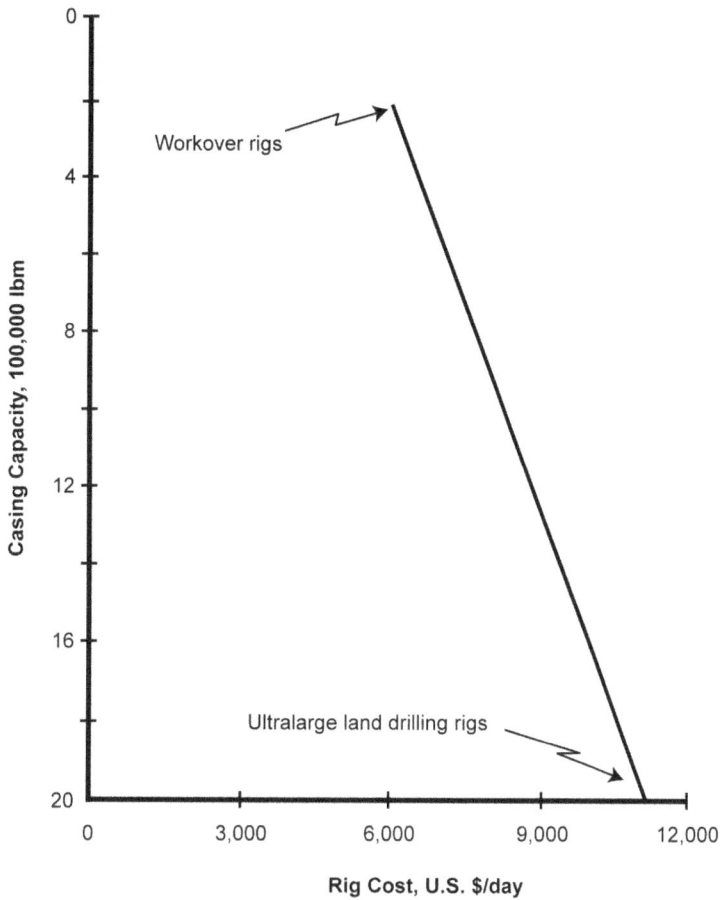

Fig. 11.27—Rig cost compared to casing capacity of the derrick and substructure.

lesser rigs. An interesting point on the illustration is that the over-supply rig costs were reasonably equal regardless of the rig size (i.e., U.S. $6,000 vs. $9,500/day for small to very large rigs).

Standby rates for drilling rigs usually range from U.S. $200 to $500/day less than the amounts shown in Fig. 11.27. The rates include crews and drillpipe. The costs are used to estimate move-in and move-out charges.

Fuel. Drilling contracts are either inclusive or exclusive of fuel on the rig. This major contract policy change occurred in the late 1970s when fuel charges increased from $0.20 to $1.20/gal.

Fuel usage is dependent on equipment type and rig. Fuel consumption rates were evaluated in the study previously described for rig cost rates. The results are shown in **Fig. 11.28**. The average consumption rate is evaluated as a function of the rig size measured by its ability to run casing.

Water. A supply of water is an important consideration. The water is used to wash the rig, mix mud and cement, and cool the engines and equipment.

Water can be supplied in three ways. A shallow water well can be drilled. This method is common in most land operations, but it is not feasible offshore or with deepwater tables on land. Water can be transported to the rig by means of truck, pipelines, barges, or boats. In addition, offshore rigs can use seawater.

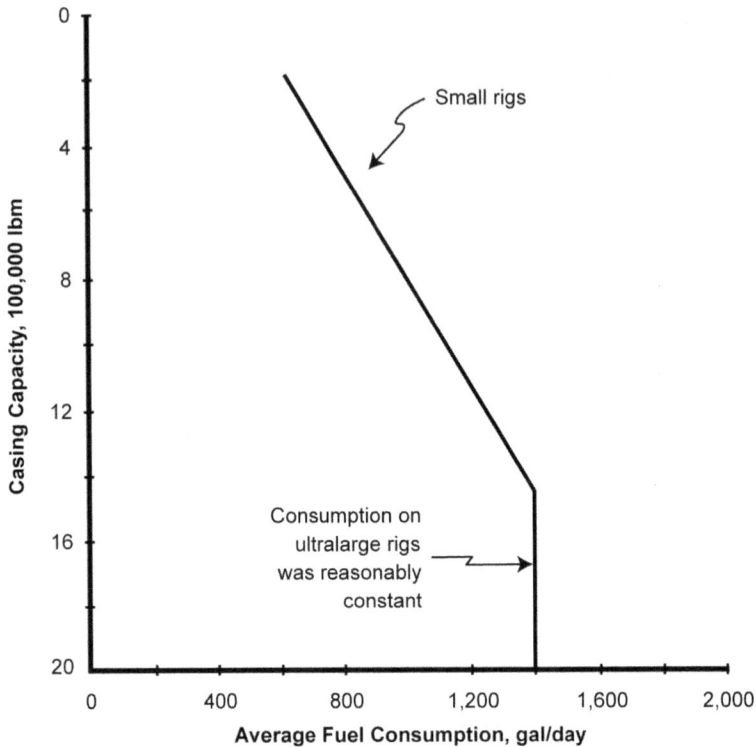

Fig. 11.28—Average fuel consumption per day for rigs with various casing capacities.

Many engineers use a value of U.S. $5,000 to $10,000 for water costs. This amount is approximately the cost to drill a shallow water well. It is also a fair estimate of the cost to lay a water line from a nearby water source. In any case, water costs are seldom considered as a major impact on the total cost estimate.

Bits. Establishing a bit cost depends on the number, size, and type of bits and their respective cost. The bit type, size, and number should have been previously defined in the well plan by the time the AFE is prepared. If the bit is a standard IADC-code bit, published prices are available. Prices are not readily available for specialty bits or for diamond and polycrystalline bits.

Diamond-bit costs depend on the bit size as well as the diamond size, spacing, and quality. In most cases, these bits are made upon demand and are not off-the-shelf items. A rule-of-thumb cost guide for diamond bits is $2,500/in. of bit diameter. For example, a 10-in. bit would cost approximately U.S. $25,000. Salvage values of up to 40% of the bit cost are often granted on used bits. From a conservative view, many engineers prefer to disregard bit salvage value when estimating bit costs, in case the bit is completely destroyed.

Polycrystalline bits are a staple in the drilling industry. Their physical structure, drilling performance, and cost are significantly different from roller-cone or diamond bits. Sample costs for these bits are shown in **Table 11.9.**

Completion Rigs. A completion rig is a small workover rig that costs considerably less than a large drilling rig. Operators often use these rigs when the completion procedures are expected to require significant amounts of time. The drilling rig is used until the production casing is run and cemented.

TABLE 11.9—POLYCRYSTALLINE-BIT COSTS	
Size, in.	Cost, U.S. dollars
6	8,750
6.25	9,000
6.5	10,000
6.75	11,000
7.875	13,750
8.5	15,250
8.75	15,500
9.875	18,500
10.625	20,500
12.25	26,000
14.75	31,000
17.50	45,000

Costs for completion rigs can be determined from Fig. 11.27. Tubing or small drillstring load requirements are used instead of casing capacity. Economic decisions to use a completion rig must also consider the cost of the rig moving onto the location, as well as the daily-rate differences between the drilling and completion rigs.

11.5.8 Drilling Fluids. Drilling fluids are an important part of the well plan and drilling program. The prices are based on build cost for a certain mud weight and a daily maintenance expense. These costs vary from different mud types and are dependent on the chemicals and weighting material required and on the base fluid phase, such as water or oil. Miscellaneous cost factors include specialty products such as hydrogen sulfide scavengers, lost-circulation materials, and hole-stability chemicals.

The build cost for a mud system (**Fig. 11.29**) is the price for the individual components and mixing requirements. Oil-based muds have a higher build cost than most water-based muds because of the expensive oil phase, the mixing and emulsion-stability chemicals, and the additional barite required to achieve comparable densities with water-based muds. **Fig. 11.30** shows a comparison of build costs for an oil-based mud (invert type) and a lignosulfonate mud. The total build cost includes purchasing the initial mud system and the expenses involved with increasing mud weight in the well as it is drilled.

The maintenance costs for deep, high-pressure wells are usually larger than the build costs. The maintenance fee includes the chemicals required daily to maintain the desired mud properties. These chemicals include fluid-loss agents, thinners, and caustic soda.

Fig. 11.30 shows an estimate of empirically derived maintenance costs for invert emulsion, oil muds, and lignosulfonate water muds. The illustration demonstrates that heavy muds can have high daily fees. A system with 1,000 bbl of 16.0-lbm/gal lignosulfonate mud would cost approximately U.S. $2,700 for daily maintenance. In addition, note that the maintenance costs for invert-emulsion muds is significantly less than that for lignosulfonate muds, even though the reverse is true for build costs.

Several additional factors affect mud costs. Small mud companies can often provide less-expensive mud systems than larger companies, although a sacrifice is made occasionally in terms of technical support and mud-problem testing capabilities. In addition, many mud companies offer mud without technical support at a price reduction over mud with engineering support.

Packer Fluids. Packer fluids are placed between the tubing and production casing above the packer. The fluid is usually a treated brine but can be an oil mud or treated water-based mud-type fluid. In some cases, a packer fluid will not be used. Although a low-density brine is commonly used, occasionally a higher-density water or mud is used for pressure control.

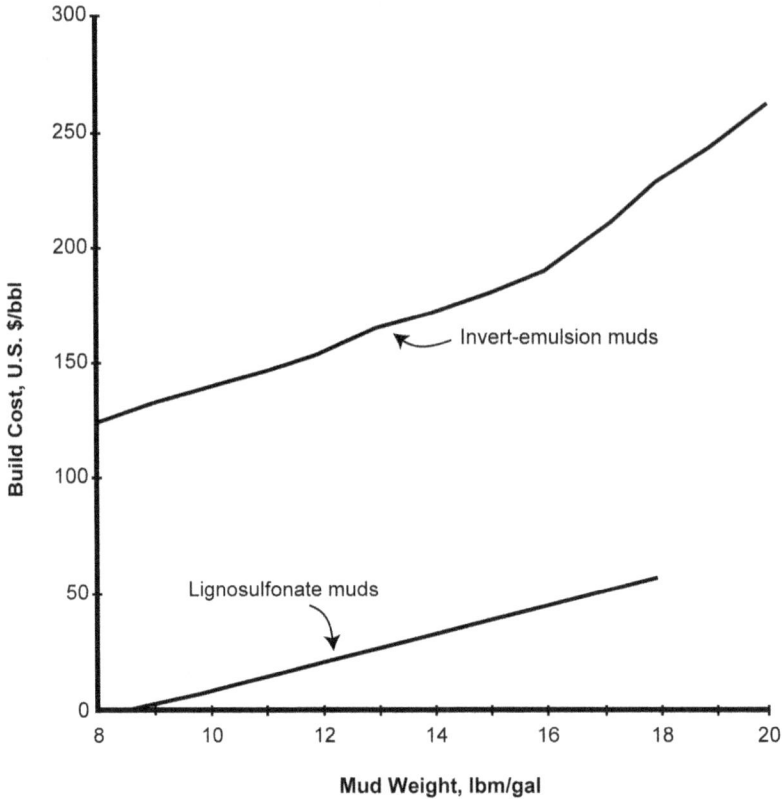

Fig. 11.29—Build costs for invert-emulsion and lignosulfonate muds.

Completion Fluids. Special fluids are occasionally used for well-completion purposes. They are usually designed to minimize formation damage. The fluids may be filtered brine, nitrogen, or oil. Costs for these fluids must be considered on a case-by-case basis.

11.5.9 Rental Equipment. Drilling equipment that is beyond the scope of the contractor-furnished items is almost always required to drill a well. These items must be rented at the expense of the contractor or operator, depending on the provisions of the contract. They can include well-control equipment, rotary tools and accessories, mud-related equipment, and casing tools. These items can represent a substantial sum in deep, high-pressure wells.

Well-Control Equipment. Drilling contractors usually furnish BOPs, chokes, choke manifolds, and, in some cases, atmospheric degasser units. However, the equipment may not be satisfactory for a particular well. In addition, some land rigs currently operate with well-control equipment that is not state of the art, such as positive chokes, manual chokes, and manifold systems that do not have centrally located drillpipe- and casing-pressure gauges.

BOP rental is expensive. High-pressure stacks range from U.S. $1,500 to $3,000/day, exclusive of chokes or manifolds. The operator must define the worst pressure case that can feasibly be attained and select preventers accordingly. Cost estimates for a complete stack must consider the spherical, multiple ram sets, spools, studs, ring gaskets, and outlet valves.

Remotely controlled, hydraulic adjustable chokes are considered state-of-the-art and are available from several sources. Contractors seldom furnish this type of choke primarily because operators have always assumed this cost responsibility. These chokes usually cost U.S. $50 to

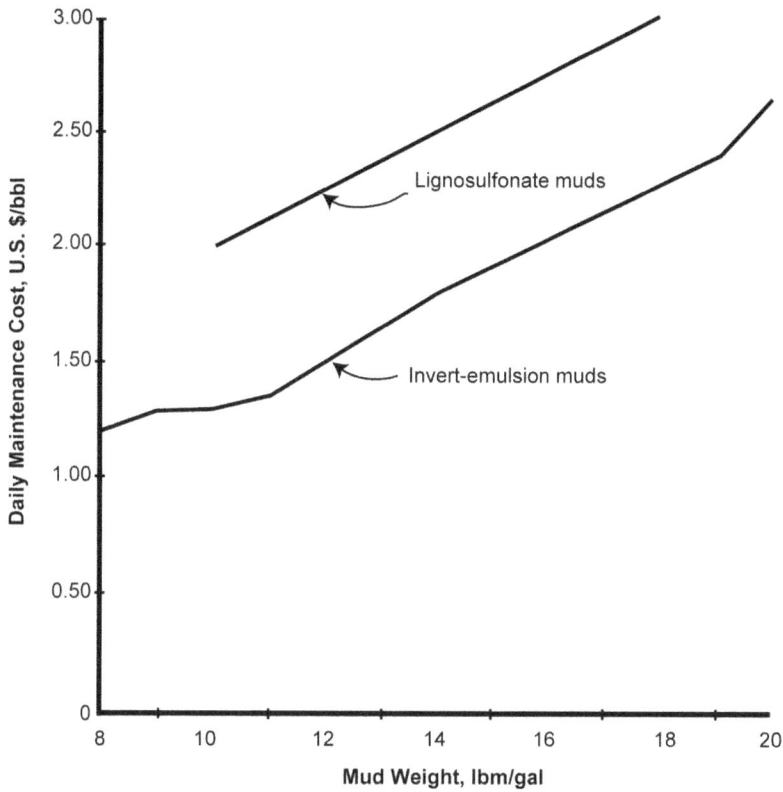

Fig. 11.30—Empirical maintenance costs for invert-emulsion and lignosulfonate muds.

$125/day with a 30-day minimum charge. Choke manifolds must be designed to withstand the maximum pressure ratings in addition to coinciding with current company philosophy.

Rotary Tools and Accessories. Rotary tools are items related to the drillstring or equipment that turns the string. The operator may be required to furnish (1) support equipment for the contractor's drillpipe, or (2) a completely different string if the contractor's drillpipe does not meet the requirements (i.e., tapered or work strings). Some of the items that may require consideration include drillpipe, drill collars, Kelly drive bushing, Kelly cock valves (upper and lower), inside BOP, full-opening safety valves (FOSV), safety clamps, elevators, slips, and pipe rubbers. The operator must evaluate the requirements for drillpipe sizes different from those offered by the contractor's rig. A recent study of U.S. rigs showed that pipe sizes on the rig could be correlated with the IADC hydraulics code (**Table 11.10**). In addition, Table 11.10 includes guides for drill-collar and casing combinations.

For example, 4.5-in. drillpipe with 6.5-in. collars would not be recommended for drilling inside of 7.625-in. casing because of the wear of the tool joints and collars on the casing. A smaller pipe- and collar-size combination would be recommended. If the 7.625-in. pipe were a drilling liner, a tapered string would be satisfactory, but an extra BOP might be required.

A work string consists of small-diameter drillpipe and collars. It is used generally during completions or workover operations. Because the pipe will be used inside production casing, the usual sizes are 2.375 to 3.5 in. Most operations require a rental string because few rigs drill with this size pipe.

Mud-Related Equipment. A properly maintained mud system offers many benefits to the operator. To achieve the desired level of system efficiency, several specialized pieces of equip-

TABLE 11.10—DRILLPIPE AND COLLAR SIZE COMBINATIONS			
IADC Code*	Drillpipe*, in.	Drill Collars**, in.	Maximum Acceptable Casing Size***, in.
1	2.875	3.25 to 3.75	4.5
2	3.5	5.0 to 5.5	6.625
3	4.5	6.0 to 6.5	8.625
4	5.0	7.0 to 7.5	>9.75

* Drillpipe sizes commonly found on rigs.
** Size ranges often used for the specified drillpipe.
*** Guide for casing/pipe/collar combinations.

ment may be required. Some of the equipment must be rented, even though the drilling rig may be well equipped with other drilling tools.

A complete suite of equipment required for the mud job usually depends on the mud type and weight. The following suite may be used for mud weights in the 8.33- to 12.0-lbm/gal range.

- Multiscreen shaker.
- Desilter (with pumps).
- Mud/gas separator.
- Degasser (vacuum).
- Pit/flow monitors.
- Drill-rate recorder.
- Gas detector.

Mud weights greater than 12.0 lbm/gal may require the use of additional equipment such as a centrifuge or mud cleaner. Oil muds need a cuttings cleaner to remove the oil from the cuttings prior to dumping.

Casing Tools. Recently, great strides have been made in running casing. Specialized equipment and crews normally handle the task rather than using the rig crew and equipment. Because most rigs are not furnished with casing-running equipment, it must be rented.

Casing tools must be selected according to size and loading requirements. A commonly used method for evaluating the load requirement is to add a design factor of 1.5 to the in-air weight of the casing string. For example, a casing string that weighed 500,000 lbf in air would require 375-ton casing tools.

The suite of equipment to run casing depends on the operator's preference. It can include elevators, slips, bales, protector rubbers, power tongs, a power-tong hydraulic unit, stabbing boards, drift gauges, a thread-cleaning unit, and safety clamps. In addition, it is usually desirable to rent several pieces of backup equipment in case of breakdowns, in most cases an inexpensive type of insurance. These items include backup tongs, a backup power unit, and a backup elevator/slip combination unit. Laydown and pickup machines were introduced to the industry in the late 1970s. These units increase the efficiency and safety of picking pipe up to the rig floor or laying it down on the pipe rack. Also, they usually minimize possible damage to pipe threads.

11.5.10 Cementing. Cost development for cementing charges requires an evaluation of the cement type and volume, spacer-fluid requirements, special additives, and pumping charges. These various charges usually apply for each primary cement job, stage slurries, squeeze slurries, plugs, and surface-casing top-outs. Cost will vary for land and offshore jobs.

Pumping Charges. Onshore and offshore pumping charges for one cementing company are shown in **Fig. 11.31.** The charges increase with depth and for the offshore case. Also, pumping charges for casing and drillpipe will vary.

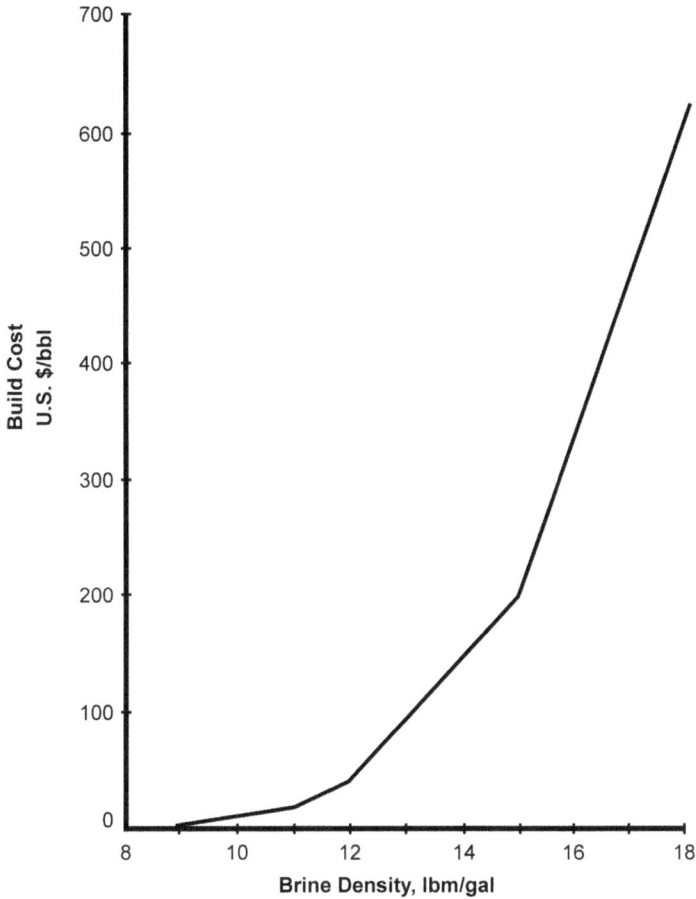

Fig. 11.31—Depth vs. pumping charge.

In addition to the primary cementing pump, most operators use a standby pump unit in case of mechanical failure on the primary unit. The ill effects of cementing-up the casing or drillpipe as a result of equipment failure overshadow the standby pumping unit charges. Rates for land-based standby pump trucks are approximately U.S. $100 to $150/hour.

Cement Spacers. A cement spacer is used to separate the cement from the drilling mud in an effort to reduce cement contamination. The chemical cost for a barrel of spacer fluid is approximately U.S. $50 to $100 depending on the amount of retarder. Barite charges or other weight materials must be added. In addition, diesel charges in the spacer must be considered when the drilling fluid has a continuous oil phase.

Cement and Additives. The major cost for large cement jobs such as surface casing is the chemical and additives charges. Typical costs are listed next.

• Cement U.S. $7.00/sack
• Barite U.S. $15.00/sack
• Gel U.S. $15.00/sack
• Mixing charges U.S. $0.95/ft^3

A reasonable rule-of-thumb for computing the cost of special additives such as water-loss agents and thinners is 75% of the charges for cement, gel, and barite.

Quick-set, top-out cement is often used on surface casing. It provides short-term strength that allows surface-equipment rigging to proceed while waiting on the other cement to cure. The slurry usually consists of 50 to 100 sacks of cement at approximately U.S. $10/sack.

11.5.11 Support Services. Drilling operations require the services of many support groups. In some cases, these groups are used because they can do a particular job more efficiently than the rig crew. An example of this efficiency is casing crews who are experienced in running large-diameter tubulars. Other support groups may provide services that cannot be performed by the rig crew or operator (i.e., well logging, pipe inspection, or specialized completions). Regardless of the reasons for using support services, their costs affect the total well cost and, as such, must be considered.

Casing Crews. During the early years of the drilling industry, the rig crews ran all casing and tubing strings into the well. However, increasing well depths and tubular sizes made the process more difficult. In addition, items such as specialized couplings and pipe torque measurements gave rise to the requirements for the use of casing crews specialized in running the tubulars. Today's industry uses not only casing crews but also groups specialized in picking up and laying down casing, tubing, and drillpipe.

Casing crew charges are dependent on crew size, pipe size, and well depth. Crew sizes usually range from 1 to 5 members. **Fig. 11.32** shows the charges for a 5-member crew. In addition, a power-tong operator is required at rates ranging from U.S. $75 to $125/hour.

Mud Logging. Monitoring services such as mud logging, cuttings interpretation, and gas monitoring are often used on deep or high-pressure wells. A variety of services at different costs are available. A few services and general cost ranges are shown in **Table 11.11.**

Well Logging. Formation-evaluation services, or well logging, are done on every well. The service may include formation evaluation, casing and cement logging, and hole-inclination surveys.

Charges for well logging vary with suppliers. However, some consistency does exist across the industry. Each logging operation will have a flat setup charge for each time the unit is rigged up (i.e., once for openhole logging call-out and once for cased-hole logging at each depth). A depth charge, usually on a per-foot basis, is applied to the deepest depth for each tool run. An operation charge is applied for each foot that the tool is operated. Estimation of the logging cost requires that a well logging program be established (**Table 11.12**). In addition, offshore logging is significantly more expensive than land operations.

Perforating. Perforating charges may not apply if the well is gravel packed or abandoned. The charges include setup, depth charge for minimum shots (usually 20), and a charge per shot over the minimum (**Table 11.13**). The total shots depend on the length of the productive zone and the shot density (e.g., 4 shots/ft). Assuming a setup charge of U.S. $750 and 20 shots as the minimum, Table 11.13 illustrates some of the costs involved with perforating.

Formation Testing. Wireline formation testing is an economical method of obtaining reliable formation information. The repeating formation tester is a device that takes samples of pressure and fluids from a zone of interest. It should be included in the cost estimate for every exploratory well.

Charges for the service are on a depth and per-sample basis. Setup charges are usually not applicable because the service is often run in conjunction with other logs. An example cost for a 15,000-ft sample would be U.S. $2,550/sample with a U.S. $0.55/ft depth charge.

Completion Logging. Various types of production logs can be run on the well if it is completed. These logs are generally run before perforation so that pre- and post-production formation evaluations can be made. Because production logging is a complex subject, the appropriate log suite must be developed jointly by the drilling and production engineers.

Tubular Inspection. Pipe inspection is an important aspect of the casing and tubing program. These support services may include magnetic particle inspection, thread and end-area

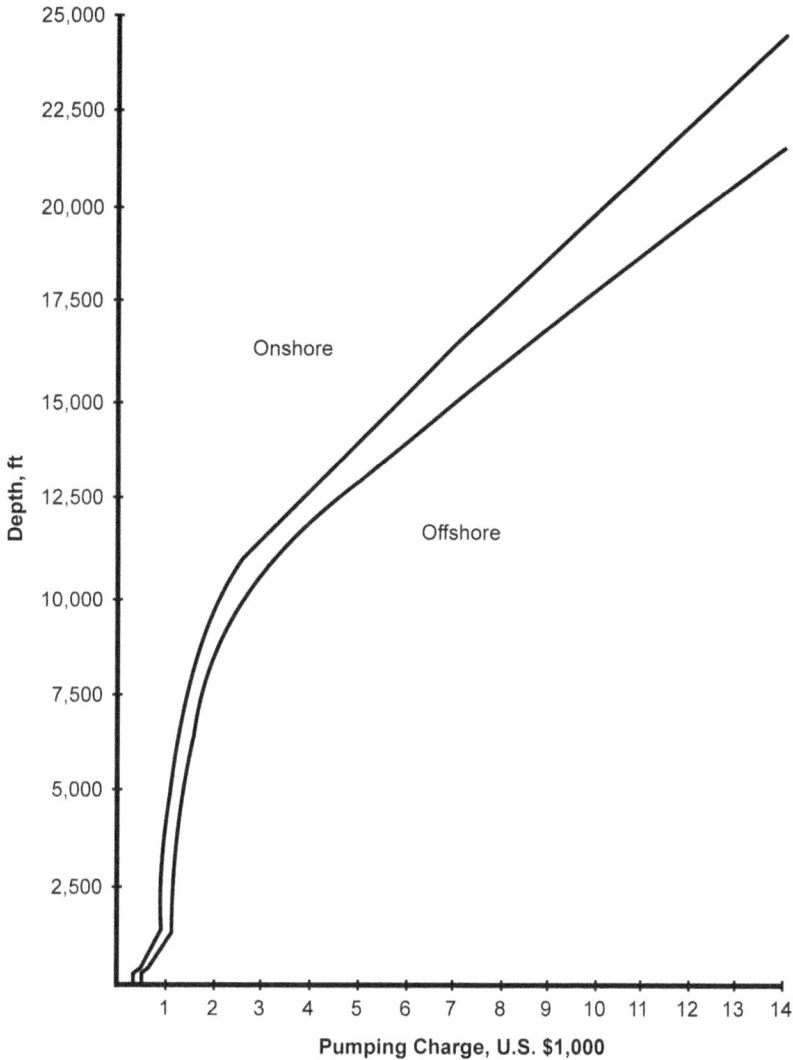

Fig. 11.32—Casing-crew costs for various depths and pipe sizes.

visual inspection, hydrostatic pressure testing, and pipe drifting. Typical charges for the services are U.S. $5 to $30/joint for each item and are service and pipe-size dependent.

Galley Services. Catering services for the galley of offshore or marsh rigs may not be included in the day-rate charges for the rig. The catering company will supply the cooks, support crews, and food for a per-man-day fee. Typical charges are U.S. $50/man-day for crews with less than 30 members and U.S. $47/man-day for crews with more than 30 members. For cost calculation purposes, average crew sizes for various rigs are given next.

Marsh barge: 30 men

Jackup: 50 men

Floater: 70 men

Special Labor. Many items used on the rig and during drilling operations require specialized labor. These services are usually on a per-hour basis and at a minimum charge (4 to 8 hours). Typical considerations are:

TABLE 11.11—MUD-LOGGING EQUIPMENT COSTS

Item	Cost, U.S. dollars/day
Gas detection, portable unit	125 to 200
Gas detection, trailer unit	250 to 300
Mud logging	800 to 1,250
Advanced mud logging	975 to 1,800
Computerized mud logging	1,250 to 3,000

TABLE 11.12—TYPICAL WELL-LOGGING PROGRAM

Item	Use	Minimum Operating Footage
ISF	Each openhole section, except surface, run over entire hole section	2,000
Sonic-BHC	Each openhole section	2,000
High-resolution dipmenter	Bottom 2,000 ft of intermediate hole and all deeper sections, includes cluster computation	2,000
Cores	Take 48 cores on each openhole section except surface	2,000
CBL	Run each casing string except surface	2,000
Caliper	Run all openhole sections except surface	2,000
Directional	Run over entire well	—
Gamma ray/neutron	Run inside of production casing	2,000
Density	Run in production casing only	2,000

TABLE 11.13—PERFORATING COSTS

Depth, ft	20-Shot Minimum Charge, U.S. dollars	Charge per Shot over 20 Minimum, U.S. dollars
5,000	1,555	40
8,000	1,695	44
12,000	2,035	56
15,000	2,535	75
20,000	4,095	136

Welding: drive pipe, casing shoes, and general construction

Rental equipment: equipment installation and repair

Service representatives: packers, wellhead equipment, and chokes

In addition to the hourly charges for this labor, mileage and expenses must be considered.

TABLE 11.14—TRANSPORTATION REQUIREMENTS

Item	Trips	Remarks
Cementing	3/Job	Two pump units, bulk truck
Logging	1/Job	—
Casing	10/String	Casing, casing crews, welders, rental items
Mud	1 (Minimum)	Add 1 trip per lbm/gal of mud weight above 9.0 lbm/gal (i.e., trips for 15.0-lbm/gal mud)
Packer fluid	2	—
Gravel pack	5	Optional

TABLE 11.15—OFFSHORE VESSEL COSTS

Boat Type	Cost/day (including fuel), U.S. dollars
100-ft crew boat	1,800
Small supply boat	3,750
Large supply boat	5,500
70-ft standby boat	1,200

TABLE 11.16—HELICOPTER COSTS

Helicopter Capacity	Base Day Rate, U.S. dollars	Flight Charge, U.S. dollars/hour	Average Time
4	500	147	2 hours/day
11	2,475	450	5 hours/week

11.5.12 Transportation. Well costs are often underestimated because of subtle items such as transportation. For example, trucking charges for cementing a casing string may exceed U.S. $3,000, which includes round-trip charges for two pump units and a bulk truck. Careful evaluation of these charges will provide a better estimate of well costs.

Transportation can include charges for land-based trucks, barges, boats, and helicopters. Long-distance crew charges via commercial or chartered airplanes may be a significant cost. Accurate estimates of transportation costs require a detailed well plan, knowledge of the distance to the rig from local stock points, and rig characteristics such as standard equipment and crew size.

Trucking charges are computed from estimates of the number of trips, the round-trip mileage, and the per-mile cost. Current trucking costs are approximately U.S. $3.50/mile. A rule of thumb for round-trip mileage is to establish a base of 100 miles from the local stock point to the rig (round trip, 200 miles). **Table 11.14** gives some guidelines for estimating the number of round trips to be considered on a well.

Marine charges are incurred for offshore operations and marshes. The costs include boats and any dock facilities. Current charges for boats operating in the Gulf of Mexico are summarized in **Table 11.15**.

Air charges occur for offshore operations and marshes. The costs include boats on a day-rate basis and begin at rig move-in. A small helicopter (3 to 4 passenger capacity) is required for day-to-day operations. A large helicopter is used for weekly crew damages (**Table 11.16**).

11.5.13 Supervision and Administration. Project management costs must be considered. These charges include well supervision and administration. Large costs can be incurred for deep wells or problem wells, such as those with H_2S incident.

Supervision includes direct management of the well, including the on-site supervisor and any members of the office staff who are dedicated to the project. Mud or completion consultants may be considered as supervision. Specialized personnel such as mud loggers are not considered in the supervisory charges.

Administration charges can be handled in several manners. Some companies prefer to apply only direct supervision charges to a given well and charge support office staff members to general company overhead. Other companies divide all overhead charges among the wells to be drilled in a fiscal year.

Regardless of the accounting method, some of the charges that must be considered are
• Staff engineering support.
• Clerical support.
• Office overhead.
• Special insurance, including blowout insurance, and bonds.
• Legal work.
• Special document preparation.

A method for computing supervision and administration costs is to assume that a consultant will handle all operations. The on-site supervisor is the drilling consultant. An office consultant performs all administrative functions on an hourly basis (e.g., 200 hours for a dry hole and 300 hours for a completed well).

11.5.14 Tubulars. Casing and tubing costs are significant factors in the well cost. In some cases, they may account for 50 to 60% of the total expenditures. The costs are dependent on well depth, size, grade requirements, and couplings.

Pipe costs are influenced heavily by several factors. Pipe size is a major consideration. **Fig. 11.33** illustrates cost variations according to pipe size for N-80 grade long-thread and coupling (LTC) pipe that exceeds a burst rating of 5,000 psi in several sizes. Although engineering considerations should have the major impact on the pipe size selection, cost considerations should have some influence.

Costs increase with higher pipe grades. **Table 11.17** shows costs for 40.0-lbm/ft., 9.625-in. pipe with LTC couplings. As in the case of the pipe sizes, however, pipe-grade selection is an engineering decision. Couplings are seldom selected as a result of costs. However, higher-price premium couplings may allow the use of smaller pipe sizes, which will reduce the overall well costs (**Table 11.18**).

Casing Equipment. Casing (or cementing) accessory equipment is used to accomplish an effective primary cement job. Although the equipment does not have a major impact on well costs, it should be considered. **Table 11.19** shows a typical suite of equipment requirements for running and cementing casing. This equipment would cost approximately U.S. $3,500 for a 7⅝-in. casing string and U.S. $25,000 for a 7⅝-in. liner.

Drive-pipe costs must be calculated for wells that utilize the pipe. The charges vary for pipe sizes and wall thickness. A drive-shoe cost must be included. Typical drive-pipe size and costs are given in **Table 11.20**.

11.5.15 Wellhead Equipment. The wellhead equipment is attached to the casing string for pressure and stability support. Its cost is dependent on the number and size of the casing and tubing strings, pressure requirements, equipment components, and special features such as H_2S stainless duty. Total equipment costs can range from U.S. $7,500 for a low-pressure set of

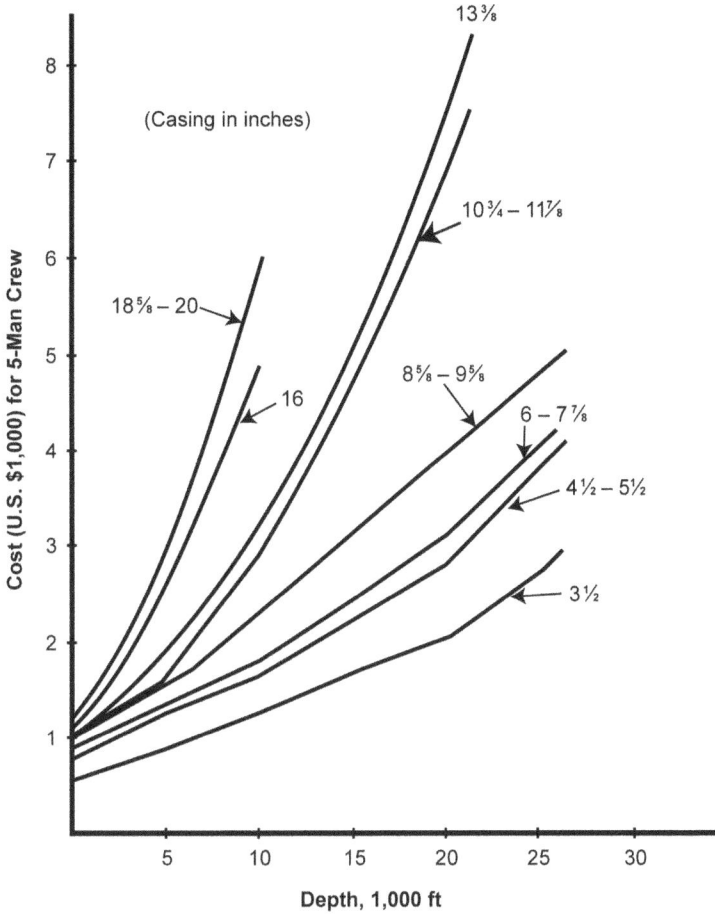

Fig. 11.33—Pipe size vs. cost.

TABLE 11.17—CASING COST COMPARISONS BY PIPE GRADE*	
Grade	Cost/ft, U.S. dollars
K-55	20.86
N-80	31.83
C-75	36.56
S-95	35.26
*9 ⅝-in. casing	

equipment to U.S. $1,500,000 for high-pressure, stainless-steel wellhead equipment and a tree. Subsea completions are even more expensive.

The wellhead equipment consists of the casing head, intermediate and tubing spools, and the production tree. The casing head is attached to the surface casing and will ultimately support all casing loads. Intermediate or production casing is hung inside the casing head. The intermediate spool supports the production casing if an intermediate string is run. The tubing

TABLE 11.18—COUPLING COST COMPARISON*

Coupling Type	Pipe Costs, U.S. dollars/ft
LTC	21.26
BTC	22.77
SFJ	34.31
FJ-P	34.31
TS	36.65
FL-4S	32.76

*7-in. casing

TABLE 11.19—EQUIPMENT REQUIREMENTS FOR RUNNING AND CEMENTING CASING

	Casing	Liner
Float shoe	1	1
Float collar	1	1
Centralizers	30	20
Scratchers	10	10
Liner hanger	0	1

TABLE 11.20—TYPICAL DRIVE-PIPE SIZE AND COSTS

Size, in.	Wall Thickness, in.	Pipe Cost, U.S. dollars/ft	Shoe Cost, U.S. dollars
14	0.375	19.54	165
16	0.375	22.53	200
20	0.500	37.49	230
24	0.500	43.92	375
26	0.500	47.66	415
30	0.500	55.14	470
36	0.750	98.82	535

spool is run only if the well is completed. It is set on the casing head or intermediate spool. The tree contains the production valves and chokes used for producing the oil or gas.

11.5.16 Completion Equipment. The completion equipment consists of downhole tools related to the tubing string. These items include: packers, seal assemblies, flow couplings, blast joints, and landing nipples. They are dependent primarily on tubing size and fluid content.

Packers. The packer is designed to divert formation fluids into the production tubing. It is selected according to production-casing size, bore size requirements, tensile loading, and seal-assembly type. In addition, H_2S-serviceable packers contain seals that are approximately 100 times more costly than the standard rubbers.

Blast Joints. Blast joints are thick-walled tubulars placed in the tubing string opposite the perforations to minimize the damage from erosion by the produced fluids. Their cost is dependent on tubing size and number of joints.

Seal Assembly. The seal assembly is an important part of the completion equipment. The cost is affected by the required number of seal units, the connection type, and the pipe size.

Nomenclature

C_B = bit cost, U.S. dollars
C_R = rig cost, U.S. dollars
d = diameter, in.
d_{DS} = drillstring diameter, in.
d_H = hole diameter, in.
D = deepest interval, ft
D_i = depth of interest, ft
D_n = deepest normal zone, ft
p_{form} = formation pressure, lbm/gal
p_h = hydrostatic pressure, psi
Q_m = mud-flow rate, gal/min
Q_p = pump output, bbl/min
T_R = rotating time, hours
T_T = trip time, hours
v = annular velocity, ft/min
V = annular volume, bbl/1,000 ft
V_a = annular velocity, ft/min
Y = footage per bit run, ft
ρ = mud weight, lbm/gal
ρ_{ekick} = equivalent mud weight at the depth of interest, lbm/gal
ρ_o = original mud weight, lbm/gal
Δp = differential pressure, psi
$\Delta\rho_{trip}$ = trip margin, lbm/gal
$\Delta\rho_{kick}$ = incremental kick mud weight increase, lbm/gal

References

1. Adams, N.J.: *Drilling Engineering: A Well Planning Approach,* PennWell Publishing Co., Tulsa (1984).

General References

Adams, N.J.: "How to Control Differential Pipe Sticking, Part 1—What is the Problem," *Petroleum Engineer* (September 1977).

Adams, N.J.: "How to Control Differential Pipe Sticking, Part 2—Procedures to Free the Drill String," *Petroleum Engineer* (October 1977).

Adams, N.J.: "How to Control Differential Pipe Sticking, Part 3—Field Study Presents New Results," *Petroleum Engineer* (November 1977).

Adams, N.J.: "How to Control Differential Pipe Sticking, Part 4—Economic Methods to Avoid or Free Stuck Pipe," *Petroleum Engineer* (January 1978).

Adams, N.J. and Hunter, D.: "Field and Laboratory Tests Indicate Muds that Resist Differential Pressure Pipe Sticking," paper presented at the 1978 Offshore Technology Conference, Houston, 8–11 May.

Adams, N.J.: "How to Estimate Well Costs," *Oil & Gas J.* (December 6, 1982).

Adams, N.J.: "Rig, Mud, Rental Tools Account For About Half the Cost of a New Well," *Oil & Gas J.* (December 13, 1982).

Adams, N.J.: "Tangible Drilling Expenses Complete Authorization-For-Expenditure Preparation," *Oil & Gas J.* (December 27, 1982).

Adams, N.J.: "Three-Step Bit Selection Can Trim Drilling Costs," *Oil & Gas J.* (June 17, 1985).

Adams, N.J.: *Well Control Problems and Solutions,* PennWell Publishing Co., Tulsa (1978).

Bourgoyne, A.T. *et al.: Applied Drilling Engineering,* Society of Petroleum Engineers, Richardson, Texas (1991).

Greenip, J.: "Care and Handling of Oilfield Tubulars," *Oil & Gas J.* Series (October 9, 1978).

Fertl, W.F.: *Abnormal Formation Pressures,* Elsevier Press, New York City (1976).

Prentice, C.M.: "Maximum Load Casing Design," *JPT* (July 1971).

SI Metric Conversion Factors

bbl	×	1.589 873	E – 01	= m^3
ft	×	3.048*	E – 01	= m
ft^3	×	2.831 685	E – 02	= m^3
gal	×	3.785 412	E – 03	= m^3
in.	×	2.54*	E + 00	= cm
lbm	×	4.535 924	E – 01	= kg
mile	×	1.609 344*	E + 00	= km
sq mile	×	2.589 988	E + 00	= km^2
psi	×	6.894 757	E + 00	= kPa
ton	×	9.071 847	E – 01	= Mg

*Conversion factor is exact.

Chapter 12
Underbalanced Drilling
Steve Nas, Weatherford Underbalanced Systems

12.1 What is Underbalanced Drilling?

In underbalanced drilling (UBD), the hydrostatic head of the drilling fluid is intentionally designed to be lower than the pressure of the formations that are being drilled. The hydrostatic head of the fluid may naturally be less than the formation pressure, or it can be induced by adding natural gas, nitrogen, or air to the liquid phase of the drilling fluid. Whether the underbalanced status is induced or natural, the result may be an influx of formation fluids that must be circulated from the well and controlled at surface.

The effective downhole circulating pressure of the drilling fluid is equal to the hydrostatic pressure of the fluid column, plus associated friction pressures, plus any pressure applied on surface.

$$\text{Overbalanced Drilling (OBD)} : P_{\text{reservoir}} < P_{\text{bottom hole}} = P_{\text{hydrostatic}} + P_{\text{friction}} + P_{\text{choke}}.$$

$$\text{UBD} : P_{\text{reservoir}} > P_{\text{bottom hole}} = P_{\text{hydrostatic}} + P_{\text{friction}} + P_{\text{choke}}.$$

Conventionally, wells are drilled overbalanced. In these wells, a column of fluid of a certain density in the hole provides the primary well-control mechanism. The pressure on the bottom of the well will always be designed to be higher than the pressure in the formation (**Fig. 12.1a**).

In underbalanced drilled wells, a lighter fluid replaces the fluid column, and the pressure on the bottom of the well is designed intentionally to be lower than the pressure in the formation (**Fig. 12.1b**).

Because the fluid no longer acts as the primary well-control mechanism, the primary well control in UBD arises from three different mechanisms:

• Hydrostatic pressure (passive) of materials in the wellbore because of the density of the fluid used (mud) and the density contribution of any drilled cuttings.

• Friction pressure (dynamic) from fluid movement because of circulating friction of the fluid used.

• Choke pressure (confining or active), which arises because of the pipe being sealed at surface, resulting in a positive pressure at surface.

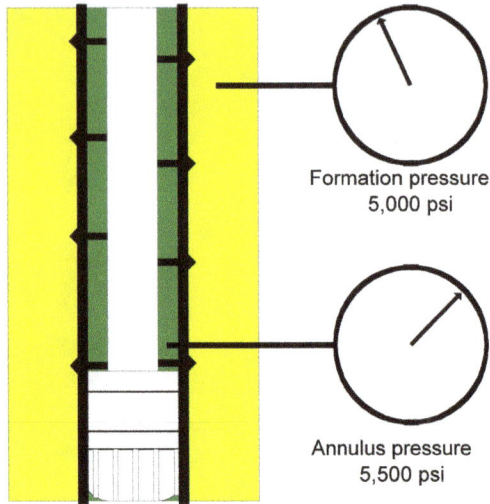

(a)

Fig. 12.1a—Pressures in conventional drilling.

Flow from any porous and permeable zones is likely to result when drilling underbalanced. This inflow of formation fluids must be controlled and any hydrocarbon fluids must be handled safely at surface.

The lower hydrostatic head avoids the buildup of filter cake on the formation as well as the invasion of mud and drilling solids into the formation. This helps to improve productivity of the well and reduce related drilling problems.

UBD produces an influx of formation fluids that must be controlled to avoid well-control problems. This is one of the main differences from conventional drilling. In conventional drilling, pressure control is the main well control principle, while in UBD, flow control is the main well-control principle. In UBD, the fluids from the well are returned to a closed system at surface to control the well. With the well flowing, the blowout preventer (BOP) system is kept closed while drilling, whereas in conventional overbalanced operations, drilling fluids are returned to an open system with the BOPs open to atmosphere (**Fig. 12.2**). Secondary well control is still provided by the BOPs, as is the case with conventional drilling operations.

12.1.1 Lowhead Drilling. Lowhead drilling is drilling with the hydrostatic head of the drilling fluid reduced to a pressure marginally higher than the pressure of the formations being drilled. The hydrostatic head of the fluid is maintained above the formation pressure, and reservoir inflow is avoided. Lowhead drilling may be undertaken in formations that would produce H_2S or would cause other issues if hydrocarbons were produced to surface.

12.1.2 Why Drill Underbalanced? The reasons for UBD can be broken down into two main categories:
  • Maximizing hydrocarbon recovery.
  • Minimizing pressure-related drilling problems.

There are also specific advantages and disadvantages of performing a drilling operation underbalanced. These are summarized in **Table 12.1**.

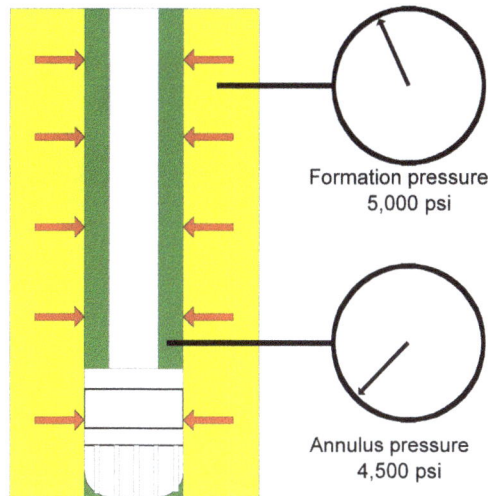

(b)

Fig. 12.1b—Pressures in underbalanced drilling.

Maximizing Hydrocarbon Recovery. *Reduced Formation Damage.* There is no invasion of solids or mud filtrate into the reservoir formation. This often eliminates the requirement for any well cleanup after drilling is completed.

Early Production. The well is producing as soon as the reservoir is penetrated with a bit. This could also be a disadvantage if hydrocarbon production cannot be handled or stored on site, or if the required export lines are not available.

Reduced Stimulation. Because there is no filtrate or solids invasion in an underbalanced drilled reservoir, the need for reservoir stimulation, such as acid washing or massive hydraulic fracture stimulation, is eliminated.

Enhanced Recovery. Because of the increased productivity of an underbalanced drilled well combined with the ability to drill infill wells in depleted fields, the recovery of bypassed hydrocarbons is possible. This can significantly extend the life of a field. The improved productivity of the wells also leads to a lower drawdown, which, in turn, can reduce water coning.

Increased Reservoir Knowledge. During an underbalanced drilling operation, reservoir productivity and the produced fluids can be measured and analyzed while drilling. This allows a well to be drilled longer or shorter, depending on production requirements. An operator is also able to determine the most productive zones in a reservoir in real time and obtain well test results while drilling.

Skin factors on most underbalanced drilled wells are negative, just as they are in wells drilled and stimulated.

Minimizing Pressure-Related Drilling Problems. *Differential Sticking.* The absence of an overburden on the formation combined with the lack of any filter cake serves to prevent the drillstring from becoming differentially stuck. This is especially useful when drilling with coiled tubing because coiled tubing lacks tool joint connections that increase the standoff in the borehole and then helps minimize sticking of conventional drillpipe.

No Losses. In general, a reduction of the hydrostatic pressure in the annulus reduces the fluid losses into a reservoir formation. In UBD, the hydrostatic pressure is reduced to a level at which losses do not occur. This is especially important in the protection of fractures in a reservoir.

Underbalanced drilling

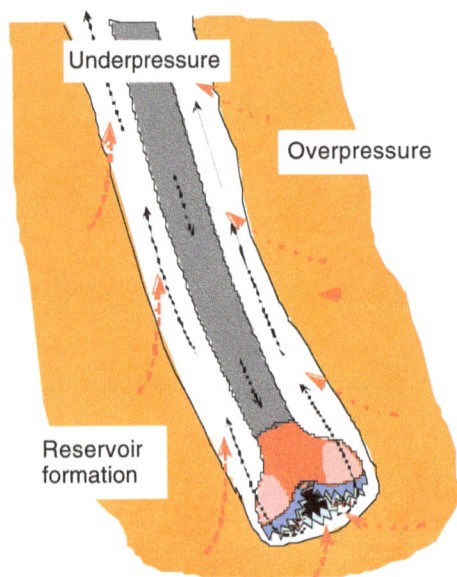

Conventional drilling

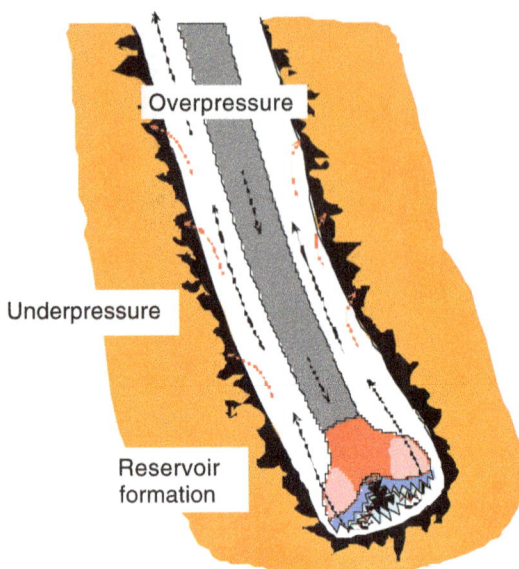

Underpressure

Overpressure

Reservoir
formation

Drilling fluid
returns to closed
circulation system

Overpressure

Underpressure

Reservoir
formation

Drilling fluid
returns to open
circulation system

Fig. 12.2—Open vs. closed circulation systems.

TABLE 12.1—ADVANTAGES VS. DISADVANTAGES OF UBD	
Advantages	Disadvantages
Increases ROP	Possible wellbore stability problems
Decreases formation damage	Increases drilling costs (depending on system used)
Eliminates risk of differential sticking	Compatibility with conventional MWD systems
Reduces risk of lost circulation	Generally higher risk with more inherent problems
Improves bit life	Possible excessive borehole erosion
Increases reservoir knowledge	Possible increased torque and drag

Improved Penetration Rate. The lowering of the wellbore pressure relative to the formation pressure has a significant effect on penetration rate. The reduction in the "chip holddown effect" also has a positive impact on bit life. The increased penetration rate combined with the effective cuttings removal from the face of the bit leads to a significant increase in bit life. In underbalanced drilled wells, sections have been drilled with only one bit where an overbalanced drilled well might need three, four, or even as many as five bits. It is normally assumed that penetration rates double when drilling underbalanced.

TABLE 12.2—RISKS ASSOCIATED WITH UNDERBALANCED DRILLED WELLS	
Level 0	Performance enhancement only—no hydrocarbon containing zones.
Level 1	Well incapable of natural flow to surface. Well is "inherently stable" and is low-level risk from a well-control point of view.
Level 2	Well capable of natural flow to surface but enabling conventional well-kill methods and limited consequences in case of catastrophic equipment failure.
Level 3	Geothermal and nonhydrocarbon production. Maximum shut-in pressures less than UBD equipment operating pressure rating. Catastrophic failure has immediate serious consequences.
Level 4	Hydrocarbon production. Maximum shut-in pressures less than UBD equipment operating pressure rating. Catastrophic failure has immediate serious consequences.
Level 5	Maximum projected surface pressures exceed UBD operating pressure rating but are below BOP stack rating. Catastrophic failure has immediate serious consequences.

TABLE 12.3—CLASSIFICATION SYSTEM FOR UNDERBALANCED TECHNIQUES												
Classification	0		1		2		3		4		5	
A = Low head	A	B	A	B	A	B	A	B	A	B	A	B
B = UBD												
Gas drilling	1	1	1	1	1	1	1	1	1	1	1	1
Mist drilling	2	2	2	2	2	2	2	2	2	2	2	2
Foam drilling	3	3	3	3	3	3	3	3	3	3	3	3
Gasified liquid drilling	4	4	4	4	4	4	4	4	4	4	4	4
Liquid drilling	5	5	5	5	5	5	5	5	5	5	5	5

12.1.3 Classification System for Underbalanced Drilling. A classification system developed by the Intl. Assn. of Drilling Contractors (IADC) is helping to establish the risks associated with underbalanced drilled wells (**Table 12.2**).

The matrix given easily classifies the majority of known underbalanced applications. This system combines the risk management categories (Levels 0 to 5) with a subclassifier to indicate either "underbalanced" or "low head" drilling using underbalanced technology. To provide a complete method of classifying the type of technology used for one or more sections of a well, or multiple wells in a particular project, a third component of the classification system addresses the underbalanced technique used, as shown in **Table 12.3**.

Example of Classification System Use. A horizontal section of a well is drilled in a known geologic area using a drilling fluid lightened with nitrogen gas to achieve an underbalanced condition through the reservoir section. The maximum predicted bottomhole pressure (BHP) is 3,000 psi with a potential surface shut-in pressure of 2,500 psi. This is classified as a 4-B-4 well indicating classification level 4 risk and UBD drilling with a gasified liquid. All wells classified as a Level 4 or Level 5 underbalanced well require significant planning to ensure safe underbalanced drilling.

12.1.4 Selecting the Right Candidate for UBD. Most reservoirs can be drilled underbalanced. Some reservoirs cannot be drilled underbalanced because of geological issues associated with rock stability. For some reservoirs, it might not be possible to drill underbalanced with the current technology because they are either prolific producers or pressures are so high that safety and environmental concerns prevent safe underbalanced drilling. These may include high-pressure or sour wells (although both types have been drilled underbalanced, but with significant engineering considerations and planning).

TABLE 12.4—UBD EFFECTS FOR RESERVOIR TYPES	
Will Benefit from UBD	Will Not Benefit from UBD
Formations that usually suffer major formation damage during drilling or completion operations. Wells with skin factors of 5 or higher.	Wells in areas of very low conventional drilling cost.
Formations that exhibit differential sticking tendencies.	Wells drilled in areas of extremely high ROP (that is, ROP≥1,000 ft/day).
Formations with zones with severe losses or fluid invasion from drilling or completion operations.	Extremely high-permeability wells.
Wells with large macroscopic fractures.	Ultralow-permeability wells.
Low-permeability wells.	Poorly consolidated formations.
Wells with massive heterogeneous or highly laminated formations characterized by differing permeabilities, porosities, and pore throat throughput.	Wells with low borehole stability.
High-production reservoirs with low to medium permeabilities.	Wells with loosely cemented laminar boundaries.
Formation, with rock fluid sensitivities.	Wells that contain multiple zones with different pressure regimes.
Formations that exhibit low ROP with OBD.	Reservoirs with interbedded shales or claystones.

Candidate selection for UBD must focus not only on the benefits of UBD but also on additional considerations. It is important that the right reservoir is selected for a UBD operation. **Table 12.4** shows reservoir types that will and will not benefit from UBD. Of course, not only the reservoir has to be evaluated, but also the well design and the possible damage mechanisms and the economic reasons for UBD. All issues must be considered carefully when choosing whether or not to drill underbalanced.

12.1.5 Reservoir Selection Issues. Appropriate reservoir screening is essential for the correct selection of a suitable reservoir application for vertical or horizontal UBD. A systematic approach, outlined in the following section, identifies the major areas of study to ascertain if sufficient information is available to initiate the design work for a viable UBD process.

Once this information is gathered and reviewed and if data show that an UBD operation is the best method for recovering hydrocarbons in an economically and technically successful manner, it is time to mobilize the team to design and execute the UBD operation. Steps in a typical UBD evaluation process are outlined in **Table 12.5**. **Fig. 12.3** shows this UBD evaluation process as a flow chart.

12.1.6 Economic Limitations. It is important not to forget the business driver behind the technology. If benefits cannot be achieved, the project must be reviewed. The improvements from UBD—increased penetration rate, increased production rate, and minimization of impairment—must offset the additional cost of undertaking a UBD project.

This is often the most difficult limitation of UBD to overcome. If the reservoir/production engineers are not convinced that there is a sound reason for drilling underbalanced for productivity reasons, most underbalanced projects will never get past the feasibility stage.

To drill a well underbalanced, extra equipment and people are required, and this adds to the drilling cost of a well. The operators must show a return for their shareholders, so they will want to know if this extra investment is worthwhile before embarking on a UBD project.

12.1.7 Costs Associated With Underbalanced Drilling. The following factors contribute to the cost increases for an underbalanced drilled well in comparison to a conventionally drilled well:
- Pre-engineering studies.
- Rotating diverter system.

TABLE 12.5—STEPS IN A TYPICAL UBD EVALUATION	
Step 1	Information gathering and a thorough review to ensure that all necessary information has been either obtained from available pre-existing data sources or directly acquired as necessary.
Step 2	Preliminary data prescreening by drilling, reservoir engineering, geology, and UBD experts to ascertain if the well meets the base criteria for optimal UBD implementation.
Step 3	Detailed review of gathered information by a cross-functional team [consisting of drilling engineers, reservoir engineers, geologists, geophysicists, petrophysicists, production engineers, UBD experts (in-house or consultants), laboratory and analytical staff (if required), regulatory and safety experts and representatives from the drilling, mud and service companies that will be involved in the execution of the operation] to commence the initial planning to drill the well. This will involve in-depth discussion and cooperation between all parties participating as members of the cross-functional team.
Step 4	Assimilation and review of the best possible services and techniques to drill and complete the reservoir in a proper underbalanced fashion.
Step 5	Selection of key personnel and equipment to execute the UBD operation.
Step 6	Detailed prespud meeting.
Step 7	Equipment procurement, transport, setup, and testing.
Step 8	UBD operation commences with capability for the acquisition of the maximum amount of useful data.
Step 9	Continuous review of the real-time data obtained during the UBD process and adjustments made, on the basis of the data, to ensure that the well is drilled properly and according to design (including contingencies for unexpected events).
Step 10	Completion of the well in an underbalanced fashion.
Step 11	Post-mortem review of the complete UBD operation by the cross-functional team.
Step 12	Production of the UBD drilled and completed view and feedback to the Step 11 review process.

• Surface separation and well-control package.
• Snubbing system to deal with pipe light.
• Data acquisition system.
• Extra downhole equipment [nonreturn valves and pressure while drilling (PWD)].
• Special drillstring connections (high-torque gas that is tight with special hardbanding).
• Additional personnel training.
• Additional operational wellsite personnel.
• Additional safety case update consistent with planned UBD operations.
• Extra time required to drill underbalanced.

From industry experience to date, we can state that underbalanced drilled wells are 20 to 30% more expensive than overbalanced drilled wells. This applies to both offshore and onshore operations in a similar area.

Cost alone is, however, not a good measure for the evaluation of UBD. The value of the well must also be recognized. The average three-fold increase in productivity of an underbalanced drilled well can add considerable value to a field development plan or a field rehabilitation program. If we add a potential increased recovery from a field to the value of an underbalanced well, even an increase as small as 1% in total hydrocarbon recovery may have a large impact on field economics.

12.1.8 Reservoir Studies. Prior to a UBD operation, some reservoir engineering work should be carried out. Not only is an accurate reservoir pressure needed, but the damage mechanism of the reservoir must be understood to ensure that the benefits of UBD can indeed be obtained. Some wells or reservoirs are suitable for underbalanced operations and result in an enhanced recovery. Other formations or fields may not be viable for a variety of reasons. If formation damage is the main driver for UBD, it is important that the reservoir and petroleum engineers

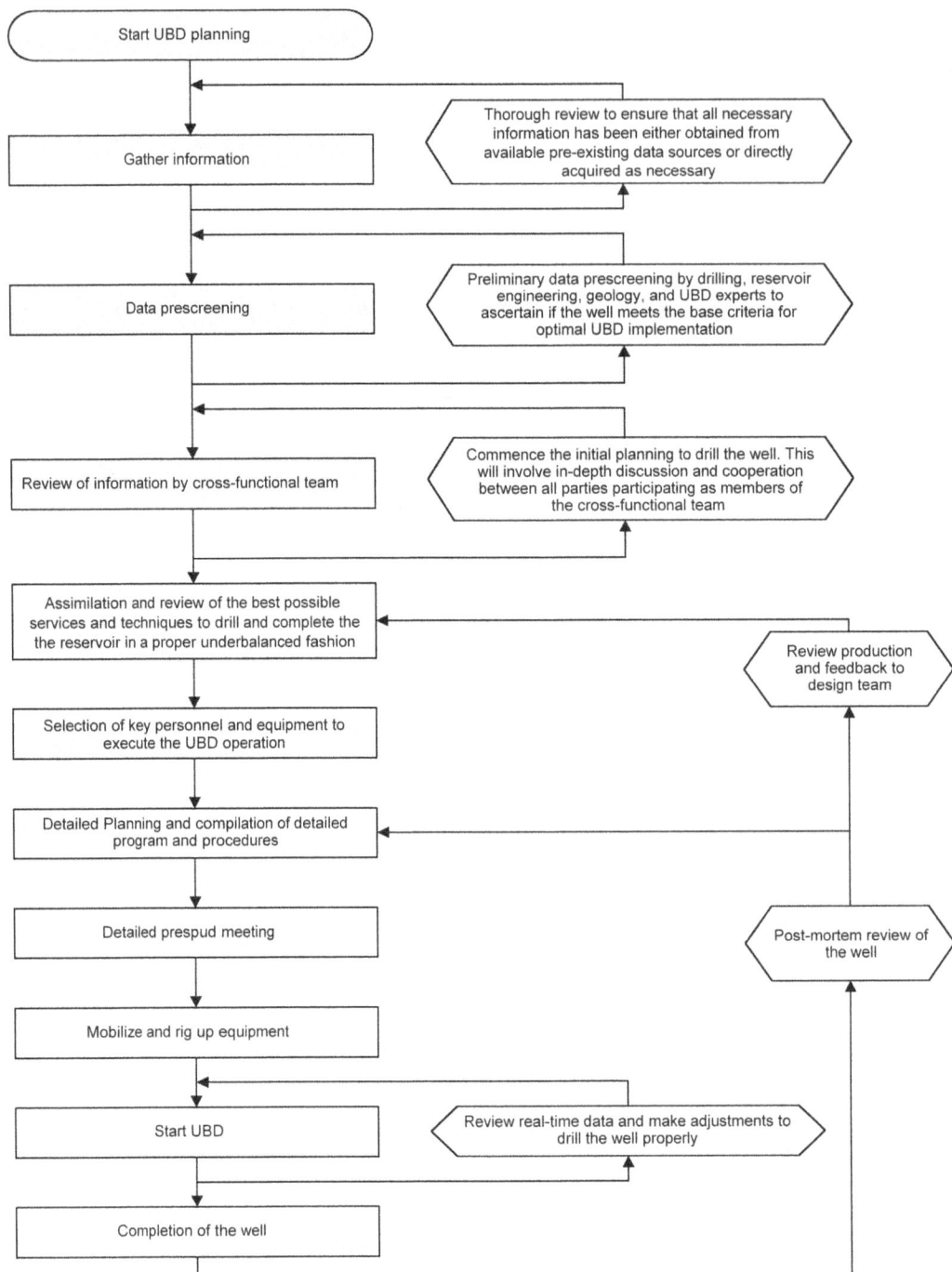

Fig. 12.3—UBD flow chart.

understand the damage mechanisms resulting from OBD. We must remember that even under-balanced drilled wells can cause formation damage.

Coreflush testing may be required to establish compatibility between the proposed drilling fluid and the produced reservoir fluids. This is critical if oil reservoirs are to be drilled under-

Technique Selection

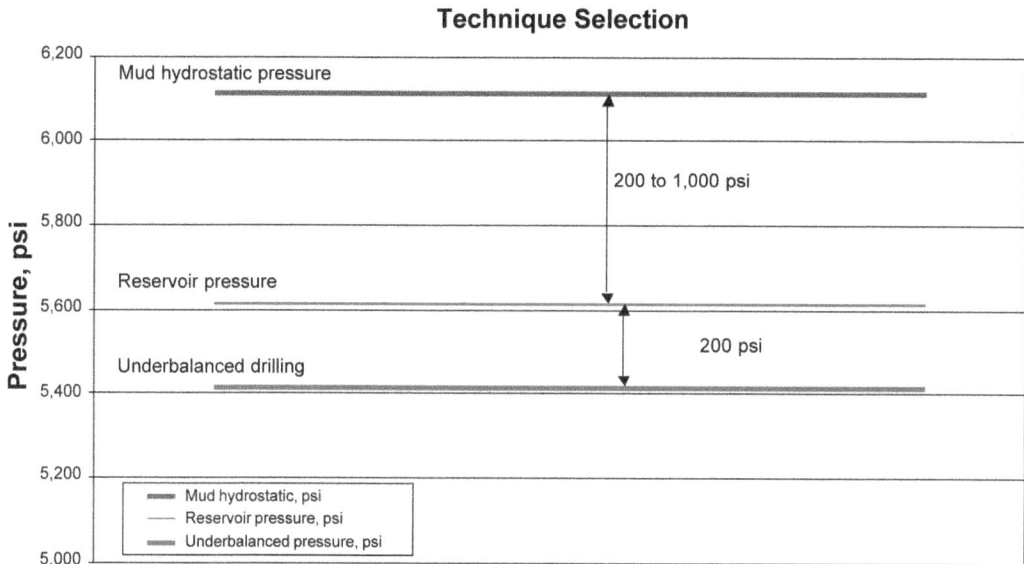

Fig. 12.4—BHP requirements

balanced. The potential for scale and emulsion forming must also be reviewed prior to starting operations. We must ascertain the stability of the zone of interest to determine if the proposed well path is structurally capable of being drilled with the anticipated formation drawdown.

Expected productivity with the proposed drawdown must be reviewed. The objective of UBD is to clean the reservoir and not to produce the well to its maximum capacity. If the reservoir is likely to produce any water, we must take this into account because water influx can have significant effects on the underbalanced process. It is important that expected productivity be analyzed with the reservoir engineers to obtain an accurate indicator as to whether UBD would be beneficial.

Once reservoir issues are fully understood, advantages to drilling underbalanced are proven, and the proposed well profile can be achieved, we can undertake the selection of the surface equipment.

12.1.9 Designing a UBD Operation. A basic four-step process can be applied to determine the options and requirements for drilling underbalanced:
1. Determine BHP requirements.
2. Identify the drilling fluid options.
3. Establish the well design and perform flowing modeling.
4. Select the surface equipment.

12.1.10 BHP Requirements. In OBD, a mud weight is selected that provides a hydrostatic pressure of 200 to 1,000 psi above the reservoir pressure. In UBD, we select a fluid that provides a hydrostatic pressure of around 200 psi below the initial reservoir pressure. This provides a good starting point for the selection of a fluid system. During the feasibility study, this drawdown is normally further refined, depending on the expected reservoir inflow and other drilling parameters. This first look provides an indication if the fluid should be foam or gasified or if the well is drilling with a single-phase fluid (**Fig. 12.4**).

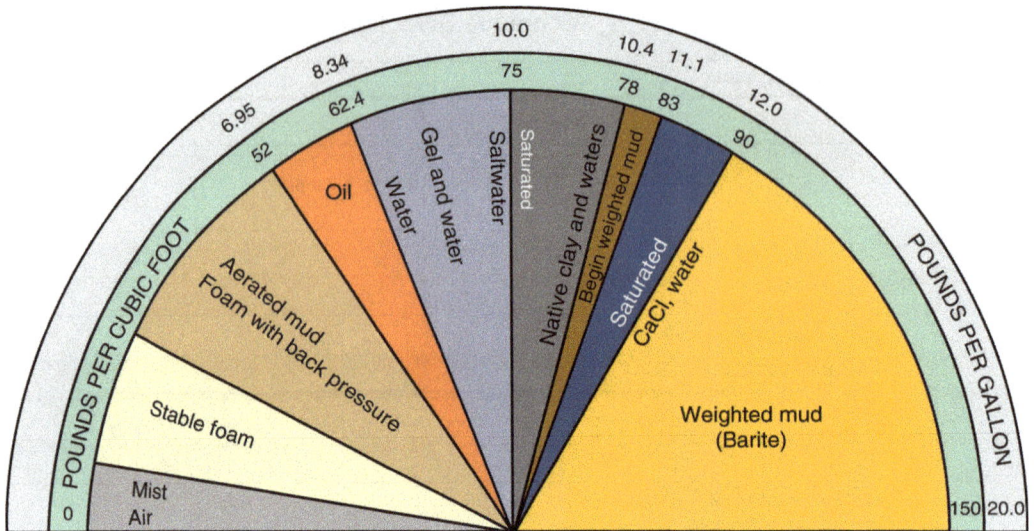

Fig. 12.5—UBD fluid density range.

12.2 Drilling Fluid Systems

Correct selection of the fluid system used in UBD is the key to a successful UBD operation (**Fig. 12.5**). Initial fluid selection for UBD operations is classified into five fluid types based primarily on equivalent circulating density: gas, mist, foam, gasified liquid, and liquid.

Final fluid selection for UBD operations can be extremely complex. Key issues such as reservoir characteristics, geophysical characteristics, well-fluid characteristics, well geometry, compatibility, hole cleaning, temperature stability, corrosion, data transmission, surface fluid handling and separation, formation lithology, health and safety, environmental impact, and fluid source availability, as well as staying below the reservoir pressure at all times, the primary objective for drilling underbalanced, must be considered before a fluid design is finalized.

12.2.1 Gaseous Fluids. Gaseous fluids are basically the gas systems. In initial UBD operations, air was used for drilling. Today, air drilling or dusting is still applied in hard rock drilling and in the drilling of water wells. The use of air in hydrocarbon-bearing formations is not recommended because the combination of oxygen and natural gas may cause an explosive mixture. There have been a number of reported cases in which downhole fires have destroyed drillstrings, with the obvious potential consequences of the rig burning down if the mixture gets to surface.

Often, nitrogen is used if hydrocarbon reservoirs are drilled with a gas. For remote or offshore locations, a nitrogen generation system can be used to reduce the logistics. Another option might be the use of natural gas, which, if available, has sometimes proved a worthy alternative in drilling operations. If a gas reservoir is being drilled underbalanced, a producing well or the export pipeline may produce sufficient gas at the right pressure to drill.

Characteristics of gas drilling are listed next:
- Fast penetration rates.
- Longer bit life.
- Greater footage per bit.
- Good cement jobs.
- Better production.
- Minimal water influx required.
- Possibility of slugging.

- Possibility of mud rings in the presence of fluid ingress.
- Relies on annular velocity to remove cuttings from the well.

12.2.2 Mist Systems. If a formation starts to produce small amounts of water when drilling with a gas system, the system is often changed to a mist system. The fluid added to the gas environment disperses into fine droplets and forms a mist system that may then be used for drilling. In general, this technique must be used in areas where some formation water exists, which prevents the use of complete "dry air" drilling. The following lists the characteristics of mist drilling:
- It is similar to gas drilling, but with addition of liquid.
- It relies on annular velocity to remove cuttings from the well.
- It reduces formation of mud rings.
- It requires high volumes (30 to 40% more than dry gas drilling).
- Its pressures are generally higher than dry gas drilling.
- Incorrect gas/liquid ratio leads to slugging with attendant pressure increase.

12.2.3 Foam Systems. Drilling with stable foam has some appeal because foam has some attractive qualities and properties at the very low hydrostatic densities that can be generated with foam systems. Foam has good rheology and excellent cuttings-transport properties. The fact that stable foam has some natural inherent viscosity, as well as fluid-loss-control properties, makes foam a very attractive drilling medium.

During foam drilling, the volumes of liquid and gas injected into the well are carefully controlled. This ensures that foam forms when the liquid enters the gas stream at the surface. The drilling fluid remains foam throughout its circulation path down the drillstring, up the annulus, and out of the well. The more stable nature of foam also results in a much more continuous downhole pressure condition because of slower fluid and gas separation when the injection is stopped.

Adding surfactant to a fluid and mixing the fluid system with a gas generates stable foam. Stable foam used for drilling has a texture not unlike shaving foam. It is a particularly good drilling fluid with a high carrying capacity and a low density. One of the problems encountered with the conventional foam systems is that the foam remains stable even when it returns to the surface, and this can cause problems on a rig if the foam cannot be broken down fast enough. In earlier foam systems, the amount of defoamer had to be tested carefully so that the foam was broken down before any fluid entered the separators. In closed-circulation drilling systems, stable foam can cause particular problems with carry-over. The recently developed stable foam systems are simpler to break, and the liquid can also be refoamed so that less foaming agent is required and a closed circulation system can be used. These systems, in general, rely on either a chemical method of breaking and making the foam or the use of an increase and decrease of pH to make and break the foam.

The foam quality at surface used for drilling is normally between 80 and 95%. This means that of the total foam, 80 to 95% of the volume is gas, with the remainder being liquid. Downhole, because of the increased hydrostatic pressure of the annular column, this ratio changes because the volume of gas is reduced. An average acceptable bottomhole foam quality (FQ) is in the region of 50 to 60%.

Characteristics of Foam Drilling.
- Extra fluid in the system reduces the influence of formation water.
- It has a very high carrying capacity.
- There are reduced pump rates because of improved cuttings transport.
- Stable foam reduces slugging tendencies of the wellbore.
- The stable foam can withstand limited circulation stoppages without affecting the cuttings removal or equivalent circulating density (ECD) to any significant degree.

- It has improved surface control and more stable downhole environment.
- The breaking down of the foam at surface must be addressed at the design stage.
- More increased surface equipment is required.

12.2.4 Gasified Systems. The next fluid system that is often used is a gasified fluid system. In these systems, gas is injected into the liquid to reduce the density. There are a number of methods that can be used to gasify a liquid system. The use of gas and liquid as a circulation system in a well significantly complicates the hydraulics program. The ratio of gas and liquid must be carefully calculated to ensure that a stable circulation system is used. If too much gas is used, slugging will occur. If not enough gas is used, the required bottomhole pressure will be exceeded, and the well will become overbalanced.

Characteristics of Gasified-Fluid Systems.
- Extra fluid in the system will almost eliminate the influence of formation fluid unless incompatibilities occur.
- The fluid properties can easily be identified prior to commencing the operation.
- Generally, less gas is required.
- Slugging of the gas and fluid must be managed correctly.
- Increased surface equipment is required to store and clean the base fluid.
- Velocities, especially at surface, are lower, reducing wear and erosion both downhole and to the surface equipment.

12.2.5 Single-Phase Fluids. If possible, the first approach used should be a single-phase fluid system with a density low enough to provide an underbalanced condition. If water can be used, then this would be the first step to take. If water is too heavy, oil can be considered. In oil reservoirs, it is not unknown to use the reservoir crude for drilling. When drilling with a crude-oil system, the rig's surface equipment must be reviewed to ensure that hydrocarbons can be handled safely with the provided rig fluid systems. On offshore rigs, a fully enclosed, vented, and nitrogen-blanketed pit system may have to be used to ensure that any gas released from the crude does not form a safety hazard.

12.2.6 Gas/Liquid Ratios. Fig. 12.6 shows fluid/gas ratios for gasified fluid systems. As we move through the various fluid systems, the amount of gas in the fluid decreases as the density of the fluid increases. This has a significant effect on the hydraulics calculations. Special hydraulics software is required to ensure that the BHP remains underbalanced when circulating.

12.2.7 Gas Lift Systems. If a fluid must be reduced in density, the use of an injection of gas into the fluid flow could be an option. This offers a choice not only of the gas used but also in the way the gas is used in the well.

Normally, natural gas or nitrogen is used as a lift gas, but both CO_2 and O_2 can also be utilized. However, gases containing oxygen are not recommended for two main reasons. First, with hydrocarbon influx, there is the danger of a downhole fire or explosion. Second, the combination of oxygen and saline fluids with the high bottomhole temperatures can cause severe corrosion to tubulars used in the well and drillstring. A number of injection methods are available to reduce the hydrostatic pressure.

12.2.8 Drillpipe Injection. Compressed gas is injected at the standpipe manifold, where it mixes with the drilling fluid. **Fig. 12.7** shows a typical drillpipe gas-injection configuration in a well. The main advantage of drillstring injection is that no special downhole equipment is required in the well. The use of reliable nonreturn valves is required to prevent flow up the drillpipe. The gas rates used when drilling with drillpipe injection systems are normally lower than with annular gas lift. Relatively low BHPs can be achieved using this system.

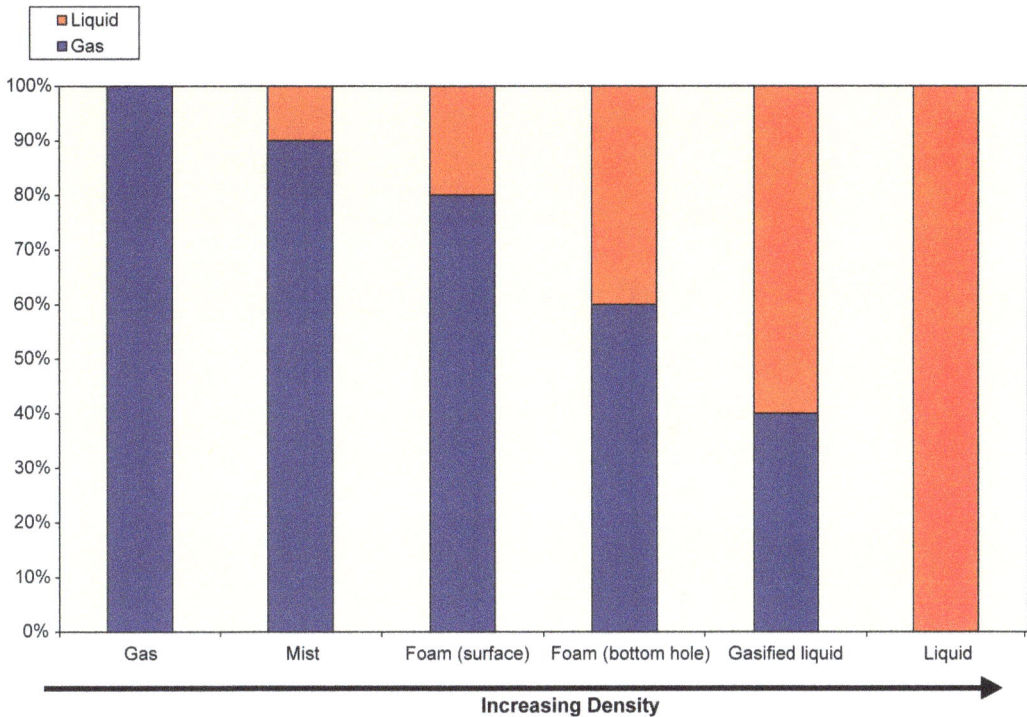

Fig. 12.6—Range of gas/liquid ratios of gasified fluids.

The disadvantages of this system include the need to stop pumping and the bleeding of any remaining trapped pressure in the drillstring every time a connection is made. This can result in an increase in BHP. It may then be difficult to obtain a stable system and avoid pressure spikes at the reservoir when using drillpipe injection.

The use of pulse-type measurement while drilling (MWD) tools is only possible with gasified fluids with up to 20% gas by volume. If higher gas volumes are used, the pulse system deployed on MWD transmission systems will no longer work. Specialist MWD tools, such as electromagnetic tools, may have to be used if high gas-injection rates are required.

A further drawback for drillstring injection is the impregnation of the gas into any downhole rubber seal. Positive displacement motors (PDMs) are especially prone to failure when rubber components are impregnated with the injection gas and then tripped back to surface.

During trips, the rubber components swell as a result of the expanding gas not being able to diffuse out of the elastomer sufficiently or quickly. This effect (explosive decompression) not only destroys downhole motors but also affects other tools with rubber seals used downhole. Special rubber compounds have been developed, and the design of motors is changing, to allow for this expansion.

The majority of motor suppliers can now provide PDMs specifically designed for use in this kind of downhole environment. Operational procedures must be written to ensure that connections can be made safely when drilling with high-pressure gas inside the drillstring.

12.2.9 Annular Injection. Annular injection through a concentric string is most commonly utilized offshore in the North Sea. In new wells, a liner is set inside the target formation. The liner is then tied back to surface using a modified tubing hanger to suspend the tieback string. Gas is injected in the casing liner annulus to facilitate the drawdown required during the drilling operation. **Fig. 12.8** shows typical annular gas-injection configuration in a well.

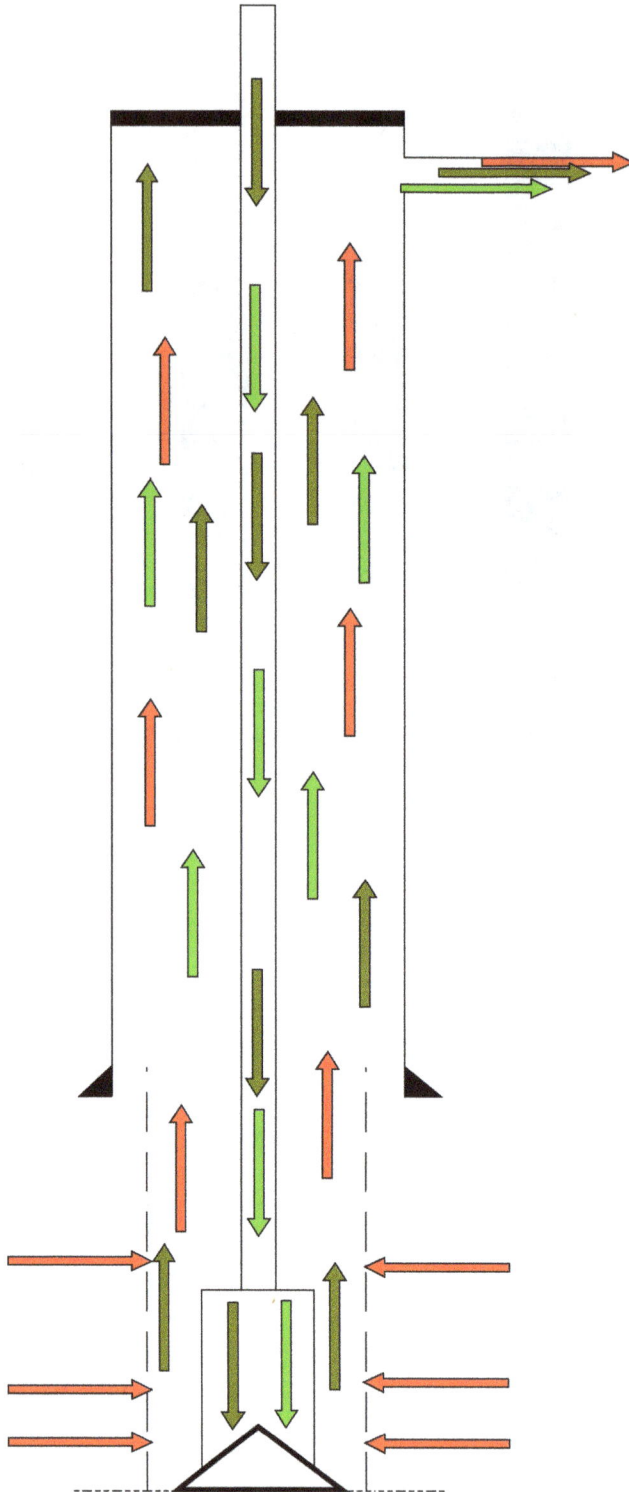

Fig. 12.7—Drillpipe gas injection.

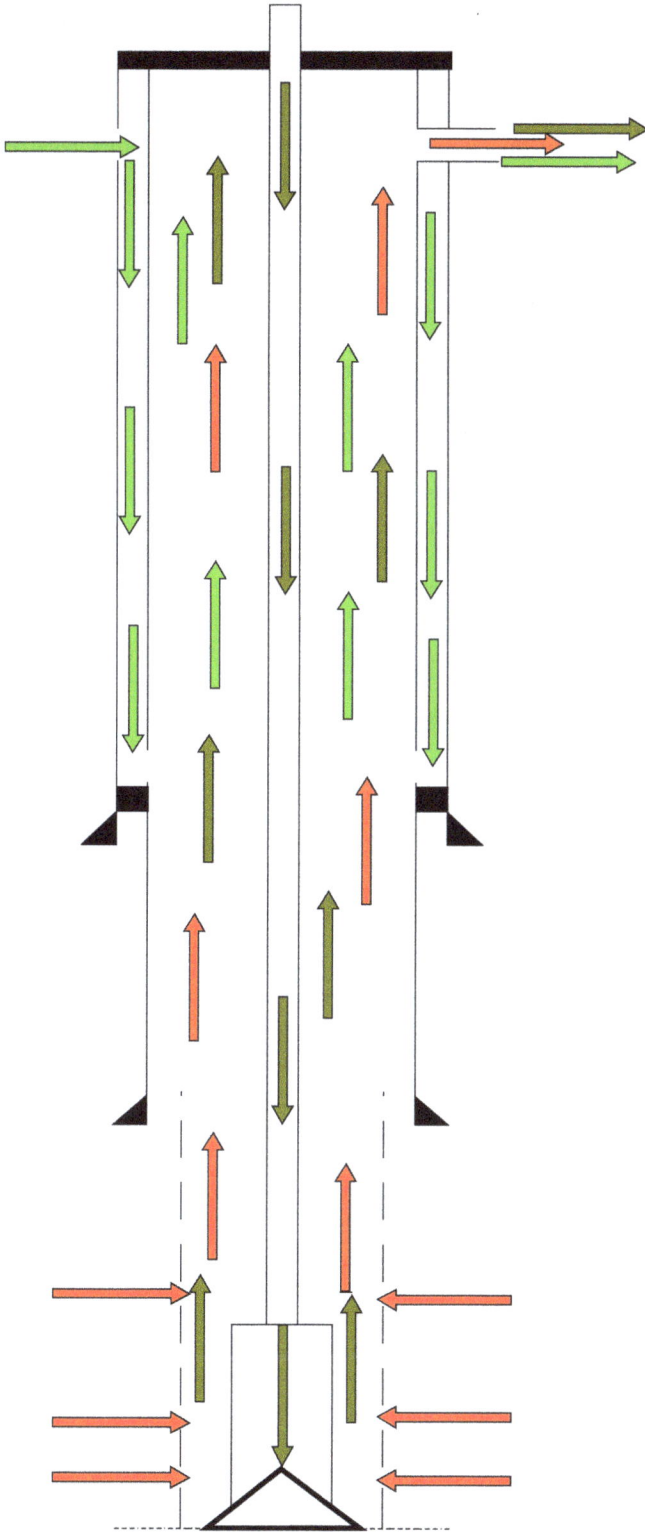

Fig. 12.8—Annular gas injection.

The tieback string is then pulled prior to installation of the final completion. The alternative is for an older well to have a completion in place incorporating gas lift mandrel pockets. These can be set up to provide the correct BHPs during the drilling operation. The drawback with this type of operation is that the hole size and tools required could be restricted by the minimum inner diameter (ID) of the completion. However, the main advantage of using an annulus to introduce gas into the system is that gas injection is continued during connections, thus creating a more stable BHP.

Because the gas is injected through the annulus, only a single-phase fluid is pumped down the drillstring. The advantage is that conventional MWD tools operate in their preferred environment, which can reduce the operational cost of a project.

However, the drawbacks of this system are that a suitable casing-completion scheme must be available and that the injection point must be low enough to obtain the required underbalanced conditions. There may also be some modifications required to the wellhead for the installation of the tieback string and the gas-injection system.

12.2.10 Parasite-String Gas Injection. Fig. 12.9 shows typical parasite-string gas-injection configuration in a well. The use of a small parasite string strapped to the outside of the casing for gas injection is used only in vertical wells. For safety reasons, two 1- or 2-in. coiled-tubing strings are strapped to the casing string above the reservoir as the casing is run in. Gas is pumped down the parasite string and injected onto the drilling annulus.

The installation of a production casing string and the running of the two parasite strings makes this a complicated operation. Wellhead modifications may be required to provide surface connections to the parasite strings.

This system is normally restricted to vertical wells to avoid damage to the parasite strings. The principles of operation and the advantages of this system are identical to the concentric gas injection system.

If natural gas is used to lighten the drilling fluid, annular injection is the preferred method. The use of natural gas through the drillstring is not recommended because gas is released on the drillfloor during connections.

12.2.11 Hydraulic Calculations. Because a compressible system is used in UBD, the annulus is always a mixture of gas and liquids. To calculate the BHPs in a gas/liquid environment, multiphase hydraulics must be used. Multiphase flow is probably some of the most complex fluid engineering known in the drilling industry. Multiphase or compressible fluids change considerably with pressures and temperatures.

12.2.12 Flow Regimes. To correctly predict friction factors and liquid holdup, the flow regime in the annulus must be known. In OBD operations, we only consider laminar or turbulent flow. In UBD, many more variations must be considered. The flow regime varies with the inclination of the well and, again, a number of methods and correlations are known to predict flow regimes.

The number of variables—fluids (gas/liquid) density, viscosity, compressibility, cuttings density, cuttings shape (or roundness), fluid composition, etc. and their interaction makes multiphase flow calculations a tasking and difficult undertaking. Because these variables are calculated over every iteration element of the well model, it is understandable that this has to be done with a computer program. Most flow models actually combine the various gas/liquid phases into the two-phase structure, as shown in **Fig. 12.10**.

Once this has been achieved, the model is now dealing with the conventional two-phase system with a liquid phase and a gas phase. The solids are combined in the liquid phase because this allows conventional fluid and cuttings transport models to be used for cuttings transport.

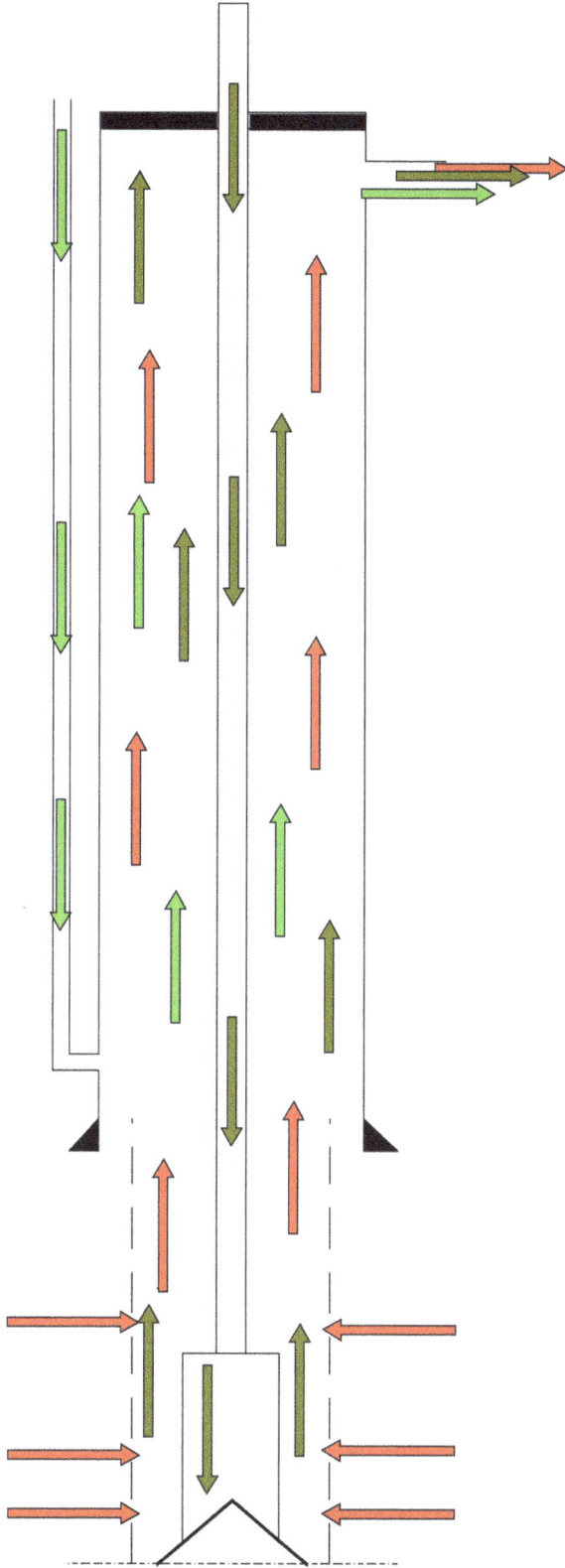

Fig. 12.9—Parasite string gas injection.

Multiphase Calculations

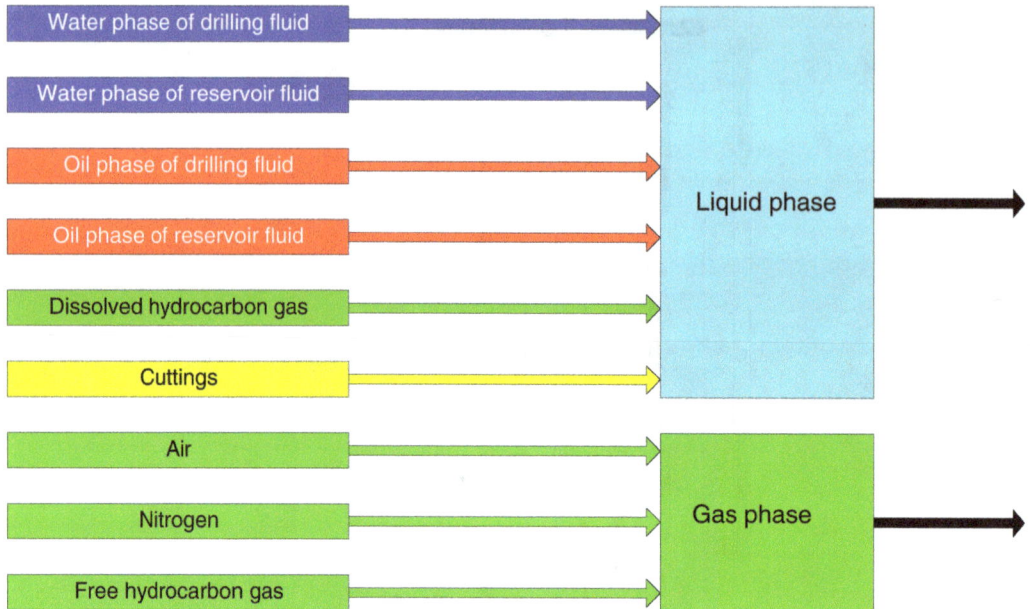

Fig. 12.10—Generating two phases from multiple components.

12.2.13 Circulation Design Calculations. In designing a UBD circulation system, the bottom-hole pressure must be maintained below the reservoir pressure. But the surface separation system must have sufficient capacity to handle the flow rates and pressures expected while drilling. The separation system must be capable of handling sudden productivity increases from the well from fractures or flush zones and retain the ability to "choke" back production if well outflow is more than what can be handled safely by the surface separation equipment. The separation system must also be able to work within the design parameters of the well. The design of a UBD circulation system must consider certain factors. These factors are discussed next.

BHP. The BHP must be less than the static reservoir pressure under static and dynamic conditions to enable reservoir fluid inflow into the wellbore. This difference creates the driving force that drives well productivity.

Reservoir Inflow Performance and Control. The productivity of the reservoir while drilling underbalanced is a function not only of BHP but also reservoir characteristics like permeability, porosity, length of reservoir exposed to the wellbore, drainage radius, and the pressure driving force. The pressure driving force (reservoir pressure—well BHP) is the most important in controlling reservoir inflow because most of the parameters are relatively fixed by the geology. Therefore, the BHP must be controlled by either hydrostatic drilling fluid or by the choke to control reservoir inflow performance.

Cuttings Transport and Hole Cleaning. Cuttings generated while drilling underbalanced must be removed from the wellbore by the hydraulic action of the drilling fluid. For hole cleaning to be effective, the fluid annular velocity has to be at least twice the cuttings' settling velocity.

Motor Performance in Multiphase-Flow Environment. While drilling with multiphase fluids, it is important that the motor performance is not compromised by the hydraulics; that is,

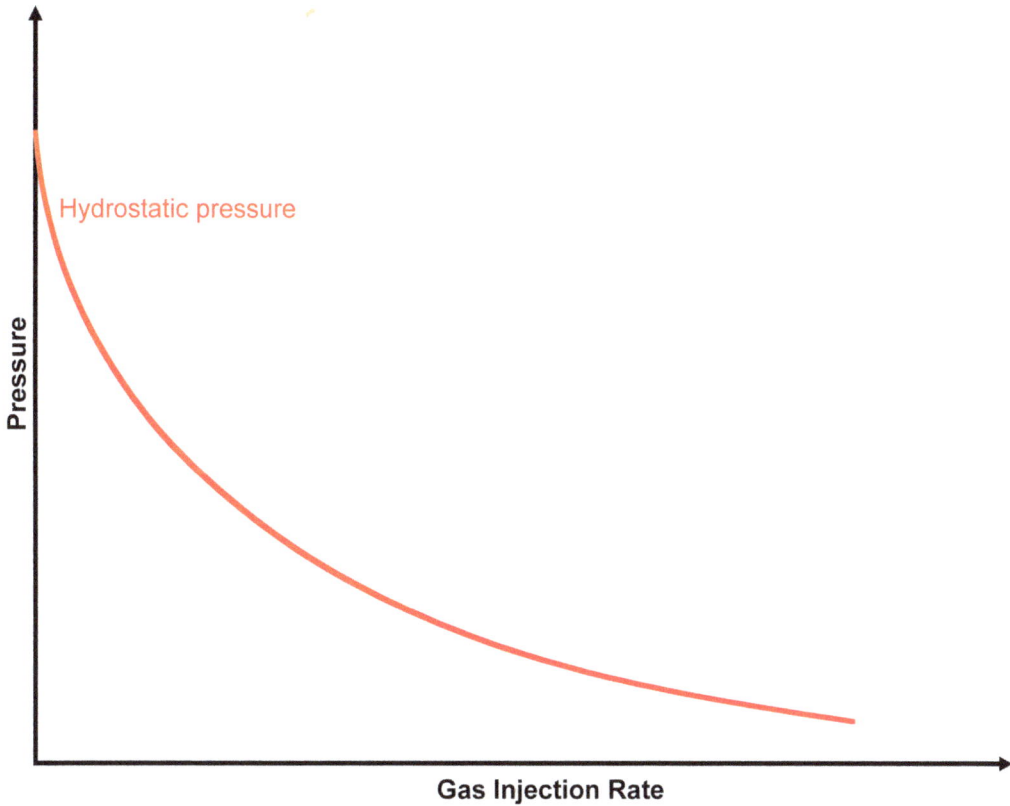

Gas Injection Rate
Fig. 12.11—Gas injection reduces BHP.

the equivalent flow rate through the motor should be sufficient to deliver the required performance and be within the motor operating envelope.

Surface Equipment Capabilities and Limitations. The productivity of the reservoir while drilling and the length of reservoir that should be exposed to the wellbore is constrained by the capacity of the surface separation facility. UBD safety systems are designed so that the surface system shuts down automatically if the rate from the well exceeds its capacity. Surface equipment capacity must always be designed to handle the maximum expected production from the well, whether instantaneous or steady-state.

Environmental Considerations. Either because of governmental legislation and/or operators' policies, UBD operations may have to be carried out with zero emissions to the environment— that is, no gas flaring. Where this is the case, the surface separation system has to be designed for total containment of the produced cuttings and reservoir fluids inflow—oil, gas, and water. Otherwise, gas re-injection will need to be considered. Gas re-injection requires a gas recompression plant so that gas can be re-injected at the right pressure.

Wellbore Stability. Exposing wellbore to pressure drawdown imposes stresses on the surrounding formation. If the stresses exceed the strength of the formation, hole collapse could occur. It is therefore important that a thorough borehole stability study be conducted in evaluating the feasibility of a reservoir as a candidate for UBD.

12.2.14 Annular Bottomhole Pressure vs. Gas Injection Rate. The graph in **Fig. 12.11** gives the first operating envelope for UBD. The operating envelope is bound by a number of curves.

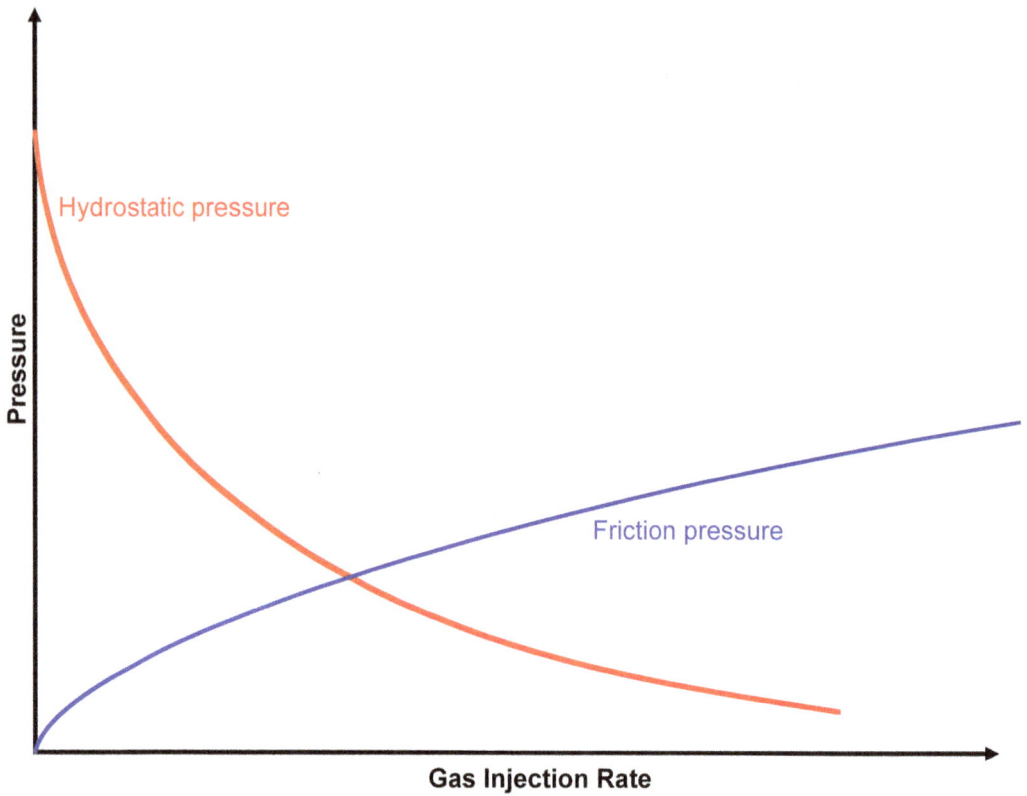

Fig. 12.12—Gas injection increases frictional pressure drop.

The annular bottomhole pressure graph is a combination chart of hydrostatic pressure vs. gas injection rate. As gas is injected into a fluid system, the hydrostatic pressure drops as more and more gas enters the system. As the amount of gas in the system increases, the gas is compressed at the bottom of the well, and the gas expands as it rises to the surface of the well. As more gas enters the system, the friction pressure in the well increases, as shown in **Fig. 12.12**. The hydrostatic pressure drops as we inject more gas, but the friction pressure starts to increase as more gas enters the well and expands on its way back to the surface.

If we combine these two effects into a single curve, then we get the typical pressure vs. gas rate curve, as shown in **Fig. 12.13**. The brown curve now shows the combined curve of hydrostatic pressure and friction pressure. In the first part of the curve, we see the rapid decline of pressure as we increase the amount of gas. This part of the curve is known as the hydrostatically dominated part of the design curve. As the amount of gas increases, the friction pressure in the well also increases as a result of the gas expansion. The flatter part of the pressure curve is known as the friction-dominated part of the curve.

As the gas-injection rate increases further, the BHP starts to increase as a result of the friction pressure.

12.2.15 BHP Stability. To design a circulation system that provides stable BHPs, the system should avoid pressure spikes as well as slugging. The operating envelope allows the drilling engineer to determine, for a particular gas-injection rate, whether the flow is dominated by hydrostatic or frictional pressure loss. Any point on the performance curve with a negative slope is dominated by hydrostatic pressure losses. These points are inherently unstable, show large pressure changes with small changes in gas flow rate, and exhibit increasing BHP with

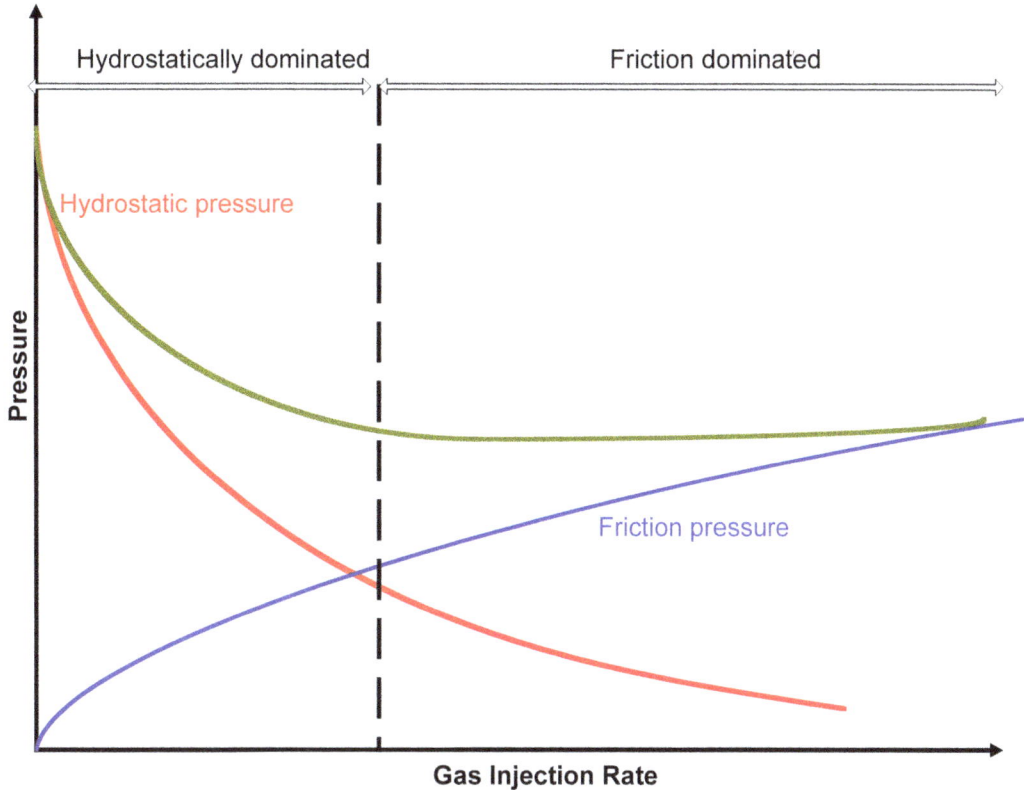

Fig. 12.13—The combined effects of gas injection.

decreasing gas flow rate. Operating on the hydrostatic-dominated slope means that severe slugging is encountered while drilling.

Points on the performance curve with a positive slope are dominated by frictional pressure loss. These points are inherently stable and exhibit increasing BHP with increasing gas flow rate.

It is important to note that "dominated by frictional pressure loss" does not necessarily imply that the frictional pressure loss is greater than the hydrostatic pressure loss. Instead, this means that the reduction in hydrostatic pressure associated with an increase in the gas-injection rate is less than the increase in frictional pressure because of the increased gas flow rate.

This information can be used in several ways. If a reduction in bottomhole pressure is required, a decrease in gas injection, the obvious answer to someone only familiar with single-phase flow, will lead to an increase in bottomhole flowing pressure if the flow is hydrostatic-dominated. Further, the cost of nitrogen (as the injection gas), if bulk liquid nitrogen is used, can be one of the most significant costs associated with UBD operations.

One of the most common misconceptions in UBD is that more nitrogen (i.e., gas) injection is better. This stems from observations of drilling operations that are hydrostatic-dominated, in which an increase in the gas-injection rate can lead to significant decreases in the bottomhole pressure. However, if the drilling operation is frictionally dominated, increasing the gas-injection rate will not only increase the bottomhole pressure but may dramatically increase the cost associated with nitrogen used while drilling. Saponja[1] recommended that UBD is carried out in the friction dominated part of the pressure curve. Operations conducted on the hydrostatic part of the curve often report that a cyclic bottomhole pressure occurs and that it is difficult to obtain a stable system. More gas is the answer here to move onto the friction-dominated part of the design curve.

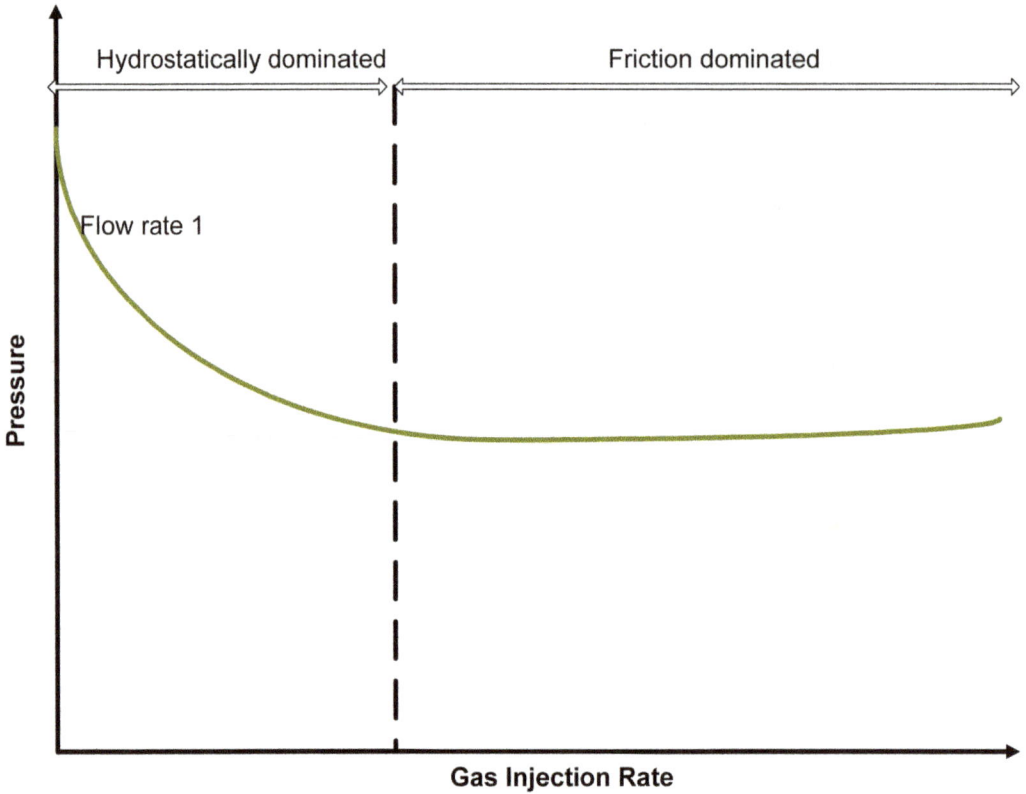

Fig. 12.14—BHP for gas injection.

Thus, for a specific design case, the operating envelope not only can confirm the feasibility of UBD but also offers valuable insights into both the acceptable and optimal gas injection rates and the influence of those rates on the bottomhole flowing pressure. Operating envelopes should be developed for a range of design parameters.

However, the operating envelope cannot tell the entire story. Each point on the operating envelope corresponds to a single wellbore calculation for a specific gas-injection rate. For all such calculations, valuable additional information can be gathered by analyzing profiles of the in-situ liquid holdup, actual gas and liquid velocities, pressures, and temperatures. At the moment, we are only concerned with the BHP. At a given flow rate, we calculate the BHP in the well for a certain fluid system, well configuration, drillstring, and surface pressure.

As we construct this first graph (**Fig. 12.14**), several other issues must be considered. The first issue is the reservoir pressure. We must establish if we can achieve a certain target pressure below the reservoir pressure. A target pressure is normally established at some 250 psi below the known reservoir pressure. **Fig. 12.15** shows liquid-flow rate and gas-injection rate vs. BHP. We now see a system that is able to achieve an underbalanced status below the reservoir pressure. We have a friction-dominated part of the design curve below the reservoir pressure and have the first operating parameters for our flow model. This curve is normally created with three or four different flow rates. Note that the shaded area is the margin between the target pressure and the predicted pressure. **Fig. 12.16** shows the margin between target pressure and actual pressure. Once we have a number of fluid rates, we continue to define the next set of operating parameters and we further define the operating window.

The next set of curves that we introduce (**Fig. 12.17**) in this curve is the minimum and maximum flow rate through the downhole motor. We now have a minimum motor speed that

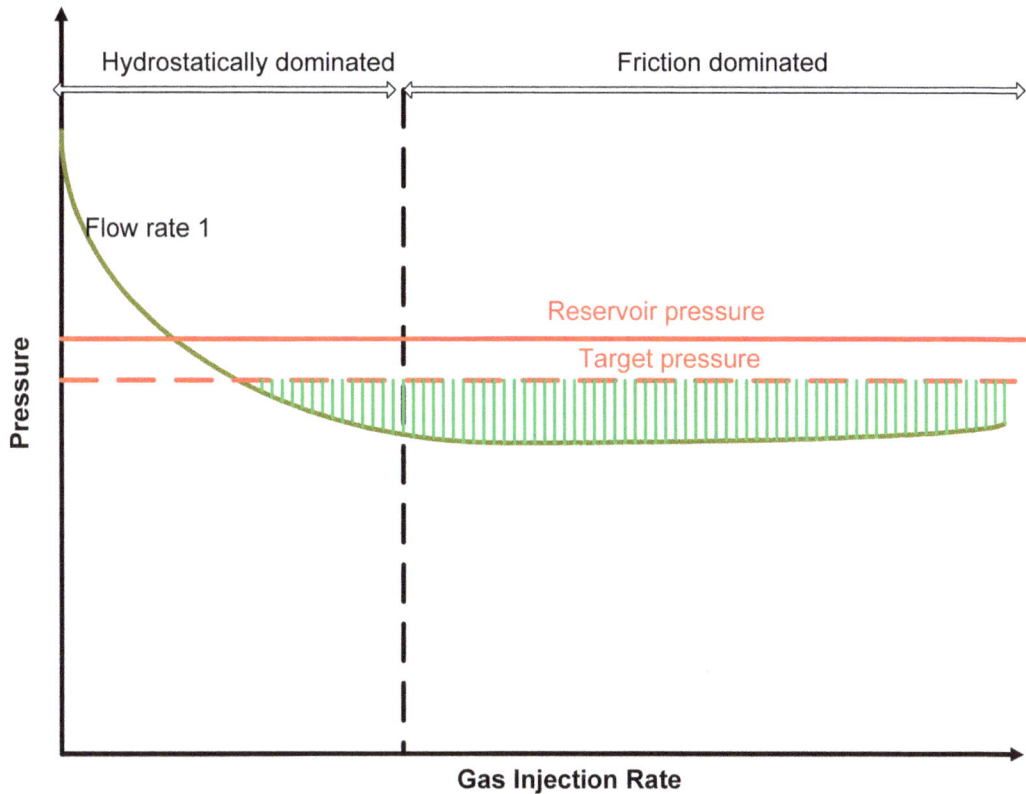

Fig. 12.15—Predicted pressure falls below target pressure at higher gas rates.

we need to drive the bit. We also have a maximum flow rate that the motor can handle without being damaged. Note that the motor limit line is slanted because the total flow rate is different for each curve at given gas rate.

It is also important to note that the maximum motor flow rate may be higher than the maximum gas-injection rate on the graph. It is not always possible to have the motor limits on the same graph.

The last information on this curve is the minimum liquid velocity for hole cleaning. Again it is sometimes impossible to show this on the design graph because the annular velocity maybe high enough without the gas injection.

12.2.16 Hole Cleaning. Fig. 12.18 shows annular liquid velocity vs. gas-injection rate and liquid-flow rate. Hole cleaning while UBD horizontally must be monitored closely. There is a reduced fluid rheology (a very thin, nonsolids-suspending mud), turbulent two-phase flow, and, normally, an increased rate of penetration (ROP). A result of two-phase flow is accelerating mud and cuttings transport velocities (because of gas expansion) as the fluid moves upward from the bit.

The main areas of concern for hole cleaning are the region where the hole angle is from 45 to 50° and the region immediately behind the bit. The area immediately behind the bit can become the critical hole-cleaning area because there is limited reservoir inflow. Liquid-phase velocity and hole cleaning in this area depend only on the fluid(s) and rate(s) being pumped or injected down the drillstring.

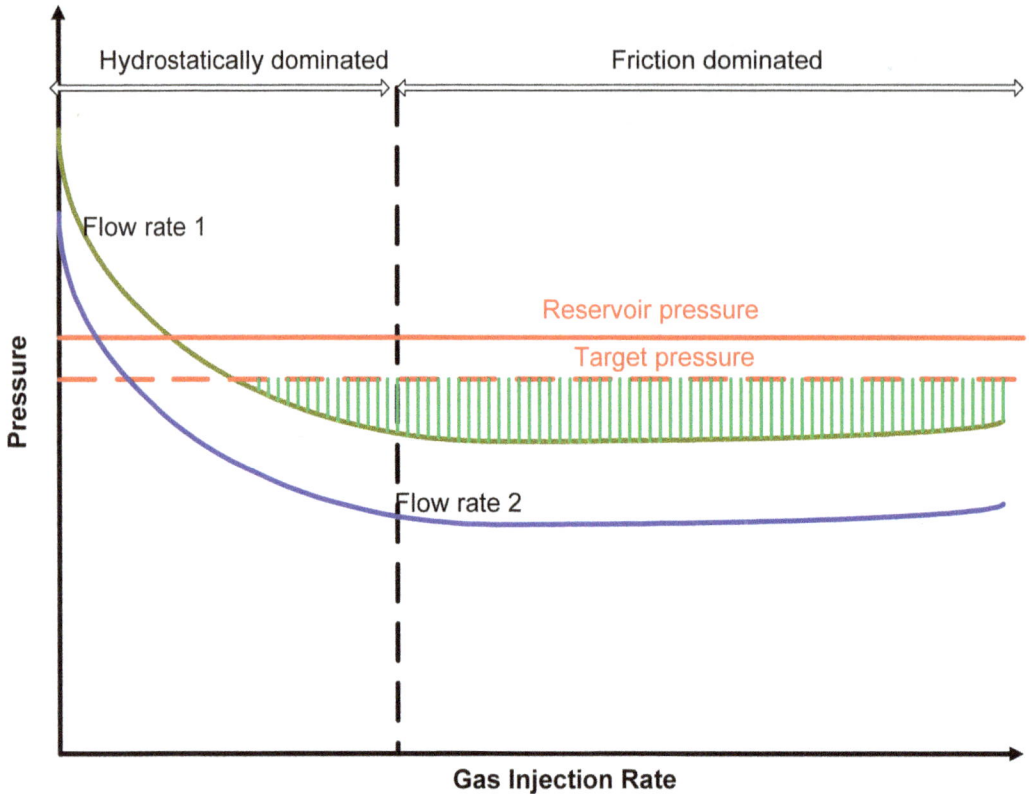

Fig. 12.16—New flow rate curve expands operating "window."

Two-phase hole cleaning is largely dependent on the same criteria as for single-phase. Hole-cleaning efficiency and solids transport are primarily controlled by liquid-phase velocities and solids concentration. Studies and field experience have shown that removal of cuttings is more efficient with two-phase fluid. The addition of a gas medium generates a turbulent flow regime, which minimizes solids bed formation. Liquid velocity is the critical parameter controlling the system's ability to transport solids. From experience, it has been concluded that a minimum liquid-phase annular velocity of 180 to 200 ft/min is required in a wellbore with a deviation greater than 10°.

12.2.17 Reservoir Inflow. In UBD, as soon as the bit penetrates the reservoir, reservoir fluids start to flow into the wellbore. At this stage, the stabilized multiphase flow regime in the well prior to reservoir fluid entry must be adjusted to account for inflow without upsetting the circulating system or moving out of the UBD window already established. The rate of reservoir fluid inflow depends, in part, on the drawdown and reservoir rock properties (the differential pressure between circulating BHP and reservoir pressure). There are a number of models that can be used to estimate the reservoir fluid inflow based on the rock and fluid parameters. However, the reservoir rock properties are fixed, and the only variable is the drawdown to control reservoir fluid inflow.

As previously defined, the inflow performance of a well represents the ability of the reservoir to produce fluids under a given condition of drawdown. The reservoir fluid inflow performance is the most important parameter in UBD, operationally and economically, because of its impact on well production and the safety operating envelope.

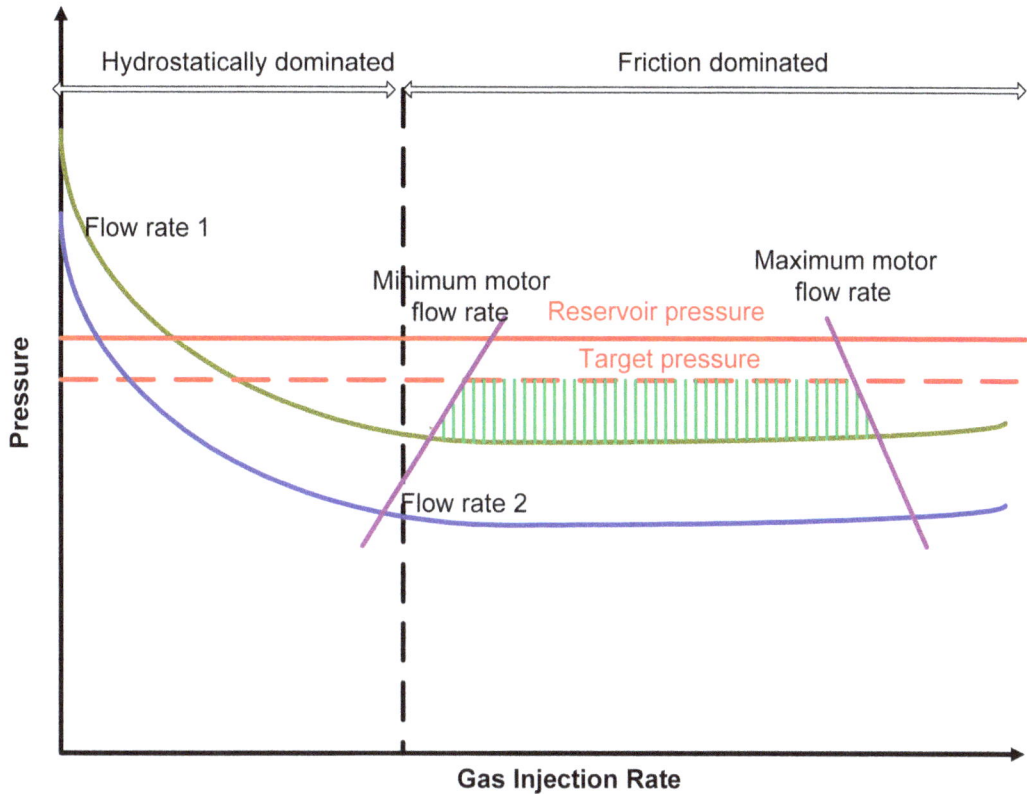

Fig. 12.17—Range of flow rate defined by motor rate limits.

The sole purpose of drilling any well underbalanced is to create conditions that induce the flow of reservoir fluid into the well while drilling, minimize reservoir damage, and optimize production of reservoir fluid from the well. Therefore, the relationship between the BHP and reservoir inflow is one of the most important parameters in UBD design and management. It is important that the BHP and reservoir inflow rate are managed and maintained within the defined operating envelope. Where the surface pressure, production rate, or BHP cannot be maintained within safe levels or underbalanced, drilling operations must cease immediately.

12.3 Downhole Equipment for UBD Operations

12.3.1 PWD Sensors. PWD sensors have proved invaluable in every UBD operation to date when they have been included in the drillstring and operated without downtime. However, quite a number of these sensors have proved problematic because of the vibration problems and fast drilling rates encountered with UBD. Adding a downhole gauge or sensor on the injection side and in the drillstring definitely enhances the UBD operation and helps the team optimize the drilling process and increase the knowledge of the reservoir.

12.3.2 Conventional MWD Tools in UBD. The most common technique for transmitting MWD data uses the drilling fluid pumped down through the drillstring as a transmission medium for acoustic waves. Mud-pulse telemetry transmits data to the surface by modifying the flow of mud in the drillpipe in such a way that there are changes in fluid pressure at surface. It involves the sequential operation of a downhole mechanism to selectively vary or modulate the dynamic flowing pressure in the drillstring and thereby sends the real-time data gathered by

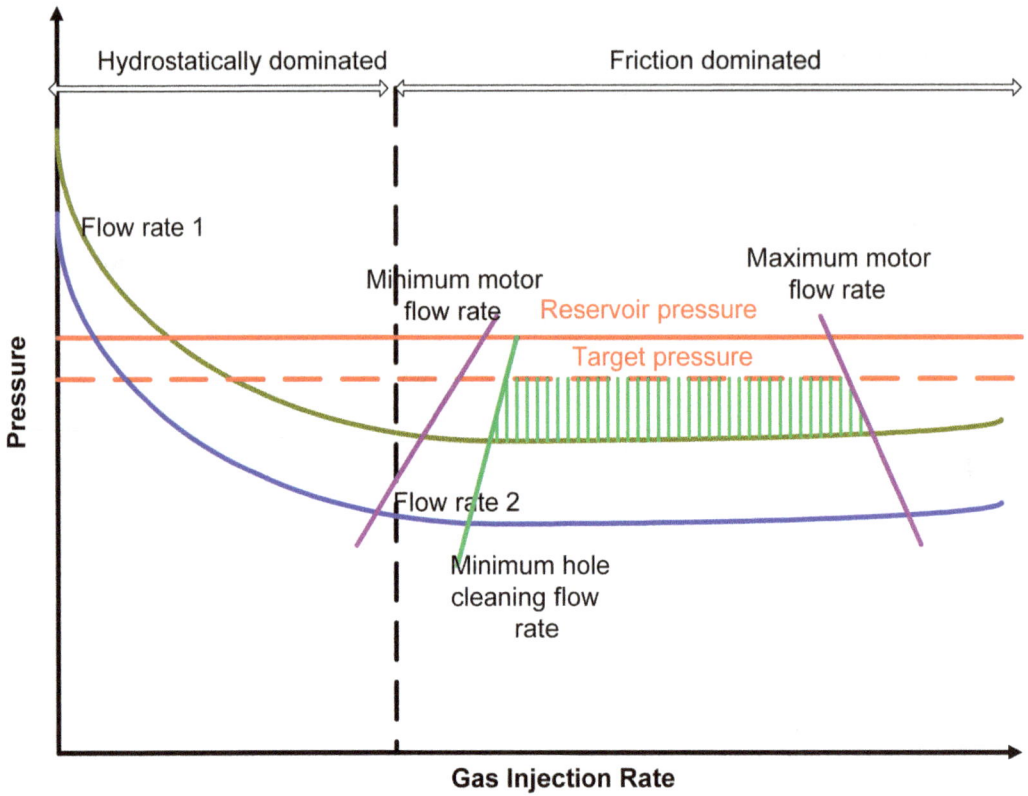

Fig. 12.18—Minimum hole cleaning rate further limits acceptable flow rates.

the downhole sensors. This variation in the dynamic pressure is detected at the surface, where it is demodulated back into the real measurements and parameters from the downhole sensors.

Signal strength at the surface depends on many factors including the mud properties, drill-string arrangement, flow rate, signal strength generated at the tool, telemetry frequency, and many others. Experience to date indicates that this enhanced mud-pulse telemetry system is best applied to scenarios with a maximum gas percentage of 20% (by volume at the standpipe), and this ratio can be extended somewhat depending on well depth, profile, liquid-phase fluid, drillstring/bottomhole assembly (BHA), pumping pressure, and flow rates. Further reductions in borehole pressure are possible with gas lift applications in which N_2 is injected into the annu-lus. A major disadvantage of the mud pulse is that it will not work if high-quality foam is needed. For such fluids, an electromagnetic method must be used.

If annular gas injection is used, we have a single-phase fluid down the drillstring, and con-ventional MWD systems can be used. If drillstring gas injection is considered, the option of using electromagnetic MWD tools must be considered.

12.3.3 Electromagnetic Measurement While Drilling (EMWD). Electromagnetic telemetry transmits data to the surface by pulsing low-frequency waves through the Earth. The first appli-cation of PWD measurements has been primarily for drilling and mud performance, kick detection, and ECD monitoring.

12.3.4 Nonreturn Valves. Float valves are necessary for UBD to prevent influx of reservoir fluids inside the drillstring either when tripping or making connections. It must be recognized that there is pressure below nonreturn valves. The positions of the float valve in the drillstring depend on the tools in the BHA and the policy of the operating philosophy underpinning the safety management of the operation. The number of float valves in the BHA and the drillstring is also a matter of company policy consistent with perceived risks and management thereof. If the drilling float valve(s) should all fail, the well may have to be circulated to kill weight fluid and a string trip undertaken to replace or repair the float valves.

It is good practice to install a float valve in the top of the drillstring, often referred to as the string float valve because it aids operational efficiency by reducing the time it takes to bleed off the pressure before making connections while also serving as an additional barrier in the event of a failure of the float valves in the BHA. This top valve is often a wireline retrievable float valve that can be retrieved, as access through the string is required. In general, a double float valve is installed just above the BHA and a further double float valve is installed above the bit so that there is redundant service. Two types of non ported drillstring floats that are commonly used are the flapper and plunger floats.

12.3.5 Deployment Valves. The underbalanced deployment valve has been designed to eliminate the need for snubbing operations or the need to kill the well to trip the drillstring during UBD operations. During UBD operations, the well is allowed to flow; this results in a flowing or shut-in pressure in the annulus at surface. With any significant pressures while tripping the drillstring, it has been necessary to either use a snubbing unit or kill the well.

The deployment valve is run as an integral part of the casing program, allowing full-bore passage for the drill bit when in the open position. When it becomes necessary to trip the drillstring, the string is tripped out until the bit is above the valve, at which time the deployment valve is closed and the annulus above the valve bled off. At this time, the drillstring can be tripped out of the well without the use of a snubbing unit and at conventional tripping speeds, thus reducing rig time requirements and providing improved personnel safety. The drillstring can then be tripped back into the well until the bit is just above the deployment valve, at which time, the deployment valve can be opened and the drillstring run in to continue drilling operations.

The deployment valve can either be run with the casing using an external casing packer for isolation or with a liner hanger and tieback. Once installed, the valve is controlled through pressure applied to the annulus, created between the intermediate and surface casing. Or the valve can be controlled through dual control lines. When using a snubbing unit, the operator not only has to consider the actual cost of the snubbing service but should also include rig-up and rig-down time together with the increased tripping times, in terms of the overall daily drilling costs.

12.3.6 Surface Equipment for UBD Operations. The surface equipment for UBD can be broken down into four categories:
- Drilling system.
- Gas-generation equipment.
- Well-control equipment.
- Surface separation equipment.

If the platform process or export equipment is used when drilling underbalanced, it is considered a separate issue and, therefore, is not included in this chapter.

Drilling Systems. Hole size and reservoir penetration, as well as directional trajectory, determine whether coiled tubing or jointed pipe is the optimal drillstring medium (**Table 12.6**). If the hole size required is larger than 6⅛ in., jointed pipe may need to be used. For hole sizes of 6⅛ in. or smaller, coiled tubing can be considered. The size of coiled tubing currently used for drilling operations is between 2 and 2⅞ in. OD. This is because of many factors, including the flow rate through the coil, pressure drop through the tubing, WOB, profile of the well, maximum pickup weight, both in-hole and surface equipment, and weight of the coiled tubing itself.

TABLE 12.6—RELATIVE MERITS OF COILED TUBING VS. JOINTED PIPE	
Coiled Tubing	Jointed Pipe
No connections made during drilling.	Connections require gas-injection shutdown, causing pressure peaks.
Higher pressure containment.	Pressure of rotating diverters limited to 3,000 psi.
Stiff wireline makes MWD systems simpler in gasified fluids.	MWD systems unreliable in gasified systems.
No snubbing system required.	Pressure deployment requires snubbing unit.
Maximum hole size, 6 in.	No hole size limit.
Hole cleaning more critical.	Hole cleaning can be assisted by rotation.
Potential for pipe collapse in high-pressure wells.	Special drillstring connections required for gas fields.
Through-tubing drilling work possible.	Through-tubing work requires special rig-floor tools on conventional rigs.
BOP stack smaller.	BOP stackup requires rotating diverter system.
Limited with drag for outreach.	Ability to drill long horizontal sections.

Occasionally, the ideal coiled tubing for an operation may be excluded because of such factors as crane or transport limitations or that the life of the coil may not be economical. Generally, coiled tubing has several advantages and disadvantages compared to jointed pipe systems. For jointed pipe systems, drillstring properties and tripping under pressure must be considered. If hole size and trajectory permit, coiled tubing is the simplest system to drill underbalanced.

Gas-Generation Equipment. Natural Gas. If natural gas is used for UBD, a natural gas compressor may be required; this would need to be reviewed once the source of the gas is known. Most production platforms have a source of high-pressure gas, and in this situation, a flow regulator and pressure regulator are required to control the amount of gas injected during the drilling process.

Cryogenic Nitrogen. The use of tanked nitrogen could be considered on onshore locations, where a large truck could be used for its supply. Cryogenic nitrogen in 2,000-gal transport tanks provides high-quality nitrogen and utilizes equipment that is generally less expensive. Liquid nitrogen is passed through the nitrogen converter, where the fluid is pumped under pressure prior to being converted to gas. The gas is then injected into the string. Generally, the requirement is for the nitrogen converter and a work tank, with additional tanks being provided as necessary. For operations in excess of 48 hours, the requirement for liquid nitrogen could be quite large, and this can result in logistical difficulties. To move away from tank transport for large nitrogen-dependent drilling operations, the use of nitrogen generators is often recommended offshore.

Nitrogen Generation. A nitrogen generator is no more than a filtering system that filters nitrogen out of the atmosphere. A nitrogen generator uses small membranes to filter the air. Oxygen-enriched air is vented to the atmosphere, and nitrogen is boosted to the required injection pressure. **Fig. 12.19** shows a nitrogen-generation system.

A nitrogen generator is 50% efficient. In real terms, if 1,500 ft²/min of nitrogen is required, then 3,000 ft²/min of air needs to be pumped into the generator. A full nitrogen system for 1,500 ft²/min would comprise of three or four large air compressors, a nitrogen generator, and a booster compressor. This equipment will take up significant deck space on an offshore rig or platform. **Fig. 12.20** shows the nitrogen generation equipment rigged up on a jackup.

Another issue associated with nitrogen generation is the purity of the nitrogen itself. Purity varies depending on the amount of nitrogen required. At 95% purity (by mole), 5% oxygen is delivered. Although this is not enough oxygen to reach explosive levels, it is sufficient oxygen to cause corrosion problems. The corrosion is further worsened when salt brine systems are used at elevated temperatures (**Fig. 12.21**).

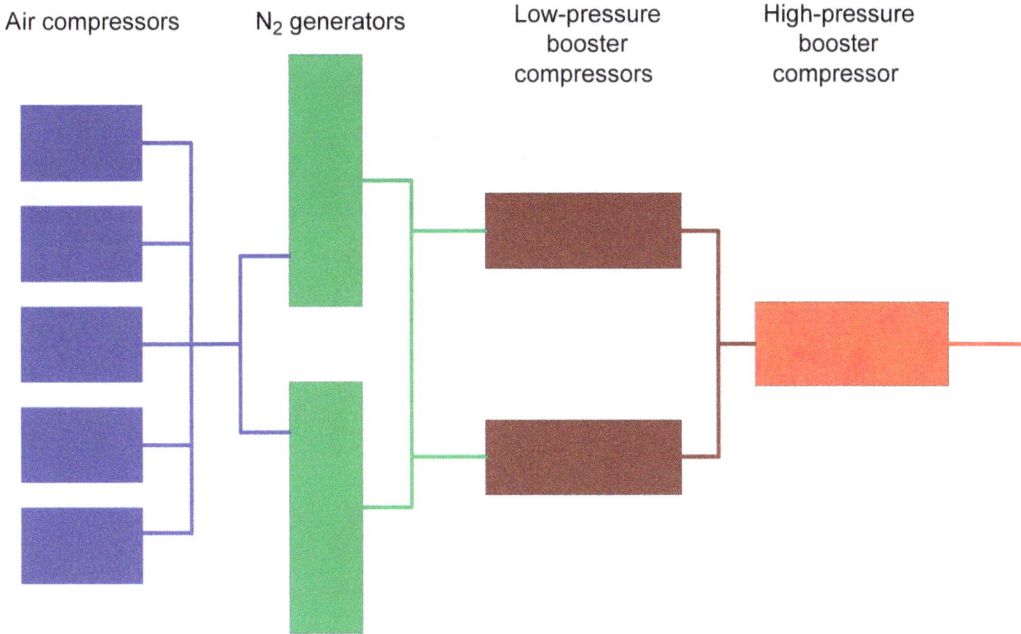

Fig. 12.19—A nitrogen generating system.

Well-Control Equipment. *Jointed-Pipe Systems.* The conventional BOP stack used for drilling is not compromised during UBD operations. The conventional BOP stack is not used for routine operations and is not used to control the well except in the case of an emergency (**Fig. 12.22**).

A rotating control-head system and primary flowline with ESD valves is installed on top of the conventional BOP. If required, a single blind ram, operated by a special Koomey unit, is installed under the BOP stack to allow the drilling BHA to be run under pressure.

Coiled-Tubing Systems. Well control is much simpler when drilling with reeled systems. A lubricator can be used to stage in the main components of the BHA, or if a suitable downhole safety valve can be used, then a surface lubricator is not required. The injector head can then be placed directly on top of the wellhead system (**Fig. 12.23**).

The reeled systems can then be tripped much faster and the rig-up is therefore much simpler. However, one consideration relating to reeled systems is the cutting strength of the shear rams. Verification is required to ascertain that the shear rams will cut the tubing and any wireline or control-line systems inside the coil. For a standalone operation on a completed well, an example stack-up is shown.

Snubbing Systems. If tripping is to be conducted underbalanced, a snubbing system must be installed on top of the rotating control-head system (**Fig. 12.24**). Current systems used offshore are called rig-assist snubbing systems. A jack with a 10-ft stroke is used to push pipe into the hole or to trip pipe out of the hole. Once the weight of the string exceeds the upward force of the well, the snubbing system is switched to standby, and the pipe is tripped in the hole using the drawworks. The ability to install a snubbing system below the rig floor allows the rig floor to be used in conventional drilling. The snubbing system is a so-called rig-assist unit. This unit needs the rig drawworks to pull and run pipe. It is designed to deal only with pipe light situations. Snubbing on an onshore rig where there is no space under the rig floor to install a

Fig. 12.20—Offshore nitrogen generating system.

snubbing unit must be conducted on the rig floor. To facilitate snubbing, so-called push/pull units are installed on the rig floor (**Fig. 12.25**).

Rotating Diverter Systems. The principle use of the rotating diverter system is to provide an effective annular seal around the drillpipe during drilling and tripping operations. The annular seal must be effective for a wide range of pressures and for a variety of equipment sizes and operational procedures. The rotating control-diverter system achieves this by packing off around the drill pipe. The rotating control-head system consists of a pressure-containing housing where packer elements are supported between roller bearings and isolated by mechanical seals.

There are currently two types of rotating diverter: active and passive. The active type uses external hydraulic pressure to activate the sealing mechanism and increase the sealing pressure as the annular pressure increases. The passive type, normally referred to as rotating control-head systems, uses a mechanical seal. All surface BOP systems have limitations in both the amount of pressure they can seal off and in the degradation of the sealing equipment from the flow and composition of the different reservoir fluids and gases over time, regardless of the type of surface BOP control system chosen.

Rotating Control Heads (Passive Systems). Rotating control heads are passive sealing systems (**Fig. 12.26**). Rotating control heads have given excellent service for more than 30 years, particularly in the air and air-foam drilling industry. The rotating control head is playing an increasingly important role in UBD, provided that its inherent pressure limitations are not being extended. The conventional, original rotating control head was developed in the 1960s. This is a low-pressure model and has been used on thousands of underbalanced and overbalanced drilled wells. It is designed to operate at 500 psi rotating and 1,000 psi static. It is capable of rotating up to 200 rpm and uses a single stripper rubber. It is currently used in

Fig. 12.21—Onshore nitrogen generator and compressors.

many underbalanced operations in the United States. The current rotating control heads are rated to a static pressure of 5,000 psi and a rotating pressure of 3,000 psi with 100 rpm.

Rotating BOPs (Active Systems). The rotating blowout preventer (RBOP), as it is commonly referred to under its trade name, is probably the most significant piece of equipment developed, with the biggest impact being its ability to drill underbalanced with jointed pipe in a variety of different reservoir and wellbore scenarios. The rotating control-head system must be sized and selected on the basis of the expected surface pressures. A well with a reservoir pressure of 1,000 psi does not need a 5,000-psi rotating control-head system. A number of companies offer rotating control-head systems for UBD (**Fig. 12.27**).

Separation Equipment. The separation system has to be tailored to the expected reservoir fluids. A separator for a dry-gas field is significantly different from a separator required for a heavy-oil field. The separation system must be designed to handle the expected influx, and it must be able to separate the drilling fluid from the return well flow so that it can be pumped down the well once again.

The surface separation system in UBD can be compared with a process plant, and there are many similarities with the process industry. Fluid streams while drilling underbalanced are often described as four-phase flow because the return flow comprises of oil, water, gas, and solids.

The challenge of separation equipment for UBD is to effectively and efficiently separate the various phases of the return fluid stream into individual streams. Several approaches in separation technology have emerged recently (**Fig. 12.28**). The chosen approach depends largely on the expected reservoir fluids.

Careful design of the surface separation system is required once the reservoir fluids are known. Dry gas is much simpler to separate than a heavy-crude or gas-condensate reservoir. However, the separation system must be tailored to reservoir and surface requirements. This requires a high degree of flexibility, and the use of a modular system helps to maintain such flexibility.

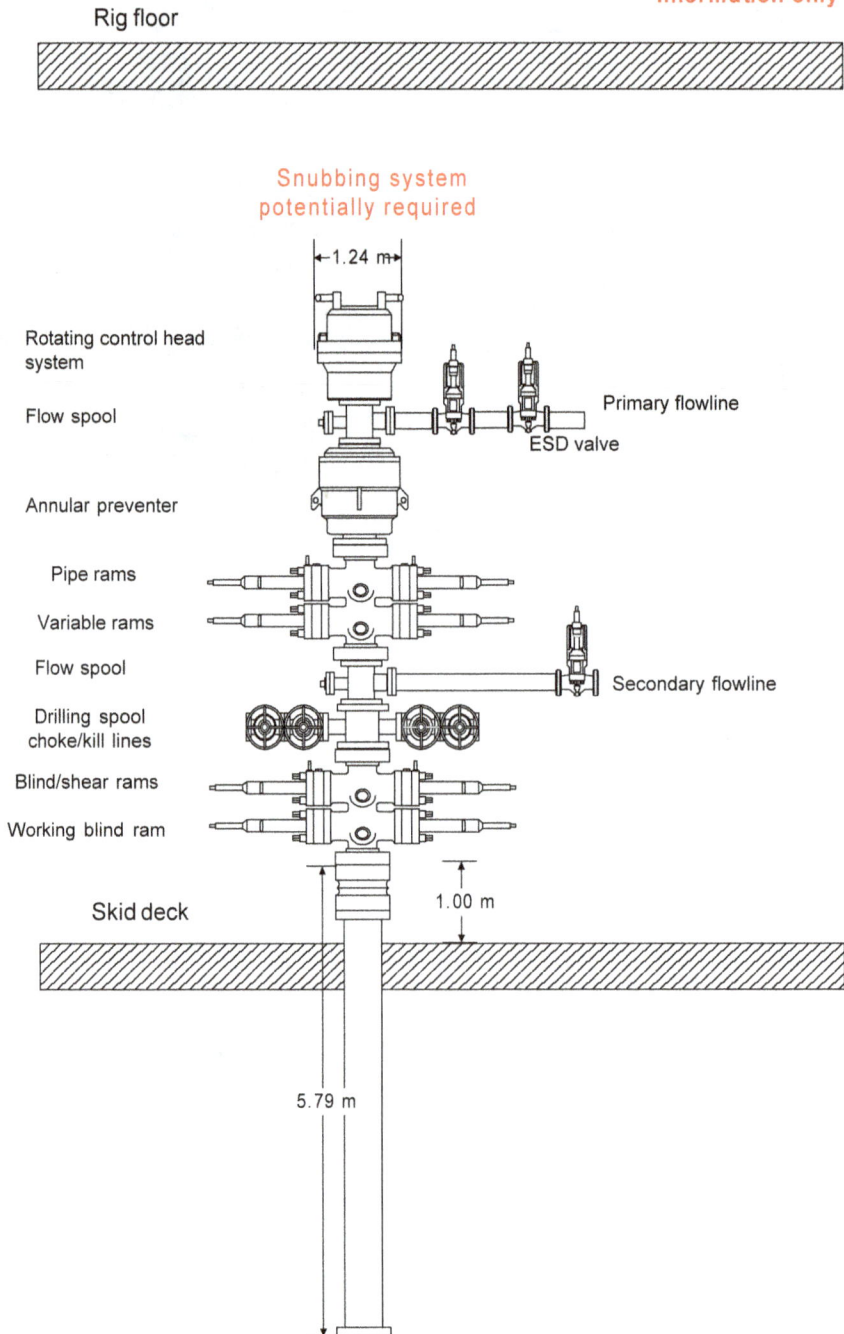

Fig. 12.22—Typical BOP stack-up.

The use of a modular system for offshore operations is often recommended because lifting capacity of platform and rig cranes is regularly limited to 15 or 20 tons. To reduce the total footprint of a separation package, vertical separators are generally used offshore as opposed to

Stuffing box and
injector head

CT riser section

Stripper assembly

Blind rams

Shear rams

Kill line

Slip rams

Pipe rams

Main flowline with ESD
valves to choke manifold

Blind ram

Tree connector

Ground level

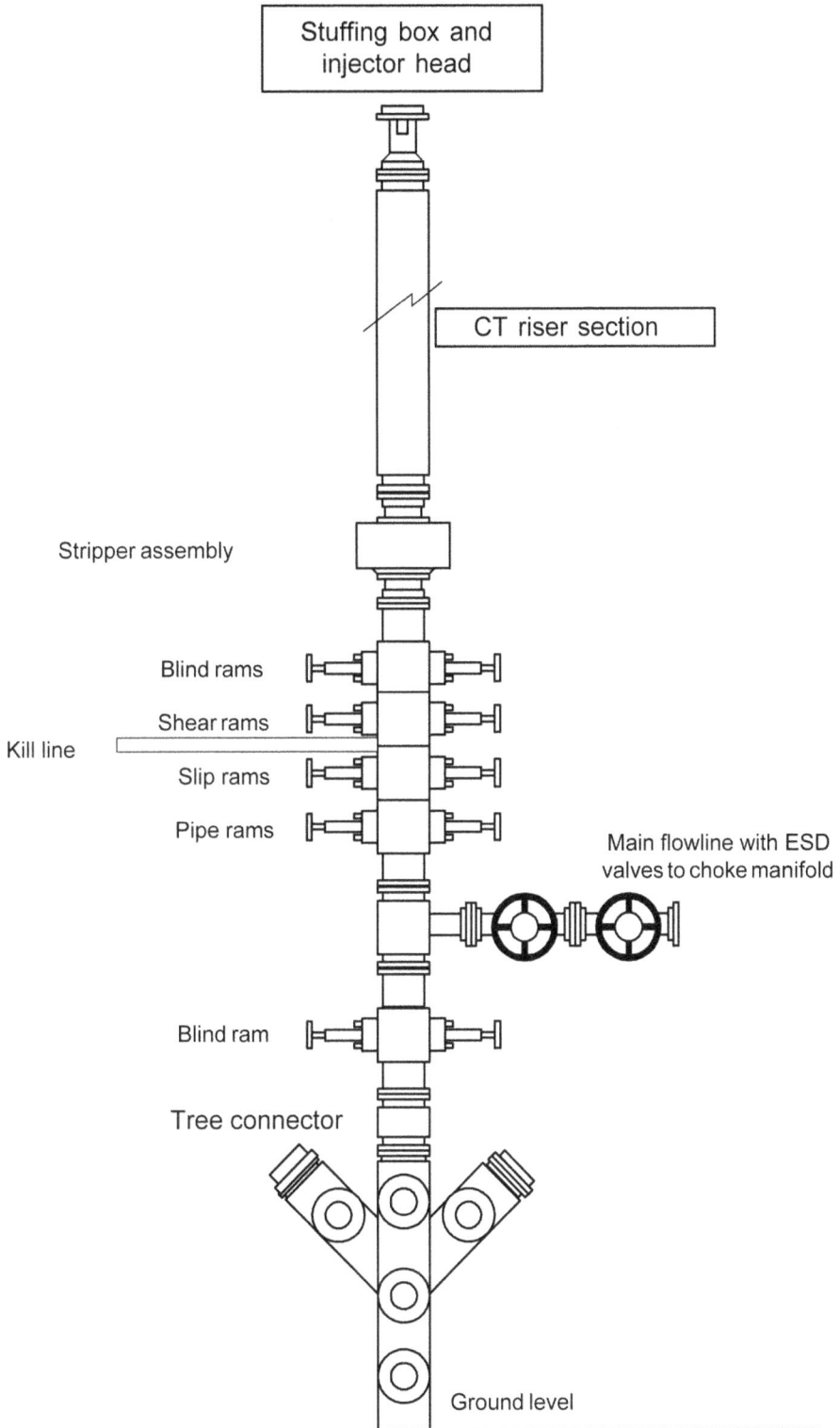

Fig. 12.23—Typical coiled-tubing stripper assembly.

Fig. 12.24—Rig-assist snubbing system.

the horizontal separators used in onshore operations. In a lot of situations, the separator is the first process equipment that receives the return flow out of a well. Separators can be classified, as shown in **Table 12.7**. Separation of liquids and gasses is achieved by relying on the density differences between liquid, gas, and solids. The rate at which gasses and solids are separated from a liquid is a function of temperature and pressure.

Horizontal and vertical separators can be used. Vertical separators are more effective when the returns are predominantly liquid, while horizontal separators have higher and more efficient gas handling capacities. In horizontal separators, well returns enter and are slowed by the velocity-reducing baffles (**Figs. 12.29 and 12.30**).

Data Acquisition. The data acquisition used on the separation system should provide the maximum amount of information about the reservoir obtainable while drilling. It should also allow for a degree of well testing during drilling. Furthermore, the safety value of data acquisition should not be overlooked because well control is related directly to the pressures and flow rates seen at surface.

Erosion Monitoring. Erosion monitoring and prediction of erosion on pipe work is essential for safe operations. The use of nondestructive testing technology has been found to be insuffi-

Down Position   **Up Position**

Fig. 12.25—Push/pull snubbing machine.

cient in erosion monitoring. An automated system using erosion probes is currently deployed, and this allows accurate prediction of erosion rates in surface pipe work.

12.3.7 Completing Underbalanced Drilled Wells. The majority of wells previously drilled underbalanced could not be completed underbalanced. The wells were displaced to an overbalanced condition with kill fluid prior to running the liner or completion. Depending on the completion fluid type, some formation damage would take place. The damage is not as severe for completion brine as with drilling mud because there are no drilled cuttings and fines in the brine. However, reductions in productivity of 20 to 50% have been encountered in underbalanced drilled wells that were killed for the installation of the completion.

If the purpose of UBD is reservoir improvement, it is important that the reservoir is never exposed to overbalanced pressure with a nonreservoir fluid. If the well has been drilled underbalanced for drilling problems and productivity improvement is not impaired, then the well can be killed and a conventional completion approach can be taken.

A number of completion methods are available for underbalanced drilled wells: liner and perforation, slotted liner, sandscreens, and barefoot. All of these options can be deployed in UBD wells. The use of cemented liners in an underbalanced drilled well is not recommended if the gains in reservoir productivity are to be maintained.

Regardless of the liner type run, the installation process for the completion is exactly the same. It is assumed that a packer-type completion is installed. The production packer and tailpipe are normally run and set on drillpipe with an isolation plug installed in the tailpipe. If

Kelly driver

Top rubber

Bearing assembly

Bottom rubber

Bowl

Fig. 12.26—Rotating control head.

the well is maintained underbalanced, well pressure will normally require the production pack-er and tailpipe to be snubbed into the well against well pressure. The use of pressure-operated setting equipment in underbalanced drilled wells is not recommended. A mechanically set pro-duction packer should be used.

Installation of a Solid Liner. Using solid pipe for the liner is no different from snubbing in drillpipe or tubing. The shoe track of the liner must be equipped with nonreturn valves to pre-vent flow up the inside of the pipe. The liner is normally run with a liner packer, and the liner can be snubbed into the live well. Once on bottom, the liner hanger and packer are set and the reservoir is now sealed. If zonal isolation is required, ECPs must be run at predetermined inter-vals. Once the liner is set, the pipe must be perforated to obtain flow. This can be achieved using the normal procedures, but it should be remembered that any fluid used must maintain the underbalanced status.

Installation of a Perforated Liner or a Sandscreen. The main disadvantage of running a slotted liner or sandscreen in an underbalanced drilled well is that isolation is not possible across the slotted section of the liner or screen with the BOPs. The use of plugged slots that dissolve once the liner is installed downhole is not deemed safe for offshore operations. The pressure integrity of each slot would have to be tested prior to running each joint, and this is not feasible.

The use of special blanking pipe in sandscreen also adds further complications to the instal-lation procedures. Running a slotted pipe or screen into a live well cannot be done safely because even if all the holes are plugged, the potential for a leak is too great. The only way to install a slotted liner in a live well is by using the well as a long lubricator and by isolating the reservoir downhole.

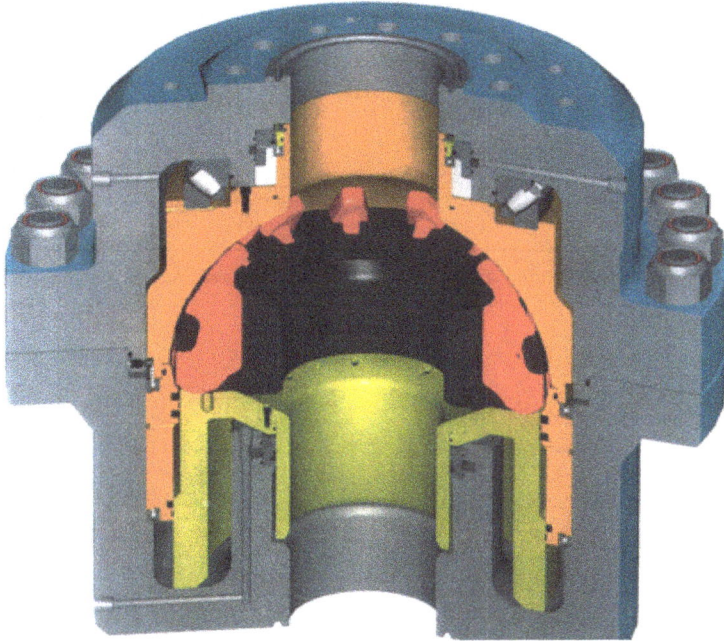

Fig. 12.27—Rotating BOP.

There are mechanical methods of downhole isolation available for the running of a slotted liner. The underbalanced liner bridge plug system is one of the systems currently on the market. This system allows a retrievable plug to be set in the last casing. This isolation plug is released by a retrieving tool that is attached to the bottom of the slotted liner. This retrieving tool unseats the isolation plug and then swallows the isolation plug or packer. The swallowing action of the retrieval tool ensures that the plug and retrieving tool are rigid and can be run to TD without hanging up in the open hole. Both the packer and retrieval tool are specifically designed to be released by the liner. If necessary, the well can be lubricated to kill fluid on top of the plug and displaced via the slotted liner when the drillstring is sealed by the rotating diverter. The procedure for running a slotted liner and the completion in an underbalanced drilled well is outlined in the following diagrams (**Fig. 12.31**).

Completion Running. The main problem with running the completion in a live well is the installation of the subsurface safety valve control line. Once the control line is connected, the BOPs no longer seal around the pipe. Once again, therefore, the simplest method is to isolate the reservoir prior to running the completion.

In the case of the completion, the production packer with a plug installed in the tailpipe is snubbed into the live well, and the production packer is set on drillpipe. The packer assembly would be lubricated into the well by utilizing the snubbing well-control system.

Once the production packer is set, the drillpipe can be used to pump completion fluid to provide an additional barrier that can be monitored if required. The completion is now run conventionally. The isolation plug in the tailpipe is retrieved during the well commissioning. Once again, before pulling this plug, the fluid must be displaced out of the completion string. This can be achieved with coiled tubing or with a sliding sleeve. Once the completion has been installed, the well is ready for production. No cleanup or stimulation is required in the case of underbalanced drilled wells.

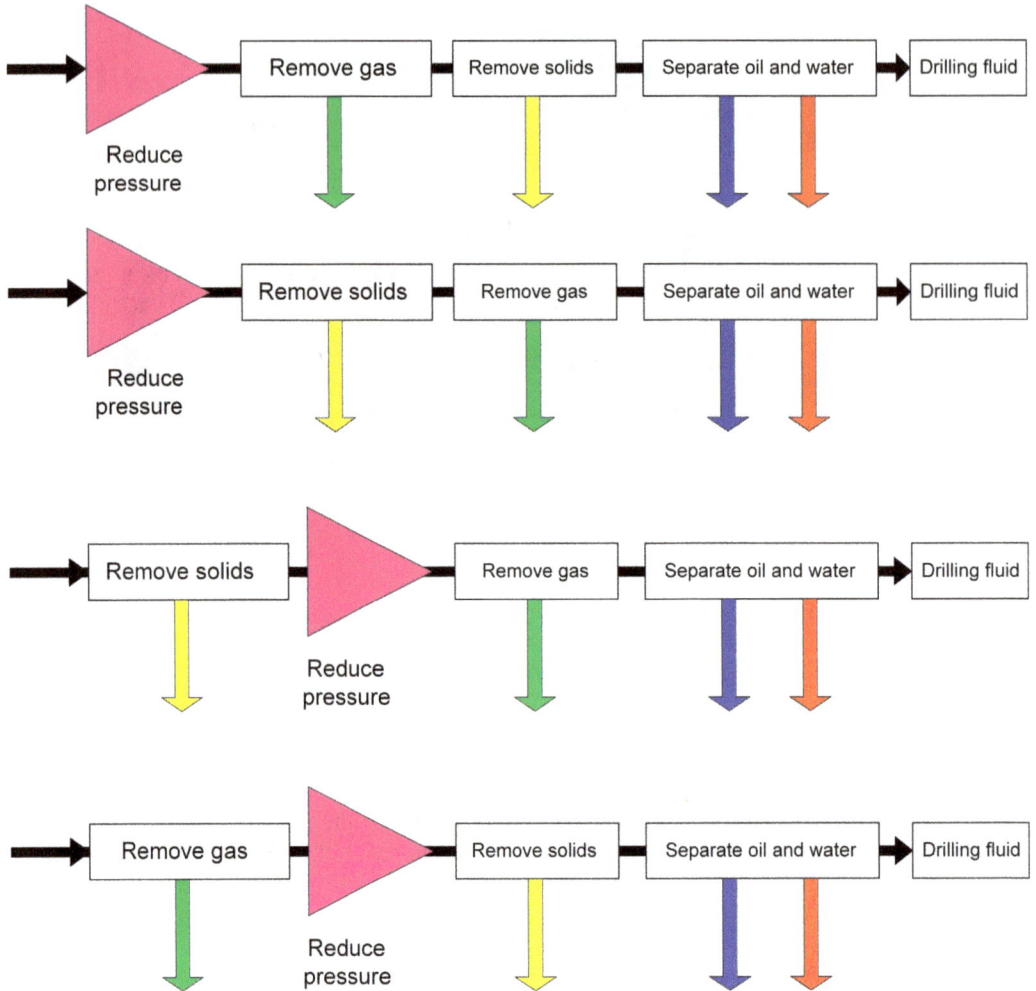

Fig. 12.28—Various solids-control and fluid-separation strategies.

TABLE 12.7—CLASSIFICATION OF SEPARATORS	
Classification	Operating Pressure
Low pressure	10 to 20 psi up to 180 to 225 psi
Medium pressure	230 to 250 up to 600 to 700 psi
High pressure	750 to 5,000 psi

12.3.8 Workover of an Underbalanced Drilled Well. The workover procedure is a reversal of the completion running (i.e., a suspension plug is installed in the production packer tailpipe, and the well is lubricated to kill fluid). After retrieving the completion, the packer-picking assembly is run to the packer depth, and the well is returned to an underbalanced condition prior to retrieving the packer. This ensures that formation-damaging kill fluid does not come into contact with the reservoir at any time.

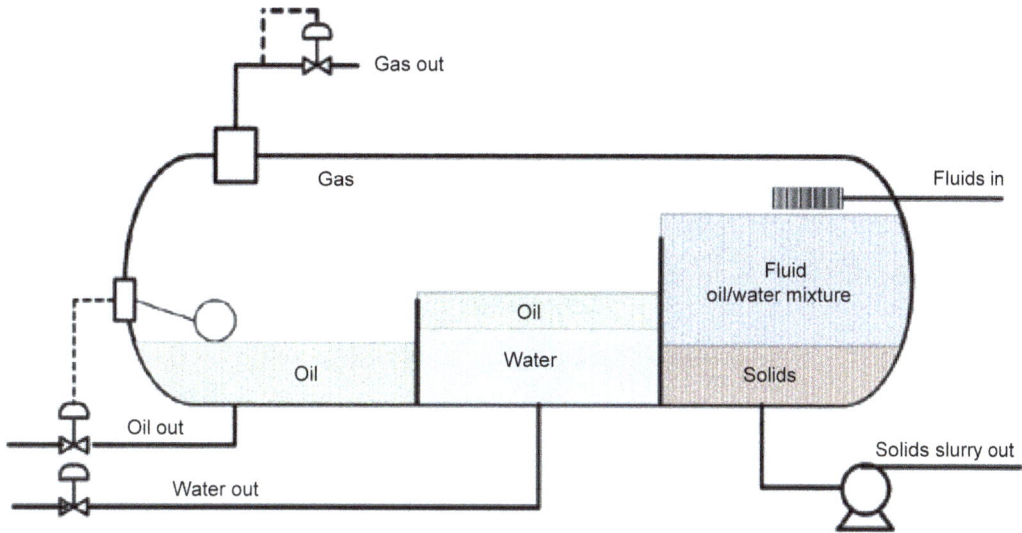

Fig. 12.29—Horizontal separator.

12.3.9 Underbalanced Drilled Multilateral Wells. The setting of the production packer with a mechanical plug allows the lower leg in a multilateral well to be isolated and remain underbalanced while the second leg is drilled. After running the liner in the second leg, the completion can be run and a second packer can be installed and stabbed into the lower packer. If leg isolation is required, a flow sleeve can be installed at the junction to allow selected stimulation or production as required. Re-entry into both legs is also possible by use of a selective system. However, more detail as to the exact requirements from a multilateral system must be reviewed.

Drilling a multilateral well underbalanced with the main bore producing can be done, but the drawdown on the reservoir is small. A further setback is that the cleaning up of the lateral is difficult if the main bore is a good producer. Getting sufficient flow through the lateral to lift fluids can be a challenge.

12.3.10 Health Safety and Environmental Issues. Because UBD involves working on a live well, a hazard operational ("hazop") analysis is required for the full process. To this end, a flow chart has been created that shows all the elements in the UBD process. Using the diagram, each element can be analyzed for input and output and the diagram has also been used to good effect to ensure that all items of an UBD system are reviewed during the hazop. It also allows procedures and documentation to be reviewed for all parts of the UBD system.

Fig. 12.32 shows an analysis path together with the interaction of the various elements. The drilling liquid system (1), the gas system (2), and the reservoir characteristics (3) specify the well system (4). The well system (4) specifies the well control system (5), which has impact on the drilling fluid system (1). This loop must be resolved before continuing to the surface separation system (6). This influences the rig fluid system (7), which must also be compatible with the drilling liquid system (1). The platform process system (8) must be consistent with the surface separation system (6) as well as the overall platform system. Multiple iterations are necessary to bring all systems into alignment.

Environmental Aspects. The UBD system is a fully enclosed system. When combined with a cuttings-injection system and an enclosed mudpit system, a sour reservoir can be drilled safely using a UBD system. The pressures and flow rates are kept as low as possible. It is not the intention to drill a reservoir and produce it to its maximum capacity. A well test can be carried

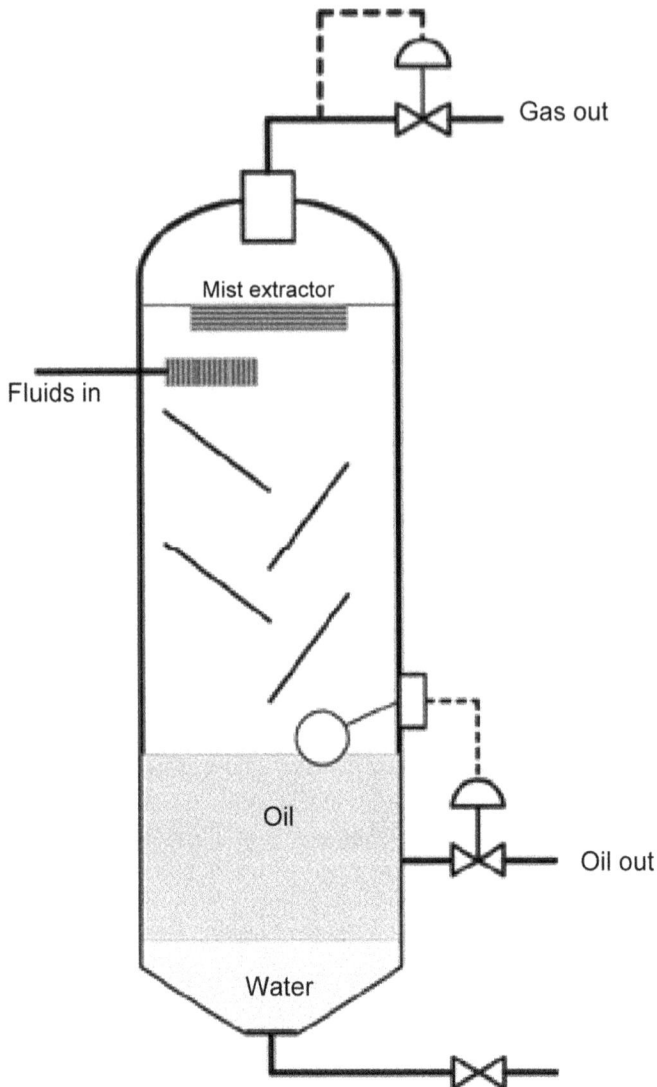

Fig. 12.30—Vertical separator.

out while drilling underbalanced to provide some productivity information. The hydrocarbons produced during the UBD process can be routed to the platform process plant, exported, or flared. There is work currently being undertaken to reduce flaring and recover the hydrocarbons for export. In a prolific well, a significant amount of gas might be flared during the drilling process. Recovering this gas provides an environmental benefit and an economic benefit. Oil and condensate recovered are normally exported via a stock tank into the process train.

Safety Aspects. Besides the full hazop, substantial crew training is required for UBD. A typical drilling crew has been instructed during its entire career that if a well kicks, it must be shut in and killed. In contrast, during UBD, the single item to be avoided is to kill the well. This may undo all the benefits of UBD. Working on a live well is not a normal operation for a drilling crew, and good training is required to ensure that accidents are avoided.

The UBD process is more complex when compared to conventional drilling operations. Gas injection, surface separation, and snubbing may be required on a well. If the hydrocarbons pro-

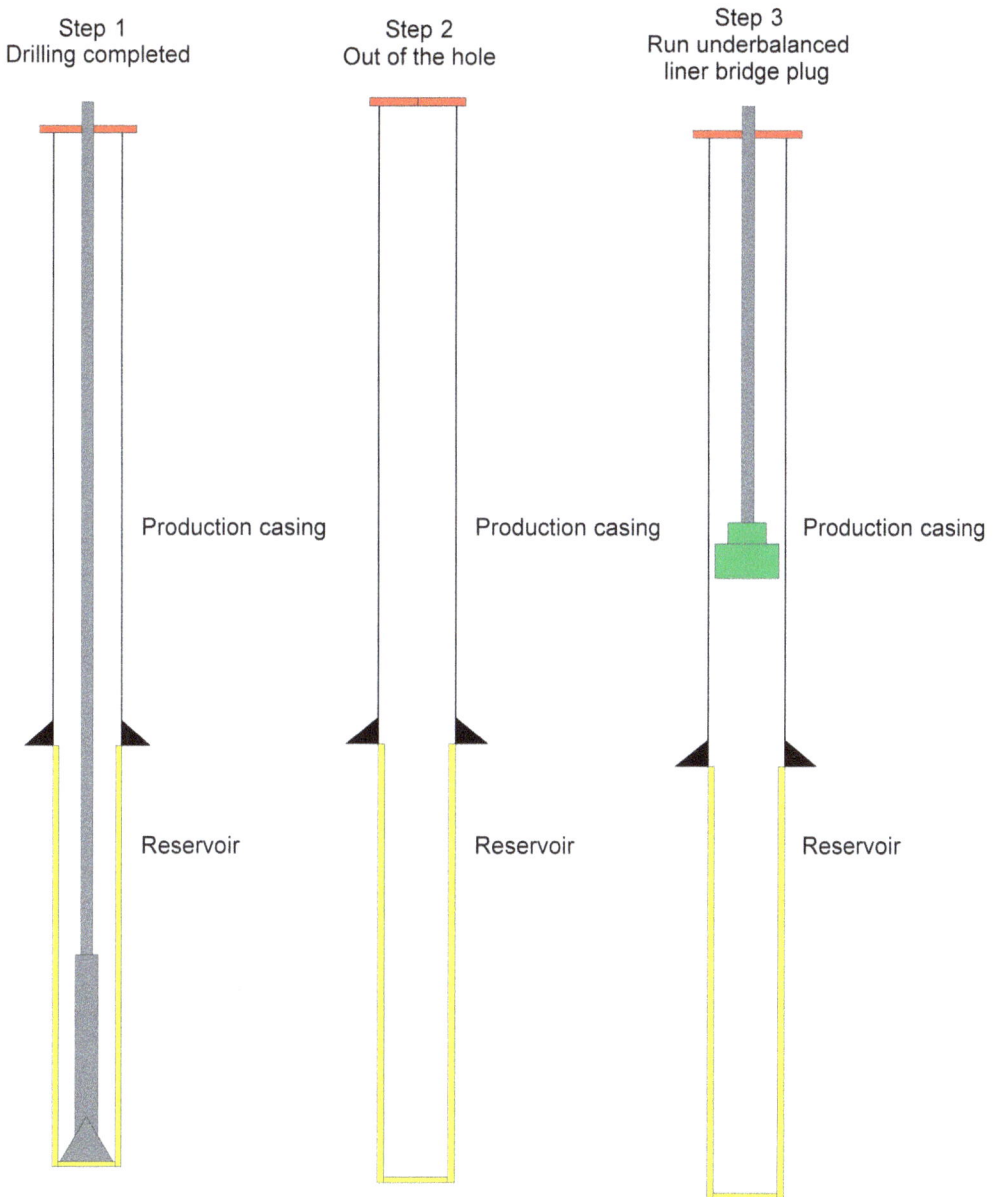

Fig. 12.31—The procedure for running a slotted liner and the completion in an underbalanced drilled well.

duced are then pumped into the process train, it is clear that drilling is no longer a standalone operation.

The reservoir is the driving force in the UBD process. The driller must understand the process and all the interaction required between the reservoir—the liquid-pump rate, the gas-injection rate, and the separation and process system—to drill the well safely. When tripping operations start, the well must remain under control. Snubbing pipe in and out of the hole is not a routine operation, and a specialized snubbing crew is normally brought on to snub the pipe in and out of the hole.

Step 4
Underbalanced liner
bridge plug set

Step 5
Run slotted liner

Step 6
Run slotted liner

Production casing

Production casing

Production casing

Reservoir

Reservoir

Reservoir

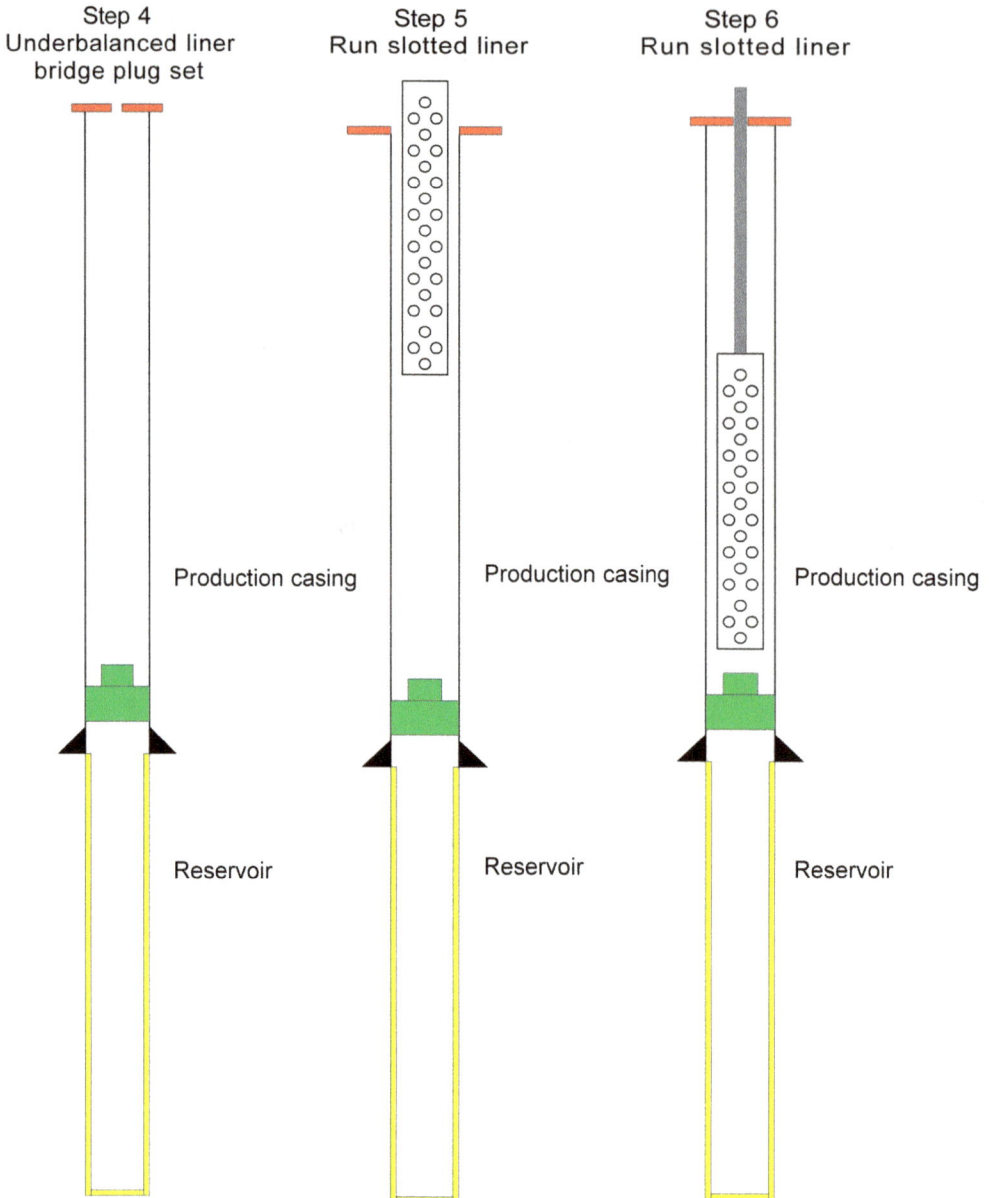

Fig. 12.31—The procedure for running a slotted liner and the completion in an underbalanced drilled well.
(Continued)

The extra equipment also brings a number of extra crewmembers to the rig. So besides a
more complex operation, a number of service hands are on the rig which now must start working with the drilling crew. Yet the drilling crew will move back to conventional drilling once
the well is completed. The drilling crew must be trained in this change of operating practice.

When a number of wells will be drilled underbalanced in a field, it may be a consideration
to batch drill the reservoir sections. This saves mobilization, and it also sets a routine with the
drilling crew. It must be stated that few accidents occur during UBD. This is mainly because
of the high emphasis on safety during live well operations.

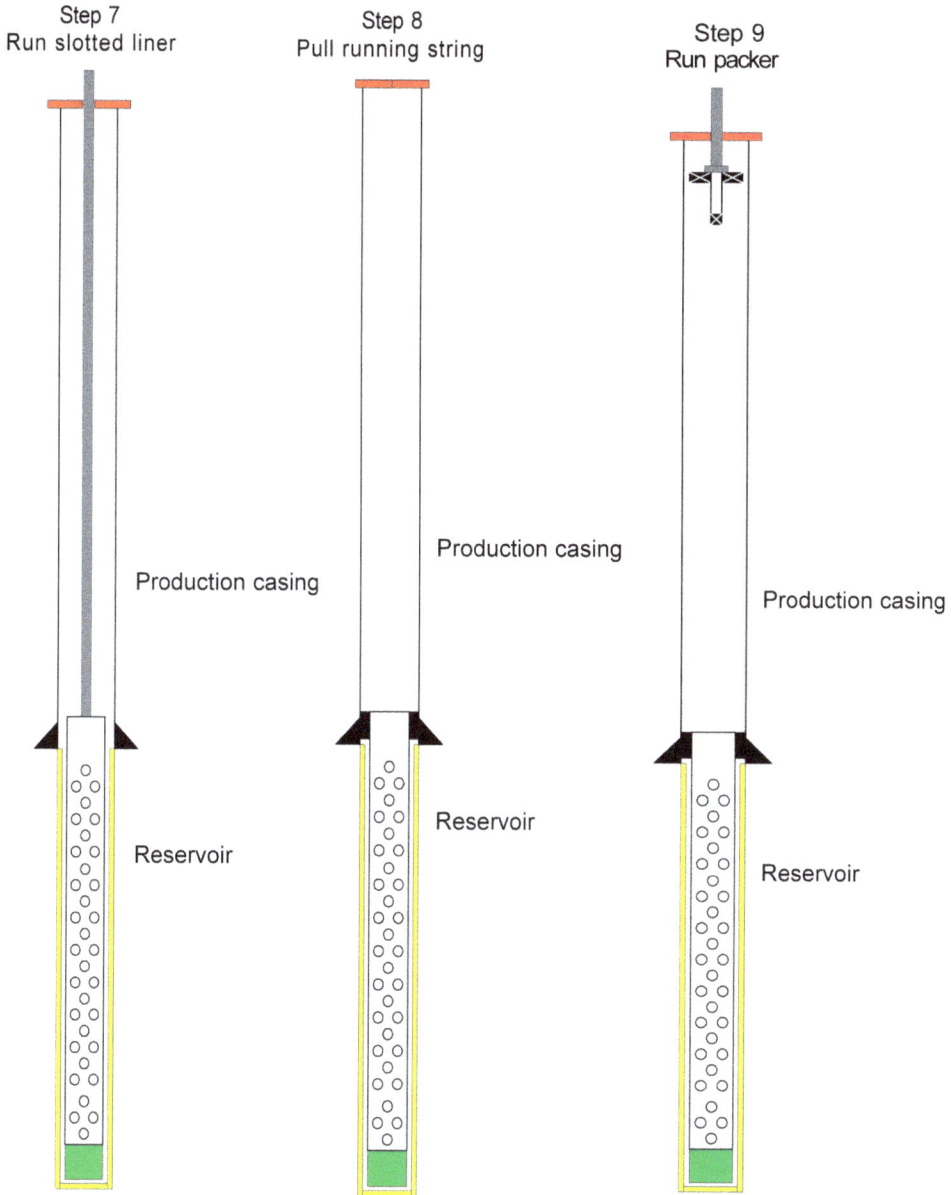

Fig. 12.31—The procedure for running a slotted liner and the completion in an underbalanced drilled well. (Continued)

12.3.11 Limitations. There are limitations, as well as advantages, to UBD. Before embarking on a UBD program, the limitations of the process must be reviewed. There are technical limitations as well as safety and economic limitations to the UBD process. The following are conditions that can adversely affect any underbalanced operation:

• Insufficient formation strength to withstand mechanical stress without collapse.

• Spontaneous imbibitions because of incompatibility between the base fluid used in the UBD fluid and the rock or reservoir fluid. Use of a nonwetting fluid can prevent or reduce this situation.

Step 10
Production packer set

Step 11
Completion

Production casing

Production casing

Reservoir

Reservoir

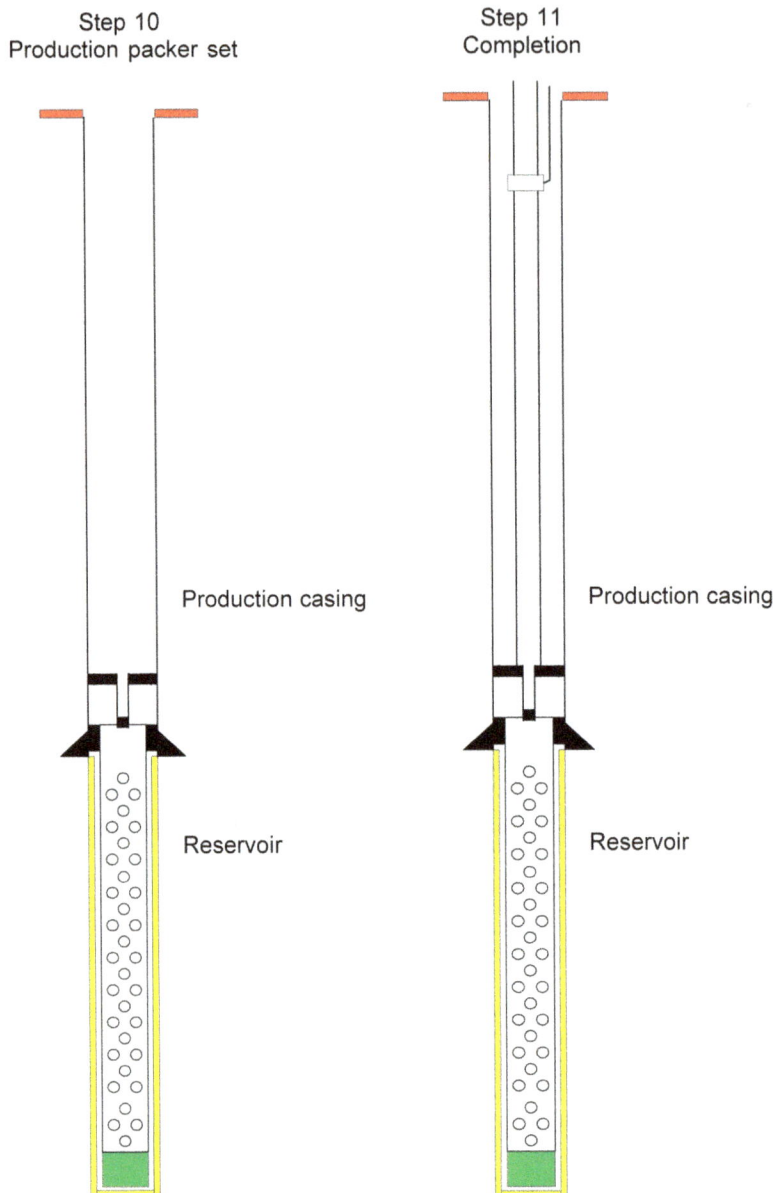

Fig. 12.31—The procedure for running a slotted liner and the completion in an underbalanced drilled well. (Continued)

  • Deep, high-pressure, highly permeable wells presently represent a technical boundary because of well control and safety issues.
  • Noncontinuous underbalanced conditions.
  • Excessive formation water.
  • High-producing zones close to the beginning of the well trajectory will adversely affect the underbalanced conditions along the borehole.
  • Wells that require hydrostatic fluid or pressure to kill the well during certain drilling or completion operations.

HSE System Elements

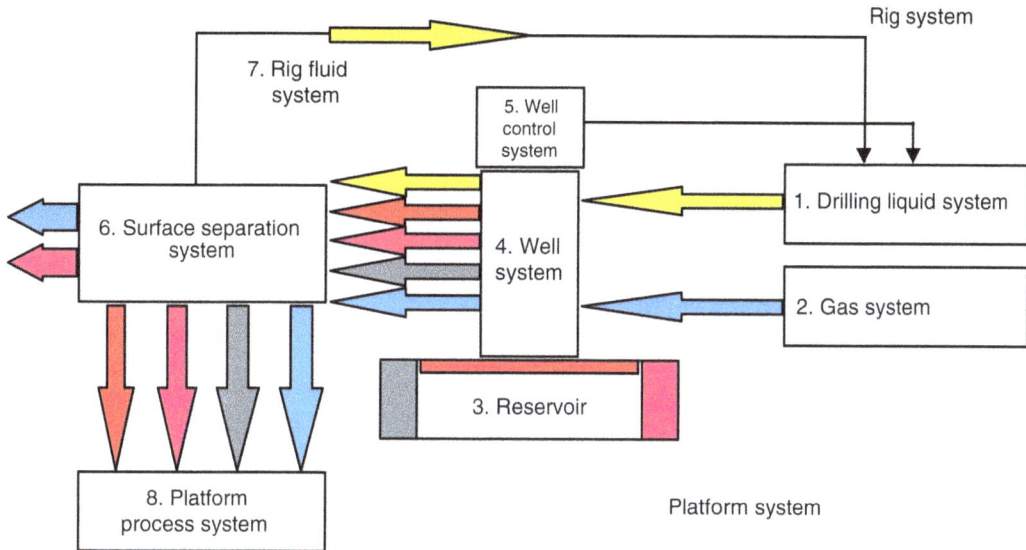

Fig. 12.32—Hazardous operation planning for UBD.

• Slimhole or drilling conditions that result in a small annulus create high backpressures because of frictional forces.

• Wells that contain targets with significant pressure or lithology variations throughout.

Technical Limitations. *Wellbore Stability.* Wellbore stability is one of the main limitations of UBD. Borehole collapse as the result of rock stresses is one issue to consider. The other issue is chemical stability, which is a problem seen in shale and claystone formations. Both these issues can have serious implications in UBD. Defining maximum drawdown and reviewing chemical compatibility with the proposed drilling fluids is a key issue in the feasibility of UBD.

Water Inflow. Water inflow in a depleted reservoir can cause severe problems in an underbalanced drilled well. If the flow rate is high enough, the well will be killed as a result of the water influx. Gas lifting a well that produces water at a high rate is almost impossible. Care must be taken that the water leg in a depleted reservoir is not penetrated when drilling underbalanced.

Directional Drilling Equipment. Directional drilling equipment can have limitations on UBD. Hydraulic operated tools cannot be used in underbalanced wells, and if a gasified system is used, the MWD pulse systems may not work. Certain motors and other directional equipment may be prone to failure as a result of the rubber components becoming impregnated with the gas used. Explosive decompression of rubber components is a consideration when selecting equipment.

The higher torque and drag seen in underbalanced wells (as much as 20 to 100%) may also prevent certain trajectories from being drilled underbalanced. The higher torque is caused by the reduced buoyancy combined with the lack of filter cake on the borehole wall.

Unsuitable Reservoir. The reservoir may not be suitable for UBD. A highly porous, high-permeability reservoir can provide too much inflow at low drawdown. It is important that the perceived benefits of UBD are kept in mind when planning for underbalanced operations.

Safety and Environment. The health, safety, and environment issues of a UBD operation may prove to be too complicated to allow UBD to proceed.

Surface Equipment. The placement of the surface equipment may prove to be impossible on some offshore locations. There can be problems with rig-floor height and with deck space or deck loading. Both the wellhead equipment and the surface separation equipment must be carefully designed to fit the platform or rig.

12.3.12 Training in UBD. The entire platform/rig crew must be trained in underbalanced techniques. Once the crew understands what is to be achieved, operations will run more smoothly and with fewer problems and accidents. Documentation, policies, and procedures should not be forgotten when considering training.

12.3.13 Personnel. The number of crew required for UBD is still considered large; 15 to 20 extra crewmembers are required for full UBD and completion.

12.3.14 Economics. The business driver behind the technology must never be forgotten. If the benefits cannot be achieved, the project must be reviewed. Improvements seen from UBD are twice the penetration rate and triple the production rate.

Nomenclature

P = pressure, m/Lt^2, psi

References

1. Saponja, J.: "Underbalanced Drilling Engineering and Well Planning," paper presented at the 1995 Intl. Underbalanced Drilling Conference and Exhibition, The Hague, 2–4 October.

General References

Adam, J. and Berry, M.: "Through Completion Underbalanced Coiled Tubing Sidetrack of Well Dalen 2," paper presented at the 1995 Intl. Underbalanced Drilling Conference and Exhibition, The Hague, 2–4 October 1995.

Adamache, I.: "A Horizontal Well Drilled Underbalanced in a Heavy Oil Unconsolidated Formation in Venezuela," paper presented at the 1997 Intl. Underbalanced Drilling Conference and Exhibition, The Hague, 8–9 October.

Atherton, G.M.: "Recent Experiences with Underbalanced Coiled Tubing Drilling," paper presented at the 1995 Intl. Underbalanced Drilling Conference and Exhibition, The Hague, 2–4 October.

Bennion, D.B.: "Underbalanced Operations Offer Pluses and Minuses," *Oil & Gas J.* (January 1996).

Bennion, D.B., Thomas, F.B., and Bietz, R.F.: "Formation Damage and Horizontal Wells—A Productivity Killer?" paper SPE 37138 presented at the 1996 International Conference on Horizontal Well Technology, Calgary, 18–20 November.

Bennion, D.B. *et al.:* "Underbalanced Drilling: Praises and Perils," *SPEDC* (December 1998) 214.

Bennion, B.D. and Thomas, F.B.: "Evaluating Reservoir Performance Improvements From Underbalanced Drilling Operations," paper presented at the 2001 IADC Underbalanced Drilling Conference and Exhibition, Aberdeen, 27–28 November.

Bern, P.A. *et al.:* "A New Downhole Tool for ECD Reduction," paper SPE 79821 presented at the 2003 SPE/IADC Drilling Conference, Amsterdam, 19–21 February.

Bieseman, T. and Emeh, V.: "An Introduction to Underbalanced Drilling," paper presented at the 1995 Intl. Underbalanced Drilling Conference and Exhibition, The Hague, 2–4 October.

Bijleveld, A.F., Koper, M., and Saponja, J.: "Development and Application of an Underbalanced Drilling Simulator," paper SPE 39303 presented at the 1998 IADC/SPE Drilling Conference, Dallas, 3–6 March.

Bourgoyne, A.T. Jr.: "Well Control Considerations for Underbalanced Drilling," paper SPE 38584 presented at the 1997 SPE Annual Technical Conference and Exhibition, San Antonio, Texas, 5–8 October.

Boyun, G: "Balance Between Formation Damage and Wellbore Damage: What Is the Controlling Factor in UBD Operations?" paper SPE 73735 presented at the 2002 SPE International Symposium and Exhibition on Formation Damage Control, Lafayette, Louisiana, 20–21 February.

Boyun, G. and Ghalambor, A.: "An Innovation in Designing Underbalanced Drilling Flow Rates: A Gas/Liquid Rate Window (GLRW) Approach," paper SPE 77237 presented at the 2002 IADC/SPE Asia Pacific Drilling Technology Conference, Jakarta, 9–11 September.

Brand, P.: "Maxus Drills First Underbalanced Well From an Offshore Structure in Southeast Asia," paper presented at the 2000 IADC Underbalanced Drilling Conference and Exhibition, Houston, 28–29 August.

Bullock, R. et al.: "New-Generation Underbalanced Drilling Four-Phase Surface Separation Technique Improves Operational Safety, Efficiency, and Data Management Capabilities," paper SPE 72153 presented at the 2001 SPE Asia Pacific Improved Oil Recovery Conference, Kuala Lumpur, 8–9 October.

Cagnolatti, E. and Curtis, F.: "Using Underbalanced Technology to Solve Traditional Drilling Problems in Argentina," paper presented at the 1995 Intl. Underbalanced Drilling Conference and Exhibition, The Hague, 2–4 October.

Chitty, G. et al.: "Case Study in UBD with Oxygenated Fluids," paper presented at the 1995 Intl. Underbalanced Drilling Conference and Exhibition, The Hague, 2–4 October.

Churcher, P.L. et al.: "Designing and Field Testing of Underbalanced Drilling Fluids to Limit Formation Damage: Examples from the Westerose Field Canada," paper presented at the 1995 Intl. Underbalanced Drilling Conference and Exhibition, The Hague, 2–4 October.

Colbert, J.: "Naturally Fractured Carbonate Drilling Technique," paper presented at the 2000 IADC Underbalanced Drilling Conference and Exhibition, Houston, 28–29 August.

Comeau, L.: "Integrating Surface Systems with Downhole Data Improves Underbalanced Drilling," Oil & Gas J. (March 1997).

Crerar, P.: "Underbalanced Re-Entry Horizontal Drilling in the Welton Field Basal Succession Reservoir Onshore U.K.," paper presented at the 1995 Intl. Underbalanced Drilling Conference and Exhibition, The Hague, 2–4 October.

Deis, P.V., Yurkiw, F.J., and Barranechea, P.J.: "The Development of an Underbalanced Drilling Process: An Operator's Experience in Western Canada," paper presented at the First Intl. Underbalanced Drilling Conference and Exhibition, The Hague 2–4 October 1995.

Doan, Q.T. et al.: "Modeling of Transient Cuttings Transport in Underbalanced Drilling," paper SPE 62742 presented at the 2000 IADC/SPE Asia Pacific Drilling Technology Conference, Kuala Lumpur, 11–13 September.

Fleck, A.R.: "Technical Challenges Encountered During Coiled Tubing Underbalanced Drilling in Oman," paper presented at the 2000 IADC Underbalanced Drilling Conference and Exhibition, Houston, 28–29 August.

Foy, J. and Brett, P.: "First North Sea Underbalanced Drilled, Jointed Pipe Well," paper presented at the 1997 Intl. Underbalanced Drilling Conference and Exhibition, The Hague, 8–9 October.

Fried, S. and McDonald, C.: "Nitrogen Supply Alternatives for Underbalanced Drilling," paper presented at the 1995 Intl. Underbalanced Drilling Conference and Exhibition, The Hague, 2–4 October.

Frink, P. *et al.:* "A Case Study: Drilling Underbalanced in Mobil Indonesia's Arun Field," paper presented at the 1999 IADC Underbalanced Drilling Conference and Exhibition, The Hague, 27–28 October.

Frink, P.J. *et al.:* "Development and Use of an Underbalanced Transient Training Simulator," paper presented at the 2001 IADC Underbalanced Drilling Conference and Exhibition, Aberdeen, 27–28 November.

Giffin, D.R. and Lyons, W.C.: "Case Histories of Design and Implementation of Underbalanced Wells," paper SPE 59166 presented at the 2000 IADC/SPE Drilling Conference, New Orleans, 23–25 February.

Guimerans, R. *et al.:* "Evaluation Criteria to Formulate Foam as Underbalanced Drilling Fluid," paper presented at the 2000 IADC Underbalanced Drilling Conference and Exhibition, Houston, 28–29 August.

Guo, B. and Ghalambor, A.: *Gas Volume Requirements for Underbalanced Drilling,* PennWell Corp., Tulsa (2002).

Hanking, T. and Rappuhn, T.F.: "Case History: Breitbrunn—Horizontal Foam Drilling Project in an Environmentally Sensitive Area in Bavaria, Germany," paper SPE 35068 presented at the 1996 IADC/SPE Drilling Conference, New Orleans, 12–15 March.

Hannegan, D. and Bourgoyne, A.T.: "Underbalanced Drilling Rotating Control Head Technology Increasing in Importance," paper presented at the 1995 Intl. Underbalanced Drilling Conference and Exhibition, The Hague, 2–4 October.

Hannegan, D. and Divine, R.: "Underbalanced Drilling—Perceptions and Realities of Today's Technology in Offshore Applications," paper SPE 74448 presented at the 2002 IADC/SPE Drilling Conference, Dallas, 26–28 February.

Hannegan, D.M. and Wanzer, G.: "Well Control Considerations—Offshore Applications of Underbalanced Drilling Technology," paper SPE 79854 presented at the 2003 SPE/IADC Drilling Conference, Amsterdam, 19–21 February.

Haselton, T.M., Pia, G., and Fuller, T.: "Underbalanced and Overbalanced Well Legs Affords Direct Comparison in the Same Reservoir Section Yielding Record Well Productivity in Lithuania," paper presented at the 2001 IADC Underbalanced Drilling Conference and Exhibition, Aberdeen, 27–28 November.

Hawkes, C.D, Smith, S.P., and McLellan, P.J.: "Coupled Modeling of Borehole Instability and Multiphase Flow for Underbalanced Drilling," paper SPE 74447 presented at the 2002 IADC/SPE Drilling Conference, Dallas, 26–28 February.

Hogg, T.: "BPX Colombia's Underbalanced Drilling Experiences," paper presented at he 1997 Intl. Underbalanced Drilling Conference and Exhibition, The Hague, 8–9 October.

Gu, H., Walton, I.C., and Stein, D.A.: "Designing Under- and Near-Balanced Coiled-Tubing Drilling by Use of Computer Simulations," *SPEDC* (June 1999) 102.

"Guideline for Underbalanced Operations," Report No. 10.18/263, *E&P Forum* (September 1997).

Jansen, S. *et al.:* "Safety Critical Learnings in Underbalanced Well Operations," paper SPE 67688 presented at the 2001 SPE/IADC Drilling Conference, Amsterdam, 27 February–1 March.

Jun, F. *et al.:* "A Comprehensive Model and Computer Simulation for Underbalanced Drilling in Oil and Gas Wells," paper SPE 68495 presented at the 2001 SPE/ICoTA Coiled Tubing Roundtable, Houston, 7–8 March.

Kindjerski, K. *et al.:* "Optimized Underbalanced Drilling Operations—A Case for Infill Drilling," paper presented at the IADC Underbalanced Drilling Conference and Exhibition, Houston, 28–29 August.

Labat, C.P., Benoit, D.J., and Vining, P.R.: "Underbalanced Drilling at its Limits Brings Life to Old Field," paper SPE 62896 presented at the 2000 SPE Annual Technical Conference and Exhibition, Dallas, 1–4 October.

Lage, A.C.V.M. *et al.*: "Full-scale Experimental Study for Improved Understanding of Transient Phenomena in Underbalanced Drilling Operations," paper SPE 52829 presented at the 1999 SPE/IADC Drilling Conference, Amsterdam, 9–11 March.

Lage, A.C.V.M., Fjelde, K.K., and Time, R.W.: "Underbalanced Drilling Dynamics: Two-Phase Flow Modeling and Experiments," paper SPE 62743 presented at the 2000 IADC/SPE Asia Pacific Drilling Technology Conference, Kuala Lumpur, 11–13 September.

Lage, A.C.V.M., Rommetveit, R., and Time, R.W.: "An Experimental and Theoretical Study of Two-Phase Flow in Horizontal or Slightly Deviated Fully Eccentric Annuli," paper presented at the 2000 IADC/SPE Asia Pacific Drilling Technology Conference, Kuala Lumpur, 11–13 September.

Lage, A.C.V.M. and Time, R.W.: "An Experimental and Theoretical Investigation of Upward Two-Phase Flow in Annuli," *SPEJ* (September 2002) 325.

Larsen, L. and Nilsen, F.: "Inflow Predictions and Testing While Underbalanced Drilling," paper SPE 56684 presented at the 1999 SPE Annual Technical Conference and Exhibition, Houston, 3–6 October.

Layne, A.W. and Yost, A.B. II: "Development of Advanced Drilling, Completion, and Stimulation Systems for Minimum Formation Damage and Improved Efficiency: A Program Overview," paper SPE 27353 presented at the 1994 SPE International Symposium on Formation Damage Control, Lafayette, Louisiana, 7–10 February.

Lorentzen, R.J. *et al.*: "Underbalanced Drilling: Real-Time Data Interpretation and Decision Support," paper SPE 67693 presented at the SPE/IADC Drilling Conference, Amsterdam, 27 February–1 March.

Monjure, N.A.: "Introduction and Review of the IADC-UBO Classification System For Underbalanced Wells," paper presented at the 1999 IADC Underbalanced Drilling Conference and Exhibition, The Hague, 27–28 October.

Moore, B.: "The Regulation of Underbalanced Drilling in the U.K. Sector," paper presented at the 1995 Intl. Underbalanced Drilling Conference and Exhibition, The Hague, 2– 4 October.

Nas, S.: "Underbalanced Drilling in a Depleted Gas Field Onshore U.K. with Coiled Tubing and Stable Foam," paper SPE 52826 presented at the 1999 IADC/SPE Drilling Conference, Amsterdam, 9–11 March.

Nas, S.: "Underbalanced Drilling in High-Pressure High-Temperature Fields," paper presented at the 1999 IADC Underbalanced Drilling Conference and Exhibition, The Hague, 27–28 October.

Nas, S. and Laird, A.: "Designing Underbalanced Through-Tubing Drilling Operations," paper SPE 69448 presented at the 2001 SPE Latin American and Caribbean Petroleum Engineering Conference, Buenos Aires, 25–28 March.

Nunes da Rosa, F.S., Santos, H., and Cunha, J.C.S.: "Underbalanced Drilling in Northeast Brazil —A Field Case History," paper presented at the IADC Underbalanced Drilling Conference and Exhibition, Houston, 28–29 August.

Parra, J.G., Celis, E., and De Gennaro, S.: "Wellbore Stability Simulations for Underbalanced Drilling Operations in Highly Depleted Reservoirs," paper SPE 65512 presented at the 2000 SPE/Petroleum Society of CIM International Conference on Horizontal Well Technology, Calgary, 6–8 November.

Pérez-Téllez, C., Smith, J.R., and Edwards, J.K.: "A New Comprehensive, Mechanistic Model for Underbalanced Drilling Improves Wellbore Pressure Predictions," paper SPE 74426 presented at the 2002 SPE International Petroleum Conference and Exhibition, Villahermosa, Mexico, 10–12 February.

Proc., First Intl. Underbalanced Drilling Conference and Exhibition, The Hague (1995).

Proc., Third Intl. Underbalanced Drilling Conference and Exhibition, The Hague (1997).

Proc., The North Sea Underbalanced Operations Forum, Aberdeen (1996).

Proc., First IADC Underbalanced Drilling Conference and Exhibition, The Hague (1998).

Proc., The IADC Underbalanced Drilling Conference and Exhibition, Houston (1999).

Proc., The IADC Underbalanced Drilling Conference and Exhibition, Aberdeen (2001).

Ramalho, J.: "Low Head Drilling in The Ommelanden Chalk," paper presented at the 1999 IADC Underbalanced Drilling Conference and Exhibition, The Hague, 27–28 October.

Rehm, B.: *Practical Underbalanced Drilling and Workover,* Petroleum Extension Service, U. of Texas at Austin, Austin, Texas (2002).

Robichaux, D.: "Successful Use of the Hydraulic Workover Unit Method for Underbalanced Drilling," paper SPE 52827 presented at the 1999 SPE/IADC Drilling Conference, Amsterdam, 9–11 March.

Robinson, S., Hazzard, V., and Leary, M.: "Redeveloping Rhourde El Baquel with Underbalanced Drilling Operations," paper presented at the 2000 IADC Underbalanced Drilling Conference and Exhibition, Houston, 28–29 August.

Rommetveit, R. *et al.:* "A Dynamic Model for Underbalanced Drilling With Coiled Tubing," paper SPE 29363 presented at the 1995 SPE/IADC Drilling Conference, Amsterdam, 28 February–2 March.

Rommetveit, R. *et al.:* "Dynamic Underbalanced Drilling Effects Are Predicted by Design Model," paper SPE 56920 presented at the 1999 SPE Offshore Europe Conference, Aberdeen, 7–9 September.

Rommetveit, R. and Lage, A.C.V.M.: "Designing Underbalanced and Lightweight Drilling Operations: Recent Technology Developments and Field Applications," paper SPE 69449 presented at the 2001 SPE Latin American and Caribbean Petroleum Engineering Conference, Buenos Aires, 25–28 March.

Roy, R. and Hay, R.: "Measuring Downhole Annular Pressure While Drilling for Optimization of Underbalanced Drilling," paper presented at the 1995 Intl. Underbalanced Drilling Conference and Exhibition, The Hague, 2–4 October.

Santos, H. *et al.:* "Offshore/Subsea Applications of Underbalanced Technology and Equipment," paper presented at the 2000 IADC Underbalanced Drilling Conference and Exhibition, Houston, 28–29 August.

Shale, L.: "Underbalanced Drilling: Formation Damage Control During High Angle or Horizontal Drilling," paper SPE 27351 presented at the 1994 SPE Symposium on Formation Damage Control, Lafayette, Louisiana, 7–10 February.

Shale, L.: "Underbalanced Drilling With Air Offers Many Pluses," *Oil & Gas J.* (June 1995).

Shi, D. and Yang, H.: "Underbalanced Drilling Practice for Lava Exploration in Daqing Oil Field," paper presented at the 2001 IADC Underbalanced Drilling Conference and Exhibition, Aberdeen, 27–28 November.

Springer, S.J., Lunan, B., and Brown, A.: "A Review of Underbalanced Drilling in the Carbonate Reservoirs of South Eastern Saskatchewan," paper presented at the 1995 Annual Meeting of the Petroleum Society of CIM in Banff, Alberta, 14–17 May.

Steiner, A.: "Production Benefits from Underbalanced Drilling—A Case Study," paper presented at the 2000 IADC Underbalanced Drilling Conference and Exhibition, Houston, 28–29 August.

Stone, R.: "The History and Development of Underbalanced Drilling in the U.S.A.," paper presented at the 1995 Intl. Underbalanced Drilling Conference and Exhibition, The Hague, 2–4 October.

Stuczynski, M.C.: "Recovery of Lost Reserves Through Application of Underbalanced Drilling Techniques in the Safah Field," paper presented at the 2001 IADC Underbalanced Drilling Conference and Exhibition, Aberdeen, 27–28 November.

Surewaard, J. *et al.:* "Underbalanced Operations in Petroleum Development Oman," paper presented at the 1995 Intl. Underbalanced Drilling Conference and Exhibition, The Hague, 2–4 October.

Tangedahl, M.J.: "Well Control: Issues of Underbalanced Drilling," paper SPE 37329 presented at the 1996 SPE Eastern Regional Meeting, Columbus, Ohio, 23–25 October.

Taylor, J., McDonald, C., and Fried, S.: "Underbalanced Drilling Total Systems Approach," paper presented at the 1995 Intl. Underbalanced Drilling Conference and Exhibition, The Hague, 2– 4 October.

Teichrob, R.R. and Manuel, J.J.: "Foam Speed Underbalanced Drilling in British Columbia," *Hart's Pet. Eng.* (October 1997).

Teichrob, R. and Baillargeon, D.: "The Changing Face of Underbalanced Drilling Technology," paper presented at the 1999 IADC Underbalanced Drilling Conference and Exhibition, The Hague, 27–28 October.

Tinkham, S.K., Meek, D.E., and Staal, T.W.: "Wired BHA Applications in Underbalanced Coiled Tubing Drilling," paper presented at the 2000 IADC Underbalanced Drilling Conference and Exhibition, Houston, 28–29 August.

"Underbalanced Drilling," Interim Directive ID 94-3, Energy Resources Conservation Board, Calgary (July 1994).

"Underbalanced Drilling and Completion Manual," Maurer Engineering Inc., DEA 101 Phase 1 (October 1996).

"Underbalanced Drilling Manual," Gas Research Inst., Chicago, GRI-97/0236.

Van Venrooy, J. *et al.:* "Underbalanced Drilling With Coiled Tubing in Oman," paper SPE 57571 presented at the 1999 SPE/IADC Middle East Drilling Technology Conference, Abu Dhabi, 6–10 November.

Vargas, A., Rodrigano, R., and Huertas, R.: "Underbalanced Drilling in Unconsolidated, Low-Gravity Oil Reservoirs in Latin America," paper presented at the 2001 IADC Underbalanced Drilling Conference and Exhibition, Aberdeen, 27–28 November.

Veira, P., Blanco, E., and Diaz, A.: "Application of Near-Balanced Drilling Technology in Eastern Venezuela," paper presented at the 1999 IADC Underbalanced Drilling Conference and Exhibition, The Hague, 27–28 October.

Walker, T. and Hopmann, M.: "Underbalanced Completions," paper SPE 30648 presented at the 1995 SPE Annual Technical Conference and Exhibition, Dallas, 22–25 October.

Wang, Z. *et al.:* "A Dynamic Underbalanced Drilling Simulator," paper presented at the 1995 IADC Underbalanced Drilling Conference and Exhibition, The Hague, 2–4 October.

Weisbeck, D. *et al.:* "Case History of First Use of Extended-Range EM MWD in Offshore, Underbalanced Drilling," paper SPE 74461 presented at the 2002 IADC/SPE Drilling Conference, Dallas, 26–28 February.

Yurkiw, F.J. *et al.:* "Optimization of Underbalanced Drilling Operations to Improve Well Productivity," presented at the 1996 Annual Technical Meeting of the Petroleum Soc., Calgary, 10–12 June.

SI Metric Conversion Factors

ft	×	3.048*	E – 01	= m
ft^2	×	9.290 304*	E – 02	= m^2
gal	×	3.785 412	E – 03	= m^3
psi	×	6.894 757	E + 00	= kPa

*Conversion factor is exact.

Chapter 13
Emerging Drilling Technologies
Roy C. Long, SPE, DOE/FE National Energy Technology Laboratory

13.1 Introduction

In a special report in the *Oil and Gas Journal*,[1] a representative of the Drilling Engineering Association's (DEA) Advisory Board (http://www.dea.main.com/) noted that "among the most important new technologies for the drilling industry are expandable tubulars, more cost-effective rotary steerable systems, and intelligent drillpipe for high-rate bottomhole data telemetry." The following discussion of emerging drilling technologies will be limited to those technologies now coming into the market, not those, such as rotary steerable and multilateral technologies, that have ready reference on service company Internet websites. Hence, this discussion is not comprehensive, but it is intended to include most of the high-impact technologies that are likely to be commercialized in the next 3 to 5 years with a brief look beyond.

The focus on drilling technology in the United States at the beginning of the 21st century is primarily in response to the fact that its remaining oil and gas resources exist in mature provinces of significantly depleted basins or in difficult drilling environments, such as the Arctic or the deepwater Gulf of Mexico (GOM). Because the United States has led the world in petroleum demand, the environment of depletion and push for further development of these mature basins will provide lessons and technology immediately applicable to the rest of the world as the world resource base continues to mature. All nations have a stake and will benefit from this development of the next redefinition of drilling state of the art.

The most basic requirement of drilling technology is that it provide safe, economic access to subsurface geologic formations to evaluate/optimize their production potential or to produce the resource existing there. The operative word is "economic." In high-cost environments, such as the deepwater offshore, technology is needed to maximize efficiency and to minimize time on location. With the advent of deepwater operations, concepts such as "parallel operations" and "flat time reduction" have become familiar technology focus areas (http://www.erch.org/workshops/FlatTime/flat_time.htm). In the onshore arena where reservoir potential is lower, the cost of accessing that potential also has to be reduced with such technologies as casing drilling. Also, with the advent of "unconventional resources" and fracture "sweet spots" as primary exploration targets, technology must provide a "smart drilling" capability to enhance finding these more difficult targets and to optimize access to the target in a manner that maximizes the production.

Advances in technology in the past decade are not all simply random evolutionary advances but represent a step change in drilling technology. They often represent a major change in drilling paradigms brought on by pressure to develop new resources in the face of existing domestic depletion and more challenging drilling environments.

Many of the technologies discussed in this chapter were presented in the keynote presentation at the DEA's Future of Well Construction workshop (http://www.dea.main.com/Future%20of %20Well%20Con/). This chapter provides an overview that includes references for more detailed information. The previously referenced website for the DEA provides an excellent summary of current industry technology focus areas (see Project Summaries). Links to other websites providing information on other key drilling technologies can be found under Information Exchange.

13.2 Offshore

Some of the most interesting near-term technologies (within 5 years of commercialization) are being developed to address the challenges of the deepwater GOM exploration. The deepwater GOM provides the high-cost environment (operating cost of U.S. $250,000/D or more) that encourages the risk-taking required to give new technologies the opportunity to demonstrate potential.

13.2.1 Dual Gradient Drilling Systems. Perhaps one of the most important ventures in the area of high-cost technologies for deepwater challenges is the development of dual gradient drilling systems (DGDSs). DGDS is often referred to as riserless drilling. It is generally accepted that DGDS is required in water depths of > 5,000 ft. There have been a number of unpublished examples, however, in which application of the technology was needed in water depths as shallow as 3,000 ft. The need for DGDS is relatively simple; it is caused by the reduced fracture gradient of formations below the mudline resulting from the reduced weight, or gradient (0.5 vs. 1.0 psi/ft), resulting from water above the mudline as viewed from a drillship operating at sea level. The various systems shown in **Fig. 13.1**, in one manner or another, isolate the borehole pressure gradient below the mudline from the drilling mud gradient above. In all but the Maurer Technology, Inc. DGDS, isolation is achieved mechanically by valves and pumping. The Maurer approach seeks to achieve the same benefit by pumping lightweight solid additives (LWSAs) from the drillship into the riser at mudline. This concept allows minimum equipment and intervention risk on the seafloor. The LWSAs investigated to date consist of hollow glass spheres and polymeric beads.

The advantage gained by these systems can be noted by comparing **Figs. 13.2 and 13.3**. For the conventional drilling case (Fig. 13.2), the gradient in the wellbore is relative to the drillship in all cases because the mud column is hydraulically continuous from the bottom of the hole up the riser to the drillship. This results in additional pressure being applied at the mudline (mud density minus seawater density times water depth times a units constant). The increased "backpressure" at the mudline has the effect of minimizing the drilling distance between casing points. The pressure at the bottom of the hole over a particular interval is usually referred to as equivalent circulating density. The equivalent circulating density from the mudline to total depth for conventional riser systems is always greater than for subsea systems in which the pressure (both circulating and static) required to get the mud from the mudline to the drillship is hydraulically isolated from the borehole or greatly reduced in density at the mudline.

Fig. 13.3 demonstrates that isolation of the pressure caused by the drilling mud above the mudline results in a borehole gradient that allows significantly longer openhole sections before reaching the depth at which casing must be set to avoid exceeding the fracture pressure.

Types of DGDS. Fig. 13.1 is an attempt to show some of the DGDS concepts being considered. The Subsea Mudlift development program produced the only prototype DGDS successful-

DeepVision

Mudlift

Shell SSPS

Nothing

Maurer

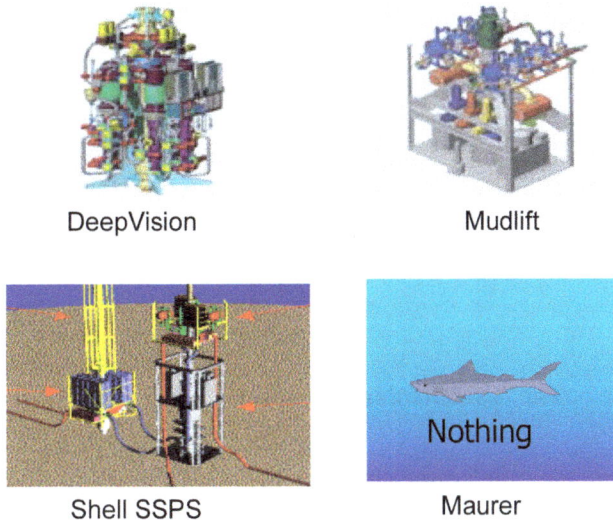

Fig. 13.1—Dual-gradient drilling systems.

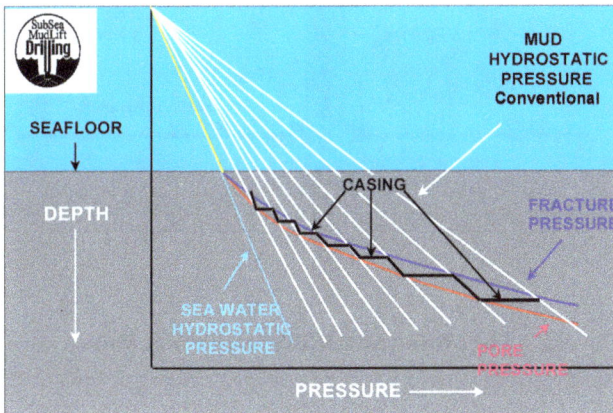

Fig. 13.2—Conventional drilling casing requirements.

ly field tested to date; however, deployment cost has proved problematic to market penetration.[2,3]

Consideration is being given to the DGDS approach (**Fig. 13.4**) proposed by Maurer to reduce the cost of achieving a dual gradient drilling capability. Maurer is leading a consortium to look into the feasibility of injecting LWSA at the mudline to control gradient in the riser (http://www.maureng.com/DGD/index-DGD.html). The strength of the approach is that it has the potential to simplify significantly the equipment installed at the mudline and hence to reduce the cost of the DGDS. In addition, the LWSAs are well behaved in the riser; they maintain constant shape and do not migrate significantly when pumping is stopped for a reasonably long period of time. To date, both hollow glass spheres and polypropylene beads have undergone testing for use as LWSAs; however, an investigation of alternatives is ongoing.

Fig. 13.5 shows a side-by-side comparison of the results of using DGDS compared with conventional single gradient drilling. As noted in the figure, the setting depth of all casing strings is significantly increased. This is achieved, in effect, by isolating the borehole pressure

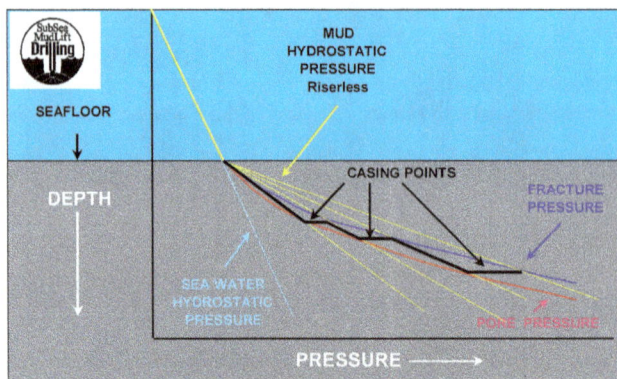

Fig. 13.3—Riserless drilling casing requirements.

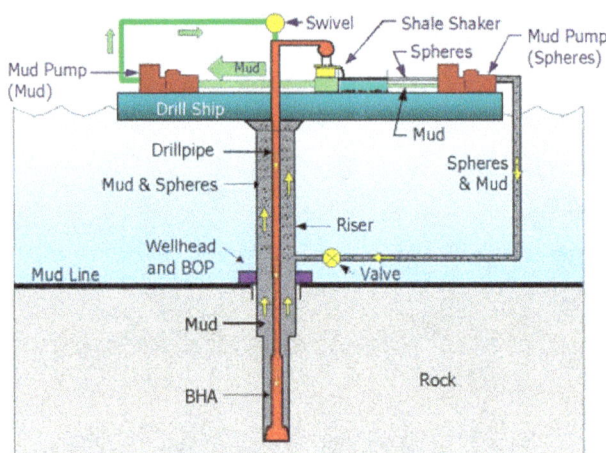

Fig. 13.4—LWSA-based DGDS proposed by Maurer.

from the weight of the seawater above the mudline. Most systems achieve this isolation through rather complex combinations of pumps and cuttings processing equipment at the mudline. The Maurer DGDS achieves a similar effect by lowering the fluid density in the riser well below 8 lbm/gal. Regardless of the system considered, the most notable benefits of DGDS technology are that it has the potential to enhance the capability of drilling to even deeper targets in ultradeep waters of the GOM and it allows active control of borehole mud gradient. It should be noted that the latter benefit could also be a significant safety consideration.

High-Speed Communications. Communication with downhole tools while drilling is currently achieved with either mud-pulse telemetry or electromagnetic-based systems. The maximum data transmission rate (correlated with bandwidth) of these systems is about 10 bits per second.[4] As a result, much of the information from measurement while drilling and logging while drilling must be processed and stored in computer memory associated with the downhole instrumentation near the drill bit. The term "real-time monitoring" can be applied in only a very limited sense with current technology.

The potential for true real-time monitoring has increased significantly with the initiation of the commercialization phase of a Dept. of Energy (DOE) technology development contract with Novatek, Inc. Novatek and its partner, Grant Prideco, have begun commercial construction

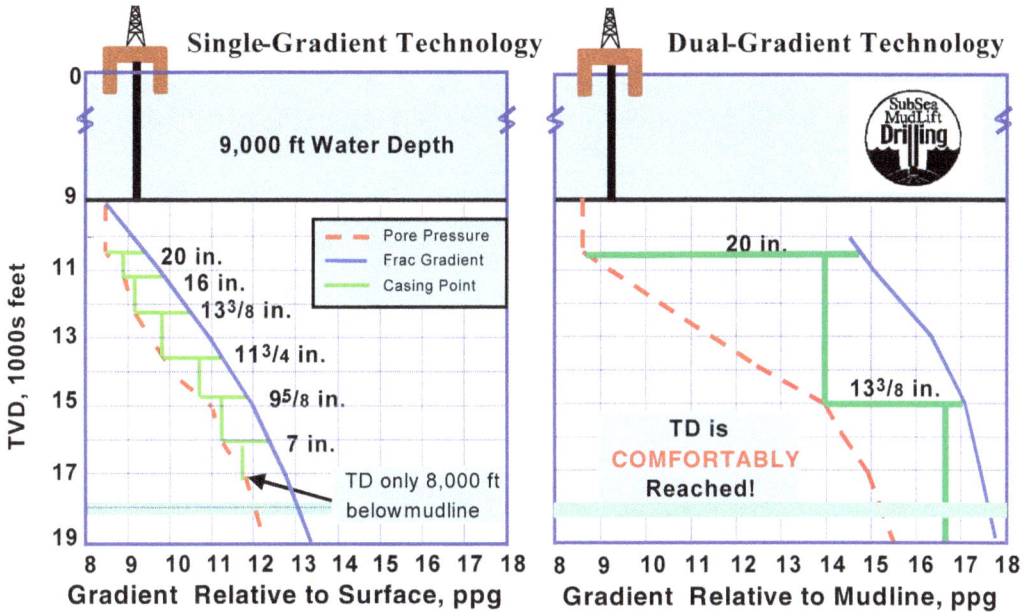

Fig. 13.5—DGDS casing point benefits comparison.

of Intellipipe® (http://www.intellipipe.com/). Intellipipe represents a novel and robust means of transmitting data up drillpipe at a transmission rate of 1 million bits per second. Key to the success of this technology was development of a high-efficiency coupling that enabled successful transmission of data across many tool joints without the need for amplification over lengths exceeding 1,000 ft. Another key feature of the system is that it will allow the drillpipe to act as a local area network within which many different tools or systems located anywhere within the drillstring can be individually addressed and/or turned on and off. **Fig. 13.6** is a concept drawing that shows the basic components of the proposed real-time monitoring and control system. **Fig. 13.7** details the components of the electromagnetic coupling across the tool joint. The recessed coil in the pin connection comes in very close, controlled proximity with the coil in the base of the box connection during makeup. The design results in a strong connection and forms the basis of a robust, reliable, efficient electromagnetic coupling for transfer of data across the connection.

Subsea Completion Systems. Drilling is not the only challenge to deepwater drilling economics. Current deepwater technology trends almost exclusively require huge discoveries and unprecedented production rates to ensure acceptable rates of return. One method to reduce the high development capital expenditures associated with deepwater environments that is being explored (**Fig. 13.8**) is a modular system designed for easy retrieval to the surface using diverless techniques for repair and maintenance, process reconfiguration, and equipment upgrade. The system reconfiguration will be accomplished by a workboat instead of a drillship. As a result, the cost of the reconfiguration, or "intervention," could be reduced as much as $200,000 per day compared with systems accomplishing similar functions. This reduction in capital expenditures associated with intervention is expected to make many smaller reservoirs economically viable.

Three-phase pumps have been used with limited success to pump deepwater production to separation facilities. However, one characteristic of the deepwater GOM is the significant topographic relief, occasionally reaching 1,000 ft in one 9-sq-mile lease. Such relief results in subsea pipelines acting as separators. As a result, severe slugging phenomena have been report-

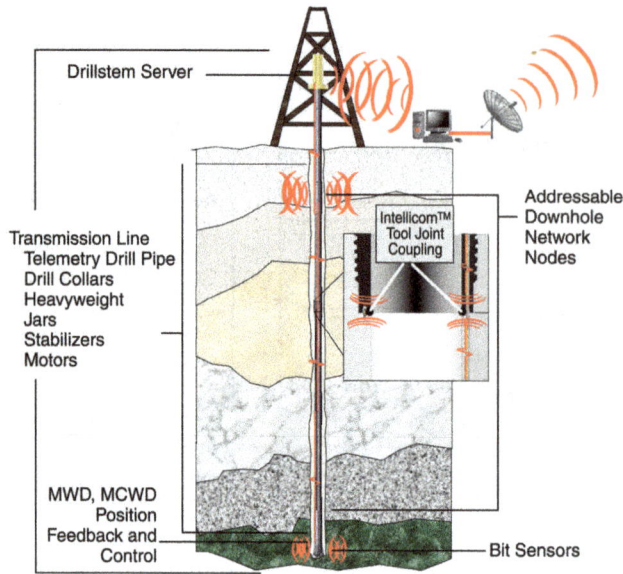

Fig. 13.6—Intellipipe local area network concept (1 million bits per second).

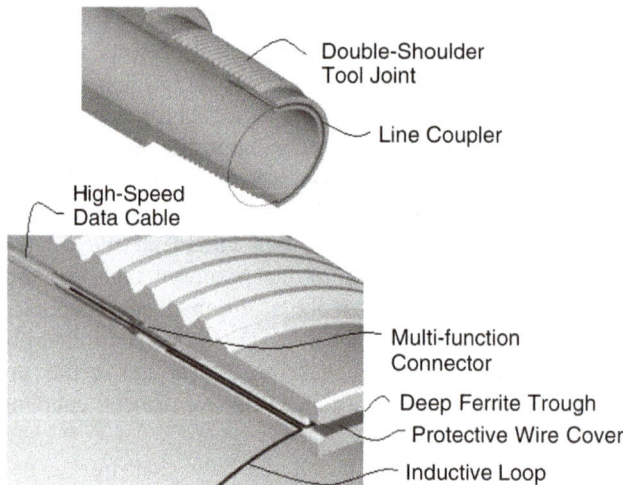

Fig. 13.7—Intellipipe coupling detail.

ed[5] in cases of extended pipeline distance and elevation change. Separating the liquid and gas very near the subsea wellheads is expected to significantly reduce pumping problems, required pumping horsepower, and many problems associated with hydrate production.

13.3 Onshore

Technologies discussed within this section are included because their development initiated with onshore field tests and because the technologies are considered essential to economics for enabling further exploration of ultradeep (> 20,000 ft) onshore petroleum resources. However, it is recognized that the need to further reduce operating cost and efficiency offshore will likely lead to expanded commercialization of these technologies more rapidly in this arena.

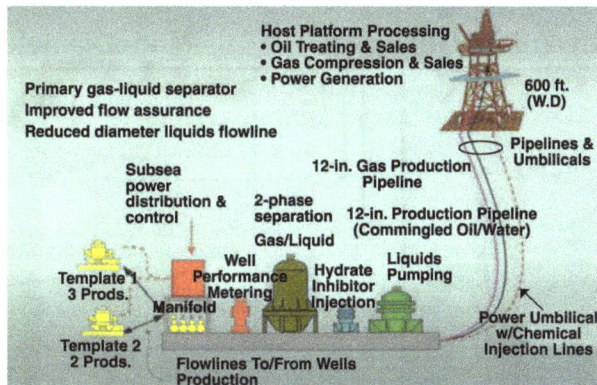

Fig. 13.8—Subsea production system.

13.3.1 Expandable Tubulars. One of the most exciting developments in the last decade has been expandable tubulars because they offer the potential for a "monoborehole" and drilling to depths no longer limited by initial hole diameter. As a result, the focus on tubulars has concentrated on expandable casing. Shell and Halliburton formed a company, Enventure, that is specializing in the commercialization of expandable casing based on earlier Shell work. A key development from that work is the concept of the monodiameter borehole (**Fig. 13.9**). Production casing can be run inside the expanded form of casing with the same diameter with this concept. It will allow, for the first time, casing to be set at will or as needed without a penalty in completed depth. Lost-circulation zones, swelling shales, and other drilling problems can be put behind pipe as necessary without jeopardizing planned total depth. Total depth limitations will now be limited primarily to mechanical capabilities of the drill rig, casing, and/or drillpipe run into and out of the borehole.

Elastomers on the exterior of the expanded casing have proved to be effective pressure seals in liner lap applications in lieu of running a conventional liner hanger seal assembly. Additional testing for sealing potential is being conducted. As of this writing, Enventure is planning a field test to investigate the potential of expandable casing to seal off against the formation without cement. If the test is successful, it will demonstrate the potential to eliminate most cementing operations, one of the costliest phases of well construction. If this becomes an accepted, safe practice, it could enable other opportunities for unprecedented reductions in exploration cost.

13.3.2 Casing Drilling. Drilling with casing is not a new concept; it has been used in the mining and water-well industries[6] for many years. However, modifying the tools and materials for oilfield use and extending drilling depth beyond a few thousand feet is new. This new approach, called Casing Drilling™, was developed[7] and field tested[8] and culminated in a successful demonstration to ≈ 9,500 ft early in 2002 in South Texas by Tesco Corp. and its partner, Conoco. The demonstration was the result of >5 years of development that included development of tools for directional drilling.[9] The demonstration resulted in an actual overall drilling time reduction of 17.5% and a potential for as much as a 33% reduction.[10] In October 2002, Tesco won World Oil's prestigious Next Generation Idea Award, which recognized the technology as a step change in drilling. **Fig. 13.10** is the comparison diagram from Tesco's website (http://www.tescocorp.com/bins/content_page.asp?cid=60) used to denote areas in which Casing Drilling has proved superior to conventional drilling. Those areas are (1) swelling formations, (2) sloughing formations, (3) washouts, (4) swabbing, (5) hole in casing or key seats, and (6) running logs and casing. One area not mentioned in the diagram is lost circulation. In the

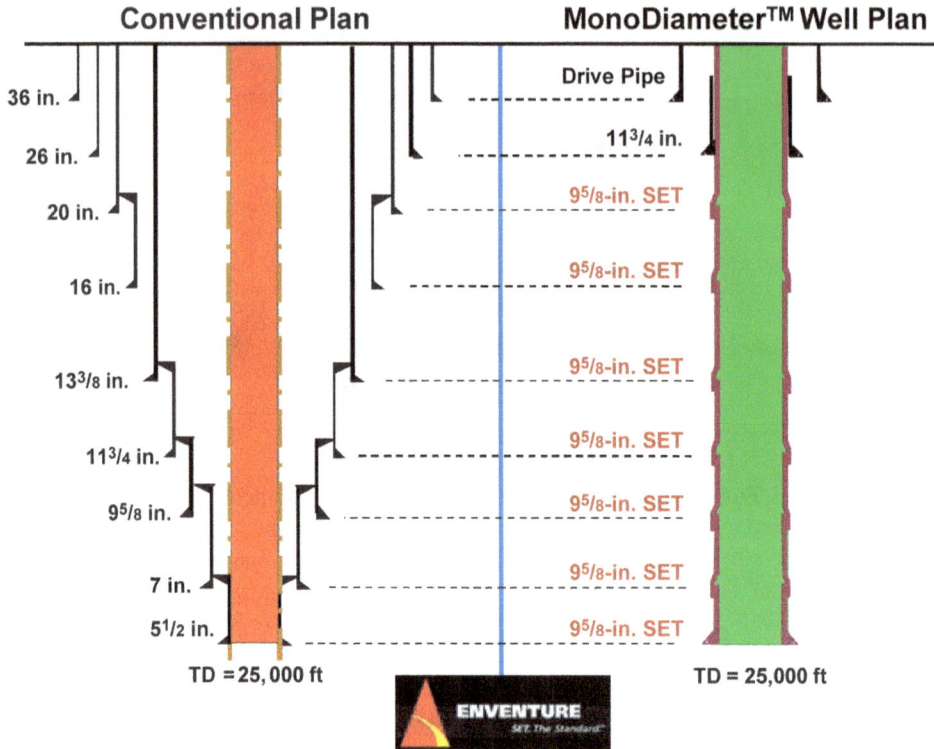

Fig. 13.9—Monodiameter borehole concept.

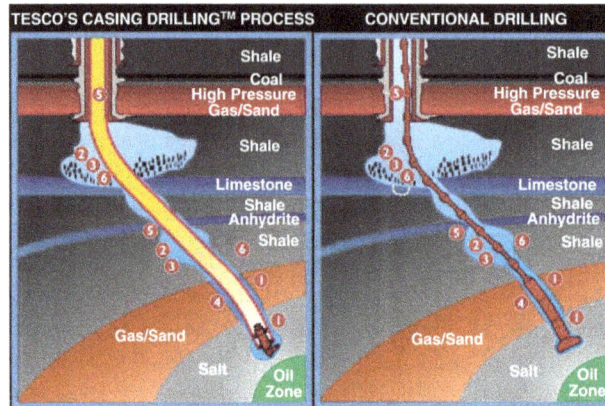

Fig. 13.10—Casing Drilling benefits.

South Texas field demonstration, conventional drilling in the area was characterized by lost circulation and stuck pipe. In fact, the offset conventional well used for comparison experienced a total of 53 hours of lost circulation and stuck pipe, whereas the Casing Drilling test had only 1 hour. Typically, stuck pipe and lost circulation accounted for 75% of the trouble time for conventionally drilled wells in the test area. The reason for fewer lost-circulation difficulties associated with Casing Drilling is not clear at this time; however, studies are currently underway to better our understanding of the phenomenon.

13.3.3 Deep Hard-Rock Drilling.
As the quest for new petroleum supplies has increased in the past few years, operators have been forced to drill deeper to find new reserves. Much of the higher cost of drilling deeper, especially onshore, is typically associated with decreased rate of penetration (ROP) caused by both harder rock and higher mud weights required to counter the overpressured reservoirs often associated with deeper drilling. The following discussion centers on technologies intended to enhance the deep drilling capability.

Mud Hammers. Industrial hammers for hard rock drilling have been around for some time, but most have been air operated and used mostly in the mining industry. Historically, hammers have been thought to have limited capability in oil and gas drilling operations, with their use limited to air drilling. Because of "chip holddown" and erosion through the hammer when drilling mud is used, hammers were not considered for drilling operations involving drilling mud. As a result, hammers have never been seriously considered for most deep drilling where hammer energy might enhance ROP by helping to overcome increased rock strength.

In an effort to develop novel drilling technologies, the DOE awarded a contract to Novatek to develop an "integrated" drilling system using a mud hammer as the primary engine (see **Fig. 13.11** and http://www.fossil.energy.gov/news/techlines/02/tl_intellipipe.html). The previously referenced high-speed communication system for drillpipe was part of that development. Another part of that development was a mud hammer that incorporated a number of revolutionary concepts, as shown in **Fig. 13.12**. Most notably, the bit was a radical departure from typical hammer bits. It was essentially a five-bladed drag bit with polycrystalline diamond cutters (PDCs) specially manufactured by Novatek to allow the aggressive drag bit profile to be used in soft formations but still allow enhanced drilling in hard formations using the high-energy impacts of the optimized industrial hammer. In addition, the hammer piston is used to energize a series of high-pressure jets ($\approx$ 5,000 psi) that exhaust directly in front of each PDC to achieve an unprecedented level of cleaning ahead of each cutter. The jets also energize fractures ahead of the bit to enhance ROP.

Directional steering is made possible by means of a directional control sub (see Fig. 13.11). The control sub causes preferential firing of the jet pulse on the side of the hole in the direction the operator wants to steer, as shown in Fig. 13.12.

The Novatek IDS hammer and other hammers were part of a test program funded by the DOE to provide a focused study program for investigation of mud hammer potential in deep hard-rock drilling environments. That program is being run by TerraTek with several industry participants. The first results of that program were published in the SPE *Journal of Petroleum Technology Online.*[11]

In summary, the current status of mud hammer investigation is still unfolding, with improvements in a number of mud hammers being driven by the testing program at TerraTek. The promise provided by mud hammers is potentially far more extensive than simply enhancing ROP in deep hard rock, although that alone would be sufficient. Mud hammers provide extremely strong seismic energy coupling into the rock. It might be possible to incorporate a mud-hammer-based seismic imaging system into the previously discussed high-speed communications system to provide a "seismic look-ahead" capability that could allow navigation directly into the desired hydrocarbon target or sweet spot. Such a system would be a significant step forward in the exploration and development of fractured, unconventional reservoirs.

Mud-Pulse Drilling. Another novel approach to enhancing ROP in deep mud drilled wells was developed by Tempress Technologies. **Fig. 13.13** shows the basic principles used in this system. Chip holddown is a well-documented phenomenon associated with mud drilling, especially in deep environments. In essence, the fluid pressure of the mud inhibits rock chips made by the drill bit from being removed from the cutting face in front of the bit. The result is regrinding of cuttings and a slowing of ROP.

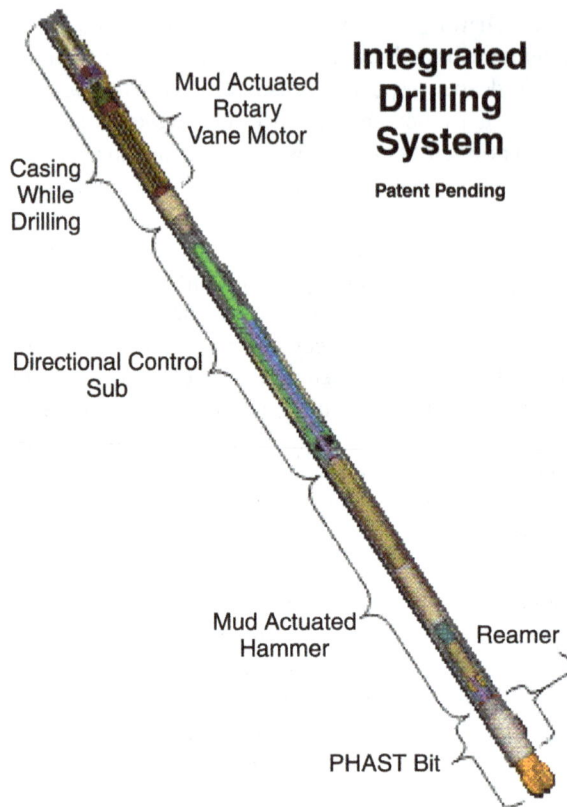

Fig. 13.11—Integrated drilling system.

The mud-pulse drilling system comprises an oscillator valve in the drillstring, which momentarily interrupts flow of the drill mud around a velocity section on the outer wall of the pipe. This interruption in flow results in extreme depressurization pulses ($> 1,500$ psi) developing below the bit. Theoretically, this causes rapid decompression of the fluids in the rock ahead of the drill bit and results in an apparent decrease in rock strength ahead of the bit, which results in increased ROP. The system can be run with almost any drill bit.

To date, development and testing are continuing. A more detailed description and the latest information can be found at Tempress' website (http://www.tempresstech.com/hydropulse.htm).

13.4 Materials

So far in this discussion, new methods have been the source of innovation. However, advancements in materials are also at the heart of the current drilling revolution. Receiving focus are both resin-based and metal composites. Resin composites have been studied extensively, with the advent of carbon-fiber-based materials showing promise for significant increases in yield strength and reductions in required weight. Metal composites have made another dramatic jump in capabilities with the commercialization of microwave-processed (MWP) diamond and tungsten carbide composites. Potential uses and combinations of MWP materials have not been fully explored.

13.4.1 Resin Composites. The term "resin" is used here to describe the family of composite materials that use a resin to bind a matrix of fibers, usually woven. Many such materials have been commercialized for coiled tubing applications because composite resiliency to cyclic

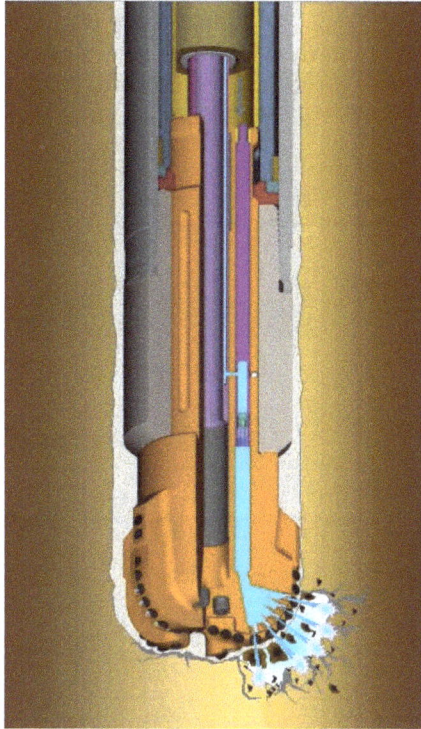

Fig. 13.12—Advanced mud hammer.

stress results in significantly longer life of the coiled tubing string than steel coiled tubing. In addition, composite coiled tubing is lighter than its steel counterpart, and communication cables can be embedded in the wall of the pipe.[12] Currently available composite coiled tubing is typically <5 in. in diameter. Some interest has been expressed in developing larger-diameter composite pipe for increased rigidity in horizontal and extended-reach drilling.

The DOE-funded project with Advanced Composite Products and Technology (ACPT) and its joint industry project partners is focused on development of a 5.5-in.-diameter composite drillpipe[13] with conventional steel connections that will be cost-competitive with steel drillpipe. Key benefits of the pipe are that it will be half the weight of steel drillpipe and will be essentially interchangeable with existing drillpipe.

Interestingly, during the development of the 5.5-in.-diameter pipe, a number of smaller-diameter test specimens were manufactured. On the basis of the results of testing the smaller-diameter drillpipe, interest was expressed in using it for developing the build section for short-radius boreholes. This interest culminated in a field test of the short-radius composite drillpipe (SR CDP). The following summary of that operation was provided from ACPT and is accompanied by photos in **Fig. 13.14**:

The field test was completed on November 6, 2002 by Grand Resources, Inc. at their Bird Creek site. Starting with an existing well that stopped producing in 1923, Grand Resources packed the bottom of the well and sealed it with concrete. Then they lowered the drill string 1208 feet and began directionally drilling a 70-ft-radius curvature through the well casing and into the strata. The SR CDP was furnished by ACPT, Inc. and DOE/NETL for the purpose of drilling the curve and lateral section that extends 1000 feet into the strata. The pipe worked flawlessly and Grand Resources was pleased with performance of the new product. Grand Resources estimates that this renewed well will produce 30 to 50 barrels of oil per day for quite

Oscillator Valve

Valve Open   Valve Closed

Flow Cycling
Valve
Poppet
High-Speed
Flow
PDC or
Tricone Bit

Exhaust
Suction
Pressure
Pulse
Hydraulic
Trust
Enhanced
Rock
Breaking

Seismic Pulse

Suction Pulse Train

Manoose Shale
Tensile
Strength

Time, milliseconds

8³/₄-in. Prototype

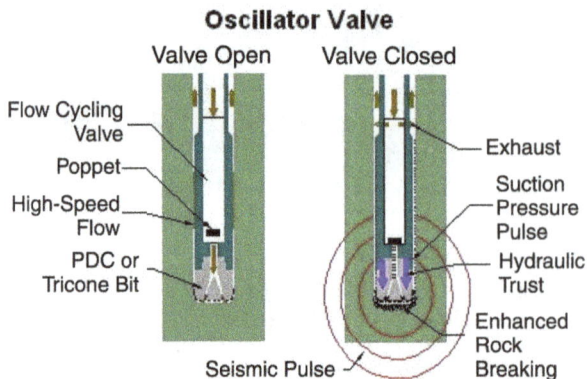

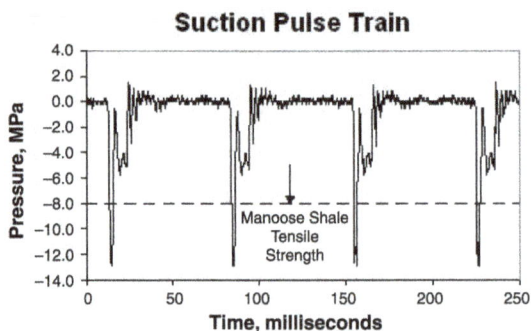

Fig. 13.13—Mud-pulse drilling system.

some time. Grand Resources plans to renew 14 additional wells in the same area in the near future and will use the new composite drill pipe in these endeavors. The CDP was not used to drill the lateral portion of this well because air hammer tools were used for this section. Grand Resources will test sections of CDP with air hammer tools in the next well. The air hammer beats at 2400 strokes per minute with a 4 to 6 inch stroke. This will be a good test of the strength and durability of the CDP. As they gain experience and confidence in the product, Grand Resources expects to extend the reach from 1000 to 2000 feet by using the CDP.

13.4.2 Metal Composites. Although there have been a number of advances in metal composite materials, much excitement has been focused on the application of microwave sintering of "green bodies" associated with powder metal technology. Powder metal technology is not new, but the sintering, or densifying, of the green body through microwave heating is a novel concept made possible only recently. The microwave energy heats relatively quickly and evenly from within the green body. This allows a process time that is a fraction of that for conventional sintering (hours vs. days). In addition, as noted in **Fig. 13.15**, the finished product is

SR CDP being readied for lifting

Condition of joint end after use—a little rust
on the steel end, but the composite has only
a few scratches.

Fig. 13.14—SR CDP field test results.

- 30% Stronger Composite Models
- Tungsten Carbide Sintered
  with Diamond Composite
- More Ductile = Better Impact
                 Resistance
- Higher Heat Conductivity

Diamond Composite
formed with Tungsten Carbide

Improved Impact Strength
and Higher Heat Conductivity

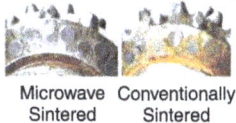

Microwave   Conventionally
Sintered   Sintered

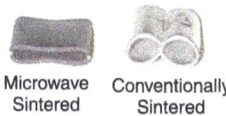

Microwave   Conventionally
Sintered   Sintered

Fig. 13.15—Advanced materials by microwave processing.

typically 30% stronger with improved impact resistance and corrosion resistance. Fig. 13.15 also shows that MWP is applicable to a number of materials (most notably tungsten carbide, diamond composite, and steels). A research program is under way at Pennsylvania State U. (with its commercialization partner, Dennis Tool Co.) to investigate the potential use of MWP for the manufacture of ceramics and hard transparent polycrystalline materials. The latter involves a Defense Advanced Research Agency-funded project to investigate the potential for using MWP to manufacture "transparent armor." If successful, such materials could find their way into hardened subsystems for MWD.

It should be noted that the ability to form diamond composite with tungsten carbide was a significant leap forward in materials development. In conventional processing/sintering, the diamond composite is turned to graphite because of the high temperatures required for sintering the tungsten carbide and the long heating and cooling times required. With microwave processing, the entire sintering process can be accomplished before the diamond composite is affected. In addition, the boundary between the diamond composite and tungsten carbide is not well defined because of a diffusion bonding process. This process is being investigated for its potential to make "functionally graded" materials. Such materials would allow entire bits and/or cutter assemblies to be manufactured in a single process in which diamond composite is bonded to tungsten carbide that is, in turn, bonded to drill steel. It might even be possible to form thread into the green body before sintering to eliminate machining.

An extension of this technology has been in the investigation of the potential for MWP to be applied to the manufacture of coiled tubing. If use of MWP technology results in a tubular that is 30% stronger and retains the same ductile character of competing steels used for coiled tubing, another leap in coiled-tubing drilling capability will be a reality. This should be known within the next 3 years.

In a related development, the above-mentioned "supermaterials" will allow even more aggressive cutting structures and drilling machines. One such drilling tool is already being evaluated by Dennis Tool Co. It is a drilling motor that uses high-speed milling concepts to abrade rock. A pilot bit is rotated counter to the direction of three high-speed cones that follow the bit and open the hole to the required diameter. The result of this action is zero torque transferred to the drillstring, a significant benefit to coiled-tubing operations. The drill appears to be capable of drilling on the order of 80 ft/hr in almost any type of rock. A final commercial version of this high-potential drilling machine could depend on proper application of the new supermaterials.

13.5 Microsystems

For purposes of this discussion, microsystems are those systems or subsystems that represent a quantum leap in the size and/or capability of currently available systems. Interest in these systems is tied to the interest in "smart" drilling and the demand for increased information and reliability. Reliability can often be defined as simplicity. Hence, smaller systems that require less power or are passive measurement devices can be of great benefit in building more complicated sensory and communication networks.

13.5.1 Fiber-Optic Devices. With the incorporation of "interferometry" technology into fiber-optic systems, it has become possible to talk about extremely small packages suitable for harsh environments,[14] such as a drilling environment. **Fig. 13.16** shows an example of a pressure sensor developed for a stationary measurement environment, such as a well completion. With the advent of composite drillpipe and the capability to embed such systems into the wall of the pipe, using much more sophisticated information systems for drilling is very possible.

13.5.2 Microdrilling. If it were possible to reduce drilling cost to the point that it could be considered a part of "predrill" prospect development, a significant capability would exist for improving the economics of developing today's fractured, unconventional resource. That feasibility is being investigated (**Fig. 13.17**) at Los Alamos Natl. Laboratory under a grant through DOE's National Gas and Oil Technology Partnership Program (http://www.sandia.gov/ngotp). The project is called microdrilling. The first enabling technologies were the microelectromechanical systems (MEMS) that made feasible a complete "rethink" of new economies possible for exploration drilling. It is possible to talk about smart drilling systems drilling 2⅜-in.-diameter and smaller boreholes with drill rigs that do not look at all like today's conventional rigs. The

Thermal bond | Protective coating

Input fiber R₂ R₁ Reflecting fiber

Glass alignment tube | Interferometric cavity

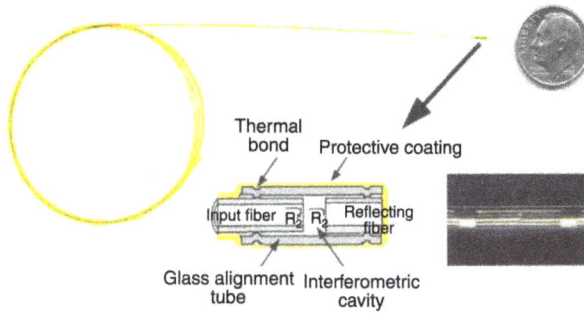

Self-Calibrated Interferometric/Intensity-Based (SCIIB) Fiber Optic Pressure Sensor

Fig. 13.16—Fiber-optic pressure sensor.

Micro-electromechanical Systems (MEMS)

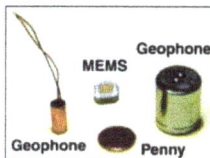

MEMS | Geophone

Geophone | Penny

Micro-Drillrig

Relative Borehole Sizes

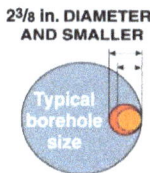

2³/₈ in. DIAMETER AND SMALLER

Typical borehole size

Downhole Systems

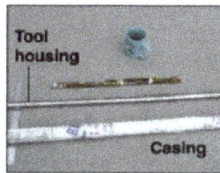

Tool housing

Casing

Fig. 13.17—Microdrilling components.

MEMS technologies also make possible downhole systems that are essential for steering and formation evaluation.

To be successful, microdrilling cannot simply be an expensive, smaller form of "slimhole" technology. The MEMS technology will allow a significant reduction in the size of drill rigs and drilling systems. However, the key will be to reduce total well cost. Such concepts as the monoborehole resulting from expandable tubular technology must be part of the complete microdrilling capability. In addition, high-ROP drilling tools will need to be investigated.

Of significant potential to all coiled-tubing drilling is the development program for a high-pressure coiled-tubing drilling system. Maurer Technology developed and tested this system with financial assistance from the DOE. Instantaneous ROPs as high as 1,400 ft/hr were recorded during surface testing in limestone. The project description can be found at http://www.netl.doe.gov/scng/projects/adv-drill/cost-reduct/dcr33063.html. The system incorporates a special moineau-type, positive-displacement motor with high-pressure (10,000 psi) housing. The motor drives a special PDC bit with high-pressure jets to etch the rock ahead of the bit. Thus, the bit only needs to break up the remaining rock not cut by the jet.

Such systems are excellent for coiled-tubing operations because they offer the potential for high ROPs without significant drilling torque.

13.6 Federally Funded Drilling Projects

Quite often, companies participate in federally cost-shared drilling technology development. The details that follow are provided to bring the reader up to date on the latest published trends in drilling technology development that could affect the drilling industry soon. It might be thought of as a long look into today's "crystal ball" for tomorrow's technologies.

On 23 September 2002, the DOE's Office of Fossil Energy announced the awards to its National Energy Technology Laboratory's Deep Trek Solicitation for initiation of development of the following technologies (if successful, commercialization anticipated within 5 to 10 years):

1. APS Technology Inc., Cromwell, Connecticut, plans to develop a two-component system that monitors and controls drilling vibrations in smart drilling technologies. Drillstring vibration causes premature failure of equipment, which reduces the depth and speed at which a well is drilled. A multiaxis active damper will be used to minimize harmful vibrations, which will extend the life of the drill bit and other components and improve the ROP. A real-time system that monitors three-axis vibrations and related measurements will be used to assess the vibration environment and adjust the damper accordingly.

2. E-Spectrum Technologies, San Antonio, Texas, proposes to develop a communications system that allows well operators to receive vital measurements while a well is being drilled, which improves drilling and consequently production. The system would directly control adjustable downhole tools and make changes in drilling in real time, greatly improving a well's future production level. E-Spectrum will build and field test a prototype of a wireless electromagnetic telemetry system for use in high-temperature (392°F) drilling beyond 20,000 ft. The system will be composed of a surface unit receiver/transmitter, downhole data-acquisition module, downhole repeater module, and a downhole receiver/transmitter module.

3. Pennsylvania State U., University Park, and Quality Tubing Inc., Houston, will develop a continuous microwave process to make seamless coiled tubing and drillpipes efficiently and economically. Improving the performance, life cycle, and ROP of these materials will allow deeper wells to be drilled. Drill mud, which contains drilling fluids, causes erosion and leaks that weaken conventionally welded drillpipes, causing them to fail.

4. Pinnacle Technologies, San Francisco, will review current and past stimulation techniques for deep-well completions to develop data that help minimize uncertainty and increase success in drilling deep formations. Information will be obtained through literature reviews; interviews with operators, service companies, and consultants; evaluations of rock mechanics and fracture growth in deep formations; and assessments of stimulation techniques in three to five gas wells. A comprehensive report will be assembled and given to the gas industry through publications and workshops.

5. Terra Tek, Salt Lake City, Utah, will develop and test prototypes of novel drill bits and advances in high-temperature, high-pressure fluids suited for slow, deep-drilling operations. With its private industry partners, Terra Tek will characterize technologies, develop and supply new bit prototypes and drilling fluids, and field test prototypes. Researchers will benchmark the performance of emerging products by conducting drilling tests in its laboratory. Joining Terra Tek will be the U. of Tulsa, Hughes Christensen, BP America, Conoco, INTEQ Drilling Fluids, Marathon Oil Co., ExxonMobil, and National Oilwell.

Find more information on the above programs at http://fossil.energy.gov/news/techlines/02/tl_deeptrek_2002sel.html.

Acknowledgments

Sincere appreciation is expressed for approval to publish graphic materials provided for this publication from the following companies: Conoco and all its Subsea Mudlift Project partners;

Maurer Technology, Inc.; Novatek and its Intellipipe partner, Grant Prideco; Conoco and its subsea completion project partner, Kvaerner; Enventure; Tesco Corp.; Tempress Technologies; ACPT and its composite drillpipe joint industry partners; Pennsylvania State U. and its partner, Dennis Tool Co.; Virginia Polytechnics Inst.; and Los Alamos Natl. Laboratory. Appreciation is also extended to NETL's Strategic Center for Natural Gas and Oil, for the information on the DOE-funded projects.

Disclaimer

A number of the technologies discussed in this chapter were sponsored by an agency of the United States government. Neither the United States government, any agency thereof, nor any of their employees make any warranty, express or implied, or assume any legal liability or responsibility for the accuracy, completeness, or usefulness of any information, apparatus, product, or process disclosed or represent that its use would not infringe privately owned rights. Reference herein to any specific commercial product, process, or service by trade name, trademark, manufacturer, or otherwise does not necessarily constitute or imply its endorsement, recommendation, or favoring by the United States government or any agency thereof. The views and opinions expressed here do not necessarily state or reflect those of the United States government or any agency thereof.

References

1. Sumrow, M.: "Point of View: Harsh Environments, Emerging Technologies, Organizational Capacity to Shape Future of Drilling," *Oil & Gas J.* (17 June 2002) 40.
2. Smith, K.L. *et al.:* "SubSea MudLift Drilling Joint Industry Project: Delivering Dual Gradient Drilling Technology to Industry," paper SPE 71357 presented at the 2001 SPE Annual Technical Conference and Exhibition, New Orleans, 30 September–3 October.
3. Eggemeyer, J.C. *et al.:* "SubSea MudLift Drilling: Design and Implementation of a Dual Gradient Drilling System," paper SPE 71359 presented at the 2001 SPE Annual Technical Conference and Exhibition, New Orleans, 30 September–3 October.
4. Jellison, M.J. *et al.:* "Telemetry Drill Pipe: Enabling Technology for the Downhole Internet," paper SPE 79885 presented at the 2003 IADC/SPE Drilling Conference, Amsterdam, 19–21 February.
5. Notes from 2001 SPE Applied Technology Workshop, "Subsea Processing," Houston, 6–7 May.
6. Driscoll, F.G.: *Groundwater and Wells,* second edition, Johnson Filtration Systems Inc., St. Paul, Minnesota (1986) 301–307.
7. Warren, T.M.: "Casing Drilling Application Design Considerations," paper IADC/SPE 59179 presented at the 2000 IADC/SPE Drilling Conference, New Orleans, 23–35 February.
8. Shepard, S.F., Reiley, R.H., and Warren, T.M.: "Casing Drilling Successfully Applied in Southern Wyoming," *World Oil* (June 2002) 33–41.
9. Warren, T., Houtchens, B., and Portas, W.: "Casing Drilling with Directional Steering in the US Gulf of Mexico," *Offshore* (January and February 2001) 50–53, 40–42.
10. Fontenot, K. *et al.:* "Casing Drilling Activity Expands in South Texas," paper SPE 79862 presented at the 2003 SPE/IADC Drilling Conference, Amsterdam, 19–21 February.
11. Tibbits, G. *et al.:* "World's First Benchmarking of Drilling Mud Hammer Performance at Depth Conditions," paper SPE 74540 presented at the 2002 SPE/IADC Drilling Conference, Dallas, 26–28 February.
12. Coats, E. and Farabee, M.: "The Hybrid Drilling System: Incorporating Composite Coiled Tubing and Hydraulic Workover Technologies Into One Integrated Drilling System," paper SPE 74538 presented at the 2002 SPE/IADC Drilling Conference, Dallas, 26–28 February.
13. Leslie, J.C. *et al.:* "Cost Effective Composite Drill Pipe: Increased ERD, Lower Cost Deepwater Drilling and Real-Time LWD/MWD Communication," paper SPE 67764 presented at the 2001 SPE/IADC Drilling Conference, Amsterdam, 27 February–1 March.

14. Ruan, H. *et al.:* "Optical Fiber Logging System for Multiphase Profile Analysis in Steam Injection Wells," paper SPE 68897 presented at the 2001 SPE Western Regional Meeting, Bakersfield, 26–30 March.

SI Metric Conversion Factors

ft	×	3.048*	E–01	= m
°F		(°F–32)/1.8		= °C
gal	×	3.785 412	E–03	= kg
in.	×	2.54*	E+00	= cm
lbm	×	4.535 924	E–01	= kg
psi	×	6.894 575	E+00	= kPa
sq mile	×	2.589 988	E+06	= m^2

*Conversion factor is exact.

Chapter 14
Offshore Drilling Units
Mark A. Childers, Atwood Oceanics

14.1 Introduction

The growth and evolution of offshore drilling units have gone from an experiment in the 1940s and 1950s with high hopes but unknown outcome to the extremely sophisticated, high-end technology and highly capable units of the 1990s and 2000s. In less than 50 years, the industry progressed from drilling in a few feet of water depth with untested equipment and procedures to the capability of drilling in more than 10,000 ft of water depth with well-conceived and highly complex units. These advances are a testament to the industry and its technical capabilities driven by the vision and courage of its engineers, crews, and management. From an all-American start to its present worldwide, multinational involvement, anyone involved can be proud to be called a "driller."

Since the beginning in the mid-1800s until today, the drilling business commercially has been very cyclic. It has been and still is truly a roller-coaster ride, with rigs being built at premium prices in good economic times and sold for pennies on the dollar in bad times. Mergers, acquisitions, fire sales, and buyouts have occurred throughout its history, yet during all these times, the drilling segment has served the oil and gas industry well. Unfortunately, all this turmoil has been hard on the people involved, but they keep coming back with enthusiasm to this very interesting and stimulating industry.

In the early days, public image, safety, and the environment took a backseat to the technical and operational challenges of offshore drilling. Today, however, these issues often drive the whole thrust of drilling activities and operations. The offshore drilling business is now a worldwide, multibillion-dollar business with high visibility that has a strong influence on the world's economic health and people of all nations.

The offshore drilling business is one of the most challenging, exciting, and rewarding businesses in which an individual can be involved. This chapter focuses on the history and evolution of offshore drilling rigs and describes the various types of offshore drilling units. The capabilities and limitations of mobile offshore drilling units (MODUs) are discussed, followed by specific subjects that have a direct bearing on their operation and use. Health, safety, environment, and security (HSE&S) also are discussed because these subjects have become a major element in the industry. The importance and reason for classification, registration, and regulation of the units are presented, along with a discussion on the relationship of the contract drillers and their customers, the operators. And for the user, an explanation is given concerning selection of the appropriate type of drilling unit for a particular job.

14.2 History and Evolution

When did offshore drilling start? If "offshore" is defined as a large body of open water generally considered an ocean or sea, in 1897, just 38 years after Col. Edwin Drake drilled the first well in 1859, H.L. Williams is credited with drilling a well off a wooden pier in the Santa Barbara Channel in California. He basically used the pier to support a land rig next to an existing field. Five years later, there were 150 "offshore" wells in the area. By 1921, steel piers were being used in Rincon and Elwood (California) to support land-type drilling rigs. In 1932, a steel-pier island (60 × 90 ft with a 25-ft air gap) was built ½ mile offshore by a small oil company, Indian Petroleum Corp., to support another onshore-type rig. Although the wells were disappointing and the island was destroyed in 1940 by a storm, it was the forerunner of the steel-jacketed platforms of today.[1]

In 1938, a field was discovered offshore Texas. Subsequently, a 9,000-ft well was drilled in 1941 in fashion similar to the California wells by use of a wooden pier; however, with the start of World War II, all offshore drilling activities halted. After the end of World War II, the state of Louisiana held an offshore state waters lease sale in 1945. This was followed in 1955 by the state of California (Cunningham-Shell Act) lease sale, which allowed exploration of oil and gas sands.[1] Before the latter act, core drilling could be done only until a show of oil and gas. At that time, all drilling had to stop and the core hole plugged with cement.

The first "on-water drilling" was born in the swamps of Louisiana in the early 1930s with the use of shallow-draft barges. These barges were rectangular with a narrow slot in the aft end of the barge for the well conductor. Canals were and still are dredged so that tugs can mobilize the barges to locations. Later, barges were "posted" on a lattice steel structure above the barge, allowing them to work in deeper water depths by submerging the barge on the bay bottoms. These barges usually required pilings around them to keep them from being moved off location by winds and waves. The first "offshore" well, defined as "out of sight of land," was started on 9 September 1947 by a tender assist drilling (TAD) unit owned by Kerr-McGee in 15 ft of water in the Gulf of Mexico (GOM). An ex-World War II 260 × 48-ft barge serviced the drilling equipment set (DES), which consisted of the drawworks, derrick, and hoisting equipment located on a wooden pile platform.[2] TADs are discussed in more detail later.

The *Breton Rig 20* (**Fig. 14.1**), designed by John T. Hayward who was with Barnsdall Refining Co. at the time, was a large "posted" submersible barge credited in 1949 with drilling some of the first wells in the open waters of Louisiana. What made it different from the Kerr-McGee barge was that all the drilling equipment was on one barge that could be towed as a complete unit. The unit, which was a conversion from an inland drilling barge, had two stability pontoons, one on each side of the barge, that hydraulically jacked up and down as the barge was submerged and pumped out. These pontoons provided the necessary stability for this operation. The *Breton Rig 20,* later known as the *Transworld Rig 40,* was a major step forward because it eliminated the cost and time required to build a wooden platform to support all or some of the offshore-type rig. Although it drilled only in predominantly protected bays in shallow water (less than 20 ft), the *Breton Rig 20* may be able to lay a qualified claim as being the first MODU.[3,4]

The first truly offshore MODU was the *Mr. Charlie,* designed and constructed from scratch by Ocean Drilling and Exploration Co. (ODECO) headed by its inventor and president, "Doc" Alden J. Laborde. The *Mr. Charlie* (**Fig. 14.2**) was truly a purpose-built submersible barge built specifically to float on its lower hull to location and, in a sequence of flooding the stern down, ended up resting on the bottom to begin drilling operations. When the *Mr. Charlie* went to its first location in June 1954, *Life* magazine wrote about the novel new idea to explore for oil and gas offshore.[5] The *Mr. Charlie*, rated for 40-ft water depth, set the tone for how most

Fig. 14.1—*Breton Rig 20,* a converted "posted" swamp drilling barge capable of drilling in open Louisiana water depths up to 20 ft in 1949. Retired in 1962.

MODUs were built in the GOM. Usually an inventor secured investors, in this case Murphy Oil, and then found a customer with a contract to drill for, in this case Shell Oil, allowing bank loans to be obtained to build the unit.

Because the shelf dropped off quickly and water depths increased rapidly offshore California, the approach there was entirely different from that in the GOM. Rigs were installed on surplus World War II ship hulls modified to drill in a floating position compared with sitting a submersible barge on the ocean bottom, as done in the GOM. Oil companies formed partnerships or proceeded independently, but MODUs were not designed and constructed by contract drilling companies in California. All design and construction was done in a highly secretive manner with little sharing of knowledge because technology was thought to give an edge in bidding for state oil and gas leases. Before the leasing of oil and gas rights in 1955, oil companies cored with small rigs cantilevered over the side midship of old World War II barges. These barges did not have well-control equipment or the ability to run a casing program. They could only drill to a designated core depth with the understanding that if they drilled into any oil and/or gas sands, they would stop, set a cement plug, and pull out of the core hole. These

Fig. 14.2—*Mr. Charlie,* **the first purpose-built (June 1954) open-water MODU rated for 40-ft water depth. Retired in late 1986 and now a museum and training rig in Morgan City, Louisiana.**

core vessels were highly susceptible to wave action, resulting in significant roll, heave, and pitch, which made them difficult to operate.

With leasing from the state of California to explore and produce oil and gas, well control and the ability to run multiple strings of casing became mandatory and required a totally new, unproven technology. The first floating drilling rig to use subsea well control was the *Western Explorer* (**Fig. 14.3**) owned by Chevron, which spudded its first well in 1955 in the Santa Barbara Channel. Others followed quickly, with all of them concerned about the marine environment and technology to allow drilling in rough weather. In 1956, the *CUSS 1* was built from another World War II barge. The unit, built by the CUSS group (Continental, Union, Shell, and Superior Oil), was 260 ft long and had a 48-ft beam. The CUSS group eventually evolved into what is now Global Santa Fe.

The original designers had no examples or experiences to go by, so novelty and innovation were the course of the day. Torque converters on the drawworks were used as heave-motion compensators; rotaries were gimbaled to compensate for roll and pitch; the derrick was placed at midship over a hole in the vessel called a "moonpool"; blowout preventers (BOPs) were run on casing to the seafloor; re-entry into the well was through a funnel above a rotating head (riserless drilling is not new); mud pits were placed in the hull with mud pumps; and living quarters were added. It was an exciting and amazing time, considering that everyone was starting with a blank sheet of paper.

Fig. 14.4 shows the *Humble SM-1* drilling barge (204 × 34 × 13 ft) built and owned by Humble Oil and Refining Co. (now ExxonMobil) in 1957. **Fig. 14.5** shows the subsea equipment used to drill the wells. Note that it has no marine riser. The *Humble SM-1* drilled 65 wells for a total cost of $11.74/ft, about double the cost of land drilling at the time, in an average water depth of 159 ft and with a maximum well depth of 5,000 ft. The unit averaged 8.93 days per well and drilled an average of 324 ft/D. Unfortunately, the unit sank in a storm in 1961 while on loan to another operator.[6] At the insistence of insurance underwriters, the

Fig. 14.3—*Western Explorer,* the first (1955) floating oil- and gas-drilling MODU that used subsea well control. Retired in 1972.

American Bureau of Shipping (ABS) wrote and implemented in 1968 the first independent codes, guidelines, and regulations concerning the design, construction, and inspections of MODU hulls.

With the *Mr. Charlie* (bottom founded) and *Western Explorer* (floating) as the first MODUs, another concept for a MODU showed up in the form of a "jackup". This type of unit floated to location on a hull with multiple legs sticking out under the hull. Once on location, the legs were electrically or hydraulically jacked down to the ocean bottom, and then the hull was jacked up out of the water. With this approach, a stable platform was available from which to drill. In World War II, the De Long spud can jacks were installed on barges for construction and/or docks. The De Long-type rigs (**Fig. 14.6** shows an example, the *Gus I)* were the first jackups built in 1954.[7] Although jackups initially were designed with 6 to 8 legs and then a few with 4 legs, the vast majority of units today have 3 legs. The *Gus I* was constructed with independent legs. The Le Tourneau Co. built for Zapata Corp. the first lattice-leg jackup, the *Scorpion* (**Fig. 14.7**), which had independent legs with spud cans. To this day, Le Tourneau continues to specialize in lattice-leg-type jackup MODUs.

A major evolution for the jackup design was the introduction of the cantilevered drill-floor substructure (**Fig. 14.8**) in the late 1970s and early 1980s. As fixed platforms got bigger, the slot jackups could not "swallow" or surround the platform with its slot containing the drilling equipment; however, the cantilever units could skid the cantilever out over the platform after

Fig. 14.4—*Humble SM-1,* a floating MODU designed and operated by Humble Oil & Refining Co. (now ExxonMobil) in 1957. One of a number of "top secret" drilling units of the mid 1950s. Courtesy of Exxon-Mobil Development Co.

jacking up next to it. Before the cantilevered substructure, all jackups had slots, usually 50 ft square, located in the aft end of the hull. During tows, the substructure was skidded to the metacenter of the hull, but during drilling operations, the substructure was skidded aft over the slot. The derrick and/or crown could be skidded port/starboard to reach wells off center just like today's units do. The water depth range for most of the early slot and cantilever designs was from 150 to just over 300 ft; cantilever drill-floor centers had a reach of 40 to 45 ft aft of the aft hull transom; and variable deck load (VDL) ratings were 3,500 to 5,000 kips. In the late 1990s, "premium" or "enhanced" jackups were designed and built that could carry much larger deck loads ($\geq$ 7,000 kips), could drill in deepwater depths ($\geq$ 400 ft), had more capable drilling machinery (7,500-psi high-pressure mud systems and 750-ton hoisting equipment), had extended cantilever reach of $\geq$ 70 ft, and had larger cantilever load ratings of double or more the earlier units (some > 2,500,000 lbm).

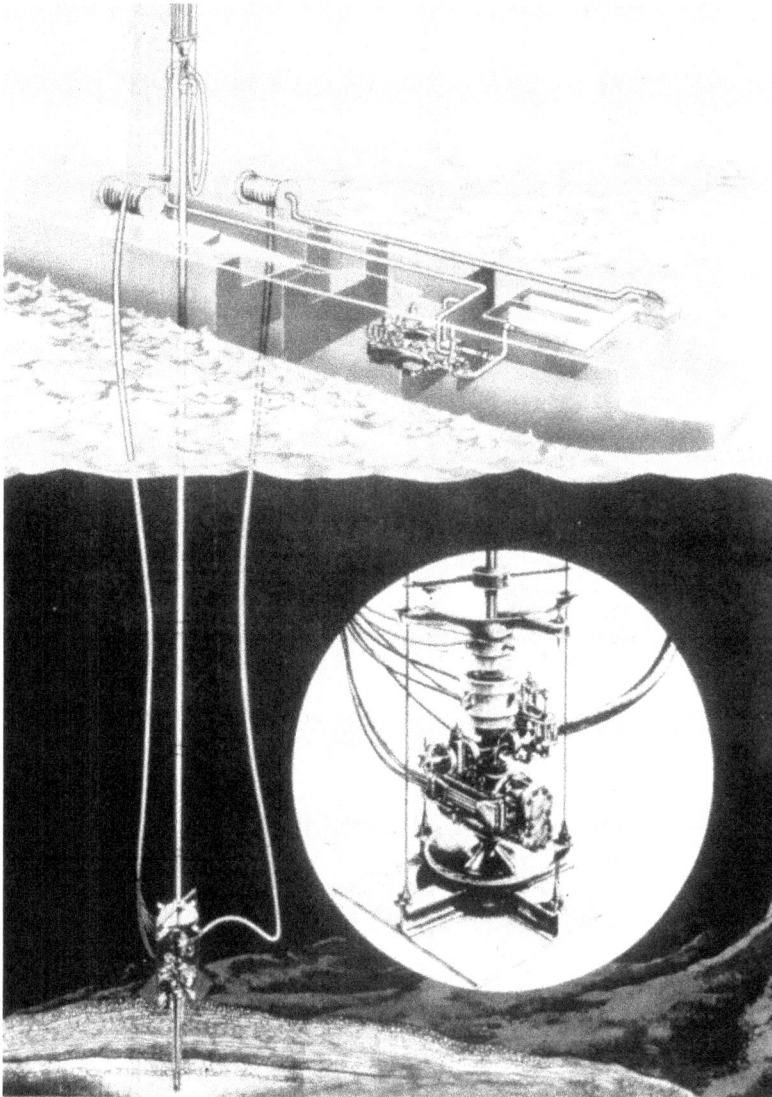

Fig. 14.5—Humble Oil and Refining Co.'s *Humble SM-1* subsea drilling system used offshore California.
Courtesy ExxonMobil Development Co.

The TAD concept was used to drill the first offshore "out of sight of land" well in the world. Initially used as an exploration method, it has evolved into a development tool. The first tenders were shaped like barges, but some are now shaped like ships for better mobilization speeds. Basically, the DES consists of the derrick, hoisting equipment, BOPs, and some mud-cleaning equipment, thus reducing the required space and weight to be placed on the fixed platform. All the rest of the rig, such as mud pits, mud pumps, power generators, tubulars and casing storage, bulk storage, accommodations, fuel, and drill water, is located on the tender hull moored next to the fixed platform. This approach turned out to be a very cost-effective way to drill from small fixed platforms. Unfortunately, in mild and especially severe weather, the mooring lines could fail, with the hull floating away, as it often did in a GOM "norther." Today, most TADs operate in benign or calm environments in the Far East and West Africa.

Fig. 14.6—With a De Long-type jacking system, the *Gus I*, built in 1954 and rated for 100-ft water depth, was the forerunner of the modern jackup. Initially, two barges that were eventually joined permanently, but the unit was lost in a storm.

In 1992, the first semisubmersible (semi) *Seahawk* TAD (**Fig. 14.9**) was converted from an old semi MODU. The semi hull offers superior station keeping and vessel motions compared with ship or barge-shaped hulls. In a semi hull, the wave train can move through the "transparent" hull without exciting it to heave, roll, and pitch, unlike a mono hull. The lower hull of the semi is below the water at a deeper draft; the columns offer a reduced area to excite the hull; and the work platform or main deck is above all wave action. TADs are seeing new use on deepwater production platforms, such as spars, tension leg platforms (TLPs), and deepwater fixed platforms, which operate beyond jackup water depths.

Things were off and running in the 1950s, with numerous operators getting into the rig ownership and operation business and new drilling contractors being formed every year. In the early 1960s, Shell Oil saw the need to have a more motion-free floating drilling platform in the deeper, stormier waters of the GOM. Shell noticed that submersibles like the *Mr. Charlie,* now numbering almost 30 units, were very motion free afloat compared with monohulls. The idea was to put anchors on a submersible, use some of the California technology for subsea equipment, and convert a submersible to what is now known as a semisubmersible or semi. Thus, in 1961, the submersible *Bluewater I* (**Fig. 14.10**) was converted to a semi amid much technological secrecy. In fact, in the mid-1960s, Shell Oil offered the industry the technology in a school priced at U.S. $100,000 per participant and had lots of takers.

Then came the *Ocean Driller,* the first semi built from the keel up (**Fig. 14.11**). The *Ocean Driller,* designed and owned by ODECO, went to work for Texaco in 1963, with the mooring and subsea equipment owned by the operator, as was common in the 1960s. The unit was designed for approximately 300 ft of water depth, with the model tests of the hull done in Doc Laborde's swimming pool. The *Ocean Driller* could also sit on bottom and act as a submersible, which it did well into the 1980s. Most of the first-generation units could sit on bottom or drill from the floating position as a hedge against unemployment. The shape and

Fig. 14.7—Le Tourneau's *Scorpion* built for Zapata (now Diamond Offshore Drilling Inc.) in 1956 for 80-ft water depth as an independent-leg jackup. Lost in 1969.

size of the first semis varied widely as designers strived to optimize vessel motion characteristics, rig layout, structural characteristics, VDL, and other considerations. The "generation" designation of semis is a very loose combination of when the unit was built or significantly upgraded, the water depth rating, and the general overall drilling capability. Generation is discussed in more detail later.

In the early 1970s, a new, second-generation semi was designed and built with newer, more sophisticated mooring and subsea equipment. This design generally was designed for 600-ft water depth, with some extending to > 1,000 ft. The *Ocean Victory* class (**Fig. 14.12**) was typical of the units of this era, which concentrated heavily on reducing motions of the platform compared with increased upper-deck VDL rating. Many were built, and in the middle to late 1980s, a number of third-generation semis were designed and built that could moor and operate in > 3,000 ft of water depth and more severe environments. Many of the third-generation units were upgraded in the 1990s to even deeper water depth ratings with more capabilities and became fourth-generation units. With a few exceptions, the operating displacement of these units went from ≈18,000 long tons in the 1970s to > 40,000 long tons in the 1980s.

In the late 1990s, the fifth-generation units, such as the *Deepwater Nautilus* shown in **Fig. 14.13**, became even larger (> 50,000-long-ton displacement) and more capable. These units can

Fig. 14.8—Le Tourneau's 116C cantilevered jackup with drill floor cantilevered over a fixed platform. Today's workhorse design of jackups. Courtesy Le Tourneau, Inc.

operate in extremely harsh environments and in > 5,000-ft water depth. Some second- and third-generation semis have been converted, given life extensions to their hulls and upgrades to their drilling equipment so as to be classed as fourth-generation units. Fig. 14.14 shows a second-generation *Ocean Victory* class unit (see Fig. 14.12) that was completely upgraded to a fifth-generation unit capable of mooring and operating in 7,000-ft water depth. Note the addition of column "blisters" for increased VDL, ≈ 50% increase in deck space, and the addition of riser

Fig. 14.9—World's first purpose-built (conversion) semi TAD unit *Seahawk.* Converted in 1992 from a semi MODU. Courtesy Atwood Oceanics.

storage and handling. A limited number of third-, fourth-, and fifth-generation semis have dynamic positioning (DP) assist or full-DP station keeping compared with a spread-mooring system.

Fifty years ago, fixed platforms had land rigs placed on them to drill and complete wells. Today's platform rigs have been repackaged so that they optimize the rig-up/load-out time, require less space, are lighter, and have more drilling capabilities; thus, they have become very sophisticated. Drilling platform rigs are still common, but today's units look far different from those of 30 or 40 years ago. Conventional platform rigs are usually loaded out with a derrick barge. Some large platforms may have two drilling units on them. To eliminate the costly derrick barge, "self-erecting" modular rigs have been built for light workovers and for drilling to moderate depths. Larger units that have the capability of a 1-million-lbm hook load have been built that are lightweight, easier to rig up/load out, and self-erecting. The advent of spars and TLPs in deep water, where space and deck load are critical, has generated even a more sophisticated modular deepwater platform rig, which is highly specialized to the structure on which it sits (**Fig. 14.15**). These platform rigs are not self-erecting, are unique to the structure they are placed on, generally are very light, and usually have limited drilling equipment capabilities.

By the mid-1960s, the jackup-designed rigs were displacing submersibles in increasing numbers. Jackups had more water depth capability than even the largest submersibles (some could operate in 175-ft water depth),[7] and they did not slide off location in severe weather. From this point on, jackup and semi designs were refined and made larger and more capable from a drilling and environmental standpoint.

Ship and barge-shaped floating MODUs, initially attractive because of their transit speed and ease in mobilizations, decreased in number as semis and jackups became more popular. One exception was the DP drillship, which held location over the wellbore by use of thrusters and main screw propulsion rather than a spread-mooring system. The first unit developed in the mid-1960s, although not an oil and gas exploration unit, was the *Glomar Challenger,*

Fig. 14.10—World's first semi MODU, *Bluewater No.1*, converted in 1961/1962 by Shell Oil from a submersible hull. Lost in 1964.

which was designed and owned by Global Marine (now Global Santa Fe) and contracted by the National Science Foundation for deep-sea coring around the world. This vessel confirmed the theory of shifting continental plates. Following the *Glomar Challenger* in the late 1960s to early 1970s were a number of first-generation DP oil and gas drillships, such as the *Sedco 445.* Subsequently, in the middle to late 1970s, the second-generation DP units were developed, such as the *Ben Ocean Lancer. The Ben Ocean Lancer* was an IHC Holland Dutch design, which also included the French rigs *Pelerin* and *Pelican,* which were owned by the French company Foramer (now Pride). These units could drill in up to ≈ 2,000- to 3,000-ft water depth, had better station-keeping ability in moderate metocean conditions, and had better overall drilling capabilities. DP ships of the late 1990s and early 2000s can operate in > 10,000-ft water depth and are two to three times larger than the earlier DP ships, with extremely complex station-keeping and dual-activity drilling systems. Dual drilling consists basically of some degree of two complete derricks and drilling systems on one hull so that simultaneous operations, such as running casing while drilling with the other derrick, can be done. These units are very expensive to build and operate but can overcome their cost with supposedly higher efficiency. For the right conditions, such as batch drilling a subsea template, large development projects over a template, deepwater short wells, and well situations in which more than one operation can benefit the overall plan, these units need to be reviewed for possible use as an alternative to standard single-operation units.

The offshore drilling industry has had spurts of construction and design improvements over its 50-year history. The first was the conception of the MODUs in the mid-1950s, followed by a mild building period in the mid-1960s. In the early 1970s, there were significant numbers of jackups and semisubmersibles built. However, the major boom of the late 1970s and early 1980s has been unmatched in numbers of rigs built. Starting in the late 1980s, a number of drilling contractors upgraded rigs built in the 1970s and early 1980s to deepwater depths, more severe environmental ratings, and better drilling abilities rather than building new units. The

Fig. 14.11—World's first purpose-built (1963) semi MODU, *Ocean Driller*. Unit could operate as a semi or submersible. Retired in 1992 and scrapped. Courtesy ODECO (now Diamond Offshore Inc.).

concept was that delivery and cost could be cut in half compared with a new build. Some drilling contractors have successfully built their entire business plan around conversion instead of new build.

Since the oil and gas bust of the mid-1980s, there has only been one spurt of new building, and that was in the late 1990s. Mergers and buyouts of drilling contractors and rigs dominated the industry from the mid-1980s to the mid-1990s. One drilling contractor, Global Santa Fe, monthly publishes a percentage number related to day rate and cost of building a new unit. A 100% rating means new units can be built profitably; however, the percentage number has lingered in the 40 to 60% range over the last 15 years or so, with spurts into 80%. By its nature, the drilling business is built on optimism for the future that may not always show proper returns on investment in terms of new builds or conversions. High on hope and the future, the contract drilling business has historically not been conservative and has not followed generally accepted rules of investment.

In the early 2000s, the average age of the fleet was > 20 years, with some units > 30 years old. Few are < 5 years old. Some have been upgraded and have had life extensions, which means that, with good care and maintenance, the basic hull, if it and/or the rig are not rendered technologically obsolete, may last > 40 years, as do units in the dredging business. "Technologically obsolete" means that the unit needs to have up-to-date top drive, mud-solids control, and pipe handling equipment, among other features, as well as enough power to run all the new equipment. The fleet in 2003 stood at approximately 390 jackups, 170 semis, 30 ships, and 7 submersibles. Fixed-platform rigs number about 50, and TADs number about 25.

The consensus is that the offshore drilling business will continue to grow, with emphasis on technical breakthroughs to reduce drilling costs. The industry has demonstrated that it can drill in water depths up to and > 10,000 ft and can operate in the most severe environments, but all at a very high cost that can run into hundreds of thousands of dollars per day. Ultra deepwater wells costing more than $50 million are common, and some wells have cost more

Fig. 14.12—ODECO's multicolumn second-generation semi Ocean Victory class of early 1970s. Unit shown is the *Ocean Voyager,* drilling in the North Sea in the early 1970s. This design proved structurally very attractive for upgrade to fourth- and fifth-generation units (see Fig. 14.14).

than $100 million. It is very difficult to justify wells that cost this much given the risks involved in drilling the unknown. The challenge to the offshore industry is to drill safely and economically, which means "technology of economics," with safety, environment, security, and personnel health all playing a large role.

14.3 Rig Types, Designs, and Capabilities

The previous section discusses the history and evolution of offshore MODUs and related offshore drilling units. This section gives more detailed technical description of today's units, their advantages and disadvantages, capabilities, and operating characteristics. One may ask why there are so many types, sizes, and capabilities of offshore units. The answer involves different technical, economic, government, and safety requirements to accomplish a specific drilling program. No one type can satisfy all the requirements for every drilling location; thus, we have to understand all types to make a correct decision on their use.

14.3.1 Fixed-Platform Rigs. As the name indicates, this type of rig is located on a fixed structure previously installed at the well location. The structure may be a fixed jacketed platform, spar, TLP, or gravity structure; whatever it is, the rig sits atop it. Fixed platforms may have as few as 3 or 4 or > 50 well conductors. Generally, the drilling rig is not a permanent part of the fixed structure; however, on some occasions, the unit is left on the platform for future

Fig. 14.13—*Deepwater Nautilus,* one of the newly built fifth-generation ultradeepwater semis that has DP assist for its spread-mooring system. Note spread columns for increased VDL and stability. Courtesy Transocean Inc.

workovers or additional drilling; sometimes, removing it is uneconomical. Most units are complete, totally self-contained units that include their own power plant, accommodations, drilling equipment, life-saving equipment, and auxiliary services. However, some do not have their own power plant and obtain power from the platform's generators, which are usually powered by produced natural gas. On large, central field platforms that have their own living quarters, the rig may not have its own accommodation facilities. In this case, the life-saving equipment (e.g., lifeboats and gas-detection, fire-fighting, and communication systems) is part of the fixed platform. Most fixed platforms have their own craneage, but usually it is not big enough to load or unload the components of a conventional platform rig. Most modern platforms are built to American Petroleum Inst. (API) standards, thus allowing movement of a standard API-configured platform rig from platform to platform with little or no modification.

There are three types of fixed-platform rigs. The first type is the conventional standard platform rig that is not self-erecting, is not particularly modular in construction, is heavy, and is built to API well spacing standards, so it can work on a wide range of platforms. This type of rig usually requires a derrick barge or a large platform crane to load and erect. Erection time may be 2 to 4 weeks, and its dry weight will probably exceed > 5,000 kips. These rigs are usually self-contained and can include up to and in a few cases over a 1 million lbm of derrick and traveling equipment.

Fig. 14.14—*Ocean Baroness,* one of the Ocean Victory class (Fig. 14.12) second-generation semis up-graded to a fifth-generation unit. Note blister additions to column, deck expansion, and much larger derrick. This semi also did surface BOP work in Malaysia in 2003, along with setting the world's record self contained spread-mooring water depth (6,152 ft). Courtesy Diamond Offshore Drilling Inc.

The second type of rig is a self-erecting, self-loading, and highly modularized rig set up to go from platform to platform quickly. Generally, they take up much less space, and their dry weight (750 to 1,250 kips) is considerably less than that of a conventional standard platform rig. Unfortunately, most of these rigs have limited hook and traveling-block capacity and some-times do not have all the auxiliary equipment, such as bulk tanks, large liquid-mud-storage capacity, and emergency power. They are particularly attractive for in-casing workovers and out-of-casing redrills. A few of the larger modular rigs have hook load ratings of 1 million lbm but also have compromised weight and ease of mobilization. Modular rigs first appeared in the late 1980s and early 1990s. They generally have no module weighing > 30 tons, have a self-erect-ing "leap frog" crane, contain modules that can be transported on any standard-sized workboat, and can be completely rigged up or down in 2 to 3 days.

A third type of the modular fixed-platform rig that has gained popularity recently is site-specifically designed and constructed to be placed on deepwater spars and TLPs. These modular rigs are very compact, lightweight, and site-specifically built (Fig. 14.15). Their mobi-lization and rig-up time is much more than that of a standard modular rig. Because they are generally not self-erecting, total rig-up and rig-down time and cost are an issue.

Fig. 14.15—Example of highly specialized and site-specific modular fixed-platform rigs used on spars, deepwater fixed platforms, and TLPs. This unit is on a TLP in the GOM. Courtesy Helmerich & Payne Intl. Drilling Co.

The first consideration in using a fixed-platform rig, usually controlled by the operator, is whether the platform is large enough and has a high enough load bearing to place and work the rig. This includes the space and dry weight of the rig itself, wet weigh (mud, operator fixed items, liquids, portable tools, etc.), live loads (hook, setback, and rotary), storage, and such expendable items as bulk casing and operator supplies. Generally, a four-pile structure is the smallest fixed structure that a conventional standard platform can be placed on and work efficiently. Usually, the second consideration is the mobilization method and cost. Numerous platform rigs when broken down for shipment cannot fit on a standard workboat, and thus a derrick barge is required. All modular rigs can usually fit on a workboat.

Why would someone want to use a fixed-platform rig? Generally, their day rate is considerably less than that of a jackup, assuming that the platform is in accessible jackup water depth and that there are enough wells to warrant the mobilization cost. The decision to use a jackup or standard platform rig is usually controlled by the number of wells to be drilled; the more wells there are to drill, the more attractive the platform rig becomes. Of course, the platform water depth, availability of a suitable jackup, metocean, and the mobilization cost and time of either unit are also factors. In shallow water, less expensive jackups are available; however, a platform rig will be more economical in deeper water. Market conditions at the time of use, like all rig types, are usually the driving economic force. Another alternative to a platform rig is a TAD; however, availability will be a problem because there are so few units, especially semi TADs. Environment may be an issue for monohull tenders. With semi TAD hulls, environment should not be an issue.

With a semi TAD, operating efficiency is higher. Studies have shown that a tender with very large load-carrying capability and space availability is operationally very attractive. A number of operators have stated that they think a semi TAD is 10 to 25% more efficient as controlled by workboat transit time, weather, specific type of wells being drilled, and space/weight limitations of the platform and platform rig.

There is no standard, easy answer for all situations to specifically recommend a specific rig type. With the appearance of extended-reach wells (ERWs), the required loads and space are becoming so great that a cantilevered jackup or TAD sometimes becomes a more attractive alternative than even a large standard platform rig.

14.3.2 Tender Assist Drilling. TADs were the rig of choice in the 1950s and early 1960s in the GOM for development drilling off fixed platforms. However, the monohull tenders tended to lose location with mooring failures during storms. This occurrence, along with severe motions of the tender, resulted in their losing favor, except for use in very mild or benign environments, such as in the Far East and West Africa. There are about 25 TADs in existence today, with most being monohull tenders. Four are semi tenders and offer the motion characteristics to drill in mild to somewhat severe environments. The TAD advantage is that its DES is relatively lightweight, one-quarter to one-fifth the weight and one-third the space of a standard platform rig. Most TADs carry the DES on the tender hull and are self-erecting, so no workboat or derrick barge is required. They are particularly attractive for situations in which there is an old platform with reduced load-carry ability and/or space, such as when a platform was drilled with a standard platform rig and then production equipment was loaded onto the platform, thus eliminating space and load-carry capacity. It is not unusual for a platform to deteriorate with age and then be unable to hold up a standard platform rig when additional wells need to be drilled. The TAD is an option for this situation. Of course, if the platform is in jackup water depth range, the jackup may also do the drilling if its cantilever can reach the well centers with adequate load capacity and if there are no incompatible spud can holes and/or a severe punch-through condition.

For spars and TLPs in deep water where weight and space are at an absolute premium, TADs, particularly semi TADs with their lightweight DES, have significant advantages in some cases over a modular platform rig. This is usually true for spars and TLPs with > 9 or 10 wells up to a maximum of ≈ 24 wells. For spars and TLPs with < 9 or 10 wells, their load and space availability are too small for any type of platform rig or DES, and those with > 24 wells are large enough to support a modular platform rig without a large weight and space penalty assuming all other factors are equal.

Semi TADs also have the advantage of acting as construction barges for platforms that are commissioning production equipment. Their large rig-up crane, open decks where the DES is stored and transported, accommodations, and general facilities offer a relatively inexpensive construction platform compared with a construction derrick barge.

Why would anyone want to use a TAD? They may be particularly attractive for standard platforms in water depths over jackup-rig rating and where space and/or load limits are a major factor, for deepwater spars and TLPs with the right number of wells, and for any platform where weight and space for long ERW are limited. Generally, a TAD costs more than a platform rig, especially the modular type, but they are a very attractive option for certain situations.

14.3.3 Conventional Ship- and Barge-Shaped Rigs. In the early days, ships were very attractive and the most common floating MODUs. They mobilized quickly and could carry a large amount of operator consumables, such as casing and bulk mud. However, their motions in weather proved to be a significant disadvantage in even mild environments. If a ship-shaped

Fig. 14.16—Typical spread-moored drillship of the 1970s. Note workboat next to rig with small ocean-going tug next to it.

unit was hit on its beam with even moderate swells, the roll could raise havoc with efficient productivity. **Fig. 14.16** shows a typical spread-moored drillship from 1970. The Offshore Co. (now Transocean) developed and patented the turret mooring system (**Fig. 14.17**). This system solved some of the motion problems, but other problems remained: decks were sometimes awash with green water, the turret could store only a limited amount of mooring wirerope because the winches were all located on the turret "plug," and the subsea BOP usually had to be stored on the drill floor. When the total number of MODUs increased, thus reducing the number of long mobilizations, and the number of semis in particular increased, the semi, with its vastly superior motion characteristics, became the MODU of choice for floating work. Another factor is that even though ships could carry large amounts of consumables, their space utility and connivance were limited by their cigar shape. The heyday of these units was the late 1950s to late 1960s, with a few being built in the early 1970s. Not until the late 1990s were more drillships built in the form of DP ultradeepwater units.

This section refers only to the spread-moored units, which were usually rated at < 1,000-ft water depth unless their mooring system was supplemented with mooring line inserts (i.e., mooring lines were inserted into the MODU's own lines by use of anchor-handling boats). For

Fig. 14.17—Discoverer II class turret-moored drillship built in 1967 and retired in 1985. The BOP stack was stored and run from the drill floor and all mooring winches were on the "plug" or turret, which was under the drill floor.

instance, ≥ 1,500 ft of mooring wirerope may be inserted into the mooring line of the drillship's own lines, thus increasing its line length and scope. With the inserts, some units have rated themselves at > 2,000 ft, but mooring MODUs in this manner is time-consuming and expensive. The alternative to moored MODUs is DP units, with their self-positioning thrusters and propulsion, are discussed later. Barges or non-self-propelled units are also not discussed here because these units, which are few in number, are used in lakes, bays, and buoys, not in offshore areas. Today, there are very few moored drillships left, and they operate only in the mild, benign environments of the Far East and West Africa. Most are > 25 years old and generally have not been upgraded technologically, which is another of their disadvantages.

Why would anyone want to use a drillship? If a location with a very benign environment is under consideration, if a conventional well is to be drilled, if the well is in a remote location where logistics is a primary consideration, and if mobilization of another type of unit is costly, then price is the driving factor.

14.3.4 Submersibles. Today, there are only seven submersibles left in existence, all located in the GOM. Their water depth range is between 9 and 85 ft, with a lesser depth rating during hurricane season. They have a narrow water depth range; however, unlike moored drillships, today they serve an important, although limited, segment of the market. Most jackup rigs cannot operate in < 18 to 25 ft of water, although a very few can move into as little as 14 ft of water. However, when they operate in very shallow water, their hull often must be placed on the ocean bottom so that their legs can be pulled. Jackup hulls are not designed for this type of service, although if no obstructions (e.g., rock outcrops, boulders, wellhead stubs, and pipelines) are present, jackups can be used. When the spud cans come out of the mud, the mud spills over onto the deck, making a huge mess. Cleaning the deck usually requires high-pressure wash-down pumps.

Submersibles are attractive in shallow water of < 14 to 20 ft and/or where the ocean bottom is very soft (< 60-psf shear strength). These soil conditions are common in river delta areas such as around the Mississippi River delta. In these areas, independent-leg jackups may drive their legs well beyond 100 ft, and then the legs may not be retrievable. Even if a mat-type jackup is used, the mat may be submerged, resulting in a loss of mat stability. In these conditions, the submersible becomes attractive.

Submersibles also have other advantages in that their VDL or well-consumable load-carrying ability is usually much higher than for comparable shallow-water jackups. They also do not leave a "footprint" like an independent-leg jackup does with its spud can holes. These footprints can cause significant structural leg problems when another jackup with different leg spacing is jacked up in the same area. Even if the second rig jacks up, it may slide into the previous spud can holes and lose its position over the platform, possibly causing significant leg damage.

The biggest disadvantage of submersible units in the past has been their susceptibility to sliding off location in even mild storms. However, one of the seven units, the *Atwood Richmond,* installed a patented station-keeping system in 2000 consisting of four 10-ft-diameter suction piles that are easily self-installable and retrievable. In 2002, the system held the unit on location in a hurricane with > 142-mile/hr winds and 30-ft seas.

14.3.5 Jackups. The jackup-type MODU has become the premier bottom-founded drilling unit, displacing submersibles and most platform units. The primary advantage of the jackup design is that it offers a steady and relatively motion-free platform in the drilling position and mobilizes relatively quickly and easily. Although they originally were designed to operate in very shallow water, some newer units, such as the "ultra-harsh environment" Maersk MSC C170-150 MC, are huge (**Fig. 14.18**) and can be operated in 550 ft in the GOM. With 673.4-ft. leg length, a hull dimension of 291×336×39 ft, and a VDL of 10,000 long tons, it is mammoth and rivals some of the larger semis. This type of unit can be commercially competitive only in the North Sea and in very special situations.

There are two basic types of jackups, the independent-leg type, usually three legs with lattice construction, and the mat type, in which the legs are attached to a very large mat that rests on the ocean bottom. Both types of jackups have a hull, float onto location, jack the legs to the ocean bottom, and then jack the hull out of the water.

For the independent-leg units, "preloading" is required to drive the legs into the ocean bottom before the hull is completely jacked out of the water. During this procedure, the jackup MODU is at risk from weather and leg "punch through"; i.e., one leg breaks through a hard crust thus putting the other legs in a large bending movement. Generally, 5-ft swells and/or a combined sea of 8 ft are the maximum seas in which these units can jack out of the water. If the hull should roll, pitch, and heave to an extent that the legs come into contact with the ocean bottom, particularly if it is hard, the legs can be severely damaged. In addition, the preload sequence is usually done in stages, with the hull never rising > 5 ft out of the water to safeguard against having a leg punch through. If the ocean bottom is soft and consists of clay, it is not uncommon to take 7 or more sequences, with each sequence taking 7 to 12 hours. The unit's pumps seawater into its preload tanks, adding weight to the hull and driving the legs. After the legs are driven and the hull goes into the water, the seawater is dumped overboard and the sequence is begun again. This process occurs until the legs no longer penetrate the ocean bottom. The concept is to load the legs to a level above that which the unit will encounter in the harshest predicted environment. The newer, enhanced premium units do a single preload in which the jacking system is strong enough to jack the unit with all the preload water onboard, the basic weight of the hull, and the full transit VDL. This is a significant advantage in that a much smaller "weather window" can be acceptable to move the unit. Jackups are most susceptible to major damage or loss when they are floating.

Fig. 14.18—Maersk's giant jackup (largest in the world) designed for deepwater use (550 ft in the GOM) and harsh North Sea environment. Courtesy Marine Structure Consultants.

The mat-type jackup also usually consists of three legs that are cylindrical and are from 8 to 12 ft in diameter (**Fig. 14.19**). The mat is carried just under the hull during mobilization, usually with ≈ 5-ft gap. When the unit comes onto location, it jacks the mat down to the ocean bottom, and because of its low bearing pressure, usually under 500 to 600 psf, the unit jacks the hull out of the water without going through the preload sequence required for independent-leg units. Bethlehem Steel Corp. built most of these units from the 1950s through the 1980s. Their key advantages are that they were relatively inexpensive to build and leave no footprint at the drilling location.

Unfortunately, the mat is also very susceptible to damage from any object on the ocean bottom. Mat-type jackups tow very slowly because the mat and hull are large and create a lot of drag. Their mats are susceptible to being gouged by workboat propellers, their upper hull has limited open deck storage space, and their legs sometimes form a wind-induced leg vibration known as vortex shedding (a form of severe vibration seen with smoke stacks without spoilers) at high winds, which can cause them to fail. Most mat rigs have cylinders for legs and are structurally limited to shallower water depths, usually < 250 to 275 ft. Only a very few units have reached 300 ft, and these units have lattice-type legs. For all these reasons, mat jackups have fallen into disfavor, although they are relatively inexpensive and for some well types are more than adequate.

Air gap, or the distance from mean water level to the bottom of the hull while the unit is jacked up in the operating condition, is a critical issue. The bottom of the hull must have a large enough air gap that the largest wave crest will not hit the hull and turn over the rig. Air gaps usually are 35 to 50 ft, with the larger air gaps in shallower water because wave heights build as water depth decreases. If a unit should work over a platform with a very high deck, air gaps of up to 100 ft are not uncommon; however, this obviously reduces the water depth

Fig. 14.19—Typical Bethlehem mat-type jackup under construction. Note size of mat and why workboats and tugs may damage it in shallow water.

rating. Jackup water depth ratings generally use a minimal leg penetration of 15 to 25 ft, which may not be the case in actual operation.

Independent- and mat-leg jackups also come in two types of drill floors, slot and cantilevered. As previously discussed, slot units were initially built in the 1950s through the late 1970s; however, with bigger platforms, the ability to cantilever the drill floor over the platform had an advantage over the slot units, which could only "swallow" minimal-size platforms. As the cantilever moves out to position itself over a well, it generally loses combined drillfloor load rating. The combined loading consists of the hook, setback, rotary, and drive-pipe tension if that tension is hung off the drill floor substructure. Generally, a minimum cantilever length ($\approx$ 14 to 20 ft) is required for moving BOPs and other items next to the hull. Full rating is usually accomplished at center positions but decreases as the cantilever moves further out and the drillfloor moves either side of center (usually ±15 ft). The rating on the extreme cantilever and extreme off-center can decrease by as much as 80%, leaving the unit capable of only light workovers.

Unlike typical earlier 1-million-lbm cantilever load units, the new premium jackups have ratings of ≥ 2 million lbm. With the advent of ERWs, deeper gas wells, and high-pressure/high-temperature requirements, the higher load ratings are required, so many older jackups have been upgraded and enhanced, although not to the extent of some of the newer premium units

Fig. 14.20—Atwood *Beacon* enhanced premium jackup typical of the large and very capable vintage units of the 2000s. Note that the rig is setting a small fixed platform by jacking up to a 161-ft air gap, at which time the platform will be righted to vertical, and then the rig will jack down and set the platform on the sea bottom. Courtesy Atwood Oceanics.

built in the late 1990s and early 2000s. The *Atwood Beacon* (**Fig. 14.20**) is shown in the process of setting a small platform. This unit has a 2-million-lbm cantilever load rating, 7,500-psi-working-pressure mud system, 70-ft cantilever, 400-ft water depth rating, accommodations for a crew of 120, and 7,500-kip VDL, which is typical of the dozen or so units like the *Atwood Beacon.*

There are more jackup-type MODUs than any other type of MODU. **Table 14.1** shows general information about the various types of major units. Marathon Le Tourneau (now Le Tourneau) has designed and built more of these units than any other designer and builder. As shown, the size and capabilities of these units vary widely, with the general trend being for them to get bigger and more expensive with higher drilling and marine capability.

Unlike platform rigs, submersibles, and ships, jackups and semis are upgradeable from a technical and commercial standpoint. Rowan Co. and Noble Drilling, both large offshore drilling contractors with large jackup fleets, have done extensive upgrades and enhancements to units built in the 1970s and 1980s. Upgrading usually consists of converting slot to cantilever units, leg strengthening and lengthening with more preload tanks, increasing environmental capability, and updating the drilling package with higher hook loads and installation of top drives.

Originally, MODUs were considered to have a life of 12 to 15 years, but through rigorous hull and equipment maintenance and technological updating, some 30-year-old units are considered "modern" and well fit for select purposes.

Why use a jackup? For water depths of 25 to 300 ft, there are many units to choose from. Some can be used in > 400-ft water depth. The jackup, with its stable work platform, relatively inexpensive mobilization costs, availability, versatility to work over a platform or to drill in

TABLE 14.1—NOTABLE JACKUP DESIGNS AND THEIR GENERAL CHARACTERISTICS

Designer (No. Built)	Type	Model or Class	Years Built	Initial Cost (M$)	Max. Water Depth (ft)	Hull Size (ft)	Leg Length (ft)	Drilling VDL (Tons)	Cantilever			Comments
									Long (ft)	Pt/Stb (+/- ft)	Load Rating (lb)	
Baker Marine (26)	3 I,M,S,C	BMC	1978–1983	21–32	148–250	191×132×16	162–336	1,621–1,975	40	15	1,000,000	Includes BMC- 150, 200 & 250 Models. Many of these models are still being used successfully at relatively inexpensive rates.
Bethlehem (30)	3 M,I	JU	1979–1982	32–35	100–200	157×132×18	162–269	1,339–2,150	25–45	10	200,000–1,000,000	Includes JU-100, 150 and 200 models. Attractive unit when spud can holes are not wanted.
Bethlehem (29)	3 M, S	JU	1974–1982	10–42	250–270	166×132×16	312	2,250	NA	NA	NA	Workhorse unit of the 1970s and 1980s and many are still in operation.
Friede – Goldman (34)	3 I, C	L-780	1981–1983	30–52	250–300	180×175×25	251–417	1,766–2,277	40	15	1,000,000	Includes L-780 and L-780 Mod II models. Still a very attractive independent leg unit.
Friede – Goldman (7)	3 I, C	Mod V B	1986–Present	62–150	350–400	228×222×31	496–540	3,707–4,166	52–70	15	1,500,000–2,500,000	Also includes KFELS Mod V B, which are upscale, enhanced units for the 2000s. Mod V A units are designed for North Sea operations.
Le Tourneau (31)	3 I, C	82	1978–1985	20–41	250	207×176×20	360	1621–1725	34–40	15	1,000,000	Includes Models 82, 82SD both slot and cantilever. Good unit for shallow water, especially in as shallow as 14 ft.
Le Tourneau (56)	3 I, S, C	116	1973–1983	9.3–43	243–350	243×200×26	410–477	2,200–2855	40–57	15	1,087,000–1,250,000	Includes Models 84, 116S and 116C. Many of the early 84S and 116S have been converted to 116C. Highly desirable and very common unit.
Le Tourneau (8)	3 I, C	Gorilla	1984–Present	85–212	328–450	306×300×36	504–605	2,594–8,125	52–75	20	1,875,000	Le Tourneau's largest, most capable and expensive unit. There are a number of different sizes for this class.
Livingston (21)	3 I, C, S	111	1971–1987	10.5–56	231–300	200×186×23	339–418	1,570–2,590	35–50	15	1,000,000–1,600,000	Some rigs have been converted to skid off units. Shipyard is out of business.
Mitsui (10)	3	300C	1981–1983	25–35	220–400	220×190×26	408–505	1,610–1,984	45	12	1,250,000–1,390,000	Few in number but a capable unit.
Offshore Co. (13)	4 I, S	Orion	1976–1988	10–50	200–307	172×134×21	305–400	2,006–2,080	NA	NA	NA	One of the older designs that is still being used in the industry. However, many have been lost, scrapped, or converted to production units.
MSC (2)	3 I, C	C-370-150 MC	2003–Present	200+	492–625	291×336×39	673	10,000	90	32.8	3,080,000	In 2003 this was the world's largest jackup, built primarily for North Sea operations.

1. The above characteristics are general. Since types are grouped, variance on listed statistics will occur.
2. Type includes independent leg (I), mat (M), slot (S), and/or cantilever (C).
3. Within designs, variations are the rule rather than the exception. When designs are grouped, the largest hull is emphasized.
4. There are many more designs that are not listed. The above listings are representative of all the designs that are available.
5. Source: Mobile Rig Register for 2000 and Offshore Data Services for 2001/02.

open water, and generally competitive day rate, lends itself as the rig type of choice in certain water depths.

14.3.6 Semisubmersible. For drilling from the floating position, the semi MODU has become the unit of choice. Sometimes referred to as a "column-stabilized" vessel, the combination of hull mass and its displacement, wave transparency of the hull because of the columns, and its deep draft enable waves to pass through the unit with minimal energy exciting it to excessive roll, pitch, sway, surge, heave, and yaw. With the work deck above the wave crests and the factors listed above, this design is a very capable work platform in severe environments. Floating units can work in very shallow water depths, < 100 ft in some cases, to the deepest water depths. The present world-record water depth for a semi is 9,472 ft set by a DP semi in Brazil in 2003 with a surface BOP, a new technique discussed later. The water depth record for a spread-moored semi is 6,152 ft, set in 2002 offshore Malaysia. A semi in 2003 set the world record for a "taut line" mooring system at 8,950 ft, also discussed later. The same rig also set the record for subsea completions in 7,571 ft in the GOM as the deepest producer.

In shallow water, the concern is the possible clashing of the lower hulls with the BOP stack if the semi moves off location. In other words, the distance between the subsea BOP stack when the lower marine-riser package is disconnected and the lower hull in the event of a move off location usually controls the minimal water depth. Heave, tidal range, slipjoint space-out, and ability to hold location are also important factors.

As with jackups, air gap is critical and is a major design consideration when the unit is rated for environmental conditions. During the design of a semi, hull motion analysis in relation to waves crashing into the upper deck is critical. Under no circumstance should a MODU be designed or rated for environmental conditions in which waves will come in contact with the upper hull. In addition, heave, roll, pitch, sway, yaw, and surge need to be analyzed in terms of the upper limits of motion in which crews and equipment can operate. For example, significant amounts of heave, if slow (long periods), may be tolerable for most operations; however, short heaves that are very fast (very short periods) are more difficult. From a crew performance standpoint, smooth predictable motions generally do not hinder performance; however, jerky unpredictable motion will have a significant negative impact. Metocean conditions throughout the world result in most semis being operated in < 8- to 10-second wave or swell periods, so motions below these periods are usually not of concern. A swell period of interest is the "resonance" or natural period in which the hull motion actually exceeds the environmental value (> 1.0 ratio) for motion (i.e., the hull heave is more than the wave height). It is generally agreed for semi designs that the resonance period for heave should be > 17 to 18 seconds in the GOM to prevent resonance. The resonance period varies in other areas.

Table 14.2 shows the relationship of common semi designs available in today's market. As seen, the size, mass or displacement, VDL, and water depth ratings vary widely. Generally, the deeper the water depth rating is, the more severe the environmental capability is, and the bigger the VDL rating is, the larger the semi displacement and dimensional size are. In the 1970s, the average semi displaced 18,000 to 21,000 long tons, whereas some of today's deepwater units displace > 50,000 long tons. Larger displacement usually means more VDL and better motion characteristics.

It is common to refer to semis as belonging to a "generation." This designation is somewhat inexact, but **Table 14.3** gives some guidance for semis. Recently, the newer ultradeep drillships have also adopted this type of designation. Many semis may start out as one generation, but an upgrade may graduate them into another one. This is particularly true of many second-generation units that are upgraded to fourth-generation units. One of the most unusual conversions and upgrades is Noble Drilling's EVA-4000 design, which originally was a shallow-water submersible. This triangular submersible was a complete redesign and turned into fourth- and fifth-generation semis. VDL and age are poor definition parameters for generation

TABLE 14.2—NOTABLE SEMISUBMERSIBLE DESIGNS AND THEIR GENERAL CHARACTERISTICS

Designer	Model (No. Built)	Year Built	Initial Cost (M$)	Water Depth (ft)	Size (ft)	Mooring	Riser Tension (kips)	Drilling Operations			Comments
								Draft (ft)	Displace (LT)	VDL (LT)	
Aker	H-3 (30)	1974–1981	28–35	600–1,500	355×221×120	Conventional chain	640	70	23,750	3,000	Most common second-generation semi. Many enhanced for more VDL. Not a common deepwater upgradeable unit.
SEDCO	700 (11)	1973–1983	30–53	600–1,500	246×229×90	Conventional chain	640	85	22,350	2,800–4,000	The SEDCO 711 Class was an enlarged version of the 700 Class with 4 units built. Most of these units have had limited water depth upgrades.
Forex Neptune	Pentagone (11)	1973–1977	20–42	660–1,500	338;326×133	Conventional wire rope	640	73.3	15,447	2,200–2350	This French design has had limited water depth upgrades of 2,000 to 2,500 ft. Two units were pushed to 4,500 ft with special equipment.
Friede Goldman	Pacesetter (5)	1974–1976	30–54	600–1650	260×218×111	Conventional chain	480	60	19373	1700	Includes Model L-900.
Friede Goldman	Enhanced Pacesetter (33)	1976–1991	40–93	660–2,200	260×200×111	Conventional chain	640	70	21,312	2,500	Includes Model L-767/907/945/1033 and 9500. Average water depth is 1,500 to 2,000 ft. Some units have had major water depth upgrades.
Korkut Engineers	New Era (5)	1974–1983	28–92	600–2,000	325×200×95	Conventional chain	800	55	22,017	2,400	There is considerable difference between various units. Three units have had major deepwater upgrades.
Noble Drilling	EVA 4000 (5)	Conv. 1998–1999	Conv. 142–175	4,500–10,000	348×328×130	Combination chain and wire rope	1,920	80	26,500	4,000–4,900	All these units are conversion from a submersible hull. One is a DP version.
ODECO	Odyssey (5)	1983–1988	65	5,000	390×259×142	Combination chain and wire rope	1,280	80	43,030	7,000	Includes one DP Semi but statistics not listed. Some models are referred to Ocean Ranger II class.
ODECO	Victory (11)	1972–1975	20–35	600–2,000	323×292×128	Conventional chain	480–640	70	23,127	2,250	Popular design for conversion to deepwater upgrade as deep as 6,500 ft.
Reading & Bates	RBS-8 (2)	2001–2002	335	10,000	374×256×136	DP assist, combination chain and wire rope	3,200	75.5	49,473	9,249	All new fifth-generation units. Very large unit with unusually high VDL but limited deck space.
SEDCO Forex	Express (3)	2000	325	6,000–7,500	349×226×111	DP	2,000	65.6	33,927	5,905	Unusual French design that has met with limited success.
Trosvik	3000 (5)	1982–1983	75–90	1,500	345×258×118	Conventional chain	640	77	27,224	2,852	Competitor to Aker H-4.2 and other third-generation units.
Trosvik	8000/9000 (3 built, 2 under construction)	1999–2001	275–480	5,000–10,000	361×246×148	DP and combination chain and wire rope	1,920–3,200	78	51,766	6,889	Most of these units are fifth-generation units.

1. The above characteristics are general, are slanted toward initial construction, and are "average." Almost all classes of rigs have been upgrade, have been modified, and/or have had major water upgrades over the years, thus increasing their mooring capability, riser tension, VDL, and displacement.
2. Source: Mobile Rig Register for 2000 and Offshore Data Services for 2001/02.

designation because some second-generation units have larger VDLs than some fourth-generation units and because age variations within a generation, especially fourth generation after
upgrade, can vary widely. The most defining qualities between generations probably are water
depth rating, the date of new build or upgrade, and the technical capability of the drilling and
subsea equipment on board the unit. Fifth-generation units usually have very large VDLs, high
marine-riser tension, hook load ratings of 1.5 million lbm, large deck space, high-pressure [7,500-
psi working pressure (WP)] mud pumps, and extensive mud-solids control systems. Floating
units require subsea well-control equipment, a marine-riser system, marine-riser tension systems, drillstring motion compensation, large mooring systems, craneage to handle all the
tubulars and marine riser, a guidance system to enter the well and to run the well-control systems, and a sophisticated management system to work all the components together. This
equipment and these procedures are discussed later.

Why would someone want to use a semi? In general, they are the most dependable, motion-
free, and capable of all the MODUs. Their cost is generally higher than that of a jackup, but in
water depths exceeding that for which jackups are rated, they are the unit of choice.

14.3.7 Ultradeepwater Units. These units, which are extremely expensive, few in number, highly capable, huge in size, and technologically advanced, are the technological forerunners and
pioneers in the offshore drilling business. **Table 14.4** gives some characteristics of these units,
most of them drillships of extraordinary size, but some are semis as listed in Table 14.3. All
were built in the late 1990s and early 2000s. Most have some degree of dual-rig activity (i.e.,
they have two drilling units on one hull). The Transocean Enterprise Class drillships (**Fig.
14.21**), for example, have the capability to run two riser and two BOP systems with one system drilling and the other completing a well on a subsea template. With this drill-and-complete
mode on a multiwell template, companies have claimed efficiency savings of 40% compared
with a single-derrick unit. For exploration wells, it is possible to run casing with one derrick
set and drill with the other, thus reducing total rig time to complete the operation. Of course,
the latter operation is accomplished before running the BOP stack. It is possible to run marine
riser and the BOP stack with one derrick set while running and cementing conductor casing
with the other. Some have the capability to produce and store crude oil, thus eliminating the
need to flare or burn the produced fluid during well testing.

The ultradeepwater drillships are the outgrowth of the second-generation DP units built in
the middle to late 1970s. These units provided technological breakthroughs in stationkeeping, re-
entry without guidelines, power management, thruster management, reliability, priority assign-
ments, and maintenance that led to the newer units shown in Table 14.4. The newer units are
"D3" rated in that they have total triple redundancy from the engines, to SCR, to electrical
switch, wiring, fuel, thruster, stationkeeping monitoring, etc. In other words, if any component
of the system should fail, another one comes online immediately; if another system fails, the
third system comes online. This approach is an effort to increase the reliability of the total
stationkeeping system.

The attractiveness of these ultradeepwater units, all of which are fifth-generation units, is
their unique ability to drill in up to 7,500 ft—and in some cases, > 10,000 ft—of water depth.
These units generally cost more than U.S. $400,000,000 to build, with some running more than
U.S. $650,000,000. The commercial viability from the contract driller's viewpoint is still ques-
tionable; however, they have proved that the industry has the ability to drill in over 10,000 ft
of water depth, a feat not imagined 15 years ago. The current world-record water depth set by
the DP drillship *Discoverer Deep Seas* in 2003 and 2004 is 10,011 ft in the GOM. The current
drill and complete for production record is 7,209 ft, also in the GOM and set in 2002 by the
sister rig of the *Discoverer Deep Seas,* the *Discoverer Spirit.*

Why use one of these units? Water depth is the primary reason. Some contract drillers believe that the dual-activity capability makes them competitive with moored units of lesser

		TABLE 14.3—DEFINITION OF SEMISUBMERSIBLE "GENERATION" DESIGNATION				
Generation	Designer/ Owner	MODU Classes (Approx. No. in Class)	MODU Names as Examples	Water Depth (ft)	Year Built/ Upgraded	Comments
1	ODECO SEDCO	Ocean Driller (2) Ocean Queen (5) SEDCO 135 (12)	*Ocean Explorer Ocean Digger SEDCO 135F*	300–600	Mid to late 1960s	Original semis built from the keel up. Only a very few of these units have escaped the scrap yard.
2	Forex Neptune & IFP, ODECO SEDCO, Aker Friede Goldman Korkut Engineers	Pentagone (11) Ocean Victory (11) SEDCO 700 (11) Aker H-3 (30) L-900 Pacesetter (5) New Era (6)	*Pentagone 87 Ocean Baroness SEDCO 702 Byford Dolphin Alaskan Star Eagle*	600–2,000	Mid to late 1970s	Many of these units are the foundation for upgrades to fourth- and fifth-generation units. Many have had moderate upgrades to 2,000 ft. The *Ocean Baroness* is an example, built second and upgraded to fifth generation.
3	ODECO Aker Friede Goldman	Odyssey (5) Aker H-4.2 (2) Enhanced Pacesetter (33)	*Ocean America Transocean Leader Global Arctic III*	1,500–5,000	Mid to late 1980s	Many of these units have had major water depth upgrades. Most were new built units rather than upgrades.
4	Diamond Offshore Atwood Oceanics Noble Drilling	Ocean Victory Upgrade (3) New Era Upgrade (3) EVA-4000 Conversion (4)	*Ocean Victory Atwood Eagle Noble Max Smith*	3,500–5,000	Late 1990s to early 2000s	These units are spread moored with a few having DP assist. Upgrades include sizeable riser tension, hook load, mud volume, mooring ability increase and other attractive features.
5	Transocean Noble Drilling Smedvig Diamond Offshore Ocean Rig ASA SEDCO Forex	R&B Falcon (2) EVA-4000 Conversion (1) Smedvig ME 5000 (1) Ocean Victory Upgrade (2) Bingo 9000 (2) Express Class (3)	*Deepwater Nautilus Noble Paul Wolff West Venture Ocean Baroness Levi Eriksson Cajun Express*	5,000+	Late 1990s to early 2000s	Most of these units are DP with a few being spread moored. They are usually designed for harsh environment and have significant riser tension, hook load, mud volume and VDL capabilities. These are ultra deep units and are compatible with Table 14.4 units.

1. The above lists the major designs, owners and class and/or designs of significance in relationship to generation classification.
2. There is overlap between generations. A good example is the technologically advanced ODECO *Ocean Ranger* built in 1975 as a second-generation unit but really a third-generation unit rated for 3,000-ft water depth and equal to many fourth-generation units in overall capability.
3. Some designs started out in one generation but after upgrade and life enhancement move into another generation. A good example is the *Ocean Victory Class* in which five units have been upgraded to fourth- or fifth-generation status. The EVA-4000 units are converted submersibles (i.e., were not even semis before conversion).
4. DP versions of some of the above generations are not applicable to water depth rating; however, the basic design is applicable to the specific generation designation.
5. Water depth ratings are very site specific and as the units were originally built; therefore, the above rating is general, and some units may claim deeper rating depths.

capability and cost. However, these units are, in general, exploration units with a "niche" development capability for large-numbered multiwell subsea templates in very deep water. They are expensive but very attractive for the right situation. Generally, for exploration wells, the deeper

TABLE 14.4—ULTRADEEPWATER DRILLSHIP DESIGNS AND THEIR GENERAL CHARACTERISTICS

Designer	Model (No. Built)	Year Built	Initial Cost (U.S. $1,000)	Maximum Water Depth (ft)	Size (ft)	Installed Power (hp)	Crude-Oil Storage (bbl)	Riser Tension (kips)	Drilling Operations			Comments
									Draft (ft)	Displacement (LT)	VDL (LT)	
Samsung Heavy Industry	Saipem 10000 (1)	2000	235	10,000	747×138×62	59,460	140,000	3,200	39.4	94,912	19,680	
Global Marine	456 Class (2)	2000	365	8,000–9,000	752×118×58	46,288	130,000	2,500	36	75,000	22,500	
Transocean	Deepwater Pathfinder Class (4)	1998–2000	355–429	7,500–10,000	835×125×62	52,118	125,000	4,800	42	92,861	19,684	Samsung involved in design.
Pride	Gusto 10000 (2)	1999–2000	235–250	6,000–10,000	671×98×62	39,600	0	1,920	32.8	41,350	17,420	
Smedvig	West Navigator MST-CAD (1)	1999	650	8,200	830×138×76	52,200	535,000	2,560	43	99,384	15,986	Hull conversion.

1. The above characteristics are general and "average."
2. Cost figures are as published, but actual cost in some cases is much higher.
3. Source: Mobile Rig Register for 2000 and Offshore Data Services for 2001/02.

Fig. 14.21—Ultradeepwater drillship Transocean Discover class *Discover Deep Seas* that currently holds the world water depth record for drilling (10,011 ft). There are three of these type of units, which are 835 ft long with a power rating of 52,000 hp and displacement of 92,800 tons. Note the dual crown block for dual activity. Courtesy Transocean.

the water depth is and the shorter the well is, the more commercially attractive they become over a standard spread-moored semi. Without them, we could not explore consistently in > 7,500 ft of water depth.

14.4 Other Considerations
Until now, we have focused on the basic hull designs and their capabilities. For any MODU to operate as designed, many associated and auxiliary factors and systems must be taken into account. Following are some major items that a driller needs to consider when selecting and operating a MODU.

14.4.1 Mobilization and the Drilling Site. Mobilizing a MODU usually falls into three categories: field, area, or long/international move. For field moves, which are short, no special preparation is done other than standard marine items. Field moves are usually defined as < 500 miles in the same environment, the same geographical area, and the availability of safe haven if required by weather conditions. In large bodies of water such as the U.S. Gulf Coast, the entire area is classified as a field move. However, if a MODU is moved from the U.S. Gulf Coast to Mexico, for example, it would be an area move. Any moves across the Atlantic or Pacific Oceans, of significant distance in Southeast Asia, from Europe to West Africa, etc., would be considered long/international moves.

Through their more favorable marine design, ships and semis have less metocean restrictions on moves than jackups and submersibles. Depending on the drilling contractor's arrangement with the insurance underwriters and third-party surveyor, a surveyor may or may not be required to be present during the move. The surveyor and underwriter are keenly interested in

Fig. 14.22—Heavy lift ship performing a dry-tow transport of a semi (*Atwood Eagle*) from Angola to Australia in 2004. Courtesy Atwood Oceanics.

the seaworthiness of the MODU. The degree of preparation is controlled by the category of move. The long/international move, which is the most restrictive, requires the most preparation. Usually, there is a long list of conditions, including mooring gear requirements, water tightness of openings, crew training and licensing, radio and communication gear, tug hookup and emergency lines, weather forecasting, class and regulator compliance, routing of the tow, post-tow inspection, and general overall condition of the MODU.

A unit may be moved in two basic ways, by wet tow with a tug or a dry tow with a heavy lift ship or barge. Tugs for wet tows come in all sizes and capabilities. Small 600- to 900-hp tugs are often used to move submersibles in shallow water near shore. For field moves in the open waters of the GOM, 4,600- to 9,000-hp units are often used, usually two to three units at a time, depending on the size of the MODU, length of tow, and type of MODU. For ocean-going wet tows, tugs with > 20,000 hp are not uncommon. In the past 15 to 20 years, a new type of tug has become popular for semis that can pull/run anchors, act as a supply boat, and tow. Some of these vessels are very large with horsepower ratings > 20,000 hp.

The second mode of transport is the use of a heavy lift ship or barge (**Fig. 14.22**). This is the most expensive transport but usually travels at > 10 knots, which, depending on the MODU, is two to three times faster than a wet tow. If collecting the MODU contract day rate is an issue for the drilling contractor, the heavy lift ship is cheaper overall because it gets to location much faster. The insurance rate is also a third to a quarter of that for a wet tow. The use of heavy lift ships has become more popular for many reasons, mainly safety and speed. There are also unpropelled submersible barges that load the same way as the heavy lift ships.

A key issue with any MODU is the site condition. This usually centers on soil characteristics, especially for jackups and submersibles, which sit on bottom, and less so for semis and drillships, which are concerned only with the anchor-holding power of the soil. The issue of punch through of a leg by an independent leg jackup is of major concern; thus, soil borings are usually required for these locations. With information on soil conditions from soil borings, punch-through conditions can usually be determined. A punch-through condition usually is associated with a hard, thin sand layer with weak soil underneath it. When the jackup preloads by filling its preload tanks with seawater and thus increases its weight, the load may become so great that the soil fails and a punch through occurs with usually just one leg-spud can. Should this occur, the other two legs will probably be quickly overstressed. If the punch-through is deep enough, the legs usually bend, and the jackup must go to the shipyard for extensive leg and hull repairs.

For submersibles, the issue is usually uneven settling or scouring under the hull. If the hull should settle unevenly because of scouring resulting from ocean currents, the hull will most likely be overstressed, resulting in possible structural damage. Although this event is very uncommon, "hogging" or bending the keel of the submersible is a very serious situation. To prevent this, some submersibles have scouring skirts around the edge of the hull. Mat jackups also may experience scouring, especially if in shallow water with high currents; therefore, a 2-ft-deep knife edge is placed all around the mat's parameter to help prevent scouring. Cement-filled sandbags have been used to prevent scouring under submersible and jackup mats. Uneven settling is a more severe condition for a mat jackup in that with misalignment over 1 to 1½ degrees vertical tilt, the cylinder legs will become wedged in the jack house, and the rig cannot jack because of friction between the leg and jack house.

Pipelines and underwater structures (e.g., natural reefs and old shipwrecks), protected underwater creatures (e.g., tube worms), and even old wells must be mapped and acknowledged. For semis, running anchor lines across and resting the anchor chains on some of the latter objects is not allowed or considered good practice. In this case, anchor patterns are altered or special mooring-line configurations are considered. Options include the use of spring buoys that lift the mooring line off the object, special vertical load anchors that do not require the anchor-line scope of a dynamically installed drag anchor, and/or special composite mooring-line makeup. Sandbags full of ready-mix cement have often been laid on pipelines in shallow water to keep mooring lines from cutting or lying on them.

Drilling next to shipping lanes and/or fairways requires planning and extra precaution. Ships sometimes stray out of their designated lanes and have on occasion collided with MODUs. Floaters may have anchors and/or anchor lines in the fairway, so coordination with the proper authorities is mandatory so that vessel traffic will not hit the MODU's mooring lines. Usually, the mooring line must be at a depth under full tension that will not threaten vessel traffic. For the GOM and other areas, notice must be given to the proper authority that a "navigation hazard" has moved into an area so that vessel traffic will be aware of the MODU and not collide with it.

The above discussion indicates that using a MODU requires the operator to plan ahead, determine conditions, and make arrangements for unusual conditions.

14.4.2 Equipment Outfitting and Capabilities. We must never lose sight of the fact that a MODU's primary goal is to drill and sometimes complete wells. Often, when concentrating on the marine aspects of the offshore drilling business, we forget this fact. Every well and site have their own requirements and demands, but following are some general comments, not always applicable to every situation, on MODU equipment and capabilities that should be considered when selecting a unit or planning a well. This list is far from complete but raises some of the most common considerations:

1. *Variable Deck Load.* VDL includes any item of weight that is not included in the light-ship of the basic vessel. Lightship is the basic weight of the MODU, including all equipment considered permanent. This includes drawworks, mud pumps, rotary, derrick, top drives, power plant, and basically all items that cannot be readily lifted off the vessel. VDL includes the

drilling contractor's drillstring, BOPs, spare parts, vertical tension to hold up a drilling marine riser, fuel, portable water, and anything loose on board. The remaining VDL is for the operator's consumables, including logging units, casing, bulk and liquid mud, cement, handling tools, subs, and anything that he may want to store on the MODU. Hook, rotary, and setback are also considered VDL and may consist of a large portion of what the MODU is required to safely carry. The depth of the well and casing program has a big impact on the amount of VDL required, as will water depth. Complicated mud programs requiring changing of mud systems will necessitate more volume and space. It is not uncommon for a development MODU to have three types of mud on board. For floating rigs, storing the entire volume of the marine riser adds significant weight and space requirements to the MODU as water depths increase. Some MODUs report large VDL capacity, but often they do not have the space to store the VDL.

In general, jackups, except for the new premium units, have the least VDL capabilities. They also do not have a lower hull or large tankage like a semi's lower hull or a drillship's auxiliary tanks or a submersible's lower hull tankage. The range of VDL for jackups (Table 14.1) runs from 1,600 to 2,600 short tons. For semis (Table 14.2), the VDL ranges from 2,500 to 4,000 short tons for older units and 4,000 to > 7,000 short tons for the newer-generation units.

2. *Stationkeeping Equipment and Marine Riser Tension.* For spread-moored MODUs, analysis must be done in relationship to the environment required for it to withstand and hold station in drilling, standby, and survival modes. Metocean data must be obtained and used in an industry standard analysis program like that published by API or other recognized authority. For DP operations, the operating limits of the system must be compared against the metocean and the return periods of major events. DP stationkeeping, unlike spread-mooring systems, functions so that the unit either maintains location or is steadily forced off location. There is no in-between when reaching the maximum capabilities of the unit. For spread-moored units, as the MODU moves off location because of increasing environmental forces, the mooring system increases in restoring force; however, the offset from the well may be too great to manage the marine-riser system safely.

The mooring and marine riser work hand-in-hand; therefore, a riser analysis in accordance with an industry standard such as API should be done. If the MODU appears to be more than adequate for the proposed location, drilling contractors can usually supply the analysis and guidance. For more challenging locations, a number of competent engineering firms can conduct studies and give guidance as to the acceptability of a specific MODU under consideration.

3. *MODU Classification and Environmental Rating.* Every MODU has design ratings approved by classification societies, country of registration, and regulations by various bodies. A unit may be able to work in the GOM but not be rated for the environment or regulatory requirements in the North Sea. Most MODUs can operate in temperate and mild environments, but such areas as the North Sea and west of the Shetlands are restrictive to many units. Some third-world countries do not have any regulatory requirements, and the regulations that do exist are loosely enforced. Pollution and environmental requirements can be major considerations. Some countries, such as Australia and Italy, have very strict rules concerning electrical, mechanical, training, staffing, and other matters. This subject is discussed in more detail later.

4. *Well-Control and Related Equipment.* The anticipated maximum surface pressure in the event of a well-control problem will determine the WP rating of the well-control equipment. Most MODUs have a 10,000-psi-WP system. A few have only 5,000-psi-WP systems, but if a 15,000-psi-WP system is required, MODU selection may be restricted. The cost of the MODU and well will increase because of a more restrictive market for high pressure and sometimes high temperature (high bottomhole temperatures). Wellheads, usually more and heavier casing and thus higher VDL requirements, more and heavier muds, and the like all drive up the required capability of the MODU. Well-control equipment is a subject in itself, and there are a

number of good references in the industry. The following text discusses subsea equipment and its relationship with stationkeeping.

5. *Accommodations Capacity.* If a simple well is to be drilled and not completed, crew and servicemen capacity requirements are far less than if a complicated well is to be drilled and completed. Most modern MODUs have capacity for at least 70 crew, many have capacity for 80 to 90, and some of the newer units have capacity for up to 120. Included are the operator's personnel, service personnel used at various stages, the contractor's crews, catering personnel, and any visitors. Room for regulatory personnel is sometimes a requirement, but as most operators will confirm, there never seems to be enough capacity. This results in a constant shuffling of personnel and crew on and off the MODU to stay within class and lifesaving allowable limits.

6. *Drilling Equipment and Power Plant Requirements.* Most drilling engineers and operations personnel will look at a MODU's drilling equipment and power plant first to see whether the unit is capable of drilling the well under consideration. Many upgraded units, even some new builds, may be short in one or more areas. If the unit has added, for example, a top drive, a third mud pump, enlarged accommodations, centrifuges, and solids-control equipment, more power and electrical equipment are required. During upgrades, this additional power is not always added. It generally is advantageous to be able to run all mud pumps, lift up a heavy load with the drawworks, back ream the hole at high torques, and have a maximum utility or "hotel" load simultaneously. Various operators have rules about what they require. Some require at least one mud pump in reserve at all times. Some want one main engine as a reserve for back-up in the event of an unexpected loss of one engine or available for routine maintenance.

Is a high-volume, high-pressure mud system (5,000- vs. 7,500-psi WP) worth the extra money required to hire an upscale MODU? All the aforementioned items should be part of the equipment evaluation. In addition, operating performance, management style of the contractor, safety performance, financial stability, honoring of contracts, and many more factors should be kept in mind.

7. *Well Testing.* If extensive well testing is anticipated, burning and/or storage of the crude must be considered as part of the MODU selection. If high production rates for a gas test are considered, cooling of the MODU is a major consideration. Piping and safety systems are of paramount importance to ensure that the operation is conducted in a safe and environmentally secure manner.

8. *Crew Capability, Training, Safety, and Overall MODU Performance.* Assuming that the "hard" or basic equipment qualifications are met, it is important to determine the capabilities, training, and safety work habits of the crews. Longevity of critical and key members of the crew, such as the offshore installation manager (OIM), tool pushers, drillers, crane operators, barge engineers, rig mechanics and electricians, is an indication of good morale, teamwork among the crew, continuity, and performance. The International Association of Drilling Contractors (IADC) has rules and guidelines to measure safety through lost-time incidents (LTIs), non-LTIs, first aid, and near misses. These statistics indicate the MODU and drilling contractor's commitment to and success in conducting a sound safety program. Overall MODU performance can be measured in downtime, for which every drilling contractor keeps records, and time-vs.-depth curves. Many receive appraisals from their customers on a well-by-well basis. If these forms are not proprietary per the drilling contract, they should be reviewed.

9. *Special Situations and Considerations.* If the well is to be drilled in an unusual area or there are atypical circumstances, MODU selection may be restricted. Very high-current areas that induce vortex shedding and thus violent vibrations of marine risers, jackup legs, mooring lines, guidelines, and BOP control lines may require special equipment. Severe cold, especially below-freezing temperatures for extended periods of time, require special winterization, which is not standard equipment. Icebergs and pack ice flows are another unusual situation that must be taken into account when selecting a unit. Extremely large tides, > 20 ft, may eliminate

some jackup MODUs because of leg length. Unusual situations and circumstances do not occur often but may have a significant impact on MODU selection when they do occur.

The above list is not all inclusive but suggests important points in reviewing a MODU for a specific drilling program. Many other factors and items, such as the MODU's maintenance records, age and condition of equipment, type of MODU (some jobs can be done by more than one type of MODU), day rate and contract conditions, mobilization/demobilization costs and distances, and timing and availability of potential MODUs, need to be considered. How to select the right unit is discussed later.

14.4.3 Subsea Equipment, Stationkeeping, and Management. For successful floating MODU operation, proper marine riser and mooring equipment and their management are critical. We have briefly discussed the two types of stationkeeping systems, spread mooring and DP. The vast majority of floating MODUs are equipped with spread-mooring systems. Some have a limited amount of dynamic thruster assist to their spread-mooring system. Almost all of today's semi and drillship MODUs have an eight-point mooring system consisting of anchor chain, wire rope, or a combination. Most of the deeper-water units have a combination of anchor chain on the anchor end and wire rope on the rig end. For a very few ultradeepwater operations, synthetic mooring line is used to increase the strength-to-weight ratio of the mooring line. However, the synthetic mooring line is not carried or deployed by the MODU, which is a distinct disadvantage from an operations standpoint.

The anchor chain used on most MODUs ranges between 2¾ in. and 3½ in., with the predominant sizes being 3 in. and 3¼ in. Anchor chain comes in various grades, the most common being oilrig quality (ORQ), followed by R3S (20% stronger than ORQ) and RQ4 (30% stronger than ORQ). Wire ropes range from 2¾-in. to 3¾-in. OD and may be as long as 15,000 ft, although the average is closer to 6,000 to 9,000 ft per line. The rated break strength of wire rope varies widely, depending on the construction and manufacturer. For example, API EIPS grade 3 in. is rated at 389-tonnes breaking strength, but Bridon Dyformed DB2K 3 in. is rated at 530 tonnes, or 36% more strength. For combination mooring systems, it is important to match the strength ratings of the wire rope and chain. Anchor chain generally performs better in shallow water depths (< 600 ft) because most of the strength is used for restoring force rather than holding up the chain's weight. Wire rope and combination wire-rope/chain systems are best for deep water (> 1,000 ft and > 2,000 ft for combination systems) because the strength-to-weight ratio is higher and more important in deeper water. Quality assurance is a critical issue for mooring lines and related equipment.

Spread-moored MODUs, depending on the metocean, can generally moor in up to ≈ 5,000 ft; however, in benign to mild metocean conditions, some MODUs can meet industry standards to moor in up to 8,000 ft. Increasingly used in ultradeepwater depths is the "taut mooring line" system, which uses synthetic mooring line and spring buoys and is prelaid, as shown in **Fig. 14.23.** The current world record for this type of mooring system was set by the *Deepwater Nautilus* (Fig. 14.13) in the GOM at 8,950 ft in 2004. This type of system is prelaid by anchor handling boats ahead of the arrival of the MODU. The taut-line systems are expensive and time-consuming to handle; however, they extend the mooring capability of some MODUs to deeper water depths and may be very economical compared with a DP unit, especially for very long wells and development projects. Anchors with very high holding power have been developed that range in dead weight from 7,500 to > 15,000 tons. The larger anchors perform best when the mooring line reaches the ocean bottom on or near tangent at full design tension; however, new vertical-load anchors have proved to be successful for special cases. These anchors are difficult to set, take special equipment, and cannot be carried by the MODU; however, they work well with some types of taut-line systems.

Deck machinery to store, deploy, and retrieve the anchors and mooring lines for a deepwater mooring system can be massive, expensive, and heavy. **Fig. 14.24** shows a typical layout

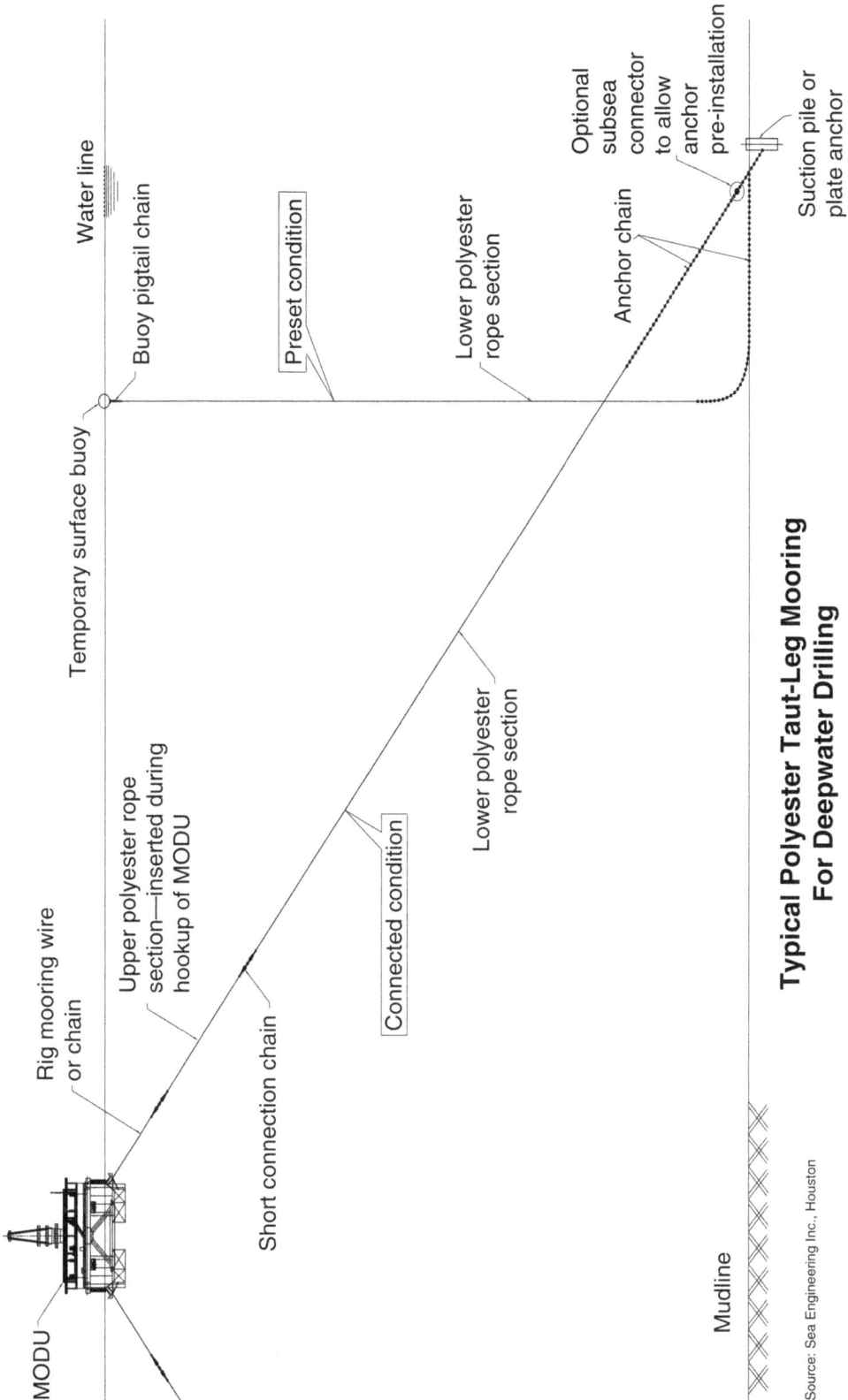

Fig. 14.23—Configuration of a taut, prelaid mooring system for ultradeep water. Shown are a unit that is installed and hooked up to a semi (left) and a unit in the prelaid condition but not hooked up to the semi (right). The world water depth record for subsea completions (7,571 ft) is held by a semi using this type mooring system. Courtesy Sea Engineering.

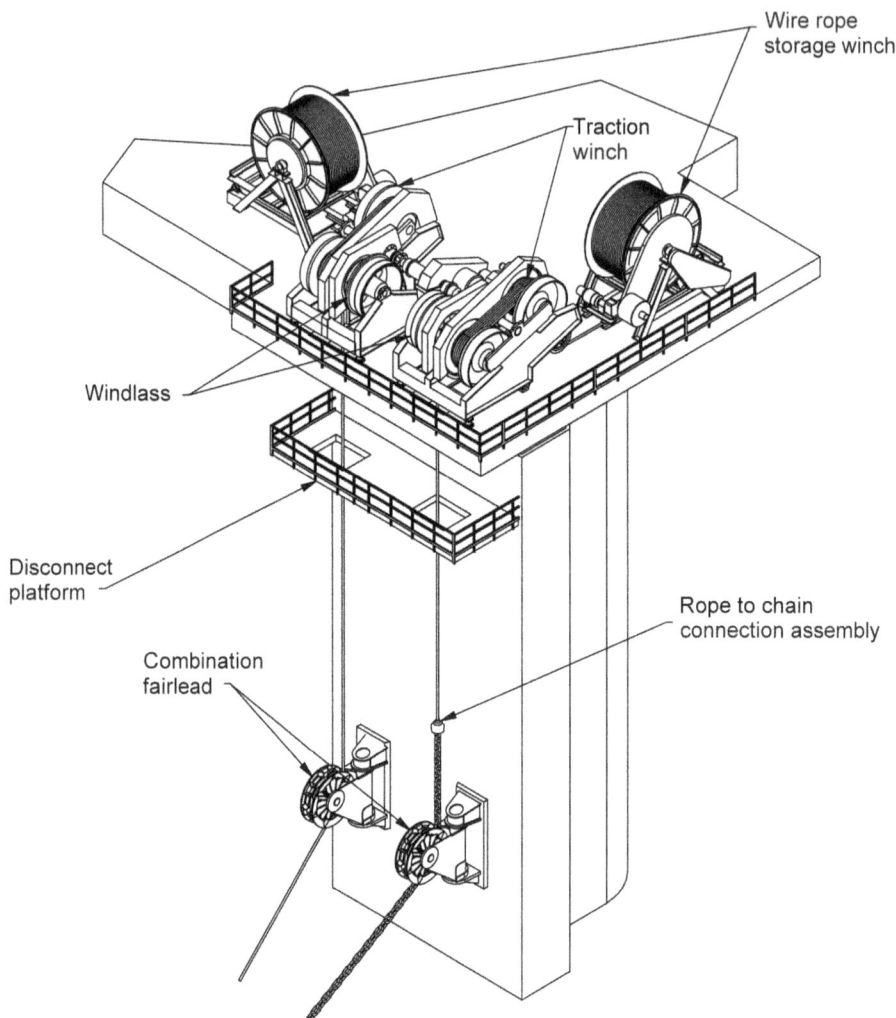

Fig. 14.24—Isometric drawing of a self-contained combination wire-rope and chain mooring system with traction winch and windlasses. Chain and wire rope are connected and disconnected on work platform. Courtesy NOV–AmClyde Products.

on one corner of a semi MODU for a deepwater combination chain/wire-rope mooring system. Chain is stored in chain lockers in the columns below the deck machinery. The chain and wire rope are connected and disconnected for storage at a platform below the deck machinery level. This operation usually takes from 20 to 40 minutes for the latter operation.

Fig. 14.25 shows all the components and their location for subsea equipment, usually defined as anything under the rotary of a floating MODU down to the ocean floor. The subsea BOP stack consists of the lower package (mostly BOPs) and the upper package (lower marine-riser package). The BOP stack is in two parts such that, in an emergency, the marine riser can be disconnected from the lower BOP at the lower marine-riser package. The BOP stack, used primarily for well control, usually consists of a minimum of four ram-type and two annular-type BOPs with three to four sets of double-outlet failsafe close valves. Valves are in sets of two with an inner and outer valve, all failsafe, with the choke side having a minimum of two sets and the kill side having one or two sets. The choke-and-kill pipeline runs are routed up past the flex joint and to the surface by use of lines attached to the marine riser. During well-

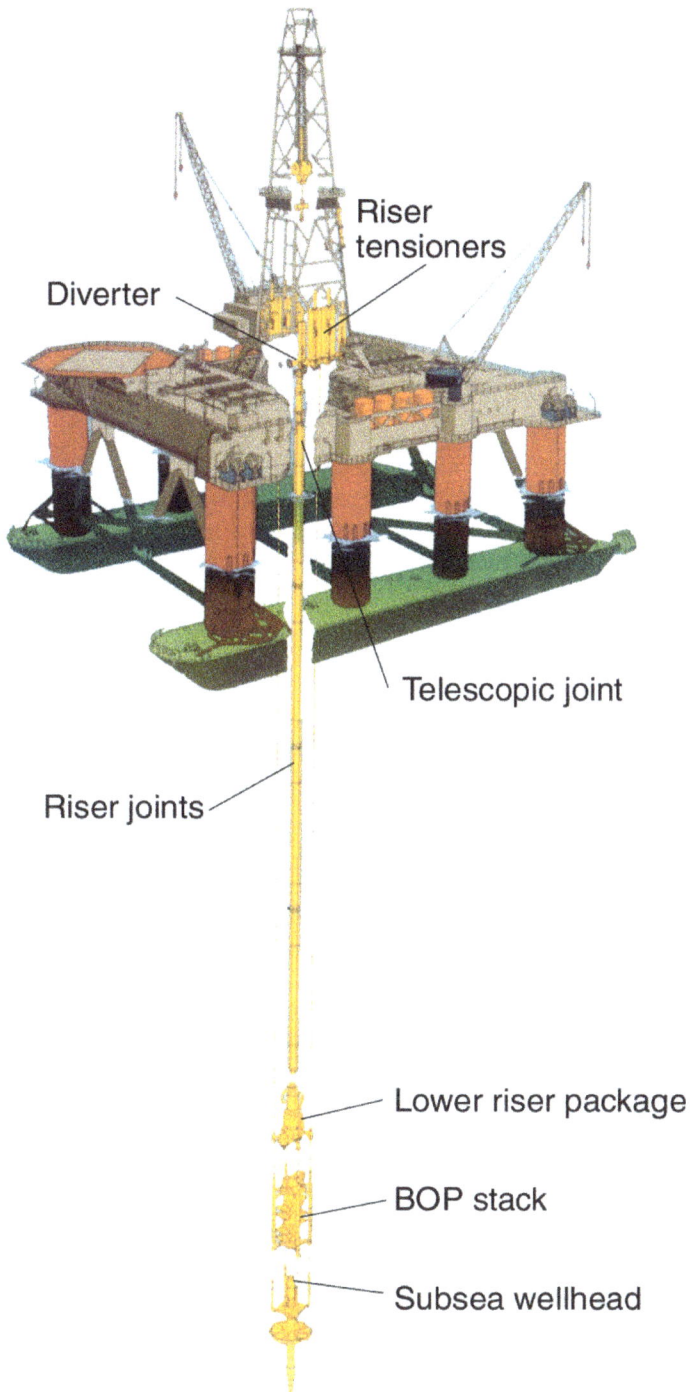

Fig. 14.25—Schematic of a typical modern subsea drilling system from underneath the rotary to the seafloor. Courtesy Vetco Gray.

control operations, the well is circulated down the drillpipe, up the choke line, and through a choke manifold in a controlled manner to pressure balance or "kill" the well.

The marine riser's primary purpose is to guide objects (bits, logging tools, casing, wellhead hangers, and seal assemblies) in and out of the wellbore while also serving as a return conduit for drilling fluids and cuttings. The marine riser also carries auxiliary lines on the outside of the main conduit for the kill (pump down to the well), choke (flow a kick back to the rig), mud-circulating line (help lift drill cuttings up the large-internal-diameter riser tube), and hydraulic conduit (hydraulic power fluid for activating the BOP stack). The flex joint at the top of the BOP stack is a pivot point to reduce stresses in the riser and acts as a hinge point. The slip or telescopic joint allows vertical motion between the floating MODU and BOP stack and marine riser, which are attached to the ocean bottom.

The outer barrel of the slip joint, attached to the BOP stack, is tensioned with strung wire rope by 6 to 16 pneumatic tensioners ranging in capacity from 80,000 to 250,000 lbf each. Riser tensioners are usually pneumatic rod/cylinder assemblies with wire rope attached to the outer barrel (the part attached to the seabed) of the slip joint. Total installed riser tension pull varies with water depth rating for the MODU, but a very-shallow-water unit will have ≈ 640,000-lbf tension and the newer MODUs will have ≈ 1.6 to 2.0 million lbf. A new type of riser-tension system consisting of large, very long hydraulic cylinders (referred to as inline tensioners) attached to the slip joint and substructure has recently been installed on some of the newer floating MODUs with tension capabilities of up to 4.8 million lbf. Total stroke for all riser tensioners usually is 50 ft, but some of the deeper-water units must have more stroke length in case the MODU moves off the well without disconnecting the lower marine-riser package from the BOP stack.

Atop the inner barrel of the slip joint, which is attached to the rig's substructure, is the diverter assembly. The diverter assembly is used to divert fluids, usually gas, that the marine riser may have in it. The diverter assembly has a low-pressure (500-psi WP) packer that may close around the drillstring and divert fluid horizontally by use of diverter lines. Diverter lines (12- to 16-in. outer diameter) are used to route well fluid away from the rig and overboard in the unlikely event that unwanted fluids should come to the surface. More detailed information is given about subsea equipment in Ref. 8.

To maintain constant weight on bit for a floating MODU, drillstring motion compensation (DSC) is required. Thus, the industry has developed inline (travels with the traveling block) and crown-block (located on top of the derrick and part of the crown assembly) motion-compensation equipment. Most drillstring motion compensators are inline and passive (the drill-string motion compensators react to MODU motion rather than sensing it, as does an active system). Drillstring motion compensator's stroke is usually 15 to 25 ft, with an average of 18 ft; however, most floating units will not operate the drillstring motion compensators with > 10 to 12 ft of heave. Active systems usually involve the drawworks motors that dissipate the energy though the rig's power-plant generators. This is one reason why DP drillships with a large power-plant system use active heave-compensation systems.

The BOP control system is critical and probably the most difficult in which to maintain the high degree of reliability required for safe offshore operations. Most floating MODUs use all hydraulic systems by use of pilot valves in a "pod" on the subsea BOP stack (Fig. 14.25) shifted by pilot lines from the surface. The power fluid is usually sent down a hydraulic conduit on the marine riser. Some deeper-water units (> 5,000 ft) use a multiplex electrically coded system as the signal medium for shifting the pilot valves in the pods. Industry standards require subsea rams to close in 45 seconds and the annulars in 60 seconds; thus, signal time is critical and very time dependent. Subsea BOP stacks differ from land BOP stacks in that they stay assembled, have remote stabbing capabilities, have hydraulic wellhead and riser connectors, have mechanical riser connectors, have BOPs and valves that are hydraulically actuated, have guidance systems, and are controlled remotely per the above description.

Operation Through BOP Stack	30-in. Conductor Angle (°)	Slip-Joint Angle (°)	Lower Flex Joint Differential Angle— Goal (°)	Lower Flex Joint Differential Angle— Maximum (°)	Hole Position (% WD)	Comment
Rotating drillpipe	<0.75	<2.0	<0.5	<1.0	±2.0	Critical wear angles.
Tripping drillpipe	<0.75	<3.0	<0.5	<3.0	±2.5	
Tripping BHA	<0.75	<2.0	<0.5	<2.0	±2.0	Tungsten carbide hard band.
Tripping completion string	<0.75	<3.0	<0.5	<3.0	±2.5	
Stationary tubulars	<0.75	<3.0	<0.5	<3.0	±2.0	
Running casing	<0.75	<3.0	<0.5	<2.0	±2.0	Drift may be a problem.
Run/pull full bore	<0.75	<2.0	<0.5	<1.75	±2.0	Hangers, bore protectors, etc.
Critical operations	<0.75	<3.0	<0.5	<3.0	±2.0	Well testing, cementing.
Miscellaneous	<0.75	<3.0	<0.5	<3.0	±2.5	BOP testing, wireline, etc.

TABLE 14.5—MARINE RISER AND MOORING SYSTEMS MANAGEMENT

1. Riser and mooring systems management is the balance of rig position and riser system angles to minimize riser component wear and to reduce the possibility of severe damage and/or failure of the equipment that could result in a rig- and personnel-threatening condition.
2. To reduce lower flex joint differential angle, the leeward 2 mooring lines should be slackened per the company mooring manual procedure. Never tension into the weather. Plan ahead with weather forecasts.
3. Differential flex joint angle is the angle between the BOP/Lower Marine Riser Package (LMRP) angle with vertical and riser-joint angle with vertical located above the flex joint. Maximum angle should be used only in an unusual situation, not routinely.
4. The acoustic riser angle riser differential angle alarm should be set at 1° above the goal or at the maximum angle toward which the rig is working.
5. The subsea television should be left at the LMRP location looking at the riser angle bull's-eye for immediate verification if the acoustic alarm sounds.
6. A 2° (not 5°) bull's-eye bubble should be used to obtain the necessary accuracy and definition.

The key to successful floating MODU operations is managing the marine-riser and mooring system together and in harmony. As stated, the mooring system objective is to restore the floating MODU within specified limits over the wellbore through varying degrees of environmental conditions and rig operations. Hole position or vessel offset from the wellbore is usually monitored with acoustic hole position indicators that work in percentage of water depth from the wellbore. Riser angle at the flex joint located on the LMRP is also measured acoustically. **Table 14.5** is an example set of criteria for allowable differential riser angle (difference between the BOP and riser angle at the flex joint, not with vertical) and hole position, depending on the rig operation being conducted. The primary purposes of these guidelines are to achieve riser angles so that tools can be run/pulled through the BOP stack and flex joint without hanging up or creating damage, to prevent damage to the subsea equipment because of drillstring key seating, and to ensure adequate structural integrity of the marine-riser system.

Recently, a new form of floating drilling has been developed in which the BOPs are located in the cellar deck rather than on the ocean bottom. With standard floating drilling, it is anticipated that if the MODU has a mooring failure, loses its station over the wellbore because of environmental conditions, or experiences a riser failure or any other mishap, the subsea BOPs can secure the well. With the surface BOP approach (**Fig. 14.26**), the loss of hole position by the MODU or a failed riser means that the well will probably be lost. The concept is that the riser is high pressure (usually 13⅜- or 16-in. casing), the metocean is very benign, and the well pressure is normal gradient, so seawater head will kill the well in the event of a riser failure. It has been very economically successful in the Far East and has cut well costs by as much as 70%; however, the risk of losing the well and/or having a blowout has deterred many operators from using the approach. One mitigating approach is to put a complete shutoff device at the ocean floor (usually at least one shear ram with hydraulic connectors top and bottom); however, this approach increases the expense and time to the point of losing all savings. However, in ultra deepwater where the well is circulated up small-ID kill and choke lines, causing significant backpressure on the formation, the surface BOP with the large high-pressure casing and BOPs at the surface eliminates the problem. In other words, there are pros and cons for every approach.[9]

Another approach similar to surface BOP is the "slim riser" approach (**Fig. 14.27**). The standard subsea system is built around an 18¾-in.-ID BOP stack and wellhead system that ordinarily uses a 21-in.-OD riser. The standard system has the capability to run up to nine casing strings by means of hangers and liners under certain conditions. In deep water where the margins between formation fracture gradient and hydrostatic head of the drilling mud to maintain well control is very close, many casing strings are often required. The GOM has this requirement, often resulting in very expensive wells costing U.S. $50 million and sometimes more than $100 million. If a more standard deepwater well is to be drilled with only two to three casing strings through the BOP stack, a 16-in.-OD riser may be used. This results in far lower mud volume requirements because of a smaller drilled hole and smaller riser ID, which in turn requires less marine-riser tension, less deck space, and thus less VDL. Most importantly, these reduced quantities allow a third- or fourth-generation MODU to be used at reduced day rate rather than a fifth-generation unit.[10] A capable third or fourth generation semi rated for 5,000 ft water depth can be increased to 7,500 ft or over.

Although not discussed in detail in this chapter, well control in deep water is much more difficult than off a jackup MODU or a land rig. With the margin of safety between the fracture gradient and mud hydrostatic pressure smaller, the shut-in point (subsea BOP) being much closer to the influx formation, the detection point still at the rotary, and long runs of kill and choke lines on the marine risers with small IDs (usually minimum of 3 in., with most being 3½ to 5 in.), detection and proper circulation is delicate and takes training, concentration, and patience. To date, the industry has an excellent deepwater well-control record.

14.4.4 Well Intervention and Remotely Operated Vehicles. In the 1980s, divers jumped in and out of saturated and pressurized systems to do almost all well and subsea equipment intervention, inspection, and repair. If the divers could not complete the repair task and/or inspection, the BOP stack or other items had to be pulled out of the water for repair. Even with the most sophisticated equipment, divers had limited capabilities because of water depth, visibility, currents, temperatures, bottom downtime, and sometimes questionable safety standards. Subsea television systems were and still are used to inspect and monitor hulls and subsea equipment by use of running down guidelines, but they can only view, not do repairs or other physical tasks.

Starting in the 1990s, coinciding with the increase in subsea completions, well intervention with highly capable remotely operated vehicles (ROVs) has developed into a common third-party addition to a floating MODU. Modern ROVs have the ability to "fly" by means of an umbilical that is attached to the transport cage (garage). Once the ROV leaves its cage, it may traverse for approximately 100 ft. The operator, or pilot, controls the flight pattern and position of the ROV so that it will not become entangled in its own umbilical or other items. Most

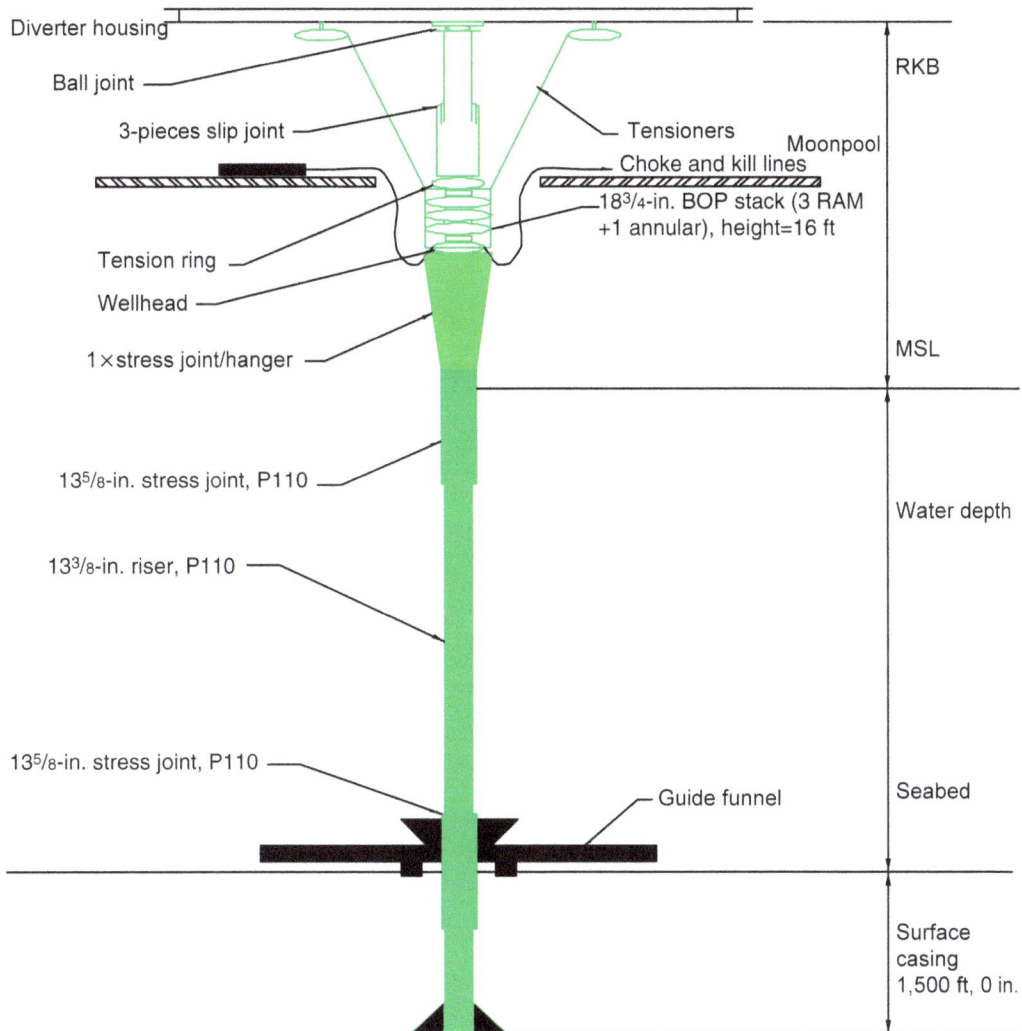

Fig. 14.26—Schematic of basic configuration and equipment for a floating surface BOP system. Courtesy Atwood Oceanics.

ROVs have visual and recording capabilities in addition to manipulator arms with various degrees of strength, feedback, and lifting capability. ROV technology has far exceeded water depth ratings of MODUs; thus, capabilities and reliability of these units have improved considerably. Changing of wellhead sealing ring gaskets, control of some functions on the BOP stack in an emergency, retrieval/installation of items on the wellhead or production hardware, and inspection are common tasks, in addition to inspections with the subsea television system. With the increase in the use of subsea completions to develop whole fields, ROVs have become an integral part of deepwater development. With subsea development, MODUs do the drilling and most of the completion, including setting trees, flying leads, jumper hoses and pipelines, umbilicals, production risers, production skids, and templates, all requiring ROV intervention. When wells need to be worked over, ROVs are required and are usually launched off MODUs or intervention vessels working in conjunction with a MODU.

As ROVs have become more important in floating MODU operations, the size and space requirements have increased dramatically. For intervention and completions, it is not uncom-

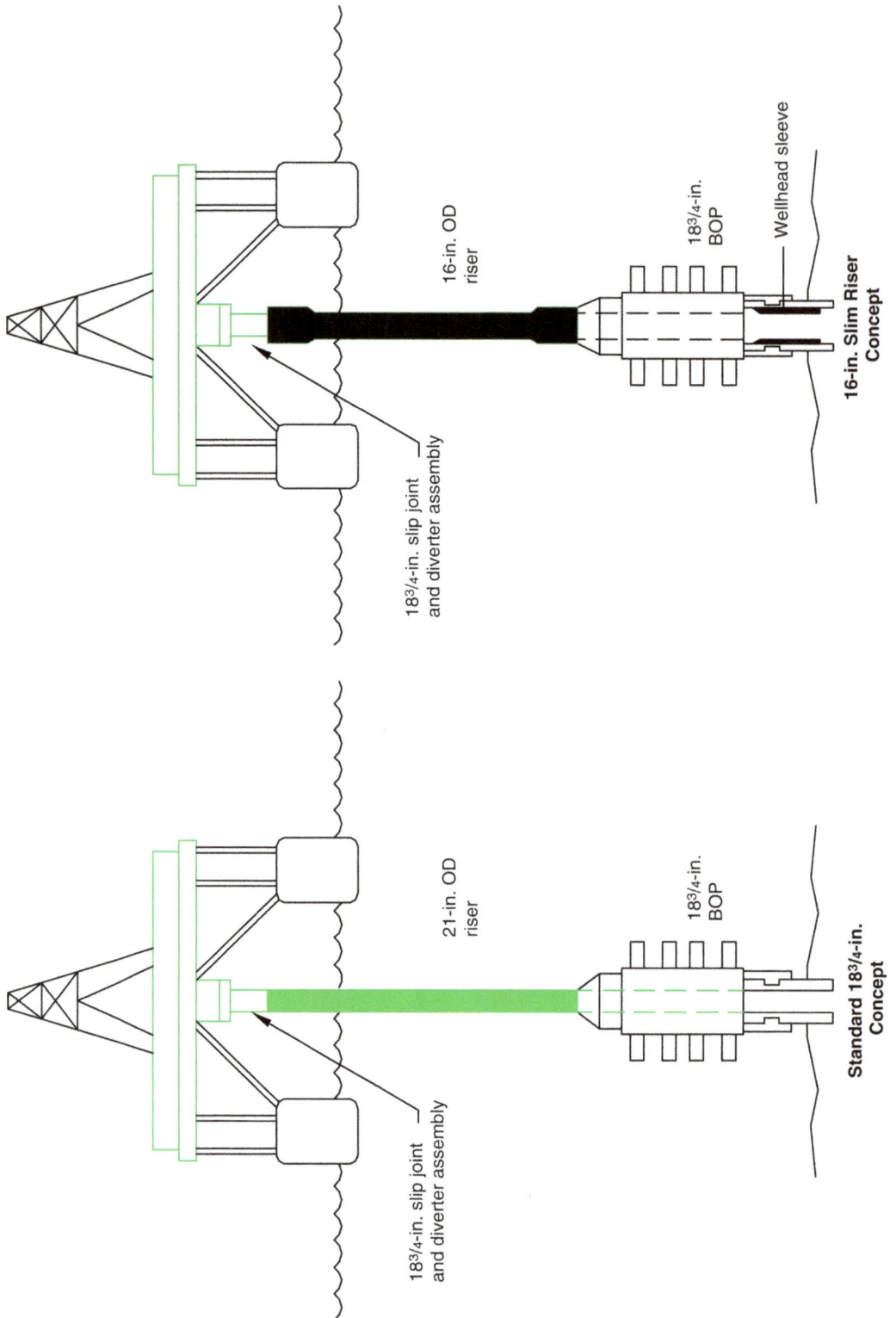

Fig. 14.27—Schematic of basic configuration and equipment for the slim-riser concept (right) and standard subsea drilling system (left). Courtesy Atwood Oceanics.

mon to have two ROV systems, requiring the storage and operating porch to be used as a work platform, structural reinforcement for the deployment winch, fendering to prevent the ROV from hitting the MODU columns/lower hull, electrical power to support the unit (can be > 200 to 300 KVA), VDL, and deck space. This can amount to a considerable support system that the MODU must accommodate, so planning ahead is important. Not every MODU can accommodate the larger ROV systems from a weight (some times over 40 tons), space (2,500 square ft or more), or power standpoint.

14.4.5 Rig Crews and Management. The importance of well-trained, motivated, skilled, safety-oriented personnel with a teamwork attitude to crew and operate offshore drilling units cannot be overstressed. No matter how well-engineered, well-equipped, and well-maintained a MODU is, it will not perform any better than the crew who manage, operate, and maintain it. The fact that the crew and management system are often the real determining factor concerning MODU performance and safety is often overlooked during the flurry of cost analysis, equipment evaluation, operating expenses assessment, and number crunching during bid analysis and MODU selection. It is often said that the low bid does not always give the best performance. A complete "hard" (equipment) and "soft" (crew, management, and safety) analysis must be done to make the best decision.

Over the last 10 to 20 years, almost every offshore drilling contactor and operator has developed very comprehensive management systems to guide and operate their companies. Management systems will normally include a mission and goal statement, a top-tier-quality control manual, and various second-tier standards and procedures manuals addressing such business functions as document control, department descriptions and responsibilities, job descriptions, bridging documents, safety and security, internal audit, contract review, purchasing, inventory control, and human resources. These policies and procedures should be reinforced from the chief executive officer to the roustabout on the rig to have a successful and well-performing organization and rig operation.

The staffing and organization of a MODU vary with each drilling contractor, operator, and country and are controlled eventually by classification and registration requirements. The most senior person on the MODU is usually the OIM who is by law the "master" or "captain" of the vessel. The OIM is responsible for all departments, including drilling, maintenance, marine, auxiliary services, and safety. The OIM works for the drilling contractor and interfaces and coordinates with the operator's (leaseholder's) representative. **Table 14.6** shows a typical MODU personnel complement for a jackup. The drilling contractor may employ the catering complement wholly or partially.

Employment contractors used by the drilling contractor are not uncommon in overseas operations. These contractors usually supply positions only from floor man down, but there are exceptions. The shore-based operation usually includes an operations manager, drilling superintendent, administrative manager, materialsman, and a secretary. Often car drivers, local agents, warehouse men, and administrative staff are included for overseas operations. If the financial and accounting functions are done on site, additional personnel may be required. With the advent of satellite communication on the MODU and local office, communication problems and time delays have been significantly reduced, resulting in a much smoother and more trouble-free operation. Procurement, inventory, and maintenance can all be monitored, directed, and recorded with ease and in a timely manner. In the early offshore days, MODUs operated like little independent companies, including their own personnel hiring/firing, procurement, accounting, materials and inventory, housing, and so on; however, with modern transportation and communications, local operations have been reduced in favor of centralized procurement, employment, accounting, and financial functions. Tax issues can be very tricky when moving from country to country; thus, outside major accounting firms are needed to interpret local laws so as to comply but not waste potentially huge sums of money.

TABLE 14.6—TYPICAL JACKUP MODU CREW COMPLEMENT AND JOB RESPONSIBILITIES

Job Classification	On Rig	Total On/Off	Job Description
Offshore Installation Manager (OIM)	1	2	In charge of all activities and legally responsible for the MODU.
Senior/Day Toolpusher	1	2	In charge of drilling activities and directing personnel that support the "hole making" activities on the MODU
Night Toolpusher	1	2	Similar duties to Day Toolpusher but works at night and usually subordinate to the Day Toolpusher
Driller	2	4	In charge of the drillfloor, drill crews, well progress and reports to the Toolpushers.
Assistant Drillers	2	4	Assists Driller, in charge of drillfloor when Driller is not present. Assists Driller in his duties.
Derrickman	2	4	On trips in and out of the well racks pipe in the derrick at the monkey board level and also assists with mud solids equipment and monitoring mud condition.
Floorman	8	16	Supervised by the Driller and works primarily on the drill floor, substructure and with drilling tools.
Crane Operator	2	4	Operates the rig's cranes and supervises the Roustabout crews.
Roustabout	4	8	Performs manual labor such as painting, unloading boats, carrying supplies to store rooms and other manual labor under the direction of the Crane Operator.
Mechanic	2	4	In charge of all rig mechanical equipment but particularly engines.
Assistant Mechanic	1	2	Splits tours between two hitches of Mechanics. Aids in Mechanic's work and is in training.
Motormen	2	4	Primary duty is to monitor and attend the engine room. Reports to the Mechanic.
Electrician/Electronics Technician	2	4	In charge of all rig electrical and electronic equipment and their maintenance.
Assistant Electrician	1	2	Splits tours between two hitches of Electricians. Aids in Electrician's work and is in training. Reports to lead Electrician.
Welder	1	2	Welds plate and pipe as necessary for drilling contractor and operator.
Materialsmen	2	4	Handles materials, data entry for maintenance, purchasing, inventory, etc.
Communications Operator	1	2	In charge of communications.
Barge Engineer	1	2	In charge of the marine equipment and its operation. In charge during rig moves and jacking. Generally the maintenance crews and specialists report to the Barge Engineer.
Catering/Camp Boss	1	2	In charge of hotel function on rig such as food, laundry, etc.
Cooks	2	4	Prepares food.
Galley Hands	10	20	Helps with food, cleans rooms, laundry, etc., under the Camp Boss
Total	**49**	**98**	Totals for basic complement

1. Crews work 12 hour shifts and usually change at 6:00 AM/PM
2. Individuals with no onboard relief are on call 24 hours per day but usually only work 12 hours per day.

The operator will have additional personnel on the MODU, such as radio operators, two or more drilling superintendents or foremen, drilling engineers, a geologist, and possibly an administrator. The operator from time to time will also have third-party service companies on board to perform and/or run mud logging, cementing, casing running, electric logging, measurement-while-drilling (MWD) and logging while drilling (LWD) drilling tools, completion and drilling tools, fishing, special downhole tools, wellheads, etc. Where rigs had accommodations of 60 to 80 personnel, including all support activities, it is now not uncommon to have a requirement for well over 100 personnel.

14.4.6 HSE&S. With operations often classified as high risk from a financial and physical standpoint and costs often in excess of a quarter of a million dollars per day, capable personnel and a defined management structure are essential. Running a drilling operation in the oil and gas business requires unique knowledge and the ability to adjust to new problems and challenges every day. It is definitely not like manufacturing widgets day in and day out.

Personal safety and health has increasingly become more of a factor and focus in offshore operations over the years. Safety statistics show that LTIs, recordable incidents, near-miss incidents, and medical treatments statistics have improved significantly over the last 10 to 15 years. Whereas the LTI rate (incidents per 200,000 hours) was commonly > 10, it is now common to be < 1 and often < 0.5. Safety offshore is no longer given mere lip service. IADC publishes statistics monthly by participating members, and the MSS gives out coveted awards in the Gulf Coast each year. From both a humanitarian and a financial standpoint, all feel that making safety a priority is the right thing to do.

All operators and drilling contractors have extensive safety programs, with the DuPont STOP program or some modification of it being the most common element. The STOP program emphasizes "observance" by everyone on the unit of the actions of each crewperson and the conditions of the surroundings. STOP cards can be written by anyone on board about anyone else, from the roustabout to the OIM, and then discussed during safety meetings. A Job Safety Analysis (JSA) is another significant program in which detailed procedures are written up for every major job and task, discussed before the job is performed, and then implemented during the job performance. The requirement to have the proper Personal Protective Equipment (PPE), such as hard hats, gloves, safety glasses with side shields, proper shirts and pants, and protective gloves, helps to provide a safe atmosphere. Drills for man overboard, firefighting, helicopter landing and takeoff, lifeboat use, first aid, entry into non-ventilated tanks, etc., contribute to improving safety in the workplace. Off and on the rig, training schools for crane operation, well control, firefighting, helicopter crash survival, team building, leadership, and other skills result in an enlightened operation and better safety and performance.

Before a crewman can be hired to go offshore, an extensive physical, including drug screening, is usually given. For newcomers, there are roustabout and roughneck schools, such as those given on the *Mr. Charlie,* now a museum and training platform in Morgan City, Louisiana. Intoxicating beverages, firearms, weapons, and illegal drugs are strictly prohibited offshore and, if discovered, usually mean instant dismissal and transport to shore for the offender. Almost every rig has a paramedic as part of the crew, with access to doctors and medical help instantly through satellite and/or other communication medium. Tens of millions of dollars and an extensive amount of time and effort continue to be spent by all trying to run a safe operation offshore, and statistics show that the industry has shown considerable improvement.

Environmental and antipollution policies and efforts have increased steadily over the last 30 to 35 years. The U.S. federal and state governments have extremely strict laws and procedures for before, during, and after leases are put up for sale, drilled, produced, and abandoned. The fear of pollution, or the potential for a spill, is so great that some areas, such as the east and west coasts of the U.S.A., have seen no drilling for years. Most of Florida is off limits, even though limited drilling has shown potential for gas. Through the International Maritime Organization (IMO), every rig has an international oil pollution plan that details the procedure to follow in the event of a spill, even a very small one. In the United States, even a very small fuel oil spill must be reported to the U.S. Coast Guard immediately. Fines of U.S. $10,000 or more can be imposed for each incident. Discharge of any toxic or potentially polluting fluid or solids overboard is strictly prohibited. Solid food waste must be ground into mulch < 1 in.[3] before discharge. Sewage waste must be treated before discharge overboard. Drill cuttings in some areas cannot be discharged overboard and must be transported to shore for disposal and/ or injected into an approved reservoir offshore, usually down a casing annulus. Some areas offshore in the GOM do not allow mooring of vessels or discharge of cuttings because of sensitive coral reefs (possibly thousands of feet underwater), tubeworms, and other protected entities. The most feared environmental event from a MODU is a blowout of crude oil. Well-control equipment capabilities, procedures, and training have improved steadily over the years to a level where a blowout of any significance is extremely rare. The industry spends billions of dollars on antipollution and environmental safeguards every year in an effort to comply with laws and the public's desire for pollution-free operations and to be just a good citizen.

A new subject in the area of personnel and equipment security has appeared in the late 1990s. The Middle East, West Africa, and radical religious sects and areas around the world have required operators and drilling contractors to take security steps not envisioned just 10 years ago. Because overseas operations usually involve air flights for personnel sometimes into hostile countries, use of security consultants and constant contact with local governments and intelligence agencies are now common. In highly sensitive areas, crews do not ride in buses to lessen the risk and to disperse the target; crew boats are searched even underwater for explosives; security personnel are stationed on board with minute-by-minute communications with army and air force support; and hotels are carefully picked for ease of escape. Contingency plans are drawn up for every conceivable event; local personnel have 24-hour evacuation plans and stay packed for quick exit; a low profile is emphasized, with advice given to stay out of native crowds; hired drivers are used to drive evasively and to lessen the risk of an expatriate getting in an accident; and in one case a 3-mile "no entrance zone" by sea or air was placed around a MODU. Security not even conceived of 10 years ago is now front and center and is a large part of offshore operations in many parts of the world.

HSE&S has become as important in offshore operations as drilling the well. Drilling contractors are taken off operator bid lists if they do not have and do not demonstrate a sound, statistically proven system for human, equipment, well, pollution, security, and operational well-being.

14.5 Classification, Registration, and Regulations

Almost every vessel, barge, or floating object, including MODUs, must have classification and registration certificates of compliance to the rules and regulations as dictated and published by the classification society and country of registration. Most insurance underwriters require classification for the vessel to qualify for marine insurance. If the vessel is not fully classified, underwriting insurance companies will not insure the property, thus leaving the owner and his financial institution "self-insured." The vessel owner may consider the risk of a financial loss resulting from self-insurance, but his bank will not. Most operators will also require a drilling contractor to have classification on the MODU to show the unit's condition and seaworthiness. In other words, MODUs must be fully "in" classification to obtain full insurance coverage, to obtain bank loans, and to comply with the operator's contract requirements.

Classification societies are usually privately owned for-profit companies that work closely with, though fully independently of, government bodies. There are twelve societies, all belonging to the International Association of Classification Societies (IACS). The primary societies are ABS (American), Det Norske Veritas (DNV, Norwegian), and Lloyd's Register of Shipping (Lloyd's Registry, English). Other members are located in France, China, Italy, Germany, Korea, Japan, Russia, Croatia, and India. It is very rare to see a MODU that is not classified by one of the three primary societies, with ABS having most of the units. Classification as an indication of seaworthiness and vessel condition was started in the late 1600s in England. ABS origin has been traced back to 1862. The first rules and regulations for MODUs appeared in 1968 and were written by ABS.

When a MODU applies for classification, usually during initial construction, it is a costly and rigorous exercise requiring months of effort by the owner, the design team or engineering company, and the classification society. The process consists of a "design review" and an on-site inspection to verify that the design is built as engineered and according to the society's published rules. Once classified, the unit will have periodic inspections that the owner and operator must plan for and schedule so that the MODU does not fall out of classification or interfere with the operator's drilling and/or well-completion program. There are "annuals" (once-per-year "walk around" unless a problem is found), yearly surveys, 2½-year surveys often called underwater inspection in lieu of drydocking (UWILD), and a "special survey" (every 5 years and usually requiring a drydock). These surveys include inspections of the steel structure and hull condition, piping, firefighting, safety at sea, corrosion protection, power and electric equipment and wiring, communication on and from the MODU, detection systems for fire and gas, crew level, mooring equipment, stability, and operating manual and emergency procedures.

In the 1940s and 1950s, it became apparent that by working together and developing common rules and regulations, the shipping industry could become safer, operate with higher principles, become more efficient, and exercise better pollution control. Thus, the Intl. Maritime Organization (IMO), an industry group that is not part of classification or registry, was assembled. Under the IMO, several regulations, guidelines, and rules have been developed and adopted by a number of countries that have become part of the requirements for vessel and MODU registration. Included under the IMO umbrella is the Safety of Life at Sea (SOLAS), which deals primarily with safety issues and communications; MODU code, which deals primarily with construction and equipment; Maritime Pollution (MARPOL), which deals with pollution control and prevention; and International Safety Management, which focuses on safety for self-propelled vessels and MODUs. Member nations of the IMO adopt the codes and enforce them through classification societies' efforts and fees charged the drilling contractor. Registration requirements often include the IMO codes.

Registration concerns the country of home port for the unit. Each country of registry has rules and regulations centered mainly on safety, communication, lifting and cargo gear, pollution, and pollution containment. Each registry has different rules and regulations, and the registry often has a working agreement with the classification society for the unit to inspect and certify on its behalf. The most popular registries are Panama, Liberia, and the Marshall Islands. The United States, England, Norway, and other industrial countries are not common registries because of their more complicated rules, regulations, and staffing requirements. Most MODU registries also fly the flag of the registry country, which controls the crewing and staffing levels and designation of crew skills. Although not common, there are some "dual registries" in which hardware and safety issues are handled by one registry and crew and staffing by another.

When a MODU has a classification and registration, it must comply with the rules and regulations of the country in which it operates. For example, a MODU may have an ABS Classification, Marshall Islands Registration for all the equipment, and registry for crewing in Germany, and it enters the United States to drill a well in the GOM. It now must comply with the U.S.A. regulations as enforced and surveyed by the U.S. Coast Guard. In addition, the operator must obtain permits to drill from the Mineral Management Service (MMS), which also inspects the MODU for MMS rules and regulations compliance. The well to be drilled and MODU must meet the MMS requirements concerning equipment, procedures, and crew training. Then, the Environmental Protection Agency (EPA), Occupational Safety and Health Agency (OSHA), and a few other agencies may enter the picture. The operator and contractor usually work together to comply with and follow the necessary rules and regulations.

Some countries, such as the United Kingdom (England), Norway, and Australia, require a "safety case" for the MODU to operate within their waters. Safety cases usually are expanded documentation, equipment and systems, and training centered on classification and registration rules but dovetailed into those particular countries' laws. Developing a safety case requires considerable time and money and should be anticipated and planned for in detail far ahead of its implementation on the MODU. It is not unusual to take 6 to 12 months or more and over half a million dollars to fully develop a documented safety case for a single MODU. The safety case is required to obtain the necessary country's approval for the MODU to drill in its waters. Fortunately, consulting companies that specialize in the development of safety cases are available.

Industry organizations, such as the American Petroleum Inst. and International Standards Organization (ISO), also have a major influence on the upstream oil and gas industry. These organizations write specifications and recommended practices (RPs) for the industry to follow. These documents usually deal with equipment, procedures, and operating systems. The documents are usually written by the industry for industry use and are widely quoted by the societies, registries, operators, and drilling contractors.

In summary, 30 years ago in the infancy of the offshore oil and gas business, none of the above was required; however, after a number of incidents and tragedies, insurance underwriters, operators, drilling contractors, and governmental bodies have developed a fairly tight system to ensure better safety and environmentally friendly systems for the benefit and health of all.

14.6 Relationship Between the Drilling Contractor and Operator

There are three separate and distinctive entities on an offshore MODU: the drilling contractor who owns and operates the MODU, the operator who contracts the drilling contractor's MODU to perform a service, and third-party contractors who work for the drilling contractor and/or the operator. In the 1950s, the relationship between these three parties was more clouded; some operators owned and operated the MODU. However, over the last 50 years, the relationship between the three entities has become standard and well defined.

There are exceptions, but the operator generally contracts the drilling contractor's MODU for a specific well or wells or a "term" contract of months or even years. The length of the contract usually is determined by the number of wells the operator wants to drill. He must decide whether a term contract or a contract for a specific number of wells is best for his program. The use of a long-term contract is usually driven by market conditions, with a tight rig market usually resulting in term contracts. This is especially true if a new rig build is involved; the drilling contractor's financial institutions may require a reasonable payback on the loan before the contractor can sign a contract and build the unit.

MODU capability, availability, mobilization, market conditions, safety, and operating performance enter the minds of the drilling contractor and operator when a potential MODU contract is at hand, but economics is generally the primary driving force. The operator will seek bids from a number of drilling contractors capable of performing the work. The operator will usually specify the type of equipment and drilling capability desired, such as the size of mud pumps, mud pit volume, hoisting load, drillpipe size and length, water depth capability, VDL, pressure rating of equipment, and size of well-control equipment. The operator will also ask for specifics about the drilling contractor's HSE&S program and request statistics showing past performance and copies of specific policies and procedures that the drilling contractor has in place. The operator will also have a preferred drilling contract. When it is a buyers' market for rigs, the operator will have a strong position to use his formulated contract with few negotiated changes; however, in a sellers' market, the drilling contractor will try to use his formulated contract, which of course favors the contractor's position. The contract, including the IADC-suggested offshore contract model, usually contains equipment and capabilities exhibit, liability clauses, payment terms, crew complement, description of work to be performed, a "menu" of who will pay for what services and materials, termination provisions, day rates and other charge items, terms for settlement of contract disputes, and numerous exhibits on customs, confidentiality, items required by law (e.g., equal opportunity), policies of both companies, etc. Sometimes a "bridging document" is required between the operator's and drilling contractor's policy and procedures manuals to eliminate confusion if the two do not agree on every item. Bridging documents are very important from a practical and legal standpoint.

A characteristic of the upstream oil and gas industry is the strong and unique cultures developed by operators and especially contract drilling companies. Although less so these days than in years gone by, egos and individualism often enter into the relationship between operators and drilling contractors, especially at the higher management levels. Some operators are very conservative and are willing to pay more for less trouble and rig downtime, greater safety, and higher-end rig capabilities. Of course, conservative drilling contractors usually work best with conservative operators. On the other hand, some operators are more freewheeling, "cut closer to the bone" so to speak, and work, for example, on a front-end cost basis, and they work best with drilling contractors who work the same way.

Generally, the most productive operations are done with good, workable, and cooperative arrangements laced with goodwill between the operator and drilling contractor. Driving the hardest bargain possible to the point of picking at every contract clause, every possible charge-back item, strongest possible indemnities, mobilization items, etc., usually results in hard feelings and a less productive operation. In other words, the relationship is very adversarial and convoluted, resulting in a difficult working relationship. Another common problem occurs when an operator decides to coordinate the drilling contractor's equipment and personnel down through the driller's position rather than communicate through the OIM. Initiative, cooperation, and a sense of responsibility and ownership by the drilling contractor personnel suffer, to the detriment of the whole operation.

If the operator has a defined drilling program for a long period (e.g., a 2- or 3-year period), the operator will generally obtain a "fit-for-purpose" MODU at a competitive price that molds itself into the operator's culture and routine during the term of the contract. This usually results in higher efficiency, a safer operation, less trouble time, less downtime for the operator and the drilling contractor, a more team-oriented effort between the parties, and overall a more cost-effective, trouble-free operation—a truly "win/win" situation. Unfortunately, not all operators can put together a drilling program of this duration, eliminating the potential for this type of relationship to develop.

14.7 Picking the Right Unit for the Job
With all the above said, one may ask, "How do I pick the right drilling rig for the job?" The answer is that often there is more than one rig type that technically can do the job. A review of previous sections of this chapter will show many items that must be considered. Following is a summary centered on the technical side of the evaluation. As stated, commercial, HSE&S, and other items need to be factored into the overall decision:

• First and foremost, and as simple as it may sound, the operator must take the time and effort to be knowledgeable about MODUs, drilling contractors, the equipment involved, and the relationship between all the parties (operator, drilling contractor, and third parties). Surprisingly, this does not always occur.

• The operator also should be aware of and obtain all the permits, and be aware of and set up logistics for boats, helicopters, ground transportation, housing, automobiles, agents, warehouses, office space, communications, contracts with third parties (e.g., bulk mud, casing, cementing services, and logging), security, drill and potable water, fuel, local supplies, and all related items.

• The operator must also be aware of any unusual requirements to drill the well. Possibilities include shipping lanes and fairways, pipelines, unusual soil conditions, strong currents and/or large ranges of tides, strict drill-cutting discharge requirements, local government requirements for use of native labor and/or professionals, and restrictions on the use of harbors and air space (military explosive dumping area or non-flyover zone). Special requirements may have a major impact on the well plan.

• With the last three points addressed, the operator must take time to engineer the exact type of performance he requires of the drilling equipment before deciding on the type of drilling rig. Sometimes the drilling rig type is obvious, such as an ultradeepwater rig; however, most of the time it is not. A checklist should include hoisting load and speed, mud volume, bulk volume (barite, bentonite, and cement), sack storage, VDL, drill-water capacity, feedstock capacities for synthetic or oil-based mud and completion fluid, metocean conditions in relationship to MODU capabilities, deck space, well-testing requirements if applicable (space, deluge for seawater and piping), and many of the items listed in the sections on equipment, outfitting, and capabilities.[11] Unfortunately, operators sometime specify a MODU and drilling equipment with not enough capabilities to drill the well with the hope that they will obtain an inexpensive unit. More often, operators specify a unit with complete overkill, eliminating very capable units that could do the job quite nicely at an attractive price. In other words, specify a unit that can do the job comfortably but do not overkill or try to squeak by.

• Is the well over a structure, such as a caisson or platform, or at an open location? If it is over a structure, then only a jackup, TAD, platform rig, and/or submersible, depending on water depth, should be considered.

• If the well is in open water then a jackup, TAD, and/or submersible, depending on water depth, should be considered. A standard moored drillship may also be evaluated if commercial issues are a key consideration.

• If the well(s) to be drilled are over a platform, some of the following questions need to be considered:

• On the subject platform, can a cantilever jackup reach the well conductor after jacking up, and does it have enough combined cantilever load rating to drill the well(s)?

• If a platform rig is being considered, is there enough fixed platform space and load-bearing capability? Older platforms sometimes weaken with age and additional production equipment is placed on them, thus reducing the space needed for a platform rig. What is the spacing for the "cap beams," or the beams the platform rig would skid and rest on? The beams may range from 30 to 62 ft; TLP, spars and large platforms may even be wider. Standard cap beam spacing usually runs from 35 to 45 ft. A jackup or a TAD should be considered if cap beam spacing, load, or space is a major issue.

• If a platform rig seems to be the best fit, required capabilities are very important when deciding between a standard and a modular unit. As a rule, modular, self-erecting units are less capable overall but offer many advantages over their larger, more expensive cousins, as discussed earlier.

• If there are weak soil conditions that increase the likelihood of a punch through or old spud-can holes that do not fit the available jackups, use of a jackup may be questionable, especially if a capable TAD, preferably a semi, is available.

• If platform space and/or load bearing are critical and the wells to be drilled are ERWs or very deep, a high-specification semi TAD will be very attractive because a TAD takes up less space, the DES is much lighter than high-specification platform rigs, the weather effect for loading and unloading consumables is generally not a factor with TADs, and the TAD can store (space and load) a considerable amount of casing, mud, cement, and operator expendables.

• If a TAD appears to be the best solution, weather, space, and VDL should be factored in when considering a monohull vs. a semi TAD.

• For spars and TLPs, modular platform rigs vs. TADs must be explored. Weight is very critical and extremely expensive to accommodate. The TAD, weighing one-fourth to one-fifth as much as a modular rig and requiring about one-third the space, is very attractive. If more than 9 to 12 long ERWs are to be drilled, a TAD spar/TLP instead of a modular platform rig "drilling" spar/TLP may be very attractive. Consumables such as mud (volume and weight), casing (weight and space), supply by boats, and the production and drilling risers will have a key impact on rig efficiency.

• Water depth of the location has a major impact on MODU selection. Following are some observations that should be kept in mind for bottom-founded units:

• In very shallow water depths (generally < 25 ft, definitely < 14 to 20 ft), submersibles offer many advantages over jackups. The smaller shallow-water jackups usually have limited drilling, deck space, and VDL capability compared with submersibles. Submersibles can operate in 10-ft water depth and generally have relatively attractive drilling capability.

• For independent-leg jackups, which most upscale jackups are, leg penetration may be critical. A 300-ft nominally rated jackup with 100-ft leg penetration becomes a unit that can drill only in ≈ 200 ft of water depth, depending on the required air gap. In addition, it will probably require many preload cycles and thus a long mob and demob period. Pulling legs may also be time consuming.

• For jackups rated for > 300 to 350 ft, a new high-specification, enhanced, premium jackup may be required, along with the additional cost. In other words, the operator should not over specify his requirements. If water depth and/or a 7,500-psi-WP mud system is thought to be required and because there are few of this class of jackup, the operator should expect to pay a premium price to obtain such a MODU.

• Selecting between a mat or an independent jackup should center on soil factors, spud-can holes (although holes can also be a problem for mat rigs if they are around a high-load-bearing area of the mat), economics, and drilling capability. Almost without exception, mat rigs are less capable than equivalent independent-leg units, but they can drill in areas where leg penetration is a major problem and/or leg punch through is of major concern. A relatively new concept for helping to prevent leg punch through, "Swiss Cheese," is being used on a limited basis. Multiple 26- to 36-in. holes are drilled through the weak load-bearing lens, allowing the spud can to penetrate the weak soil easily through to the stronger soil below the zone in question. However, it is very expensive and not always a sure solution.

• If the well under consideration is in jackup water depth but the soil conditions are very unsuitable, shallow gas flows are likely, and a jackup is not available, a shallow-water semi may be able to drill the well very economically.[12]

• If the water depth exceeds jackup capability, a moored MODU should be considered. Once again, the operator should not generally specify a unit with a lot more capability than required. Following are some observations:

• Semis generally can be grouped into three broad categories of water depth, which usually follows their generation designation: second-generation units work in < 1,500 to 2,000 ft, third- and fourth-generation units in 2,000 to 5,000 ft, and fifth-generation units in 4,000 to 6,500 ft and beyond. Costs generally increase with water depth, but so do the capabilities of the unit. Again, a sledgehammer is not needed to drive a tack.

• A prelaid taut or semi-taut mooring system can extend the depths of some units, but the prelaid systems are very expensive to purchase, deploy, and maintain. In addition, other requirements, such as VDL, deck space, marine riser tension, and liquid volume capacity may not be adequate.

• A second- or third-generation unit can be "stretched" beyond its normal water depth rating by mooring line inserts, but as pointed out earlier, other requirements may be limited.

• In some limited cases in which day rates for second-generation semis are reasonably competitive with those of deepwater jackups, a semi can drill a well faster and more economically than a jackup. This is usually in water depths of 275 to 300 ft or more and wells of short duration. The reasons are the longer time to preload/pull legs, eliminating all the casing strings that must be run and pulled between the rotary and seafloor, and potential moving delays, all of which the semi does not contend with.

• Generally, a DP MODU will not be commercially competitive with a moored vessel; however, in deep water and short-duration wells, they can be commercially competitive even with much higher day rates.

• Ultradeepwater water depths are generally the domain of DP fifth-generation drillships and a limited number of semis. There usually is no valid substitute for their use other than in some limited cases when slim riser and surface BOP technology and/or a prelaid taut or semi-taut mooring system can be used.

• Environment and metocean have a critical impact on MODU selection. There are three general metocean categories that MODUs fall into. Most can operate any place in the world except the North Sea, in arctic conditions (< 32°F), and in select areas (e.g., the southwest coast of West Australia and New Zealand). The second category of rig can operate in the most severe, hostile, and usually artic conditions. These very-high-end units are very costly. The third category can operate only in the benign to very calm environments of West Africa and the Far East. There are exceptions to these categories, such as some mat jackups, so it is very important to specify the environment and then compare it with the classification ratings of the unit. Mooring and riser analyses for floating units also need to be performed.

The above points are provided for guidance, but other factors may be the determining ones. Most important is the understanding that many unit types may be able to perform the work. The operators should do their homework and the evaluation in a knowledgeable, methodical manner. Once the technical side has been evaluated, HSE&S, the drilling contractor's reputation, crews, management style, drilling contract issues, price, etc., need to be factored into the final selection. Finally, such intangible issues as mutual confidence and respect, perception of ability to work out problems with anticipation of an equitable solution, political influence with local governments, and agent's impact and help need to be weighed.

14.8 The Future
In the offshore drilling business, predicting the future has been difficult at best. Through the transitions of the last 50 years, a few things have been constant:

• The industry is amazingly resilient. One way or another, the industry has moved forward in good and bad times. It seems to find ways to do things better, more efficiently, more safely, and in some cases more profitably. Mistakes and wrong courses are common, but a service with improving quality has resulted.

• The need for oil and gas over the long term continues to increase, although with some ups and downs. The services required to produce petroleum products will be needed for the foreseeable future.

• Technologically, the rigs of today are vastly superior to the rigs of just 10 to 15 years ago. Today's technologically superior machines can now drill more efficiently and safely in up to and over 10,000-ft water depth. It has been said that the technology required to function in the offshore drilling business is more complex and demanding than the National Aeronautics and Space Administration requirements to go to the moon.

• The industry has matured from the rough-and-ready, full-steam-ahead, damn-the-torpedoes approach of the early years to a more methodical business approach that emphasizes performance and HSE&S.

• Consolidation of drilling contactors, service companies, and operators has not stymied innovation and improvement, as has been the rule rather than the exception in other industries.

So what can be expected in the next 5, possibly the next 10, years? Following are some thoughts:

• The demand for petroleum products continues to increase, as does the need to drill to deeper depths offshore, in deeper water, and in more remote areas. The need for MODUs will still be there; however, it is unlikely that a new rig-building boom like that in the late 1970s to early 1980s or late 1990s will occur anytime in the near future. Improvement in existing units to do more for less cost is the order of the day. Possible exceptions to more units being built are enhanced, premium jackups and semi TADs.

• As stated previously, we are entering a stage of "technology of economics" concerning MODUs and their use. We have drilled in over 10,000 ft of water depth and are producing in > 7,000 ft; however, we must do it more economically. Following are some developments under way for MODUs to accomplish that goal:

• Continue to develop "dual-activity" technology in which a single MODU does some degree of two well operations simultaneously.

• Focus on better, more efficient use of MODUs, machinery, and crews. This would benefit the industry from many standpoints.

• The ability to drill ERWs and subsalt wells with relative ease and efficiency will require a shakeout among MODUs of high-pressure mud requirements, fluid storage, cuttings disposal, setback loads, storage VDL and associated space, and automatic pipe-handling systems. ERWs, especially in deep and ultradeepwater, are an attractive approach to more cost-effective development. MODUs need to fine tune themselves to drill these wells cost-effectively.

• Use of less expensive or lower-generation MODU for ultradeepwater exploration and certain types of development is a must. SBOPs and slim risers show promise under certain conditions. These approaches allow a third- or fourth-generation rig to drill in ultradeep water, thus reducing the cost of drilling the well.

• The concept of "dual gradient" shows great promise; however, it needs considerably more development to become practical and commercially viable. To eliminate three or four casing strings, skating on the edge of fracture gradients with confidence that the well will reach the planned depth is the goal of dual-gradient technology.

• Taut and semi-taut mooring systems for exploration and development MODUs will be refined and become more economical.

• Ultradeepwater units will be upgraded and modified for more efficient deepwater development. This process has already started.

• Drilling in > 10,000 ft, especially > 12,000 ft, where some think we will hit a technological roadblock, will be worked on, but economics and cost at this point are of major concern.

The issue for the next 5 to 10 years will not be whether we can drill in ultradeepwater depths or drill difficult ERWs or wells > 30,000 ft, but can it be done in a cost-effective manner.

Acknowledgments

My thanks to SPE for asking me to write this chapter, which has turned into a bigger project than anticipated but also a labor of love. I have been involved with offshore drilling units for > 35 years, and I cannot find a more interesting subject in our industry. My thanks also to Atwood Oceanics for allowing me the time to write this chapter in hopes that it will enhance the knowledge and enthusiasm for this subject that so many of us have developed during our careers. Also thanks to all those people, from chief executive officers to roustabouts, who helped me gain the knowledge needed to produce this chapter. And finally, thanks go to my family for riding this roller coaster with me that has given us such enjoyment and a solid livelihood.

Acronyms and Definitions

Throughout this chapter, words have been used that are common vernacular in the offshore drilling business. Following is a listing of terms with definitions used in the text that may not be familiar to all readers:

- API: American Petroleum Institute. An industry organization that, among other functions, publishes recommended practices, specifications, and procedures.
- BOP: Blowout preventers. Large wellbore-sized valves placed on top of the well to close it in to control high pressures and wellbore fluid flows.
- DES: Drilling equipment set. The portion of the drilling unit consisting of the derrick, drawworks, traveling equipment, and substructure that sits on a platform, with the remaining equipment moored on a tender next to the platform.
- HT/HP: High temperature/high pressure. Wells that have unusually high wellbore temperatures and pressures.
- IADC: International Association of Drilling Contractors. An industry association that represents onshore and offshore drilling contractors on many issues.
- IMO: International Maritime Organization. A private industry group that functions outside classification societies, country registration, flag state, and governmental regulatory bodies.
- DP: Dynamic Positioning. A means of stationkeeping over a location of a MODU by means of computer-controlled thrusters.
- Drilling Contractor or "Contractor": The company that owns the MODU, staffs it, and operates under contract to the operator.
- ERW: Extended-reach well. A well that has a very long horizontal length and is more challenging than a standard directional or straight vertical well.
- Floater or Floating Unit: Commonly referred to as a MODU that drills from the floating position, such as semisubmersibles and ships.
- Generation: An industry practice to categorize semisubmersibles into five categories. This categorization centers on a combination of when the unit was built or upgraded, its water depth rating, and its drilling equipment outfitting
- GOM: Gulf of Mexico. Large body of water on the southeastern coast of the United States.
- Kips: Unit of weight or force equivalent to 1,000 lbf or lbm.
- LT: Long tons or 2,240 lbm.
- LTI: Lost-time incident. This is an accident after which the individual cannot return to duties within the specified time period as defined by the IADC.
- Metocean: The wind, ocean current, and sea condition data and statistics for various return periods, i.e., 1, 10, or 50 years.
- M Tons: Metric Tonnes or 2,204 lbm.
- MODU: Mobile Offshore Drilling Unit. A MODU is any offshore drilling unit that can be moved from location to location.
- OIM: Offshore installation manager. This individual is the highest authority on the MODU, similar to a captain or master of a ship.
- Operator: The oil and gas exploration and producing company that hires the drilling contractor and MODU. The operator directs the contractor in drilling the well.
- Registration: The formal legality of a MODU for flagging and staffing of the unit.
- Regulations: Usually associated with the laws of a nation that controls the operation of a MODU.
- HSE&S: Health, safety, environment, and security. These four functions are usually grouped together into one department on and off the rig.
- Spars: These are production-type platforms that float. They are used in deepwater, are moored to the seafloor with a spread-moored system, and are shaped like a long cylinder from

72 ft in diameter to over 120 ft and up to 705 ft long. One is currently moored in 5,610 ft in the GOM.

• Stons: Short tons or 2,000 lbm.

• TAD: Tender-assist drilling. A concept in which the derrick and associated equipment sit on a platform (fixed, spar, or TLP) and the rest of the equipment is on a tender barge moored next to the platform.

• TLP: Tension-leg platform. A floating platform held on location by use of long steel vertical tendons fixed to the seafloor. TLPs are used in deep to ultra-deepwater depths.

• Tonnes: Metric weight equivalent to 2,204 lbm.

• VDL: Variable deck load. This is the drilling consumables and items that can be readily offloaded from the main deck of a jackup, semi, submersible, etc.

References

1. Silcox, W.H., *et al.:* "Offshore Operations," *Petroleum Engineering Handbook,* second edition, SPE, Richardson, Texas (1987) Chapter 18.
2. Barnes, K.B., and McCaslin, L.S. Jr.: "Gulf of Mexico Discovery," *Oil & Gas J.* (March 18, 1948) 96.
3. *Mobile Rig Register,* eighth edition, ODS-Petrodata, Houston (2002).
4. Howe, R.J.: "Evolution of Offshore Mobile Drilling Units," *Ocean Industry* (April 1966) 10.
5. Laborde, A.J.: *My Life and Times,* Laborde Publishing Co., New Orleans (1996).
6. Harris, L.M.: "Humble SM-1 Offshore Exploration Vessel, Petroleum Engineering Project Report," Humble Oil and Refining Co., Production Department California Area, Los Angeles, California (1957).
7. Howe, R.J.: "Evolution of Offshore and Production Technology," paper OTC 5354 presented at the 1986 Offshore Technology Conference, Houston, 5–8 May.
8. *Marine Riser Systems and Subsea Blowout Preventers,* Unit V, Lesson 10, first edition, Petroleum Extension Service, University of Texas, Austin (2003).
9. Childers, M.: "Surface BOP, Slim Riser or Conventional 21-Inch Riser: What is the Best Concept to Use," paper SPE/IADC 92762 presented at the 2005 IADC/SPE Drilling Conference, Amsterdam, 23–25 February.
10. Childers, M. and Quintero, A.: "Slim Riser: Effective Tool for Ultra Deepwater Drilling," paper SPE/IADC 87982 presented at the 2004 Asia Pacific Drilling Technology Conference and Exhibition, Kuala Lumpur, 13–15 September.
11. Sheffield, R.: "Floating Drilling: Equipment and Its Use," *Practical Drilling Technology,* volume 2, Gulf Publishing Co., Houston (1980).
12. Childers, M.A.: "Operating Efficiency Comparison Between a Deepwater Jackup and a Semisubmersible in the Gulf of Mexico," paper SPE/IADC 18623 presented at the 1989 SPE/IADC Drilling Conference, New Orleans, 28 February–3 March.

SI Metric Conversion Factors

bbl	×	1.589 873	E–01	= m^3
ft	×	3.048*	E–01	= m
°F		(°F-32)/1.8		= °C
hp	×	7.460 43	E–01	= kW
in.	×	2.54*	E+00	= cm
in.^{3}	×	1.638 706	E+01	= cm^3
kip	×	4.448 222	E+03	= N
knot	×	5.144 444	E–01	= m/s
lbf	×	4.448 222	E+00	= N
lbm	×	4.535 924	E–01	= kg
long ton	×	1.016 047	E+00	= Mg
mile	×	1.609 344*	E+00	= km

| psi | × | 6.894 757 | E+03 | = Pa |
| tonne | × | 1.0* | E+00 | = Mg |

*Conversion factor is exact.

Chapter 15
Drilling-Data Acquisition

Iain Dowell, Halliburton Energy Services, **Andrew Mills,** Esso Australia Ltd., **Marcus Ridgway**, and **Matt Lora**, Landmark Graphics Corp.

15.1 Introduction

The prototype data-collection system for drilling wells previously consisted of paper reports from data collected and recorded by hand, culminating in the daily "morning report" of well progress. Because of the progress in computer hardware and software over the past 20 years, spurred by the increased use of measurement-while-drilling (MWD) and logging-while-drilling (LWD) tools, wellsite data collection, storage, and use have increased many times above the meager data available only a few years ago.

15.2 Surface-Data Sensors

By analyzing cuttings, drilling mud, and drilling parameters for hydrocarbon-associated phenomena, we can develop a great deal of information and understanding concerning the physical properties of a well from the surface to final depth. A critical function in data analysis is familiarity with the different sensors used for gathering surface data. These sensors can be grouped as follows:

- Depth Tracking.
- Flow-In Tracking.
- Pressure Tracking.
- Flow-Out Tracking.
- Drill Monitoring.
- Pit Monitoring.
- Gas Detection.

15.2.1 Depth-Tracking Sensors. Current depth-tracking sensors digitally count the amount of rotational movement as the draw-works drum turns when the drilling line moves up or down. Each count represents a fixed amount of distance traveled, which can be related directly to depth movement (increasing or decreasing depth). Moreover, the amount of movement also can be tied into a time-based counter, which will give either an instantaneous or an average rate of penetration (ROP).

Alternatively, some companies still use a pressurized depth-tracking/ROP sensor. The pressurized ROP system works on the principle of the change in hydrostatic pressure in a column

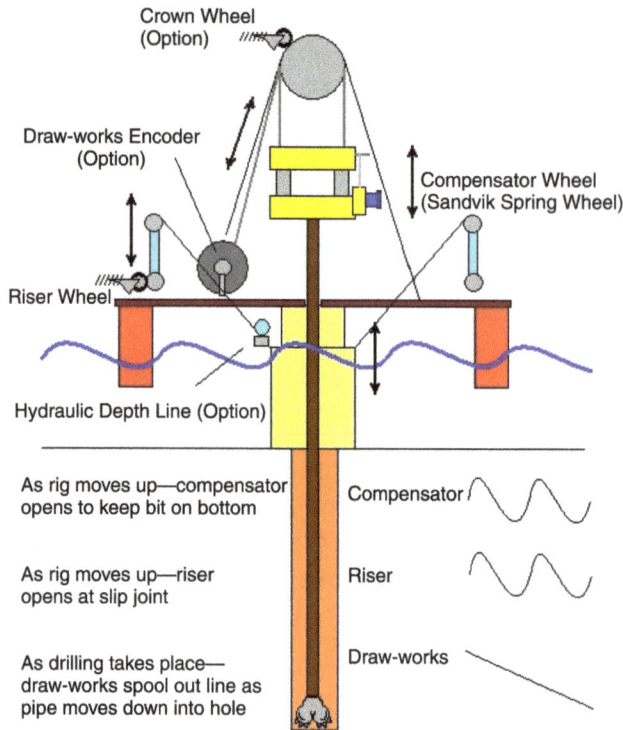

Fig. 15.1—Example of a typical depth-tracking system.

of water as the height of that column is varied. This change can then be indirectly related to a depth measurement. Again, a time-based counter is used to calculate an instantaneous or average ROP.

Additionally, accurate depth measurement on offshore rigs such as semisubmersibles, submersibles, and drill ships is affected by both lateral (tidal movement) and axial (the up-and-down motion of the rig, also called "rig heave") effects. To properly compensate for this, most of these rigs have a rig-compensator system installed on their traveling block. As the rig moves up, the compensator opens, thereby allowing the bit to stay on bottom. Similarly, as the rig moves down, the compensator must shut to keep the same relative bit position and weight on the bit.

The same digital sensors are attached to the compensators so that any change in movement can be taken into account, allowing accurate depth measurement (**Fig. 15.1**).

15.2.2 Flow-In Tracking Sensors. Flow-tracking sensors are used to monitor fluid-flow rate being applied downhole as well as the pump strokes required to achieve this flow rate. Data gathered from these sensors are essential inputs to calculating drilling-fluid hydraulics, well control, and cuttings lag. Monitoring changes in trends may also indicate potential downhole problems such as kicks or loss of circulation.

Two commonly used types are proximity and/or whisker switches. A proximity switch, activated either by an electromagnet (coil) or a permanent magnet, acts as a digital relay switch when it incorporates electrical continuity. A whisker switch, on the other hand, is a microswitch that is activated only when an external rod (called a whisker) forces a piston to raise a ball bearing to initiate contact against it (**Fig. 15.2**). Both types are digital counters; an increase in counts will correspond to a specific increase in both flow rate and pump rate.

Fig. 15.2—Example of proximity and whisker switch.

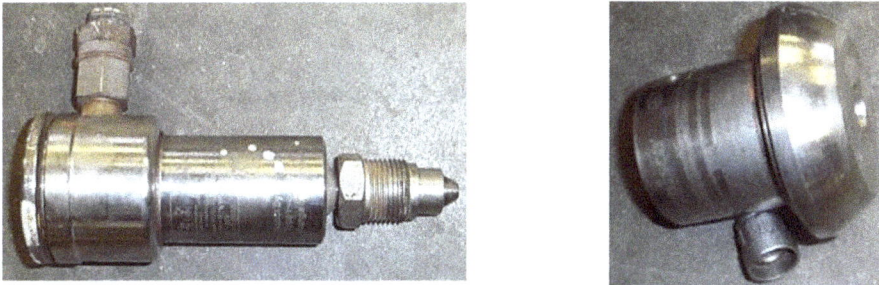
Fig. 15.3—Example of pressure transducers.

15.2.3 Pressure-Tracking Sensors. Pressure-tracking sensors are used mainly to monitor surface pressure being applied downhole. Data gathered from these sensors are used either to validate calculated values or to confirm potential downhole problems such as washouts, kicks, or loss of circulation.

Two types of sensors are available, and both monitor pressure from a high-pressure diaphragm unit (knock-on head) located on either the standpipe or the pump manifold. The first sensor type derives its physical input from mud pressure expanding a rubber (or viton when high temperature is involved) diaphragm within the knock-on head. This expansion proportionally increases the pressure in the hydraulic-oil-filled system and, in doing so, relays the mud pressure to the appropriate transducer. The second sensor type makes a direct connection with the standpipe manifold itself (i.e., the transducer face is in contact with the mud; see **Fig. 15.3**).

15.2.4 Flow-Out Tracking Sensor. Commonly called a "flow paddle," this sensor measures flow rate coming out of the annulus using a strain-gauge analog transducer (**Fig. 15.4**). Changes in resistance values are directly related to either an increase or a decrease in mud-flow rate. This sensor provides an early warning of either a kick condition (sudden increase in flow rate) or a loss of circulation (sudden decrease in flow rate).

15.2.5 Drill-Monitor Sensors. Drill-monitor sensors monitor surface revolutions-per-minute (RPM) values, rotary torque, and hook load. The torque sensor is a clamp (**Fig. 15.5**) that sits around the main power cable to the top-drive system (TDS). It works on the principle of the deformation of Hall-effect chips by the magnetic field produced around the cable owing to the current being drawn through it (i.e., the greater the torque being produced as the pipe rotates, the greater the current drawn by the TDS and therefore the greater the Hall effect). (Note: the Hall effect is a transverse voltage caused by electric current flow in a magnetic field.) Torque

Fig. 15.4—Schematic of a flow-paddle sensor.

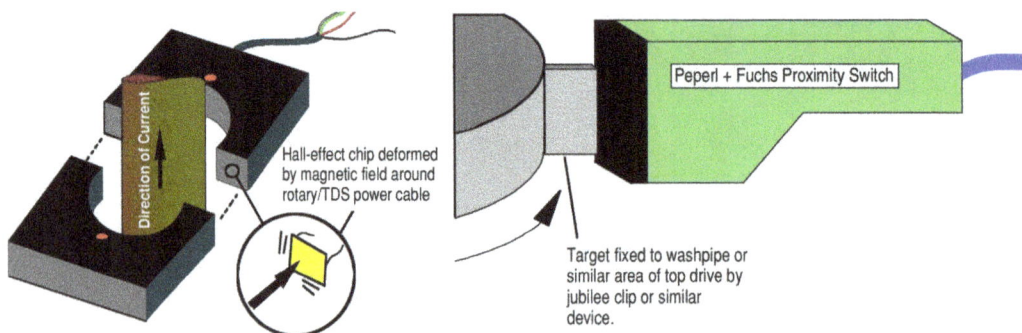

Fig. 15.5—Example of a torque (left) and an RPM sensor (right).

changes can then be related to either formation lithology or downhole drilling problems such as pipe stick/slip or motor stalling.

A digital rotary sensor is similar to a proximity sensor used in a pump. It is shaped differently but acts on the same principle. RPM changes are used to drill the well efficiently and minimize downhole vibration effects.

The combined weight of the bit, bottomhole assembly (BHA), drillpipe, etc., is called the string weight (SW). The block weight (BW) is the weight of the lines and blocks (including top drive or kelly). When the bit is on bottom (i.e., drilling), the hook load is seen to reduce. The amount of weight suspended by the bottom of the hole is the amount of weight on bit (WOB), as shown below:

$$\text{Hook Load} = \text{SW} + \text{WOB}.$$

This hook-load sensor uses the same transducer type as in a pressure-tracking sensor. As the deadline experiences strain, the reservoir has load applied across it, which pressures the hydraulic fluid. This pressure increase is translated to a measurement value (**Fig. 15.6**). These

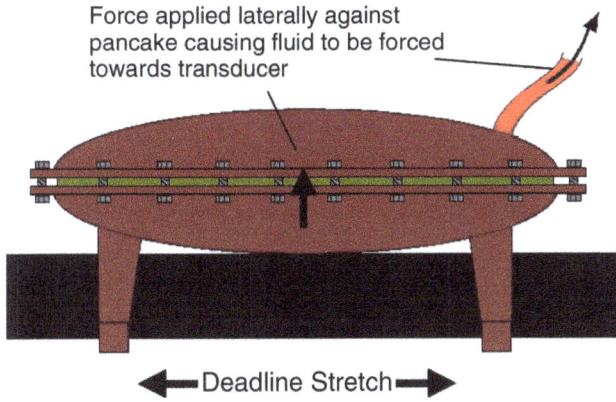

Force applied laterally against pancake causing fluid to be forced towards transducer

Deadline Stretch

Fig. 15.6—Example of a hook-load sensor.

measurement values are then correlated to potential downhole problems such as kicks or stuck pipe.

15.2.6 Pit-Monitor Sensor. Most pit-monitor sensors use ultrasonic transit time to measure mud level. The sensor is mounted over the pit above the maximum mud level and continuously sends a sonic wave that is reflected back to the receiver (**Fig. 15.7**). The transit-time measurement is then directly transformed to a volume measurement. This critical measurement is actively used to monitor potential kicks (rapid increase in pit volume) or loss of circulation (rapid decrease in pit volume).

15.2.7 Gas-Detection Sensors. The gas-detection sensors consist mainly of a gas trap, a pneumatic line linking the gas trap to the gas-detection equipment (which is found inside a mud-logging unit), and the gas-detection instruments (chromatograph and total-gas detectors).

The gas trap is basically a floating chamber with a rotating "agitator" inside. It works on the principle that mud flowing through the gas trap is agitated, thereby releasing the vast majority of any gases contained within the mud. This gas is then extracted from the trap through the unit sample line to be analyzed in the unit (**Fig. 15.8**).

The principle behind gas chromatography is simple. The gas from an oil well consists of several hydrocarbon components, ranging from light gases (methane) to oil. A gas chromatograph then takes a sample of gas and separates out some of these components for individual analysis. Typically, methane (C_1) through pentane (C_5) are the gases of interest. These can be plotted individually, or they may be used in gas-ratio analysis for reservoir characterization.

Most logging companies currently use a flame ionization detector (FID) gas chromatograph and total-gas detector (**Fig. 15.9**). The FID responds primarily to hydrocarbons and has the widest linear range of any detector in common use. The output signal is linear for a given component when concentrations vary from less than one part per million (ppm) to percent levels, and with care, resolution can be obtained in the low part-per-billion (ppb) range. The total-gas detector samples gas in a manner similar to that of a chromatograph, the only difference being that there is no column in the detector and, hence, no separation of components (i.e., it burns the "total" hydrocarbon gas sample as one). This also means that there is no injection time and, therefore, the gas is being sampled continuously (Fig. 15.8).

15.2.8 Additional Sensors. In addition, exploration and production companies may require specialized services such as formation-pressure monitoring and drilling optimization. To effective-

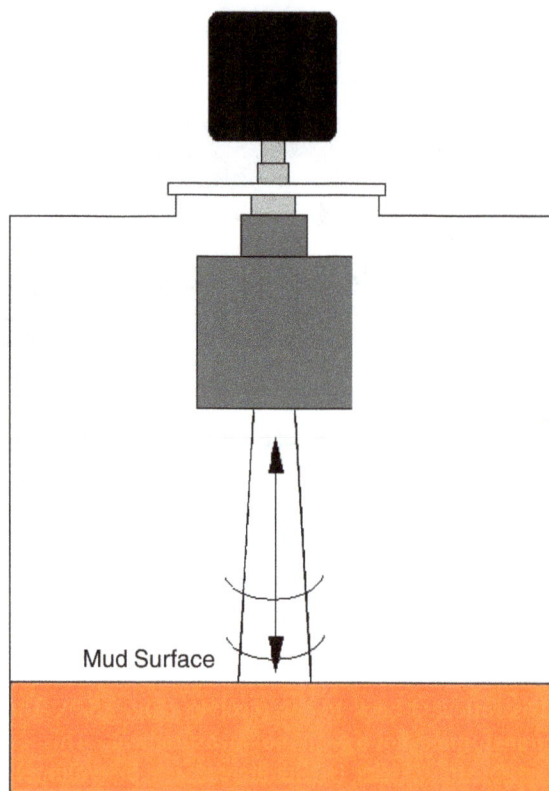

Fig. 15.7—Example of a sonic pit-volume sensor.

ly support these services, additional sensors may be required such as fluid temperature, density, and conductivity. In areas of high H_2S or CO_2 gas, corresponding sensors that exclusively monitor these gases may be required as well.

15.3 MWD and LWD Applications

15.3.1 Introduction. No other technology used in petroleum-well construction has evolved more rapidly than MWD and LWD. Early in the history of the oil field, drillers and geologists often debated conditions at the drillbit. With advances in electronic components, materials science, and battery technology, it became technically feasible to make measurements at the bit and transmit them to the surface so that the questions could begin to be answered.

Directional measurements were the first measurements to have commercial application, with almost all use in offshore, directionally drilled wells. As long as MWD achieved certain minimum-reliability targets, it was less costly than single shots, and it gained popularity accordingly. The dual challenges of MWD and LWD technology were reliable operation in the harsh downhole environment and achievement of wireline-quality measurements.

In the early 1980s, qualitative measurements of formation parameters were introduced, often based on early wireline technology. Coring points and casing points were selected using short normal-resistivity and natural gamma ray measurements, but limitations in these measurements kept them from replacing wireline for quantitative formation evaluation. In the late 1980s, the first rigorously quantitative measurements of formation parameters were made. Initially, the measurements were stored in tool memory, but soon the 2-MHz resistivity, neutron

The following diagram shows a schematic of the floating gas trap.

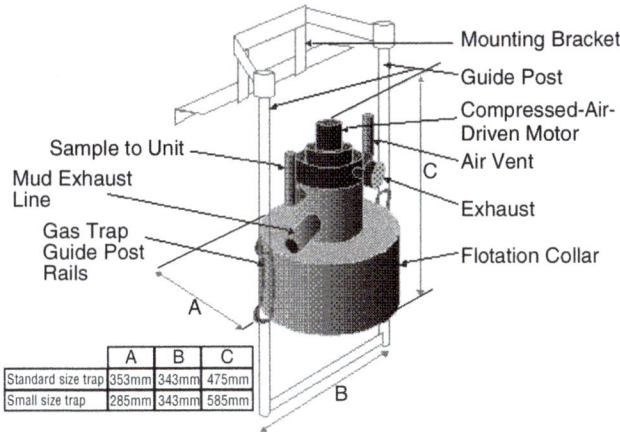

	A	B	C
Standard size trap	353mm	343mm	475mm
Small size trap	285mm	343mm	585mm

Fig. 15.8—Example of a gas trap.

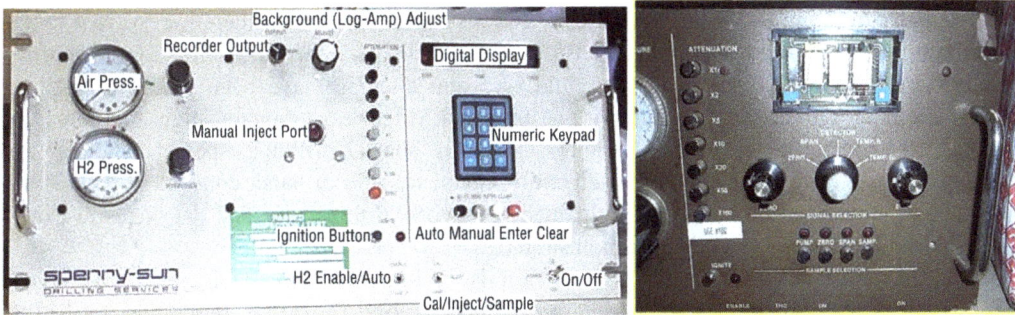

Fig. 15.9—Example of an FID chromatograph and a total-gas detector.

porosity, and gamma density measurements were transmitted to the surface in real time. By the early years of the new millennium, there was a rapid expansion of the types of measurement available while drilling, including acoustic, formation pressure, imaging, and seismic.

The terms MWD and LWD are not used consistently throughout the industry. Within the context of this section, the term MWD refers to directional-drilling measurements, and LWD refers to wireline-quality formation measurements made while drilling.

15.3.2 MWD. Although many measurements are taken while drilling, the term MWD refers to measurements taken downhole with an electromechanical device located in the BHA. Telemetry methods had difficulty in coping with the large volumes of downhole data, so the definition of MWD was broadened to include data that were stored in tool memory and recovered when the tool was returned to the surface. All MWD systems typically have three major subcomponents: a power system, a telemetry system, and a directional sensor.

Power Systems. Power systems in MWD generally may be classified as one of two types: battery or turbine. Both types of power systems have inherent advantages and liabilities. In many MWD systems, a combination of these two types of power systems is used to provide power to the MWD tool so power will not be interrupted during intermittent drilling-fluid flow conditions. Batteries can provide this power independent of drilling-fluid circulation, and they are necessary if logging will occur during tripping in or out of the hole.

Lithium-thionyl chloride batteries are commonly used in MWD systems because of their excellent combination of high-energy density and superior performance at MWD service temperatures. They provide a stable voltage source until very near the end of their service life, and they do not require complex electronics to condition the supply. These batteries, however, have limited instantaneous energy output, and they may be unsuitable for applications that require a high current drain. Although these batteries are safe at lower temperatures, if heated above 180°C, they can undergo a violent, accelerated reaction and explode with a significant force. As a result, there are restrictions on shipping lithium-thionyl chloride batteries in passenger aircraft. Even though these batteries are very efficient over their service life, they are not rechargeable, and their disposal is subject to strict environmental regulations.

The second source of abundant power generation, turbine power, uses the rig's drilling-fluid flow. Rotational force is transmitted by a turbine rotor to an alternator through a common shaft, generating a three-phase alternating current (AC) of variable frequency. Electronic circuitry rectifies the AC into usable direct current (DC). Turbine rotors for this equipment must accept a wide range of flow rates to accommodate all possible mud-pumping conditions. Similarly, rotors must be capable of tolerating considerable debris and lost-circulation material (LCM) entrained in the drilling fluid.

Telemetry Systems. Mud-pulse telemetry is the standard method in commercial MWD and LWD systems. Acoustic systems that transmit up the drillpipe suffer an attenuation of approximately 150 dB per 1000 m in drilling fluid.[1] Several attempts have been made to construct special drillpipe with an integral hardwire. Although it offers exceptionally high data rates, the integral hardwire telemetry method requires expensive special drillpipe, special handling, and hundreds of electrical connections that must all remain reliable in harsh conditions. The explosion of downhole measurements has stimulated new work in this area,[2] and data rates in excess of 2,000,000 bits/second have been demonstrated.

Low-frequency electromagnetic transmission is in limited commercial use in MWD and LWD systems. It is sometimes used when air or foam is used as drilling fluid. The depth from which electromagnetic telemetry can be transmitted is limited by the conductivity and thickness of the overlying formations. Repeaters or signal boosters positioned in the drillstring extend the depth from which electromagnetic systems can transmit reliably.

Three mud-pulse telemetry systems are available: positive-pulse, negative-pulse, and continuous-wave systems. These systems are named for the ways in which their pulses are propagated in the mud volume. Negative-pulse systems create a pressure pulse lower than that of the mud volume by venting a small amount of high-pressure drillstring mud from the drillpipe to the annulus. Positive-pulse systems create a momentary flow restriction (higher pressure than the drilling-mud volume) in the drillpipe. Continuous-wave systems create a carrier frequency that is transmitted through the mud, and they encode data using the phase shifts of the carrier. Many different data-coding systems are used, which are often designed to optimize the life and reliability of the pulser because it must survive direct contact with the abrasive, high-pressure mud flow.

Telemetry-signal detection is performed by one or more transducers located on the rig standpipe. Data are extracted from the signals by surface computer equipment housed either in a skid unit or on the drill floor. Successful data decoding is highly dependent on the signal-to-noise ratio.

A close correlation exists between the signal size and the telemetry data rate; the higher the data rate, the smaller the pulse size becomes. Most modern systems have the ability to reprogram the tool's telemetry parameters and slow down data-transmission speed without tripping out of the hole; however, slowing the data rate adversely affects log-data density.

The most notable sources of signal noise are the mud pumps, which often create a relatively high-frequency noise. Interference among pump frequencies leads to harmonics, but these background noises can be filtered out with analog techniques. Pump-speed sensors can be a very effective method of identifying and removing pump noise from the raw telemetry signal. Lower-frequency noise in the mud volume is often generated by drilling motors. Well depth and mud type also affect the received-signal amplitude and width. In general, oil-based muds (OBMs) and pseudo-oil-based muds are more compressible than water-based muds; therefore, they result in the greatest signal losses. Nevertheless, signals have been retrieved without significant problems from depths of almost 9144 m (30,000 ft) in compressible fluids.

Directional Sensors. The state of the art in directional-sensor technology is an array of three orthogonal fluxgate magnetometers and three accelerometers. Although in normal circumstances, standard directional sensors provide acceptable surveys, any application in which uncertainty in the bottomhole location exists can be troublesome. Recent trends to drill longer and more complex wells focused attention on the need for a standard error model.

Work carried out by the Industry Steering Committee on Wellbore Accuracy (ISCWA) aimed to provide a standard method of quantifying positional uncertainties with associated confidence levels. The key sources of error were classified as sensor errors, magnetic interference from the BHA, tool misalignment, and magnetic-field uncertainty.

Along with uncertainties in the measured depth, bottomhole survey uncertainties are one contributor to errors in the absolute depth. Note that all methods of real-time azimuth correction require raw data to be transmitted to the surface, which imposes load on the telemetry channel.

The development of gyroscope (gyro)-navigated MWD offers significant benefits over existing navigation sensors. In addition to greater accuracy, gyros are not susceptible to interference from magnetic fields. Current gyro technology centers upon incorporating mechanical robustness, minimizing external diameter, and overcoming temperature sensitivity. The main application of the technology is in saving the rig time used by wireline gyros when carrying out kickoffs from areas affected by magnetic interference.

MWD and LWD System Architecture. As MWD and LWD systems have evolved, the importance of customized measurement solutions has increased. The ability to add and remove measurement sections of the logging assembly as wellsite needs change is valuable, thus prompting the design of modular MWD/LWD systems. Operational issues, such as fault tolerance, power sharing, data sharing across tool joints, and memory management, have become increasingly important in LWD systems. The introduction of 3D rotary-steerable systems, which often use the same telemetry channel as the LWD systems, has reinforced the links between directional drilling and LWD.

A natural division in system architecture exists for drill-collar outside diameters (ODs) of 4¾ in. or less. Smaller-diameter tool systems tend to use positive-pulse telemetry systems and battery-power systems and are encased in a probe-type pressure housing. The pressure housing and internal components are centered on elastomer standoffs and mounted inside a drill collar. Some MWD/LWD systems are retrievable and replaceable, in case tool failure or tool sticking occurs. Retrievability from the drill collar while in the hole often compromises the system's mounting scheme; therefore, these types of systems are typically less reliable. Because the MWD string can be changed without tripping the entire drillstring, retrievable systems can be less-reliable, but still cost-effective, solutions.

For collar ODs greater than 6¾ in., LWD systems are often turbine-powered. When used with other modules, interchangeable power systems and measurement modules must supply power and transmit data across tool joints. Often, a central stinger assembly protrudes from the lower collar joint and mates with an upward-looking electrical connection as the collar-joint threads are made up on the drillfloor. These electrical and telemetry connections can be compromised by factors such as high build rates in the drillstring and electrically conductive muds. Recent MWD/LWD designs ensure that each module contains an independent battery and memory so that logging can continue even if central power and telemetry are interrupted. Battery power and memory also enable logging to be performed while tripping out of the hole. As the quantities of data gathered downhole increase, time spent dumping data on the rig floor becomes a significant factor affecting the economics of wireline replacement. Increasing efforts will be made to make the data-download process more rapid over the coming years.

Drilling Dynamics. The aim of drilling-dynamics measurement is to make drilling the well more efficient and to minimize nonproductive time (NPT). Approximately 75% of all lost-time incidents of more than 6 hours are caused by drilling-mechanics failures.[3] Therefore, extensive effort is made to ensure that the drilling-mechanics information acquired is converted to a format usable by the driller and that usable data are provided to the rig floor.

The most frequently measured downhole drilling-mechanics parameters are downhole mud pressures (PWD), WOB, torque on bit, shock, temperature, and caliper. Formation testing while drilling (FTWD) provides key formation pressures for drilling optimization. The data provided by these measurements are intended to enable informed, timely decisions by the drilling staff and thereby improve drilling efficiency. The two main causes of NPT are hole problems (addressed by hydraulics measurement and wellbore-integrity measurement) and drillstring and tool failure (addressed by drillstring-integrity measurement).

To have a positive effect on drilling efficiency, drilling dynamics must have a quick feedback loop to the driller. Recent advances have made it possible to observe the cyclic oscillations in WOB.[4] If the oscillations exceed a predetermined threshold, they can be diagnosed as bit bounce, and a warning is transmitted to the surface. The driller can then take corrective action (such as altering WOB) and observe whether the bit has stopped bouncing on the next data transmission. Other conditions, such as "stick-slip" (intermittent sticking of the bit and drillstring with rig torque applied, followed by damaging release or slip) and torsional shocks, also can be diagnosed and corrected.

Another application is the use of downhole shock sensors, which count the number of shocks that exceed a preset force threshold over a specific period. This number of occurrences is then transmitted to the surface. Downhole shock levels can be correlated with the design specification of the MWD tool. If the tool is operated above design thresholds for a period, the likelihood of tool failure increases proportionally. Of course, a strong correlation exists between continuous shocking of the BHA and the mechanical failure that causes the drillstring to part. In most cases, lateral-shock readings have been observed at significantly higher levels than axial (along the tool axis) shock.

Hydraulics management with PWD has proved a key enabling technology in extended-reach wells where long tangent sections may have been drilled. Studies performed on such wells have shown that hole cleaning can be difficult and that cuttings can build up on the lower side of the borehole. If this buildup is not identified early enough, loss of ROP and sticking problems can result. A downhole annulus-pressure measurement can monitor backpressure while circulating the mud volume, and, assuming that flow rates are unchanged, it can identify precisely if a wiper trip should be performed to clean the hole. **Fig. 15.10** shows an example in which cuttings have fallen out of suspension in the annulus during a period of sliding. Once rotation is resumed, the cuttings are agitated and suspended once more in the mudstream with a consequent increase in equivalent circulating density (ECD).

In wells in which there is a narrow window between pore pressure and fracture gradient (e.g., deep water), the uncertainties can be reduced greatly through the use of PWD and FTWD technology. Downhole measurement and transmission of leakoff tests eliminate errors associated with surface measurements. Real-time ECD measurements pinpoint key pressure parameters

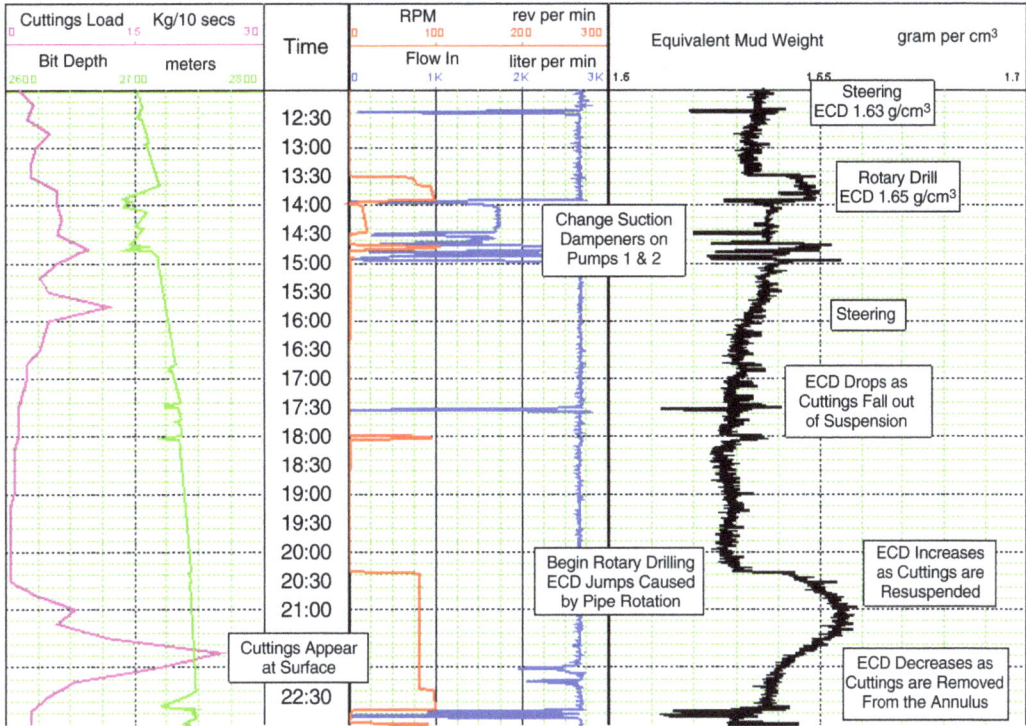

Fig. 15.10—Downhole sensors provide useful drilling measurements.

frequently and accurately. Finally, real-time measurement of pore pressure identifies exactly the mud weight required.

Tool Operating Environment and Tool Reliability. MWD systems are used in the harshest operating environments. Obvious conditions such as high pressure and temperature are all too familiar to engineers and designers. The wireline industry has a long history of successfully overcoming these conditions.

Most MWD tools can operate continuously at temperatures up to 150°C, with some sensors available with ratings up to 175°C. MWD-tool temperatures may be 20°C lower than formation temperatures measured by wireline logs, owing to the cooling effect of mud circulation, so the highest temperatures encountered by MWD tools are those measured while running into a hole in which the drilling-fluid volume has not been circulated for an extended period. In such cases, it is advisable to break circulation periodically while running in the hole. Using a Dewar flask to protect sensors and electronics from high temperatures is common in wireline, where downhole exposure times are usually short, but using flasks for temperature protection is not practical in MWD because of the long exposure times at high temperatures that must be endured.

Downhole pressure is less a problem than temperature for MWD systems. Most tools are designed to withstand up to 20,000 psi, with specialist tools rated to 25,000 psi. The combination of hydrostatic pressure and system backpressure rarely approaches this limit.

However, it is downhole shock and vibration that present MWD systems with their most severe challenges. Contrary to expectation, early tests using instrumented downhole systems showed that the magnitudes of lateral (side-to-side) shocks are dramatically greater than axial shocks during normal drilling. Modem MWD tools are generally designed to withstand shocks of approximately 500 G for 0.5 ms over a life of 100,000 cycles. Torsional shock, produced by stick/slip torsional accelerations, may also be significant. If subjected to repeated stick/slip, tools can be expected to fail.

Early work done to standardize the measurement and reporting of MWD-tool reliability statistics focused on defining a failure and dividing the aggregate number of successful circulating hours by the aggregate number of failures. This work resulted in a mean-time-between-failure (MTBF) number. If the data were accumulated over a statistically significant period (typically 2,000 hours), meaningful failure-analysis trends could be derived. As downhole tools became more complex, however, the Intl. Assn. of Drilling Contractors (IADC) published recommendations on the acquisition and calculation of MTBF statistics.[5]

15.3.3 LWD. *Electromagnetic Logging.* The electromagnetic-wave resistivity (EWR) tool has become the standard of the LWD environment. The nature of the electromagnetic measurement requires that the tool typically be equipped with a loop antenna that fits around the OD of the drill collar and emits electromagnetic waves. The waves travel through the immediate wellbore environment and are detected by a pair of receivers. Two types of wave measurements are performed at the receivers. The attenuation of the wave amplitude as it arrives at the two receivers yields the attenuation ratio. The phase difference in the wave between the two receivers is measured, yielding the phase-difference measurement. Typically, these measurements are then converted back to resistivity values through the use of a conversion derived from computer-modeling or test-tank data.

The primary purpose of resistivity-measurement systems is to obtain a value of true formation resistivity (R_t) and to quantify the depth of invasion of the drilling-fluid filtrate into the formation. A critical parameter in MWD measurements is formation exposure time (FET), the time difference between the drillbit disturbing in-situ conditions and sensors measuring the formation. MWD systems have the advantage of measuring R_t after a relatively short FET, typically 30 to 300 minutes. Interpretation difficulties sometimes can be caused by variable FET, and logs should always contain at least one formation exposure curve.

Knowledge of FET does not, however, rule out other effects. **Fig. 15.11** shows a comparison between phase and attenuation resistivity with an FET of less than 15 minutes and a wireline laterolog run several days later. Even the attenuation resistivity has been affected dramatically by invasion, reading about 10 Ω·m, whereas the true resistivity is in the region of 200 Ω·m.

Another example, shown in **Fig. 15.12**, illustrates invasion effects in the interval from 2995 to 3025 m. Very deep invasion by conductive muds in the reservoir has caused the 2-MHz tool to read less than 10 Ω·m in a 200-Ω·m zone. Between 3058 and 3070 m, the deep invasion has caused the hydrocarbon-bearing zone to be almost completely obscured. Only by comparison with the overlying, deeply invaded zone from 2995 to 3025 m was this productive interval identified.

Similarly, LWD data density is dependent upon ROP. Good-quality logs typically have graduations or "tick" marks in each track to give a quick-look indication of measurement-density variations with respect to depth.

Early resistivity systems emphasized the difference between the phase and attenuation curves and suggested that one curve was a "deep" (radius of investigation) curve and another was a "medium" curve. Difficulties with this interpretation in practice[6] led to the development of a generation of tools that derive their differences in investigation depth from additional physical spacings. Identification and presentation of invasion profiles, particularly in horizontal holes, can lead to a greater understanding of reservoir mechanisms. Many of the applications in which LWD logs have replaced wireline logs occur in high-angle wells. This trend leads to an emphasis on LWD for certain specialist-interpretation issues.

The depth of investigation of 2-MHz-wave resistivity devices is dependent on the resistivity of the formation being investigated. The measurement response of a device (both phase and attenuation) with four different receiver spacings is shown in **Fig. 15.13**. The region measured by the 25-in. sensor (R25P) is based on a 25-in. diameter of investigation in a formation

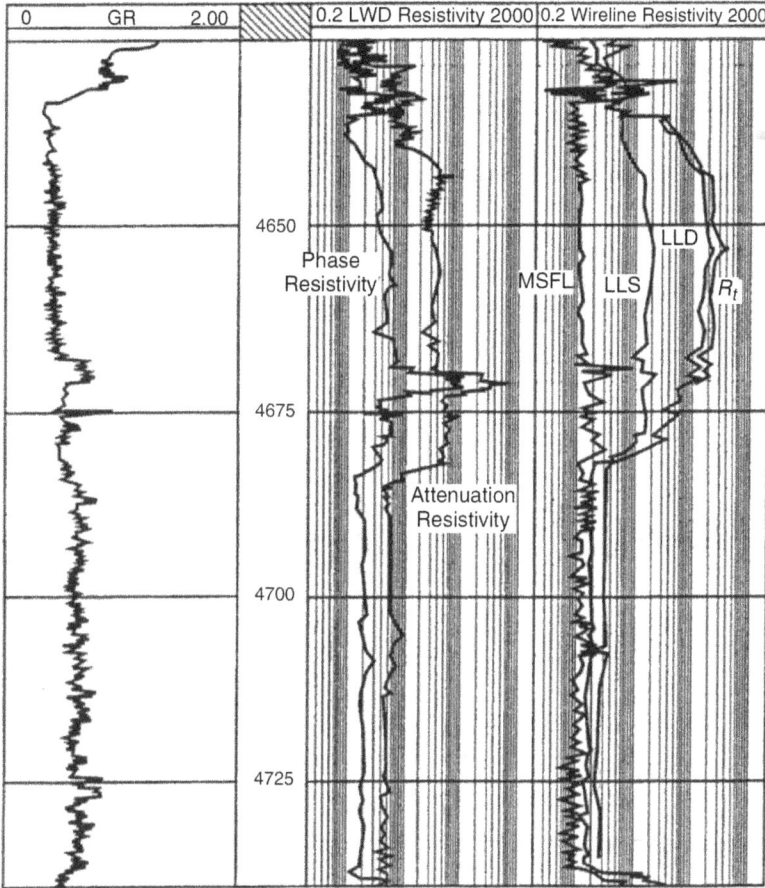

Fig. 15.11—Comparison of EWR and wireline resistivity in a deeply invaded, high-permeability sandstone; LLD = dual laterolog, LLS = single laterolog, and MSFL = micro spherically focused log. (From Economides, Watters, and Dunn-Norman, *Petroleum Well Construction,* © 1998; reproduced by permission of John Wiley & Sons Ltd.)

known to have a resistivity of 1 $\Omega \cdot$m. The phase measurement looks deeper (away from the borehole) and loses vertical resolution as the charts progress to greater resistivities. In contrast, the amplitude ratio at first looks deeper than the phase measurement, and the expected penalty of poorer vertical resolution is paid. In the most resistive case, the attenuation measurement shows a 129-in. diameter of investigation. Many electromagnetic tools transmit at variable frequencies (2MHz, 1MHz, and 400kHz) to capture the benefit of variable depths of investigation and to minimize eccentering effects. Some systems can yield more than 20 resistivity measurements per data point, from which a greater understanding of reservoir characteristics can be derived.

Dielectric effects are responsible for some discrepancies between phase and attenuation resistivity measurements. Errors are greatest in the most resistive formations. Different approaches to this issue have been taken by vendors, with some opting to assume a set dielectric-constant value such as "10," whereas others have chosen to vary the dielectric constant as a function of formation resistivity.

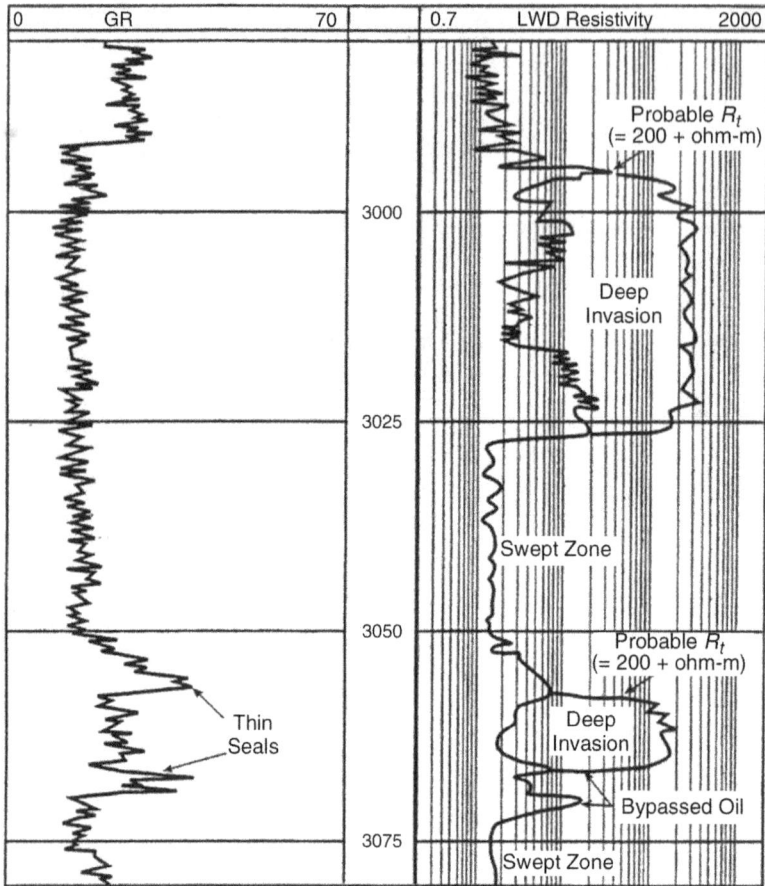

Fig. 15.12—Effects of very deep invasion by conductive muds. (From Economides, Watters, and Dunn-Norman, *Petroleum Well Construction,* © 1998; reproduced by permission of John Wiley & Sons Ltd.)

Further discrepancies between phase and attenuation resistivity measurements also may be attributed to the effects of formation anisotropy. Anisotropy may also be responsible for the separation of measurements taken at different spacings or at different frequencies. This can be easily misinterpreted as an invasion effect. Anisotropy effects are caused by differences in the resistance of the formation when measured across bedding planes (R_v) or along bedding planes (R_h). An assumption is generally made that R_h is independent of orientation. As borehole inclination increases, the angle between the borehole and formation dip typically increases. When this relative angle exceeds approximately 40°, resultant effects become significant. Anisotropy has the effect of increasing the observed resistivity above R_h. Effects are greater on the phase measurements than the attenuation measurements and greater on longer receiver spacings than short ones. It is important to understand that separation between resistivity curves caused by conductive invasion will result in the deep-resistivity-curve reading less than or equal to the true formation resistivity, whereas resistivity-curve separation caused by anisotropy will lead to measured deep resistivities being greater than true formation resistivity. The importance of trying to resolve these effects has led to a substantial and ongoing effort by the industry to develop robust, fast resistivity-modeling packages.

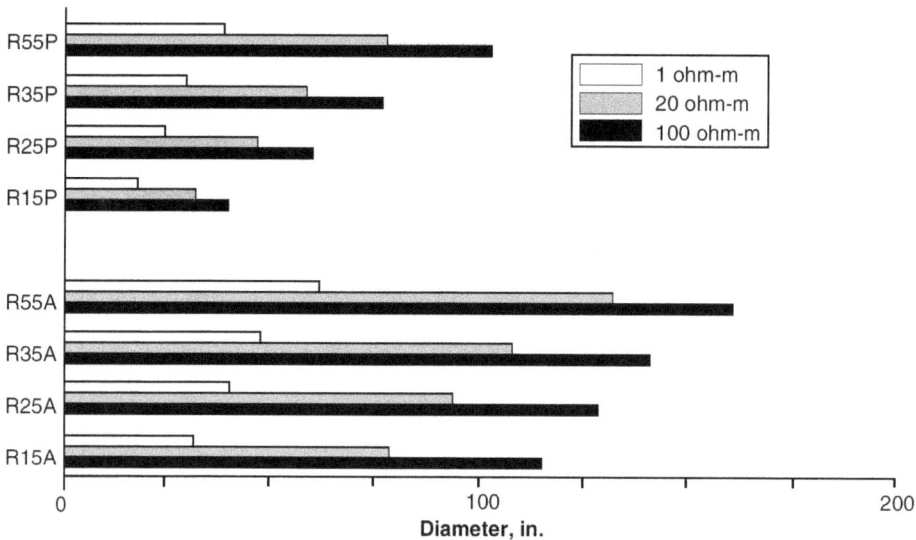

Fig. 15.13—Depths of investigation for various sensor spacings. (From Economides, Watters, and Dunn-Norman, *Petroleum Well Construction,* © 1998; reproduced by permission of John Wiley & Sons Ltd.)

Wave resistivity tools are run in most instances in which LWD systems are used, but toroidal resistivity measurements are desirable under some circumstances.[7] Toroidal resistivity tools typically consist of a transmitter that is excited by an AC, which induces a current in the BHA. Two receivers are placed below the transmitter, and the amount of current measured exiting the tool to the formation between the receivers is the lateral (or ring) resistivity. The amount of current passing through the lower measuring point is the bit resistivity (**Fig. 15.14**). Because of the large number of variables involved, bit resistivity measurements have been difficult to quantify, but measurements from current-generation tools now compare favorably with wireline laterolog measurements. In formations with high resistivities (greater than 100 $\Omega \cdot$m), measurements with a toroidal resistivity tool may be more appropriate than measurements with other tool types. An important side benefit of this technology is its insensitivity to anisotropic effects.

The log example in **Fig. 15.15** shows a case in which 2-MHz measurements have saturated because of the high salinity of the mud. If the drilling fluid is conductive or if conductive invasion is expected, then toroidal resistivity measurement is preferred. If early identification of a coring or casing point is crucial, then bit resistivity measurements give a good first look. In geosteering applications, toroidal bit resistivity measurements are an immediate indicator of a fault crossing.

The first formation images while drilling were acquired through the use of toroidal resistivity tools. When a small-button electrode is placed on the OD of a stabilizer, the current flowing through that electrode can be monitored. The current is proportional to the formation resistivity in the immediate proximity. Effective measurements are best taken in salty muds with resistive formations. Vertical resolution is 2 to 3 in., and azimuthal resolution is less than 1 in.[8] With the tool rotating at least 30 RPM, internal magnetometer readings are taken, and resistivity values are scanned and stored appropriately. A sample of the data is pulsed to the surface in real time to provide a low-resolution measurement. At the surface, tool memory is dumped, and the data are related to the correct depth. Quality checks are made to ensure that poor microdepth measurements are not affecting the reading.

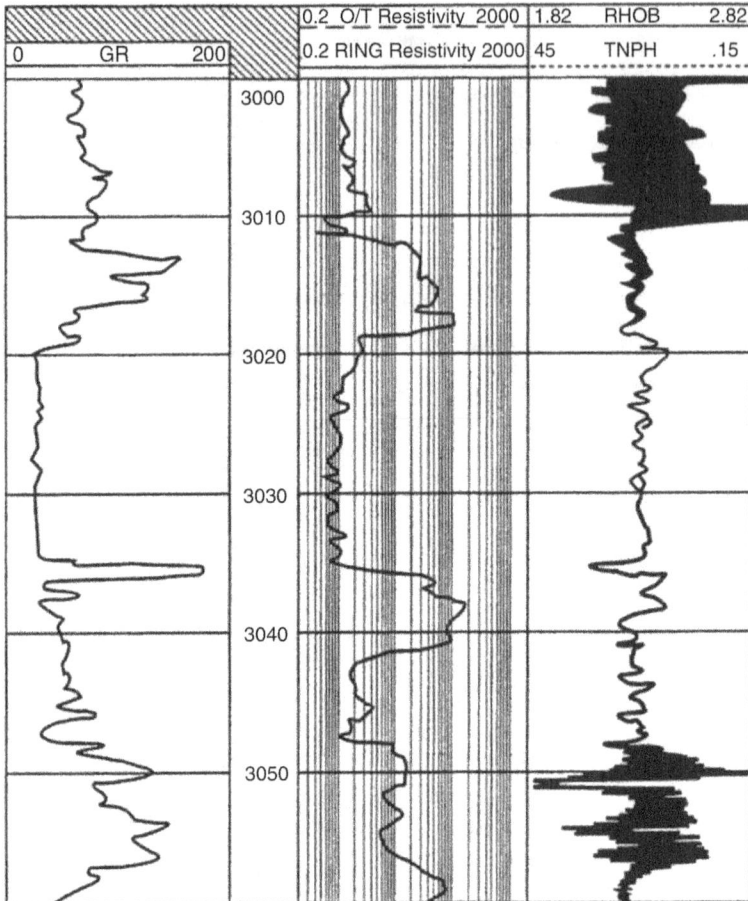

Fig. 15.14—Ring and bit resistivity measurements show good corroboration; RHOB = bulk density, and TNPH = Φ_{nt}, porosity from neutron log, thermal. (From Economides, Watters, and Dunn-Norman, *Petroleum Well Construction*, © 1998; reproduced by permission of John Wiley & Sons Ltd.)

Imaging while drilling can provide a picture of formation structure, nonconformities, large fractures, and other visible formation features. Azimuthal-density devices may also be processed to provide dip information. Imaging is increasingly used as in geosteering applications. Real-time dip calculations can be carried out in structures with relatively high apparent dips.

Nuclear Logging. Gamma ray measurements have been made while drilling since the late 1970s. These measurements are relatively inexpensive, although they require a more sophisticated surface system than is needed for directional measurements. Log plotting requires a depth-tracking system and additional surface computer hardware.

Applications have been made in both reconnaissance mode, where qualitative readings are used to locate a casing or coring point, and evaluation mode. Verification of proper MWD gamma ray detector function is normally performed in the field with a thorium blanket or an annular calibrator.[9]

The main differences between MWD and wireline gamma ray curves are caused by spectral biasing of the formation gamma rays and logging speeds.[10]

Neutron porosity (Φ) and bulk-density (ρ_b) measurements in LWD tools are often combined in one sub or measurement module. Reproducing wireline-density accuracy has proven to be

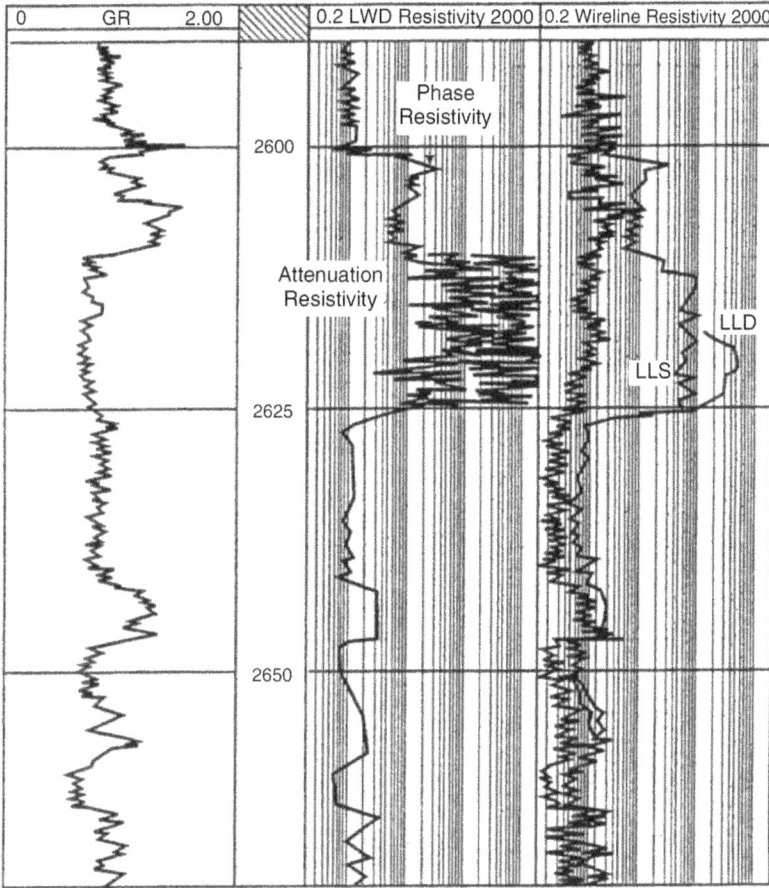

Fig. 15.15—EWR phase and attenuation resistivity measurements saturate when run in high-salinity water-based muds. (From Economides, Watters, and Dunn-Norman, *Petroleum Well Construction*, © 1998; reproduced by permission of John Wiley & Sons Ltd.)

one of the most difficult challenges facing LWD tool designers. Tool geometry typically consists of a cesium gamma ray source (located in the drill collar) and two detectors, one at a short spacing from the source and one at a long spacing from the source. Gamma counts arriving at each of the detectors are measured. Count rates at the receivers depend upon the density of the media between them. Density measurements are severely affected by the presence of drilling mud between the detectors and the formation. If more than 1 in. of standoff exists, the tendency of the gamma rays to travel the (normally less dense) mud path and "short circuit" the formation-measurement path becomes overwhelming. The gamma ray short-circuit problem is solved by placing the gamma detectors behind a drilling stabilizer. With the detector mounted in the stabilizer, in gauge holes, the maximum mud thickness is 0.25 in., and the mean mud thickness is 0.125 in. Response of the tool is characterized for various standoffs in various mud weights, and various formations and corrections are applied.

Placing the gamma detector in the stabilizer does have some drawbacks. Detector placement can affect the directional tendency of the BHA. In horizontal and high-angle wells, in which the density measurement is most frequently run, the stabilizer can sometimes hang up and prevent weight from being properly transferred to the bit. It is important to note that in

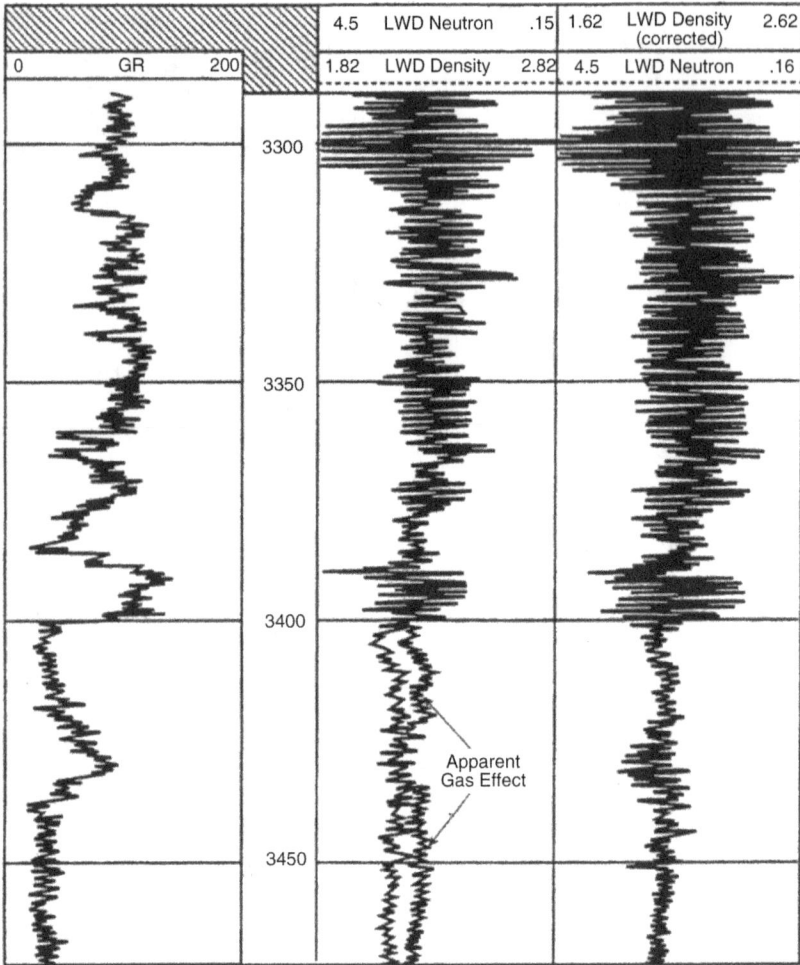

Fig. 15.16—Density logs run with undergauge stabilizers in high-angle wells can be severely affected. (From Economides, Watters, and Dunn-Norman, *Petroleum Well Construction,* **© 1998; reproduced by permission of John Wiley & Sons Ltd.)**

enlarged boreholes, gamma detectors deployed in the drilling stabilizer may not accurately measure density.

Assuming that an 8½-in. bit and an 8¼-in. density sleeve are used and the tool is rotating slowly in the hole, the average standoff is 0.125 in., and the maximum standoff is 0.25 in. If, however, the borehole enlarges to 10 in., the average standoff increases to 0.92 in., and the maximum standoff increases to 1.75 in. In big hole conditions, very large corrections are required to obtain an accurate density reading. An example of an erroneous gas effect using older-generation neutron density devices in an enlarged 9⅞-in. hole is shown in **Fig. 15.16.**

Varying approaches have been developed to obtain accurate density measurements in enlarged boreholes. Most widely accepted are the azimuthal density method, the rapid-sampling method, and the constant-standoff method. Azimuthal density links the counts to an orientation of the borehole by taking regular readings from a magnetometer.[11] When this method is used, the wellbore (which is generally inclined) is divided into multiple segments (often 4, 8, or 16). Incoming gamma counts are placed into one of the bins. From this, the segment densities and an average density are obtained. A coarse image of the borehole can be obtained when beds of

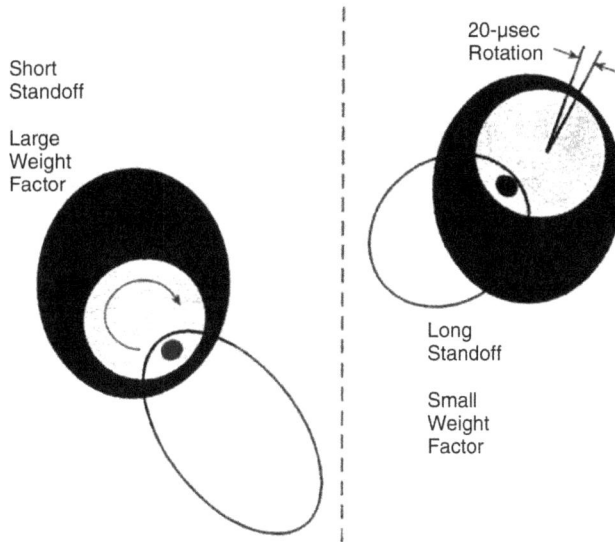

Fig. 15.17—Use of ultrasonic measurements to compensate for mud effects. (From Economides, Watters, and Dunn-Norman, *Petroleum Well Construction*, © 1998; reproduced by permission of John Wiley & Sons Ltd.)

varying density arrive in one segment before another. Azimuthal density can be run without stabilization, but it relies on the assumption that standoff is minimal in the bottom quadrant of the wellbore.

Another method is referred to as rapid sampling. In this method, statistical techniques are applied to rapid samples taken on incoming gamma counts. When the tool is rotating and there is a significant difference between mud weight and formation density, there will be an unexpectedly high standard deviation. This is used to create limits for a high- and low-count rate bin. The total counts arriving in the low-count rate bin are used to calculate a rapid sample density.

Another method of obtaining density in enlarged boreholes relies on the constant measurement of standoff using a series of ultrasonic calipers.[12] A standoff measurement is made at frequent intervals, and a weighted average is calculated. High weight is given to gamma rays arriving at the detector when the standoff is low, and low weight is given to those gamma rays that arrive when the standoff is high (**Fig. 15.17**). This method attempts to replicate the wireline technique of dragging a tool pad up the side of the borehole. The constant-standoff method can also be applied to neutron porosity tools.

All density measurements suffer if the drillstring is sliding in a high-angle or horizontal borehole with the gamma detectors pointing up (away from the bottom of the wellbore). To overcome this problem, orientation devices are often inserted in the toolstring. As the BHA is being made up, the offset between the density sleeve and the tool face is measured. Adjusting the location of the orientation device allows the density measurement to be set to the desired offset. While the drillstring is sliding to build angle, the density detectors can be oriented downward by setting the offset to 180°.

LWD porosity measurements use a source (typically americium beryllium) that emits neutrons into the formation. Neutrons arrive at the two detectors (near and far) in proportion to the amount they are moderated and captured by the media between the source and detectors. The best natural capture medium is hydrogen, generally found in the water, oil, and gas in the pore spaces of the formation. The ratio of neutron counts arriving at the detectors is calculated and stored in memory or transmitted to the surface. A high near/far ratio implies a high concentration of hydrogen in the formation and, hence, high porosity.

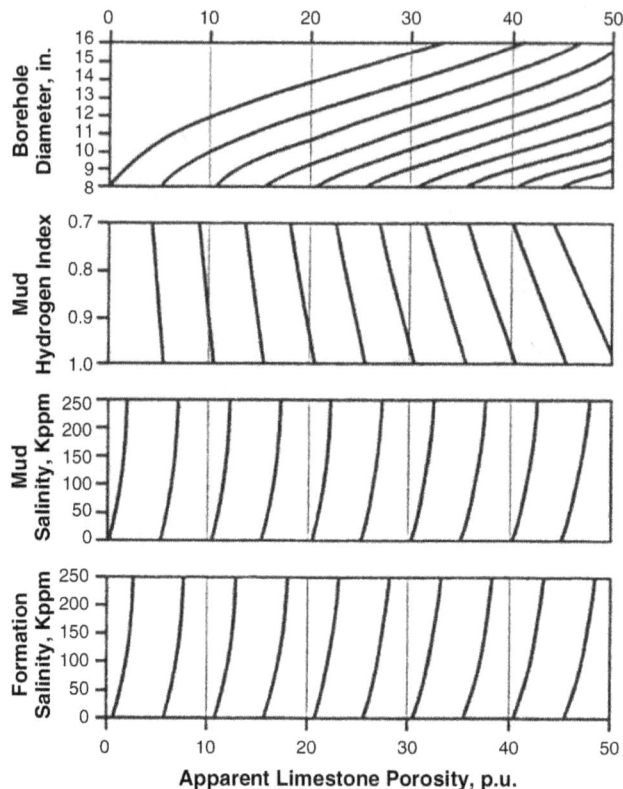

Fig. 15.18—Effects of tool centering. (From Economides, Watters, and Dunn-Norman, *Petroleum Well Construction,* © 1998; reproduced by permission of John Wiley & Sons Ltd.)

Neutron measurements are susceptible to a large number of environmental effects. Unlike wireline or LWD density measurements, the neutron measurement has minimal protection from mud effects. Neutron source/detector arrays are often built into a section of the tool that has a slightly larger OD than the rest of the string. The effect of centering the tool has been shown[13] to have a dramatic influence on corrections required compared to wireline (**Fig. 15.18**). Stand-off between the tool and the formation requires corrections of approximately 5 to 7 porosity units (p.u.) per inch. Borehole-diameter corrections can range from 1 to 7 p.u./in. depending on tool design. Neutron porosity measurements are also affected by mud salinity, hydrogen index, formation salinity, temperature, and pressure. However, these effects are generally much smaller, requiring corrections of approximately 0.5 to 2.0 p.u.

Statistical effects on nuclear measurements are quite significant. Uncertainties increase as ROP increases. LWD nuclear measurements can be performed either while drilling or while tripping. LWD rates vary because of ROP changes, but they typically range from 15 to 200 ft/ hr, whereas instantaneous logging rates can be significantly higher. Tripping rates can range from 1,500 to 3,000 ft/hr. Typical wireline rates are approximately 1,800 ft/hr and constant. Statistical uncertainty in LWD nuclear logging also varies with formation type. In general, log quality begins to suffer increased statistical uncertainties at logging rates above 100 ft/hr. This limits the value of logging while tripping to repeating formation intervals of particular interest.

Acoustic Logging. Ultrasonic caliper measurements while drilling were introduced principally for improving neutron and density measurements. Caliper transducers consist of two or more piezoelectric-crystal stacks placed in the wall of the drill collar. These transducers generate a high-frequency acoustic signal, which is reflected by a nearby surface (ideally, the borehole wall). The quality of the reflection is determined by the acoustic-impedance mismatch between the original and reflected signals. Often, there are difficulties in obtaining caliper measurement in wells with high drilling-fluid weights. Compared to the wireline mechanical caliper, the ultrasonic caliper provides readings with much higher resolution.

Acoustic-velocity data are important in many lithologies for correlation with seismic information. These data also can be a useful porosity indicator in certain areas. Shear-wave velocity also can be measured and used to calculate rock mechanical properties. Four main challenges in constructing an LWD acoustic tool are described as follows[14]:

• Preventing the compressional wave from traveling down the drill collar and obscuring the formation arrival. Unlike wireline tools, the bodies of LWD tools must be rigid structural members that can withstand and transmit drilling forces down the BHA. Therefore, it is impractical to adopt the wireline solution of cutting intricate patterns into the body of the tool to delay the arrival of the compressional wave. Isolator design is crucial and is still implemented to enable successful signal processing in a wide variety of formations, particularly the slower ones [those having a compressional delta time (Δt_C) slower than approximately 100 μsec].

• Mounting transmitters and receivers on the OD of the drill collar without compromising their reliability.

• Eliminating the effect of drilling noise from the measurement.

• Processing the data so that they can be synthesized into a single Δt_C and that this data point can be transmitted by mud pulse. This is particularly challenging given the large quantity of raw data that must be acquired and processed.

In its most basic form, an acoustic-logging device consists of a transmitter with at least two receivers mounted several feet away. Additional receivers and transmitters enhance the measurement quality and reliability. The transmitters and receivers are piston-type piezoelectric stacks that operate at a higher frequency than typical drilling noise. Drilling noise has been shown to be concentrated in the lower frequencies (**Fig. 15.19**). A data-acquisition cycle is performed as the transmitter fires and the waveforms are measured and stored. Arrival time is measured from the time the transmitter fires until the wave arrives at each receiver. From this acoustic-velocity information, the tool's downhole data-processing electronics, using digital signal-processing techniques, calculate the formation slowness or Δt_C. This value is the reciprocal of velocity and is expressed in units of μsec/ft. Waveforms also are stored in tool memory for later processing at the surface when the memory is dumped. Developments in acoustic LWD have focused on increasing the array of transmitters and receivers and operating with dual frequencies. These have shown much better ability to provide shear measurements when the shear velocity is greater than the mud velocity. When the converse condition exists, there is no shear-wave arrival, and corrections have to be applied to other modes to derive shear. The processing required both at surface and downhole has become ever more sophisticated.[15]

The log in **Fig. 15.20** shows an example of a log processed at the surface from waveforms stored downhole. Here, the Δt_C values have been reprocessed from the stored waveforms. When compared with a wireline log, this log is clearly less affected by the washout below the shoe and in the shale at X235 measured depth. LWD acoustic devices, by nature of their size, fill a much larger portion of the borehole than wireline devices and are less susceptible to the effects of borehole washout. Synthetic seismograms can be produced when acoustic and density data are combined, which yield valuable correlations with seismic information. Nevertheless, synthetic seismograms derived from LWD suffer from the same frequency-dispersion issues as wireline when making comparisons with data acquired from surface seismic.

To deal with this issue, LWD-tool designers have made progress in developing seismic-while-drilling systems that can be used to provide seismic checkshots. In this system, sensitive instruments are placed in a downhole sub connected to the telemetry system. A surface gun is located on the surface. If the well is vertical or near vertical, this might be on the rig; other-

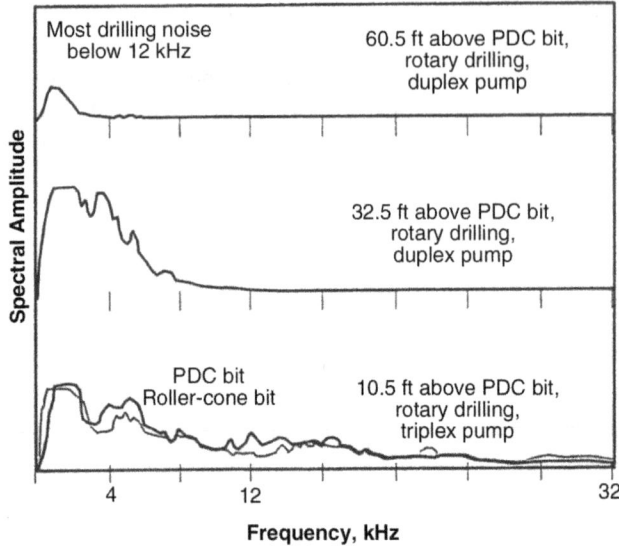

Fig. 15.19—Drilling noise is concentrated at low frequencies. (Note: PDC = polycrystalline diamond compact.) (From Economides, Watters, and Dunn-Norman, *Petroleum Well Construction,* © 1998; reproduced by permission of John Wiley & Sons Ltd.)

wise, it will be on a boat located above the receivers. When the gun is fired, which is typically at a connection to ensure quiet conditions, the arriving waveform is detected by the instrumentation and stored in memory. Processing is carried out, and information is sent to the surface, from which the one-way seismic travel time can be derived. One of the key challenges in seismic while drilling is overcoming the lack of an electrical link between surface guns and downhole receivers.

Seismic while drilling has the potential to reduce the positional uncertainty in the earth model. The main applications are in exploration wells or where there is limited confidence in the velocity model. Data can be acquired at connections either while drilling or while pulling out the hole. The cost of locating a boat on a station with guns for the duration of drilling may be an impediment to routine operations of this sort in deviated wells. Data quality is not currently thought to be adequate for processing for vertical-seismic-profile purposes.

Magnetic-Resonance-Image (MRI) Logging. Another measurement that is in the process of making the transition to a while-drilling environment is magnetic resonance. The use of chemical nuclear sources downhole has been a logistical and management headache. MRI, by measuring in real time the free-fluid, capillary-bound-water, and clay-based-water volumes, offers an alternative, lithology-independent porosity measurement in complex lithologies. It can be used for geosteering and geostopping when sufficient productive formation has been exposed to the wellbore.

Like most measurements, at an initial phase there are specialist applications that are more susceptible to realizing the value of magnetic-resonance logging. In this case, applications of interest are the evaluation of shaly sands and low-resistivity pays, particularly in deepwater and exploration wells.

Tool designers have had to meet a number of challenges in converting the measurement to a drilling environment. Shock, vibration, rotation, and general tool movement mitigate against the use of the T_2 measurement, which is sensitive to excessive motion while drilling. As a result, the T_1 measurement has been adopted as a *de facto* standard in real-time (reconnaissance) applications. This is supplemented by T_2 measurements when a more-detailed characteri-

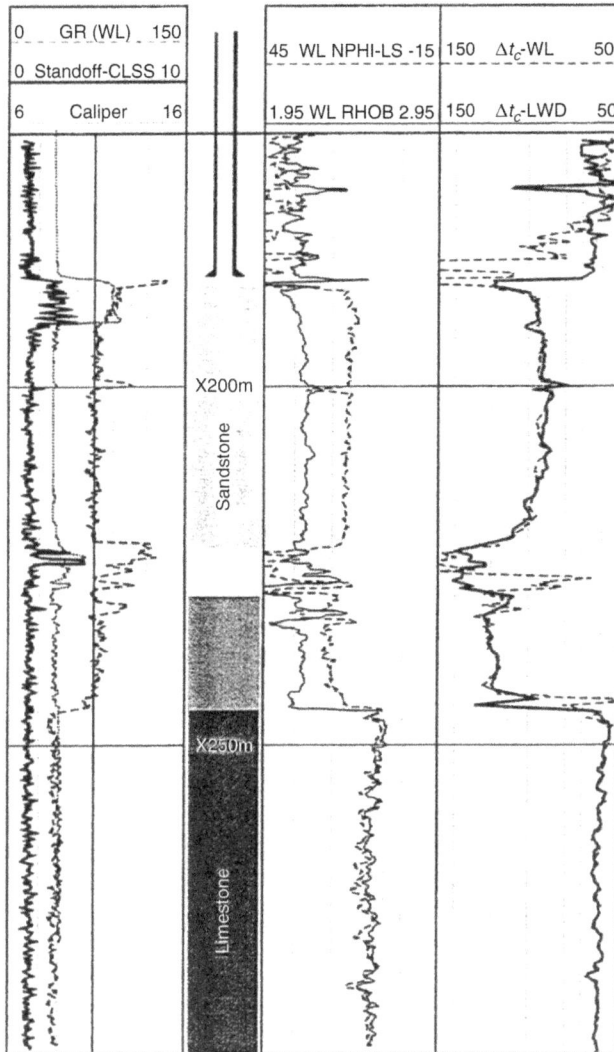

Fig. 15.20—Comparison of wireline and LWD acoustic measurements; NPHI = Φ_m, neutron porosity, GR = gamma ray log, and CLSS = caliper log. (From Economides, Watters, and Dunn-Norman, *Petroleum Well Construction*, © 1998; reproduced by permission of John Wiley & Sons Ltd.)

zation of the formation is required. Devices in use investigate a rotationally symmetric volume with a diameter of 14 in. They benefit from the generally lower ROP experienced in the drilling environment. Some care needs to be taken in the relative position of the large permanent magnets in the magnetic-resonance device and the magnetometers in the directional module, although correction algorithms can be used to eliminate interference.

15.3.4 Formation Testing While Drilling. FTWD has a broad interest in all the different disciplines involved in drilling and evaluating the well. For the drilling engineer and the geologist, a number of different approaches to the problem of acquiring formation-pressure data while drilling have been tried. A sophisticated subindustry has evolved aimed at pore-pressure prediction using proven methods such as "D exponent" (see the chapter on Drilling Geology in this section of the Handbook), connection gas, and cuttings analysis. Real-time formation-pressure data will, at a minimum, allow more-frequent calibration of pressure models. For the reservoir engineer, it opens the possibility of "barosteering"; where there is doubt in mature fields about whether a compartment has been drained, immediate measurements can be taken and a decision reached about whether to geostop or geosteer for a more-promising compartment. It allows immediate testing to verify whether geological barriers are sealing, and it opens the possibility of pressure profiling to identify (from gradient information) types of fluids present and contact points. For the drilling engineer, the precise identification of mud weight needed offers potential for improvements in ROP. For all, particularly in high-angle wells, it offers the prospect of eliminating the need to acquire costly pressure measurements by pipe-conveyed wireline techniques. All will have concerns about the time required to take a test, especially if no circulation is permitted because those conditions increase the likelihood of tool sticking.

Two different approaches have been taken to the problem of acquiring the data. The first adopts the traditional testing approach associated with drillstem tests (DSTs). In this manifestation, dual inflatable packers are mounted on the outside of the tester. When a zone of interest is reached, a command is issued from the surface, and the packers are inflated to isolate the zone of interest. The drawdown pump is activated to remove a controlled volume of fluid from the annulus between the packers. Circulation above the tool can be maintained with a diverter sub, and pressure data continue to be pumped to the surface until sufficient data are acquired. One advantage of an approach of this sort is that it investigates a greater depth in the formation and is not susceptible to seating problems in laminated formations in which there may be a chance of landing a probe on a hard streak. Conversely, the exposure of relatively large areas of inflatable packers to the wellbore environment calls for careful design and handling to avoid damage.

An alternative approach to acquiring data follows the traditional wireline approach.[16] In this, a small extendable probe with an elastomeric seal is applied to the formation on command from the surface. An internal piston is then actuated to draw down the pressure by as much as 8,000 psi below hydrostatic pressure. Formation fluids then flow into the probe and build up the pressure in the probe to the formation pore pressure. Pressure measurements are taken both with fast-acting strain gauges and high-accuracy quartz gauges. Tests can be acquired either with the pumps on or off. The drawdown and buildup profiles also provide information used in the determination of formation permeability. The reduced area associated with the probe should reduce the chances of drilling damage, and the smaller volumes involved in the test should provide reasonable data in a shorter time period (although from a shallower depth) than the DST-type design.

Early indications are that FTWD tools will be adopted quite rapidly by the industry provided that they can be shown to provide high-quality, reliable measurements without significantly increasing loss-in-hole risks.

Depth Measurement. Good, consistent knowledge of the absolute depth of critical bed boundaries is important for geological models. Knowledge of the relative depth from the top of a reservoir to the oil/water contact is vital for reserves estimates. Nevertheless, of all the measurements made by wireline and LWD, depth is the one most taken for granted (despite being one of the most critical). Depth discrepancies between LWD and wireline have plagued the industry.

LWD depth measurements have evolved from mud-logging methods. Depth readings are tied, on a daily basis, to the driller's depth. Driller's depths are based on measurements of the length of drillpipe going in the hole and are referenced to a device for measuring the height of the kelly or top drive with respect to a fixed point. These instantaneous measurements of depth are stored with respect to time for later merging with LWD downhole-memory data. The final log is constructed from this depth merge. On fixed installations, such as land rigs or jackup rigs, a number of well-documented sources exist that describe environmental error being introduced in the driller's depth method. Floating rigs can introduce additional errors. One study suggested that the following environmental errors would be introduced in a 3000-m well[17]:

- Drillpipe stretch: 5- to 6-m increase.
- Thermal expansion: 3- to 4-m increase.
- Pressure effects: 1- to 2-m increase.

Floating rigs have special problems associated with depth measurements. Errors are caused principally by rig heave and tidal action. In LWD, these effects are sufficiently overcome by the placement of compensation transducers in locations fixed with respect to the seabed.

Wireline measurements are also significantly affected by depth errors, as shown by the amount of depth shifting required between logging runs, which are often performed only hours apart. Given the errors inherent to depth measurement, if wireline and LWD ever tagged a marker bed at the same depth, it would be sheer coincidence.

Environmentally corrected depth would be a relatively simple measure to implement in LWD. Although this measure would certainly reduce depth errors, it probably would not eliminate them. The "cost" of corrected depth is an additional depth measurement that must be monitored. Driven by the increasing availability of wireline-quality measurements while drilling, the industry is beginning to realize the need to adopt a new process for measuring depth accurately. Running a cased-hole gamma ray during completion operations is a practice adopted by many operators as a check against LWD depth errors and lost-data zones.

15.4 Drilling-Data Management and Reporting

15.4.1 Overview. From the late 1960s and early 1970s to the present, oilfield drilling and well-services rigs and work units have seen an increase in electronic data-recording, monitoring, engineering, and reporting systems that have replaced manual or mechanical recording systems and hard-copy paper reports completed by rigsite personnel. Implementation of service-company, operator, and rig-contractor software systems has enabled the electronic capture of drilling and well-services operations and equipment data that provide significant value to engineers involved in operations monitoring, data analysis, well planning, and external reporting. Live capture of real-time data fed into engineering and geoscience systems has enabled asset-team members to make more-informed timely decisions that positively affect wellbore placement, resulting in more-profitable wells for the operator.

Advancement of rigsite software systems has seen applications evolve from early mainframe to mini-computer systems to UNIX multitasking systems, Microsoft DOS applications, Microsoft Windows applications, and the current emergence of Intranet or Internet applications. Early systems used by single operators developed in-house have now been replaced by customizable commercial systems shared by a large number of operators.

15.4.2 Rigsite Software Systems. *Service Company.* The most comprehensive data-acquisition systems present at the rigsite are provided by service companies such as mud-logging, MWD/ LWD, and wireline vendors. Real-time data-acquisition systems typically are connected to a suite of surface and downhole sensors that enable live monitoring of the rig-equipment operation and the well-construction process. Service-company systems are typically capable of accepting Wellsite Information Transfer Specification (WITS) inputs from other vendors so that sensor readings from all data-acquisition systems may be collated into a single real-time data set that may be provided to the operator at the end of the well. In addition to collating sensor readings, service-company software systems also enable various interpretative reports to be entered into the system depending on the service provided, such as mud logs, drilling-data logs, pressure logs, wellsite geology, mud, and cementing. The combination of surface and downhole sensors with networked graphical data logs and text outputs enables the operator's supervisory staff, service company, and rig contractor to maintain an accurate picture of the drilling or well-services operation and track well progress to ensure that the new-wellbore placement or completion meets the operator's safety, geologic, and production requirements.

Rig Contractor. Rig-contractor personnel may use any number of commercially available electronic tour-sheet applications that enable them to complete their Intl. Assn. of Drilling Contractors (IADC)/Canadian Assn. of Oilwell Drilling Contractors (CAODC) report electronically on a PC rather than fill in traditional paper-based forms. These electronic tour-sheet applications may be hooked up to the rig's own data-acquisition system, which records surface-sensor readings from all rig equipment, such as hookload, WOB, ROP, kelly or stand height, surface torque and RPM, pump pressure, pump flow rate, pump speed, and pit volumes, all in an electronic drilling recorder (EDR) system.

Increasingly, data from rig-contractor EDR systems and service-company systems are being supplied live back to the beach or office and made available as a service to operators through commercial Website offerings that provide online or offline logs of drilling and well-services data.

Operator. From an operator's perspective, rigsite data acquisition typically consists of daily operations morning reporting systems, survey-data management, and well-engineering software systems.

Operations Reporting. The daily operations report is the operator's record of the construction, completion, workover, or abandonment operation occurring on the well. The daily operations report is a comprehensive record of all daily activity and equipment operations that occur over a reporting interval. Current operations status, progress and current formation/lithology information, time summary information, and daily cost, survey, drilling fluids, bit, BHA, mud-cleaning-equipment, safety, personnel, support-craft, and weather information are typically entered. Rigsite supervisors or field engineers enter a number of associated reports depending on the type of well operation, rig equipment used, or operator and regional government reporting requirements.

For the drilling process, reports are typically entered for daily operations, pipe tallies, casing, cementing, wellsite geology, coring, logging, and DSTs when these operations occur. For completion and workover operations, engineers enter reports for downhole wellbore equipment, wellhead installations, perforation, stimulation, remedial cementing, production tests, and pressure surveys. For artificial-lift completions, engineers will enter detailed report information for conventional pumps, gas lift, electrical submersible pumps, progressing-cavity pumps, and hydraulic-lift completions. For all types of operations, performance is measured through detailed, planned (vs. actual) activity tracking, NPT analysis, and equipment-failure analysis. Operational learnings are recorded and collated in lessons-learned systems associated with key data parameters so that this information may be shared across an organization and used for future well-performance assessments or well-planning operations. Health, safety, and environmental assessment and monitoring of the well operation and fluids/chemicals used are an increasingly important part of the well-operations reporting process.

Survey-Data Management. Correct placement of the wellbore to meet geological and production requirements is the primary goal of any drilling operation. In the office, directional-well planners will use a survey-data-management solution to design the well trajectory to intersect one or more drilling targets, avoid adjacent wellbores within safe collision-avoidance tolerances, and not exceed other well-design criteria. At the rigsite, the system is used to record survey-station data for specific survey-tool runs. Survey-tool error models are used to calculate positional uncertainty down the wellbore. The definitive wellpath is updated continuously to calculate the most accurate well trajectory, compare planned vs. actual well trajectory, and perform anticollision risk assessment for any nearby wellbores. Tools are also available to quality assess the survey data to ensure that survey-station data are within acceptable tolerances.

Well Planning/Drilling Engineering. Many commercial software vendors provide a suite of drilling-engineering applications that enable casing/tubing design, torque/drag, hydraulics, hole cleaning, swab/surge, well control, cementing, drillstring-vibration/directional-performance, and wellbore-stability analysis to be performed. These engineering systems enable well planners to design the well within concise engineering constraints. These planned models are updated during the drilling process to monitor the well and to ensure that design constraints are not exceeded.

Drilling/Rigsite Simulators. The electronic capture of real-time rig-operations information into rig or drilling simulations or modeling systems enables the users of these systems to "play back" the well operation so that detailed research or analysis may be performed. This enables researchers to simulate the use of new technologies or monitoring systems before their actual use at the rigsite. The increased availability of usable data sets provided by various rigsite data-acquisition vendors in WITS or Wellsite Information Transfer Standard Markup Language (WITSML) format is enabling operators to store this information consistently within their own data stores. Previously, service companies could provide real-time information only in proprietary or other nonstandard formats, making consistent storage of this data for reuse much more difficult.

Other Software Systems. Associated rigsite systems used by operators include site construction and reclamation software and environmental-assessment and -monitoring systems. The rig contractor and/or the operator may also be using human resources systems and materials/inventory-tracking software systems to manage the flow of personnel and materials to and from the rigsite. A new software area at the rigsite and in the office is e-invoicing, where service-company and materials/equipment vendors invoice the operator electronically using Extensible Markup Language (XML) -based systems instead of traditional paper invoices or field tickets.

15.4.3 Enter It Once! The Value of Integration. Historically, all these types of rigsite software systems have been separate applications or application suites hosted on separate data stores and IT infrastructures with little to no connectivity between them. These software systems did not integrate because they were used by different companies, teams of users, or single users who did not expect integration because they were using their software to perform specific tasks. With increasingly complex and costly drilling and well-services operations and technologies, all office rigsite personnel who use well-information management systems today expect to use innovative suites of applications that integrate across the geoscience, well-engineering, and rigsite-management disciplines.

The current trend in oilfield software development is to provide integrated systems used by multiple well-engineering disciplines that support numerous engineering workflows that meet rigsite monitoring requirements. These systems use a single common repository of well data that covers an ever-increasing extent of the well life cycle from initial wellsite environmental surveys, initial well construction, and completion to production field-data capture, accounting, economics, workover, abandonment, site reclamation, and follow-up environmental monitoring. Engineers expect to see efficiencies resulting from shared use of a common data store that enables them to more efficiently perform their specific tasks or perform analysis without having to duplicate or transfer information entered elsewhere.

Where systems do not share the same data store, field users expect to be able to import or exchange data between systems with no loss of content or data quality. To meet this requirement, electronic data-exchange systems have evolved from the 1980s WITS standard and various system-specific methods to modern XML-based systems. Additionally, standardization of software systems on the Windows operating system enables rigsite systems to exchange information through Microsoft OLE and ODBC standard methods.

15.4.4 Value From Data. The shared use of information at the rigsite or data transmitted in real time or offline to the office is used for a variety of purposes that provide real value to the operator. Operators implement corporate stores of this information to realize several goals:

• Enabling an open database to reliably store historical drilling, completion, and well-services information in a common data store.

- Providing instant access to data across the organization.
- Supporting consistent rigsite data capture and reporting across all operations.
- Supporting the implementation of consistent data-quality methods and procedures.
- Providing consistent output reports and electronic output formats.
- Supporting multiple units of measure.
- Enabling operations engineers to remotely oversee drilling and well-services operations.
- Enabling operations statistics and performance benchmarks to be performed so that procedures requiring improvement can be identified.
- Providing well planners with accurate historical operations-performance data with which to perform statistical risk analysis for future well operations.
- Making informed decisions with greater effectiveness at the time they have to be taken.

Output Reporting. From an operator's perspective, the most immediate benefit of rigsite software systems that collate information is to enable consistent output reporting through all types of well operations and across all geographic areas. Daily well-operations information is required by operations engineers supervising well progress, fellow asset-team members, senior managers, and members of associated disciplines such as materials management, accountancy, and health, safety, and environment. Traditionally, operations reports have been faxed in from the rig or completed in the office using information provided from the rig by telephone. Increasingly, information-management systems enable operations-report data sets to be sent electronically from the rig to the office so that the data may be used in town. Hard-copy reports can then be distributed from the office, often generated through automated systems that filter data. Increasingly, electronic output report formats such as Adobe Acrobat Reader (PDF) and dynamically populated Websites are used to disseminate well-operations information across the various disciplines.

In many regions, local or federal government agencies require well-operations and equipment information to be submitted as hard-copy or electronic reports so that the government has an accurate record of the well operation and completion. Hard-copy reports in government-required formats are easily generated from electronic information systems. Digital data-submission files also can be extracted from electronic data stores and formatted to the government requirement so that they may be uploaded directly into government master data stores. Examples of digital data submissions include the Norwegian Petroleum Directorate DDRS system for daily operations data and the Alberta Energy & Utilities Board Guide 59 Standard for event-summary data for each phase of well operations.

Wellbore Schematics. Historical wellbore-equipment visualization based on field-entered data is a key requirement for many operators, who demand accurate wellbore-equipment schematic diagrams and reports to be automatically drawn from well-operations data. Some systems enable wellbore drawings to be generated directly from the operations reporting system data store for any phase of the wellbore life history. Other products enable detailed wellbore-equipment schematic diagrams to be constructed manually and associated to planned or actual equipment parameters. A completion manager enables slick wellbore drawings to be manually constructed. Well-services engineers in the office and in the field about to go on a job require the ability to quickly generate an up-to-date wellbore drawing that enables them to plan their next job.

Data Analysis. The primary function of a well-operations database is to enable analysis of the captured data so that they may be used to improve future well operations. This enables the operator to use the information as a real asset that provides value. A well-organized well-operations data model should easily facilitate analysis through use of simple Structured Query Language (SQL) queries, summary output reports, and sophisticated data-analysis tools. This enables operator engineers to perform any kind of structured query for a variety of analyses, performance benchmarking, research, or collation of statistical information for corporate or government reporting. Typically, commercial software systems now provide data-analysis tools with which queries and analyses can be shared across the network. These systems store queries with the data so that they can be reused at any time.

Performance Benchmarking. Well planners and operations engineers are often required to analyze the cost or operations performance of their drilling and/or well-services operations. These analyses may be performed to identify areas for improvement, as well as to identify operators or operations that are performing above or below standard, or they may be performed to compare various operator or contractor performances. Analyses also may be performed to compare different well-construction methods or technologies to evaluate their effectiveness. The electronic capture of data at the rigsite integrated into corporate reporting systems or data stores enables the operator to perform these types of analysis.

Technical-Limit Well Planning and Operations. A high-profile well-planning and operations-monitoring method used by an increasing number of operators is technical-limit drilling (TLD) or well services. The technical limit is defined as the most optimal well-construction process that enables the well to be drilled or serviced safely in as short a time as possible. The method is used to challenge well-construction teams to reach their objective safely while identifying performance bottlenecks or procedures that may be performed more quickly with other methods or technologies while achieving the same result. Many operators have formal technical-limit initiatives in place that enable the entire well-construction team to improve operational performance. A significant part of the TLD process is the historical analysis of comparable offset-well data, which enables the well-planning team to identify the most efficient procedures and best performance for each phase of the well operation. This analysis of historical data is enabled through the capture of operations or activity information at the rigsite. Without offset-well data, identification of the desired "gold medal" performance is difficult if not impossible.

With a technical-limit operation plan defined for the new well operation, the actual execution of the well program may be compared to the technical-limit performance identified for each phase of the well operation. Deviations of actual performance from the technical-limit plan may be recorded for both improved or degraded performance to identify more-efficient procedures or technologies, as well as reasons why targeted performance was not achieved. Recording this information enables future well-planning teams to incorporate these findings into future well designs.

Knowledge Management. With historically inadequate replacement of employees leaving the industry, the oil field is currently witnessing decreased availability of experienced knowledge workers. The result is that fewer people are available to perform the same level of activity that has been performed previously. Additionally, other factors, such as a reduced occurrence of easy-to-find, accessible hydrocarbon reservoirs and an increased demand for hydrocarbons caused by an increasing population and more energy-demanding industries and technologies, have forced the industry to use increasingly more-complex operations methods, equipment, and technologies to replace existing hydrocarbon reserves. With the reduced availability of experienced knowledge workers, operators are looking at various technologies to enable their workforce to more effectively leverage the knowledge and experience retained within the corporation so that new or existing technologies, methods, and equipment may be used more efficiently.

Many operators and service companies are looking at knowledge-management best practices as a framework for capturing engineering experience, lessons-learned information, and results for various procedures and technologies. Storing this knowledge in an information-management system enables operators to distribute it more effectively across the organization to maximize its value. Different types of knowledge-capture systems are being implemented across the industry, including the rigsite, where immediate operational knowledge or experience may be recorded or referenced to improve operations. This enables service companies to more easily disseminate operations experience across the organization for their various product service lines and equipment. Operators are able to more easily share well-construction and well-planning experience across various operating regions. Information systems and other information technologies increasingly are being used to bring together experts of the same domain or discipline to form "networks of excellence" in which experience or other knowledge may be shared.

15.4.5 Data-Management Systems. *Overview.* "E&P project data-management" systems are database-management systems designed to support integrated suites of exploration and production applications. Applications share data through a single common data store. Data are administered with a single set of tools for importing, exporting, viewing, and editing and for performing database administration. By centralizing all data available for an E&P asset in a single integrated data store, project data-management systems greatly reduce the amount of time spent moving data between applications. Data-management systems serve as readily accessible repositories for the knowledge about an asset. This knowledge comes from a variety of studies and continues to grow over the life of the asset.

Before integrated project data-management systems, each fit-for-purpose E&P application had its own private, proprietary data store. Each individual data store supported different data-exchange formats and procedures. Moving data between applications involved exporting files, reformatting them manually, and importing them into the target application. This process was often so cumbersome that data were not exchanged at all, or they could be exchanged only by manually retyping the data into each application. Project data-management systems allow applications to share data without moving them from application to application. Outputs created by one application are automatically available in all applications connected to the project data-management system.

15.4.6 Key Features and Functions of Project Data-Management Systems. *Broad Application Support.* Project data-management systems should support a rich set of E&P applications that solve a broad range of technical problems. Key workflows should be completed entirely within a system, without the use of external tools for data manipulation.

Open Extensible Environment. Given the diversity of the oil and gas industry, no software vendor can offer a solution to all problems. Instead, project-management systems should provide an open-development environment, allowing niche application vendors to plug in "best of breed" applications.

Technology Based on a Standard Database. Many requirements of an E&P project data-management system are similar to those of database systems in other industries. Systems based on common horizontal-market technologies allow the use of relatively cheap and powerful horizontal-market database tools.

The most mature database-management systems are "relational databases." Relational databases have been used by many industries to store mission-critical data for more than 25 years. Researchers at IBM performed much of the early research on the relational model in the late 1960s and early 1970s.[18] A relational model views data logically as a series of tables and columns, with a mathematical model for operations on these structures. The physical arrangement of data is hidden; instead, one depends only on a simple, logical view of tabular data. All data are reduced to the simple "flat" tabular form. Relational databases support SQL, which allows users to build queries that filter the rows of a single table. SQL queries can also combine data from one table with data from another on the basis of shared foreign key fields. This allows SQL statements to join data from multiple tables and build powerful ad hoc reports.

Relational database-management systems range from desktop databases to enterprise data-management systems. Robust database systems typically provide:

• Network access to data; flexible and powerful data security; tools for "hot backups," allowing a system to be backed up without shutting down; and recovery tools in case of a system crash.

• A rich set of utilities that allow administrators to configure, control, and monitor the system.

Several very powerful tools have been developed for working with relational databases. These include systems that generate flexible ad hoc reports with rich format control and systems to build queries graphically without the use of SQL. Many tools that were originally designed for the horizontal market can be used when working with E&P data. These tools expose the data model of the underlying database. For a project database to be readily accessible by these generic tools, it must use relational technology and have a relatively simple and well-documented data model.

Object-oriented databases are a newer trend in database-management systems. They have the flexibility to store complex, structured data and to associate software logic with that data. Object databases allow data to be stored in the form needed by today's object-oriented applications. This can create a performance advantage, but the flexibility offered by object databases makes it difficult to write generic tools or evaluate arbitrary user-defined queries against object databases. Many large database vendors are moving toward a "hybrid database." Hybrid databases contain most features of relational databases, but they extend the table/column view of a relational database to allow a column to contain rich, complex user-defined data types. If used carefully, this allows developers the best of both worlds. Most data are modeled relationally for flexibility and ease of query. When performance becomes critical, however, certain columns are modeled with optimized object structures. This is particularly important for E&P data because this flexibility is important for managing certain E&P data types.

Efficient Handling of Bulk Data. Many E&P data types lend themselves readily to representation as relational rows and columns. However, a few critically important data types in the petroleum industry have special performance concerns because of their size. Well logs, seismic surveys, and continuous sensor readings are examples of data types that can produce very large amounts of data. Large data items cannot be stored efficiently using the row/column abstraction of relational databases. These data are stored more efficiently in unstructured "blob" data types, either within the database or within operating-system files outside of the database. An E&P project-management system should blend these specialized data types seamlessly with more traditional relational data in integrated user presentations and displays. The location of data should be irrelevant to the user.

Rich Suite of Project Data-Management Tools. Although many key data-management functions are provided by generic database-management systems, it is the responsibility of the E&P project-management system software to provide both data-type and domain-specific functionality to manage E&P technical data. Project-management systems provide rich data-management utility applications that allow:

• Flexible importing/exporting of data.
• Data browsing/querying/editing.
• Simple project database administration.

Project data-management tools should isolate end users from the complexity of working directly with the database when performing common data-management tasks.

Data import/export routines support a wide variety of common data-exchange formats. They should provide management for units of measure and unit conversion, if necessary. They also should convert surface locations between different map-projection systems. Other domain-specific functionality is provided when importing particular data types (e.g., the computation of wellbore paths from directional survey information).

Data browsing, querying, and editing tools should fill in the gaps left by horizontal-market query and browse tools, offering industry-specific displays for key data types such as well logs, production plots, and seismic displays. These tools should allow the updating and editing of project data and should enforce standard business rules and data integrity.

Project-administration tools should allow users with relatively little database knowledge to perform the following:

- Project database creation.
- Control of user access to a project database.
- Backup and restoration of a project database.
- Allocation of disk space and other database resources to a project database.

Support Industrywide Standards. Several industry consortia, including the Petrotechnical Open Software Corp. (POSC; www.posc.org) and the Public Petroleum Data Model Assn. (PPDM; www.ppdm.org) offer standard data models for many common E&P data types. As vendors move to support these standards, it should become easier to integrate data between project data-management systems from different vendors.

Integrate User Experience Across Applications. Project data-management systems should provide a framework for applications to work together seamlessly using the same data. This requires more than sharing the same database. Applications should be notified when another application changes data in the shared database, allowing them to refresh their display to reflect changes made in other applications. In addition, a data selection made in one application should be available in another application. Consider selecting a well in map view to view in a utility that provides cross-sectional views of downhole well equipment. It is more efficient for a user to select a well in map view and send it to the utility than for a user to type in the name of the well in each utility.

This integrated functionality typically requires an interprocess communication scheme, allowing different applications in a user session to communicate. In addition, "session management" is needed so that users may select parameters that apply to all applications in a session (e.g., the active project or units of measure for display).

Nomenclature

R_h = formation resistivity along bedding planes

R_t = true formation resistivity

R_v = formation resistivity across bedding planes

T_1 = longitudinal, or spin-lattice, relaxation time; this time constant characterizes the alignment of spins with the external static magnetic field. Refer to the chapter on Logging in this section of the *Handbook* for more information.

T_2 = transverse, or spin-spin, relaxation time; this time constant characterizes the loss of phase coherence that occurs among spins oriented at an angle to the main magnetic field, caused by interactions between spins

Δt_c = formation slowness (reciprocal velocity), μsec/ft

Φ = neutron porosity

ρ = density

ρ_b = bulk density

References

1. Spinnler, R.F. and Stone, F.A.: "Mud Pulse Logging While Drilling Telemetry System Design, Development and Demonstrations," paper presented at the 1978 IADC Drilling Technology Conference, Houston, 7 March.
2. Jellison, M.J. *et al.*: "Telemetry Drill Pipe: Enabling Technology for the Downhole Internet," paper SPE 79885 presented at the 2003 SPE/IADC Drilling Conference, Amsterdam, 19–21 February.

3. Burgess, T.M. and Martin, C.A.: "Wellsite Action on Drilling Mechanics Information Improves Economics," paper SPE 29431 presented at the 1995 SPE/IADC Drilling Conference, Amsterdam, 28 February–2 March.

4. Hutchinson, M., Dubinsky, V., and Henneuse, H.: "An MWD Assistant Driller," paper SPE 30523 presented at the 1995 SPE Annual Technical Conference and Exhibition, Dallas, 22–25 October.

5. Ng, F.: "Recommendations for MWD Tool Reliability Statistics," paper SPE 19862 presented at the 1989 SPE Annual Technical Conference and Exhibition, San Antonio, Texas, 8–11 October.

6. Shen, L.C.: "Theory of a Coil-Type Resistivity Sensor for MWD Application," *The Log Analyst* (September–October 1991) **32,** No. 5, 603.

7. Gianzero, S. *et al.:* "A New Resistivity Tool for Measurement-While-Drilling," paper A presented at the 1985 SPWLA Annual Logging Symposium, Dallas, 17–20 June.

8. Rosthal, R.A. *et al.:* "Formation Evaluation and Geological Interpretation From the Resistivity-at-the-Bit Tool," paper SPE 30550 presented at the 1995 SPE Annual Technical Conference and Exhibition, Dallas, 22–25 October.

9. Brami, J.B.: "Current Calibration and Quality Control Practices for Selected Measurement-While-Drilling Tools," paper SPE 22540 presented at the 1991 SPE Annual Technical Conference and Exhibition, Dallas, 6–9 October.

10. Coope, D.F.: "Gamma Ray Measurements While Drilling," *The Log Analyst* (January–February 1983) **24,** No. 1, 3.

11. Holenka, J. *et al.:* "Azimuthal Porosity While Drilling," *Trans.,* 1995 SPWLA Annual Logging Symposium, Paris (26–29 June 1995) paper BB.

12. Moake, G., Beals, R., and Schultz, W.: "Reduction of Standoff Effects on LWD Density and Neutron Measurements," *Trans.,* 1996 SPWLA Annual Logging Symposium, New Orleans (16–19 June 1996) paper V.

13. Allen, D.F. *et al.:* "The Effect of Wellbore Condition on Wireline and MWD Neutron Density Logs," *SPEFE* (March 1993) 50.

14. Aron, J. *et al.:* "Sonic Compressional Measurements While Drilling," *Trans.,* 1994 SPWLA Annual Logging Symposium, paper SS.

15. Market, J. *et al.:* "Processing and Quality Control of LWD Dipole Sonic Measurements," *Trans.,* 2002 SPWLA Annual Logging Symposium, Oiso, Japan (2–5 June 2002) paper PP.

16. Proett, M. *et al.:* "Formation Testing While Drilling, A New Era in Formation Testing," paper SPE 84087 presented at the 2003 SPE Annual Technical Conference and Exhibition, Denver, 5–8 October.

17. Kirkman, M. and Seim, P.: "Depth Measurement with Wireline and MWD Logs," in *Measurement While Drilling,* Reprint Series, SPE, Richardson, Texas (1989) **40,** 27.

18. Codd, E.F.: "A Relational Model of Data for Large Shared Data Banks," *Communications Assn. of Computing Machinery* (1970) **13,** No. 6.

General References

Ahmed, U., Bordelon, D., and Allen, D.: "MWD Rock Mechanical Properties To Avoid Drilling Related Problems," paper SPE 25692 presented at the 1993 SPE/IADC Drilling Conference, Amsterdam, 23–25 February.

Allison, J.L., Rezvani, M., and Leake, R.E.: "Time and Cost Reductions Through a Database-Designed Directional Drilling Program," paper SPE 17218 presented at the 1988 IADC/SPE Drilling Conference, Dallas, 28 February—2 March.

Amar, Z.H.B.T.: "The Benefits of Logging While Drilling (LWD) for Formation Evaluation in the Dulang West Field," *SPEREE* (December 1998) 496.

Bates, T.R. Jr. and Martin, C.A.: "Multi-sensor Measurements-While-Drilling Tool Improves Drilling Economics," *Oil & Gas J.* (19 March 1984) **82,** No. 11, 119.

Bernasconi, G. *et al.:* "Compression of Downhole Data," paper SPE 52806 presented at the 1999 SPE/IADC Drilling Conference, Amsterdam, 9–11 March.

Bonner, S.D. *et al.*: "New 2-MHz Multiarray Borehole-Compensated Resistivity Tool Developed for MWD in Slim Holes," paper SPE 30547 prepared for the 1995 SPE Annual Technical Conference and Exhibition, Dallas, 22–25 October.

Booth, J.E. and Hebert, J.W. II: "Support of Drilling Operations Using a Central Computer and Communications Facility With Real-Time MWD Capability and Networked Personal Computers," paper SPE 19127 presented at the 1989 SPE Petroleum Computer Conference, San Antonio, Texas, 26–28 June.

Brandon, T.L., Mintchev, M.P., and Tabler, H.: "Adaptive Compensation of the Mud Pump Noise in a Measurement-While-Drilling System," *SPEJ* (June 1999) 128.

Bryant, T.M., Grosso, D.S., and Wallace, S.N.: "Gas-Influx Detection With MWD Technology," *SPEDE* (December 1991) 273.

Bryant, T.M. and Wallace, S.N.: "Field Results of an MWD Acoustic Gas Influx Detection Technique," paper SPE 21963 presented at the 1991 SPE/IADC Drilling Conference, Amsterdam, 11–13 March.

Cantrell, L.A. *et al.*: "Case Histories of MWD as Wireline Replacement: An Evolution of Formation Evaluation Philosophy," paper SPE 24673 presented at the 1992 SPE Annual Technical Conference and Exhibition, Washington, DC, 4–7 October.

Carman, G. J. and Hardwick, P.: "Geology and Regional Setting of the Kuparuk Oil Field, Alaska." *Oil & Gas J.* (22 November 1982) **79,** No. 47, 153.

Chunduru, R. *et al.*: "Joint Inversion of MWD and Wireline Resistivity Measurements," paper SPE 59004 presented at the 2000 SPE International Petroleum Conference and Exhibition in Mexico, Villahermosa, Mexico, 1–3 February.

Dalton, C.L., Paulk, M.D., and Bittar, M.: "Real-Time, Time-Lapse Resistivity Logging With a Wired Composite Tubing," paper SPE 74380 presented at the 2002 SPE International Petroleum Conference and Exhibition in Mexico, Villahermosa, Mexico, 10–12 February.

Clark, D.D. and Barth. J.W.: "Calculator Programs Guide Directionally Drilled Wells Through Tangled Thums Lease," *Oil & Gas J.* (10 October 1983) **80,** No. 41, 100.

Close, D.A., Owens, S.C., and MacPherson, J.D.: "Measurement of BHA Vibration Using MWD," paper SPE 17273 presented at the 1988 IADC/SPE Drilling Conference, Dallas, 28 February–2 March.

Coope, D., Shen, L.C., and Huang, F.S.C.: "The Theory of 2 MHz Resistivity Tool and Its Application to Measurement-While-Drllling," *The Log Analyst* (May–June 1984) **25,** No. 3, 35.

Courteille, J.-M. and Boutrolle, P.: "Drilling and Completion Computerized Information System: PINFOR," paper SPE 20338 presented at the 1990 SPE Petroleum Computer Conference, Denver, 25–28 June.

da Bruijn, H.J., Kemp, A.J., and van Dongen, J.C.M.: "The Use of MWD for Turbodrill Performance Optimization as a Means to Improve Rate of Penetration," paper SPE 13000 presented at the 1984 European Petroleum Conference, London, 25–28 October.

Date, C.J.: *An Introduction to Database Systems,* Addison Wesley, New York City (1999).

Date, C.J. and Darwin, H.: *Foundation for Future Database Systems: The Third Manifesto,* second edition, Addison-Wesley, New York City (2000).

Denison, E.B.: "High Data-Rate Drilling Telemetry System," *JPT* (February 1979) 155.

Dewan, J.T.: *Essentials of Modern Open-Hole Log Interpretation,* PennWell, Tulsa (1983).

Domangue, P.M. and Peressini, R.J.: "A Novel Application of Open Systems at the Wellsite," paper SPE 24424 presented at the 1992 SPE Petroleum Computer Conference, Houston, 19–22 July.

Dumont, J.: "What Strategy for a Technical Drilling Information System?," paper SPE 30211 presented at the 1995 SPE Petroleum Computer Conference, Houston, 11–14 June.

Efnik, M.S. *et al.:* "Using New Advances in LWD Technology for Geosteering and Geologic Modeling," paper SPE 57537 presented at the 1999 SPE/IADC Middle East Drilling Technology Conference, Abu Dhabi, UAE, 8–10 November.

Farruggio, G.: "Innovative Use of BHAs and LWD Measurements to Optimize Placement of Horizontal Laterals," paper SPE 52825 presented at the 1999 SPE/IADC Drilling Conference, Amsterdam, 9–11 March.

Fay, J.B., Fay, H., and Couturier, A.: "Wired Pipes for a High-Data-Rate MWD System," paper SPE 24971 presented at the 1992 SPE European Petroleum Conference, Cannes, France, 16–18 November.

Fontenot, J.E.: "Measurement While Drilling—A New Tool," *JPT* (February 1986) 128.

Ford, G. *et al.:* "Dip Interpretation from Resistivity at Bit Images (RAB) Provides a New and Efficient Method for Evaluating Structurally Complex Areas in the Cook Inlet, Alaska," paper SPE 54611 presented at the 1999 SPE Western Regional Meeting, Anchorage, 26–28 May.

Foreman, R.D.: "The Drilling Command and Control System," paper SPE 14387 presented at the 1985 SPE Annual Technical Conference and Exhibition, Las Vegas, Nevada, 22–25 September.

Fredericks, P.D., Hearn, F.P., and Wisler, M.M.: "Formation Evaluation While Drilling With a Dual Propagation Resistivity Tool," paper SPE 19622 presented at the 1989 SPE Annual Technical Conference and Exhibition, San Antonio, Texas, 8–11 October.

Gaudin, D.B. and Beasley, J.C.: "A Comparison of MWD and Wireline, Steering Tool Guidance Systems in Horizontal Drilling," paper SPE 22536 presented at the 1991 SPE Annual Technical Conference and Exhibition, Dallas, 6–9 October.

Gearhart, M., Ziemer, K.A., and Knight, O.M.: "Mud Pulse MWD Systems Report," *JPT* (December 1981) 2301.

Gearhart, M., Moseley, L.M., and Foste, M.: "Current State of the Art of MWD and Its Application in Exploration and Development Drilling," paper SPE 14071 presented at the 1986 SPE International Meeting on Petroleum Engineering, Beijing, 17–20 March.

Gianzero, S.C., Chemali, R.E., and Su, S-M.: "Induction, Resistivity, and MWD Tools in Horizontal Wells," paper SPE 22347 presented at the 1992 SPE International Meeting on Petroleum Engineering, Beijing, 24–27 March.

Hache, J.-M. and Till, P.: "New-Generation Retrievable MWD Tool Delivers Superior Performance in Harsh Drilling Environments," paper SPE 67718 presented at the 2001 SPE/IADC Drilling Conference, Amsterdam, 27 February–1 March.

Hansen, R.R. and White, J.: "Features of Logging-While-Drilling (LWD) in Horizontal Wells," paper SPE 21989 presented at the 1991 SPE/IADC Drilling Conference, Amsterdam, 11–13 March.

Hendricks, W.E., Coope, D.F., and Yearsley, E.N.: "MWD: Formation Evaluation Case Histories in the Gulf of Mexico," paper SPE 13187 presented at the 1984 SPE Annual Technical Conference and Exhibition, Houston, 16–19 September.

Honeybourne, W.: "Formation MWD Benefits, Evaluation and Efficiency," *Oil & Gas J.* (25 February 1985) **83,** No. 9.

Hutchinson, M.W.: "Comparisons of MWD, Wireline, and Core Data From a Borehole Test Facility," paper SPE 22735 presented at the 1991 SPE Annual Technical Conference and Exhibition, Dallas, 6–9 October.

Hutin, R., Tennent, R.W., and Kashikar, S.V.: "New Mud Pulse Telemetry Techniques for Deepwater Applications and Improved Real-Time Data Capabilities," paper SPE 67762 presented at the 2001 SPE/IADC Drilling Conference, Amsterdam, 27 February–1 March.

Jackson, C.E. and Heysse, D.R.: "Improving Formation Evaluation by Resolving Differences Between LWD and Wireline Log Data," paper SPE 28428 presented at the 1994 SPE Annual Technical Conference and Exhibition, New Orleans, 25–28 September.

Jan, Y.-M. and Campbell, R.L. Jr.: "Borehole Correction of MWD Gamma Ray and Resistivity Logs," paper PP presented at the 1984 SPWLA Annual Logging Symposium, New Orleans, 10–13 June.

Jantsen, H. *et al.:* "Format, Content and Information Transfer Standards for Digital Rigsite Data," paper SPE 16141 presented at the 1987 SPE/IADC Drilling Conference, New Orleans.

Johancsik, C.A. *et al.:* "Application of Measurement While Drilling in a Shallow, Highly Deviated Drilling Program," paper 84-35-116 presented at the 1984 Petroleum Soc. of CIM Annual Technical Meeting, Calgary, 10–13 June.

Kashikar, S.V. and Lesso, W.G. Jr.: "The Principles and Procedures of Geosteering," paper SPE 35051 presented at the 1996 IADC/SPE Drilling Conference, New Orleans, 12–15 March.

Kline, K. and Kline, D.: *SQL in a Nutshell,* O'Reilly & Assocs., Cambridge, Massachusetts (2000).

Knox, D.J.W. and Milne, J.M.: "Measurement-While-Drilling Tool Performance in the North Sea," paper SPE 16523 presented at the 1987 SPE Offshore Europe Conference, Aberdeen, 8–11 September.

Koopersmith, C.A. and Barnett, W.C.: "Environmental Parameters Affecting Neutron Porosity, Gamma Ray, and Resistivity Measurements Made While Drilling," paper SPE 16758 presented at the 1987 SPE Annual Technical Conference and Exhibition, Dallas, 27–30 September.

Lah, M. *et al.:* "Real-Time Data Analysis While Drilling Provides Risk Management for Both Geological and Geometric Uncertainties in the Sotong K2.0 Reservoir," paper SPE 64477 presented at the 2000 SPE Asia Pacific Oil and Gas Conference and Exhibition, Brisbane, Australia, 16–18 October.

Lassoued, C., Dowla, N., and Wendt, B.: "Deepwater Improvements Using Real-Time Formation Evaluation," paper SPE 74397 presented at the 2002 SPE International Petroleum Conference and Exhibition in Mexico, Villahermosa, Mexico, 10–12 February.

Leggett, J.V. *et al.:* "Field Test Results and Processing Methods for Remote Acoustic Sensing of Stratigraphic Bed Boundaries," paper SPE 50593 presented at the 1998 SPE European Petroleum Conference, The Hague, 20–22 October.

Lesage, M., Casso, C.G., and Zanker, K.J.: "A New Approach to Rig Sensors," paper SPE 19999 presented at the 1990 IADC/SPE Drilling Conference, Houston, 27 February–2 March.

Lesso, W.G. Jr. and Burgess, T.M.: "Pore Pressure and Porosity From MWD Measurements," paper SPE 14801 presented at the 1986 IADC/SPE Drilling Conference, Dallas, 10–12 February.

Lesso, W.G. Jr., Rezmer-Cooper, I.M., and Chau, M.: "Continuous Direction and Inclination Measurements Revolutionize Real-Time Directional Drilling Decision-Making," paper SPE 67752 presented at the 2001 SPE/IADC Drilling Conference, Amsterdam, 27 February–1 March.

MacDonald, R.R.: "Drilling the Cold Lake Horizontal Well Pilot No. 2," *SPEDE* (September 1987) 193; *Trans.,* AIME, **283.**

Mallary, C.R., Varco, M., and Quinn, D.: "Pressure-While-Drilling Measurements To Solve Extended-Reach Drilling Problems on Alaska's North Slope," *SPEDC* (June 2002).

Maranuk, C.: "Development of an MWD Hole Caliper for Drilling and Formation Evaluation Applications," paper SPE 38585 presented at the 1997 SPE Annual Technical Conference and Exhibition, San Antonio, Texas, 5–8 October.

Marsh, J.L., Fraser, E.C., and Holt, A.L. Jr.: "Measurement-While-Drilling Mud Pulse Detection Process: An Investigation of Matched Filter Responses to Simulated and Real Mud Pressure Pulses," paper SPE 17787 presented at the 1989 SPE Symposium on Petroleum Industry Applications of Microcomputers, San Jose, California, 27–29 June.

Martin, C.A. *et al.:* "Innovative Advances in MWD," paper SPE 27516 presented at the 1994 IADC/SPE Drilling Conference, Dallas, 15–18 February.

Minear, J. *et al.:* "Compressional Slowness Measurements While Drilling," *Trans.,* 1995 SPWLA Annual Logging Symposium, Paris (26–29 June 1995) paper VV.

Mitrou, T.J. *et al.:* "Comparison of Magnetic Single-Shot Instruments With a Directional MWD System," *SPEDE* (April 1986) 163.

Monroe, S.P.: "Applying Digital Data-Encoding Techniques to Mud Pulse Telemetry," paper SPE 20326 presented at the 1990 SPE Petroleum Computer Conference, Denver, 25–28 June.

Montaron, B.A., Hache, J.-M.D., and Voisin, B.: "Improvements in MWD Telemetry: 'The Right Data at the Right Time'," paper SPE 25356 presented at the 1993 SPE Asia Pacific Oil and Gas Conference and Exhibition, Singapore, 8–10 February.

Morley, J. *et al.:* "Field Testing of a New Nuclear Magnetic Resonance Logging-While-Drilling Tool," paper SPE 77477 presented at the 2002 SPE Annual Technical Conference and Exhibition, San Antonio, Texas, 29 September–2 October.

Naville, C. *et al.:* "Drillbit SWD Reverse Walkaway Using Downhole Reference Signal," paper SPE 65115 presented at the 2000 SPE European Petroleum Conference, Paris, 24–25 October.

Noureldin, A. *et al.:* "A New Borehole Surveying Technique for Horizontal Drilling Processes Using One Fiber Optic Gyroscope and Three Accelerometers," paper SPE 59198 presented at the 2000 IADC/SPE Drilling Conference, New Orleans, 23–25 February.

Ohlinger, J.J., Gantt, L.L., and McCarty, T.M.: "A Comparison of Mud Pulse and E-Line Telemetry in Alaska CTD Operations," paper SPE 74842 presented at the 2002 SPE/ICoTA Coiled Tubing Conference and Exhibition, Houston, 9–10 April.

Orban, J.J. *et al.:* "New Ultrasonic Caliper for MWD Operations," paper SPE 21947 presented at the 1991 SPE/IADC Drilling Conference, Amsterdam, 11–14 March.

Parrish, R., Fielder, C., and Ishmael, R.: "Bi-Center Drill Bits and MWD/LWD Tools in a Horizontal Application Prove Effective in Reducing Well Costs and Increasing Liner-Size Capability," paper SPE 37640 presented at the 1997 SPE/IADC Drilling Conference, Amsterdam, 4–6 March.

Patrick, J.J.: *SQL Fundamentals,* Prentice Hall, New York City (2002).

Patton, B.J. *et al.:* "Development and Successful Testing of a Continuous-Wave, Logging-While-Drilling Telemetry System," *JPT* (October 1977) 1215.

Peach, S.R. and Kloss, P.J.C.: "A New Generation of Instrumented Steerable Motors Improves Geosteering in North Sea Horizontal Wells," paper SPE 27482 presented at the 1994 IADC/SPE Drilling Conference, Dallas, 15–18 February.

Prammer, M.G. *et al.:* "The Magnetic-Resonance While-Drilling Tool: Theory and Operation," paper SPE 62981 presented at the 2000 SPE Annual Technical Conference and Exhibition, Dallas, 1–4 October.

Prammer, M.G. *et al.:* "The Magnetic-Resonance While-Drilling Tool: Theory and Operation," *SPEREE* (August 2001) 270.

Rasmus, J. *et al.:* "Optimizing Horizontal Laterals in a Heavy Oil Reservoir Using LWD Azimuthal Measurements," paper SPE 56697 presented at the 1999 SPE Annual Technical Conference and Exhibition, Houston, 3–6 October.

Rasmus, J.C. and Gray Stephens, D.M.R.: "Real-Time Pore-Pressure Evaluation From MWD/LWD Measurements and Drilling-Derived Formation Strength," *SPEDE* (December 1991) 264.

Reiss, L.H. *et al.:* "Offshore and Onshore European Horizontal Wells," paper OTC 4791 presented at the 1984 Offshore Technology Conference, Houston, 7–9 May.

Roberts, A., Newton, R., and Stone, F.: "MWD Field Use and Results in the Gulf of Mexico," paper SPE 11226 presented at the 1982 SPE Annual Technical Conference and Exhibition, New Orleans, 26–29 September.

Robinson, L.H. Jr. *et al.:* "Exxon Completes Wireline Drilling Data Telemetry System," *Oil & Gas J.* (14 April 1980) **77,** No. 17, 137.

Robnett E.W. *et al.*: "Real-Time Downhole Drilling Process Data Complement Surface Data in Drilling Optimization," paper SPE 77248 presented at the 2002 IADC/SPE Asia Pacific Drilling Technology Conference, Jakarta, 9–11 September.

Rodney. P.F. and Wisler, M.M.: "Electromagnetic Wave Resistivity MWD Tool," *SPEDE* (October 1986) 337; *Trans.,* AIME, **281.**

Roesler, R.F. and Paske, W.C.: "Theory and Application of a Measurement-While-Drilling Neutron Porosity Sensor," paper SPE 16057 presented at the 1987 SPE/IADC Drilling Conference, New Orleans, 15–18 March.

Sawaryn, S.: "A Drilling Information System—Its Application and Future Development," paper SPE 19044 available from SPE, Richardson, Texas (1989).

Sawaryn, S. and Grayson, H.: "Industry Participation in the Development of an Integrated Drilling Engineering Applications Platform," paper SPE 23892 presented at the 1992 IADC/SPE Drilling Conference, New Orleans, 18–21 February.

Schenato, A. *et al.*: "ADIS: Advanced Drilling Information System Project," paper SPE 22317 presented at the 1991 SPE Petroleum Computer Conference, Dallas, 17–20 June.

Serra, O.: *Fundamentals of Well Log Interpretation: Vol. 1—The Acquisition of Logging Data,* Elsevier, Amsterdam (1984).

Skillingstad, T.: "At-Bit Inclination Measurement Improves Directional Drilling Efficiency and Control," paper SPE 59194 presented at the 2000 IADC/SPE Drilling Conference, New Orleans, 23–25 February.

Simms, G.J. and Koopersmith, C.A.: "Hydrocarbon Type Identification With MWD Neutron Porosity Logging: A Case Study," *SPEFE* (September 1991) 343.

Soulier, L. and Lemaitre, M.: "E.M. MWD Data Transmission Status and Perspectives," paper SPE 25686 presented at the 1993 SPE/IADC Drilling Conference, Amsterdam, 23–25 February.

Spoerker, H.F. and Kroell, E.: "Rigsite Information Management—Are We Making Use of Modern System's Full Potential?" paper SPE 37476 presented at the 1997 SPE Production Operations Symposium, Oklahoma City, Oklahoma, 9–11 March.

Stockhausen, E.J. and Lesso, W.G. Jr.: "Continuous Direction and Inclination Measurements Lead to an Improvement in Wellbore Positioning," paper SPE 79917 presented at the 2003 SPE/IADC Drilling Conference, Amsterdam, 19–21 February.

Sugimura, Y. *et al.*: "Development for Acoustic Sensing System Ahead of the Bit for Detection of Abnormally High-Pressured Formation," paper SPE 56700 presented at the 1999 SPE Annual Technical Conference and Exhibition, Houston, 3–6 October.

Tait, C.A. and Hamlin, K.H.: "Solution to Depth-Measurement Problems in LWD," paper presented at the 1996 Energy Week Conference and Exhibition, Houston, 29 January–2 February.

Tanguy, D.R. and Zoeller, W.A.: "Applications of Measurements While Drilling," paper SPE 10324 presented at the 1981 SPE Annual Technical Conference and Exhibition, San Antonio, Texas, 5–7 October.

Teige, T.G., Undersrud, E., and Rees, M.: "MWD: A Case Study in Applying New Technology in Norwegian Block 34/10," *SPEDE* (December 1986) 426.

Tinkham, S.K., Meek, D.E., and Staal, T.W.: "Wired BHA Applications in Underbalanced Coiled Tubing Drilling," paper SPE 59161 presented at the 2000 IADC/SPE Drilling Conference, New Orleans, 23–25 February.

Tochikawa, T. *et al.*: "Acoustic Telemetry: The New MWD System," paper SPE 36433 presented at the 1996 SPE Annual Technical Conference and Exhibition, Denver, 6–9 October.

Tsai, C.R.: "Improve Drilling Safety and Efficiency With MWD Sensors," paper SPE 22386 presented at the 1992 SPE International Meeting on Petroleum Engineering, Beijing, 24–27 March.

Turvill, J.A., Evans, H.B., and Hebel, J.B.: "Optimizing Design and Performance of an MWD Resistivity Sensor," paper SPE 19621 presented at the 1989 SPE Annual Technical Conference and Exhibition, San Antonio, Texas, 8–11 October.

Tare, U.A., Mese, A.I., and Mody, F.K.: "Interpretation and Application of Acoustic and Transient Pressure Response to Enhance Shale (In)Stability Predictions," paper SPE 63052 presented at the 2000 SPE Annual Technical Conference and Exhibition, Dallas, 1–4 October.

Underhill, W. *et al.:* "Demonstrations of Real-Time Borehole Seismic From an LWD Tool," paper SPE 71365 presented at the 2001 SPE Annual Technical Conference and Exhibition, New Orleans, 30 September–3 October.

Vang, S.: *SQL and Relational Databases,* Microtrend Books, San Marcos, California (1991).

Weisbeck, D. *et al.:* "Case History of First Use of Extended-Range EM MWD in Offshore, Underbalanced Drilling," paper SPE 74461 presented at the 2002 IADC/SPE Drilling Conference, Dallas, 26–28 February.

Wu, P.T., Tabanou, J.R., and Bonner, S.D.: "Petrophysical Interpretation of a Multispacing 2-MHz MWD Resistivity Tool in Vertical and Horizontal Wells," paper SPE 36547 presented at the 1996 SPE Annual Technical Conference and Exhibition, Denver, 6–9 October.

Zoeller, W.A.: "Pore Pressure Detection From the MWD Gamma Ray," paper SPE 12166 presented at the 1983 SPE Annual Technical Conference and Exhibition, San Francisco, 5–8 October.

SI Metric Conversion Factors

$$
\begin{array}{lll}
\text{cycles/sec} \times 1.0^* & \text{E+00} & = \text{Hz} \\
\text{ft} \times 3.048^* & \text{E–01} & = \text{m} \\
°\text{F} \quad (°\text{F} - 32)/1.8 & & = °\text{C} \\
\text{in.} \times 2.54^* & \text{E+00} & = \text{cm} \\
\text{in.}^3 \times 1.638\ 706 & \text{E+01} & = \text{cm}^3 \\
\text{psi} \times 6.894\ 757 & \text{E+00} & = \text{kPa}
\end{array}
$$

*Conversion factor is exact.

Chapter 16
Coiled-Tubing Well Intervention and Drilling Operations

Alex Sas-Jaworsky, II, SAS Industries Inc., **Curtis Blount,** Conoco-Phillips, and **Steve M. Tipton,** U. of Tulsa

16.1 Birth of Coiled-Tubing (CT) Technology

Numerous continuous-length tubular service concept trials and inventions paved the way for the creation of present day CT technology. The following discussion outlines some of the inventions and major milestones that directly contributed to the evolution of the continuous-length tubular products used in modern CT services.

The origins of continuous-length, steel-tubing technology can be traced to engineering and fabrication work pioneered by Allied engineering teams during the Second World War. Project 99, code named "PLUTO" (an acronym for Pipe Lines Under The Ocean), was a top-secret Allied invasion enterprise involving the deployment of pipelines from the coast of England to several points along the coast of France. The 3-in. inside diameter (ID) continuous-length pipelines were wound upon massive hollow conundrums, which were used to spool up the entire length of individual pipeline segments. The reported dimensions of the conundrums were 60 ft in width (flange-to-flange), a core diameter of 40 ft, and a flange diameter of 80 ft. These conundrums were designed to be sufficiently buoyant with a full spool of pipeline to enable deployment when towed behind cable-laying ships. Six of the 17 pipelines deployed across the English Channel were constructed of 3-in.-ID steel pipe (0.212-in. wall thickness). The 3-in.-ID steel pipelines, described as "Hamel Pipe," were fabricated by butt-welding 40-ft lengths of pipe into approximately 4,000-ft segments of pipeline. These 4,000-ft segments were then butt-welded together and spooled onto the conundrums. A total of 172,000,000 gallons of petrol was reported to have been delivered to the allied armies through PLUTO pipelines at a rate of more than 1 million gal/D.[1]

Although the initial development effort of spoolable steel tubulars was reported to have occurred in the early 1940s, the first concept developed for use of continuous-length tubing in oil/gas wellbore services can be found in U.S. Patent 1,965,563, "Well Boring Machine," awarded on 10 July 1934 to Clyde E. Bannister.[2] This approach utilized "reelable drillpipe," which was flexible enough to be coiled within a basket for storage when it was run into or out of the borehole (**Fig. 16.1**). This original concept used a rubber hose as the drillpipe, with the hose couplings designed to accommodate the attachment of two steel cables to provide the axial load support for the weight of the hose and bottomhole drilling assembly. The hose-coupling-

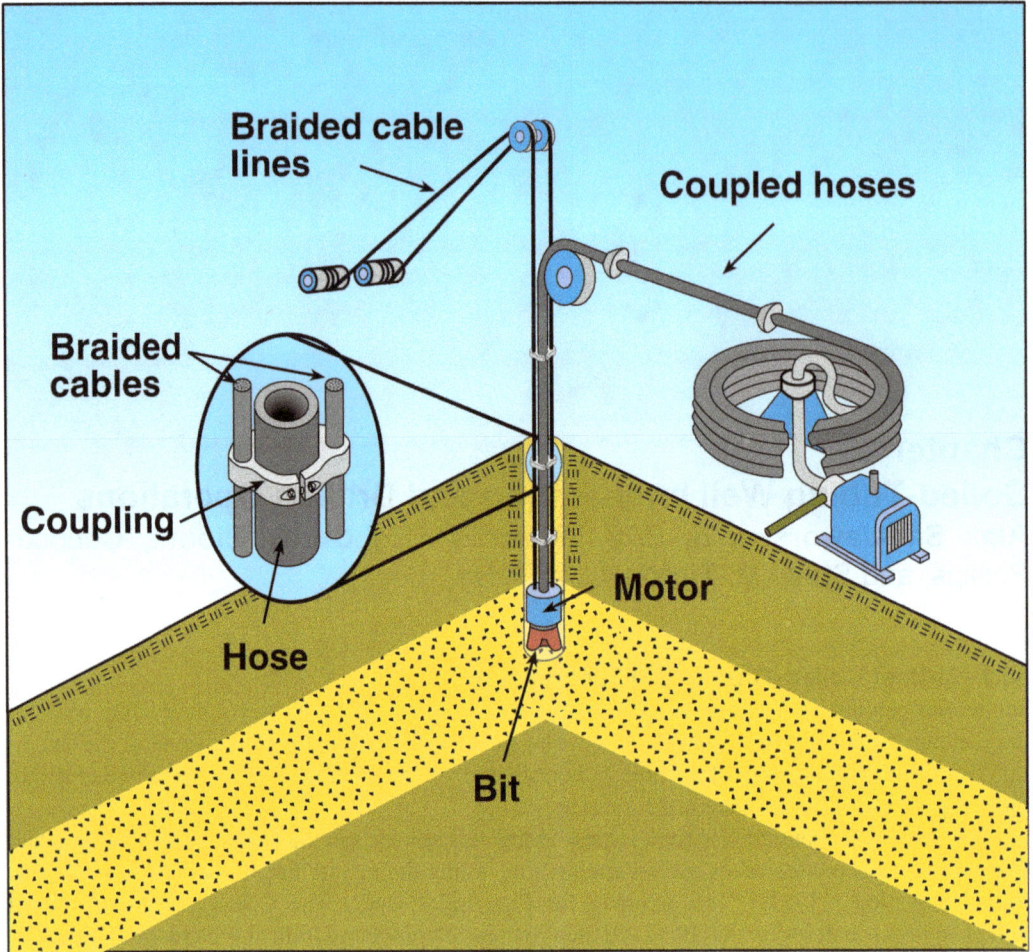

Fig. 16.1—Illustration of Bannister well-boring machine (courtesy of SAS Industries Inc.).

cable-attachment clamps were also designed to allow removal of the steel cables as the flexible drillstring was removed from the wellbore. When pulling the flexible drillstring out of the wellbore, the separate cable lines were spooled onto drums for storage.

The reeled drillstring system repeatedly used a bottomhole mud percussor and an oscillator (at different times) to drive the bit. In 1935, a total of 4,000 ft of borehole was drilled with this system, with a maximum single borehole depth of 2,000 ft.[3] The Bannister reelable drillpipe system reportedly became inactive in or about 1940 because of the lack of suitable downhole motors available.

The first concept on record for use of continuous-length steel tubing in well-service work was proposed by George D. Priestman and Gerald Priestman, as seen in U.S. Patent 2,548,616, "Well Drilling," awarded 10 April 1951.[4] The patent claimed the invention of a reeled, rigid-pipe drilling system in which the steel pipe is spooled onto a carrier reel. The reeled drillpipe was proposed to be deployed into the wellbore through a series of rollers mounted above the carrier reel, which was also fashioned to serve as a pipe bender (**Fig. 16.2**). Once the steel pipe was run through the pipe-bending device, the pipe was oriented vertically and entered a pipe straightener mounted on the wellhead. This straightener was proposed as a series of motor-powered rollers and also served as the drive mechanism for deploying and retrieving the drillpipe.

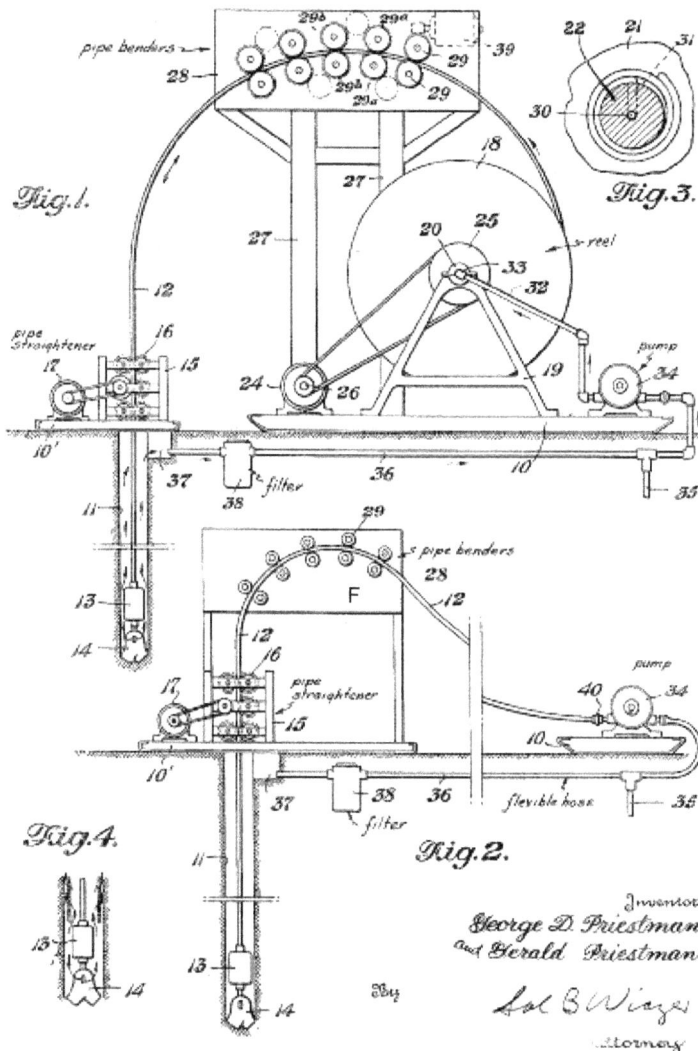

April 10, 1951   G. D. PRIESTMAN ET AL   2,548,616

WELL DRILLING

Filed Feb. 2, 1948

Fig. 16.2—Well drilling apparatus (U.S. Patent No. 2,548,616) (after Priestman *et al.*[4]).

16.2 Modern CT Technology

The chronology of modern-day steel CT technology development appears to begin in the early 1950s with U.S. Patent 2,567,009, "Equipment for Inserting Small Flexible Tubing into High Pressure Wells," awarded to George H. Calhoun and Herbert Allen on 4 September 1951. The fundamental concepts developed and claimed by Calhoun and Allen[5] served as the basis for the vertical, counter-rotating chain tractor device, which was upscaled to serve as the design for the first CT injector placed in operation.

This apparatus provided the ability to insert, suspend, and extract strings of elongated cylindrical elements (such as tubing) for well-intervention services with surface pressure present. A modified version of this device was originally developed to enable submarine vessels to deploy a radio communications antenna up to the ocean surface while still submerged. Using the Calhoun and Allen concept, Bowen Tools developed a vertical, counter-rotating chain tractor device called the "A/N Bra-18 Antenna Transfer System," which was designed to deploy a ⅝-in. outside diameter (OD) polyethylene encapsulated brass antenna from as deep as 600 ft beneath the water level. Fabric-reinforced phenolic "saddle blocks" grooved to match the OD of the tube were installed as the middle section of the drive chain sets, securing the antenna during operations. The antenna was stored on a carrier reel located beneath the antenna transfer system for ease of deployment and retrieval. The pressure seal was provided by a stripper-type element, which allowed the antenna to penetrate the hull of the vessel. The basic principles of this design concept aided in the development of the prototype Bowen Tools CT injector system.

In 1962, the California Oil Co. and Bowen Tools developed the first working prototype "continuous-string light workover unit" for use in washing out sand bridges in U.S. Gulf of Mexico oil/gas wells (**Fig. 16.3**). The original "Unit No. 1" injector was designed as a vertical, counter-rotating, chaindrive system built to run a string of 1.315-in.-OD tubing and operate with surface loads of up to 30,000 lbf. The core diameter of the tubing reel was 9 ft and was equipped with a rotating swivel mounted on the reel axle to allow continuous pumping down the tubing throughout the workover operation.

The first full-scale continuous length of CT was fabricated from highly ductile 40-ksi-yield, low-alloy Columbium steel. This low-alloy Columbium "skelp" was reportedly rolled to a thickness of 0.125 in. by the Great Lakes Steel Co. (Detroit, Michigan) and then milled into 1.315-in.-OD tubing by Standard Tube Co. (Detroit, Michigan). The 50-ft milled tube lengths were butt-welded together using a combination tungsten inert gas (TIG) and metal inert gas (MIG) process. The assembled tubing string was spooled onto a reel with a 9-ft core diameter to a total length of 15,000 ft and then subjected to numerous bending and loading cycles. The performance of this tubing string and CT unit was tested in several wells (located inland and offshore of south Louisiana) in the early to mid-1960s. The services performed by this original CT unit included sand washing and fishing a storm choke out of the existing completion tubing.[6]

16.3 CT Equipment Design

There are several CT equipment manufacturers presently marketing various designs of CT injectors, service tubing reels, and related well-control equipment in the industry today. The injector designs available within the industry include the opposed counter-rotating, chaindrive system, arched-chain roller drive, single-chain opposed-gripper-drive system, and the sheavedrive system. At present, the predominant equipment design for CT well-intervention and drilling services incorporates the vertically mounted, counter-rotating chaindrive type of injector. For purposes of practical demonstration, the following descriptions of CT equipment focus on the specific unit components supporting the vertical, counter-rotating chaindrive type of injector.

The CT unit is a portable, hydraulically powered service system that is designed to inject and retrieve a continuous string of tubing concentric to larger-ID production tubing or casing strings. At the present time, CT manufactured for well intervention and drilling application is available in sizes ranging from 0.750 to 3.500 in. OD. A simplified illustration of a CT unit is shown in **Fig. 16.4**.

The basic components of a CT unit are listed next.
• Injector and tubing guide arch.
• Service reel with CT.
• Power supply/prime mover.
• Control console.
• Control and monitoring equipment.
• Downhole CT connectors and bottomhole assembly (BHA) components.
• Well-control equipment.

Fig. 16.3—Bowen Tool Injector No. 1 (courtesy of Damon Slator).

16.3.1 Tubing Injector. The CT injector is the equipment component used to grip the continuous-length tubing and provide the forces needed for deployment and retrieval of the tube into and out of the wellbore. The injector assembly is designed to perform three basic functions:

• Provide the thrust required to snub the tubing into the well against surface pressure and/or to overcome wellbore friction forces.

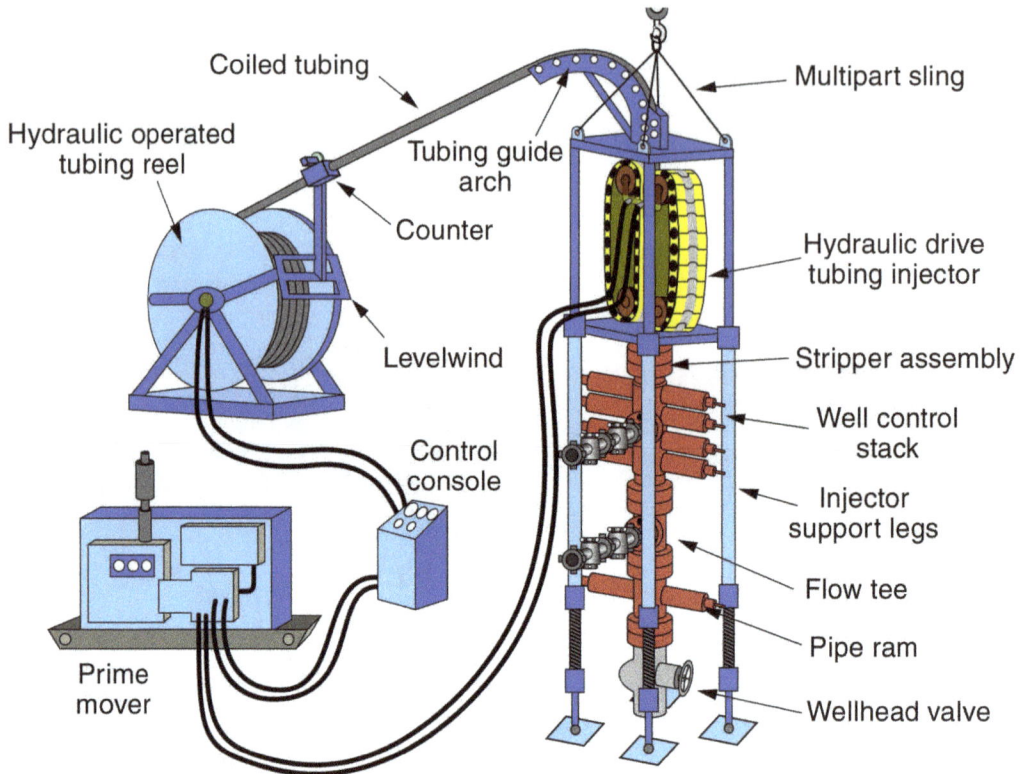

Fig. 16.4—Mechanical elements of a hydraulic CT unit (courtesy of SAS Industries Inc.).

• Control the rate of lowering the tubing into the well under various well conditions.

• Support the full weight of the tubing and accelerate it to operating speed when extracting it from the well.

Fig. 16.5 illustrates a typical rig-up of a CT injector and well-control stack on a wellhead. There are several types of counter-rotating, chaindrive injectors working within the industry, and the manner in which the gripper blocks are loaded onto the tubing varies depending on design. These types of injectors manipulate the continuous tubing string using two opposed sprocketdrive traction chains, which are powered by counter-rotating hydraulic motors.

The fundamental operating concept of the counter-rotating, opposed-chain injector is one that utilizes drive chains fabricated with interlocking gripper blocks mounted between the chain links (**Fig. 16.6**). These types of gripper blocks are designed to minimize damage to the CT and may be machined to fit the circumference of the CT string or formed in a "V" shape to accommodate variable OD sizes of CT (**Fig. 16.7**). The chaindrive assembly operates on the principle of frictional restraint, in that the CT is loaded by the opposing gripper blocks with sufficient magnitude of applied normal force that the resulting tangential friction forces are greater than the axial tubing loads (tension or compression).

In all traction-loading systems, hydraulic cylinders are used to supply the traction pressure and subsequent normal force applied to the CT (**Fig. 16.8**). The primary means of applying hydraulic pressure to this circuit may be through pumps on the prime mover, air-over-hydraulic pumps, or manual hand pumps. In addition, chain-loading systems require an emergency pressure source to maintain traction in case of a loss of hydraulic pressure supply. Typically, this

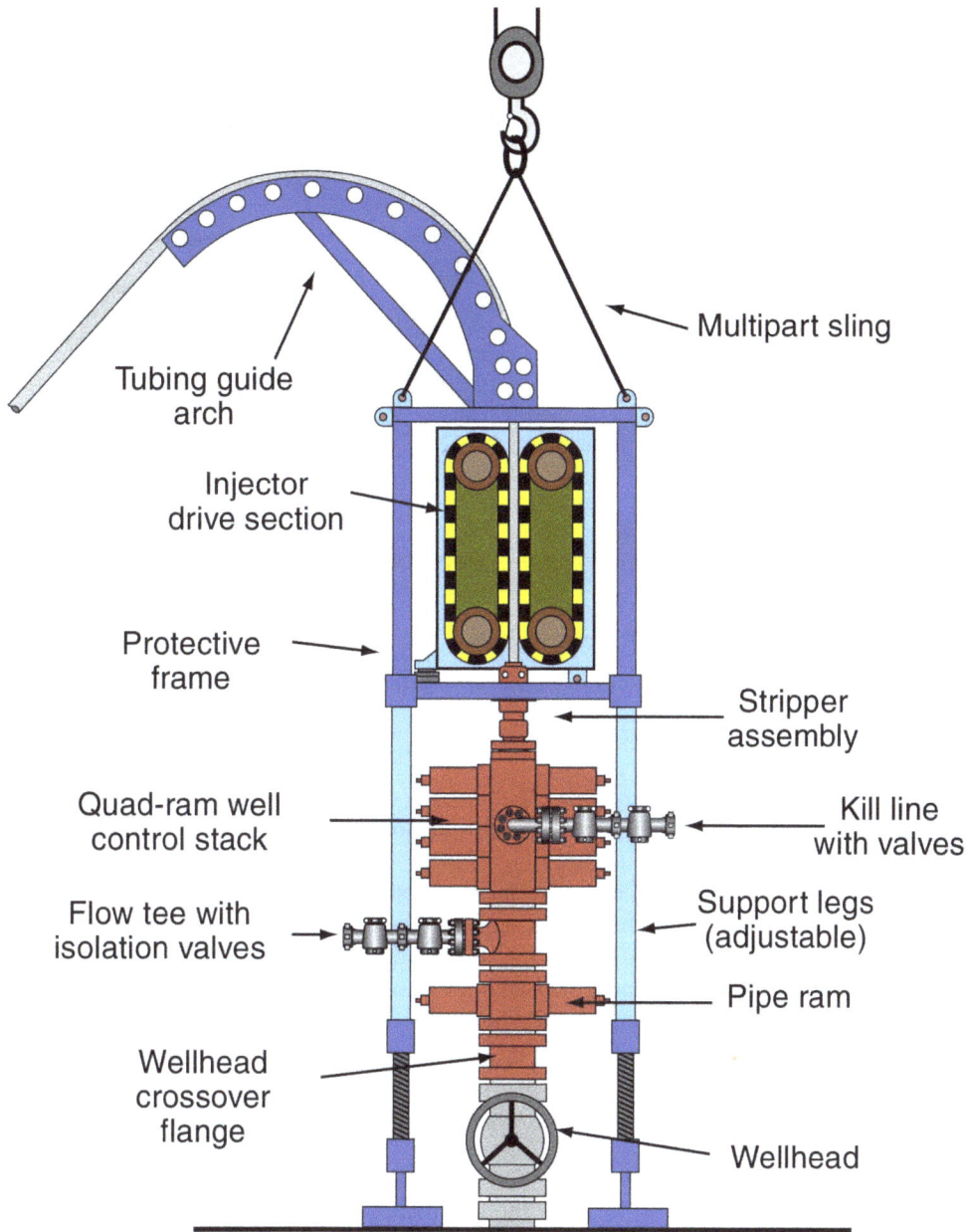

Fig. 16.5—CT injector and typical well-control stack rig-up (courtesy of SAS Industries Inc.).

system consists of an accumulator and a manual hydraulic pump or air-over-hydraulic pump located in the control cabin.

It is critical that the injector be equipped with a weight indicator that measures the tensile load in the CT (above the stripper), with the weight measurement displayed to the equipment operator during well intervention or drilling services. There should also be a weight indicator that measures the compressive force in the tubing below the injector when CT is being thrust into the well (often referred to as negative weight). Some weight indicators are capable of measuring a limited amount of negative weight—typically equal to the weight of the chaindrive

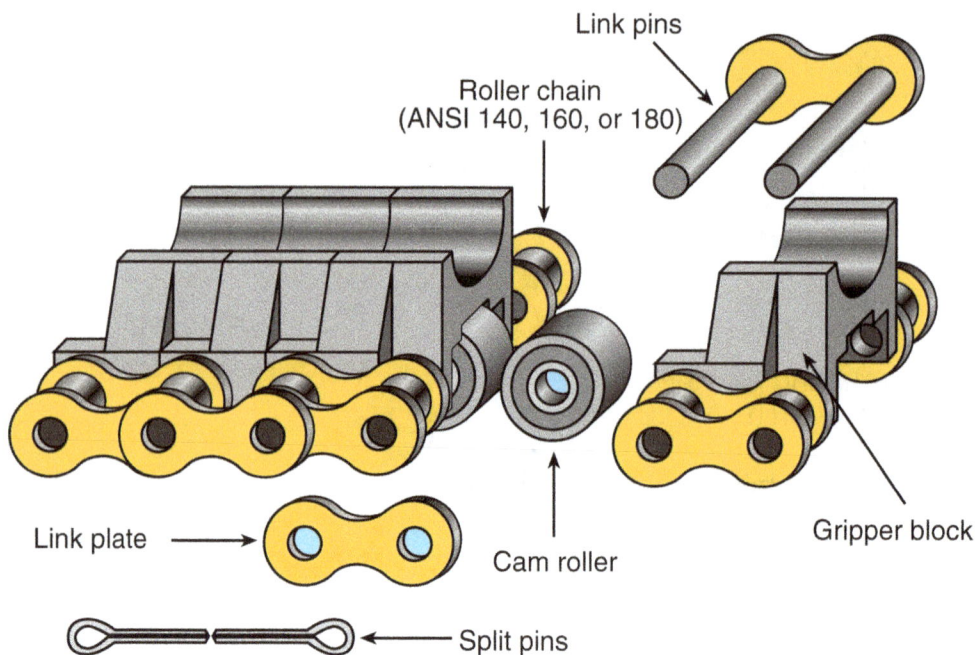

Fig. 16.6—Typical CT injector-drive-chain system (courtesy of SAS Industries Inc.).

assembly mounted in the injector frame. If this type of weight indicator is being used, the thrust force applied during the CT operation should not exceed the weight of the chaindrive assembly.

The counter-rotating, opposed-chaindrive injectors used in well intervention and drilling operations utilize a tubing guide arch, located directly above the injector. The tubing guide arch supports the tubing through the 90°+ bending radius and guides the CT from the service reel into the injector chains. The tubing guide arch assembly may incorporate a series of rollers along the arch to support the tubing or may be equipped with a fluoropolymer-type slide pad run along the length of the arch. The tubing guide arch should also include a series of secondary rollers mounted above the CT to center the tubing as it travels over the guide arch. The number, size, material, and spacing of the rollers can vary significantly with different tubing guide arch designs.

For CT used repeatedly in well intervention and drilling applications, the radius of the tubing guide arch should be at least 30 times the specified OD of the CT in service. This factor may be less for CT that will be bend-cycled only a few times, such as in permanent installations. The continuous-length tubing should enter and exit the tubing guide arch tangent to the curve formed by the guide arch. Any abrupt bending angle over which the CT passes causes increased bending strains, dramatically increasing the fatigue damage applied to the tubing. During normal CT operations, the reel tension applies a bending moment to the base of the tubing guide arch. Therefore, the tubing guide arch must be designed to be strong enough to withstand the bending caused by the required reel back tension for the applicable tubing size.

The injector should be stabilized when rigged up to minimize the potential for applying damaging bending loads to the well-control stack and surface wellhead during the well-intervention program. The injector may be stabilized above the wellhead using telescoping legs, an elevating frame, or a mast or rig-type structure. The injector support is the means provided to the injector to prevent a bending moment (such as reel back tension) from being applied to the

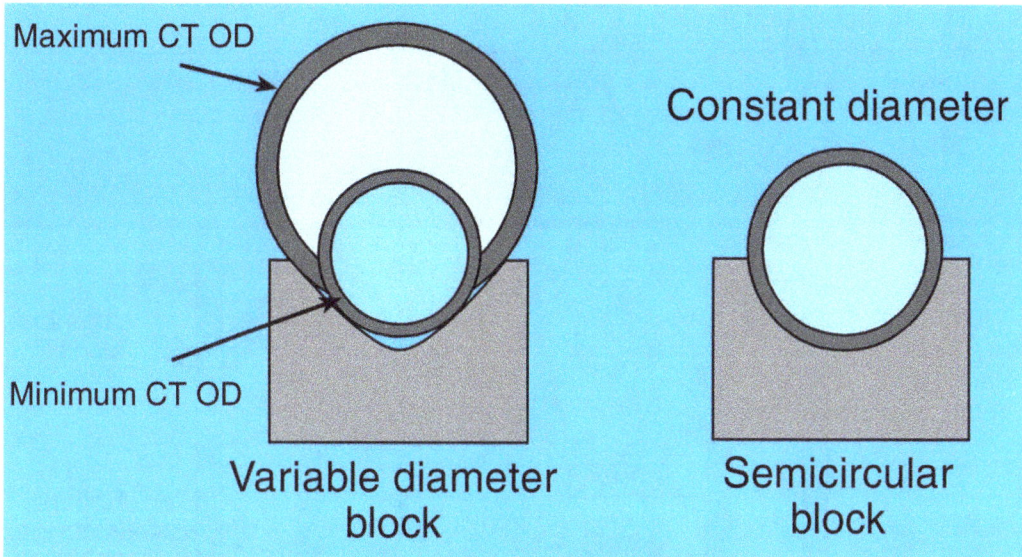

Fig. 16.7—Typical CT gripper-block configurations (courtesy of SAS Industries Inc.).

wellhead of such magnitude as to cause damage to the wellhead or well-control stack under normal planned operating conditions. Precautions should be taken to minimize the transfer of loads resulting from the weight of the injector, well-control equipment, and the hanging weight of the CT into the tree along the axis of the wellhead.

Telescoping legs are generally used in rig-ups where the height of the injector or wellhead does not permit the use of an elevating frame. When telescoping legs are used, the top sections are inserted into the four cylinders located on the corners of the injector frame and then secured with pins at the required height.

Footpads are placed beneath each telescoping leg to distribute the weight of the injector to the surface grade. Further stiffness of the legs is achieved by tightening the turnbuckles mounted beneath the leg sections. When telescoping legs are used, the weight and operating forces of the injector and well-control stack assembly are transferred directly to the wellhead, requiring that the rig-up load be supported with a crane or traveling block to minimize the load applied onto the wellhead.

In rig-up scenarios where an unobstructed surface is available (e.g., offshore platforms), it is recommended to support the injector using a hydraulically or mechanically controlled elevating frame structure. Once the desired height of the stand is achieved, the four legs on the perimeter of the stand are pinned and secured in place. The base of the elevating frame distributes the weight of the injector evenly around the perimeter of the frame. The benefits of using an elevating frame over the telescoping legs include greater stability, latitude in releasing the overhead crane support in noncritical service, and safety.

In rig-up scenarios in which a mast or derrick is required, precautions must be taken to minimize the axial load placed on the wellhead by the injector and well-control stack. In addition, the injector should be secured in some fashion within the mast or derrick to minimize the pitch and yaw motion of the injector during service.

16.3.2 Service Reel. The service reel serves as the CT storage apparatus during transport and as the spooling device during CT well-intervention and drilling operations. **Figs. 16.9 and 16.10** show the side view and front view of a typical service reel.

Twin hydraulic motors
with integral fail-safe
braking

Gripper
block
chain

Handling
frame

Rollers
integral to
chains

Gripper block
loading
cylinders

Inside skate

Chain tension
cylinders

Compression
load cell

Tension
load cell

Drive section
fulcrum point

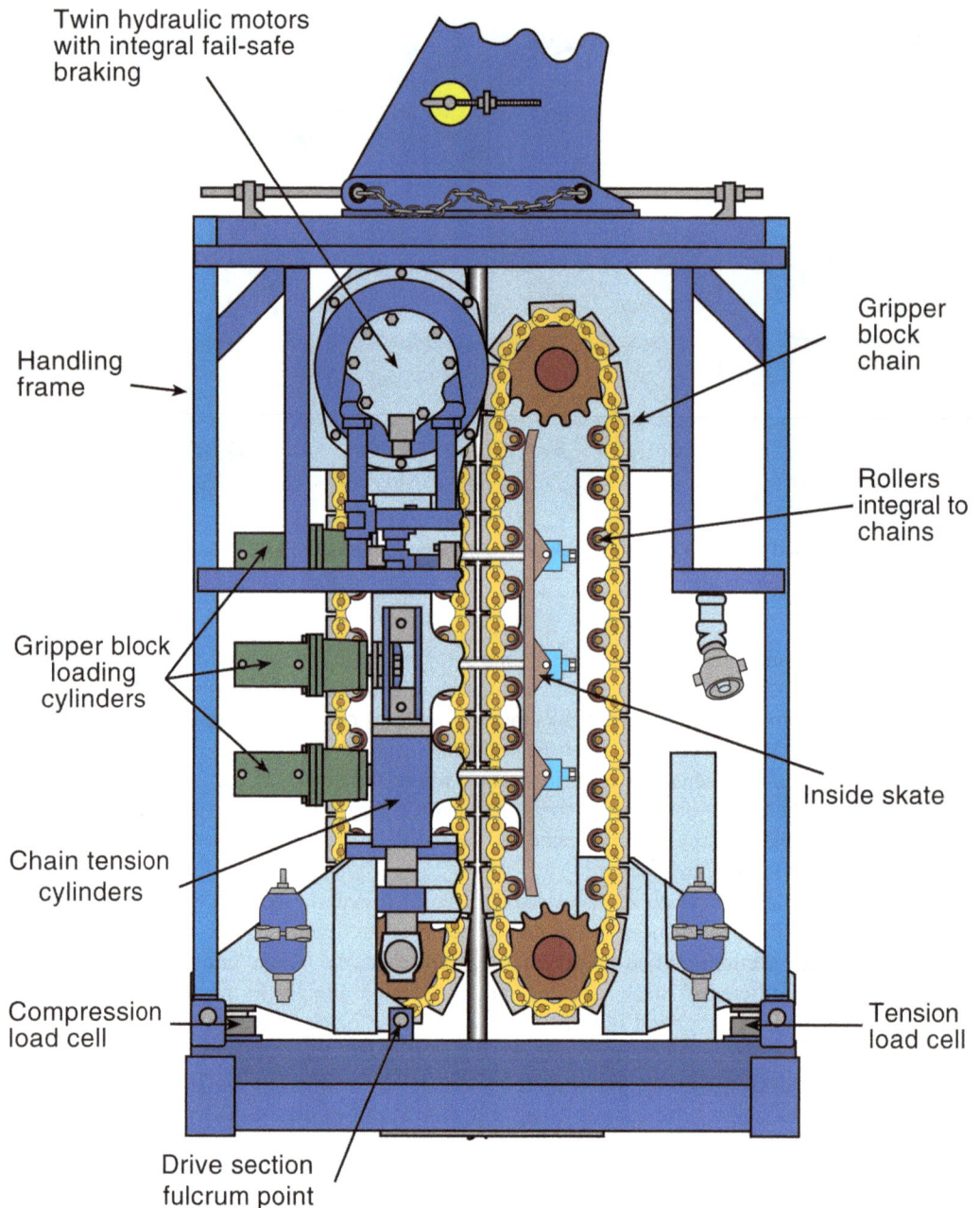

Fig. 16.8—Cutaway view of injector-drive assembly (courtesy of SAS Industries Inc.).

The inboard end of the CT may be connected either to the hollow segment of the reel shaft (spoke and axle design) or to a high-pressure piping segment (concave flange plates), both of which are then connected to a high-pressure rotating swivel. This high-pressure fluid swivel is secured to a stationary piping manifold, which provides connection to the treatment-fluid pumping system. As a result, continuous pumping and circulation can be maintained throughout the job. A high-pressure shutoff valve should be installed between the CT and reel shaft swivel for emergency use in isolating the tubing from the surface pump lines. The reel should also have a

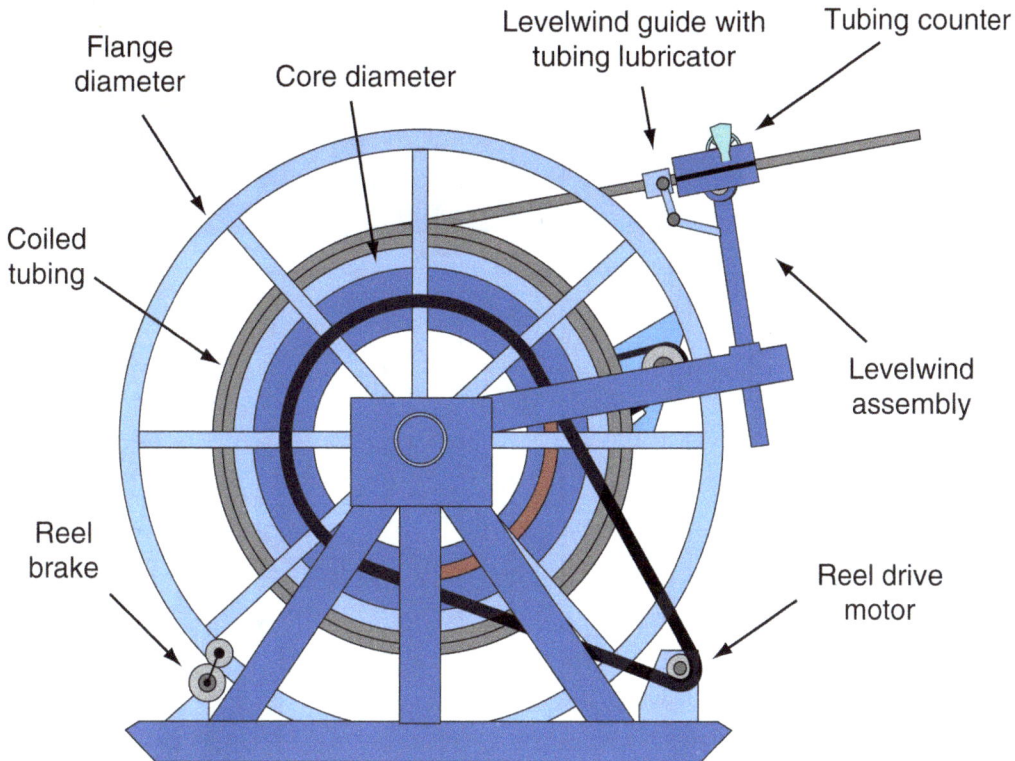

Fig. 16.9—Side view of typical CT service reel (courtesy of SAS Industries Inc.).

mechanism to prevent accidental rotational movement of the drum when it is required to remain stationary. In any event, the reel supporting structure should be secured to the deck or surface grade on location to prevent movement during operations.

In addition to the fluid-pumping service of the reel, electric wireline may be installed within the CT string to provide a means for conducting logging and downhole tool manipulation operations. The wireline is run inside the CT and is terminated at the reel shaft within a pressure bulkhead on the CT manifold. The single or multiconductor cable is then run from the pressure bulkhead to a rotating electric connection (slip collector ring) similar to that found on electric wireline units. On reels equipped for electric-line service, this electric connection may be located on the reel shaft opposite the rotating fluid swivel or at the pressure bulkhead adjacent to the inboard swivel piping.

In preparation for initial installation, a wing union is typically welded onto the end of the CT to be hooked up to the high-pressure piping within the reel (typically referred to as the "reference" end). The mechanical connection is then inserted through a slot in the reel core drum and made up to the high-pressure piping. Once the connection has been properly terminated, the tube is then bent over a preset guide to create a reasonably smooth bend transition to the outer surface of the core drum.

The initial layer of the tubing is spooled across the core drum until the tubing wrap reaches the opposing flange. The tubing is then spooled back over the base layer, resting in the recesses between the tubes on the previous layer. This wrapping process is continued through the remaining successive layers until the desired amount of tubing is spooled onto the reel. The manner in which the tubing is wrapped onto the reel allows the tube to be supported within the space formed by the previously wrapped tubing and offers a unique stacking geometry.

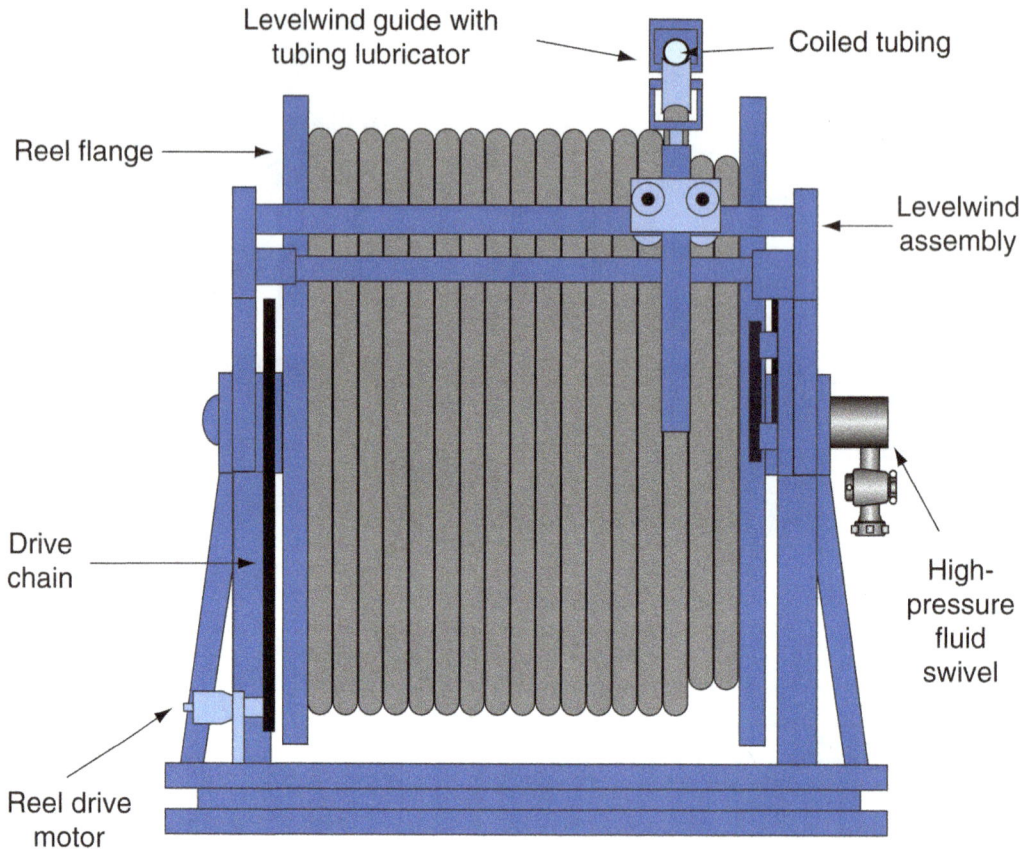

Fig. 16.10—Front view of a typical CT reel (courtesy of SAS Industries Inc.).

The core radius of the service reel defines the smallest bending radius for the tubing. For CT used repeatedly in well intervention and drilling applications, the core radius should be at least 20 times the specified OD of the CT. This factor may be less for CT that will be bend-cycled only a few times, such as for permanent installations.

The rotation of the service reel is controlled by a hydraulic motor, which may be mounted as a direct drive on the reel shaft or operated by a chain-and-sprocket drive assembly. This motor is used to provide a given tension on the tubing, thereby maintaining the pipe tightly wrapped on the reel. Back-pressure is kept on the reel motor during deployment, keeping tension on the tubing between the injector and service reel. This tensile load applied to the tubing by the reel motor is commonly called "reel back tension," requiring the injector to pull the tubing off the reel. The amount of reel back tension required increases with an increase in CT OD, yield strength (increased bending stiffness of the tubing), and distance between the service reel and injector. In addition, the required load on the reel drive system increases as the size of the core radius increases. Note that this tension results in an axial load imposed onto the tubing guide arch and creates a bending moment that is applied to the top of the injector. Therefore, it is critical that the injector is secured properly so that the bending moment is not translated to the well-control stack components or wellhead.

During operations, the reel back tension also prevents the tubing from "springing." Although the CT stored on a service reel has been plastically deformed during the spooling process, the tubing still has internal residual stresses that create a condition for potential unwrapping and outward springing of the tubing from the reel if the back tension is released. To prevent the CT from "springing," the free end of the tubing must always be kept in tension. When not in operation, the free end of the CT must be restrained to prevent springing.

The reel drive system must produce the tension required to bend the CT over the tubing guide arch and onto the reel. When CT is retrieved from the wellbore, the hydraulic pressure in the reel motor circuit is increased, providing the torque needed to allow reel rotation to keep up with the extraction rate of the tubing injector. Also, the reel drive system should have enough torque to accelerate the reel drum from stop to maximum injector speed at an acceptable rate. The torque should be capable of handling a fully loaded reel drum with the tubing full of fluid.

Additional safety items should also be included in the reel package to provide for an ancillary remote-activated braking system. The primary function of the reel brake is to stop drum rotation if the tubing accidentally parts between the reel and injector and limit tubing-reel rotation if a runaway condition develops. This braking system is not intended to halt the uncontrolled dispensing or retrieval of tubing in a runaway mode but only to offer resistance to slow down the reel rotation. The brake can also minimize tubing on the reel from springing in the case of loss of hydraulic pressure and, thus, the loss in reel back tension. When the reel is being transported, the brake should be engaged to prevent reel rotation. Many units incorporate a device in their hydraulic power systems to impose backpressure at the motor to slow the reel down. Other units employ a caliper-type or friction-pad braking system, which is hydraulically or mechanically applied onto the outer diameter of the reel flange to aid in slowing the reel rotation down.

The tubing is typically guided between the service reel and injector using a mechanism called the "levelwind assembly," which properly aligns the tubing as it is wrapped onto or spooled off the reel. The levelwind assembly spans across the width of the service reel drum and can be raised to any height, which will line up the CT between the tubing guide arch and the reel. Generally, a mechanical depth counter is mounted on the levelwind assembly, which typically incorporates a series of roller wheels placed in contact with the CT and geared to mechanically measure the footage of the tubing dispensed through it. The levelwind must be strong enough to handle the bending and side loads of the CT. During transportation, the free end of the CT is usually clamped to the levelwind to prevent springing. The levelwind may also be equipped with a hydraulically or pneumatically operated clamp, which can be manipulated to secure the CT at the crossbar of the levelwind frame.

In many cases, the service reel is equipped with a system for lubricating the outside of the CT. This tube lubricating system acts to protect against atmospheric corrosion and reduces the frictional loads encountered when deploying the tubing through an energized stripper assembly.

The high-pressure rotating swivel and treatment fluid plumbing must have a working pressure rating greater than the maximum anticipated pressure for the specified job. Special consideration should be given to cases in which the swivel and piping may come in contact with native wellbore fluids. These components must be suitable for the type of service and fluids encountered (e.g., H_2S, high temperature, etc.). At least one high-pressure isolation valve should be incorporated between the high-pressure swivel and the surface-treatment piping.

16.3.3 Prime Mover.
CT power supply units are built in many different configurations depending on the operating environment. Most are hydraulic-pressure pump systems powered by diesel engines, though a limited few employ electrical power. In general, the prime mover packages used on CT units are equipped with diesel engines and multistage hydraulic pumps that are typically rated for operating pressures of 3,000 to 5,000 psig. The hydraulic drive unit is supplied in the size necessary to operate all of the CT components in use and will vary with the needs of the hydraulic circuits employed.

The most common hydraulic power pack system is described as an "open loop" circuit, in which the fluid is discharged from the prescribed motor and returned to the hydraulic reservoir at atmospheric pressure. In general, open-loop power packs are equipped with vane-type hydraulic pumps and are rated for a maximum 3,000 psig service pressure applied to the hydraulic circuit. The pumps in these power packs provide source power for the injector, service reel, levelwind, well-control stack accumulators, console priority, and auxiliary panels as needed.

Where additional power to the injector circuit is needed, the hydraulic power pack may be designed as a "high-pressure, open-loop" system or as a "closed-loop" system. In both of these enhanced hydraulic power systems, the high-pressure circuit is limited to the injector hydraulics, with the remaining circuits powered by the vane-type pumps. The increased pressure in the hydraulic circuit for the injector provides the means for generating higher force loads within the injector motors as compared to the vane pumps, which are limited to 3,000 psig service. The high-pressure open-loop system typically uses a piston pump to provide hydraulic pressure as high as 5,000 psig to the injector circuit. The hydraulic fluid is discharged from the injector motors to the hydraulic reservoir tank at atmospheric pressure. The closed-loop hydraulic system also provides injector pressure to a maximum of 5,000 psig, with the distinction being that the hydraulic fluid is recirculated to the injector without returning to the hydraulic reservoir. The hydraulic fluid losses experienced through the injector motors are compensated by a charge pump incorporated into the closed loop circuit.

In general, the hydraulic pumps on the power pack are equipped with pressure-relief valves or unloader valves that limit the amount of hydraulic pressure the pump can deliver to the prescribed circuit. These pressure-relief or unloader valves are set at the desired pressure for the respective circuit and must be checked periodically to ensure that they are functioning properly.

Specifically, the pressure-relief or unloader valve on the injector circuit should be set at a pressure that limits the amount of force that can be applied to the tubing in tension (pulling) and compression (thrust). Before dispatch of CT service equipment from the vendor facility, the unloader valve on the injector circuit (either on the power pack or in the console) should be set to a pressure which does not exceed the safe load limit of the CT in service. Tests should be performed before equipment load-out to verify the sustained pressure output and fluid flow rate for the hydraulic pumps.

In current power pack design, an accumulator circuit is typically included to provide fluid volume and pressure for the well-control stack operation. The number of accumulator bottles typically ranges from one to six, depending on the size and pressure rating of the well-control stack in service. The accumulator package for well-control operation must have sufficient volume and pressure capacity to complete three complete function cycles of all the rams incorporated within the well-control stack without recharge from the power pack. These function cycles are typically described as "close-open-close" cycles and should be performed periodically to ensure that the accumulators are precharged to the appropriate pressure and that the circuit is free of hydraulic leaks.

16.3.4 Control Console. The control-console design for the CT unit may vary with manufacturers, but normally, all controls are positioned on one remote console panel. A diagram of a typical well-intervention unit control panel is seen in **Fig. 16.11.** The console assembly is complete with all controls and gauges required to operate and monitor all of the components in use and may be skid-mounted for offshore use or permanently mounted as with the land units. The skid-mounted console may be placed where needed at the wellsite as desired by the operator. The reel and injector motors are activated from the control panel through valves that determine the direction of tubing motion and operating speed. Also located on the console are the control systems that regulate the pressure for the drive chain, stripper assembly, and various well-control components.

Well control stack pressure

#2 Stripper pressure

Stripper system pressure

Chain tension pressure

Traction pressure drain

Wellhead pressure

Circulating pressure

Engine RPM

System air

Well control stack system pressure

#1 Stripper pressure

Stripper pressure bleedoff

Injector tension controls

Traction supply pressure

Thrust/weight load indicator

Emergency injector brake

Upper Panel Section

Well control stack function controls

#2 Stripper function control

Stripper selection valve

Emergency traction supply

Injector chain traction pressure adjust

Injector speed control

Injector speed selection

Reel brake

Horn

Wiper control

Engine throttle

On Off

Well control stack pressure supply valve

#1 Stripper function control

Stripper pressure adjust

Chain traction valves

Injector direction control

Injector pressure adjust

Reel pressure adjust

Reel function controls

Engine emergency kill

Engine kill switch

Lower Panel Section

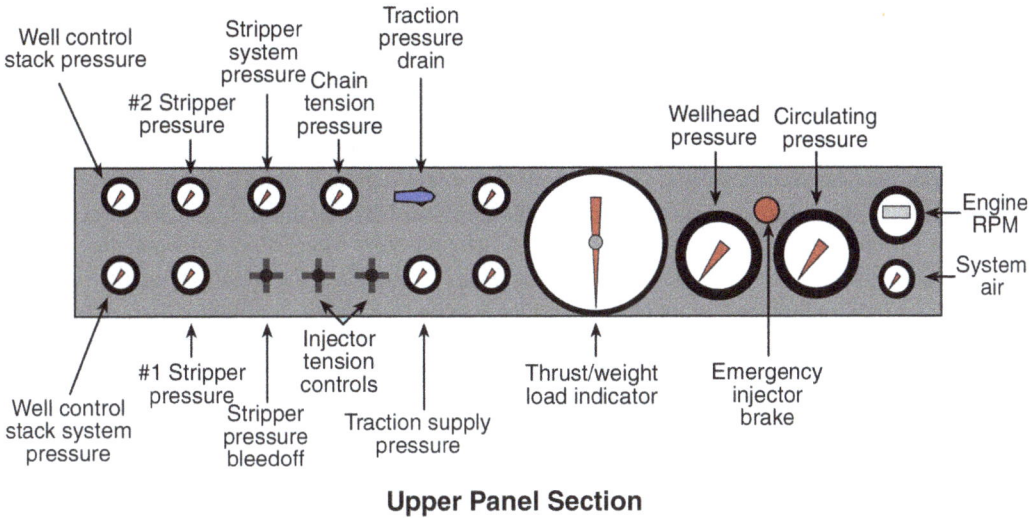

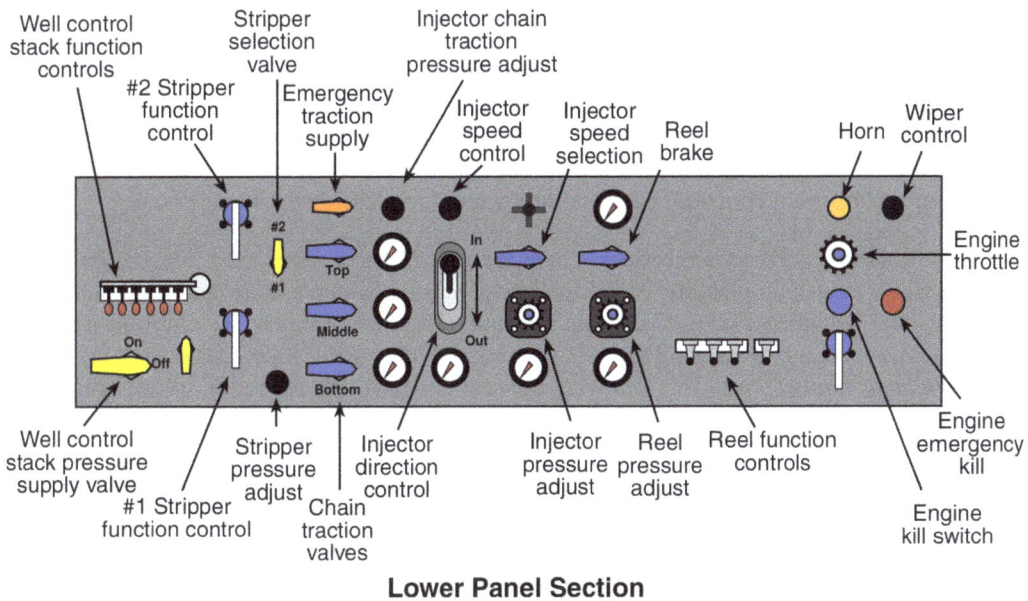

Fig. 16.11—Simplified layout of a console control panel (courtesy of SAS Industries Inc.).

16.3.5 Control and Monitoring Equipment. The CT equipment-related parameters that should be monitored to ensure the equipment is functioning correctly include traction force, chain tension, well-control system hydraulic pressure, reel motor pressure, injector motor pressure, and stripper hydraulic pressure. The critical job parameters that must be monitored throughout the job are discussed next.

Load Measurement. Load may be defined as the tensile or compressive force in the CT just above the stripper and is one of the most important measurements needed for proper operation of the prescribed service. Load may be affected by several parameters other than the hang weight of the CT and include wellhead pressure, stripper friction, reel back tension, and the density of the fluids inside and outside the tubing. Load should be measured directly using a load cell that measures the tensile and compressive forces applied to the CT by the injector. A secondary load measurement may be obtained indirectly by measuring the hydraulic pressure applied to the injector motors where the specified hydraulic pressure-to-load ratio is known.

Measured Depth. Measured depth is the length of CT deployed through the injector. Measured depth may be significantly different from the actual depth of the CT in the well because of stretch, thermal expansion, mechanical elongation, etc. Measured depth can be directly observed at several places on a CT unit using a friction-type wheel that contacts the tubing. Measured depth may also be obtained indirectly by measuring the rotation of the injector shafts. A CT unit should not be operated without a dedicated depth measurement system being displayed to the CT operator. Measured depth should be recorded as a function of time and in relation to internal pressure applied to the CT string for use in bend-cycle fatigue calculations.

Speed Measurement. Speed may be calculated from the change in measured depth over a specified time period.

CT Inlet Pressure. Pumping pressure at the inlet to the CT should be monitored and displayed to the CT operator, as well as recorded for use in bend-cycle fatigue calculations or for post-job reviews. This pressure-measurement system must incorporate a method of isolating the pumped-fluid circuit, eliminating the possibility for pumped fluid to discharge into the control cabin if gauge failure occurs. It is recommended that a pressure recorder be incorporated in the CT pressure-monitoring package to record pump pressure throughout the prescribed service.

Wellhead Pressure. Well pressure around the outside of the CT at the wellhead should be monitored and displayed to the CT operator, as well as recorded for use in post-job reviews. This pressure-measurement system must incorporate a method of isolating the wellbore fluid circuit, eliminating the possibility for well fluids to discharge into the control cabin if gauge failure occurs. It is recommended that a pressure recorder be incorporated in the CT pressure-monitoring package to record well pressure throughout the prescribed service.

16.3.6 Downhole CT Tool Connections. There are several connections used in CT services for the purpose of isolating pressure and transferring tension, compression, and torsional loads from tools and bottomhole assemblies onto the tube. These connections are typically designed to be field installed and reusable. The most common CT connections are discussed next.

Nonyielding Connections. The following connections have the capability of securing loads and pressure to the end of the CT in a manner that, during makeup, does not result in yielding of the tube body.

External Slip Type. The external slip-type connection requires the use of a slip or grapple-type load ferrule placed on the OD of the tube body. The load ferrule is typically constructed with sharp "spiraled" teeth that secure the ferrule onto the CT. The tool connection mechanically wedges the load ferrule onto the CT OD during connection makeup.

Pressure integrity of this connection is typically maintained with the use of O-rings or other types of elastomeric seals on the OD of the CT body. The external CT surface must be prepared to allow for an effective seal.

Internal Slip Type. The internal slip-type connection requires the use of a slip or grapple-type load ferrule placed within the ID of the tube body. The load ferrule is typically constructed with sharp "spiraled" teeth that secure the ferrule onto the ID wall of the CT. The tool connection mechanically wedges the load ferrule into the CT ID during connection makeup. Pressure integrity of this connection is typically maintained with the use of O-rings or other types of elastomeric seals applied against the ID of the CT body. The internal CT seam must be removed and the ID prepared for an effective seal by smoothing and buffing the ID surface scars and imperfections.

It should be noted that these nonyielding connections require that the terminated end of the CT be reasonably round, with OD/ID dimensions within the connector size tolerance. Problems often arise when using these connections on older, used CT that have become oval (distortion of tube roundness) or have experienced diametral growth (see 16.5.3).

The changes in CT geometry caused by ovality and diametral growth make sealing difficult when using common O-ring technology. Other elastomeric seals are often employed in larger OD CT when the diametric clearance may exceed O-ring sealing parameters.

Changes in CT OD geometry also present problems when trying to install the connectors onto the CT. Where CT geometry changes, resulting from ovality, are present (typical in larger OD CT sizes), a swaging tool may be used to return the tube body to a "near round" condition. The swaging tool is constructed similar to a muffler-pipe expander, with a hollow-core hydraulic jack pulling a short swaging cone up through a longer split skirt expander. The diameters of the swaging cone and skirt expander are chosen for a given CT size to yield the tube body OD a few thousandths larger than desired. After swaging, the CT body springs back, providing a CT OD with minimal ovality.

Yielding Connections. The following connections have the capability of securing loads and pressure at the end of the CT in a manner that, during makeup, results in yielding of the tube body.

External Dimple Type. The external-dimple-type connection is secured onto the tube body through the use of numerous mechanical screws. Forces exceeding the CT material yield strength create "dimples" in the tubing. These "dimples" serve as recessed receptacles for the mechanical screws that secure the connection to the tube body OD. Other variations of this basic connection method include pressure actuated dimpling tools used in conjunction with a template. Pressure integrity of this connection is typically maintained with the use of O-rings or other types of elastomeric seals on the OD of the CT body.

Internal Dimple Type. The internal-dimple-type connection is secured onto the CT body through the use of numerous mechanical indentations into recesses on an internal mandrel insert. Forces exceeding the CT material yield strength create "dimples" in the tubing, serving as the load transfer mechanism that secures the connection to the CT body ID. Pressure integrity of this connection is typically maintained with the use of O-rings or other types of elastomeric seals on the ID of the coiled tube body.

Roll-On Type. The roll-on type connection incorporates a machined insert mandrel designed to fit inside the CT. The mandrel is machined with circular recesses or "furrows." The connection is secured to the tube by means of mechanically yielding the tube body into the machined recesses on the mandrel. Pressure integrity of this connection is typically maintained with the use of O-rings or other types of elastomeric seals on the ID of the tube body.

Weld-On Connectors. Weld-on connectors are used in special applications such as coiled tubing drilling (CTD) in which larger CT is used and the aforementioned connectors impose operating limitations. These limitations include an excessively large OD, reduced torque or other load ratings, vibratory and oscillating load suitability, or a restricted ID that reduces the size of pump-down darts or ball.

Properly designed weld-on connectors will exhibit 100% of the torque and yield ratings of the CT material.

Important parameters to proper weld-on connector design and application include:

• A gradually tapered insertion neck several inches up from a straight-wall section that continues to the weld-bead location (**Fig. 16.12**). This taper provides a bend support to prevent concentrating bending loads at a single point where the CT meets the weld-on connector. The straight-wall section of the connector is used to eliminate all bending to the CT heat-affected zone.

• "Chill blocks" aid in limiting the heat-affected zone from welding operations. The connector's straight-wall OD exceeds the ID of the CT by several thousands of an inch. The

2⁵/₈-in.×0.203-in. CT Weld-On Connector

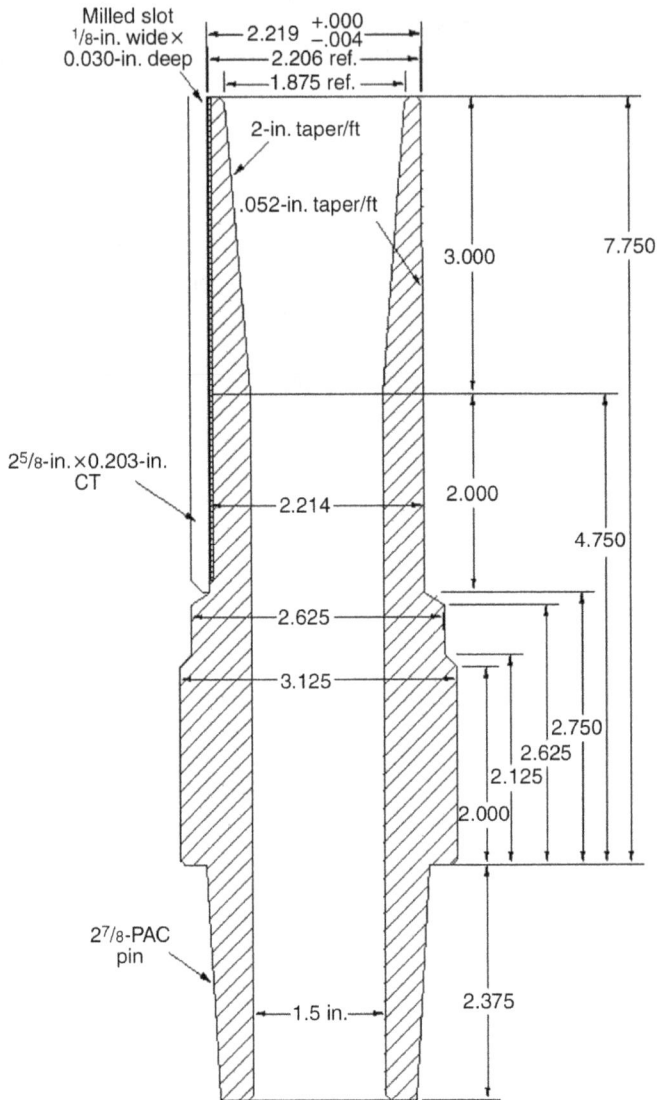

Fig. 16.12—Weld-on connector.

weld-on connection is therefore inserted after the CT is preheated in preparation for welding operations.

16.3.7 Well-Control Stack. The well-control stack system is a critical part of the CT unit pressure containment package and is composed of a stripper assembly and hydraulically operated rams, which perform the functions described next.

For typical well-intervention service, the four ram compartments are equipped (from top down) with blind rams, tubing shear rams, slip rams, and pipe rams. (**Fig. 16.13**). The blind rams are used to seal the wellbore off at the surface when well control is lost. Sealing of the blind rams occurs when the elastomer elements in the rams are compressed against each other.

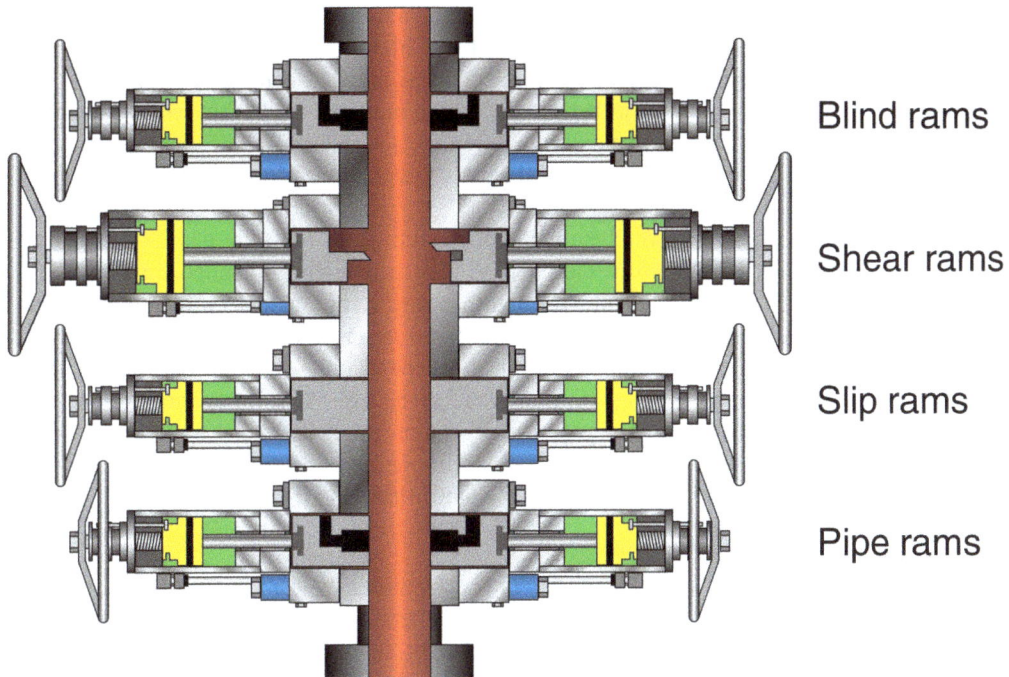

Fig. 16.13—Typical quad-ram well-control stack configuration (courtesy of SAS Industries Inc.).

For the blind rams to work properly, the tubing or other obstructions across the ram bonnets must be removed.

The tubing shear rams are used to mechanically break the CT in the event the pipe gets stuck within the well-control stack or whenever it is required to cut the tube and remove the surface equipment from the well. As the shearing blades are closed onto the CT, the forces imparted will mechanically yield the body of the tube to failure. The cut is deformed and typically must be dressed to return to the proper geometry.

The slip rams should be equipped with bidirectional teeth, which, when activated, secure against the tubing and support the weight of the CT and BHA below. An additional utility of the slip rams is the ability to close onto the tube and secure movement in the event that well-pressure risks blowing the tubing out of the borehole. The slip rams are outfitted with guide sleeves that properly center the CT into the grooved recesses of the ram body as the slips are being closed.

The pipe rams are equipped with elastomer seals preformed to the specified OD size of CT in service. When closed against the CT, the pipe rams are used to isolate the wellbore annulus pressure below the rams. These rams are also outfitted with guide sleeves that properly center the CT into the preformed recess as the rams are being closed.

Typically, a kill-line flange inlet is positioned directly below the tubing shear ram set and above the slip ram set in the well-control stack. Two valves rated to the maximum allowable working pressure (MAWP) of the well-control stack are mounted onto the kill-line flange, which typically includes a high-pressure check valve installed in the high-pressure chicksan line run to the high-pressure pump. The practice of taking returns through the kill line is not recommended because it exposes the lower sets of rams and bonnets to accumulation of solids, debris, and other return fluids that may adversely affect the performance of the rams.

On all well-intervention services that require the circulation of wellbore returns to surface (solids, debris, spent acid, etc.), the use of a separate "flow-tee" or "flow-cross" mounted directly below the primary well-control stack rams is recommended. This flow-tee or flow-cross connection should be equipped with a minimum of two high-pressure isolation valves and rated to the same working pressure and NACE classification as the well-control stack rams.

On most well-control stack assemblies, the blind ram and pipe ram compartments are equipped with ports that, when activated, allow pressure to equalize within the ram body. This allows for differential pressure to be equalized across the ram compartments before opening the rams.

The union positioned at the top of the well-control ram stack typically connects to the stripper assembly located on the bottom of the injector. The recommended connection at the bottom of the well-control ram stack is an integral high-pressure flange assembly or another suitable metal-to-metal seal connection. The pressure rating and arrangement of the well-control stack components for a given CT operation will typically depend upon the type of application employed and the maximum anticipated surface pressure in the well. When preparing for a CT well-intervention or drilling operation, the well-control equipment must be in compliance with the local regulatory authority and should reference applicable industry best practices [e.g., American Petroleum Inst. (API), Intl. Organization for Standardization (ISO), etc.].

16.4 Coiled Tubing

CT is an electric-welded tube manufactured with one longitudinal seam formed by high-frequency induction welding without the addition of filler metal. The first step in the typical CT manufacturing process involves the acquisition of steel stock supplied in 40- to 48-in.-wide sheets that are wrapped onto a "master coil" to a nominal weight of approximately 40,000 lbm. As a result, the lengths of sheet steel will vary depending upon the wall thickness. For example, a 40-in.-wide sheet of steel having a wall thickness of 0.109 in. rolled to a length of 2,700 ft will weigh approximately 40,000 lbm. If the 40-in.-wide sheet steel has a wall thickness of 0.156 in., a 40,000 lbm-master coil will have a length of approximately 1,900 ft.

When the diameter of the CT is selected, the sheet steel on the master coil is "slit" into a continuous strip of a specific width to form the circumference of the specified tube. The flat skelp is then welded to another segment of skelp to form a continuous length of steel. The welded area is dressed off smooth, cleaned, and then x-ray inspected to ensure that the weld is free from defects. Once a sufficient length of the continuous skelp steel is rolled onto the skelp take-up reel, the tube milling process can begin.

The skelp is then run through a series of roller dies that mechanically work the flat steel into the shape of a tube. At a point immediately ahead of the last set of forming rollers, the edges of the tube walls are positioned very close to each other. These edges are then joined together by an electric welding process described as "high-frequency induction" (HFI) welding. The HFI coil generates the heat for welding by the resistance to flow of electric current. As the tube is run through the high-frequency induction coil positioned inches ahead of the forming rollers, the edges of the walls are heated to the temperature needed to create the seam weld when pressed together within the last set of forming rollers.

The weld flash exposed on the outside of the tube is removed, and the welded seam is annealed. The tube is allowed to cool in air and then within a liquid bath before passing through a nondestructive inspection station to inspect the tube body. The inspection is typically performed with an eddy-current device that creates a magnetic field around the tube body and looks for distortions in the field created by surface defects in the tube body.

The manufacturing process continues as the tube is run through a sizing mill that slightly reduces the diameter after welding and works the tubing to the required OD and roundness tolerances. At this time, the tubing undergoes full-body heat treatment using induction coils. The purpose of the heat treatment is to stress-relieve the entire tube, increasing the ductility of the steel. The tube is allowed to cool, first gradually in air and then within a liquid bath. This process results in the development of pearlite and ferrite grain sizes within the steel microstructure. The final product is a high-strength CT string with ductility and physical properties appropriate for the specified yield range. The tube is then spooled onto a steel or wooden take-up reel and subjected to a hydrostatic pressure test using water treated with corrosion inhibitors.

Alternative CT manufacturing processes may require that a string be constructed by butt-welding sections of tube together. The tube-to-tube welding technique may be performed using TIG or MIG welding practices, and each weld should be inspected using radiographic inspection (x-ray) or ultrasonic inspection to evaluate the quality of the weld. Note that the exterior surface of the tube-to-tube weld may or may not be dressed off and that the weld bead on the ID surface is not disturbed in any way. The string of tubing is then spooled onto the service reel or shipping reel as required.

All manufactured spools of CT are given a unique identification number that is assigned at the time of manufacture. Documentation for each spool of CT should include the identification number, OD of the tubing, material grade, wall thickness(es), weld positions, and total length. A spool of CT may be manufactured from one heat or a combination of heats that are selected according to a documented procedure provided by the manufacturer. However, the steel used to fabricate the string must have a uniform material yield strength throughout. The manufacturer should maintain traceability of the CT product throughout the manufacturing and testing process. The requirements of the purchaser often include traceability to the heat of steel.

16.4.1 Tapered Wall Thickness String Design. In general, tapered CT strings can be manufactured by changing the wall thickness of the tubing within the length of a spool while maintaining a constant OD. The changes in wall thickness along the string length are intended to increase the performance properties of the CT in the selected sections. The construction of a tapered CT string may be achieved in one of the following ways:

• A continuously milled string incorporating multiple single-wall-thickness skelp segments joined using skelp-end welds.

• A continuously milled string incorporating single-wall-thickness skelp segments with continuously tapered skelp segments joined using skelp-end welds.

• Continuously milled, single-wall-thickness CT segments joined to another finished tube segment of a different wall thickness using the tube-to-tube welding process.

Continuously-tapered skelp is milled having a specified wall thickness at the leading end of the steel skelp, progressively increasing in wall thickness along the length of the skelp to a second specified wall thickness at the trailing end of the skelp.

The construction of tapered CT strings conforms to the previously described manufacturing processes. Although tube segments of different wall thickness can be assembled within the string construction, it is critical that all of the segments have a uniform material yield strength. The change in specified wall thickness, t, between the adjoining CT segments should not exceed the following specified values:

• 0.008 in. where the specified wall thickness of the thicker of the adjoining segments is less than 0.110 in.

• 0.020 in. where the specified wall thickness of the thicker of the adjoining segments is between 0.110 and 0.223 in.

• 0.031 in. where the specified wall thickness of the thicker of the adjoining segments is 0.224 in. and greater.

For tapered-wall string designs, the transition point may be defined as the points along the string having a change in the specified wall thickness. For single-wall-thickness segments, the transition points are defined as the points where the different specified wall thicknesses are mechanically joined together. For continuously tapered skelp segments, the transition points are defined as the points where the ramping of the skelp occurs. The typical design criteria for selecting transition points (desired lengths of specified-wall thickness segments) within the string includes weight and overpull loading, wellbore condition, and combined pressure loading. Note that the overpull load rating increases for the adjoining tube segment with an increased wall thickness. The effect of changes in overpull load should be applied across the entire tapered string to ensure that a given overpull load does not exceed the limits of any other tube segment within the string. Therefore, the maximum length of each CT wall thickness segment should be evaluated using overpull and combined pressure loading to confirm that the stress applied to the CT string below any point on the segment does not exceed the triaxial stress load at that point for the given safety factor.

16.5 CT Performance

CT well intervention and drilling operations require that the continuous-length tube be subjected to repeated deployment and retrieval cycles during its working life. The tubing stored on a service reel is deployed into the wellbore to the designated depth and then retrieved back onto the service reel. The working life of the CT may be defined as the duration of service for the continuous-tubing string when subjected to the following conditions: bend-cycle fatigue, internal pressure loading, applied axial loading, corrosion, and mechanical damage.

All of the aforementioned items act on the tube body to some degree during any CT service and contribute to the eventual mechanical failure of the tubing. To ensure safe and reliable well intervention and drilling operations, the user must understand the unique behavior of CT to minimize the possibility of tubing failure. Numerous decisions must be made throughout the working life of a CT string to maximize the remaining life. From this approach, the decision to retire the tubing must be made on the basis of current tube conditions, service history, and the anticipated service loading.

16.5.1 Description of Fatigue. Fatigue is generally considered to be the single major factor in determining the working life of CT. The deployment and retrieval of the continuous-length tubing string require that the tube be subjected to repeated bending and straightening events, commonly referred to as "bend-cycling." The amount of strain imposed upon the tube body during the bend-cycling process is considered to be enormous, in many cases on the order of 2 to 3%. When subjecting the CT to this type of fatigue cycling, the stress and/or strain fluctuations to failure may be estimated using conventional axial fatigue life prediction approaches.[7]

However, when the bend-cycling process is coupled with internal tube pressure loading, conventional multiaxial life prediction approaches cannot accurately predict CT behavior. Numerous tests performed have confirmed the fact that bend-cycling CT with internal pressure loading dramatically reduces the fatigue life of the tube when compared to the cycle life of unpressurized tubing. This happens despite the fact that the tangential (hoop) stresses imposed by typical "high" pressures in CT service are on the order of only 50 to 60% of nominal yield strength. However, when combined with fluctuating axial strain ranges on the order of 2 to 3%, significant cyclic ratcheting (ballooning) occurs, and fatigue damage development is accelerated. Outside of the CT industry, there are essentially no other applications involving steel alloys in which cyclic loading of this magnitude is intentionally imposed and expected to survive the prescribed service.[8]

To better understand the abuse imposed upon CT during normal service operations, a brief review of the relationship between stress and strain in high-strength low-alloy (HSLA) steel is given next. **Fig. 16.14** is a typical stress-strain curve for HSLA steels. The axial stress, α, is plotted along the Y-axis, and the corresponding axial strain, ε, is plotted along the X-axis. As stress is applied to the steel material, a corresponding strain develops as defined by Hooke's law that states that the amount of stress is equal to the amount of strain, multiplied by the modulus of elasticity of the material. This relationship is graphically represented as the line segment O-A in Fig. 16.14, where the modulus of elasticity defines the slope. The stress at Point A is referred to as the "proportionality limit," also referred to as the "elastic limit." As

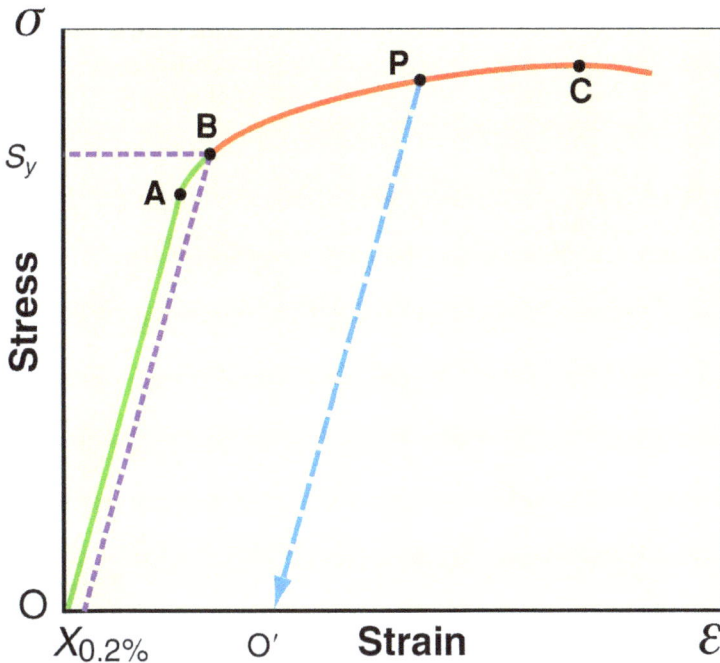

Fig. 16.14—Stress/strain relationship representing behavior of HSLA material (courtesy of Dr. Steve Tipton, U. Tulsa).

long as the stress levels within the steel are held below the elastic limit, the strains are also considered to be elastic, and no permanent deformation will occur.

However, as increased stress is applied to the steel, the elastic limit is exceeded and will reach Point B, which is termed the "yield point." The yield point defines the yield strength, S_y, or the stress that causes the initiation of plastic strain in the material. Once the yield point is reached, permanent deformation occurs, and plastic strain is developed as the material begins to elongate permanently. To determine the yield strength of alloy steel materials in a consistent manner, a standard 0.2% offset plastic strain value was adopted to locate the yield point on the stress-strain curve. This is shown as the dotted line "AB-X$_{.2\%}$." Loading into the plastic region to Point P, followed by unloading to Point O, Line P-O is usually believed to also define the slope of the modulus of elasticity and intersect the X-axis to represent the amount of plastic strain resulting from the deformation. However, it has been shown that plastic deformation tends to reduce the modulus of the elastic unloading curves by as much as 15 to 20% for loading typically seen in CT well intervention and drilling services.

By applying additional stress, we will reach Point D, which is referred to as the ultimate tensile strength of the material. Once Point D is reached, the material will suffer a separation failure.

The amount of stress experienced by the HSLA material can be fully appreciated when considering the degree of bending the tube must undergo during conventional service activities. The radii of typical CT service reels and tubing guide arches are significantly less than the yield radius of curvature (R_Y) for any given size of CT listed. When the tubing is bent and wound onto the service reel or storage spool, the HSLA steel is yielded and becomes plastically deformed. This plastic deformation event is similar to the graphical illustration seen in Fig. 16.14 as Curve O-P. When the amount of stress applied to the tubing is relaxed, the strain remaining is permanent and has a value represented by the extension of Line P-O$'$.

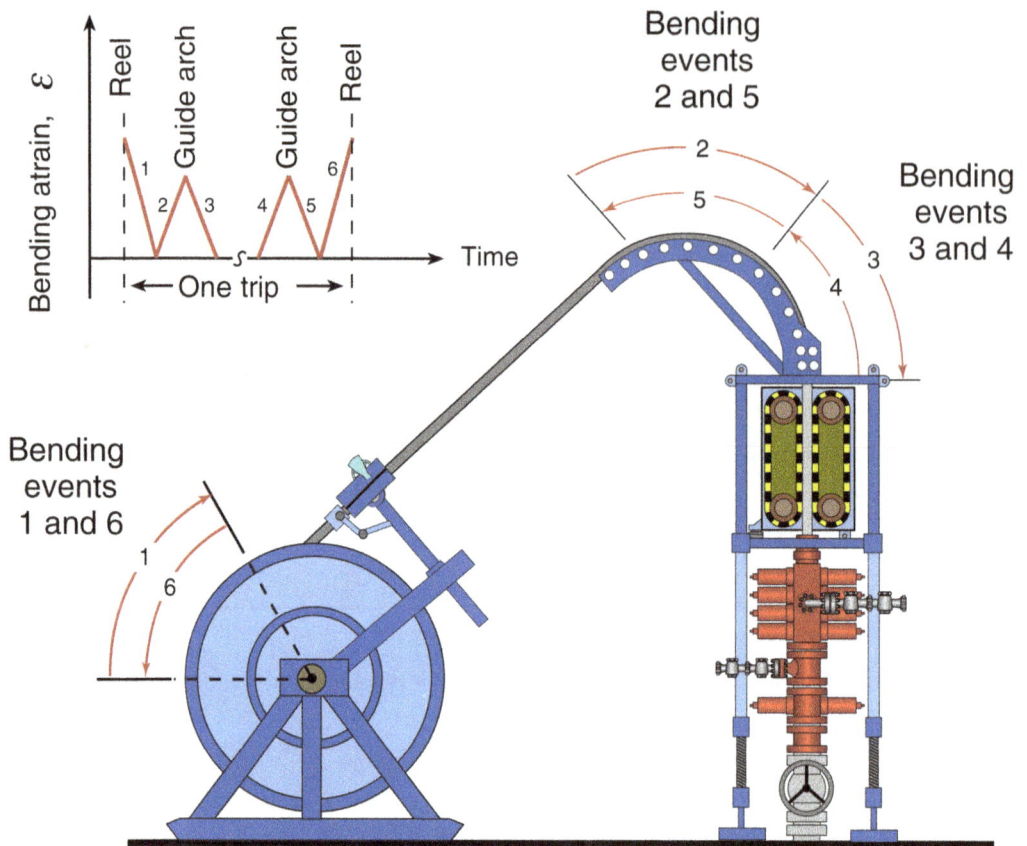

Fig. 16.15—Bending events occurring during CT operations (courtesy of Dr. Steve Tipton, U. Tulsa).

The fatigue imparted to the CT material during normal service operations is the result of bending the continuous-length tubing beyond its elastic limit and forcing the material into plastic deformation. **Fig. 16.15** illustrates the typical operating sequence whereby bend cycles are imposed on the CT during deployment and retrieval. For this illustration, the initial state of the tubing will be in the "as wrapped" condition on the service reel. The bend event sequence is described next.

• The tubing is pulled off the service reel by the injector. The hydraulic motor on the reel provides resistance to the pull of the injector, placing the tubing in tension. The tension applied is typically limited to the amount needed to bend the tubing over the tubing guide arch and maintain control of the tube during deployment. Therefore, the tubing is straightened out somewhat, constituting the first bend event.

• Once the CT reaches the tubing guide arch, the tubing is bent over the prescribed radius. This event constitutes the second bend event experienced during deployment.

• As the tubing travels over the tubing guide arch, the tubing is returned to the straightened orientation before entering the gripper blocks in the injector (third bending event).

These three bending and straightening events are repeated in reverse as the tubing is extracted from the well, resulting in a total of six bending events commonly described as a "trip." In relative terms, the smaller the bending radius, the larger the value of bending strain induced into the tube segment. This is seen as the more severe bending strains imposed when the CT is cycled over the service reel, as compared to the less severe bending strains imposed when bending occurs over the tubing guide arch. Therefore, during a complete service trip, the tubing will experience two bending events inducing relatively higher bending strain and four bending events inducing relatively lower bending strain. Note that the pairing of two bending events with equivalent strain magnitudes is defined as a "bend cycle." Therefore, within each CT service trip, there are two relatively low-strain bend cycles and one relatively high-strain bend cycle.

In general, for CT well-intervention and drilling operations, the number of trips assigned to the tubing is not the same over the entire length of the tubing string. For example, during routine service work, it is common to periodically stop deployment and reverse the motion of the CT string, retrieving a prescribed length of tubing back out of the wellbore to check weight and drag of the tubing. The intermediate segments of the tubing string, which were subjected to these "drag checks," will have a greater number of trips allocated than that which is allocated to the remainder of the tubing string. In addition, where a prescribed service program requires repeated bend cycling over a specified segment of the tubing string, this segment will have an accumulation of fatigue damage that is significantly greater than the fatigue damage subjected to the remainder of the tubing-string length.

The discussion on CT bend cycling makes apparent that most of the material fatigue occurs at the reel and tubing guide arch, with very little, if any, within the wellbore. In CT services, the plastic deformation of the tube material that results in the cumulative damage is defined as "ultralow cycle fatigue." The loading in ultralow cycle fatigue is plastic, and failure of the material generally occurs within 2,000 bend cycles.

A consequence of ultralow cycle fatigue in CT operations is the eventual formation of microcracks in the tube body. With continued bend cycling, the cracks propagate through the tubing wall until the crack establishes full penetration from one side of the wall to the other. Once the crack has fully translated through the tube body wall, pressure integrity is lost. Typically, the initial size of the crack is very small, making detection of this type of body damage very difficult. Furthermore, cracks tend to initiate on the inner wall of the tubing, despite the fact that the bending strain is greater on the outer surface. In conditions in which high internal pressure is present, the hoop stress may cause the crack to instantaneously propagate along the circumference of the tube. In this condition, a major transverse crack will occur and may result in the mechanical separation of the tubing. For this reason, crack initiation is typically the criterion used to define "failure" within a CT section.

Loss of pressure integrity in any condition makes the CT string unfit for service. At present, the fatigue condition of a segment of tubing with an unknown bending history cannot be accurately measured nondestructively. Because fatigue appears to be a statistical phenomenon, variations can be expected in observed fatigue life for any sample, regardless of comparable bending history records.

Various types of bend-cycle fatigue test machines (**Fig. 16.16**) have been developed in an attempt to simulate the wellsite bendcycling of CT over a constant radius with internal pressure loading. These bend-cycle fatigue fixtures generate a statistically significant quantity of fatigue data and provide a means to estimate bend-cycle fatigue over a wide range of test conditions.

Numerous analyses of the trends recorded from the ever-increasing volumes of CT fatigue tests suggest that bend-cycle events imposed onto a given tubing specimen with high internal pressure (causing hoop stresses on the order of 30 to 60% of the yield strength) accumulate fatigue damage at a much greater rate than bendcycles imposed with low internal pressure loading. In addition, the magnitude of fatigue damage realized from a given bend-cycle event cannot be applied to working-life predictions in a linear fashion. Evidence from the volume of testing suggests that a given bend-cycle load applied later in the tube working life may cause greater fatigue damage than the equivalent bend-cycle load applied earlier in the tube working life. The bend-cycle fatigue data obtained from the test fixtures has also provided insights into several other areas of CT service behavior. These items are discussed next.

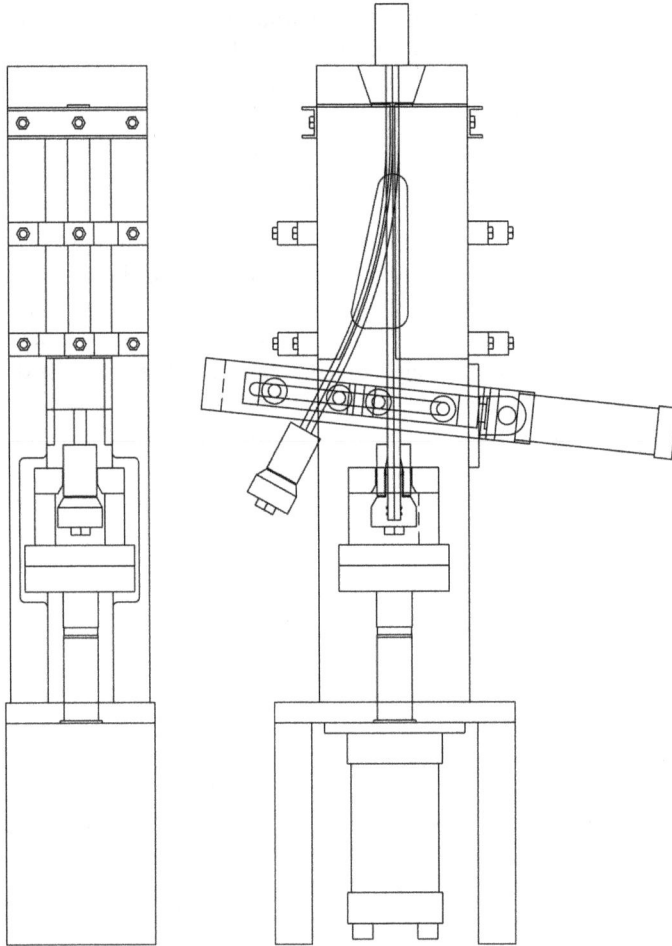

Fig. 16.16—Illustration of CT fatigue testing machine (courtesy of Dr. Steve Tipton, U. Tulsa).

16.5.2 "Cyclic Softening" of CT Material. One significant consequence of repeated bend cycling of CT material is a phenomenon referred to as "cyclic softening," which results in a reduction in material yield strength as the CT performs its prescribed service.[9] The magnitude of the cyclic softening realized in CT materials can be directly related to the amount of material strain imposed during service.

Standard strain-controlled, low-cycle fatigue testing is commonly performed on axial-test coupons of a specified material to evaluate its behavior when subjected to fluctuating strain conditions. These tests attempt to simulate the magnitude and intensity of the strain fluctuations anticipated during the prescribed service and are used to help predict the fatigue strength of the material at the test conditions. For steel CT products, similar types of stress-cycle tests have been performed to provide insight into fatigue life and performance when subjected to multiaxial plastic deformation events.

The strain-controlled fatigue-testing process is illustrated in **Fig. 16.17**, in which the CT material is cut into the designated test coupon geometry and installed into a fatigue test machine that subjects the material specimen to strain-controlled, axial load cycles between a maximum and minimum strain, γ_{max} and γ_{min}. In these tests, cyclic softening was observed because the peak stresses 3, 5, and 7 continually diminish from the initial peak at Point 1.

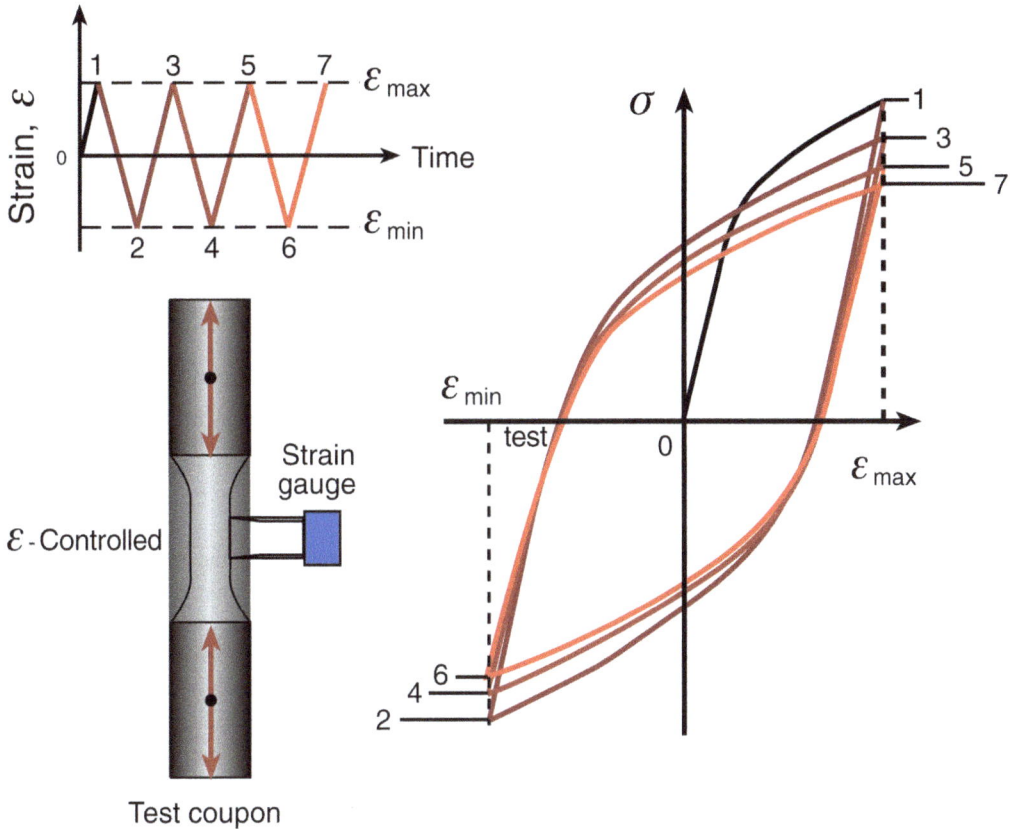

Fig. 16.17—Cyclic softening of coupon during strain-controlled fatigue test (courtesy of Dr. Steve Tipton, U. Tulsa).

The stress/strain diagram shown in Fig. 16.17 illustrates the hysteresis effect observed during the strain-controlled axial-stress load-cycle tests. If plotted in terms of stress vs. time as shown schematically in **Fig. 16.18**, the coupons exhibit a transient stress response, in which the recorded peak stress values diminish incrementally relative to the previous value. This cyclic softening corresponds to a reduction in the material yield strength. Eventually, stable hysteresis loops form, reaching a condition described as "cyclic stabilization" of the material. In the conventional low-cycle fatigue tests conducted on the material coupons, cyclic stabilization appears to occur when the accumulated fatigue damage is within the range of 20 to 40% of expended cycle life.

The material strain imposed during bend cycling of full-body CT specimens is considered to be at the upper limit of the low-cycle fatigue regime, especially because this bending can combine with pressure to cause lives on the order of fewer than 20 cycles. As previously discussed, the category that best describes the behavior of CT bend-cycling is that of "ultralow cycle fatigue." At the strain levels realized during full-body CT bendcycling over typical bend radii, the bulk of the material softening can be seen within the first few load cycles, with continued material softening occurring over the remainder of the tube cycle life. For full-body CT bendcycling, truly stable stress responses may not be achieved, and cyclic stabilization may never be truly realized.

The importance of this discussion becomes apparent when considering that the strain range experienced in CT services varies significantly from point to point around the circumference of

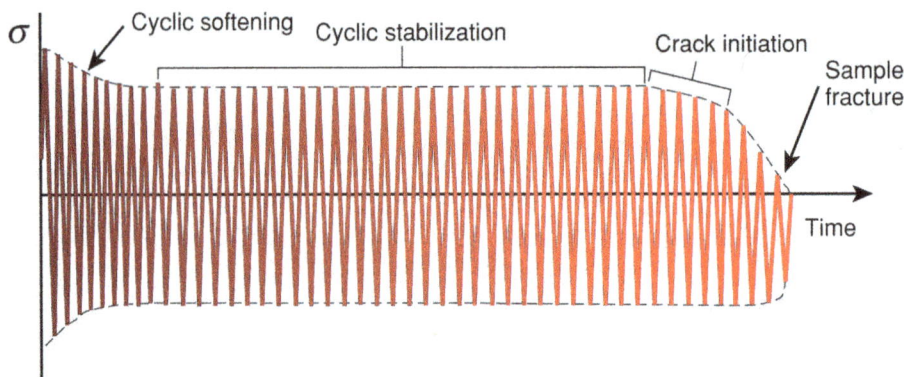

Fig. 16.18—Stress history of coupon during strain-controlled fatigue test (courtesy of Dr. Steve Tipton, U. Tulsa).

the tube body. At positions between the points of maximum stress (i.e., the top and bottom of the tube body) and the neutral axis, the intermediate strain ranges and transient material softening will most likely occur throughout the entire working life of the tube. As a result, material strength and performance properties can be expected to vary relative to the position around the circumference of the tube body. In some CT material samples tested, the reduction in yield strength was found to be as much as 10 to 20% of the parent material yield strength.

In actual service, the neutral bending axis on the CT when deployed into the wellbore will most likely be in a different orientation on the tube circumference from when it is retrieved from the wellbore. This change in neutral bending axis may be the result of relaxation of residual stresses in the tube when axial tensile forces are applied during deployment into the well or because of some other phenomena. During the extraction process, the tubing is retrieved into the injector gripper blocks with the previous neutral axis rotated slightly off alignment. Once the tube is secured into its "new" orientation relative to the gripper blocks, a new neutral axis will be defined as the tubing is bent over the tubing guide arch. This repositioning of the neutral axis during well intervention and drilling operations is believed to distribute the high-strain bend cycles around the circumference, creating a more uniform accumulation of fatigue damage within the tube.

16.5.3 Diametral Growth. When subjected to plastic deformation because of bendcycling with internal-pressure loading, the diameter of the CT tends to grow or "balloon." Even when the internal pressure loading is well below the yielding stress of the material, the tube body is subjected to hoop and radial stresses that cause the material to grow macroscopically in diameter and to decrease in wall thickness (**Fig. 16.19**). In uncontrolled fatigue tests, diametral growth exceeding 30% has been observed. The primary factors influencing diametral growth are material properties, bend radius, internal-pressure loading, tube OD, and tube wall thickness.

One major concern for diametral growth is the interaction with surface handling and pressure-control equipment. The injector gripper block loading on CT usually has an impact on the tube geometry, and the effect tends to vary according to the magnitude of gripper block normal force, block geometry and wear, and CT geometry, internal pressure, and material type. Most conventional counter-rotating chain injectors have gripper blocks that are machined to fit the OD of the specified size of tubing. When CT experiences diametral growth, the increase in tube size creates a nonsymmetrical loading condition, concentrating the normal force load at

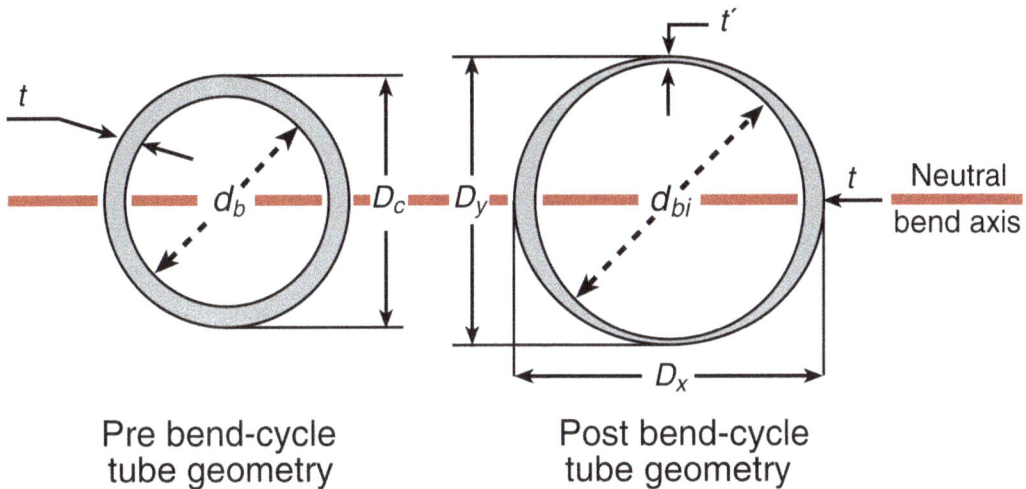

Fig. 16.19—Illustration of typical diametral growth for CT (courtesy of Dr. Steve Tipton, U. Tulsa).

contact points on the edges of the gripper block. These focused stress concentrations induce additional damage into the tube body and result in added tube deformation.

Some injector manufacturers have made adjustments to the design of the gripper blocks in an attempt to minimize this additional damage to the tube. The segmented-type and "variable"-type gripper-block designs appear to better accommodate increases in tube body growth. However, as the diametral growth of the tube increases, the two-point contact loading of the gripper blocks will induce lines of concentrated stress on the tube body. With this focused stress concentration, the gripper blocks will impart greater damage onto the tube at the normal force loads typical for properly fitted gripper blocks.

Another concern for diametral growth relates to the interaction with the pressure-control equipment. The stripper assembly contains brass bushings that are used to prevent extrusion of the elastomeric elements. These bushings have an internal diameter that is slightly larger than the specified OD of the tubing. If the actual CT diameter on any axis reaches or exceeds the internal diameter of the brass bushings, the CT will bind up within the bushings, resulting in surface damage to the tubing. Once this condition is reached, the CT may no longer pass through the stripper. To prevent this situation, limitations should be placed on the maximum allowable CT diameter.

The recommended method of defining maximum allowable growth of CT is "absolute diameter," at which the CT is retired from service when the diameter reaches a given value greater than its specified dimension. The limit typically used is 0.050 in. and can be considered valid for all CT sizes when the aforementioned type of stripper apparatus is used.

Observations of bend-cycle fatigue testing directly relating to diametral growth in CT are listed next.

• The growth rate of the OD increases with increasing internal-pressure loading.

• The diametral growth of larger-diameter CT as a percentage of specified diameter is greater than that of smaller-diameter CT.

• CT specimens with higher material-yield-strength demonstrate less diametral growth than lower material-yield-strength specimens.

• Mechanical limitations of surface handling and pressure-control equipment tolerances for allowable diametral growth restrict the effective working life of CT in high-pressure service to only a fraction of the projected fatigue life.

16.5.4 Differential Wall Thinning. As a consequence of diametral growth, CT experiences differential wall thinning. If we assume that the cross-sectional area of the tube body is constant, then as the diameter grows, the redistribution of material causes the wall of the tube to get thinner. The concept of differential thinning is also illustrated in Fig. 16.19. As the tube is bend cycled about the neutral axis, the top and bottom of the tube are subjected to the highest stress concentrations and subsequently experience the greatest amount of thinning. Note that the wall thickness at the neutral axis (where no bending stress occurs) remains at the initial wall thickness. Therefore, the wall-thinning process is not uniform. Although the severe ballooning and localized wall-thickness reduction are typically observed on samples tested in the laboratory, such gross geometric changes are rarely seen in the field. This is because of factors such as tubing rotation (which causes the wall thinning to occur more uniformly) and the ID of the stripper brass bushings (which limit the use of tubing having excessive diametral growth).

Although extensive diametral growth and wall thinning are observed in the lab, empirical data obtained from extensive full-scale field cycle testing conducted by Walker *et al.*[10] has shown that for a 3% increase in tube body diameter, the wall thickness tends to decrease approximately 7.5%, resulting in a loss of burst and collapse pressure rating of as much as 10%.

The mechanics behind the phenomena of diametral growth and wall thinning is referred to as transverse cyclic ratcheting, whereby material deforms permanently with each cycle of loading in directions transverse to the major fluctuating load directions. Conventional incremental plasticity theory has been shown to overpredict cyclic ratcheting in the typical CT loading regime. Therefore, refined plasticity models have been developed and used successfully to predict the behavior of CT samples in the lab and the field.

16.5.5 Tubing Length Extension. As a result of the stress/strain response exhibited by CT material (as seen in Figs. 16.14 and 16.17), the tubing enters the well with a significant distribution of residual stress. The residual stresses vary from tension on one side to compression on the other, at magnitudes equal to the material's cyclic yield strength. This residual stress profile has a first-order influence on the load- deflection behavior of the tubing and tends to induce secondary deformation mechanisms which contribute to permanent increases in tube length. This condition is commonly described as "elongation," but should not be confused with the API definition of elongation, which represents the change in specimen gauge length once ultimate tensile strength is exceeded during a destructive tensile test. To avoid confusion, the term "extension" is used in this text to refer to the observed permanent increase in CT length resulting from realignment of residual stresses where axial force loading is applied.

As with diametral growth, extension occurs despite the fact that the applied axial stress (in terms of load over cross-sectional area) is maintained to a value substantially below the material yield stress. The investigation of extension was prompted by observations in CTD operations where the BHA could not return to maximum deployed depth after the string was retrieved to surface and then redeployed to depth (referenced by painted points on the reel). The reported "stack-out" depth was several feet higher than the previously reached depth, in excess of calculated corrections for tube helix reorientation within the well. Extraction and subsequent redeployment of this CT string found the stack-out depth several feet higher than the previous "stackout" depth, prompting concern for accurate depth control. Other onsite observations found that the amount of permanent tube length increase was greater for new tubing strings as compared to "seasoned" tubing strings.

Detailed studies have documented that extension occurs during high-pressure/constant-pressure bend cycling because of the constant axial stress caused by pressure loading on the end caps. When higher tensile loads are applied to the tube immediately after bend cycling has occurred, the section of the tube wall with residual tensile stresses equal to the yield strength cannot sustain any more axial stress. Therefore, the residual stresses within the tube body are redistributed to reach equilibrium, resulting in a permanent extension of tube length. The intensity of the residual stress distribution in the tube body is directly related to the imposed strain, which suggests that the amount of extension for a given axial load increases as the bend radius is reduced. Therefore, the location of maximum tube-length extension within the string will typically be in the region where the smallest bend radii and highest axial tensile loads are imposed. This region of the tubing string is typically located near the core diameter of the service reel (reference end), with high axial loading resulting from the weight of the tubing suspended within the wellbore.

Multiaxial plasticity models have been developed to compute stresses and strains induced by the surface handling equipment (tubing guide arch and service reel), internal pressure loading, and axial loads experienced by the tube. These complex and sophisticated algorithms handle highly nonproportional loading and must accommodate changes within the tube cross section as well as axially along the entire length of the string. Therefore, the accuracy of the plasticity model predictions depends upon the proper mapping of imposed strains throughout the service life of the tubing string.

16.5.6 Mechanical Damage. The bend-cycle fatigue life of a CT string is also sensitive to surface damage, such as scars, scratches, gouges and dents. These types of mechanical damage serve as localized stress-strain concentrations where repeated bend cycling can cause cracks to develop in the tube body. Testing was performed by Quality Tubing to quantify the reduction in bend-cycle fatigue life for 1.750-in.-OD, 0.134-in.-wall-thickness tube samples, with and without surface damage present.[11] The 80-ksi-yield tubing samples were subjected to bend-cycle testing over an equivalent 72-in. bend radius with a 3,000-psig internal pressure. A baseline test was performed with an undamaged tube and survived approximately 725 bend cycles before loss of internal pressure (failure). Two samples were prepared with transverse notches cut to an approximate depth of 10% of wall thickness. One sample located the notch across the longitudinal weld seam, with the other sample locating the notch 180° from the longitudinal weld seam. Both of these samples survived approximately 120 bend cycles, yielding an 83% reduction in fatigue cycle life relative to baseline. Two additional samples were prepared with the 10% wall thickness notches located at the same positions on each tube body, but these notches were ground out and polished smooth before being subjected to bend cycling. By removing the transverse notches, the tube samples survived approximately 590 bend cycles, yielding a 19% reduction in fatigue-cycle life relative to baseline. With the removal of the surface damage, the bend-cycle fatigue life increased from 120 to 590 cycles, a 390% improvement in survival life.

A recent study has noted that not only is the geometry of the defect important (length, width, depth and shape of the defect), but also the mechanism by which the defect was imposed into the tubing.[12] For instance, ball-nosed end mills were used to cut hemispherical defects into a set of CT samples. In another set of tubes, defects with identical geometries were imposed by pressing hard balls into the surface, rather than removing material by machining. When tested under identical constant amplitude bend cycling, the samples with the milled defects reduced the life of the samples by as much as 60% (relative to defect-free baseline samples), while the impressed defects caused virtually no life reduction.

Significant research is currently under way to quantify the influence of defects on CT life, as well as the effectiveness of repair strategies, such as simple removal by surface grinding. The research currently under way is addressing the influence of loading parameters (bending radii, pressure, tube geometry) and material grade.

An important component of the research is the incorporation of CT inspection technology. The most prevalent CT inspection technique is magnetic flux leakage. Although this technique has proven effective in detecting the presence of a flaw, existing technology provides no information about the severity of the flaw. The research currently under way has made important steps[13] toward extracting quantitative information about the geometry of the defect, as well as its influence on fatigue. Over time, a continually growing database will facilitate the development of refined signal-processing algorithms that provide important feedback directly from magnetic flux leakage results.

16.5.7 Surface Rippling. An additional phenomenon that occurs as a result of CT bend-cycle fatigue is described as "surface rippling." It is common to find "ripples" developing on the top surface of the CT (relative to the neutral axis) when subjected to bend cycling in combination with high internal-pressure loading. On many samples examined, the period of the ripple formation has typically been observed to be approximately twice the tube diameter. This condition has been observed to occur both in field service operations and with tube specimens tested in fatigue cycle fixtures. This consequence of bend cycling is attributed to the fact that CT does not yield continuously along its length as it is deformed. What occurs is more of a buckling phenomenon, where the outer fibers simply lengthen to a point where they buckle in compression when straightened and never quite return to a perfectly straight condition upon rewrapping. Rippling typically occurs relatively late in the fatigue life of the tubing. Therefore, if surface rippling is observed in the CT, it is recommended that the tube be immediately withdrawn from service.

16.6 Commonly Used Bend-Cycle Fatigue Derating Methods

Over the years, attempts have been made to track the working history of CT strings in service to maximize the service utility of the tube while minimizing fatigue failures. As a result, three commonly used methodologies for predicting the fatigue condition of the CT were developed.

16.6.1 The "Running-Feet" Method. A relatively simplistic approach used to predict the working life of CT is commonly described as the "running-feet" method, in which the footage of tubing deployed into a wellbore is recorded for each job performed. This deployed footage is then added to the existing record of footage deployed in service for any given string. Depending upon the service environment, type of commonly performed services, and local field history, the CT string is retired when the total number of running feet reaches a predetermined amount.

The running-feet method offers the service vendor relative simplicity of use, requiring only that the maximum depth of CT deployed into the wellbore be recorded. However, there are numerous limitations of this fatigue-tracking method as a reliable means of determining ultimate working life of a CT service string. Several limitations are described next.

• The value of maximum footage to retirement for any CT string is based on the service vendor's previous experience with the same type of tubing, performing wellsite operations with similar well depths and types of service. In this method, there is generally no consideration given to duration of corrosive services performed or effects of exposure to atmospheric corrosion.

• The running-feet method typically focuses on the specified OD of the CT string in service, with minimal consideration for tubing wall thickness, tube material type, and yield strength.

• The running-feet method does not have a means of accounting for variations in tubing guide arch radius, service reel core radius, internal pressure loading, or identification of specific tube segments where additional bending cycles are applied.

• The working life-derating method used in the running-feet approach cannot be extended to different tubing sizes or operating conditions. This method can be used only where working history for the specific tube material, geometry, and surface handling equipment has been gathered and analyzed to yield the prescribed maximum running-feet value.

16.6.2 The "Trip" or "Empirical" Method. A natural extension of the running-feet fatigue derating approach can be found in what is commonly described as the "trip" method. In the trip method, numerous improvements have been incorporated to the running-feet approach, providing greater reliability in predicting working life of the CT string. One major improvement entails evaluating the CT string as a series of partitioned segment lengths that can range from 100 to 500 ft long. This approach applies a greater sensitivity to the working life analysis by identifying sections of the CT that are subjected to more bending cycles than others during a specified service. The number of trips over the service reel and tubing guide arch for each discrete segment can then be tracked and recorded. When employing this method, a reduction in the length of the section increment increases the accuracy of the bend-cycle record. This type of analysis makes it possible to identify the CT string segments that have experienced the most bend-cycle fatigue damage.

Another major improvement with the trip method incorporates the effects of internal-pressure loading. For a given tubing guide arch and service reel core radius, CT bend-cycle fatigue life decreases significantly with increased internal pressure loading. The evolution of the trip method incorporated extensive CT bend-cycle fatigue testing using full-scale service equipment (injector, tubing guide arch, and service reel) and varying amounts of internal-pressure loading. In this scenario, numerous bend-cycle fatigue tests are performed for a given size of CT at specified amounts of internal pressure.

Data recorded in these tests were initially used to create a database for statistical projection of CT working life. From these types of tests, a segment of the CT string that had accumulated a considerable amount of bend-cycle fatigue damage can be identified, thereby providing the user with options for removing the heavily damaged segment of tubing from service.

As more full-scale fatigue cycle tests were performed, trends in CT fatigue were identified for various pipe sizes, tube geometry, and internal-pressure load conditions. Analysis of these trends provided the service vendor with the ability to "curve-fit" the data points and derive empirical coefficients that were incorporated into conventional multiaxial fatigue-life prediction approaches, yielding the early CT fatigue prediction models.

The aforementioned improvements in fatigue damage tracking realized by the trip method offer enhanced accounting of operating conditions present when the bend-cycling events occurred, along with a greater sensitivity of identifying tubing-string segments subjected to bend cycling. The limitations with the trip method of empirical modeling include:

• The derived empirical coefficients for fatigue-damage are generally different for each combination of CT material, OD, wall thickness, and bending radius.

• Bend-cycle testing using full-scale equipment is required to obtain the fatigue coefficients experimentally (expensive and time consuming).

• The trip method does not incorporate tube body damage incurred as a result of well-servicing operations. This type of damage includes exterior tube body wear, interior and exterior corrosion (atmospheric and industrial), or nicks, cuts, or scarring resulting from contact with surface handling equipment.

• The test data obtained from fatigue bend-cycling machines is usually at a constant internal pressure. In well-servicing operations in which fluid pumping is required, the amount of internal pressure present in the CT varies along the entire length of the string. Therefore, as the tubing is deployed and retrieved, each section of the string has a different internal pressure at the point where bend cycling occurs.

• The varying internal-pressure loading at the point of bend cycling requires a complicated record and prediction procedure to provide a realistic working-life prediction. This requires investment in surface recording instrumentation and sophisticated data collection systems, such as portable computers, as well as complicated tubing management software systems for tracking and maintaining up-to-date records of the compiled tubing working life.

16.6.3 The "Theoretical" Method. A third method for predicting bend-cycle fatigue incorporates the same approach developed in the "trip/empirical" method for estimating the bend-straighten-pressure history for each segment of tubing along a string. However, a theoretical model based on the fundamental principles of mechanics and fatigue is used to estimate the stress, strain, and fatigue behavior of each section in the string.

The theoretical modeling of fatigue typically involves use of "plasticity" algorithms and "damage" algorithms. The plasticity algorithm is used to estimate the stress and strain history of the CT material as it is bent or straightened over a particular bending radius at a particular internal pressure. The damage algorithm uses the concept of cumulative fatigue damage to quantify the reduction in working tube life caused by each bend or straighten event. The approach is very mechanistic, for instance, taking into account the fact that the pressure during each bending or straightening event can be different. The fatigue damage computed for each event is summed throughout life and is usually expressed as a percentage of the predicted working life. Since each section of tubing along the length of a string can endure differing bend-straighten-pressure histories, the damage profile can (and usually does) vary along the length of a typical working string.

The plasticity algorithm in the theoretical model requires input of the specific material properties. These properties come from two types of testing. First, the aforementioned low-cycle fatigue testing conducted on axial coupons, and second, full-scale data typically taken from a CT fatigue testing fixture, or from full-scale equipment.

The low-cycle fatigue data are used to compute both elastic material properties and the cyclic stress-strain curve for the particular CT alloy. Although these properties are generated for axial loading only, they serve as the "constitutive relations" (i.e., the relations between stress and strain) for a multiaxial plasticity algorithm, which is capable of estimating the history of all stress and strain components (axial, hoop and radial) in the tubing. Since the plastic deformation caused by bend cycling is so severe, it was determined that conventional plasticity theory was inadequate to describe the behavior of CT accurately. Conventional theories tended to overpredict phenomena such as ballooning and wall thickness reduction. To overcome this, new theories were developed specifically for CT. These models are effectively "tuned" to specific alloys by collecting data from constant pressure bend-cycling tests conducted on laboratory testing fixtures (although data from full-scale equipment can also be utilized) to supplement the low-cycle fatigue data. The empirical parameters derived from these test results cause the algorithms to do an excellent job of estimating ballooning and wall thickness reduction, as well as fatigue under complex loading histories.

The use of empirically derived data in this approach assures that the model can be mapped back to realistic behavior exhibited by real CT sections. In reality, CT mechanical properties must be allowed to vary within a particular grade. For this reason, it is important to collect as many experimental data points as possible to characterize the scatter caused by typical material variation. The greater the number of experimental data points, the stronger the statistical validity of the model.

The advantages to the use of theoretical models include greater accuracy of bend-cycle fatigue life prediction with the capability to predict fatigue life for variable loading conditions. The use of such an algorithm in the field is dependent upon the use of a reliable string-management routine that keeps track of the depth and pressure history of the string throughout its use and is capable of computing the bend-straighten-pressure history for each section of tubing, based on that depth-pressure log. Fortunately, software is available commercially to implement the approach either in real time or following the job.

The advantage of this model is its ability to make quantitative predictions that are based on statistically significant quantities of empirical data. **Fig. 16.20** shows estimated trips to failure vs. internal pressure for an 80-ksi tubing material with three different diameters and a 0.134-in. wall thickness, run over a 96-in. reel core diameter and 72-in. tubing guide arch. (The life

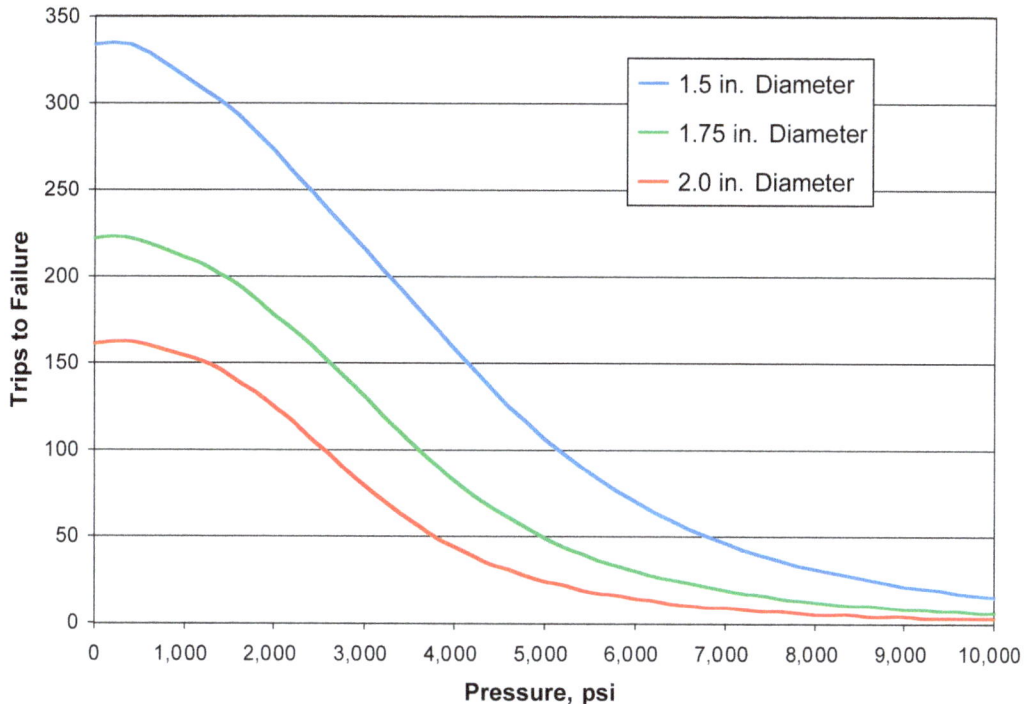

Fig. 16.20—Bend-cycle fatigue-life comparison for various CT OD cases.

prediction algorithm used to make these predictions comes from the Flexor TU4 Life Prediction Model).

In the field, the model is capable of monitoring the fatigue life profile along the length of the tubing. But, more importantly, it can predict the effect of impending jobs on that profile. This allows the engineer to modify the use of each string to effectively maximize its life, avoiding potentially dangerous situations and reducing costs.

The limitations of theoretical models include:

• Two sets of input data are required to characterize a particular material: (a) low-cycle fatigue data taken from axial coupons and (b) constant pressure bend-cycle data taken from CT fatigue fixtures or full-scale equipment. The greater the number of data points from the latter set, the stronger the statistical validity of the model.

• Current theoretical models do not incorporate tube body mechanical damage incurred as a result of well intervention or drilling operations. This type of damage includes exterior tube body wear, interior and exterior corrosion (atmospheric and industrial), or nicks, cuts, or scarring resulting from contact with surface handling equipment. However, research is under way to develop models that can estimate the influence of surface defects. Such a model exists and is currently being refined.

• The implementation of a theoretical approach must be in concert with a routine capable of estimating the varying internal-pressure loading at the point of bend cycling. This requires a reliable record of not only the depth-pressure history of a string throughout its use, but also a record of any string modification (section splicing and/or removal). This requires investment in surface recording instrumentation and sophisticated data-collection systems, such as portable computers, as well as complicated tubing management software systems for tracking and maintaining up-to-date records of the compiled tubing working life.

16.7 CT Management

As discussed in the previous section, the service vendor must maintain a history of the various services for which each CT string has been employed to ensure prudent management of CT strings. This tubing-string record should include the following information (as a minimum):

• Pressure/bending fatigue-cycle history and locations of repeated bending cycling. The data for these records should be obtained from daily service activity reports or through electronic record-keeping devices.

• If pressure bend cycling is not recorded, then the service vendor should provide the "total running feet" (records must reflect footage into and out of the well).

• Maximum pumping pressures through the CT string when stationary.

• Exposure of tubing string to acid service. This record should list the number of acid jobs performed, type and volume of the acid system pumped, duration of the acid-pumping pro-gram, and vendor-recommended derating factors for the string.

• Locations of welds, identification of type of weld, and observations of deformity, ovality, or surface damage.

• Locations in the CT string where tensile loads exceeding 80% minimum yield were placed upon the pipe.

• A detailed record of any splicing or section removal that takes place along the length of the string.

16.7.1 Effects of Bend-Cycle Fatigue on Welds. Welds of several types are fundamental to the manufacture of steel CT. Welds are a common concern because the bend-cycle fatigue life of certain welds can be significantly less than that of the parent tube material. As a result, the performance of welds is of critical importance to the working condition of the tubing string as a whole.

The longitudinal seam weld runs the entire length of the coiled tube and is created during the manufacturing process as the base skelp is formed into tubing in the mill. Problems with the longitudinal seam weld are usually detected at the mill during either the manufacturing process or the subsequent hydrostatic pressure testing procedure. However, failures along the seam have occurred during field services in high-pressure applications because of excessive triaxial stress loading. During laboratory CT fatigue testing, it has become common practice to place the longitudinal weld seam along the curved bending mandrel, or on the compressive side of the tube wall. In general, cracking occurs on the compressive side of the tubing at least as often as (if not more often than) it does on the tensile side of the tubing. This observation holds even for testing when the weld seam is placed along the neutral axis of the tubing. In general, tests performed on the orientation of the longitudinal weld, with respect to the bending axis, have found no correlation of cycle fatigue life reduction because of longitudinal seam-weld damage. In the process of derating the CT because of bend cycling, the longitudinal weld is generally disregarded.

The welding technique that joins base metal skelp is referred to as the skelp-end weld. The flat skelp is cut and welded at an approximate 45° angle, causing the weld to form a helical wrap around the tube girth when the strip is subsequently formed into a cylinder. The weld is typically made in ideal conditions (with full control over the geometry and weld penetration), resulting in a weld of high and consistent quality. Because of the quality control, treatment of the weld and the subsequent helical distribution of stress, the skelp-end weld performs favor-ably in bend-cycle fatigue tests. Experimental results taken from bend-cycle fatigue tests comparing base tube samples to skelp-end welded samples found that the skelp-end weld sur-vival can range from 80 to 90% of the base tube working life with relatively little variation from one weld to the next. This is well within the typical range of scatter exhibited by CT fatigue lives tested in the laboratory and the field.

A butt-weld or tube-to-tube weld is a means for joining lengths of previously formed tubing. The two ends of the finished tube to be joined are cut square, slightly beveled, carefully aligned, and TIG- or MIG-welded around the circumference. The resulting weld is oriented perpendicular to the CT axis, focusing the bending stress perpendicular to the tube axis (directly through the weldment). Because the weld is made from the tubing exterior, the quality of the weld penetration through the tubing wall is paramount. There are currently two types of tube-to-tube welds in use in CT services, which are discussed next.

Factory Butt-Weld. Before the development of the skelp-end weld manufacturing process, all CT strings were constructed by butt-welding several lengths of finished tubing together using an automated welder at the tubing mill. These welds were made in carefully controlled conditions, and the quality of these welds was generally very good. However, these types of welds remain significantly inferior to skelp-end welds, with bend-cycle fatigue life observed to be in the range of 40% of base pipe working life.

Field Butt-Weld. Once the CT is delivered to the field location, tubing sections which need to be joined are typically butt-welded by hand using a TIG or MIG technique. This "field repair" is frequently the only option available at the service vendor location. Manual butt-welds require a very high level of skill to have acceptable survival rates. As a result, field butt-welds are the most problematic of all CT welds. With the high likelihood of significant variations in quality of welds, the recommended practice based on experimental tests is to derate field butt-welds to approximately 20% of base tube working life.

16.8 CT Applications

There are numerous well-intervention applications that are performed using CT services. The advantages of CT include:

• Deployment and retrievability while continuously circulating fluids.

• Ability to work with surface pressure present (no need to kill the well).

• Minimized formation damage when operation is performed without killing the well.

• Reduced service time as compared to jointed tubing rigs because the CT string has no connections to make or break.

• Increased personnel safety because of reduced pipe handling needs.

• Highly mobile and compact. Fewer service personnel are needed.

• Existing completion tubulars remaining in place, minimizing replacement expense for tubing and components.

• Ability to perform continuous well-control operations, especially while pipe is in motion.

However, there are several disadvantages to CT operations.

• CT is subjected to plastic deformation during bend-cycling operations, causing it to accumulate fatigue damage and reduce service life of the tubing string.

• Only a limited length of CT can be spooled onto a given service reel because of reel transport limitations of height and weight.

• High pressure losses are typical when pumping fluids through CT because of small diameters and long string lengths. Allowable circulation rates through CT are typically low when compared to similar sizes of jointed tubing.

• CT cannot be rotated at the surface to date. However, interest in rotating CT has been high in recent years, and several companies are actively designing equipment that will allow rotating of CT.

The most common CT well-intervention and drilling applications involve issues related to sand cleanouts or solids-transport efficiency. The process of cleaning sand or solids out of a wellbore requires pumping a fluid down into the well, entraining the solids into the wash fluid, and subsequently carrying the solids to the surface. In most cases, the wash fluids and solids are captured in surface return tanks with sufficient volume to allow the solids to settle out of the fluid. Where practical, the cleanout fluids are recirculated in the wellbore, thereby optimizing the cleanout program economics. One of the most important concerns in designing a solids cleanout program is the correct selection of the circulated fluid system. An overview of the two types of cleanout fluids used in CT services, categorized as "compressible" and "incompressible," is offered next.

Incompressible cleanout fluids are, for this discussion, limited to aqueous and hydrocarbon liquids. This type of cleanout program is the less complicated of the two categories. The cleanout fluid selected should be one that provides for solids removal in a "piston displacement" manner. The desired cleanout fluid is one that adequately transports solids out of the annulus. If circulation pump rates achieve annular velocities sufficient to exceed the terminal particle settling velocity, then Newtonian fluids can be used. Depending on the CT OD size, Newtonian fluids are generally adequate when performing a cleanout inside of production tubing. However, when circulating cleanout fluids within large-ID bore tubing or casing, the reduced annular velocities are typically insufficient to transport the solids out of the wellbore. In these cases, the cleanout fluid should be gelled to a higher viscosity. The non-Newtonian fluids used in this situation are generally sheared biopolymer gels or gelled oil systems.

Compressible fluid cleanout programs are more difficult to design and implement than incompressible fluid cleanout programs. Compressible fluids incorporate various fractions of gas in their composition and are selected to compensate for underpressured formations or where liquid cleanout fluid annular velocities are insufficient to lift solids. Fluid volumes change relative to temperature and pressure in a compressible system; therefore, the annular velocities of these fluid returns do not travel at the same rates throughout the length of the annulus.

Once circulation is established in a compressible fluid cleanout program, unit volumes of fluid are pumped down the CT at pressures needed to overcome the total system friction pressure losses. In this condition, the compressible fluid is experiencing high pressure and occupies minimal volume. As the unit volume of compressible fluid exits the end of the CT, it begins its rise in the annulus. The decreasing hydrostatic pressure of the fluid in the annulus, coupled with a reduction in system friction pressure loss, allows the gas within the fluid to expand. The expanding gas within the fluid causes the velocity of the unit volume to increase. The expansion of the compressible fluid and subsequent increase in unit volume velocity creates an environment of high frictional pressure losses.

As a result, annular velocities and solids removal capabilities require complex mathematical calculations to predict. In these cases, it is recommended to obtain computer-generated cleanout-program predictions from the CT service companies to evaluate the performance of a compressible fluid procedure. The fluids that fall under the compressible fluid category are dry nitrogen and foam (aqueous or oil-based).

Nitrogen is an inert gas and, therefore, cannot react with hydrocarbons to form a combustible mixture. In addition, nitrogen is only slightly soluble in water and other liquids that allow it to remain in bubble form when commingled with wash liquids. Nitrogen is a nontoxic, colorless, and odorless gas that is typically brought to location in liquid form in cryogenic bottles at temperatures below –320°F. The liquid nitrogen is pumped through a triple-stage cryogenic pump at a specified rate into an expansion chamber that allows the nitrogen to absorb heat from the environment and vaporize into a dry gas. The gas is then displaced out of the expansion chamber and into the treatment piping at the required surface pressure to perform the prescribed job. Although crogenic nitrogen does not contain oxygen, several other nitrogen sources such as pulse swing adsorption or membrane units can contain significant percentages of oxygen. This oxygen content can exceed 3% and represents a potential corrosion problem in some applications such as CT drilling.

In completed wellbores that are critically underpressured or liquid-sensitive, nitrogen pumped at high rates can be used to transport solids up the annulus and out of the wellbore. The solids-removal mechanism within the wellbore is directly dependent upon the annular velocity of the nitrogen returns. If the nitrogen pump rate is interrupted during the cleanout program, all solids being transported up the annulus will immediately fall back. Of equal concern are the tremendous erosional effects on the production tube, CT, and surface flow tee or flow cross that will occur at the rates needed to maintain solids transport up the annulus. Because of the difficulty to safely execute this type of cleanout program, solids removal programs using nitrogen should be considered as a "last resort" option.

Foam may be defined as a fluid that is an emulsion of gas and liquid. For this discussion, the liquid can be aqueous or oil-based, but the gas will always be nitrogen. In a stable foam, the liquid is the continuous phase, and the nitrogen is the discontinuous phase. In order to homogeneously disperse the nitrogen gas into the cleanout liquid, a small amount of surfactant is used to reduce the surface tension and create a "wet" liquid phase. The surfactant is usually mixed into the liquid phase in concentrations ranging from 1 to 5% of liquid volume. The "wet" liquid is then pumped down the treatment line and commingled with nitrogen in a "foam generating tee." The turbulent action created by the nitrogen intermixing with the "wet" liquid provides sufficient dispersion for the formation of a homogeneous, emulsified fluid.

Foam is generally selected as the preferred fluid media when performing solids removal programs in underpressured wellbores. Foam can be generated in hydrostatic pressure gradients ranging from 0.350 psi/ft through 0.057 psi/ft, depending upon the wellbore pressures and temperatures. The rheology of stable foam most closely resembles that of a Bingham plastic fluid, where the yield stress must be overcome to initiate movement of the fluid.

The industry-accepted term for describing the volumetric gas content of a foam fluid regime is "quality," which is arithmetically defined as

$$Q_f = \frac{\text{volume of nitrogen}}{\text{volume of liquid } + \text{ volume of nitrogen}} \quad\text{.................................. (16.1)}$$

A stable foam regime possesses two significantly unique wash-fluid properties. The first is a solids suspension capability as high as 10 times that of liquids or gels. The second is the ability to act as a diverting system, withstanding up to 1,000 psig applied pressure with a minimal loss of wash fluids to the completion. However, if the foam quality exceeds the stable regime limits, the solids-suspension characteristics of the foam are reduced. At this point, the gas in the foam has expanded significantly, and the velocity of the gas in the annulus is maintaining suspension of the solids particles.

Note that because foam is compressible, the quality of this fluid regime is temperature- and pressure-dependent. As a result, the quality of the system is not uniform throughout the entire wellbore annulus. At surface treatment temperatures and pressures, the foam regime occupies a specific volume, thus defining the initial quality of the system. As the unit volume of foam is pumped down the CT and back up the annulus, the total frictional pressure loss acting against this unit volume decreases. Along with the reduction in annular hydrostatic pressure, the nitrogen gas in the foam expands as it approaches the surface. The result is a dynamic profile of foam quality in which the effects of friction pressure losses, viscosity, and fluid velocity are in constant flux.

Where CT solids-cleanout services are performed to re-establish communication with an open completion interval, it is a common practice to underbalance the pressure within the annular fluid system relative to the bottomhole pressure. This minimizes the loss of circulated cleanout fluids to the formation and the damage associated with deposited solids. As the annular fluid velocities increase, the frictional pressure loss and equivalent hydrostatic pressure acting against the open formation correspondingly increase. If the formation is open to take fluids, then the volume of cleanout fluids returning to the surface decreases to a rate that maintains the proper balance of friction pressure and annular hydrostatic pressure acting on the open completion.

If the cleanout fluid was designed to hydrostatically balance the bottomhole completion pressure, then any additional pressure applied to the circulating system will cause an overbalance condition to occur. If the formation is highly permeable, then it is likely that a portion of the circulated cleanout fluids will be lost to the open completion once communication with the wash system is established. In effect, if the wellbore circulating system is balanced at a specific rate, the incremental increase in surface pump rate intended to increase circulation rates will most likely be diverted into the completion.

Note that the annular pressure losses because of friction for the circulating system are for "clean" circulated fluids. If the solids concentration within the cleanout fluids are maintained below 2 ppg, the effect of frictional pressure loss because of an increase in solids concentration in the annular wash fluids is considered to be minimal. However, a cleanout-fluid solids concentration in excess of 2 ppg is likely to cause a change in fluid rheology and a noticeable increase in annular friction pressure loss.

The rate of penetration of CT into a column of packed solids (wellbore cleanout) or drilled hole, coupled with a constant circulated fluid annular velocity, directly determines the concentration of solids captured within the cleanout fluid. The dispersion of the solids in the fluid media causes an increase in effective weight of the annular returns fluid. As a result, the hydrostatic pressure differential increases between the "clean" fluids pumped down the CT and the "dirty" fluids circulated up the annulus.

The type of formation fluids produced can also determine the effectiveness of the solids-removal program. In a liquid-producing wellbore (oil and water), the fluids in the wellbore are slightly compressible and can, therefore, support a "piston"-type displacement of the captured solids back up the annulus. If the produced fluid is a gas, then caution must be taken to prepare for "gas influx surges" or lost returns when breaking through sand bridges or drilled gas pockets. In addition, the difference in fluid densities between gas and liquids causes the gas to override the circulated cleanout fluid. When in communication with a permeable gas zone or completion interval, liquids are likely to be lost to the gas zone, regardless of the bottomhole pressure.

When performing a solids removal program in an underpressured oil-producing wellbore with an aqueous foam, precautions must be taken for foam degradation when commingled into the oil. The oil rapidly destabilizes the foam regime at the contact interface and breaks down into a gasified, oil/water emulsion. As this gasified oil/water emulsion continues to degenerate and move up the annulus, the solids-laden foam in the returns becomes compromised, and fallback of the solids can occur.

For wellbore cleanout applications, the selection of a wash tool should define the hydrodynamic action of the cleanout program. In other words, the wash tool should provide additional downhole turbulent action as needed. Several wash tools available within the industry are designed with ported jet nozzles for imparting hydraulic energy on packed solids or mechanical assistance in breaking up bridged solids. Many times, these wash tools can be constructed to serve as mandrel bypass tools, further extending their utility. Depending on the number and size of nozzle ports, along with the cleanout fluid system selected, frictional pressure losses can be significant.

With the evaluation of the aforementioned criteria for selecting a cleanout fluid system completed, the cleanout program can be implemented using either the "conventional-circulation" or "reverse-circulation" techniques. These two techniques are discussed next.

Conventional circulation is the process of pumping a fluid down the CT and allowing the fluid to travel back up the wellbore annulus to the surface. Conventional circulation is by far the most common CT service technique used for removing solids out of wellbores. Along with all of the aforementioned criteria used to determine the cleanout fluid system, the maximum tensile stress loads to be placed on the CT string should be estimated to ensure that the loads do not approach the minimum yield rating of the tube.

Both compressible and incompressible fluids can be used with the conventional-circulation cleanout technique. The selection of appropriate CT size depends on the minimum pump rates needed, total circulation system pressure losses, and the minimum yield load rating required to safely retrieve pipe from the wellbore. The use of downhole flow check devices (check valves) and ported wash tools should not inhibit the intended execution of conventional circulation wash programs.

Reverse circulation is the process of pumping the "clean" circulated fluid down the concentric tube annulus and forcing the "dirty" fluids to travel up the CT ID to the surface. In general, reverse-circulating solids cleanout programs are used where annular velocities are insufficient to lift solids out of normally pressured or geopressured wellbores. The cleanout program is designed to pump the clean fluids down the tubing annulus and use the higher fluid velocities within the CT to lift the solids out of the wellbore. This technique is more complicated to plan and execute than the conventional circulation cleanout program.

The planning of a reverse-circulation cleanout program requires that a minimum effective fluid pump rate be established and the frictional pressure loss through the CT and annulus be calculated for that rate with a high degree of accuracy. Information on the particle size, geometry, adhesive tendencies, and settling velocity must be obtained to ensure that no settling or plugging of the CT string is likely to occur. In a reverse-circulation cleanout program, the highest pump pressures act against the OD of the CT directly below the stripper assembly. Depending on the amount of differential pressure between the annulus and the CT ID, coupled with the condition of the CT (tensile forces, ovality, wall thickness, etc.), plastic collapse of the coiled tube can occur.

Reverse-circulating cleanout programs are generally limited to incompressible fluid applications. The selection of an appropriate CT string is limited to larger-ID tube sizes that minimize friction pressure losses. However, the larger OD of the CT causes higher annular pressure losses. In addition, reverse-circulating programs cannot be performed with downhole flow check devices or restrictive wash tools installed on the CT string.

16.9 Coiled Tubing Drilling

16.9.1 Introduction. CTD has a rather extensive history and received a large amount of press and hype from the 1990s to date, not an insignificant amount being less than positive. There have been numerous highly successful applications of CTD technology in such regions as Alaska and the United Arab Emirates, yet CTD is still considered an immature new technology. Reasons for this are numerous, ranging from lack of understanding of CTD technology, to misapplication, to exaggerated expectations. One example is the knowledge that advantages to CT services include small footprint, high mobility, and quick operations. The aforementioned advantages may be true for conventional CT services and simple, short CTD jobs where directional control is not required and the hole can be left uncased. However, when more complex CTD services are planned, including directional drilling and cased completions, these advantages may no longer apply.

The complex drilling operations routinely require pipe handling equipment, provisions for handling long BHAs, large diameter CT, larger blowout preventer (BOP) stacks, and fluid-handling equipment for cleaning, mixing, and recirculating fluids, which are typically not required for conventional CT services. When including the additional separators and nitrogen-pumping equipment required for underbalanced drilling (UBD), the advantages related to small footprint and high mobility may no longer be the case. Numerous truckloads of equipment can take days to rig up in preparation to drill with CT.

Fig. 16.21—Purpose-built CTD rig working in Oman (courtesy of Baker Hughes Inteq).

Fig. 16.21 shows a purpose-built CTD rig working in Oman. When considering all the equipment necessary to handle completion pipe, allow fluid recirculation and provide for UBD operations. The small footprint and high mobility commonly associated with CT may no longer be a valid assumption.

However, even considering the challenges to CTD, there are certain applications in which the unique aspects and capabilities of CTD technology clearly demonstrate that it is the best tool for the job. The most common applications for directionally controlled CTD technology are re-entry drilling/sidetracking from existing wellbores (often through the existing wellbore's production tubing) and underbalanced, managed-pressure, or low-bottomhole-pressure drilling.

Another niche market for CTD technology includes the combination of a CTD unit with a low-cost conventional rotary drilling rig. In this application, the rotary rig is used to drill a quick and simple wellbore and sets casing just above the desired zone. CT is then used to drill a small, clean penetration into the desired zone and is used to run any required completion. The following sections will further discuss CTD technology.

16.9.2 Brief History. As previously mentioned, drilling with CT was one of the first ideas for application of continuous workstrings dating back to the 1926 Bannister concept for a flexible hose drillstring and the 1948 G.D. Priestman patent application work for the more conventional reeled rigid pipe. The Bannister work involved using hose for fluid circulation with support cables attached to the sides to carry the weight. The system was reported to be technically successful, but marginally reliable, and development work ceased in 1940 reportedly because of the "lack of a suitable downhole motor" for the new technology.

The G.D. Priestman patent conceived what is today considered modern CTD technology as far as the spooled tubing and operation is concerned. However, it was 25 years before the first actual steel coiled tube drilling found practical application with Flex Tube Ltd. and the Uni-Flex Rig Co. Ltd. by drilling numerous shallow gas wells in Canada. This initial rigid CTD effort was pioneered by Ben Gray through the drilling of approximately 18 wells over a 14-year period in Canada.

The flexible-hose Bannister work was followed up by R.H. Cullen in the late 1950s and early 1960s. R.H. Cullen Research came up with an armor-wrapped flexible-string drilling system that used off-the-rack types of motors and drill bits. The Cullen work improved on the original by braiding the hose to carry the "drillstring" weight.

This flexible braided hose had a 2⅝-in. OD and electric-powered cable running internal to the pipe. The BHA comprised an electric motor and drill collars. R.H. Cullen drilled two separate boreholes, approximately 4¾ in. in diameter, to a depth greater than 1,000 ft.

At approximately the same time period as the Cullen work in the late 1950s and early 1960s, Inst. Français du Pétrol (IFP) also showed interest in continuous-string flexible-hose drilling technology. The IFP drilling hose was spooled up onto a reel roughly 5 in. in diameter. A four-skate injector was used to translate the spooled tubing into and out of the wellbore.

The IFP spearheaded work ran turbodrills and electric drilling motors in the BHA. In this test program, over 20,000 ft of borehole was drilled, with hole sizes ranging from 6¾ up to 12¼ in. This development effort was tested both onshore and offshore. The maximum depth reportedly drilled in the IFP project was 3,380 ft because of length restrictions of the pipe on the reel.

Advantages to the IFP flexible drillstring technology of the era included:
• Reduced trip time.
• No connections.
• Continuous circulation.
• Improved well condition.
• Improved safety.
• Optimized bit performance, directional control, early kick detection, and bottomhole parameter monitoring.

Many of these same advantages are still touted today when discussing advantages of CTD technology when compared with conventional rotary drilling. What one mentioned IFP spooled-drillstring advantage, "better working conditions," was based on is somewhat difficult to understand. The project was abandoned because of lack of support.

From 1964 to 1969, there was an additional spooled-drillstring development effort in the form of a consortium of different companies to develop a longer string of spooled hose. This consortium developed a larger-diameter flexible drillstring up to 12,000 ft long. Again, the power cable was run internal to the pipe, and the BHA comprised electric motors and drill collars. One borehole was drilled to 4,500 ft, but there was insufficient support for further development of this type of drilling concept in the industry at the time. Like all previous projects based on this new technology, the project was soon abandoned.

Application of spooled rigid pipe drilling (PD) systems followed the early flexible reinforced hose work. From 1976 through 1978, Ben Gray with Flex Tube Ltd. assembled Rig No.

11, which used 3,000 ft of 2⅜-in. OD butt-welded X-42 line pipe. This coiled-steel drillstring was spooled onto a 13-ft diameter reel, and this apparatus was used to drill shallow gas wells in Canada. The BHA reportedly comprised three 4¾-in. drill collars, a 5-in. downhole PD motor and a 6⅝-in. tricone bit. A sixteen near-vertical, nonsteered wells were drilled, with the deepest being approximately 1,700 ft. It is important to note that Flex Tube Ltd. had also constructed a string of spooled aluminum drillpipe for this new drilling technology toward the end of this pioneering development period, but it was not placed in service.

The reasons cited for the need for this continuous drillstring included:

• Escalating pipe prices.
• Expensive handling equipment.
• Eliminating need to handle heavy 30-ft pipe joints.
• Eliminating two men per shift at a time when the industry had problems attaining people "capable and willing to do a good job" and because "long hours requiring physical and mental endurance make [the oil industry] an unattractive career."[14]

The first two aforementioned reasons do not appear to fit the current CT market in that CT is an expendable that typically costs more than oil country tubular goods (OCTGs) and that CT handling equipment prices have come up similar to other options. However, it is hard to disagree with the logic of the last two reasons.

The roughly one-half century of spooled-drillstring technology development reportedly ended because of

• Lack of petroleum industry sponsorship for high-tech ventures.
• Competitive markets.
• Fully depreciated rigs.
• Proven rig technology.
• CTD benefits did not translate into immediate cost savings.

These reasons are easy to believe even today.[15]

16.9.3 The New Era of CTD. Following the initial Canadian spooled rigid pipe work, little activity occurred until 1991, when interest in the technology was again piqued and CTD began anew in France and west Texas. This renewed interest continues today in niche markets throughout the world.

As of 2003, approximately 12 years have passed from this renaissance of CTD. Out of a fleet of approximately 1,100 CT units in the world today, approximately 60 to 100 of the CT units are considered applicable for CTD, depending on reel capacity and numerous other needs and logistics limiting parameters.

The total 2002 CT-drilling-based revenues are estimated to be approximately U.S. $43,000,000, while drilling revenues are estimated at approximately U.S. $4,000,000,000. The CTD market is then estimated to be about ½ to 3% (depending on using only the CTD portion of revenue or entire job costs) of rotary-drilling-based on revenues.

Interestingly, a market survey in 1994 put the market share somewhat less than 1%. As can be seen, the revenue growth in this industry has been flat over the last decade. However, CTD does offer some unique advantages to other options, and it does come with distinct disadvantages as well.

16.9.4 Advantages to CTD. *UBD.* The ability to work with surface pressure while flowing produced fluids and continuously pumping when tripping into and out of the hole clearly represent the most important advantage to CTD. This unique ability allows for maintaining underbalanced conditions on the formation to minimize the potential for formation damage and increase drilling penetration rate. Maintaining underbalanced conditions on the reservoir at all times is critical in reducing the potential for formation damage in sensitive reservoirs. The majority of CTD operations performed in Canada are primarily for this reason.

Managed Pressure Drilling. Again, the ability to work with surface pressure gives a unique advantage to the CTD process. Experienced coiled tubing unit (CTU) crews are well trained in working with surface pressure, and the CT equipment is designed to work with significant surface pressure. Once the BHA is pressure deployed, commonly done with a lubricated wireline rig up and deployment BOPs, there is no need to snub or strip connections through a rotary BOP stripper. This capability, combined with reduced pipe handling, helps increase the safety of the operation and minimizes the risk of spills.

CT Provides Continuous Use of Hardwired Telemetry and Conduits. As previously mentioned, CT can have electric logging line or other signal telemetry options installed that are fully operational even while tripping. These power and signal paths significantly increase the communication bandwidth available for bidirection telemetry. The hardwired telemetry data transmission rates surpass any mud pulse telemetry, allowing greater data acquisition while drilling. Hardwired telemetry also allows deeper attainable communications than other technologies, such as electromagnetic telemetry. These power and signal paths significantly increase the communication bandwidth available for bidirectional telemetry. The hardwired telemetry surpasses any mud pulse telemetry, allowing greater data acquisition while drilling. Other pressure conduit(s) such as small capillary tubing are often installed in CTD reels, which enable the unique capabilities for operating downhole tools.

Fully Contained Well Pressure. CTD operations are most often performed with fully contained well pressure via the well-control stack including a lubricator and upper stripper of hydraulic packoff. This mechanical pressure-control system is often considered a part of the primary well control as opposed to the drilling fluid in most conventional rotary-drilling operations. In properly designed and engineered jobs, taking a kick is not as much of a threat to manpower and equipment as in common rotary drilling operations.

Small Footprint and Greater Mobility. Many of the more recent CTD programs ranging from the McKittrick work in California during 1994 to the Cerro Dragon work in Argentina during 2001 chose this method over more conventional rotary equipment because of the smaller footprint and ease of mobility of CTD equipment.

Quicker Trip Times. The CTD program in Alaska is arguably the most successful continuous CTD program in the world to date. The CTD program has been operating uninterrupted for over 10 years, and yet the vast majority of the jobs are performed overbalanced. The reasons for this are simple. Many formations will not support underbalanced conditions, and CT drilled boreholes have been shown to be less expensive than rotary-drilled wells, both from a cost-per-well and cost-per-barrel perspective. The quicker trip times allow for lower-cost penetrations when multiple trips and encountering unexpected geological formation changes require operational flexibility and increase the potential for changing the target and trajectory.

Potentially Fewer Service Personnel Are Needed. This is not always the case, but generally speaking, CTD operations require fewer service personnel because of the reduction in pipe-handling requirements. This again, helps lower the cost of the well on a daily basis.

16.9.5 Disadvantages to CTD Techniques. *Inability to Rotate.* The inability to rotate the pipe accounts for the largest single disadvantage to CTD technology. Running drilling operations in 100% slide mode would be the closest analogy to understanding CTD limitations. The ability to prevent cuttings beds uphole, achievable depths, and tolerance of solids in the drilling fluid are all reduced with this inability to rotate.

The buildup of solids beds requires numerous short trips to stir the cuttings bed back into the drilling fluid. In Alaska, short trips to prevent "duning" in the high angle to horizontal sections of the well's account for more than nine times the drilling penetration measured depths. On-bottom testing has confirmed that rotation and short tripping are virtually the only two ways to effectively remove solids beds once they have been deposited in the wellbore above the BHA. Some work has been applied to designing CT equipment that can be rotated, but the CT will be able to withstand the abrasive environment typical in many rotary-drilling operations. Maximum depths achievable in high-angle to horizontal holes are reduced in a large part because of the increased friction of being in essentially static rather than dynamic mode as when the drillpipe is rotated.

When drilling overbalanced, differential pressure can increase the chance of differentially sticking the drillstring or BHA. This is particularly true for CTD for a number of reasons. First, CT is run in essentially buckled mode because of residual stresses in the CT, even when low to moderate tensile loads are present in the CT string. This, coupled with the lack of stand-off normally provided by the drillpipe connections, increases the surface area of the drillstring to differential sticking. Solids accumulation within the drilling fluid system further exasperate this sticking tendency. Field data have shown that drilling efficiency is greatly reduced as solids loading in the drilling fluid approached 1%.

Cost of Consumables. Jointed drillpipe can be maintained for a relatively long life by having connections recut and resurfaced, or damaged joints may simply be replaced by another 30-ft joint. CT, on the other hand, is a consumable commodity. Unlike drillpipe, CT is plastically yielded 6 times every round trip in the hole. After a finite number of trips into the hole, the entire CT string is scrapped or sold for less severe applications. This price differential can be compounded by the fact that CT typically cost more per foot than OCTG products of similar size and weight. Because the probability of having a pinhole or parted CT is higher than in a properly maintained drillpipe, a well-defined contingency plan for such an occurrence is essential. A downhole motor is required for all CTD operations because no current method of rotating CT has been applied in the field. This adds to the cost per foot.

Limited Drilling-Fluids Life. As previously mentioned, CTD requires a low-solids loading in the drilling fluid to provide the highest weight on bit (WOB), assure adequate rate of penetration (ROP), and to maximize the potential reach. Relatively low achievable CTD pump rates often mandate relatively high viscosity to assure adequate hole cleaning. This high viscosity often exceeds a low shear-rate viscosity (LSRV) of 40,000 or more and tasks the ability of solids-control equipment to efficiently remove solids. Finally, the high friction losses and associated turbulence degrade many common biopolymers used in CTD applications. All these factors result in higher costs to maintain a drilling-fluid system.

Limited Equipment and Limited Experienced Manpower Base. As previously discussed, the limited equipment base and lack of widespread application of CTD technology limit the availability of equipment and experienced manpower. These factors often result in higher-cost operations, and because the experience base is not nearly as high as that for rotary technology, the potential also exists for reduced chance factor of success in some instances.

Logistics of Getting Equipment to the Work Location. Drilling requires a conduit to carry drilling fluids at a sufficient rate to lubricate and cool the bit and remove the cuttings at the depth required to reach the desired targets. The higher the achievable pump rate, the more efficient the cutting-transport back to the surface. Generally, a relatively large-diameter conduit is more desirable. The target depth is fixed. Almost without exception, all the CT required to drill a well is spooled up onto a single drum. The needs for hole cleaning and reaching the required depth often results in relatively large reels of CT, which make the logistics of getting the equipment to many potential drillsites problematic. Meeting road restrictions is a challenge for many on-site locations, and the offshore arena has its own set of equal or more challenging logistic problems. Not only is room at a premium, but, also, cranes needed to lift the spools of CT are often inadequate. These and other problems require more preplanning, engineered solutions, and often butt-welds in the CT. As previously mentioned, butt-welds significantly reduce the available useful life of a reel of CT that is already a consumable.

Reduced Pump Rates, Torque, and WOB. This is not unique to CTD operations. Drilling operations encounter the same limits when they are performed with small-diameter drillstring similar to most common CTD operations.

More Tortuous Path. Currently, CT cannot be rotated to drill the reach and horizontal sections of horizontal and high-angle holes. However, new technology is currently under development and is expected to offer some relief with the ability to continuously rotate portions of the BHA to provide a smoother trajectory.

Newer Technology with Lack of Operator Experience Base. Rotary drilling is a proven technology with reasonably well understood capabilities and limitations. This can be said for CTD only in a few geographic locations where CTD is continuously used, such as Alaska and Canada. Selling CTD technology in a new location is a difficult proposition owing in part to the truth of the following quote: "Bad memories die hard in the oil field, and many remain suspicious of the technology. Because the reputation of a project engineer or manager is always on the line, it is natural to choose the proven over the new, potentially risky technique, regardless of potential cost savings."[16] Despite these limitations, the unique advantages to CTD technology often outweigh the disadvantages.

16.9.6 General Discussion of CTD Equipment. There is the common belief that CTD equipment is compact and highly mobile. In reality, the equipment required to provide drilling functions can make this a misconception.

CTD ancillary equipment needs do not significantly differ from those of jointed-pipe or rotary-drilling operations. CTD mechanics and limitations are the same as when slide drilling with rotary-drilling units. The similarities between CTD and jointed-pipe drilling in needed equipment far outweigh the differences.

Current CTD is based on the same equipment used in rotary drilling with small-diameter drillstrings. Other than common CT ancillary equipment, such as connectors, flapper check valves, disconnects, and circulation subs, the only unique CTD equipment required for directional drilling is the orienter in the BHA. With the exception of the CT itself, all other drilling physics and required equipment is the same as slide drilling with jointed pipe. **Fig. 16.22** shows an example of mud pulse telemetry CTD BHA.

Fig. 16.23 shows an example of a modular CTD BHA that relies on an electric line installed within the CT for telemetry. Although differential pressure drop through the CT is higher than CT without wire installed, this type of BHA has proven to be efficient in CTD applications.

The orienter, as the name suggests, provides a method to orient the toolface of the bottomhole drilling assembly. Early orienters included a design with a lead screw that would provide a rotational torque to the BHA when the pumps were off and slackoff weight was varied—much like the operation of a small child's mechanical top. The new orientation was locked in once circulation was started again. The next improvement in orienting tools was indexing tools. These tools were operated by alternating the pumps on and off to index the orienter in typically 30 to 60° steps. These early orienters were often slow and less than optimally reliable, but they did and still do provide a method to directionally steer a mud-pulse BHA where no wires or umbilicals are installed inside the CT. The most recent design of an orienter for mud-pulse-directional drilling included use of a smart sub that can recognize words pumped downhole by varying the pump rate, decipher the words, and then rotate the BHA to the requested tool face.

The other category of orienters includes those that are wireline or hydraulically operated. Using these tools requires the inclusion of wire and/or one or more small-diameter umbilicals installed inside the CT. The wires and/or umbilicals reduce the effective CT ID, causing increased frictional pressure losses in the CT, and add more weight per foot. Typically, these electric or hydraulically operated orienters have a range of rotation of approximately 400°. How-

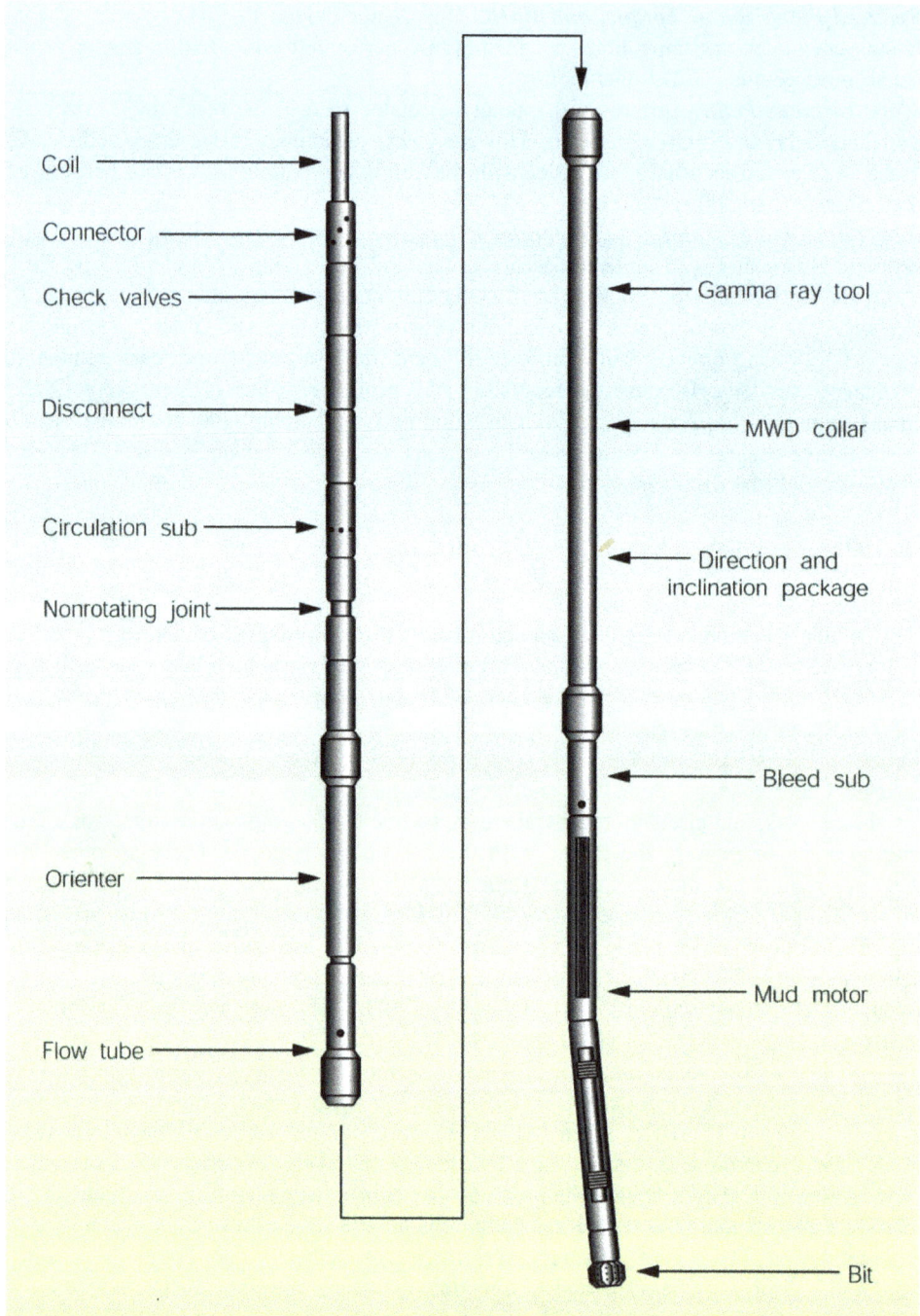

Fig. 16.22—Mud-pulse telemetry CTD BHA. Frictional pressure drop through the CT can be significantly reduced and problem contingencies are increased when compared to wire- or umbilical-containing CT. Data transmission, however, is significantly slower than wireline telemetry options and is not compatible with compressible gas within the CT.

ever, these tools have provided much more predictable results, and the high-rate telemetry path of installed wireline has proven to outweigh the disadvantages to their use. Often overlooked

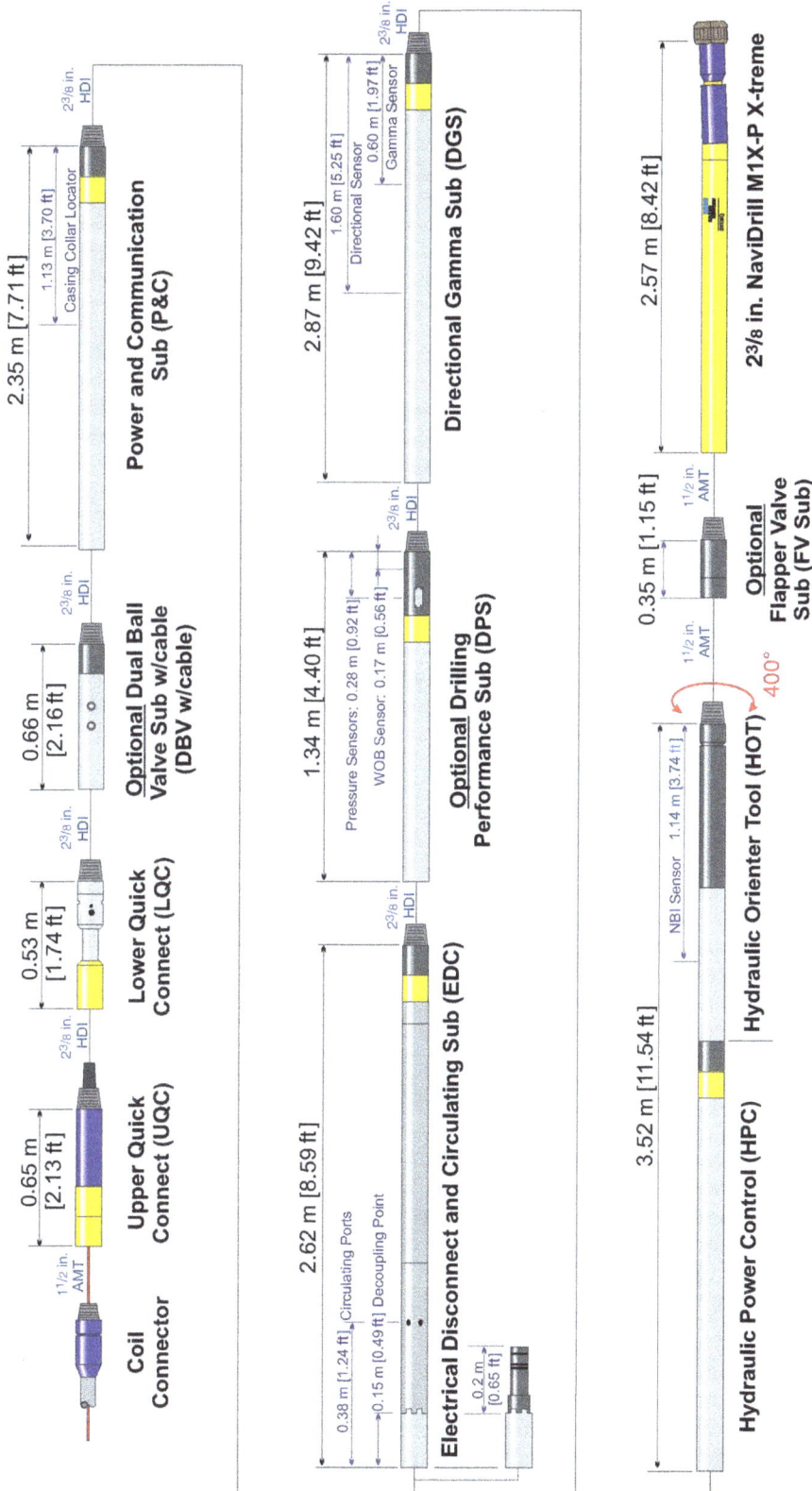

Fig. 16.23—Modular CTD BHA that relies on an electric line installed within the CT for telemetry. Although differential pressure drop through the CT is higher than CT without wire installed, this type of BHA has proven to be efficient in CTD applications. Electric line telemetry will operate with gas phases within the CT as commonly used in underbalanced drilling applications (courtesy of Baker Hughes Inteq.)

potential problems with CT containing umbilical or wire include the contingency to cut the CT should the BHA or CT become stuck and restrictions to pumping balls or darts through the CT.

Many, if not most, drilling applications require cleaning and recirculation of viscosified drilling fluid and call for completing the zone drilled with some type of tubulars. The vast majority of common CT units are designed to effectively run CT into and out of live wells. They are not designed to handle jointed pipe and typically carry no fluid-handling equipment.

There are numerous methods used to provide needed drilling and completion capabilities during CTD operations. Many CTD service providers combine CT equipment with common drilling components including substructures, pipe-handling equipment, drilling-fluid handling and solids-removal equipment, and some sort of mast or vertical support.

Occasionally, CTD service providers simply rig up a coiled tubing drilling unit (CTDU) with a workover or drilling rig. Others build new equipment custom designed to meet the restrictions and needs of areas in which they operate. (See **Fig. 16.24**).

Hybrid CTDUs are available in several locations worldwide that combine CT equipment with a drilling or completion rig in an integrated package. These hybrid CTDUs are as efficient drilling with CT as they are in handling pipe or performing other common drilling, completion, and workover functions. However, with the versatility often come higher day rates and a larger and heavier equipment spread. **Fig. 16.25** shows an example of a hybrid CTDU that provides all common drilling rig functions and efficient CTD. These hybrid units are extremely efficient and operationally flexible but can weigh well over 1.5 million pounds.

16.9.7 Guidelines for Successfully Applying CTD Technology. Although a number of operators ranging in several geographical areas have successfully applied CTD technology, its widespread use still has not been accepted. Reasons for this are numerous, with the most common probably being lack of commitment to get over the learning curve and into the exploitation mode. The following is a list of things an engineer can do to help assure a successful CTD program:
 • Have the correct target and keep it simple: (a) Proper reservoir, (b) low difficulty, (c) lower-challenge drilling especially for wells early in the program, and (d) drillability considering the confines of slide drilling with small tubing.
 • Have the correct program size; have enough candidate wells to get over the learning curve and into exploitation.
 • Have the correct equipment for the effort.
 • Have the commitment: management and technology resources from both operator and service provider.

When bringing CTD into a new area, plan on using sound engineering concepts combined with extensive preparation and planning. It is imperative to remember that CTD drilling technology is still a drilling function that relies on the same best practices for drilling. Pull together a multiexperienced crew that knows CT and drilling practices. Prepare and train for numerous contingencies. Read all available technical papers on CTD.

The following is a list of items to consider and parameters that are within the abilities of CTD technology. None of these are "records." Instead, they fall in the middle of what can be accomplished using proper CTD techniques.
 • WOB is a challenge; CTD currently is 100% slide drilling.
 • Depths to about 17,000 ft measured depths with reasonable geometry.
 • 6 ⅛ in. or smaller hole sizes.
 • Done 13 in. shallow.
 • 3,000 ft or less measured-depth laterals kicking off at 10,000 ft measured depths.
 • Plan on less than 55° per 100 ft build sections.
 • When drilling in 4½-in. casing, the most common CT size used is 2⅜ in. or 2⅝ in. that hydraulically fits the bit for the casing size.

Fig. 16.24—Purpose built CTD rig working in Canada. These rigs are often built considering local area needs and regulations. This rig includes equipment for deploying BHAs into a pressurized wellbore (courtesy of BJ CT Services).

• In 5½-in. casing, the most commonly used CT sizes range from 2⅜ in. to 3$\frac{1}{16}$ in., again with the 3$\frac{1}{16}$ in. CT optimized for hydraulic requirements.
• In 3½ in. casing, the most common CT size is 2 in. OD.
The following is a list of where CTD technology may be applicable:
• Underbalanced (UB) drilling to minimize potential damage to the formation.
• Where costs to mobilize rotary rigs are high.
• Areas with campaign number of candidates.
• Where logistics/area for rig may be tight.
• Managed pressure drilling candidates.
• Through faults with high differential pressure between zones.
• In stable wells where ROP dramatically increases during UB drilling.

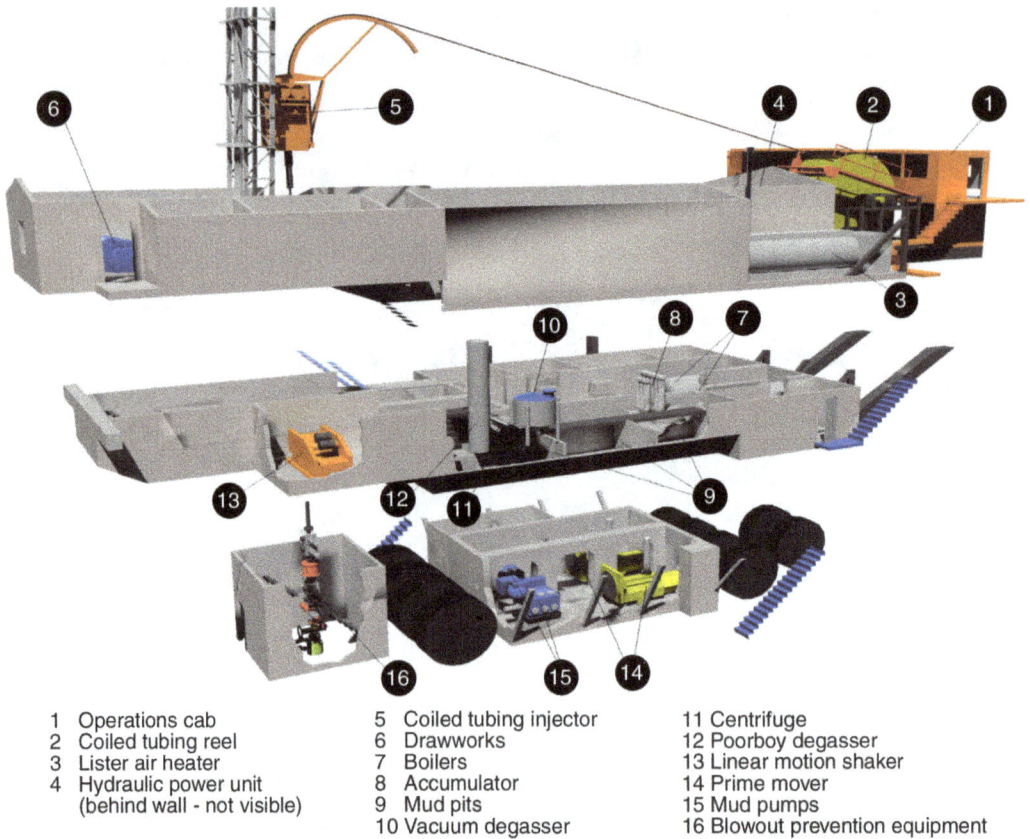

1 Operations cab
2 Coiled tubing reel
3 Lister air heater
4 Hydraulic power unit
   (behind wall - not visible)

5 Coiled tubing injector
6 Drawworks
7 Boilers
8 Accumulator
9 Mud pits
10 Vacuum degasser

11 Centrifuge
12 Poorboy degasser
13 Linear motion shaker
14 Prime mover
15 Mud pumps
16 Blowout prevention equipment

Fig. 16.25—Example of hybrid CTD unit that provides all common drilling rig functions and efficient CT drilling (courtesy of Schlumberger).

• Areas with access to: (a) right well candidates, (b) drillable formations within CT range, (c) personnel, (d) right people, with the right skill sets, doing the right thing, and (e) right equipment available.

Properly applied, CTD's unique capabilities can be used to provide reliable and repeatable drilling solutions, even in demanding circumstances.

16.9.8 CTD Tools, Techniques, and Equipment Under Development. Currently there are numerous ideas in various stages of development that may extend the utilization of CT for drilling operations. These new developments in technology, techniques, and/or equipment target special needs:

1. Allow rotating the entire string of CT to extend attainable measured depths and improve hole-cleaning efficiency.

2. Include telemetry built into the CT string such as used in the Halliburton Anaconda Project. The Anaconda CT was made of composites and contained numerous conductors wound into the composite body. The composite material reduced needed crane load capacities and extended the CT fatigue life. The built-in conductors allowed for power and telemetry without intrusion into the inner diameter of the CT.

3. Special designed equipment and techniques to reduce costs for exploiting existing brownfield assets.

4. More-compact and lighter units for more flexible movement on existing roadways.

5. Equipment to efficiently drill smaller-diameter wells, for both directional and nonsteered applications.

6. More-efficient managed-pressure-drilling operations, especially in extremely low or high BHP applications

7. Offshore packages for CTD intervention. This equipment is designed to address problems common to offshore environments including limited deck space, limited crane capacities, heave and swell problems, and time required to rig up and test equipment.

Only time will tell if this equipment can be successfully developed and applied to extend the utility of CTD or if these new ideas will disappear into obscurity, as have many of the early innovations in continuous conduit drilling.

Nomenclature

d_b = inside diameter
d_{bi} = inside diameter increase owing to diametral growth
D_c = specified outside diameter
Dx = outside diameter (measured along the x-axis)
D_y = outside diameter (measured along the y-axis)
E = modulus of elasticity
Q_f = foam quality
R_y = yield-stress radius of curvature
S_y = material yield strength
t' = reduced wall thickness owing to thinning
t = specified wall thickness
v = Poisson's ratio
γ_{max} = maximum strain
γ_{min} = minimum strain
ε = axial strain
σ = axial stress

References

1. Hartley, A.C.: "Operation Pluto," *The Civil Engineer in War,* Institution of Civil Engineers, London (1948) **3**, 201–210.
2. Bannister, C.E.: "Well Boring Machine," U.S. Patent No. 1,965,563 (July 10, 1934).
3. Ledgerwood, L.W. Jr.: "Efforts to Develop Improved Oilwell Drilling Methods," *Trans.,* AIME (1960) **219**, 61–74.
4. Priestman, G.D. *et al.:* "Well Drilling,", U.S. Patent No. 2,548,616 (10 April 1951).
5. Calhoun, G.H. and Allen, H.: "Equipment for Inserting Small Flexible Tubing into High-Pressure Wells," U.S. Patent No. 2,567,009 (4 September 1951).
6. Slator, D.T. and Hanson, W.E. Jr.: "Continuous String Light Workover Unit," *JPT* (January 1965) 39.
7. Tipton, S.M. and Newburn, D.A.: "Plasticity and Fatigue Damage Modeling of Severely Loaded Tubing," *Advances in Fatigue Lifetime Predictive Techniques, STP 1122*, American Soc. for Testing and Materials (1992) 369–382.
8. Tipton, S.M. and Brown, P.A.: "Monitoring Coiled Tubing Fatigue Life," paper presented at the 1994 Intl. Conference on Coiled Tubing Technology, Houston, 29–31 March.
9. Tipton, S.M.: "Low-Cycle Fatigue Testing of Coiled Tubing Materials," paper presented at the 1997 World Oil Coiled Tubing and Well Intervention Technology Conference, Houston, 4–6 February.
10. Walker, E.J. and Mason, C.M.: "Collapse Tests Expand Coiled Tubing Uses," *Oil & Gas J.* (March 1990) **88**, No. 10, 56.

11. Stanley, R.K.: "Results from NDE Inspections of Coiled Tubing," paper SPE 46023 presented at the 1998 SPE/ICoTA Coiled Tubing Roundtable, Houston, 15–16 April.

12. Tipton, S.M. *et al.:* "Quantifying the Influence of Surface Defects on Coiled Tubing Fatigue Resistance," paper SPE 74827 presented at the 2002 SPE/ICoTA Coiled Tubing Conference, Houston, 9–10 April.

13. Moran, D.W. *et al.:* "Challenges Facing the Development of Coiled Tubing Inspection Technology," paper SPE 74835 presented at the 2002 SPE/ICoTA Coiled Tubing Conference, Houston, 9–10 April.

14. "New Rig Concept Uses Continuous Drill String," (*World Oil* eds.) *World Oil* (March 1977) 94.

15. Hatala, R., Olanson, M., and Davis, P.: "Canadian CT Horizontal Drilling Technology and Applications," *Cdn. CT Horizontal Drill. Tech. & Applications* (1994)

16. "Coiled Tubing Technology: Changing Well-Servicing Techniques and Economics," Morgan Stanley U.S. Investment Research (2 February 1994) 6.

General References

Brown, A.D.F., Merrett, S.J., and Putnam, J.S.: "Coiled-Tubing Milling/Underreaming of Barium Sulphate Scale and Scale Control in the Forties Field," paper SPE 23106 presented at the 1991 SPE Offshore Europe Conference, Aberdeen, 3–6 September.

Cobb, C.C., Headworth, C.S., and Wharton, W.: "A Subsea Reeled Tubing Service Unit," paper SPE 19277 presented at the 1989 SPE Offshore Europe Conference, Aberdeen, 5–8 September.

Crouse, P.C. and Lunan, W.B.: "Coiled-Tubing Drilling—Expanding Application Key to Future," paper SPE 60706 presented at the 2000 SPE/ICoTA Coiled Tubing Roundtable, Houston, 5–6 April.

Cruise, D.S., Davis, D.L., and Elliott, R.H.: "Use of Continuous Coiled Tubing for Subsurface Scale and Corrosion Treating Rangely Weber Sand Unit," paper SPE 11853 presented at the 1983 SPE Rocky Mountain Regional Meeting, Salt Lake City, Utah, 23–25 May.

Donald, D. *et al.:* "Planning, Execution and Review of Brent's First Coiled-Tubing-Drilled Well," paper SPE 37655 presented at the 1997 SPE/IADC Drilling Conference, Amsterdam, 4–6 March.

Elsborg, C., Carter, J., and Cox, R.: "High Penetration Rate Drilling with Coiled Tubing," paper SPE 37074 presented at the 1996 SPE International Conference on Horizontal Well Technology, Calgary, 18–20 November.

Feechan, M., Makselon, C., and Nolet, S.: "Field Experience With Composite Coiled Tubing," paper SPE 82045 presented at the 2003 SPE/ICoTA Coiled Tubing Conference, Houston, 8–9 April.

Flowers, J.K. and Nessim, A.E.: "Solutions to Coiled-Tubing Depth Control," paper SPE 74833 presented at the 2002 SPE/ICoTA Coiled Tubing Conference and Exhibition, Houston, 9–10 April.

Flynn, T.S. and Jahn, B.: "Vibration Analysis for Coiled-Tubing Drilling in Prudhoe Bay," paper SPE 60751 presented at the 2000 SPE/ICoTA Coiled Tubing Roundtable, Houston, 5–6 April.

Forgenie, V.H. *et al.:* "Coiled-Tubing Fishing Operations Utilize a First-Time Technique To Strip Over and Recover 9,500 Feet of Stuck Slickline Wire," paper SPE 30678 presented at the 1995 SPE Annual Technical Conference and Exhibition, Dallas, 22–25 October.

Going, W.S.: "Inhibitor Treatment by Coiled-Tubing Unit Can Now Be Performed While Maintaining Production," paper SPE 18891 presented at the 1989 SPE Production Operations Symposium, Oklahoma City, Oklahoma, 13–14 March.

Hatzignatiou, D.G. and Olsen, T.N.: "Innovative Production Enhancement Interventions Through Existing Wellbores," paper SPE 54632 presented at the 1999 SPE Western Regional Meeting, Anchorage, 26–28 May.

Higdon, A. *et al.:* "Mechanics of Materials," third edition, John Wiley and Sons Inc. (1976) 87–106. "High Hopes for CTD," *Petroleum Economist* (October 1994) 16.

Kirk, A. and Sembiring, T.: "Application of C.T.D. Offshore, Indonesia Phase One Pilot Project," paper SPE 54502 presented at the 1999 SPE/ICoTA Coiled Tubing Roundtable, Houston, 25–26 May.

Kumar, M. *et al.:* "Turbo Drilling Through Coiled Tubing Using Foam In Sub-hydrostatic Well: A Case History for Jotana Field in Cambay Basin, India," paper SPE 47840 presented at the 1998 IADC/SPE Asia Pacific Drilling Technology, Jakarta, 7–9 September.

Laun, L. *et al.:* "Improved CT Operational Efficiency by Use of Detailed Planning," paper SPE 74821 presented at the 2002 SPE/ICoTA Coiled Tubing Conference and Exhibition, Houston, 9–10 April.

Loughlin, M.J. and Plante, M.: "History of and Applications for a Coil-Tubing-Conveyed, Inflatable, Selective Injection Straddle Packer," paper SPE 50655 presented at the 1998 SPE European Petroleum Conference, The Hague, 20–22 October.

McCarty, T.M., Stanley, M.J., and Gantt, L.L.: "Coiled-Tubing Drilling: Continued Performance Improvement in Alaska," paper SPE 67824 presented at the 2001 SPE/IADC Drilling Conference, Amsterdam, 27 February–1 March.

Mascarà, S. *et al.:* "Acidizing Deep Openhole Horizontal Wells: A Case History on Selective Stimulation and Coiled-Tubing-Deployed Jetting System," paper SPE 54738 presented at the 1999 SPE European Formation Damage Conference, The Hague, 31 May–1 June.

Nirider, H.L. *et al.:* "Coiled Tubing as Initial Production Tubing: An Overview of Case Histories," paper SPE 29188 presented at the 1994 SPE Eastern Regional Conference and Exhibition, Charleston, West Virginia, 8–10 November.

Rixse, M. and Johnson, M.O.: "High-Performance Coiled-Tubing Drilling in Shallow North Slope Heavy Oil," paper SPE 74553 presented at the 2002 IADC/SPE Drilling Conference, Dallas, 26–28 February.

RP 5C7, Recommended Practice for Coiled Tubing Operations in Oil and Gas Well Services, first edition, API, Washington, DC (1996).

Sas-Jaworsky, A. II: "Coiled Tubing Operations and Services, Part 1—The Evolution of Coiled Tubing Equipment," *World Oil* (November 1991) 41.

Sas-Jaworsky, A. II: "Tube Technology and Capabilities," *Coiled Tubing Handbook*, third edition, Gulf Publishing Co. (1998) Houston, 19–24.

Sas-Jaworsky, A. II: *Coiled Tubing—Design and Application of Concentric Solutions Manual,* SAS Industries Inc. (2002) Houston.

Stanley, M.J. and Stoltz, D.S.: "The Evolution of Profitable Development Drilling in Prudhoe Bay: A Case of Adapting To Survive," paper SPE 37613 presented at the 1997 SPE/IADC Drilling Conference, Amsterdam, 4–6 March.

Stiles, E.K. *et al.:* "Coiled-Tubing Ultrashort-Radius Horizontal Drilling in a Gas Storage Reservoir: A Case Study," paper SPE 57459 presented at the 1999 SPE Eastern Regional Meeting, Charleston, West Virginia, 20–22 October.

Svendsen, Ø. *et al.:* "Optimum Fluid Design for Drilling and Cementing a Well Drilled with Coiled-Tubing Technology," paper SPE 50405 presented at the 1998 SPE International Conference on Horizontal Well Technology, Calgary, 1–4 November.

SI Metric Conversion Factors

ft	× 3.048*	E – 01	= m
°F	(°F – 32)/1.8		= °C
in.	× 2.54*	E + 00	= cm
lbf	× 4.448 222	E + 00	= N
lbm	× 4.535 924	E – 01	= kg

*Conversion factor is exact.

AUTHOR INDEX

A

Adams, N., 185–219, 455–518
Allen, D.F., 666
Allen, H., 689, 690
Anderson, E.M., 3, 7, 10, 12, 33
Angel, R.R., 166, 168
Arafa, H., 111
Aron, J., 667
Athy, L.F., 40
Azar, J.J., 163, 433–454
Aziz, K., 141, 142

B

Bannister, C.E., 687
Barnes, K.B., 590
Barr, J.D., 283
Bassal, A.A., 164
Becker, T.E., 163
Beique, M., 49
Bell, J.S., 24
Bentson, H.G., 223
Bern, P.A., 164
Beyer, A.H., 169
Bickham, K.L., 159, 164
Biot, M.A., 6, 7, 37, 68
Blauer, R.D., 169
Blount, C., 687–742
Bosworth, S., 269
Bourgoyne, A.T., 127, 128, 158
Bowers, G., 45
Brami, J.B., 662
Brandon, B.D., 253, 255
Brice, J., 382
Bruce, B., 45
Burgess, T.M., 268, 656
Burrows, K., 104, 105

C

Calhoun, G.H., 689, 690
Calvert, D.G., 382, 383
Cameron, C., 103
Campos, W., 164
Castillo, D.A., 6
Chang, C., 14
Cheatham, C.A., 277
Chen, D., 265–286
Chia, R., 230
Chien, S.F., 156, 159, 165
Childers, M., 630
Childers, M.A., 589–646, 641
Clark, R.K., 159, 164
Coats, E., 581
Codd, E.F., 676
Colebrook, C.F., 139
Coope, D.F., 662
Crandall, S.H., 295
Crook, R., 369–431
Cullen, R.H., 729

D

Dodge, D.W., 140, 150, 151
Dowell, I., 647–685
Driscoll, F.G., 577
Dutta, N.C., 41

E

Eaton, L., 106
Economides, M.J., 47

F

Finkbeiner, T., 34
Fisher, E.K., 268
Fontenot, K., 577
Ford, J.T., 164

G

Gaarenstroom, L., 47, 83
Gavignet, A., 164
Gianzero, S., 661
Gough, D.I., 24
Govier, G.W., 139, 141, 142, 169
Gray, G.R., 139
Gray, K.E., 167
Gregory, A.R., 31
Gruenhagen, H., 283
Guild, G.J., 165, 270
Gust, D.A., 268

H

Haimson, B.C., 51
Hale, A.H., 73
Hall, J., 89–118
Hanson, K.E., 31
Hanson, P., 164
Harris, L.M., 592
Hartley, A.C., 687
Hatala, R., 730
Hauck, M., 266
Hemphill, T., 164
Herrick, C.G., 51
Hickman, S.H., 25
Hill, T.H., 165
Hodgeson, H., 266
Holenka, J., 664
Howe, R.J., 593, 599
Hsia, R., 104
Hutchinson, M., 656
Hzouz, I., 450

I

Ikoku, C.U., 169
Isambourg, P., 101

J

Jalukar, L.S., 164
Jellison, M.J., 574, 654

K

Kapur, K.C., 332
Kenny, P., 165
Keshavan, M.K., 228
Kirkman, M., 670

L

Laborde, A.J., 590
Lamberson, L.R., 332
Larsen, T.I., 164, 165
Lawrence, L., 283
Ledgerwood, L.W. Jr., 688

Lee, B., 113
Leslie, J.C., 581
Long, R.C., 571–588
Lora, M., 647–685
Lubinski, A., 129, 153
Luo, Y., 164
Lyons, E.P., 266
Lyons, W.C., 95

M

Market, J., 667
Martin, M., 164
Mason, W., 92
McGehee, D.Y., 238, 253, 255
McLean, M.R., 444
Metzner, A.B., 140, 150, 151
Mills, A., 647–685
Mitchell, A., 39, 40
Mitchell, R.F., 119–183, 129, 153, 287–342
Moake, G., 665
Modi, S., 270
Mody, F.K., 73
Moore, P.L., 156, 159
Moos, D., 1–87
Morales, L., 105
Moran, D.W., 718
Mouchet, J.P., 39, 40

N

Nas, S., 519–569
Negrao, A.F., 96
Ng, F., 658
Nolte, K.G., 47

O

Okrajni, S.S., 163
Oraskar, A.D., 164

P

Pasicznyk, A., 269
Patel, A., 98
Payne, M., 270
Pigott, R.J.S., 156, 162
Pilehvari, A.A., 104
Plumb, R.A., 25
Portwood G., 228
Press, W.H., 126
Priestman, G., 688
Priestman, G.D., 688, 689
Proett, M., 670

R

Rai, C.S., 31
Ravi, K., 119–183
Ridgway, M., 647–685
Rosthal, R.A., 661
Ruan, H., 584
Russell, A.W., 277
Russell, J., 277

S

Salesky, W.J., 228
Sanghani, V., 169
Saponja, J., 539
Sas-Jaworsky, A. II, 687–742
Savins, J.G., 138
Schaaf, S., 283
Schutjens, P.M.T.M., 15
Scott, P.W., 270

Seaton, S., 89–118
Seeberger, M.H., 163
Sewell, M., 104
Sheffield, R., 640
Shen, L.C., 658
Shepard, S.F., 577
Sifferman, T.R., 156, 158, 163
Silcox, W.H., 590
Slator, D.T., 690, 691
Smith, D.K., 384, 385, 394
Smith, R.C., 268, 269
Sobey, I., 164
Speer, M., 343–363
Spinnler, R.F., 654
Stanley, R.K., 717
Stokes, G.G., 156
Streeter, V.L., 155
Sumrow, M., 571
Swarbrick, R., 41
Sweatman, R., 100

T

Terzaghi, K., 7
Thorogood, J.L., 266
Tibbits, G., 579
Tipton, S.M., 687–742, 708, 717
Tomren, P.H., 162
Townend, J., 11

U

Uner, D., 149

W

Walker, E.J., 716
Walker, R.E., 156
Walstrom, J.E., 277
Ward, C., 49
Warren, T., 577
Warren, T.M., 577
West, G., 89–118
Whitfill, D., 104
Whitfill, D.L., 100, 107
Whitmore, R.L., 164
Wilcox, R.D., 99
Williamson, H.S., 278
Wilson, K.C., 164
Wiprut, D.J., 29, 54
Wolff, C.J.M., 277

Y

Yonezawa, T., 283

Z

Zamora, M., 164
Zoback, M.D., 6, 11, 24, 54
Zucrow, M.J., 125
Zwart, G., 43

SUBJECT INDEX

A

accelerators, 398
acoustic logging, 666–668
acoustic systems, 654
2-acrylamido- 2-methylpropanesulfonic acid (AMPS), 399
additives, 396
    accelerators and, 398
    antifoam, 410
    chemical extenders, 403
    crystalline-growth, 409
    dispersants and, 404–405
    fluid-loss-control additives (FLAs), 405–406
    foamed cement, 403–404
    lightweight, 400–401
    mud-decontaminant, 410
    pozzolanic extenders, 401–403
    retarders and, 398–400
    weighting agents and, 404
adjustable-gauge stabilizers (AGSs), 283
Adobe Acrobat Reader (PDF), 674
aft hull transom, 594
air drilling, 166
alkali hydroxides, 398
alkalis, 393
all-oil fluids, 94
American Bureau of Shipping (ABS), 593
American Petroleum Institute (API), 288, 290, 349, 366, 644
    buttress casing joint strength and, 301
    cement classifications and, 385
    cement properties specifications and, 385
    connection ratings, 299–303
    coupling internal yield pressure and, 299–300
    couplings, 492
    extreme-line casing-joint strength and, 301–302
    proprietary connections and, 302–303
    round-thread casing-joint strength and, 300–301
    standards, 603
American Society for Testing and Materials (ASTM), 383, 385, 401–402
americium beryllium, 665
Anderson's faulting theory, 7, 10
anisotropic strength, 65–68
annular bottomhole pressure, 537–538
annular flow
    of Bingham plastic fluids, 144–145
    of Newtonian fluids, 143
    of power law fluid, 146–147
annular-fluid velocity, 448
annular gas injection, 533
annulus cementing, 372–373
annulus seals, 347–349
    slip-and-seal assembly and, 347–348
    wear bushings and, 348
APS Technology Inc., 586
aqueous-filtrate blockage, 447
arctic well completions, 326
    external freeze-back and, 328–329
    internal freeze-back and, 327
    permafrost cementing, 327–328
    thaw subsidence and, 329–330
asymmetric nozzle configurations, 230–231
attapulgite (salt gel) and physical extenders, 401
Atwood Beacon, 612

Authority for Expenditures (AFE)
    administration and, 514
    cementing and, 508–510
    completion equipment and, 516
    cost categories and, 498
    drilling fluids and, 505–506
    drilling tools and, 501–505
    location preparation and, 499–501
    project drilling time and, 493–496
    rental equipment and, 506–508
    support services and, 510–513
    tangible and intangible costs, 498–499
    time categories and, 496
    time considerations and, 496–498
    transportation and, 513
    tubular and, 514
    wellhead equipment and, 514–516
axial loads, 322
    air weight of casing and, 315
    bending loads and, 316
    buoyed weight and, 315
    changes in, 314
    green cement pressure test and, 315
    incremental, 315
    overpull while running and, 315
    running in hole and, 314–315
    service load and, 316
    shock loads and, 315–316
axial strength, 294–295
azimuth, 274, 276–278
azimuthal density, 662–665

B

back rake, 245
balance equations, 120–121
ballooning, 46–48, 298, 304–305, 314, 316, 323
barge-shaped rigs, 606–608
barite, 93
    sag and drilling challenges, 104
    weighting agents and, 404
Barlow equation, 290
bearing(s)
    grading, 257–258
    journal, 232–233
    open, 233
    primary, 232
    roller, 232–233
    roller-cone bit and, 231–233
    seals, 234–235
    secondary, 232
    section, 281
bending
    loads, 316
    moment, 307
    stress, 307
Ben Ocean Lancer, 600
bent-housing motor, 280, 282–283
bentonite
    microsilica and, 406
    polyvinyl alcohol (PVA), 406
    synthetic latex and, 406
bentonite gel, 400–401. *See also* bentonite
Bethlehem Steel Corp., 610
big bore subsea wellhead systems, 364

Bingham plastic fluids, 135
  annular flow and, 144–145
  frictional pressure drop and, 144–145
  plastic flow and, 144
  rheology of, 144–146
  slit flow and, 145–146
biocides, 95–96
biopolymers, 408
bit deviating forces, 440
bit-grading data, 463
bit manufacturers, 463
bit records
  bit-grading data, 463
  cost-per-foot, 465–468
  drilling analysis, 463
bit-size selection, 485, 492
blast joints, 516
block weight (BW), 650
blowout preventers (BOPs), 502, 520
  control system, 628
  isolation test tool, 364
  stack-up, 550
blowout prevention (BOP), 287, 312, 318, 343, 345, 354,
  358, 362
blowouts, 185
borehole expansion, 153
borehole fluids, 443, 445
borehole instability
  causes of, 441
  collapse and, 442
  fracturing and, 442
  hole closure and, 442
  hole enlargement and, 442
  mechanical rock-failure mechanisms and, 443
  pipe sticking and, 435
  prevention of, 444
  principles of, 443
  shale instability and, 443–444
  wellbore-stability analysis and, 444
bore protectors, 362–363
bottomhole assembly (BHA), 376, 435–436, 439–440, 650
bottomhole assembly (BHA) design
  building assemblies, 279
  deviation tools, 279–284
  dropping assemblies, 279
  fulcrum principle, 279
  holding assemblies, 279
  hydraulics and, 278
  packed hole, 279
  pendulum principle, 279
  principles-bit side force and tilt, 278
  rotary assemblies, 278–279
bottomhole circulating temperature (BHCT), 370–371, 394,
  399, 407, 411, 417
bottomhole pressure (BHP), 523, 536
  gas injection and, 537
  requirements, 527
  stability, 538–541
bottomhole static temperature (BHST), 370, 394
bottom-up design, 319–320
bradenhead placement method, 378–379
bradenhead squeeze, 376
breakeven analysis, 261–262
breathing. *See* ballooning
Breton Rig 20, 590
bridging document, 639
British Geological Survey (BGS), 273
brittle failure, 243

buckling
  dogleg, 306
  strain, 307–308
build rate, 268–269, 280
bulk-density, 662–663
buoyancy, 295, 314–315
  effect, 42–43
burst strength, 289–291

C

calcium aluminate, 390–391
calcium aluminate cements (CACs)
  cement composition and, 389
  well cementing and, 389
calcium aluminoferrite, 391–392
calcium chloride, 398
calcium-ions contamination, 445
calcium silicate, 390
calcium sulfates, 393
capillary pressure, 443
capillary suction time (CST) test, 99
carbonate contamination, 445
carbon dioxide foams, 168
carboxymethyl hydroxyethyl cellulose (CMHEC), 407
casing
  conductor, 287, 473
  design, 487
  equipment, 514
  intermediate, 474–475
  liners, 475
  mechanical properties of, 289–299
  production, 475
  properties of, 288
  running, 498
  setting-depth selection and, 473–485
  strings, 287
  structural, 473
  surface, 473–474
  tieback string, 475
  types of, 473–475
casing buckling
  calculations for, 309–310
  corkscrewing and, 309–310
  correlations for, 306–309
  length change, 307–308
  models, 304–306
  in oilfield operations, 304
casing centralizers, 420–423
  restraining device and, 422
  types of, 421
casing design, 317
  API connection ratings, 299–303
  arctic well completions, 326–330
  and buckling, 303–310
  casing strings, 286–288
  connection design limits, 303
  connection failures, 303
  critique of risk-based, 335
  detailed, 322–323
  external pressure loads and, 311
  internal pressure loads and, 311–314
  mechanical loads and, 314–316
  method, 317
  objectives, 317
  pipe strength, 288–299
  preliminary, 318–322
  properties, 288
  required information, 318
  risk-based, 330–334

sample design calculations, 323–326
temperature effects and, 316–317
thermal loads and, 316–317
tubing, 288
casing drilling, 577
benefits of, 578
test, 578
casing hangers, 360–362
mandrel type, 354–357, 360
casing joint strength, 300–302
casing selection, 59
casing-size selection, 485, 491–492
casing strings, 286
conductor casing, 287
intermediate casing, 287–288
liner, 288
loads on, 310–311
production casing, 288
surface casing, 287
tieback string, 288
casing-to-hole annulus, 487
cement composition, 382–390
API and, 385
cement classification and, 383–385
cement manufacture and, 382–383
cement hydration
of cement phases, 392–393
of pure mineral phases, 390–392
sulfate attack and, 394
temperature and, 393
cementing
additives and, 509–510
annulus, 372–373
API and, 385
cement classification and, 383–385
cement composition and, 382–390
cement hydration and, 390–394
cement manufacture and, 382–383
cement-placement design and, 371–374
cement spacers, 509
delayed-set, 373
formation characteristics and, 371
formation pressures and, 371
hardware, 412–427
of high-pressure/high-temperature wells, 373–374
hole preparation and, 379–382
inner-string, 372
multiple-string, 373
operations (*see* cementing operations)
placement techniques and, 377–379
plug, 376–377
primary, 369, 371
pumping charges, 508–509
remedial, 369, 374–379
reverse-circulation, 373
slurry design and, 394–396
slurry-design testing and, 411–412
squeeze, 374–375
temperature and, 370–371
through pipe and casing, 372
water-insoluble materials and, 406
water-soluble materials and, 407–411
wellbore geometry and, 370
well depth and, 370
well parameters and, 370–371
cementing hardware
casing centralizers and, 420–423
cementing plugs, 415–419
cement mixing and, 426–427

floating equipment and, 412–414
multiple-stage cementing tools, 419–420
plug containers and, 414–415
pumping equipment and, 424–426
scratchers and, 423
special equipment and, 423–424
cementing operations, 369
cementing plugs, 415–419
cement mixer, 426–427
cement-placement design
annulus cementing and, 372–373
cementing of high-pressure/high-temperature wells and,
373–374
delayed-set cementing and, 373
inner-string cementing and, 372
multiple-string cementing and, 373
primary cementing and, 371
reverse-circulation cementing and, 373
stage cementing and, 372
cement spacers, 509
centroid effect, 42–43
chain/wire-rope mooring system, 626
characteristic equations, 153–155
chemical extenders, 403
chip holddown, 579
effect, 522
choke pressure, 519
circulatability treatment, 379–381
circulatability treatment and hole preparation, 379–381
clay-particle swelling, 446
closed-circulation drilling systems, 529
coiled-tubing
drilling, 585
stripper assembly, 551
systems, 547
collapse
above packer, 313
below packer, 313
cementing, 313
and D/t ratio, 291–294
elastic, 292–293
gas migration, 313–314
lost returns with mud drop, 313–314
other load cases, 313
plastic, 291–292
pressure, 324
salt loads, 314
strength, 291–294
transition, 292
column-stabilized vessel, 614
combined stress effects, 295–299
compaction, 11–12, 15–16
completion equipment
blast joints, 516
packers, 516
seal assembly, 516
completion fluids, 506
compressible fluid flow
air drilling and, 166
static wellbore pressure solutions and, 124–124
compressible gas, 123
compressive strength, 411
compressive wellbore failure, 23
concurrent method, 201–202
conductor casing, 287, 473
cone offset, 224
constant-bottomhole-pressure concept, 194
contact force, 308–309
contact time, 382

continuous-wave systems, 654
core-based stress analysis, 34
    strain measurement while reloading and, 33
    strain relaxation measurement and, 32
    velocity measurement and, 33
coreflush testing, 526–527
CO_2 resistance, 390
corrosion, 440–441
    inhibitors, 95–96
coupling internal pressure leak resistance, 300
coupling selection, 492
crewboats, 502
cross-axial interference, 277
crossed-dipole sonic logs, 32
crossflow, 230–231
crushed coal, 401
crushing, 222–223
crustal anomalies, 273
cryogenic nitrogen, 546
cut mud, 188
cutter density
    IADC fixed-cutter bit classification system and, 252–254
    polycrystalline diamond compact (PDC) cutter and, 244–245
    polycrystalline diamond compact (PDC) drill bit and, 245–247
cutter dull characteristic (D), 257
cutters, PDC, 247–250
    back rake, 245
    cutter density of, 244–245
    cutter orientation of, 245
    cutter wear and, 255–257
    diamond table and, 242
    diamond table bonds and, 247–249
    optimization, 249
    synthetic diamond and, 241–242
    thermally stable PDC (TSP) and, 247–249
cutter size, 252–254
cutter wear, 255–257
cuttings concentration(s), 161
    vertical wells and, 159–162
cuttings erosion testing, 98
cuttings transport, 155
    air drilling and, 166
    in deviated wells, 156, 162–166
    drillpipe rotation and, 164
    equations and single-phase flow, 167–168
    foam drilling and, 168–169
    hole cleaning and, 169–170
    mechanistic modeling and, 164–166
    mist drilling and, 166–168
    models, 164–166
    particle slip velocity and, 156–159
    in vertical wells, 156, 159–162
cutting structure
    IADC dull grading system and, 255–257
    polycrystalline diamond compact (PDC) drill bit and, 243, 245, 252
    roller-cone drill bit and, 222, 224–226, 228, 231
cyclic stresses, 441

D

damage mechanisms, 445–447
    aqueous-filtrate blockage, 447
    clay-particle swelling, 446
    emulsion blockage, 447
    precipitation of soluble salts and, 447
    saturation change, 446
    solids plugging, 446

    wettability reversal, 446–447
data-acquisition cycle, 667
data-management systems, 676
3D Earth models, 270
deep hard-rock drilling, 579–580
    mud hammer, 579
    mud pulse drilling, 579–580
deepwater fixed platforms, 596
deepwater Gulf of Mexico (GOM), 382, 571–572, 590–591
Deepwater Nautilus, 597
deepwater operations
    hydrates and, 106
    PP/FG and, 106
    shallow-water flow (SWF) and, 105–106
    temperature variation and, 105
delayed-set cementing, 373
Dennis Tool Co., 584
deployment valves, 545
depth-tracking sensors, 647–648
derivatized cellulose, 399
    carboxymethyl hydroxyethyl cellulose (CMHEC), 407
    hydroxyethyl cellulose (HEC), 407
derrickman, 193
design
    polycrystalline diamond compact (PDC) drill bit, 239–245
    roller-cone drill bit, 221–231, 236–237
design check equation (DCE), 333–334
design factors (DF), 322–323
design objectives
    design strings and, 317
    well mechanical integrity and, 317
design process
    detailed design and, 317
    phases of, 317
    preliminary design and, 317
design wells
    3D Earth models and, 270–271
    3D visualization and, 270–271
detailed design, 317
    design factors and, 322–323
    load cases and, 322
    sound engineering judgment and, 322
deviated wellbores, 49–51
deviated wells, 156
deviation angle, 370
deviation tools
    adjustable-gauge stabilizers (AGSs), 283
    intermediate radius applications and, 282–283
    jetting bits, 284
    medium-radius applications and, 282
    positive-displacement motors (PDM) applications and, 281–282
    rotary-steerable systems (RSS) and, 283
    short-radius applications and, 282–283
    steerable motor assemblies, 280–281 (*see also* positive-displacement motors)
    turbines, 284
    whipstocks, 283–284
diamond bits, 250
diamond-enhanced inserts, 228–229
diamond grit, 241–242
diamond table bonds, 247–249
diamond tables, 242, 247–249
diatomaceous Earth (DE), 385, 401, 403
dielectric effects, 659
differential pressure, 472
differential-pressure pipe sticking
    drillstring rotation and, 435
    mudcake and, 433–435

pull force and, 434
surge pressures and, 435
differential-sticking potential, 477
directional drilling, 498
applications, 265–266
bottomhole assembly (BHA) design, 278–284
directional survey, 272–278
directional-well profiles, 266–271
environmental impact of, 265
equipment, 563
fault drilling, 265
field developments and, 265
inaccessible surface locations and, 265
multiple target zones, 265
positive-displacement motors (PDM) applications in,
    281–282
relief-well drilling, 266
river-crossing applications, 266
salt-dome exploration, 266
sidetrack, 265
directional plan
bottomhole assembly (BHA) and, 322
geological targets and, 322
directional sensors, 655
directional survey
average angle method and, 275
balanced tangential method and, 275
calculation methods, 275
cross-axial interference, 277
curvature radius method and, 275
geomagnetic field, 273–274
gyroscopic sensors, 274–275
hole direction and, 272–273, 275
inclination and, 272, 274–275
magnetic interference, 276–277
magnetic sensors, 272–273
measured depth (MD) and, 275
minimum curvature method and, 275
sources of errors in, 276–277
survey instruments, 272–277
survey quality control, 278
tangential method and, 275
tool misalignment, 276
and true vertical depth (TVD), 272
wellbore position error, 277–278
directional wells, 266
drilling, 435, 447–449
directional-well profiles
design wells, 270–271
extended-reach wells, 265, 269–270
horizontal wells, 268–269
multilateral wells, 269
overburden section, 266–268
reservoir-penetration section, 268
surface-hole section, 266
dispersants, 404
hydroxycarboxylic acids, 405
polysulfonated naphthalene (PNS), 405
dispersed water-based fluids (WBFs), 92
diverter lines, 193
dogleg severity (DLS), 269
double layer model, 164
downhole shock sensors, 656
downhole stresses, 205
downhole survey techniques, 266
drill bits
economic analysis and, 261–262
fixed-cutter, 221, 239 (see also polycrystalline diamond
    compact (PDC) drill bit)

hydraulic performance and, 259–261
milled-tooth drill bits, 221–222, 228, 238, 261
polycrystalline diamond compact (PDC), 239–259
roller-cone, 221–239
selection and operating practices for, 262–263
drill-bit performance, 222
driller, 193, 589
driller's method. See two-circulation method
drill-in fluids (DIF), 93
drilling
air, 166
break, 189
contractor, 638–639
costs, 462
foam, 168–169
liners, 475
mist, 166–168
mud records, 468–469
noise, 667
operator, 638–639
platform rigs, 599
rig, 501
shut-in procedures and, 191–193
drilling-data acquisition
additional sensors, 651–652
depth measurement and, 670–671
depth-tracking sensors and, 647–648
drill-monitor sensors and, 649–651
flow-in tracking sensors and, 648
flow-out tracking sensor and, 649
formation testing while drilling (FTWD) and, 669–670
gas-detection sensors and, 651
logging-while-drilling (LWD) (see logging-while-drilling)
management and (see drilling-data management)
measurement-while-drilling (MWD) and
    (see measurement-while-drilling)
pit-monitor sensor and, 651
pressure-tracking sensor and, 649
surface data sensors and, 648
drilling-data management
data-management systems and, 676
integration value and, 673
project data-management systems and, 676–678
rigsite software systems and, 671–673
value from data and, 671–676
drilling dynamics, 656–657
drilling engineering, 1–82
Drilling Engineering Association's (DEA), 571
drilling equipment, 623
drilling equipment set (DES), 590
drilling exponent (D_c), 36–37
drilling fluid(s)
barite sag and, 104
basic functions of, 90–92
capillary suction time (CST) test and, 99
cementing and, 370–374, 379–382, 420
cementing plugs and, 415
challenges and, 99–105
completion fluids, 506
contamination sources and, 111
costs and, 90
cuttings erosion testing and, 98
cuttings transport and, 155–170
deepwater operations and, 105–106
drillstrings lubrication and, 91
dynamic high-angle sag test (DHAST), 98
environmental considerations and, 107–113
environment protection and, 107–111
field tests, 96

fluid mechanics of, 119–174
formation damage and, 90
formation integrity test (FIT) and, 101
Gulf-of-Mexico Compliance-Testing profile, 111–113
high-angle sag test (HAST), 98
high-temperature fluid aging test, 98
hole cleaning and, 103
hole-cleaning sweeps and, 103–104
HP/HT wells and, 106–107
kick indicators and, 187–189
laboratory tests, 96–99
leakoff test (LOT) and, 101
loss, 97–98
lost-circulation and, 100
packer fluids, 505
particle-plugging test (PPT), 98
particle size and, 113–114
particle-size distributions (PSD), 99
pneumatic, 92, 94–95
return-permeability test and, 99
rheology, 97, 134–151
riserless interval and, 105
risk minimization and, 91–92
rubber zones and, 104–105
salt formations and, 104–105
saltwater, 93
selection and, 116
settling velocity and, 114–115
shale instability and, 102–103
shale-shaker screens and, 115
shale stability and, 98–99
situations for, 105–107
slake-durability tester and, 98
solids concentration, 113
solids-control and, 113–116
stuck pipe and, 101–102
surface area and, 113–114
testing, 96–99
total fluids management and, 115–116
toxicity of, 97
transporting drilled cuttings to surface by, 90
types of, 92–96
waste management and, 113–116
waste volume of cuttings and, 115
wellbore information by, 91
wellbore stability and, 90
well-construction and, 89–90
well-control issues and, 90
X-ray diffraction and, 99
drilling fluids, types
    all-oil, 94
    biocides, 95–96
    corrosion inhibitors, 95–96
    drill-in (DIF), 93
    invert-emulsion systems, 92
    lost-circulation materials (LCM), 95
    lubricants, 95
    oil-based fluids (OBF), 91, 93–94
    pneumatic, 92, 94–95
    scavengers, 95–96
    specialty products, 95–96
    spotting fluids, 95
    synthetic-based (SBF), 91, 94
    water-based, 91–93
drilling fluid systems
    annular bottomhole pressure vs. gas injection rate and,
        537–538
    annular injection and, 531–534
    bottomhole pressure (BHP) stability and, 538–541

circulation design calculations and, 536–537
drillpipe injection and, 530–532
flow regimes and, 534–536
foam systems and, 529–530
gaseous fluids and, 528–529
gasified systems and, 530
gas lift systems and, 530
gas/liquid ratios and, 530
hole cleaning and, 541–542
hydraulic calculations and, 534
mist systems and, 529
parasite-string gas injection and, 534
reservoir inflow and, 542–543
single-phase fluids and, 530
drilling materials
    metal composites, 582–584
    resin composites, 580–582
drilling-mechanics failures, 656
drilling-mud report, 91–92
drilling problems
    borehole instability and, 441–444
    drillpipe failures and, 440–441
    equipment and, 450–451
    hole deviation and, 439–440
    hydrogen-sulfide-bearing zones and, 449–450
    loss of circulation and, 437–439
    mud contamination and, 444–445
    personnel, 451
    pipe sticking and, 433–437
    producing formation damage and, 445–447
    shallow gas and, 449–450
drilling rig, 501–505, 639–642
drilling rig and tools, 501
    bits, 504
    completion rigs, 504–505
    day-work bid, 502–503
    footage bid, 502
    fuel, 503
    move-in and move-out, 502
    water, 503–504
drilling solutions, 447–449
drilling technologies
    deepwater offshore and, 571–572
    economic, 571
    requirement of, 571
    United States and, 571
drilling time, 593–594
    drilling-time information, 594–596
drilling tools, 501–505
drill-monitor sensors, 649–651
drill mud, 586
drillpipe failures, 440–441
    burst, 441
    collapse, 441
    fatigue, 441
    parting, 441
    prevention of, 441
    twistoff, 441
drillpipe floats, 196–197
drillpipe injection, 530–532
drillpipe rotation, 164
drill rates, 496–497
drillstem tests (DSTs), 670
drillstrings
    drilling challenges and, 99–100
    drilling fluids and, 91
    kicks and, 188–189
    lubrication, 91
    rotation, 435, 448–449

shut-in procedures and, 192
vibration, 586
drillstring motion compensation (DSC), 628
dual-diameter bits, 250–252
dual drilling, 600
dual gradient drilling systems (DGDS)
   high-speed communications and, 574–575
   isolation, 572
   subsea completion systems, 575–576
   types of, 572–574
dyes, 410
dynamic high-angle sag test (DHAST), 98
dynamic pressure, 151–152
   borehole expansion and, 153
   governing equations and, 152–153
   method of characteristics and, 153–155
dynamic seals, 233–234

E

Earth's magnetic field, 272
   harmonic expansion of, 273
Eaton's method, 38–39
   effective stress, 7–8
   methods, 39–40
elasticity and rock properties, 12–15
elastic modulus. *See* elasticity
elastic-wave velocities, 30–31
elastic wellbore stress concentration
   mud weight effect on, 27–28
   stress-induced wellbore breakouts and, 23
   tensile fracture detection and, 24–27
   tensile wellbore failure and, 23–24
   thermal effects and, 27–28
   in vertical well, 19–23
   wellbore breakout detection and, 24–27
elastomers, 577
electromagnetic logging, 658–662
electromagnetic measurement while drilling (EMWD), 544
emulsion blockage, 447
engineer's method. *See* one-circulation method
environmental considerations
   contamination sources and, 111
   and drilling fluids, 107–113
   environment protection, 107–111
   Gulf-of-Mexico Compliance-Testing profile, 111–113
environmental factors, 222
Environmental Protection Agency (EPA), 638
equations
   characteristic, for dynamic pressure, 153–155
   cuttings transport and single-phase flow, 167
   density, and wellbore flow, 125–127
   dynamic pressure, 152–153
   energy, and wellbore flow, 120–122, 125–127
   mass, and wellbore flow, 120–122, 152–153
   mass conservation, 124
   momentum, and wellbore flow, 120–122, 152–153
   pressure, and wellbore flow, 125–127
   pressure drop, 124
equipment outfitting
   accommodations capacity and, 623
   crew capability and, 623
   drilling equipment and, 623
   marine riser tension and, 622
   MODU performance and, 623
   power plant requirements and, 623
   safety and, 623
   stationkeeping equipment and, 622
   training and, 623
   variable deck load (VDL) and, 621–622

well-control equipment and, 622–623
   well testing and, 623
equivalent circulating density (ECD), 319, 437–438, 444, 448, 656–657
equivalent depth method, 37–38
equivalent mud weight (EMW), 319
E-Spectrum Technologies, 586
ettringite, 391
expandable tubulars, 577
expanded perlite, 401
expanding cements, 389
   crystalline-growth additives, 409
   in-situ gas generation and, 409
extended-reach drilling (ERD), 103
extended-reach wells, 265, 267
   measured depth (MD), 269–270
   true vertical depth (TVD), 269–270
extended-reach wells (ERWs), 606
Extensible Markup Language (XML), 673
external freeze-back, 328
external pressure loads
   openhole pore pressure and, 311
   permeable zones and, 311
   pressure distributions and, 311
   top-of-cement (TOC) and, 311

F

Fanning friction factor, 121
   correlations and fluid rheology, 139–141
fault drilling, 265
faulting regime, 3–4, 6
federally funded drilling projects
   APS Technology Inc., 586
   E-Spectrum Technologies, 586
   Pennsylvania State U., University Park, and Quality Tubing Inc., 586
   Pinnacle Technologies, 586
   Terra Tek, 586
fiber-optic cables, 266
fiber-optic device, 584
field tests, 96
fishtail PDC bits, 254
five percent maximum concentration model, 159–162
fixed-cutter bits, 221, 239. *See also* polycrystalline diamond compact (PDC) drill bit
fixed-platform rigs, 602–606
flame ionization detector (FID) gas chromatography, 651
floating equipment, 412–414
floating rigs, 192–193
floorhand (roughneck), 193
flowing wellbore pressure solutions, 123
   compressible fluid and, 124–125
   constant density and, 124
   linearly varying density and, 124
flow-in tracking sensors, 648
flow-out tracking sensor, 649
flow paddle, 649
flow tubes, 230
fluid
   gradient, 313–314
   loss, 97–98
   rheology (*see* rheology)
   velocity, 259–261
fluid mechanics
   cuttings transport and, 155–170
   dynamic pressure prediction and, 151–155
   flowing wellbore pressure solutions and, 123–125
   fluid rheology and, 134–151
   general steady flow wellbore pressure solutions and, 125–127

governing equations and, 120–122
method of characteristics and, 153–155
problems, 119
single-phase flow and, 120–122
static wellbore pressure solutions and, 123
surge pressure prediction and, 128–134
wellbore hydraulic simulation and, 122–123
wellbore pressure and, 127–128
fluid-separation strategies, 556
fluxgate magnetometers, 273
fly ash, 401–402
foam
    quality, 529
    systems, 529–530
foam drilling, 529–530
    cuttings transport and, 168–169
foamed cement, 403–404
formation damage
    drill-in fluids (DIFs) and, 93
    drilling fluids and, 90
formation exposure time (FET), 658
formation integrity test (FIT), 101
formation pressure, 458
formation testing while drilling (FTWD), 656, 669–670
    depth measurement and, 670–671
fracture mud weight, 186
fracture strength, 300
free gyros, 274
free-water control
    biopolymers and, 408
    sodium silicate and, 408
    synthetic polymers and, 409
frictional pressure drop
    Bingham plastic fluids, 144–145
    in eccentric annulus, 149–151
    Newtonian fluids, 141, 143
    power law fluids, 146–147
    yield power law fluids, 147–148
frictional strength, 8–11
friction bearings. See journal bearings
friction pressure, 519
fulcrum principle, 279
full evacuation, 313
functionally graded materials, 584

G

galley services, 511
gas chromatography, 651
gas-detection sensors, 651
gas drilling, 528–529
gas generation equipment
    cryogenic nitrogen, 546
    natural gas, 546
    nitrogen generation, 546
gasified fluid systems, 530
gas kicks, 186, 206–208, 311
gas migration, 313
gas trap, 651
gauge grading (G), 258–259
gauge inserts, 228
gauge row, 228
general steady flow wellbore pressure solutions, 125–127
geomagnetic field, 273
geomechanics
    effective stress and, 7–8
    elastic wellbore stress concentration and, 19–28
    faulting regime and, 3–4, 6
    horizontal stresses and, 3–6
    model, 35–53

pore pressure and, 4, 6
principal stresses and, 2–3
real-time wellbore stability analysis and, 77–80
rock properties and, 12–19
stress constraints and, 8–12
stresses and, 2–4
stress orientation determination and, 28–34
stress tensors and, 2–3, 23
vertical stress and, 3–4
wellbore stability models and, 65–77
wellbore stability prediction and, 53–64
geomechanical model
    casing seat selection and, 59
    least principal stress and, 43–46
    least principal stress from ballooning and, 46–48
    overburden pressure (S_v) and, 35
    pore pressure and, 35–43
    validation of, 59–61
    wellbore failure to constrain stress magnitude and, 48–53
    wellbore stability and, 53–65
gilsonite, 401
Glomar Challenger, 599–600
gouging, 222–223
governing equations, 120, 152–153
gravitational loading, 3
gravity accelerometers, 273, 275
ground rubber, 401
Gulf-of-Mexico Compliance-Testing profile, 111–113
gypsum, 444–445
gypsum cements
    cement composition and, 388–389
    thixotropy and, 388
gyroscopic sensors
    azimuth and, 274
    speeds of, 274
gyroscopic systems (gyros)
    free gyros, 274
    inertial navigation systems, 274–275
    rate gyros, 274
gyro surveys, 266

H

Hall-effect chips, 649
hard-facing materials, 229–230
hard-formation bits, 224–226
hard-rock drilling, 496–497
hausmannite, 404
health, safety, environment, and security (HSE&S), 635–636, 639
hematite, 404
Herschel-Bulkley fluids. See yield power law fluid
hesitation squeeze, 375
high-angle sag test (HAST), 98
high-pressure/high-temperature wells cementing, 373–374
high-pressure squeeze, 375
Hoek and Brown (HB) criterion, 17–18
hole cleaning, 487, 490, 541–542
    annular-fluid velocity and, 448
    cuttings characteristics and, 449
    drilling challenges and, 103
    drillstring rotation and, 448–449
    factors in, 448–449
    hole inclination angle and, 448
    hydraulics-modeling software and, 103
    mud properties and, 449
    pipe eccentricity and, 449
    rate of penetration (ROP) and, 449
    sweeps, 103–104
hole deviation, 439–440

bit deviating forces and, 440
  bottomhole assembly (BHA) and, 439–440
  rock/bit interaction and, 440
  weight on bit (WOB) and, 440
hole-geometry selection
  casing- and bit-size selection, 485, 491–492
  design procedure and, 486–487
  size-selection problems, 487–491
hole inclination angle, 448
hole preparation
  circulatability treatment and, 380–381
  contact time and, 382
  flushes and, 381
  spacers and, 381
  standoff and, 379–380
  sweep pill design and analysis, 382
  velocity and, 379
hole problems, 497–498
hole stability, 321
horizontal stresses, 3–6
horizontal wells, 268–269
  Austin Chalk play and, 268
  build rate and, 268–269
  drop rate and, 268–269
HP/HT wells, 106–107
Humble SM-1 drilling barge, 592
hybrid databases, 677
hydrates, 106
hydraulic diameter, 160
hydraulic energy, 259–261
hydraulic erosion, 442, 448
hydraulic performance
  dual-diameter bit and, 252
  roller-cone drill bit and, 230–231
hydraulic pressure, 283
hydraulics management, 656
hydraulics-modeling software, 103
hydrocarbon-bearing formation, 265
hydrocarbon-bearing formations, 528
hydrocarbon recovery, 521–523
hydrogen sulfide
  contamination, 445
  cracking, 288
hydrogen-sulfide-bearing zones, 449–450
hydrostatic pressure, 4, 6, 371, 394, 409–410, 412–413,
  421, 435, 519
hydroxycarboxylic acids, 399, 405
hydroxyethyl cellulose (HEC), 407

I

ilmenite, 404
impregnated bits, 250
improper hole fill-up during trips, 187–188
inaccessible surface locations, 265
inclined wellbores. See deviated wells
incremental forces, 314
independent jackups, 609–611
Indian Petroleum Corp., 590
Industry Steering Committee on Wellbore Accuracy
  (ISCWA), 655
Industry Steering Committee on Wellbore Survey Accuracy
  (ISCWSA), 278
inertial navigation systems, 274–275
influx, 206–208
inner-string cementing, 372, 413
inserts
  diamond-enhanced tungsten carbide, 228–229
  gauge, 228

roller-cone bit design and, 224–225
  tungsten carbide, 227–228
in-situ gas generation, 409
in-situ stresses, 441, 443–444, 449
instantaneous shutin pressure (ISIP), 45
insufficient mud weight, 186–187
integral-blade stabilizers, 278
Intellipipe®, 574–575
Intellipipe coupling, 576
intermediate casing, 287–288
intermediate-pipe depth, 480
internal freeze-back, 327
internal pressure loads
  annular pressure buildup and, 314
  cementing and, 312
  collapse and, 313–314
  displacement to gas and, 312
  gas kick and, 311
  gas migration and, 313
  injection down casing and, 313
  lost returns with water and, 312
  maximum load concept and, 312
  mud hydrostatic column and, 312
  pressure distributions and, 311–314
  pressure test and, 312
  surface protection and, 312
International Assn. of Oil and Gas Producers (OGP), 111
International Assn. of Drilling Contractors (IADC), 237
  bit dull grading system, 254–257
  fixed-cutter bit classification system, 252–254
  reports, 469–470
  rig, 502
  rig hydraulics code, 498
  roller-cone bit classification method, 238–239
International Maritime Organization (IMO), 635–637
International Organization for Standardization (ISO), 349
invert-emulsion systems, 92

J

jackups, 609
  characteristics of, 613
  designs, 613
  independent-leg type, 609–611
  mat-type, 610–611
  MODUs and, 612
jackup drilling rig
  mudline hangers and, 352
  mudline suspension equipment and, 349–354
  reconnection and, 353–354
  well offshore drilling and, 349–354
jetting bits, 279, 284
job safety analysis (JSA), 635
jointed-pipe systems, 547
journal angle, 224
journal bearings, 232–233

K

kicks
  concurrent method and, 201–202
  cut mud and, 188
  detection and monitoring for, 190
  drilling break and, 189
  flowing well with pumps off and, 189
  flow rate increase and, 188
  gas, 186, 206–208
  identification and, 197
  improper hole fill-up during trips and, 187–188
  improper hole fill-up on trips and, 189
  indicators, 188–189

influx and, 206–208
insufficient mud weight and, 186–187
kill-weight increment and, 208–210
labels, 186
lost circulation and, 188
one-circulation method and, 200–201, 210–217
pit volume increase and, 189
pump pressure decrease and, 189
pump stroke increase and, 189
saltwater, 208
string weight change and, 189
swabbing and, 188
two-circulation method and, 201
kick-imposed pressures, 481
kill sheet, 217
kill-weight increment, 208–210
kill-weight mud, 198–200
concurrent method and, 201–202
knowledge management, 675–678

L

laboratory tests, 96–99
Lamé equation, 290, 295
laminar flow, 134
land or bottom-supported offshore rigs, 191–192
land well drilling
American Petroleum Inst. (API) and, 349
annulus access and, 348–349
annulus seals and, 347–348
blowout preventer (BOP) stack and, 343
International Organization for Standardization (ISO)
and, 349
load-carrying components and, 345–347
pressure containment and, 343–345
product material specifications and, 349
latex cement, 389
latitude/longitude mapping system, 462
"lat/long" system, 462
leakoff tests (LOT), 101
extended, least principal stress and, 43–46
least principal stress (S3)
from ballooning, 46–48
geomechanical model and, 43–46
lightweight additives, 400–401
lightweight solid additives (LWSAs), 572
lignosulfonates, 399
lime cements, 387–388
limit-state function (LSF), 332
linearly varying density, 124
liner casing string, 288
lithium-thionyl chloride batteries, 654
load and resistance factor design (LRFD), 331–334
load-carrying components, 345–347
mandrel-style casing hanger and, 345–346
slip-and-seal casing-hanger assembly and, 345–346
location costs, 500
logging-while-drilling (LWD), 270–271
acoustic logging, 666–668
electromagnetic logging, 658–662
magnetic-resonance-image (MRI) logging, 668–669
nuclear logging, 662–666
porosity, 665
tool operating environment, 657–658
log headers, 463, 472
log libraries, 463
lost circulation
additives, 407–408
control, 376
drilling challenges and, 100

kick causes and, 188
lost-circulation zones, 437–438
mud equivalent circulating density (ECD) and, 438
mud window and, 57–59
prevention of, 438–439
remedial measures and, 439
lost-circulation materials (LCMs), 95, 100, 439, 654
low-frequency electromagnetic transmission, 654
lowhead drilling, 520
low-pressure squeeze, 375
low-solids, nondispersed (LSND) polymers, 92
Lubinski's solution, 304, 307
lubricants
roller-cone drill bit and, 235–236
specialty products and, 95
for water-based fluids (WBFs), 101–102
lubrication systems, 235–236

M

magnetic interference, 276–277
magnetic-resonance-image (MRI) logging, 668–669
magnetic sensors
electronic compasses, 272
fluxgate magnetometers, 273
gravity accelerometers, 273
mechanical compasses, 272
maleic anhydride, 399
mandrel casing hangers, 345–346, 354–357, 360.
See also casing hangers
marine riser, 628
tension, 622
Maritime Pollution (MARPOL), 637
matrix-body PDC bit, 239–241
mat-type jackups, 610–611
Maurer Technology Inc., 572, 585
maximum load concept, 312
MD error, 276
mean-time-between failure (MTBF) number, 658
measured depth (MD), 269–270, 272
measurement-while-drilling (MWD), 272–273, 277, 282
directional sensors and, 655
drilling dynamics and, 656–657
logging-while-drilling (LWD) and, system architecture,
655–656
power systems and, 653–654
telemetry systems and, 654–655
tool operating environment, 657–658
tool reliability and, 657–658
tools, 543–544
measurement-while-drilling (MWD) systems
kick detection and, 190
mechanical loads. *See* axial loads
mechanical pipe sticking
borehole instability and, 435
bottomhole assembly (BHA) and, 435–436
directional well drilling and, 435
drilled cuttings and, 435
drillstring rotation and, 435
key seating and, 435–437
metal composites, 582–584
method of characteristics, 153–155
metocean, 630
microdrilling, 584–586
microelectromechanical systems (MEMS), 584–585
microfine cements, 389
microsilica, 402–403, 406
microspheres, 402

microsystems
    fiber-optic device, 584
    microdrilling, 584–586
milled-tooth drill bits, 221–222, 228, 238, 261
Mineral Management Service (MMS), 272, 637
minifrac tests. *See* leakoff tests
mist drilling, 529
    cuttings transport and, 166–168
mist systems, 529
mobile offshore drilling units (MODUs)
    anchor chain and, 624
    barge-shaped rigs and, 606–608
    capabilities, 621–624
    classification and, 622, 636–638
    conventional ship-shaped rigs and, 606–608
    drilling contractor and, 638–639
    drilling operator and, 638–639
    drilling site and, 619–621
    drillstring motion compensation (DSC) and, 628
    equipment outfitting and, 621–624
    evolution of, 590–602
    fixed-platform rigs and, 602–606
    future of, 642–643
    health, safety, environment, and security (HSE&S) and,
        635–636, 639
    history of, 590–602
    jackups and, 609–614
    management and, 624–630
    mobilization and, 619–621
    performance, 623
    registration and, 636–638
    regulations and, 636–638
    remotely operated vehicles and, 630–633
    rig crews and management, 633–635
    semisubmersible and, 614–616
    spread-moored, 624
    stationkeeping and, 624–630
    submersibles and, 608–609
    subsea equipment and, 624–630
    tender assist drilling (TAD) and, 606
    ultradeepwater units and, 616–619
    unit selection and, 639–642
    well intervention and, 630–633
model(s)
    Angel's, 168
    Bingham plastic, 169
    buckling, 304–306
    Coulomb, 18
    cuttings transport and, 164–166
    earth, 270–271, 668
    end-cap, 12
    failure, 16–18, 76, 335
    flexor TU4 life prediction, 721
    fluid, 122
    geomagnetic, 277
    geomechanical, 2, 8, 35–53, 59–61, 78
    hydraulics, 103
    mechanical, 332
    multiaxial plasticity, 717
    openhole, 153
    pseudoplastic, 169
    rheological, 141–149
    transport, 164
modern well-planning software, 266
Moineau principle, 280
mono-calcium aluminate and accelerators, 398
monocone bits, 236
monodiameter borehole, 578
Monte Carlo simulation, 332–333

moonpool, 592
mudcake, 433–435
mud contamination
    bicarbonate contaminant and, 445
    calcium-ions contaminant and, 445
    carbonate contaminant and, 445
    hydrogen sulfide contaminant and, 445
    saltwater flows and, 445
    solids contaminant and, 444–445
    treatments of, 444–445
mud-decontaminant additives, 410
mud hammer, 579
mudline hanger(s), 352
    system, 350–351
mudline suspension equipment
    jackup drilling rig and, 349–354
    mudline hangers and, 352
    reconnection and, 353–354
    well offshore drilling and, 349–354
mud logging, 510
    drilling parameters, 471–472
    gamma ray curve, 471
    records, 468–469
mud program, 318
mud pulse drilling, 579–580
mud-pulse telemetry, 654
    system, 544
mud/rock interactions, 72–75
mud system. *See* drilling fluids
mud weight
    geomechanical design with little data and, 61–62
    wellbore failure and, 55–57
    wellbore stress concentration and, 27–28
mud window, 57–59
multilateral wells, 269
multiple-stage cementing tools, 419–420
multiple-string cementing, 373
multiple target zones, 265

N
natural gas, 530, 546
near-bit stabilizer, 279
negative-pulse systems, 654
neutron porosity, 662–663
Newtonian fluids
    annular flow and, 143
    pipe flow and, 141–142
nitrogen foams, 168
nitrogen gas generation, 546
Noble Drilling's EVA-4000 design, 614
nondispersed water-based fluids (WBFs), 92
non-Newtonian fluids, 135
nonproductive time (NPT), 94, 656
nonreturn valves, 544–545
normal-pressure wells, 458
Novatek IDS hammer, 579
nozzles, 230–231
nuclear logging, 662–666

O
Occupational Safety and Health Agency (OSHA), 638
Ocean Driller, 596
Ocean Drilling and Exploration Co. (ODECO), 590
offset-well selection, 460–462
offshore
    deepwater, 571
    dual gradient drilling systems (DGDS), 573–576
offshore drilling units. *See* mobile offshore drilling units
oil-based fluids (OBFs), 91, 93–94

oil-based muds (OBMs), 655
one-circulation method, 200–201
    implementation of, 210–217
one-dimensional (1D) fluid flow, 120–122
    balance equations and, 120–121
onshore, 576
    casing drilling, 577–578
    deep hard-rock drilling, 579–580
    expandable tubulars, 577
open bearing systems, 233
organophosphonates, 399
O-seals, 235
osmotic pressure, 443–444
outer diameter (OD), 283
outside cementing, 372–373
overbalanced drilling (OBD), 519
overburden pressure (Sv), 35
overburden section, 266–268
    build-and-hold, 267
    deflection angles and, 267
    stratigraphic trap, 267–268
    surface casing and, 267

P

packed hole, 279
packer fluids, 505
packers, 516
packer squeeze, 375–376
parasite-string gas injection, 534–535
partial evacuation, 313
particle-plugging test (PPT), 98
particle-size distributions (PSD), 99
particle slip velocity, 156–159
Paslay buckling force, 304–305, 309
PDC. See polycrystalline diamond compact (PDC) drill bit
PDC cutters, 244–245, 247–250
    back rake, 245
    cutter density of, 244–245
    cutter orientation of, 245
    cutter wear and, 255–257
    diamond table and, 242
    diamond table bonds and, 247–249
    optimization, 249
    synthetic diamond and, 241–242
    thermally stable PDC (TSP) and, 247–249
Pelerin, 600
Pelican, 600
pendulum device, 272
pendulum principle, 279
penetrometer testing, 18–19
Pennsylvania State U., University Park, and
    Quality Tubing Inc., 586
perforating, 510
permafrost, 326
    cement, 389–390
    cementing, 327–328
    loading mechanism, 326
    mechanical response of, 326
permeability, 186
personal protective equipment (PPE), 635
physical extenders, 400–401
Pinnacle Technologies, 586
pipe couplings, 492
pipe eccentricity, 449
pipe flow
    of Newtonian fluids, 141–142
    of power law fluid, 146
    of yield power law (YPL) fluid, 147–148
pipe inspection, 510–511

pipe selection, 491–492
pipe sticking
    borehole instability and, 435
    bottomhole assembly (BHA) and, 435–436
    differential-pressure, 433–435
    drilled cuttings and, 435
    key seating and, 435–437
    mechanical, 435–437
pipe strength
    axial strength, 294–295
    burst strength, 289–291
    collapse strength, 291–294
    combined stress effects, 295–299
pipe thread strength, 301
placement techniques
    balanced plugs and, 379
    bullheading and, 378–379
    coiled tubing and, 377–378
    dump bailer and, 377
plastic cements, 390
plastic failure, 243
plastic flow of Bingham plastic fluids, 144
plug cementing
    abandonment and, 376
    conformance and, 377
    directional drilling and, 376
    formation testing and, 377
    lost-circulation control and, 376
    sidetracking and, 376
    wellbore stability and, 377
    well control and, 376
    zonal isolation and, 377
pneumatic drilling fluids, 92, 94–95
polycrystalline diamond compact (PDC) drill bit, 417
    cutter density and, 245–247
    cutting structure of, 243, 245, 252, 255–258
    design, 239–245
    diamond bits and, 250
    diamond grit and, 241–242
    dual-diameter bits and, 250–252
    durability and, 245–247
    economic aspect of, 261–262
    hydraulic performance and, 259–261
    IADC bit dull grading system and, 254–257
    IADC fixed-cutter bit classification system and, 252–254
    impregnated bits and, 250
    inserts, 228
    long parabolic profiles and, 246–247
    materials for, 239–243
    matrix-body, 239–241
    medium parabolic profiles and, 246–247
    operating practices of, 262–263
    parabolic profiles and, 246–247
    PDC cutters and, 241–242, 244–245, 247–250
    profile of, 245–247
    ROP and, 245–247
    short parabolic profiles and, 246–247
    stability and, 245–247
    steel-body, 239–241
    steerability and, 245–247
polycrystalline-diamond (PCD), 458
    cutters, 579
polyester taut-leg mooring, 625
polypropylene glycol, 410
polysulfonated naphthalene (PNS), 405
polyvinyl alcohol (PVA), 406
polyvinyl pyrrolidone (PVP), 407
pore fluid properties, 41

pore pressure, 4, 6
  buoyancy effect and, 42–43
  centroid effect and, 42–43
  complications in determination of, 40–42
  drilling exponent (D_c), 36–37
  Eaton's method and, 38–39
  effective stress and, 7
  effective stress methods and, 39–40
  equivalent depth method and, 37–38
  geomechanical model and, 35–43
  pore fluid properties and, 41
  ratio method and, 37–38
  undercompacted shales and, 42
poroelasticity, 68–72
poroelastic-plastic analysis, 5–76
porosity, 186
ported drillstring floats, 545
portland cement, 382–387, 389–390, 394
  strength-retrogression inhibitors, 408
positive displacement motors (PDMs), 279–281, 531
potassium chloride, 398
power
  plants, 623
  systems, 653–654
  unit, 281
power law fluid
  annular flow and, 146–147
  pipe flow and, 146
  slit flow and, 147
pozzolanic cements
  cement composition and, 386–387
  diatomaceous earth (DE) and, 386
pozzolanic extenders
  diatomaceous earth (DE) and, 402–403
  fly ash and, 401–402
  microsilica and, 402–403
  microspheres and, 402
preliminary design, 317
  bottom-up design and, 319–320
  differential sticking and, 321
  directional drilling and, 321
  directional plan, 322
  drill-bit sizes and, 318
  drilling and, 318
  equivalent circulating density (ECD) and, 319
  equivalent mud weight (EMW) and, 319
  hole and pipe diameters and, 318–322
  mud program and, 318
  number of strings and, 318–321
  slimhole drilling and, 318
  top-down design and, 320
  top-of-cement (TOC) depths, 321–322
  well-depth and, 319
  zonal isolation and, 321
pressure differential, 186
pressure diffusion, 444
pressure losses, 259–261
pressure while drilling (PWD)
  sensors, 543
  tool, 77–78
primary bearings, 232
principal stresses, 2–4
  and stress magnitudes, 8, 10
producing formation damage
  aqueous-filtrate blockage and, 447
  borehole fluids and, 445
  clay-particle swelling and, 446
  damage mechanisms and, 445–447
  emulsion blockage and, 447

  precipitation of soluble salts and, 447
  saturation change and, 446
  solids plugging and, 446
  wettability reversal and, 446–447
production casing, 288
project-administration tools, 678
project data-management systems
  broad application support, 676
  handling of bulk data, 677
  integrate user experience across applications, 678
  open extensible environment, 676
  suit of, 677–678
  support industrywide standards, 678
  technology based on a standard database, 676–677
proprietary connections, 302–303
  horizontal well, 302
Prudhoe Bay, 327
pseudo-oil-based muds, 655
pull force, 434
pulse-type measurement while drilling (MWD) tools, 531
pumping equipment, 424–425
pump-stroke measurement method, 187–188
push-the-bit system, 283

Q
quantitative risk assessment (QRA), 62–64, 331–333

R
radioactive tracers, 410
rate gyros, 274
rate of penetration (ROP), 449
  bit economics and, 262
  polycrystalline diamond compact (PDC) bit design and, 243, 245, 252
  roller-cone bit design and, 226
ratio method, 37–38
real-time monitoring, 574–575
real-time wellbore stability analysis, 77–78
reliability-based design approaches, 332–334
  load and resistance factor, 333–334
  quantitative risk assessment and, 332–333
relief-well drilling, 266
remotely operated vehicles (ROVs), 630–633
rental equipment
  accessories, 507
  casing tools, 508
  mud-related equipment, 507–508
  rotary tools, 507
  well-control equipment, 506–507
reservoir fluid inflow, 542
reservoir inflow, 542–543
reservoir-penetration section
  aspects of, 268
  production efficiency and, 268
  realization and, 268
  target location and, 268
resin
  cements, 390
  composites, 580–582
retainer squeeze, 375–376
retarders, 398–400
  borax and, 400
  cellulose derivatives, 399
  hydroxycarboxylic acids, 399
  inorganic compounds, 399–400
  lignosulfonates and, 399
  organophosphonates and, 399
  salt and, 400
  synthetic, 399

return-permeability test, 99
reverse-circulation cementing, 373
rheology
 Bingham plastic fluids and, 135, 144–146
 dilatant fluid and, 137
 drilling fluid test and, 97
 Fanning friction factor correlations and, 139–141
 laminar flow and, 134
 Newtonian fluids and, 141–143
 non-Newtonian fluids and, 135
 power law fluid and, 146–147
 pseudoplastic fluid and, 136
 shear stress and, 135
 turbulent flow and, 134
 viscometry and, 137–139
 viscosity and, 135
 yield power law fluid and, 136–137, 147–149
rig(s), 501
 Breton Rig 20, 590
 costs and, 502
 crews management, 633–635
 drilling, 501–505, 639–642
 fixed-platform, 602–606
 floating, 192–193
 heave, 648
 hydraulics, 450
 hydraulics code, 498
 jackup drilling, 349–354
 land or bottom-supported offshore, 191–192
 move-in, 498
 move-out, 498
 platform, 599
 ship-shaped and barge-shaped, 606–608
 Transworld Rig 40, 590
rigsite software systems, 671
 drilling/rigsite simulators and, 673
 operations reporting and, 672
 operator and, 672
 others, 673
rig-contractor and, 672
 service company and, 671
 survey-data management and, 672
 well planning/drilling engineering and, 672–673
risk-based casing design, 330
 background of, 331
 critique of, 335
 design check equation (DCE) and, 333–334
 limit-state function (LSF) and, 332
 load and resistance factor design (LRFD) and, 331–334
 quantitative risk assessment (QRA) and, 331–333
 reliability-based design approaches, 332–334
 safety factors (SFs) and, 330–332, 334
 and working stress design (WSD), 330–332, 334–335
river-crossing applications, 266
rock/bit interaction, 440
rock properties
 compaction, 15–16
 elasticity, 12–15
 end-cap plasticity, 15–16
 failure models and, 16–18
 Hoek and Brown (HB) criterion and, 17–18
 penetrometer testing and, 18–19
 rock strength models and, 16–18
 scratch testing and, 18–19
 single-sample testing and, 18
 2D linear Mohr-Coulomb criterion and, 16–17
rock strength models, 16–18
roller bearings, 232–233

roller-cone drill bit
 asymmetric nozzle configurations and, 230–231
 bearing seals and, 234–235
 bearings system of, 231–233
 bit diameter/available space, 223
 components, 231–236
 cone offset and, 224
 crossflow and, 230
 cutting structure of, 222, 224–226, 228, 231
 design, 221–231, 236–237
 drill-bit performance and, 222
 drilling action of, 222–223
 dynamic seals and, 233–234
 economic aspect of, 261–262
 environmental factors and, 222
 flow tubes and, 230
 hydraulic features and, 230–231, 259–261
 IADC roller-cone bit classification method and, 238–239
 insert design of, 224–225
 journal angle and, 224
 journal bearings and, 232–233
 lubricants and, 235–236
 lubrication systems and, 235–236
 materials design and, 226–227
 monocone bit design and, 236
 nozzles and, 230–231
 open bearing systems and, 233
 operating practices of, 262–263
 O-seals and, 235
 primary bearings and, 232
 roller bearings and, 232–233
 seal system of, 233–235
 secondary bearings and, 232
 selection, 262–263
 tooth design of, 224–225
 tungsten carbide hard facing and, 229–230
 tungsten carbide inserts (TCI) and, 227–229
 two-cone bit design and, 236–237
rotary assemblies, 278–279
 building assemblies, 279
 dropping assemblies, 279
 holding assemblies, 279
rotary-steerable systems (RSSs), 269–270, 283
rotating control heads, 548–549
rotating diverter systems, 548
rotational force, 654
rubber zones, 104–105
running tools
 casing-hanger seal-assembly, 363
 conductor wellhead, 363
 high-pressure wellhead, 363
 multipurpose, 364
 seal-assembly, 364

S
Safety of Life at Sea (SOLAS), 637
salt-dome exploration, 266
salt domes, 34
salt formations, 104–105
saltwater
 drilling fluids, 93
 kicks, 208
sample design calculations
 collapse and, 324–325
 triaxial comparison and, 323–324
 uniaxial tension and, 325–326
satellite surveys, 501
scavengers, 95–96
Scorpion, 593

scout tickets, 463, 470, 495–496
scratchers, 423
scratch testing, 18–19
seal assembly, 357–358, 516
    metal-to-metal, 357–358, 360, 362
seal grading, 257–258
seal system, 233–235
seawater, 398
secondary bearings, 232
seismic anisotropy, 30–32
seismic work, 463
semisubmersible, 614
    characteristics of, 615
    design of, 615
    generation designation, 617
semisubmersible (semi) Seahawk tender assist drilling
    (TAD), 596
separator
    horizontal, 557
    vertical, 558
service loads, 316, 322
setting-depth design procedures, 475–485
settling velocity, 114–115
shale
    inhibition, 93
    instability, 102–103, 443–444
    stability, 98–99
shale instability
    borehole-fluid invasion and, 444
    capillary pressure and, 443
    chemical, 443
    mechanical, 443
    osmotic pressure and, 443–444
    pressure diffusion and, 444
shale-shaker screens, 115
shallow casing strings, 481
shallow gas, 438, 449–450
shallow-water flow (SWF), 105–106
shear-enhanced compaction, 11–12, 15–16
shear stress, 135, 441, 443
shock loads, 295, 310, 315–316
shut-in drillpipe pressure, 193, 215
    constant-bottomhole-pressure concept and, 194
    drillpipe floats and, 196–197
    reading, 194
    time and, 194–195
    trapped pressure and, 195–196
shut-in procedures
    crewmember responsibilities for, 193
    initial, 190–193
    one-circulation method and, 213–215
sidetrack, 265, 268, 281
single-phase flow, 120–122
    balance of energy equation for, 121
    balance of mass equation for, 120
    balance of momentum equation for, 120–121
    Fanning friction factor and, 121
single-phase fluids, 530
single-sample testing, 18
size selection
    casing design, 487
    casing-to-hole annulus, 487
    drillstring/hole annulus, 487
skidding, 222–223
sleeve stabilizers, 278
sliding mode, 283
slip-and-seal casing-hanger assembly, 343–345
slit flow
    of Bingham plastic fluids, 145–146
    of power law fluid, 147
    of yield power law (YPL) fluid, 148–149
slow drilling rates, 498
slurry design, 394–396
slurry-design testing
    diagnostic, 411–412
    mixing effects and, 411
    performance and, 411
smart drilling, 571
snubbing systems, 547–548
sodium chloride, 398
sodium silicate, 403, 408
    accelerators and, 398
soft-formation bits and roller-cone bit design, 224–226
solids contamination, 444–445
solids-control strategies, 556
solids plugging, 446
sonic pit-volume sensor, 652
sour gas well design, 317
spars, 596
spool system
    inside diameter (ID) and, 343–345
    intermediate casing spools and, 345
    outside diameter (OD) and, 345
    slip-and-seal assembly, 343–345
    starting casing head and, 343–345
    tubing spool and, 345
spotting fluids, 95, 102
squeeze cementing, 375–376
S-shaped well, 267
stability
    dual-diameter bit and, 252
    polycrystalline diamond compact (PDC) drill bit and,
        245–247
stabilizing effect, 305
stage cementing, 372
static wellbore pressure solutions
    compressible gas and, 123
    constant density, 123
stationkeeping, 624–630
    equipment, 622
steel-body PDC bit, 239–241
steel grade, 288, 290
steerability, 245–247
strain relaxation and core-based stress analysis, 32
stress(es), 2–4
    constraints (see stress magnitude constraints)
    limit, 11
    and pore pressure, 7–8
    tensors, 2–3, 23
stress magnitude constraints, 2–4
    deviated wellbores and, 49–51
    frictional strength and, 8–11
    shear-enhanced compaction and, 11–12
    Visund field and, 51–53
    wellbore failure and, 48–53
stress orientation
    core-based analysis of, 32–34
    crossed-dipole sonic logs and, 32
    geological indicators of, 34
    salt domes and, 34
    seismic anisotropy and, 30–32
    wellbore failure and, 28–30
string weight (SW), 650
Structured Query Language (SQL), 674
stuck pipe, 463
    drilling challenges and, 101–102
    spotting fluids and, 102
submersibles, 608–609

subnormal-pressure wells, 458
subsea equipment, 624–630
Subsea Mudlift development program, 572–573
subsea production system, 577
subsea well drilling
    BOP stack and, 358
    bore protectors and, 362
    casing hangers and, 360–362
    drilling guide base and, 358–360
    high-pressure housing and, 360
    low-pressure housing and, 360
    metal-to-metal annulus seal assembly and, 362
    running tools and, 363
    test tools and, 363
substrates and PDC cutter design, 247–249
sulfate attack, 394
support services
    casing crews, 510
    completion logging, 510
    formation testing, 510
    galley services, 511
    mud logging, 510
    perforating, 510
    pipe inspection, 510–511
    special labor, 511–513
    well logging, 510
surface casing, 287
surface-casing string, 481
surface-hole section
    collisions and, 266
    directional steering and, 266
    directional wells and, 266
    downhole survey techniques and, 266
    environmental footprint and, 266
    gyro surveys, 266
    traveling-cylinder diagram (TCD) and, 266
surface pressures, 203–204
surface protection, 312
surge pressures, 435, 438
    analysis, 128–129
    boundary conditions and, 129–134
    solution of and, 134
survey instruments, 272–277
    azimuth and, 276
    calibration performance and, 276
    Earth's spin rate and, 276
swabbing, 188
sweep pill design and analysis, 382
synthetic-based fluids (SBFs), 91, 94
synthetic diamond, 241–242
synthetic fibers, 410–411
synthetic latex, 406
synthetic polymers, 409
    polyvinyl pyrrolidone (PVP), 407
synthetic retarders, 399

T
target zone, 267–268
taut line mooring system, 614, 624, 641–642
2D linear Mohr-Coulomb criterion, 16–17
technical-limit well planning and operations, 675
Technology Advancement for Multi-Laterals (TAML), 269
technology of economics, 602, 643
tectonic stresses, 2–4
    sources of, 4
telecommunication companies, 266
telemetry-signal detection, 654
telemetry systems, 654–655
temperature and deepwater, 105

temperature effects
    sour gas well design and, 317
    temperature dependent yield and, 317
    tubing thermal expansion and, 317
    on tubular design, 316–317
tender assist drilling (TAD), 595, 605–606
tensile fracture. See tensile wellbore failure
tensile stress, 441, 443
tensile wellbore failure, 23–24
    detection, 24–27
    Visund field and, 51–53
tension-leg platforms (TLPs), 366–367, 596
Terra Tek, 579, 586
Tesco Corp., 577
thaw subsidence, 329
thermal effects, 27–28
thermal expansion, 304, 314, 316. See also ballooning
    coefficient of, 325
    tubing, 317
thermal loads
    annular fluid expansion pressure, 316–317
    sour gas well design and, 317
    and temperature dependent yield, 317
    tubing thermal expansion and, 317
thermally stable PDC (TSP), 247–249
thermoporoelasticity, 68–72
thief zone, 437–438
thixotropy
    gypsum cement and, 388
    slurries and, 379, 403
3D visualization, 270–271
3D well trajectory, 267
three-phase pumps, 575–576
tieback casing string, 288
tieback tools
    rotation-lock design and, 353
    types of, 353
time
    consideration and well control, 203
    shut-in drillpipe pressure and, 194–195
tool
    operating environment, 657–658
    reliability, 657–658
tooth design, 224–225
top-down design, 320
top-of-cement (TOC), 311, 314
total depth (TD), 311, 313
toxicity, 97
Transworld Rig 40, 590
trapezoidal thread, 299–300
trapped pressure, 195–196
traveling-cylinder diagram (TCD), 266
trend-line method, 37, 39
triaxial stress, 295–296, 298–299, 323–324.
    See also von Mises stress
tripping, 191–193
trip-tank method, 187–188
trip time, 497
true vertical depth (TVD), 269–270, 272
tubing, 473–475
tubing buckling
    calculations for, 309–310
    corkscrewing and, 304
    correlations for, 306–309
    models, 304–306
    in oilfield operations, 304
tubing properties, 288
    mechanical, 289–299
tubing strings, 310–311

tubing thermal expansion, 317
tungsten carbide
    hard facing, 229–230
    inserts (TCI), 227–229
tungsten carbide inserts (TCI) bits. *See* roller-cone drill bit
turbine(s), 279, 284
    rotors, 654
two-circulation method, 201
two-cone bits, 236–237
2D rotary systems, 279, 283. *See also* adjustable-gauge
    stabilizers

U

UBD circulation design calculation
    bottomhole pressure (BHP) and, 536
    cuttings transport and, 536
    environmental considerations and, 537
    hole cleaning and, 536
    motor performance in multiphase-flow environment,
        536–537
    reservoir inflow performance and, 536
    wellbore stability and, 537
UBD downhole equipment
    conventional MWD tools in, 543–544
    deployment valves, 545
    economics and, 564
    electromagnetic measurement while drilling (EMWD), 544
    environmental aspects and, 557–558
    health safety issues and, 558–561
    limitations of, 561–564
    nonreturn valves, 544–545
    personnel for, 564
    pressure while drilling (PWD) sensors, 543
    surface equipment for, 545–553
    training and, 564
    underbalanced drilled multilateral wells, 557
    underbalanced drilled wells, 553–556
UBD surface equipment
    drilling systems, 545–546
    gas-generation equipment, 546
    and limitations, 563
    well-controlled equipment, 547–553
ultimate tensile strength (UTS), 290, 301, 303
ultradeepwater drillships, 616
    characteristics, 616
    design, 616
ultradeepwater units, 616–619
ultra-harsh environment, 609
ultrasonic calipers, 665
ultrasonic sensors, 190
ultrasonic transit time, 651
underbalanced drilled multilateral wells, 557
underbalanced drilled wells, 553–556, 559
    risk associated with, 523
underbalanced drilling (UBD)
    advantages of, 522
    bottomhole pressure (BHP) requirements and, 527–528
    candidate for, 523–524
    circulation design calculation (*see* UBD circulation
        design calculation)
    classification system for, 523
    costs associated with, 524–525
    definition of, 519–520
    disadvantages of, 522
    downhole equipment (*see* UBD downhole equipment)
    economic limitations and, 524
    effects for reservoir types, 524
    evaluation, 525
    fluid density range, 528

    fluid selection for, 528
    lowhead drilling and, 520
    maximizing hydrocarbon recovery and, 520–521
    minimizing pressure-related drilling problems and, 520–523
    operation designing and, 527
    operation planning for, 563
    planning, 526
    reasons for, 520–523
    reservoir selection issues and, 524
    reservoir studies and, 525–527
    surface equipment for (*see* UBD surface equipment)
    unsuitable reservoir and, 563
undercompacted shales, 42
underreaming, 491
underwater inspection in lieu of drydocking (UWILD), 637
United Kingdom Offshore Operators Assn. (UKOOA), 111
unitized wellhead
    body of, 354–355
    mandrel casing hangers and, 354–357
    seal assembly and, 357–358
usable holes, 456
U-tube effect, 209

V

variable deck load (VDL), 594, 621–622
velocity
    hole preparation and, 379
    measurement and core-based stress analysis, 33
vertical stresses, 3–4
vertical wells
    cuttings slip velocity and, 159
    cuttings transport in, 156
    elastic wellbore stress concentration in, 19–23
    five percent maximum concentration model and, 159–162
    mist drilling and cuttings transport in, 166–168
viscometry, 137–139
viscosity, 444, 449
    apparent, 136–137
    bit hydraulics and, 259–261
    drilling fluid and, 114–115, 374, 381
    fluid rheology and, 134–135
    HP/HT wells and, 106–107
    of slurry, 401–402, 406–407
    synthetic-based drilling fluids and, 94
    wellbore hydraulic simulation and, 122–123
von Mises stress, 295, 298, 324

W

wait-and-weight method. *See* one-circulation method
washouts, 442
water-based fluids (WBFs), 91–93
    lubricants for, 101–102
    shale formation and, 103
water-based mud (WBM), 382
water-hammer effect, 191
water insoluble materials
    bentonite and, 406
    microsilica and, 406
    polyvinyl alcohol (PVA) and, 406
    synthetic latex and, 406
water-soluble materials
    antifoam additives and, 410
    derivatized cellulose and, 407
    dyes and, 410
    expansive cements and, 409
    fibers and, 410–411
    free-water control and, 408–409
    lost-circulation additives and, 407–408
    miscellaneous additives and, 409

mud-decontaminant additives and, 410
radioactive tracers and, 410
strength-retrogression inhibitors and, 408
synthetic polymers and, 407, 409
wear bushings, 362–363
weather window, 609
weighting agents, 404
weight on bit (WOB), 279, 281, 440, 545, 656, 672
roller-cone bit design and, 221–231, 236–237
weight-set elastomeric seal assembly, 346
weight-up procedures, 213–215
welded-blade stabilizers, 278
well
intervention, 630–633
logging, 510
testing, 623
wellbore, 90–91
flow, 122–123
geometry, 370
pressure, 127–128
schematics, 674
stress concentration (*see* elastic wellbore stress
concentration)
wellbore failure
compressive, 23
detection, 24–27
in plastic rock, 75–76
stress magnitude constraints and, 48–53
stress orientation determination and, 28–30
tensile, 23–27
wellbore position error, 277–278
2D ellipses, 277
wellbore pressure solutions
flowing, 123–125
general steady flow, 125–127
static, 123
wellbore stability, 377, 563
analysis, 444
anisotropic strength and, 65–68
casing seat selection and, 59
drilling fluids and, 90
geomechanical design and, 61–62
geomechanical model and, 53–65
models of, 65–76
mud/rock interactions and, 72–75
mud window and, 57–59
real-time analysis of, 77–8
thermoporoelasticity and, 68–72
uncertainties and, 62–65
wellbore failure in plastic rock and, 75–76
wells of all orientations and, 55–57
well cementing and calcium aluminate cements (CACs), 390
well completion systems, 498
well construction and drilling fluid, 89–90
well control, 376
best method for, 202–206
concurrent method and, 201–202
downhole stresses and, 205
drilling fluids and, 90
fracture mud weight and, 186
kick and, 185–190
kick detection and monitoring for, 190
kick identification and, 197
kill-weight mud calculation and, 198–200
nonconventional procedures for, 217
one-circulation method and, 200–201, 210–217
procedure complexity and, 204–205
procedures, 200–202
shut-in pressures and, 193–197

shut-in procedures and, 190–193
surface pressures and, 203–204
time and, 202
time consideration and, 203
two-circulation method and, 201
variables affecting, 206–210
well controlled equipment, 622–623
coiled-tubing systems, 547
data acquisition, 552
erosion monitoring, 552–553
jointed-pipe systems, 547
rotating blowout preventers (BOPs), 549
rotating control heads, 548–549
rotating diverter systems, 548
separation equipment, 551–552
snubbing systems, 547–548
well design procedures
bottom-to-top approach, 486
flow-string sizing, 486–487
planning for problems, 487
well head equipment, 415–416
wellhead systems
annulus access and, 348–349
annulus seals and, 347–348
big bore subsea, 364
bore protectors and, 362
casing hangers and, 360–362
casing strings and, 364
drilling guide base and, 358–360
jackup drilling rig and (*see* jackup drilling rig)
land well drilling and, 343–349
load-carrying components and, 345–347
low-pressure housing and, 360
mandrel casing hangers and, 345, 354–357
metal-to-metal annulus seal assembly and, 362
mudline hangers and, 352
mudline suspension equipment and (*see* mudline
supension equipment)
oil and gas industry and, 343, 364
pressure containment and, 343–345
product material specifications and, 349
running tools and, 363–364
seal assembly and, 357
spool system and (*see* spool system)
subsea well drilling and, 358–364
test tools and, 363–364
unitized, 354–358
wear bushings and, 362
well offshore drilling and (*see* well offshore drilling)
well offshore drilling
from jackup drilling rig, 349–354
mudline hangers and, 352
mudline suspension equipment and, 349–354
reconnection and, 353–354
total depth (TD) and, 352
well parameters
depth and, 370
formation characteristics and, 371
formation pressures and, 371
temperature and, 370–371
wellbore geometry and, 370
well planning
Authority for Expenditures (AFE), 493–516
bit records and, 463–468
casing, types of, 473–475
casing setting-depth selection and, 473–485
costs and, 456–458
data collection and, 459–460
data sources and, 462–463

drilling engineering and, 455
flow path for, 459
formation pressure and, 458
hole-geometry selection and, 485–493
IADC reports and, 469–470
log headers and, 463, 472
mud-logging records and, 471–472
mud records and, 468–469
objective of, 455–456
offset-well selection and, 460–462
process and, 458–459
production history of, 472
safety, 455–456
scout tickets and, 463, 470, 495–496
seismic studies and, 473
setting-depth design procedures and, 475–485
tubing, types of, 473–475
usable holes and, 456
weather and, 498
well-type classification and, 456–457
wellsite information transfer specification (WITS), 671
Wellsite Information Transfer Standard Markup Language
(WITSML), 673
well-type characteristics, 457
wettability reversal, 446–447
whipstocks, 279, 283–284
whisker, 648
Wildcats, 456, 462–463, 473
wildcatter, 456–457
wireline-quality measurements, 652
working stress design (WSD), 330–332, 334–335

X
X-ray diffraction, 99

Y
yield power law (YPL) fluid, 136
pipe flow and, 147–148
slit flow and, 148–149
yield strength (YS), 289–292, 294–295, 301, 303

Z
zonal isolation, 321, 377

www.ingramcontent.com/pod-product-compliance
Lightning Source LLC
Chambersburg PA
CBHW080338220326
41598CB00030B/4536